A TEXT-BOOK OF

ELECTRICAL TECHNOLOGY

IN S.I. UNITS

Volume II

AC & DC MACHINES

B.L. THERAJA
A.K. THERAJA

A TEXTBOOK OF

ELECTRICAL TECHNOLOGY

IN
S.I. SYSTEM OF UNITS

(Including rationalized M.K.S.A. System)
for the Examinations of B.Sc. (Egg.) ; Sec. A & B of A.M.I.E. (I) ; A.M.I.E.E. (London); City & Guilds (London) ; I.E.R.E. (London) ; Grad. I.E.T.E. (I)

VOLUME II
AC & DC MACHINES

2002
S. CHAND & COMPANY LTD.
RAM NAGAR, NEW DELHI-110 055

S. CHAND & COMPANY LTD.

(An ISO 9002 Company)

Head Office : 7361, RAM NAGAR, NEW DELHI - 110 055
Phones : 3672080-81-82; Fax : 91-11-3677446
Shop at: **schandgroup.com**
E-mail: **schand@vsnl.com**

Branches :

No. 6, Ahuja Chambers, 1st Cross, Kumara Krupa Road, **Bangalore**-560 001. Ph : 2268048
152, Anna Salai, **Chennai**-600 002. Ph : 8460026
Pan Bazar, **Guwahati**-781 001. Ph : 522155
Sultan Bazar, **Hyderabad**-500 195. Ph : 4651135, 4744815
Mai Hiran Gate, **Jalandhar** - 144008. Ph. 401630
613-7, M.G. Road, Ernakulam, **Kochi**-682 035. Ph : 381740
285/J, Bipin Bihari Ganguli Street, **Kolkata**-700 012. Ph : 2367459, 2373914
Mahabeer Market, 25 Gwynne Road, Aminabad, **Lucknow**-226 018. Ph : 226801, 284815
Blackie House, 103/5, Walchand Hirachand Marg , Opp. G.P.O., **Mumbai**-400 001. Ph : 2690881, 2610885
3, Gandhi Sagar East, **Nagpur**-440 002. Ph : 723901
104, Citicentre Ashok, Govind Mitra Road, **Patna**-800 004. Ph : 671366

First Edition 1959
Subsequent Editions and Reprints 1960, 61, 62, 64, 65, 66, 67, 68, 69, 70, 71, 72, 1973 (Twice), 74, 75, 76, 77, 78, 79, 80, 81, 82, 83, 84 (Twice), 85, 86, 89, 91, 93, 94, 95, 96, 97 (Twice), 98 (Twice), 99, 2000, 2001
Reprint 2002

Other Parts Available :
Volume I Basic Electrical Engineering
Volume II A.C. & D.C. Machines
Volume III Transmission, Distribution & Utilization
Volume IV Electronic Devices & Circuits

ISBN : 81-219-1142-7

PRINTED IN INDIA

By Rajendra Ravindra Printers (Pvt.) Ltd., 7361, Ram Nagar, New Delhi-110 055 and published by S. Chand & Company Ltd., 7361, Ram Nagar, New Delhi-110 055.

PREFACE TO THE TWENTY-SECOND EDITION

Authors feel happy to present to their esteemed readers a thoroughly revised twenty-second edition of Vol. II of Electrical Technology whose popularity seems to be growing day by day. Each chapter has been properly pruned and updated with the help of subject material selected from the latest curricula of various engineering colleges and technical institutions in India and abroad. Special attention has been given to the chapters dealing with three-phase induction motors, single-phase motors and synchronous motors. Additions comprise Operation of Salient-pole synchronous Machines, Power Developed by a Synchronous Generator and Motor and detailed discussion of Salient-pole Synchronous Motor etc.

A new chapter entitled **Special Machines** has been added to this volume for the first time. It deals with various types of **Stepper Motors, Permanent-magnet DC Motors, Printed-Circuit DC Motors, PM Synchronous Motors, Resolvers** and **Servomotors** both DC and AC. Stepper motors are widely used for operation control in computer peripherals, textile industry, IC fabrication and robotics. They are also used for applications requiring incremental motion. PM synchronous motors are used where precise speed must be maintained to ensure consistent product. The switched-reluctance motor is still in its infancy but has been successfully applied to a wide range of applications both industrial and domestic.

Many redundant solved as well as unsolved examples have been deleted and selectively replaced by the latest numerical examples set in various engineering examination papers throughout India. Similarly, all Objective Tests have been recast and extended where required. It is hoped that with these new additions and deletions, the present volume would prove more useful to our esteemed readers than the earlier one.

As ever before, we are thankful to our publishers particularly Shri Ravindra Kumar Gupta for the personal interest he took in the expeditious printing of this book and for the highly attractive cover design suggested by him. Our sincere thanks go to their hyperactive and result-oriented overseas manager for his globe-trotting efforts to popularise the book from one corner of the globe to the other. Lastly, we would love to record our sincere thanks to two brilliant ladies, nimbled-fingered mini-sized Mrs. Uma Arora and calm and quiet Ms. Shweta Bhardwaj for the secretarial support they provided us during the preparation of this book.

AUTHORS

CONTENTS

24 D.C. GENERATORS

24.1. Generator Principle

An electrical generator is a machine which converts mechanical energy (or power) into electrical energy (or power).

The energy conversion is based on the principle of the production of dynamically (or motionally) induced e.m.f. As seen from Fig. 7.3., whenever a conductor cuts magnetic flux, dynamically induced e.m.f. is produced in it according to Faraday's Laws of Electromagnetic Induction. This e.m.f. causes a current to flow if the conductor circuit is closed.

Hence, two basic essential parts of an electrical generator are (*i*) a magnetic field and (*ii*) a conductor or conductors which can so move as to cut the flux.

24.2. Simple Loop Generator

Construction

In Fig. 24.1 is shown a single-turn rectangular copper coil *ABCD* rotating about its own axis in a magnetic field provided by either permanent magnets or electromagnets. The two ends of the coil

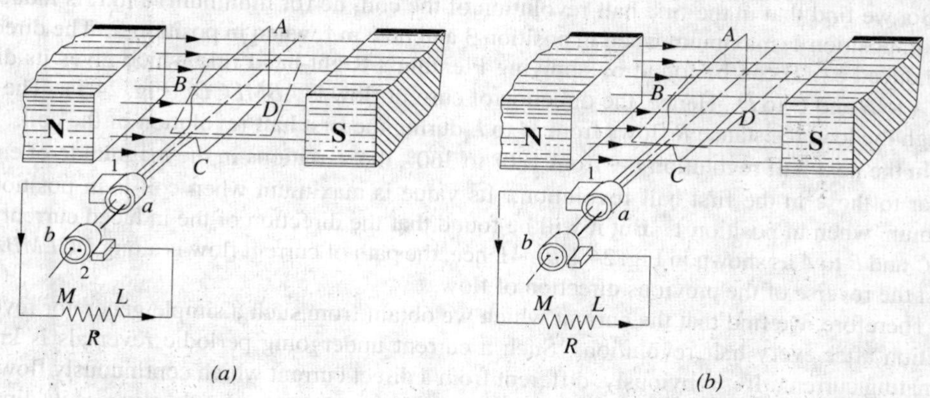

Fig. 24.1

are joined to two slip-rings '*a*' and '*b*' which are insulated from each other and from the central shaft. Two collecting brushes (of carbon or copper) press against the slip-rings. Their function is to collect the current induced in the coil and to convey it to the external load resistance *R*.

The rotating coil may be called 'armature' and the magnets as 'field magnets'.

Working

Imagine the coil to be rotating in clock-wise direction (Fig. 24.2). As the coil assumes successive positions in the field, the flux linked with it changes. Hence, an e.m.f. is induced in it which is proportional to the rate of change of flux linkages ($e = Nd\Phi dt$). When the plane of the coil is at right angles to lines of flux *i.e.* when it is in position, 1, then flux linked with the coil is maximum but *rate of change of flux linkages is minimum.*

It is so because in this position, the coil sides *AB* and *CD* do not cut or shear the flux, rather they slide along them *i.e.* they move parallel to them. Hence, there is no induced e.m.f. in the coil. Let

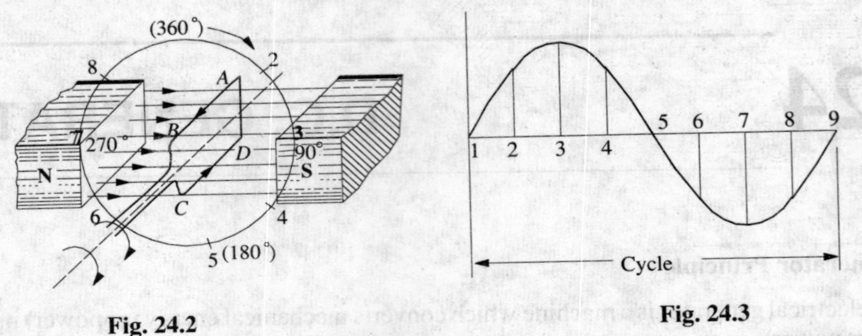

Fig. 24.2

Fig. 24.3

us take this no-e.m.f. or vertical position of the coil as the starting position. The angle of rotation or time will be measured from this position.

As the coil continues rotating further, the rate of change of flux linkages (and hence induced e.m.f. in it) increases, till position 3 is reached where $\theta = 90°$. Here, the coil plane is horizontal *i.e.* parallel to the lines of flux. As seen, the flux linked with the coil is minimum but *rate of change of flux linkages is maximum*. Hence, maximum e.m.f. is induced in the coil when in this position (Fig. 24.3).

In the next quarter revolution *i.e.* from 90° to 180°, the flux linked with the coil gradually *increases* but the rate of change of flux linkages *decreases*. Hence, the induced e.m.f. decreases gradually till in position 5 of the coil, it is reduced to zero value.

So, we find that in the first half revolution of the coil, no (or minimum) e.m.f. is induced in it when in position 1, maximum when in position 3 and no e.m.f. when in position 5. The direction of this induced e.m.f. can be found by applying Fleming's Right-hand rule which gives its direction from *A* to *B* and *C* to *D*. Hence, the direction of current flow is *ABMLCD* (Fig. 24.1). The current through the load resistance *R* flows from *M* to *L* during the first half revolution of the coil.

In the next half revolution *i.e.* from 180° to 360°, the variations in the magnitude of e.m.f. are similar to those in the first half revolution. Its value is maximum when coil is in position 7 and minimum when in position 1. But it will be found that the direction of the induced current is from *D* to *C* and *B* to *A* as shown in Fig. 24.1(*b*). Hence, the path of current flow is along *DCLMBA* which is just the reverse of the previous direction of flow.

Therefore, we find that the current which we obtain from such a simple generator reverses its direction after every half revolution. Such a current undergoing periodic reversals is known as alternating current. It is, obviously, different from a direct current which continuously flows in one and the same direction. It should be noted that alternating current not only reverses its direction, it does not even keep its magnitude constant while flowing in any one direction. The two half-cycles may be called positive and negative half-cycles respectively (Fig. 24.3).

For making the flow of current unidirectional in the *external* circuit, the slip-rings are replaced by split-rings (Fig. 24-4). The split-rings are made out of a conducting cylinder which is cut into two halves or segments insulated from each other by a thin sheet of mica or some other insulating material (Fig. 24.5).

As before, the coil ends are joined to these segments on which rest the carbon or copper brushes.

It is seen [Fig. 24.6(*a*)] that in the first half revolution current flows along *ABMLCD i.e.* the brush No. 1 in contact with segment '*a*' acts as the positive end of the supply and '*b*' as the negative end. In the next half revolution [Fig. 24.6 (*b*)], the direction of the induced current in the coil has reversed. But at the same time, the positions of segments '*a*' and '*b*' have also reversed with the result that brush No. 1 comes in touch with that segment which is positive *i.e.* segment '*b*' in this case. Hence, current in the load resistance again flows from *M* to *L*. The waveform of the current

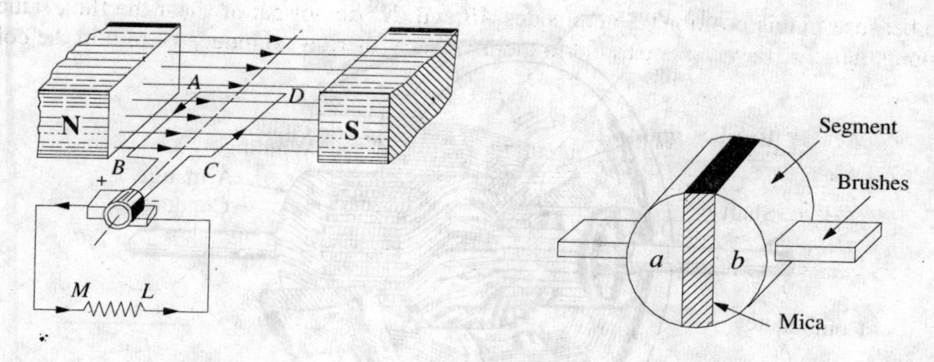

Fig. 24.4

Fig. 24.5

through the external circuit is as shown in Fig. 24.7. *This current is unidirectional but not continuous like pure direct current.*

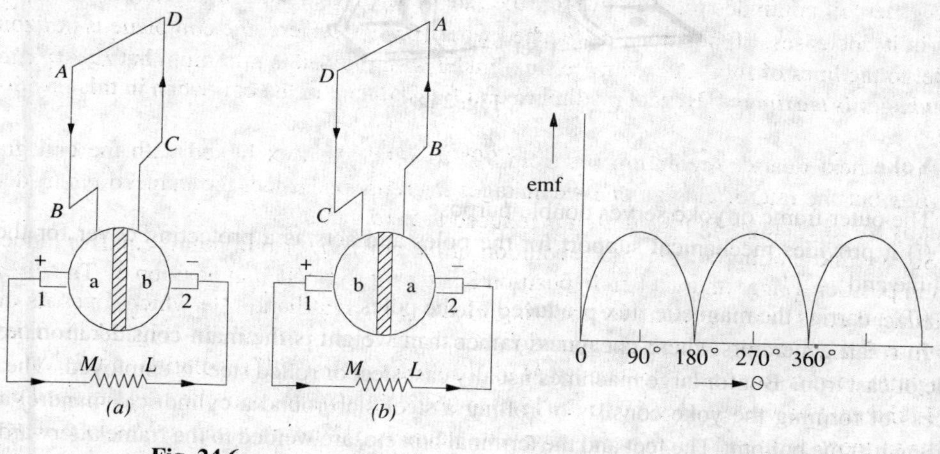

Fig. 24.6

Fig. 24.7

It should be noted that the position of brushes is so arranged that the changeover of segments '*a*' and '*b*' from one brush to the other takes place when the plane of the rotating coil is at right angles to the plane of the lines of flux. It is so because in that position, the induced e.m.f. in the coil is zero.

Another important point worth remembering is that even now the current induced in the coil sides is alternating as before. It is only due to the rectifying action of the split-rings (also called commutator) that it becomes unidirectional in the external circuit. Hence, it should be clearly understood that even in the armature of a d.c. generator, the induced voltage is alternating.

24.3. Practical Generator

The simple loop generator has been considered in detail merely to bring out the basic principle underlying construction and working of an actual generator illustrated in Fig. 24.8 which consists of the following essential parts :

1. Magnetic Frame or Yoke
2. Pole-cores and Pole-shoes
3. Pole Coils or Field Coils
4. Armature Core
5. Armature windings or Conductors
6. Commutator
7. Brushes and Bearings

Of these, the yoke, the pole cores, the armature core and air gaps between the poles and the armature core or the magnetic circuit whereas the rest form the electrical circuit.

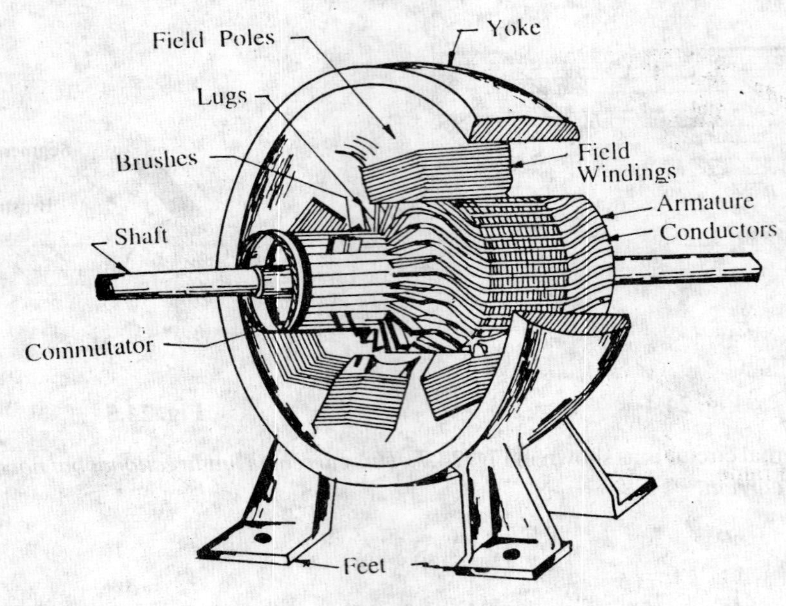

Fig. 24.8

24.4 Yoke

The outer frame or yoke serves double purpose :

(*i*) it provides mechanical support for the poles and acts as a protecting cover for the whole machine and

(*ii*) it carries the magnetic flux produced by the poles.

In small generators where cheapness rather than weight is the main consideration, yokes are made of cast iron. But for large machines usually cast steel or rolled steel is employed. The modern process of forming the yoke consists of rolling a steel slab round a cylindrical mandrel and then welding it at the bottom. The feet and the terminal box etc. are welded to the frame afterwards. Such yokes possess sufficient mechanical strength and have high permeability.

24.5. Pole Cores and Pole Shoes

The field magnets consist of pole cores and pole shoes. The pole shoes serve two purposes (*i*) they spread out the flux in the air gap and also, being of larger cross-section, reduce the reluctance of the magnetic path (*ii*) they support the exciting coils (or field coils) as shown in Fig. 24.14.

There are two main types of pole construction.

(*a*) The pole core itself may be a solid piece made out of either cast iron or cast steel but the pole shoe is laminated and is fastened to the pole face by means of counter sunk screws as shown in Fig. 24.10.

(*b*) In modern design, the complete pole cores and pole shoes are built of thin laminations of annealed steel which are rivetted together under hydraulic pressure (Fig. 24.11). The thickness of laminations varies from 1 mm to 0.25 mm. The laminated poles may be secured to the yoke in any of the following two ways :

(*i*) either the pole is secured to the yoke by means of screws bolted through the yoke and into the pole body or

(*ii*) the holding screws are bolted into a steel bar which passes through the pole across the plane of laminations (Fig. 24.12).

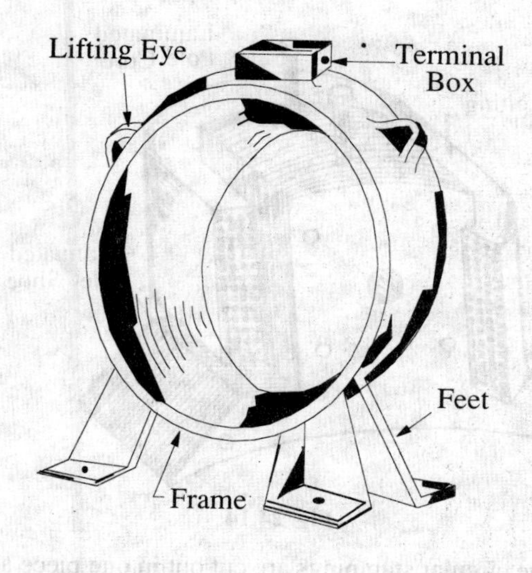

Fig. 24.9

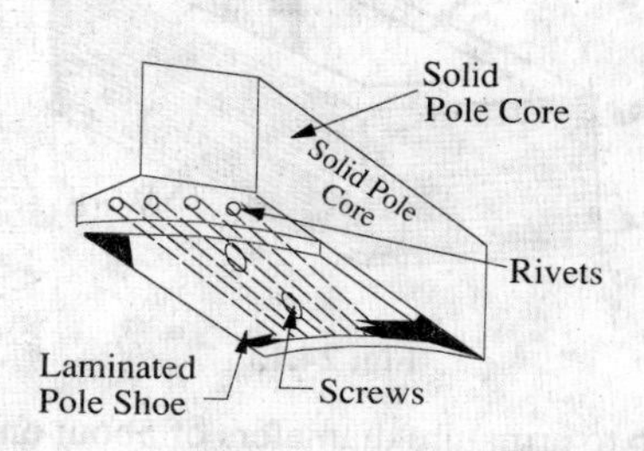

Fig. 24.10

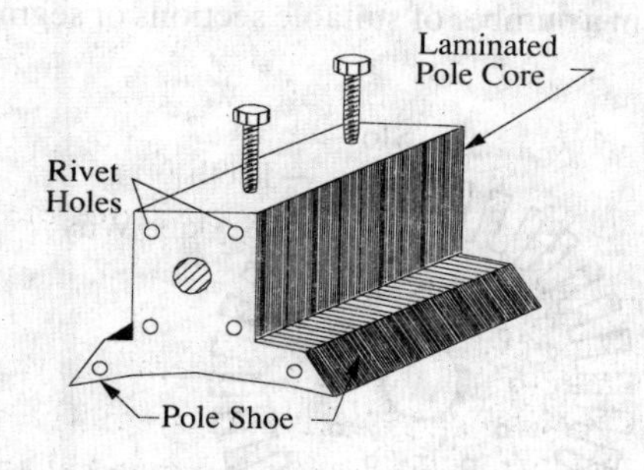

Fig. 24.11

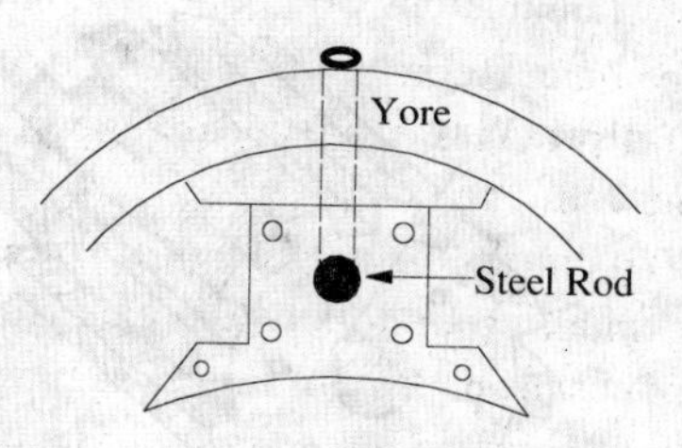

Fig. 24.12

24.6. Pole Coils

The field coils or pole coils, which consist of copper wire or strip, are former-wound for the correct dimension (Fig. 24.13). Then, the former is removed and wound coil is put into place over the core as shown in Fig. 24.14.

When current is passed through these coils, they electromagnetise the poles which produce the necessary flux that is cut by revolving armature conductors.

24.7. Armature Core

It houses the armature conductors or coils and causes them to rotate and hence cut the magnetic flux of the field magnets. In addition to this, its most important function is to provide a path of very low reluctance to the flux through the armature from a *N*-pole to a *S*-pole.

It is cylindrical or drum-shaped and is built up of usually circular sheet steel discs or laminations approximately 0.5 mm thick (Fig. 24.15). It is keyed to the shaft.

The slots are either die-cut or punched on the outer periphery of the disc and the keyway is located on the inner diameter as shown. In small machines, the armature stampings are keyed directly to the shaft. Usually, these laminations are perforated for air ducts which permits axial flow of air through the armature for cooling purposes. Such ventilating channels are clearly visible in the laminations shown in Fig. 24.16 and Fig. 24.17.

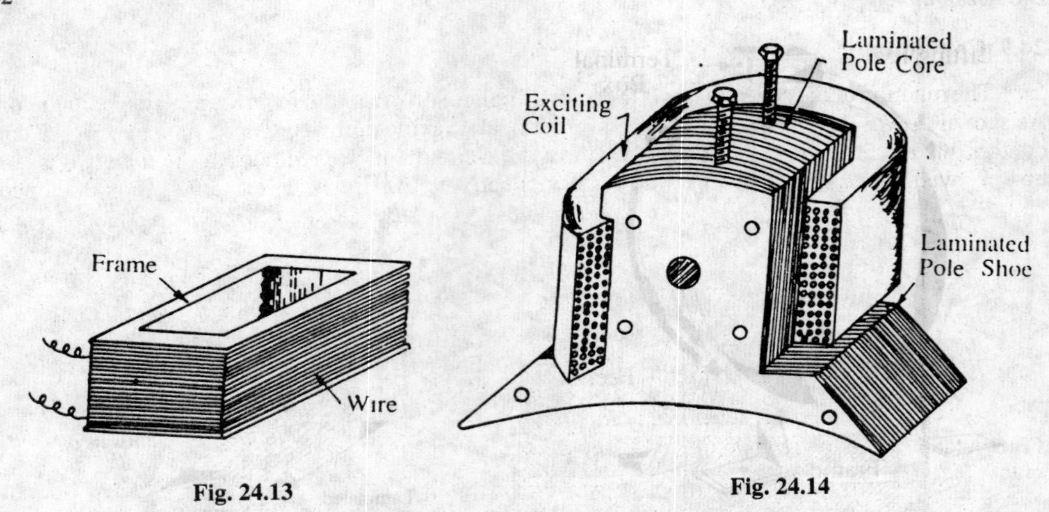

Fig. 24.13 **Fig. 24.14**

Up to armature diameters of about one metre, the circular stampings are cut out in one piece as shown in Fig. 24.16. But above this size, these circles, especially of such thin sections, are difficult to handle because they tend to distort and become wavy when assembled together. Hence, the circular laminations, instead of being cut out in one piece, are cut in a number of suitable sections or segments which form part of a complete ring (Fig. 24.17).

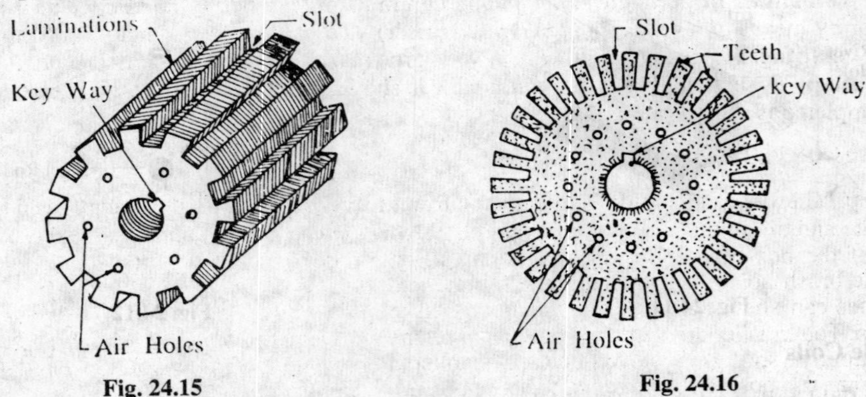

Fig. 24.15 **Fig. 24.16**

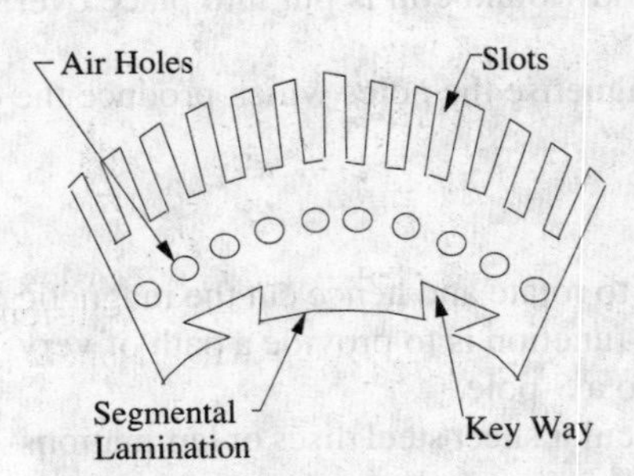

Fig. 24.17.

A complete circular lamination is made up of four or six or even eight segmental laminations. Usually, two keyways are notched in each segment and are dove-tailed or wedge-shaped to make the laminations self-locking in position.

The purpose of using laminations is to reduce the loss due to eddy currents. Thinner the laminations, greater is the resistance offered to the induced e.m.f., smaller the current and hence lesser the $I^2 R$ loss in the core.

24.8. Armature Windings

The armature windings are usually former-wound. These are first wound in the form of flat rectangular coils and are then pulled into their proper shape in a coil puller. Various conductors of the coils are insulated from each other. The conductors are placed in the armature slots which are lined with tough insulating material. This slot insulation is folded over above the armature conductors placed in the slot and is secured in place by special hard wooden or fibre wedges.

24.9. Commutator

The function of the commutator is to facilitate collection of current from the armature conductors. As shown in Art. 24.2, it rectifies *i.e.* converts the alternating current induced in the armature conductors into unidirectional current in the external load circuit. It is of cylindrical structure and is built up of wedge-shaped segments of high-conductivity hard-drawn or drop forged copper.

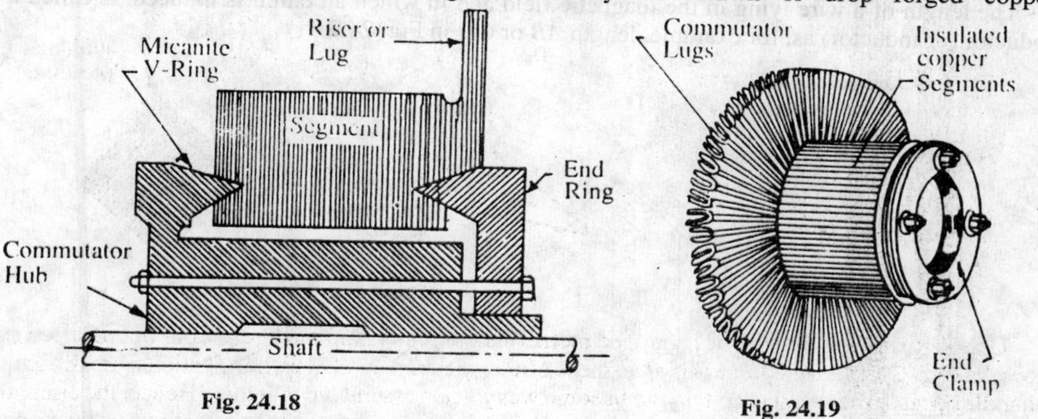

Fig. 24.18

Fig. 24.19

These segments are insulated from each other by thin layers of mica. The number of segments is equal to the number of armature coils. Each commutator segment is connected to the armature conductor by means of a copper lug or strip (or riser). To prevent them from flying out under the action of centrifugal forces, the segments have *V*-grooves, these grooves being insulated by conical micanite rings. A sectional view of commutator is shown in Fig. 24.18 whose general appearance when completed is shown in Fig. 24.19.

21.10 Brushes and Bearings

The brushes whose function is to collect current from commutator, are usually made of carbon or graphite and are in the shape of a rectangular block. These brushes are housed in brush-holders usually of the box-type variety. As shown in Fig. 24.20, the brush-holder is mounted on a spindle and the brushes can slide in the rectangular box open at both ends. The brushes are made to bear down on the commutator by a spring whose tension can be adjusted by changing the position of lever in the notches. A flexible copper pigtail mounted at the top of the brush conveys current from the brushes to the holder. The number of brushes per spindle depends on the magnitude of the current to be collected from the commutator.

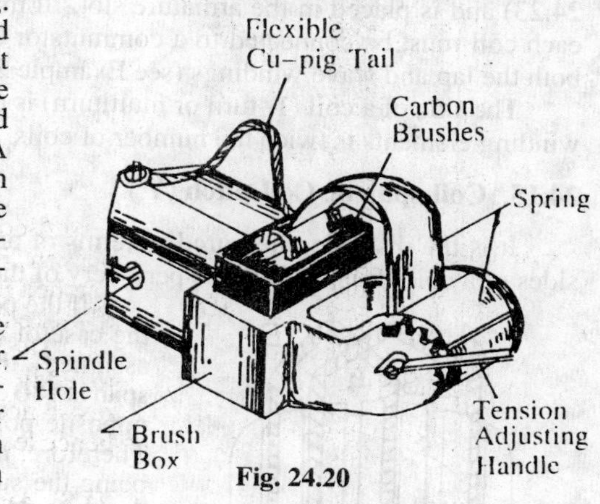

Fig. 24.20

Because of their reliability, ball-bearings are frequently employed, though for heavy duties, roller bearings are preferable. The ball and rollers are generally packed in hard oil for quieter operation and for reduced bearing wear, sleeve bearings are used which are lubricated by ring oilers fed from oil reservoir in the bearing bracket.

24.11. Armature Windings

Now, we will discuss the winding of an actual armature. But before doing this, the meaning of the following terms used in connection with armature winding should be clearly kept in mind.

24.12. Pole-pitch

It may be variously defined as :

(*i*) The periphery of the armature divided by the number of poles of the generator *i.e.* the distance between two adjacent poles.

(*ii*) It is equal to the number of armature conductors (or armature slots) per pole. If there are 48 conductors and 4 poles, the pole pitch is 48/4 = 12.

24.13. Conductor

The length of a wire lying in the magnetic field and in which an e.m.f. is induced, is called a conductor (or inductor) as, for example, length *AB* or *CD* in Fig. 24.21.

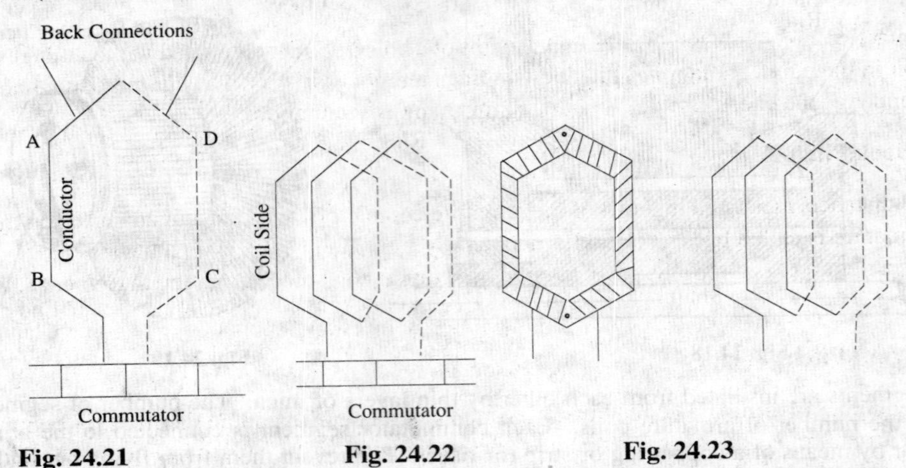

Fig. 24.21 Fig. 24.22 Fig. 24.23

24.14. Coil and Winding Element

With reference to Fig. 24.21, the two conductors *AB* and *CD* along with their end connections constitute one coil of the armature winding. The coil may be single-turn coil (Fig. 24.21) or multi-turn coil (Fig. 24.22). A single-turn coil will have two conductors. But a multi-turn coil may have many conductors per coil side. In Fig. 24.22, for example, each coil side has 3 conductors. The group of wires or conductors constituting a coil side of a multi-turn coil is wrapped with a tape as a unit (Fig. 24.23) and is placed in the armature slot. It may be noted that since the beginning and the end of each coil must be connected to a commutator bar, there are as many commutator bars as coils for both the lap and wave windings (see Example 24.1).

The side of a coil (1–turn or multiturn) is called a winding element. Obviously, the number of winding elements is twice the number of coils.

24.15. Coil-span or Coil-pitch (Y_s)

It is the distance, measured in terms of armature slots (or armature conductors) between two sides of a coil. It is, in fact, the periphery of the armature spanned by the two sides of the coil.

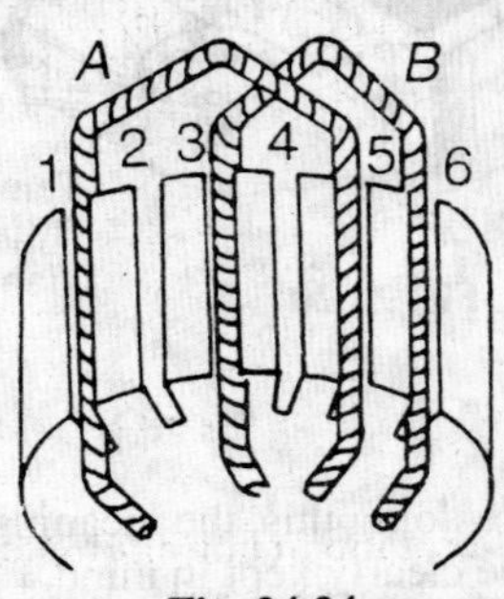

Fig. 24.24

If the pole span or coil pitch is equal to the pole pitch (as in the case of coil *A* in Fig. 24.24 where pole-pitch of 4 has been assumed), then winding is called *full-pitched*. It means that coil span is 180 electrical degrees. In this case, the coil sides lie under opposite poles, hence the induced e.m.fs. in them are additive. Therefore, maximum e.m.f. is induced in the coil as a whole, it being the sum of the e.m.fs. induced in the two coil sides. For example, if there are 36 slots and 4 poles, then coil span is 36/4 = 9 slots. If number of slots in 35, then Y_S = 35/4 = 8 because it is customary to drop fractions.

If the coil span is less than the pole pitch (as in coil *B* where coil pitch is 3/4th of the pole pitch), then the winding is fractional-pitched. In this case, there is a phase difference between the e.m.fs. in the two sides of the coil. Hence, the total e.m.f. round the coil which is the vector sum of e.m.fs. in the two coil sides, is less in this case as compared to that in the first case.

24.16. Pitch of a Winding (Y)

In general, it may be defined as the distance round the armature between two successive conductors which are directly connected together. Or, it is the distance between the beginnings of two consecutive turns.

$$Y = Y_B - Y_F \quad \text{..........for lap winding}$$

$$= Y_B + Y_F \quad \text{........for wave winding}$$

In practice, coil-pitches as low as eight-tenths of a pole pitch are employed without much serious reduction in the e.m.f. Fractional-pitched windings are purposely used to effect substantial saving in the copper of the end connections and for improving commutation.

24.17. Back Pitch (Y_B)

The distance, measured in terms of the armature conductors, which a coil advances on the back of the armature is called back pitch and is denoted by Y_B.

As seen from Fig. 24.28, element 1 is connected on the back of the armature to element 8. Hence, Y_B =- (8−1) = 7.

24.18. Front Pitch (Y_F)

The number of armature conductors or elements spanned by a coil on the front (or commutator end of an armature) is called the front pitch and is designated by Y_F. Again in Fig. 24.28, element 8 is connected to element 3 on the front of the armature, the connections being made at the commutator segment. Hence, $Y_F = 8 - 3 = 5$.

Alternatively, the front pitch may be defined as the distance (in terms of armature conductors) between the second conductor of one coil and the first conductor of the next coil which are connected together at the front *i.e.* commutator end of the armature. Both front and back pitches for lap and wave-winding are shown in Fig. 24-25 and 24.26.

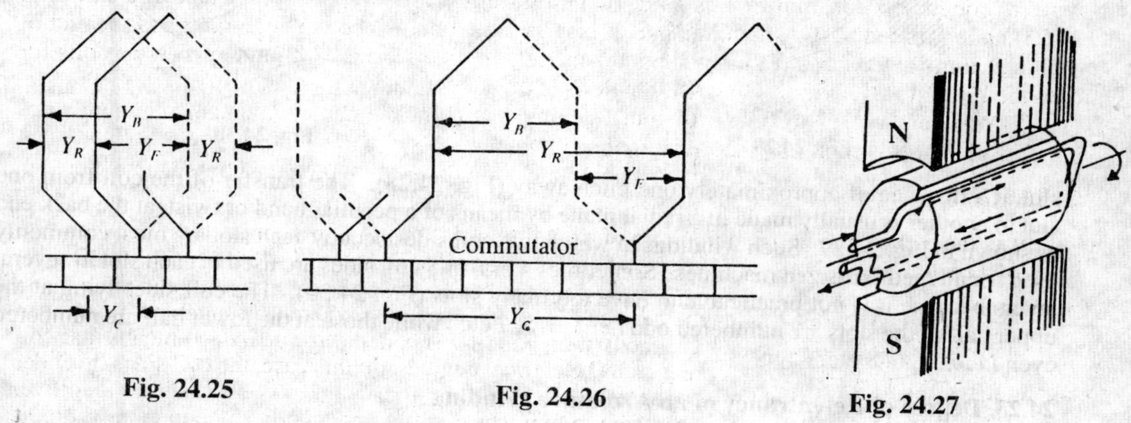

Fig. 24.25 Fig. 24.26 Fig. 24.27

24.19. Resultant Pitch (Y_R)

It is the distance between the beginning of one coil and the beginning of the next coil to which it is connected (Fig. 24.25 and 24.26).

As a matter of precaution, it should be kept in mind that all these pitches, though normally stated in terms of armature conductors, are also sometimes given in terms of armature slots or commutator bars because commutator is, after all, an image of the winding.

24.20. Commutator Pitch (Y_G)

It is the distance (measured in commutator bars or segments) between the segments to which the two ends of a coil are connected. From Fig. 24.25 and 24.26 it is clear that for lap winding, Y_C is the *difference* of Y_B and Y_F whereas for wave-winding it is the *sum* of Y_B and Y_F. Obviously, commutator pitch is equal to the number of bars between coil leads. In general, Y_C equals the 'plex' of the lap-wound armature. Hence, it is equal to 1, 2, 3, 4 etc. for simplex –, duplex –, triplex – and quadruplex – etc. lap-windings.

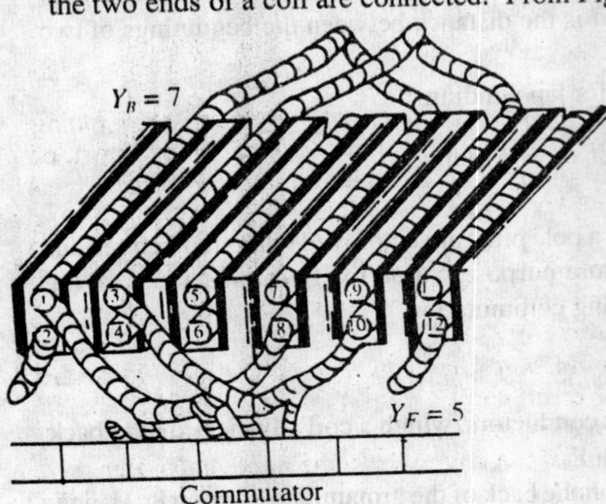

Fig. 24.28

24.21. Single-layer Winding

It is that winding in which one conductor or one coil side is placed in each armature slot as shown in Fig. 24.27. Such a winding is not much used.

24.22. Two-layer Winding

In this type of winding, there are two conductors or coil sides per slot arranged in two layers. Usually, one side of every coil lies in the upper half of one slot and other side lies in the lower of half of some other

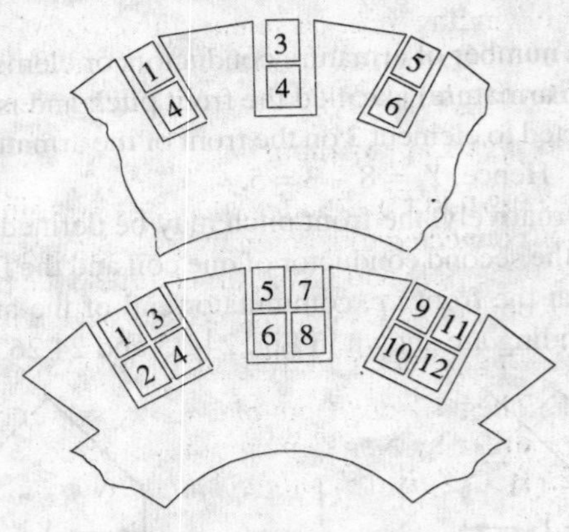

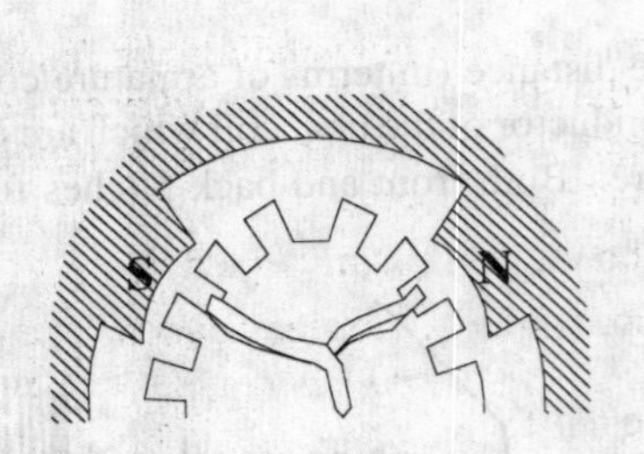

Fig. 24.29

Fig. 24.30

slot at a distance of approximately one pitch away (Fig. 24.28). The transfer of the coil from one slot to another is usually made in a radial plane by means of a peculiar bend or twist at the back end as shown in Fig. 24.29. Such windings in which two coil sides occupy each slot are most commonly used for all medium-sized machines. Sometimes 4 or 6 or 8 coil sides are used in each slot in several layers because it is not practicable to have too many slots (Fig. 24.30). The coil sides lying at the upper half of the slots are numbered odd *i.e.* 1, 3, 5, 7 etc. while those at the lower half are numbered even *i.e.* 2, 4, 6, 8 etc.

24.23. Degree of Re-entrancy of an Armature Winding

A winding is said to be singly re-entrant if on tracing through it once, all armature conductors are included on returning to the starting point. It is doubly re-entrant if only half the conductors are included in tracing through the winding once and so on.

24.24. Multiplex Winding

In such windings, there are several sets of completely closed and independent windings. If there is only one set of closed winding, it is called simplex wave winding. If there are two such windings on the same armature, it is called duplex winding and so on. The multiplicity affects a number of parallel paths in the armature. For a given number of armature slots and coils, as the multiplicity increases, the number of parallel paths in the armature increases thereby increasing the current rating but decreasing the voltage rating.

24.25. Lap and Wave Windings

Two types of windings mostly employed for drum-type armatures are known as Lap Winding and Wave Winding. The difference between the two is merely due to the different arrangement of the end connections at the front or commutator end of armature. Each winding can be arranged progressively or retrogressively and connected in simplex, duplex and triplex. The following rules, however, apply to both types of the windings :

(*i*) The front pitch and back pitch are each approximately equal to the pole-pitch *i.e.* windings should be full-pitched. This results in increased e.m.f. round the coils. For special purposes, fractional-pitched windings are deliberately used (Art. 24.15).

(*ii*) Both pitches should be odd, otherwise it would be difficult to place the coils (which are former-wound) properly on the armature. For example, if Y_B and Y_F were both even, the *all* the coil sides and conductors would lie either in the upper half of the slots or in the lower half. Hence, it would become impossible for one side of the coil to lie in the upper half. Hence, it would become impossible for one side of the coil to lie in the upper half of one slot and the other side of the same coil to lie in the lower half of some other slot.

(*iii*) The number of commutator segments is equal to the number of slots or coils (or half the number of conductors) because the front ends of conductors are joined to the segments in pairs.

(*iv*) The winding must close upon itself *i.e.* if we start from a given point and move from one coil to another, then all conductors should be traversed and we should reach the same point again without a break or discontinuity in between.

24.26. Simplex Lap-winding*

It is shown in Fig. 24.25 which employs single-turn coils. In lap winding, the finishing end of one coil is connected to a commutator segment and to the starting end of the adjacent coil situated under the same pole and so on, till and the coils have been connected. This type of winding derives its name from the fact it doubles or laps back with its succeeding coils.

Following points regarding simplex lap winding should be carefully noted :

1. The back and front pitches are odd and of opposite sign. But they cannot be equal. They differ by 2 or some multiple thereof.
2. Both Y_B and Y_F should be nearly equal to a pole pitch.
3. The average pitch $Y_A = \dfrac{Y_B + Y_F}{2}$. It equals pole pitch $= \dfrac{Z}{P}$
4. Commutator pitch $Y_C = \pm 1$. (In general, $Y_C = \pm m$)
5. Resultant pitch Y_R is even, being the arithmetical difference of two odd numbers. *i.e.* $Y_R = Y_B - Y_F$
6. The number of slots for a 2-layer winding is equal to the number of coils (*i.e.* half the number of coil sides). The number of commutator segments is also the same.
7. The number of parallel paths in the armature = mP where m is the multiplicity of the winding and P the number of poles.

Taking the first condition, we have $Y_B = Y_F \pm 2$.**

---

*However, where heavy currents are necessary, duplex or triplex lap windings are used. The duplex lap winding is obtained by placing two similar windings on the same armature and connecting the even-numbered commutator bars to one winding and the odd-numbered ones to the second winding. Similarly, in triplex lap winding, there would be three windings, each connected to one third of the commutator bars.

**In general, $Y_B = Y_F \pm 2m$ where $m = 1$ for simplex lap winding and $m = 2$ for duplex lap winding etc.

(a) If $Y_B > Y_F$ *i.e.* $Y_B = Y_F + 2$, then we get a progressive or right-handed winding *i.e.* a winding which progresses in the clockwise direction as seen from the commutator end. In this case, obviously, $Y_C = +1$.

(b) If $Y_B < Y_F$ *i.e.* $Y_B = Y_F - 2$, then we get a retrogressive or left-handed winding *i.e.* one which advances in the anti-clockwise direction when seen from the commutator side. In this case, $Y_C = -1$.

(c) Hence, it is obvious that

$$\left.\begin{aligned} Y_F &= \frac{Z}{P} - 1 \\ Y_B &= \frac{Z}{P} + 1 \end{aligned}\right] \text{ for progressive winding and } \left.\begin{aligned} Y_F &= \frac{Z}{P} + 1 \\ Y_B &= \frac{Z}{P} - 1 \end{aligned}\right] \text{ for retrogressive winding}$$

Obviously, Z / P must be even to make the winding possible.

24.27. Numbering of Coils and Commutator Segments

In the d.c. winding diagrams to follow, we will number the coils only (not individual turns). The upper side of the coil will be shown by a firm continuous line whereas the lower side will be shown by a broken line. The numbering of coil sides will be consecutive *i.e.* 1, 2, 3 ...... etc. and such that odd numbers are assigned to the top conductors and even numbers to the lower sides for a two-layer winding. The commutator segments will also be numbered consecutively, the number of the segments will be the same as that of the upper side connected to it.

Example 24.1. *Draw a developed diagram of a simple 2-layer lap-winding for a 4-pole generator with 16 coils. Hence, point out the characteristics of a lap-winding.*

(Elect. Engineering, Madras Univ. 1981)

Solution. The number of commutator segments = 16

Number of conductors or coil sides 16 × 2 = 32; pole pitch= 32/4 = 8

Now remembering that (*i*) Y_B and Y_F have to be odd and (*ii*) have to differ by 2, we get for a progressive winding $Y_B = 9$; $Y_F = -7$ (retrogressive winding will result if $Y_B = 7$ and $Y_F = -9$). Obviously, commutator pitch $Y_C = -1$.

[Otherwise, as shown in Art. 24.26, for progressive winding

$$Y_F = \frac{Z}{P} - 1 = \frac{32}{4} - 1 = 7 \text{ and } Y_B = \frac{Z}{P} - 1 = \frac{32}{4} + 1 = 9]$$

The simple winding table is given as under :

Back Connections		*Front Connections*
1 to (1 + 9) = 10	⟶	10 to (10 − 7) = 3
3 to (3 + 9) = 12	⟶	12 to (12 − 7) = 5
5 to (5 + 9) = 14	⟶	14 to (14 − 7) = 7
7 to (7 + 9) = 16	⟶	16 to (16 − 7) = 9
9 to (9 + 9) = 18	⟶	18 to (18 − 7) = 11
11 to (11 + 9) = 20	⟶	20 to (20 − 7) = 13
13 to (13 + 9) = 22	⟶	22 to (22 − 7) = 15
15 to (15 + 9) = 24	⟶	24 to (24 − 7) = 17
17 to (17 + 9) = 26	⟶	26 to (26 − 7) = 19
19 to (19 + 9) = 28	⟶	28 to (28 − 7) = 21
21 to (21 + 9) = 30	⟶	30 to (30 − 7) = 23
23 to (23 + 9) = 32	⟶	32 to (32 − 7) = 25
25 to (25 + 9) = 34 = (34 − 32) = 2	⟶	2 to (34 − 7) = 27
27 to (27 + 9) = 36 = (36 − 32) = 4	⟶	4 to (36 − 7) = 29
29 to (29 + 9) = 38 = (38 − 32) = 6	⟶	6 to (38 − 7) = 31
31 to (31 + 9) = 40 = (40 − 32) = 8	⟶	8 to (40 − 7) = 33 = (33 − 32) = 1

The winding ends here because we come back to the conductor from where we started.

We will now discuss the developed diagram which is one that is obtained by imagining the armature surface to be removed and then laid out flat so that the slots and conductors can be viewed without the necessity of turning round the armature in order to trace out the armature windings. Such a developed diagram is shown in Fig. 24.31.

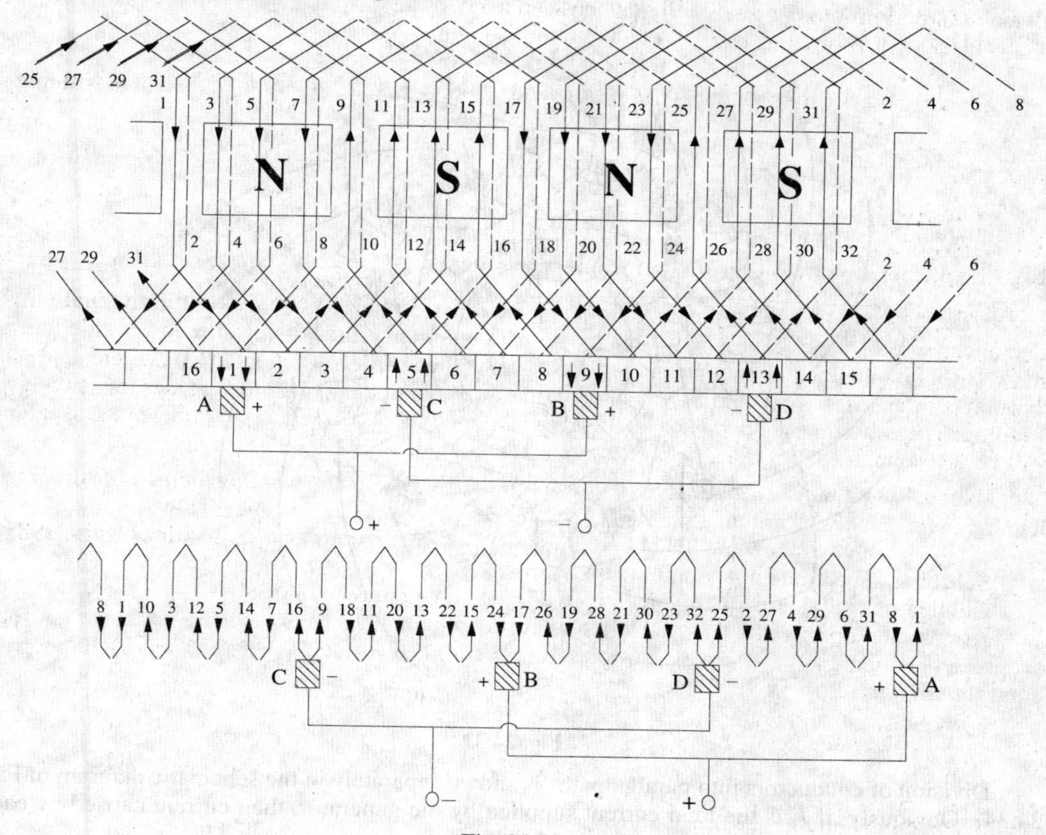

Fig. 24.31

The procedure of developing the winding in this :

Front end of the upper side of coil No. 1 is connected to a commutator segment (whose number is also 1). The back end is joined at the back to the 1 + 9 = 10th coil side in the *lower* half of 5th slot. The front end of coil side 10 is joined to commutator segment 2 to which is connected the front end of 10 − 7 = 3 *i.e.* 3rd coil side lying in the upper half of second armature slot. In this way, by travelling 9 coil sides to the right at the back and 7 to the left at the front we complete the winding, thus including every coil side once till we reach the coil side 1 from where we started. Incidentally, it should be noted that all upper coil sides have been given odd numbers, whereas lower ones have been given even numbers as shown in the polar diagram (Fig. 24.32) of the winding of Fig. 24.31.

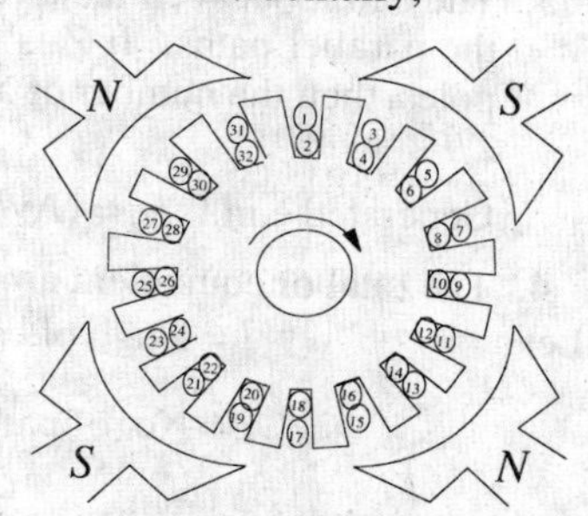

Fig. 24.32

Brush positions can be located by finding the direction of currents flowing in the various conductors. If currents in the conductors under the influence of a *N*-pole are assumed to flow downwards (as shown), then these will flow upwards in conductors under the influence of *S*-pole. By putting proper arrows on the conductors (shown separately in the equivalent ring diagram), it is found that commutator bars No. 1 and 9 are the meeting points of e.m.fs. and hence currents are flowing out of these conductors. The positive brushes should, therefore, be placed at these commutator bars. Similarly, commutator bars No. 5 and 13 are the separating points of e.m.fs, hence negative brushes are placed there. In all, there are four brushes, two positive and two negative. If brushes of the same polarity are connected together, then all the armature conductors are divided into four parallel paths.

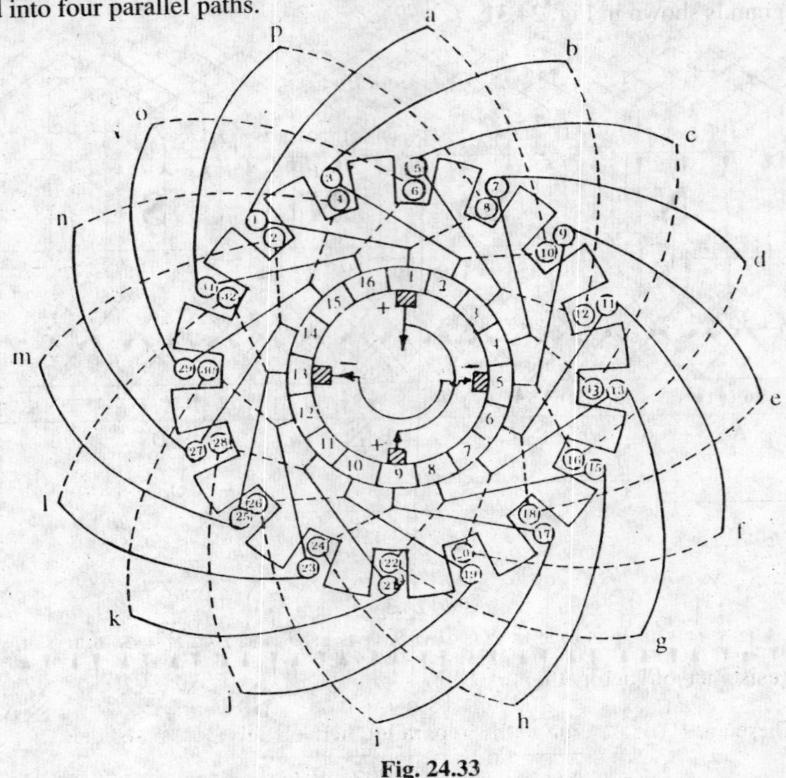

Fig. 24.33

Division of conductors into parallel paths is shown separately in the schematic diagram of Fig. 24.34. Obviously, if I_a is the total current supplied by the generator, then current carried by each parallel path is $I_a/4$.

Summarizing these conclusions, we have

1. The total number of brushes is equal to the number of poles.
2. There are as many parallel paths in the armature as the number of poles. That is why such a winding is sometimes known as 'multiple circuit' or 'parallel' winding. In general, number of parallel paths in armature = mP where m is the multiplicity (plex) of the lap winding. For example, a 6-pole duplex lap winding has $(6 \times 2) = 12$ parallel paths in its armature.
3. The e.m.f. between the +ve and –ve brushes is equal to the e.m.f. generated in any one of the parallel paths. If Z is the total number of armature conductors and P the number of poles, then the number of armature conductors (connected in series) in any parallel path is Z/P.

$\therefore$ Generated e.m.f. E_g = (Average e.m.f./conductor) $\times \frac{Z}{P} = e_{av} \times \frac{Z}{P}$

4. The total or equivalent armature resistance can be found as follows :

Let l = lengh of each armature conductor; S = its cross-section

A = No. of parallel paths in armature $= P$ – for simplex lap winding

R = resistance of the whole windingthen $R = \frac{\rho l}{S} \times Z$

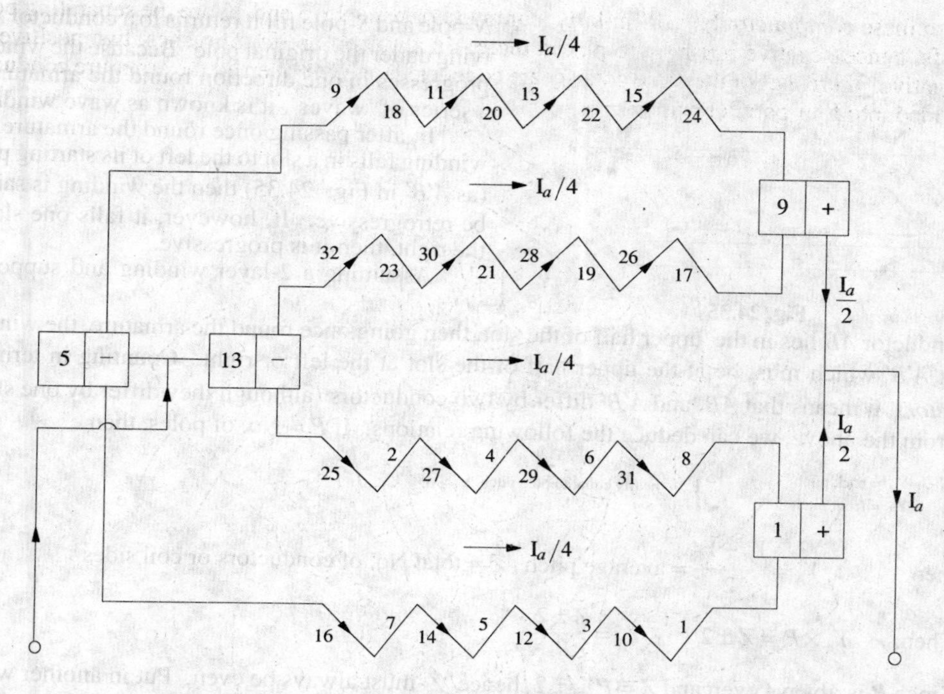

Fig. 24.34

$$\text{Resistance of each path} = \frac{\rho l Z}{S \times A}$$

There are P (or A) such paths in parallel, hence equivalent resistance

$$= \frac{1}{A} \times \frac{\rho l Z}{SA} = \frac{\rho l Z}{SA^2}$$

5. If I_a is the total armature current, then current per parallel path (or carried by each conductor) is I_a/P.

24.28. Simplex Wave Winding*

From Fig. 24.31, it is clear that in lap winding, a conductor (or coil side) under one pole is connected at the back to a conductor which occupies an almost corresponding position under the *next* pole of *opposite* polarity (as conductors 3 and 12). Conductor No. 12 is then connected to conductor No. 5 under the *original* pole but which is a little removed from the initial conductor No. 3. If, instead of returning to the same N-pole, the conductor No. 12 were taken *forward* to the next N-pole, it would make no difference so far as the direction and magnitude of the e.m.f. induced in the circuit are concerned.

As shown in Fig. 24.35, conductor *AB* is connected to *CD* lying under *S*-pole and then to *EF* under the next *N*-pole. In this way, the winding progresses, passing successively under every

*Like lap winding, a wave winding may be duplex, triplex or may have any degree of multiplicity. A simplex wave winding has two paths, a duplex wave winding four paths and a triplex one six paths etc.

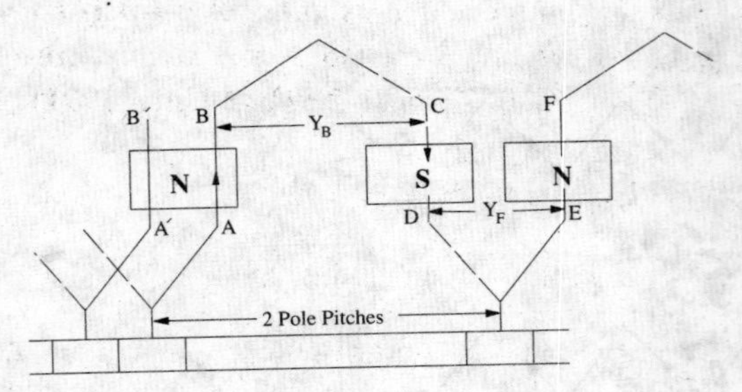

Fig. 24.35

N-pole and S-pole till it returns to a conductor $A'B'$ lying under the original pole. Because the winding progresses in one direction round the armature in a series of 'waves', it is known as wave winding.

If, after passing once round the armature, the winding falls in a slot to the left of its starting point (as $A'B'$ in Fig. 24.35) then the winding is said to be retrogressive. If, however, it falls one slot to the right, then it is progressive.

Assuming a 2-layer winding and supposing that conductor AB lies in the upper half of the slot, then going once round the armature, the winding ends at $A'B'$ which must be at the upper half of the slot at the left or right. Counting in terms of *conductors*, it means that AB and $A'B'$ differ by two conductors (although they differ by one slot).

From the above, we can deduce the following relations. If P = No. of poles, then

Y_B = back pitch, Y_F = front pitch } nearly equal to pole pitch

then $$Y_A = \frac{Y_B + Y_F}{2} = \text{average pitch}$$; Z = total No. of conductors or coil sides

Then, $$Y_A \times P = Z \pm 2 \qquad Y_A = \frac{Z \pm 2}{P}$$

Since P is always even and $Z = PY_A \pm 2$, hence Z must always be even. Put in another way, it means that $\frac{Z \pm 2}{P}$ must be an even integer.

The plus sign will give a progressive winding and the negative sign a retrogressive winding.

Points to Note :

1. Both pitches Y_B and Y_F are odd and of the same sign.
2. Back and front pitches are nearly equal to the pole pitch and may be equal or differ by 2, in which case, they are respectively one more or one less than the average pitch.
3. Resultant pitch $Y_R = Y_F + Y_B$.
4. Commutator pitch, $Y_C = Y_A$ (in lap winding $Y_C = \pm 1$).

   Also, $$Y_C = \frac{\text{No. of commutator bars} \pm 1}{\text{No. of pair of poles}}$$

5. The average pitch which must be an integer is given by

   $$Y_A = \frac{Z \pm 2}{P} = \frac{\frac{Z}{2} \pm 1}{P/2} = \frac{\text{No. of commutator bars} \pm 1}{\text{No. of pair of poles}}$$

   It is clear that for Y_A to be an integer, there is a restriction on the value of Z. With $Z = 32$, this winding is impossible for a 4-pole machine (though lap winding is possible). Values of $Z = 30$ or 34 would be perfectly alright.

6. The number of *coils i.e.* N_C can be found from the relation.

   $$N_C = \frac{PY_A \pm 2}{2}$$

   This relation has been found by rearranging the relation given in (5) above.

7. It is obvious from (5) that for a wave winding, the number of armature conductors with 2 either added or subtracted must be a multiple of the number of poles of the generator. This restriction eliminates many even numbers which are unsuitable for this winding.
8. The number of armature parallel paths = $2m$ where m is the multiplicity of the winding

Example 24.2. *Draw a developed diagram of a simplex 2-layer wave-winding for a 4-pole d.c. generator with 30 armature conductors. Hence, point out the characteristics of a simple wave winding.* **(Elect. Engg-I, Nagpur Univ. 1991)**

Solution. Here, $Y_A = \frac{30 \pm 2}{4} = 8^*$ or 7. Taking $Y_A = 7$, we have $Y_B = Y_F = 7$

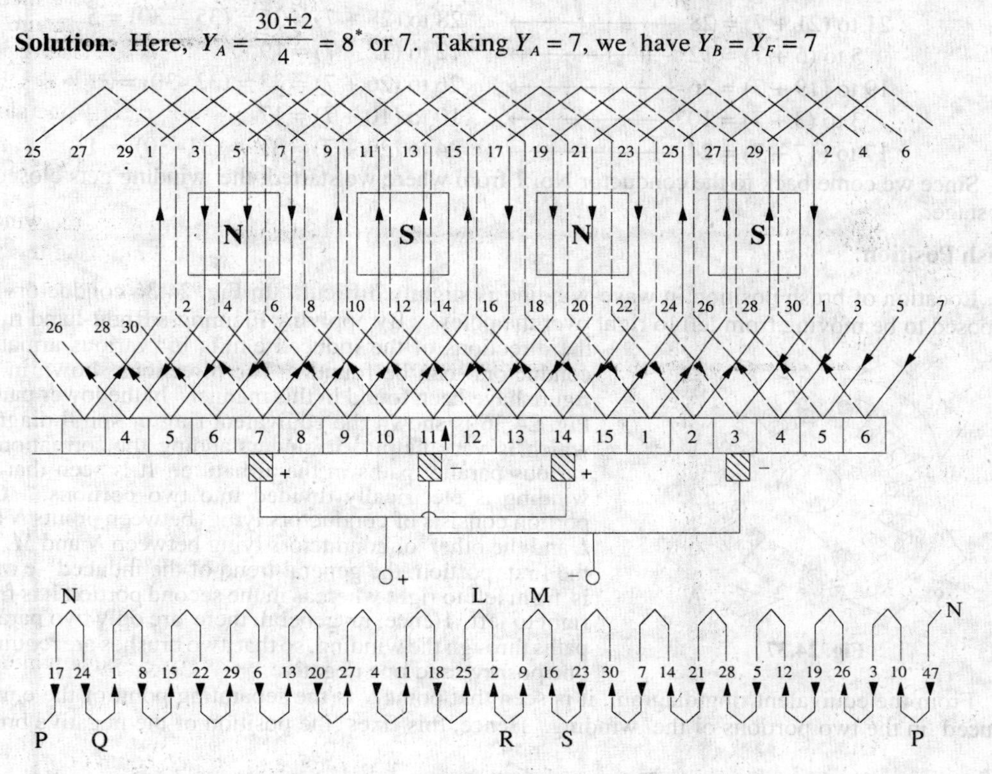

Fig. 24.36

As shown in Fig. 24.36 and 24.37, conductor No. 5 is taken to conductor No. 5 + 7 = 12 at the back and is joined to commutator segment 5 at the front. Next, the conductor No. 12 is joined to commutator segment 5 + 7 = 12 (∵ Y_C = 7) to which is joined conductor No. 12 + 7 = 19. Continuing this way, we come back to conductor No. 5 from where we started. Hence, the winding closes upon itself.

The simple winding table is as under :

Back Connections		Front Connections
1 to (1 + 7) = 8	⟶	8 to (8 + 7) = 15
15 to (15 + 7) = 22	⟶	22 to (22 + 7) = 29
29 to (29 + 7) = 36 = (36–30) = 6	⟶	6 to (6 + 7) = 13
13 to (13 + 7) = 20	⟶	20 to (20 + 7) = 27
27 to (27 + 7) = 34 = (34–30) = 4	⟶	4 to (4 + 7) = 11
11 to (11 + 7) = 18	⟶	18 to (18 + 7) = 25
25 to (25 + 7) = 32 = (32–30) = 2	⟶	2 to (2 + 7) = 9

*If we take 8, then the pitches would be : $Y_B = 9$ and $Y_F = 7$ or $Y_B = 7$ and $Y_F = 9$. Incidentally, if $Y_A = Y_C$ is taken as 7, armature will rotate in one direction and if $Y_C = 8$, it will rotate in the opposite direction.

9 to (9 + 7) = 16 ⟶ 16 to (16 + 7) = 23

23 to (23 + 7) = 30 ⟶ 30 to (30 + 7) = 37 = (37−30) = 7

7 to (7 + 7) = 14 ⟶ 14 to (14 + 7) = 21

21 to (21 + 7) = 28 ⟶ 28 to (28 + 7) = 35 = (35 − 30) = 5

5 to (5 + 7) = 12 ⟶ 12 to (12 + 7) = 19

19 to (19 + 7) = 26 ⟶ 26 to (26 + 7) = 33 = (33−30) = 3

3 to (3 + 7) = 10 ⟶ 10 to (10 + 7) = 17

17 to (17 + 7) = 24 ⟶ 24 to (24 + 7) = 31 = (31−30) = 1

Since we come back to the conductor No. 1 from where we started, the winding gets closed at this stage.

Brush Position

Location of brush position in wave-winding is slightly difficult. In Fig. 24.36 conductors are supposed to be moving from left to right over the poles. By applying Fleming's Right-hand rule, the directions of the induced e.m.fs in various armature conductors can be found. The directions shown in the figure have been found in this manner. In the lower part of Fig. 24.36 is shown the equivalent ring or spiral diagram which is very helpful in understanding the formation of various parallel paths in the armature. It is seen that the winding is electrically divided into two portions. One portion consists of conductors lying between points *N* and *L* and the other of conductors lying between *N* and *M*. In the first portion, the general trend of the induced e.m.fs. is from left to right whereas in the second portion it is from right to left. Hence, in general, there are only two parallel paths through the winding, so that two brushes are required, one positive and one negative.

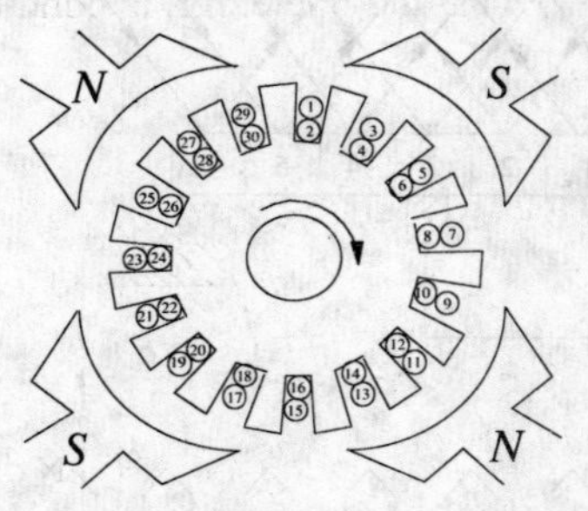

Fig. 24.37

From the equivalent ring diagram, it is seen that point *N* is the separating point of the e.m.fs. induced in the two portions of the winding. Hence, this fixes the position of the negative brush.

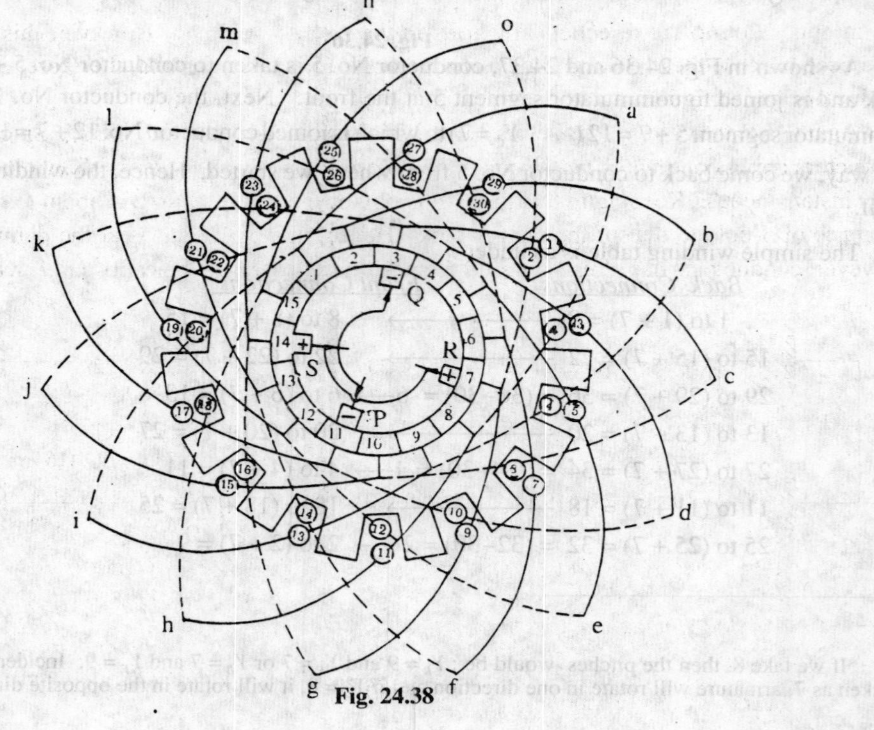

Fig. 24.38

But as it is at the back and not at the commutator end of the armature, the negative brush has two alternative positions *i.e.* either at point P or Q. These points on the equivalent diagram correspond to commutator segments No. 3 and 11.

Now, we will find the position of the positive brush. It is found that there are two meeting points of the induced e.m.fs. *i.e.* points L and M but both these points are at the back or non-commutator end of the armature. These two points are separated by one loop only, namely, the loop composed of conductors 2 and 9, hence the middle point R of this loop fixes the position of the positive brush, which should be placed in touch with commutator segment No. 7. We find that for one position of the +ve brush, there are two alternative positions for the −ve brush.

Taking the +ve brush at point R and negative brush at point P, the winding is seen to be divided into the following two paths.

In path 1 (Fig. 24.36) it is found that e.m.f. in conductor 9 is in opposition to the general trend of e.m.fs. in the other conductors comprising this path. Similarly, in path 2, the e.m.f. in conductor

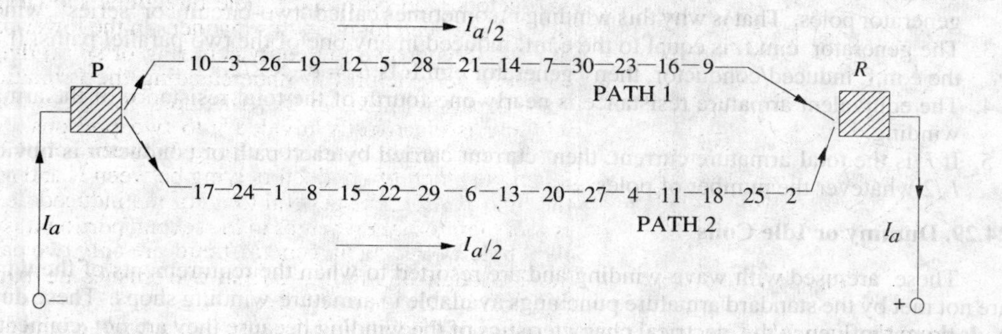

Fig. 24.39

2 is in opposition to the direction of e.m.fs. in the path as a whole. However, this will make no difference because these conductors lie almost in the interpolar gap and, therefore, e.m.fs. in these conductors are negligible.

Again, take the case of conductors 2 and 9 situated between points L and M. Since the armature conductors are in continuous motion over the pole faces, their positions as shown in the figure are only instantaneous. Keeping in this mind, it is obvious that conductor 2 is about to move from the influence of S-pole to that of the next N-pole. Hence, the e.m.f. in it is at the point of reversing. However, conductor 9 has already passed the position of reversal, hence its e.m.f. will not reverse,

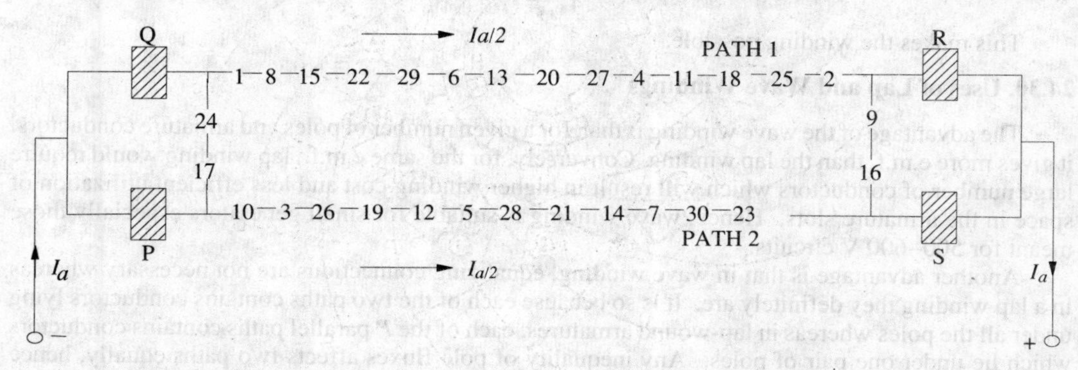

Fig. 24.40

rather it will increase in magnitude gradually. It means that in a very short interval, point M will become the meeting point of the e.m.fs. But as it lies at the back of the armature, there are two alternative positions for the +ve brush *i.e.* either point R which has already been considered or point S which corresponds to commutator segment 14. This is the second alternative position of the positive brush. Arguing in the same way, it can be shown that after another short interval of time, the alternative position of the positive brush will shift from segment 14 to segment 15. Therefore, if one positive brush is in contact with segment 7, then the second positive brush if used, should be in touch with both segments 14 and 15.

It may be noted that if brushes are placed in both alternative positions for both positive and negative (*i.e.* if in all, 4 brushes are used, two +ve and two −ve), then the effect is merely to short-circuit the loop lying between brushes of the same polarity. This is shown in Fig. 24.40 where it will also be noted that irrespective of whether only two or four brushes are used, the number of parallel paths through the armature winding is still two.

Summarizing the above facts, we get

1. Only two brushes are necessary, though their number may be equal to the number of poles.
2. The number of parallel paths through the armature winding is two irrespective of the number of generator poles. That is why this winding is sometimes called 'two-circuit' or 'series' winding.
3. The generator e.m.f. is equal to the e.mf. induced in any one of the two parallel paths. If e_{ev} is the e.m.f. induced/conductor, then generator e.m.f. Is $E_g = e_{av} \times Z/2$.
4. The equivalent armature resistance is nearly one-fourth of the total resistance of the armature winding.
5. If I_a is the total armature current, then current carried by each path or conductor is obviously $I_a/2$ whatever the number of poles.

24.29. Dummy or Idle Coils

These are used with wave-winding and are resorted to when the requirements of the winding are not met by the standard armature punchings available in armature-winding shops. These dummy coils do not influence the electrical characteristics of the winding because they are not connected to the commutator. They are exactly similar to the other coils except that their ends are cut short and taped. They are there simply to provide mechanical balance for the armature because an armature having some slots without windings would be out of balance mechanically. For example, suppose number of armature slots is 15, each containing 4 sides and the number of poles is 4. For a simplex wave-winding.

$$Y_A = \frac{Z \pm 2}{P} = \frac{60 \pm 2}{4}$$

which does not come out to be an integer (Art. 24.28) as required by this winding. However, if we make one coil dummy so that we have go 58 active conductors, then

$$Y_A = \frac{58 \pm 2}{4} = 14 \text{ or } 15$$

This makes the winding possible.

24.30. Uses of Lap and Wave Windings

The advantage of the wave winding is that, for a given number of poles and armature conductors, it gives more e.m.f. than the lap winding. Conversely, for the same e.m.f., lap winding would require large number of conductors which will result in higher winding cost and less efficient utilization of space in the armature slots. Hence, wave winding is suitable for small generators especially those meant for 500–600 V circuits.

Another advantage is that in wave winding, equalizing connections are not necessary whereas in a lap winding they definitely are. It is so because each of the two paths contains conductors lying under all the poles whereas in lap-wound armatures, each of the P parallel paths contains conductors which lie under one pair of poles. Any inequality of pole fluxes affects two paths equally, hence their induced e.m.fs. are equal. In lap-wound armatures, unequal voltages are produced which set up a circulating current that produces sparking at brushes.

However, when large currents are required, it is necessary to use lap winding, because it gives more parallel paths.

Hence, lap winding is suitable for comparatively low-voltage but high-current generators whereas wave-winding is used for high-voltage, low-current machines.

Tutorial Problem No. 24.1

1. Write down the winding table for a 2-layer simplex lap-winding for a 4-pole d.c. generator having (*a*) 20 slots and (*b*) 13 slots. What are the back and front pitches as measured in terms of armature conductors ?

[**Hint** : (*a*) No. of conductors = 40 ; $Y_B = 11$ and $Y_F = -9$] (**Elect. Engineering, Madras Univ. 1978**)

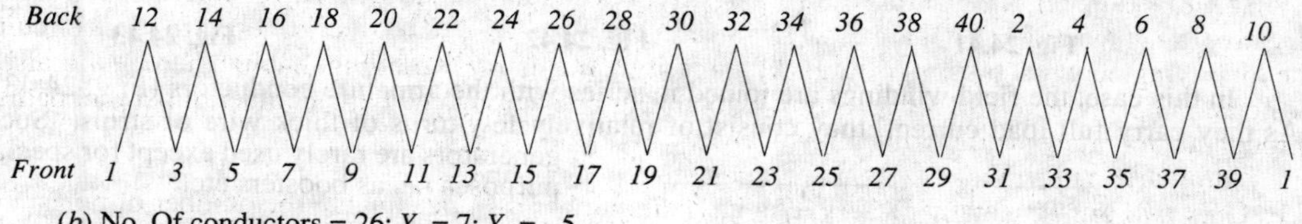

(*b*) No. Of conductors = 26; $Y_B = 7$; $Y_F = -5$

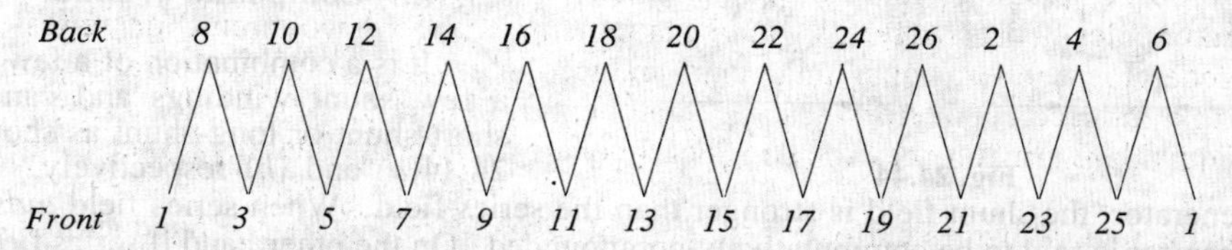

2. With a simplex 2-layer wave winding having 26 conductors and 4-poles, write down the winding table. What will be the front and back pitches of the winding ?

[**Hint.** $Y_F = 7$ and $Y_B = 5$] (**Electric Machinery-I, Madras Univ. Nov. 1979**)

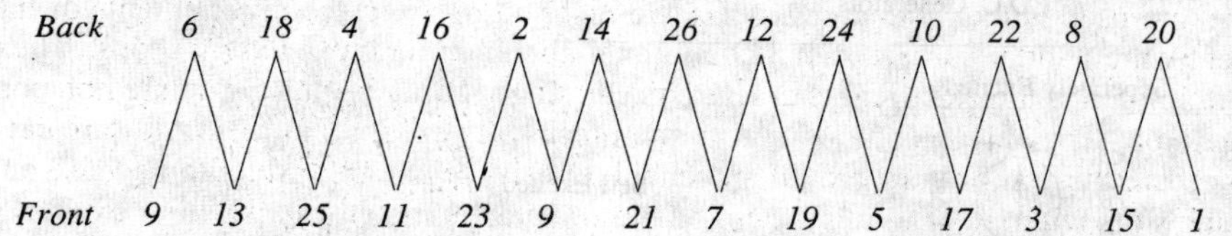

3. Is it possible to get simplex wave winding for a 4-pole d.c. machine with 28 conductors ? Explain the reason for your answer. [**No, it would contain only 4 conductors**]

4. State for what type of winding each of the following armatures could be used and whether the winding must be four or six-pole if no dummy coils are to be used (*a*) 33 slots, 165 commutator segments (*b*) 64 solts, 256 commutator segments (*c*) 65 slots, 260 commutator segments.

[(*a*) **4-pole lap with commutator pitch 82 or 83 or 6-pole lap.**
(*b*) **4-pole lap or 6-pole wave with commutator pitch 85.**
(*c*) **6-pole wave with commutator pitch 87.**]

24.31. Types of Generators

Generators are usually classified according to the way in which their fields are excited. Generators may be divided into (*a*) separately-excited generators and (*b*) self-excited generators.

(a) Separately-excited generators are those whose field magnets are energised from an independent external source of d.c. current. It is shown diagramatically in Fig. 24.41.

(b) Self-excited generators are those whose field magnets are energised by the current produced by the generators themselves. Due to residual magnetism, there is always present some flux in the poles. When the armature is rotated, some e.m.f. and hence some induced current is produced which is partly or fully passed through the field coils thereby strengthening the residual pole flux.

There are three types of self-excited generators named according to the manner in which their field coils (or windings) are connected to the armature.

(i) Shunt wound

The field windings are connected across or in parallel with the armature conductors and have the full voltage of the generator applied across them (Fig. 24.42).

(ii) Series Wound

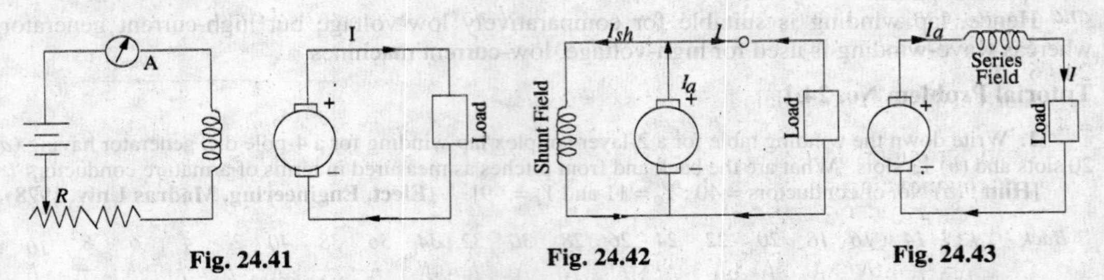

Fig. 24.41 Fig. 24.42 Fig. 24.43

In this case, the field windings are joined in series with the armature conductors (Fig. 24.43). As they carry full load current, they consist of relatively few turns of thick wire or strips. Such generators are rarely used except for special purposes *i.e.* as boosters etc.

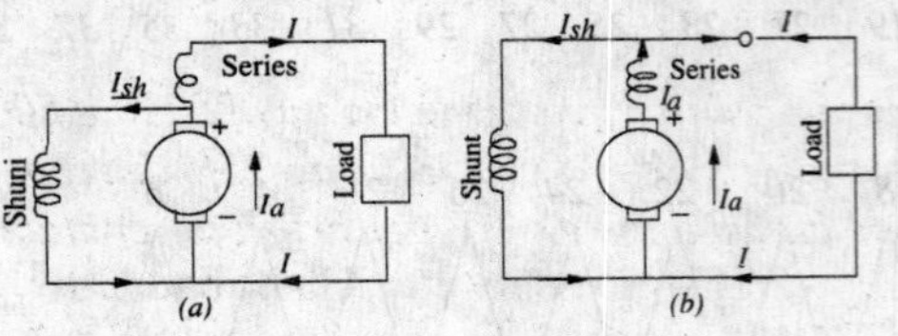

Fig. 24.44

(iii) Compound Wound

It is a combination of a few series and a few shunt windings and can be either short-shunt or long-shunt as shown in Fig. 24.44(*a*) and (*b*) respectively. In a compound generator, the shunt field is stronger than the series field. When series field *aids* the shunt field, generator is said to be commutatively-compounded. On the other hand if series field *opposes* the shunt field, the generator is said to be differentially compounded. Various types of d.c. generators have been shown separately in Fig. 24.45.

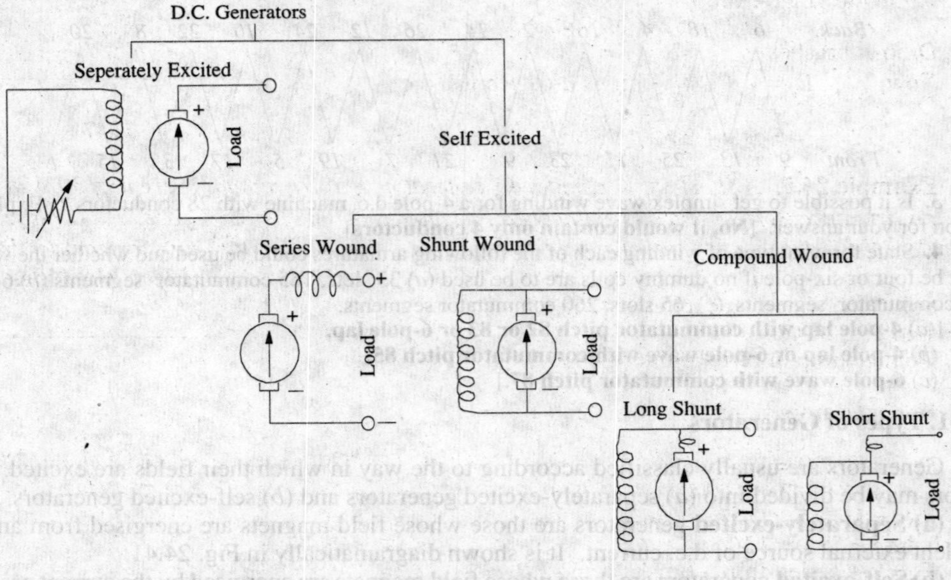

Fig. 24.45

24.32. Brush Contact Drop

It is the voltage drop over the brush contact resistance when current passes from commutator segments to brushes and finally to the external load. Its value depends on the amount of current and the value of contact resistance. This drop is usually small and includes brushes of both polarities. However, in practice, the brush contact drop is assumed to have following constant values for all loads.

0.5 V for metal-graphite brushes.

2.0 V for carbon brushes.

Example 24.3. *A shunt generator delivers 450 A at 230 V and the resistance of the shunt field and armature are 50 Ω and 0.03 Ω respectively. Calculate the generated e.m.f.*

Solution. Generator circuit is shown in Fig. 24.46.

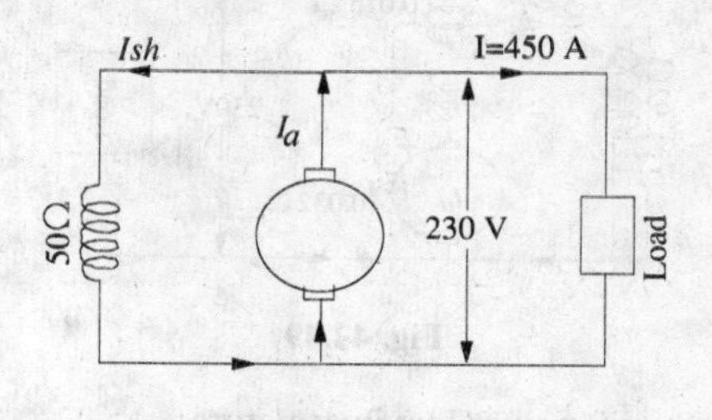

Fig. 24.46

Current through shunt field winding is $I_{sh} = 230/50 = 4.6$ A

Load current $I = 450$ A

∴ Armature current $I_a = I + I_{sh}$
$= 450 + 4.6 = 454.6$ A

Armature voltage drop

$I_aR_a = 454.6 \times 0.03 =$ **13.6 V**

Now E_g = terminal voltage + armature drop
$= V + I_aR_a$

∴ e.m.f. generated in the armature

$E_g = 230 + 13.6 =$ **243.6 V**

Example 24.4. *A long-shunt compound generator delivers a load current of 50 A at 500 V and has armature, series field and shunt field resistances of 0.05 Ω, 0.03 Ω ad 250 Ω respectively. Calculate the generated voltage and the armature current. Allow 1 V per brush for contact drop.*

(Elect. Science 1, Allahabad Univ. 1992)

Solution. Generator circuit is shown in Fig. 24.47.

$I_{sh} = 500/250 = 2$ A

Current through armature and series winding is
$= 50 + 2 = 52$ A

Voltage drop on series field winding
$= 52 \times 0.03 = 1.56$ V

Armature voltage drop
$I_aR_a = 52 \times 0.05 = 2.6$ V

Drop at brushes $= 2 \times 1 = 2$ V

Now, $E_g = V + I_aR_a +$ series drop + brush drop

$= 500 + 2.6 + 1.56 + 2 =$ **506.16 V**

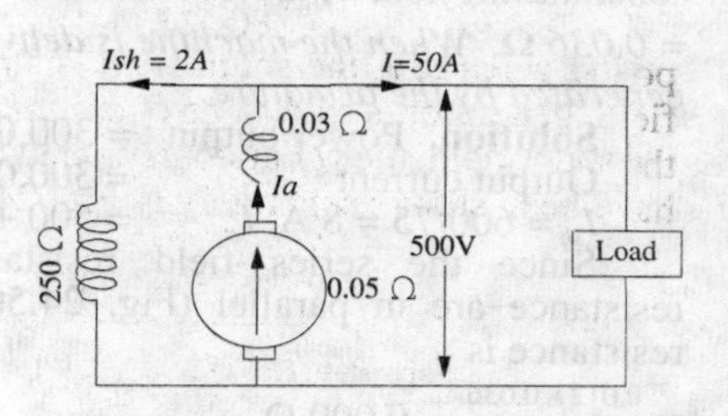

Fig. 24.47

Example 24.5. *A short-shunt compound generator delivers a load current of 30 A at 220 V, and has armature, series-field and shunt-field resistances of 0.05 Ω 0.30 Ω and 200 Ω respectively. Calculate the induced e.m.f. and the armature current. Allow 1.0 V per brush for contact drop.*

(AMIE Sec. B. Elect. Machines 1991)

Solution. Generator circuit diagram is shown in Fig. 24.48.

Voltage drop in series winding $= 30 \times 0.3 = 9$ V

Voltage across shunt winding $= 220 + 9 = 229$ V

$I_{sh} = 229/200 = 1.145$ A

$I_a = 30 + 1.145 = 31.145$ A

$I_aR_a = 31.145 \times 0.05 = 1.56$ V

Brush drop $= 2 \times 1 = 2$ V

$E_g = V +$ series drop + brush drop $+ I_aR_a = 220 + 9 + 2 + 1.56 = 232.56$ V

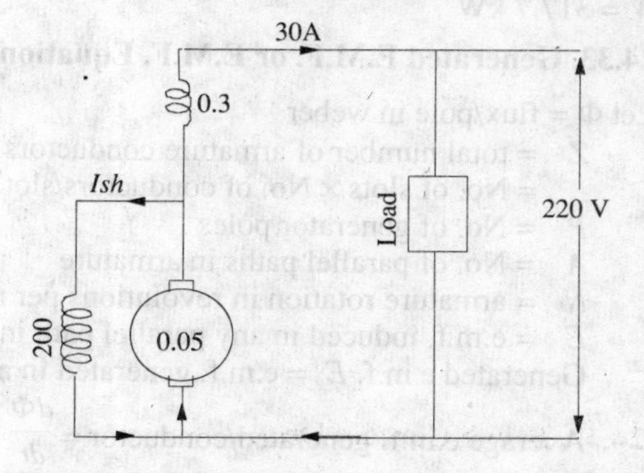

Fig. 24.48

Example 24.6. *In a long-shunt compound generator, the terminal voltage is 230 V when generator delivers 150 A. Determine (i) induced e.m.f. (ii) total power generated and (iii) distribution of this power. Given that shunt field, series field, divertor and armature resistances are 92 Ω, 0.015 Ω, 0.03 Ω and 0.032 Ω respectively.*

(Elect. Technology-II, Gwalior Univ. 1987)

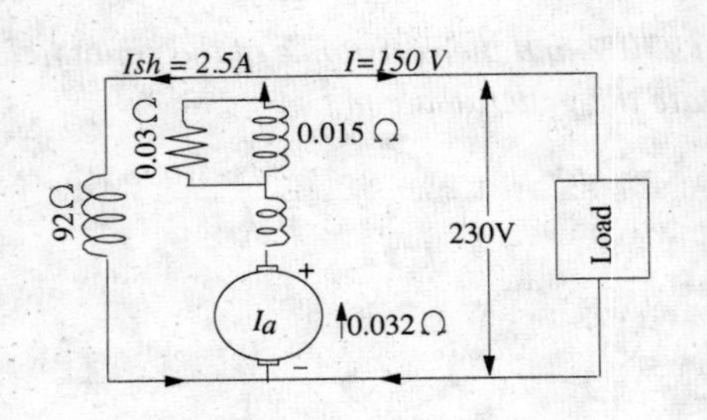

Fig. 42.49

Solution. $I_{sh} = 230/92 = 2.5$ A ; $I_a = 150 + 2.5 = 152.5$ A

Since series field resistance and divertor resistance are in parallel (Fig. 24.49) their combined resistance is

$= 0.03 \times 0.015/0.045 = 0.01\ \Omega$

Total armature circuit resistance is

$= 0.032 + 0.01 = 0.042\ \Omega$

Voltage drop $= 152.5 \times 0.042 = 6.4$ V

(*i*) voltage generated by armature

$E_g = 230 + 6.4 =$ **236.4 V**

(*ii*) total power generated in armature

$= E_g I_a = 236.4 \times 152.5 =$ **36,051 W**

(*iii*) power lost in armature $= I_a^2 R_a = 152.5^2 \times 0.032 = 744$ W

Power lost in series field and divertor $= 152.5^2 \times 0.01 = 232$ W

Power dissipated in shunt winding $= VI_{sh} = 230 \times 0.01 = 575$ W

Power delivered to load $= 230 \times 150 = 34500$ W

Total $= 36,051$ W

Example 24.7. *The following information is given for a 300-kW, 600-V, long-shunt compound generator : Shunt field resistance = 75 Ω, armature resistance including brush resistance = 0.03 Ω, commutating field winding resistance = 0.011 Ω, series field resistance = 0.012 Ω, divertor resistance = 0.036 Ω. When the machine is delivering full load, calculate the voltage and power generated by the armature.* **(Elect. Engg -II, Pune Univ. No. 1989)**

Solution. Power output $= 300{,}000\ W$

Output current $= 300{,}000/600 = 500$ A

$I_{sh} = 600/75 = 8$ A, $I_a = 500 + 8 = 508$ A

Since the series field resistance and divertor resistance are in parallel (Fig. 24.50) their combined resistance is

$$= \frac{0.012 \times 0.036}{0.048} = 0.009\ \Omega$$

Total armature circuit resistance

$= 0.03 – 0.011 + 0.009 = 0.05\ \Omega$

Voltage drop $= 508 \times 0.05 = 25.4$ V

Voltage generated by armature

$= 600 + 25.4 = 625.4$ V

Power generated $= 625.4 \times 508 = 317{,}700$ W $= 317.7$ kW

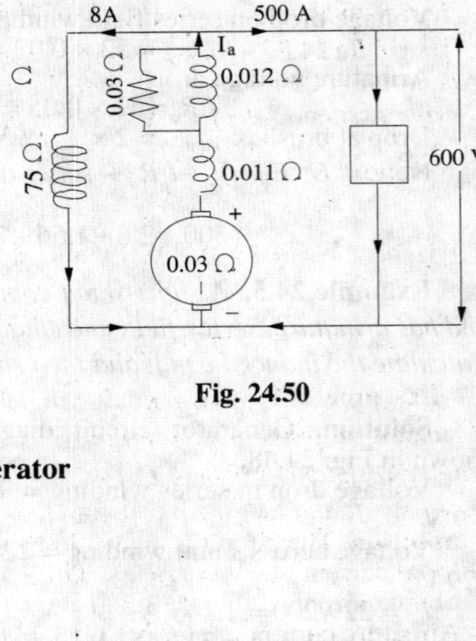

Fig. 24.50

24.33. Generated E.M.F. or E.M.F. Equation of a Generator

Let Φ = flux/pole in weber

Z = total number of armature conductors
= No. of slots × No. of conductors/slot

P = No. of generator poles

A = No. of parallel paths in armature

N = armature rotation in revolutions per minute (r.p.m.)

E = e.m.f. induced in any parallel path in armature

Generated e.m.f. E_g = e.m.f. generated in any one of the parallel paths *i.e.* E.

Average e.m.f. generated/conductor $= \dfrac{d\Phi}{dt}$ volt ($\because\ n = 1$)

Now, flux cut/conductor in one revolution $d\Phi = \Phi P\ Wb$

No. of revolutions/second $= N/60$ ∴ time for one revolution, $dt = 60/N$ second

Hence, according to Faraday's Laws of Electromagnetic Induction,

E.M.F. generated/conductor $= \dfrac{d\Phi}{dt} = \dfrac{\Phi PN}{60}$ volt

For a simplex wave-wound generator

No. of parallel paths = 2

No. of conductors (in series) in one path = $Z/2$

$$\therefore \text{ E.M.F. generated/path} = \frac{\Phi PN}{60} \times \frac{Z}{2} = \frac{\Phi ZPN}{120} \text{ volt}$$

For a simplex lap-wound generator

No. of parallel paths = P

No. of conductors (in series) in one path = Z/P

$$\therefore \text{ E.M.F #r[generated/path6} = \frac{\Phi PN}{60} \times \frac{Z}{P} = \frac{\Phi ZN}{60} \text{ volt}$$

In general, generated e.m.f. $E_g = \frac{\Phi ZN}{60} \times \left(\frac{P}{A}\right)$ volt

where $A = 2$- for simplex wave-winding

$= P$-for simplex lap-winding

Also, $E_g = \frac{1}{2\pi} \cdot \left(\frac{2\pi N}{60}\right) \Phi Z \left(\frac{P}{A}\right) = \frac{\omega \Phi Z}{2\pi}\left(\frac{P}{A}\right)$ volt $-\omega$ in rad/s

For a given d.c machine, Z, P and A are constant. Hence, putting $K_a = ZP/A$, we get $E_g = K_a \Phi N$ volts ---- where N is in r.p.s.

Example 24.8. *A four-pole generator, having wave-wound armature winding has 51 slots, each slot containing 20 conductors. What will be the voltage generated in the machine when driven at 1500 rpm assuming the flux per pole to be 7.0 mWb* **(Elect. Machines-I, Allahabad Univ. 1993)**

Solution. $E_g = \frac{\Phi ZN}{60}\left(\frac{P}{A}\right)$ volts

Here, $\Phi = 7 \times 10^{-3}$ Wb, $Z = 51 \times 20 = 1020$, $A = P = 4$, $N = 1500$ r.p.m.

$$\therefore \quad E_g = \frac{7 \times 10^{-3} \times 1020 \times 1500}{60}\left(\frac{4}{2}\right) = \mathbf{178.5\ V}$$

Example 24.9. *An 8-pole d.c. shunt generator with 778 wave-connected armature conductors and running at 500 r.p.m. supplies a load of 12.5 Ω resistance at terminal voltage of 50 V. The armature resistance is 0.24 Ω and the field resistance is 250 Ω. Find the armature current, the induced e.m.f. and the flux per pole.* **(Electrical Engg-I, Bombay Univ. 1988)**

Solution. The circuit is shown in Fig. 24.51.

Load current $= V/R = 250/12.5 = 20$ A

Shunt current $= 250/250 = 1$ A

Armature current $= 20 + 1 =$ **21 A**

Induced e.m.f. $= 250 + (21 \times 0.24) =$ **255.04 V**

Now $E_g = \frac{\Phi ZN}{60} - \left(\frac{P}{A}\right)$

$$\therefore \quad 255.04 = \frac{\Phi \times 778 \times 500}{60}\left(\frac{8}{2}\right)$$

$$\therefore \quad \Phi = \mathbf{9.83\ mWb}$$

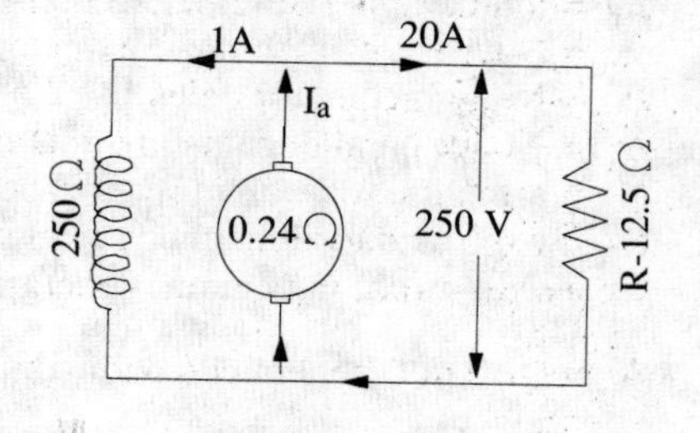

Fig. 24.51

Example 24.10 *A separately excited generator, when running at 100 r.p.m supplied 200 A at 125 V. What will be the load current when the speed drops to 800 r.p.m if. I_f is unchanged? Given that the armature resistance = 0.04 ohm and brush drop = 2 V.*

(Elect. Machines Nagpur Univ. 1993)

Solution. The load resistance $R = 125/200 = 0.625\ \Omega$

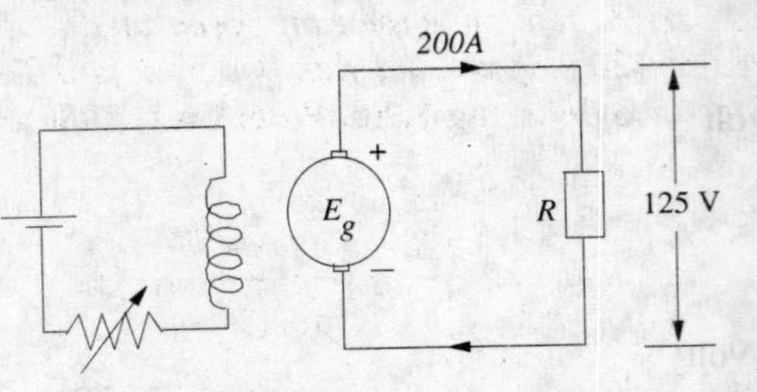

Fig. 24.52

$E_{g1} = 125 + 200 \times 0.04 + 2 = 135$ V ; $N_1 = 1000$ r.p.m

At 800 r.p.m. $E_{g2} = 135 \times 800/1000 = 108$ V

If I is the new load current, then terminal voltage V is given by

$V = 108 - 0.04\ I - 2 = 106 - 0.04\ I$

$\therefore I = V/R = (106 - 0.4\ I)/0.625$; $I = 159.4$ A

Example 24.11. *A 4–pole, 900 r.p.m d.c machine has a terminal voltage of 220 V and an induced voltage of 240 V at rated speed. The armature circuit resistance is 0.2 Ω. Is the machine operating as a generator or a motor ? Compute the armature current and the number of armature coils if the air-gap flux /pole is 10 mWb and the armature turns per coil are 8. The armature is wave-wound.*

(Elect. Machines AMIE Sec. B 1990)

Solution. Since the induced voltage E is more than the terminal voltage v, the machine is working as a generator.

$E - V = I_a R_a$ or $240 - 220 = I_a \times 0.2$; $I_a = 100$ A

Now, $E_b = ZN\ (P/A)$ or $240 = 10 \times 10^{-3} \times z \times (900/600)(4/2)$; $Z = 8000$

Since there are 8 turns in a coil, it means there are 16 active conductors/coil. Hence, the number of coils = 8000/16 = 500.

Example 24.12. *In a 120 V compound generator, the resistances of the armature, shunt and series windings are 0.06 Ω, 25 Ω and 0.04 Ω respectively. The load current is 100 A at 120 V. Find the induced e.m.f. and the armature current when the machine is connected as (i) long-shunt and as (ii) short-shunt. How will the ampere-turns of the series field be changed in (i) if a diverter of 0.1 ohm be connected in parallel with the series winding ? Neglect brush contact drop and ignore armature reaction.*

(Elect. Machines AMIE Sec. B, 1992)

Solution. (*i*) Long Shunt [Fig. 24.53(*a*)]

$I_{sh} = 120/125 = 4.8$ A; $I = 100$ A; $I_a = 104.8$ A

Voltage drop in series winding = $104.8 \times 0.04 = 4.19$ V

Armature voltage drop = $104.8 \times 0.06 = 6.29$ V

$\therefore E_g = 120 + 4.19 + 6.29 = 130.5$ V

(*ii*) *Short Shunt* [Fig. 24.53(*c*)]

Voltage drop in series winding = $100 \times 0.04 = 4$ V

Voltage across shunt winding = $120 + 4 = 124$ V

$\therefore I_{sh} = 124/25 = 5$ A; $\therefore I_a = 100 + 5 = 105$ A

Armature voltage drop = $105 \times 0.06 = 6.3$ V

$E_g = 120 + 5 + 4 = 129$ V

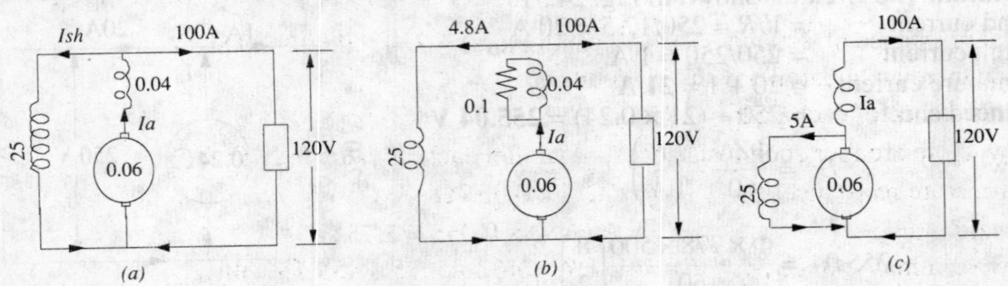

Fig. 24.53

When a diverter of 0.1 Ω is connected in parallel with the series winding, the diagram becomes as shown in Fig. 24.53(*b*). As per current-divider rule, the current through the series winding is = $104.8 \times 0.1/(0.1 + 0.04) = 74.86$ A. It means that the series field current has decreased from an original value of 104.8 A to 74.86 A. Since No. of turns in the series winding remains the same, the change in series field ampere-turns would be the same as the change in the field current. Hence, the percentage decrease in the series field ampere-turns = $(74.86 - 104.8) \times 100/104.8 = -28.6\%$.

Example 24.13. *A 4-pole, long-shunt lap-wound generator supplies 25 kW at a terminal voltage of 500 V. The armature resistance is 0.03 ohm, series field resistance is 0.04 ohm and shunt field resistance is 200 ohm. The brush drop may be taken as 1.0 V. Determine the e.m.f. generated.*

Calculate also the No. of conductors if the speed is 1200 r.p.m. and flux per pole is 0.02 weber. Neglect armature reaction. **(Elect. Engineering-I, Sd. Patel Univ. 1986)**

Solution. $I = 25,000/500 = 50$ A, $I_{sh} = 500/200 = 2.5$ A (Fig. 24.54)

$I_a = I + I_{sh} = 50 + 2.5 = 52.5$ A

series field drop $= 52.5 \times 0.04 = 2.1$ V

armature drop $= 52.5 \times 0.03 = 1.575$ V

brush drop $= 2 \times 1 = 2$ V

Generated e.m.f. $E_g = 500 + 2.1 + 1.575 + 2 = \mathbf{505.67\ V}$

Now, $E_g = \dfrac{\Phi ZN}{60}\left(\dfrac{P}{A}\right)$

or $505.67 = \dfrac{0.02 \times Z \times 1200}{60}\left(\dfrac{4}{4}\right), Z = \mathbf{1264}$

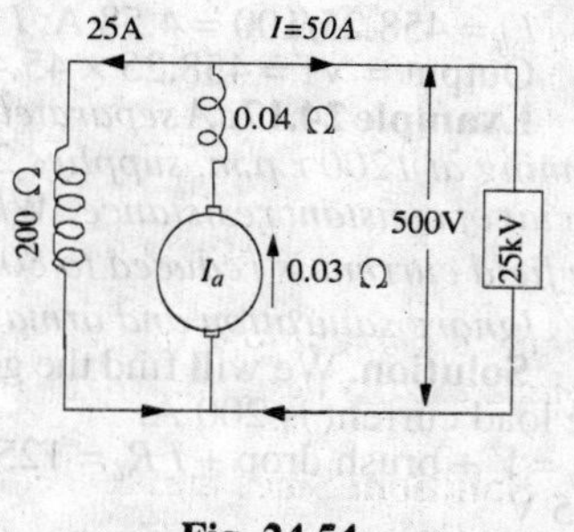

Fig. 24.54

Example 24.14. *A 4-pole d.c. generator runs at 750 r.p.m. and generates an e.m.f. of 240 V. The armature is wave-wound and has 792 conductros. If the total flux from each pole is 0.0145 Wb, what is the leakage coefficient ?*

Solution. Formula used :

$$E = \frac{\Phi ZN}{60}\left(\frac{P}{A}\right) \text{volt} \;\therefore\; 240 = \frac{\Phi \times 750 \times 792}{60} \times \frac{4}{2}$$

$\therefore$ Working flux/pole, $\Phi = 0.0121$ Wb ; Total flux/pole $= 0.0145$ Wb

$$\therefore \text{ Leakage coefficient } \lambda = \frac{\text{total flux/pole}}{\text{working flux/pole}} = \frac{0.0145}{0.0121} = \mathbf{1.2}$$

Example 24.15. *A 4-pole, lap-wound, d.c. shunt generator has a useful flux per pole of 0.07 Wb. The armature winding consists of 220 turns each of 0.004 Ω resistance. Calculate the terminal voltage when running at 900 r.p.m. if the armature current is 50 A.*

Solution. Since each turn has two sides,

$$Z = 220 \times 2 = 440\,; N = 900 \text{ r.p.m.}\,;\ \Phi = 0.07\ Wb;\ P = A = 4$$

$$\therefore \quad E_g = \frac{\Phi ZN}{60}\cdot\left(\frac{P}{A}\right) = \frac{0.07 \times 440 \times 900}{60} \times \left(\frac{4}{4}\right) = 462 \text{ volt}$$

Total resistance of 220 turns (or 440 conductors) $= 220 \times 0.004 = 0.88\ \Omega$

Since there are 4 parallel paths in armature,

$\therefore$ resistance of each path $= 0.88/4 = 0.22\ \Omega$

Now, there are four such resistances in parallel each of value 0.22 Ω.

$\therefore$ armature resistance, $R_a = 0.22/4 = 0.055\ \Omega$

Armature drop $= I_a R_a = 50 \times 0.055 = 2.75\ \Omega$.

Now, terminal voltage $V = E_g - I_a R_a = 462 - 2.75 = \mathbf{459.25\ volt.}$

Example 24.16. *A 4-pole, lap-wound, long-shunt, d.c. compound generator has useful flux per pole of 0.07 Wb. The armature winding consists of 220 turns and the resistance per turn is 0.004 ohms. Calculate the terminal voltage if the resistance of shunt and series field are 100 ohms and 0.02 ohms respectively ; when the generator is running at 900 r.p.m with armature current of 50 A. Also calculate the power output in kW for the generator.*

(Basic Elect. Machine Nagpur Univ. 1993)

Solution. $E_b = \dfrac{0.07\times(220\times2)\times900}{60}\times\left(\dfrac{4}{4}\right)=462$ V

As found in Ex. 24.15, $R_a = 0.055\ \Omega$
Arm. circuit resistance $= R_a + R_{se} = 0.055 + 0.02 = 0.075\ \Omega$
Arm. Circuit drop = $50 \times 0.075 = 3.75$ V;
$V = 462 - 3.75 = 458.25$ V
$I_{sh} = 458.25/100 = 4.58$ A; $I = 50 - 4.58 = 45.42$ A
Output = VI = 458.25×45.42, = 20,814 W = 20.814 kW

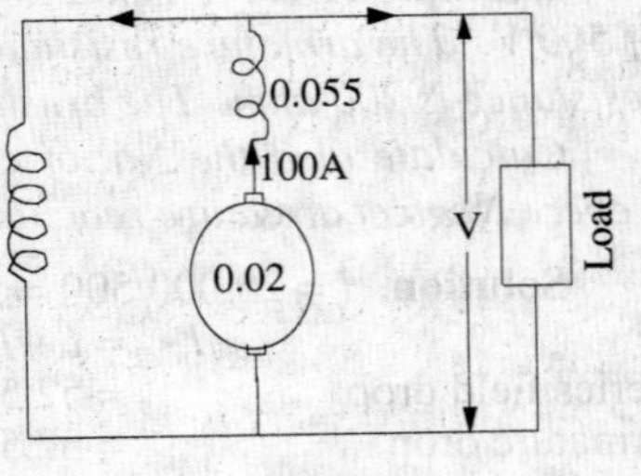

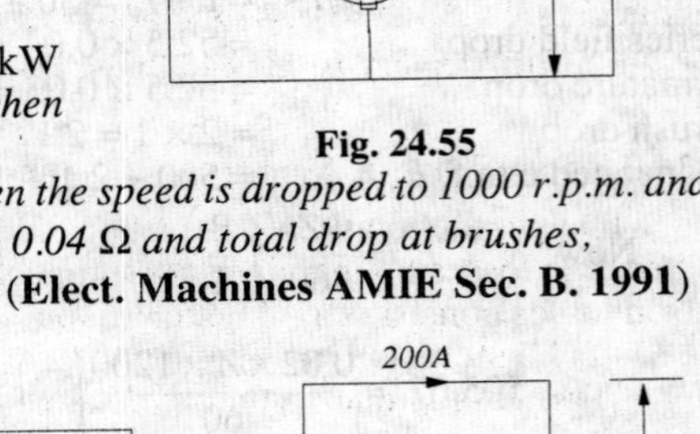

Fig. 24.55

Example 24.17. *A separately excited d.c. generator, when running at 1200 r.p.m. supplies 200 A at 125 V to a circuit of constant resistance. What will be the current when the speed is dropped to 1000 r.p.m. and the field current is reduced to 80% ? Armature resistance, 0.04 Ω and total drop at brushes, 2 V. Ignore saturation and armature reaction.*

(Elect. Machines AMIE Sec. B. 1991)

Solution. We will find the generated emf. when the load current is 200 A.

$E_{g1} = V$ + brush drop + $I_aR_a = 125 + 2 + 200 \times 0.04 = 135$ V

Now, $E_{g1} \times \propto \Phi_1 N_1$ and $E_{g2} \propto \Phi_2 N_2$

$$\therefore \quad \frac{E_{g2}}{E_{g1}} = \frac{\Phi_2 N_2}{\Phi_1 N_1} \quad \text{or} \quad \frac{E_{g2}}{135} = 0.8 \times \frac{1000}{1200} = 90 \text{ V}$$

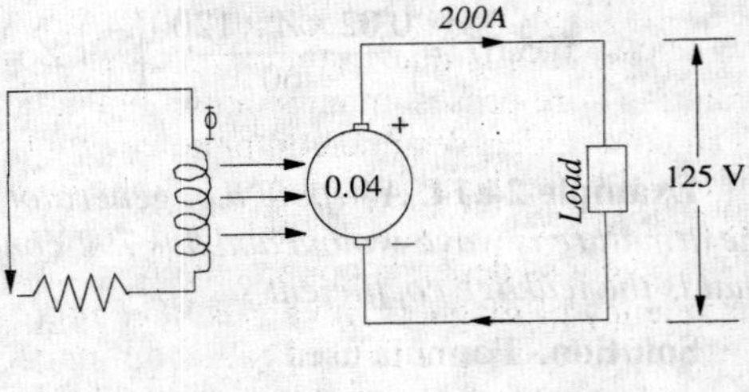

Fig. 24.56

Example **24.18.** *A 4-pole, d.c. shunt generator with a shunt field resistance of 100 Ω and an armature resistance of 1 Ω has 378 wave-connected conductors in its armature. The flux per pole is 0.02 Wb. If a load resistance of 10 Ω is connected across the armature terminals and the generator is driven at 1000 r.p.m., calculate the power absorbed by the load.*

(Elect. Technology, Hyderabad Univ. 1991)

Solution. Induced e.m.f. In the generator is

$$E_g = \frac{\Phi ZN}{60}\left(\frac{P}{A}\right) \text{ volt}$$

$$= \frac{0.02\times378\times1000}{60}\left(\frac{4}{2}\right) = 252 \text{ volt}$$

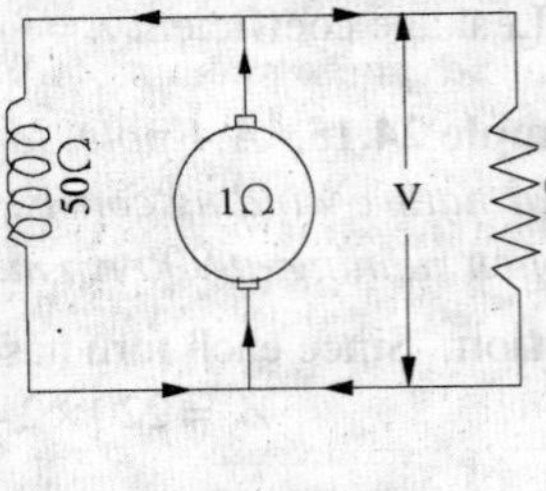

Fig. 24.57

Now, let V be the terminal voltage *i.e.* the voltage available across the load as well as the shunt resistance (Fig. 24.57)

Load current = $V/10$ A and Shunt current = $V/100$ A

Armature current $= \dfrac{V}{10} + \dfrac{V}{100} = \dfrac{11V}{100}$ Now, $V = E_g$ – armature drop

$$\therefore \qquad V = 252 - 1 \times \frac{11V}{100} \quad \therefore \quad V = 227 \text{ Volt}$$

Load current = 227/10 = 22.7 A Power absorbed by the load is = 227×22.7 = **5,153W**

Example 24.19. *A 6-pole dc generator runs at 1200 r.p.m on no-load and has a generated e.m.f of 250 V. Its armature diameter is 350 mm and the radial air-gap between the field poles and the armature is 3 mm. The axial length of the field poles is 260 mm and the field pole effective coverage is 80% including fringing. If the armature has 96 coils having 3 turns per coil and is wound duplex lap, calculate (a) flux per pole (b) effective pole arc length and (c) average air-gap flux density.*

Solution. (*a*) Z = (96 × 3) × 2 = 576, *P* = 6, *A* = P × plex = 6 × 2 = 12, N = 1200 r.p.m.

$$\therefore\ 250 = \frac{\Phi\times576\times1200}{6}\left(\frac{6}{12}\right); \quad \therefore\ \Phi = 0.0434 \text{ Wb}$$

(*b*) Inner diameter of the pole shoe circle is = 350 + 6 = 356 mm.

Since there are 6 poles, the net field pole flux coverage is 80% of one-sixth of the pole shoe circle. Hence, the effective pole arc length is

$$= \frac{1}{6} \times \pi d \times 0.8 = \frac{1}{6} \times \pi \times 356 \times 0.8 = 149 \text{ mm} = 0.149 \text{ m.}$$

(*c*) Pole surface area = pole shoe arc × axial length of the pole (Fig. 24.58). $= 0.149 \times 0.260 = 0.03874 \text{ mm}^2$

$\therefore$ flux density $B = 0.0434/0.03874 = 1.12$ T

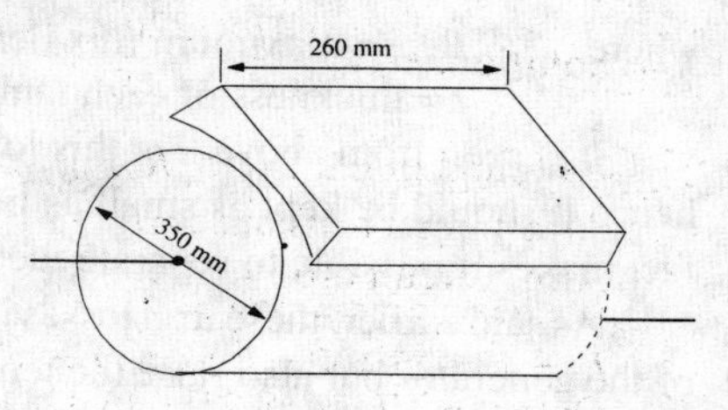

Fig. 24.58

24.34. Iron Loss in Armature

Due to the rotation of the iron core of the armature in the magnetic flux of the field poles, there are some losses taking place continuously in the core and are known as Iron Losses or Core Losses. Iron losses consist of (*i*) **Hysteresis** loss and (*ii*) **Eddy Current** loss.

(*i*) **Hysteresis Loss** (W_h)

This loss is due to the reversal of magnetisation of the armature core. Every portion of the rotating core passes under *N* and *S* pole alternately, thereby attaining *S* and *N* polarity respectively. The core undergoes one complete cycle of magnetic reversal after passing under one *pair* of poles. If *P* is the number of poles and *N*, the armature speed in r.p.m., then frequency of magnetic reversals is $f = PN/120$.

The loss depends upon the volume and grade of iron, maximum value of flux density B_{max} and frequency of magnetic reversals. For normal flux densities (*i.e.* upto 1.5 Wb/m²), hysteresis loss is given by **Steinmetz** formula. According to this formula,

$$W_h = \eta B_{max}^{1.6} f V \text{ watt}$$

where V = volume of the core in m³

η = Steinmetz hysteresis coefficient.

Value of η for :

Good dynamo sheet steel = 502 J/m³, Silicon steel = 191 J/ m³, Hard cast steel = 7040 J/m³, Cast steel = 750–3000 J/m³ and Cast iron = 2700–4000 J/m³.

(ii) Eddy Current Loss (W_e)

When the armature core rotates, it also cuts the magnetic flux. Hence, an e.m.f. is induced in the body of the core according to the laws of electromagnetic induction. This e.m.f. though small, sets up large current in the body of the core due to its small resistance. This current is known as eddy current. The power loss due to the flow of this current is known as eddy current loss. This loss would be considerable if solid iron core were used. In order to reduce this loss and the consequent heating of the core to a small value, the core is built up of thin laminations, which are stacked and then riveted at right angles to the path of the eddy currents. These core laminations are insulated from each other by a thin coating of varnish. The effect of laminations is shown in Fig. 24.59. Due to the core body being one continuous solid iron piece (Fig. 24.59(*a*)], the magnitude of eddy currents is large. As armature cross-sectional area is large, its resistance is very small, hence eddy current loss is large. In Fig. 24.59(*b*), the same core has been split up into thin circular discs insulated from each other. It is seen that now each current path, being of much less cross-section, has a very high resistance. Hence, magnitude of eddy currents is reduced considerably thereby drastically reducing eddy current loss.

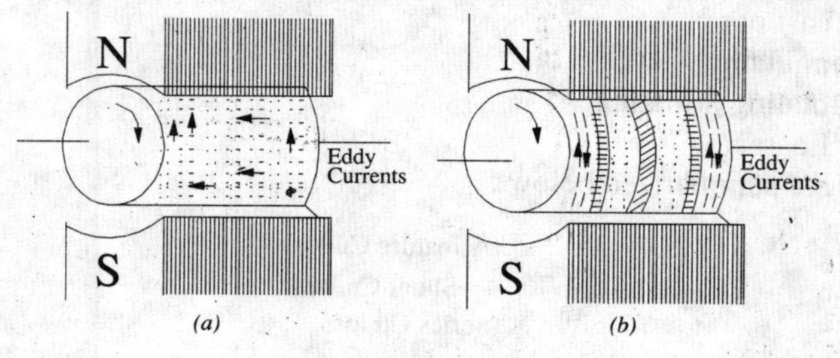

Fig. 24.59

It is found that eddy current loss W_e is given by the following relation :

$$We = KB^2_{max} f^2 t^2 V^2 \text{ watt}$$

where B_{max} = maximum flux density f = frequency of magnetic reversals
t = thickness of each lamination V = volume of armature core.

It is seen from above that this loss varies directly as the square of the thickness of laminations, hence it should be kept as small as possible. Another point to note is that $W_h \propto f$ but $W_e \propto f^2$. This fact makes it possible to separate the two losses experimentally if so desired.

As said earlier, these iron losses if allowed to take place unchecked not only reduce the efficiency of the generator but also raise the temperature of the core. As the output of the machines is limited, in most cases, by the temperature rise, these losses have to be kept as small as is economically possible.

Eddy current loss is reduced by using laminated core but hysteresis loss cannot be reduced this way. For reducing the hysteresis loss, those metals are chosen for the armature core which have a low hysteresis coefficient. Generally, special silicon steels such as stalloys are used which not only have a low hysteresis coefficient but which also possess high electrical resistivity.

24.35 Total Loss in a D.C. Generator

The various losses occurring in a generator can be sub-divided as follows :

(a) Copper Losses

(*i*) Armature copper loss = $I_a^2 R_a$ (not $E_g I_a$)

where R_a = resistance of armature and interpoles and series field winding etc.

This loss is about 30 to 40% of full-load losses.

(*ii*) Field copper loss. In the case of shunt generators, it is practically constant and = $I_{sh}^2 R_{sh}$ (or VI_{sh}). In the case of series generator, it is = $I_{Se}^2 R_{Se}$ where R_{Se} is resistance of the series field winding.

This loss is about 20 to 30% of F.L. losses.

(*iii*) The loss due to brush contact resistance. It is usually included in the armature copper loss.

(b) Magnetic Losses (also known as iron or core losses).

(*i*) hysteresis loss, $W_h \propto B_{max}^{1-6} f$ and (*ii*) eddy current loss, $W_e \propto B_{max}^2 f^2$

These losses are practically constant for shunt and compound-wound generators, because in their case, field current is approximately constant.

Both these losses total up to about 20 to 30% of F.L. losses

(c) Mechanical Losses. These consist of :

(*i*) friction loss at bearings and commutator.

(*ii*) air-friction or windage loss of rotating armature.

These are about 10 to 20% of F.L. Losses.

The total losses in a d.c. generator are summarized below :

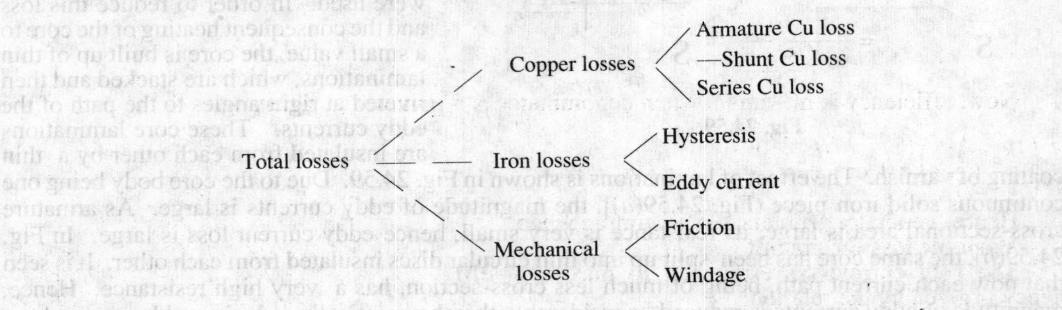

24.36. Stray Losses

Usually, magnetic and mechanical losses are collectively known as **Stray Losses.** These are also known as rotational losses for obvious reasons.

24.37. Constant or Standing Losses

As said above, field Cu loss is constant for shunt and compound generators. Hence, stray losses and shunt Cu loss are constant in their case. These losses are together known as **standing or constant** losses W_c.

Hence, for shunt and compound generators,

Total loss = armature copper loss + $W_c = I_a^2R_a + W_c = (I + I_{sh})^2R_a + W_c$.

Armature Cu loss $I_a^2R_a$ is known as variable loss because it varies with the load current.

Total loss = variable loss + constant losses W_c

24.38. Power Stages

Various power stages in the case of a d.c. generator are shown below :

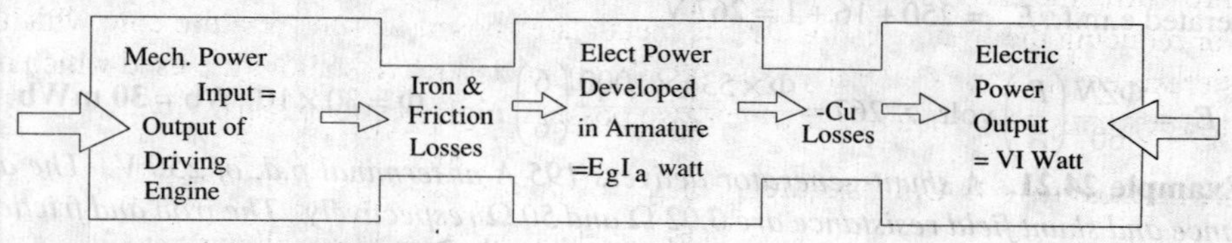

Following are the three generator efficiencies :

1. Mechanical Efficiency

$$\eta_m = \frac{B}{A} = \frac{\text{total watts generated in armature}}{\text{mechanical power supplied}} = \frac{E_g I_a}{\text{output of driving engine}}$$

2. Electrical Efficiency

$$\eta_e = \frac{C}{B} = \frac{\text{watts avaialble in load circuit}}{\text{total watts generated}} = \frac{VI}{E_g I_a}$$

3. Overall or Commercial Efficiency

$$\eta_c = \frac{C}{A} = \frac{\text{watts available in load circuit}}{\text{mechanical power supplied}}$$

It is obvious that overall efficiency $\eta_c = \eta_m \times \eta_e$. For good generators, its value may be as high as 95%.

Note. Unless specified otherwise, commercial efficiency is always to be understood.

24.39. Condition for Maximum Efficiency

Generator output $= VI$

Generator input = output + losses

$$= VI + I_a^2R_a + W_c = VI + (I + I_{sh})^2R_a + W_c \qquad (\because \; I_a = I + I_{sh})$$

However, if I_{sh} is negligible as compared to load current, then $I_a = I$ (approx.)

$$\therefore \qquad \eta = \frac{\text{output}}{\text{input}} = \frac{VI}{VI + I_a^2R_a + W_c} = \frac{VI}{VI + I^2R_a + W_c} \qquad (\because \; I_a = I)$$

$$= \frac{1}{1 + \left(\frac{IR_a}{V} + \frac{W_c}{VI}\right)}$$

Now, efficiency is maximum when denominator is minimum *i.e.* when

$$\frac{d}{dI}\left(\frac{IR_a}{V} + \frac{W_c}{VI}\right) = 0 \text{ or } \frac{R_a}{V} - \frac{W_c}{VI^2} = \text{ or } I^2R_a = W_c$$

Hence, generator efficiency is maximum when

Variable loss = constant loss

The load current corresponding to maximum efficiency is given by the relation.

$$I^2R_a = W_C \text{ or } I = \sqrt{\frac{W_c}{R_a}}$$

Variation of η with load current is shown in Fig. 24.60.

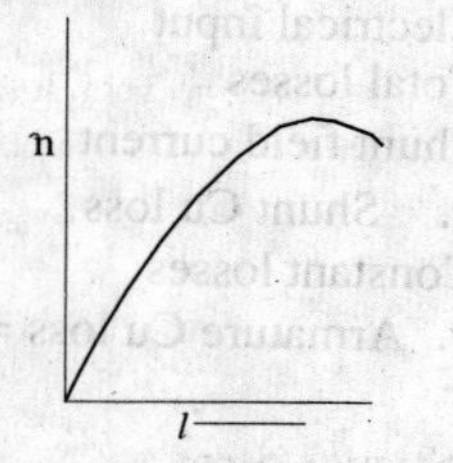

Fig. 24.60

Example 24.20. *A 10-kW, 250-V, d.c., 6-pole shunt generator runs at 1000 r.p.m. when delivering full-load. The armature has 534 lap-connected conductors. Full-load Cu loss is 0.64 kW. The total brush drop is 1 volt. Determine the flux per pole. Neglect shunt current.*

(Elect. Engg. & Electronics, M.S. Univ./ Baroda 1987)

Solution. Since shunt current is negligible, there is no shunt Cu loss. The copper loss occurs in armature only.

$I = I_a = 10,000/250 = 40$ A ; $I_a^2 R_a =$ Arm. Cu loss or $40^2 \times R_a = 0.64 \times 10^3; R_a = 0.4\Omega$

$I_a R_a$ drop $= 0.4 \times 40 = 16$ V ; Brush drop $= 2 \times 1 = 2$ V

$\therefore$ generated e.m.f. $E_g = 250 + 16 + 1 = 267$ V

Now, $E_g = \frac{\Phi ZN}{60}\left(\frac{P}{A}\right)$ volt $\therefore 267 = \frac{\Phi \times 534 \times 1000}{60}\left(\frac{6}{6}\right)$ $\therefore$ $\Phi = 30 \times 10^{-3}$Wb $=$ **30 mWb**

Example 24.21. *A shunt generator delivers 195 A at terminal p.d. of 250 V. The armature resistance and shunt field resistance are 0.02 Ω and 50 Ω respectively. The iron and friction losses equal 950 W. Find*

(a) E.M.F. generated (b) Cu losses (c) output of the prime motor

(d) commercial, mechanical and electrical efficiencies.

(Elect. Machines-I, Nagpur Univ. 1991)

Solution. (*a*) $I_{sh} = 250/50 = 5$A ; $I_a = 195 + 5 = 200$ A

Armature voltage drop $= I_a R_a = 200 \times 0.02 = 4$ V

$\therefore$ generated e.m.f. $= 250 + 4 =$ **254 V**

(*b*) Armature Cu loss $= I_a^2 R_a = 200^2 \times 0.02 = 800$ W

Shunt Cu loss $= V.I_{sh} = 250 \times 5 = 1250$ W

$\therefore$ Total Cu loss $= 1250 + 800 =$ **2050 W**

(*c*) Stray losses $= 950$ W ; Total losses $= 2050 + 950 = 3000$ W

Output $= 250 \times 195 = 48,750$ W; Input - $48,750 + 3,000 = 51750$ W

$\therefore$ Output of prime mover $=$ **51.750 W**

(*d*) generator input $= 51,750$ W ; Stray losses $= 950$ W

Electrical power produced in armature $= 51,750 - 950 = 50,800$

$\eta_m = (51,800/51,750) \times 100 =$ **98.2%**

Electrical or Cu losses $= 2050$ **W**

$\therefore$ $\eta_e = \frac{48,750}{48,750 + 2,050} \times 100 =$ **95.9%**; $\eta_c = (48,750/51,750) \times 100 =$ **94.2%**

Example 24.22. *A shunt generator has a F.L. current of 196 A at 220 V. The stray losses are 720 W and the shunt field coil resistance is 55 Ω. If it has a F.L. efficiency of 88 %, find the armature resistance. Also, find the load current corresponding to maximum efficiency.*

(Electrical Technology Punjab Univ. Nov. 1988)

Solution. Output $= 220 \times 196 = 43,120$ W ; $\eta = 88\%$ (overall efficiency)

Electrical input $= 43,120/0.88 = 49,000$ W

Total losses $= 49,000 - 43,120 = 5,880$ W

Shunt field current $= 220/55 = 4$A $\therefore$ $I_a = 196 + 4 = 200$ A

$\therefore$ Shunt Cu loss $= 220 \times 4 = 880$ W; Stray losses $= 720$ W

Constant losses $= 880 + 720 = 1,600$

$\therefore$ Armature Cu loss $= 5,880 - 1,600 = 4,280$ W

$\therefore$ $I_a^2 R_a = 4,280$ W

$200^2 R_a = 4,280$ or $R_a = 4,280/200 \times 200 =$ **0.107 Ω**

For maximum efficiency,

$I^2 R_a =$ constant losses $= 1,600$ W ; $I = \sqrt{1,600/0.107} =$ **122.34 A**

Example 24.23. *A long-shunt dynamo running at 1000 r.p.m. supplies 22 kW at a terminal voltage of 220 V. The resistances of armature, shunt field and the series field are 0.05, 110 and 0.06 Ω respectively. The overall efficiency at the above load is 88%. Find (a) Cu losses (b) iron and friction losses (c) the torque exerted by the prime mover.*

(Elect. Machinery-I, Bangalore Univ. 1987)

Solution. The generator is shown in Fig. 24.61

$I_{sh} = 220/110 = 2$ A

$I = 22{,}000/220 = 100$ A, $I_a = 102$ A

Drop in series field winding

$= 102 \times 0.06 = 6.12$ V

(*a*) $I_a^2 R_a = 102^2 \times 0.05 = 520.2$ W

Series field loss $= 102^2 \times 0.06 = 624.3$ W

Shunt field loss $= 4 \times 110 = 440$ W

Total Cu losses $= 520.2 + 624.3 + 440 =$ **1584.5 W**

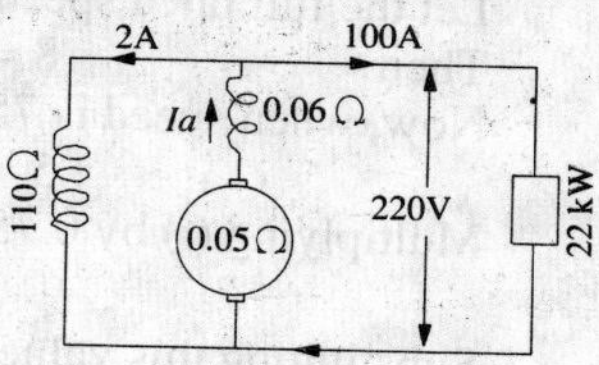

(*b*) Output $= 22.000$ W : Input $= 22{,}000/0.88 = 25{,}000$ W

$\therefore$ Total losses $= 25{,}000 - 22{,}000 = 3{,}000$ W

$\therefore$ Iron and friction losses $= 3.000 - 1.584.5 =$ **1,415.5 W**

Fig. 24.61

$$\text{Now } T \times \frac{2\pi N}{60} = 25{,}000; \quad T = \frac{25{,}000 \times 60}{1{,}000 \times 6.284} = \textbf{238.74 N-m}$$

Example 24.24. *A 4-pole d.c. generator is delivering 20 A to a load of 10 Ω. If the armature resistance is 0.5 Ω and the shunt field resistance is 50 Ω, calculate the induced e.m.f. and the efficiency of the machine. Allow a drop of 1V per brush.*

(Electrical Technology-I, Osmania Univ., 1990)

Solution. Terminal voltage $= 20 \times 10 = 200$ V

$I_{sh} = 200/50 = 4$ A; $I_a = 20 + 4 = 24$ A

$I_a R_a = 24 \times 0.5 = 12$ V; Brush drop $= 2 \times 1 = 2$V

$\therefore E_g = 200 + 12 + 2 =$ **214 V**

Since iron and friction losses are not given, only electrical efficiency of the machine can be found out

Total power generated in the armature

$= 214 \times 24 = 5{,}136$ W

Useful output $= 200 \times 20 = 4{,}000$ W

$\therefore \eta_e = 4{,}000/5{,}136 = 0.779$ or **77.9%**

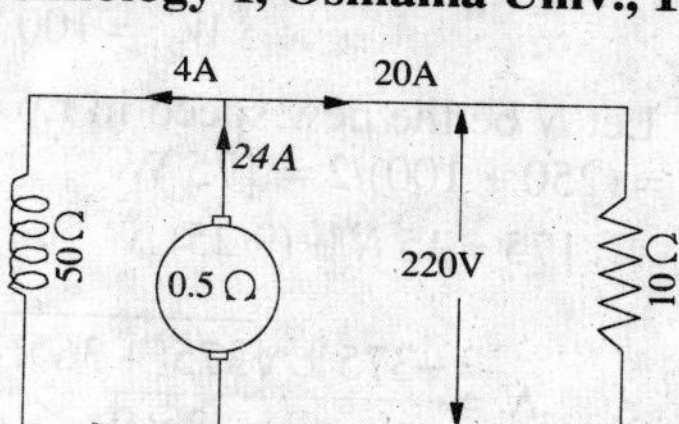

Fig. 24.62

Example 24.25. *A long-shunt compound-wound generator gives 240 volts at F.L. output of 100 A. The resistances of various windings of the machine are : armature (including brush contact) 0.1 Ω, series field 0.02 Ω, interpole field 0.025 Ω, shunt field (including regulating resistance) 100 Ω. The iron loss at F.L. is 1000 W; windage and friction losses total 500 W. Calculate F.L. efficiency of the machine.*

(Electrical Machinery-I, Indore Univ. 1989)

Solution. Output $= 240 \times 100 = 24{,}000$ W

Total armature circuit resistance $= 0.1 + 0.02 + 0.025 = 0.145$ Ω

$I_{sh} = 240/100 = 2.4$ A $\quad \therefore I_a = 100 + 2.4 = 102.4$ A

$\therefore$ Armature circuit copper loss $= 102.4^2 \times 0.145 = 1{,}521$ W

Shunt field copper loss $= 2.4 \times 240 = 576$ W

Iron loss $= 1000$ W ; Frinction loss $= 500$ W

$$\text{Total loss} = 1{,}521 + 1{,}500 + 576 = 3{,}597 \text{ W}; \ \eta = \frac{24{,}000}{24{,}000 + 3{,}597} = 0.87 = \textbf{87\%}$$

Example 24.26. *In a.d.c. machine the total iron loss is 8 kW at its rated speed and excitation. If excitation remains the same, but speed is reduced by 25%, the total iron loss is found to be 5 kW. Calculate the hysteresis and eddy current losses at (i) full speed (ii) half the rated speed.*

Solution. We have seen in Art. 24.32 that

$$W_h \propto f \text{ and } W_e \propto f^2$$

Since *f*, the frequency of reversal of magnetization, is directly proportional to the armature speed,

$$W_h \propto A \times N \quad \text{and} \quad W_e \propto N^2$$

$\therefore$ $W_h =$ and $W_e = BN^2$ where A and B are constants

Total loss $W = W_h + W_e = AN + BN^2$

Let the full rated speed be 1.

Then $8 = A \times 1 + B \times 1^2$ or $8 = A + B$ ...(*i*)

Now, when speed is 75% of full rated speed, then

$5 = A \times (0.75) + B(0.75)^2$ ...(*ii*)

Multiplying (*i*) by 0.75 and subtracting (*ii*) from it, we get

$0.1875\ B = 1 \therefore\ B = 1/0.1875 =$ **5.33 kW**

Substituting this value in (*i*) above

$8 = 5.33 + A \qquad \therefore A =$ **2.67 kW**

(*i*) W_h at rated speed = **2.67 kW** $\qquad W_e$ at rated speed = **5.33 kW**

(*ii*) W_h at half the rated speed $= 2.67 \times 0.5 =$ **1.335 kW**

W_e at half the rated speed $= 5.33 \times 0.5^2 =$ **1.3325 kW**

Example 24.27. *The hysteresis and eddy current losses in a d.c. machine running at 1000 r.p.m. are 250 W and 100 W respectively. If the flux remains constant, at what speed will be total iron losses be halved ?* **(Electrical Machines-I, Gujarat Univ. 1989)**

Solution. Total loss $W = W_h + W_e = AN + BN^2$

Now, $W_h = 250\ W \quad \therefore \quad A \times (1000/60) = 250\ ; \quad A = 15$

$W_e = 100\ W \quad \therefore \quad B \times (1000/60)^2 = 100; \quad B = 9/25$

Let N be the new speed in r.p.s. at which total loss is one half of the loss at 1000 r.p.m. New loss = (250 + 100)/2 = 175 W

$\therefore\ 175 = 15\,N + (9/25)\,N^2$ or $9N^2 + 375\,N - 4,375 = 0$

$$\therefore \quad N = \frac{-375 \pm \sqrt{375^2 + 36 \times 4,375}}{2 \times 9} = \frac{-375 \pm 546}{18} = 9.5 \text{ r.p.s} = \textbf{570 r.p.m.}^*$$

Note. It may be noted that at the new speed, $W_h = 250 \times (570/100) = 142.5$ W and $W_e = 100 \times (570/1000)^2 = 32.5$ W. Total loss = 142.5 + 32.5 = 175 W.

Tutorial Problem No. 24.2

1. A 4-pole, d.c. generator has a wave-wound armature with 792 conductors. The flux per pole is 0.0121 Wb. Determine the speed at which it should be run to generate 240 V on no-load. **[751.3 r.p.m.]**

2. A 20-kW compound generator works on full-load with a terminal voltage of 230 V. The armature, series and shunt field resistances are 0.1, 0.05 and 115 Ω respectively. Calculate the generated e.m.f. when the generator is connected short-shunt. **[243.25 V]**

(Elect. Engg. Madras Univ. April. 1978)

3. A d.c. generator generates an e.m.f. of 520 V. It has 2,000 armature conductors, flux per pole of 0.013 Wb, speed of 1200 r.p.m. and the armature winding has four parallel paths. Find the number of poles. **[4]** *(Elect. Technology, Aligarh Univ. 1978)*

4. When driven at 1000 r.p.m. with a flux per pole of 0.02 Wb, a.d.c. generator has an e.m.f. of 200 V. If the speed is increased to 1100 r.p.m. and at the same time the flux per pole is reduced to 0.019 Wb per pole, what is then the induced e.m.f. ? **[209 V]**

5. Calculate the flux per pole required on full-load for a 50-kW, 400, V, 8 pole, 600 r.p.m. d.c. shunt generator with 256 conductors arranged in a lap-connected winding. The armature winding resistances is 0.1 Ω, the shunt field resistance is 200 Ω and there is a brush contact voltage drop of 1 V at each brush on full-load. **[0.162 Wb]**

---

*The negative value has been rejected-being mathematically absurd.

6. Calculate the flux in a 4-pole dynamo with 722 armature conductors generating 500 V when running at 1000 r.p.m. when the armature is (*a*) lap connected (*b*) wave connected.

[(*a*) **41.56 mWb** (*b*) **20.78 mWb**] (*City & Guilds, London*)

7. A 4-pole machine running at 1500 r.p.m. has an armature with 90 slots and 6 conductors per slot. The flux per pole is 10 mWb. Determine the terminal e.m.f. as d.c. Generator if the coils are lap-connected. If the current per conductor is 100 A, determine the electrical power. [**810 V, 324 kW**] (*London Univ.*)

8. An 8-pole lap-wound d.c. generator has 120 slots having 4 conductors per slot. If each conductor can carry 250 A and if flux /pole is 0.05 Wb, calculate the speed of the generator for giving 240 V on open circuit. If the voltage drops to 220 V on full load, find the rated output of the machine.

[**600 V, 440 kW**]

9. A 110-V shunt generator has a full-load current of 100 A, shunt field resistance of 55 Ω and constant losses of 500 W. If F.L. efficiency is 88%, find armature resistance. Assuming voltage to be constant at 110 V, calculate the efficiency at half F.L. And at 50% overload. Find the load current corresponding to maximum efficiency. [**0.078 Ω ; 85-8% ; 96.2 A**]

10. A short-shunt compound d.c. Generator supplies a current of 100 A at a voltage of 220 V. If the resistance of the shunt field is 50 Ω, of the series field 0.025 Ω, of the armature 0.05 Ω, the total brush drop is 2 V and the iron and friction losses amount to 1 kW, find

(*a*) the generated e.m.f. (*b*) the copper losses (*c*) the output power of the prime-mover driving the generator and (*d*) the generator efficiency.

[(*a*) **229.7 V** (*b*) **1.995 kW** (*c*) **24.99 kW** (*d*) **88%**]

11. A 20 kW, 440 V, short-shunt, compound d.c. Generator has a full-load efficiency of 87%. If the resistance of the armature and interpoles is 0.4 Ω and that of the series and shunt fields 0.25 Ω and 240 Ω respectively, calculate the combined bearing friction, windage and core-loss of the machine.

[**725 W**]

12. A long-shunt, compound generator delivers a load current of 50 A at 500 V and the resistances of armature, series field and shunt field are 0.05 ohm and 250 ohm respectively. Calculate the generated electromotive force and the armature current. Allow 1.0 V per brush for contact drop.

[**506.2 V ; 52 A**] (*Elect. Engg. Banaras Hindu Univ. 1977*)

13. In a 110-V compound generator, the resistances of the armature, shunt and the series windings are 0.06 Ω, 25 Ω and 0.04 Ω respectively. The load consists of 200 lamps each rated at 55 W, 110 V.

Find the total electromotive force and armature current when the machine is connected (*i*) long shunt (*ii*) short shunt. Ignore armature reaction and brush drop.

[(*a*) **1200-4, 104.4 A** (*b*) **120.3 V, 104.6 A**] (*Electrical Machines-I, Bombay Univ. 1979*)

Objective Tests 24

1. The basic requirement of a d.c. armature winding is that it must be
(*a*) a closed one
(*b*) a lap winding
(*c*) a wave winding
(*d*) either (*b*) or (*c*)

2. A wave winding must go at least —— around the armature before it closes back where it started.
(*a*) once
(*b*) twice
(*c*) thrice
(*d*) four times

3. The d.c. armature winding in which coil sides are a pole pitch apart is called —— winding.
(*a*) multiplex
(*b*) fractional-pitch
(*c*) full-pitch
(*d*) pole-pitch

4. For making coil span equal to a pole pitch in the armature winding of a d.c. generator, the back pitch of the winding must equal the number of
(*a*) commutator bars per pole
(*b*) winding elements
(*c*) armature conductors per path
(*d*) armature parallel paths.

5. The primary reason for making the coil span of a d.c. armature winding equal to a pole pitch is to
(*a*) obtain a coil span of 180° (electrical)
(*b*) ensure the addition of e.m.fs. of consecutive turns
(*c*) distribute the winding uniformly under different poles
(*d*) obtain a full-pitch winding

6. In a 4-pole, 35-slot d.c. armature, 180 electrical-degree coil span will be obtained when coils occupy —— slots.
(*a*) 1 and 10
(*b*) 1 and 9
(*c*) 2 and 11
(*d*) 3 and 12

7. The armature of a dc generator has a 2-layer lap-winding housed in 72 slots with six conductors/slot. What is the minimum number of commutator bars required for the armature ?
(*a*) 72
(*b*) 432
(*c*) 216
(*d*) 36

8. The sole purpose of a commutator in a.d.c. Generator is to
(*a*) increase output voltage
(*b*) reduce sparking at brushes
(*c*) provide smoother output
(*d*) convert the induced a.c. into d.c.

9. For a 4-pole, 2-layer, d.c., lap-winding with 20 slots and one conductor per layer, the number of commutator bars is
(*a*) 80
(*b*) 20
(*c*) 40
(*d*) 160

10. A 4-pole, 12-slot lap-wound dc armature has two coil-sides/slot. Assuming single turn coils and progressive winding, the back pitch would be
(*a*) 5
(*b*) 7
(*c*) 3
(*d*) 6

11. If in the case of a certain d.c. armature, the number of commutator segments is found either one less or more than the number of slots, the armature must be having a simplex —— winding.
(*a*) wave
(*b*) lap
(*c*) frog leg
(*d*) multielement

12. Lap winding is suitable for —— current, —— voltage d.c. generators.
(*a*) high, low
(*b*) low, high
(*c*) low, low
(*d*) high, high

13. The series field of a short-shunt d.c. generator is excited by —— currents.
(*a*) shunt
(*b*) armature
(*c*) load
(*d*) external

14. In a d.c. generator, the generated e.m.f. is directly proportional to the
(*a*) field current
(*b*) pole flux
(*c*) number of armature parallel paths
(*d*) number of dummy coils

15. In a 12-pole triplex lap-wound d.c. armature, each conductor can carry a current of 100 A. The rated current of this armature is —— ampere
(*a*) 600
(*b*) 1200
(*c*) 2400
(*d*) 3600

16. The commercial efficiency of a shunt generator is maximum when its variable loss equals —— loss.
(*a*) constant
(*b*) stray
(*c*) iron
(*d*) friction and windage

17. In small d.c. machines, armature slots are sometimes not made axial but are skewed. Though skewing makes winding a little more difficult, yet it results in
(*a*) quieter operation
(*b*) slight decrease in losses
(*c*) saving of copper
(*d*) both (*a*) and (*b*)

18. The critical resistance of the d.c. generator is the resistance of
(*a*) armature
(*b*) field
(*c*) load
(*d*) brushes

(Grad. I.E.T.E Dec. 1985)

Answers

1. *a* **2.** *b* **3.** *c* **4.** *a* **5.** *b* **6.** *b* **7.** *c* **8.** *d* **9.** *b* **10.** *b*
11. *a* **12.** *a* **13.** *c* **14.** *b* **15** *d* **16.** *a* **17.** *d* **18.** *b*

25 ARMATURE REACTION AND COMMUTATION

25.1. Armature Reaction

By armature reaction is meant the effect of magnetic field set up by armature current on the distribution of flux under main poles of a generator. The armature magnetic field has two effects :

(i) *it demagnetises* or *weakens the main flux* and
(ii) *it cross-magnetises* or *distorts it.*

The first effect leads to reduced generated voltage and the second to the sparking at the brushes.

These effects are well illustrated in Fig. 25.1 which shows the flux distribution of a bipolar generator when there is no current in the armature conductors. For convenience, only two poles have been considered, though the following remarks apply to multipolar fields as well. Moreover, the brushes are shown touching the armature conductors directly, although in practice, they touch commutator segments, It is seen that

(a) the flux is distributed symmetrically with respect to the polar axis, which is the line joining the centres of *NS* poles.

(b) The magnetic neutral axis or plane (*M.N.A.*) coincides with the geometrical neutral axis or plane (*G.N.A.*)

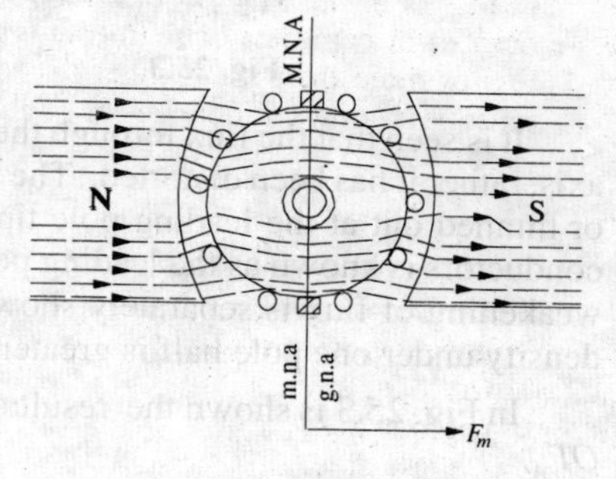

Fig. 25.1

Magnetic neutral axis may be defined as the axis along which no e.m.f. is produced in the armature conductors because they then move parallel to the lines of flux.

Or *M.N.A.* is the axis which is perpendicular to the flux passing through the armature.

As hinted in Art. 25.2, brushes are always placed along *M.N.A.* Hence, *M.N.A.* is also called 'axis of commutation' because reversal of current in armature conductors takes place across this axis. In Fig. 25.1 is shown vector OF_m which represents, both in magnitude and direction, the m.m.f. producing the main flux and also *M.N.A.* which is perpendicular to OF_m.

In Fig. 25.2 is shown the field (or flux) set up by the armature conductors *alone* when carrying current, the field coils being unexcited. The direction of the armature current is the same as it would actually be when the generator is loaded. It may even be found by applying Fleming's Right-hand Rule. The current direction is downwards in conductors under *N*-pole and upwards in those under *S*-pole. The downward flow is represented by crosses and upward flow by dots.

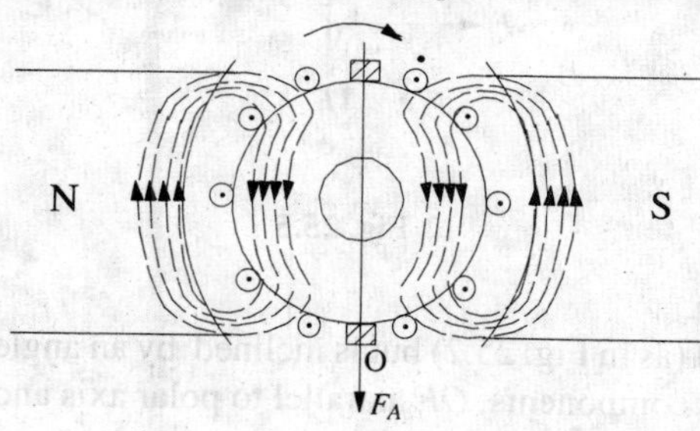

Fig. 25.2

As shown in Fig. 25.2, the m.m.fs. of the armature conductors combine to send flux downwards through the armature. The direction of the lines of force can be found by applying cork-screw rule. The armature m.m.f. (depending on the strength of the armature current) is shown separately both in magnitude and direction by the vector OF_A which is parallel to the brush axis.

So far, we considered the main m.m.f. and armature m.m.f. separately as if they existed independently, which is not the case in practice. Under actual load conditions, the two exist simultaneously in the generator as shown in Fig. 25.3.

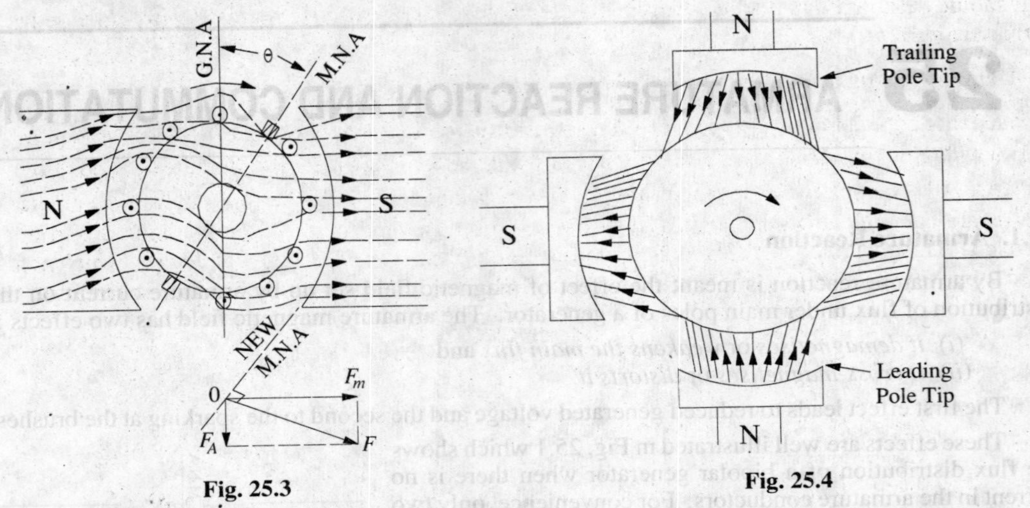

Fig. 25.3

Fig. 25.4

It is seen that the flux through the armature is no longer uniform and symmetrical about the pole axis, rather it has been distorted. The flux is seen to be crowded at the trailing pole tips but weakened or thinned out at the leading pole tips (the pole tip which is first met during rotation by armature conductors is known as the leading pole tip and the other as trailing pole tip). The strengthening and weakening of flux is separately shown for a four-pole machine in Fig. 25.4. As seen, air-gap flux density under one pole half is greater than that under the other pole half.

In Fig. 25.3 is shown the resultant m.m.f. OF which is found by vectorially combining OF_m and OF_A.

The new position of *M.N.A.*, which is always perpendicular to the resultant m.m.f. vector OF, is also shown in the figure. With the shift of *M.N.A.*, say through an angle θ, brushes are also shifted so as to lie along the new position of *M.N.A.* Due to this brush shift (or forward lead), the armature conductors and hence armature current is redistributed. Some armature conductors which were earlier under the influence of *N*-pole come under the influence of *S*-pole and *vice-versa*. This regrouping is shown in Fig. 25.5, which also shows the flux due to armature conductors. Incidentally, brush position shifts in the *same* direction as the direction of armature rotation.

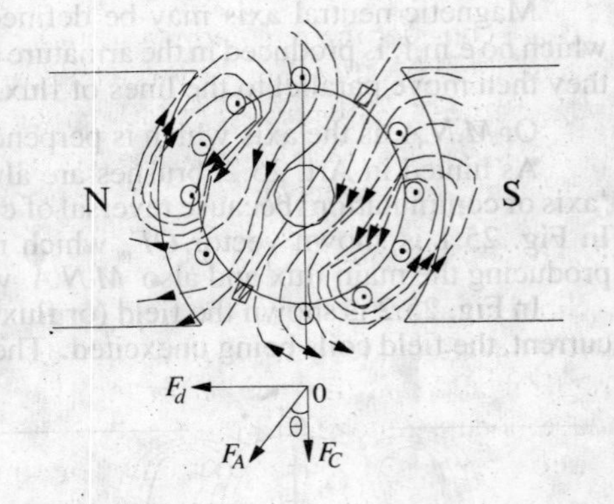

Fig. 25.5

All conductors to the left of new position of *M.N.A.* but between the two brushes, carry current downwards and those to the right carry current upwards. The armature m.m.f. is found to lie in the direction of the new position of *M.N.A.* (or brush axis). The armature m.m.f. is now represented by the vector OF_A which is not vertical (as in Fig. 25.2) but is inclined by an angle θ to the left. It can now be resolved into two rectangular components, OF_d parallel to polar axis and OF_C perpendicular to this axis. We find that

(*i*) component OF_C is at right angles to the vector OF_m (of Fig. 25.1) representing the main m.m.f. It produces distortion in the main field and is hence called the cross-magnetising or distorting component of the armature reaction.

(*ii*) The component OF_d is in direct opposition to OF_m which represents the main m.m.f. It exerts a demagnetising influence on the main pole flux. Hence, it is called the demagnetising or weakening component of the armature reaction.

It should be noted that both distorting and demagnetising effects will increase with increase in the armature current.

25.2. Demagnetising and Cross-magnetising Conductors

The exact conductors which produce these distorting and demagnetising effects are shown in Fig. 25.6 where the brush axis has been given a forward lead of θ so as to lie along the new position of *M.N.A.* All conductors lying within angles $AOC = BOD = 2\theta$ at the top and bottom of the armature, are carrying current in such a direction as to send the flux *through* the armature from right to left. This fact may be checked by applying crockscrew rule. It is these conductors which act in direct opposition to the main field and are hence called the demagnetising armature conductors.

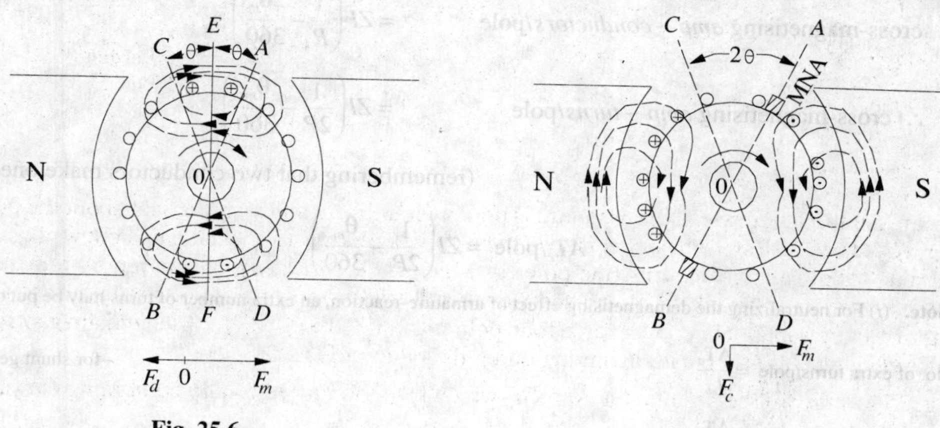

Fig. 25.6 **Fig. 25.7**

Now, consider the remaining armature conductors lying between angles *AOD* and *COB*. As shown in Fig. 25.7, these conductors carry current in such a direction as to produce a combined flux pointing vertically downwards *i.e.* at right angles to the main flux. This results in distortion of the main field. Hence, these conductors are known as cross-magnetising conductors and constitute distorting ampere-conductors.

25.3. Demagnetising AT per Pole

Since armature demagnetising ampere-turns are neutralized by adding extra ampere-turns to the main field winding, it is essential to calculate their number. But before proceeding further, it should be remembered that the number of turns is equal to half the number of conductors because two conductors-constitute one turn.

Let z = total number of armature conductors

I = current in each armature conductor

$= I_a/2$ .... for simplex wave winding

$= I_a/P$ .... for simplex lap winding

θ_m = forward lead in mechanical or geometrical or angular degrees.

Total number of armature conductors in angles *AOC* and *BOD* $\dfrac{4\theta\, m}{360} \times Z$

As two conductors constituted one turn,

$\therefore$ total number of turns in these angles = $\dfrac{2\theta\, m}{360} \times Z\,I$

$\therefore$ demagnetising amp-turns per pair of poles = $\dfrac{2\theta\, m}{360} \times Z\,I$

$\therefore$ demagnetising amp – tuns/pole $= \frac{\theta_m}{360} \times ZI$ $\therefore$ AT_d per pole $= ZI \times \frac{\theta m}{360}$

25.4. Cross-magnetising AT per pole

The conductors lying between angles *AOD* and *BOC* constitute what are known as distorting or cross-magnetising conductors. Their number is found as under :

Total armature-*conductors*/pole both cross and demagnetising = *Z*/*P*

Demagnetising *conductors*/pole $= Z . \frac{2\theta_m}{360}$ (found above)

$\therefore$ cross-magnetising *conductors*/pole $= \frac{Z}{P} - Z \times \frac{2\theta_m}{360} = Z\left(\frac{1}{P} - \frac{2\theta_m}{360}\right)$

$\therefore$ cross-magnetising *amp – conductors*/pole $= ZI\left(\frac{1}{P} - \frac{2\theta_m}{360}\right)$

$\therefore$ cross-magnetising *amp – turns*/pole $= ZI\left(\frac{1}{2P} - \frac{\theta_m}{360}\right)$

(remembering that two conductors make one turn)

$$\therefore \quad AT_c/\text{pole} = ZI\left(\frac{1}{2P} - \frac{\theta_m}{360}\right)$$

Note. (*i*) For neutralizing the demagnetising effect of armature-reaction, an extra number of turns may be put on each pole.

No. of extra turns/pole $= \frac{AT_d}{I_{sh}}$ –for shunt generator

$= \frac{AT_d}{I_a}$ –for series generator

If the leakage coefficient λ is given, then multiply each of the above expressions by it.

(*ii*) If lead angle is given in electrical degrees, it should be converted into mechanical degrees by the following relation.

$$\theta\ (\text{mechanical}) = \frac{\theta\ (\text{electrical})}{\text{pair of poles}} \quad \text{or} \quad \theta_m = \frac{\theta_e}{P/2} = \frac{2\theta_e}{P}$$

25.5. Compensating Windings

These are used for large direct current machines which are subjected to large fluctuations in load *i.e.* rolling mill motors and turbo-generators etc. Their function is to neutralize the cross-magnetizing effect of armature reaction. In the absence of compensating windings, the flux will be suddenly shifting backward and forward with every change in load. This shifting of flux will induce *statically* induced e.m.f. in the armature coils. The magnitude of this e.m.f. will depend upon the rapidity of changes in load and the amount of change. It may be so high as to strike an arc between the consecutive commutator segments across the top of the mica sheets separating them. This may further develop into a flash-over around the whole commutator thereby short-circuiting the whole armature.

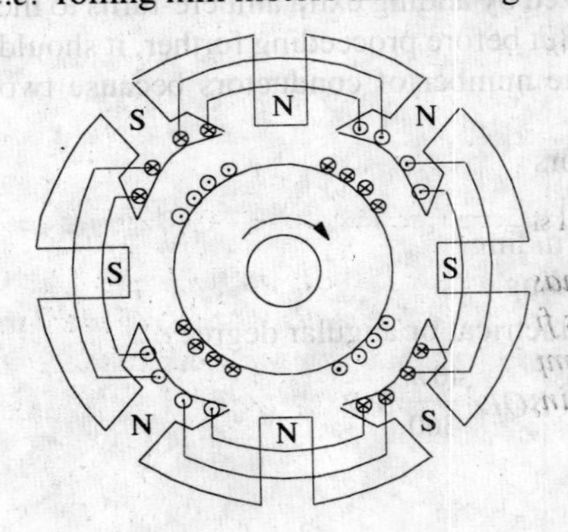

Fig. 25.8

These windings are embedded in slots in the pole shoes and are connected in series with armature in such a way that the current in them flows in opposite direction to that flowing in armature conductors directly below the pole shoes. An elementary scheme of compensating winding is shown in Fig. 25.8.

It should be carefully noted that compensating winding must provide sufficient m.m.f so as to counterbalance the armature m.m.f. Let

Z_c = No. of compensating conductors/pole face
Z_a = No. of active armature conductors/pole, I_a = total armature current
I_a/A = current/armature conductor

$\therefore\ Z_c I_a = Z_a (I_a/A)$ or $Z_c = Z_a/A$

Owing to their cost and the room taken up by them, the compensating windings are used in the case of large machines which are subject to violent fluctuations in load and also for generators which have to deliver their full-load output at considerably low induced voltage as in the Ward-Leonard set.

25.6. No. of Compensating Windings

No. of armature *conductors*/pole $= \dfrac{Z}{P}$ No. of armature *turns*/pole $= \dfrac{Z}{2P}$

$\therefore$ No. of armature-*turns* immediately under one pole

$$= \frac{Z}{2P} \times \frac{\text{Pole are}}{\text{Pole pitch}} = 0.7\frac{Z}{2P} \text{ (approx.)}$$

$\therefore$ No. of armature amp-*turns*/pole for compensating winding

$$= 0.7 \times \frac{ZI}{2P} = 0.7 \times \text{armature amp-turns/pole}$$

Example 25.1. *A 4-pole generator has a wave-wound armature with 722 conductors, and it delivers 100* A *on full load. If the brush lead is 8°, calculate the armature demagnetising and cross-magnetising ampere turns per pole.* **(Advanced Elect. Machines AMIE Sec. B 1991)**

Solution. $I = I_a/2 = 100/2 = 50\text{A};\ Z = 722;\ \theta_m = 8°$

$$\text{AT}_d\text{/pole} = ZI.\frac{\theta_m}{360} = 722 \times 50 \times \frac{8}{360} = 802$$

$$\text{AT}_c\text{/pole} = ZI\left(\frac{1}{2P} - \frac{\theta_m}{360}\right) = 722 \times 50\left(\frac{1}{2 \times 4} - \frac{8}{360}\right) = 37/8$$

Example 25.2. *An 8-pole generator has an output of 200 A at 500 V, the lap-connected armature has 1280 conductors, 160 commutator segments. If the brushes are advanced 4-segments from the no-load neutral axis, estimate the armature demagnetizing and crossmagnetizing ampere-turns per pole.* **(Electrical Machines-I, South Gujarat Univ. 1986)**

Solution. $I = 200/8 = 25\text{ A}, Z = 1280, \theta_m = 4 \times 360/160 = 9°\ ; P = 8$

$$\text{AT}_d\text{/pole} = ZI\theta_m/360 = 1280 \times 25 \times 9/360 = \mathbf{800}$$

$$\text{AT}_c\text{/pole} = ZI\left(\frac{1}{2p} - \frac{\theta_m}{360}\right) = 1280 \times 25\left(\frac{1}{2 \times 8} - \frac{9}{360}\right) = \mathbf{1200}$$

Note. No. of coils = 160, No. of conductors = 1280. Hence, each coil side contains 1280/160 = 8 conductors.

Example 25.3. *A 4-pole wave-wound motor armature has 880 conductors and delivers 120 A. The brushes have been displaced through 3 angular degrees from the geometrical axis. Calculate (a) demagnetising amp-turns/pole (b) cross-magnetising amp-turns/pole (c) the additional field current for neutralizing the demagnetisation if the field winding has 1100 turns/pole.*

Solution. $Z = 880;\ I = 120/2 = 60\text{ A};\ \theta = 3°$ angular

(*a*) $\therefore\ \text{AT}_d = 880 \times 60 \times \dfrac{3}{360} = \mathbf{440\ AT}$

(b) $\therefore AT_C = 880 \times 60\left(\frac{1}{8} - \frac{3}{360}\right) = 880 \times \frac{7}{60} \times 60 = \mathbf{6{,}160}$

or Total AT/pole = 440 × 60/4 = 6600

Hence, AT_C /pole = Total AT/pole – AT_d/pole = 6600 – 440 = 6160

(c) Additional field current = 440/1100 = **0.4 A**

Example 25.4. *A 4-pole generator supplies a current of 143 A. It has 492 armature conductors (a) wave-wound (b) lap-wound. When delivering full load, the brushes are given an actual lead of 10°. Calculate the demagnetising amp-turns/pole. This field winding is shunt connected and takes 10 A. Find the number of extra shunt field turns necessary to neutralize this demagnetisation.*

(Elect. Machines, Nagpur Univ. 1993)

Solution. $Z = 492$; $\theta_m = 10°$; $AT_d/\text{pole} = ZI \times \frac{\theta_m}{360}$

$I_a = 143 + 10 = 153$ A ; $I = 153/2$ ... when wave-wound

$= 153/4$ ... when lap-wound

(a) $\therefore AT_d\text{pole} = 492 \times \frac{153}{2} \times \frac{10}{360} = \mathbf{1046\ AT}$

Extra shunt field turns = 1046/10 = **105 approx**

(b) $AT_d/\text{pole} = 492 \times \frac{153}{4} \times \frac{10}{360} = \mathbf{523}$

Extra shunt field turns = 523/10 = **52 (approx.)**

Example 25.5. *A 4-pole, 50-kW, 250-V wave-wound shunt generator has 400 armature conductors. Brushes are given a lead of 4 commutator segments. Calculate the demagnetisation amp-turns/pole if shunt field resistance is 50 Ω. Also, calculate extra shunt field turns/pole to neutralize the demagnetisation.*

Solution. Load current supplied = 50,000/250 = 200 A

$I_{sh} = 250/50 = 5$ A $\therefore I_a = 200 + 5 = 205$ A

Current in each conductor $I = 205/2$ A

No. of commutator segments $= N/A$ where $A = 2$ ...for wave-winding

$\therefore$ No. of segments = 400/2 = 200 ; $\theta = \frac{4}{200} \times 360 = \frac{36}{5}$ degrees

$\therefore AT_d/\text{pole} = 400 \times \frac{205}{2} \times \frac{36}{5 \times 360} = \mathbf{820\ AT}$

Extra shunt turns/pole $= \frac{AT_d}{I_{sh}} = \frac{820}{5} = \mathbf{164}$

Example 25.6. *Determine per pole the number (i) of cross-magnetising ampere-turns (ii) of back ampere-turns and (iii) of series turns to balance the back ampere-turns in the case of a.d.c. generator having the following data.*

500 conductors, total current 200 A, 6 poles, 2-circuit wave winding, angle of lead = 10°, leakage coefficient = 1.3 **(Electrical Machines-I, Bombay University, 1986)**

Solution. Current/path, $I = 200/2 = 100$ A, $\theta = 10°$ (mech), $Z = 500$

(a) $AT_c/\text{pole} = ZI\left(\frac{1}{2P} - \frac{\theta_m}{360}\right) = 500 \times 100\left(\frac{1}{2 \times 6} - \frac{10}{360}\right) = \mathbf{2{,}778}$

(b) $AT_d/\text{pole} = 500 \times 100 \times 10/360 = \mathbf{1390}$

(c) Series turns required to balance the demagnetising ampere-turns are

$$= \lambda \times \frac{\text{AT}_d}{I_a} = 1.3 \times \frac{1390}{200} = \mathbf{9}$$

Example 25.7. *A 22.38 kW, 440-V, 4-pole wave-wound d.c. shunt motor has 840 armature conductors and 140 commutator segments. Its full-load efficiency is 88% and the shunt field current is 1.8 A. If brushes are shifted backwards through 1.5 segments from the geometrical neutral axis, find the demagnetising and distorting amp-turns/pole.* **(Elect. Engg. Punjab Univ. 1991)**

Solution. The shunt motor is shown diagrammatically in Fig. 25.9.

Motor output $= 22,380\, W;\ \eta = 0.88$

$\therefore$ Motor input $= 22,380/0.88$ W

Motor input current $= \dfrac{22,380}{0.88 \times 440} = 57.8$ A

$I_{sh} = 1.8\, A;\quad I_a = 57.8 - 1.8 = 56$ A

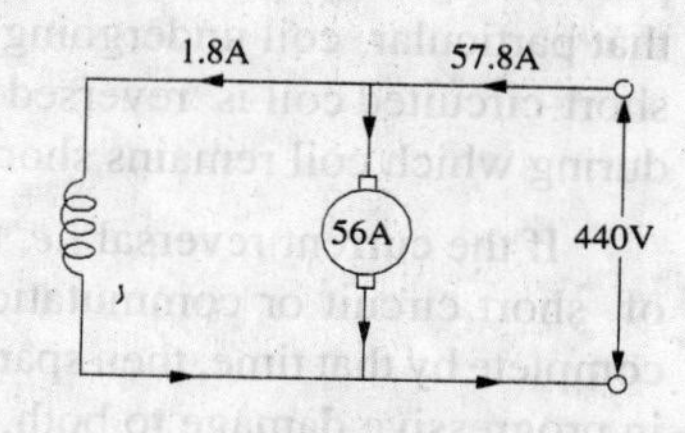

Fig. 25.9

Current in each conductor $= 56/2 = 28$ A

$$\theta = 1.5 \times 360/140 = 27/7 \text{ degrees}$$

$$\therefore \quad \text{AT}_d/\text{pole} = 840 \times 28 \times \frac{27}{7 \times 360} = \mathbf{252}$$

$$\text{AT}_c/\text{pole} = Z\,I\left(\frac{1}{2P} - \frac{\theta_m}{360}\right) = 840 \times 28\left(\frac{1}{8} - \frac{27}{7 \times 360}\right) = \mathbf{2{,}688}$$

Example 25.8. *A 400-V, 1,000-A, lap-wound d.c. machine has 10 poles and 860 armature conductors. Calculate the number of conductors in the pole face to give full compensation if the pole face covers 70% of pole span.*

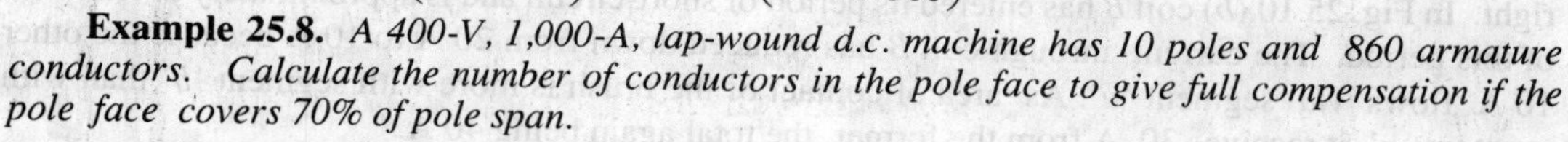

Solution. AT/pole for compensating winding

$$= \text{armature amp-turns/pole} \times \frac{\text{pole arc}}{\text{pole pitch}} = 0.7\,\frac{ZI}{2P}$$

Here $\quad I$ = current in each armature conductor = 1,000/10 = 100 A

$Z = 860;\quad P = 10$

$\therefore$ AT/pole for compensating winding = $0.7 \times 860 \times 100/2 \times 10 = \mathbf{3{,}010}$

Tutorial Problem No. 25.1

1. Calculate the demagnetising amp-turns of a 4-pole, lap-wound generator with 720 turns, giving 50 A, if the brush lead is 10° (mechanical). **[250 AT/pole]**

2. A 250-V, 25-kW, 4-pole d.c. generator has 328 wave-connected armature conductors. When the machine is delivering full load, the brushes are given a lead of 7.2 electrical degrees. Calculate the cross-magnetising amp-turns/pole. **[1886, 164]**

3. An 8-pole lap-connected d.c. shunt generator delivers an output of 240 A at 500 V. The armature has 1408 conductors and 160 commutator segments. If the brushes are given a lead of 4 segments from the no-load neutral axis, estimate the de-magnetising and cross-magnetising AT/pole. **[1056, 1,584]** (*Electrical Engineering, Bombay Univ. 1978*)

4. A 500-V, wave-wound, 750 r.p.m. shunt generator supplies a load of 195 A. The armature has 720 conductors and shunt field resistance is 100 Ω. Find the demagnetising amp-turns/pole if bruhses are advanced through 3 commutator segments at this load. Also, calculate the extra shunt field turns required to neutralize this demagnetisation. **[600, 4,800, 120]**

5. A 4-pole, wave-wound generator has 320 armature conductors and carries an armature current of 400 A. If the pole arc/pole pitch ratio is 0.68, calculate the AT/pole for a compensating winding to give uniform flux density in the air gap. **[5440]**

6. A 500-kW, 500-V, 10 pole d.c. generator has a lap-wound armature with 800 conductors. Calculate the number of pole-face conductors in each pole of a compensating winding if the pole face covers 75 per cent of the pitch. **[6 conductors/pole]**

25.7. Commutation

It was shown in Art 24.2 that currents induced in armature conductors of a d.c. generator are alternating. To make their flow unidirectional in the external circuit, we need a commutator. Moreover, these currents flow in one direction when armature conductors are under *N*-pole and in the opposite direction when they are under *S*-pole. As conductors pass out of the influence of a *N*-pole and enter that of *S*-pole, the current in them is reversed. This reversal of current takes place along magnetic neutral axis or brush axis *i.e.* when the brush spans and hence short-circuits that particular coil undergoing reversal of current through it. This process by which current in the short-circuited coil is reversed while it crosses the *M.N.A.* is called commutation. The brief period during which coil remains short-circuited is known as commutation period T_c.

If the current reversal *i.e.* the change from $+I$ to zero and then to $-I$ is completed by the end of short circuit or commutation period, then the commutation is ideal. If current reversal is not complete by that time, then sparking is produced between the brush and the commutator which results in progressive damage to both.

Let us discuss the process of commutation or current reversal in more detail with the help of Fig. 25.10 where ring winding has been used for simplicity. The brush width is equal to the width of one commutator segment and one mica insulation. In Fig. 25.10 (*a*) coil *B* is about to be short-circuited because brush is about to come in touch with commutator segment '*a*'. It is assumed that each coil carries 20 A, so that brush current is 40 A. It is so because every coil meeting at the brush supplies half the brush current whether lap wound or wave wound. Prior to the beginning of short circuit, coil *B* belongs to the group of coils lying to the left of the brush and carries 20 A from left to right. In Fig. 25.10 (*b*) coil *B* has entered its period of short-circuit and is approximately at one-third of this period. The current through coil *B* has reduced down from 20 A to 10 A because the other 10 A flows *via* segment '*a*'. As area of contact of the brush is more with segment '*b*' than with segment '*a*', it receives 30 A from the former, the total again being 40 A.

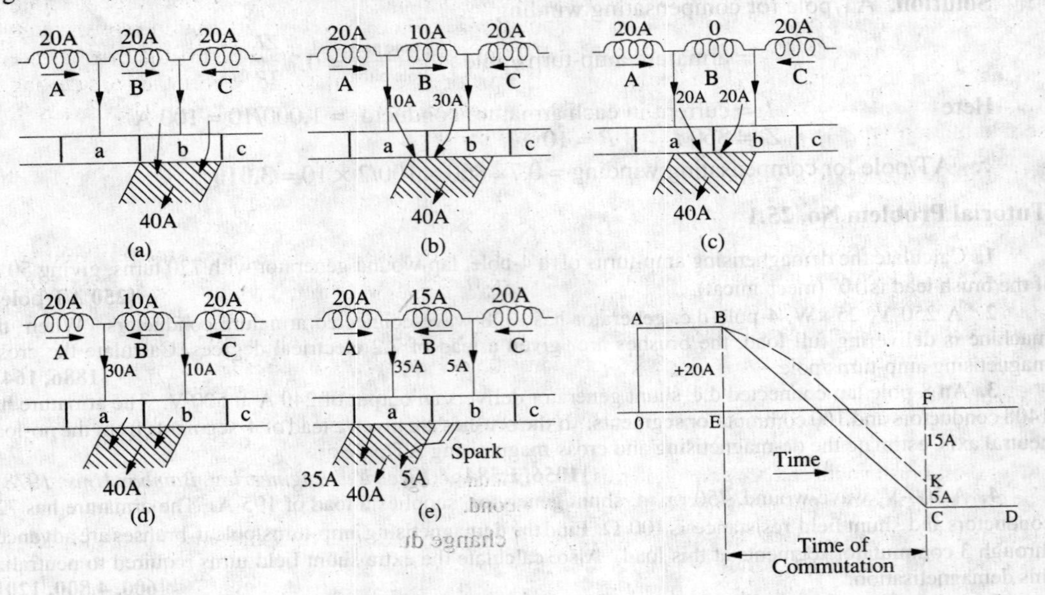

Fig. 25.10

Fig. 25.11

Fig. 25.10 (*c*) shows the coil *B* in the middle of its short-circuit period. The current through it has decreased to zero. The two currents of value 20 A each, pass to the brush directly from coil *A* and *C* as shown. The brush contact areas with the two segments '*b*' and '*a*' are equal.

In Fig. 25.10 (*d*), coil *B* has become part of the group of coils lying to the right of the brush. It is seen that brush contact area with segment '*b*' is decreasing rapidly whereas that with segment '*a*' is increasing. Coil *B* now carries 10 A in the reverse direction which combines with 20 A supplied by coil *A* to make up 30 A that passes from segment '*a*' to the brush. The other 10 A is supplied by coil *C* and passes from segment '*b*' to the brush, again giving a total of 40 A at the brush.

Fig. 25.10 (*e*) depicts the moment when coil *B* is almost at the end of commutation or short-circuit period. For ideal commutation, current through it should have reversed by now but, as shown, it is carrying 15 A only (instead of 20 A). The difference of current between coils *C* and *B i.e.* 20–15 = 5 A in this case, jumps directly from segment *b* to the brush through air thus producing spark.

If the changes of current through coil *B* are plotted on a time base (as in Fig. 25.11) it will be represented by a horizontal line *AB i.e.* a constant current of 20 A up to the time of beginning of commutation. From the finish of commutation, the current will be represented by another horizontal line *CD*. Now, again the current value is *FC* = 20 A, although in the reversed direction. The way in which current changes from its positive value of 20 A (= *BE*) to zero and then to its negative value of 20 A (= *CF*) depends on the conditions under which the coil *B* undergoes commutation. If the current varies at a uniform rate *i.e.* if *BC* is a straight line, then it is referred to as linear commutation. However, due to the production of self-induced e.m.f. in the coil (discussed below) the variations follow the dotted curve. It is seen that, in that case, current in coil *B* has reached only a value of *KF* = 15 A in the reversed direction, hence the difference of 5 A (20 A – 15 A) passes as a spark.

So, we conclude that sparking at the brushes, which results in poor commutation is due to the inability of the current in the short-circuited coil to reverse completely by the end of short-circuit period (which is usually of the order of 1/500 second).

At this stage, the reader might ask for the reasons which make this current reversal impossible in the specified period *i.e.* what factors stand in the way of our achieving ideal commutation. The main cause which retards or delays this quick reversal is the production of self-induced e.m.f. in the coil undergoing commutation. It may be pointed out that the coil possesses appreciable amount of self inductance because it lies embedded in the armature which is built up of a material of high magnetic permeability. This self-induced e.m.f. is known as *reactance voltage* whose value is found as given below. This voltage, even though of a small magnitude, produces a large current through the coil whose resistance is very low due to short circuit. It should be noted that if the brushes are set so that the coils undergoing short-circuit are in the magnetic neutral plane, where they are cutting no flux and hence have no e.m.f. induced in them due to armature rotation, there will still be the e.m.f. of self-induction which causes severe sparking at the brushes.

25.8. Value of Reactance Voltage

Reactance voltage = coefficient of self-inductance × rate of change of current.

It should be remembered that the time of short-circuit or commutation is the time required by the commutator to move a distance equal to the circumferential thickness of the brush minus the thickness of one insulating plate or strip of mica.

Let W_b = brush width in cm; W_m width of mica insulation in cm

v = peripheral velocity of commutator segments in cm/second

Then T_c = time of commutation or short-circuit = $\dfrac{W_b - W_m}{v}$ second

Note. If brush width etc. are given in terms of commutator segments, then commutator velocity should also be converted in terms of commutator segments per second.

If I is the current through a conductor, then total change during commutation = $I - (-I) = 2I$.

∴ Self-induced or reactance voltage

$$= L \times \frac{2I}{T_c} \quad \text{– if commutation is linear}$$

$$= 1.11\, L \times \frac{2I}{T_c} \quad \text{– if commutation is sinusodial}$$

As said earlier, the reactance e.m.f. hinders the reversal of current. This means that there would be sparking at the brushes due to the failure of the current in short-circuited coil to reach its full value in the reversed direction by the end of short-circuit. This sparking will not only damage the brush and the commutator but this being a cumulative process, it may worsen and eventually lead to the short-circuit of the whole machine by the formation of an arc round the commutator from brush to brush.

Example 25.9. *The armature of a certain dynamo runs at 800 r.p.m. The commutator consists of 123 segments and the thickness of each brush is such that the brush spans three segments. Find the time during which the coil of an armature remains short-circuited.*

Solution. As W_m is not given, it is considered negligible.

$W_b = 3$ segments and $v = (800/60) \times 123$ segments/second

$\therefore$ commutation time $= \dfrac{3 \times 60}{800 \times 123} = 0.00183$ second $=$ **1.83 millisecond**

Example 25.10. *A 4-pole, wave-wound, d.c. machine running at 1500 r.p.m. has a commutator of 30 cm diameter. If armature current is 150 A, thickness of brush 1.25 cm and the self-inductance of each armature coil is 0.07 mH, calculate the average e.m.f. induced in each coil during commutation. Assume linear commutation.*

Solution. Formula : $E = L\dfrac{2I}{T_c}$

Here $L = 0.07 \times 10^{-3}$ H, $I = 150/2 = 75$ A ($\because$ it is wave-wound)

$W_b = 1.25$ cm $W_m = 0$ ...considered negligible

$v = \pi \times 30 \times (1500/60) = 2356$ cm/s ; $T_c = W_b/v = 1.25/2356 = 5.3 \times 10^{-4}$ second

$E = L \times 2I/T_c = 0.07 \times 10^{-3} \times 2 \times 75/5.3 \times 10^{-4} =$ **19.8 V**

Example 25.11. *Calculate the reactance voltage for a machine having the following particulars. Number of commutator segments = 55, Revolutions per minute = 900, Brush width in commutator segments = 1.74, Coefficient of self-induction = 153×10^{-6} henry, Current per coil = 27 A.*

(Advanced Elect. Machines. AMIE Sec. Winter 1991)

Solution. Current per coil, $I = 27$ A ; $L = 153 \times 10^{-6}$ H

$v = 55 \times (900/60) = 825$ segments/second; $T_c = W_b/v = 1.74/825 = 2.11 \times 10^{-3}$ second

Assuming linear commutation, $E = L \times 2I/T_C$

$\therefore E = 153 \times 10^{-6} \times 2 \times 27/2.11 \times 10^{-3} =$ **3.91 V**

Example 25.12. *A 4-pole, lap-wound armature running at 1500 r.p.m. delivers a current of 150 A and has 64 commutator segments. The brush spans 1.2 segments and inductance of each armature coil is 0.05 mH. Calculate the value of reactance voltage assuming (i) linear commutation (ii) sinusoidal commutation. Neglect mica thickness.*

Solution. Formula : $E = L.\dfrac{2I}{T_c}$ Now, $L = 0.05 \times 10^{-3}$ H ; $W_b = 1.2$ segments

$$v = \frac{1500}{60} \times 64 = 1600 \text{ segments/second}$$

$$\therefore \quad T_c = \frac{1.2}{1600} = 7.5 \times 10^{-4} \text{ seocnd} ; I = \frac{150}{4} \text{ A} = 37.5 \text{ A}$$

$$\therefore \quad \frac{2I}{T_c} = \frac{2 \times 37.5}{7.5 \times 10^{-4}} = 10^5 \text{ A/s}$$

For linear commutation, $E = 0.05 \times 10^{-3} \times 10^5 =$ **5 V**

For sinusoidal commutation, $E = 1.11 \times 5$ = **5.55 V**

25.9. Methods of Improving Commutation

There are two practical ways of improving commutation *i.e.* of making current reversal in the short-circuited coil as sparkless as possible. These methods are known as (*i*) resistance commutation and (*ii*) e.m.f. commutation (which is done with the help of either brush lead or interpoles, usually the later).

25.10. Resistance Commutation

This method of improving commutation consists of replacing low-resistance Cu brushes by comparatively high-resistance carbon brushes.

From Fig. 25.12, it is seen that when current *I* from coil' *C* reaches the commutator segment *b*, it has two parallel paths open to it. The first path is straight from bar '*b*' to the brush and the other parallel path is *via* the short-circuited coil *B* to bar '*a*' and then to the brush. If the Cu brushes (which have low contact resistance) are used, then there is no inducement for the current to follow the second longer path, it would preferably follow the first path. But when carbon brushes having high resistance are used, then current *I* coming from *C* will prefer to pass through the second path because (*i*) the resistance r_1 of the first path will increase due to the diminishing area of contact of bar '*b*' with the brush and because (*ii*) resistance r_2 of second path will decrease due to rapidly increasing contact area of bar '*a*' with the brush.

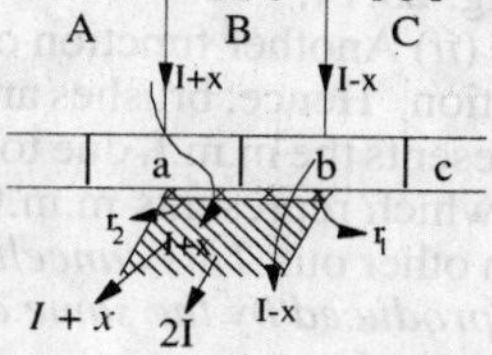

Fig. 25.12

Hence, carbon brushes have, usually, replaced Cu brushes. However, it should be clearly understood that the main cause of sparking commutation is the self-induced e.m.f. (*i.e.* reactance voltage), so brushes alone do not give a sparkless commutation; though they do help in obtaining it.

The additional advantages of carbon brushes are that (*i*) they are to some degree self-lubricating and polish the commutator and (*ii*) should sparking occur, they would damage the commutator less than when Cu brushes are used.

But some of their minor disadvantages are : (*i*) due to their high contact resistance (which is beneficial to sparkless commutation) a loss of approximately 2 volt is caused. Hence, they are not much suitable for small machines where this voltage forms an appreciable percentage loss (*ii*) owing to this large loss, the commutator has to be made somewhat larger than with Cu brushes in order to dissipate heat efficiently without greater rise of temperature (*iii*) because of their lower current density (about 7–8 A/cm^2 as compared to 25–30 A/cm^2 for Cu brushes) they need larger brush holders.

25.11. E.M.F. Commutation

In this method, arrangement is made to neutralize the reactance voltage by producing a reversing e.m.f. in the short-circuited coil under commutation. This reversing e.m.f., as the name shows, is an e.m.f. in opposition to the reactance voltage and if its value is made equal to the latter, it will completely wipe it off, thereby producing quick reversal of current in the short-circuited coil which will result in sparkless commutation. The reversing e.m.f. may be produced in two ways : (*i*) either by giving the brushes a forward lead sufficient enough to bring the short-circuited coil under the influence of next pole of opposite polarity or (*ii*) by using interpoles.

The first method was used in the early machines but has now been abandoned due to many other difficulties it brings along with.

25.12. Interpoles or Compoles

These are small poles fixed to the yoke and spaced in between the main poles. They are wound with comparatively few heavy gauge Cu wire turns and are connected in series with the armature so that they carry full armature current. Their polarity, in the case of a generator, is the same as that of the main pole *ahead* in the direction of rotation (Fig. 25.13).

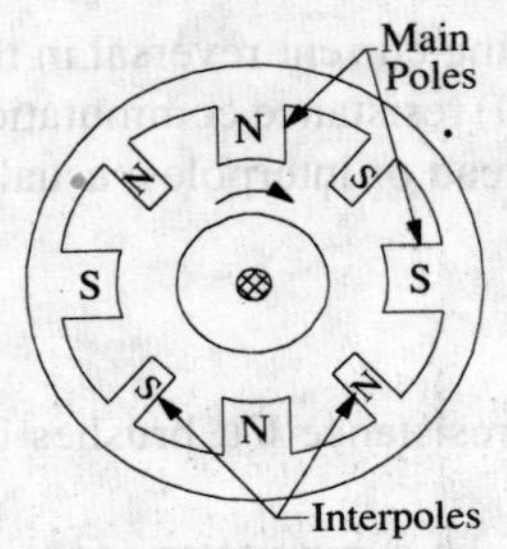

Fig. 25.13

The function of interpoles is two-fold :

(*i*) As their polarity is the same as that of the main pole ahead, they induce an e.m.f. in the coil (under commutation) which helps the reversal of current. The e.m.f. induced by the compoles is known as *comutating* or *reversing e.m.f.* The commutating e.m.f. neutralizes the reactance e.m.f. thereby making commutation sparkless. With interpoles, sparkless commutation can be obtained up to 20 to 30% overload with fixed brush position. In fact, interpoles raise sparking limit of a machine to almost the same value as heating limit. Hence, for a given output, an interpole machine can be made smaller and, therefore, cheaper than a non-interpolar machine.

As interpoles carry armature current, their commutating e.m.f. is proportional to the armature current. *This ensures automatic neutralization of reactance voltage which is also due to armature current.* Connections for a shunt generator with interpoles are shown in Fig. 25.14.

(*ii*) Another function of the interpoles is to neutralize the cross-magnetising effect of armature reaction. Hence, brushes are not to be shifted from the original position. In Fig. 25.15, *OF* as before, represents the m.m.f. due to main poles. *OA* represents the cross-magnetising m.m.f. due to armature. *BC* which represents m.m.f. due to interpoles, is obviously in opposition to *OA*, hence they cancel each other out. *This cancellation of cross-magnetisation is automatic and for all loads because both are produced by the same armature current.*

The distinction between the interpoles and compensating windings should be clearly understood. Both are connected in series and their m.m.fs. are such as to neutralize armature reaction. But compoles additionally supply m.m.f. for counteracting the reactance voltage induced in the coil undergoing commutation. Moreover, the action of the compoles is localized, they have negligible effect on the armature reaction occurring on the remainder of the armature periphery.

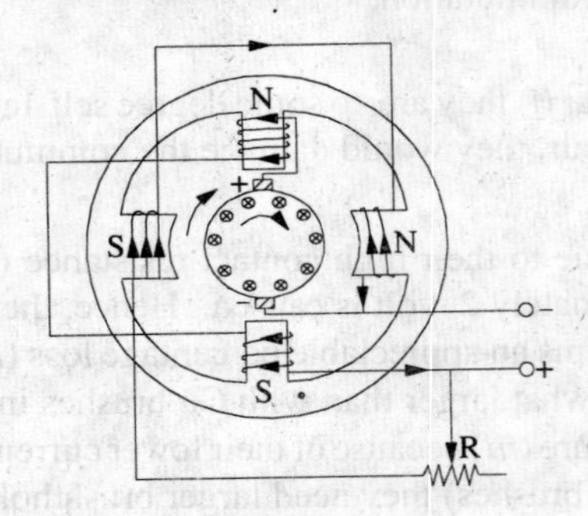

Fig. 25.14

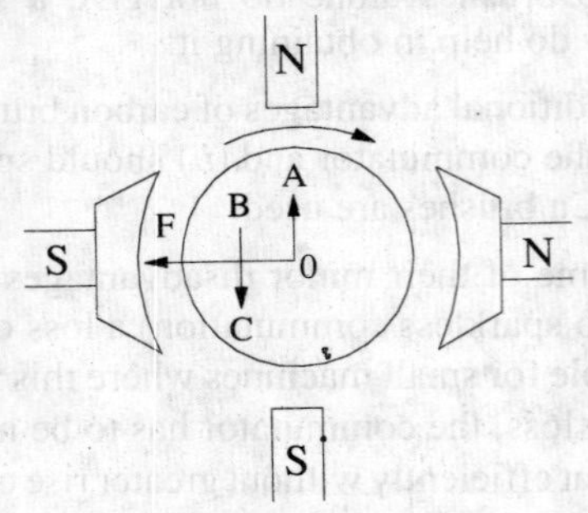

Fig. 25.15

Example 25.13. *Determine the number of turns on each commutating pole of a 6-pole machine, if the flux density in the air-gap of the commutating pole = 0.5 Wb/m² at full load and the effective length of the air-gap is 4 mm. The full-load current is 500 A and the armature is lap-wound with 540 conductors. Assume the ampere turns required for the remainder of the magnetic circuit to be one-tenth of that of the air gap.* **(Advanced Elect. Machines AMIE Sec.B, 1991)**

Solution. It should be kept in mind that compole winding must be sufficient to oppose the armature m.m.f. (which is directed along compole axis) and to provide the m.m.f. for compole air-gap and its magnetic circuit.

$\therefore\ N_{CP} I_a = ZI_C/2P + B_g l_g/\mu_0$

where N_{cp} = No. of turns on the compole ; I_a = armature current

Z = No. of armature conductors ; I_c = coil current

P = No. of poles ; l_g = air-gap length under the compole

$Z = 540$; $I_a = 500$ A; $I_c = 500/6$ A; $P = 6$

$\therefore$ Arm. m.m.f. $= 540 \times (500/6)/2 \times 6 = 3750$

Compole air-gap m.m.f. = $B_g \times l_g / \mu_0 = 0.5 \times 4 \times 10^{-3}/4\pi \times 10^{-7} = 1591$

m.m.f. reqd. for the rest of the magnetic circuit = 10% of 1591 = 159

$\therefore$ total compole air-gap m.m.f. = 1591 + 159 = 1750

Total m.m.f. reqd. = 3750 + 1750 = 5500

$\therefore N_{cp} I_a = 5500$ or $N_{cp} = 5500/500 = \mathbf{11}$

25.13. Equalizing Connections

It is characteristic of lap-winding that all conductors in any parallel path lie under one pair of poles. If fluxes from all poles are exactly the same, then e.m.f. induced in each parallel path is the same and each path carries the same current. But despite best efforts, some inequalities in flux inevitably occur due either to slight variations in air-gap length or in the magnetic properties of steel. Hence, there is always a slight imbalance of e.m.f. in the various parallel paths. The result is that conductors under stronger poles generate greater e.m.f. and hence carry larger current. The current distribution at the brushes becomes unequal. Some brushes are overloaded *i.e.* carry more than their normal current whereas others carry less. Overloaded brushes spark badly whatever their position may be. This results in poor commutation and may even limit the output of the machine.

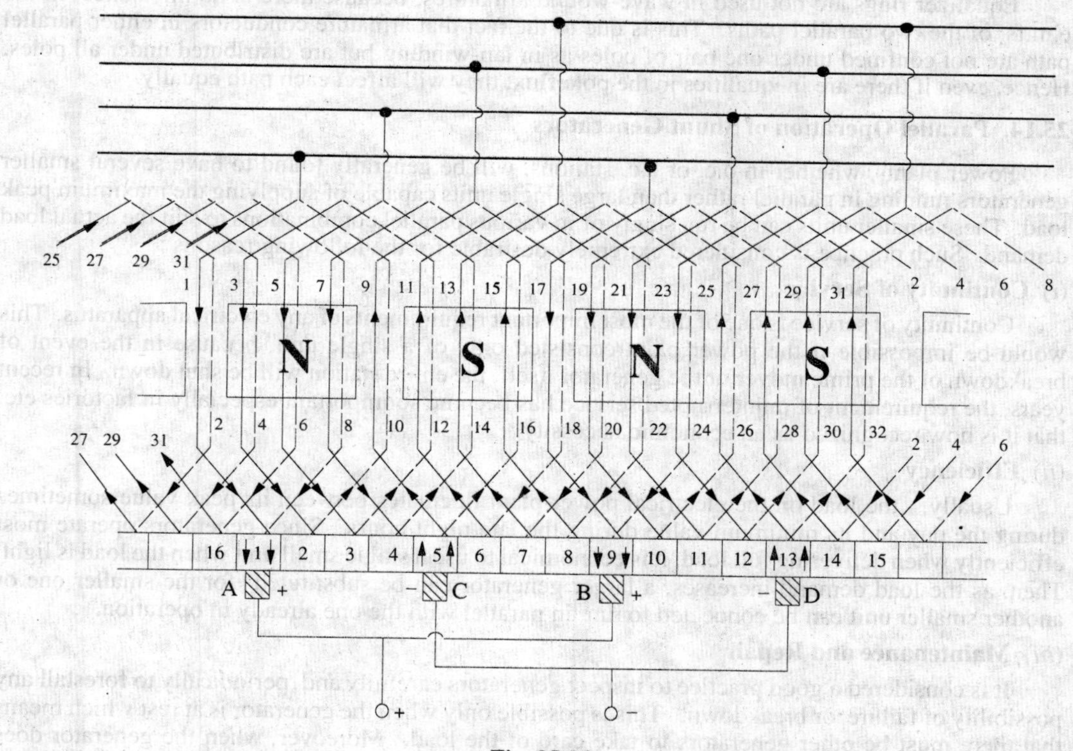

Fig. 25.16

By connecting together a number of symmetrical points on armature winding which would be at equal potential if the pole fluxes were equal, the difference in brush currents is diminished. This requires that there should be a whole number of slots per pair of poles so that, for example, if there is a slot under the centre of a *N*-pole, at some instant, then there would be one slot under the centre of every other *N*-pole. The equalizing conductors, which are in the form of Cu rings at the armature back and which connect such points are called *Equalizer Rings*. The circulating current due to the slight difference in the e.m.fs. of various parallel paths, passes through these *equalizer* rings instead of passing through the brushes.

Hence, the function of equalizer rings is to avoid unequal distribution of current at the brushes thereby helping to get sparkless commutation.

One equalizer ring is connected to all conductors in the armature which are two poles apart (Fig. 25.17). For example, if the number of poles is 6, then the number of connections for each equalizer ring is 3 *i.e.* equal to the number of pair of poles. Maximum number of equalizer rings is equal to the number of conductors under one pair of poles. Hence, number of rings is

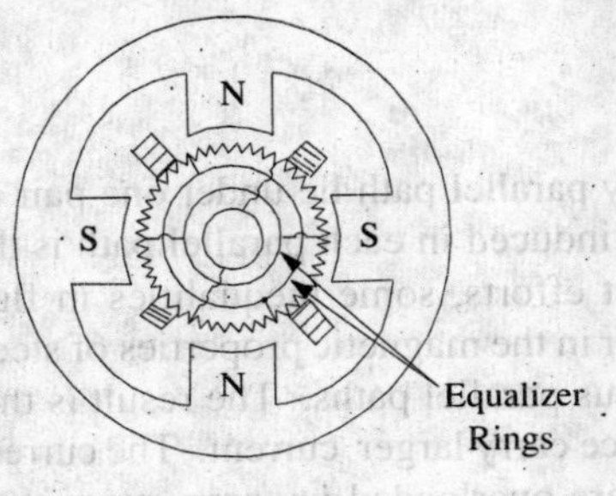

Fig. 25.17

$$= \frac{\text{No. of conductors}}{\text{No. of pair of poles}}$$

In practice, however, the number of rings is limited to 20 on the largest machines and less on smaller machines. In Fig. 25.16 is shown a developed armature winding. Here, only 4 equalizing bars have been used. It will be seen that the number of equalizing connections to each bar is two *i.e.* half the number of poles. Each alternate coil has been connected to the bar. In this case, the winding is said to be 50% equalized. If all conductors were connected to the equalizer rings, then the winding would have been 100% equalized.

Equalizer rings are not used in wave-wound armatures, because there is no imbalance in the e.m.fs. of the two parallel paths. This is due to the fact that armature conductors in either parallel path are not confined under one pair of poles as in lap-winding but are distributed under all poles. Hence, even if there are inequalities in the pole flux, they will affect each path equally.

25.14. Parallel Operation of Shunt Generators

Power plants, whether in d.c. or a.c. stations, will be generally found to have several smaller generators running in parallel rather than large single units capable of supplying the maximum peak load. These smaller units can be run singly or in various parallel combinations to suit the actual load demand. Such practice is considered extremely desirable for the following reasons :

(*i*) Continuity of Service

Continuity of service is one of the most important requirements of any electrical apparatus. This would be impossible if the power plant consisted only of a single unit, because in the event of breakdown of the prime mover or the generator itself, the entire station will be shut down. In recent years, the requirement of uninterrupted service has become so important especially in factories etc. that it is now recognized as an economic necessity.

(*ii*) Efficiency

Usually, the load on the electrical power plant fluctuates between its peak value sometimes during the day and its minimum value during the late night hours. Since generators operate most efficiently when delivering full load, it is economical to use a single small unit when the load is light. Then, as the load demand increases, a larger generator can be substituted for the smaller one or another smaller unit can be connected to run in parallel with the one already in operation.

(*iii*) Maintenance and Repair

It is considered a good practice to inspect generators carefully and periodically to forestall any possibility of failure or breakdown. This is possible only when the generator is at rest which means that there must be other generators to take care of the load. Moreover, when the generator does actually breakdown, it can be repaired with more care and not in a rush, provided there are other generators available to maintain service.

(*iv*) Additions to Plant

Additions to power plants are frequently made in order to deliver increasingly greater loads. Provision for future extension is, in fact, made by the design engineers right from the beginning. It becomes easy to add other generators for parallel operation as the load demand increases.

25.15. Paralleling DC Generator

Whenever generators are in parallel, their +ve and – ve terminals are respectively connected to the +ve and – ve sides of the bus-bars. These bus-bars are heavy thick copper bars and they act as

+ve and –ve terminals for the whole power station. If polarity of the incoming generator is not the same as the line polarity, as serious short-circuit will occur when S_1, is closed.

Moreover, paralleling a generator with reverse polarity effectively short-circuits it and results in damaged brushes, a damaged commutator and a blacked-out plant. Generators that have been tripped off the bus-because of a heavy fault current should always be checked for reversed polarity before paralleling.

In Fig. 25.18 is shown a shunt generator No. 1 connected across the bus-bars *BB* and supplying some of the load. For putting generator No. 2 in parallel with it the following procedure is adopted.

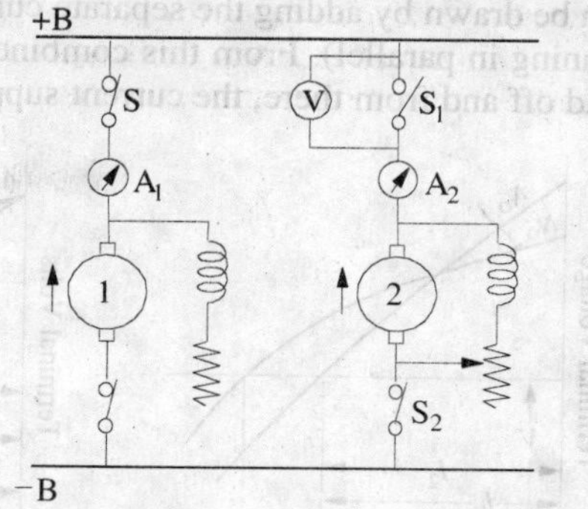

Fig. 25.18

The armature of generator No. 2 is speeded by the prime-mover up to its rated value and then switch S_2 is closed and circuit is completed by putting a voltmeter *V* across the open switch S_1. The excitation of the incoming generator No. 2 is changed till *V* reads zero. Then it means that its terminal voltage is the same as that of generator No. 1 or bus-bar voltage. After this, switch S_1 is closed and so the incoming machine is paralleled to the system. Under these conditions, however, generator No. 2 is not taking any load, because its *induced* e.m.f. is the same as bus-bar voltage and there can be no flow of current between two points at the same potential. The generator is said to be 'floating' on the bus-bar. If generator No. 2 is to deliver any current, then its induced e.m.f. *E* should be greater than the bus-bar voltage *V*. In that case, current supplied by it is $I = (E - V)/R_a$ where R_a is the resistance of the armature circuit. The induced e.m.f. of the incoming generator is increased by strengthening its field till it takes its proper share of load. At the same time, it may be found necessary to weaken the field of generator No. 1 to maintain the bus-bar voltage *V* constant.

25.16. Load Sharing

Because of their slightly drooping voltage characteristics, shunt generators are most suited for stable parallel operation. Their satisfactory operation is due to the fact that any tendency on the part of a generator to take more or less than its proper share of load results in certain changes of voltage in the system which immediately oppose this tendency thereby restoring the original division of load. Hence, once paralleled, they are automatically held in parallel.

Similarly, for taking a generator out of service, its field is weakened and that of the other generator is increased till the ammeter of the generator to be cleared reads zero. After that, its breaker and then the switch are opened thus removing the generator out of service. This method of connecting in and removing a generator from service helps in preventing any shock or sudden disturbance to the prime-mover or to the system itself.

It is obvious that if the field of one generator is weakened too much, then power will be delivered to it and it will run in its original direction as a motor, thus driving its prime-mover.

In Fig. 25.19 and 25.20 are shown the voltage characteristics of two shunt generators. It is seen that for a common terminal voltage *V*, the generator No. 1 delivers I_1 amperes and generator No. 2, I_2 amperes. It is seen that generator No. 1, having more drooping characteristic, delivers less current. It is found that two shunt generators will divide the load properly at all points if their characteristics are similar in form and each has the same voltage drop from no-load to full-load.

If it is desired that two generators of different kW ratings automatically share a load in proportion to their ratings, then their external characteristics when plotted in terms of their percentage full-load currents (not actual currents) must be identical as shown in Fig. 25.21. If, for example, a 100-kW generator is working in parallel with a 200-kW generator to supply a total load of 240-kW, then first generator will supply 80 kW and the other 160 kW.

When the individual characteristics of the generators are known, their combined characteristics can be drawn by adding the separate currents at a number of equal voltages (because generators are running in parallel). From this combined characteristic, the voltage for any combined load can be read off and from there, the current supplied by each generator can be found (Fig. 25.20).

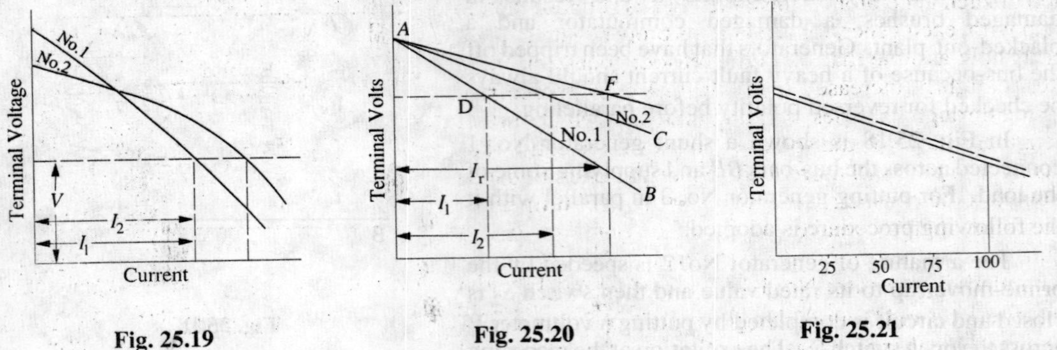

Fig. 25.19 Fig. 25.20 Fig. 25.21

If the generators have straight line characteristics, then the above result can be obtained by simple calculations instead of graphically.

Let us discuss the load sharing of two generators which have unequal no-load voltages.

Let E_1, E_2 = no-load voltages of the two generators

R_1, R_2 = their armature resistances

V = common terminal voltage

Then $$I_1 = \frac{E_1 - V}{R_1} \text{ and } I_2 = \frac{E_2 - V}{R_2}$$

$$\therefore \quad \frac{I_2}{I_1} = \frac{E_2 - V}{E_1 - V} \cdot \frac{R_1}{R_2} = \frac{K_2 N_2 \Phi_2 - V}{K_1 N_1 \Phi_1 - V} \cdot \frac{R_1}{R_2}$$

From the above equation, it is clear that bus-bar voltage can be kept constant (and load can be transferred from 1 to 2) by increasing Φ_2 or N_2 or by reducing N_1 and Φ_1. N_2 and N_1 are changed by changing the speed of driving engines and Φ_1 and Φ_2 are changed with the help of regulating shunt field resistances.

It should be kept in mind that

(*i*) Two parallel shunt generators having equal no-load voltages share the load in such a ratio that the load current of each machine produces the same drop in each generator.

(*ii*) In the case of two parallel generators having unequal no-load voltages, the load currents produce sufficient voltage drops in each so as to keep their terminal voltage the same.

(*iii*) The generator with the least droop assumes greater share of the change in bus load.

(*iv*) Paralleled generators with different power ratings but the same voltage regulation will divide any oncoming bus load in direct proportion to their respective power ratings (Ex. 25.14).

25.17. Procedure for Paralleling D.C. Generators

(*i*) Close the disconnect switch of the incoming generator

(*ii*) Start the prime-mover and adjust it to the rated speed of the machine

(*iii*) Adjust the voltage of the incoming machine a few volts higher than the bus voltage

(*iv*) Close the breaker of the incoming generator

(*v*) Turn the shunt field rheostat of the incoming machine in the raise-voltage direction and that of the other machine(s) already connected to the bus in the lower-voltage direction till the desired load distribution (as indicated by the ammeters) is achieved.

25.18. Compound Generators in Parallel

In Fig. 25.22 are shown two compound generators (designated as No. 1 and No. 2) running in parallel. Because of the rising characteristics of the usual compounded generators, it is obvious that in the absence of any corrective devices, the parallel operation of such generators is unstable. Let us suppose that, to begin with, each generator is taking its proper share of load. Let us now assume that for some reason, generator No. 1 takes a slightly increased load. In that case, the current passing through its series winding increases which further strengthens its field and so raises its generated e.m.f. thus causing it to take still more load. Since the system load is assumed to be constant, generator No. 2 will drop some of its load, thereby weakening its series field which will result in its further dropping off its load. Since this effect is cumulative, generator No. 1 will, therefore, tend to take the entire load and finally drive generator No. 2 as a motor. The circuit breaker of at least one of the two generators will open, thus stopping their parallel operation.

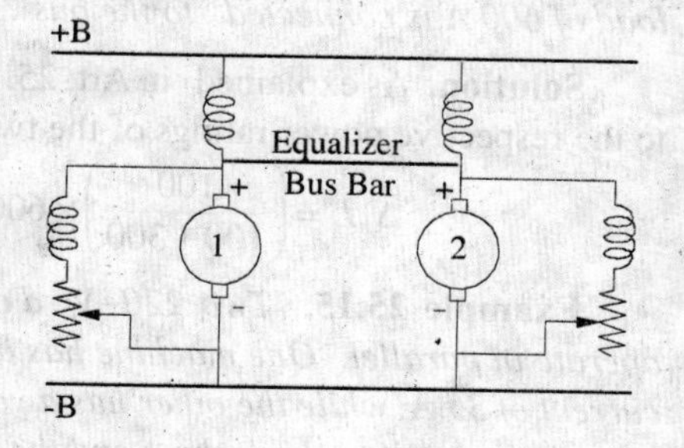

Fig. 25.22

For making the parallel operation of over-compound and level-compound generators stable,* they are always used with an equalizer bar (Fig. 25.22) connected to the armature ends of the series coils of the generators. The equalizer bar is a conductor of low resistance and its operation is as follows :

Suppose that generator No. 1 starts taking more than its proper share of load. Its series field current is increased. But now this increased current passes partly through the series field coil of generator. No. 1 and partly it flows via the equalizer bar through the series field winding of generator No. 2. Hence, the generators are affected in a similar manner with the result the generator No. 1 cannot take the entire load. For maintaining proper division of load from no-load to full-load, it is essential that

(*i*) the regulation of each generator is the same.

(*ii*) the series field resistances are inversely proportional to the generator rating.

25.19. Series Generators in Parallel

Fig. 25.23 shows two identical series generators connected in parallel. Suppose E_1 and E_2 are initially equal, generators supply equal currents and have equal shunt resistances. Suppose E_1 increases slightly so that $E_1 > E_2$. In that case, I_1 becomes greater than I_2. Consequently, field of machine 1 is strengthened thus increasing E_1 further whilst the field of machine 2 is weakened thus decreasing E_2 further. A final stage is reached when machine 1 supplies not only the whole load but also supplies power to machine 2 which starts running as a motor. Obviously, the two machines will form a short-circuited loop and the current will rise indefinitely. This condition can be prevented by using equalizing bar because of which two similar machines pass approximately equal currents to the load, the slight difference between the two currents being conf ined to the loop made by the armatures and the equalizer bar.

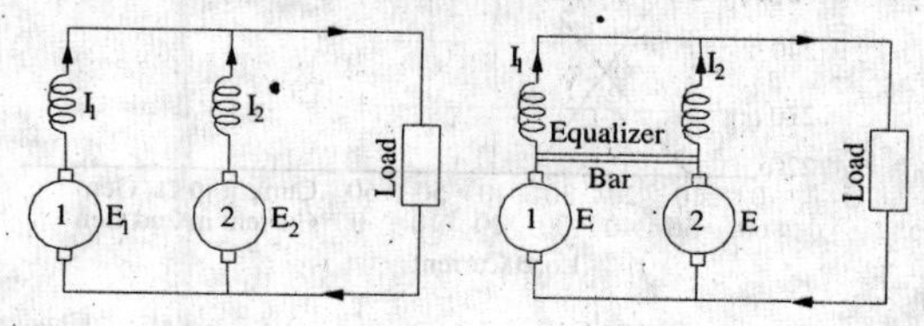

Fig. 25.23

---

*Like shunt generators, the under-compound generators also do not need equalizers for satisfactory parallel operation.

Example 25.14. *A 100-kW, 250-V generator is paralleled with a 300 kW, 250-V generator. Both generators have the same voltage regulation. The first generator is supplying a current of 200 A and the other 500 A. What would be the current supplied by each generator if an additional load of 600 A is connected to the bus ?*

Solution. As explained in Art. 25.17, this additional load would be divided in direct proportion to the respective power ratings of the two generators.

$$\Delta I_1 = \left(\frac{100}{100+300}\right)\times 600 = 150 \text{ A} ; \Delta I_2 = \left(\frac{300}{100+300}\right)\times 600 = 450 \text{ A}$$

Example 25.15. *Two 220–V, d.c. generators, each having linear external characteristics, operate in parallel. One machine has a terminal voltage of 270 V on no-load and 220 V at a load current of 35 A, while the other has a voltage of 280 V at no-load and 220 V at 50 A. Calculate the output current of each machine and the bus-bar voltage when the total load is 60 A. What is the kW output of each machine under this condition ?*

Solution. Generator No. 1.

Voltage drop for 35 A = 270–220 = 50 V

∴ Voltage drop/ampere = 50/35 = 10/7 V/A

Generator No. 2

Voltage drop/ampere = (280–220)/50 = 1.2 V/A

Let V = bus-bar voltage

I_1 = current output of generator No.1

I_2 = current output of generator No. 2

then $V = 270-(10/7)\, I_1$ ...for generator No. 1

$= 280-1.2\, I_2$ ...for generator No. 2

Since bus – bar voltage is the same

$270-10\, I_1/7 = 280-1.2\, I_2$ or $4.2\, I_2 - 5\, I_1 = 35$ ...(*i*)

Also $I_1 + I_2 = 60$ ...(*ii*)

Solving the two equations, we get I_1 = **23.6 A**;

I_2 = **36.4 A**

Now $V = 280-1.2\, I_2$

$= 280-1.2\times 36.4$

= **236.3 V**

Output of Ist machine

$= 236.3\times 23.6/1000$

= **5.577 kW**

Output of 2nd machine

$= 236.3\times 36.4/1000$

= **8.602 kW**

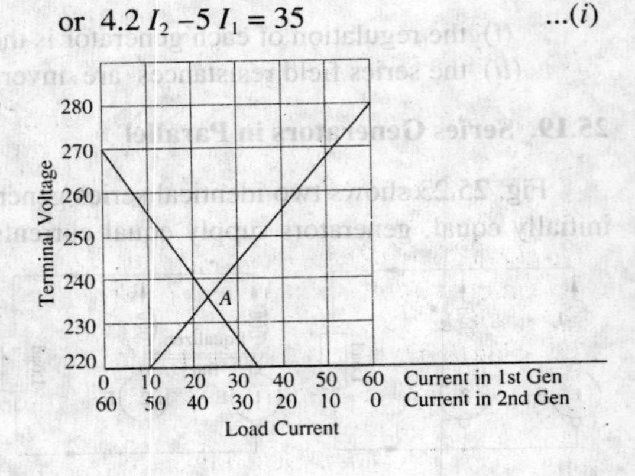

Fig. 25.24

Graphical Solution.

In Fig. 25.24, total load current of 60 A has been plotted along *X*-axis and the terminal voltage along *Y*-axis. The linear characteristics of the two generators are drawn from the given data. The common bus-bar voltage is given by the point of intersection of the two graphs. From the graph, it is seen that V = 236.3 V ; I_1 = 23.6 A ; I_2 = 36.4 A.

Example 25.16. *Two shunt generators each with an armature resistance of 0.01 Ω and field resistance of 20 Ω run in parallel and supply a total load of 4000 A. The e.m.f.s are respectively 210 V and 220 V. Calculate the bus-bar voltage and output of each machine.*

(Electrical Machines-1, South Gujarat Univ. 1988)

Solution. Generators are shown in Fig. 25.25.

Let V = bus-bar voltage
I_1 = output current of G_1
I_2 = output current of G_2

Now, $I_1 + I_2 = 4000$ A, $I_{sh} = V/20$.

$I_{a1} = (I_1 + V/20)$; $I_{a2} = (I_2 + V/20)$

In each machine,

V + armature drop = induced e.m.f.

$\therefore V + I_{a1} R_a = E_1$

or $V + (I_1 + V/20) \times 0.01 = 210$ ...1st machine

Also $V + I_{a2}R_a = E_2$

or $V + (I_2 + V/20) \times 0.01 = 220$...2nd machine

Subtracting, we have $0.01 (I_1 - I_2) = 10$ or $I_1 - I_2 = 1000$

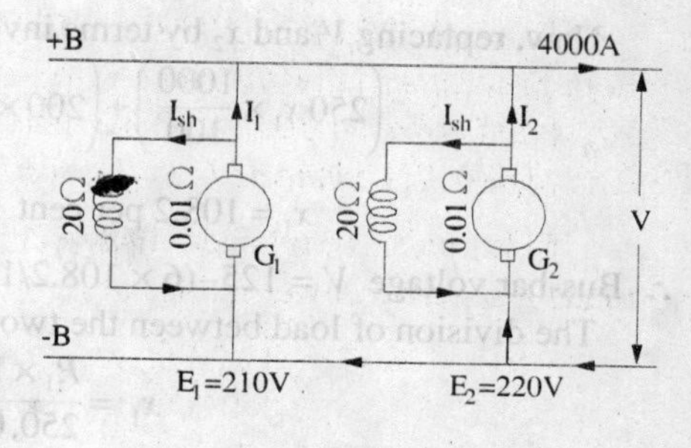

Fig. 25.25

Also, $I_1 + I_2 = 4000$ A $\therefore I_1 =$ **2500 A** ; $I_2 =$ **1500 A**

Substituting the value of I_1 above, we get

$V + (2500 + V/20) \times 0.01 = 210 \quad \therefore V =$ **184.9 V**

Output of Ist generator $= 184.9 \times 2500/1000 =$ **462.25 kW**

Output of 2nd generator $= 184.49 \times 1500/1000 =$ **277.35 kW**

Example 25.17. *Two shunt generators operating in parallel deliver a total current of 250 A. One of the generators is rated 50 kW and the other 100 kW. The voltage rating of both machines is 500 V and have regulations of 6 per cent (smaller one) and 4 per cent. Assuming linear characteristics, determine (a) the current delivered by each machine (b) terminal voltage.*

(Elect. Machines, Nagpur Univ. 1991)

Solution. 50 kW generator

F.L. voltage drop $= 500 \times 0.06 = 30$ V ; F.L. current = 50,000/500 = 100 A

Drop per ampere = 30/100 = 3/10 V/A

100 kW generator

F.L. drop $= 500 \times 0.04 = 20$ V ; F.L. current = 100,000/5000 = 200 A

Drop per ampere = 20/200 = 1/10 V/A

If I_1 and I_2 are currents supplied by the two generators and V the terminal voltage, then

$V = 500 - (3I_1/10)$ –1st generator

$= 500 - (I_2/10)$ –2nd generator

$\therefore \quad 3I_1/10 = I_2/10$ or $3I_1 = I_2$; Also $I_1 + I_2 = 250$ –given

(*a*) Solving the above two equations, we get $I_1 =$ **62.5 A** ; $I_2 =$ **187.5 A**

(*b*) $V = 500 - (3 \times 62.5/10) =$ **481.25 V**

Example 25.18. *Two shunt generators, each with a no–load voltage of 125 V are run in parallel. Their external characteristics can be taken as straight lines over this operating ranges. Generator No. 1 is rated at 250 kW and its full-load voltage is 119 V, Generator No. 2 is rated at 200 kW at 116 V. Calculate the bus-bar voltage when the total load is 3500 A. How is the load divided between the two ?*

(Elect. Machinery - I, Mysore Univ. 1988)

Solution. Let V = bus-bar voltage

x_1, x_2 = load carried by each generator in terms of percentage of rated load

P_1, P_2 = load carried by each generator in watt

$V = 125 - [(125 - 119)(x_1/100)$ ...Generator No. 1

$V = 125 - [(125 - 116)(x_2/100)]$ ...Generator No. 2

$$\therefore \quad 125 - \frac{6x_1}{100} = 125 - \frac{9x_2}{100} \qquad x_2 = \frac{6x_1}{9} = \frac{2x_1}{3}$$

Since in d.c. circuits, power delivered is given by VI watt, the load on both generators is

$$\left(250\,x_1 \times \frac{1000}{100}\right) + \left(200\,x_2 \times \frac{1000}{100}\right) = V \times 3500$$

Now, replacing V and x_2 by terms involving x_1, we get as a result

$$\left(250\,x_1 \times \frac{1000}{100}\right) + \left(200 \times \frac{2x_1}{3} \times \frac{1000}{100}\right) = \left(125 - \frac{6x_1}{100}\right) \times 3500$$

$$x_1 = 108.2 \text{ per cent}$$

$\therefore$ Bus-bar voltage $V = 125-(6 \times 108.2/100) =$ **118.5 V**

The division of load between the two generators can be found thus :

$$x_1 = \frac{P_1 \times 100}{250,000} \quad \text{and} \quad x_2 = \frac{P_2 \times 1000}{200,000}$$

$$\therefore \quad \frac{x_1}{x_2} = \frac{P_1 \times 200,000}{P_2 \times 250,000} = \frac{4P_1}{5P_2} = \frac{3}{2} \quad \therefore \quad \frac{3}{2} = \frac{4P_1}{5P_2} = \frac{4VI_1}{5VI_2} = \frac{4I_1}{5I_2} \quad \ldots(i)$$

Since $\quad I_1 + I_2 = 3500 \quad \therefore \quad I_2 = 3500 - I_1$

Hence (*i*) above becomes, $\dfrac{3}{2} = \dfrac{4I_1}{5(3500 - I_1)}$

$\therefore \quad I_1 =$ **2,283 A** and $\quad I_2 =$ **1,217 A**

Example 25.19. *In a certain sub-station, there are 5 d.c. shunt generators in parallel, each having an armature resistance of 0.1 Ω, running at the same speed and excited to give equal induced e.m.f.s. Each generator supplies an equal share of a total load of 250 kW at a terminal voltage of 500 V into a load of fixed resistance. If the field current of one generator is raised by 4%, the others remaining unchanged, calculate the power output of each machine and their terminal voltages under these conditions. Assume that the speeds remain constant and flux is proportional to field current.*

(Elect. Technology, Allahabad Univ. 1991]

Solution. Generator connections are shown in Fig. 25.26.

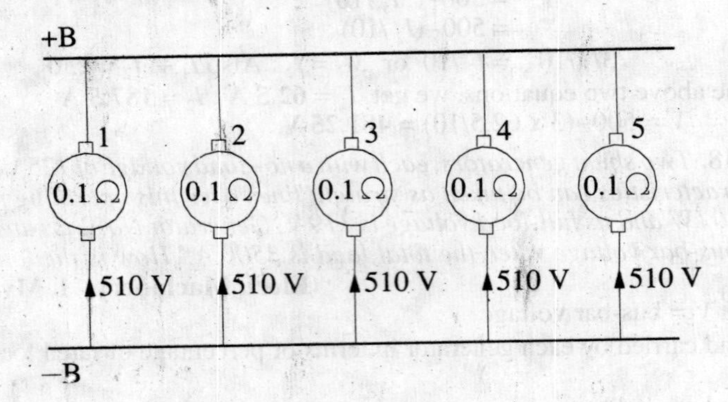

Fig. 25.26

Load supplied by each $= 250/5 = 50$ kW $\therefore$ Output of each $= 50,000/500 = 100$ A

Terminal voltage of each $= 500$ V Armature drop of each $= 0.1 \times 100 = 10$ V

Hence, induced e.m.f. of each $= 510$ V

When field current of one is increased, its flux and hence its generated e.m.f. is increased by 4%. Now, 4% of 510 V = 20.4 V

$\therefore$ Induced e.m.f. of one $= 510 + 20.4 = 530.4$ V

Let $\quad I_1$ = current supplied by one generator after increased excitation

I_2 = current supplied by each of the other 4 generators

V = new terminal or bus-bar voltage

$\therefore$
$$530.4 - 0.1\,I_1 = V \qquad ...(i)$$
$$510 - 0.1\,I_2 = V \qquad ...(ii)$$

Now, fixed resistance of load = 500/500 = 1 Ω ; Total load current = $I_1 + 4I_2$

$\therefore$
$$1 \times (I_1 + 4I_2) = V \quad \text{or} \quad I_1 + 4I_2 = V \qquad ...(iii)$$

Subtracting (*ii*) from (*i*), we get, $I_1 - I_2 = 204$ ...(*iv*)

Subtracting (*iii*) from (*ii*), we have $I_1 + 4.1\,I_2 = 510$ ...(*v*)

From (*iv*) and (*v*), we get I_2 = 3060/51 = 59/99 = 60 A (approx.)

From (*iv*) I_1 = 204 + 60 = 264 A

From (*iii*) V = 264 + 240 = **504 Volt**

Output of Ist machine = 504 × 264 watt = **133 kW**

Output of *each* of other four generators = 504 × 60 W = **30.24 kW**

Example 25.20. *Two d.c. generators are connected in parallel to supply a load of 1500 A. One generator has an armature resistance of 0.5 Ω and an e.m.f. of 400 V while the other has an armature resistance of 0.04 Ω and an e.m.f. of 440 V. The resistances of shunt fields are 100 Ω and 80 Ω respectively. Calculate the currents I_1 and I_2 supplied by individual generator and terminal voltage V of the combination.* **(Power Apparatus-I, Delhi Univ. Dec. 1987).**

Solution. Generator connecion diagram is shown in Fig. 25.27.

Let V = bust–bar voltage

I_1 = output current of one generator

I_2 = output current of other generator = $(1500 - I_1)$]

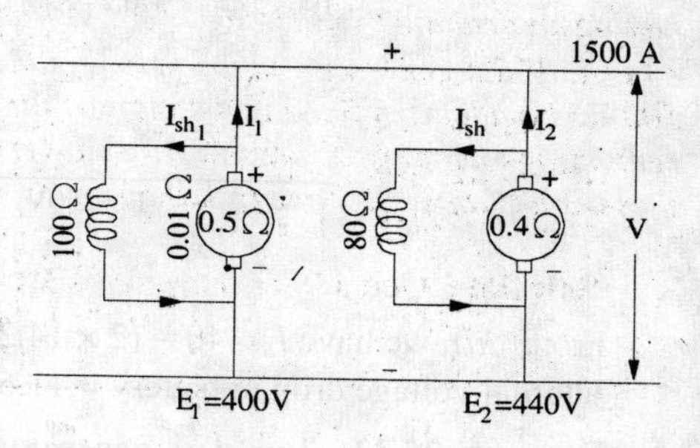

Fig. 25.27

Now, $I_{sh1} = V/100$ A ; $I_{sh2} = V/80$ A

$$I_{a1} = \left(I_1 + \frac{V}{100}\right) \text{ and } I_{a2} = \left(I_2 + \frac{V}{80}\right)$$

or
$$I_{a2} = \left(1500 - I_1 + \frac{V}{80}\right)$$

For each machine

E – armature drop = V, $\therefore$ $400 - \left(I_1 + \frac{V}{100}\right) \times 0.5 = V$

or $400 - 0.5\,I_1 - 0.005\,V = V$ or $0.5\,I_1 = 400 - 1.0005 = V$ ...(*i*)

Also $440 - \left(1500 - I_1 + \frac{V}{80}\right) \times 0.04 = V$ or $0.04\,I_1 = 1.0005\,V - 380$ ...(*ii*)

Dividing Eq. (*i*) by (*ii*), we get

$$\frac{0.5\,I_1}{0.04\,I_1} = \frac{400 - 1.005\,V}{1.0005\,V - 380} \qquad \therefore\ V = \mathbf{381.2\ V}$$

Substituting this value of V in Eq. (*i*), we get $0.5\,I_1 = 400 - 1.005 \times 381.2$

$\therefore$ I_1 = **33.8 A** ; I_2 – 1500 – 33.8 = **1466.2 A**

Output of Ist generator = 381.2 × 33.8 × 10^{-3} = **12.88 kW**

Output of 2nd generator = 381.2 × 1466.2 × 10^{-3} = **558.9 kW**

Example 25.21. *Two shunt generators and a battery are working in parallel. The open circuit voltage, armature and field resistances of generators are 250 V, 0.24 Ω, 100 Ω are 248 V, 0.12 Ω and 100 Ω respectively. If the generators supply the same current when the load on the bus-bars is 40 A, calculate the e.m.f. of the battery if its internal resistance is 0.172 Ω.*

Solution. Parallel combination is shown in Fig. 25.28.

Values of currents and induced e.m.fs. are shown in the diagram.

$$V + \left(I + \frac{V}{100}\right) \times 0.24 = 250 \qquad \ldots(i)$$

$$V + \left(I + \frac{V}{100}\right) \times 0.12 = 248 \qquad \ldots(ii)$$

Also

$$I + I + I_b = 40 \qquad \ldots(iii)$$

$$I_b + 2I = 40$$

Subtracting (*ii*) from (*i*), we get $\left(I + \frac{V}{100}\right) \times 0.12 = 2$ ...(*iv*)

Putting this value in (*ii*) above, V = 246 volt.

Putting this value of V in (*iv*), $\left(1 + \frac{246}{100}\right) \times 0.12 = 2$ ...(*v*)

$\therefore \quad I = 50/3 - 2.46 = 14.2$ A

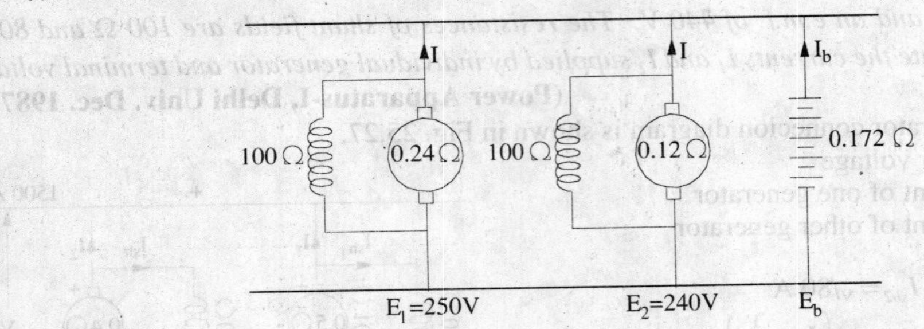

Fig. 25.28

From (*iii*), we have I_b = 40 − (2 × 14.2) = 11.6 A

Internal voltage drop in battery = 11.6 x 0.172 = 2 V $\therefore$ E_b = 246 + 2 = **248 V**

Example 25.22. *Two d.c. generators A and B are connected to a common load. A has a constant e.m.f. of 400 V and internal resistance of 0.25 Ω while B has a constant e.m.f. of 410 V and an internal resistance of 0.4 Ω. Calculate the current and power output from each generator if the load voltage is 390 V. What would be the current and power from each and the terminal voltage if the load was open-circuited ?* **(Elect. Engg; I, Bangalore Univ. 1987)**

Solution. The generator connections are shown in Fig. 25.29(*a*).

Since the terminal or output voltage is 390V, hence

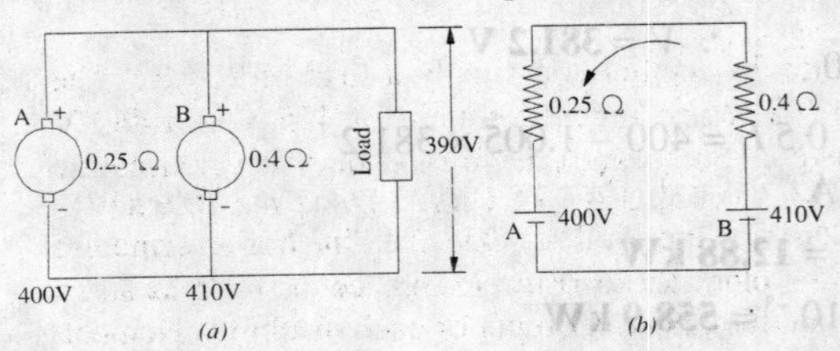

Fig. 25.29.

Load supplied by A = (400−390)/0.25 = **40 A**

Load supplied by **B** = (410−390)/0.4 = **50 A**

$\therefore$ Power output from A = 40 × 390 = **15.6 kW**

Power output from B = 50 × 390 = **19.5 kW**

If the load is open−circuited as shown in Fig. 25.29(*b*), then the two generators are put in series with each other and a circulatory current is set up between them.

Net voltage in the circuit	= 410−400 = 10 V
Total resistance	= 0.4 + 0.25 = 0.65 Ω
∴ circulatory current	= 10/0.65 = 15.4 A
The terminal voltage	= 400 + (15.4 × 0.25) = **403.8 V**

Obviously, machine B with a higher e.m.f. acts as a generator and drives machine A as a motor.

Power taken by A from B = 403.8 × 15.4 = **6,219 W**

Part of this appears as mechanical output and the rest is dissipated as armature Cu loss.

Mechanical output = 400 × 15.4 = **6.16 kW** ; Armature Cu loss = 3.8 × 15.4 = **59 W**

Power supplied by B to A = **6,219 W** ; Armature Cu loss = 6.16 × 15.4 = **95 W**

Example 25.23. *Two compound generators A and B, fitted with an equalizing bar, supply a total load current of 500 A. The data regarding the machines are :-*

	A	*B*
Armature resistance (ohm)	*0.01*	*0.02*
Series field winding (ohm)	*0.004*	*0.006*
Generated e.m.fs. (volt)	*240*	*244*

Calculate (a) current in each armature (b) current in each series winding (c) the current flowing in the equalizer bar and (d) the bus-bar voltage. Shunt currents may be neglected.

Solution. The two generators (with their shunt windings omitted) are shown in Fig. 25.30.

Let V = bus-bar voltage ; v = voltage between equalizer bus-bar and the negative

i_1, i_2 = armature currents of the two generators

Now , $i_1 + i_2 = 500$

or $$\frac{240 - v}{0.01} + \frac{244 - v}{0.01} = 500$$

Multiplying both sides by $\frac{1}{100}$ we get

$240 - v + 122 - (v/2) = 5$

$\therefore\ v = 238$ volt

(*a*) $\therefore\ i_1 = \frac{240 - 238}{0.01} =$ **200 A**

$i_2 = \frac{244 - 238}{0.02} =$ **300 A**

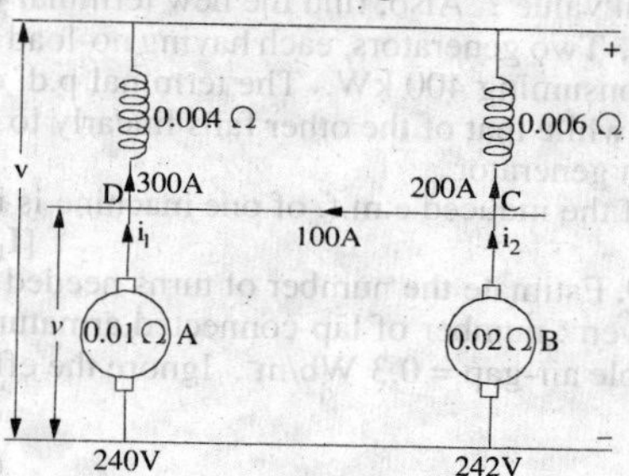

Fig. 25.31

(*b*) The total current of 500 A divides between the series windings in the inverse ratio of their resistance *i.e.* in the ratio of $\frac{1}{0.004} : \frac{1}{0.006}$ or in the ratio 3 : 2

Hence, current in the series winding of generator **A** = 500 × 3/5 = **300 A**

Similarly, current in the series winding of generator **B** = 500 × 2/5 = **200 A**

(*c*) It is obvious that a current of 100 A flows in the equalizing bar from C to D. It is so because the armature current of generator A is 200 A only. It means that 100 A comes from the armature of generator, B, thus making 300 A for the series field winding of generator A.

(*d*) $V = v$ – voltage drop in one series winding = 238 – (300 × 0.004) = **236.8 V**

Tutorial Problem No. 25.2

1. Two separately-excited d.c. generators are connected in parallel and supply a load of 200 A. The machines have armature circuit resistances of 0.05 Ω and 0.1 Ω and induced e.m.fs. of 425 V and 440 V respectively. Determine the terminal voltage, current and power output of each machine. The effect of armature reaction is to be neglected. **[423.3 V ; 33.3 A ; 14.1 kW ; 166.7 A ; 70.6 kW]**

2. Two shunt generators operating in parallel give a total output of 600 A. One machine has an armature resistance of 0.02 Ω and a generated voltage of 455 V and the other an armature resistance of 0.025 Ω and a generated voltage of 460 V. Calculate the terminal voltage and the kilowatt output of each machine. Neglect field currents. **[450.56 V; 100 kW ; 170.2 kW]**

3. The external characteristics of two d.c. shunt generators A and B are straight lines over the working range between no-load and full-load.

	Generator A		*Generator B*	
	Nos-load	*Full-load*	*No-load*	*Full-load*
Terminal P.D. (V)	400	360	420	370
Load current (A)	0	80	0	70

Determine the common terminal voltage and output current of each generator when sharing a total load of 100 A. **[57.7 A ; 42.3 A ; 378.8 V]**

4. Two shunt generators operating in parallel have each an armature resistance of 0.02 Ω. The combined external load current is 2500 A. If the generated e.mfs. of the machines are 560 V and 550 V respectively, calculate the bus–bar voltage and output in kW of each machine. **[530 V ; 795 kW ; 530 kW]**

5. Two shunt generators *A* and *B* operate in parallel and their load characteristics may be taken as straight lines. The voltage of *A* falls from 240 *V* at no-load to 220 V at 200 A, while that of *B* falls from 245 V at no-load to 220 V at 150 A. Determine the current which each machine supplies to a common load of 300 A and the bus-bar voltage at this load. **[169 A; 131 A ; 223.1 V]**

6. Two shunt-wound d.c. generators are connected in parallel to supply a load of 5,000 A. Each machine has an armature resistance of 0.03 Ω and a field resistance of 60 Ω, but the e.m.f. of one machine is 600 V and that of the other is 640 V. What power does each machine supply ? **[1,004 kW ; 1,730 kW including the fields]**

7. Two shunt generators running in parallel share a load of 100 kW equally at a terminal voltage of 230 V. On no-load, their voltages rise to 240 V and 245 V respectively. Assuming that their volt-ampere characteristics are rectilinear, find how would they share the load when the total current is reduced to half its original value ? Also, find the new terminal voltage. **[20 kW ; 30 kW, 236 V]**

8. Two generators, each having no-load voltage of 500 V, are connected in parallel to a constant resistance load consuming 400 kW. The terminal p.d. of one machine falls linearly to 470 V as the load is increased to 850 A while that of the other falls linearly to 460 V when the load is 600 A. Find the load current and voltage of each generator.

If the induced e.m.f. of one machine is increased to share load equally, find the new current and voltage. **[I_1 = 626 A ; I_2 = 313 A ; V = 479 V ; I = 469.5 A ; V = 484.4 V]**

9. Estimate the number of turns needed on each interpole of a 6-pole generator delivering 200 kW at 200 V ; given : number of lap-connected armature conductors = 540 ; interpole air gap = 1.0 cm ; flux-density in interpole air-gap = 0.3 Wb/m^2. Ignore the effect of iron parts of the circuit and of leakage. **[10]** (*Electrical Machines, B.H.U. 1980*)

Objective Tests –25

1. In d.c. generators, armature reaction is produced actually by
(*a*) its field current
(*b*) armature conductors
(*c*) field pole winding
(*d*) load current in armature

2. In a d.c generator, the effect of armature reaction on the main pole flux is to
(*a*) reduce it
(*b*) distort it
(*c*) reverse it
(*d*) both (*a*) and (b)

3. In a clockwise-rotating loaded dc generator, brushes have to be shifted
(*a*) clockwise
(*b*) counterclockwise
(*c*) either (*a*) or (*b*)
(*d*) neither (*a*) nor (*b*).

4. The primary reason for providing compensating windings in a d.c. generator is to
(*a*) compensate for decrease in main flux
(*b*) neutralize armature mmf
(*c*) neutralize cross-magnetising flux
(*d*) maintain uniform flux distribution.

5. The main function of interpoles is to minimize —— between the brushes and the commutator when the d.c. machine is loaded.
(*a*) friction (*b*) sparking
(*c*) current (*d*) wear and tear

6. In a 6-pole d.c. machine, 90 mechanical degrees correspond to —— electrical degrees.
(*a*) 30 (*b*) 180
(*c*) 45 (*d*) 270

7. The most likely cause/s of sparking at the brushes in a d.c. machine is/are
(*a*) open coil in the armature
(*b*) defective interpoles
(*c*) incorrect brush-spring pressure
(*d*) all of the above

8. In a 10-pole, lap-wound d.c. generator, the number of active armature conductors per pole is 50. The number of compensating conductors per pole required is
(*a*) 5 (*b*) 50
(*c*) 500 (*d*) 10

9. The commutation process in a d.c. generator basically involves
(*a*) passage of current from moving armature to a stationary load
(*b*) reversal of current in an armature coil as it crosses MNA
(*c*) conversion of a.c. to d.c.
(*d*) suppression of reactance voltage

10. Point out the WRONG statement. In d.c. generators, commutation can be improved by
(*a*) using interpoles
(*b*) using carbon brushes in place of Cu brushes
(*c*) shifting brush axis in the direction of armature rotation
(*d*) non of the above

11. Each of the following statements regarding interpoles is true **except**
(*a*) they are small yoke-fixed poles spaced in between the main poles
(*b*) they are connected in parallel with the armature so that they carry part of the armature current

(*c*) their polarity, in the case of generators is the same as that of the main pole ahead
(*d*) they automatically neutralize not only reactance voltage but cross-magnetisation as well

12. Shunt generators are most suited for stable parallel operation because of their–voltage characteristics.
(*a*) identical
(*b*) drooping
(*c*) linear
(*d*) rising

13. Two parallel shunt generators will divide the total load equally in proportion to their kilowatt output ratings only when they have the same
(*a*) rated voltage
(*b*) voltage regulation
(*c*) internal I_aR_a drops
(*d*) both (*a*) and (*b*)

14. The main function of an equalizer bar is to make the parallel operation of two over-compounded dc generators
(*a*) stable
(*b*) possible
(*c*) regular
(*d*) smooth

15. The *essential* condition for stable parallel operation A Two d.c. gerators having similar characteristics is that they should have
(*a*) same kilowatt output rattings
(*b*) drooping voltage characterisitcs
(*c*) same percentage reulation
(*d*) same no-load and full-load speed

16. The main factor which leads to unstable parallel operation of flat-and over-compound d.c. generators is
(*a*) unequal number of turns in their series field windings
(*b*) unequal series field resistances
(*c*) their rising voltage characteristics
(*d*) unequal speed regulation of their prime movers

17. The simplest way to shift load from one dc shunt generator running in parallel with another is to
(*a*) adjust their field rheostats
(*b*) insert resistance in their armature circuits
(*c*) adjust speeds of their prime movers
(*d*) use equalizer connections

18. Which one of the following types of generators does NOT need equalizers for satisfactory parallel operation ?
(*a*) series
(*b*) over-compound
(*c*) flat-compound
(*d*) under-compound.

Answers

1. *d* **2.** *d* **3.** *a* **4.** *c* **5.** *b* **6.** *d* **7.** *d* **8.** *a* **9.** *b* **10.** *d*
11. *b* **12.** *b* **13** *d* **14.** *a* **15** *b* **16.** *c* **17.** *a* **18.** *d*

26 GENERATOR CHARACTERISTICS

26.1. Characteristics of D.C. Generators

Following are the three *most* important characteristics or curves of a d.c. generator :

1. No–load saturation Characteristic (E_o/I_f)

It is also known as Magnetic Characteristic or Open-circuit Characteristic (*O.C.C.*). It shows the relation between the no-load generated e.m.f. in armature, E_0 and the field or exciting current I_f at a given fixed speed. It is just the magnetisation curve for the material of the electromagnets. Its shape is practically the same for **all** generators whether separately-excited or self-excited.

2. Internal or Total Characteristic (E/I_a)

It gives the relation between the e.m.f. *E actually* induced in the armature (after allowing for the demagnetising effect of armature reaction) and the armature current I_a. This characteristic is of interest mainly to the designer and can be obtained as explained in Art. 26-12.

3. External Characteristic (V/I)

It is also referred to as performance characteristic or sometimes voltage-regulating curve.

It gives relation between that terminal voltage *V* and the load current *I*. This curve lies below the internal characteristic because it takes into account the voltage drop over the armature circuit resistance. The values of *V* are obtained by subtracting I_aR_a from corresponding values of *E*. This characteristic is of great importance in judging the suitability of a generator for a particular purpose. It may be obtained in two ways (*i*) by making simultaneous measurements with a suitable voltmeter and an ammeter on a loaded generator (Art. 26.10) or (*ii*) graphically from the O.C.C. provided the armature and field resistances are known and also if the demagnetising effect (under rated load conditions) or the armature reaction (from the short-circuit test) is known.

26.2. Separately-excited Generator

(*a*) (*i*) No-load Saturation Characteristic (E_0/I_f)

The arrangement for obtaining the necessary data to plot this curve is shown in Fig. 26.1. The exciting or field current I_f is obtained from an external independent d.c. source. It can be varied from zero upwards by a potentiometer and its value read by an ammeter *A* connected in the field circuit as shown.

Now, the voltage equation of a d.c. generator is, $E_g = \frac{\Phi ZN}{60} \times \left(\frac{P}{A}\right)$ volt

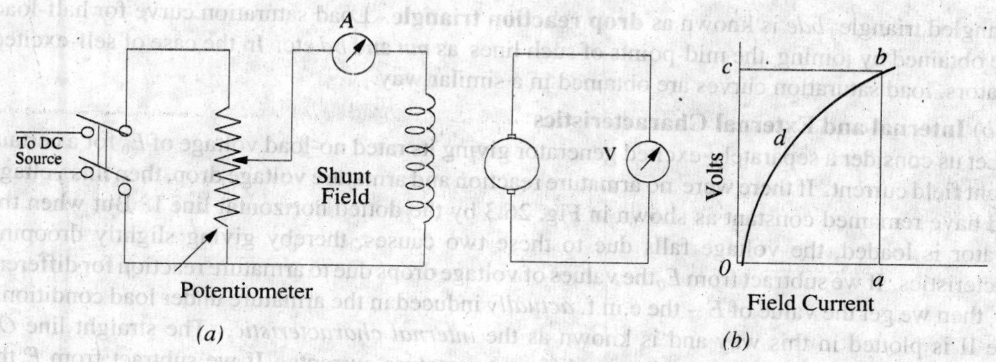

Fig. 26.1

Hence, if speed is constant, the above relation becomes $E = k\,\Phi$

It is obvious that when I_f is increased from its initial small value, the flux Φ and hence generated e.m.f. E_g increases directly as current so long as the poles are unsaturated. This is represented by the straight portion *od* in Fig. 26.1 (*b*). But as the flux density increases, the poles become saturated, so a greater increase in I_f is required to produce a given increase in voltage than on the lower part of the curve. That is why the upper portion *db* of the curve *odb* bends over as shown.

(*ii*) **Load Saturation Curve (V/I_f)**

The curve showing relation between the terminal voltage V and field current I_f when the generator is loaded, is known as Load Saturation Curve.

The curve can be deduced from the no-load saturation curve provided the values of armature reaction and armature resistance are known. While considering this curve, account is taken of the

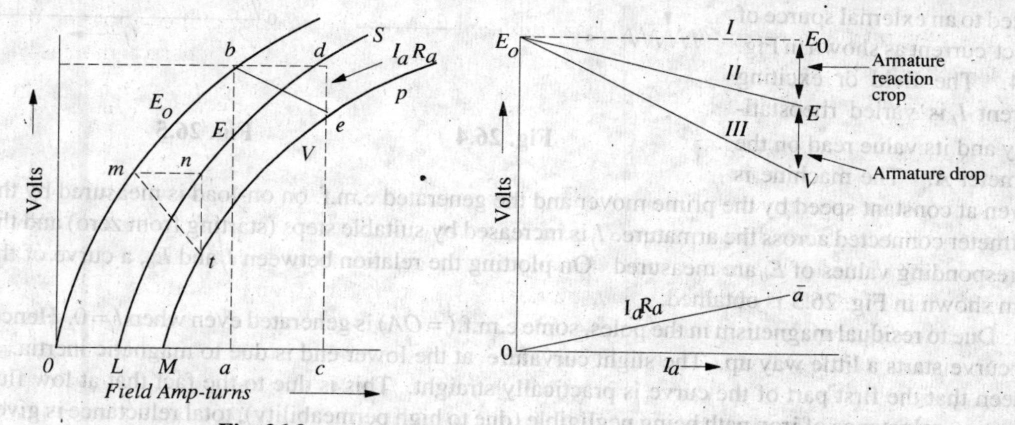

Fig. 26.2

Fig. 26.3

demagnetising effect of armature reaction and the voltage drop in armature which are practically absent under no-load conditions. The no-load saturation curve of Fig. 26.1 has been repeated in Fig. 26.2 on a base of field amp-turns (and not current) and it is seen that at no-load, the field amp-turns required for rated no-load voltage are given by *Oa*. Under load conditions, the voltage will decrease due to demagnetising effect of armature reaction. This decrease can be made up by suitably increasing the field amp-turns. Let *ac* represent the equivalent demagnetising amp-turns per pole. Then, it means that in order to generate the same e.m.f. on load as at no-load, the field amp-turns/pole must be increased by an amount *ac* = *bd*. The point *d* lies on the curve *LS* which shows relation between the voltage E generated under *load* conditions and the field amp-turns. The curve *LS* is practically parallel to curve *ob*. The terminal voltage V will be less than this generated voltage E by an amount $= I_a R_a$ where R_a is the resistance of the armature circuit. From point *d*, a vertical line *de* $= I_a R_a$ is drawn. The point *e* lies on the full-load saturation curve for the generator. Similarly, other points are obtained in the same manner and the full-load saturation curve *Mp* is drawn. The right-angled triangle *bde* is known as **drop reaction triangle**. Load saturation curve for half-load can be obtained by joining the mid-points of such lines as *mn* and *bd* etc. In the case of self-excited generators, load saturation curves are obtained in a similar way.

(*b*) **Internal and External Characteristics**

Let us consider a separately-excited generator giving its rated no-load voltage of E_0 for a certain constant field current. If there were no armature reaction and armature voltage drop, then this voltage would have remained constant as shown in Fig. 26.3 by the dotted horizontal line I. But when the generator is loaded, the voltage falls due to these two causes, thereby giving slightly drooping characteristics. If we subtract from E_0 the values of voltage drops due to armature reaction for different loads, then we get the value of E – the e.m.f. *actually* induced in the armature under load conditions. Curve II is plotted in this way and is known as the *internal characteristic*. The straight line *Oa* represents the I_aR_a drops corresponding to different armature currents. If we subtract from E the armature drop I_aR_a, we get terminal voltage V. Curve *III* represents the external characteristic and is obtained by subtracting ordinates of line *Oa* from those of curve *II*.

26.3. No-load Curve for Self-excited Generator

The *O.C.C.* or no-load saturated curves for self-excited generators whether shunt or series-connected, are obtained in a similar way.

The field winding of the generator (whether *shunt* or *series* wound) is disconnected from the machine and connected to an external source of direct current as shown in Fig. 26.4. The field or exciting current I_f is varied rheostatically and its value read on the ammeter *A*. The machine is driven at constant speed by the prime mover and the generated e.m.f. on on-load is measured by the voltmeter connected across the armature. I_f is increased by suitable steps (starting from zero) and the corresponding values of E_0 are measured. On plotting the relation between I_f and E_0, a curve of the form shown in Fig. 26.5 is obtained.

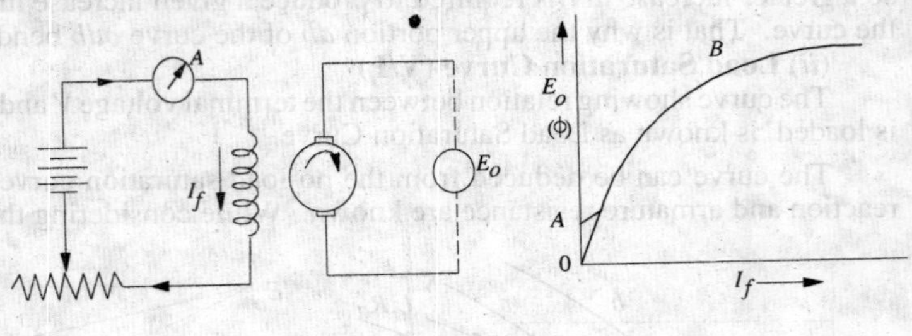

Fig. 26.4 Fig. 26.5

Due to residual magnetism in the poles, some e.m.f.($= OA$) is generated even when $I_f = 0$. Hence, the curve starts a little way up. The slight curvature at the lower end is due to magnetic inertia. It is seen that the first part of the curve is practically straight. This is due to the fact that at low flux densities, reluctance of iron path being negligible (due to high permeability), total reluctance is given by the air-gap reluctance which is constant. Hence, the flux and consequently, the generated e.m.f. is directly proportional to the exciting current. However, at high flux densities, – where μ is small, iron path reluctance becomes appreciable and straight relation between E and I_f no longer holds good. In other words, after point B, saturation of poles starts. However, the initial slope of the curve is determined by air-gap width.

It should be noted that *O.C.C.* for a higher speed would lie above this curve and for a lower speed, would lie below it.

It should also be noted that the load-saturation curve for a shunt generator is similar to the one shown in Fig. 26.2.

Critical Resistance for Shunt Generator

Now, connect the field windings back to the armature and run the machine as a shunt generator.

Due to residual magnetism, some initial e.m.f. and hence current, would be generated. This current while passing through the field coils will strengthen the magnetism of the poles (provided field coils are properly connected as regards polarity). This will increase the pole flux which will further increase the generated e.m.f. Increased e.m.f. means more current which further increases the flux and so on. This mutual reinforcement of e.m.f. and flux proceeds on till equilibrium is reached at some point like P (Fig. 26.6). The point lies on the resistance line OA of the field winding. Let R be the resistance of the field winding. Line OA is drawn such that its slope equals the field winding resistance *i.e.* every point on this curve is such that volt/ampere = R

The voltage OL corresponding to point P represents the maximum voltage to which the machine will build up with R as field resistance. OB represents smaller resistance and the corresponding voltage OM is slightly greater than OL. If field resistance is increased, then slope of the resistance line increased, and hence the maximum voltage to which the generator will build up at a given speed, decreases. If R is increased so much that the resistance line does not cut the $O.C.C.$ at all (like OT), then obviously the machine will fail to excite *i.e.* there will be no 'build up' of the voltage. If the resistance line just lies along the slope, then with that value of field resistance, the machine will *just* excite. The value of the resistance represented by the tangent to the curve, is known as **critical resistance** R_c for a *given speed.*

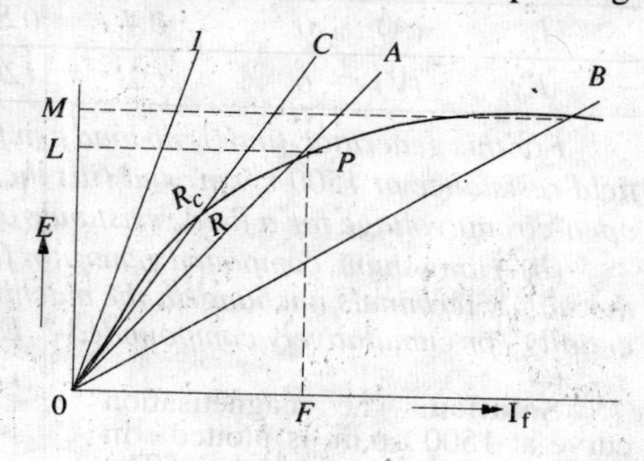

Fig. 26.6

26.4. How to Find Critical Resistance R_c ?

First, $O.C.C.$ is plotted from the given data. Then, tangent is drawn to its initial portion. The slope of this curve gives the critical resistance for the speed at which the data was obtained.

26.5. How to Draw *O.C.C.* at Different Speeds ?

Suppose we are given the data for $O.C.C.$ of a generator run at a fixed speed, say, N_1. It will be shown that $O.C.C.$ at any other constant speed N_2 can be deduced from the $O.C.C.$ for N_1. In Fig. 26.7 is shown the $O.C.C.$ for speed N_1.

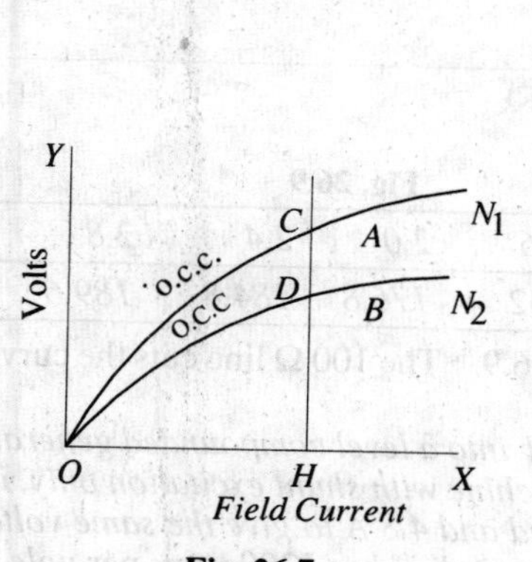

Fig. 26.7

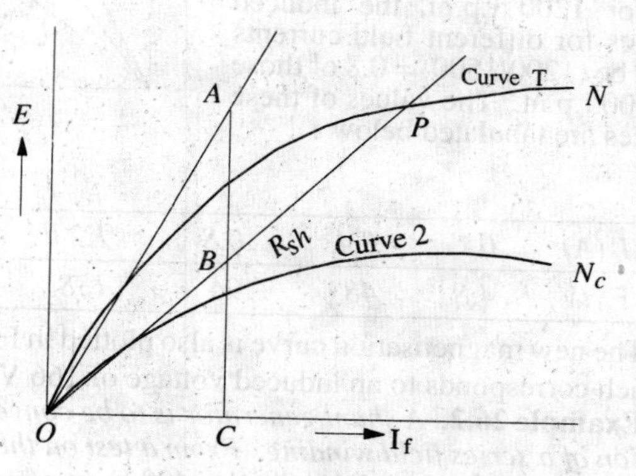

Fig. 26.8

Since $E \propto N$ for any fixed excitation, hence, $\frac{E_2}{E_1} = \frac{N_2}{N_1}$ or $E_2 = E_1 \times \frac{N_2}{N_1}$

As seen, for $I_f = OH$, $E_1 = HC$. The value of new voltage for the same I_f but at N_2

$$E_2 = HC \times \frac{N_2}{N_1} = HD$$

In this way, point D is located. In a similar way, other such points can be found and the new $O.C.C.$ at N_2 drawn.

26.6. Critical Speed N_C

Critical speed of a shunt generator is that speed for which the given shunt field resistance represents critical resistance. In Fig. 26.8, curve 2 corresponds to critical speed because R_{sh} line is tangential to it. Obviously

$$\frac{BC}{AC} = \frac{N_c}{\text{Full speed}} = \frac{N_c}{N} \quad \therefore \quad N_c = \frac{BC}{AC} \times \text{full speed } N$$

Example 26.1. *The magnetization curve of a d.c. shunt generator at 1500 r.p.m. is:*

I_f	(A) :	0	0.4	0.8	1.2	1.6	2.0	2.4	2.8	3.0
E_0	(V) :	6	60	120	172.5	202.5	221	231	237	240

For this generator find (i) no load e.m.f. for a total shunt field resistance of 100 Ω (ii) the critical field resistance at 1500 r.p.m. and (iii) the magnetization curve at 1200 r.p.m. and therefrom the open-circuit voltage for a field resistance of 100 Ω

(b) A long shunt, compound generator fitted with interpoles is cummutatively-compounded. With the supply terminals unchanged, the machine is now run as compound motor. Is the motor differentially or cumulatively compounded ? Explain. **(Elect, Machines, A.M.I.E. Sec B, 1990)**

Solution. The magnetisation curve at 1500 r.p.m. is plotted in Fig. 26.9 from the given data. The 100 Ω resistance line OA is obtained by joining the origin (0, 0) with the point (1A, 100 V). The voltage corresponding to point A is 227.5 V. Hence, no-load voltage to which the generator will build-up is 227.5 V.

The tangent OT represents the critical resistance at 1500 r.p.m. considering point B, R_C = 225/1.5 = 150 Ω

For 1200 r.p.m, the induced voltages for different field currents would be (1200/1500) = 0.8 of those for 1500 r.p.m. The values of these voltages are tabulated below :

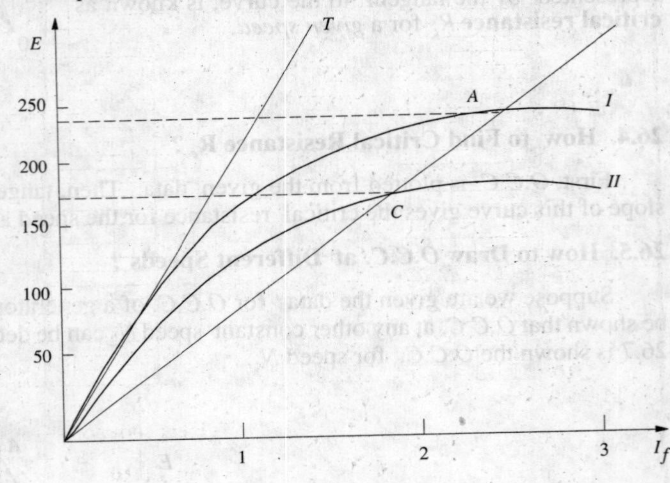

Fig. 26.9

I_f (A)	0	0.4	0.8	1.2	1.6	2.0	2.4	2.8	3.0
E_0 (V)	4.8	48	96	138	162	176.8	184.8	189.6	192

The new magnetisation curve is also plotted in Fig. 26.9. The 100 Ω line cuts the curve at point C which corresponds to an induced voltage of 166 V.

Example 26.2. *A shunt generator is to be converted into a level compounded generator by the addition of a series field winding. From a test on the machine with shunt excitation only, it is found that the shunt current is 3.1 A to give 400 V on no = load and 4.8 A to give the same voltage when the machine is supplying its full load of 200 A. The shunt winding has 1200 turns per pole. Find the number of series turns required per pole.* **(Elect. Machines, A.M.I.E. Sec. B, 1989)**

Solution. At no-load the ampere turns required to produce 400 V = 3.1 × 1200 = 3720

On full-load ampere turns required to produce the same voltage = 4.8 × 1200 = 5760

Additional ampere turns required due to de-magnetising effect of load current = 5760 − 3720 = 2040

If N is a number of series turns required when load current is 200 A, then

$$N \times 200 = 2040, N = 10.2$$

Example 26.3. *The open-circuit characteristic of a.d.c. shunt generator driven at rated speed is as follows :-*

Field Amperes	0.5	1.0	1.5	2.0	2.5	3.0	3.5 A
Induced Voltage	60	120	138	145	149	151	152 V

If resistance of field circuit is adjusted to 53 Ω, calculate the open circuit voltage and load current when the terminal voltage is 100 V. Neglect armature reaction and assume an armature resistance of 0.1 Ω. **(Electrical Technology Punjab Univ. Dec. 1989)**

Solution. Take $I_f = 3$ A, $R_{sh} = 53$ Ω, ∴

Point A is (3A, 159V) point. Line *OA* is the 53 Ω line in Fig. 26.1. It cut drop = 3 × 53 = 159 V. *O.C.C.* at *B*. Line *BM* is drawn parallel to the base. *OM* represents the *O.C.* voltage which equals **150 V.**

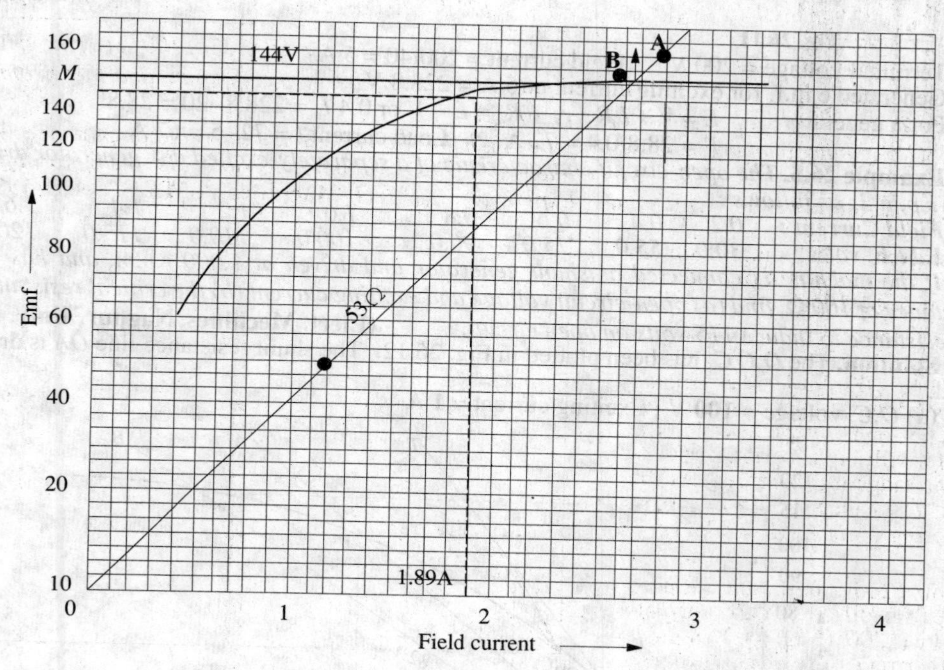

Fig. 26.10

Now, when $V = 100$ V, $I_{sh} = I_f = 100/53 = 1.89$ A

Generated or *O.C.* voltage corresponding to this exciting current as seen from graph of Fig. 26.10 is 144 V.

Now $\quad E = V + I_a R_a \quad$ or $I_a R_a = 144 - 100 = 44$ V

∴ $\quad 0.1\, I_a = 44$ or $I_a = 44/0.1 =$ **440 A**

Example 26.4. *The following figures give the O.C.C. of a.d.c. shunt generator at 300 r.p.m.*

Field amperes :	0	2	3	4	5	6	7
Armature volt :	7.5	92	132	162	183	190	212

Plot the O.C.C. for 375 r.p.m. and determine the voltage to which the machine will excite if field circuit resistance is 40 Ω

(a) What additional resistance would have to be inserted in the field circuit to reduce the voltage to 200 volts at 375 r.p.m.

(b) Without this additional resistance, determine the load current supplied by the generator, when its terminal voltage is 200 V. Ignore armature reaction and assume speed to be constant. Armature resistance is 0.4 Ω. **(Elect. Machines-I, South Gujarat Univ. 1986)**

Solution. The e.m.f. induced at 375 r.p.m. would be increased in the ratio 375/300 corresponding to different shunt field current values. A new table in given with the voltages multiplied by the above ratio.

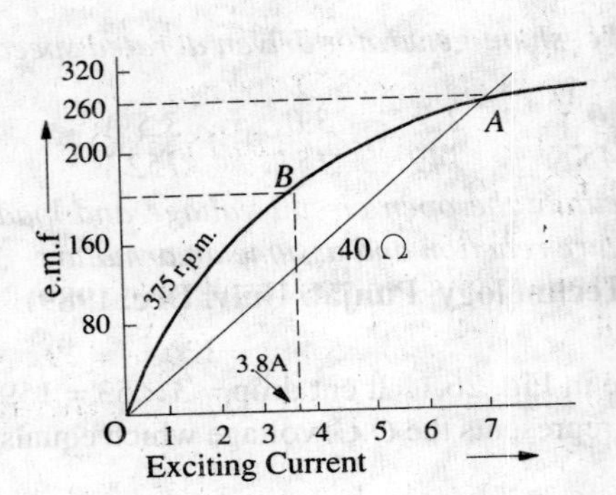

Fig. 26.11

Field amp :	0	2	3	4	5	6	7
Arm. volt :	9.4	115	165	202.5	228.8	248.8	265

The new *O.C.C.* at 375 r.p.m. is shown in Fig. 26.11. Line *OA* represents 40-Ω line.

The voltage corresponding to point *A* is 260 V. Hence machine will excite to 260 volt with 40 Ω shunt field resistance

(*a*) From Fig. 26.11, it is clear that for exciting the generator to 200 V, exciting current should be 3.8 A.

∴ Field circuit resistance = 200/3.8 = 52.6 Ω

∴ Additional resistance required = 52.6 −40 = **12.6 Ω**

(*b*) In this case, shunt field resistance = 40 Ω ...(as above)

Terminal voltage = 200 V ∴Field current = 200/40 = 5 A

Generated e.m.f. for exciting current of 5 A = 228.8 V

For a generator $E = V + I_a R_a$ ∴ $I_a R_a = E - V$ or $0.4\, I_a = 228.8–20 = 28.8$

∴ $I_a = 28.8/0.4 = 72$ A ∴ Load current $I = 72–5 =$ **67 A**

Example 26.5. *The open-circuit characteristic of a separately-excited d.c. generator driven at 1000 r.p.m. is as follows :*

Field Current :	*0.2*	*0.4*	*0.6*	*0.8*	*1.0*	*1.2*	*1.4*	*1.6*
E.M.F. volts :	*30.0*	*55.0*	*75.0*	*90.0*	*100.0*	*110.0*	*115.0*	*120.0*

If the machine is connected as shunt generator and driven at 1,000 r.p.m. and has a field resistance of 100 Ω, find (a) open-circuit voltage and exciting current (b) the critical resistance and (c) resistance to induce 115 volts on open circuit. **(Elect. Mechines, Nagpur, Univ. 1993)**

Solution. The *O.C.C.* has been plotted in Fig. 26.12. The shunt resistance line *OA* is drawn as usual.

(*a*) *O.C.* voltage = **100 V** ; Exciting current = **1 A**

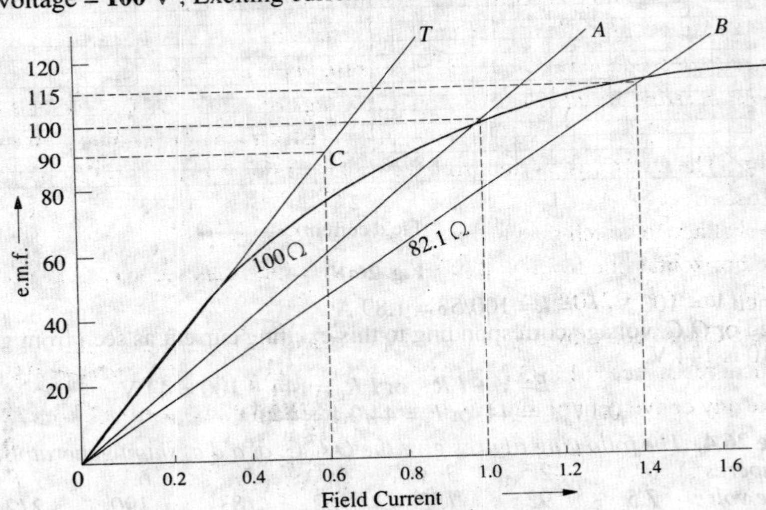

Fig. 26.12

(*b*) Line *OT* is tangent to the initial part of the *O.C.C.* It represents critical resistance. As seen from point *C*, value of critical resistance is 90/0.6 = **150 Ω**

(*c*) Line *OB* represents shunt resistance for getting 115 V on open-circuit. Its resistance = 115/1.4 = **82.1 Ω**

Example 26.6. *A.d.c. generator has the following magnetisation characteristics.*

Field current (A) :	*1*	*2*	*3*	*4*	*5*	*6*	*7*	*8*
Generated e.m.f. (V) :	*23*	*45*	*67*	*85*	*100*	*112*	*121*	*126*

If the generator is shunt excited, determine the load current.

(a) when terminal p.d. is 120 V, the field resistance is 15 Ω at a speed of 600 r.p.m. And

(b) when terminal p.d. is 144 V, the field resistance is 18 Ω at a speed of 700 r.p.m.

Solution. (*a*) When terminal p.d. = 120 V, then field current $I_f = V/R_{sh} = 120/15 = 8\ \Omega$

From the given data, it is seen that the generated e.m.f. = 126 V*

Voltage drop = 126 − 120 = 6 V

Since drop due to armature reaction is neglected, this represents the armature drop.

$\therefore \quad I_a R_a = 6$ or $I_a = 6/0.02 = 300$ A Load current = 300−8 = **292 A**

(*b*) The *O.C.* data at 700 r.p.m. can be obtained by multiplying the given values of generated e.m.f. by a factor of 700/600 = 7/6. Hence, the new data is :-

Field current (A) :	1	2	3	4	5	6	7	8
Generated e.m.f. (V) :	26.8	52.5	78.2	99.2	116.6	131	141	146

When $V = 144$ V and $R_{sh} = 18\ \Omega$, field current = 144/18 = 8.A

From the given data, the corresponding generated e.m.f. is = 146 V

Voltage drop $I_a R_a = 146 - 144 = 2V$

$\therefore \quad I_a$ = 2/0.02 = 100 A ∴ Load current = 100−8 = **92 A**

Example 26.7. *The O.C.C. of a d.c. generator driven at 400 rev/min is a s follows :*

Field current (A) :	*2*	*3*	*4*	*5*	*6*	*7*	*8*	*9*
Terminal volts :	*110*	*155*	*186*	*212*	*230*	*246*	*260*	*271*

Find :-

(a) voltage to which the machine will excite when run as a shunt generator at 400 rev/min with shunt field resistance equal to 34 Ω

(b) resistance of shunt circuit to reduce the O.C. voltage to 220 V

(c) critical value of the shunt field circuit resistance

(d) the critical speed when the field circuit resistance is 34 Ω

(e) lowest possible speed at which an O.C. voltage of 225 V can be obtained.

(Electrical Technology, Bombay Univ. 1987)

Solution. The *O.C.C.* as plotted from the given data is shown in Fig. 26.13. The 34-Ω line *OA* is drawn as usual.

(*a*) The voltage to which machine will excite = *OM* = **255 V**

(*b*) The horizontal line from *N* (220 V) is drawn which cuts the *O.C.C.* at point *B*. Resistance represented by line *OB* = 220/5.4 = **40.7 Ω**

(*c*) Line *OC* has been drawn which is tangential at the origin to the *O.C.C.* This represents the value of critical resistance = 140/2.25 = **62.2 Ω.**

(*d*) Take any convenient point *D* and erect a perpendicular which cuts both *OA* and *OC*.

$$\frac{DE}{DF} = \frac{N_C}{400} \quad \text{or} \quad \frac{110}{202} = \frac{N_C}{400}, N_C = \mathbf{218}\ rp.m.$$

(*e*) From point *P* (225 *V*) draw a horizontal line cutting *OA* at point *G*. From *G*, draw a perpendicular line *GK* cutting the *O.C.C.* at point *H*. If *N′* is the lowest speed possible for getting 225 volt with 34 Ω shunt circuit resistance, then

$$\frac{GK}{HK} = \frac{N'}{400} \quad \text{or} \quad \frac{225}{241} \quad \text{or} \quad N' = \mathbf{375\ r.p.m.}$$

---

*We need not plot the *O.C.C.* in this particular case.

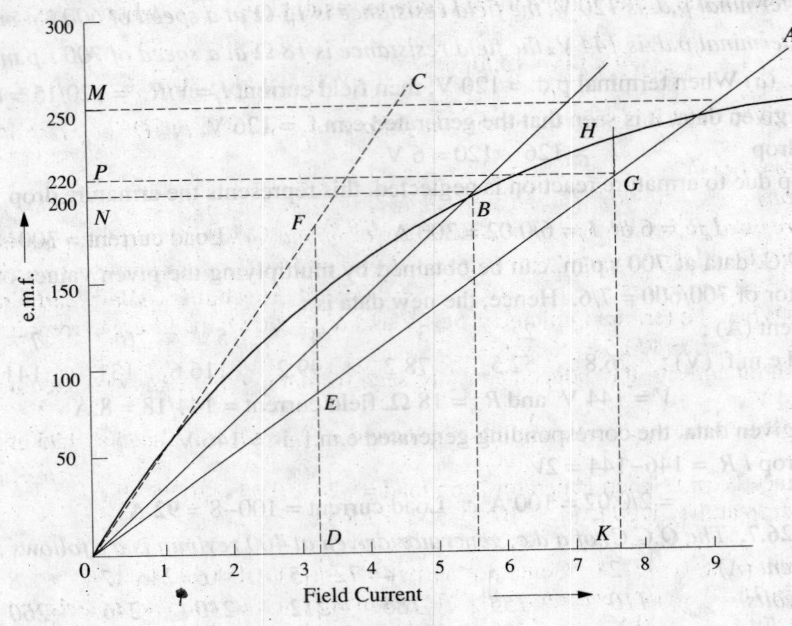

Fig 26.13

Example 26.8. *The magnetization characteristic for a 4-pole, 110-V, 1000 r.p.m. shunt generator is as follows :*

field current	*0*	*0.5*	*1*	*1.5*	*2*	*2.5*	*3 A*
O.C. voltage	*5*	*50*	*85*	*102*	*112*	*116*	*120 V*

Armature is lap-connected with 144 conductors. Field resistance is 45 ohms. Determine

(i) voltage the machine will build up at no-load. (ii) the critical resistance.

(iii) the speed at which the machine just fails to excite (iv) residual flux per pole.

(Electrical Machinery-I, Mysore Univ. 1988)

Solution. In Fig. 26.14, *OA* represents the 45-Ω line which is drawn as usual.

(*i*) The voltage to which machine will build up = *OM* = **118 V**

(*ii*) *OT* is tangent to the initial part of the *O.C.C.* If represents critical resistance. Take point *B* lying on this line. Voltage and exciting current corresponding to this point are 110 V and 1.1 A respectively.

$\therefore R_C = 110/1.1 = \mathbf{100\ \Omega}$

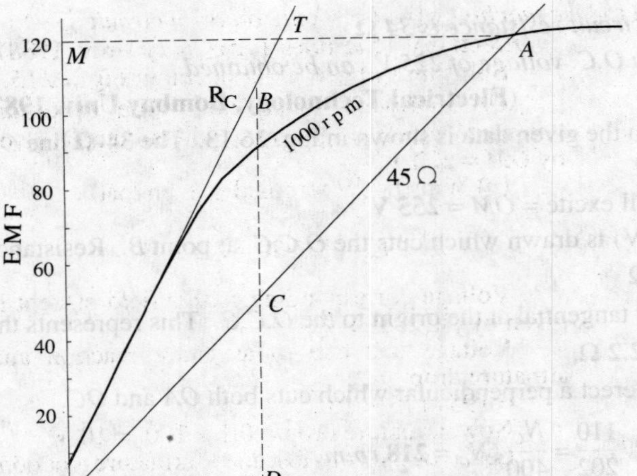

Fig. 26.14

(*iii*) From any point on *OT*, say point *B*, drop the perpendicular *BD* on *X*-axis.

$$\frac{CD}{BD} = \frac{N_C}{1000} \quad \text{or} \quad \frac{49}{110} = \frac{N_C}{1000}$$

$$\therefore \quad N_C = \mathbf{445\ r.p.m}$$

(*iv*) As given in the table, induced e.m.f. due to residual flux (*i.e.* when there is no exciting current) is 5 V.

$$\therefore \quad 5 = \frac{\Phi \times 144 \times 1000}{60}\left(\frac{4}{4}\right) \quad \therefore \quad \Phi = \mathbf{2.08\ mWb.}$$

Example 26.9. *A shunt generator gave the following results in the O.C.C. test at a speed of r.p.m.*

Field current (A) :	*1*	*2*	*3*	*4*	*6*	*8*	*10*
E.M.F. (volt) :	*90*	*185*	*251*	*290*	*324*	*345*	*360*

The field resistance is adjusted to 50 Ω and the terminal voltage is 300 V on load. Armature resistance is 0.1 Ω and assuming that the flux is reduced by 5% due to armature reaction, find the load supplied by the generator. **(Electromechanic, Allahabad Univ, 1992)**

Solution. When the terminal voltage is 300 V and R_{sh} = 50 Ω, then field current is = 300/50 = 6 A

With this shunt current, the induced e.m.f. as seen from the given table (we need not draw the *O.C.C.*) is = 324 *V*

Due to armature reaction, the flux and hence the induced e.m.f. is reduced to 0.95 of its no-load value.

Hence, induced e.m.f. when generator is on load = 324 × 0.95 = 307.8 V

Armature drop at the given load = 307.8 −300 = 7.8 V

$$I_aR_a = 7.8, \quad I_a = 7.8/0.1 = 78 \text{ A}$$

Load current = 78−6 = 72 A ; Generator output = 72 × 300/1000 = **21.6 kW**

Example 26.10. *A shunt generator gave the following open-circuit characteristic :*

Field current :	*0.5*	*1.0*	*1.5*	*2.0*	*2.5*	*3.0*	*3.5 A*
O.C. e.m.f. :	*54*	*107*	*152*	*185*	*210*	*230*	*245 V*

The armature and field resistances are 0.1 Ω and 80Ωrespectively. Calculate (a) the voltage to which the machine will excite when run as a shunt generator at the same speed.

(b) the volts lost due to armature reaction when 100 A are passing in the armature at a terminal voltage of 175 V.

(c) the percentage reduction in speed for the machine to fail to excite on open circuit.

(Electrical Machines-I, Bombay Univ. 1988)

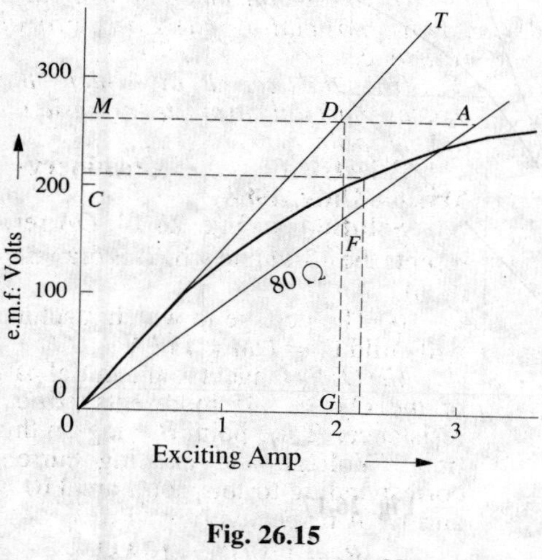

Solution. (*a*) *O.C.C.* is shown in Fig. 26.15.

OA represents 80 Ω line. The maximum voltage to which the generator will build up is given by *OM* = **222 V**.

(*b*) With 175 V terminal p.d. on load

$$I_{sh} = 175/80 = 2.2 \text{ A}$$

Voltage corresponding to this field current is given by *OC* = 195 V.

Voltage lost due to armature reaction and armature drop

= 195−175 = 20 V

Now, armature drop = 0.1 × 100 = 10 V

Let '*x*' be the volts lost due to armature reaction.

Then $\quad 10 + x = 20 \quad \therefore x = \mathbf{10\ V}$

(*c*) Line *OT* is drawn tangential to the curve. *DFG* is perpendicular to the base line.

Fig. 26.15

$$\frac{N_c}{N} = \frac{FG}{DG} = \frac{160}{220} \quad \text{or} \quad \frac{N_e - N}{N} = \frac{-60}{220}$$

Percentage reduction in speed = $\frac{-60}{220} \times \mathbf{100} = \mathbf{-27.3\%}$

Example 26.11. *The O.C.C. of a shunt generator running at 800 r.p.m. is as follows :*

Field current (amp) :	*1*	*2*	*3*	*4*	*5*	*6*
Induced e.m.f. (volt) :	*82.5*	*180*	*225*	*252*	*273*	*282*

If the shunt field resistance is 60 Ω, find

(i) the voltage to which the machine will build up running at the same speed (ii) the value of field regulating resistance if the machine is to build up to 120 V when its field coils are grouped in two parallel circuits and generator is run at half the speed.

(Electrical Technology, Hyderabad Univ. 1991)

Solution. *O.C.C.* is drawn from the given data and is shown in Fig. 26.16. *OA* represents 60 Ω. The voltage corresponding to point *A* is *OM* = 260 V.

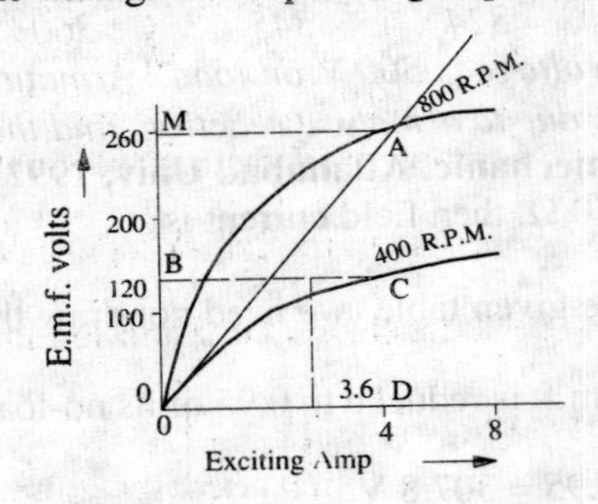

Fig. 26.16

(*i*) Generator will build up to **260 volts.**

(*ii*) Lower curve represents the induced e.m.f. for different exciting current values at 400 r.p.m. From point *B* which represents 120 V, draw *BC*. From *C* draw a perpendicular *CD* which gives the exciting current of 3.6A. It means that current through *each* parallel path of shunt field coils is 3.6 A. Total current which passes through the field regulator resistance is = 3.6 × 2 = 7.2 A because it is in series with the field coils.

Hence, total shunt field resistance = 120/7.2 = 16.67 Ω

Now, resistance of each shunt parallel path = 60/2 = 30 Ω

Joint resistance of two parallel paths = 30/2 = 15 Ω

∴ Shunt field regulator resistance = 16.67 − 15 = **1.67 Ω**

26.7 Voltage Build up of a Shunt Generator

Before loading a shunt generator, it is allowed to build up its voltage. Usually, there is always present some residual magnetism in the poles, hence a small e.m.f. is produced initially. This e.m.f. circulates a small current in the field circuit which increases the pole flux (provided field circuit is properly connected to armature, otherwise this current may wipe off the residual magnetism). When flux is increased, generated e.m.f. is increased which further increases the flux and so on. As shown in Fig. 26.17, *Oa* is the induced e.m.f. due to residual magnetism which appears across the field circuit and causes a field current *Ob* to flow. This current aids residual flux and hence produces a larger induced e.m.f. *Oc*. In turn, this increased e.m.f. *Oc* causes an even larger current Od which creates more flux for a still larger e.m.f. and so on.

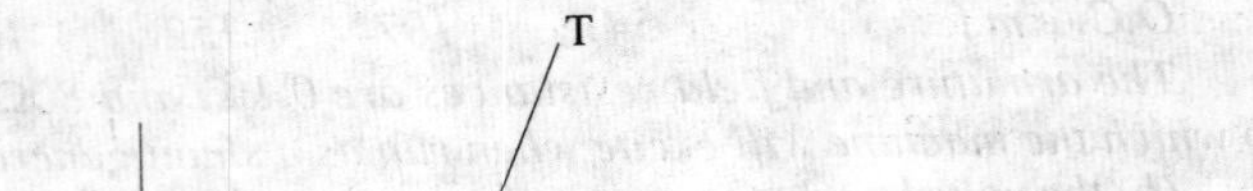

Fig. 26.17

Now, the generated e.m.f. in the armature has

(*a*) to supply the ohmic drop $I_f R_{sh}$ in the winding and (*b*) to overcome the opposing self-induced e.m.f. in the field coil *i.e.* $L dI_f/d_t$ because field coils have appreciable self-inductance.

$$e_g = I_f R_{sh} + L - d\,I_f/d_t$$

If (and so long as), the generated e.m.f. is in excess of the ohmic drop $I_f R_{sh}$, energy would continue being stored in the pole fields. For example, as shown in Fig. 26.17, corresponding to field current *OA*, the generated e.m.f. is *AC*. Out of this, *AB* goes to supply ohmic drop $I_f R_{sh}$ and *BC* goes to overcome self-induced e.m.f. in the coil. Corresponding to $I_f = OF$, whole of the generated e.m.f. is used to overcome the ohmic drop. None is left to overcome $L.\ dI_f/dt$. Hence no energy is stored in the pole fields. Consequently, there is no further increase in pole flux and the generated e.m.f. With the given shunt field resistance represented by line OP, the maximum voltage to which the machine will build up is OE. If resistance is decreased, it will build up to a somewhat higher voltage. OR represents the resistance known as *critical resistance.* If shunt field resistance is greater than this value, the generator will fail to excite.

26.8. Conditions for Build-up of a Shunt Generator

We may summarize the conditions necessary for the build-up of a (self-excited) short generator as follows :-

1. There must be some residual magnetism in the generator poles.
2. For the given direction of rotation, the shunt field coils should be correctly connected to the armature *i.e.* they should be so connected that the induced current reinforces the e.m.f. produced initially due to residual magnetism.
3. If excited on open circuit, its shunt field resistance should be less than the critical resistance (which can be found from its *O.C.C.*)
4. If excited on load, then its shunt field resistance should be more than a certain minimum value of resistance which is given by internal characteristic (Art. 26.11)

26.9. Other Factors Affecting Voltage Building of a DC Generator

In addition to the factors mentioned above, there are some other factors which affect the voltage building of a self-excited d.c. generator. These factors are (*i*) reversed shunt field connection (*ii*) reversed rotation and (*iii*) reversed residual magnetism. These adverse effects would be explained with the help of Fig. 26.18 and the right-hand rule for finding the direction of the coil flux. For the sake of simplicity, only one field pole has been shown in the figure.

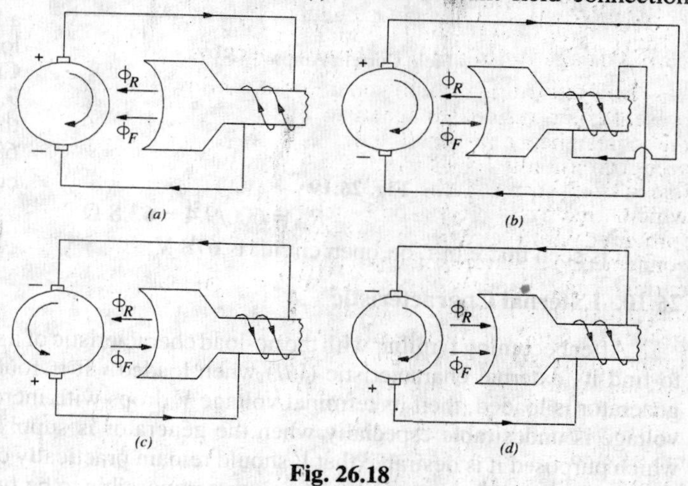

Fig. 26.18

Fig. 26.18(*a*) represents the normal operation, the prime mover rotation is clockwise and both the residual flux Φ_R and the field flux Φ_F are directed to the left.

Fig. 26.18(*b*) shows reversed connection of the field circuit which causes Φ_F to oppose Φ_R. Consequently, the generator voltage builds down from its original residual value.

In Fig. 26.18(*c*), reversed armature rotation causes the reversal of the voltage produced by the residual magnetism. Even though the field coil connections are correct, the reversed field current flow causes Φ_F to oppose Φ_R so that the voltage builds down from its original residual value.

Fig. 26.18 (*d*) shows the case when due to some reason the residual magnetism gets reversed. Hence, the armature voltage is also reversed which further reverses the field current. Consequently, both Φ_F and Φ_R are reversed but are directed to the right as shown. Under this condition, the voltage buildup is in the reversed direction. Obviously, the generator will operate at rated voltage but with reversed polarity.

If desired, the reversed polarity can be corrected by using an external d.c. source to remagnetise the field poles in the correct direction. This procedure is known as field flashing.

Example 26.12. *The O.C.C. of a generator is given by the following :*

Field current	*1.5*	*3*	*4.5*	*6*	*7.5*	*9*	*10.5*
E.M.F.	*168*	*330*	*450*	*490*	*600*	*645*	*675*

The speed at which data is obtained is 1000 r.p.m. Find the value of the shunt field resistance that will give a p.d. of 600 V with an armature current of 300 A at the same speed. Due to armature reaction, the shunt field current is given by I_{sh} *(eff.)* $= I_{sh} - 0.003\ I_a$. *Armature resistance, including brush contact resistance, is 0.1 Ω. What will be the p.d. on open circuit at the same speed ?*

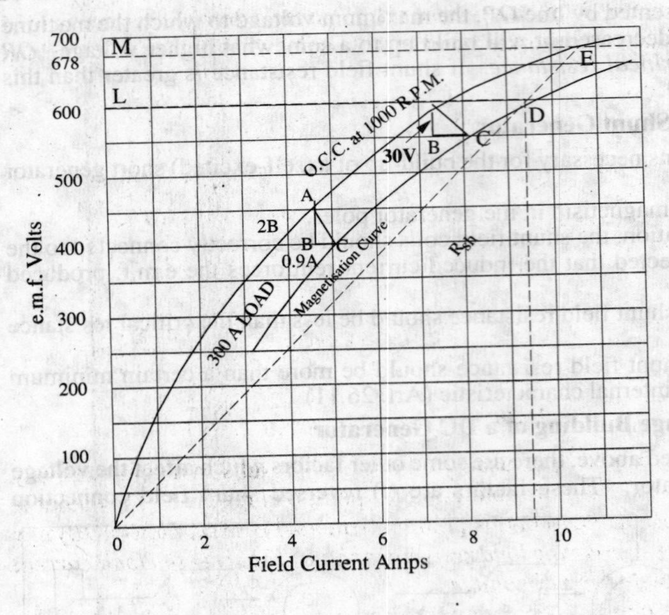

Fig. 26.19

Solution. As shown in Fig. 26.19, the *O.C.C.* has been plotted from the given data.

Voltage drop due to armature resistance

$= 300 \times 0.1 = 30$ V

Reduction of field current due to armature reaction $= 0.003 \times 300 = 0.9$ A

Any point *A* is taken on the *O.C.C.* A vertical distance $AB = 30$ V is taken and then the horizontal line $BC = 0.9$ A is drawn thus completing triangle *ABC* which is known as drop reaction triangle. Then, point *C* lies on the 300 ampere load saturation curve. This curve can be drawn by finding such similar points like *C′* etc.

From point *L* representing 600 *V*, a horizontal line is drawn cutting the load saturation curve at *D*. Join *OD*. Current corresponding to point *D* is 9.4 A. The slope of I_f the line *OD* gives the value of shunt resistance to give 600 V with 300 amperes of armature current.

$\therefore \quad R_{sh} = 600/9.4 = \mathbf{63.8\ \Omega}$

It is seen that e.m.f. on open circuit is **678 V**

26.10. External Characteristic

After becoming familiar with the no-load characteristic of a shunt generator, we will now proceed to find its external characteristic (*V*/*I*) when loaded. It is found that if after building up, a shunt generator is loaded, then its terminal voltage *V* drops with increase in load current. Such a drop in voltage is undesirable especially when the generator is supplying current for light and power for which purposed it is desirable that *V* should remain practically constant and independent of the load. This condition of constant voltage is almost impossible to be fulfilled with a shunt generator unless the field current is being automatically adjusted by an automatic regulator. Without such regulation terminal voltage drops considerably as the load on the generator is increased. These are three main reasons for the drop in terminal voltage of a shunt generator when under load.

(*i*) **Armature resistance drop**

As the load current increases, more and more voltage is consumed in the ohmic resistance of the armature circuit. Hence, the terminal voltage $V = E - I_aR_a$ is decreased where *E* is the induced e.m.f. in the armature under load condition.

(*ii*) **Armature reaction drop**

Due to the demagnetising effect of armature reaction, pole flux is weakened and so the induced e.m.f. in the armature is decreased.

(*iii*) The drop in terminal voltage *V* due to (*i*) and (*ii*) results in a decreased field current I_f which further reduces the induced e.m.f.

For obtaining the relation between the terminal voltage and load current, the generator is connected as shown in Fig. 26.20(*a*).

The shunt generator is first excited on no-load so that it gives its full open circuit voltage = *Oa* [Fig. 26.20(*b*)]. Then, the load is gradually applied and, at suitable intervals, the terminal voltage *V* (as read by the voltmeter) and the load current *I* (as read by the ammeter A_2) are noted. The field current as recorded by ammeter A_1, is kept constant by a rheostat (because during the test, due to heating, shunt field resistance is increased). By plotting these readings, the external characteristic of Fig. 26.20(*b*) is obtained. The portion *ab* is the working part of this curve. Over this part, if the load

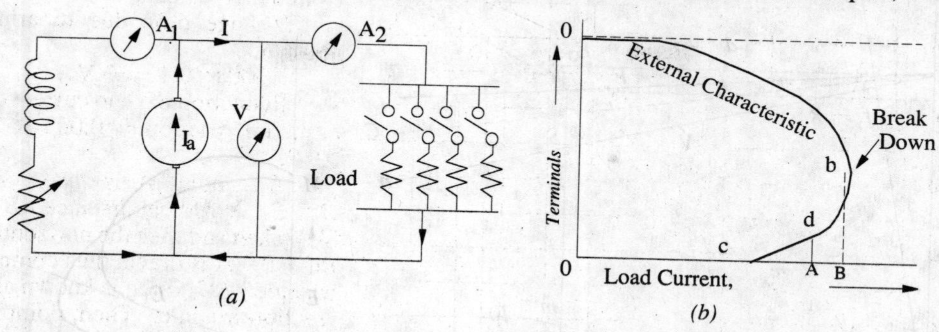

Fig. 26.20

resistance is decreased, load current is increased as usual, although this results in a comparatively small additional drop in voltage. These conditions hold good till point *b* is reached. This point is known as breakdown point. It is found that beyond this point (where load is maximum = *OB*) *any effort to increase load current by further decreasing load resistance results in decreased load current (like OA) due to a very rapid decrease in terminal voltage.*

We will discuss the reason for this unusual behavior of the generator in more details. Over the earlier portion *ab* [Fig. 26.20(*b*)] where the load current is comparatively small, when external load resistance is decreased, it results in increased load current as might be expected keeping Ohm's law in mind. It should not, however, be forgotten that due to increase in load current, *V* is also decreased somewhat due to the cause (*iii*) given above. But over the portion *ab*, the effect of decrease in load resistance predominates the effect of decrease in *V* because load current is relatively small.

At point *b*, generator is delivering a very large current *i.e.* current which is many times greater than its normal current. If load resistance is decreased at this point so as to be able to draw a load current greater than *OB*, the current is *increased* momentarily. But due to the severe armature reaction for this heavy current and increased I_aR_a drop, the terminal voltage *V* is drastically reduced. The effect of this drastic reduction in *V* results in less load current (= *OA*). In other words, over the portion *bdc* of the curve, the terminal voltage *V* decreases more rapidly than the load resistance. Hence, any, further decrease in load resistance actually causes a *decrease* in load current (it may seem to contravene Ohm's law but this law is not applicable here since *V* is not constant). As load resistance is decreased beyond point *b*, the curve turns back till when the generator is actually short-circuited, it cuts the current axis at point *c*. Here, terminal voltage *V* in reduced to zero, though there would be some value of *E* due to residual magnetism (Fig. 26.22).

26.11. Voltage Regulation

By voltage regulation of a generator is meant the change in its terminal voltage with the change in load current when it is run at a constant speed. If the change in voltage between no-load and full load is small, then the generator is said to have good regulation but if the change in voltage is large, then it has poor regulation. The *voltage regulation of a d.c. generator is the change in voltage when the load is reduced from rated value to zero, expressed as percentage of the rated load voltage.*

If no-load voltage of a certain generator is 240 V and rated-load voltage is 220 V, then,

$$\text{regn.} = (240-220)/220 = 0.091 \text{ or } 9.1\ \%$$

26.12. Internal or Total Characteristic

As defined before, internal characteristic gives the relation between *E* and I_a. Now in a shunt generator

$$I_a = I + I_f \quad \text{and} \quad E = V + I_aR_a$$

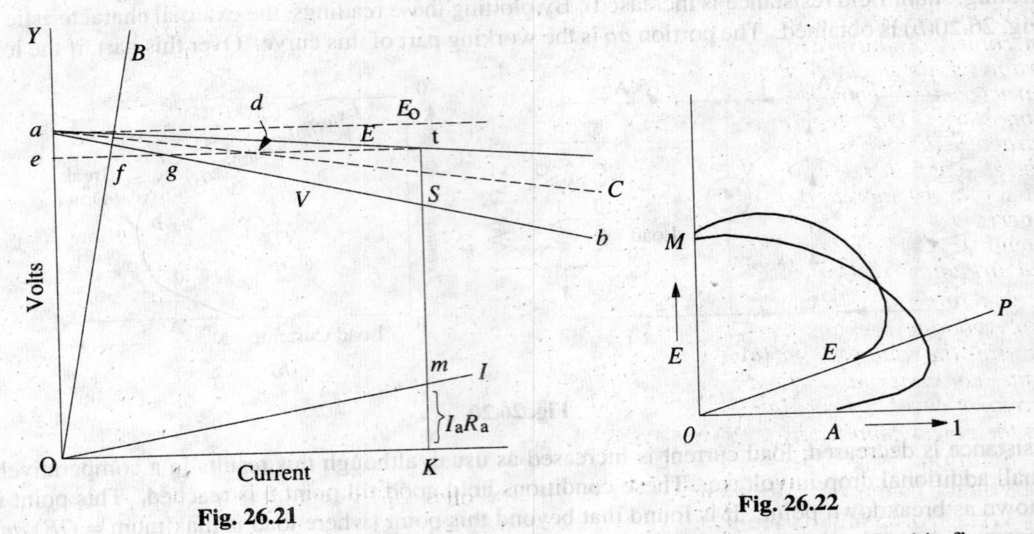

Fig. 26.21

Fig. 26.22

Hence, E/I_a curve can be obtained from V/I curve as shown in Fig. 26.21 . In this figure, *ab* represents the external characteristic as discussed above. The field resistance line *OB* is drawn as usual. The horizontal distances from *OY* line to the line *OB* give the values of field currents for different terminal voltages. If we add these distances horizontally to the external characteristic *ab*, then we get the curve for the total armature current *i.e.* dotted curve *ac*. For example, point *d* on *ac* is obtained by making $gd = ef$. The armature resistance drop line *Or* is then plotted as usual. If brush contact resistance is assumed constant, then armature voltage drop is proportional to the armature current. For any armature current = *OK*, armature voltage drop $I_aR_a = mK$. If we add these drops to the ordinates of curve *ac*, we get the internal characteristic. For example, $St = mK$. The point *t* lies on the internal characteristic. Other points like *t* can be found similarly at different armature currents as the total characteristic can be drawn.

It may be noted here, in passing, that product EI_a gives the total power developed within the armature. Some of this power goes to meet I^2R losses in armature and shunt field windings and the rest appears as output.

As explained in Art. 26.10 if load resistance is decreased, the armature current increases up to a certain load current value. After that, any decrease in load resistance is not accompanied by increase in load current. Rather, it is decreased and the curve turns back as shown in Fig. 26.22. If the load resistance is too small, then the generator is short-circuited and there is no generated e.m.f. due to heavy demagnetisation of main poles.

Line *OP* is tangential to the internal characteristic *MB* and its slope gives the value of the *minimum* resistance with which the generator will excite if excited *on load.*

26.13. Series Generator

In this generator, because field windings are in series with the arinature [Fig. 26.23(*a*)], they carry full armature current I_a. As I_a, is increased, flux and hence generated e.m.f. is also increased as shown by the curve. Curve *Oa* is the *O.C.C.* The extra exciting current necessary to neutralize the weakening effect of armature reaction at full load is given by the horizontal distance *ab*. Hence, point *b* is on the internal characteristic. If the ordinate $bc = gh$ = armature voltage drop, then point *C* lies on the external characteristic [Fig. 26.23(*b*)].

It will be noticed that a series generator has rising voltage characteristic *i.e.* with increase in load, its voltage is also increased. But it is seen that at high loads, the voltage starts decreasing due to excessive demagnetising effects of armature reaction. In fact, terminal voltage starts decreasing as load current is increased as shown by the dotted curve. For a load current *OC′*, the terminal voltage is reduced to zero as shown.

Example 26.13. *In a 220-V supply system, a series generator, working on a linear portion of its magnetisation characteristic is employed as a booster. The generator characteristic is such that induced e.m.f. increases by 1 volt for every increase of 6 amperes of load current through the generator. The total armature resistance of the generator is 0.02Ω. If supply voltage remains constant, find the voltage supplied to the consumer at a load current of 96 A. Calculate also the power supplied by the booster itself.*

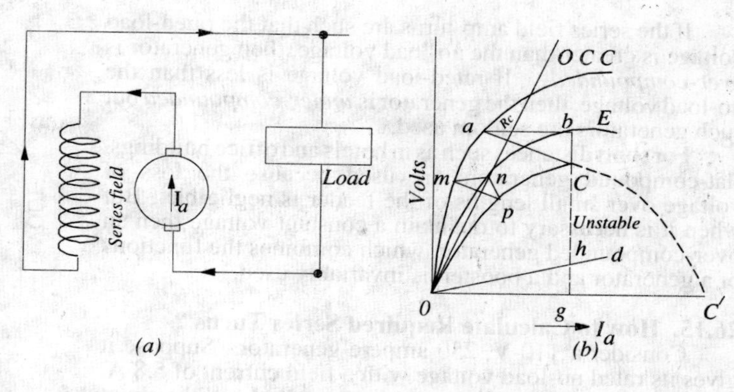

Fig. 26.23

Solution.

Voltage increase for 6 amperes	= 1V ∴ Voltage increase for 96 A = 96/6 = 16V
Voltage drop in series coils	= 96 × 0.02 = 1.9 V
Net voltage rise due to booster	= 16–1.9 = 14.1 V
Voltage at consumer end	= 220 + 14.1 = **234.1 V**
Power supplied by booster itself	= 14.1 × 96 = 1354 W = **1.354 kW**

Example 26.14. *A.d.c. series generator, having an external characteristic which is a straight line through zero to 50 V at 200 A is connected as a booster between a station bus bar and a feeder of 0.3 ohm resistance. Calculate the voltage difference between the station bus-bar and the far end of the feeder at a current of (i) 200 A and (ii) 50 A.*

(AIME Sec. B Elect. Machine Summer 1991)

Solution. (*i*) Voltage drop on feeder = 200 x 0.3 = 60 V

Booster voltage provided by series generator for 200 A current as given = 50 V.

∴ net voltage decrease = 60–50 = 10 V

(*ii*) Feeder drop = 50 × 0.3 = 15 V

Booster voltage provided by series generator (by proportion) is

= 50 × 50/200 = 12.5 V

∴ net decrease in voltage = 15 – 12.5 = 2.5 V

26.14. Compound-wound Generator

A shunt generator is unsuitable where constancy of terminal voltage is essential, because its terminal voltage decreases as the load on it increases. This decrease in *V* is particularly objectionable for lighting circuit where even slight change in the voltage makes an appreciable change in the candle power of the incandescent lamps. A shunt generator may be made to supply substantially constant voltage (or even a rise in voltage as the load increases) by adding to it a few turns joined in series with either the armature or the load (Fig. 26.24). These turns are so connected as to aid the shunt turns when the generator supplies load. As the load current increasess, the current through the series windings also increase thereby increasing the flux. Due to the increase in flux, induced e.m.f. is also increased. By adjusting the number of series turns (or series amp-turns), this increase in e.m.f. can be made to balance the combined voltage drop in the generator due to armature reaction and the armature drop. Hence, *V* remains practically constant which means that field current is also almost unchanged. We have already discussed the three causes which decrease the terminal voltage of a shunt generator (Art. 26.10). Out of these three, the first two are neutralized by the series field amp-turns and the third one, therefore, does not occur.

If the series field amp-turns are such as to produce the same voltage at rated load as at no-load, then the generator is *flat-compounded.* It should be noted, however, that even in the case of a flat-compounded generator, the voltage is not constant from no-load to rated-load. At half the load, the voltage is actually greater than the rated voltage as seen from Fig. 26.24.

If the series field amp-turns are such that the rated-load voltage is greater than the no-load voltage, then generator is *over-compounded.* If rated-load voltage is less than the no-load voltage, then the generator is *under-compounded* but such generators are seldom used.

For short distances such as in hotels and office buildings, flat-compound generators are used because the loss of voltage over small lengths of the feeder is negligible. But when it is necessary to maintain a constant voltage then an over-compounded generator, which combines the functions of a generator and a booster, is invariably used.

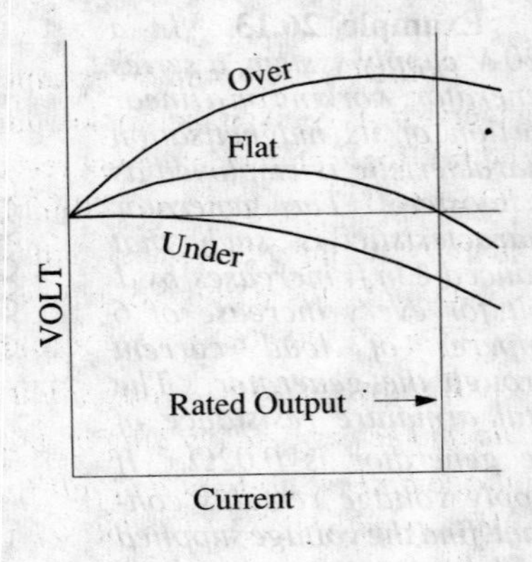

Fig. 26.24

26.15. How to Calculate Required Series Turns ?

Consider a 110-V, 250-ampere generator. Suppose it gives its rated no-load voltage with a field current of 5.8 A. If, now, the series windings are disconnected and the shunt field rheostat is left unchanged them the machine will act as shunt generator, hence its voltage will fall with increase in load current. Further, suppose that the field current has to be increased to 6.3 A in order to maintain the rated terminal voltage at full load. If the number of turns of the shunt field winding is 2000, then 2000 × (6.3−5.8) = 1000 amp-turns represent the additional excitation that has to be supplied by the series windings. As series turns will be carrying a full load current of 250 A, hence number of series turns = 1000/250 = 4.

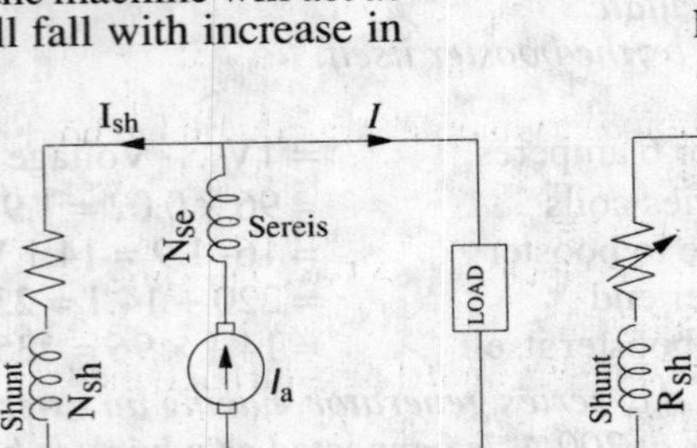

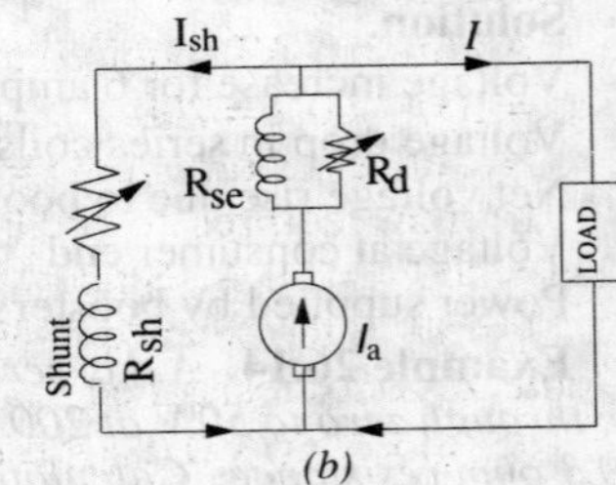

Fig. 26.25

In general, let

ΔI_{sh} = increase in shunt field current required to keep voltage constant from no-load to full-load

N_{sh} = No. of shunt field turns per pole (or the total number of turns)

N_{se} = No. of series turns per pole (or the total number of turns)

I_{se} = current through series winding

= armature current I_a —for long-shunt

= load current I —for short-shunt

It is seen that while running as a simple shunt generator, the increase in shunt field ampere-turns necessary for keeping its voltage constant from no-load to full-load is = N_{sh}. Δ_{sh}. This increase in field excitation can be alternatively achieved by adding a few series turns to the shunt generator [Fig. 26.25(*a*)] thereby converting it into a compound generator.

$$\therefore \quad N_{sh} \cdot \Delta I_{sh} = N_{se} I_{se}$$

If other things are known, N_{se} may be found from the above equation.

In practice, a few extra series amp-turns are taken in order to allow for the drop in armature. Any surplus amp-turns can be changed with the help of a divertor across the series winding as shown in Fig. 26.25(*b*).

As said above, the degree of compounding can be adjusted with the help of a variable-resistance, divertor as shown in Fig. 26.25(*b*). If I_d is the current through the divertor of resistance R_d, then remembering that series windings and divertor are in parallel,

$$\therefore \quad I_{se} \cdot R_{se} = I_d R_d \quad \text{or} \quad R_d = I_{se} R_{se} / I_d$$

Example 26.15. *A shunt generator is converted into a compound generator by addition of a series field winding. From the test on the machine with shunt excitation only, it is found that a field current of 5 A gives 440 V on no-load and that 6 A gives 440 V at full load current of 200 A. The shunt winding has 1600 turns per pole. Find the number of series turns required.*

(Elect. Machines, A.M.I.E., Sec., B, 1991)

Solution. It would be assumed that shunt generator is converted into a short shunt compound generator. It is given that for keeping the voltage of shunt generator constant at 440 V both at no-load and full-load, shunt field ampere-turns per pole have to be increased from $1600 \times 5 = 8000$ to $(1600 \times 6) = 9600$ *i.e.* an increase of $(9600–8000) = 1600$ AT. The same increase in field AT can be brought about by adding a few series turns.

Let n be the number of series turns required per pole. Since they carry 200 A,

$\therefore n \times 200 = 1600$; $n =$ **8 turns/pole**

Example 26.16. *A long shunt compound generator has a shunt field winding of 1000 turns per pole and series field winding of 4 turns per pole and resistance 0.05 Ω. In order to obtain the rated voltage both at no-load and full-load for operation as shunt generator, it is necessary to increase field current by 0.2 A. The full-load armature current of the compound generator is 80 A. Calculate the divertor resistance connected in parallel with series field to obtain flat compound operation.*

(Elect. Machines A.M.I.E. Sec B, 1993)

Solution.

Additional AT required to maintain rated voltage both at no-load and full-load (Fig. 26.26).

$= 1000 \times 0.2 = 200$

No. of series turns/pole = 4

Current required to produce 200 AT by the series field = 200/4 = 50 A. Since

$I_a = 80$ A, the balance of 30 A must pass through the parallel divertor resistance

$\therefore$ $30\,R = 50 \times 0.05$, $R = 0.0833\ \Omega$

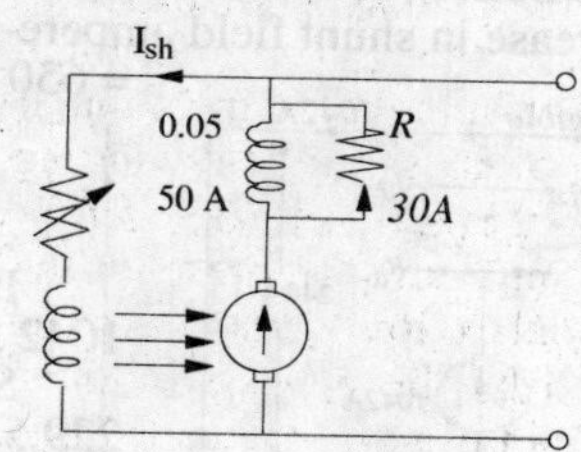

Fig. 26.26

Example 26.17. *A 220-V compound generator is supplying a load of 100 A at 220 V. The resistances of its armature, shunt and series windings are 0.1 Ω, 50 Ω and 0.06 Ω respectively. Find the induced e.m.f. and the armature current when the machine is connected (a) short shunt (b) long shunt (c) how will the series amp-turns be changed in (b) if a divertor of 0.14 Ω is connected in parallel with the series windings ? Neglect armature reaction and brush contact drop.*

Solution. (*a*) **Short-shunt** (Fig. 26.27).

Voltage drop in series winding = $100 \times 0.06 = 6$ V ; $I_{sh} = 226/50 = 4.5$ A.

$$\therefore \quad I_a = 100 + 4.5 = \mathbf{104.5}$$

Armature drop = $104.5 \times 0.1 = 10.5$ V $\therefore$ Induced e.m.f. = 220 + 6 + 10.5 = **236.5 V**

(*b*) **Long-shunt** (Fig. 26.28)

$$I_{sh} = 220/50 = 4.4 \text{ A} \quad \therefore \quad I_a = 100 + 4.4 = \mathbf{104.4\ A}$$

Voltage drop over armature and series field winding = $104.4 \times 0.16 = 16.7$ V

$\therefore$ Induced e.m.f. = 220 + 16.7 = **236.7 V**

(*c*) As shown in Fig. 26.29, a divertor of resistance 0.14 Ω is connected in parallel with the series field winding. Let n be the number of series turns.

Number of series amp-turns without divertor = $n \times 104.4 = 104.4\,n$

When divertor is applied, then current through series field is

$$= \frac{104.4 \times 0.14}{(0.14 + 0.06)} = 73.1 \text{ A}$$

$\therefore$ series amp-turns = $73.1 \times n$ $\therefore$ series amp-turns are reduced to $\frac{73.1n}{104.4n} \times \mathbf{100 = 70\%}$

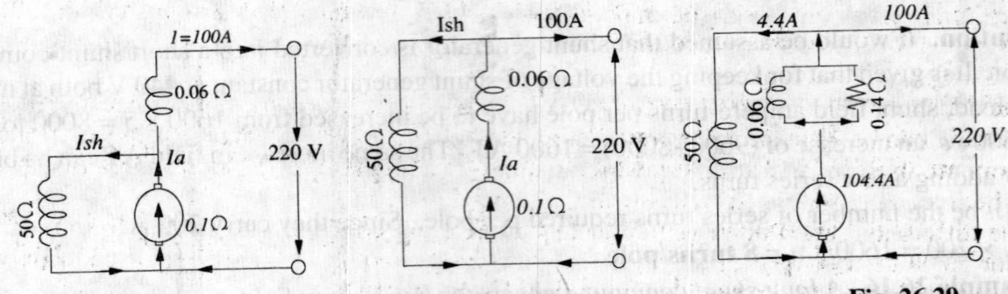

Fig. 26.27 Fig. 26.28 Fig. 26.29

Example 26.18. *A 250-kW, 240-V generator is to be compounded such that its voltage rises from 220 volts at no-load to 240 V at full load. When series field is cut out and shunt field is excited from an external source, then from the load test it is found that this rise in voltage can be obtained by increasing the exciting current from 7 A at no-load to 12 A at full-load. Given shunt turns/pole = 650, series turns/pole = 4 and resistance of series winding, 0.006 Ω. If the machine is connected long-shunt, find the resistance of the series amp-turns at no-load and drop in series winding resistance at full-load.*

Solution. Full-load current = $250 \times 10^3/240 = 1042$ A

Increase in shunt field ampere-turns to over-compound the shunt generator

$= 650\ (12-7) = 3{,}250$. As seen from Fig. 23.35

$4 \times I_{se} = 3250$

$I_{se} = 3250/4 = 812.5$ A

$\therefore \quad I_d = 1042 - 812.5 = 229.5$ A

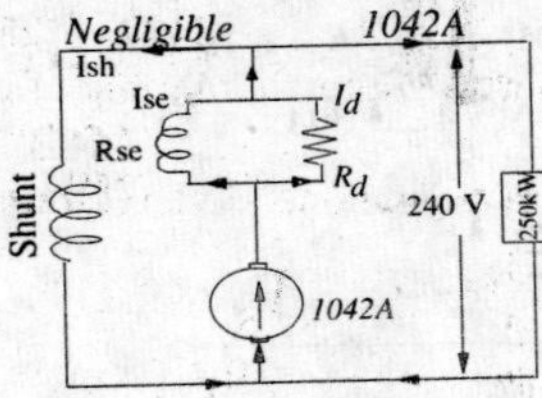

It is so because no-load shunt current being negligible, $I_a = I =$ 1042 A

Since series winding and divertor are in parallel, $I_d R_d = I_{se} R_{se}$ or

$229.5\ R_d = 812.5 \times 0.006$

$\therefore \quad R_d = \mathbf{0.0212\ \Omega}$

Fig. 26.30

Example 26.19. *A 60-kW d.c. shunt generator has 1600 turns/pole in its shunt winding. A shunt field current of 1.25 A is required to generate 125 V at no-load and 1.75 A to generate 150 V at full load. Calculate*

(i) the minimum number of series turns/pole needed to produce the required no-load and full-load voltages as a short-shunt compound generator.

(ii) if the generator is equipped with 3 series turns/pole having a resistance of 0.02 Ω, calculate divertor resistance required to produce the desired compounding.

(iii) voltage regulation of the compound generator.

Solution. (*i*) Extra excitation ampere-turns required = 1600 (1.75−1.25) = 800

$I_{se} = I = 60{,}000/150 = 400$ A

$\therefore$ No. of series turns/pole required = 800/400 = **2**

Hence, minimum number of series turns/pole required for producing the desired compound generator terminal voltage is **2**.

(ii) Now, actual No. of series turns/pole is 3. Hence, current passing through it can be found from

$3 \times I_{se} = 800;\ I_{se} = 800/3$ A

As shown in Fig. 26.31, $I_d = 400 - (800/3) = 400/3$ A

Also $(800/3) \times 0.02 = (400/3) \times R_d$; $R_d = \mathbf{0.04\ \Omega}$

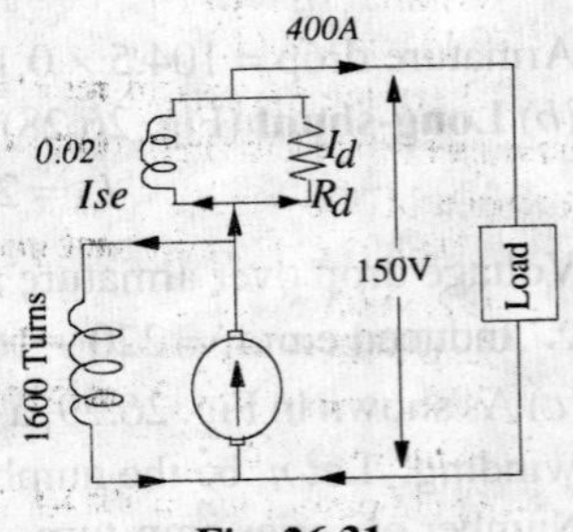

Fig. 26.31

(*iii*) % regn. = (125−150) × 100/150 = **−16.7%**

26.16. Uses of D.C. Generators

1. Shunt generators with field regulators are used for ordinary lighting and power supply purposes. They are also used for charging batteries because their terminal voltages are almost constant or can be kept constant.

2. **Series generators** are not used for power supply because of their rising characteristics. However, their rising characteristic makes them suitable for being used as boosters (Ex. 26.15) in certain types of distribution systems particularly in railway service.

3. **Compound generators**

The cumulatively-compound generator is the most widely used dc generator because its external characteristic can be adjusted for compensating the voltage drop in the line resistance. Hence, such generators are used for motor drives which require d.c. supply at constant voltage, for lamp loads and for heavy power service such as electric railways.

The differential-compound generator has an external characteristic similar to that of a shunt generator but with large demagnetization armature reaction. Hence, it is widely used in arc welding where larger voltage drop is desirable with increase in current.

Tutorial Problem No. 26.1

1. The *O.C.* curve of a d.c. shunt generator for a speed of 1000 r.p.m. is given by the following table.

Field current (A)	2.0	3.0	4.0	5.0	6.0	7.0
E.M.F. (V)	102	150	188	215	232	245

The shunt has a resistance of 37 Ω. Find the speed at which excitation may be expected to build up. The armature resistance is 0.04 Ω. Neglecting the effects of brush drop and armature reaction, estimate the p.d. when the speed is 1000 r.p.m. and the armature delivers a current of 100 A

[725 r.p.m. ; 231 V]

2. A. d.c. shunt generator running at 850 r.p.m. gave the following O.C.C. data :

Field current (A)	0	0.5	1	2	3	4	5
Armature e.m.f. (V)	10	60	120	199	232	248	258

If the resistance of the shunt field is 50 Ω, determine the additional resistance required in the shunt field circuit to give 240 V at a speed of 1000 r.p.m. **(64.3 Ω]**

3. Sketch the load characteristic of a d.c. generator with (*i*) shunt (*ii*) series excitation. Give reasons for the particular shape in each case.

The *O.C.C.* at 700 r.p.m. of a series generator with separately-excited field is as follows :

Field current (A)	20	40	50	60	75
Armature e.m.f. (V)	190	360	410	450	480

Determine the current and terminal voltage as a self-excited series machine when running at 600 r.p.m. with a load of 6 Ω connected to the terminals. Resistance of armature and series winding is 0.3 Ω. Ignore effect of armature reaction. **[369 V; 61.5A]**

4. The O.C.C. data for separately-excited generator when run at 130 r.p.m. on open circuit is

E.M.F. (V)	12	44	73	98	113	122	127
Exciting current (A)	0	0.2	0.4	0.6	0.8	1.0	1.2

Deduce the curve of e.m.f. and excitation when the generator is running separately-excited at 1000 r.p.m. To what voltage will the generator build up on no-load when running at 1000 r.p.m., To what voltage will the generator build up on no-load when running at 100 r.p.m. if the total field resistance is 100 Ω **[91V]**

5. The following figures give the *O.C.C.* of d.c. Shunt generator driven at a constant speed of 700 r.p.m.

Terminal voltage (V)	10	20	40	80	120	160	200	240	260
Field current (A)	0	0.1	0.24	0.5	0.77	1.2	1.92	3.43	5.2

Determine the critical resistance at (*a*) 700 r.p.m. (*b*) 850 r.p.m. If resistance of field coils is 50 Ω, find the range of the field rheostat required to vary the voltage between the limits of 180 V and 250 V on open circuit at a speed of 700 r.p.m. **[160 Ω ; 194 Ω ; 70 Ω to 10 Ω]**

6. The *O.C.C.* of a shunt generator when separately-excited and running at 1000 r.p.m. is given by :

O.C. volt	56	112	150	180	200	216	230
Field amp.	0.5	1.0	1.5	2.0	2.5	3.0	3.5

If the generator is shunt-connected and runs at 1100 r.p.m. with a total field resistance of 80 Ω, determine

(*a*) no-load e.m.f.

(*b*) the output when the terminal voltage is 200 V if the armature resistance is 0.1 Ω.

(*c*) the terminal voltage of the generator when giving the maximum output current

Neglect the effect of armature reaction and of brush contact drop.

[236 V ; 200 V ; 460 V ; 150 V (approx)]

7. A long-shunt compound d.c. generator with armature, series field and shunt field resistances of 0.5, 0.4 and 250 Ω respectively gave the following readings when run at constant speed :-

Load current (A)	0	10	20	30	40
Terminal p.d. (V)	480	478	475	471	467

Plot the curve of internally generated e.m.f. against load current. Explain fully the steps by which this curve is obtained and tabulate the values from which it is plotted. **(For 40 A ; E = 504.7 V (approx)]**

8. A shunt generator has the following open-circuit characteristic at 800 r.p.m.

Field amperes :	0.5	1.0	1.5	2.0	2.5	3.0	3.5
E.M.F. volt :	54	107	152	185	210	230	245

Armature and shunt field resistances are respectively 0.1 Ω and 80 Ω. The terminal p.d. falls to 175 V when the armature current is 100 A. Find the *O.C.* volts and the volts lost due to (*i*) reduction in the field current (*ii*) armature resistance (*iii*) armature reaction. **(220 V (*i*) 27 V (*ii*) 10 V (*iii*) 8 V]**

9. The open-circuit characteristic of a shunt generator when driven at normal speed is as follows :

Field current :	0.5	1.0	1.5	2.0	2.5	3.0	3.5 A
O.C. volts :	54	107	152	185	210	230	240 V

The resistance of armature circuit is 0.1 Ω. Due to armature reaction the effective field current is given by the relation I_{sh} (eff.) = I_{sh}−0.003 I_a. Find the shunt field circuit resistance that will give a terminal voltage of 220 V with normal speed (*a*) on open circuit (*b*) at a load current of 100 A. Also find (*c*) number of series turns for level compounding at 220 V with 100 A armature current ; take number of shunt turns per pole as 1200 and (*d*) No. of series turns for over-compounding giving a terminal voltage of 220 V at no-load and 230 V with 100 A armature current. **[(*a*) 80 Ω (*b*) 66 Ω (*d*) 6.8 Ω (*d*) 9.1 turns]**

10. Find how many series turns per pole are needed on a 500-kW compound generator required to give 450 V on no-load and 500 V on full-load, the requisite number of ampere-turns per pole being 9,000 and 6,500 respectively. The shunt winding is designed to give 450 V at no-load when its temperature is 20°C. The final temperature is 60°C. Take α_0 = 1/234.5 per °C.

[2.76] (Electrical Technology, Allahabad Univ. 1977)

Objective Tests – 26

1. The external characteristic of a shunt generator can be obtained directly from its —— characteristic.
   (*a*) internal (*b*) open-circuit
   (*c*) load-saturation (*d*) performance
2. Load saturation characteristic of a d.c. generator gives relation between
   (*a*) V and I_a (*b*) E and I_a
   (*c*) E_0 and I_f (*d*) V and I_f
3. The slight curvature at the lower end of the *O.C.C.* of a self-excited dc generator is due to
   (*a*) residual pole flux
   (*b*) high armature speed
   (*c*) magnetic inertia
   (*d*) high field circuit resistance
4. For the voltage built-up of a self-excited d.c. generator, which of the following is not an essential condition ?
   (*a*) there must be some residual flux
   (*b*) field winding mmf must aid the residual flux
   (*c*) total field circuit resistance must be less than the critical value
   (*d*) armature speed must be very high
5. The voltage build-up process of a d.c. generator is
   (*a*) difficult (*b*) delayed
   (*c*) cumulative (*d*) infinite
6. Which of the following d.c. generator cannot build up on open-circuit ?
   (*a*) shunt (*b*) series
   (*c*) short shunt (*d*) long shunt
7. If a self-excited dc generator after being installed, fails to build up on its first trial run, the first thing to do is to
   (*a*) increase the field resistance
   (*b*) check armature insulation
   (*c*) reverse field connections
   (*d*) increase the speed of prime mover
8. If residual magnetism of a shunt generates is destroyed accidentally, it may be restored by connecting its shunt field
   (*a*) to earth (*b*) to an a.c. source
   (*c*)_ in reverse (*d*) to a d.c. source
9. The three factors which cause decrease in the terminal voltage of a shunt generator are
   (*a*) armature reactancex (*b*) armature resistance
   (*c*) armature leakagex (*d*) armature reaction
   (*e*) reduction in field current
10. If field resistance of a d.c. shunt generator is increased beyond its critical value, the generator

(*a*) output voltage will exceed its name-plate rating
(*b*) will not build up
(*c*) may burn out if loaded to its name-plate rating
(*d*) power output may exceed its name-plate rating

11. An ideal d.c. generator is one that has —— voltage regulation.
(*a*) low (*b*) zero
(*c*) positive (*d*) negative

12. The —— generator has poorest voltage regulation.
(*a*) series (*b*) shunt
(*c*) compound (*d*) overcompound

13. The voltage regulation of an overcompound d.c. generator is always ——
(*a*) positive (*b*) negative
(*c*) zero (*d*) high

14. Most commercial compound d.c. generators are normally supplied by the manufactures as over compound machines because
(*a*) they are ideally suited for transmission of d.c. energy to remotely-located loads
(*b*) degree of compounding can be adjusted by using a diverter across series field.
(*c*) they are more cost effective than shunt generators
(*d*) they have zero percent regulatio

ANSWER

1. *b* **2.** *d* **3.** *c* **4.** *d* **5.** *c* **6.** *b* **7.** *c* **8.** *d* **9.** *b,d*
10. *b* **11.** *b* **12.** *a* **13.** *b* **14** *b*.

QUESTIONS & ANSWERS ON D.C. GENERATORS

Q.1. How may the number of parallel paths in an armature be increased ?
Ans. By increasing the number of magnetic poles.
Q.2. How are brushes connected in a d.c. generator ?
Ans. Usually, all positive brushes are connected together and all the negative brushes together (Fig. 26.32).
Q.3. What is meant by armature reaction ?
Ans. It is the effect of armature magnetic field on the distribution of flux under main poles of a generator. The armature magnetic field has two effects :
(*i*) it demagnetises or weakens the main flux and
(*ii*) it cross-magnetises or distorts it.

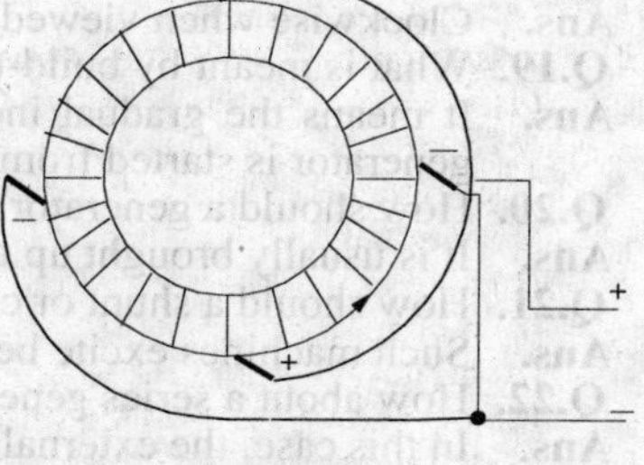

Fig. 26.32

Q.4. What is the effect of this distortion on the operation of the machine ?
Ans. It acts as a magnetic drag on the armature which consequently requires more power to turn it.
Q.5. How can field distortion be remedied ?
Ans. By using compensating windings which are embedded in the slots in the pole-shoes and are connected in series with the armature.
Q.6. What is meant by normal neutral plane ?
Ans. It is a plane which passes through the axis of the armature perpendicular to the magnetic field of the generator when there is no flow of current through the armature.
Q.7. What is the importance of this plane in the working of the machine ?

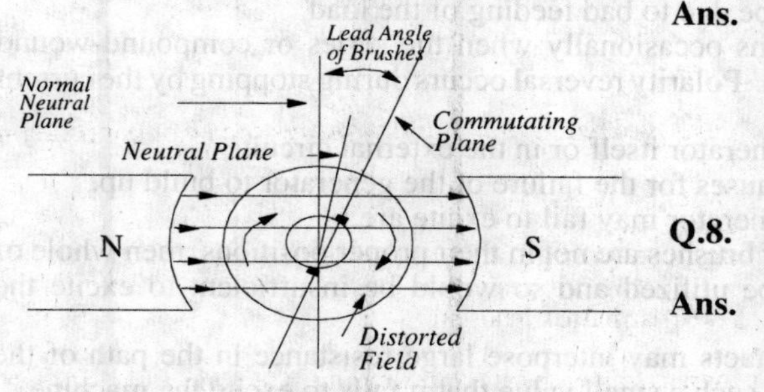

Fig. 26.33

Ans. It gives the position where brushes would be placed to prevent sparking during the operation of the generator were the main pole field not distorted by armature field and were there no self-induction in the coils.
Q.8. How do you differentiate between normal neutral plane and the neutral plane ?
Ans. The *NNP* is the position of zero induction and hence minimum sparking assuming no field distortion *i.e.* on no-load. It is perpendicular to magnetic axis. *NA* is the

position of zero induction and hence minimum sparking with distorted field *i.e. when generator is on load.*

Q.9 How do you define 'commutating plane'?

Ans. It is the plane which passes through the axis of the armature and through centre of contact of the brushes as shown in Fig. 26.33.

Q.10. What is the angle of lead ?

Ans. It is the angle between the *NNP* and the commutating plane.

Q.11. What affects this angle ?

Ans. For sparkless commutation, the angle of lead varies directly with load. Its value can be kept small by making main pole field considerably more powerful than the armature field.

Q.12. What is the best way of minimizing eddy currents in an armature ?

Ans. Lamination.

Q.13. How should the armature be laminated for the purpose ?

Ans. It should be laminated at right angle to its axis.

Q.14. How does field distortion affect commutation ?

Ans. The neutral plane no longer coincides with the normal neutral plane but is advanced by a certain angle in the direction of rotation of the armature.

Q.15. Should the brushes of a loaded generator be placed in the neutral plane ?

Ans. No.

Q.16. Why not ?

Ans. The brushes must be advanced by a certain angle (called brush lead) beyond the neutral plane to prevent sparking.

Q.17. What causes sparking at the brushes ?

Ans. It is due to the self-induction of the coil undergoing commutation.

Q.18. What is the standard direction of rotation of the d.c. generators ?

Ans. Clockwise when viewed from the end opposite the driven end.

Q.19. What is meant by build-up of a generator ?

Ans. It means the gradual increase in the generator voltage to its maximum value after the generator is started from rest.

Q.20. How should a generator be started ?

Ans. It is usually brought up to speed with the help of the driving engine called prime-mover.

Q.21. How should a shunt or compound generator be started ?

Ans. Such machines excite best when all switches controlling the external circuit are open.

Q.22. How about a series generator ?

Ans. In this case, the external circuit must be closed otherwise the generator will not build-up.

Q.23. What is the procedure for shutting down a generator ?

Ans. First, the load should be gradually reduced, if possible, by easing down the driving engine, then when the generator is supplying little or no current, the main switch should be opened. When the voltmeter reads almost zero, then brushes should be raised from the commutator.

Q.24. What are the indications and causes of an overloaded generator ?

Ans. A generator is said to be overloaded if a greater output is taken from it that it can safely carry. Overloading is indicated by (*i*) excessive sparking at brushes and (*ii*) overheating of the armature and other parts of the generator. Most likely causes of overloading are :

1. excessive voltage–as indicated by the voltmeter or the increased brilliancy of the pilot lamp. This could be due to over-excitation of field magnets or too high speed of the engine.
2. excessive current–which could be due to bad feeding of the load.
3. reversal of polarity–this happens occasionally when the series or compound-wound generators are running in parallel. Polarity reversal occurs during stopping by the current from the machines at work.
4. short-circuit or ground in the generator itself or in the external circuit.

Q.25. Mention and explain the various causes for the failure of the generator to build up.

Ans Principal causes due to which a generator may fail to excite are :

1. brushes not properly adjusted–if brushes are not in their proper positions, then whole of the armature voltage will not be utilized and so would be insufficient to excite the machine.
2. defective contacts–unclean contacts may interpose large resistance in the path of the exciting current and reduce it to such a small value that it fails to excite the machine.

3. incorrect adjustment of regulators–in the case of shunt and compound generators, it is possibly that the resistance of field regulator may be too high to permit the passage of sufficient current through the field windings.
4. speed too low–in the case of shunt-and compound-wound generators, there is certain critical armature speed below which they will not excite.
5. open-circuit–in the case of series machines.
6. short circuit–in the generator or external circuit.
7. reversed field polarity–usually caused by the reversed connections of the field coils.
8. insufficient residual magnetism–the trouble normally occurs when the generator is new. It can be remedied by passing a strong direct current through the field coils.

Q.26. How do we conclude that connections between field coils and armature are correct ?

Ans. If the generator builds up when brought to full speed. If it does not; then connections are reversed.

Q.27. When a generator loses its residual magnetism either due to lighting or short circuit, how can it be made to build up ?

Ans. By temporarily magnetising the main poles with the help of current from an external battery.

Q.28. Can a generator be reversed by reversing the connections between the armature and field coils ?

Ans. No, because if these connections are reversed, the generator will not build up at all.

Q.29. Will a generator build up if it becomes reversed ?

Ans. Yes.

Q.30. Then, what is the objection to a reversed generator ?

Ans. Since the current of such a reversed generator is also reversed, serious trouble can occur if attempt is made to connect it in parallel with other machines which are not reversed.

Q.31. What are the two kinds of sparking produced in a generator ?

Ans. One kind of sparking is due to bad adjustment of brushes and the other due to bad condition of the commutator. The sparks of the first are bluish whereas those of the other are reddish in colour.

Q.32. What is the probable reason if sparking does not disappear in any position when brushes are rocked around the commutator.

Ans. (*i*) the brushes may not be separated at correct distance.
(*ii*) the neutral plane may not be situated in the true theoretical position on the commutator due to faulty winding.

Q.33. What is the permissible rise of temperature in a well-designed generator ?

Ans. 27°C above the surrounding air.

Q.34. What are the causes of hot bearings ?

Ans. (*i*) lack of oil (*ii*) belt too tight (*iii*) armature not centered with respect to pole pieces (*iv*) bearing too tight or not in line.

Q.35. What causes heating of armature ?

Ans. 1. eddy currents.
2. moisture which almost short-circuits the armature.
3. unequal strength of magnetic poles.
4. operation above rated voltage and below normal speed.

Q.36. What is the commutator pitch of a 4-pole d.c. armature having 49 commutator bars ?

Ans. $Y_c = (49 \pm 1)/2 = 24$ or 25.

Q.37. Will it make any difference if lower figure of 24 is selected in preference to other.

Ans. Yes. Direction of armature rotation would be reversed.

27 D.C. MOTOR

27.1. Motor Principle

An Electric motor is a machine which converts electric energy into mechanical energy. Its action is based on the principle that when a current-carrying conductor is placed in a magnetic field, it experiences a mechanical force whose direction is given by Fleming's Left-hand Rule and whose magnitude is given by $F = BIl$ newton.

Constructionally, there is no basic difference between a d.c. generator and a d.c. motor. In fact, the same d.c. machine can be used interchangeably as a generator or as a motor. D.C. motors are also like generators, shunt-wound or series-wound or compound-wound.

In Fig. 27.1 is shown a part of a multipolar d.c. motor. When its field magnets are excited and its armature conductors are supplied with current from the supply mains, they experience a force tending to rotate the armature. Armature conductors under *N*-pole are assumed to carry current downwards (crosses) and those under *S*-poles, to carry current upwards (dots). By applying Fleming's Left-hand Rule, the direction of the force on each conductor can be found. It is shown by small arrows placed above each conductor. It will be seen that each conductor can be found. It will be seen that each conductor experiences a force *F* which tends to rotate the armature in anticlockwise direction. These forces collectively produce a driving torque which sets the armature rotating.

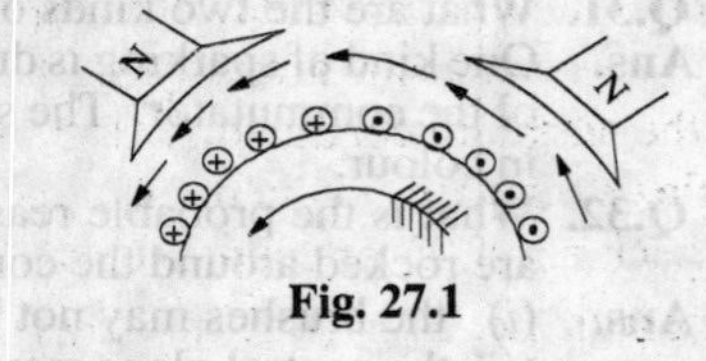

Fig. 27.1

It should be noted that the function of a commutator in the motor is the same as in a generator. By reversing current in each conductor as it passes from one pole to another, it helps to develop a continuous and unidirectional torque.

27.2. Comparison of Generator and Motor Action

As said above, the same d.c. machine can be used, at least theoretically, interchangeably as a generator or as a motor. When operating as a generator, it is driven by a mechanical machine and it develops voltage which in turn produces a current flow in an electric circuit. When operating as a motor, it is supplied by electric current and it develops torque which in turn produces mechanical rotation.

Let us first consider its operation as a generator and see how exactly and through which agency, mechanical power is converted into electric power.

In Fig. 27.2 is shown part of a generator whose armature is being driven clockwise by its prime mover.

Fig. 27.2 (*a*) represents the fields set up independently by the main poles and the armature conductors like *A* in the figure. The resultant field or magnetic lines on flux are shown in Fig. 27.2 (*b*). It is seen that there is a crowding of lines of flux on the right-hand side of *A*. These magnetic lines of flux may be likened to the rubber bands under tension. Hence, the bent lines of flux set up a mechanical force on *A* much

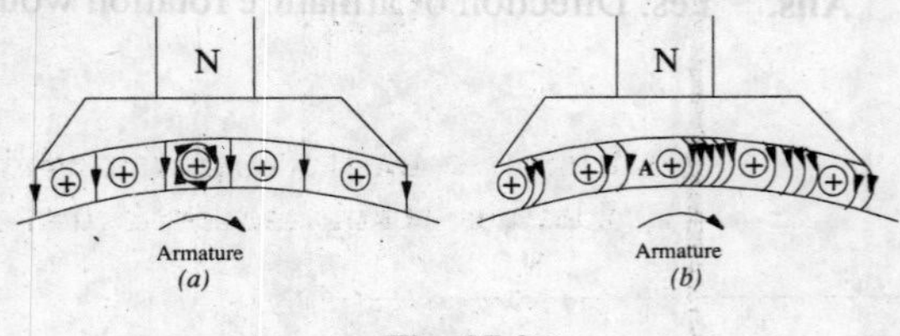

Fig. 27.2

in the same way as the bent elastic rubber band of a catapult produces a mechanical force on the stone piece. It will be seen that this force is in a direction opposite to that of armature rotation. Hence, it is known as backward force or magnetic drag on the conductors. It is against this drag action on all armature conductor that the prime mover has to work. The work done in overcoming this opposition is converted into electric energy. Therefore, it should be clearly understood that it is only through the instrumentality of this magnetic drag that energy conversion is possible in a d.c. generator.*

Next, suppose that the above d.c. machine is uncoupled from its prime mover and that current is sent through the armature conductors under a N-pole in the downward direction as shown in Fig. 27.3. The conductors will again experience a force in the anticlockwise direction (Fleming's Left-hand Rule). Hence, the machine will start rotating anticlockwise, thereby developing a torque which can produce mechanical rotation. The machine is then said to be motoring.

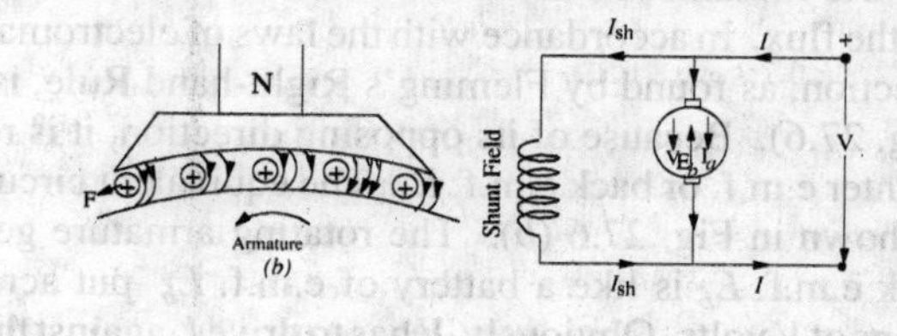

Fig. 27.3 Fig. 27.4

As said above, energy conversion is not possible unless there is some opposition whose overcoming provides the necessary means for such conversion. In the case of a generator, it was the magnetic drag which provided the necessary opposition. But what is the equivalent of that drag in the case of a motor ? Well, it is the back e.m.f. It is explained in this manner :

As soon as the armature starts rotating, dynamically (or motionally) induced e.m.f. is produced in the armature conductors. The direction of this induced e.m.f. as found by Fleming's Ring-hand Rule, is outwards *i.e.* in direct opposition to the applied voltage (Fig. 27.4). This is why it is known as back e.m.f. E_b or counter e.m.f. Its value is the same as for the motionally induced e.m.f. in the generator *i.e.* $E_b = (\Phi ZN) \times (P/A)$ volts. The applied voltage V has to force current through the armature conductors against this back e.m.f. E_b. The electric work done in overcoming this opposition is converted into mechanical energy developed in the armature. Therefore, it is obvious that but for the production of this opposing e.m.f. energy conversion would not have been possible.

Now, before leaving this topic, let it be pointed out that in an actual motor with slotted armature, the torque is not due to mechanical force on the conductors themselves, but due to tangential pull on the armature teeth as shown in Fig. 27.5.

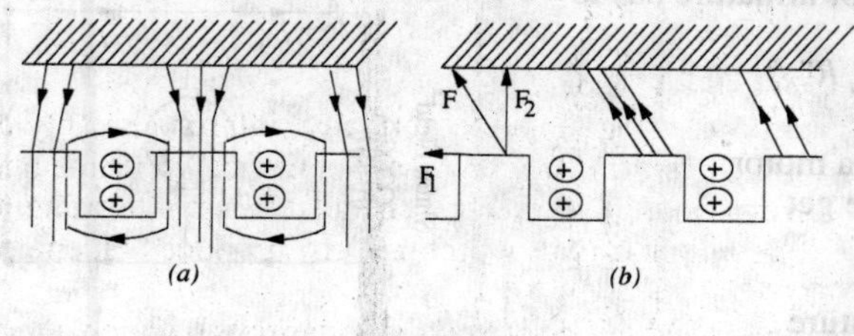

Fig. 27.5

It is seen from Fig. 27.5(*a*) that the main flux is concentrated in the form of tufts at the armature teeth while the armature flux is shown by the dotted lines embracing the armature slots. The effect of armature flux on the main flux, as shown in Fig. 27.5(*b*), is two-fold :

(*i*) It increases the flux on the left-hand side of the teeth and decreases it on the right-hand side, thus making the distribution of flux density across the tooth section unequal.

*In fact, it seems to be one of the fundamental laws of Nature that no energy conversion from one form to another is possible until there is some one to oppose the conversion. But for the presence of this opposition, there would simply be no energy conversion. In generators, opposition is provided by magnetic drag whereas in motors, back e.m.f. does this job. Moreover, it is only that part of the input energy which is used for overcoming this opposition that is converted into the other form.

(*ii*) It inclines the direction of lines of force in the air-gap so that they are not radial but are disposed in a manner shown in Fig. 27.5(*b*). The pull exerted by the poles on the teeth can now be resolved into two components. One is the tangential component F_1 and the other vertical component F_2. The vertical component F_2, when considered for all the teeth round the armature, adds up to zero. But the component F_1 is not cancelled and it is this tangential component which, acting on all the teeth, gives rise to the armature torque.

27.3. Significance of the Back e.m.f.

As explained in Art. 27.2, when the motor armature rotates, the conductors also rotate and hence cut the flux. In accordance with the laws of electromagnetic induction, e.m.f. is indued in them whose direction, as found by Fleming's Right-hand Rule, is in opposition to the applied voltage (Fig. 27.6). Because of its opposing direction, it is referred to as counter e.m.f. or back e.m.f. E_b. The equivalent circuit of a motor is shown in Fig. 27.6 (*b*). The rotating armature generating the back e.m.f. E_b is like a battery of e.m.f. E_b put across a supply mains of V volts. Obviously, V has to drive I_a against the opposition of E_b. The power required to overcome this opposition is E_bI_a.

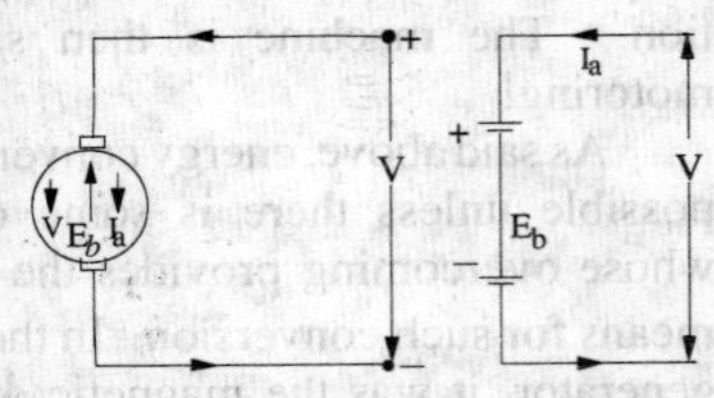

Fig. 27.6

In the case of a cell, this power over an interval of time is converted into chemical energy, but in the present case, it is converted into mechanical energy.

It will be seen that $I_a = \dfrac{\text{net voltage}}{\text{resistance}} = \dfrac{V - V_b}{R_a}$ where R_a is the resistance of the armature circuit. As pointed out above, $E_b = \Phi ZN \times (P/A)$ volt where N is in r.p.s.

Back e.m.f. depends, among other factors, upon the armature speed. If speed is high, E_b is large, hence armature current I_a, as seen from the above equation, is small. If the speed is less, then E_b is less, hence more current flows which develops motor torque (Art. 27.7). So, we find that E_b acts like a governor *i.e.* it makes a motor self-regulating so that it draws as much current as is just necessary.

27.4. Voltage Equation of a Motor

The voltage V applied across the motor armature has to
(*i*) overcome the back e.m.f. E_b and
(*ii*) supply the armature ohmic drop $I_a R_a$

$\therefore \quad V = E_b + I_a^2 R_a$

This is known as voltage equation of a motor:
Now, multiplying both sides by I_a, we get

$VI_a = E_bI_a + I_a^2R_a$

As shown in Fig. 27.7,

VI_a = electrical input to the armature
$E_b I_a$ = electrical equivalent of mechanical power developed in the armature
$I_a^2 R_a$ = Cu loss in the armature

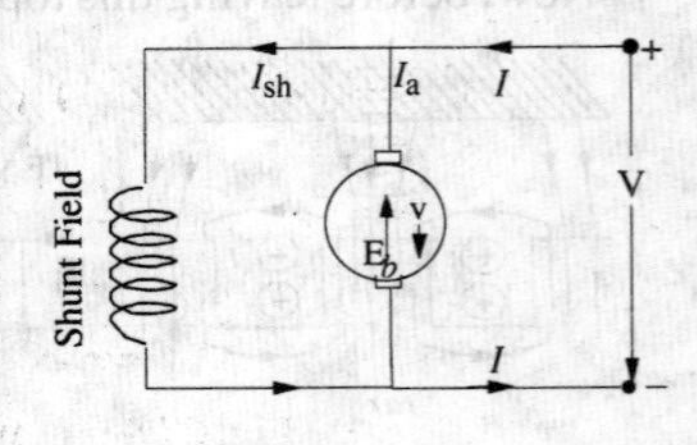

Fig. 27.7

Hence, out of the armature input, some is wasted in I^2R loss and the rest is converted into mechanical power *within* the armature.

It may also be noted that motor efficiency is given by the ratio of power developed by the armature to its input *i.e.* $E_b I_a / VI_a = E_b / V$. Obviously, higher the value of E_b as compared to V, higher the motor efficiency.

27.5. Condition for Maximum Power

The gross mechanical power developed by a motor is $P_m = VI_a - I_a^2 R_a$

Differentiating both sides with respect to I_a and equating the result to zero, we get

$$dp_m/dI_a = V - 2I_aR_a = 0 \quad \therefore \quad I_aR_a = V/2$$

As $\quad V = E_b + I_aR_a$ and $I_aR_a = V/2 \quad \therefore \quad E_b = V/2$

Thus gross mechanical power developed by a motor is maximum when back e.m.f. is equal to half the applied voltage. This condition is, however, not realized in practice, because in that case current would be much beyond the normal current of the motor. Moreover, half the input would be wasted in the form of heat and taking other losses (mechanical and magnetic) into consideration, the motor efficiency will be well below 50 per cent.

Example 27.1. *A 220-V d.c. machine has an armature resistance of 0.5 Ω. If the full-load armature current is 20A, find the induced e.m.f. when the machine acts as (i) generator (ii) motor.*

(Electrical Technology-I, Bombay Univ. 1987)

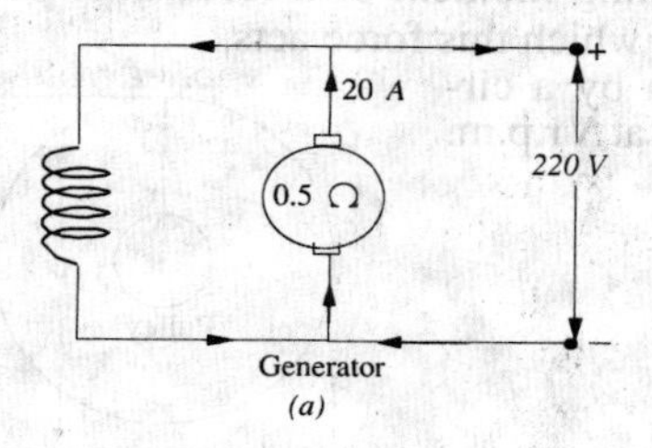

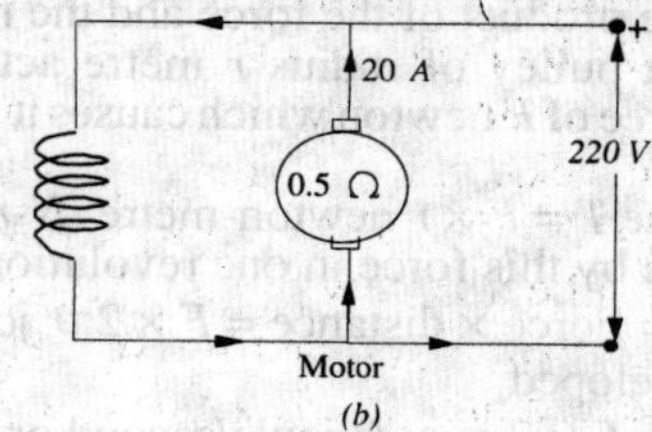

Fig. 27.8

Solution. As shown in Fig. 27.8, the d.c. machine is assumed to be shunt-connected. In each case, shunt current is considered negligible because its value is not given.

(*a*) As Generator [Fig. 27.8 (*a*)] $\quad E_g = V + I_aR_a = 220 + 0.5 \times 20 =$ **230 V**

(*b*) As Motor [Fig. 27.8 (*b*)] $\quad E_b = V - I_a = R_a = 220 - 0.5 \times 20 =$ **210 V**

Example 27.2. *A 440-V, shunt motor has armature resistance of 0.8 Ω and field resistance of 200 Ω. Determine the back e.m.f. when giving an output of 7.46 kW at 85 per cent efficiency.*

Solution. Motor input power = $7.46 \times 10^3/0.85$ W

Motor input current = $7460/0.85 \times 440 = 19.95$ A ; $I_{sh} = 440/200 = 2.2$ A

$I_a = 19.95 - 2.2 = 17.75$ A ; Now, $E_b = V - I_a R_a$

$\therefore \quad E_b = 440 - (17.75 \times 0.8) =$ **425.8 V**

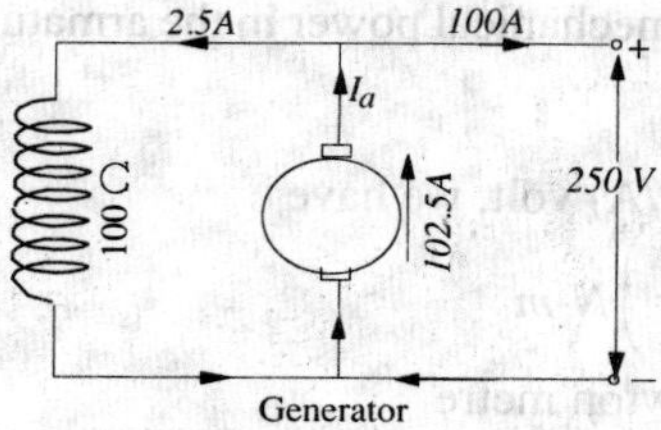

Generator

Fig. 27.9

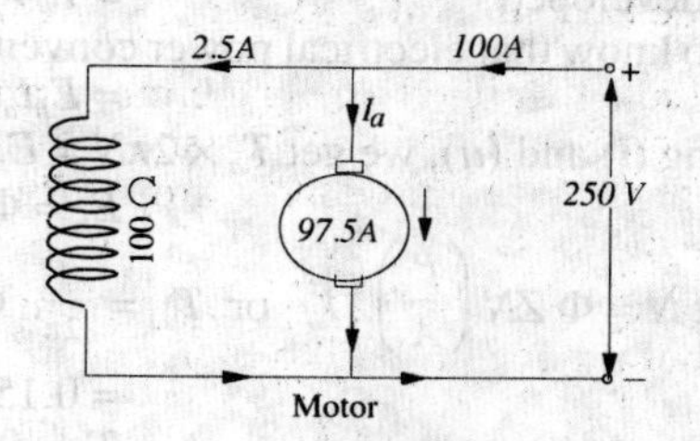

Fig. 27.10

Example 27.3. *A 25-kW, 250-V, d.c. shunt generator has armature and field resistances of 0.06 Ω and 100 Ωrespectively. Determine the total armature power developed when working (i) as a generator delivering 25 kW output and (ii) as a motor taking 25 kW input*

(Electrical Technology, Punjab Univ. June 1991)

Solution. As Generator (Fig. 27.9).

Output current $\quad = 25{,}000/250 = 100$ A; $I_{sh} = 250/100 = 2.5$ A; $I_a = 102.5$ A

Generated e.m.f. $\quad = 250 + I_aR_a = 250 + 102.5 \times 0.06 = 256.15$ V

Power developed in armature = $E_bI_a = \dfrac{256.15 \times 102.5}{1000} =$ **26.25 kW**

As Motor (Fig. 27.10)

Motor input current = 100 A ; $I_{sh} = 2.5$ A, $I_a = 97.5$ A

$E_b = 250 - (97.5 \times 0.06) = 250 - 5.85 = 244.15$ V

Power developed in armature = $E_b I_a = 244.15 \times 97.5/1000 =$ **23.8 kW**

Tutorial Problem No. 27.1

1. What do you understand by the term 'back e.m.f.'? A d.c. motor connected to a 460-V supply has an armature resistance of 0.15 Ω. Calculate.

(*a*) The value of back e.m.f. when the armature current is 120 A.

(*b*) The value of armature current when the back e.mf. is 447.4 V. [(*a*) **442 V** (*b*) **84** A]

2. A d.c. motor connected to a 460–V supply takes an armature current of 120 A on full load. If the armature circuit has a resistance of 0.25 Ω, calculate the value of the back e.m.f. at this load. **[430 V]**

3. A 4-pole d.c. motor takes an armature current of 150 A at 440 V. If its armature circuit has a resistance of 0.15 Ω, what will be the value of back e.m.f. at this load ? **[417.5 V]**

27.6. Torque

By the term torque is meant the turning or twisting moment of a force about an axis. It is measured by the product of the force and the radius at which this force acts.

Consider a pulley of radius *r* metre acted upon by a circumferential force of *F* newton which causes it to rotate at *N* r.p.m. (Fig. 27.11).

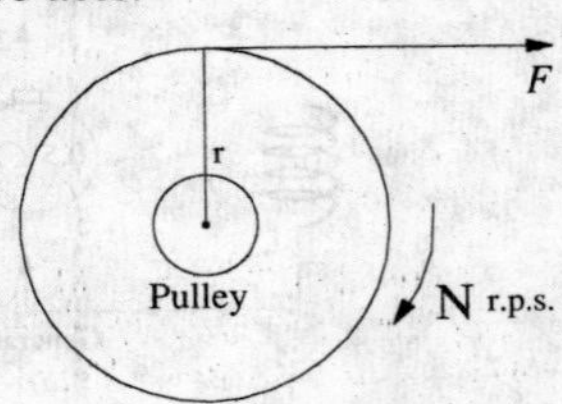

Fig. 27.11

Then torque $T = F \times r$ newton-metre (*N-m*)

Work done by this force in one revolution

$= \text{Force} \times \text{distance} = F \times 2\pi r$ joule

Power developed

$= F \times 2\,\pi r \times N$ joule/second or watt

$= (F \times r) \times 2\pi\, N$ watt

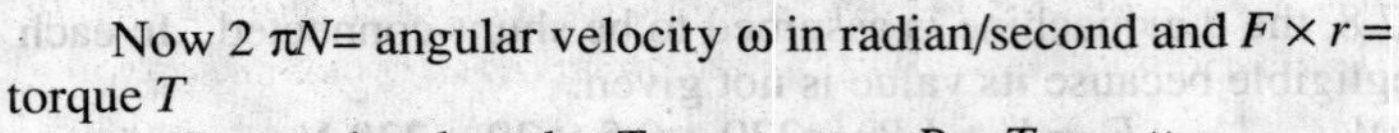

Now $2\,\pi N$= angular velocity ω in radian/second and $F \times r$ = torque *T*

∴ Power developed = $T \times \omega$ watt or $P = T\omega$ watt

Moreover, if *N* is in rpm, then

$\omega = 2\,\pi N/60$ rad/s

$$\therefore \quad P = \frac{2\,\pi N}{60} \times T \text{ or } P = \frac{2\,\pi}{60} \cdot NT = \frac{NT}{9.55}$$

27.7. Armature Torque of a Motor

Let T_a be the torque developed by the armature of a motor running at N r.p.s. If T_a is in *N* /m, then power developed $= T_a \times 2\pi\, N$ watt ...(*i*)

We also know that electrical power converted into mechanical power in the armature (Art. 27.4) is $= E_b I_a$ watt ...(*ii*)

Equating (*i*) and (*ii*), we get $T_a \times 2\pi N = E_b\, I_a$ ...(*iii*)

Since $E_b = \Phi ZN \times (P/A)$ volt, we have

$$T_a \times 2\pi\, N = \Phi\, ZN \left(\frac{P}{A}\right) .\ I_a \quad \text{or} \quad T_a = \frac{1}{2\pi} .\ \Phi\, ZI_0 \left(\frac{P}{A}\right) N\text{-}m$$

$= 0.159$ N newton metre

$\therefore \quad T_a = 0.159\ \Phi Z I_a \times (P/A)$ N-m

Note. From the above equation for the torque, we find that $T_a \propto \Phi I_a$

(*a*) In the case of a series motor, Φ is directly proportional to I_a (before saturation) because field windings carry full armature current $\therefore\ T_a \propto I_a^2$

(*b*) For shunt motors, Φ is practically constant, hence $T_a \propto I_a$

As seen from (*iii*) above

$$T_a = \frac{E_b\, I_a}{2\pi\, N} \text{ N - m - N in rps}$$

If *N* is in rpm, then

$$T_a = \frac{E_b\, I_a}{2\pi\, N/60} = 60 \frac{E_b\, I_a}{2\pi N} = \frac{60}{2\pi} \frac{E_b I_a}{N} = 9.55 \frac{E_b I_a}{N} \text{ N-m}$$

27.8. Shaft Torque (T_{sh})

The whole of the armature torque, as calculated above, is not available for doing useful work, because a certain percentage of it is required for supplying iron and friction losses in the motor.

The torque which is available for doing useful work is known as shaft torque T_{sh}. It is so called because it is available at the shaft. The motor output is given by

output $= T_{sh} \times 2\pi N$ Watt provided T_{sh} is in N-m and N in r.p.s.

$$\therefore \quad T_{sh} = \frac{\text{output in wattts}}{2\pi N} \text{ N-m } -N \text{ in r.p.s.}$$

$$= \frac{\text{output in watts}}{2\pi N/60} \text{ N-m } -N \text{ in r.p.m.}$$

$$= \frac{60}{2\pi}\frac{\text{output}}{N} = 9.55\frac{\text{output}}{N} \text{ N-m.}$$

The difference $(T_a - T_{sh})$ is known as lost torque and is due to iron and friction losses of the motor.

Note. The value of back e.m.f. E_b can be found from

(*i*) the equation, $E_b = V - I_a R_a$ (*ii*) the formula

$E_b = \Phi ZN \times (P/A)$ volt

Example 27.4. *A. d.c. motor takes an armature current of 110 A at 480 V. The armature circuit resistance is 0.2 Ω. The machine has 6-poles and the armature is lap-connected with 864 conductors. The flux per pole is 0.05 Wb. Calculate (i), the speed and (ii) the gross torque developed by the armature.* **(Elect. Machines, A.M.I.E. Sec B, 1989)**

Solution. $E_b = 480 - 110 \times 0.2 = 458$ V, $\Phi = 0.05$ W, $Z = 864$

Now, $E_b = \frac{\Phi ZN}{60}\left(\frac{P}{A}\right)$ or $458 = \frac{0.05 \times 864 \times N}{60} \times \left(\frac{6}{6}\right)$

$\therefore$ N = 636 r.p.m.

$T_a = 0.159 \times 0.05 \times 864 \times 110(6/6) = 756.3$ N-m

Example 27.5. *A 250-V, 4-pole, wave-wound d.c. erie motor has 782 conductors on its armature. It has armature and series field resistance of 0.75 ohm. The motor takes a current of 40 A. Estimate its speed and gross torque developed if it has a flux per pole of 25 mWb.* **(Elect. Engg.-II, Pune Univ. 1991)**

Solution. (*i*) $E_b = \Phi ZN\ (P/A)$

Now, $E_b = V - I_a R_a = 50 - 40 \times 0.75 = 220$ V

$\therefore \quad 220 = 25 \times 10^{-3} \times 782 \times N \times 0.75 = 220$ V

$\therefore$ $220 = 0.159\ \Phi\ ZI_a\ (P/A)$

$= 0.159 \times 25 \times 10^{-3} \times 782 \times 40 \times (4/2) =$ **249 N–m**

Example 27.6. *A d.c. shunt machine develops an o.c e.m.f. of 250 V at 1500 r.p.m. Find its torque and mechanical power developed for an armature current of 50A. State the simplifying assumptions.* **(Basic Elect. Machine Nagpur Univ. 1993)**

Solution. A given d.c. machine develops the same e.m.f. in its armature conductors whether running as a generator or as a motor. Only difference is that this armature e.m.f. is known as back e.m.f. when the machine is run as a motor.

Mechanical power developed in the arm. $= E_b I_a = 250 \times 50 = 12{,}500\ \omega$

$T_a = 9.55\ \ E_b I_a/N = 9.55 \times 250 \times 50/1500 = 79.6$ N-m.

Example 27.7. *Determine developed torque and shaft torque of 220-V, 4-pole series motor with 800 conductors wave-connected supplying a load of 8.2 kW by taking 45 A from the mains. The flux per pole is 25 mWb and its armature circuit resistance is 0.6 Ω.*

(Elect. Machine AMIE Sec. B Winter 1991)

Solution. Developed torque or gross torque is the same thing as armature torque.

$$\therefore \quad T_a = 0.159\, \Phi ZA(P/A)$$

$$= 0.159 \times 25 \times 10^{-3} \times 800 \times 45(4/2) = 286.2 \text{ N-m}$$

$$E_b = V - I_a R_a = 220 - 45 \times 0.6 = 193 \text{ V}$$

Now, $E_b = \Phi ZN(P/A)$ or $193 = 25 \times 10^{-3} \times 800 \times N \pi \times (4/2)$

$$\therefore \quad N = 4.825 \text{ rps}$$

Also, $2\pi N T_{sh}$ = output or $2\pi \times 4.825\, T_{sh} = 8200 \quad \therefore T_{sh} = 270.5$ N-m

Example 27.8. *A 220-V d.c. shunt motor runs at 500 r.p.m. when the armature current is 50 A. Calculate the speed if the torque is doubled. Given that $R_a = 0.2\ \Omega$.*

(Electrical Technology-II, Gwalior Univ. 1985)

Solution. As seen from Art. 27.7, $T_a \propto \Phi I_a$. Since Φ is constant, $T_a \propto I_a$

$\therefore\ T_{a1} \propto I_{a1}$ and $T_{a2} \propto I_{a2} \quad \therefore\ T_{a2}/T_{a1} = I_{a2}/I_{a1}$

$\therefore\ 2 = I_{a2}/50$ or $I_{a2} = 100$ A

Now, $N_2/N_1 = E_{b2}/E_{b1}$ – since Φ remains constant

$F_{b1} = 220 - (50 \times 0.2) - 210\ V \qquad E_{b2} = 220 - (100 \times 0.2) = 200$ V

$\therefore\ N_2/500 = 200/210 \qquad \therefore\ N_2 =$ **476 r.p.m.**

Example 27.9. *A 500-V, 37.3 kW, 1000 r.p.m. d.c. shunt motor has on full-load an efficiency of 90 per cent. The armature circuit resistance is 0.24 Ω and there is total voltage drop of 2 V at the brushes. The field current is 1.8 A. Determine (i) full-load line current (ii) full load shaft torque in N-m and (iii) total resistance in motor starter to limit the starting current to 1.5 times the full-load current.*

(Elect. Engg.I.; M.S. Univ. Baroda 1987)

Solution. (*i*) Motor input = 37,300/0.9 = 41,444 W

F.L. line current = 41,444/500 = **82.9 A**

(*ii*) $T_{sh} = 9.55 \dfrac{\text{output}}{N} = 9.55 \times \dfrac{37,300}{1000} = 356$ N-m

(*iii*) Starting line current = 1.5 × 82.9 = 124.3 A

Arm. current at starting = 124.3 − 1.8 = 122.5 A

If R is the starter resistance (which is in series with armature), then

$122.5\,(R + 0.24) + 2 = 500 \quad \therefore\ R =$ **3.825 Ω**

Example 27.10. *A 4-pole, 220-V shunt motor has 540 lap-wound conductor. It takes 32 A from the supply mains and develops output power of 5.595 kW. The field winding takes 1 A. The armature resistance is 0.9 Ω and the flux per pole is 30 mWb. Calculate (i) the speed and (ii) the torque developed in newton-metre.*

(Electrical Technology, Nagpur Univ. 1992)

Solution. $I_a = 32 - 1 = 31$ A; $E_b = V - I_a R_a = 220 - (0.9 \times 31) = 217.2$ V

Now, $E_b = \dfrac{\Phi ZN}{60}\left(\dfrac{P}{A}\right) \quad \therefore\ 217.2 = \dfrac{30 \times 10^{-3} \times 540 \times N}{60}\left(\dfrac{4}{4}\right)$

(*i*) $\therefore\ N =$ **804.4 r.p.m.**

(*ii*) $T_{sh} = 9.55 \times \dfrac{\text{output in watts}}{N} = 9.55 \times \dfrac{5,595}{804.4} =$ **66.5 N-m**

Example 27.11. *A d.c. series motor takes 40 A at 220 V and runs at 800 r.p.m. If the armature and field resistance are 0.2 Ω and 0.1 Ω respectively and the iron and friction losses are 0.5 kW, find the torque developed in the armature. What will be the output of the motor ?*

Solution. Armature torque is given by $T_a = 9.55 \dfrac{E_b I_a}{N}$ N-m

Now $E_b = V - I_a(R_a + R_{se}) = 220 - 40\,(0.2 + 0.1\,) = 208$ V

$\therefore$ $T_a = 9.55 \times 208 \times 40/800 =$ **99.3 N-m**

Cu loss in armature and series-field resistance $= 40^2 \times 0.3 = 480$ W

Iron and friction losses = 500 W ; Total looses = 480 + 500 = 980 W

Motor power input = 220 × 40 = 8,800 W

Motor output = 8,800–980 = 7,820 *W* = **7.82 kW**

Example 27.12. *A cutting tool exerts a tangential force of 400 N on a steel bar of diameter 10 cm. which is being turned in a simple lathe. The lathe is driven by a chain at 840 r.p.m. from a 220 V d.c. Motor which runs at 1800 r.p.m. Calculate the current taken by the motor if its efficiency is 80%. What size is the motor pulley if the lathe pulley has a diameter of 24 cm ?*

(Elect. Technology-II, Gwalior Univ. 1985)

Solution.

Torque T_{sh} = tangential force × radius = 400 × 0.05 = 20 N-m

Output power = $T_{sh} \times 2\pi N$ watt = $20 \times 2\pi \times (840/60)$ watt = 1,760 W

Motor $\eta = 0.8$ $\therefore$ Motor input = 1,760/0.8 = 2,200 W

Current drawn by motor = 2200/220 = **10 A**

Let N_1 and D_1 be the speed and diameter of the driver pulley respectively and N_2 and D_2 the respective speed and diameter of the lathe pulley

Then $N_1 \times D_1 = N_2 \times D_2$ or $1{,}800 \times D_1 = 840 \times 0.24$

$\therefore$ $D_1 = 840 \times 0.24/1{,}800 = 0.112$ m = **11.2 cm**

Example 27.13. *The armature winding of a 200-V, 4-pole, series motor is lap-connected. There are 280 slots and each slot has 4 conductors. The current is 45 A and the flux per pole is 18 mWb. The field resistance is 0.3 Ω ; the armature resistance 0.5 Ω and the iron and friction losses total 800 W. The pulley diameter is 0.41 m. Find the pull in newton at the rim of the pulley*

(Elect. Engg. AMIETE Sec A. 1991)

Solution. $E_b = V - I_a R_a = 200 - 45(0.5 + 0.3) = 164$ V

Now $E_b = \dfrac{\Phi Z N}{60} \cdot \left(\dfrac{P}{A}\right)$ volt

$\therefore$ $164 = \dfrac{18 \times 10^{-3} \times 280 \times 4) \times N}{60} \times \dfrac{4}{4}$ $\therefore N = 488$ r.p.m.

Total input $= 200 \times 45 = 9{,}000$ W ; Cu loss $= I_a^2 R_a = 45^2 \times 0.8 = 1{,}620$ W

Iron + friction losses = 800 W ; Total losses = 1,620 + 800 = 2,420 W

Output = 9,000 – 2,420 = 6,580 W

$\therefore$ $T_{sh} = 9 \times 55 \times \dfrac{6580}{488} = 128.$N-m

Let *F* be the pull in newtons at the rim of the pulley.

Then $F \times 0.205 = 128.8$ $\therefore$ $F = 128.8/0.205\ N =$ **634 N**

Example 27.14. *A 4-pole, 240 V, wave connected shunt motor gives 11.19 kW when running at 1000 r.p.m. and drawing armature and field currents of 50 A and 1.0 A respectively. It has 540 conductors. Its resistance is 0.1 Ω. Assuming a drop of 1 volt per brush, find (a) total torque (b) useful torque (c) useful flux/pole (d) rotational losses and (e) efficiency.*

Solution. $E_b = V - I_a R_a$ - brush drop = 240 – (50 × 0.1) – 2 = 233 V

Also $I_a = 50$ A

(*a*) Armature torque $T_a = 9.55\dfrac{E_bI_a}{N}$ N-m $= 9.55 \times \dfrac{233 \times 50}{1000}$ = **111 N -m**

(*b*) $T_{sh} = 9.55\dfrac{\text{output}}{N} = 9.55 \times \dfrac{11,190}{1000}$ = **106.9 N-m**

(*c*) $E_b = \dfrac{\Phi ZN}{60} \times \left(\dfrac{P}{A}\right)$ volt

$\therefore \quad 233 = \dfrac{\Phi \times 540 \times 1000}{60} \times \left(\dfrac{4}{2}\right) \quad \therefore$ **Φ = 12.9 mWb**

(*d*) Armature input $= VI_a = 240 \times 50$ = **12,000 W**

Armature Cu loss $= I_a^2R_a = 50^2 \times 0.1 = 250\ W$; Brush contact loss $= 50 \times 2 = 100$ W

$\therefore$ power developed = 12,000 − 350 = 11,6850 W ; Output = 11.19 kW = 11,190 W

$\therefore$ rotational losses = 11,650 − 11,190 = **460 W**

(*e*) Total motor input $= VI = 240 \times 51$ = 12,340 W; Motor output = 11,190 W

$\therefore$ efficiency $= \dfrac{11,190}{12,240} \times 100$ = **91.4%**

Example 27.15. *A 460-V series motor runs at 500 r.p.m. taking a current of 40 A. Calculate the speed and percentage change in torque if the load is reduced so that the motor is taking 30 A. Total resistance of the armature and field circuits is 0.8 Ω. Assume flux is proportional to the field current.* **(Elect. Engg.-II, Kerala Univ. 1988)**

Solution. Since $\Phi \propto I_a$, hence $T \propto \Phi I_a \propto I_a^2$

$\therefore \quad T_1 \propto 40^2$ and $T_2 \propto 30^2 \quad \therefore \quad \dfrac{T_2}{T_1} = \dfrac{9}{16}$

$\therefore$ percentage change in torque is

$= \dfrac{T_1 - T_2}{T_1} \times 100 = \dfrac{7}{16} \times 100$ = **43.75%**

Now $E_{b1} = 460 - (40 \times 0.8) = 428$ V; $E_{b2} = 460 - (30 \times 0.8) = 436$ V

$\dfrac{N_2}{N_1} = \dfrac{E_{b2}}{E_{b1}} \times \dfrac{I_{a1}}{I_{a2}} \quad \therefore \quad \dfrac{N_2}{500} = \dfrac{436}{428} \times \dfrac{40}{30} \quad \therefore \quad N_2$ = **679 r.p.m.**

Example 27.16. *A 460-V, 55.95 kW, 750 r.p.m. shunt motor drives a load having a moment of inertia of 252.8 kg-m². Find approximate time to attain full speed when starting from est against full-load torque if starting current varies between 1.4 and 1.8 times full-load current.*

Solution. Let us suppose that the starting current has a steady value of (14 + 1.8)/2 = 1.6 times full-load value.

Full-load output = 55.95 kW = 55,950 W ; Speed = 750 r.p.m. = 12.5 r.p.s.

F.L. shaft torque T = power/ω = power/2π N = 55,950π × (750/60) = 712.4 N-m

During starting period, average available torque

$= 1.6\ T - T = 0.6\ T = 0.6 \times 712.4 = 427.4$ N-m

This torque acts on the moment of inertial $I = 252.8$ km-m²

$\therefore \quad 427.4 = 252.8 \times \dfrac{d\omega}{dt} = 252.8 \times \dfrac{2\pi \times 12.5}{dt}, \quad \therefore \quad dt$ = **46.4 s**

Example 27.17. *A 14.92 kW, 400V, 400-r.p.m. d.c. shunt motor draws a current of 40 A when running at full-load. The moment of inertia of the rotating system is 7.5 kg-m2. If the starting current is 1.2 times full-load current, calculate*

(a) full-load torque

(b) the time required for the motor to attain the rated speed against full-load.

(Electrical Technology, Gujarat Univ. 1988)

Solution. (*a*) F.L output 14.92 kW = 14,920 *W* ; Speed = 400 r.p.m. = 20/3 r.p.s.
Now, $T\omega$ = output $\therefore$ $T = 14{,}920/2\pi \times (20/3) =$ **356 N-m**
(*b*) During the starting period, the torque available for accelerating the motor armature is

$$= 1.2\,T - T = 0.2\,T = 0.2 \times 356 = 71.2 \text{ N-m}$$

Now, torque $= I\dfrac{d\omega}{dt}$ $\quad\therefore\quad 71.2 = 7.5 \times \dfrac{2\pi \times (20/3)}{dt}$ $\quad\therefore\ dt =$ **4.41 second**

27.9. Speed of a D.C. Motor

From the voltage equation of a motor (Art. 27.4), we get

$$E_b = V - I_a R_a \quad \text{or} \quad \frac{\Phi Z N}{60}\left(\frac{P}{A}\right) = V - I_a R_a$$

$$\therefore \quad N = \frac{V - I_a R_a}{\Phi} \times \left(\frac{60A}{ZP}\right) \text{ r.p.m.}$$

Now $\quad V - I_a R_a = E_b \quad \therefore \quad N = \dfrac{E_b}{\Phi} \times \left(\dfrac{60A}{ZP}\right) \text{ r.p.m. or } N = K\dfrac{E_b}{\Phi}$

It shows that speed is directly proportional to back e.m.f. E_b and inversely to the flux Φ or $N \propto E_b/\Phi$.

For Series Motor

Let $\quad N_1$ = Speed in the 1st case ; I_{a1} = armature current in the 1st case
Φ_1 = flux/pole in the first case
N_2, I_{a2}, Φ_2 = corresponding quantities in the 2nd case.

Then, using the above relation, we get

$$N_1 \propto \frac{E_{b1}}{\Phi_1} \text{ where } E_{b1} = V - I_{a1}R_a;\ N_2 \propto \frac{E_{b2}}{\Phi_2} \text{ where } E_{b2} = V - I_{a2}R_a$$

$$\therefore \quad \frac{N_2}{N_1} = \frac{E_{b2}}{E_{b1}} \times \frac{\Phi_1}{\Phi_2}$$

Prior to saturation of magnetic poles ; $\Phi \propto I_a$ $\therefore \dfrac{N_2}{N_1} = \dfrac{E_{b2}}{E_{b1}} \times \dfrac{I_{a1}}{I_{a2}}$

For Shunt Motor

In this case the same equation applies,

i.e. $\quad \dfrac{N_2}{N_1} = \dfrac{E_{b2}}{E_{b1}} \times \dfrac{\Phi_1}{\Phi_2}$ If $\Phi_2 = \Phi_1$, then $\dfrac{N_2}{N_1} = \dfrac{E_{b2}}{E_{b1}}$

27.10. Speed Regulation

The term speed regulation refers to the change in speed of a motor with change in applied load torque, other conditions remaining constant. By change in speed here is meant the change which occurs under these conditions due to inherent properties of the motor itself and not those changes which are affected through manipulation of rheostats or other speed-controlling devices.

The speed regulation is defined as *the change in speed when the load on the motor is reduced from rated value to zero, expressed as per cent of the rated load speed.*

$$\therefore \quad \%\text{speed regulation} = \frac{\text{N.L. speed - F.L. speed}}{\text{F.L. speed}} \times 100 = \frac{dN}{N} \times 100$$

27.11. Torque and Speed of a DC Motor

It will be proved that though torque of a motor is admittedly a function of flux and armature current, *yet it is independent of speed.* In fact, it is the speed which depends on torque and not nice versa. It has been proved earlier that

$$N = K\frac{V - I_a R_a}{\Phi} = \frac{KE_b}{\Phi} \quad \text{–Art. 27.9}$$

Also, $$T_a \propto \Phi I_a \quad \text{–Art. 27.7}$$

It is seen from above that increase in flux would decrease the speed but increase the armature torque. It cannot be so because torque always tends to produce rotation. If torque increases, motor speed must *increase* rather than *decrease*. The apparent inconsistency between the above two equations can be reconciled in the following way:

Suppose that the flux of a motor is decreased by decreasing the field current. Then, following sequence of events takes place :

1. back e.m.f. E_b (= NΦ/K) drops instantly (the speed remains constant because of inertia of the heavy armature).
2. due to decrease in E_b,I_a is increased because $I_a = (V - E_b)/R_a$. Moreover, a small reduction in flux produces a proportionately large increase in armature current.
3. hence, in equation $T_a \propto \Phi I_a$, a small decrease in ϕ is more than counterbalanced by a *large* increase in I_a with the result that there is a net *increase* in T_a
4. this increase in T_a produces an increase in motor speed.

It is seen from above that with the applied voltage *V* held constant, motor speed varies inversely as the flux. However, it is possible to increase flux and, at the same time, increase the speed provided I_a is held constant as is actually done in a d.c. servomotor.

Example 27.18. *A 4-pole series motor has 944 wave-connected armature conductors. At a certain load, the flux per pole is 34.6 m Wb and the total mechanical torque developed is 209 N-m Calculate the line current taken by the motor and the speed at which it will run with an applied voltage of 500 V. Total motor resistance is 3 ohm.*

(Elect. Engg. AMIETE Sec. A Part II June 1991)

Solution. $T_a = 0.159\ \phi\ ZI_a$ (P/A) N-m

$\therefore$ $209 = 0.159 \times 34.6 \times 10^{-3} \times 944 \times I_a$ (4/2) ; I_a= 20.1 A

$E_a = V - I_a R_a = 500 - 20.1 \times 3 = 439.7$ V

Now, speed may be found either by using the relation for E_b or T_a as given in Art.

$E_b = \Phi$ ZN × (P/A) or $439.7 = 34.6 \times 10^{-3} \times 944 \times N \times 2$

$\therefore$ N = 6.73 rps or **382.2 r.p.m**

Example 27.19. *A 250-V shunt motor runs at 1000 r.p.m. at no-load and takes 8 A. The total armature and shunt field resistances are respectively 0.2 Ω and 250 Ω. Calculate the speed when loaded and taking 50 A. As sume the flux to be constant* **(Elect. Engg. A.M.Ae. S.I. June 1991)**

Solution. Formula used : $\frac{N}{N_0} = \frac{E_b}{E_{bo}} \times \frac{\Phi_0}{\Phi}$; Since $\Phi_0 = \Phi$ (given); $\frac{N}{N_0} = \frac{E_b}{E_{bo}}$

I_{sh} = 250/250 = 1 A $\qquad I_{ao} = 8 - 1 = 7$ A ; $I_a = 50 - 1 = 49$ A

$E_{bo} = V - I_{ao} R_a = 250 - (7 \times 0.2) = 248.6\,V$; $E_b = V - I_a R_a = 250 - (49 \times 0.2) = 240.2V$

$\therefore \quad \frac{N}{1000} = \frac{240.2}{248.6} \qquad N =$ **966.1 r.p.m.**

Example 27.20. *A d.c. series motor operates at 800 r.p.m. with a line current of 100 A from 230-V mains. Its armature circuit resistance is 0.15 Ω and its field resistance 0.1 Ω. Find the speed at which the motor runs at a line current of 25 A, assuming that the flux at this current is 45 per cent of the flux at 100 A.* **(Electrical Machinery-I, Bangalore Univ. 1986)**

Solution. $\dfrac{N_2}{N_1}=\dfrac{E_{b2}}{E_{b1}}\times\dfrac{\Phi_1}{\Phi_2}$; $\Phi_2 = 0.45\,\Phi_1$ or $\dfrac{\Phi_1}{\Phi_2}=\dfrac{1}{0.45}$

$$E_{b1} = 230-(0.15+0.1)\times 100 = 205\text{V} ; E_{b2} = 230-25\times 0.25 = 223.75 \text{ V}$$

$$\therefore \quad \frac{N_2}{800}=\frac{223.75}{205}\times\frac{1}{0.45}; N_2 = \mathbf{1940\ r.p.m.}$$

Example 27.21. *A 230-V d.c. shunt motor has an armature resistance of 0.5 Ω and field resistance of 115 Ω. At no load, the speed is 1,200 r.p.m. and the armature current 2.5 A. On application of rated load, the speed drops to 1,120 r.p.m. Determine the line current and power input when the motor delivers rated load.* **(Elect. Technology, Kerala Univ. 1988)**

Solution. $N_1 = 1200$ r.p.m., $E_{b1} = 230-(0.5\times 2.5) = 228.75$ V
$N_2 = 1120$ r.p.m., $E_{b2} = 230-0.5\,I_{a2}$

Now, $\dfrac{N_2}{N_1}=\dfrac{E_{b2}}{E_{b1}}$ $\therefore$ $\dfrac{1120}{1200}=\dfrac{230-0.5I_{a2}}{228.75}$; $I_{a2} = \mathbf{33\ A}$

Line current drawn by motor = $I_{a2} + I_{sh} = 33 + (230/115) =$ **35 A**
Power input at rated load = 230 × 35 = **8,050 W**

Example 27.22. *A belt-driven 100-kW, shunt generator running at 300 r.p.m. on 220-V bus-bars continues to run as a motor when the belt breaks, then taking 10 kW. What will be its speed ? Given armature resistance = 0.025 Ω, field resistance = 60 Ω and contact drop under each brush = 1 V, Ignore armature reaction.* **(Elect. Machines (E-3) AMIEC Sec. Winter 1991)**

Solution. As Generator [Fig. 27.12 (*a*)]
Load current, I = 100,000/220 = 454.55 A; I_{sh} = 220/60 = 3.67 A
$I_a = I + I_{sh}$ = 458.2 A ; $I_a R_a$ = 458.2 × 0.025 = 11.45
E_b = 220 + 11.45 + 2×1 = 233.45 V; N_1 = 300 r.p.m.

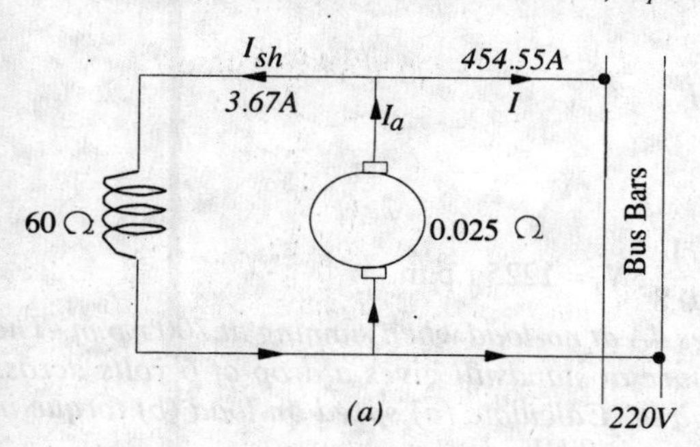

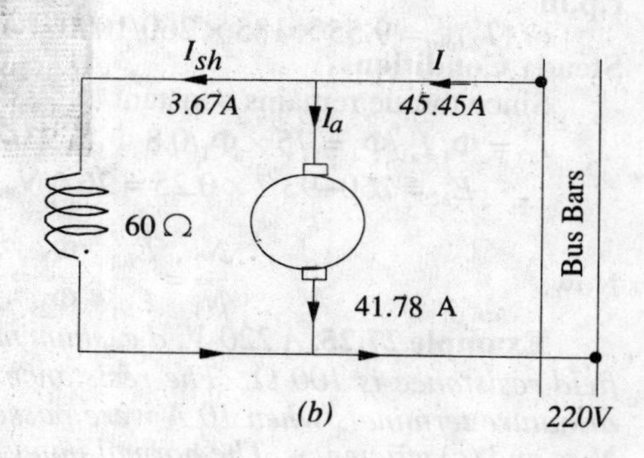

Fig. 27.12

As Motor [Fig. 27.12 (b).
Input line current = 100,000/220 = 45.45 A ; I_{sh} = 220/60 = 3.67 A
I_a = 45.45−3.67 = 41.78 A; $I_a R_a$ = 41.78 × 0.025 = 1.04 V; E_{b2} = 220-1.04-2 × 1 = 216.96 V

$$\frac{N_2}{N_1}=\frac{E_{b2}}{E_{b1}}\times\frac{\Phi_1}{\Phi_2}; \text{ since } \Phi_1=\Phi_2 \text{ because } I_{sh} \text{ is constant.}$$

$$\therefore \quad \frac{N_2}{300}=\frac{216.96}{233.45}; N_2 = 279 \text{ r.p.m.}$$

Example 27.23. *A d.c. shunt machine generates 250-V on open circuit at 1000 r.p.m. Effective armature resistance is 0.5 Ω, field resistance is 250 Ω, input to machine running as a motor on no-load is 4 A at 250 V. Calculate speed of machine as a motor taking 40 A at 250 V. Armature reaction weakens field by 4%.* **(Electrical Machines-I, Gujarat Univ. 1987)**

Solution. Consider the case when the machine runs as a motor on no-load.

Now, $I_{sh} = 250/250 = 1$ A; Hence, $I_{ao} = 4 - 1 =$ A ; $E_{bo} = 250 - 0.5 \times 3 = 248.5$ V

It is given that when armature runs at 1000 r.p.m, it generates 250 V. When it generates 248.5 V, it must be running at a speed = 1000 × 248.5/250 = 994 r.p.m.

Hence, $N_o = 994$ r.p.m.

When Loaded

$I_a = 40 - I = 39$ A; $E_b = 250 - 39 \times 0.5 = 230.5$ V Also, $\Phi_0/\Phi = 1/0.96$

$\frac{N}{E} = \frac{E_b}{E_{bo}}$ ∴ $\frac{N}{994} = \frac{230.5}{248.5} \times \frac{1}{0.96}$, **N = 960 r.p.m.**

Example 27.24. *A 250-V shunt motor giving 14.92 kW at 1000 rpm takes an armature current of 75 A. The armature resistance is 0.25 ohm and the load torque remains constant. If the flux is reduced by 20 per cent of its normal value before the speed changes, find the instantaneous value of the armature current and the torque. Determine the final value of the armature current and speed.* **(Elect. Engg. AMIETE (New Scheme) 1990)**

Solution. $E_{b1} = 250 - 75 \times 0.25 = 231.25$ V

When flux is reduced by 20%, the back e.m.f. is also reduced instantly by 20% because speed remains constant due to inertia of the heavy armature (Art. 27.11).

∴ instantaneous value of back e.m.f. $(E_b)_{inst} = 231,25 \times 0.8 = 185$ V

$(I_a)_{inst} = [V - (E_b)_{inst}]/R_a = (250 - 185)/0.25 = 260$ A

instantaneous value of the torque $= 9.55 \times \frac{(E_b)_{inst} \times (I_a)_{inst}}{N}$ r.p.m

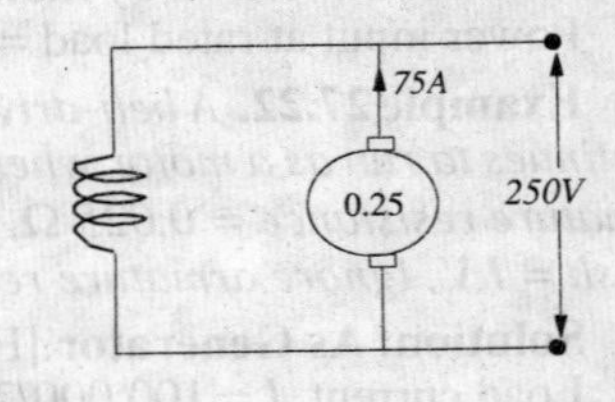

Fig. 27.13

or $(T_a)_{inst} = 9.55 \times 185 \times 260/1000 = 459$ N-m

Steady Conditions

Since torque remains constant ∴ $\Phi_1 I_{a1} = \Phi_2 I_{a2}$

$I_{a2} = \Phi_1 I_{a1}/\Phi_2 = 75 \times \Phi_1/0.8\,\Phi_1 =$ **93.7 A**

∴ $E_{b2} = 250 - 93.7 \times 0.25 = 26.6$ V

Now, $\frac{N_2}{N_1} = \frac{E_{b2}}{E_{b1}} \times \frac{\Phi_1}{\Phi_2} = \frac{226.6}{231.25} \times \frac{1}{0.8}$; $N_2 = 1225$ r.p.m.

Example 27.25. *A 220-V, d.c. shunt motor takes 4A at no-load when running at 700 r.p.m. The field resistance is 100 Ω. The resistance of armature at standstill gives a drop of 6 volts across armature terminals when 10 A were passed through it. Calculate (a) speed on load (b) torque in N-m and (c) efficiency. The normal input, of the motor is 8 kW.* **(Electrotechnics-II; M.S. Univ. Baroda 1988)**

Solution. (*a*) $I_{sh} = 200/100 = 2$A

F.L. Power input = 8,000 W   F.L. line current = 8,000/200 = 40 A

$I_a = 40 - 2 = 38$ A $R_a = 6/10 = 0.6\ \Omega$

$E_{bo} = 200 - 2 \times 0.6 = 198.8$ V ; $E_b = 200 - 38 \times 0.6 = 177.2$ V

Now, $\frac{N}{N_0} = \frac{E_b}{E_{bo}}$ or or $\frac{N}{700} = \frac{177.2}{198.8}$; N = **623.9 r.p.m.**

(*b*) $T_a = 9.55\, E_b I_a/N = 9.55 \times 177.2 \times 38/623.9 =$ **103 N-m**

(*c*) N.L. power input = 200 × 4 = 800 W ; Arm. Cu loss $= I_a^2 R_a = 2^2 \times 0.6 = 2.4$ W

Constant losses = 800 − 2.4 = 797.6 W; F.L. arm. Cu loss $= 38^2 \times 0.6 = 866.4$ W

Total F.L. losses = 797.6 + 866.4 = 1664 W; F.L. output = 8,000 − 1664 = 6336 W

F.L. Motor efficiency = 6336/8,000 = 0.72 or **79.2%**

Example 27.26. *The input to 230-V, d.c. shunt motor is 11 kW. calculate (a) the torque developed (b) the efficiency (c) the speed at this load. The particulars of the motor are as follows :*

No-load current = 5 A; No-load speed = 1150 r.p.m.

Arm. resistance = 0.5 Ω; shunt field resistance = 110 Ω

(Elect Technology; Bombay University 1988)

Solution. No-load input = 220 × 5 = 1,100 W ; I_{sh} = 220/110 = 2A; I_{ao} = 5 − 2 = 3 A

No-load armature Cu loss = $3^2 \times 0.5$ = 4.5 W

∴ constant losses = 1,100 − 4.5 = 1,095.5 W

When input is 11 kW

Input current = 11,000/220 = 50 A ; Armature current = 50 −2 = **48 A**

Arm. Cu loss = $48^2 \times 0.5$ = 1,152 W; Total loss = arm. Cu loss + constant losses

= 1152 + 1095.5 = 2248 W

Output = 11,000 − 2,248 = 8,752 W

(*b*) Efficiency = 8,752 × 100/11,000 = **79.6%**

(*c*) Back e.m.f. at no-load = 220−(3 × 0.5) = 218.5 V

Back e.m.f. at given load = 220 − (48 × 0.5) = 196 V

∴ speed *N* = 1,150 × 196/218.5 = **1,031 r.p.m.**

(*a*) $$T_a = 9.55 \times \frac{196 \times 48}{1031} = \mathbf{87.1\ N\text{-}m}$$

Example 27.27. *The armature circuit resistance of a 18.65 kW 250-V series motor is 0.1 Ω, the brush voltage drop is 3V, and the series field resistance is 0.05. When the motor takes 80 A, speed is 600 r.p.m. Calculate the speed when the current is 100 A.*

(Elect. Machines, A.M.I.E. Sec B, 1993)

Solution. E_{b1} = 250 − 80(0.1 + 0.05) − 3 = 235 V. E_{b2} = 250 − 100(0.1 + 0.05) − 3 = 232 V

Since $\Phi \propto I_a$, hence, $\Phi \propto 80$, $\Phi_2 \propto 100$, $\Phi_1/\Phi_2 = 80/100$

Now $\frac{N_2}{N_1} = \frac{E_{b2}}{E_{b1}} \times \frac{\Phi_1}{\Phi_2}$ or $\frac{N_2}{600} = \frac{232}{235} \times \frac{80}{100}$; N_2 = **474 r.p.m**

Example 27.28. *A 220-volt d.c. series motor is running at a speed of 800 r.p.m. and draws 100 A. Calculate at what speed the motor will run when developing half the torque. Total resistance of the armature and field is 0.1 ohm. Assume that the magnetic circuit is unsaturated*

(Elect Machines ; A.M.I.E. Sec. B, 1991)

Solution. $$\frac{N_2}{N_1} = \frac{E_{b2}}{E_{b1}} \times \frac{\Phi_1}{\Phi_2} = \frac{E_{b2}}{E_{b1}} \times \frac{I_{a1}}{I_{a2}} \quad (\therefore \ \Phi \propto I_a)$$

Since field is unsaturated, $T_a \propto \Phi I_a \propto I_a^{\ 2}$ ∴ $T_1 \propto I_{a1}^{\ 2}$ and $T_2 \propto I_{a2}^{\ 2}$

or $T_2/T_1 = (I_{a2}/I_{a1})^2$ or $1/2 = (I_{a2}/I_{a1})^2$; $I_{a2} = I_{a2}/\sqrt{2}$ = 70.7 A

E_{b1} = 220 − 100 × 0.1 = 210 V ; E_{b2} = 220 − 0.1 × 70.7 = 212.9 V

∴ $\frac{N_2}{800} = \frac{212.9}{210} \times \frac{100}{70.7}$; N_2 = **1147 r.p.m.**

Example 27.29. *A 4-pole d.c. motor runs at 600 r.p.m. on full load taking 25 A at 450 V. The armature is lap-wound with 500 conductors and flux per pole is expressed by the relation.*

$$\Phi = (1.7 \times 10^{-2} \times I^{0.5})\textit{weber}$$

where 1 is the motor current. If supply voltage and torque are both halved, calculate the speed at which the motor will run. Ignore stray losses. **(Elect. Machines, Nagpur Univ. 1993)**

Solution. Let us first find R_a

Now $$N = \frac{E_b}{Z\Phi}\left(\frac{60\,A}{P}\right) \text{ r.p.m.}$$

$\therefore$ $$600 = \frac{E_b}{1.7 \times 10^{-2} \times 25^{0.5}} \times \frac{60 \times 4}{500 \times 4}$$

$\therefore$ $$E_b = 10 \times 1.7 \times 10^{-2} \times 5 \times 500 = 425 \text{ V}$$

$$I_a R_a = 450 - 425 = 25 \text{ V} ;\ R_a = 25/25 = 1.0\ \Omega$$

Now in the Ist Case

$T_1 \propto \Phi_1 I_1 \quad \therefore\ T_1 \propto 1.7 \times 10^{-2} \times \sqrt{25} \times 25.$

Similarly $T_2 \propto 1.7 \times 10^{-2} \times \sqrt{1} \times I$ Now $T_1 = 2T_2$

$\therefore$ $1.7 \times 10^{-2} \times 125 = 1.7 \times 10^{-2} \times I^{3/2} \times 2 \quad \therefore\ I = (125/2)^{2/3} = 15.75$ A

$E_{b1} = 425$ V ; $E_{b2} = 225 - (15.75 \times 1) = 209.3$ V

Using the relation $\dfrac{N_2}{N_1} = \dfrac{E_{b2}}{E_{b1}} \times \dfrac{\Phi_1}{\Phi_2}$; we have

$$\frac{N_2}{600} = \frac{209.3}{425} \times \frac{1.7 \times 10^{-2} \times 5}{1.7 \times 10^{-2} \times \sqrt{15.75}} ;\ N_2 = \mathbf{372 \text{ r.p.m.}}$$

Tutorial Problem No. 27.2

1. Calculate the torque in newton-metre developed by a 440-V d..c motor having an armature resistance of 0.25 Ω and running at 750 r.p.m. when taking a current of 60 A. **[325 N-m]**

2. A 4-pole, lap-connected d.c. motor has 576 conductors and draws an armature current of 10 A. If the flux per pole is 0.02 Wb, calculate the armature torque developed. **[18.3 N-m]**

3. (*a*) A d.c. shunt machine has armature and field resistances of 0.025 Ω and 80 Ω respectively. When connected to constant 400-V bus-bars and driven as a generator at 450 r.p.m., it delivers 120 kW. Calculate its speed when running as a motor and absorbing 120 kW from the same bus-bars.

(*b*) Deduce the direction of rotation of this machine when it is working as a motor assuming a clockwise rotation as a generator. **[(*a*) 435 r.p.m. (*b*) Clockwise]**

4. The armature current of a series motor is 60 A when on full-load. If the load is adjusted to that this current decreases to 40-A, find the new torque expressed as a percentage of the full-load torque. The flux for a current of 40 A is 70% of that when current is 60 A. **[46%]**

5. A 4-pole, d.c. shunt motor has a flux per pole of 0.04 Wb and the armature is lap-wound with 720 conductors. The shunt field resistance is 240 Ω and the armature resistance is 0.2 Ω. Brush contact drop is 1 *V* per brush. Determine the speed of the machine when running (*a*) as a motor taking 60 *A* and (*b*) as a generator supplying 120 A. The terminal voltage in each case is 480 V **[972 r.p.m. ; 1055 r.p.m]**

6. A 25-kW shunt generator is delivering full output to 400-V hus-bars and is driven at 950 r.p.m. by belt drive. The belt breaks suddenly but the machine continues to run as a motor taking 25 kW from the bus-bars. At what speed does it run ? Take armature resistance including brush contact resistance as 0.5 Ω and field resistance as 160 Ω **[812.7 r.p.m.]** [*Elect. Technology, Andhra Univ. Apr. 1977*)

7. A 4-pole, d.c. motor has a wave-wound armature with 65 slots each containing 6 conductors. The *flux* per pole is 20 mwb and the armature has a resistance of 0.15 Ω. Calculate the motor speed when the machine is operating from a 250-V supply and taking a current of 60 A. **[927 r.p.m.]**

8. A 500-V, d.c. shunt motor has armature and field resistances of 0.5 Ω and 200 Ω respectively. When loaded and taking a total input of 25 kW, it runs at 400 r.p.m. Find the speed at which it must be driven as a shunt generator to supply a power output of 25 kW at a terminal voltage of 500 V. **[442 r.p.m.]**

9. A d.c. shunt motor runs at 900 r.p.m. from a 400 V supply when taking an armature current of 25 A. Calculate the speed at which it will run from a 230 V supply when taking an armature current of 15 A. The resistance of the armature circuit is 0.8 Ω. Assume the flux per pole at 230 V to have decreased to 75% of its value at 400 V. **[595 r.p.m.]**

10. A shunt machine connected to 250-A mains has an armature resistance of 0.12 Ω and field resistance of 100 Ω. Find the ratio of the speed of the machine as a generator to the speed as a motor, if line current is 80 A in both cases. **[1.08]** (*Electrical Engineering-II, Bombay Univ. April. 1977, Madras Univ. Nov. 1978*)

11. A 20-kW d..c shunt generator delivering rated output at 1000 r.p.m. has a terminal voltage of 500 V. The armature resistance is 0.1 Ω, voltage drop per brush is 1 volt and the field resistance is 500 Ω.

Calculate the speed at which the machine will run as a motor taking an input of 20 kW from a 500 V d.c. supply **[976.1 r.p.m]** (*Elect. Engg-I, Bombay Univ. 1975*)

12. A 4-pole, 250-V, d.c. shunt motor has a lap-connected armature with 960 conductors. The flux per pole is 2×10^{-2} Wb. Calculate the torque developed by the armature and the useful torque in newton-metre when the current taken by the motor is 30 A. The armature resistance is 0.12 ohm and the field resistance is 125 Ω. The rotational losses amount to 825 W.

[85.5 N-m ; 75.3 N-m] (*Electric Machinery-I, Madras Univ. Nov. 1979*)

27.12. Motor Characteristics

The characteristic curves of a motor are those curves which show relationships between the following quantities.

1. **Torque and armature current** *i.e* T_a/I_a characteristic. It is known as *electrical characteristic.*
2. **Speed and armature current** *i.e.* N/I_a characteristic.
3. **Speed and torque** *i.e.* N/T_a characteristic. It is also known as *mechanical characteristic*. It can be found from (1) and (2) above.

While discussing motor characteristics, the following two relations should always be kept in mind :

$$T_a \propto \Phi I_a \quad \text{and} \quad N \propto \frac{E_b}{\Phi}$$

27.13. Characterisics of Series Motors

1. T_a/I_a Characteristic. We have seen that $T_a \propto \Phi I_a$. In this case, as field windings also carry the armature current, $\Phi \propto I_a$ up to the point of magnetic saturation. Hence, before saturation, $T_a \propto \Phi I_a$ and $\therefore\ T_a \propto I_a^2$

At light loads, I_a and hence Φ is small. But as I_a increases, T_a increases as the square of the current. Hence, T_a/I_a curve is a parabola as shown in Fig. 27.14. After saturation, Φ is almost independent of I_a hence $T_a \propto I_a$ only. So the characteristic becomes a straight line. The shaft torque T_{sh} is less than armature torque due to stray losses. It is shown dotted in the figure. So we conclude that (prior to magnetic saturation) on heavy loads, a series motor exerts a torque proportional to the square of armature current. Hence, in cases where huge starting torque is required for accelerating heavy masses quickly as in hoists and electric trains etc., series motors are used.

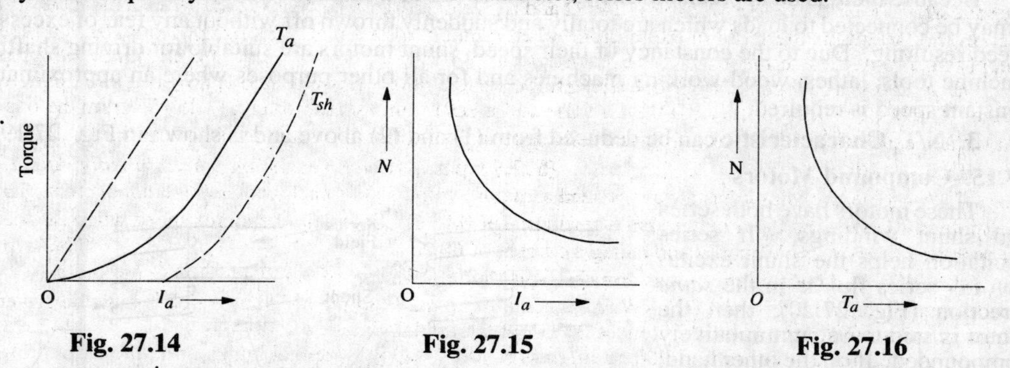

Fig. 27.14 **Fig. 27.15** **Fig. 27.16**

2. N/I_a Characteristics. : Variations of speed can be deduced from the formula :

$$N \propto \frac{E_b}{\Phi}$$

Change in E_b, for various load currents is small and hence may be neglected for the time being. With increased I_a, Φ also increases. Hence, speed varies inversely as armature current as shown in Fig. 27.15.

When load is heavy, I_a is large. Hence, speed is low (this decreases E_b and allows more armature current to flow). But when load current and hence I_a falls to a small value, speed becomes dangerously high. Hence, a series motor should never be started without some mechanical (not belt-driven) load on it otherwise it may develop excessive speed and get damaged due to heavy centrifugal forces so produced. It should be noted that series motor is a variable speed motor.

3. N/T$_a$ or mechanical characteristic. It is found from above that when speed is high, torque is low and *vice-versa*. The relation between the two is as shown in Fig. 27.16.

27.14. Characteristics of Shunt Motors

1. T$_a$/I$_a$ Characteristic

Assuming Φ to be practically constant (though at heavy loads, Φ decreases somewhat due to increased armature reaction) we find that $T_a \propto I_a$

Hence, the electrical characteristic as shown in Fig. 27.17, is practically a straight line through the origin. Shaft torque is shown dotted. Since a heavy starting load will need a heavy starting current, shunt motor should never be started on (heavy) load.

2. N/I$_a$ Characteristic

If Φ is assumed constant, then $N \propto E_b$. As E_b is also practically constant, speed is, for most purposes, constant (Fig. 27.18).

But strictly speaking, both E_b and Φ decrease with increasing load. However, E_b decreases slightly more than Φ so that on the whole, there is some decrease in speed. The drop varies from 5 to 15% of full-load speed, being dependent on saturation, armature reaction and brush position. Hence, the actual speed curve is slightly drooping as shown by the dotted line in Fig. 27.18. But, for all practical purposes, shunt motor is taken as a constant-speed motor.

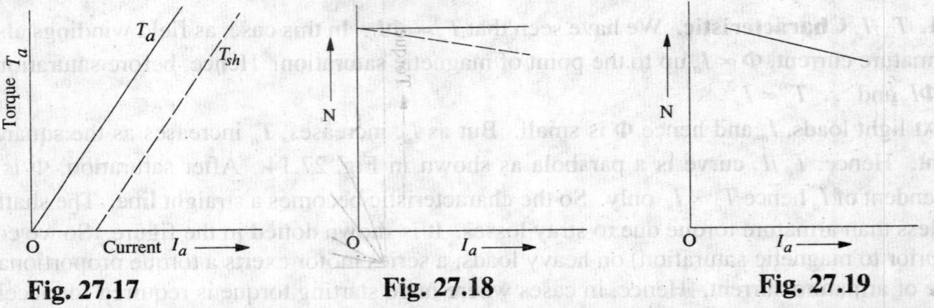

Fig. 27.17 Fig. 27.18 Fig. 27.19

Because there is no appreciable change in the speed of a shunt motor from no-load to full-load, it may be connected to loads which are totally and suddenly thrown off without any fear of excessive speed resulting. Due to the constancy of their speed, shunt motors are suitable for driving shafting, machine tools, lathes, wood-working machines and for all other purposes where an approximately constant speed is required.

3. N/T$_a$ Characteristic can be deduced from (1) and (2) above and is shown in Fig. 27.19.

27.15. Compound Motors

These motors have both series and shunt windings. If series excitation helps the shunt excitation *i.e.* series flux is in the *same* direction (Fig. 27.20); then the motor is said to be cummulatively compounded. If on the other hand, series field opposes the shunt field, then the motor is said to be differentially compounded.

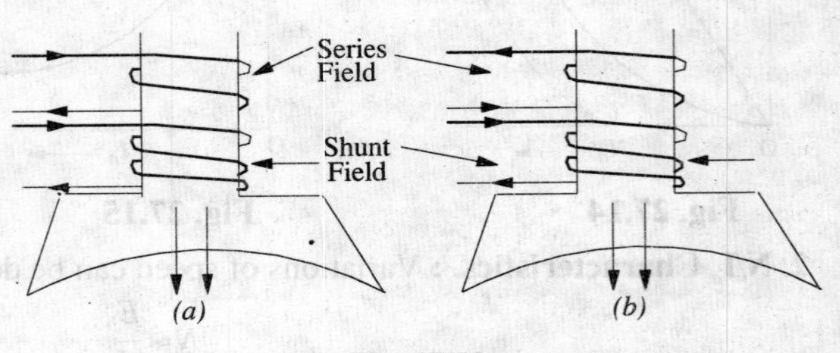

Fig. 27.20

The characteristics of such motors lie in between those of shunt and series motors as shown in Fig. 27.21.

(*a*) Cumulative-compound Motors

Such machines are used where series characteristics are required and where, in addition, the load is likely to be removed totally such as in some types of coal cutting machines or for driving heavy machine tools which have to take sudden cuts quite often. Due to shunt windings, speed will not become excessively high but due to series windings, it will be able to take heavy loads. In conjunction with fly-wheel (functioning as load equalizer), it is employed where there are sudden temporary loads as in rolling mills. The fly-wheel supplies its stored kinetic energy when motor slows down due to sudden heavy load. And when due to the removal of load motor speeds up, it gathers up its kinetic energy.

Compound-wound motors have greatest application with loads that require high starting torques or pulsating loads (because such motors smooth out the energy demand required of a pulsating load). They are used to drive electric shovels, metal-stamping machines, reciprocating pumps, hoists and compressors etc.

(b) Differential-compound Motors

Since series field opposes the shunt field, the flux is decreased as load is applied to the motor. This results in the motor speed remaining almost constant or even increasing with increase in load (because, $N \propto E_b / (\Phi)$). Due to this reason, there is a decrease in the rate at which the motor torque increases with load. Such motors are not in common use. But because they can be designed to give an accurately constant speed under all load conditions, they find limited application for experimental and research work.

One of the biggest drawback of such a motor is that due to weakening of flux with increase in load, there is a tendency towards speed instability and motor running away unless designed properly.

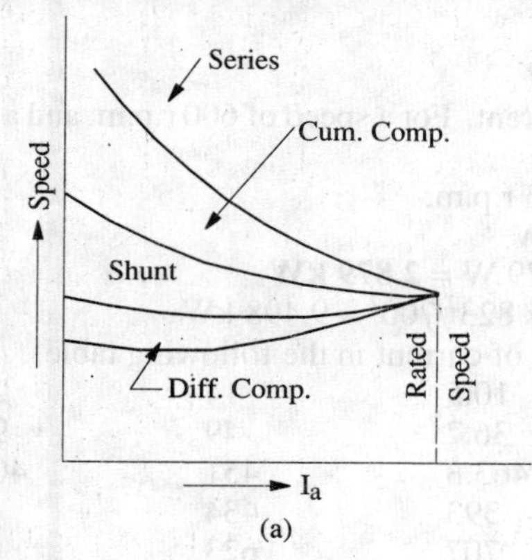

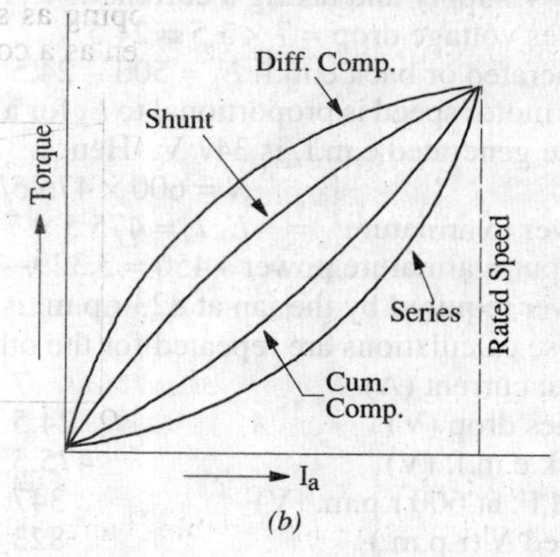

Fig. 27.21

Example 27.30. *The following result were obtained from a static torque test on a series motor :*

Current (A) :	*20*	*30*	*40*	*50*
Torque (N-m) :	*128.8*	*230.5*	*349.8*	*46.2*

Deduce the speed/torque curve for the machine when supplied at a constant voltage of 460V. Resistance of armature and field winding is 0.5 Ω. Ignore iron and friction losses.

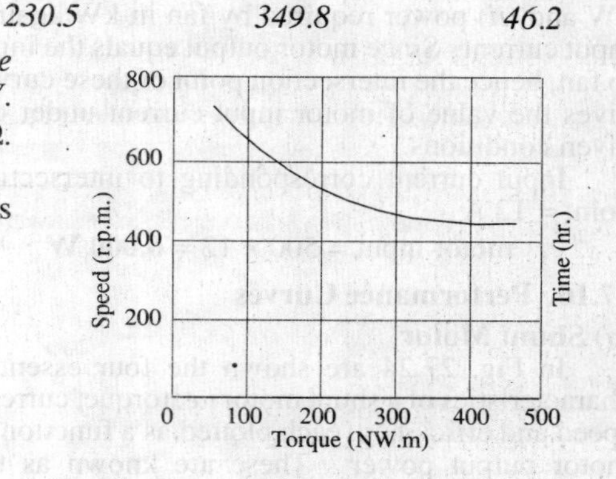

Fig. 27.22

Solution. Taking the case when input current is 20 A, we have

Motor input $= 460 \times 20 = 9{,}200$ W

Field and armature Cu loss

$= 20^2 \times 0.5 = 200$ W

Ignoring iron and friction losses,

output $= 9{,}200 - 200 = 9{,}000$ W

Now, $T_{sh} \times 2\pi N =$ output in watts

$\therefore \; 128.8 \times 2\pi \times N = 9{,}000$

$\therefore \; N = 9{,}000 / 2\pi \times 128.8$

= 11.12 r.p.s. = 667 r.p.m.

Similar calculations for other values of current are tabulated below :

Current (A)	20	30	40	50
Input (W)	9,200	13,800	18,400	23,000
I^2R loss (W)	200	450	800	1,250
Output (W)	9,200	13,350	17,600	21,850
Speed (r.p.m.)	667	551	480	445
Torque (N-m)	128.8	230.5	349.8	469.2

From these values, the speed/torque curve can be drawn as shown in Fig. 27.22.

Example 27.31. *A fan which requires 8 h.p. (5.968 kW) at 700 r.p.m. is coupled directly to a d.c. series motor. Calculate the input to the motor when the supply voltage is 500 V, assuming that power required for fan varies as the cube of the speed. For the purpose of obtaining the magnetisation characteristics, the motor was run as a self-excited generator at 600 r.p.m. and the relationship between the terminal voltage and the load current was found to be as follows :*

Load current (A)	*7*	*10.5*	*14*	*27.5*
Terminal voltage (V)	*347*	*393*	*434*	*458*

The resistance of both the armature and field windings of the motor is 3.5 Ω and the core, friction and other losses may be assumed to be constant at 450 W for the speeds corresponding to the above range of currents at normal voltage. **(I.E.E. London)**

Solution. Let us, by way of illustration, calculate the speed and output when motor is running off a 500-V supply and taking a current of 7 A.

Series voltage drop = 7 × 3.5 = 24.5 V

Generated or back e.m.f. E_b = 500 − 24.5 = 475.5 V

The motor speed is proportional to E_b for a given current. For a speed of 600 r.p.m. and a current of 7A, the generated e.m.f. is 347 V. Hence.

$$N = 600 \times 475.5/347 = 823 \text{ r.p.m.}$$

Power to armature = $E_b I_a$ = 475.5 × 7 = 3,329 W

Output = armature power −450 = 3,329−450 = 2.879 W = **2.879 kW**

Power required by the fan at 823 r.p.m. is = $5.968 \times 823^2/700^2$ = 9.498 kW

These calculations are repeated for the other values of current in the following table.

Input current (A)	7	10.5	14	27.5
Series drop (V)	24.5	36.7	49	96.4
Back e.m.f. (V)	475.5	463.3	451	403.6
E.M.F. at 600 r.p.m. (V)	347	393	434	458
Speed N (r.p.m.)	823	707	623	528
Armature power (W)	3329	4870	6310	11,100
Motor output (kW)	2.879	4.420	5.860	10.65
Power required by fan (kW)	9.698	6.146	4.222	2.566

In Fig. 27.23 is plotted (*i*) the motor output in kW and (*ii*) power required by fan in kW against input current. Since motor output equals the input to fan, hence the intersection point of these curves gives the value of motor input current under the given conditions.

Input current corresponding to intersection point = 12 A

∴ motor input = 500 × 12 = **6,000 W**

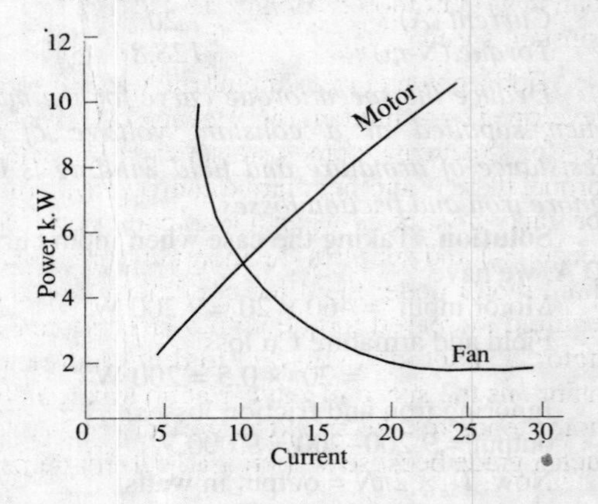

Fig. 27.23

27.16. Performance Curves

(*a*) Shunt Motor

In Fig. 27.24 are shown the four essential characteristics of a shunt motor *i.e.* torque, current, speed and efficiency, each plotted as a function of motor output power These are known as the *performance curves* of a motor.

It is seen that shunt motor has a definite no-load speed. Hence, it does not 'run away' when load is suddenly thrown off provided the field circuit remains closed. The drop in speed from no-load to full-load is small, hence this motor is usually referred to as *constant speed* motor. The speed for any load within the operating range of the motor can be readily obtained by vaying the field current by means of a field rheostat.

The efficiency curve is usually of the same shape for all electric motors and generators. The shape of efficiency curve and the point of maximum efficiency can be varied considerably by the designer, though it is advantageous to have an efficiency curve which is farily flat, so that there is little change in efficiency between load and 25% overload and to have the maximum efficiency as near to the full load as possible.

It will be seen from the curves, that a certain value of current is required even when output is zero. The motor input under no-load conditions goes to meet the various losses occurring within the machine.

As compared to other motors, a shunt motor is said to have a lower starting torque. But this should not be taken of mean that a shunt motor is incapable of starting a heavy load. Actually, it means that series and compound motors are capable of starting heavy loads with less excess of current inputs over normal values than the shunt motors and that consequently the depreciation on the motor will be relatively less. For example, if twice full-load torque is required at start, then shunt motor draws twice the full-load current ($T_a \propto I_a$ or $I_a \propto \sqrt{T_a}$) whereas series motor draws only approximately one and a half times the full load current ($T_a \propto I_a^2$ or $I_a \propto \sqrt{T_a}$).

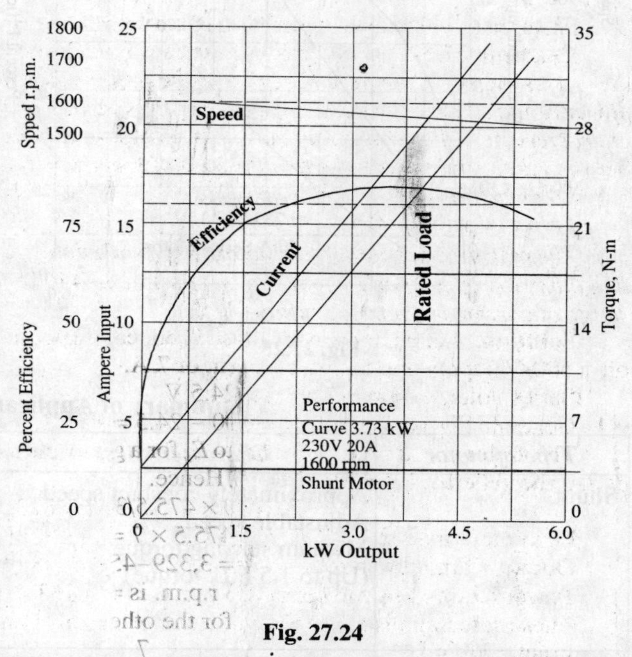

Fig. 27.24

The shunt motor is widely used with loads that require essentially constant speed but where high starting torques are not needed. Such loads include centrifugal pumps, fans, winding reels, conveyors and machine tools etc.

(*b*) Series Motor

The typical performance curves for a series motor are shown in Fig. 27.25.

It will be seen that drop in speed with increased load is much more prominent in series motor than in a shunt motor. Hence, a series motor is not suitable for applications requiring a substantially constant speed.

For a given current input, the starting torque developed by a series motor is greater than that developed by a shunt motor. Hence, series motors are used where huge starting torques are necessary *i.e.* for street cars, cranes, hoists and for electric-railway operation. In addition to the huge starting torque, there is another unique characteristic of series motors which makes them especially desirable for traction work *i.e* when a load comes on a series motor, it responds by decreasing its speed (and hence, E_b) and supplies the increased torque with a small increase in current. On the other hand a shunt motor under the same conditions would hold its speed nearly constant and would supply the required increased torque with a large increase of input current. Suppose that instead of a series motor, a shunt motor is used to drive a street car. When the car ascends a grade, the shunt motor maintains the speed of the car at approximately the same value it had on the level ground, but the motor tends to take an excessive current. A series motor, however, automatically slows down on such a grade because of increased current demand, and so it develops more torque at reduced speed.

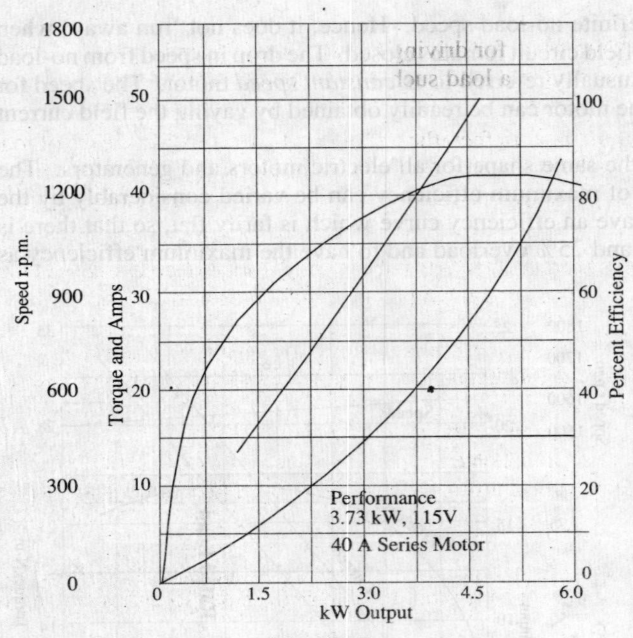

Fig. 27.25

The drop in speed permits the motor to develop a large torque with but a moderate increase of power. Hence, under the same load conditions, rating of the series motor would be less than for a shunt motor.

27.17. Comparison of Shunt and Series Motors

(*a*) Shunt Motors

The different characteristics have been discussed in Art. 27.14. It is clear that

(*a*) speed of a shunt motor is sufficiently constant.

(*b*) for the same current input, its starting torque is not as high as that of series motor. Hence, it is used.

(*i*) when the speed has to be maintained approximately constant from N.L. to F.L. *i.e.* for driving a line of shafting etc.

(*ii*) when it is required to drive the load at various speeds, any one speed being kept constant for a relatively long period *i.e.* for individual driving of such machines as lathes. The shunt regulator enables the required speed control to be obtained easily and economically.

Summary of Applications

Type of motor	*Characteristics*	*Applications*
Shunt	Approximately constant speed Adjustable speed Medium starring torque (Up to 1.5 F.L. torque)	For driving constant speed line shafting Lathes Centrifugal pumps Machine tools Blowers and fans Reciprocating pumps
Series	Variable speed Adjustable variying speed High Starting torque	For traction work *i.e.* Electric locomotives Rapid transit systems Trolley cars etc. Cranes and hoists Conveyors
Commulative Compound	Variable speed Adjustable varying speed High starting torque	For intermittent high torque loads For shears and punches Elevators Conveyors Heavy planers Heavy planers Rolling mills; Ice machines; Printing presses; Air compressors

(*b*) Series Motors

The operating characteristics have been discussed in Art. 27.13. These motors

1. have a relatively huge starting torques.
2. have good accelerating torque
3. have low speed at high loads and dangerously high speed at low loads.

Hence, such motors are used

1. when a large starting torque is required *i.e.* for driving hoists, cranes, trams etc.

2. when the motor can be directly coupled to a load such as a fan whose torque increases with speed.

3. if constancy of speed is not essential, then, in fact, the decrease of speed with increase of load has the advantage that the power absorbed by the motor does not increase as rapidly as the torque. For instance, when torque is doubled, the power approximately increases by about 50 to 60% only. ($\therefore\ I_a \propto \sqrt{T_a}$).

4. a series motor should not be used where there is a possibility of the load decreasing to a very small value. Thus, it should not be used for driving centrifugal pumps or for a belt-drive of any kind.

27.18. Losses and Efficiency

The losses taking place in the motor are the same as in generators. These are (*i*) Copper losses (*ii*) Magnetic losses and (*iii*) Mechanical losses.

The condition for maximum *power* developed by the motor is

$$I_a R_a = V/2 = E_b$$

The condition for maximum *efficiency* is that armature Cu losses are equal to constant losses (Art. 24.33).

27.19. Power Stages

The various stages of energy transformation in a motor and also the various losses occurring in it are shown in the flow diagram of Fig. 27.26.

Overall or commercial efficiency $\eta_c = \frac{C}{A}$; Electrical efficiency $\eta_e = \frac{B}{A}$; Mechanical efficiency $\eta_m = \frac{C}{B}$

The efficiency curve for a motor is similar in shape to that for a generator (Art. 24.35).

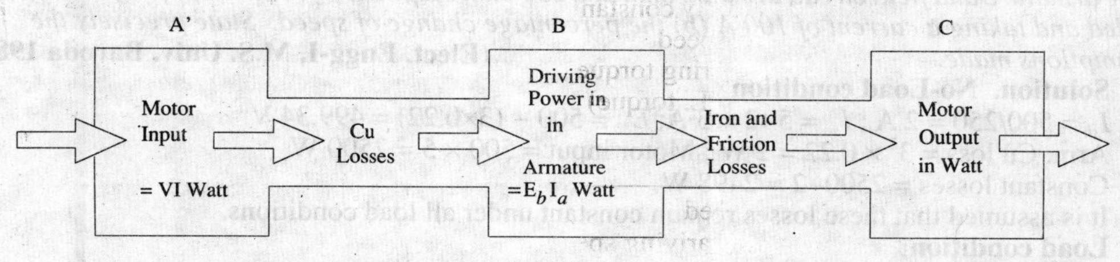

Fig. 27.26

It is seen that $A-B$ = copper losses and $B-C$ = iron and friction losses.

Example 27.32. *One of the two similar 500-V shunt machines A and B running light takes 3A. When A is mechanically coupled to B, the input to A is 3.5 A with B unexcited and 4.5 A when B is separately-excited to generate 500 V. Calculate the friction and windage loss and core loss of each machine.* **(Electric Machinery-I, Madras Univ. 1985)**

Solution. When running light, machine input is used to meet the following losses (*i*) armature Cu loss (*ii*) shunt Cu loss (*iii*) iron loss and (*iv*) mechanical losses *i.e* friction and windage losses. Obviously, these no-load losses for each machine equal 500 × 3 = 1500 W.

(*a*) **With B unexcited**

In this case, only mechanical losses take place in *B*, there being neither Cu loss nor iron-loss because *B* is unexcited. Since machine *A* draws 0.5 A more current.

Friction and windage loss of *B* = 500 × 0.5 = **250 W**

(*b*) **With B excited**

In this case, both iron losses as well as mechanical losses take place in machine *B*. Now, machine *A* draws 4.5−3 = 1.5 A more current.

Iron and mechanical losses of *B* = 1.5 × 500 = 750 W

Iron losses of *B* = 750 − 250 = **500 W**

Example 27.33. *A 220 V shunt motor has an armature resistance of 0.2 ohm and field resistance of 110 ohm. The motor draws 5 A at 1500 r.p.m. at no load. Calculate the speed and shaft torque if the motor draws 52 A at rated voltage.* **(Elect. Machines Nagpur Univ. 1993)**

Solution. $I_{sh} = 220/110 = 2A$; $I_{a1} = 5-2 = 3$ A ; $I_{a2} = 52 - 2 = 50$ A
$E_{b1} = 220 - 3 \times 0.2 = 219.4$ V ; $E_{b2} = 220 - 50 \times 0.2 = 210$ V

$$\frac{N_2}{1500} = \frac{210}{219.4}; N_2 = 1436 \text{ r.p.m.} \quad (\because \quad \Phi_1 = \Phi_2)$$

For finding the shaft torque, we will find the motor output when it draws a current of 52 A. First we will use the no-load data for finding the constant losses of the motor.

No load motor input = 220 × 5 = 1100 W; Arm. Cu loss = $3^2 \times 0.2 = 2$ W
∴ Constant or standing losses of the motor = 1100–2 = 1098
When loaded, arm. Cu loss = $50^2 \times 0.2 = 500$ W
Hence, total motor losses = 1098 + 500 = 1598 W
Motor input on load = 220 × 52 = 11,440 W; output = 11,440–1598 = 9842 W
∴ $T_{sh} = 9.55 \times \text{output}/N = 9.55 \times 9842/1436 =$ **65.5 N-m**

Example 27.34. *250 V shunt motor on no load runs at 1000 r.p.m. and takes 5 amperes. Armature and shunt field resistances are 0.2 and 250 ohms respectively. Calculate the speed when loaded taking a current of 50 A. The armature reaction weakens the field by 3%.*
(Elect. Engg.I Nagpur Univ. 1993)

Solution. $I_{sh} = 250/250 = 1$ A; $I_{a1} = 5-1 = 4$ A ; $I_{a2} = 50 - 1 = 49$ A
$E_{b1} = 250-4 \times 0.2 = 249.2$V ; $E_{b2} = 250\text{-}49 \times 0.2 = 240.2$ V

$$\frac{N_2}{1000} = \frac{240.2}{249.2} \times \frac{\Phi_1}{0.97\Phi_1}; N_2 = 944 \text{ r.p.m.}$$

Example 27.35. *A 500-V d.c. shunt motor takes a current of 5 A on no-load. The resistances of the armature and field circuit are 0.22 ohm and 250 ohm respectively. Find (a) the efficiency when loaded and taking a current of 100 A (b) the percentage change of speed. State precisely the assumptions made.* **(Elect. Engg-I, M.S. Univ. Baroda 1987)**

Solution. No-Load condition
$I_{sh} = 500/250 = 2$ A ; $I_{ao} = 5-2 = 3$ A; $E_{bo} = 500 - (3{\times}0.22) = 499.34$ V
Arm. Cu loss = $3^2 \times 0.22 = 2$ W ; Motor input = 500 × 5 = 2500 W
Constant losses = 2500–2 = 2498 W
It is assumed that these losses remain constant under all load conditions.
Load condition
(*a*) Motor current = 100 A ; $I_a = 100 - 2 = 98$ A ; $E_b = 500 - (98 \times 0.22) = 478.44$ V
Arm. Cu loss = $98^2 \times 0.22 = 2110$ W, Total losses = 2110 + 2498 = 4608 W
Motor input = 500 × 100 = 50,000 W, Motor output = 50,000 – 4,608 = 45,392 W
Motor η = 45,392/50,000 = 0.908 or **90.8%**

$$\frac{N}{N_0} = \frac{E_b}{E_{b0}} = \frac{478.44}{499.34} \text{ or } \frac{N - N_0}{N_0} = \frac{-20.9}{499.34} = -0.0418 \text{ or } \mathbf{4.18\%}$$

Example 27.36. *A 250 V d.c. shunt motor runs at 1000 r.p.m. while taking a current of 25 A. Calculate the speed when the load current is 50 A if armature reaction weakens the field by 3%. Determine torques in both cases.*
R_a = 0.2 ohm ; R_f = 250 ohms
Voltage drop per brush is 1 V. **(Elect. Machines Nagpur Univ. 1993)**

Solution. $I_{sh} = 250/250 = 1$A; $I_{al} = 25-1 = 24$ A
E_{bh} = 250 – arm. drop–brush drop
= 250 – 24 × 0.2 – 2 = 243.2V
$I_{a2} = 50-1 = 49$ *A*; $E_{b2} = 250 - 49 \times 0.2 - 2 = 238.2$ V

$$\frac{N_2}{1000} = \frac{238.2}{243.2} \times \frac{\Phi_1}{0.97\Phi_1}; N_2 = 1010 \text{ r.p.m.}$$

$T_{a1} = 9.55\, E_{b1} I_{a1}/N_1 = 9.55 \times 243.2 \times 24/1000 = 55.7$ N-m

$T_{a2} = 9.55 \times 238.2 \times 49/1010 = 110.4$ r.p.m.

Example 27.37. *A d.c. shunt machine while running as generator develops a voltage of 250 V at 1000 r.p.m. on no-load. It has armature resistance of 0.5 Ω and field resistance of 250 Ω. When the machine runs as motor, input to it at no-load is 4 A at 250 V. Calculate the speed and efficiency of the machine when it runs as a motor taking 40 A at 250 V. Armature reaction weakens the field by 4%.* **(Electrical Technology, Aligarh Muslim Univ. 1989)**

Solution. $\dfrac{N_2}{N_1} = \dfrac{E_{b2}}{E_{b1}} \times \dfrac{\Phi_1}{\Phi_2}$

Now, when running as a generator, the machine gives 250 V at 1000 r.p.m. If this machine were run as motor at 1000 r.p.m., it will, obviously, have a back e.m.f. of 250 V produced in its armature. Hence $N_1 = 1000$ r.p.m. and $E_{b1} = 250$ V.

When it runs as a motor, drawing 40 A, the back e.m.f. induced in its armature is

$E_{b2} = 250 - (40-1) \times 0.5 = 230.5$ V; Also, $\Phi_2 = 0.96\,\Phi_1$, $N_2 = ?$

Using the above equation we have

$$\frac{N_2}{1000} = \frac{230.5}{250} \times \frac{\Phi_1}{0.96\Phi_1};\ N_2 = \mathbf{960\ r.p.m.}$$

Efficiency

No-load input represents motor losses which consists of

(*a*) armature Cu loss = $I_a^2 R_a$ which is variable.

(*b*) constant losses W_c which consists of (*i*) Shunt Cu loss (*ii*) magnetic losses and (*iii*) mechanical losses.

No-load input or total losses = 250 × 4 = 1000 W

Arm. Cu loss = $I_a^2 R_a = 3^2 \times 0.5 = 4.5$ W $\therefore W_c = 1000 - 4.5 = 995.5$ W

When motor draws a line current of 40 A, its armature current is (40–1) = 39 A

Arm. Cu loss = $39^2 \times 0.5 = 760.5$ W; Total losses = 760.5 + 955.5 = 1756 W

input = 250 × 40 = 10,000 W ; output = 10,000 – 1756 = 8,244 W

$\therefore$ $\eta = 8{,}244 \times 100/10{,}000 = \mathbf{82.44\ \%}$

Example 27.38. *The armature winding of a 4-pole, 250 V d.c. shunt motor is lap connected. There are 120 slots, each slot containing 8 conductors. The flux per pole is 20 mWb and current taken by the motor is 25 A. The resistance of armature and field circuit are 0.1 and 125 Ω respectively. If the rotational losses amount to 810 W find.*

(i) gross torque (ii) useful torque and (iii) efficiency **(Elect. Machines Nagpur Univ. 1993)**

Solution. $I_{sh} = 250/125 = 2$ A ; $I_a = 25 - 2 = 23$ A; $E_b = 250 - (23 \times 0.1) = 247.7$ V

Now, $E_b = \dfrac{\Phi ZN}{60}\left(\dfrac{P}{A}\right)$ $\therefore 247.7 = \dfrac{20 \times 10^{-3} \times 960 \times N}{60}\left(\dfrac{4}{4}\right)$; $N = 773$ r.p.m.

(*i*) Gross torque or armature torque $T_a = 9.55\,\dfrac{E_b I_a}{N}$ 9.55 × $9.55 \times \dfrac{247.7 \times 223}{773}$ = **70.4 N-m**

(*ii*) Arm. Cu loss = $23^2 \times 0.1 = 53$ W; Shunt Cu loss = 250 × 2 = 500 W

Rotational losses = 810 W; Total motor losses = 810 + 500 + 53 = 1363 W

Motor input = 250 × 25 = 6250 W ; Motor output = 6250 – 1363 = 4887 W

$T_{sh} = 9.55 \times \text{output}/N = 9.55 \times 4887/773 = 60.4$ N-m

(*iii*) Efficiency = 4887/6250 = 0.782 = 78.2 %

Example 27.39. *A 20-hp (14.92 kW); 230-V, 1150-r.p.m. 4-pole, d.c. shunt motor has a total of 620 conductors arranged in two parallel paths and yielding an armature circuit resistance of 0.2 Ω. When it delivers rated power at rated speed, it draws a line current of 74.8 A and a field current of 3A. Calculate (i) the flux per pole (ii) the torque developed (iii) the rotational losses (iv) total losses expressed as a percentage of power.* **(Electrical Machinery-I, Bangalore Univ. 1987)**

Solution. $I_a = 74.8 - 3 = 71.8$ A ; $E_b = 230 - 71.8 \times 0.2 = 215.64$ V

(*i*) Now, $E_b = \dfrac{\Phi ZN}{60}\left(\dfrac{P}{A}\right)$; $215.64 = \dfrac{\Phi \times 620 \times 1150}{60}\left(\dfrac{4}{2}\right)$; $\mathbf{\Phi = 9\ mWb}$

(*ii*) Armature torque, $T_a = 9.55 \times 215.64 \times 71.8 / 1150 =$ **128.8 N-m**
(*iii*) Driving power in armature $= E_b I_a = 215.64 \times 71.8 = 15{,}483$ W
Output = 14,920 W ; rotational losses = 15,483 − 14,920 = **563 W**
(*iv*) Motor input = $VI = 230 \times 74.8 = 17{,}204$ W ; Total loss = 17,204–14,920 = 2,284 W
Losses expressed as percentage of power input = 2284/17,204 = 0.133 or **13.3%**

Example 27.40. *A 7.46 kW, 250-V shunt motor takes a line current of 5 A when running light. Calculate the efficiency as a motor when delivering full load output, if the armature and field resistances are 0.5 Ω and 250 Ω respectively. At what output power will the efficiency be maximum ? Is it possible to obtain this output from the machine ?*

(Electrotechnics-II, M.S. Univ. Baroda 1985)

Solution. When loaded lightly

Total motor input (or total no-load losses) = 250 × 5 = 1,250 W
$I_{sh} = 250/250 = 1$A ∴ $I_a = 5 - 1 = 4$ A
Field Cu loss = 250 × 1 = 250 W; Armature Cu loss = $4^2 \times 0.5 = 8$ W
∴ Iron losses and friction losses = 1250 − 250 − 8 = 992 W
These losses would be assumed constant
Let I_a be the full-load armature current, then armature input is = $(250 \times I_a)$ W
F.L. output = 7.46 × 1000 = 7,460 W
The losses in the armature are :
(*i*) Iron and friction losses = 992 W
(*ii*) Armature Cu loss = $I_a^2 \times 0.5$ W ∴ $250\, I_a = 7{,}460 + 922 + I_a^2 \times 0.5$
or $0.5\, I_a^2 - 250 I_a + 8{,}452 = 0$ ∴ $I_a = 36.5$ A
∴ F.L. input current = 36.5 + 1 = 37.5 A ; Motor input = 250 × 37.5 W
F.L. output = 7,460 W ∴ F.L. efficiency = 7460 × 100/250 × 37.5 = 79.6%
Now, efficiency is maximum when armature Cu loss equals constant loss.
i.e $I_a^2 R_a = I_a^2 \times 0.5 = (1{,}250 - 8) = 1{,}242$ W or $I_a = 49.84$ A
∴ Armature input = 250 × 49.84 = 12,460 W.
Armature Cu loss = $49.84^2 \times 0.5 = 1242$ W; Iron and friction losses = 992 W
∴ Armature output = 12,460 − (1,242 + 992) = 10,226 W
∴ output power = 10,226 W = **10.226 kW**
As the input current for maximum efficiency is beyond the full-load motor current, it is never realised in practice.

Example 27.41. *A d.c. series motor drives a load, the torque of which various as the square of the speed. Assuming the magnetic circuit to remain unsaturated and the motor resistance to be negligible, estimate the percentage reduction in the motor terminal voltage which will reduce the motor speed to half the value it has on full voltage. What is then the percentage fall in the motor current and efficiency ? Stray losses of the motor may be ignored.*

(Electrical Engineering-III, Pune Univ. 1987)

Solution. $T_a \propto \Phi I_a \propto I_a^2$ Also, $T_a \propto N^2$. Hence, $N^2 \propto I_a^2$ or $N \propto I_a$
∴ $N_1 \propto I_{a1}$ and $N_2 \propto I_{a2}$ or $N_2/N_1 = I_{a2}/I_{a1}$
Since, $N_2/N_1 = 1/2$ ∴ $I_{a2}/I_{a1} = 1/2$ or $I_{a2} = I_{a1}/2$

Let V_1 and V_2 be the voltages across the motor in the two cases. Since motor resistance is negligible, $E_{b1} = V_1$ and $E_{b2} = V_2$. Also $\Phi_1 \propto I_{a1}$ and $\Phi_2 \propto I_{a2}$ or $\Phi_1/\Phi_2 = I_{a1}/I_{a2} = I_{a1} \times 2/I_{a1} = 2$

Now, $$\frac{N_2}{N_1} = \frac{E_{b2}}{E_{b1}} \times \frac{\Phi_1}{\Phi_2} \quad \text{or} \quad \frac{1}{2} = \frac{V_2}{V_1} \times 2 \quad \text{or} \quad \frac{V_2}{V_1} = \frac{1}{4}$$

∴ $$\frac{V_1 - V_2}{V_1} = \frac{4-1}{4} = 0.75$$

$\therefore$ percentage reduction in voltage $= \dfrac{V_1 - V_2}{V_1} \times 100 = 0.75 \times 100 = \mathbf{75\%}$

Percentage change in motor current $= \dfrac{I_{a1} - I_{a2}}{I_{a1}} \times 100 = \dfrac{I_{a1} - I_{a1}/2}{I_{a1}} \times 100 = \mathbf{50\ \%}$

Example 27.42. *A 6-pole, 500-V wave-connected shunt motor has 1200 armature conductors and useful flux/pole of 20 mWb. The armature and field resistances are 0.5 Ω and 250 Ω respectively. What will be the speed and torque developed by the motor when it draws 20 A from the supply mains ? Neglect armature reaction. If magnetic and mechanical losses amount to 900 W, find (i) useful torque (ii) output in kW and (iii) efficiency at this load.*

Solution. (*i*) $I_{sh} = 500/250 = 2\text{A}$ $\therefore$ $I_a = 20 - 2 = 18$ A

$\therefore$ $E_b = 500 - (18 \times 0.5) = 491$ V ; Now, $E_b = \dfrac{\Phi ZN}{60} \times \left(\dfrac{P}{A}\right)$ volt

$\therefore$ $491 = \dfrac{20 \times 10^{-3} \times 1200 \times N}{60} \times \left(\dfrac{6}{2}\right)$; $N = \mathbf{410\ r.p.m.\ (approx.)}$

Now $T_a = 9.55 \dfrac{E_b I_a}{N} = 9.55 \dfrac{491 \times 18}{410} = \mathbf{206\ N\text{-}m}$

Armature Cu loss = $18^2 \times 0.5 = 162$ W ; Field Cu loss = $500 \times 2 = 1000$ W

Iron and friction loss = 900 W. Total loss = 162 + 1000 + 900 = 2,062 W

Motor input = $500 \times 20 = 10{,}000$ W

(*i*) $T_{sh} = 9.55$ 7,$\dfrac{938}{410}$ = **184.8 N-m**

(*ii*) output = 10,000 – 2062 = **7.938 kW**

(*iii*) % $\eta = \dfrac{\text{output}}{\text{input}} \times 100 = \dfrac{7{,}938 \times 100}{10{,}000} = 0.794 = \mathbf{79.4\%}$

Example 27.43. *A 50-h.p. (37.3 kW), 460-V, d.c. shunt motor running light takes a current of 4 A and runs at a speed of 660 r.p.m. The resistance of the armature circuit (including brushes) is 0.3 Ω and that of the shunt field circuit 270 Ω.*

Determine when the motor is running at full load.

(i) the current input (ii) the speed. Determine the armature current at which efficiency is maximum. Ignore the effect of armature reaction.

(Elect. Technology Punjab, Univ. 1991)

Solution. $I_{sh} = 460 \times 270 = 1.7$ A; Field Cu loss = $460 \times 1.7 = 783$ W

When running light

$I_a = 4 - 1.7 = 2.3$ A; Armature Cu loss = $2.3^2 \times 0.3 = 1.5$ W (negligible)

No-load armature input = $460 \times 2.3 = 1{,}058$ W

As armature Cu loss is negligible, hence 1,058 W represents iron, friction and windage losses which will be assumed to be constant.

Let full-load armature input current be I_a. Then

Armature input $= 460\, I_a$ W ; Armature Cu loss = $I_a^2 \times 0.3$ W

Output = 37.3 kW = 37,300 W

$\therefore$ $460\, I_a = 37{,}300 + 1{,}058 + 0.3\, I_a^2$ or $0.3\, I_a^2 - 460\, I_a + 38{,}358 = 0$

$\therefore$ $I_a = 88.5$ A

(*i*) current input = 88.5 + 1.7 = **90.2 A**

(*ii*) $E_{b1} = 460 - (2.3 \times 0.3) = 459.3$ V ; $E_{b2} = 460 - (88.5 \times 0.3) = 433.5$ V

$\therefore$ $N_2 = 660 \times 433.5/459.3 = \mathbf{624\ r.p.m.}$

For maximum efficiency, $I_a^2 R_a$ = constant losses (Art. 24.37)

$\therefore$ $I_a^2 \times 0.3 = 1.058 + 783 = 1{,}841$ $\therefore$ $I_a = (1841/0.3)^{1/2} = \mathbf{78.33\ A}$

Tutorial Problem No. 27.3.

1 A 4-pole, 250-V, d.c. series motor has a wave-wound armature with 496 conductors. Calculate
(*a*) the gross torque (*b*) the speed
(*c*) the output torque and (*d*) the efficiency, if the motor current is 50 A
The value of flux per pole under these conditions is 22 mWb and the corresponding iron, friction and windage losses total 810W. Armature resistance = 0.19 Ω, field resistance = 0.14 Ω.
[(*a*) **173.5 N-m** (*b*) **642 r.p.m.** (*c*) **161.4 N-m** (*d*) **86.9%**]

2. On no-load, a shunt motor takes 5 A at 250 V, the resistances of the field and armature circuits are 250 Ω and 0.1 Ω respectively. Calculate the output power and efficiency of the motor when the total supply current is 81 A at the same supply voltage. State any assumptions made.
18.5 kW; 91%. It is assumed that windage, friction and eddy current losses are independent of the current and speed]

3. A 230-V series motor is taking 50 A. Resistance of armature and series field windings is 0.2 Ω and 0.1 Ω respectively. Calculate.
(*a*) brush voltage (*b*) back e.m.f.
(*b*) power wasted in armature (*d*) mechanical power developed
[(*a*) **215 V** (*b*) **205 V** (*c*) **500 W** (*d*) **13.74 h.p.** **(10.25 kW)**

4. Calculate the shaft power of a seires motor having the following data; overall efficiency 83.5%, speed 550 r.p.m. when taking 65 A; motor resistance 0.2 Ω, flux per pole 25 mWb, armature winding lap with 1200 conductor. **(15.66 kW)**

5. A shunt motor running on no-load takes 5 A at 200 V. The resistance of the field circuit is 150 Ω and of the armature 0.1 Ω. Determine the output and efficinecy of motor when the input current is 120 A at 200 V. State any conditions assumed. **[89.8%]**

6. A d.c. shunt motor with interpoles has the following particulars :-
Output power ; 8.952 kW, 440-V, armature resistance 1.1 Ω, brush contact drop 2 V, interpole winding resistance 0.4 Ω shunt resistance 650 Ω, resistance in the shunt regulator 50 Ω. Iron and friction losses on full-load 450 W. Calculate the efficinecy when taking the full rated current of 24 A. **[85%]**

7. A d.c. series motor on full-load takes 50 A from 230 V d.c. mains. The total resistance of the motor is 0.22 Ω. If the iron and friction losses together amount to 5% of the input, calculate the power delivered by the motor shaft. Total voltage drop due to the brush contact is 2 A **[10.275 kW]**

8. A 2-pole d.c. shunt motor operating from a 200 V supply takes a full-load current of 35 A, the no-load current being 2 A. The field resistance is 500 Ω and the armature has a resistance of 0.6 Ω. Calculate the efficiency of the motor on full-load. Take the brush dorp as being equal to 1.5 V per brush arm. Neglect temperature rise. **[82.63%]**

Objective Tests – 27

1. In a d.c. motor, undirectional torque is produced with the help of
(*a*) brushes (*b*) commutator
(*c*) end-plates (*d*) both (*a*) and (*b*)

2. The counter e.m.f. of a d.c. motor
(*a*) often exceeds the supply voltage
(*b*) aids the applied voltage
(*c*) helps in energy conversion
(*d*) regulates its armature voltage

3. The normal value of the armature resistance of a d.c. motor is
(*a*) 0.005 (*b*) 0.5
(*c*) 10 (*d*) 100
(Grad. I.E.T.E. June 1987)

4. The E_b/V ratio of a d.c. motor is an indication of its
(*a*) efficiency (*b*) speed regulation
(*c*) starting torque (*d*) Running Torque
(Grade. I.E.T.E. June 1987)

5. The mechanical power developed by the armature of a d.c. motor is equal to
(*a*) armature current multiplied by back e.m.f.
(*b*) power input minus losses
(*c*) power output multiplied by efficiency
(*d*) power output plus iron losses

6. The induced e.m.f. in the armature conductors of a d.c. motor is
(*a*) sinusoidal (*b*) trapezoidal
(*c*) rectangular (*d*) alternating

7. A d.c. motor can be looked upon as d.c. generator with the power flow
(*a*) reduced (*b*) reversed
(*c*) increased (*d*) modified

8. In a d.c. motor, the mechanical output power actually comes from
(*a*) field system (*b*) air-gap flux
(*c*) back e.m.f. (*d*) electrical input power

9. The maximum torque of d.c. motors is limited by
(*a*) commutation (*b*) heating
(*c*) speed (*d*) armature current

10. Which of the following quantity maintains the same direction whether a d.c. machine runs as a generator or as a motor ?
(*a*) induced e.m.f. (*b*) armature current
(*c*) field current (*d*) supply current

11. Under constant load conditions, the speed of a d.c. motor is affected by
(*a*) field flux (*b*) armature current
(*c*) back e.m.f. (*d*) both (*b*) and (*c*)

12. It is possible to increase the field flux and, at the same time, increase the speed of a d.c. motor provided its —— is held constant.
(*a*) applied voltage (*b*) torque
(*c*) Armature circuit resistance
(*d*) armature current

13. The current drawn by a 120 - V d.c. motor of armature resistance 0.5 Ω and back e.m.f. 110 *V* is —— ampere
(*a*) 20 (*b*) 240
(*c*) 220 (*d*) 5

14. The shaft torque of a d.c. motor is less than its armature torque because of —— losses.
(*a*) copper (*b*) mechanical
(*c*) iron (*d*) rotational

15. A d.c. motor develops a torque of 200 N-m at 25 rps. At 20 rps, it will develop a torque of —— N-m.
(*a*) 200 (*b*) 160
(*c*) 250 (*d*) 128

16. Neglecting saturation, if current taken by a series motor is increased from 10 A to 12 A, the percentage increase in its torque is —— percent
(*a*) 20 (*b*) 44
(*c*) 30.5 (*d*) 16.6

17. If load on a d.c. shunt motor is increased, its speed is decreased due primarily to
(*a*) increase in its flux
(*b*) decrease in back e.m.f.
(*c*) increase in armature current
(*d*) increase in brush drop

18. If the load current and flux of a d.c. motor are held constant and voltage applied across its armature is increased by 10 per cent, its speed will
(*a*) decrease by about 10 per cent
(*b*) remain unchanged
(*c*) increase by about 10 per cent
(*d*) increase by 20 per cent

19. If the pole flux of a d.c. motor approaches zero, its speed will
(*a*) approach zero
(*b*) approach infinity
(*c*) no change due to corresponding change in back e.m.f.
(*d*) approach a stable value somewhere between zero and infinity.

20. If the field circuit of a loaded shunt motor is suddenly opened
(*a*) it would race to almost infinite speed
(*b*) it would draw abnormally high armature current
(*c*) circuit breaker or fuse will open the circuit before too much damage is done to the motor
(*d*) torque developed by the motor would be reduced to zero.

21. Which of the following d.c. motor would be suitable for drives requiring high starting torque but only fairly constant speed such as crushers ?
(*a*) shunt (*b*) series
(*c*) compound (*d*) permanent magnet

22. A d.c. shunt motor is found suitable to drive fans because they require
(*a*) small torque at start up
(*b*) large torque at high speeds
(*c*) practically constant voltage
(*d*) both (*a*) and (*b*)

23. Which of the following load would be best driven by a d.c. compound motor ?
(*a*) reciprocating pump
(*b*) centrifugal pump
(*c*) electric locomotive
(*d*) fan

24. As the load is increased, the speed of a d.c. shunt motor
(*a*) increases proportionately
(*b*) remains constant
(*c*) increases slightly
(*d*) reduces slightly

25. Between no-load and full-load, —— motor develops the *least* torque
(*a*) series (*b*) shunt
(*c*) cumulative compound
(*d*) differential compound

26. The T_a/I_a graph of a d.c. series motor is a
(*a*) parabola from no-load to overload
(*b*) straight line throughout
(*c*) parabola throughout
(*d*) parabola upto full-load and a straight line at over loads

27. As compared to shunt and compound motors, series motor has the highest torque because of its comparatively —— at the start.
(*a*) lower armature resistance
(*b*) stronger series field
(*c*) fewer series turns
(*d*) larger armature current

28. Unlike a shunt motor, it is difficult for a series motor to stall under heavy loading because
(*a*) it develops high overload torque
(*b*) its flux remains constant
(*c*) it slows down considerably
(*d*) its back e.m.f. is reduced to almost zero.

29. When load is removed, —— motor will run at the *highest* speed.
(*a*) shunt
(*b*) cumulative-compound
(*c*) differential compound
(*d*) series

30. A series motor is best suited for driving
(*a*) lathes (*b*) cranes and hoists
(*c*) shears and punches (*d*) machine tools

31. A 220 V shunt motor develops a torque of 54 N-m at armature current of 10A. The torque produced when the armature current is 20 A, is
(*a*) 54 N-m (*b*) 81 N-m
(*c*) 108 N-m (*d*) None of the above
(*Elect. Machines, A.M.I.E. Sec B, 1993*)

32. The d.c. series motor should never be switched on at no load because
(*a*) the field current is zero
(*b*) The machine does not pickup
(*c*) The speed becomes dangerously high
(*d*) It will take too long to accelerate.
(*Grad. I.E.T.E. June 1988*)

33. A shunt d.c. motor works on a.c. mains
(*a*) unsatisfactorily
(*b*) satisfactorily
(*c*) not at all
(*d*) none of the above
(*Elect. Machines, A.M.I.E. Sec. B, 1993*)

34. A 200V, 10 A motor could be rewound for 100 V, 20 A by using —— as many turns per coil of wire having —— the cross-sectional area.
(*a*) twice, half
(*b*) thrice, one third
(*c*) half, twice
(*d*) four times, one-fourth.

Answers

1. (*d*) **2.** (*c*) **3.** (*b*) **4.** (*a*) **5.** (*a*) **6.** (*a*) **7.** (*b*) **8.** (*d*) **9.** (*a*) **10.** (*a*)
11. (*a*) **12.** (*d*) **13** (*a*) **14.** (*d*) **15** (*a*) **16.** (*b*) **17.** (*d*) **18.** (*c*) **19.** (*b*) **20.** (*c*)
21 (*c*) **22.** (*d*) **23** (*a*) **24.** (*d*) **25.** (*a*) **26.** (*d*) **27.** (*b*) **28** (*a*)
29. (*d*) **30** (*b*) **31.** (*c*) **32.** (*c*) **33.** (*a*) **34.** (*c*)

28 SPEED CONTROL OF D.C. MOTORS

28.1. Factors Controlling Motor Speed

It has been shown earlier that the speed of a motor is given by the relation

$$N = \frac{V - I_a R_a}{Z\Phi}\left(\frac{A}{P}\right) = K\frac{V - I_a R_a}{\Phi} \text{ r.p.s.}$$

where R_a = armature circuit resistance.

It is obvious that the speed can be controlled by varying (*i*) flux/pole, Φ (Flux Control) (*ii*) resistance R_a of armature circuit (Rheostatic Control) and (*iii*) applied voltage V (Voltage Control). These methods as applied to shunt, compound and series motors will be discussed below.

28.2. Speed Control of Shunt Motors

(i) Variation of Flux or Flux Control Method

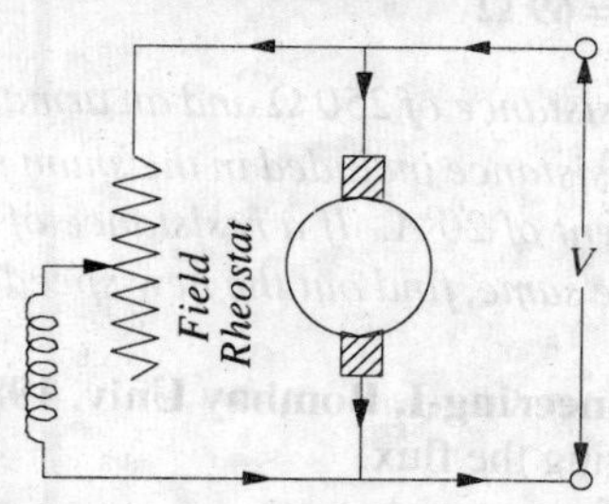

Fig. 28.1

It is seen from above that $N \propto 1/\Phi$. By decreasing the flux, the speed can be increased and *vice versa*. Hence, the name *flux* or *field control* method. The flux of a d.c. motor can be changed by changing I_{sh} with help of a shunt field rheostat (Fig. 28.1). Since I_{sh} is relatively small, shunt field rheostat has to carry only a small current, which means I^2R loss is small, so that rheostat is small in size. This method is, therefore, very efficient. In non-interpolar machines, the speed can be increased by this method in the ratio 2 : 1. Any further weakening of flux Φ adversely affects the commutation and hence puts a limit to the maximum speed obtainable with this method. In machines fitted with interpoles, a ratio of maximum to minimum speeds of 6 : 1 is fairly common.

Example 28.1. *A 500-V shunt motor runs at its normal speed of 250 rpm when the armature current is 200 A. The resistance of armature is 0.12 ohm. Calculate the speed when a resistance is inserted in the field reducing the shunt field to 80% of normal value and the armature current is 100 ampere.*

(Elect. Engg. A.M.A.e. S.I. June 1992)

Solution. $E_{b1} = 500 - 200 \times 0.12 = 476$ V; $E_{b2} = 500 - 100 \times 0.12 = 488$ V

$\Phi_2 = 0.8\ \Phi_1$; $N_1 = 250$ rpm; $N_2 = ?$

Now, $\dfrac{N_2}{N_1} = \dfrac{E_{b2}}{E_{b1}} \times \dfrac{\Phi_1}{\Phi_2}$ or $\dfrac{N_2}{250} = \dfrac{488}{476} \times \dfrac{\Phi_1}{0.8\Phi_1}$, $N_2 = 320.4$ rpm

Example 28.2. *A 250-volt d.c. shunt motor has armature resistance of 0.25 ohm, on load it takes an armature current of 50 A and runs at 750 r.p.m. If the flux of motor is reduced by 10% without changing the load torque, find the new speed of the motor.*

(Elect. Engg-II, Pune Univ. 1987)

Solution. $\frac{N_2}{N_1} = \frac{E_{b2}}{E_{b1}} \times \frac{\Phi_1}{\Phi_2}$

Now, $T_a \propto \Phi I_a$. Hence, $T_{al} \propto \Phi_1 I_{al}$ and $T_{a2} \propto \Phi_2 I_{a2}$.

Since $T_{a1} = T_{a2}$ $\therefore$ $\Phi_1 I_{cl} = \Phi_2 I_{a2}$

Now, $\Phi_2 = 0.9\,\Phi_1$ $\therefore$ $50\ \Phi_1 = 0.9\,\Phi, I_{a2}, I_{a2} = 55.6$ A

$\therefore$ $E_{bl} = 250-(50 \times 0.25) = 237.5$ V ; $E_{b2} = 250 - (55.6 \times 0.25) = 231.1$ V

$\therefore$ $\frac{N_2}{750} = \frac{231.1}{237.5} \times \frac{\Phi}{0.9\Phi_1}$; $N_2 =$ **811 r.p.m**

Example 28.3. *Describe briefly the method of speed control available for dc motors.*
A 230-V d.c. shunt motor runs at 800 r.p.m. and takes armature current of 50 A. Find resistance be added to the field circuit to increase speed to 1000 rpm at an armature current of 80 A. Assume flux proportional to field current. Armature resistance = 0.15 Ω and field winding resistance = 250 Ω.

(Elect. Technology, Hyderabad Univ. 1991)

Solution. $\frac{N_2}{N_1} = \frac{E_{b2}}{E_{b1}} \times \frac{\Phi_1}{\Phi_2} = \frac{E_{b2}}{E_{b1}} \times \frac{I_{sh1}}{I_{sh2}}$ since flux $\propto$ field current

$E_{b1} = 230-(50 \times 0.15) = 222.5$ V ; $E_{b2} = 230-(80 \times 0.15) = 218$ V

Let R_t = total shunt resistance = $(250 + R)$ where R is the additional resistance

$I_{sh1} = 230/250 = 0.92$ A, $I_{sh2} = 230/R_t$; $N_1 = 800$ rpm ; $N = 1000$ rpm

$\therefore$ $\frac{1000}{800} = \frac{218}{222.5} \times \frac{0.92}{230/R_t}$; $R_t = 319\ \Omega$ $\therefore$ $R = 319 - 250 =$ **69 Ω**

Example 28.4. *A 250-V, d.c. shunt motor has a shunt field resistance of 250 Ω and an armature resistance of 0.25 Ω. For a given load torque and no additional resistance included in the shunt field circuit, the motor runs at 1500 r.p.m. drawing an armature current of 20 A. If a resistance of 250 Ω is inserted in series with the field, the load torque remaining the same, find out the new speed and armature current. Assume the magnetisation curve to be linear.*

(Electrical Engineering-I, Bombay Univ. 1987)

Solution. In this case, the motor speed is changed by changing the flux.

Now, $\frac{N_2}{N_1} = \frac{E_{b2}}{E_{b1}} \times \frac{\Phi_1}{\Phi_2}$

Since it is given that magnetisation curve is linear, it means that flux is directly proportional to shunt current, hence $\frac{N_2}{N_1} = \frac{E_{b2}}{E_{b1}} \times \frac{I_{sh1}}{I_{sh2}}$ where $E_{b2} = V - I_{a2}R_a$ and $E_{bl} = V - I_{al}R_a$

Since load torque remains the same $\therefore T_a \propto \Phi_1 I_{a1} \propto \Phi_2 I_{a2}$ or $\Phi_1 I_{a1} = \Phi_2 I_{a2}$

$\therefore$ $I_{a2} = I_{a1} \times \frac{\Phi_1}{\Phi_2} = I_{a1} \times \frac{I_{sh1}}{I_{sh2}}$

Now $I_{sh1} = 250/250 = 1A$; $I_{sh2} = 250/(250+250) = 1/2\ A$

$\therefore$ $I_{a2} = 20 \times \frac{1}{1/2} =$ **40 A** $\therefore$ $E_{b2} = 250 - (40 \times 0.25) = 240$ V and

$E_{b1} = 250 - (20 \times 0.25) = 245$ V $\therefore$ $\frac{N_2}{1500} = \frac{240}{245} \times \frac{1}{1/2}$ $\therefore$ $N_2 =$ **2,930 r.p.m.**

Example 28.5. *A 250-V, d.c. shunt motor has an armature resistance of 0.5 Ω and a field resistance of 250 Ω. When driving a load of constant torque at 600 r.p.m., the armature current is 20 A. If it is desired to raise the speed from 600 to 800 r.p.m., what resistance should be inserted in the shunt field circuit ? Assume that the magnetic circuit is unsaturated.*

(Elect. Engg.. AMIETE, June 1992)

Solution. $\frac{N_2}{N_1} = \frac{E_{b2}}{E_{b1}} x \frac{\Phi_1}{\Phi_2}$

Since the magnetic circuit is unsaturated, it means that flux is directly proportional to the shunt current.

$$\therefore \quad \frac{N_2}{N_1} = \frac{E_{b2}}{E_{b1}} \times \frac{I_{sh1}}{I_{sh2}} \qquad \text{where } E_{b2} = V - I_{a2}R_a \text{ and } E_{b1} = V - I_{a1}R_a$$

Since motor is driving a load of constant torque,

$$T_a \propto \Phi_1 I_{a1} \propto \Phi_2 I_{a2} \quad \therefore \Phi_2 I_{a2} = \Phi_1 I_{a1} \text{ or } I_{a2} = I_{a1} \times \frac{\Phi_1}{\Phi_2} = I_{a1} \times \frac{I_{sh1}}{I_{sh3}}$$

Now, $I_{sh1} = 250/250 = 1$ A ; $I_{sh2} = 250/R_t$

where R_t is the total resistance of the shunt field circuit.

$$\therefore I_{a2} = 20 \times \frac{1}{250/R_t} = \frac{2R_t}{25}; \; E_{b1} = 250 - (20 \times 0.5) = 240 \text{ V}$$

$$E_{b2} = 250 - \left(\frac{2R_t}{25} \times 0.5\right) = 250 - (R_t/25) \quad \therefore \frac{800}{600} = \frac{250 - (R_t/25)}{240} \times \frac{1}{250/R_t}$$

$$0.04\,R_t^2 - 250\,R_t + 80,000 = 0$$

$$\text{or} \quad R_t = \frac{250 \pm \sqrt{62,500 - 12,800}}{0.08} = \frac{27}{0.08} = 337.5\ \Omega$$

Additional resistance required in the shunt field circuit is = 337.5−250 = **87.5 Ω**

Example 28.6. *A 220-V shunt motor has an armature resistance of 0.5 Ω and takes a current of 40 A on full-load. By how much must the main flux be reduced to raise the speed by 50% if the developed torque is constant.*

(Elect. Machines, AMIE, Sec B, 1991)

Solution. Formula used is $\frac{N_2}{N_1} = \frac{E_{b2}}{E_{b1}} \times \frac{\Phi_1}{\Phi_2}$. Since torque remains constant,

$$\text{Hence} \quad \Phi_1 I_{a1} = \Phi_2 I_{a2} \quad \therefore \; I_{a2} = I_{a1} \cdot \frac{\Phi_1}{\Phi_2} = 40\,x \text{ where } x = \frac{\Phi_1}{\Phi_2}$$

$$E_{b1} = 220 - (40 \times 0.5) = 200 \text{ V} \; ; E_{b2} = 220 - (40x \times 0.5) = (220 - 20\,x) \text{ V}$$

$$\frac{N_2}{N_1} = \frac{3}{2} \text{ ...given} \; \therefore \; \frac{3}{2} = \frac{(220 - 20x)}{200} \times x \quad \therefore \; x^2 - 11x + 15 = 0$$

$$\text{or } x = \frac{11 \pm \sqrt{121 - 60}}{2} = \frac{11 \pm 7.81}{2} = 9.4^* \text{ or } 1.6 \; \therefore \; \frac{\Phi_1}{\Phi_2} = 1.6 \text{ or } \frac{\Phi_2}{\Phi_1} = \frac{1}{1.6}$$

$$\therefore \quad \frac{\Phi_1 - \Phi_2}{\Phi_1} = \frac{1.6 - 1}{1.6} = \frac{3}{8} \; \therefore \text{ percentage change in flux } = \frac{3}{8} \times 100 = \mathbf{37.5\%}$$

* This figure is rejected as it does not give the necessary increase in speed.

Example 28.7. *A 220-V, 10-kW, 2500 rpm shunt motor draws 41 A when operating at rated conditions. The resistances of the armature, compensating winding, interpole winding and shunt field winding are respectively 0.2 Ω, 0.05 Ω, 0.1 Ω and 110 Ω. Calculate the steady-state values of armature current and motor speed if pole flux is reduced by 25%, a 1 Ω resistance is placed in series with the armature and the load torque is reduced by 50%.*

Solution. $I_{sh} = 220/110 = 2$ A; $I_{al} = 41 - 2 = 39$ A (Fig. 28.2)

$T_1 \propto \Phi_1 I_{a1}$ and $I_2 \propto \Phi_2 I_{a2}$

$\therefore \quad \frac{T_2}{T_1} = \frac{\Phi_2}{\Phi_1} \times \frac{I_{a2}}{I_{a1}}$ or $\frac{1}{2} = \frac{3}{4} \times \frac{I_{a2}}{39} \quad \therefore \quad I_{a2} = 26$ A

$E_{b1} = 220 - 39\,(0.2 + 0.1 + 0.05) = 206.35$ V

$E_{b2} = 220 - 26\,(1 + 0.35) = 184.9$ V

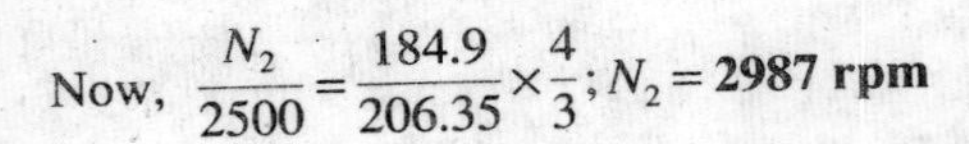

Now, $\frac{N_2}{2500} = \frac{184.9}{206.35} \times \frac{4}{3}; N_2 = \mathbf{2987\ rpm}$

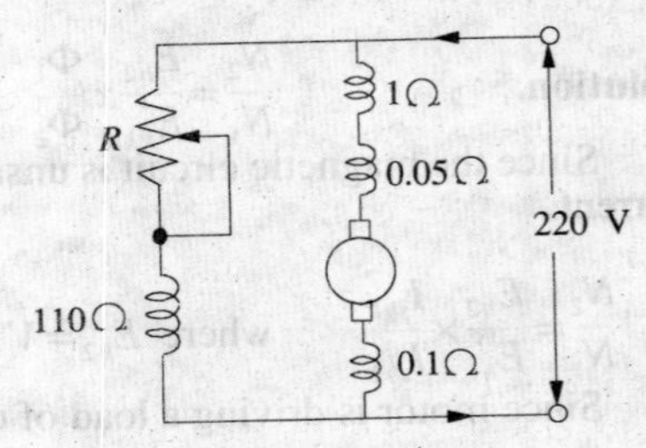

Fig. 28.2

Example 28.8. *A 220-V, 15-kW, 850 rpm shunt motor draws 72.2 A when operating at rated condition. The resistances of the armature and shunt field are 0.25 Ω and 100 Ω respectively. Determine the percentage reduction in field flux in order to obtain a speed of 1650 rpm when armature current drawn is 40 A.*

Solution. $I_{sh} = 220/100 = 2.2$ A; $I_{a1} = 72.2 - 2.2 = 70$ A

$E_{b1} = 220 - 70 \times 0.25 = 202.5$ V, $E_{b2} = 220 - 40 \times 0.25 = 210$ V.

Now $\frac{N_2}{N_1} = \frac{E_{b2}}{E_{b1}} \times \frac{\Phi_1}{\Phi_2}$ or $\frac{1650}{850} = \frac{210}{202.5} \times \frac{\Phi_1}{\Phi_2}$

$\therefore \quad \Phi_2 = 0.534\,\Phi_1; \quad \therefore$ reduction in field flux $= \frac{\Phi_1 - 0.534\,\Phi_1}{\Phi_1} \times 100 = \mathbf{46.6\%}$

Example 28.9. *A 220-V shunt motor has an armature resistance of 0.5 ohm and takes an armature current of 40 A on a certain load. By how much must the main flux be reduced to raise the speed by 50% if the developed torque is constant ? Neglect saturation and armature reaction.*

(Elect. Machines, A.M.I.E., Sec B, 1991)

Solution. $T_1 \propto \Phi_1 I_{a1}$, and $T_2 \propto \Phi_2 I_{a2}$ Since, $T_1 = T_2$

$\therefore \quad \Phi_2 I_{a2} = \Phi_1 I_{a2}$ or $\Phi_2/\Phi_1 = I_{a1}/I_{a1} = 40/I_{a2}$

$E_{b1} = 220 - 40 \times 0.5 = 200$ V; $E_{b2} = 220 - 0.5\,I_{a2}$

Now, $\frac{E_{b2}}{E_{b1}} = \frac{N_2 \Phi_2}{N_1 \Phi_1}$ or $\frac{220 - 0.5 I_{a2}}{200} = \frac{1.5 N_1}{N_1} \times \frac{40}{I_{a2}}$

$\therefore \quad I_{a2} - 40 I_{a2} + 24{,}000 = 0$ or $I_{a2} = 63.8$ A

$\therefore \quad \frac{\Phi_2}{\Phi_1} = \frac{40}{63.8} = 0.627$ or $\Phi_2 = 0.627\,\Phi_1 = 62.7\%$ of Φ_1

Tutorial Problem No. 28.1

1. A d.c. shunt motor runs at 900 r.p.m. from a 460-V supply when taking an armature current of 25 A. Calculate the speed at which it will run from a 230-V supply when taking an armature current of 15 A. The resistance of the armature circuit is 0.8 Ω. Assume the flux per pole at 230 V to have decreased to 75% of its value at 460 V. **[595 r.p.m.]**

2. A 250-V shunt motor has an armature resistance of 0.5 Ω and runs at 1200 r.p.m. when the armature current is 80 A. If the torque remains unchanged, find the speed and armature current when the field is strengthened by 25%. **[998 r.p.m. ; 64 A]**

3. When on normal full-load, a 500-V, d.c. shunt motor runs at 800 r.p.m. and takes an armature current of 42 A. The flux per pole is reduced to 75% of its normal value by suitably increasing the field circuit resistance. Calculate the speed of the motor if the total torque exerted on the armature is (*a*) unchanged (*b*) reduced by 20%. The armature resistance is 0.6 Ω and the total voltage loss at the brushes is 2 V.

[(*a*) 1,042 r.p.m. (*b*) 1,061 r.p.m.]

4. The following data apply to a d.c. shunt motor.

Supply voltage = 460 V ; armature current = 28 A; speed = 1000 r.p.m. ; armature resistance = 0.72 Ω. Calculate (*i*) the armature current and (*ii*) the speed when the flux per pole is increased to 120% of the initial value, given that the total torque developed by the armature is unchanged. **[(*i*) 23.33 A (*ii*) 840 r.p.m.]**

5. A 100-V shunt motor, with a field resistance of 50 Ω and armature resistance of 0.5 Ω runs at a speed of 1,000 r.p.m. and takes a current of 10 A from the supply. If the total resistance of the field circuit is reduced to three quarters of its original value, find the new speed and the current taken from the supply. Assume that flux is directly proportional to field current. **[1,089 r.p.m. ; 8.33 A]**

6. A 250-V d.c. shunt motor has armature circuit resistance of 0.5 Ω and a field circuit resistance of 125 Ω. It drives a load at 1000 r.p.m. and takes 30 A. The field circuit resistance is then slowly increased to 150 Ω. If the flux and field current can be assumed to be proportional and if the load torque remains constant, calculate the final speed and armature current. **[1186 r.p.m. 33.6 A]**

7. A 250-V, shunt motor with an armature resistance of 0.5 Ω and a shunt field resistance of 250 Ω drives a load the torque of which remains constant. The motor draws from the supply a line current of 21 A when the speed is 600 r.p.m. If the speed is to be raised to 800 r.p.m., what change must be affected in the shunt field resistance ? Assume that the magnetisation curve of the motor is a straight line. **[88 Ω]**

8. A 240-V, d.c. shunt motor runs at 800 r.p.m. with no extra resistance in the field or armature circuit, on no-load. Determine the resistance to be placed in series with the field so that the motor may run at 950 r.p.m. when taking an armature current of 20 A. Field resistance = 160 Ω. Armature resistance = 0.4 Ω. It may be assumed that flux per pole is proportional to field current. **[33.6 Ω]**

9. A shunt-wound motor has a field resistance of 400 Ω and an armature resistance of 0.1 Ω and runs off 240-V supply. The armature current is 60 A and the motor speed is 900 r.p.m. Assuming a straight line magnetisation curve, calculate (*a*) the additional resistance in the field to increase the speed to 1000 r.p.m. for the same armature current and (*b*) the speed with the original field current of 200 A. **[(*a*) 44.4 Ω (*b*) 842.5 r.p.m.]**

10. A 230-V d.c. shunt motor has an armature resistance of 0.5 Ω and a field resistance of $76\frac{2}{3}$Ω. The motor draws a line current of 13 A while running light at 1000 r.p.m. At a certain load, the field circuit resistance is increased by $38\frac{1}{3}$ Ω. What is the new speed of the motor if line current at this load is 42 A.**[1400 r.p.m.]**

(*Electrical Engineering ; Grad. I.E.T.E. Dec. 1986*)

(*ii*) Armature or Rheostatic Control Method

This method is used when speeds below the no-load speed are required. As the supply voltage is normally constant, the voltage across the armature is varied by inserting a variable rheostat or resistance (called controller resistance) in series with the armature circuit as shown in Fig. 28.3 (*a*). As controller resistance is increased, p.d. across the armature is decreased, thereby decreasing the armature speed. For a load of constant torque, speed is approximately proportional to the p.d. across the armature. From the speed/armature current characteristic [Fig. 28.3 (*b*)], it is seen that greater the resistance in the armature circuit, greater is the fall in speed.

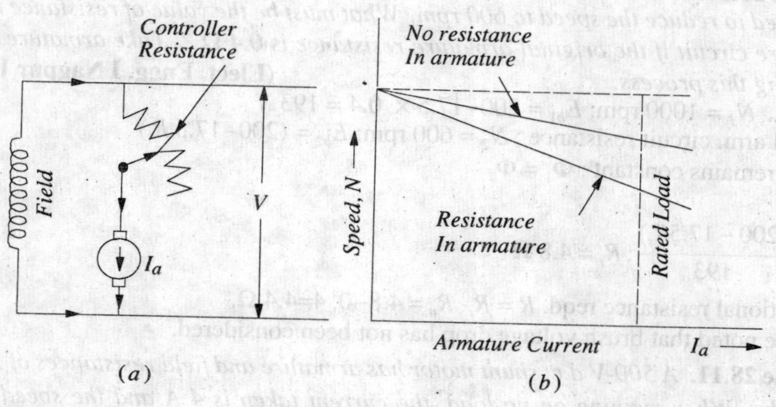

Fig. 28.3

Let I_{a1} = armature current in the first case

I_{a2} = armature current in the second case

(If $I_{a1} = I_{a2}$, then the load is of constant torque)

N_1, N_2 = corresponding speeds, V = supply voltage

Then $N_1 \propto V - I_{a1}R_a \propto E_{b1}$

Let some controller resistance of value R be added to the armature circuit resistance so that its value becomes $(R + R_a) = R_t$

Then $N_2 \propto V - I_{a2} Rt \propto E_{b2}$ $\therefore \dfrac{N_2}{N_1} = \dfrac{E_{b2}}{E_{b1}}$

(In fact, it is a simplified form of relation given in Art. 27.9 because here $\Phi_1 = \Phi_2$)

Considering no-load speed, we have $\dfrac{N}{N_0} = \dfrac{V - I_a R_t}{V - I_{ao} R_a}$

Neglecting $I_{a0}R_a$ with respect to V, we get

$$N = N_0\left(1 - \frac{I_a R_t}{V}\right)$$

It is seen that for a given resistance R_t the speed is a linear function of armature current I_a as shown in Fig. 28.4.

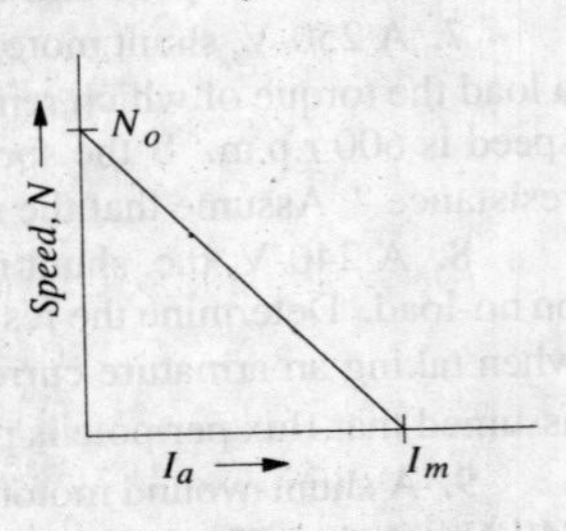

Fig. 28.4

The load current for which the speed would be zero is found by putting $N = 0$ in the above relation.

$$\therefore\ 0 = N_0\left(1 - \frac{I_a R_t}{V}\right) \text{ or } I_a = \frac{V}{R_t}$$

This is the maximum current and is known as *stalling* current.

As will be shown in Art. 28.4, this method is very wasteful, expensive and unsuitable for rapidly changing loads because for a given value of R_t, speed will change with load. A more stable operation can be obtained by using a divertor across the armature in addition to armature control resistance (Fig. 28.5). Now, the changes in armature current (due to changes in the load torque) will not be so effective in changing the p.d. across the armature (and hence the armature speed).

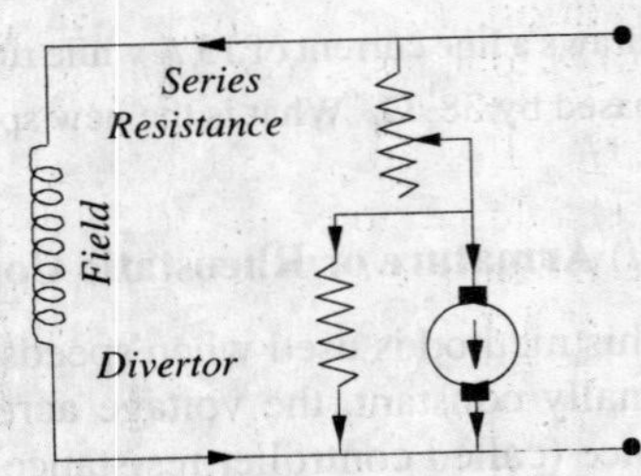

Fig. 28.5

Example 28.10. *A 200 d.c. shunt motor running at 1000 rpm takes an armature current of 17.5 A. It is required to reduce the speed to 600 rpm. What must be the value of resistance to be inserted in the armature circuit if the original armature resistance is 0.4 Ω ? Take armature current to be constant during this process.* **(Elect. Engg. I Nagpur Univ. 1993)**

Solution. $N_1 = 1000$ rpm; $E_{b1} = 200 - 17.5 \times 0.4 = 193$ V

R_t = total arm. circuit resistance ; $N_2 = 600$ rpm; $E_{b2} = (200 - 17.5\, R_t)$

Since I_{sh} remains constant ; $\Phi_1 = \Phi_2$

$$\therefore\ \frac{600}{1000} = \frac{(200 - 17.5R_t)}{193};\ R_t = 4.8\ \Omega$$

$\therefore$ additional resistance reqd. $R = R_t - R_a = 4.8 - 0.4 = 4.4\ \Omega$.

It may be noted that brush voltage drop has not been considered.

Example 28.11. *A 500-V d.c. shunt motor has armature and field resistances of 1.2 Ω and 500 Ω respectively. When running on no-load, the current taken is 4 A and the speed is 1000 rpm. Calculate the speed when motor is fully loaded and the total current drawn from the supply is 26 A.*

Estimate the speed at this load if (a) a resistance of 2.3 Ω is connected in series with the armature and (b) the shunt field current is reduced by 15%.

(Electrical Engineering-I, Sd. Patel Univ. 1985)

Solution. $I_{sh} = 500/500 = 1$ A ; $I_{a1} = 4-1 = 3$ A

$E_{b1} = 500-(3 \times 1.2) = 496.4$ V $\quad I_{a2} = 26-1 = 25$ A

$E_{b2} = 500-(25 \times 1.2) = 470$ V $\therefore \dfrac{N_2}{1000} = \dfrac{470}{496.4}$; $N_2 =$ **947 r.p.m.**

(*a*) In this case, total armature circuit resistance = 1.2 + 2.3 = 3.5 Ω

$\therefore E_{b2} = 500-(25 \times 3.5) = 412.5$ V $\therefore \dfrac{N_2}{1000} = \dfrac{412.5}{496.4}$; $N_2 =$ **831 r.p.m.**

(*b*) When shunt field is reduced by 15%, $\Phi_2 = 0.85\ \Phi_1$ assuming straight magnetisation curve.

$$\frac{N_2}{1000} = \frac{412.5}{496.4} \times \frac{1}{0.85};\ N_2 = \mathbf{977.6\ r.p.m.}$$

Example 28.12. *A 250-V shunt motor (Fig. 28.6) has an armature current of 20 A when running at 1000 rpm against full load torque. The armature resistance is 0.5 Ω. What resistance must be inserted in series with the armature to reduce the speed to 500 rpm at the same torque and what will be the speed if the load torque is halved with this resistance in the circuit ? Assume the flux to remain constant throughout and neglect brush contact drop.*

(Elect. Machines AMIE Sec. B Summar 1991)

Solution. $E_{b1} = V - I_{a1}R_a = 250-20 \times 0.5 = 240$ V

Let R_t to be total resistance in the armature circuit *i.e.* $R_t = R_a + R$ where R is the additional resistance.

$\therefore E_{b2} = V - I_{a2}R_t = 250-20\ R_t$

It should be noted that $I_{a1} = I_{a2} = 20$ A because torque remains the same and $\Phi_1 = \Phi_2$ in both cases.

$$\therefore \frac{N_2}{N_1} = \frac{E_{b2}}{E_{b1}} \times \frac{\Phi_1}{\Phi_2} = \frac{E_{b2}}{E_{b1}} \text{ or } \frac{500}{1000} = \frac{250-20\,R_t}{240}$$

$\therefore R_t = 6.5$ Ω ; hence, R = 6.5–0.5 = 6 Ω

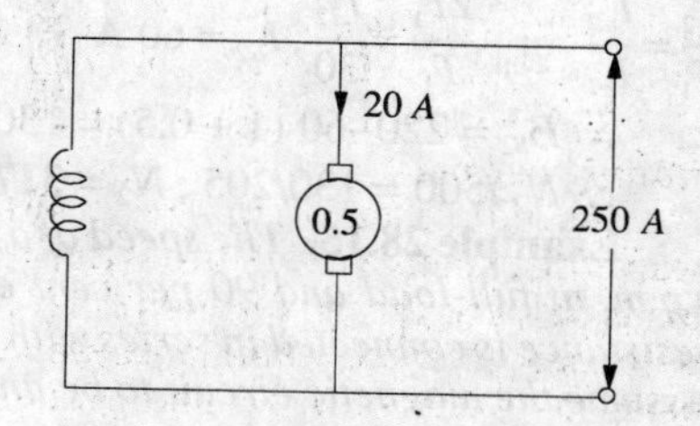

Fig. 28.6

Since the load is halved, armature current is also halved because flux remains constant. Hence, $I_{a3} = 10$ A.

$$\therefore \frac{N_3}{1000} = \frac{250-10 \times 6.5}{240} \text{ or } N_3 = 771 \text{ rpm}$$

Example 28.13. *A 250-V shunt motor with armature resistance of 0.5 ohm runs at 600 r.p.m. on full-load and takes an armature current of 20 A. If resistance of 1.0 ohm is placed in the armature circuit, find the speed at (i) full-load torque (ii) half full-load torque.*

(Electrical Machines-II, Punjab Univ. May 1991)

Solution. Since flux remains constant, the speed formula becomes $\dfrac{N_2}{N_1} = \dfrac{E_{b2}}{E_{b1}}$

(*i*) In the first, case, full-load torque is developed.

$N_1 = 600$ rpm; $E_{b1} = V - I_{a1}R_{a1} = 250-20 \times 0.5 = 240$ V

Now, $T \propto \Phi I_a \propto I_a$ ($\because$ Φ is constant)

$$\therefore \frac{T_2}{T_1} = \frac{I_{a2}}{I_{a1}} \quad \text{Since } T_2 = T_1;\ I_{a2} = I_{a1} = 20 \text{ A}$$

$$E_{b2} = V - I_{a2}R_{a2} = 250 - 20 \times 1.5 = 220 \text{ V}, N_2 = ;\ \frac{N_2}{600} = \frac{220}{240};\ N_2 = \frac{600 \times 220}{240} = \mathbf{550\ rpm}$$

(*ii*) In this case, the torque developed is half the full-load torque

$$\frac{T_2}{T_1}=\frac{I_{a2}}{I_{a1}} \text{ or } \frac{T_1/2}{T_1}=\frac{I_{a2}}{20}; I_{a2}=10\text{ A}; E_{b2}=250-10\times 1.5=235\text{ V}$$

$$\therefore \frac{N_2}{600}=\frac{235}{240}; \; N_2=600\times 235/240=\mathbf{587.5\ rpm}$$

Example 28.14. *A 220 V shunt motor with an armature resistance of 0.5 ohm is excited to give constant main field. At full load the motor runs at 500 rev. per minute and takes an armature current of 30 A. If a resistance of 1.0 ohm is placed in the armature circuit, find the speed at (a) full load torque (b) double-full load torque.* **(Elect. Machines-I, Nagpur Univ. 1993)**

Solution. Since flux remains constant, the speed formula becomes $N_2/N_1 = E_{b2}/E_{b1}$.

(a) Full – load Torque

With no additional resistance in the armature circuit,

$N_1 = 500$ rpm; $I_{a1} = 30$ A; $E_{b1} = 220–30 \times 0.5 = 205$ V

Now, $T \propto I_a$ (since Φ is constant) $\therefore \dfrac{T_2}{T_1}=\dfrac{I_{a2}}{I_{a1}}$ Since $T_2 = T_1$; $I_{a2} = I_{a1} = 30$ A

When additional resistance of 1 Ω is introduced in the armature circuit,

$E_{b2} = 220–30\ (1 + 0.5) = 175$ V; $N_2 = ?$ $\therefore N_2/500 = 175/205$; $N_2 = 427$ rpm

(b) Double Full-load Torque

$$\frac{T_2}{T_1}=\frac{I_{a2}}{I_{a1}} \text{ or } \frac{2T_1}{T_1}=\frac{I_{a2}}{30}; I_{a2}=60\text{ A}$$

$\therefore E_{b2} = 220–60\ (1 + 0.5) = 130$ V

$\therefore N_2/500 = 130/205$; $N_2 =$ **317 r.p.m**

Example 28.15. *The speed of a 50 h.p (37.3 kW) series motor working on 500-V supply is 750 r.p.m. at full-load and 90 per cent efficiency. If the load torque is made 350 N-m and a 5 ohm resistance is connected in series with the machine, calculate the speed at which the machine will run. Assume the magnetic circuit to be unsaturated and the armature and field resistance to be 0.5 ohm.* **(Electrical Machinery I, Madras Univ. 1986)**

Solution. Load torque in the first case is given by

$T_1 = 37{,}300/2\pi(750/60) = 474.6$ N-m

Input current, $I_{a1} = 37{,}300/0.9 \times 500 = 82.9$ A

Now, $T_2 = 250$ N-m ; $I_{a2} = ?$

In a series motor, before magnetic saturation,

$$T \propto \Phi I_a \propto I_{a2} \quad \therefore \quad T_1 \propto I_{a1}^{\;2} \text{ and } T_2 \propto I_{a2}^{\;2}$$

$$\therefore \left(\frac{I_{a2}}{I_{a1}}\right)^2=\frac{T_2}{T_1} \quad \therefore \; I_{a2}=82.9\times\sqrt{250/474.6}=60.2\text{ A}$$

Now, $E_{b1}=500-(82.9\times 0.5)=458.5\text{ V}; E_{b2}=500-60.2(5+0.5)=168.9\text{ V}$

Using, $\dfrac{N_2}{N_1}=\dfrac{E_{b2}}{E_{b1}}\times\dfrac{I_{a1}}{I_{a2}}$, we get $\dfrac{N_2}{750}=\dfrac{168.9}{458.5}\times\dfrac{82.9}{60.2}$ $\therefore N_2=$ **381 r.p.m.**

Example 28.16. *A 7.46 kW, 220-V, 900 r.p.m. shunt motor has a full-load efficiency of 88 per cent, an armature resistance of 0.08 Ω and shunt field current of 2 A. If the speed of this motor is reduced to 450 r.p.m. by inserting a resistance in the armature circuit, the load torque remaining constant, find the motor output, the armature current, the external resistance and the overall efficiency.* **(Elect. Machines, Nagpur Univ. 1993)**

Solution. Full-load motor input current $I = 7460/220 \times 0.88 = 38.5$ A

$\therefore$ $I_{a1} = 38.5 - 2 = 36.5$ A

Now, $T \propto \Phi I_a$. Since flux remains constant

$\therefore$ $T \propto I_a \therefore T_{a1} \propto I_{a1}$ and $T_{a2} \propto I_{a2}$ or $T_{a2}/T_{a1} = I_{a2}/I_{a1}$

It is given that $T_{a1} = T_{a2}$; hence $I_{a1} = I_{a2} = \mathbf{36.5\ A}$

$E_{b1} = 220 - (36.5 \times 0.08) = 217.1\ \Omega$

$E_{b2} = 220 - 36.5\ R_t$; $N_1 = 900$ r.p.m.; $N_2 = 450$ r.p.m.

Now, $$\frac{N_2}{N_1} = \frac{E_{b2}}{E_{b1}} \times \frac{\Phi_1}{\Phi_2} = \frac{E_{b2}}{E_{b1}} \quad (\because \ \Phi_1 = \Phi_2)$$

$$\therefore \quad \frac{450}{900} = \frac{220 - 36.5\,R_t}{217.1}; R_t = 3.05\ \Omega$$

$\therefore$ external resistance $R = 3.05 - 0.08 = \mathbf{2.97\ \Omega}$

For calculating the motor output, it will be assumed that all losses except copper losses vary directly with speed.

Since motor speed is halved, stray losses are also halved in the second case. Let us find their value.

In the first case, motor input = 200×38.5 = 8,470 W; Motor output = 7,460 W

Total Cu losses + stray losses = 8470 − 7460 = 1010 W

Arm. Cu loss = $I_{a1}^2 R_a = 36.5^2 \times 0.08 = 107$ W; Field Cu loss = $220 \times 2 = 440$ W

Total Cu loss = 107 + 440 = 547 W $\therefore$ Stray losses in first case = 1010−547 = 463 W

Stray losses in the second case = $463 \times 450/900 = 231$ W

Field Cu loss = 440 W − as before; Arm. Cu loss = $36.5^2 \times 3.05 = 4{,}064$ W

Total losses in the 2nd case = 231 + 440 + 4,064 = 4,735 W

Input = 8,470 W − as before

Output in the second case = 8,470−4,735 = **3,735 W**

$\therefore$ overall η = 3,735/8,470 = 0.441 or **44.1 per cent***

Example 28.17. *A 240 V shunt motor has an armature current of 15A when running at 800 rpm against F.L. torque. The arm. resistance is 0.6 ohms. What resistance must be inserted in series with the armature to reduce the speed to 400 rpm, at the same torque ?*

What will be the speed if the load torque is halved with this resistance in the circuit ? Assume the flux to remain constant throughout. **(Elect. Machines-I Nagpur Univ. 1993)**

Solution. Here, $N_1 = 800$ rpm; $E_{b1} = 240 - 15 \times 0.6 = 231$ V

Flux remaining constant, $T \propto I_a$. Since torque is the same in both cases, $I_{a2} = I_{a1} = 15$ A. Let R be the additional resistance inserted in series with the armature. $E_{b2} = 240 - 15\,(R + 0.6)$; $N_2 = 400$ rpm

$$\therefore \quad \frac{400}{800} = \frac{240 - 15(R + 0.6)}{231}; R = 7.7\ \Omega$$

When load torque is halved

With constant flux when load torque is halved, I_a is also halved. Hence, $I_{a3} = I_{a1}/2 = 15/2 = 7.5$ A

$\therefore$ $E_{b3} = 240 - 7.5\,(7.7 + 0.6) = 177.75$ V ; $N_3 = ?$

$N_3/N_1 = E_{b3}/E_{b1}$ or $N_3/800 = 177.75/231$; $N_3 = 614.7$ rpm

Example. 28.18. *(a) A 400 V shunt connected d.c. motor takes a total current of 3.5 A on no load and 59.5 A at full load. The field circuit resistance is 267 ohms and the armature circuit resistance is 0.2 ohms (excluding brushes where the drop may be taken as 2V). If the armature reaction effect at 'full load' weakens the flux per pole by 2 percentage change in speed from no-load to full load.*

(b) What resistant must be placed in series with the armature in the machine of (a) if the full-load speed is to be reduced by 50 per cent with the gross torque remaining constant. Assume no change in the flux. **(Electrical Machines, A.M.I.E. Sec. B, 1989)**

*It may be noted that efficiency is reduced almost in the ratio of the two speeds

Solution. (*a*) Shunt current $I_{sh} = 400/267 = 1.5$ A. At no load, $I_{a1} = 3.5 - 1.5 = 2$ A, $E_{b1} = V - I_{a1}$ R_{ai} – brush drop = 400 – 2 × 0.2 – 2 = 397.6 V. On full load, $I_{a2} = 59.5 - 1.5 = 58$ A, $E_{b2} = 400 -$ 58 × 0.2 – 2 = 386.4 V.

$$\therefore \frac{E_{b1}}{E_{b2}} = \frac{\Phi_1 N_1}{\Phi_2 N_2} \text{ or } \frac{397.6}{386.4} = \frac{\Phi_1 N_1}{0.98\Phi N_2}; \frac{N_1}{N_2} = 1.084$$

$$\% \text{ change in speed } = \frac{N_1 - N_2}{N_1} \times 100$$

$$\left(\frac{1 - N_2}{N_1}\right) \times 100 = \left(1 - \frac{1}{1.0084}\right) \times 100 = 0.833$$

(*b*) Since torque remains the same, I_a remains the same, hence $I_{a3} = I_{a2}$. Let R be the resistance connected in series with the armature.

$E_{b3} = V - I_{a2}\ (R_a + R)$ – brush drop = 400 – 58 (0.2 + R) – 2 = 386.4 – 58 R

$$\therefore \frac{E_{b2}}{E_{b3}} = \frac{\Phi_2 N_2}{\Phi_3 N_3} = \frac{N_2}{N_3} \quad \therefore \quad \Phi_2 = \Phi_3$$

$$\frac{386.4}{386.4 - 58R} = \frac{1}{0.5}; \quad R = \mathbf{3.338\Omega}$$

Example 28.19. *A d.c. shunt motor drives a centrifugal pump whose torque varies as the square of the speed. The motor is fed from a 200 V supply and takes 50 A when running at 1000 r.p.m. What resistance must be inserted in the armature circuit in order to reduce the speed to 800 r.p.m. ? The armature and field resistances of the motor are 0.1 Ω and 100 Ω respectively.*

(Elect. Machines, Allahabad Univ. 1992)

Solution. In general, $T \propto \Phi I_a$

For shunt motors whose excitation is constant,

$T \propto I_a \propto N^2$ ... given

$\therefore$ $I_a \propto N^2$. Now $I_{sh} = 200/100 = 2$ A $\therefore$ $I_{al} = 50 - 2 = 48$ A

Let I_{a2} = new armature current at 800 r.p.m.

then $48 \propto N_1^2 \propto 1000^2$ and $I_{a2} \propto N_2^2 \propto 800^2$

$$\therefore \frac{I_{a2}}{48} = \left(\frac{800}{1000}\right)^2 = 0.8^2 \therefore I_{a2} = 48 \times 0.64 = 30.72 \text{ A}$$

$$\therefore E_{b1} = 200 - (48 \times 0.1) = 195.2V; E_{12} = (200 - 30.72\, Rt)\text{ V}$$

Now $$\frac{N_2}{N_1} = \frac{E_{b2}}{E_{b1}} \therefore \frac{800}{1000} = \frac{200 - 30.72 R_t}{195.2}, Rt = 14.2\ \Omega$$

Additional resistance = 1.42 – 0.2 = **1.32 Ω**

Example 28.20. *A 250-V, 50 h.p. (37.3 kW) d.c. shunt motor has an efficiency of 90% when running at 1,000 r.p.m. on full-load. The armature and field resistances are 0.1 Ω and 115 Ω respectively. Find*

(a) the net and developed torque on full-load.

(b) the starting resistance to have the line start current equal to 1.5 times the full-load current.

(c) the torque developed at starting. **(Elect. Machinery-I, Kerala Univ. 1987)**

Solution. (*a*) $T_{sh} = 9.55 \times 37.300/1000 = \mathbf{356.2\ N\text{-}m}$

Input current $= \dfrac{37,300}{250 \times 0.9} = 165.8A\ ;\ I_{sh} = \dfrac{250}{125} = 2\ \text{A}$

$\therefore \quad I_a = 165.8 - 2 = 163.8\ \text{A}; E_b = 250 - (163.8 \times 0.1) = 233.6\ \text{V}$

$\therefore \quad T_a = 9.55 \dfrac{233.6 \times 163.8}{1000} = \mathbf{365.4\ N\text{-}m}$

(*b*) F.L. input line I = 165.8 A ; Permissible input I = 165.8 × 1.5 = 248.7 A
Permissible armature current= 248.7−2 = 246.7 A
Total armature resistance = 250/246.7 = 1.014 Ω
∴ starting resistance required = 1.014 −0.1 = **0.914 Ω**
(*c*) Torque developed with 1.5 times the F.. current would be practically 1.5 times the F.L. torque *i.e.* 1.5 × 365.4 = **548.1 N-m**

Example 28.21. *A 200-V shunt motor with a shunt resistance of 40 Ω and armature resistance of 0.02 Ω takes a current of 55 A and runs at 595 r.p.m. when there is a resistance of 0.58 Ω in series with armature. Torque remaining the same, what change should be made in the armature circuit resistance to raise the speed to 630 r.p.m. ? Also find*

(*i*) *AT what speed will the motor run if the load torque is reduced such that armature current is 15 A.*
(*ii*) *Now, suppose that a divertor of resistance 5 Ω is connected across the armature and series resistance is so adjusted that motor speed is again 595 r.p.m., when armature current is 50 A. What is the value of this series resistance ? Also, find the speed when motor current falls of 15 A again.*

Solution. The circuit is shown in Fig. 28.7

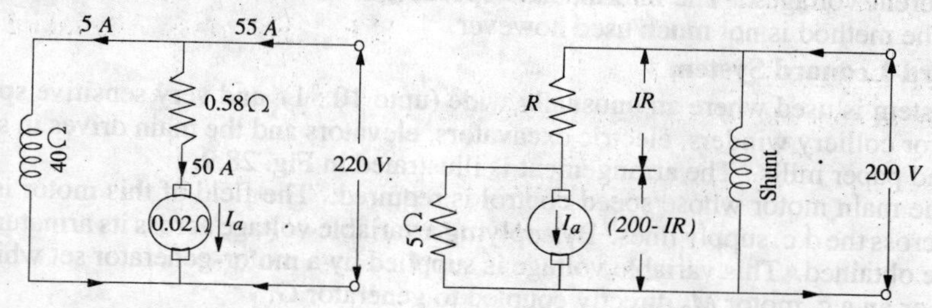

Fig. 28.7. **Fig. 28.8.**

$I_{sh} = 200/40 = 5\ \text{A} \quad \therefore I_{a1} = 55 - 5 = 50\ \text{A}$
Armature circuit resistance = 0.58 + 0.02 = 0.6 Ω
$\therefore\ E_{b1} = 200 - (50 \times 0.6) = 170\ \text{V}$
Since torque is the same in both cases, $\therefore\ I_{a1}\ \Phi_1 = I_{a2}\ \Phi_2$
Moreover, $\Phi_1 = \Phi_2 \quad \therefore\ I_{a1} = I_{a2} \quad \therefore\ I_{a2} = 50\ \text{A}$
Now $E_{b1} = 170\ \text{V}$, $N_1 = 595$ r.p.m. $N_2 = 630$ r.p.m. $E_{b2} = ?$

Using $\dfrac{N_2}{N_1} = \dfrac{E_{b2}}{E_{b1}},\ (\because\ \Phi_1 = \Phi_2)$ we get $E_{b2} = 170 \times (630/595) = 180\ \text{V}$

Let R_2 be the new value of armature circuit resistance for raising the speed from 595 r.p.m. to 630 r.p.m.

$\therefore \quad 180 = 200 - 50\,R_2 \quad \therefore\ R_2 = 0.4\ \Omega$

Hence, armature circuit resistance should be reduced by 0.6−0.4 = **0.2 Ω**

(*i*) We have seen above that

$I_{a1} = 50$ A, $E_{b1} = 170$ V, $N_1 = 595$ r.p.m.

If $I_{a2} = 15$ A, $E_{b2} = 200-(15 \times 0.6) = 191$ V

$\therefore \quad \dfrac{N_2}{595} = \dfrac{191}{170}$ $\quad \therefore$ N_2 = **668.5 r.p.m.**

(*ii*) When armature divertor is used (Fig. 28.8)

Let R be the new value of series resistance

$\therefore \quad E_{b3} = 200-IR-(50 \times 0.2) = 199-IR$

Since speed is 595 r.p.m., E_{bs} must be equal to 170 V

$\therefore \quad 170 = 199-IR \quad \therefore \; IR = 29$ V ; P.D. across divertor $= 200-29 = 171$ V

Current through divertor $I_d = 171/5 = 34.2$ A $\therefore \; I = 50 + 34.2 = 84.2$ A

As $IR = 29$ V $\quad \therefore \; R = 29/84.2 =$ **0.344 Ω**

When $I_a = 15$ A, then $\quad I_d = (I-15)$ A

P.D. across divertor $= 5(I-15) = 200 - 0.344\,I \;\therefore\; I = 51.46$ A

$E_{b4} = 200-0.344\,I-(15 \times 0.02) = 200 - (0.344 \times 51.46) - 0.3 = 182$ V

$\therefore \quad \dfrac{N_4}{N_1} = \dfrac{E_{b4}}{E_{b1}}$ or $\dfrac{N_4}{595} = \dfrac{182}{170}$ $\quad \therefore \; N_4$ = **637 r.p.m.**

The effect of armature divertor is obvious. The speed without divertor is 668.5 r.p.m. and with armature divertor, it is 637 r.p.m.

(*iii*) **Voltage Control Method**

(*a*) **Multiple Voltage Control**

In this method, the shunt field of the motor is connected permanently to a fixed exciting voltage, but the armature is supplied with different voltages by connecting it across one of the several different voltages by means of suitable switchgear. The armature speed will be approximately proportional to these different voltages. The intermediate speeds can be obtained by adjusting the shunt field regulator. The method is not much used however.

(*b*) **Ward-Leonard System**

This system is used where an unusually wide (upto 10 : 1) and very sensitive speed control is required as for colliery winders, electric excavators, elevators and the main drives in steel mills and blooming and paper mills. The arrangement is illustrated in Fig. 28.9.

M_1 is the main motor whose speed control is required. The field of this motor is permanently connected across the d.c. supply lines. By applying a variable voltage across its armature, any desired speed can be obtained. This variable voltage is supplied by a motor-generator set which consists of either a d.c. or an a.c. motor M_2 directly coupled to generator G.

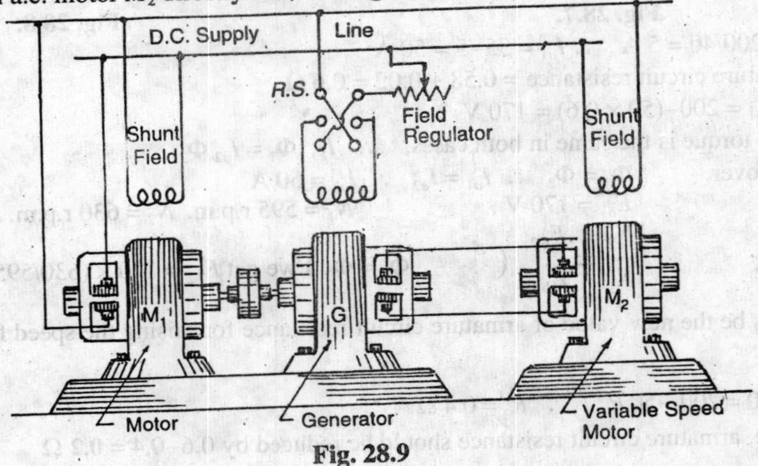

Fig. 28.9

The motor M_2 runs at an approximately constant speed. The output voltage of G is directly fed to the main motor M_1. The voltage of the generator can be varied from zero up to its maximum value by means of its field regulator. By reversing the direction of the field current of G by means of the reversing switch RS, generated voltage can be reversed and hence the direction of rotation of M_1. It should be remembered that motor generator set always runs in the same direction.

Despite the fact that capital outlay involved in this system is high because (*i*) a large output machine must be used for the motor generator set and (*ii*) that two extra machines are employed, still it is used extensively for elevators, hoist control and for main drive in steel mills where motor of ratings 750 kW to 3750 kW are required. The reason for this is that the almost unlimited speed control in either direction of rotation can be achieved entirely by field control of the generator and the resultant economies in steel production outweigh the extra expenditure on the motor generator set.

A modification of the Ward-Lenoard system is known as Ward-Leonard-Ilgner gner system which uses a smaller motor-generator set with the addition of a flywheel whose function is to reduce fluctuations in the power demand from the supply circuit. When main motor M_1 becomes suddenly overloaded, the driving motor M_2 of the motor generator set slows down, thus allowing the inertia of the flywheel to supply a part of the overload. However, when the load is suddenly thrown off the main motor M_1, then M_2 speds up thereby again storing energy in the flywheel.

When the Ilgner system is driven by means of an a.c. motor (whether induction or synchronous) another refinement in the form of a 'slip regulator' can be usefully employed thus giving an additional control.

The chief disadvantage of this system is its low overall efficiency especially at light loads. But as said earlier, it has the outstanding merit of giving wide speed control from maximum in one direction through zero to the maximum in the opposite direction and of giving a smooth acceleration.

Example 28.22. *The O.C.C. of the generator of a Ward-Leonard set is*

Field amps :	*1.4*	*2.2*	*3*	*4*	*5*	*6*	*7*	*8*
Armature volts :	*212*	*320*	*397*	*472*	*522*	*560*	*586*	*609*

The generator is connected to a shunt motor, the field of which is separately-excited at 550 V. If the speed of motor is 300 r.p.m. at 550 V, when giving 485 kW at 95.5% efficiency, determine the excitation of the generator to give a speed of 180 r.p.m. at the same torque. Resistance of the motor armature circuit = 0.01 Ω, resistance of the motor field = 60 Ω, resistance of generator armature circuit = 0.01 Ω. Ignore the effects of armature reaction and variation of the core factor and the windage losses of the motor.

Solution. Motor input $= 485 \times 10^3/0.955 = 509{,}300$ W

Motor to motor field $= 550/60 = 55/6$ A

Input to motor field $= 550 \times 55/6 = 5{,}040$ W

$\therefore$ motor armature input $= 509{,}300 - 5{,}040 = 504{,}260$ W

$\therefore$ armature current $= 504{,}260/550 = 917$ A

Back e.m.f. E_{b1} at 300 r.p.m. $= 550 - (917 \times 0.01) = 540.83$ V

Back e.m.f. E_{b2} at 180 r.p.m. $= 540.83 \times 180/300 = 324.5$ V

Since torque is the same, the armature current of the main motor is also the same *i.e.* 917 A because its excitation is independent of its speed.

$\therefore \quad V = 324.5 + (917 \times 0.01) = 333.67$ V

Generated e.m.f. $= V + I_a R_a$ ...for generator

$= 333.67 + (917 \times 0.011) = 343.77$ V

If O.C.C. is plotted from the above given data, then it would be found that the excitation required to give 343.77 V is 2.42 A.

∴ Generator exciting current = **2.42 A**

28.3. Speed Control of Series Motors

1. Flux Control Method

Variations in the flux of a series motor can be brought about in any one of the following ways :

(a) Field Divertors

The series windings are shunted by a variable resistance known as field divertor (Fig. 28.10). Any desired amount of current can be passed through the divertor by adjusting its resistance. Hence the flux can be decreased and, consequently, the speed of the motor increased.

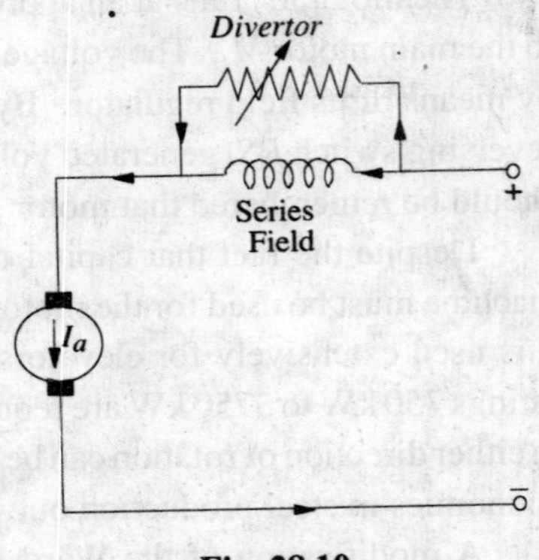

Fig. 28.10

(b) Armature Divertor

A divertor across the armature can be used for giving speeds lower than the normal speed (Fig. 28.11). For a *given constant load torque,* if I_a is reduced due to armature divertor, then Φ must increase

($\because$ $T_a \propto \Phi\, I_a$). This results in an increase in current taken from the supply (which increases the flux and a fall in speed ($N \propto I/\Phi$). The variation in speed can be controlled by varying the divertor resistance.

(c) Tapped Field Control

This method is often used in electric traction and is shown in Fig. 28.12.

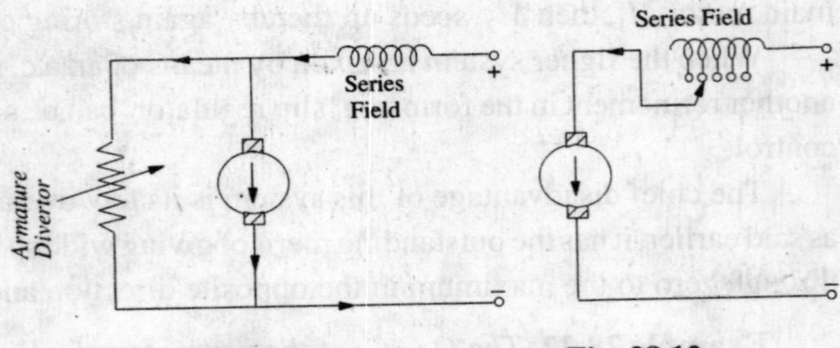

Fig. 28.11 Fig. 28.12

The number of series field turns in the circuit can be changed at will as shown. With full field, the motor runs at its minimum speed which can be raised in steps by cutting out some of the series turns.

(d) Paralleling Field coils

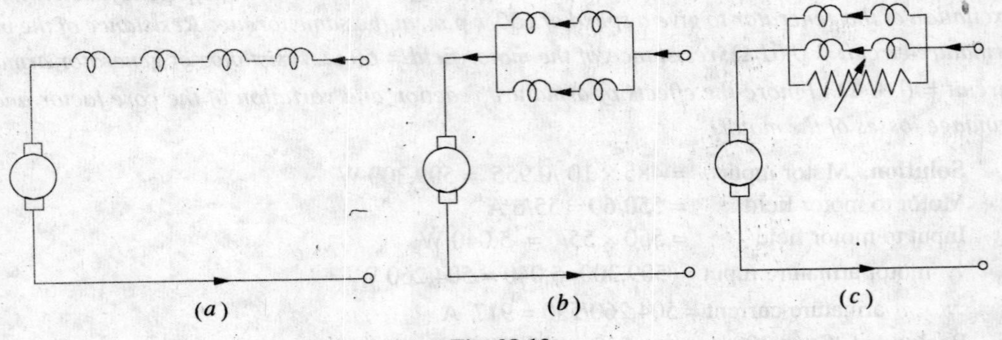

Fig. 28.13

In this method, used for fan motors, several speeds can be obtained by regrouping the field coils as shown in Fig. 28.13. It is seen that for a 4-pole motor, three speeds can be obtained easily. (Ex. 28.27).

2. Variable Resistance in Series with Motor

By increasing the resistance in series with the armature (Fig. 28.14) the voltage applied across the armature terminals can be decreased.

With reduced voltage across the armature, the speed is reduced. However, it will be noted that since full motor current passes through this resistance, there is a considerable loss of power in it.

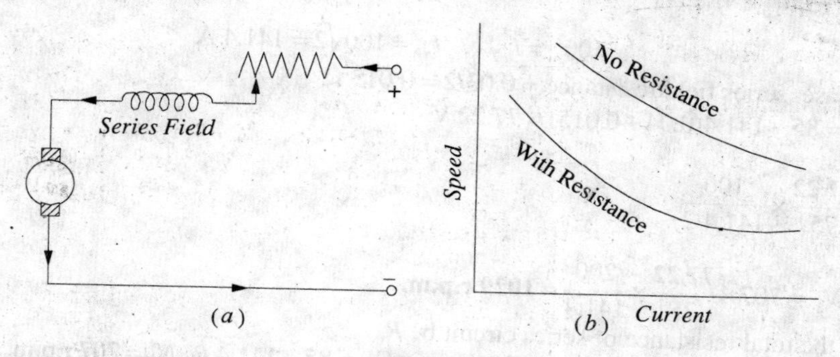

Fig. 28.14

Example 28.23. *A d.c. series motor drives a load the torque of which varies as the square of the speed. The motor takes a current of 15 A when the speed is 600 r.p.m. Calculate the speed and the current when the motor field winding is shunted by a diverter of the same resistance as that of the field winding. Mention the assumptions made, if any.*

(Elect. Mechines, A.M.I.E. Sec B, 1993)

Solution.

$$T_{a1} \propto N_1^2, T_{a2} \propto N_1^2 \therefore T_{a2}/T_{a1} = N_2^2/N_1^2$$

Also, $$T_{a1} \propto \Phi_1 I_{a1} \propto I_{a1}^2, T_{a2} \propto \Phi_2 I_{a2} \propto (I_{a2}/2) I_{a2} \propto I_{a2}^2/2$$

It is so because in the second case, field current is half the armature current

$$\therefore \quad N_2^2 \frac{2}{N_1^2} = \frac{I_{a2}^2/2}{I_{a1}^2} \quad \text{or} \quad \frac{N_2}{N_1} = \frac{I_{a2}}{\sqrt{2} I_{a1}} \quad \ldots(1)$$

Now, $$\frac{N_2}{N_1} = \frac{E_{b2}}{E_{b1}} \times \frac{\Phi_1}{\Phi_2}$$

If we neglect the armature and series winding drops as well as brush drop, then $E_{b1} = E_{b2} = V$

$$\therefore \quad \frac{N_2}{N_1} = \frac{\Phi_1}{\Phi_2} = \frac{I_{a1}}{I_{a2}/\sqrt{2}} = \frac{2I_{a1}}{I_{a2}} \quad \ldots(ii)$$

From (*i*) and (*ii*), $\dfrac{I_{a2}}{\sqrt{2} I_{a1}} = \dfrac{2I_{a1}}{I_{a2}}$ or $I_{a2}^2 = 2\sqrt{2} I_{a1}^2 = 2\sqrt{2} \times 15^2$ or $I_{a2} = 25.2$ A

From (*ii*), we get, $N_2 = 600 \times 2 \times 15/252 =$ **714 rpm**

Example 28.24. *A 2-pole series motor runs at 707 r.p.m. when taking 100 A at 85 V and with the field coils in series. The resistance of each field coil is 0.03 Ω and that of the armature 0.04 Ω. If the field coils are connected in parallel and load torque remains constant, find (a) speed (b) the additional resistance to be inserted in series with the motor to restore the speed to 707 r.p.m.*

Solution. Total armature circuit resistance = 0.04 + (2 × 0.03) = 0.1 Ω

$I_{a1} = 100$ A ; $E_{b1} = 85 - (100 \times 0.1) = 75$ V

When series field windings are placed in parallel, the current through each is half the armature current.

If I_{a2} = new armature current ; then $\Phi_2 \propto I_{a2}/2$

As torque is the same in the two cases

$$\therefore \quad \Phi_1 I_{a1} = \Phi_2 I_{a2} \text{ or } I_{a1}^2 = \frac{I_{a2}}{2} \times I_{a2} = \frac{I_{a2}^2}{2}$$

$$\therefore \quad 100^2 = I_{a2}^2 \therefore \; I_{a2} = 100\sqrt{2} = 141.4 \text{ A}$$

In this case, series field resistance = 0.03/2 = 0.015 Ω

$\therefore \; E_{b2} = 85 - 141.4(0.04 + 0.015) = 77.22$ V

$$\frac{N_2}{707} = \frac{77.22}{75} \times \frac{100}{141.4/2} \qquad (\because \; \Phi_2 \propto I_{a2}/2)$$

(a) $\therefore \quad N_2 = 707 \times \frac{77.22}{75} \times \frac{200}{141.4} =$ **1029 r.p.m.**

(b) Let the total resistance of series circuit be R_t.

Now, $E_{b1} = 77.22$ V, $N_1 = 1029$ r.p.m. ; $E_{b2} = 85 - 141.4\,R_t$, $N_2 = 707$ r.p.m.

$$\frac{707}{1029} = \frac{85 - 141.4R_t}{77.22} \quad \therefore \quad R_t = 0.226 \; \Omega$$

$\therefore$ additional resistance = 0.226–0.04–0.015 = **0.171 Ω**

Example 28.25. *A 240-V series motor takes 40 amperes when giving its rated output at 1500 rpm. Its resistance is 0.3 ohm. Find what resistance must be added to obtain rated torque (i) at starting (ii) at 1000 r.p.m.* **(Elect. Engg., Madras Univ. 1987)**

Solution. Since torque remains the same in both cases, it is obvious that current drawn by the motor remains constant at 40 A $\quad (\because \; T_a \propto I_a^2)$

(*i*) If R is the series resistance added, then $40 = 240/(R + 0.3) \; \therefore \; R = 5.7 \; \Omega$

(*ii*) Current remaining constant, $T_a \propto E_b/N$ –Art. 24.7

$$\frac{E_{b1}}{N_1} = \frac{E_{b2}}{N_2}$$

Now, $E_{b1} = 240 - 40 \times 0.3 = 228 \; V;\; N_1 = 1500$ r.p.m

$E_{b2} = 240 - 40\,(R + 0.3)\; V;\; N_2 = 1000$ r.p.m.

$$\therefore \quad \frac{228}{1500} = \frac{240 - 40(R + 0.3)}{1000};\; R = \mathbf{1.9 \; \Omega}$$

Example 28.26. *A 4-pole, series-wound fan motor runs normally at 600 r.p.m. on a 250 V d.c. supply taking 20 A. The field coils are connected all in series. Estimate the speed and current taken by the motor if the coils are reconnected in two parallel groups of two in series. The load torque increases as the square of the speed. Assume that the flux is directly proportional to the current and ignore losses.* **(Elect. Machines, AMIE, Sec, B. 1990)**

Solution. When coils are connected in two parallel groups, current through each becomes $I_{a2}/2$ where I_{a2} is the new armature current.

Hence, $\Phi_2 \propto I_{a2}/2$

Now $T_a \propto \Phi I_a$ –Art. 25.7

$\propto N^2$ –given

$$\therefore \quad \Phi_1 I_{a1} \propto N_1^2 \text{ and } \Phi_2 I_{a2} \propto N_2^2 \therefore \left(\frac{N_2}{N_1}\right)^2 = \frac{\Phi_2 I_{a2}}{\Phi_1 I_{a1}} \quad \ldots(i)$$

Since losses are negligible, field coil resistance as well as armature resistance are negligible. It means that armature and series field voltage drops are negligible. Hence, back e.m.f. in each case equals the supply voltage.

$$\therefore \quad \frac{N_2}{N_1} = \frac{E_{b2}}{E_{b1}} \times \frac{\Phi_1}{\Phi_2} \text{ becomes } \frac{N_2}{N_1} = \frac{\Phi_1}{\Phi_2} \quad ...(ii)$$

Putting this value in (*i*) above, we get

$$\left(\frac{\Phi_1}{\Phi_2}\right)^2 = \frac{\Phi_2 I_{a2}}{\Phi_1 I_{a1}} \text{ or } \frac{I_{a2}}{I_{a1}} = \left(\frac{\Phi_1}{\Phi_2}\right)^3$$

Now, $\Phi_1 \propto 20$ and $\Phi_2 \propto I_{a2}/2 \;\therefore\; \frac{I_{a2}}{20} = \left(\frac{20}{I_{a2}/2}\right)^3$ or $I_{a2} = 20 \times 2^{3/4} = \mathbf{33.64\ A}$

From (*ii*) above, we get

$$\frac{N_2}{N_1} = \frac{\Phi_1}{\Phi_2} = \frac{I_{a1}}{I_{a2}/2} = \frac{2I_{a1}}{I_{a2}};\; N_2 = 600 \times 2 \times 20/33.64 = \mathbf{714\ r.p.m.}$$

Example 28.27. *A d.c. series motor having a resistance of 1 Ω drives a fan for which the torque varies as the square of the speed. At 220-V, the set runs at 350 rpm and takes 25 A. The speed is to be raised to 500 rpm by increasing the voltage. Determine the necessary voltage and the corresponding current assuming the field to be unsaturated.*

(Electrical Engineering, Banaras Hindu Univ. 1988)

Solution. Since $\Phi \propto I_a$, hence $T_a \propto \Phi I_a \propto I_a^2$. Also $T_a \propto N^2$ ...(given)

$\therefore I_a^2 \propto N^2$ or $I_a \propto N$ or $I_{a1} \propto N_1$ and $I_{a2} \propto N_2$

$$\therefore \frac{I_{a2}}{I_{a1}} = \frac{N_2}{N_1} = \frac{500}{350};\; I_{a2} = 25 \times \frac{500}{350} = \mathbf{\frac{250}{7}\ A}$$

$$E_{b1} = 220 - 25 \times 1 = 195\text{ V};\; E_{b1} = V - (250/7) \times 1,\; \frac{\Phi_1}{\Phi_2} = \frac{25}{250/7} = \frac{7}{10}$$

$$\text{Now, } \frac{N_2}{N_1} = \frac{E_{b2}}{E_{b1}} \times \frac{\Phi_1}{\Phi_2};\; \frac{500}{350} = \frac{V - (250/7)}{195} \times \frac{7}{10};\; V = \mathbf{433.7\ V}$$

Example 28.28. *A d.c. series motor runs at 1000 rpm when taking 20 A at 200 V. Armature resistance is 0.5 Ω. Series field resistance is 0.2 Ω. Find the speed for a total current of 20 A when a divertor of 0.2 Ω resistance is used across the series field. Flux for a field current of 10 A is 70 per cent of that for 20 A.*

Solution. $E_{b1} = 200 - (0.5 + 0.2) \times 20 = 186\text{ V};\; N_1 = 1000$ r.p.m.

Since divertor resistance equals series field resistance, the motor current of 20 A is divided equally between the two. Hence, a current of 10 A flows through series field and produces flux which is 70% of that corresponding to 20 A. In other words, $\Phi_2 = 0.7\,\Phi_1$ or $\Phi_1/\Phi_2 = 1/0.7$

Moreover, their combined resistance = 0.2/2 = 0.1 Ω

Total motor resistance becomes = 0.5 + 0.1 = 0.6 Ω

$$\therefore \quad E_{b2} = 200 - 0.6 \times 20 = 188\text{ V};\; N_2 = ?$$

$$\therefore \quad \frac{N_2}{1000} = \frac{188}{186} \times \frac{1}{0.7};\quad N_2 = \mathbf{1444\ rpm}$$

Example 28.29. *A 220-V, d.c. series motor takes 40 A when running at 700 rpm. Calculate the speed at which the motor will run and the current taken from the supply if the field is shunted by a resistance equal to the field resistance and the load torque is increased by 50%.*

Armature resistance = 0.15 Ω, field resistance = 0.1 Ω

It may be assumed that flux per pole is proportional to the field.

Solution. In a series motor, prior to magnetic saturation

$$T \propto \Phi I_a \propto I_a^2 \quad \therefore T_1 \propto I_{a1}^2 \propto 40^2 \quad ...(i)$$

If I_{a2} is the armature current (or motor current) in the second case when divertor is used, then only $I_{a2}/2$ passes through the series field winding.

$\therefore\ \Phi_2 \propto I_{a2}/2$ and $T_2 \propto \Phi_2 I_{a2} \propto (I_{a2}/2) \times I_{a2} \propto I_{a2}^2/2$ ...(ii)

From (i) and (ii), we get $\dfrac{T_2}{T_1} = \dfrac{I_{a2}^2}{2 \times 40^2}$

Also $T_2/T_1 = 1.5 \quad \therefore\ 1.5 = I_a^2/2 \times a^2$

$\therefore \quad I_{a2} = \sqrt{1.5 \times 2 \times 40^2} = 69.3$ A ; Now $E_{b1} = 220 - (40 \times 0.25) = 210$ V

$E_{b2} = 220 - (69.3 \times 0.2^*) = 206.14$ V ; $N_1 = 700$ r.p.m. ; $N_2 = ?$

$\dfrac{N_2}{N_1} = \dfrac{E_{b2}}{E_{b1}} \times \dfrac{\Phi_1}{\Phi_2} \quad \therefore \quad \dfrac{N_2}{700} = \dfrac{206.14}{210} \times \dfrac{40}{69.3/2} \quad \therefore\ N_2 = \mathbf{794\ r.p.m.}$

Example 28.30. *A 4-pole, 250 V d.c. series motor takes 20 A and runs at 900 r.p.m. Each field coil has resistance of 0.025 ohm and the resistance of the armature is 0.1 ohm. At what speed will the motor run developing the same torque if :*

(i) a diverter of 0.2 ohm is connected in parallel with the series field.

(ii) Rearranging the field coils in two series and parallel groups.

Assume unsaturated magnetic operation.

(Electric Drives and their Control, Nagpur Univ. 1993)

Solution. The motor with its field coils all connected in series is shown in Fig. 28.15 (*a*). Here, $N_1 = 900$ rpm. $E_{b1} = 250 - 20 \times (0.1 + 4 \times 0.025) = 246$ V.

In Fig. 28.15. (*b*), a diverter of resistance 0.2 Ω has been connected in parallel with the series field coils. Let I_{a2} be the current drawn by the motor under this condition. As per current-divider rule, part of the current passing through the series field is $= I_{a2} \times 0.2/(0.1 + 0.2) = 2\,I_{a2}/3$. Obviously, $\Phi_2 \propto 2\,I_{a2}/3$.

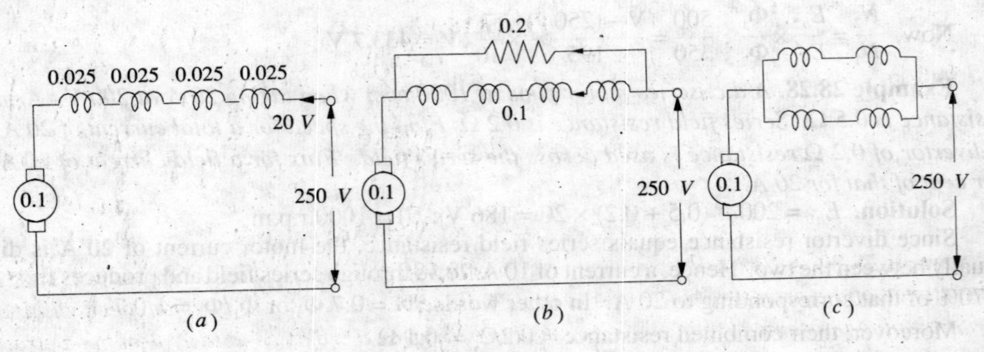

Fig. 28.15

Now, $T_1 \propto \Phi_1 I_{a1} \propto I_{a1}^2$; $T_2 \propto \Phi_2 I_{a2} \propto (2I_{a2}/3) I_{a2} \propto 2I_{a2}^2/3$.

Since $T_1 = T_2$; $\therefore\ I_{a1}^2 = 2I_{a2}^2/3$ or $20^2 = 2I_{a2}^2/3$; $\therefore\ I_{a2} = 24.5$ A.

Combined resistance of the field and divertor $= 0.2 \times 0.1/0.3 = 0.667\ \Omega$; Arm. circuit resistance $= 0.1 + 0.0667 = 0.1667\ \Omega$ $E_{b2} = 250 - 24.5 \times 0.1667 = 250 - 4.1 = 245.9$ V

$\dfrac{N_2}{N_2} = \dfrac{E_{b1}}{E_{b1}} \times \dfrac{\Phi_1}{\Phi_2}$; $\dfrac{N_2}{900} = \dfrac{245.9}{246} \times \dfrac{20}{(2/3)24.5}$; $N_2 = 1102$ rpm ...($\because\ \Phi_2 \propto 2I_{a2}/3$)

---

*The combined resistance of series field winding and the divertor is $0.1/2 = 0.05\ \Omega$. Hence, the total resistance $= 0.15 + 0.05 = 0.2\ \Omega$

(*ii*) In Fig. 28.15(*c*), the series field coils have been arranged in two parallel groups. If the motor current is I_{a2}, then it is divided equally between the two parallel paths. Hence, $\Phi_2 \propto I_{a2}/2$.

Since torque remains the same,

$T_1 \propto \Phi_1 I_{a1} \propto I_{a1}^2 \propto 20^2$; $T_2 \propto \Phi_2 I_{a2} \propto (I_{a2}/2) I_{a2} \propto I_{a2}^2/2$

Since $T_1 = T_2$; $\therefore\ 20^2 = I_{a2}^2/2$; $I_{a2} = 28.28$ A

Combined resistance of the two parallel paths = 0.05/2 = 0.025 Ω

Total arm. circuit resistance = 0. 1 + 0.025 = 0.125 Ω

$\therefore\ E_{b2} = 250 - 28.28 \times 0.125 = 246.5$ V

$$\frac{N_2}{900} = \frac{246.5}{246} \times \frac{20}{28.28/2}; \ N_2 = \mathbf{1275\ rpm}$$

Example 28.31. *A 4-pole, 230-V series motor runs at 1000 rpm when the load current is 12 A. The series field resistance is 0.8 Ω and the armature resistance is 1.0 Ω. The series field coils are now regrouped from all in series to two in series with two parallel paths. The line current is now 20 A. If the corresponding weakening of field is 15% calculate the speed of the motor.*

(Electrotechnology-I, Gauhati Univ. 1987)

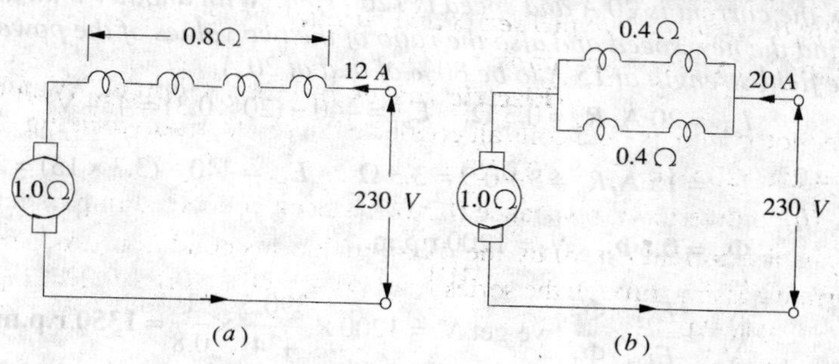

Fig. 28.16

Solution.
$$\frac{N_2}{N_1} = \frac{E_{b2}}{E_{b1}} \times \frac{\Phi_1}{\Phi_2}; \ E_{b1} = 230 - 12 \times 1.8 = 208.4 \text{ V}$$

$$E_{b2} = 230 - 20(1 + 0.4/2) = 206 \text{ V}; \ \Phi_2 = 0.85; \Phi_1 \text{ or } \Phi_1/\Phi_2 = 1/0.85$$

$$\therefore \quad \frac{N_2}{1000} = \frac{206}{208.4} \times \frac{1}{0.85} \qquad \therefore\ N_2 = \mathbf{1163\ r.p.m}$$

Example 28.32. *A 200-V, d.c. series motor runs at 500 rpm when taking a line current of 25 A., The resistance of the armature is 0.2 Ω and that of the series field 0.6 Ω. At what speed will it run when developing the same torque when an armature divertor of 10 Ω is used ? Assume a straight line magnetisation curve.*

(DC. Machines, Jadavpur Univ. 1988)

Solution. Resistance of motor 0.2 + 0.6 = 0.8 Ω

$\therefore\ E_{b1} = 200 - (25 \times 0.8) = 180$ V

Let the motor input current be I_2 when armature divertor is used.

Series field voltage drop = 0.6 I_2

$\therefore$ P.D. at brushes = 200−0.6 I_2

$$\therefore \quad \text{Arm. divertor current} = \left(\frac{200 - 0.6I_2}{10}\right) \text{A}$$

$$\therefore \quad \text{Armature current} = I_2 \left(\frac{200 - 0.6I_2}{10}\right)$$

$$\therefore \quad I_{a2} = \frac{10.6I_2 - 200}{10}$$

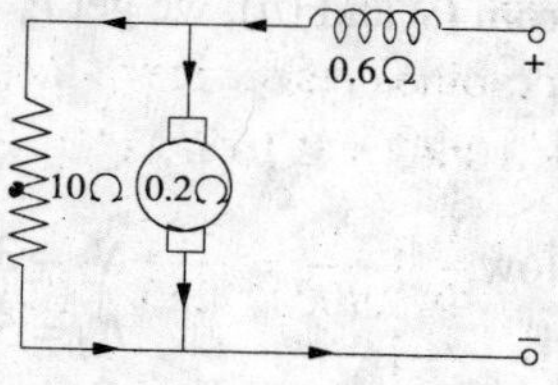

Fig. 28.17

As torque in both cases is the same, $\therefore\ \Phi_1 I_1 = \Phi_2 I_2$

$$\therefore \quad 25 \times 25 = I_2\left(\frac{10.6 I_2 - 200}{10}\right) \text{ or } 6,250 = 10.6 I_2^{\,2} - 200 I_2$$

or $$10.6\, I_2^2 - 200\, I_2 - 6250 = 0 \text{ or } I_2 = 35.6 \text{ A}$$

P.D. at brushes in this case = 200−(35.6 × 0.6) = 178.6 V

$$\therefore \quad I_{a2} = \frac{10.6 \times 35.6 - 200}{10} = 17.74 \text{ A} ; E_{b2} = 178.6 - (17.74 \times 0.2) = 175 \text{ V}$$

Now $$\frac{N_2}{N_1} = \frac{E_{b2}}{E_{b1}} \times \frac{\Phi_1}{\Phi_2} \text{ or } \frac{N_2}{N_1} = \frac{E_{b2}}{E_{b1}} \times \frac{I_1}{I_2} \quad (\because \quad \Phi \propto I)$$

$$\therefore \quad \frac{N_2}{500} = \frac{175}{180} \times \frac{25}{35.6} \quad \therefore N_2 = \mathbf{314 \text{ r.p.m.}}$$

Example 28.33. *A series motor is run on a 440-V circuit with a regulating resistance of R Ω for speed adjustment. The armature and field coils have a total resistance of 0.3 Ω. On a certain load with R = zero, the current is 20 A and speed is 1200 rpm. With another load and R = 3 Ω, the current is 15 A. Find the new speed and also the ratio of the two values of the power outputs of the motor. Assume the field strength at 15 A to be 80% of that at 20 A.*

Solution. $I_{a1} = 20 \text{ A}, R_a = 0.3\ \Omega; \quad E_{b1} = 440 - (20 \times 0.3) = 434 \text{ V}$

$I_{a2} = 15 \text{ A}, R_a = 3 + 0.3 = 3.3\ \Omega \therefore\ E_{b2} = 440 - (3.3 \times 15) = 390.5 \text{ V}$

$\Phi_2 = 0.8 \Phi_1, \quad N_1 = 1200 \text{ r.p.m.}$

Using $$\frac{N_2}{N_1} = \frac{E_{b2}}{E_{b1}} \times \frac{\Phi_1}{\Phi_2}, \text{ we get } N_2 = 1200 \times \frac{390.5}{434} \times \frac{1}{0.8} = \mathbf{1350 \text{ r.p.m.}}$$

Now, in a series motor,

$$T \propto \Phi\, I_a \text{ and power } P \propto T \times N \text{ or } P \propto \Phi\, N I_a$$

$$\therefore \quad P_1 \propto \Phi_1 \times 1200 \times 20 \text{ and } P_2 \propto 0.8\, \Phi_1 \times 1350 \times 15$$

$$\therefore \quad \frac{P_1}{P_2} = \frac{1200 \times 20\, \Phi_1}{1350 \times 15 \times 0.8 \Phi_1} = \mathbf{1.48}$$

Hence, power in the first case is 1.48 times the power in the second case.

Example 28.34. *A series motor, with an unsaturated field and negligible resistance when running at a certain speed on a given load takes 50 A at 460 V. If the load torque varies as the cube of speed, calculate the resistance required to reduce the speed by 25%.*

(Elect. Machinery- I, Madras Univ. 1987)

Solution. Since the field is unsaturated, $\Phi \propto I_a$

$$\therefore \quad T \propto \Phi I_a \propto I_a^2 \quad \text{...(i)}$$

$$T \propto N^3 \quad \text{(given)} \quad \text{...(ii)}$$

From (*i*) and (*ii*), we get $I_a^2 \propto N^3$

$$\therefore \quad I_{a1}^{\,2} \propto N_1^{\,3} \quad \text{...Ist case}$$

$$I_{a2}^{\,2} \propto N_2^{\,3} \quad \text{...2nd case}$$

or $$(I_{a2}/I_{a1})^2 = (N_2/N_1)^3$$

Now, $$N_2 = 3N_1/4 \quad \therefore\ N_2/N_1 = 3/4 \quad \therefore\ (I_{a2}/I_{a1})^2 = (3/4)^3 = 27/64$$

$$\therefore \quad I_{a2} = 50\sqrt{27/64} = 32.48 \text{ A}$$

$$\therefore \quad E_{b1} = 460 - I_{a1} \times 0 = 460 \text{ V} ; E_{b2} = 460 - (32.48\, R)$$

where R is the required series resistance.

Also, $$\frac{N_2}{N_1} = \frac{E_{b2}}{E_{b1}} \times \frac{I_{a1}}{I_{a2}} \quad \therefore \quad \frac{3}{4} = \frac{460 - 32.48\, R}{460} \times \frac{50}{32.48} \quad \therefore R = \mathbf{7.26\ \Omega}$$

28.4. Merits and Demerits of Rheostatic Control Method

1. Speed changes with every change in load, because speed variations depend not only on controlling resistance but on load current also. This double dependence makes it impossible to keep the speed sensibly constant on rapidly changing loads.
2. A large amount of power is wasted in the controller resistance. Loss of power is directly proportional to the reduction in speed. Hence, efficiency is decreased.
3. Maximum power developed is diminished in the same ratio as speed.
4. It needs expensive arrangement for dissipation of heat produced in the controller resistance.
5. It gives speeds *below* the normal, not above it because armature voltage can be decreased (not increased) by the controller resistance.

This method is, therefore, employed when low speeds are required for a short period only and that too occasionally as in printing machines and for cranes and hoists where motor is continually started and stopped.

Advantages of Field Control Method

This method is economical, more efficient and convenient though it can give speeds *above* (not below) the normal speed. The only limitation of this method is that commutation becomes unsatisfactory, because the effect of armature reaction is greater on a weaker field.

It should, however, be noted that by combining the two methods, speeds above and below the normal may be obtained.

28.5. Series-parallel Control

In this system of speed control, which is widely used in electric traction, two or more similar mechanically-coupled series motors are employed. At low speeds, the motors are joined in series (Fig. 28.18) and for high speeds, are joined in parallel (Fig. 28.19).

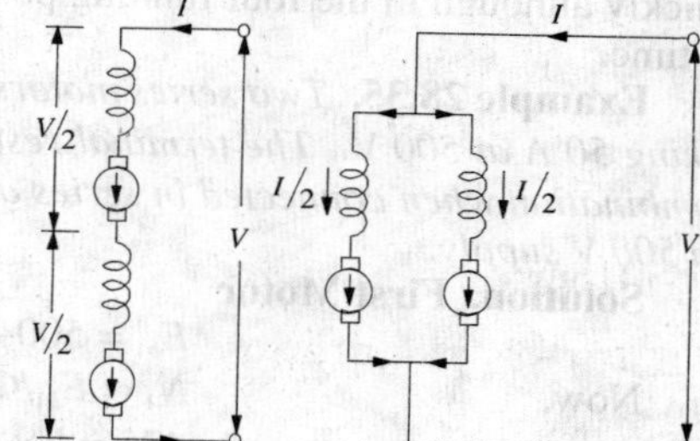

Fig. 28.18 **Fig. 28.19**

When in series, the two motors have the same current passing through them, although the voltage across each motor is $V/2$ *i.e.* half the supply voltage. When joined in parallel, voltage across each machine is V, though current drawn by each motor is $I/2$.

When in Parallel

Now speed $\propto E_b/\Phi \propto E_b/\text{current}$ (being series motors)

Since E_b is approximately equal to the applied voltage V

$$\therefore \quad \text{speed} \propto \frac{V}{I/2} \propto \frac{2V}{I}$$

Also, $$\text{torque} \propto \Phi I \propto I^2 \quad (\because \quad \Phi \propto I) \qquad \ldots(i)$$

$$\therefore \quad T \propto (I/2)^2 \propto I^2/4 \qquad \ldots(ii)$$

When in Series

Here $$\text{speed} \propto \frac{E_b}{\Phi} \propto \frac{V/2}{I} \propto \frac{V}{2I} \qquad \ldots(iii)$$

This speed is one-fourth of the speed of the motors when in parallel.

Similarly $$T \propto \Phi I \propto I^2$$

The torque is four times that produced by motors when in parallel.

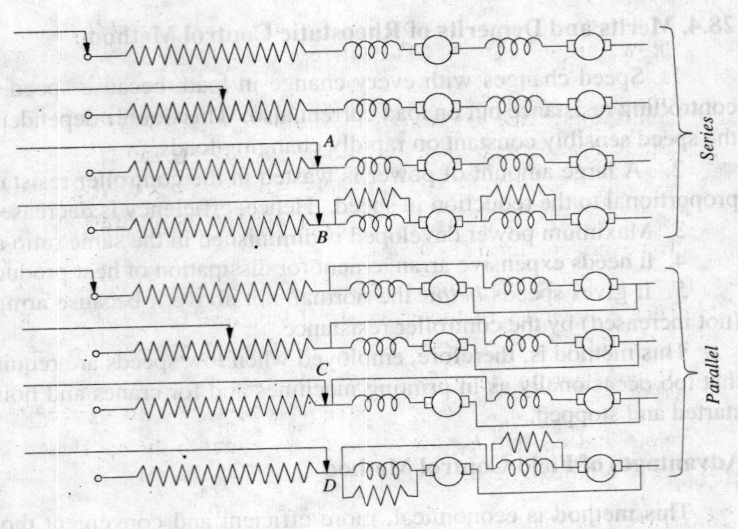

Fig. 28.20

This system of speed control is usually combined with the variable resistance method of control described in Art. 28.3(2).

The two motors are started up in series with each other and with variable resistance which is cut out in sections to increase the speed. When all the variable series resistance is cut out, the motors are connected in parallel and at the same time, the series resistance is reinserted. The resistance is again reduced gradually till full speed is attained by the motors. The switching sequence is shown in Fig. 28.20. As the variable series controller resistance is not continuously rated, it has to be cut out of the circuit fairly quickly although in the four running positions *A,B,C* and *D*, it may be left in circuit for any length of time.

Example 28.35. *Two series motors run at a speed of 500 rpm and 550 rpm respectively when taking 50 A at 500 V. The terminal resistance of each motor is 0.5 Ω. Calculate the speed of the combination when connected in series and coupled mechanically. The combination is taking 50 A on 500 V supply.* **(Electrical Machinery-I, Mysore Univ. 1985)**

Solution. First Motor

$E_{b1} = 500-(50 \times 0.5) = 475$ V ; $I = 50$ A

Now, $N_1 \propto E_{b1}/\Phi_1$ or $E_{b1} \propto N_1\Phi_1$ or $E_{b1} = kN_1\ \Phi_1$

$\therefore$ $475 = k \times 500 \times \Phi_1 \ \therefore\ k\Phi_1 = 475/500$

Second Motor $E_{b2} = 500-(50 \times 0.5) = 475$ V. Similarly, $k\Phi_2 = 475/550$

When both motors are in series

$E_b' = 500-(50 \times 2 \times 0.5) = 450$ V

Now, $E_b' = E_{b1} + E_{b2} = k\Phi_1 N + k\Phi_2 N$

where N is the common speed when joined in series.

$$\therefore \quad 450 = \frac{475}{500}N + \frac{475}{550}N \quad \therefore \quad N = \mathbf{248\ r.p.m.}$$

Example 28.36. *Two similar 20 h.p. (14.92 kW), 250-V, 1000-rpm series motors are connected in series with each other across a 250-V supply. The two motors drive the same shaft through a reduction gearing 5 : 1 and 4 : 1 respectively. If the total load torque on the shaft is 882 N-m, calculate* (*i*) *the current taken from the supply main* (*ii*) *the speed of the shaft and* (*iii*) *the voltage across each motor. Neglect all losses and assume the magnetic circuits to be unsaturated.* **(Elect. Machines, Punjab Univ., 1991)**

Solution. (*i*) Rated current of each motor = 14,920/250 = 59.68 A

Back e.m.f. $E_b = 250$ V (neglecting I_aR_a drop)

Now, $E_b \propto N\Phi$ As $\Phi \propto I \ \therefore\ E_b \propto NI$ or $E_b = kNI$

$250 = k \times (1000/60) \times 59.68 \quad \therefore \quad k = 0.25$

Let N_{sh} be the speed of the shaft.

Speed of the first motor $N_1 = 5\,N_{sh}$; Speed of the second motor $N_2 = 4\,N_{sh}$

Let I be the new current drawn by the series set, then

$E_b' = E_{b1} + E_{b2} = kI \times 5\,N_1 + kI \times N_2 = kI \times 5N_{sh} + kI \times 4N_{sh}$

$250 = 9 \times kI\,N_{sh}$ ...(*i*)

Now, torque $\quad T = 0.159 \dfrac{E_b I}{N} = 0.159 \times \dfrac{kI N_{sh} \times I}{N_{sh}} = 0.159\ kI^2$

Shaft torque due to gears of 1st motor = $5 \times 0.159\ kI^2$

Shaft torque due to gears of 2nd motor = $4 \times 0.159\ kI^2$

$\therefore \quad 882 = kI^2 (5 \times 0.159 + 4 \times 0\text{-}.159) = 1.431\ kI^2$

$\therefore \quad I^2 = 882/1.431 \times 0.25 = 2{,}449$ A $\therefore$ I = **49.5 A**

(*ii*) From equation (*i*), we get

$250 = 9 \times 0.25 \times 49.5 \times N_{sh} \therefore N_{sh} = 2.233$ r.p.s. = **134 rpm.**

(*iii*) Voltage across the armature of Ist motor is

$E_{b1}' = 5kI\ N_{sh} = 5 \times 0.25 \times 49.5 \times 2.233 =$ **139 V**

Voltage across the armature of 2nd motor

$E_{b2} = 4kIN_{sh} = 4 \times 0.25 \times 49.5 \times 2.233 =$ **111 V**

Note that E_{b1} and E_{b2} are respectively equal to the applied voltage across each motor because $I_a R_a$ drops are negligible.

28.6. Electric Braking

A motor and its load may be brought to rest quickly by using either (*i*) Friction Braking or (*ii*) Electric Braking. The commonly-used mechanical brake has one drawback : it is difficult to achieve a smooth stop because it depends on the condition of the braking surface as well as on the skill of the operator.

Three excellent electric braking methods are available which eliminate the need of brake lining, levers and other mechanical gadgets. Electric braking, both for shunt and series motors, is of the following three types (*i*) rheostatic or dynamic braking (*ii*) plugging *i.e.* reversal of torque so that armature tends to rotate in the opposite direction and (*iii*) regenerative braking.

Obviously, friction brake is necessary for holding the motor even after it has been brought to rest.

28.7. Electric Braking of Shunt Motors

(*a*) Rheostatic or Dynamic Braking

In this method, the armature of the shunt motor is disconnected from the supply and is connected across a variable resistance R as shown in Fig. 28.21(*b*). The field winding is, however, left connected across the supply undisturbed. The braking effect is controlled by varying the series resistance R. Obviously, this method makes use of generator action in a motor to bring it to rest.* As seen from Fig. 28.21(*b*), armature current is given by

$$I_a = \frac{E_b}{R + R_a} = \frac{\Phi ZN(P/A)}{R + R_a} = \frac{k_1 \Phi N}{R + R_a}$$

Braking torque is given by

$$T_B = \frac{1}{2\pi} \Phi Z I_a \left(\frac{P}{A}\right) \text{N-m}$$

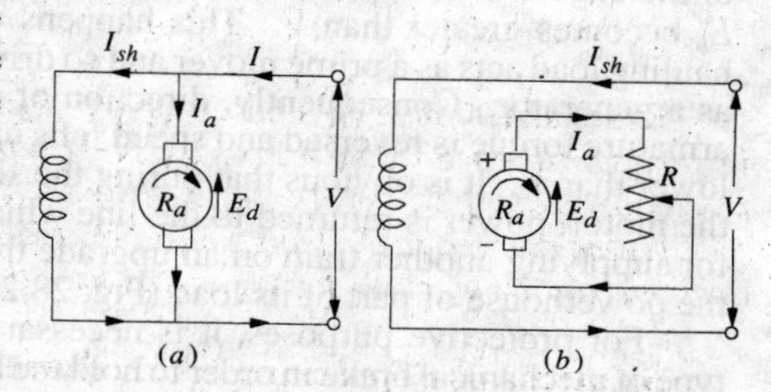

Fig. 28.21

*The motor while acting as a generator feeds current to the resistor dissipating heat at the rate of I^2R. The current I_a produced by dynamic braking flows in the opposite direction thereby producing a counter torque that slows down the machine.

$$= \frac{1}{2\pi}\Phi Z\left(\frac{P}{A}\right)\cdot\frac{\Phi ZN(P/A)}{R+R_a} = \frac{1}{2\pi}\left(\frac{ZP}{A}\right)^2\frac{\Phi^2 N}{R+R_a} = k_2\Phi^2 N \quad \therefore\ T_B \propto N$$

Obviously, T_B decreases as motor slows down and disappears altogether when it comes to a stop.

(b) Plugging or Reverse Current Braking

This method is commonly used in controlling elevators, rolling mills, printing presses and machine tools etc.

In this method, connections to the armature terminals are reversed so that motor tends to run in the opposite direction (Fig. 28.22). Due to the reversal of armature connections, applied voltage V and E_b start acting *in the same direction* around the circuit. In order to limit the armature current to a reasonable value, it is necessary to insert a resistor in the circuit while reversing armature connections.

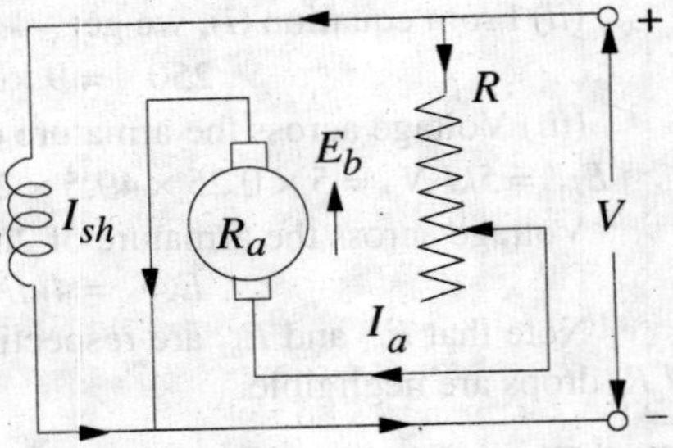

$$I_a = \frac{V+E_b}{R+R_a} = \frac{V}{R+R_a} + \frac{E_b}{R+R_a}$$

$$= \frac{V}{R+R_a} + \frac{\Phi ZN(P/A)}{R+R_a} = \frac{V}{R+R_a} + \frac{k_1\Phi N}{R+R_a}$$

Fig. 28.22

$$T_B = \frac{1}{2\pi}.\Phi ZI_a\left(\frac{P}{A}\right) = \frac{1}{2\pi}\cdot\left(\frac{\Phi ZP}{A}\right)\cdot I_a = \frac{1}{2\pi}\left(\frac{\Phi ZP}{A}\right)\left[\frac{V}{R+R_a} + \frac{\Phi ZN(P/A)}{R+R_a}\right]$$

$$= \frac{1}{2\pi}\left(\frac{ZP}{A}\right)\left(\frac{V}{R+R_a}\right)\cdot\Phi + \frac{1}{2\pi}\left(\frac{ZP}{A}\right)^2\frac{\Phi^2 N}{R+R_a} = k_2\Phi + k_3\Phi^2 N$$

or $\quad T_b = k_4 + k_5 N$ where $k_4 = \frac{1}{2\pi}\left(\frac{\Phi ZP}{A}\right)\left(\frac{V}{R+R_a}\right)$ and $k_5 = \frac{1}{2\pi}\left(\frac{\Phi ZP}{A}\right)^2 \times \frac{1}{(R+R_a)}$

Plugging gives greater braking torque than rheostatic braking. Obviously, during plugging, power is drawn from the supply and is dissipated by R in the form of heat. It may be noted that even when motor is reaching zero speed, there is some braking torque $T_B = k_4$ (see Ex. 28.38)

(c) Regenerative Braking

This method is used when the load on the motor has overhauling characteristic as in the lowering of the cage of a hoist or the downgrade motion of an electric train. Regeneration takes place when E_b becomes greater than V. This happens when the overhauling load acts as a prime mover and so drives the machine as a generator. Consequently, direction of I_a and hence of armature torque is reversed and speed falls until E_b becomes lower than V. It is obvious that during the slowing down of the motor, power is returned to the line which may be used for supplying another train on an upgrade thereby relieving the powerhouse of part of its load (Fig. 28.23).

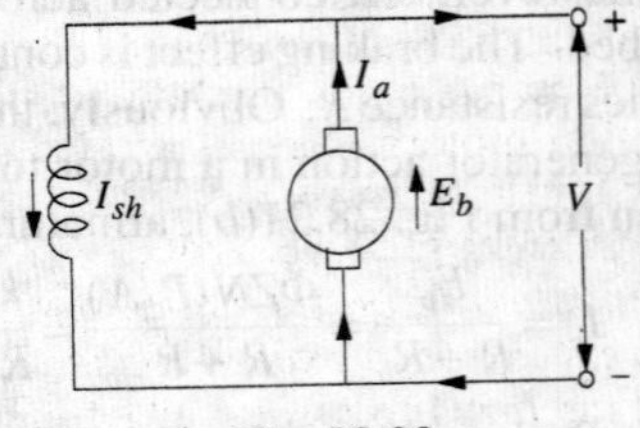

Fig. 28.23

For protective purposes, it is necessary to have some type of mechanical brake in order to hold the load in the event of a power failure.

Example 28.37. *A 220-V compensated shunt motor drives a 700 N-m torque load when running at 1200 rpm. The combined armature compensating winding and interpole resistance is 0.008 Ω and shunt field resistance is 55 Ω. The motor efficiency is 90%. Calculate the value of the dynamic braking resistor that will be capable of 375 N-m torque at 1050 rpm. The windage and friction losses may be assumed to remain constant at both speeds.*

Solution. Motor output = ωT = 2π NT = 2π (1200/60) × 700 = 87,965 W

Power drawn by the motor = 87,965/0.9 = 97,740 W

Current drawn by the motor = 97,740/220 = 444 A.

I_{sh} = 220/55 = 4 A; I_{a1} = 444−4 = 440 A

E_{b1} = 220−440 × 0.008 = 216.5 V

Since field flux remains constant, T_1 is proportional to I_{a1} and T_2 to I_{a2}.

$$\therefore \quad \frac{T_2}{T_1} = \frac{I_{a2}}{I_{a1}} \text{ or } I_{a2} = 440 \times \frac{375}{700} = 236 \text{ A}$$

$$\frac{N_2}{N_1} = \frac{E_{b2}}{E_{b1}} \text{ or } \frac{1050}{1200} = \frac{E_{b2}}{216.5}; E_{b2} = 189.4 \text{ V}$$

With reference to Fig. 28.23, we have

$$189.4 = 236(0.008 + R); R = 0.794\ \Omega$$

28.8. Electric Braking of Series Motor

The above-discussed three methods as applied to series motors are as follows :

(*a*) **Rheostatic (or Dynamic) Braking**

The motor is disconnected from the supply, the field connections *are reversed* and the motor is connected in series with a variable resistance R as shown in Fig. 28.24. Obviously, now, the machine is running as a generator. The field connections are reversed to make sure that current through field winding flows in the *same* direction as before (*i.e.* from M to N) in order to assist residual magnetism. In practice, the variable resistance employed for starting purpose is itself used for braking purposes. As in the case of shunt motors.

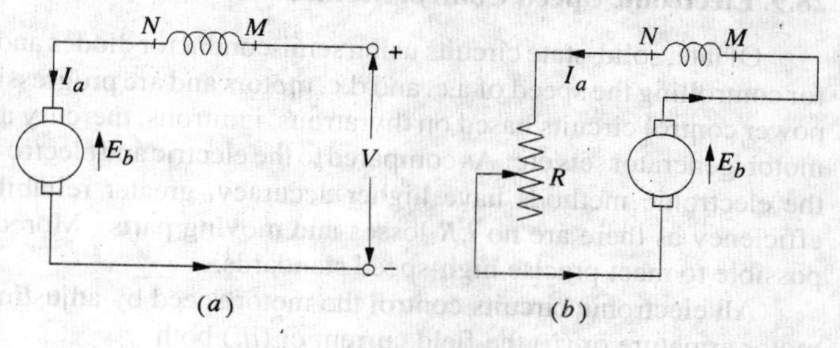

Fig. 28.24

$$T_B = k_2\,\Phi^2 N = k_3 I_a^2\, N \qquad \text{.....prior to saturation}$$

(*b*) **Plugging or Reverse Current Braking**

As in the case of shunt motors, in this case also the connections of the armature are reversed and a variable resistance R is put in series with the armature as shown in Fig. 28.25. As found in Art. 28.7 (*b*),

$$T_B = k_2\,\Phi + k_3\Phi^2 N$$

(*c*) **Regenerative Braking**

This type of braking of a series motor is not possible without modification because reversal of I_a would also mean reversal of the field and hence of E_b. However, this method is sometimes used with traction motors, special arrangements being necessary for the purpose.

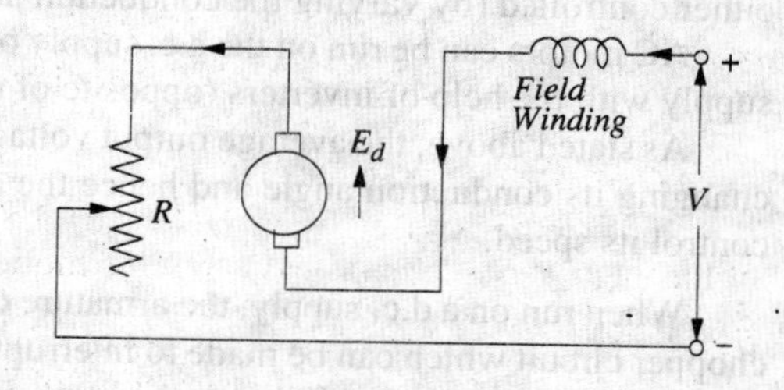

Fig. 28.25

Example 28.38. *A 400-V, 25-h.p. (18.65 kW), 450-r.p.m., d.c. shunt motor is braked by plugging when running on full load. Determine the braking resistance necessary if the maximum braking current is not to exceed twice the full-load current. Determine also the maximum braking torque and the braking torque when the motor is just reaching zero speed. The efficiency of the motor is 74.6% and the armature resistance is 0.2 Ω.*

(Electrical Technology, M.S. Univ. Baroda, 1988)

Solution.

F.L. Motor input current I = 18,650/0.746 × 400 = 62.5 A

I_a = 62.5 A (neglecting I_{sh}); E_b = 400−62.5 × 0.2 = 387.5 V

Total voltage around the circuit is= 400 + 387.5 = 787.5 V

Max. braking current = 2 × 62.5 = 125 A

Total resistance required in the circuit= 787.5/125 = 6.3 Ω

Braking resistance R = 6.3−0.2 = **6.1 Ω**

Maximum braking torque will be produced initially when motor speed is maximum *i.e.* 450 r.p.m. or 7.5 r.p.s.

Maximum value of $T_B = k_4 + k_5 N$ --Art. 25.7(*b*)

Now, $$k_4 = \frac{1}{2\pi}\left(\frac{\Phi ZP}{A}\right)\left(\frac{V}{R+R_a}\right) \text{ and } k_5 = \frac{1}{2\pi}\left(\frac{\Phi ZP}{A}\right)^2 \cdot \frac{1}{(R+R_a)}$$

Now, $E_b = \Phi ZN(P/A)$; also $N = 450/60 = 7.5$ r.p.s.

$\therefore$ $387.5 = 7.5(\Phi ZP/A)$ or $(\Phi ZP/A) = 51.66$

$$\therefore \quad k_4 = \frac{1}{2\pi} \times 51.66 \times \frac{400}{6.3} = 522 \text{ and } k_5 = \frac{1}{2\pi} \times (51.66)^2 \times \frac{1}{6.3} = 67.4$$

$\therefore$ Maximum $T_B = 522 + 67.4 \times 7.5 =$ **1028 N-m**

When speed is also zero *i.e.* $N = 0$, the value of torque is $T_B = K_4 =$ **522 N-m**

28.9. Electronic Speed Control Methods for DC Motors

Of late, solid-state circuits using semi-conductor diodes and thyristors have become very popular for controlling the speed of a.c. and d.c. motors and are progressively replacing the traditional electric power control circuits based on thyratrons, ignitrons, mercury arc rectifiers, magnetic amplifiers and motor-generator sets etc. As compared to the electric and electro-mechanical systems of speed control, the electronic methods have higher accuracy, greater reliability, quick response and also higher efficiency as there are no I^2R losses and moving parts. Moreover, full 4-quadrant speed control is possible to meet precise high-speed standards.

All electronic circuits control the motor speed by adjusting either (*i*) the voltage applied to the motor armature or (*ii*) the field current or (*iii*) both.

DC motors can be run from the d.c. supply if available or from a.c. supply after it has been converted into dc supply with the help of rectifiers which can be either half-wave or full-wave and either controlled (by varying the conduction angle of the thyristors used) or uncontrolled.

AC motors can be run on the a.c. supply or from d.c. supply after it has been converted into a.c. supply with the help of inverters (opposite of rectifiers).

As stated above, the average output voltage of a thyristor-controlled rectifier can be changed by changing its conduction angle and hence the armature voltage of the d.c. motor can be adjusted to control its speed.

When run on a d.c. supply, the armature d.c. voltage can be changed with the help of a thyristor chopper circuit which can be made to interrupt d.c. supply at different rates to give different average values of the d.c. voltage. If dc supply is not available, it can be obtained from the available ac supply with the help of uncontrolled rectifiers (using only diodes and not thyristors). The dc voltage so obtained can be then chopped with the help of a thyristor chopper circuit.

A brief description of rectifiers, inverters* and d.c. choppers would now be given before discussing the motor speed control circuits.

28.10. Uncontrolled Rectifiers

As stated earlier, rectifiers are used for a.c. to dc conversion *i.e.* when the supply is alternating but the motor to be controlled is a d.c. machine.

---

*Rectifiers convert a.c. power into d.c. power whereas inverters convert d.c. power into a.c. power. However, converter is a general term embracing both rectifiers and inverters.

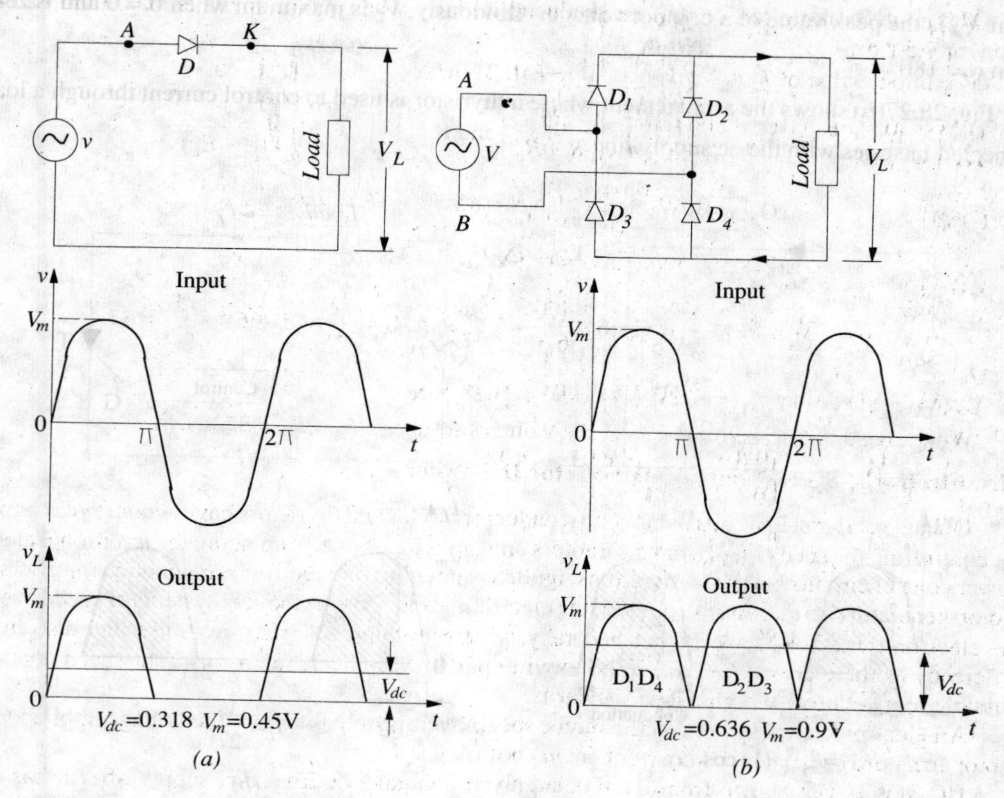

Fig. 28.26

Fig. 28.26(*a*) shows a half-wave uncontrolled rectifier. The diode *D* conducts only during positive half-cycles of the single-phase a.c. input *i.e.* when its anode *A* is positive with respect to its cathode *K*. As shown, the average voltage available across the load (or motor) is 0.45 *V* where *V* is the r.m.s. value of the ac voltage (in fact, $V = V_m/\sqrt{2}$). As seen, it is a pulsating d.c. voltage.

In Fig. 28.26 (*b*) is shown a single-phase, full-wave bridge rectifier which uses four semiconductor diodes and provides double the voltage *i.e.* 0.9 V. During positive input half-cycles when end *A* is positive with respect to end *B*, diodes D_1 and D_4 conduct (*i.e.* opposite diodes) whereas during negative input half-cycles, D_2 and D_3 conduct. Hence, current flows through the load during both half-cycles in the *same* direction. As seen, the d.c. voltage supplied by a bridge rectifier is much less pulsating than the one supplied by the half-wave rectifier.

28.11. Controlled Rectifiers

In these rectifiers, output load current (or voltage) can be varied by controlling the point in the input a.c. cycle at which the thyristor is turned ON with the application of a suitable low-power gate pulse. Once triggered (or fired) into conduction, the thyristor remains in the conducting state for the rest of the half-cycle *i.e.* upto 180°. The firing angle α can be adjusted with the help of a control circuit. When conducting, it offers no resistance *i.e.* it acts like a short-circuit.

Fig. 28.27(*a*) shows an elementary half-wave rectifier in which thyristor triggering is delayed by angle α with the help of a phase-control circuit. As shown, the thyristor starts conducting at point A and not at point O because its gate pulse is applied after a delay of α. Obviously, the conduction angle is reduced from 180° to (180 − α) with a consequent decrease in output voltage whose value is given by

$$V_L = \frac{V_m}{2\pi}(1+\cos\alpha) = 0.16\,V_m(1+\cos\alpha) = 0.32\,V_m\cos^2\frac{\alpha}{2}$$

where V_m is the peak value of a.c. input voltage. Obviously, V_L is maximum when $\alpha = 0$ and is zero when $\alpha = 180°$

Fig. 28.27(*b*) shows the arrangement where a thyristor is used to control current through a load connected in series with the ac supply line.

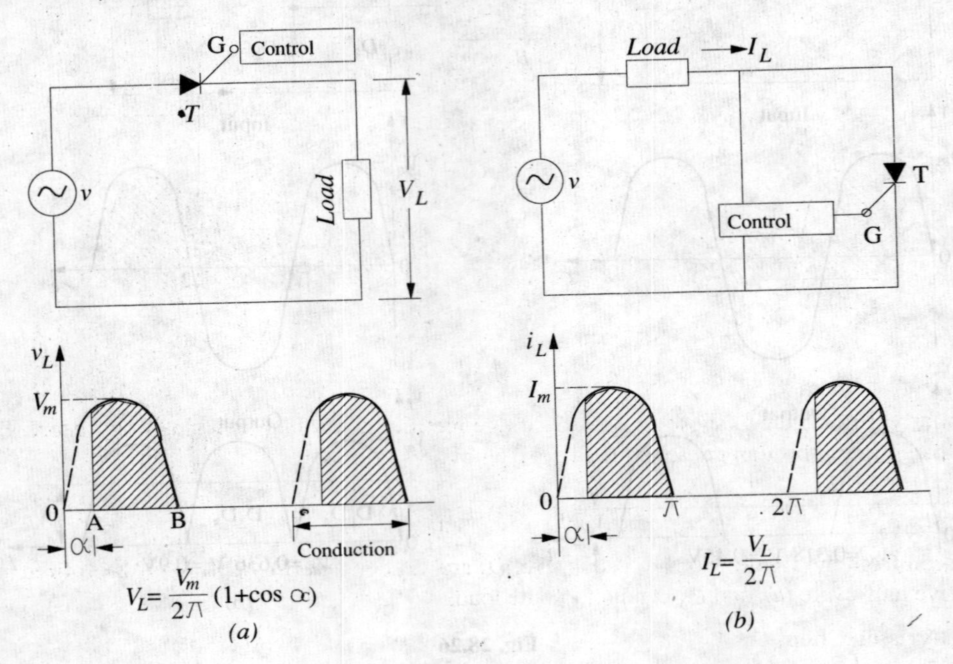

Fig. 28.27

The load current is given by

$$I_L = \frac{V_L}{R_L} = \frac{V_m}{2\pi R_L}(1+\cos\alpha) = \frac{V_m}{\pi R_L}\cos^2\frac{\alpha}{2}$$

Fig. 28.28(*a*) shows a single-phase, full-wave half-controlled rectifier. It is called half-controlled because it uses two thyristors and two diodes instead of four thyristors. During positive input half-cycle when *A* is positive, conduction takes place via T_1, load and D_1. During the negative half-cycle when *B* becomes positive, conduction route is via T_2, load and D_2.

The average output voltage V_L or V_{dc} is given by $V_L = 2 \times$ half-wave rectifier output

$$\therefore\ V_L = 2 \times \frac{V_m}{2\pi}(1+\cos\alpha) = \frac{V_m}{\pi}(1+\cos\alpha)^* = \frac{2V_m}{\pi}\cos^2\frac{\alpha}{2}$$

*For a fully-controlled bridge rectifier, its value is

$$V_L = \frac{2V_m}{\pi}\cos\alpha$$

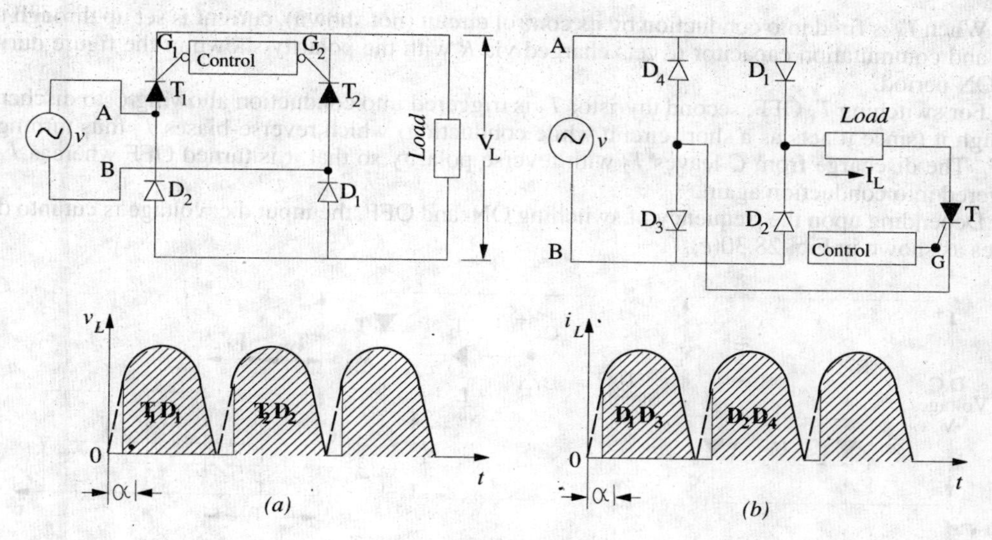

Fig. 28.28

Similarly, Fig. 28.28(*b*) shows a 4-diode bridge rectifier controlled by a single thyristor. The average load current through the series load is given by

$$I_L = \frac{V_m}{\pi R_L}(1 + \cos \alpha) = \frac{2V_m}{\pi R_L}\cos^2\frac{\alpha}{2}$$

As seen from the figure, when A is positive, D_1 and D_3 conduct provided T has been fired. In the negative half-cycle, D_2 and D_4 conduct via the load.

28.12. Thyristor Choppers

Since thyristors can be switched *ON* and *OFF* very rapidly, they are used to interrupt a d.c. supply at a regular frequency in order to produce a lower (mean) d.c. voltage supply. In simple words, they can produce low-level d.c. voltage from a high-voltage d.c. supply as shown in Fig. 28.29.

The mean value of the output voltage is given by

$$V_{dc} = V_L = V\frac{T_{ON}}{T_{ON} + T_{OFF}} = V\frac{T_{ON}}{T}$$

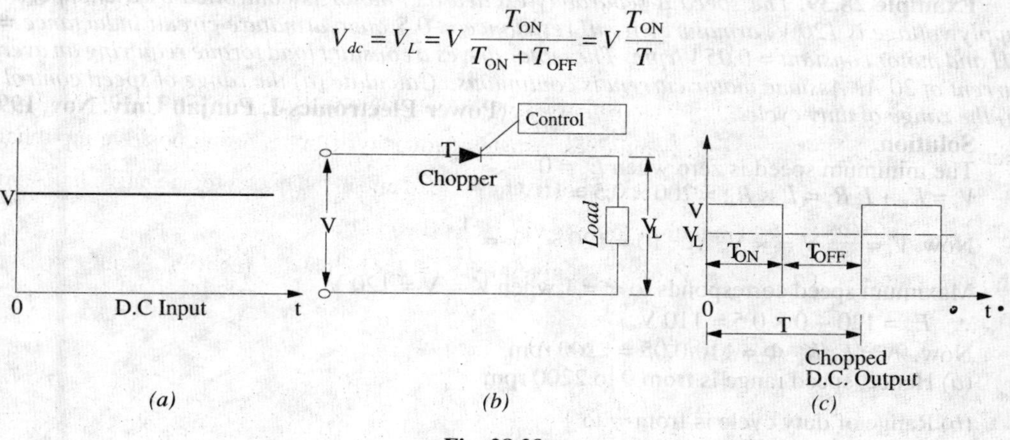

Fig. 28.29

Fig. 28.28(*a*) shows a simple thyristor chopper circuit alongwith extra commutating circuitry for switching T_1 OFF. As seen, T_1 is used for d.c. chopping whereas R, T_2 and C are used for commutation purposes as explained below.

When T_1 is fired into conduction by its control circuit (not shown), current is set up through the load and commutation capacitor C gets charged via R with the polarity shown in the figure during this ON period.

For switching T_1 OFF, second thyristor T_2 is triggered into conduction allowing C to discharge through it (since it acts as a short-circuit while conducting) which reverse-biases T_1 thus turning it OFF. The discharge from C leaves T_2 with reverse polarity so that it is turned OFF whereas T_1 is triggered into conduction again.

Depending upon the frequency of switching ON and OFF, the input d.c. voltage is cut into d.c. pulses as shown in Fig. 28.30(c).

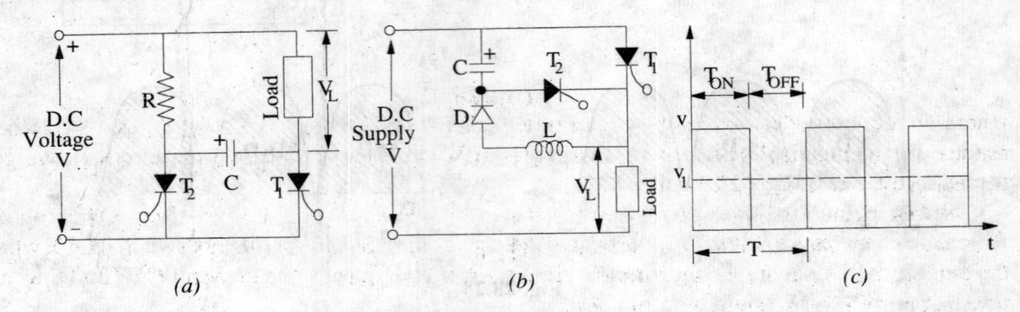

Fig. 28.30

In Fig. 28.30(b), T_1 is the chopping thyristor whereas C, D, T_2 and L constitute the commutation circuitry for switching T_1 OFF and ON at regular intervals.

When T_2 is fired, C becomes charged via the load with the polarity as shown. Next, when T_1 is fired, C reverse-biases T_2 to OFF by discharging via T_1, L and D and then recharges with reverse polarity. T_2 is again fired and the charge on C reverse-biases T_1 to non-conducting state.

It is seen that output (or load) voltage is present only when T_1 is ON and is absent during the interval it is OFF. The mean value of output d.c. voltage depends on the relative values of T_{ON} and T_{OFF}. In fact, output d.c. voltage is given by

$$V_{dc} = V_L \frac{T_{ON}}{T_{ON} + T_{OFF}} = V\frac{T_{ON}}{T}$$

Obviously, by varying thyristor ON/OFF ratio, V_L can be made any percentage of the input d.c. voltage V.

Example 28.39. *The speed a separately-excited d.c. motor is controlled by a chopper. The supply voltage is 120 V, armature circuit resistance = 0.5 ohm, armature circuit inductance = 20 mH and motor constant = 0.05 V/rpm. The motor drives a constant load torque requiring an average current of 20 A. Assume motor current is continuous. Calculate (a) the range of speed control (b) the range of duty cycle.* **(Power Electronics-I, Punjab Univ. Nov. 1990)**

Solution.

The minimum speed is zero when $E_b = 0$

$V_t = E_b + I_a R_a = I_a \times R_a, = 200 \times 0.5 = 10\text{V}$

Now, $V_t = \frac{T_{ON}}{T} \text{V} = \propto \text{V}$, $\therefore$ $10 = 120 \propto$, $\propto = \frac{1}{12}$

Maximum speed corresponds to $\propto = 1$ when $V_t = \text{V} = 120$ V

$\therefore$ $E_b = 120 - 0 \times 0.5 = 110$ V

Now, $N = E_b/K_a\ \Phi = 110/0.05 = 2200$ rpm

(a) Hence, speed range is from 0 to 2200 rpm

(b) Range of duty cycle is from $\frac{1}{12}$ to 1

28.13. Thyristor Inverters

Such inverters provide a very efficient and economical way of converting direct current (or voltage) into alternating current (or voltage). In this application, a thyristor serves as a controlled switch alternately opening and closing a d.c. circuit. Fig. 28.31(a), shows a basic inverter circuit

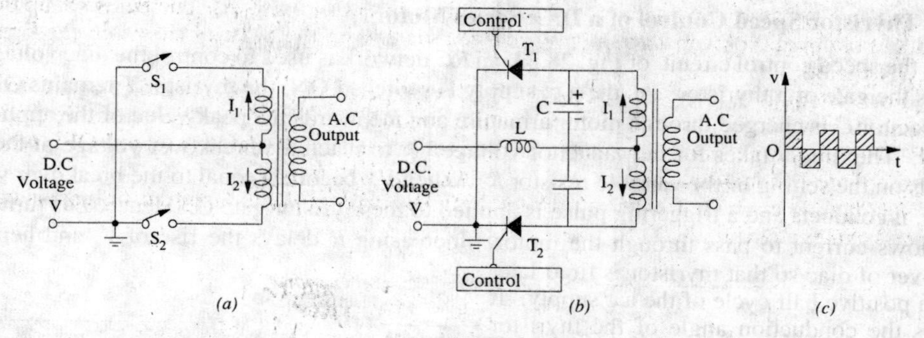

Fig. 28.31

where an a.c. output is obtained by alternately opening and closing switches S_1 and S_2. When we replace the mechanical switches by two thyristors (with their gate triggering circuits), we get the thyristor inverter of Fig. 28.31(*b*).

Before discussing the actual circuit, it is worthwhile to recall that thyristor is a latching devi which means that once it starts conducting, gate loses control over it and cannot switch it OFF whateve the gate signal. A separate *commutating* circuitry is used to switch the thyristor OFF and thus enable it to perform ON-OFF switching function.

Suppose T_1 is fired while T_2 is still OFF. Immediately I_1 is set up which flows through L, one half of transformer primary and T_1. At the same time, C is charged with the polarity as shown.

Next when T_2 is fired into condition, I_2 is set up and C starts discharging through T_1 thereby reverse-biasing it to CUT-OFF.

When T_1 is again pulsed into conduction, I_1 is set up and C starts discharging thereby reverse-biasing T_2 to OFF and the process just described repeats. As shown in Fig. 28.31(*c*), the output is an alternating voltage whose frequency depends on the switching frequency of thyristors T_1 and T_2.

28.14. Thyristor Speed Control of Separately-excited D.C. Motor

In Fig. 28.32, the bridge rectifier circuit converts a voltage into d.c. voltage which is then applied to the armature of the separately-excited d.c. motor M.

As we know, speed of a motor is given by

$$N = \frac{V - I_a R_a}{\Phi}\left(\frac{A}{ZP}\right)$$

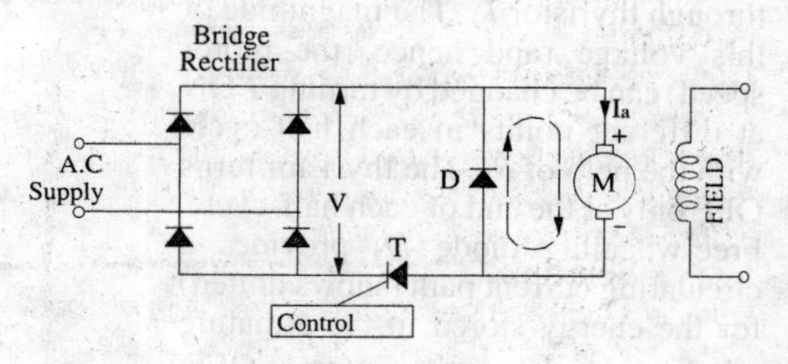

Fig. 28.32

If Φ is kept constant and also if $I_a R_a$ is neglected, then, $N \propto V \propto$ voltage across the armature. The value of this voltage furnished by the rectifier can be changed by varying the firing angle $\propto$ of the thyristor T with the help of its control circuit. As α is increased *i.e.* thyristor firing is delayed more, its conduction period is reduced and, hence, armature voltage is decreased which, in turn, decreases the motor speed. When α is decreased i.e. thyristor is fired earlier, conduction period is increased which increases the mean value of the voltage applied across the motor armature. Consequently, motor speed is increased. In short, as α increases, V decreases and hence N decreases. Conversely, as α decreases, V increases and so, N increases. The free-wheeling diode D connected across the motor provides a circulating current path (shown dotted) for the energy stored in the inductance of the armature winding at the time T turns OFF. Without D, current will flow through T and bridge rectifier, prohibiting T from turning OFF.

28.15. Thyristor Speed Control of a D.C. Series Motor

In the speed control circuit of Fig. 28.33, an RC network is used to control the diac voltage that triggers the gate of a thyristor. As the a.c. supply is switched ON, the thyristor T remains OFF but the capacitor C is charged through motor armature and R towards the peak value of the applied a.c. voltage. The time it takes for the capacitor voltage V_C to reach the breakover voltage of the diac* depends on the setting of the variable resistor R. When V_C becomes equal to the breakover voltage of diac, it conducts and a triggering pulse is applied to the thyristor gate G. Hence, T is turned ON and allows current to pass through the motor. Increasing R delays the rise of V_C and hence the breakover of diac so that thyristor is fired later in each positive half cycle of the a.c supply. It reduces the conduction angle of the thyristor which, consequently, delivers less power to the motor. Hence, motor speed is reduced.

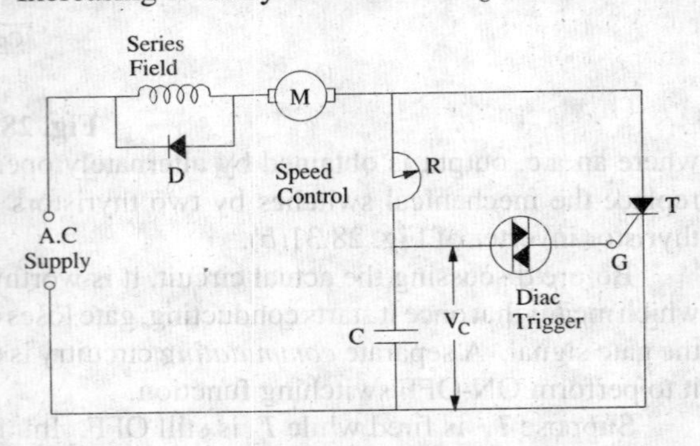

Fig. 28.33

If R is reduced, time-constant of the RC network is decreased which allows V_C to rise to the breakover voltage of diac more quickly. Hence, it makes the thyristor fire early in each positive input half-cycle of the supply. Due to increase in the conduction angle of the thyristor, power delivered to the motor is increased with a subsequent increase in its speed. As before, D is the free-wheeling diode which provides circulating current path for the energy stored in the inductance of the armature winding.

28.16. Full-wave Speed Control of a Shunt Motor

Fig 28.34 shows a circuit which provides a wide range of speed control for a fractional kW shunt d.c. motor. The circuit uses a bridge circuit for full-wave rectification of the ac supply. The shunt field winding is permanently connected across the d.c. output of the bridge circuit. The armature voltage is applied through thyristor T. The magnitude of this voltage (and hence, the motor speed) can be changed by turning T ON at different points in each half-cycle with the help of R. The thyristor turns OFF only at the end of each half-cycle. Free-wheeling diode D_3 provides a circulating current path (shows dotted) for the energy stored in the armature winding at the time T turns OFF. Without D_3, this current would circulate through T and the bridge rectifier thereby prohibiting T from turning OFF.

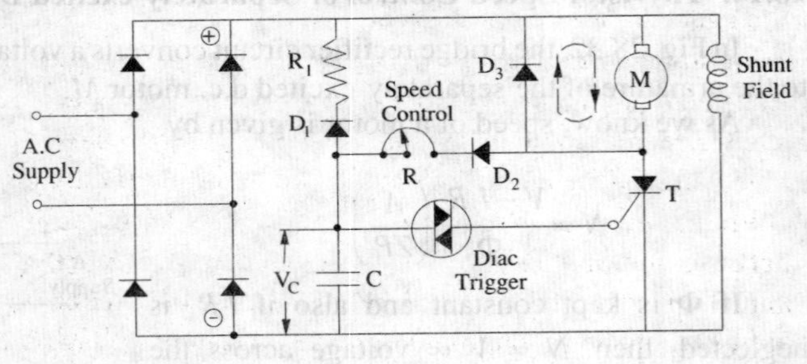

Fig. 28.34

At the beginning of each half-cycle, T is the *OFF* state and C starts charging up via motor armature, diode D_2 and speed-control variable resistor R (it cannot charge through R_1 because of reverse-biased diode D_1). When voltage across C *i.e.* V_C builds up to the breakover voltage of diac, diac conducts and applies a sudden pulse to T thereby turning it ON. Hence, power is supplied to the motor armature for the remainder of that half-cycle. At the end of each half-cycle, C is discharged through D_1, R_1 and shunt field winding. The delay angle α depends on the time it takes V_C to become

*Like a triac, it has directional switching characteristics.

equal to the breakover voltage of the diac. This time, in turn, depends on the time-constant of the *R-C* circuit and the voltage available at point *A*. By changing *R*, V_C can be made to build-up either slowly or quickly and thus change the angle α at will. In this way, the average value of the d.c. voltage across the motor armature can be controlled. It further helps to control the motor speed because it is directly proportional to the armature voltage.

Now, when load is increased, motor tends to slow down. Hence, E_b is reduced. The voltage of point *A* is increased because it is equal to the d.c. output voltage of the bridge rectifier *minus* back emf E_b. Since V_A increases *i.e.* voltage across the *R-C* charging circuit increases, it builds up V_C more quickly thereby decreasing which leads to early switching ON of *T* in each half-cycle. As a result, power supplied to the armature is increased which increases motor speed thereby compensating for the motor loading.

28.17. Thyristor Speed Control of a Shunt Motor

The speed of a shunt d.c motor (upto 5 kW) may be regulated over a wide range with the help of the full-wave rectifier using only one main thyristor (or *SCR*) *T* as shown in Fig. 28.35. The firing angle α of *T* is adjusted by R_1 thereby controlling the motor speed. The thyristor and *SUS* (silicon unilateral switch*) are reset (*i.e.* stop conduction) when each half wave of voltage drops to zero. Before switching on the supply, R_1 is increased by turning it in the counter-clockwise direction. Next, when supply is switched ON, *C* gets charged via motor armature and diode D_1 (being forward biased). It means that it takes much longer for V_C to reach the breakdown voltage of SUS due to large time constant of R_1–*C* network. Once V_C reaches that value, SUS conducts suddenly and triggers *T* into conduction. Since thyristor starts conducting late (*i.e.* its α is large), it furnishes low voltage to start the motor. As speed selector R_1 is turned clockwise (for less resistance), *C* charges up more rapidly (since time constant is decreased) to the breakover voltage of SUS thereby firing *T* into conduction earlier. Hence, average value of the dc voltage across the motor armature increases thereby increasing its speed.

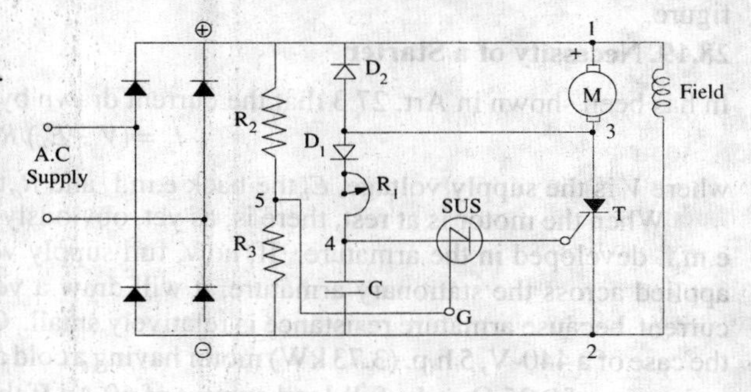

Fig. 28.35

While the motor is running at the speed set by R_1, suppose that load on the motor is increased. In that case, motor will tend to slow down thereby decreasing armature back emf. Hence, potential of point 3 will rise which will charge *C* faster to the breakover voltage of SUS. Hence, thyristor will be fired earlier thereby applying greater armature voltage which will return the motor speed to its desired value. As seen, the speed is automatically regulated to offset changes in load.

The function of free-wheeling diode D_2 is to allow dissipation of energy stored in motor armature during the time the full-wave rectified voltage drops to zero between half-cycles. If D_2 is not there, then decreasing armature current during those intervals would be forced to flow through *T* thereby preventing its being reset. In that case, *T* would not be ready to be fired in the next half-cycle.

Similarly, towards the end of each half cycle as points 1 and 5 decrease towards zero potential, the negative going gate *G* turns SUS on thereby allowing *C* to discharge completely through SUS and thyristor gate-cathode circuit so that it can get ready to be charged again in the next half-cycle.

---

*It is a four-layer semiconductor diode with a gate terminal. Unlike diac, it conducts in one direction only.

28.18. Thyristor Speed Control of a Series D.C. Motor

Fig. 28.36 shows a simple circuit for regulating the speed of a d.c. motor by changing the average value of the voltage applied across the motor armature by changing the thyristor firing angle ∝. The trigger circuit R_1–R_2 can give a fring range of almost 180°. As the supply is switched on, full d.c. voltage is applied across $R_1 - R_2$. By changing the variable resistance R_2, drop across it can be made large enough to fire the SCR at any desired angle from 0° – 180°. In this way, output voltage of the bridge rectifier can be changed considerably, thus enabling a wide-range control of the motor speed. The speed control can be made somewhat smoother by joining a capacitor C across R_2 as shown in the figure.

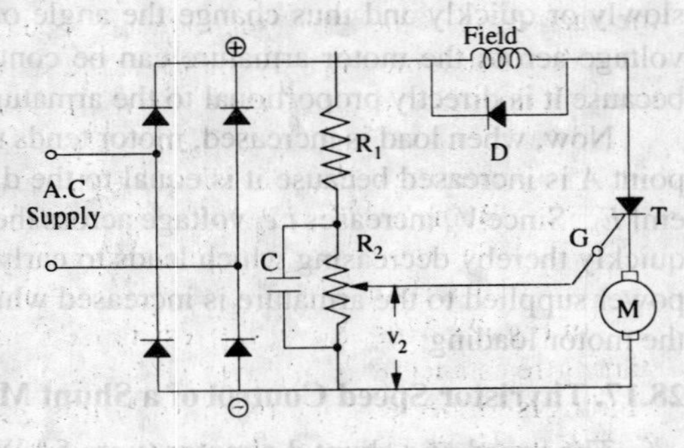

Fig. 28.36

28.19. Necessity of a Starter

In has been shown in Art. 27.3 that the current drawn by a motor armature is given by the relation

$$I_a = (V - E_b)/R_a$$

where V is the supply voltage, E_b the back e.m.f. and R_a the armature resistance.

When the motor is at rest, there is, as yet, obviously no back e.m.f. developed in the armature. If, now, full supply voltage is applied across the stationary armature, it will draw a very large current because armature resistance is relatively small. Consider the case of a 440-V, 5 h.p. (3.73 kW) motor having a cold armature resistance of 0.25 Ω and a full-load current of 50 A. If this motor is started from the line directly, it will draw a starting currant of 400/0.25 = 1760 A which is 1760/50 = 35.2 times its full-load current. This excessive current will blow out the fuses and, prior to that, it will damage the commutator and brushes etc. To avoid this happening, a resistance is introduced in series with the

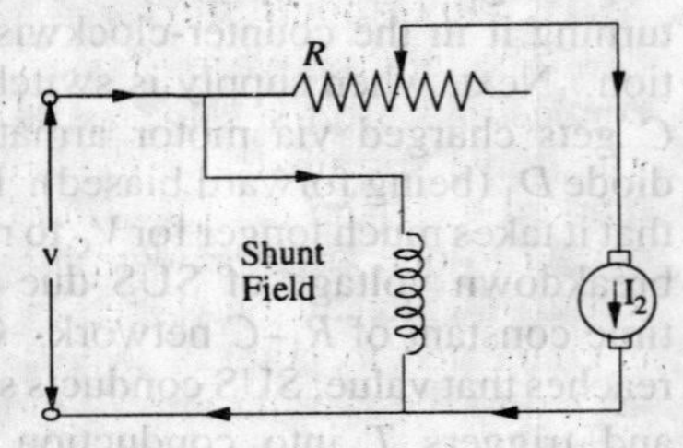

Fig. 28.37

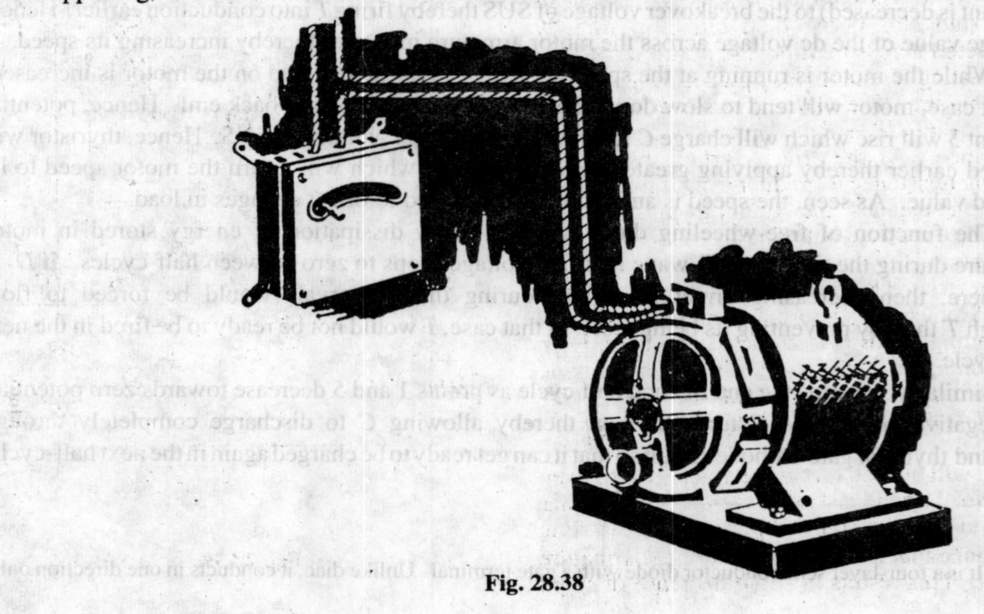

Fig. 28.38

armature (for the duration of starting period only, say 5 to 10 seconds) which limits the starting current to a safe value. The starting resistance is gradually cut out as the motor gains speed and develops the back e.m.f. which then regulates its speed.

Very small motors may, however, be started from rest by connecting them directly to the supply lines. It does not result in any harm to the motor for the following reasons :

1. Such motors have a relatively higher armature resistance than large motors, hence their starting current is not so high.
2. Being small, they have low moment of inertia, hence they speed up quickly.
3. The momentary large starting current taken by them is not sufficient to produce a large disturbance in the voltage regulation of the supply lines.

In Fig. 28.37 is shown the resistance R used for starting a shunt motor. It will be seen that the starting resistance R is in series with the *armature* and not with the *motor* as a whole. The field winding is connected directly across the lines, hence shunt field current is independent of the resistance R. If R were introduced in the motor circuit, then I_{sh} will be small at the start, hence starting torque T_{st} would be small ($\because$ $T_a \propto \Phi I_a$) and there would be experienced some difficulty in starting the motor.

Such a simple starter is shown diagrammatically in Fig. 38.38.

28.20. Shunt Motor Starter

The face-plate box type starters used for starting shunt and compound motors of ordinary industrial capacity are of two kinds known as three-point and four-point starters respectively.

28.21. Three-point Starter

The internal wiring for such a starter is shown in Fig. 28.39 and it is seen that basically the connections are the same as in Fig. 28.37 except for the additional protective devices used here. The three terminals of the starting box are marked A, B and C. One line is directly connected to one armature terminal and one field terminal which are tied together. The other line is connected to point A which is further connected to the starting arm L, through the over-current (or over-load) release M.

To start the motor, the main switch is first closed and then the starting arm is slowly moved to the right. As soon as the arm makes contact with stud No. 1, the field circuit is directly connected across the line and at the same time full starting resistance R_s is placed in series with the armature. The starting current drawn by the armature $= V/(R_a + R_s)$ where R_s is the starting resistance. As the arm is further moved, the starting resistance is gradually cut out till, when the arm reaches the running position, the resistance is all cut out. The arm moves over the various studs against a strong spring which tends to restore it to OFF position. There is a soft iron piece S attached to the arm which in the full 'ON' or running position is attracted and held by an electromagnet E energised by the shunt current. It is variously known as 'HOLD-ON' coil, LOW VOLTAGE (or NO-VOLTAGE) release.

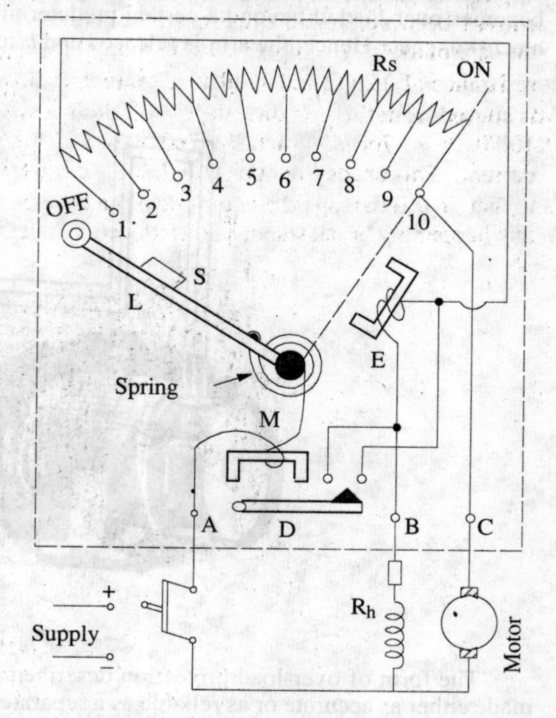

Fig. 28.39

It will be seen that as the arm is moved from stud No. 1 to the last stud, the field current has to travel back through that portion of the starting resistance that has been cut out of the armature circuit. This results in slight decrease of shunt

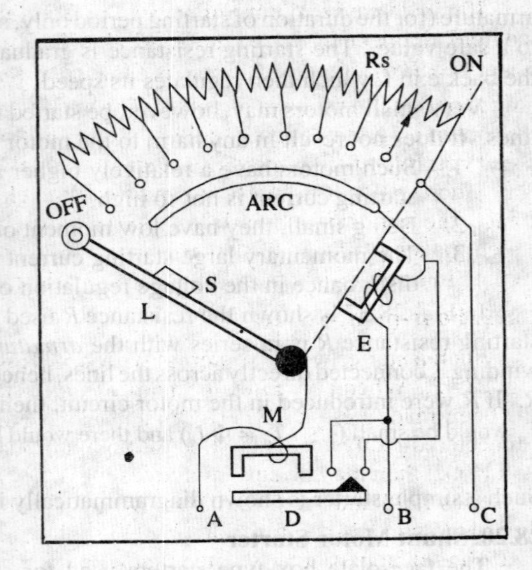

Fig. 28.40

current. But as the value of starting resistance is very small as compared to shunt field resistance, this slight decrease in I_{sh} is negligible. This defect can, however, be remedied by using a brass arc which is connected to stud No. 1 (Fig. 28.40). The field circuit is completed through this arc so that current does not now pass through the starting resistance as it did in Fig. 28.39.

Now, we will discuss the action of the two protective devices shown in Fig. 28.39. The normal function of the HOLD-ON coil is to hold on the arm in the full running position when the motor is in normal operation. But, in the case of failure or disconnection of the supply or a break in the field circuit, it is de-energised thereby releasing the arm which is pulled back by the spring to the OFF position. This prevents the stationary armature from being put across the lines again when the supply is restored after temporary shunt down. This would have happened if the arm were left in the full ON position. One great advantage of connecting the HOLD-ON coil in series with the shunt field is that, should the field circuit become open, the starting arm immediately springs back to the OFF position thereby preventing the motor from running away.

The over-current release consists of an electromagnet connected in the supply line. If the motor becomes over-loaded beyond a certain predetermined value, then D is lifted and short-circuits the electromagnet. Hence, the arm is released and returns to OFF position.

Fig. 28.41

The form of over-load protection described above is becoming obsolete, because it cannot be made either as accurate or as reliable as a separate well-designed circuit breaker with a suitable time element attachment. Many a times a separate magnetic contactor with an over-load relay is also used.

Often the motors are protected by thermal over-load relays in which a bimetallic strip is heated by the motor current at approximately the same rate at which the motor is itself heating up. Above a certain temperature, this relay trips and opens the line contactor thereby isolating the motor from the supply.

If it is desired to control the speed of the motor in addition, then a field rheostat is connected in the field circuit as indicated in Fig. 28.39. The motor speed can be increased by weakening the flux ($\because$ $N \propto I/\Phi$). Obviously, there is a limit to the speed increase obtained in this way, although speed ranges of three to four are possible. The connections of a starter and speed regulator with the motor are shown diagrammatically in Fig. 28.41. But there is one difficulty with such an arrangement for speed control. If too much resistance is 'cut in' by the field rheostat, then field current is reduced very much so that it is unable to create enough electromagnetic pull to overcome the spring tension. Hence, the arm is pulled back to OFF position. It is this undesirable feature of a three-point starter which makes it unsuitable for use with variable-speed motors. This has resulted in wide-spread application of four-point starter discussed below.

28.22. Four-point Starter

Such a starter with its internal wiring is shown connected to a long-shunt compound motor in Fig. 28.42. When compared to the three-point starter, it will be noticed that one important change has been made *i.e.* the HOLD ON coil has been taken out of the shunt field circuit and has been connected directly across the line through a protecting resistance as shown. When the arm touches stud No. 1, then the line current divides into three parts (*i*) one part passes through starting resistance R_S, series field and motor armature (*ii*) the second part passes through the shunt field and its field rheostat R_h and (*iii*) the third part passes through the HOLD-ON coil and current-protecting resistance *R*. It should be particularly noted that with this arrangement any change of current in the shunt field circuit does not at all affect the current passing through the HOLD-ON coil because the two circuits

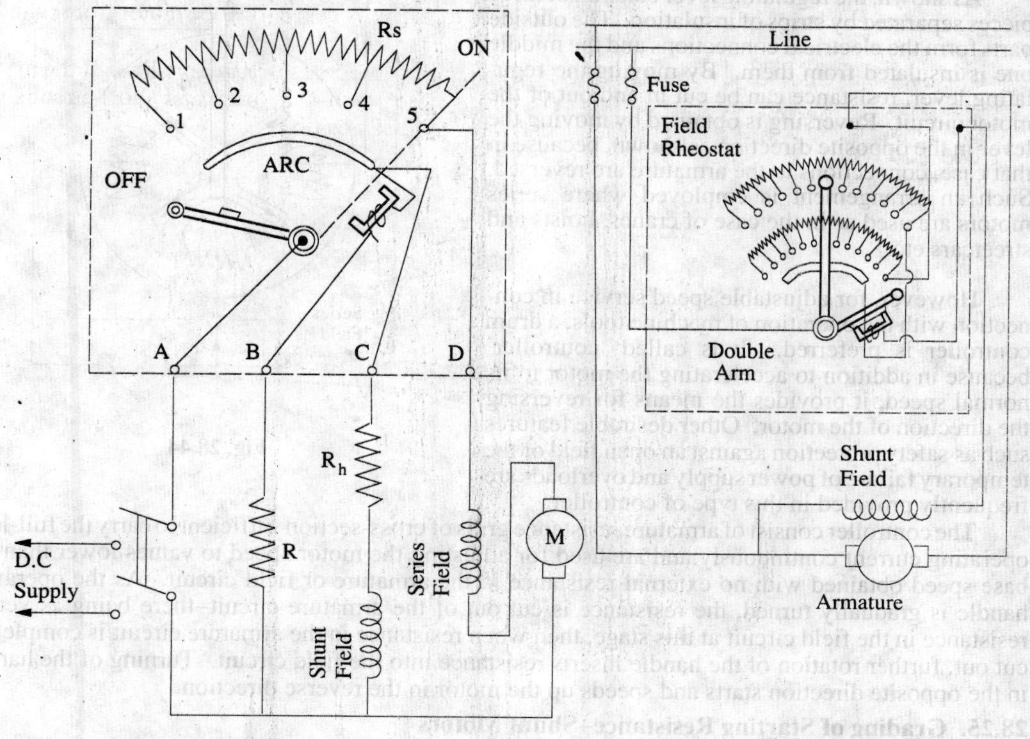

Fig. 28.42

Fig. 28.43

are independent of each other. It means that the electromagnetic pull exerted by the HOLD-ON coil will always be sufficient and will prevent the spring from restoring the starting arm to OFF position no matter how the field rheostat or regulator is adjusted.

28.23. Starter and Speed-control Rheostats

Sometimes, for convenience, the field rheostat is also contained within the starting box as shown in Fig. 28.43. In this case, two arms are used. There are two rows of studs, the lower ones being connected to the armature. The inside starting arm moves over the lower studs on the starting resistor whereas the outside field lever moves over the upper ones on the field rheostat. Only the outside field arm is provided with an operating handle. While starting the motor, the two arms are moved together, but field lever is electrically inoperative because the field current flows directly from the starting arm through the brass arc to HOLD-ON coil and finally to the shunt field winding. At the end of the starting period, the starting arm is attracted and held in FULL-ON position by the HOLD-ON coil, and the contact between the starting arm and brass arc is broken thus forcing field current to pass through the field rheostat. The field lever can be moved back to increase the motor speed. It will be seen that now the upper row of contacts is operative because starting arm no longer touches the brass arc.

When motor is stopped by opening the main switch, the starting arm is released and on its way back it strikes the field lever so that both arms are returned simultaneously to OFF position.

28.24. Starting and Speed Control of Series Motor

For starting and speed control of series motor either a face-plate type or drum-type controller is used which usually has the reversing feature also. A face-plate type of reversing controller is shown in Fig. 28.44.

Except for a separate over-load circuit, no inter-locking or automatic features are required because the operator watches the performance continuously.

As shown, the regulating lever consists of three pieces separated by strips of insulation. The outside parts form the electrical connections and the middle one is insulated from them. By moving the regulating lever, resistance can be cut in and out of the motor circuit. Reversing is obtained by moving the lever in the opposite direction as shown, because in that case, connections to the armature are reversed. Such an arrangement is employed where series motors are used as in the case of cranes, hoists and streetcars etc.

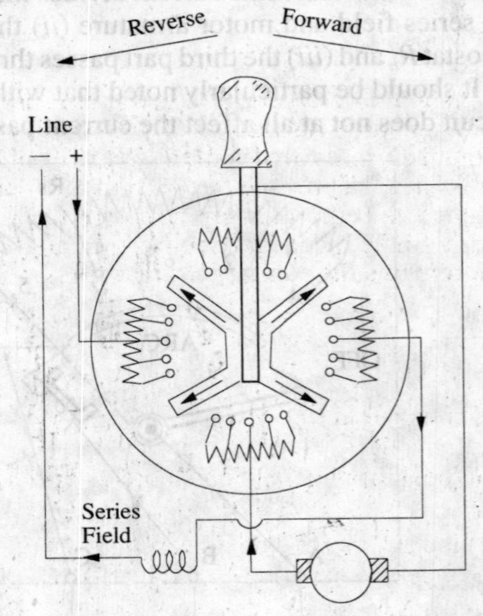

Fig. 28.44

However, for adjustable speed service in connection with the operation of machine tools, a drum controller is preferred. It is called 'controller' because in addition to accelerating the motor to its normal speed, it provides the means for reversing the direction of the motor. Other desirable features such as safety protection against an open field or the temporary failure of power supply and overloads are frequently provided in this type of controller.

The controller consist of armature resistance grids of cross-section sufficient to carry the full-load operating current continuously and are used for adjusting the motor speed to values lower than the base speed obtained with no external resistance in the armature or field circuit. As the operating handle is gradually turned, the resistance is cut out of the armature circuit–there being as yet no resistance in the field circuit at this stage, then when resistance in the armature circuit is completely cut out, further rotation of the handle inserts resistance into the field circuit. Turning of the handle in the opposite direction starts and speeds up the motor in the reverse direction.

28.25. Grading of Starting Resistance–Shunt Motors

T_{st} would be small In designing shunt motor starters, it is usual to allow an overload of 50% for starting and to advance the starter a step when armature current has fallen to a definite lower value. Either this lower current limit may be fixed or the number of starter steps may be fixed. In the former case, the number of steps are so chosen as to suit the upper and lower current limits whereas in the latter case, the lower current limit will depend on the number of steps specified. It can be shown that the resistances in the circuit on successive studs form geometrical progression, having a common ratio equal to lower current limit/upper current limit *i.e.* I_2/I_1.

In Fig. 28.45 is shown the starter connected to a shunt motor. For the sake of simplicity, four live studs have been taken. When arm A makes contact with stud No. 1, full shunt field is established and at the same time the armature current immediately jumps to a maximum value I_1 given by $I_1 = V/R_1$ where R_1 = armature and starter resistance (Fig. 28.45).

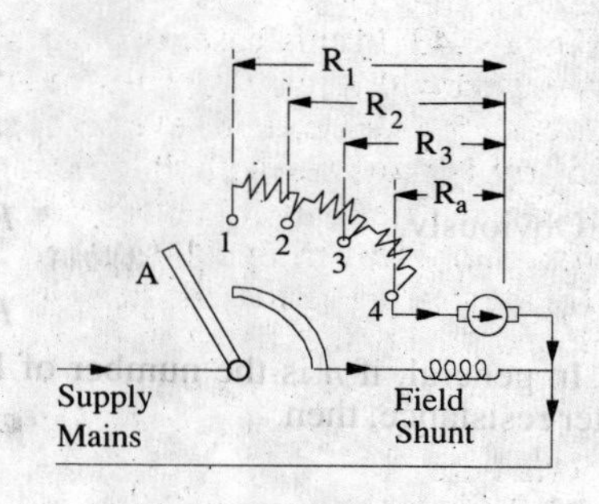

Fig. 28.45

I_1 the maximum permissible armature current at the start (I_{max}) and is, as said above, usually limited to 1.5 times the full-load current of the motor. Hence, the motor develops 1.5 times its full-load torque and accelerates very rapidly. As the motors speeds up, its back e.m.f. grows and hence decreases the armature current as shown by curve *ab* in Fig. 28.46.

When the armature current has fallen to some predetermined value I_2 (also called I_{min}) arm A is moved to stud No. 2. Let the value of back e.m.f. be E_{b1} at the time of leaving stud No. 1. Then

$$I_2 = \frac{V - E_{b1}}{R_1} \quad ...(i)$$

It should be carefully noted that I_1 and I_2 (I_{max}) and (I_{min}) are respectively the maximum and minimum currents of the motor. When arm A touches stud No. 2, then due to diminution of circuit resistance, the current again jumps up to its previous value I_1. Since speed had no time to change, the back e.m.f. remains the same as initially.

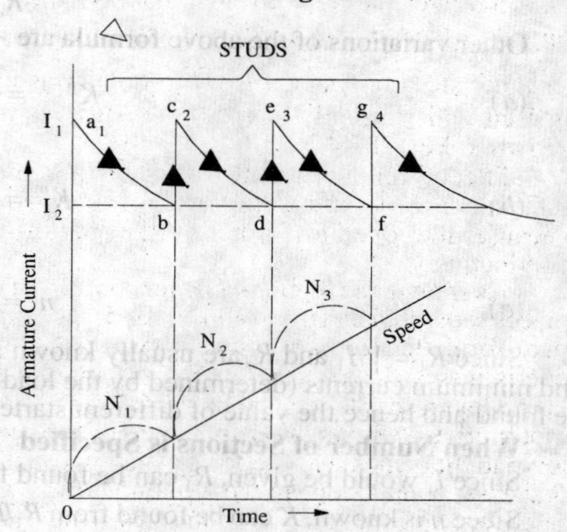

Fig. 28.46

$$\therefore \quad I_1 = \frac{V - E_{b1}}{R_2} \quad ...(ii)$$

From (*i*) and (*ii*), we get $\dfrac{I_1}{I_2} = \dfrac{R_1}{R_2}$ ...(*iii*)

When arm A is held on stud No. 2 for some time, then speed and hence the back e.m.f. increases to a value E_{b2} thereby decreasing the current to previous value I_2, so that $I_2 = \dfrac{V - E_{b2}}{R_2}$ ....(*iv*)

Similarly, on first making contact with stud No. 3, the current is

$$I_1 = \frac{V - E_{b2}}{R_3} \quad ...(v)$$

From (*iv*) and (*v*), we again get $\dfrac{I_1}{I_2} = \dfrac{R_2}{R_3}$ ...(*vi*)

When arm A is held on stud No. 3 for some time, the speed and hence back e.m.f. increases to a new value E_{b3}, thereby decreasing the armature current to a value I_2 such that

$$I_2 = \frac{V - E_{b3}}{R_3} \quad ...(vii)$$

On making contact with stud No. 4, current jumps to I_1 given by

$$I_1 = \frac{V - E_{b3}}{R_a} \quad ...(viii)$$

From (*vii*) and (*viii*), we get $\dfrac{I_1}{I_2} = \dfrac{R_3}{R_a}$ ...(*ix*)

From (*iii*), (*vi*) and (*ix*), it is seen that

$$\frac{I_1}{I_2} = \frac{R_1}{R_2} = \frac{R_2}{R_3} = \frac{R_3}{R_a} = K \text{ (say)} \qquad \ldots(x)$$

Obviously,
$$R_3 = KR_a;\ R_2 = KR_3 = K^2R_a$$
$$R_1 = KR_2 = K.K^2R_a = K^3R_a$$

In general, if n is the number of live studs and therefore $(n-1)$ the number of sections in the starter resistance, then

$$R_1 = K^{n-1}.R_a \text{ or } \frac{R_1}{R_a} = K^{n-1} \text{ or } \left(\frac{I_1}{I_2}\right)^{n-1} = \frac{R_1}{R_a} \qquad \ldots\text{from } (x)$$

Other variations of the above formula are

(a) $$K^{n-1} = \frac{R_1}{R_a} = \frac{V}{I_1R_a} = \frac{V}{I_{max}R_a}$$

(b) $$K^{n} = \frac{V}{I_1R_a}.\frac{I_1}{I_2} = \frac{V}{I_2R_a} = \frac{V}{I_{min}.R_a} \text{ and}$$

(c) $$n = 1 + \frac{\log (V/R_a.I_{max})}{\log K} \text{ from } (a) \text{ above.}$$

Since $R_1 = V/I_1$ and R_a are usually known and K is known from the given values of maximum and minimum currents (determined by the load against which motor has to start), the value of n can be found and hence the value of different starter sections.

When Number of Sections is Specified

Since I_1 would be given, R_1 can be found from $R_1 = V/I_1$.

Since n is known, K can be found from $R_1/R_a = K^{n-1}$ and the lower current limit I_2 from $I_1/I_2 = K$.

Example 28.39. *A 10 b.h.p. (7.46 kW) 200-V shunt motor has full-load efficiency of 85%. The armature has a resistance of 0.25 Ω. Calculate the value of the starting resistance necessary to limit the starting current to 1.5 times the full-load current at the moment of first switching on. The shunt current may be neglected. Find also the back e.m.f. of the motor, when the current has fallen to its full-load value, assuming that the whole of the starting resistance is still in circuit.*

Solution. Full-load motor current= 7,460/200 × 0.85 = 43.88 A

Starting current $I_1 = 1.5 \times 43.88 = 65.83$ A

$R_1 = V/I_1 = 200/65.83 = 3.038\ \Omega$; $R_a = 0.25\ \Omega$

∴ starting resistance $= R_1 - R_a = 3.038 - 0.25 = 2.788\ \Omega$

Now, full-load current $I_2 = 43.88$ A

Now, $I_2 = \dfrac{V - E_{b1}}{R_1}$ ∴ $E_{b1} = V - I_2R_1 = 200 - (43.88 \times 3.038) =$ **67 V**

Example 28.40. *A 220-V shunt motor has an armature resistance of 0.5 Ω. The armature current at starting must not exceed 40 A. If the number of sections is 6, calculate the values of the resistor steps to be used in this starter.* **(Elect. Machines, AMIE Sec. B, 1992)**

Solution. Since the number of starter sections is specified, we will use the relation.

$R_1/R_a = K^{n-1}$ or $R_1 = R_aK^{n-1}$

Now, $R_1 = 220/40 = 5.5\ \Omega$, $R_a = 0.4\ \Omega$; $n-1 = 6$, $n = 7$

∴ $5.5 = 0.4K^6$ or $K^6 = 5.5/0.4 = 13.75$

$6 \log_{10} K = \log_{10} 13.75 = 1.1383$; $K = 1.548$

Now, $R_2 = R_1/K = 5.5/1.548 = 3.553\ \Omega$, $R_3 = R_2/K = 3.553/1.548 = 2.295\ \Omega$

$R_4 = 2.295/1.548 = 1.482\ \Omega$; $R_5 = 1.482/1.548 = 0.958\ \Omega$

$R_6 = 0.958/1.548 = 0.619\ \Omega$

Resistance of Ist section $= R_1 - R_2 = 5.5 - 3.553 = \mathbf{1.947\ \Omega}$
" 2nd" $= R_2 - R_3 = 3.553 - 2.295 = \mathbf{1.258\ \Omega}$
" 3rd" $= R_3 - R_4 = 2.295 - 1.482 = \mathbf{0.813\ \Omega}$
" 4th" $= R_4 - R_5 = 1.482 - 0.958 = \mathbf{0.524\ \Omega}$
" 5th" $= R_5 - R_6 = 0.958 - 0.619 = \mathbf{0.339\ \Omega}$
" 6th" $= R_6 - R_a = 0.619 - 0.4 = \mathbf{0.219\ \Omega}$

Example 28.41. *Find the values of the step resistances in a 6-stud starter for a 5 h.p. (3.73 kW), 200-V shunt motor. The maximum current in the line is limited to twice the full-load value. The total Cu loss is 50% of the total loss. The normal field current is 0.6 A and the full-load efficiency is found to be 88%.* **(D.C. Machines, Jadavpur Univ. 1988)**

Solution. Output $= 3{,}730$ W Input $= 3{,}730/0.88 = 4{,}238$ W
Total loss $= 4{,}238 - 3{,}730 = 508$ W
Armature Cu loss alone $= 508/2 = 254$ W
Input current $= 4{,}238/200 = 21.19$ A
armature current $= 21.19 - 0.6 = 20.59$ A

$\therefore \quad 20.59^2 R_a = 254 \quad \therefore\ R_a = 254/20.59^2 = 0.5989\ \Omega$

Permissible input current $= 21.19 \times 2 = 42.38$ A
Permissible armature current $= 42.38 - 0.6 = 41.78$ A

$\therefore \quad R_1 = 200/41.78 = 4.787\ \Omega\ ;\ n = 6\ ;\ n-1 = \mathbf{5}$

$\therefore \quad 4.787 = K^5 \times 0.5989 \qquad \therefore\ K^5 = 4.787/0.5989 = 7.993$

or $\quad 5 \log K = \log 7.993 = 0.9027 \therefore\ \log K = 0.1805\ ;\ K = 1.516$

Now $\quad R_1 = R_l/K = 4.789/1.516 = 3.159\ \Omega$

$R_2 = 2.084\ \Omega\ ;\ R_4 = 1.376\ \Omega\ ;\ R_5 = 0.908\ \Omega$.

Resistance in 1st step $= R_1 - R_2 = 4.787 - 3.159 = \mathbf{1.628\ \Omega}$
Resistance in 2nd step $= R_2 - R_3 = 3.159 - 2.084 = \mathbf{1.075\ \Omega}$
Resistance in 3rd step $= R_3 - R_4 = 2.084 - 1.376 = \mathbf{0.708\ \Omega}$
Resistance in 4th step $= R_4 - R_5 = 1.376 - 0.908 = \mathbf{0.468\ \Omega}$
Resistance in 5th step $= R_5 - R_a = 0.908 - 0.5989 = \mathbf{0.309\ \Omega}$

The various sections are shown in Fig. 28.47.

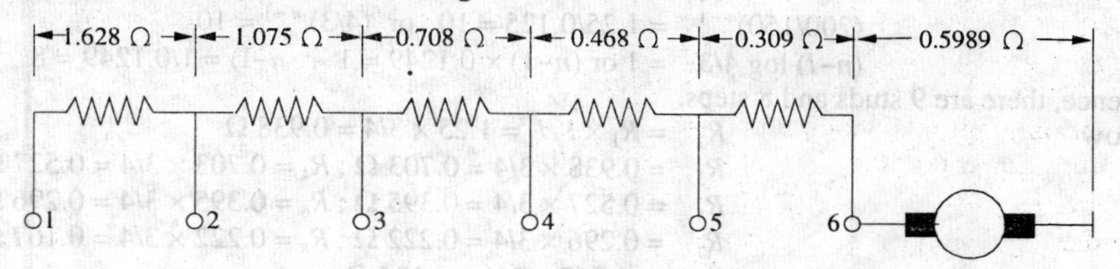

Fig. 28.47

Example 28.42. *Design the resistance sections of a seven-stud starter for 36.775 kW, 400 V, d.c. shunt motor. Full-load efficiency is 92%, total Cu losses are 5% of the input. Shunt field resistance is 200 Ω. The lower limit of the current through the armature is to be full-load value.* **(Elect. Machines, Gujarat Univ. 1987)**

Solution. Output $= 36{,}775$ W ; Input $= 36{,}775/0.92 = 39{,}980$ W
Total *Cu* loss $= 0.05 \times 39{,}980 = 1{,}999$ W
Shunt Cu loss $= V^2/R_{sh} = 400^2/200 = 800$ W
Armature Cu loss $= 1{,}999 - 800 = 1199$ W
F.L. input current $= 39{,}980/400 = 99.95$ A
$I_{sh} = 400/200 = 2\text{A};\ I_a = 99.95 - 2 = 97.95$ A

$\therefore \quad 97.95^2\, R_a = 1199\text{d} \text{ or } R_a = 0.125\text{n}$

Now, minimum armature current equals full-load current *i.e.* $I_a = 97.95$ A. As seen from Art. 28.25 (*b*), we have

$$K^n = \frac{V}{I_2 R_a} \text{ or } K^7 = 400/97.95 \times 0.125 = 32.68 \therefore K = 32.68^{1/7} = 1.645$$

I_1 = maximum permissible armature current

$= KI_2 = 1.645 \times 97.94 = 161$ A

$\therefore R_1 = V/I_1 = 400/161 = 2.483\ \Omega$

$R_2 = R_1/K = 2.483/1.645 = 1.51\ \Omega$

$R_3 = 1.51/1.645 = 0.917\ \Omega$

$R_4 = 0.917/1.645 = 0.557\ \Omega$

$R_5 = 0.557/1.645 = 0.339\ \Omega$

$R_6 = 0.339/1.645 = 0.206\ \Omega$

$R_7 = 0.206/1.645 = 0.125\ \Omega$

Resistance in Ist step $= R_1 - R_2 =$ **0.973 Ω**

Resistance in 2nd step $= R_2 - R_3 =$ **0.593 Ω**

Resistance in 3rd step $= R_3 - R_4 =$ **0.36 Ω** ;

Resistance in 4th step $= R_4 - R_5 =$ **0.218 Ω**

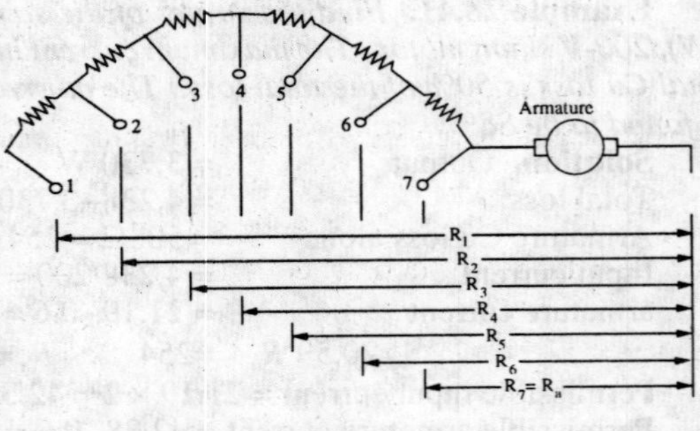

Fig. 28.48

Resistance in 5th step $= R_5 - R_6 =$ **0.133 Ω**; Resistance in 6th step $= R_6 - R_a =$ **0.081 Ω**

The various starter sections are shown in Fig. 28.48.

Example 28.43. *Calculate the resistance steps for the starter of a 250-V, d.c. shunt motor having an armature resistance of 0.125 Ω and a full-load current of 150 A. The motor is to start against full-load and maximum current is not to exceed 200 A.*

(Electrical Engineering-I, Bombay Univ. 1989)

Solution. As the motor is to start against its full-load, the minimum current is its F.L. current *i.e.* 150 A. We will use the formula given in Art. 28.25.

$(I_1/I_2^{n-1}) = R_1/R_a$

Here $I_1 = 200$ A $\quad I_2 = 150$ A $\quad R_1 = 250/200 =$ **1.25 Ω**

$R_a = 0.125\ \Omega$; n = No of live studs

$\therefore (200/150)^{n-1} = 1.25/0.125 = 10$ or $(4/3)^{n-1} = 10$

$(n-I) \log 4/3 = 1$ or $(n-1) \times 0.1249 = 1 \therefore n - I) = 1/0.1249 = 8$

Hence, there are 9 studs and 8 steps.

Now $R_2 = R_1 \times I_2/I_1 = 1.25 \times 3/4 = 0.938\ \Omega$

$R_3 = 0.938 \times 3/4 = 0.703\ \Omega$; $R_4 = 0.703 \times 3/4 = 0.527\ \Omega$

$R_5 = 0.527 \times 3/4 = 0.395\ \Omega$; $R_6 = 0.395 \times 3/4 = 0.296\ \Omega$

$R_7 = 0.296 \times 3/4 = 0.222\ \Omega$; $R_8 = 0.222 \times 3/4 = 0.167\ \Omega$

$R_a = 0.167 \times 3/4 = 0.125\ \Omega$

$\therefore$ resistance of Ist element $= 1.25 - 0.938 =$ **0.312 Ω**

" 2nd " $= 0.938 - 0.703 =$ **0.235 Ω**

" 3rd " $= 0.703 - 0.527 =$ **0.176 Ω**

" 4th " $= 0.527 - 0.395 =$ **0.132 Ω**

" 5th " $= 0.395 - 0.296 =$ **0.099 Ω**

" 6th " $= 0.296 - 0.222 =$ **0.074 Ω**

" 7th " $= 0.222 - 0.167 =$ **0.055 Ω**

" 8th " $= 0.167 - 0.125 =$ **0.042 Ω**

Example 28.44. *The 4-pole, lap-wound armature winding of a 500-V, d.c. shunt motor is housed in a total number of 60 slots each slot containing 20 conductors. The armature resistance is 1.31 Ω. If during the period of starting, the minimum torque is required to be 218 N-m and the maximum torque 1.5 times the minimum torque, find out how many sections the starter should have and calculate the resistances of these sections. Take the useful flux per pole to be 23 mWb.*

(Elect. Machinery-II, Bangalore Univ. 1991)

Solution. From the given minimum torque, we will be able to find the minimum current required during starting. Now

$$T_a = 0.159\ \Phi\ ZI_a\ (P/A)$$

$\therefore$ $218 = 0.159 \times 23 \times 10^{-3} \times (60 \times 20)\, I_a \times (4/4)$ $\therefore$ $I_a = 50$ A (approx.)

Maximum current = 50 × 1.5 = 75 A

$\therefore$ $I_1 = 75$ A ; $I_2 = 50$ A $\therefore$ $I_1/I_2 = 75/50 = 1.5$; $R_1 = 500/75 = 6.667\ \Omega$

If n is the number of starter studs, then

$(I_1/I_2)^{n-1} = R_1/R_a$ or $1.5^{n-1} = 6.667/1.31 = 5.09$

$\therefore$ $(n-1)\log_{10}1.5 = \log_{10}5.09$ $\therefore$ $(n-1) \times 0.1761 = 0.7067$ $\therefore$ $(n-1) = 4$ or $n = 5$

Hence, there are five studs and four sections.

$R_2 = R_1 \times I_2/I_1 = 6.667 \times 2/3 = 4.44\ \Omega$

$R_3 = 4.44 \times 2/3 = 2.96\ \Omega$; $R_4 = 2.96 \times 2/3 = \mathbf{1.98\ \Omega}$

Resistance of the first sections $= R_1 - R_2 = 6.67 - 4.44 = \mathbf{2.23\ \Omega}$

" " 2nd " $= R_2 - R_3 = 4.44 - 2.96 = \mathbf{1.48\ \Omega}$

" " 3rd " $= R_3 - R_4 = 2.96 - 1.98 = \mathbf{0.98\ \Omega}$

" " 4th " $= R_4 - R_a = 1.98 - 1.31 = \mathbf{0.67\ \Omega}$

28.26. Series Motor Starters

The basic principle employed in the design of a starter for series motor is the same as for a shunt motor *i.e.* the motor current is not allowed to exceed a certain upper limit as the starter arm moves from one stud to another. However, there is one significant difference. In the case of a series motor, the flux does not remain constant but varies with the current because armature current is also the exciting current. The determination of the number of steps is rather complicated as illustrated in Example 28.45. It may however, be noted that the section resistances form a geometrical progression.

The face-plate type of starter formerly used for d.c. series motor has been almost entirely replaced by automatic starter in which the resistance steps arc cut out automatically by means of a contactor operated by electromagnets. Such starters are well-suited for remote control.

However, for winch and crane motors where frequent starting, stopping, reversing and speed variations are necessary, drum type controllers are used. They are called controllers because they can be left in the circuit for any length of time. In addition to serving their normal function of starters, they also act as speed controllers.

Example 28.45. *(a) Show that, in general, individual resistances between the studs for a rheostat starter for a series d.c. motor with constant ratio of maximum to minimum current at starting, are in geometrical progression, stating any assumptions made.*

(b) Assuming that for a certain d.c. series motor the flux per pole is proportional to the starting current, calculate the resistance of each rheostat section in the case of a 50 b.h.p (37.3 kW) 440-V motor with six sections.

The total armature and field voltage drop at full-load is 2% of the applied voltage, the full-load efficiency is 95% and the maximum starting current is 130% of full-load current.

Solution. (*a*) Let I_1 = maximum current, I_2 = minimum current

Φ_1 = flux/pole for I_1 ; Φ_2 = flux/pole for I_2

$$\frac{I_1}{I_2} = \text{K and } \frac{\Phi_1}{\Phi_2} = \alpha$$

Let us now consider the conditions when the starter arm is on the nth and $(n + 1)$th stud. When the current is I_2, then $E_b = V - I_2R_n$

If, now, the starter is moved up to the $(n + 1)$th stud, then

$$E_b' = \frac{\Phi_1}{\Phi_2} \cdot E_b = \alpha E_b$$

$$\therefore \quad R_{n+1} = \frac{V - E_{b1}}{I_1} = \frac{V - \alpha E_b}{I_1} = \frac{V - \alpha(V - I_2R_n)}{I_1} = \frac{V}{I_1}(1 - \alpha) + \alpha\frac{I_2}{I_1} \cdot R_b$$

Now, $V/I_1 = R_1$–the total resistance in the circuit when the starter arm is on the first stud.

$$\therefore \quad R_{n+1} = R_1(1-\alpha) + \frac{\alpha}{K}R_n$$

Similarly, by substituting $(n-1)$ for n, we get $R_n = R_1(1-\alpha) + \frac{\alpha}{K}R_{n-1}$

Therefore, the resistance between the nth and $(n+1)$ th studs is

$$r_n = R_n - R_{n+1} = \frac{\alpha}{K}R_{n-1} - \frac{\alpha}{K}R_n = \frac{\alpha}{K}(R_{n-1} - R_n) = \frac{\alpha}{K}r_{n-1}$$

$$\therefore \quad \frac{r_n}{r_{n-1}} = \frac{\alpha}{K} = \frac{\Phi_1}{\Phi_2} \times \frac{I_2}{I_1} = b \quad \text{– constant}$$

Obviously, the resistance elements form a geometrical progression series.

(*b*) Full-load input current $= \dfrac{37,300}{440 \times 0.95} = 89.2$ A

Max. starting current $I_1 = 1.3 \times 89.2 = 116$ A

Arm. and field voltage drop on full-load = 2% of 440 = $0.02 \times 440 = 8.8$ V

Resistance of motor = 8.8/89.2 = 0.0896 Ω

Total circuit resistance on starting, $R_1 = V/I_1 = 440/116 = 3.79\ \Omega$

Assuming straight line magnetisation, we have $I_1 \propto \Phi_1$ and $I_2 \propto \Phi_2$

$\therefore \quad I_1/I_2 = \Phi_1/\Phi_2 \;\therefore\; \alpha = K$ and $b = \alpha/K = 1 \;\therefore\; r_n = b \times r_{n-1} = r_{n-1}$

In other words, all sections have the same resistance

$$\therefore \quad r = \frac{R_1 - R_{motor}}{\text{No. of sections}} = \frac{3.79 - 0.0896}{6} = \mathbf{0.6176\ \Omega}$$

Example 28.46. *A 75 h.p. (55.95 kW) 650-V, d.c. series traction motor has a total resistance of 0.51 Ω. The starting current is to be allowed to fluctuate between 140 A and 100 A, the flux at 140 A being 20% greater than at 100 A. Determine the number of steps required in the controller and the resistance of each step.*

Solution. Let R_1 = total resistance on the first stud = 650/140 = 4.65 Ω

When motor speeds up, then back e.m.f. is produced and current falls to I_2.

$$V = E_{b1} + I_2R_1 \qquad \ldots(i)$$

When the starter moves to the next stud, the speed is still the same, but since current rises to I_1 for which flux is 1.2 times greater than for I_2, hence back e.m.f. becomes 1.2 E_{b1}. Since resistance in the circuit is now R_2.

$$\therefore \quad V = I_1R_2 + 1.2\,E_{b1} \qquad \ldots(ii)$$

From (*i*) and (*ii*) we get, $0.2\,V = 1.2\,I_2R_1 - I_1R_2$

$\therefore \quad R_2 = (1.2\,I_2/I_1)R_1 - 0.2\,V/I_1 = (1.2\,I_2/I_1)\,R_1 - 0.2\,R_1 = (1.2\,I_2/I_1 - 0.2)R_1$

Similarly $\quad R_3 = (1.2\,I_2/I_1)R_2 - 0.2\,R_1$

In this way, we continue till we reach the value of resistance equal to the armature resistance. Hence, we obtain R_1, R_2 etc. and also the number of steps.

In the present case, $I_1 = 140$ A, $I_2 = 100$ A, $V = 650$ V and $I_2/I_1 = 100/140 = 1.14$

$$\therefore \quad R_1 = 4.65\Omega\,; R_2 = (1.2/1.4 - 0.2)4.65 = 3.07\ \Omega$$

$$R_3 = (1.2/1.4) \times 3.07 - 0.2 \times 4.65 = 1.70\ \Omega$$

$$R_4 = (1.2/1.4) \times 1.7 - 0.2 \times 4.65 = 0.53\ \Omega$$

We will stop here because R_4 is very near the value of the motor resistance. Hence, there are 4 studs and 3 sections or steps.

$R_1 - R_2 = 4.65 - 3.07 = 1.58\ \Omega$, $R_2 - R_3 = 3.07 - 1.7 = 1.37\ \Omega$

$R_3 - R_4 = 1.70 - 0.53 = 1.17\ \Omega$

Note. It will be seen that

$R_2 - R_3 = (1.2\ \ I_2/I_1)\,(R_1 - R_2)$ and $R_3 - R_4 = (1.2\,I_2/I_1)\,(R_2 - R_3)$ and so on.

or $$\frac{R_2-R_3}{R_1-R_2}=\frac{R_3-R_4}{R_2-R_3}=1.2\frac{I_2}{I_1}=\frac{\Phi_1}{\Phi_2}\times\frac{I_2}{I_1}=\frac{\alpha}{K}=b$$

It is seen that individual resistances of various sections decrease in the ratio of $\alpha/K = b$.

28.27 Thyristor Controller Starters

The moving parts and metal contacts etc. of the resistance starters discussed in Art 28.21 can be eliminated by using thyristors which can short circuit the resistance sections one after another. A thyristor can be switched on to the conducting state by applying a suitable signal to its gate terminal. While conducting, it offers zero resistance in the forward (*i.e.* anode-to-cathode) direction and thus acts a short-circuit for the starter resistance section across which it is connected. It can be switched off (*i.e.* brought back to the non-conducting state) by reversing the polarity of its anode-cathode voltage. A typical thyristor-controlled starter for d.c. motors is shown in Fig. 28.49.

After switching on the main supply, when switch S_1 is pressed, positive signal is applied to gate G of thyristor T_1 which is, therefore, turned ON. At the same time, shunt field gets established since it is directly connected across the d.c. supply. Consequently, motor armature current I_a flows via T_1, R_2, R_3 and R_4 because T_2, T_3 and T_4 are, as yet in the non-conducting state. From now onwards, the starting procedure is automatic as detailed below :

1. as S_1 is closed, capacitor C starts charging up with the polarity as shown when I_a starts flowing.
2. the armature current and field flux together produce torque which accelerates the motor and load.

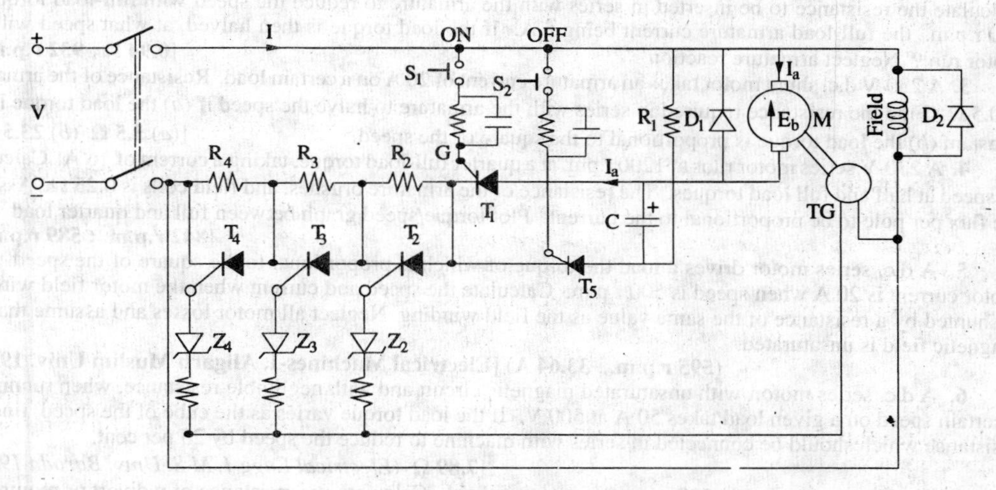

Fig. 28.49

3. as motor speeds up, voltage provided by techogenerator (*TG*) is proportionately increased because it is coupled to the motor.
4. at some motor speed, the voltage provided by *TG* becomes large enough to breakdown Zener diode Z_2 and hence trigger T_2 into conduction. Consequently, R_2 is shorted out and now I_a flows via motor armature, T_1, T_2, R_3 and R_4 and back to the negative supply terminal.
5. as R_2 is cut out, I_a increases, armature torque increases, motor speed increases which further increases the voltage output of the tachogenerator. At some speed, Z_3 breaks down thereby triggering T_3 into conduction which cuts out R_3.
6. after sometime, R_4 is cut out as Z_4 breaks down and triggers T_4 into conduction. In fact, Zener diodes Z_2, Z_3 and Z_4 can be rated for 1/3, 1/2 and 3/4 full speed respectively.

For stopping the motor, switch S_2 is closed which triggers T_5 into conduction thereby establishing current flow via R_1. Consequently, capacitor C starts discharging thereby reverse-biasing T_1 which stops conducting. Hence, I_a ceases and, at the same time, T_2, T_3 and T_4 also revert back to their non-conducting state.

Incidentally, it may be noted that the function of C is to switch T_1 ON and OFF. Hence, it is usually called *commutating capacitor*.

The function of the diodes D_1 and D_2 is to allow the decay of inductive energy stored in the motor armature and field when supply is disconnected. Supply failure will cause the thyristors to block because of this current decay thereby providing protection usually given by no-voltage release coil.

Recently, thyristor starting circuits have been introduced which use no starting resistance at all thereby making the entire system quite efficient and optimized as regards starting time. These are based on the principle of 'voltage chopping' (Art. 28.12). By varying the chopping frequency, the ratio of the time the voltage is ON to the time it is OFF can be varied. By varying this ratio, the average voltage applied to the motor can be changed. A low average voltage is needed to limit the armature current while the motor is being started and gradually the ratio is increased to reach the maximum at the rated speed of the motor.

Tutorial Problem No. 28.2

1. A shunt-wound motor runs at 600 r.p.m. from a 230-V supply when taking a line current of 50 A. Its armature and field resistances are 0.4 Ω and 104.5 Ω respectively. Neglecting the effects of armature reaction and allowing 2 V brush drop, calculate (*a*) the no-load speed if the no-load line current is 5A (*b*) the resistance to be placed in armature circuit in order to reduce the speed to 500 r.p.m. when motor is taking a line current of 50 A (*c*) the percentage reduction in the flux per pole in order that the speed may be 750 r.p.m. when the armature current is 30 A with no added resistance in the armature circuit. [(*a*) **652 r.p.m.** (*b*) **0.73 Ω** (*c*) **1.73%**]

2. The resistance of the armature of a 250-V shunt motor is 0.3 Ω and its full-load speed is 1000 r.p.m. Calculate the resistance to be inserted in series with the armature to reduce the speed with full-load torque to 800 r.p.m., the full-load armature current being 5 A. If the load torque is then halved, at what speed will the motor run ? Neglect armature reaction. [**0.94 Ω ; 932 r.p.m.**]

3. A 230-V d.c. shunt motor takes an armature current of 20 A on a certain load. Resistance of the armature is 0.5 Ω. Find the resistance required in series with the armature to halve the speed if (*a*) the load torque is constant (*b*) the load torque is proportional to the square of the speed. [(*a*) **5.5 Ω** (*b*) **23.5 Ω**]

4. A 230-V series motor runs at 1200 r.pm. at a quarter full-load torque, taking a current of 16 A. Calculate its speed at half and full load torques. The resistance of the armature brushes, and field coils is 0.25 Ω. Assume the flux per pole to be proportional to the current. Plot torque/speed graph between full and quarter load. [**842 r.p.m. ; 589 r.p.m.**]

5. A d.c. series motor drives a load the torque of which is proportional to the square of the speed. The motor current is 20 A when speed is 500 r.p.m. Calculate the speed and current when the motor field winding is shunted by a resistance of the same value as the field winding. Neglect all motor losses and assume that the magnetic field is unsaturated.

(595 r.p.m. ; 33.64 A) [Electrical Machines-I, Aligarh Muslim Univ. 1979)

6. A d.c. series motor, with unsaturated magnetic circuit and with negligible resistance, when running at a certain speed on a given load takes 50 A at 500 V. If the load torque varies as the cube of the speed, find the resistance which should be connected in series with machine to reduce the speed by 25 per cent.

[**7.89 Ω** (*Electrical Engg-I, M.S. Univ. Baroda 1980*)

7. A series motor runs at 500 r.p.m. on a certain load. Calculate the resistance of a divert or required to rise the speed to 650 r.p.m. with the same load current, given that the series field resistance is 0.05 Ω and the field is unsaturated. Assume the ohmic drop in the field and armature to be negligible. [**0.1665 Ω**]

8. A 230-V d.c. series motor has armature and field resistances of 0.5 Ω and 0.3 Ω respectively. he motor draws a line current of 40 A while running at 400 rpm. If a divertor of resistance 0.15 Ω is used, find the new speed of the motor for the same armature current.

It may be assumed that flux per pole is directly proportional to the field current.

[**1204 r.p.m.**] (*Electrical Engineering Grad. I.E.T.E. June 1986*)

9. A 250-V, d.c. shunt motor runs at 700 r.p.m. on no-load with no extra resistance in the field and armature circuit. Determine :

(*i*) the resistance to be placed in series with the armature for a speed of 400 r.p.m. when taking a total current of 16 A.

(*ii*) the resistance to be placed in series with the field to produce a speed of 1,000 r.p.m. when taking an armature current of 18 A.

Assume that the useful flux is proportional to the field. Armature resistance = 0.35 Ω, Field resistance = 125 Ω. [(*i*) **7.3 Ω** (*ii*) **113 Ω**] (*Elect. Engg. Grad. I.E.T.E., June 1984*)

10. A d.c. series motor is operating from a 220-V supply. It takes 50 A and runs at 1000 r.p.m. The resistance of the motor is 0.1 Ω. If a resistance of 2 Ω is placed in series with the motor, calculate the resultant speed if the load torque is constant. [**534 r.p.m.**]

11. A d.c. shunt motor takes 25 A when running at 1000 r.p.m. from a 220-V supply.

Calculate the current taken from the supply and the speed if the load torque is halved, a resistance of 5 Ω is placed in the armature circuit and a resistance of 50 Ω is placed in the field circuit.

Armature resistance = 0.1 Ω ; field resistance = 100 Ω.

Assume that the field flux per pole is directly proportional to the field current.

[17.1 A; 915 r.p.m.] (*Elect. Technology, Gwalior Univ. Nov. 1977*)

12. A 440-V shunt motor takes an armature current of 50 A and has a flux/pole of 50 mWb. If the flux is suddenly decreased to 45 mWb, calculate (*a*) instantaneous increase in armature current (*b*) percentage increase in the motor torque due to increase in current (*c*) value of steady current which motor will take eventually (*d*) the final percentage increase in motor speed. Neglect brush contact drop and armature reaction and assume an armature resistance of 0.6 Ω.

[(*a*) **118 A** (*b*) **112%** (*c*) **5.55 A** (*d*) **10%**]

13. A 440-V shunt motor while running at 1500 r.p.m. takes an armature current of 30 A and delivers a mechanical output of 15 hp. (11.19 kW). The load torque varies as the square of the speed. Calculate the value of resistance to be connected in series with the armature for reducing the motor speed to 1300 r.p.m. and the armature current at that speed.

[2.97 Ω, 22.5 A]

14. A 460-V series motor has a resistance of 0.4 Ω and takes a current of 25 A when there is no additional controller resistance in the armature circuit. Its speed is 1000 r.p.m. The control resistance is so adjusted as to reduce the field flux by 5%. Calculate the new current drawn by the motor and its speed. Assume that the load torque varies as the square of the speed and the same motor efficiency under the two conditions of operation.

[22.6 A; 926 r.p.m.] (*Elect. Machines, South Gujarat Univ. Oct. 1977*)

15. A 460-V, series motor runs at 500 r.p.m. taking a current of 40 A. Calculate the speed and percentage change and torque if the load is reduced so that the motor is taking 30 A. Total resistance of armature and field circuit is 0.8 Ω. Assume flux proportional to the field current.

[680 r.p.m. 43.75%]

16. A 440-V, 25 h.p (18.65 kW) motor has an armature resistance of 1.2 Ω and full-load efficiency of 85%. Calculate the number and value of resistance elements of a starter for the motor if maximum permissible current is 1.5 times the full-load current.

[1.92 Ω, 1.30 Ω, 0.86 Ω ; 0.59 Ω]

17. A 230-V, d.c. shunt motor has an armature resistance of 0.3 Ω. Calculate (*a*) the resistance to be connected in series with the armature to limit the armature current to 75 A at starting and (*b*) value of the generated e.m.f. when the armature current has fallen to 50 A with this value of resistance still in circuit.

[(*a*) **2.767 Ω**(*b*) **76.7A**]

18. A 200-V, d.c. shunt motor takes full-load current of 12 A. The armature circuit resistance is 0.3 Ω and the field circuit resistance is 100 Ω. Calculate the value of 5 steps in the 6-stud starter for the motor. The maximum starting current is not to exceed 1.5 times the full-load current.

[6.57 Ω, 3.12 Ω, 1.48 Ω, 0.7 Ω, 0.33Ω]

19. The resistance of a starter for a 200-V, shunt motor is such that maximum starting current is 30 A. When the current has decreased to 24 A, the starter arm is moved from the first to the second stud.

Calculate the resistance between these two studs if the maximum current in the second stud is 34 A. The armature resistance of the motor is 0.4 Ω.

[1.334 Ω]

20. A totally-enclosed motor has thermal time constant of 2 hr. and final temperature rise of at no load and 40° on full load.

Determine the limits between which the temperature fluctuates when the motor operates on a load cycle consisting of alternate period of 1 hr. on full-load and 1 hr. on no-load, steady state conditions having been established.

[28.7°C, 21.3°C]

21. A motor with a thermal time constant of 45 min. has a final temperature rise of 75°C on continuous rating (*a*) What is the temperature rise after one hour at this load ? (*b*) If the temperature rise on one-hour rating is 75°C, find the maximum steady temperature at this rating (*c*) When working at its one-hour rating, how long does it take the temperature to increase from 60°C to 75°C ?

[(*a*) **55°C** (*b*) **102°C** (*c*) **20 min**] (*Electrical Technology, M.S. Univ. Baroda. 1976*)

Objective Tests–28

1. The speed of a d.c. motor can be controlled by varying
(*a*) its flux per pole
(*b*) resistance of armature circuit
(*c*) applied voltage
(*d*) all of the above

2. The most efficient method of increasing the speed of a 3.75 kW d.c. shunt motor would be the —— method.
(*a*) armature control
(*b*) flux control
(*c*) Ward-Leonard
(*d*) tapped-field control

3. Regarding Ward-Leonard system of speed control which statement is **false?**
(*a*) it is usually used where wide and very sensitive speed control is required
(*b*) it is used for motors having ratings from 750 kW to 4000 kW
(*c*) capital outlay involved in the system is right since it uses two extra machines.
(d) it gives a speed range of 10 : 1 but in one direction only
(*e*) it has low overall efficiency especially at light loads

4. In the rheostatic method of speed control for a d.c. shunt motor, use of armature divertor makes the method
(*a*) less wasteful (*b*) less expensive
(*c*) unsuitable for changing loads
(*d*) suitable for rapidly changing loads

5. The chief advantage of Ward-Lenoard system of d.c. motor speed control is that it
(*a*) can be used even for small motors
(*b*) has high overall efficiency at all speeds
(*c*) gives smooth, sensitive and wide speed control
(*d*) uses a flywheel to reduce fluctuations in power demand

6. The flux control method using paralleling of field coils when applied to a 4-pole series d.c. motor can give —— speeds.
(*a*) 2 (*b*) 3
(*c*) 4 (*d*) 6

7. The series-parallel system of speed control of series motors widely used in traction work gives a speed range of about
(*a*) 1 : 2 (*b*) 1 : 3
(*c*) 1 : 4 (*d*) 1 : 6

8. In practice, regenerative braking is used when
(*a*) quick motor reversal is desired
(*b*) load has overhauling characteristics
(*c*) controlling elevators, rolling mills and printing presses etc.
(*d*) other methods cannot be used.

9. Statement 1. A direct-on-line (**DOL**) starter is used to start a small d.c. motor because
Statement 2. it limits initial current drawn by the armature circuit.
(*a*) both statements 1 and 2 are incorrect
(*b*) both statements 1 and 2 are correct
(*c*) statement 1 is correct but 2 is wrong
(*d*) statement 2 is correct but 1 is wrong

10. Ward-Lenoard system of speed control is **NOT** recommended for
(*a*) wide speed range
(*c*) constant-speed operation
(*c*) frequent motor reversals
(*d*) very low speeds

11. Thyristor chopper circuits are employed for
(*a*) lowering the level of a d.c. voltage
(*b*) rectifying the a.c. voltage
(*c*) frequency conversion
(*d*) providing commutation circuitry

12. An inverter circuit is employed to convert
(*a*) a.c. voltage into d.c. voltage
(*b*) d.c. voltage into a.c. voltage
(*c*) high frequency into low frequency
(*d*) low frequency into high frequency

13. The phase-control rectifiers used for speed of d.c. motors convert fixed a.c. supply voltage into
(*a*) variable d.c. supply voltage
(*b*) variable a.c. supply voltage
(*c*) full-rectified a.c. voltage
(*d*) half-rectified a.c. voltage

14. If some of the switching devices in a converter are controlled devices and some are diodes, the converter is called
(*a*) full converter (*b*) semiconverter
(*c*) solid-state chopper (*d*) d.c. converter

15. A solid-state chopper convertes a fixed-voltage d.c. supply into a
(*a*) variable-voltage a.c. supply
(*b*) variable-voltage d.c. supply
(*c*) higher-voltage d.c. supply
(*d*) lower-voltage a.c. supply

16. The d.c. motor terminal voltage supplied by a solid-state chopper for speed control purposes varies —— with the duty ratio of the chopper
(*a*) inversely (*b*) indirectly
(*c*) lineraly (*d*) parabolically

Answers

1. *d* **2.** *b* **3.** *d* **4.** *d* **5.** *c* **6.** *b* **7.** *c* **8.** *b* **9.** *c* **10.** *b* **11.** *a* **12.** *b* **13.** *a* **14.** *b* **15.** *b* **16.** *c*

29 TESTING OF D.C. MACHINES

29.1. Brake Test

It is a direct method and consists of applying a brake to a water-cooled pulley mounted on the motor shaft as shown in Fig. 29.1. The brake band is fixed with the help of wooden blocks gripping the pulley. One end of the band is fixed to earth via a spring balance S and the other is connected to a suspended weight W_1. The motor is run and the load on the motor is adjusted till it carries its full load current.

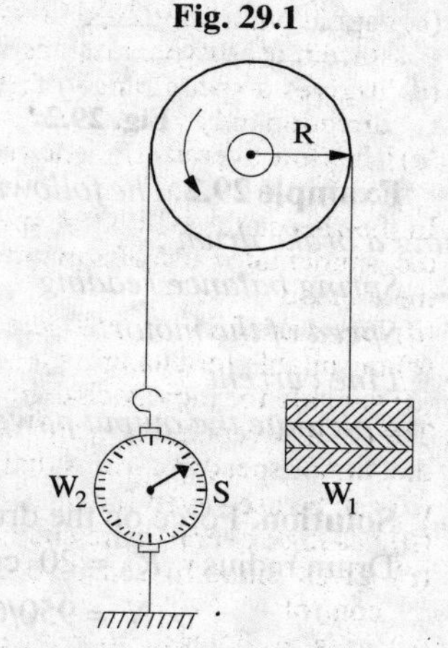

Fig. 29.1

Let W_1 = suspended weight in kg
W_2 = reading on spring balance in kg-wt

The net pull on the band due to friction at the pulley is (W_1-W_2) kg. wt. or 9.81 (W_1-W_2) newton

If R = radius of the pulley in metre
and N= motor or pulley speed in r.p.s.

Then, shaft torque T_{sh} developed by the motor
$= (W_1-W_2)\,R$ kg-m $= 9.81\,(W_1-W_2)\,R\;N\text{-}m$

Motor output power $= T_{sh} \times 2\pi\;N$ watt
$= 2\pi \times 9.81\;\;N(W_1-W_2)\,R$ watt
$= 61.68\;N(W_1-W_2)\,R$ watt

Let, V = supply voltage; I = full-load currrent taken by the motor.

Then, input power = VI watt

$$\therefore\;\; \eta = \frac{\text{output}}{\text{input}} = \frac{61.68\,N\,(W_1 - W_2)\,R}{VI}$$

The simple brake test described above can be used for small motors only, because in the case of large motos, it is difficult to dissipate the large amount of heat generated at the brake.

Another simple method of measuring motor output is by the use of poney brake one form of which is shown in Fig. 29.2. A rope is wound round the pulley and its two ends are attached to two spring balances S_1 and S_2. The tension of the rope can be adjusted with the help of swivels. Obviously, the force acting tangentially on the pulley is equal to the difference between the readings of the two spring balances. If R is the pulley radius, the torque at the pulley is $T_{sh} = (S_1-S_2)\,R$. If $\omega\;(= 2\pi N)$ is the angular velocity of the pulley, then

motor output $= T_{sh} \times \omega = 2\,\pi\,N(S_1-S_2)\,R$ m-kg. wt. $= 9.81 \times 2\pi N\;S_1-S_2)\,R$ watt.

The motor input may be measured as shown in Fig. 29.3. Efficiency may, as usual, be found by using the relation η = output/input.

Example 29.1. *In a brake test the effective load on the branch pulley was 38.1 kg, the effective diameter of the pulley 63.5 cm and speed 12 r.p.s. The motor took 49 A at 220 V. Calculate the output power and the efficiency at this load.*

Solution. Effective load (W_1-W_2) = 38.1 kg. wt ; radius = 0.635/2 = 0.3175 m
Shaft torque = 38.1 × 0.3175 kg-m = 9.81 × 38.1 × 0.3175 = 118.6 N-m
Power output = torque × angular velocity in rad/s = 118.6 × 2π × 12 = **8,945 W**

Now, motor input = 49 × 220 W ∴ motor $\eta = \frac{8,945}{49 \times 220} =$ **0.83** or **83%**

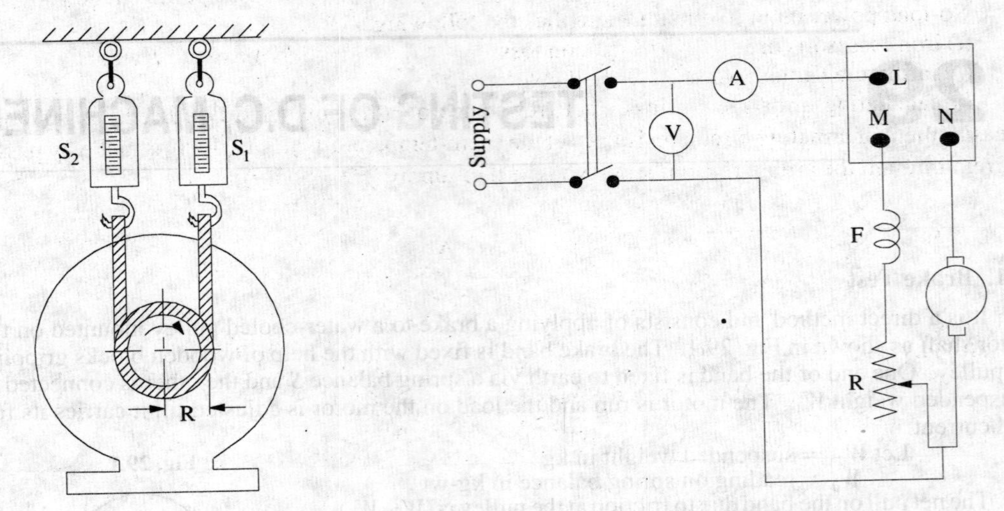

Fig. 29.2

Fig. 29.3

Example 29.2. *The following readings are obtained when doing a load test on a d.c. shunt motor using a brake drum :-*

Spring balance reading	*10 kg and 35 kg*	*Diameter of the drum*	*40 cm*
Speed of the motor	*950 r.p.m.*	*Applied voltage*	*200 V*
Line current	*30 A*		

Calculate the output power and the efficiency,

(Electrical Engineering, Madras Univ. 1986)

Solution. Force on the drum surface $F = (35-10) = 25$ kg wt $= 25 \times 9.8$ N

Drum radius $R = 20$ cm $= 0.2$m ; Torque $T_{sh} = F \times R = 25 \times 9.8 \times 0.2 = 49\ N$

$N = 950/60 = 95/6$ r.p.s.; $\omega = 2\pi(95/6) = 99.5$ rad/s

Motor output $= T_{sh} \times \omega$ watt $= 49 \times 99.5 = 4,876$ W

Motor input $= 200 \times 30 = 6000$ W; $\eta = 4876/6000 = 0.813$ or **81.3%**

29.2. Swinburnes* Test (or No-load Test or Losses Method)

It is a simple method in which losses are measured separately and from their knowledge, efficiency at any desired load can be predetermined in advance. The only running test needed is no-load test. *However, this test is applicable to those machines in which flux is practically constant i.e. shunt and compound-wound machines*.

The machine is run as a motor on no-load at its rated voltage *i.e.* voltage stamped on the name-plate. The speed is adjusted to the rated speed with the help of shunt regulator as shown in Fig. 29.4.

The no-load current I_0 is measured by the ammeter A_1 whereas shunt field currrent I_{sh} is given by ammeter A_2. The no-load armature current is $(I_0 - I_{sh})$ or I_{a0}

---

*Sir James Swinburne (1858-1958) made outstanding contributions to the development of electric lamps, electric machines and synthetic resins.

Let, supply voltage $= V$ no-load input $= VI_0$ watt
∴ power input to armature $= V(I_0 - I_{sh})$; power input to shunt $= VI_{sh}$
No-load power input to armature supplies the following :-
(*i*) iron losses in core (*ii*) friction loss (*iii*) windage loss and
(*iv*) armature Cu loss, $(I_0 - I_{sh})^2 R_a$ or $I_{a0}^2 R_a$

In calculating armature Cu loss, 'hot' resistance of armature should be used. A stationary measurement of armature circuit resistance at the room-temperature of, say, 15°C is made by passing current through the armature from a low voltage d.c. supply [Fig. 29.5(*a*)].

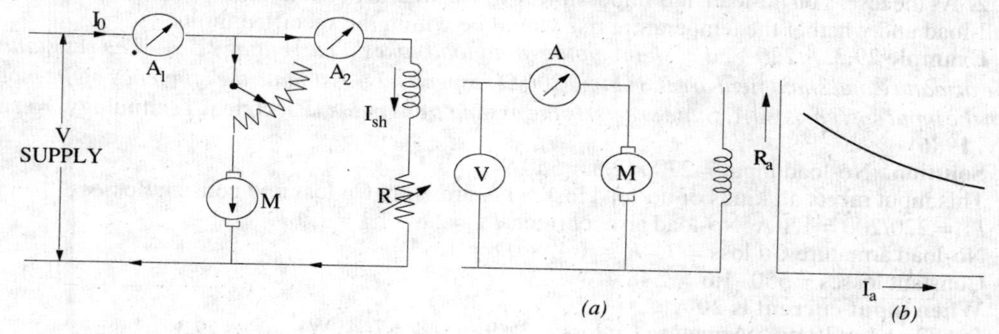

Fig. 29.4 **Fig. 29.5**

Then, the 'hot' resistance, allowing a tempeature rise of 50°C is found thus :

$$R_{15} = R_0(1 + 15\alpha_0) : R_{65} = (1 + 65\alpha_0), R_{65} = R_{15} \times \frac{1 + 65\alpha_0}{1 + 15\alpha_0}$$

Taking $\alpha_0 = 1/234.5$, we have $R_{65} = R_{15} \times \frac{234.5 + 65}{234.5 + 15} = 1.2\, R_{15}$(approx.*)

If we subtract from the total input the no-load armature Cu loss, then we get constant losses. ∴ constant losses $W_C = VI_0 - (I_0 - I_{sh})^2 R_a$

Knowing the constant losses of the machine, its efficiency at any other load can be determined as given below. Let I = load current at which efficiency is required.

Then, armature current is $I_a = I - I_{sh}$ ...if machine is motorting
$= I + I_{sh}$ ...if machine is generating

Efficiency when running as a motor

Input $= VI$ Armature Cu loss $= I_a^2 R_a = (I - I_{sh})^2 R_a$

Constant losses $= W_c$ ...found above

$$\therefore \text{ Total losses } = (I - I_{sh})^2 R_a + W_C \; ; \; \eta_m = \frac{\text{input - losses}}{\text{input}} = \frac{VI - (I - I_{sh})^2 R_a - W_e}{VI}$$

Efficiency when running as a generator :

Output $= VI$; Armature, Cu loss $= (I + I_{sh})^2 R_a$; Constant loss $= W_c$ ...found above

$$\therefore \text{ total losses } = (I + I_{sh})^2 R + W_c \; ; \; 2_g = \frac{\text{output}}{\text{output + losses}} = \frac{VI}{VI + (I + I_{sh})^2 R_a + W_c}$$

*The armature resistance is found to decreases slightly with increasing armature current as shown n Fig. 26.5(*b*). This is due to the fact that brush contact resistance is approximately proportional to the armature current.

29.3. Advantages of Swinburne's Test

1. It is convenient and economical because power required to test a large machine is small *i.e.* only no-load input power.

2. The efficiency can be predetermined at any load because constant - losses are known.

29.4. Main Disadvantages

1. No account is taken of the change in iron losses from no-load to full load. At full-load, due to armature reaction, flux is distorted which increases the iron losses in some cases by as much as 50%.

2. As the test i s on no-load, it is impossible to know whether commutation would be satisfactory at full-load and whether the temperature rise would be within the specified limits.

Example 29.3. *A 220 V, d.c. shunt motor at no load takes a current of 2.5 A. The resistances of the armature and shunt field are 0.8 Ω and 200 Ω respectively. Estimate the efficiency of the motor when the input current is 20 A. State precisely the assumptions made.* **(Electrical Technology, Kerala Univ. 1986)**

Solution. No-load input = 220 × 2.5 = 550 W

This input meets all kinds of no-load losses *i.e.* armature Cu loss and constant losses

I_{sh} = 220/200 = 1.1 A No-load arm. current, I_{a0} = 2.5 −1.1 = 1.4 A

No-load armature Cu loss = $I_{a0}^2 R_a = 1.4^2 \times 0.8 = 1.6$ W

Constant losses = 550−1.6 = 548.4 W

When input current is 20 A

I_a = 32−1.1 = 30.9 A; Armature Cu loss = $30.9^2 \times 0.8$ = 764 W

Total loss = 764 + 548.4 = 1315 W (approx.) ; Input = 220 × 20 = 4,400 W

Output = 4,400−1,312 = 3,088 W ; efficiency = (3088/4400) × 100 = **70.2%**

In the above calculations, it has been assumed that :-

1. mechanical losses remain constant even through motor speed changes from no-load to the given load.

2. effect of armature reaction on main pole flux with a consequent change in iron losses has been neglected.

3. decrease in flux due to increase in shunt resistance by heating has been neglected.

Example 29.4. *When running on no-load, a 400-V shunt motor takes 5 A. Armature resistance is 0.5 Ω and field resistance 200 Ω. Find the output of the motor and efficiency when running on full-load and taking a current of 50 A. Also, find the percentage change in speed from no-load to full-load* *(Electro Mechanics, Allahabad Univ. 1991)*

Solution. No-load input = 400 × 5 = 2,000 W

This input goes to meet all kinds of no-load losses *i.e.* armature Cu loss and constant losses.

I_{sh} = 400/200 = 2 A ; No-load I_a = 5−2 = 3 A

No-load arm Cu loss = $3^2 \times 0.5$ = 4.5 W ; Constant losses = 2,000−4.5 = 1,995.5 W

When line current is 50 A

Ia = 50−2 = 48 A; Arm. Cu loss = $48^2 \times 0.5$ = 1,152 W

Total loss on F.L = 1,152 + 1,995.5 = 3,147.5 W ; Input = 50 × 400 = 20,000 W

Output= 20,000−3,147.5 = **16,852.5 W = 16.8 kW**

F.L. efficiency = 16,852.5/20,000 = 0.8426 or **84.26%**

Now, E_{b1} = 400−(3 × 0.5) = 398.5 V ; E_{b2} = 400−(48 × 0.5) = 376 V

$$\frac{N_1}{N_2} = \frac{E_{b1}}{E_{b2}} = \frac{398.5}{376} \quad \therefore \quad \frac{N_1 - N_2}{N_2} = \frac{22.5}{376} = 0.0598$$

∴ percentage change in speed = **5.98**

Example 29.5. *The no-load test of a 44.76 kW, 220-V, d.c. shunt motor gave the following figures :*

Input current = 13.25 A; field current = 2.55 A; resistance of armature at 75°C = 0.032 Ω and brush drop = 2V. Estimate the full-load current and efficiency.

(Electrical Engineering, Madras Univ, 1987)

Solution. No-load Condition

No-load input = 220 × 13.25 = 2915 W; Armature current = 13.25 − 2.55 = 10.7 A

Armature Cu loss = $10.7^2 \times 0.032 = 3.6$ W
Loss due to brush drop = $2 \times 10.7 = 21.4$ W
Variable loss $= 21.4 + 3.6 = 25\ W_e$ Constant losses $W_C = 2915–25 = 2890$ W

Full-load Condition

If I_a is the full-load armature current, then full-load motor input current is $(I_a + 2.55)$ A.
F.L. motor power input = $220\ (I_a + 2.55)$ W
This input must be equal to the sum of
(*i*) output = 44.76 kW = 44,760 W (*ii*) W_C = 2,890 W
(*iii*) brush loss = $2I_a$ watt (*iv*) Arm. Cu loss = $0.032\ I_a^2$

$\therefore\ 220\ (I_a + 2.55) = 44{,}760 + 2{,}890 + 2I_a + 0.032\ I_a^2$

or $$0.032\ I_a^2 - 218\ I_a + 47{,}090 = 0$$

or $$I_a = \frac{218 \pm \sqrt{218^2 - 4 \times 0.032 \times 47{,}090}}{2 \times 0.032} = 223.5 \text{ A}$$

Line input current $I = I_a + I_{sh} = 223.5 + 2.55 =$ **226 A**
F.L. power input $= 226 \times 220 = 49{,}720$ W
$\therefore$ F.L. efficiency = 44,760/49,720 = 0.9 or **90%.**

Example 29.6. *A 200-V, shunt motor develops an output of 17.158 kW when taking 20.2 kW. The field resistance is 50 Ω and armature resistance 0.06 Ω. What is the efficiency and power input when the output is 7.46 kW ?*

(Elect. Machines-I, Aligarh Muslim Univ. 1989)

Solution. In the first case:

Output = 17,158 W Input = 20,200 W
Total losses = 20,200 – 17,158 = 3,042 W; Input current = 20,200/200=101 A
$I_{sh} = 200/50 = 4$ A, $I_a = 101 - 4 = 97$ A
$\therefore$ armature Cu loss $= 97^2 \times 0.06 = 564.5$ W
$\therefore$ constant losses = 3,042 – 564.5 = 2,477.5 = 2478 W (approx.)

In the second case:

Let, I_a = armature current Input current = $(I_a + 4)$ A
Now, input power = output + $I_a{}^2 R_a$ + constant losses

$\therefore\ 200(I_a + 4) = 7{,}460 + 0.06\ I_a{}^2 + 2{,}478$

or $0.06\ I_a{}^2 - 200\ I_a + 9{,}138 = 0$

$$\therefore \quad I_a = \frac{200 \pm \sqrt{200^2 - 4 \times 0.06 \times 9{,}138}}{2 \times 0.06} = \frac{200 \pm 194}{0.12} = 3{,}283.3 \text{ A or } 46 \text{ A}$$

We will reject the larger value because it corresponds to unstable operation of the motor. Hence, take $I_a = 46$ A

$\therefore$ input current $I = I_a + I_{sh} = 46 + 4 = 50$ A

Power input $= \dfrac{50 \times 200}{1000} =$ **10 kW** $\therefore\ \eta = \dfrac{7{,}460 \times 100}{10{,}000} =$ **74.6%**

Example 29.7. *A 200-V, 14.92 kW dc shunt motor when tested by the Swinburne method gave the following results :*

Running light : armature current was 6.5 A and field current 2.2 A. With the armature locked, the current was 70 A when a potential difference of 3 V was applied to the brushes. Estimate the efficiency of the motor when working under full-load conditions.

(Electrical Engg.-I, Bombay Univ. 1985)

Solution. No-load input current= 6.5 + 2.2 = 8.7 A
No-load power input = $200 \times 8.7 = 1{,}740$ W
No-load input equals Cu losses and stray losses.
Field Cu loss = $200 \times 2.2 = 440$ W

Armature Cu loss $= 6.5^2 \times 0.04286 = 1.8$ W ($\because R_a = 3/70 = 0.04286\ \Omega$)

$\therefore$ constant losses $= 1{,}740 - 1.8 = 1738$ W

We will assume that constant losses are the same at full-load also.

Let, I_a = full-load armature current

F.L. armature Cu loss $= 0.04286\ I_a^2$ W; Constant losses = 1,738 W

F.L. total loss $= 1{,}738 + 0.04286\ I_a^2$

F.L. output = 14,920 W; F.L. input $= 200\ (I_a + 2.2)$ W

We know, input = output + losses

or $200\ I_a + 440 = 14{,}920 + 1{,}738 + 0.04286\ I_a^2$

or $0.04286\ I_a^2 - 200\ I_a + 16{,}218 = 0 \quad \therefore I_a = 82.5$ A

$\therefore$ input current $= 82.5 + 2.2 = 84.7$ A

F.L. power input $= 200 \times 84.7$ A = **16,940 Ω**

$\therefore \quad \eta = 14{,}920 \times 100/16{,}940 =$ **88%**.

Example 29.8. *In a test on a d.c. shunt generator whose full-load output is 200 kW at 250 V, the following figures were obtained:*

(a) When running light as a motor at full speed, the line current was 36 A, the field current 12 A and the supply voltage 250.

(b) With the machine at rest, a p.d. of 6 V produced a current of 400 A through the armature circuit. Explain how these results may be utilised to obtain the efficiency of generator at full-load and half-load. Neglect brush voltage drop.

Solution. At no-load :

$I_a = 36 - 12 = 24$ A ; $R_a = 6/400 = 0.015\ \Omega$

$\therefore$ armature Cu loss $= 24^2 \times 0.015 = 8.64$ watt

No-load input = total losses in machine $= 250 \times 36 = 9{,}000$ W

Constant losses $= 9{,}000 - 8.64 =$ **8.991.4 W**

At full-load:

Output = 200,000 W ; Output current = 200,000/250 = 800 A ; $I_{sh} = 12$ A

$\therefore$ F.L. armature current = 800 + 12 = 812 A

$\therefore$ F.L. armature Cu losses $= 812^2 \times 0.015 = 9.893$ W

$\therefore$ F.L total losses $= 9.893 + 8{,}991.4 = 18{,}884$ W $\therefore \eta = \dfrac{200{,}000 \times 100}{200{,}000 + 18{,}884} =$ **91.4%**

At half-load:

Output = 100,000 W; Output current = 100,000/250 = 400 A

$I_a = 400 + 12 = 412$ A $\quad \therefore \quad I_a^2 R_a = 412^2 \times 0.015 = 2{,}546$ W

Total losses $= 8{,}991.4 + 2{,}546 = 11{,}545$ W $\therefore \eta = \dfrac{100{,}000 \times 100}{111{,}545} =$ **89.6%**

Example 29.9. *A 250-V, 14.92 kW shunt motor has a maximum efficiency of 88% and a speed of 700 r.p.m. when delivering 80% of its rated output. The resistance of its shunt field is 100 ω. Determine the efficiency and speed when the motor draws a current of 78 A from the mains.*

Solution. Full-load output = 14,920 W

80% of F.L. output $= 0.8 \times 14{,}920 = 11{,}936$ W; $\eta = 0.88$

Input $= 11{,}936/0.88 = 13{,}564$ W

Total losses $= 13{,}564 - 11{,}936 = 1{,}628$ W

As efficiency is maximum at this load, the variable loss is equal to constant losses.

$\therefore \quad W_c = I_a^2 R_a = 1{,}628/2 \quad \therefore \quad I_a^2 R_a = 814$ W

Now, input current $= 13{,}564/250 = 54.25$ A

$\therefore \quad I_{sh} = 250/100 = 2.5$ A $\quad \therefore \quad I_a = 54.25 - 2.5 = 51.75$ A

$\therefore \quad 51.75^2\ R_a = 814 \quad \therefore \quad R_a = 814/51.75^2 = 0.3045\ \Omega$

When input current is 78 A

$I_a = 78 - 2.5 = 75.5 \text{ A} \quad \therefore \quad I_a^2 R_a = 75.5^2 \times 0.3045 = 1{,}736 \text{ W}$

Total losses $= 1{,}736 + 814 = 2{,}550 \text{ W}$; Input $= 250 \times 78 = 19{,}500$ W

$$\eta = \frac{19{,}500 - 2{,}550}{19{,}550} \times 100 = \mathbf{86.9\%}$$

Speed :

$$\frac{N_2}{N_1} = \frac{E_{b2}}{E_{b1}} \text{ or } \frac{N_2}{700} = \frac{250 - (75.5 \times 0.3045)}{250 - (51.75 \times 0.3045)} = \frac{227}{234.25};\ N_2 = \mathbf{680 \text{ r.p.m.}}$$

29.5. Regenerative or Hopkinson's Test (Block-to-Back Test)

By this method, *full-load test* can be carried out on two shunt machines, preferably identical ones, without wasting their outputs. The two machines are mechanically coupled and are so adjusted electrically that one of them runs as a motor and the other as a generator. The mechanical output of the motor drives the generator and the electrical output of generator is used in supplying the greater part of input to the motor. If there were no losses in the machines, they would have run without any external power supply. But due to these losses, generator output is not sufficient to drive the motor and *vice-versa*. The losses are supplied either by an extra motor which is belt-connected to the motor-generator set or as suggested by Kapp, electrically from the supply mains.

Essential connections for the test are shown in Fig. 29.6. The two shunt machines are connected in parallel. They are, to begin with, started as unloaded motors. Then, the field of one is weakened and that of the other is strengthened so that the former runs as a motor and the latter as a generator. The usual method of procedure is as follows:

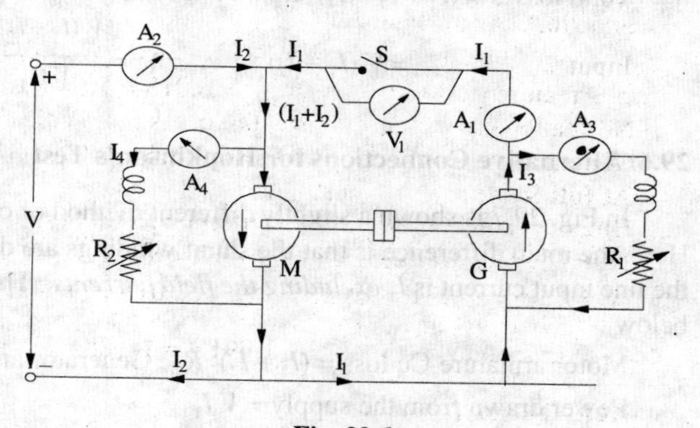

Fig. 29.6

Machine M is started up from the supply mains with the help of a starter (not shown) whereas main switch S of the other machine is kept open. Its speed is adjusted to normal value by means of its shield regulator. Machine M drives machine G as a generator and its voltage is read on voltmeter V_1. The voltage of G is adjusted by its field regulator until voltmeter V_1 reads zero, thereby showing that its voltage is the same, both in polarity and magnitude as that of the main supply. Thereafter, S is closed to parallel the machines. By adjusting the respective field regulators, any load can now be thrown on to the machines. Generator current I_1 can be adjusted to any desired value by increasing the excitation of G or by reducing the excitation of M and the corresponding values of different ammeters are read.

The electrical output of the generator *plus* the small power taken from the supply, is taken by the motor and is given out as a mechanical power *after* supplying the motor losses.

If supply voltage is V, then

Motor input $= V(I_1 + I_2)$ where I_2 is the current taken from the supply.

Generator output $= V I_1$ ...(*i*)

Assuming that both machines have the same efficiency η,

Output of motor $= \eta \times \text{input} = \eta\, V(I_1 + I_2) = $ generator input

Output of generator $= \eta \times \text{input} = \eta \times \eta\, V(I_1 + I_2) = \eta^2 V(I_1 + I_2)$ ...(*ii*)

Hence, from (*i*) and (*ii*), we get

$$\eta^2 V (I_1 + I_2) = VI_1 \quad \text{or} \quad \eta = \sqrt{\frac{I_1}{I_1 + I_2}}$$

However, it is not quite correct to assume equal efficiencies for two machines because their armature currents as well as excitations are different. We will now find the efficiencies separately :

Let R_a = armature resistance of each machine
I_3 = exciting current of the generator
I_4 = exciting current of the motor

Armature Cu loss in generator = $(I_1 + I_3)^2 R_a$; Armature Cu loss in motor = $(I_1 + I_2 - I_4)^2 R_a$

Shunt Cu loss in generator = $V I_3$; Shunt Cu loss in motor = $V I_4$

But total motor and generator losses are equal to the power supplied by the mains.

Power drawn from supply = $V I_2$

If we subtract the armature and shunt Cu losses from this, we get the stray losses of *both* machines.

$\therefore$ Total stray losses for the set

$$= V I_2 - [(I_1 + I_3)^2 R_a + (I_1 + I_2 - I_4)^2 R_a + V I_3 + V I_4] = W \text{ (say)}$$

Making one assumption that stray losses are equally divided between the two machines, we have

Stray loss per machine = W/2

For Generator

Total losses = $(I_1 + I_3)^2 R_a + VI_3 + W/2 = W_g$ (say)

Output = VI_1 $\therefore$ $\eta_g = \dfrac{VI_1}{VI_1 + W_g}$

Total losses = $(I_1 + I_2 - I_4)^2 R_a + VI_d + W/2 = W_m$ (say)

Input = $V(I_1 + I_2)$ $\therefore$ $\eta_m = \dfrac{V(I_1 + I_2) - W_m}{V(I_1 + I_2)}$

29.6. Alternative Connections for Hopkinson's Test

In Fig. 29.7 is shown a slightly different method of connecting the two machines to the supply. Here, the main difference is that the shunt windings are directly connected across the lines. Hence, the line input current is I_1 *excluding the field currents*. The efficiencies can be calculated as detailed below :-

Motor armature Cu loss = $(I_1 + I_2)^2 R_a$: Generator armature Cu loss = $I_2^2 R_a$

Power drawn from the supply = $V I_1$

$\therefore$ Total stray losses *i.e.* iron, friction and windage losses for the two machines are

$$= VI_1 - [(I_1 + I_2)^2 R_a - I_2^2 R_a] = W \text{ (say)}$$

$\therefore$ stray loss for each machine = $W/2$

Motor Efficiency

Motor input = armature input + shunt field input = $V(I_1 + I_2) + VI_3 = W_{input}$

Motor losses = armature Cu loss + shunt Cu loss + stray losses
= $(I_1 + I_2)^2 R_a + VI_3 + W/2 = W_m$ (say)

Motor $\eta_1 = \dfrac{W_{input} - W_m}{W_{input}} \times 100$

Generator Efficiency

Generator output = $V I_2$; Generator losses = $I_2^2 R_a + V I_4 + W/2 = W_g$ (say)

$\therefore$ generator $\eta = \dfrac{VI_2}{VI_2 + W_g}$

29.7 Merits of Hopkinson's Test

1. Power required for the test is small as compared to the full-load powers of the two machines.

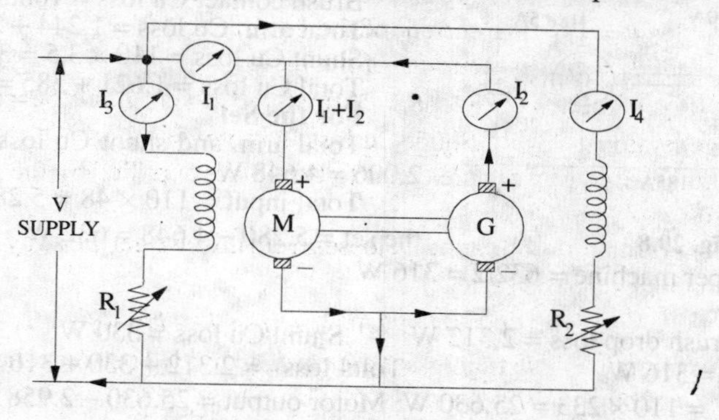

Fig. 29.7

2. As machines are being tested under full-load conditions, the temperature rise and the commutation qualities of the machines can be observed.

3. Because of full-load conditions, any change in iron loss due to flux distortion at full-load, is being taken into account.

The only disadvantage is with regard to the availability of two identical machines.

Example 29.10. *In a Hopkinson's test on two 220-V, 100-kW generators, the circulating current is equal to the full-load current and, in addition, 90 A are taken from the supply. Obtain the efficiency of each machine.*

Solution. Output current of the generator

$$I_1 = \frac{100,000}{220} = \frac{5,000}{11} = 454.5 \text{ A}, I_2 = 90 \text{ A}$$

Assuming equal efficiencies, from Art. 29.5, we have

$$\eta = \sqrt{\frac{I_1}{I_1 + I_2}} = \sqrt{\frac{454.5}{454.5 + 90}} = 0.914 \text{ or } \mathbf{91.4\%}$$

Example 26.11. *The Hopkinson's test on two similar shunt machines gave the following full-load data :*

Line voltage	*= 110 V*	*Field currents are 3 A and 3.5 A.*
Line current	*= 48 A*	*Arm. resistance of each is = 0.035 Ω.*
Motor arm. current	*= 230 A*	

Calculate the efficiency of each machine assuming a brush contact drop of 1 volt per brush.

(Electrical Machines, Nagpur Univ, 1992)

Solution. The motor-generator set is shown in Fig. 29.8. It should also be noted that the machine with lesser excitation is motoring. We will find the total armature Cu losses and brush contact loss for both machines.

Motor

Arm. Cu loss	$= 230^2 \times 0.035 = 1,851.5$ W
Brush contact loss	$= 230 \times 2 = 460$ W
Total arm. Cu loss	$= 1,8515 + 460 = 2,312$ W
Shunt Cu loss	$= 110 \times 3 = 330$ W
Total Cu loss	$= 2,312 + 330 = 2,642$ W

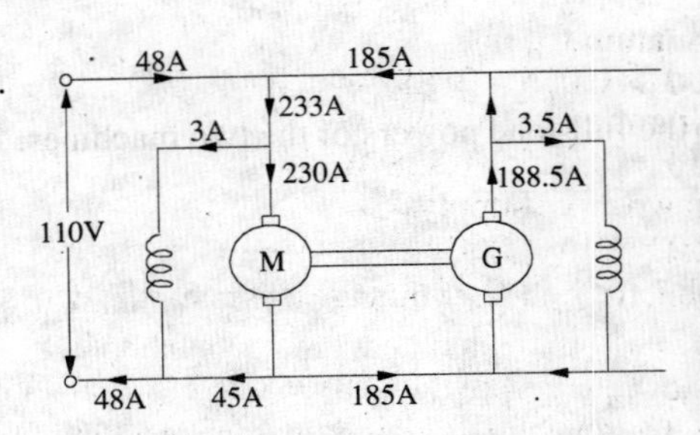

Fig. 29.8

Generator

Generator arm. current = 233 – 48 + 3.5 = 188.5 W

Arm. Cu loss = $188.5^2 \times 0.035$ = 1,244 W

Brush contact Cu loss = 188.5×2 = 377 W

Total arm. Cu loss = 1,244 + 377 = 1,621 W

Shunt Cu loss = 110×3.5 = 385 W;

Total Cu loss = 1,621 + 385 = 2,006 W

For the Set

Total arm. and shunt Cu loss for the set = 2,642 + 2,006 = 4,648 W

Total input = 110×48 = 5,280 W ; Stray losses for the set = 5,280 – 4,648 = 632 W

Stray losses per machine = 632/2 = 316 W

Motor Efficiency

Arm. Cu + brush drop loss = 2,312 W Shunt Cu loss = 330 W

Stray losses = 316 W Total loss = 2,312 + 330 + 316 = 2,958 W

Motor input = 110×233 = 25,630 W; Motor output = 25,630 – 2,958 = 22,672 W

$\therefore$ $\eta = 22,672 \times 100/25,630 =$ **88.8%**.

Generator Efficiency

Total losses = 2,006 + 316 = 2,322 W; Output = 110×185 = 20,350 W

Generator input = 20,350 + 2,322 = 22,672 W = motor input

η = 20,350/22,672 = 0.894 or **89.4%**

Example 29.12. *In a Hopkinson's test on a pair of 500-V, 100-kW shunt generators, the following data was obtained :*

Auxiliary supply, 30 A at 500 V : Generator output current, 200 A

Field currents, 3.5 A and 1.8 A

Armature circuit resistances, 0.075 Ω each machine. Voltage drop at brushes, 2 V (each machine).

Calculate the efficiency of the machine acting as a generator.

(Elect. Technology-1, Gwalior Univ. 1986)

Solution. Motor arm. current = 200 + 30 = 230 A

Motor arm. Cu loss = $230^2 \times 0.075 + 230 \times 2$ = 4,428 W

Motor field Cu loss = 500×1.8 = 900 W

Generator arm. Cu loss = $200^2 \times 0.075 + 200 \times 2$ = 3,400 W

Generator field Cu loss = 500×3.5 = 1,750 W

Total Cu loss for two machines

= 4,428 + 900 + 3400 + 1750 = 10,478 W

Power taken from auxiliary supply

= 500×30 = 15,000 W

Stray losses for the two machines

= 15,000 – 10,478 = 4,522 W

Stray loss per machine = 4,522/2 = 2,261 W

Total losses in generator = 3400 + 1750 + 2261

= 7,411 W

Generator output = 500×200

= 100,000 W

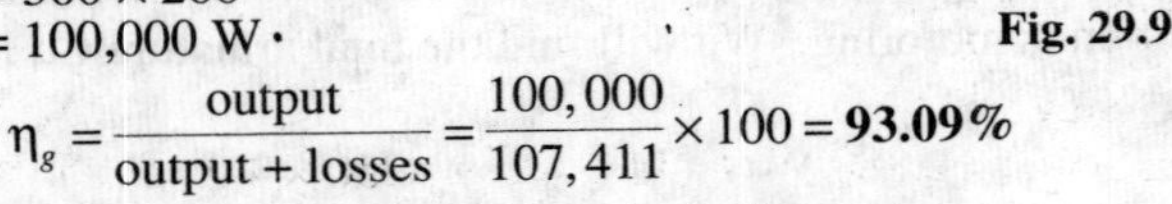

Fig. 29.9

$$\therefore \quad \eta_g = \frac{\text{output}}{\text{output + losses}} = \frac{100,000}{107,411} \times 100 = \mathbf{93.09\%}$$

Example 29.13. *Explain the Hopkinson test on a pair of shunt motors.*

In such a test on 250-V machines, the line current was 50 A and the motor current 400 A not including the field currents of 6 and 5 A. The armature resistance of each machine was 0.015 Ω. Calculate the efficiency of each machine.

(Adv. Elect. Machines, A.M.I.E. Sec. B, 1991)

Solution. The connections are shown in Fig. 29.10.

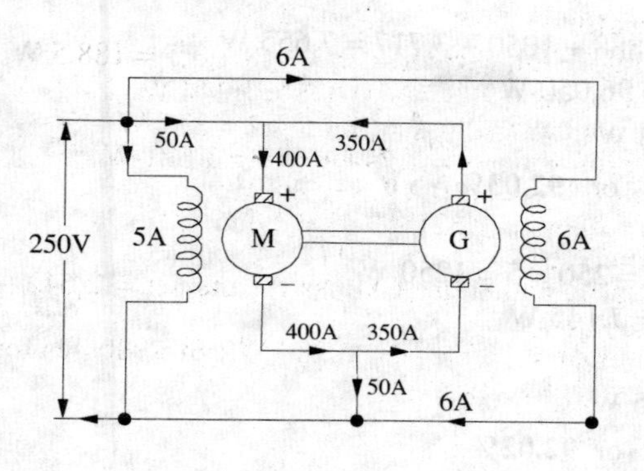

Fig. 29.10

Motor armature Cu loss
$= 400^2 \times 0.015 = 2,400$ W
Generator armature Cu loss
$= 350^2 \times 0.015 = 1,838$ W
Power drawn from supply
$= 250 \times 50 = 12,500$ W
$\therefore$ iron, friction and windage losses for the two machines
$= 12,500 - (2,400 + 1,838) = 8,262$ W
Iron, friction and windage loss per machine
$= 8,262/2 = 4,130$ W* (approx.)

Motor Losses and Efficiency

Motor arm. Cu loss = 2,400 W ; Motor field Cu loss $= 250 \times 5 = 1,250$ W
Iron, friction and windage losses = 4,130 W
Total motor losses $= 2,400 + 1250 + 4,130 = 7,780$ W
Motor input $= 250 \times 400 + 250 \times 5 = 101,250$ W
$\therefore$ motor efficiency $= (101,250 - 7,780)/101,250 = 0.923$ or **92.3%**

Generator Losses and Efficiency

Generator arm. Cu loss = 1,838 W; Generator field Cu loss $= 250 \times 6 = 1,500$ W
Iron, friction and windage loss= 4,130 W
Total losses $= 1,838 + 1,500 + 4,130 = 7,468$ W
Generator output $= 250 \times 350 = 87,500$ W
$\therefore$ generator efficiency $= (87,500 - 7,468)/87,500 = 0.915$ or **91.5%**

Example 29.14. *The Hopkinson's test on two shunt machines gave the following results for full-load :*

Line voltage = 250 V ; current taken from supply system excluding field currents = 50 A; motor armature current = 380 A ; field currents 5 A and 4.2 A. Calculate the efficiency of the machine working as a generator. Armature resistance of each machine is 0.02 Ω.

(Electrical Machinery-I Mysore Univ. 1988)

Solution. The connections are shown in 29.11.

Motor arm. Cu loss $= 380^2 \times 0.02 = 2,888$ W
Generator arm. Cu loss $= 330^2 \times 0.02 = 2,178$ W
Power drawn from supply $= 250 \times 50 = 12,500$ W
Stray losses for the two machines
$= 12,500 - (2,888 + 2,178) = 7,434$ W
Stray losses per machine
$= 7,434/2 = 3,717$ W

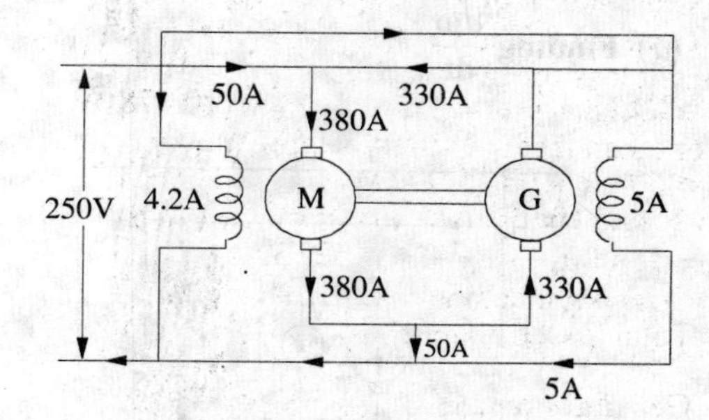

Fig. 29.11

Motor Efficiency

Arm. Cu loss = 2,888 W Field Cu loss $= 250 \times 4.2 = 1050$ W

---

*We could also get this value as follows :
Total supply input $= 250 \times 61 = 15,250$ W ; Gen. and motor field Cu loss $= 250 \times 6 + 250 \times 5 = 2,750$ W
Iron, friction and windage losses for both machines
$= 15,250 - (2,400 + 1,838 + 2,750) =$ **8,262 W** –as before

Stray losses = 3,717 W Total loss = 2,888 + 1050 + 3,717 = 7,655 W
Motor input = 250 × 380 + 250 × 4.2 = 96,050 W
Motor output = 96,050 − 7,655 = 88,395 W
∴ η = 88,395/96,050 = 0.9203 or **92.03%**

Generator Efficiency

Arm. Cu loss = 2,178 W; Field Cu loss = 250 × 5 = 1250 W
Stray losses = 3,717 W; Total losses = 7,145 W
Generator output = 250 × 330 = 82,500 W
Generator input = 82,500 + 7,145 = 89,645 W
∴ η = 82,500/89,645 = 0.9202 or **92.02%**

29.8 Retardation or Running Down Test

This method is applicable to shunt motors and generators and is *used for finding stray losses.* Then, knowing the armature and shunt Cu losses at a given load current, efficiency can be calculated.

The machine under test is speeded up slightly beyond its normal speed and then supply is cut off from the armature while keeping the field excited. Consequently, the armature slows down and its kinetic energy is used to meet the *rotational losses i.e.* friction, windage and iron losses.*

Kinetic energy of the armature is $K.E. = \frac{1}{2} I\omega^2$

where I = moment of inertia of the armature and ω = angular velocity

∴ rotational losses, W = rate of loss of $K.E.$

$$\therefore \; W = \frac{d}{dt}\left(\frac{1}{2} I\omega^2\right) = I\omega \cdot \frac{d\omega}{dt}$$

Two quantities need be known (*i*) moment of inertia (I) of the armature and (*ii*) $\frac{d\omega}{dt}$ or $\frac{dN}{dt}$ because $\omega \propto N$. These are found as follows:

(*a*) **Finding $\frac{d\omega}{dt}$**

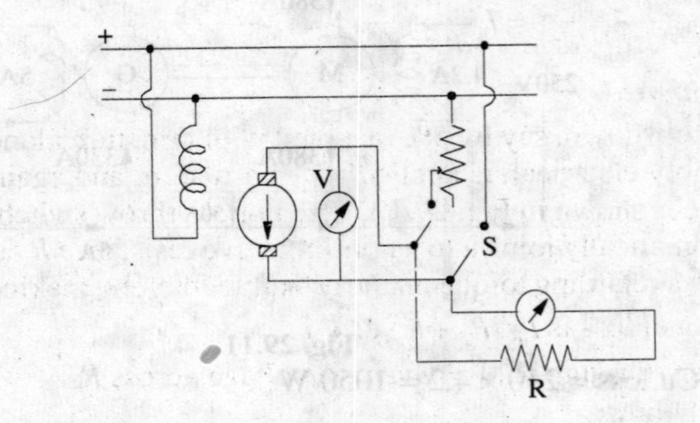

Fig. 29.12

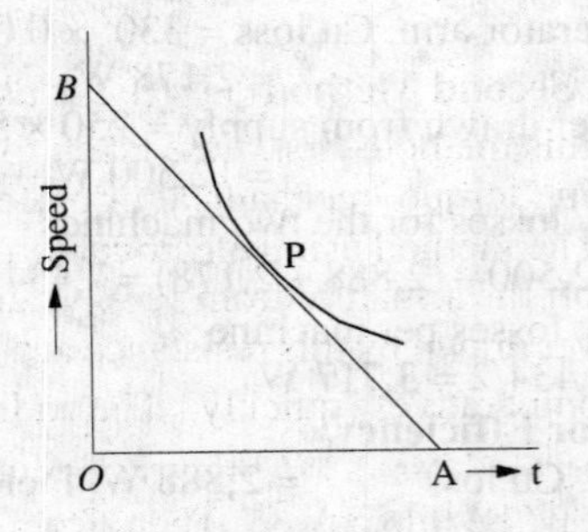

Fig. 29.13

*If armature slows down with no excitation, then energy of the armature is used to overcome mechanical losses only, there being no iron losses (see Ex. 29.18).

As shown in Fig. 29.12, a voltmeter V is connected across the armature. This voltmeter is used as a speed indicator by suitably graduating it, because $E \propto N$. When supply is cut off, the armature speed and hence voltmeter reading falls. By noting different amounts of voltage fall in different amounts of time, a curve is drawn between time and the speed (obtained from voltage values) as shown in Fig. 29.13.

From any point P which corresponds to normal speed, a tangent AB is drawn.

Then $$\frac{dN}{dt} = \frac{OB \text{ (in r.p.m.)}}{OA \text{ (in seconds)}}$$

From (*i*), above, $$W = i\,\omega\frac{d\omega}{dt}$$

Now $$\omega = \frac{2\pi N}{60} \quad \ldots(N \text{ in r.p.m.})$$

$$W = I\left(\frac{2\pi N}{60}\right)\times\frac{d}{dt}\left(\frac{2\pi N}{60}\right);\; W = \left(\frac{2\pi}{60}\right)^2 I.N.\frac{dN}{dt} = 0.011\,I.N.\frac{dN}{dt} \qquad \ldots(ii)$$

(*ii*) **Finding Moment of Inertia (I)**

(*a*) **First Method**–*where I is calculated.*

First, slowing down curve is drawn with armature alone. Next, a fly-wheel of known moment of inertia I_1 is keyed onto the shaft and slowing down curve is drawn again. Obviously, slowing down time will be longer due to combined increased moment of inertia of the two. For any given speed, (dN/dt_1) and (dN/dt_2) are determined as before. It should be noted that the losses in both cases would be almost the same, because addition of a fly-wheel will not make much difference to the losses.

Hence, from equation (*ii*) above

In the first case, $W = \left(\frac{2\pi}{60}\right)^2 IN\left(\frac{dN}{dt_1}\right)$

In the second case, $W = \left(\frac{2\pi}{60}\right)^2 (I + I_1)N \times \left(\frac{dN}{dt_2}\right)$

$$\therefore \quad (I + I_1)\left(\frac{dN}{dt_2}\right) = I\left(\frac{dN}{dt_1}\right) \text{ or } \left(\frac{I + I_1}{I}\right) = \left(\frac{dN}{dt_1}\right) = \left(\frac{dN}{dt_1}\right)\Big/\left(\frac{dN}{dt_2}\right)$$

$$\therefore \quad I = I_1 \times \frac{(dN/dt_2)}{(dN/dt_1) - (dN/dt_2)} = I_1\frac{dt_1}{dt_2 - dt_1};\; I = I_1\frac{t_1}{t_2 - t_1}$$

(*b*) **Second Method**–*where I is eliminated.*

In this method, first, time taken to slow down, say by 5%, is noted with armature alone. Next, a retarding torque–mechanical or preferably electrical, is applied to the armature and again time is noted. The method using electrical torque is shown in Fig. 29.12. The double-throw switch S while cutting off the armature from supply, automatically joins it to a non-inductive resistance R as shown. The power drawn by this resistance acts as a retarding torque on the armature, thereby making it slow down comparatively quickly. The additional loss is $I_a{}^2(R_a + R)$ or $V I_a$

where I_a = *average* current through R; V = *average* voltage across R

Let W' be this power. Then, from (*i*) above

$$\left.\begin{aligned} W &= \left(\frac{2\pi}{60}\right)^2 I.N.\frac{dN}{dt_1} \\ W + W' &= \left(\frac{2\pi}{60}\right)^2 I.N.\frac{dN}{dt_2} \end{aligned}\right\} \text{ if } dN \text{ is the same}$$

$$\therefore \quad \frac{W + W'}{W} = \frac{dt_1}{dt_2} \qquad \therefore W = W' \times \frac{dt_2}{dt_1 - dt_2} \text{ or } W = W' \times \frac{t_2}{t_1 - t_2}$$

where $\frac{dN}{dt_1}$ = rate of change of speed *without* extra load

$\frac{dN}{dt_2}$ = rate of change of speed *with* extra electrical load.

Example 29.15. *In a retardation test on a separately-excited motor, the induced e.m.f. in the armature falls from 220 V to 190 V in 30 seconds on disconnecting the armature from the supply. The same fall takes place in 20 seconds if, immediately after disconnection, armature is connected to a resistance which takes 10 A (average) during this fall. Find stray losses of the m or.*

(Adv. Elect. Machines, A.M.I.E. Sec. B, 1992)

Solution.

Let W = stray losses (mechanical and magnetic losses)

Average voltage across resistance = (200 + 190)/2 = 195 V, Average current = 10 A

$\therefore$ power absorbed W' = 1950 W

Using the relation $\frac{W}{W'} = \frac{t_2}{t_1 - t_2}$; we get $W = 1950 \times \frac{20}{30 - 20}$ = **3,900 watt**

Example 29.16. *In a retardation test on a d.c. motor, with its field normally excited, the speed fell from 1525 to 1475 r.p.m. in 25 seconds. With an average load of 1.0 kW supplied by the armature, the same speed drop occurred in 20 seconds. Find out the moment of inertia of the rotating parts in kg.m².*

(Electrical Machines-III, Gujarat Univ. 1984)

Solution. As seen from Art. 29.8 (*ii*) (*b*),

$$W' = \left(\frac{2\pi}{60}\right)^2 I.N.\frac{dN}{dt} \qquad \text{Also } W = W' \times \frac{t_2}{t_1 - t_2}$$

Here, W' = 1 kW = 1000 W, t_1 = 25 second, t_2 = 20 second

$\therefore$ $W = 1000 \times 20/(25 - 20) = 4000$ W

Now, N = 1500 r.p.m. (average speed); dN = 1525 – 1475 = 50 r.p.m. ; dt = 25

$\therefore$ 4000 = $(2\pi/60)^2 I.1500 \times 50/25$ $\therefore$ I = **121.8 kg.m².**

Example 29.17. *A retardation test is made on a separately-excited d.c. machine as a motor. The induced voltage falls from 240 V to 225 V in 25 seconds on opening the armature circuit and 6 seconds on suddenly changing the armature connection from supply to a load resistance taking 10 A (average). Find the efficiency of the machine when running as a motor and taking a current of 25 A on a supply of 250 V. The resistance of its armature is 0.4 Ω and that of its field winding is 250 Ω.*

(Elect. Technology, Allahabad Univ. 1991)

Solution. Average voltage across load

= (240 + 225)/2 = 232.5 V; I_{av} = 10 A

$\therefore$ power absorbed W' = 232.5 × 10 = 2,325 W

and t_1 = 30 second, t_2 = 6 second ; W = stray loss

Using $\frac{W}{W'} = \frac{t_2}{t_1 - t_2}$ = **734.1 W**, we get

Stray losses $W = 2325 \times \frac{6}{25 - 6}$ = **734.1 W**

Input current = 25 A; I_{sh} = 250/250 = 1A; I_a = 25 – 1 = 24 A

Armature Cu loss = $24^2 \times 0.4$ = 230.4 W ; Shunt Cu loss = 250 × 1 = 250 W

$\therefore$ Total losses = 734.1 + 230.4 + 250 = 1,215 W (approx.)

Input = 250 × 25 = 6,250 W ; Output = 6,250 – 1,215 = 5,035 W

$\therefore$ η = 5,035/6,250 = 0.806 or **80.6%**

Example 29.18. *A retardation test is carried out on a 1000 r.p.m. d.c. machine. The time taken for the speed to fall from 1030 r.p.m. to 970 r.p.m. is :-*

(a) 36 seconds with no excitation

(b) 15 seconds with full excitation and

(c) 9 seconds with full excitation and the armature supplying an extra load of 10 A at 219 V.

Calculate (i) the moment of inertia of the armature in kg.m² (ii) iron losses and (iii) the mechanical losses at the mean speed of 1000 r.p.m.

Solution. It should be noted that

(*i*) when armature slows down with no excitation, its kinetic energy is used to overcome *mechanical* losses only ; because due to the absence of flux, there is no iron loss.

(*ii*) with excitation, ion kinetic energy is used to supply mechanical and iron losses collectively known as stray losses.

(*iii*) If I is taken in kg-m² unit, then rate of loss of energy is in watts.

Mechanical loss $$W_m = \left(\frac{2\pi}{60}\right)^2 I.N.\frac{dN}{dt} \qquad \text{...Art. 29.8}$$

Here $dN = 1030 - 970 = 60$ r.p.m., $dt = 36$ seconds, $N = 1000$ r.p.m.

$$W_m = \left(\frac{2\pi}{60}\right)^2 I.N.\frac{60}{36} \qquad ...(i)$$

Similarly $$W_s = \left(\frac{2\pi}{60}\right)^2 I.N.\frac{60}{15} \qquad ...(ii)$$

Also $$W_s = W'\frac{t_2}{t_1 - t_2} = 219 \times 10 \times \frac{9}{15-9} = 3,285 \text{ W}$$

Using equation (*ii*), we get

$$3,285 = \left(\frac{2\pi}{60}\right)^2 \times I \times 1000 \times \frac{60}{15}$$

(*i*) ∴ $\mathbf{I = 75\ kg.m^2}$

Dividing (*i*) by (*ii*), we get $\frac{W_m}{W_s} = \frac{15}{36}$

(*ii*) ∴ $W_m = 3,285 \times 15/36 = \mathbf{1,369\ W}$

(*iii*) ∴ iron losses $= W_s - W_m = 3,285 - 1,369 = \mathbf{1,916\ W}$

29.9. Field's Test for Series Motor

This test is applicable to two similar *series motors*. Series motors which are mainly used for traction work are easily available in pairs. The two machines are coupled mechanically.

One machine runs normally as a motor and drives generator whose output is wasted in a variable load R (Fig. 29.14). Iron and friction losses of two machines are made equal (*i*) by joining the series

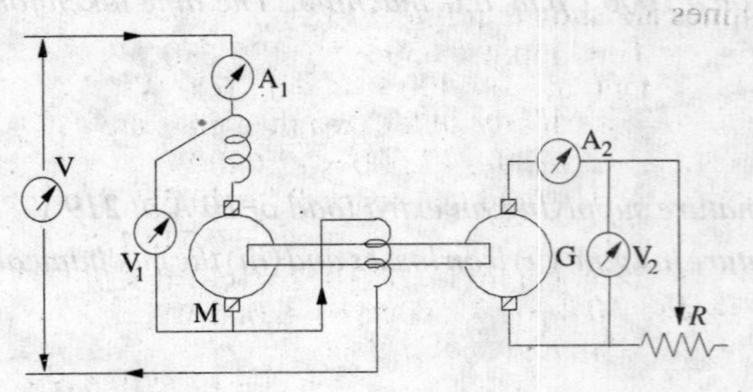

Fig. 29.14

field winding of the generator in the motor armature circuit so that both machines are equally excited and (*ii*) by running them at equal speed. Load resistance R is varied till the motor current reaches its full-load value indicated by ammeter A_1. After this adjustment for full load current, different ammeter and voltmeter readings are noted.

Let V = supply voltage; I_1 = motor current; V_2 = terminal p.d. of generator; I_2 = load current.

∴ intake of the whole set $= VI_1$; output $= V_2I_2$.

Total losses in the set, $W_t = VI_1 - V_2I_2$

Armature and field Cu losses $W_{cu} = (R_a + 2R_{se})I_1^2 + I_2^2R_a$

where R_a = hot armature resistance of each machine

R_{se} = hot series field resistance of each machine

∴ stray losses for the set $= W_t - W_{cu}$

Stray losses per machine $W_s = \dfrac{W_t - W_{cu}}{2}$

Stray losses are equally divided between the machines because of their equal excitation and speed.

Motor Efficiency

Motor input $= V_1I_1$

Motor losses = armature + field Cu losses + stray losses

$= (R_a + R_{se})I_1^2 + W_s = W_m$ (say)

$$\eta_m = \frac{V_1I_1 - W_m}{V_1I_1}$$

Generator Efficiency

The generator efficiency will be of little use because it is running under abnormal conditions of *separate* excitation. However, the efficiency under these unusual conditions can be found if desired.

Generator output $= V_2I_2$

Field Cu loss $= I_1^2R_{se}$ (∵ motor current is passing through it)

Armature Cu loss $= I_2^2R_a$; Stray losses $= W_s$

Total losses $= I_1^2R_{se} + I_2^2R_a + W_s = W_g$ (say)

$$\eta_g = \frac{V_2I_2}{V_2I_2 + W_g}$$

It should be noted that although the two machines are mechanically coupled yet it is not a regenerative method, because the generator output is wasted instead of being fed back into the motor as in Hopkinson's (back-to back) test.

Example 29.19. *A test on two coupled similar tramway motors, with their fields connected in series, gave the following results when one machine acted as a motor and the other as a generator.*

Motor : Armature current $= 56$ *A ; Armature voltage* $= 590$ *V*

Voltage drop across field winding $= 40$ *V*

Generator :

Armature current $= 44$ *A ; Armature voltage* $= 400$ *V*

Field voltage drop $= 40$ *V ; Resistance of each armature* $= 0.3\ \Omega$

Calculate the efficiency of the motor and gearing at this load.

(Elect. Machinery-II, Nagpur Univ. 1992)

Solution. The connections for the two machines are shown in Fig. 29.15.

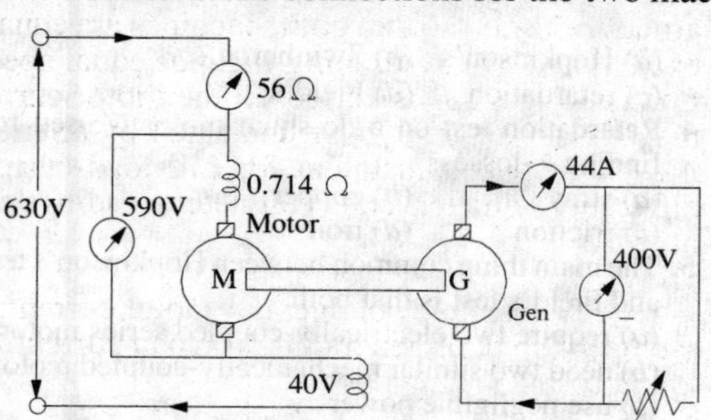

Fig. 29.15

Total input $= 630 \times 56 = 35{,}280$ W

Output $= 400 \times 44 = 17{,}600$ W

Total losses in the two machines are

$= 35{,}280 - 17{,}600 = 17{,}680$ W

Series field resistance $R_{se} = 40/56$

$= 0.714\ \Omega$

Total Cu loss $= (0.3 + 2 \times 0.714) \times 56^2$

$+ 44^2 \times 0.3 = 5{,}425 + 581 = 6{,}006$ W

Stray losses of the set

$= 17{,}680 – 6{,}006$

$= 11{,}674$ W

$\therefore$ stray losses/machine $= 11{,}674/2 = 5{,}837$ W

Motor Efficiency

Motor armature input $=$ arm. voltage $\times$ motor current $= 590 \times 56 = 33{,}040$ W

Armature circuit Cu loss $= (0.3 + 0.714) \times 56^2 = 3{,}180$ W

Stray loss $= 5{,}837$ W —found above

Total losses $= 3{,}180 + 5{,}837 = 9{,}017$ W, Output $= 33{,}040 - 9{,}017 = 24{,}023$ W

$\therefore \quad \eta_m = 24{,}023/33{,}040 = 0.727$ or **72.7%**

Generator Efficiency

Armature Cu loss $= 44^2 \times 0.3 = 581$ W, Series field Cu loss $= 40 \times 56 = 2{,}240$ W

Stray losses $=$ 5,837 W; Total losses $= 581 + 2{,}240 + 5{,}837 = 8{,}658$ W

Output $= 400 \times 44 = 17{,}600$ W

$\therefore \quad \eta_g = 17{,}600/(17{,}600 + 8{,}658) = 0.67$ or **67%**

Tutorial Problem No. 29.1

1. A 500-V, shunt motor takes a total current of 5 A when running unloaded. The resistance of armature circuit is 0.25 Ω and the field resistance is 125 Ω. Calculate the efficiency and output when the motor is loaded and taking a current of 100 A. **[90.4%; 45.2 kW]**

2. A d.c. shunt motor rated at 12.5 kW output runs at no-load at 1000 r.p.m. from a 250-V supply consuming an input current of 4 A. The armature resistance is 0.5 Ω and shunt field resistance is 250 Ω. Calculate the efficiency of the machine when delivering full-load output of 12.5 kW while operating at 250 V.

[**81.57%**] (*Elect. Technology-I Madras Univ. 1979*)

3. The following results were obtained during Hopkinson's test on two similar 230-V machines; armature currents 37 A and 30 A; field currents 0.85 A and 0.8 A. Calculate the efficiencies of machines if each has an armature resistance of 0.33 Ω. **[Generator 87.9%, Motor 87.7%]**

4. In a Field's test on two 230-V, 1.492 kW mechanically-coupled similar series motors, the following figures were obtained. Each had armature and compole resistance of 2.4 Ω, series field resistance of 1.45 Ω and total brush drop of 2 V. The p.d. across armature and field was 230 V with a motor current of 10.1 A. The generator supplied a current of 8.9 A at a terminal p.d. of 161 V. Calculate the efficiency and output of the motor for this load. **[76.45%, 1.775 kW]**

5. Describe the Hopkinson test for obtaining the efficiency of two similar shunt motors. The readings obtained in such a test were as follows; line voltage 100 V; motor current 30 A; generator current 25 A; armature resistance of each machine 0.25 Ω. Calculate the efficiency of each machine from these results, ignoring the field currents and assuming that their iron and mechanical losses are the same.

[Motor 90.05%; Generator 92.5%]

6. The Hopkinson's test on two similar d.c. shunt machines gave the following results:

Line voltage = 220 V; line current excluding field currents = 40 A; the armature current of motoring machine = 200 A; field currents 6 A and 7 A. Calculate the efficiency of each of the machines at the given load conditions. The armature resistance of each of the machines is 0.05 Ω.

[η_m = **86.58%** ; η_g = **86.3%**] (*Electrical Engg-I, M.S. Univ. Baroda 1980*)

Objective Tests–29

1. One of the main advantages of Swinburne's test is that it
   (*a*) is applicable both to shunt and compound motors
   (*b*) needs one running test
   (*c*) is very economical and convenient
   (*d*) ignores any change in iron loss
2. The main disadvantage of Hopkinson's test for finding efficiency of shunt d.c. motors is that it
   (*a*) requires full-load power
   (*b*) ignores any change in iron loss
   (*c*) needs one motor and one generator
   (*d*) requires two identical shunt machines
3. The most economical method of finding no-load losses of a large d.c. shunt motor is—test
   (*a*) Hopkinson's (*b*) Swinburne's
   (*c*) retardation (*d*) Field's
4. Retardation test on a dc shunt motor is used for finding—losses.
   (*a*) stray (*b*) copper
   (*c*) friction (*d*) iron
5. The main thing common between Hopkinson's test and Field's test is that both
   (*a*) require two electrically-coupled series motors
   (*b*) need two similar mechanically-coupled motors
   (*c*) use negligible power
   (*d*) are regenerative tests
6. The usual test for determining the efficiency of a traction motor is the______test.
   (*a*) Field's (*b*) retardation
   (*c*) Hopkinson's (*d*) Swinburne's

Answers

1. *c* **2.** *d* **3.** *b* **4.** *a* **5.** *b* **6.** *a*

QUESTIONS AND ANSWERS ON D.C. MOTORS

Q. 1. How may the direction of rotation of a d.c. motor be reversed ?
Ans. By reversing either the field current or current through the armature. Usually, reversal of current through the armature is adopted.
Q. 2. What will happen if both currents are reversed ?
Ans. The motor will run in the original direction.
Q. 3. What will happen if the field of a d.c. shunt motor is opened ?
Ans. The motor will achieve dangerously high speed and may destroy itself.
Q. 4. What happens if the direction of current at the terminals of a series motor is reversed ?
Ans. It does not reverse the direction of rotation of motor because current flows through the armature in the same direction as through the field.
Q. 5. Explain what happens when a d.c. motor is connected across an a.c. supply ?
Ans. 1. since on a.c. supply, reactance will come into the picture, the a.c. supply will be offered impedance (not resistance) by the armature winding. Consequently, with a.c. supply, current will be much less. The motor will run but it would not carry the same load as it would on d.c. supply.
2. there would be more sparking at the brushes.
3. though motor armature is laminated as a rule, the field poles are not. Consequently, eddy currents will cause the motor to heat up and eventually burn on a.c. supply.
Q. 6. What will happen if a shunt motor is directly connected to the supply line.
Ans. Small motors up to 1 kW rating may be line-started without any adverse results being produced. High rating motors must be started through a suitable starter in order to avoid the huge starting current which will
(*i*) damage the motor itself and (*ii*) badly affect the voltage regulation of the supply line.
Q. 7. What is the function of interpoles and how are interpole windings connected ?
Ans. Interpoles are small poles placed in between the main poles. Their function is to assist commutation by producing the auxiliary or commutating flux. Consequently, brush sparking is practically eliminated. Interpole windings are connected in series with the armature winding.
Q. 8. In rewinding the armature of a d.c. motor, progressive connections are changed to retrogressive ones. Will it affect the operation in any way ?
Ans. Yes. Now, the armature will rotate in the opposite direction.
Q. 9. A d.c. motor fails to start when switched on. What could be the possible reasons and remedies ?

Ans. Any one of the following reasons could be responsible :-

1. open-circuit in controller–should be checked for open starting resistance or open switch or open fuse.
2. low terminal voltage–should be adjusted to name-plate value.
3. overload–should be reduced if possible otherwise larger motor should be installed.
4. excessive friction–bearing lubrication should be checked.

Q. 10. A d.c. motor is found to stop running after a short period of time. What do you think could be the reasons? How would you remedy each ?

Ans. Possible causes are as under :-

1. motor not getting enough power–check voltage at motor terminals as well as fuses, clups and overload relay.
2. weak or no field–in the case of adjustable-speed motors, check if rheostat is correctly set. Also, check field winding for any 'open'. Additionally, look for any loose winding or broken connection.
3. motor torque insufficient for driving the given load–check line voltage with name-plate voltage. If necessary, use larger motor to match the load.

Q. 11. What are the likely causes if a d.c. motor is found to run too slow under load ? And the remedy ?

Ans. 1. supply line voltage too low–remove any excessive resistance in supply line, connections or controller
2. brushes ahead of neutral–set them on neutral
3. overload–reduce it to allowable value or use larger motor.

Q. 12. Why does a d.c. motor sometime run too fast when under load ? Give different possible causes and their remedies.

Ans. Different possible causes are as under :-

1. weak field–remove any extra resistance in shunt field circuit. Also, check for 'grounds'
2. line voltage too high–reduce it to name-plate value
3. brushes back of neutral–set them on neutral,

Q. 13. Under what conditions is sparking produced at the brushes of a d.c. motor ? How would you remedy it ?

Ans. 1. commutator in bad condition–clean and reset brushes
2. commutator either eccentric or rough–grind and true the commutator. Also, undercut mica.
3. excessive vibration–balance armature. Make sure that brushes ride freely in holders
4. brush-holding spring broken or sluggish–replace spring and adjust pressure to recommended value
5. motor overloaded–reduce load or install motor of proper rating
6. short-circuit in armature circuit–remove any metallic particles between commutator segments and check for short between adjacent commutator risers. Locate and repair internal armature short if any.

Q. 14. Sometimes a hissing noise (or brush chatter) is heard to emanate from the commutator end of a running d.c. motor. What could it be due to and how could it be removed ?

Ans. Any one of the following causes could produce brush chatter :-

1. excessive clearance of brush holders–adjust properly
2. incorrect angle of brushes–adjust to correct value
3. unsuitable brushes–replace them
4. high mica–undercut it
5. wrong brush spring pressure–adjust to correct value.

Q. 15. What are the possible causes of excessive sparking at brushes in a d.c. motor ?

Ans. 1. poor brush fit on commutator–sand-in the brushes and polish commutator.
2. brushes binding in the brush holders–clean holders and brushes and remove any irregularities on surfaces of brush holders or rough spots on brushes.
3. excessive or insufficient pressure on brushes–adjust pressure
4. brushes off neutral–set them on neutral.

Q. 16. Why does a d.c. motor sometime spark on light load ?

Ans. Due to the presence of paint spray, chemical, oil or grease etc. on commutator.

Q. 17. When is the armature of a d.c. motor likely to get over-heated ?

Ans. 1. when motor is over-loaded
2. when it is installed at a place having restricted ventilation
3. when armature winding is shorted.

Q. 18. What causes are responsible for over-heating of commutator in a d.c. motor ?

Ans. It could be due either to the brushes being off neutral or being under excessive spring pressure. Accordingly, brushes should be adjusted properly and the spring pressure should be reduced but not to the point where sparking is introduced.

30 TRANSFORMER

30.1. Working Principle of a Transformer

A transformer is a static (or stationary) piece of apparatus by means of which electric power in one circuit is transformed into electric power of the same frequency in another circuit. It can raise or lower the voltage in a circuit but with a corresponding decrease or increase in current. The physical basis of a transformer is *mutual induction* between two circuits linked by a common magnetic flux. In its simplest form, it consists of two inductive coils which are electrically separated but magnetically linked through a path of low reluctance as shown in Fig. 30.1. The two coils possess high mutual inductance. If one coil is connected to a source of alternating voltage, an alternating flux is set up in the laminated core, most of which is linked with the other coil in which it produces mutually-induced e.m.f. (according to Faraday's Laws of Electromagnetic Induction $e = MdI/dt$). If the second coil circuit is closed, a current flows in it and so electric energy is transferred (entirely magnetically) from the first coil to the second coil. The first coil, in which electric energy is fed from the a.c. supply mains, is called *primary* winding and the other from which energy is drawn out, is called *secondary* winding. In brief, a transformer is a device that

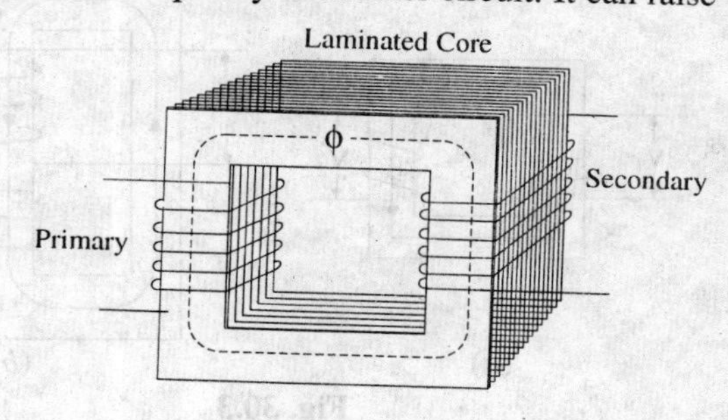

Fig. 30.1

1. transfers electric power from one circuit to another
2. it does so without a change of frequency
3. it accomplishes this by electromagnetic induction and
4. where the two electric circuits are in mutual inductive influence of each other.

30.2. Transformer Construction

The simple elements of a transformer consist of two coils having mutual inductance and a laminated steel core. The two coils are insulated from each other and the steel core. Other necessary parts are : some suitable container for assembled core and windings ; a suitable medium for insulating the core and its windings from its container ; suitable bushings (either of porcelain, oil-filled or capacitor-type) for insulating and bringing out the terminals of windings from the tank.

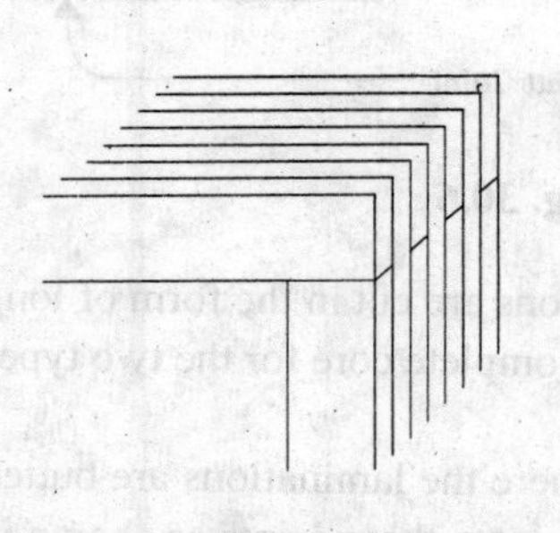
Fig. 30.2

In all types of transformers, the core is constructed of transformer sheet steel laminations assembled to provide a continuous magnetic path with a minimum of air-gap included. The steel used is of high silicon content, sometimes heat treated to produce a high permeability and a low hysteresis loss at the usual operating flux densities. The eddy current loss is minimised by laminating the core, the laminations being insulated from each other by a light coat of core-plate varnish or by an oxide layer on the surface. The thickness of laminations varies from 0.35 mm for a frequency of 50 Hz to 0.5 mm for a frequency of 25 Hz. The core laminations (in the form of strips) are joined as shown in Fig. 30.2. It is seen that the joints in the alternate layers are staggered in order to avoid the presence of narrow gaps right through the cross-section of the core. Such staggered joints are said to be 'imbricated'.

Constructionally, the transformers are of two general types, distinguished from each other merely by the manner in which the primary and secondary coils are placed around the laminated core. The two types are known as (*i*) **core-type** and (*ii*) **shell-type**. Another recent development is *spiral-core* or *wound-core type*, the trade name being *spirakore* transformer.

In the so-called core type transformers, the *windings surround a considerable part of the core* whereas in shell-type transformers, the *core surrounds a considerable portion of the windings* as shown schematically in Fig. 30.3 (*a*) and (*b*) respectively.

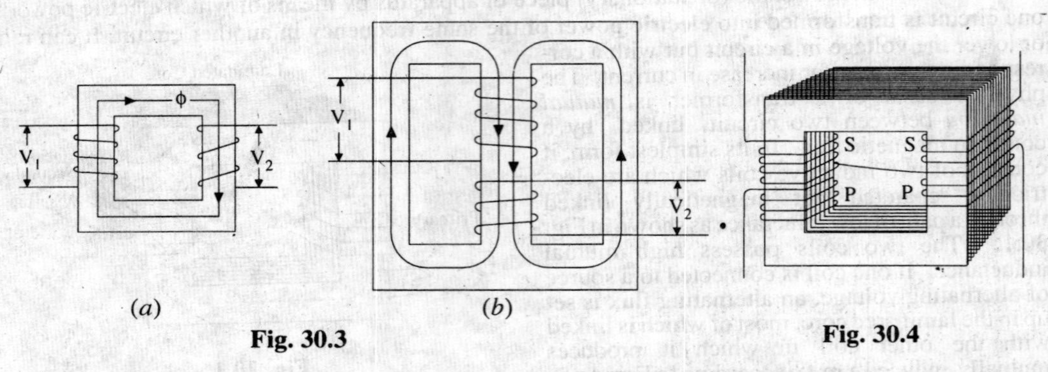

Fig. 30.3

Fig. 30.4

In the simplified diagram for the core type transformers [Fig. 30.3(*a*)], the primary and secondary winding are shown located on the opposite legs (or limbs) of the core, but in actual construction, these are always interleaved to reduce leakage flux. As shown in Fig. 30.4, half the primary and half the secondary winding have been placed side by side or concentrically on each limb, not primary on one limb (or leg) and the secondary on the other.

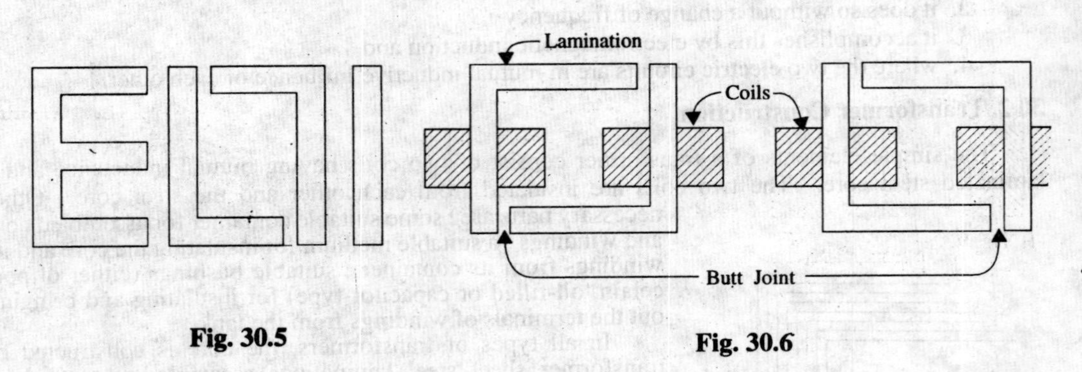

Fig. 30.5

Fig. 30.6

In both core and shell-type transformers, the individual laminations are cut in the form of long strips of *L*'s, *E*'s and *I*'s as shown in Fig. 30.5. The assembly of the complete core for the two types of transformers is shown in Fig. 30.6 and Fig. 30.7.

As said above, in order to avoid high reluctance at the joints where the laminations are butted against each other, the alternate layers are stacked differently to eliminate these joints as shown in Fig. 30.6 and 30.7.

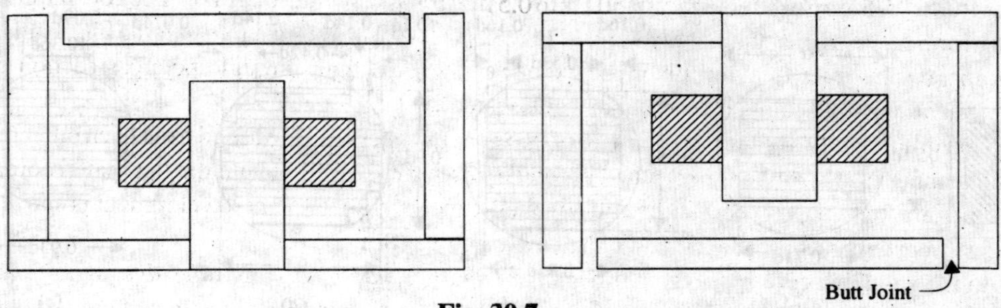

Fig. 30.7

30.3. Core-type Transformers

The coils used are form-wound and are of the cylindrical type. The general form of these coils may be circular or oval or rectangular. In small size core-type transformers, a simple rectangular core is used with cylindrical coils which are either circular or rectangular in form. But for large-size core-type transformers, round or circular cylindrical coils are used which are so wound as to fit over a cruciform core section as shown in Fig. 30-8(*a*). The circular cylindrical coils are used in most of the core-type transformers because of their mechanical strength. Such cylindrical coils are wound in helical layers with the different layers insulated from each other by paper, cloth, micarta board or cooling ducts. Fig. 30-8(*c*) shows the general arrangement of these coils with respect to the core. Insulating cylinders of fuller board are used to separate the cylindrical windings from the core and from each other. Since the low-voltage (*LV*) winding is easiest to insulate, it is placed nearest to the core (Fig. 30.8).

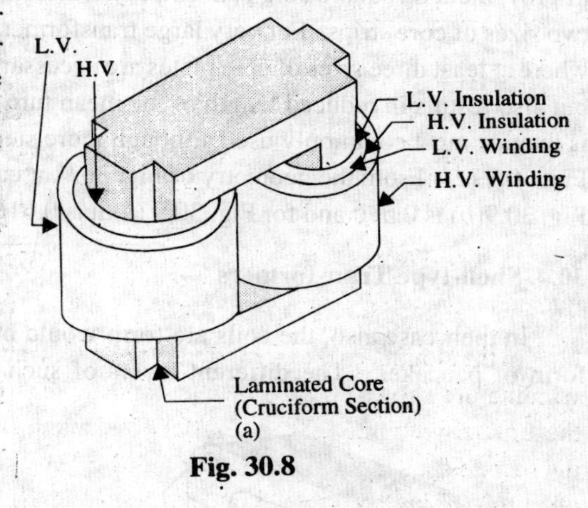

Fig. 30.8

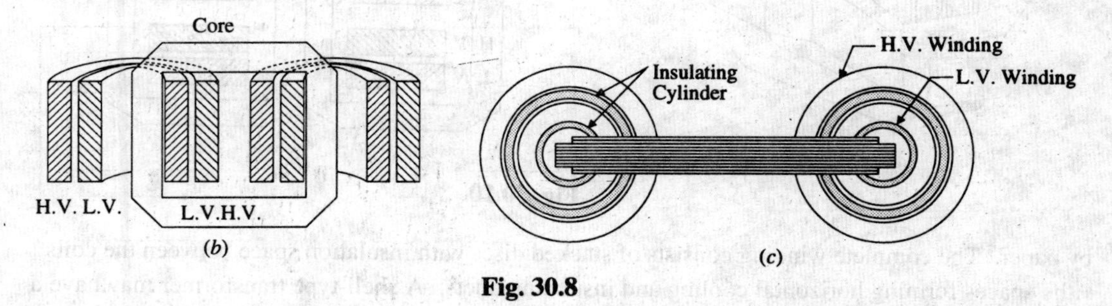

Fig. 30.8

Because of laminations and insulation, the net or effective core area is reduced, due allowance for which has to be made (Ex. 30.6). It is found that, in general, the reduction in core sectional area due to the presence of paper, surface oxide etc. is of the order of 10% approximately.

As pointed out above, rectangular cores with rectangular cylindrical coils can be used for small-size core-type transformers as shown in Fig. 30.9(*a*) but for large-sized transformers, it becomes wasteful to use rectangular cylindrical coils and so circular cylindrical coils are preferred. For such purposes, square cores may be used as shown in Fig. 30.9 (*b*) where circles represent the tubular

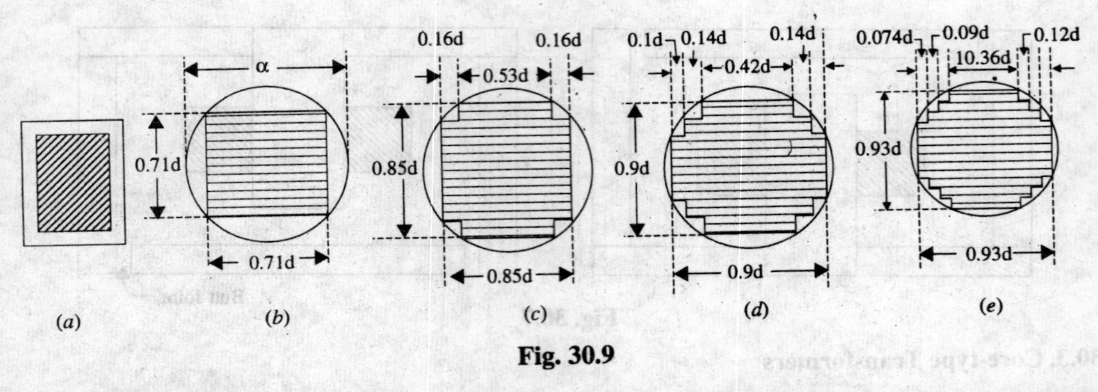

Fig. 30.9

former carrying the coils. Obviously, a considerable amount of useful space is still wasted. A common improvement on square core is to employ cruciform core as in Fig. 30.9 (*c*) which demands, at least, two sizes of core strips. For very large transformers, further core-stepping is done as in Fig. 30.9 (*d*) where at least three sizes of core plates are necessary. Core-stepping not only gives high space factor but also results in reduced length of the mean turn and the consequent I^2R loss. Three stepped core is the one most commonly used although more steps may be used for very large transsformers as in Fig. 30.9 (*e*). From the geometry of Fig. 30.9, it can be shown that maximum gross core section for Fig. 30.9(*b*) is 0.5 d^2 and for Fig. 30.9 (*c*) it is 0.616d^2 where *d* is the diameter of the cylindrical coil.

30.4. Shell-type Transformers

In their case also, the coils are form-would but are multi-layer disc type usually wound in the form of pancakes. The different layers of such multi-layer discs are insulated from each other

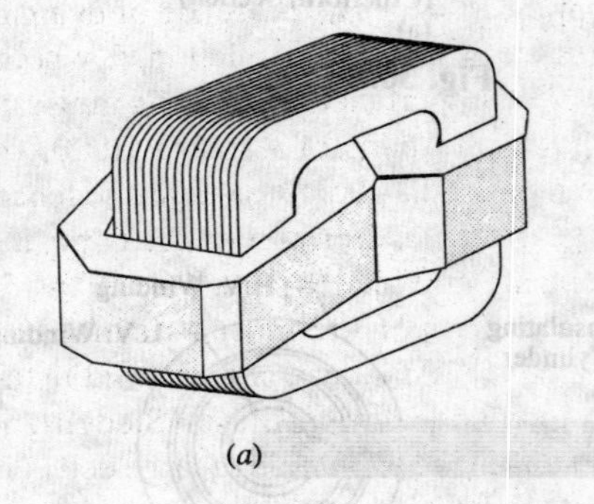

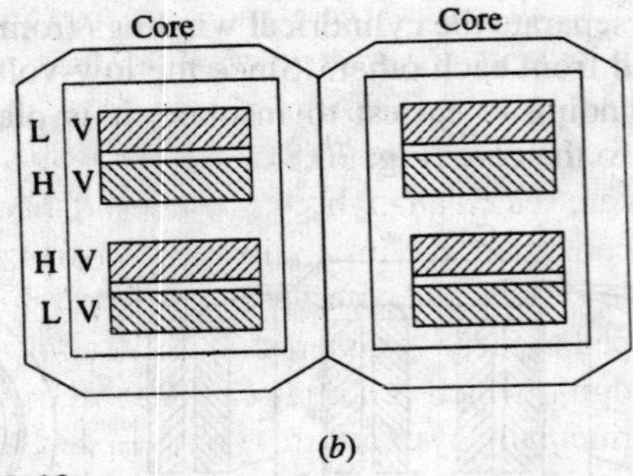

Fig. 30.10

by paper. The complete winding consists of stacked discs with insulation space between the coils —the spaces forming horizontal cooling and insulating ducts. A shell-type transformer may have a simple rectangular form as shown in Fig. 30.10 or it may have distributed form as shown in Fig. 30.11.

A very commonly-used shell-type transformer is the one known as Berry Transformer–so called after the name of its designer and is cylindrical in form. The transformer core consists of laminations arranged in groups which radiate out from the centre as shown in section in Fig. 30.12.

It may be pointed out that cores and coils of transformers must be provided with rigid mechanical bracing in order to prevent movement and possible insulation damage. Good bracing reduces vibration and the objectionable noise–a humming sound–during operation.

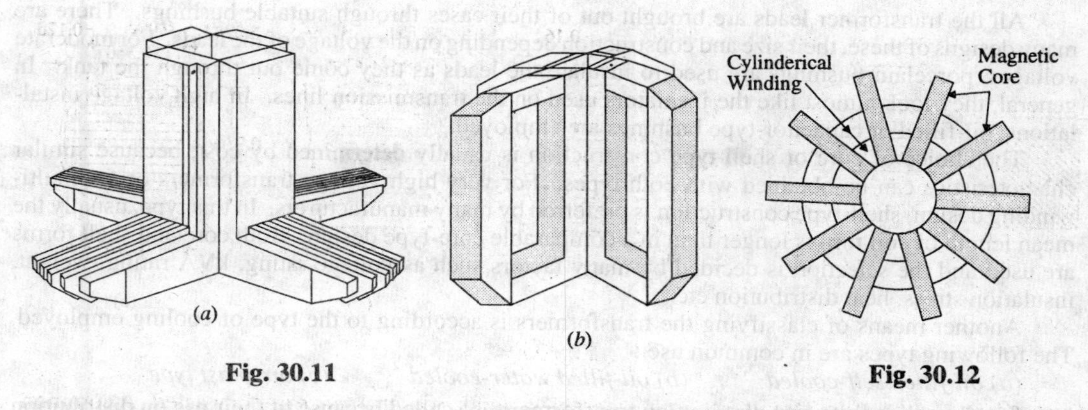

Fig. 30.11

Fig. 30.12

The spiral-core transformer employs the newest development in core construction. The core is assembled of a continuous strip or ribbon of transformer steel wound in the form of a circular or elliptical cylinder. Such construction allows the core flux to follow the grain of the iron. Cold-rolled steel of high silicon content enables the designer to use considerably higher operating flux densities with lower loss per kg. The use of higher flux density reduces the weight per kVA. Hence, the advantages of such construction are (*i*) a relatively more rigid core (*ii*) lesser weight and size per kVA rating (*iii*) lower iron losses at higher operating flux densities and (*iv*) lower cost of manufacture.

Transformers are generally housed in tightly-fitted sheet-metal; tanks filled with special insulating oil*. This oil has been highly developed and its function is two-fold. By circulation, it not only keeps the coils reasonably cool, but also provides the transformer with additional insulation not obtainable when the transformer is left in the air.

In cases where a smooth tank surface does not provide sufficient cooling area, the sides of the tank are corrugated or provided with radiators mounted on the sides. Good transformer oil should be absolutely free from alkalies, sulphur and particularly from moisture. The presence of even an extremely small percentage of moisture in the oil is highly detrimental from the insulation viewpoint because it lowers the dielectric strength of the oil considerably. The importance of avoiding moisutre in the transformer oil is clear from the fact that even an addition of 8 parts of water in 1,000,000 reduces the insulating quality of the oil to a value generally recognized as below standard. Hence, the tanks are sealed air-tight in smaller units. In the case of large-sized transformers where complete air-tight construction is impossible, chambers known as *breathers* are provided to permit the oil inside the tank to expand and contract as its temperature increases or decreases. The atmospheric moisture is entrapped in these breathers and is not allowed to pass on to the oil. Another thing to avoid in the oil is sledging which is simply the decomposition of oil with long and continued use. Sledging is caused principally by exposure to oxygen during heating and results in the formation of large deposits of dark and heavy matter that eventually clogs the cooling ducts in the transformer.

No other feature in the construction of a transformer is given more attention and care than the insulating materials, because the life of the unit almost solely depends on the quality, durability and handling of these materials. All the insulating materials are selected on the basis of their high quality and ability to preserve high quality even after many years of normal use.

---

*Instead of natural mineral oil, now-a-days synthetic insulating fluids known as ASKARELS (trade name) are used. They are non-inflammable and, under the influence of an electric arc, do not decompose to produce inflammable gases. One such fluid commercially known as PYROCLOR is being extensively used because it possesses remarkable stability as a dielectric and even after long service shows no deterioration through sledging, oxidation, acid or moisture formation. Unlike mineral oil, it shows no rapid burning.

All the transformer leads are brought out of their cases through suitable bushings. There are many designs of these, their size and construction depending on the voltage of the leads. For moderate voltages, porcelain bushings are used to insulate the leads as they come out through the tank. In general, they look almost like the insulators used on the transmission lines. In high voltage installations, oil-filled or capacitor-type bushings are employed.

The choice of core or shell-type construction is usually determined by cost, because similar characteristics can be obtained with both types. For very high-voltage transformers or for multi-winding design, shell-type construction is preferred by many manufacturers. In this type, usually the mean length of coil turn is longer than in a comparable core-type design. Both core and shell forms are used and the selection is decided by many factors such as voltage rating, kVA rating, weight, insulation stress, heat distribution etc.

Another means of classifying the transformers is according to the type of cooling employed. The following types are in common use :

(a) oil-filled self-cooled *(b) oil-filled water-cooled* *(c) air-blast type*

Small and medium size distribution transformers—so called because of their use on distribution systems as distinguished from line transmission—are of type (*a*). The assembled windings and cores of such transformers are mounted in a welded, oil-tight steel tank provided with steel cover. After putting the core at its proper place, the tank is filled with purified, high quality insulating oil. The oil serves to convey the heat from the core and the windings to the case from where it is radiated out to the surroundings. For small size, the tanks are usually smooth-surfaced, but for larger sizes, the cases are frequently corrugated or fluted to get greater heat radiation area without increasing the cubical capacity of the tank. Still larger sizes are provided with radiatiors or pipes.

Construction of very large self-cooled transformers is expensive, a more economical form of construction for such large transformers is provided in the oil-immersed, water-cooled type. As before, the windings and the core are immersed in the oil, but there is mounted near the surface of oil, a cooling coil through which cold water is kept circulating. The heat is carried away by this water. The largest transformers such as those used with high-voltage transmission lines, are constructed in this manner.

Oil-filled transformers are built for outdoor duty and as these require no housing oher than their own, a great saving is thereby effected. These transformers require only periodic inspection.

For voltages below 25,000 V transformers can be built for cooling by means of an air-blast. The transformer is not immersed in oil, but is housed in a thin sheet-metal box open at both ends through which air is blown from the bottom to the top by means of a fan or blower.

30.5. Elementary Theory of an Ideal Transformer

An ideal transformer is one which has no losses *i.e.* its windings have no ohmic resistance, there is no magnetic leakage and hence which has no I^2R and core losses. In other words, an ideal transformer consists of two purely inductive coils wound on a loss-free core. It may, however, be noted *that it is impossible to realize such a transformer in practice, yet for convenience, we will start with such a transformer and step by step approach an actual transformer.*

Consider an ideal transformer [Fig. 30.13 (*a*)] whose secondary is open and whose primary is connected to sinusoidal altenating voltage V_1. This potential difference causes an alternating current

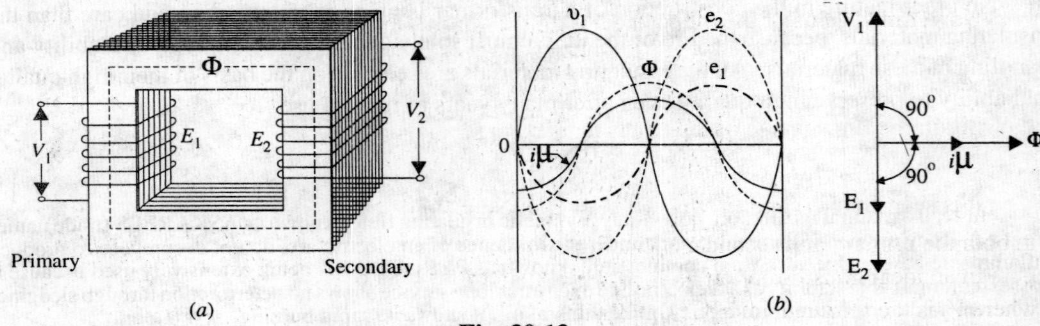

Fig. 30.13

to flow in the primary. Since the primary coil is purely inductive and there is no output (secondary being open) the primary draws the magnetisting current I_μ only. The function of this current is merely to magnetise the core, it is small in magnitude and lags V_1 by 90°. This alternating current I_μ produces an alternating flux ϕ which is, at all times, proportional to the current (assuming permeability of the magnetic circuit to be constant) and, hence, is in phase with it. This changing flux is linked both with the primary and the secondary windings. Therefore, it produces self-induced e.m.f. in the primary. This *self-induced* e.m.f. E_1 is, at every instant, equal to and in opposition to V_1. It is also known as counter e.m.f. or back e.m.f. of the primary.

Similarly, there is produced in the secondary an induced e.m.f. E_2 which is known as *mutually* induced e.m.f. This e.m.f. is antiphase with V_1 and its magnitude is proportional to the rate of change of flux and the number of secondary turns.

The instantaneous values of applied voltage, induced e.m.fs, flux and magnetising current are shown by sinusoidal waves in Fig. 30.13(*b*). Fig. 30.13 (*c*) shows the vectorial representation of the effective values of the above quantities.

30.6. E.M.F. Equation of a Transformer

Let N_1 = No. of turns in primary

N_2 = No. of turns in secondary

Φ_m = maximum flux in core in webers $= B_m \times A$

f = frequency of a.c. input in Hz

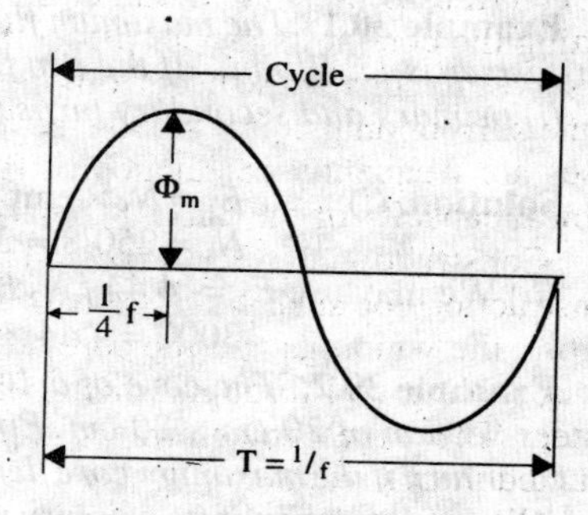

Fig. 30.14

As shown in Fig. 30.14, flux increases from its zero value to maximum value Φ_m in one quarter of the cycle *i.e.* in 1/4 *f* second

$$\therefore \quad \text{average rate of change of flux} = \frac{\Phi_m}{i/4f}$$

$$= 4f\Phi_m \text{ Wb/s or volt}$$

Now, rate of change of flux per turn means induced e.m.f. in volts.

$\therefore$ average e.m.f./turn $= 4f\Phi_m$ volt

If flux Φ varies *sinusoidally*, then r.m.s. value of induced e.m.f. is obtained by multiplying the average value with form factor.

$$\text{Form factor} = \frac{\text{r.m.s. value}}{\text{average value}} = 1.11$$

$\therefore$ r.m.s. value of e.m.f./turn $= 1.11 \times 4f\Phi_m = 4.44f\Phi_m$ volt

Now, r.m.s. value of the induced e.m.f. in the whole of primary winding

= (induced e.mf./turn) × No. of primary turns

$$E_1 = 4.44f\,N_1\,\Phi_m = 4.44\,f N_1\,B_m\,A \qquad ...(i)$$

Similarly, r.m.s. value of the e.m.f. induced in secondary is,

$$E_2 = 4.44f\,N_2\Phi_m = 4.44f\,N_2 B_m\,A \qquad ...(ii)$$

It is seen from (*i*) and (*ii*) that $E_1/N_1 = E_2/N_2 = 4.44\,f\Phi_m$. It means that e.m.f./turn is the *same* in both the primary and secondary windings.

In an ideal transformer on no-load, $V_1 = E_1$ and $E_2 = V_2$ where V_2 is the terminal voltage (Fig. 30-15).

30.7. Voltage Transofrmation Ratio (K)

From eqations (*i*) and (*ii*), we get

$$\frac{E_2}{E_1} = \frac{N_2}{N_1} = K$$

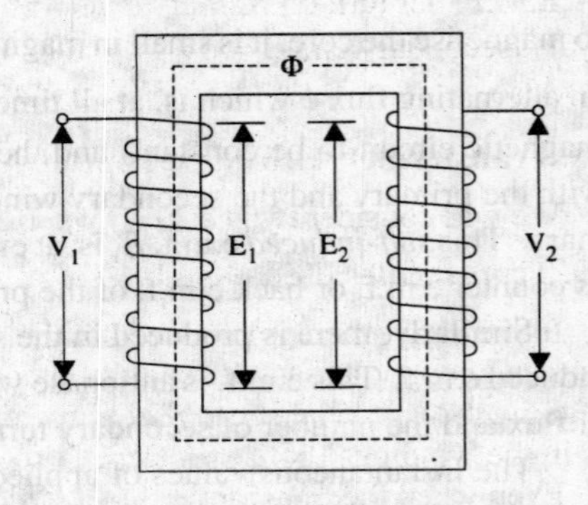

Fig. 30.15

This constant *K* is known as voltage transformation ratio.

(*i*) If $N_2 > N_1$ *i.e.* $K > 1$, then transformer is called *step-up* transformer.

(*ii*) If $N_2 < N_1$ *i.e.* $K < 1$, then transformer is known as *step-down* transformer.

Again, for an *ideal* transformer,
input VA = output VA

$$V_1 I_1 = V_2 I_2 \text{ or } \frac{I_2}{I_1} = \frac{V_1}{V_2} = \frac{1}{K}$$

Hence, currents are in the inverse ratio of the (voltage) transformation ratio.

Example 30.1. *The maximum flux density in the core of a 250/3000-volts, 50-Hz single-phase transformer is 1.2 Wb/m². If the e.m.f. per turn is 8 volt, determine*
(i) primary and secondary turns (ii) area of the core.

(Electrical Engg-I, Nagpur Univ. 1991)

Solution. (*i*) $E_1 = N_1 \times$ emf induced/turn
$N_1 = 250/8 = \mathbf{32}$; $N_2 = 3000/8 = \mathbf{375}$

(*ii*) We may use $E_2 =- 4.44 f N_2 B_m A$

$\therefore$ $3000 = 4.44 \times 50 \times 375 \times 1.2 \times A$; $\mathbf{A = 0.03\ m^2}$.

Example 30.2. *The core of a 100-kVA, 11000/550 V, 50-Hz, 1-ph, core type transformer has a cross-section of 20 cm × 20 cm. Find (i) the number of h.v. and 1.v. turns per phase and (ii) the e.m.f. per turn if the maximum core density is not to exceed 1.3 Tesla. Assume a stacking factor of 0.9. What will happen if its primary voltage is increased by 10% on no-load ?*

(Elect. Machines, A.M.I.E. Sec. B, 1991)

Solution. (*i*) $B_m = 1.3\ T$, $A = (0.2 \times 0.2) \times 0.9 = 0.036\ m^2$

$\therefore$ $11{,}000 = 4.44 \times 50 \times N_1 \times 1.3 \times 0.036$, $N_1 = 1060$

$550 = 4.44 \times 50 \times N_2 \times 1.3 \times 0.036$; $N_2 = 53$

or $N_2 = KN_1 = (550/11{,}000) \times 1060 = 53$

(*ii*) e.m.f./turn = 11,000/1060 = 10.4 V or 550/53 = 10.4 V

Keeping supply frequency constant, if primary voltage is increased by 10%, magnetising current will increase by much more than 10%. However, due to saturation, flux density will increase only marginally and so will the eddy current and hysteresis losses.

Example 30.3. *A single-phase transformer has 400 primary and 1000 secondary turns. The net cross-sectional area of the core is 60 cm². If the primary winding be connected to a 50-Hz supply at 520 V, calculate (i) the peak value of flux density in the core (ii) the voltage induced in the secondary winding.*

(Elect. Engg-I, Pune Univ. 1989)

Solution. $K = N_2/N_1 = 1000/400 = 2.5$

(*i*) $E_2/E_1 = K$ $\therefore E_2 = KE_1 = 2.5 \times 520 = \mathbf{1300\ V}$

(*ii*) $E_1 = 4.44\ f\ N_1 B_m A$

or $520 = 4.44 \times 50 \times 400 \times B_m \times (60 \times 10^{-4})$ $\therefore B_m = \mathbf{0.976\ Wb/m^2}$

Example 30.4. *A 25-kVA transformer has 500 turns on the primary and 50 turns on the secondary winding. The primary is connected to 3000-V, 50-Hz supply. Find the full-load primary and secondary currents, the secondry e.m.f. and the maximum flux in the core. Neglect leakage drops and no-load primary current.*

(Elect. & Electronic Engg., Madras Univ. 1985)

Solution. $K = N_2/N_1 = 50/500 = 1/10$

Now, full-load $I_1 = 25{,}000/3000 = \mathbf{8.33\ A}$. F.L. $I_2 = I_1/K = 10 \times 8.33 = \mathbf{83.3\ A}$.

E.M.F. per turn on primary side = 3000/500 = 6 V

$\therefore$ secondary e.m.f. = 6 × 50 = **300 V** (or $E_2 = KE_1 = 3000 \times 1/10 = 300\ V$)

Also, $E_1 = 4.44\, f N_1\, \Phi_m$; $3000 = 4.44 \times 50 \times 500 \times \Phi_m$ $\therefore$ $\Phi_m = \mathbf{27\ m\ Wb}$

30.8. Transformer with Losses but no Magnetic Leakage

We will consider two cases (*i*) when such a transformer is on no-load and (*ii*) when it is loaded.

30.9. Transformer on No-load

In the above discussion, we assumed an ideal transformer *i.e.* one in which there were no core losses and copper losses. But practical conditions require that certain modifications be made in the foregoing theory. When an *actual* transformer is put on load, there is iron loss in the core and copper loss in the windings (both primary and secondary) and these losses are not entirely negligible.

Even when the transformer is on no-load, the primary input current is not wholly reactive. The primary input current under no-load conditions has to supply (*i*) iron losses in the core *i.e.* hysteresis loss and eddy current loss and (*ii*) a very small amount of copper loss in primary (there being no Cu loss in secondary as it is open). Hence, the no-load primary input current I_0 is not at 90° behind V_1 but lags it by an angle $\phi_0 < 90°$. No-load input power

$$W_0 = V_1 I_0 \cos \phi_0$$

where cos ϕ_0 is primary power factor under no-load conditions. No-load condition of an actual transformer is shown vectorially in Fig. 30.16.

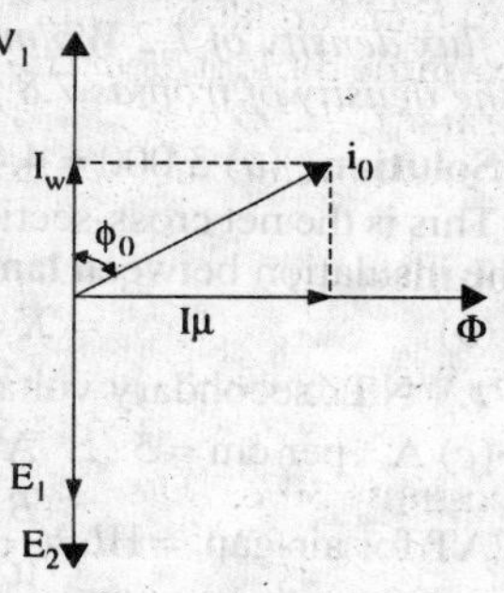

Fig. 30.16

As seen from Fig. 30.16, primary current I_0 has two components :

(*i*) one in phase with V_1. This is known as *active* or *working* or *iron* loss component I_w because it mainly supplies the iron loss plus small quantity of primary Cu loss.

$$I_w = I_0 \cos \phi_0$$

(*ii*) the other component is in quadrature with V_1 and is known as *magnetising* component I_μ because its function is to sustain the alternating flux in the core. It is wattless.

$$I_\mu = I_0 \sin \phi_0$$

Obviously, I_0 is the vector sum of I_w and I_μ, hence $I_0 = (I_\mu^2 + I_\omega^2)$.

The following points should be noted carefully :

1. The no-load primary current I_0 is very small as compared to the full-load primary current. It is about 1 per cent of the full-load current.

2. Owing to the fact that the permeability of the core varies with the instantaneous value of the exciting current, the wave of the exciting or magnetising current is not truly sinusoidal. As such, it should not be represented by a vector because only sinusoidally varying quantities are represented by rotating vectors. But, in practice, it makes no appreciable difference.

3. As I_0 is very small, the no-load primary Cu loss is negligibly small which means *that no-load primary input is practically equal to the iron loss in the transformer.*

4. As it is principally the core-loss which is responsible for shift in the current vector, angle ϕ_0 is known as *hysteresis angle of advance.*

Example 30.5. *(a) A 2,200/200-V transforr er draws a no-load primary current of 0.6 A and absorbs 400 watts. Find the magnetising and iron loss currents.*

(b) A 2,200/250-V transformer takes 0.5 A at a p.f. of 0.3 on open circuit. Find magnetising and working components of no-load primary current.

Solution. (*a*) Iron-loss current

$$= \frac{\text{no-load input in watts}}{\text{primary voltage}} = \frac{400}{2,200} = \mathbf{0.182\ A}$$

Now $$I_0^{\ 2} = I_w^{\ 2} + I_\mu^{\ 2}$$

Magnetising component $I_\mu = \sqrt{(0.6^2 - 0.182^2)} = \mathbf{0.572\ A}$

The two components are shown in Fig. 27.16.

(*b*) $I_0 = 0.5$ A, $\cos\phi_0 = 0.3$ $\therefore$ $I_w = I_0 \cos\phi_0 = 0.5 \times 0.3 = 0.15$ A

$I_\mu = \sqrt{(0.5^2 - 0.15)} = 0.476$ A

Example 30.6. *A single-phase transformer has 500 turns on the primary and 40 turns on the secondary winding. The mean length of the magnetic path in the iron core is 150 cm and the joints are equivalent to an air-gap of 0.1 mm. When a p.d. of 3,000 V is applied to the primary, maximum flux density is 1.2 Wb/m². Calculate (a) the cross-sectional area of the core (b) no-load secondary voltage (c) the no-load current drawn by the prin ry (d) power factor on no-load. Given that AT/cm for a flux density of 1.2 Wb/m² in iron to be 5, the corresponding iron loss to be 2 watt/kg at 50 Hz and the density of iron as 7.8 gram/cm³*

Solution. (*a*) $3,000 = 4.44 \times 50 \times 500 \times 1.2 \times A$ $\therefore$ $A = 0.0225$ m² = **225 cm²**

This is the net cross-sectional area. However, the gross area would be about 10% more to allow for the insulation between laminations.

(*b*) $K = N_2/N_1 = 40/500 = 4/50$

$\therefore$ N.L. secondary voltage = KE_1 = (4/50) × 3000 = **240 V**

(*c*) A per cm = 5 $\therefore$ AT for iron core = 150 × 5 = **750**

$$\text{AT for air-gap} = Hl = \frac{B}{\mu_0} \times l = \frac{1.2}{4\pi \times 10^{-7}} \times 0.0001 = 95.5$$

Total AT for given B_{max} = 750 + 95.5 = 845.5

Max. value of magnetising current drawn by primary = 845.5/500 = 1.691 A

Assuming this current to be sinusoidal, its r.m.s. value is $I_\mu = 1.691/\sqrt{2} = 1.196$ A

Volume of iron = length × area = 150 × 225 = 33,750 cm³

Density = 7.8 gram/cm³ $\therefore$ mass of iron = 33,750 x 7.8/1000 = 263.25 kg

Total iron loss = 263.25 × 2 = 526.5 W

Iron loss component of no-load primary current I_0 is I_w = 526.5/3000 = 0.176 A

$$I_0 = \sqrt{I_u^2 + I_w^2} = \sqrt{1.196^2 + 0.176^2} = \mathbf{1.208\ A}$$

(*d*) Power factor, $\cos\phi_0 = I_w/I_0 = 0.176/1.208 = \mathbf{0.1457}$

Tutorial Problem No. 30-1

1. The number of turns on the primary and secondary windings of a 1−φ transformer are 350 and 35 respectively. If the primary is connected to a 2.2 kV, 50-Hz supply, determine the secondary voltage on no-load. **[220 V]** (*Elect. Engg.-II, Kerala Univ. 1980*)

2. A 3000/200-V, 50-Hz, 1-phase transformer is built on a core having an effective cross-sectional area of 150 cm² and has 80 turns in the low-voltage winding. Calculate
(*a*) the value of the maximum flux density in the core
(*b*) the number of turns in the high-voltage winding. [(*a*) **0.75 Wb/m²** (*b*) **1200**]

3. A 3,300/230-V, 50-Hz, 1-phase transformer is to be worked at a maximum flux density of 1.2 Wb/m² in the core. The effective cross-sectional area of the transformer core is 150 cm². Calculate suitable values of primary and secondary turns. **[830 ; 58]**

4. A 40-kVA, 3,300/240-V, 50 Hz, 1-phase transformer has 660 turns on the primary. Determine
(*a*) the number of turns on the secondary (*b*) the maximum value of flux in the core
(*c*) the approximate value of primary and secondary full-load currents.
Internal drops in the windings are to be ignored [(*a*) **48** (*b*) **22.5 mWb** (*c*) **12.1 A ; 166.7 A**]

5. A double-wound, 1-phase transformer is required to step down from 1900 V to 240 V, 50-Hz. It is to have 1.5 V per turn. Calculate the required number of turns on the primary and secondary windings respectively.

The peak value of flux density is required to be not more than 1.2 Wb/m^2. Calculate the required cross-sectional area of the steel core. If the output is 10 kVA, calculate the secondary current.

[1,267 ; 160 ; 56.4 cm^2; 41.75 A]

6. The no-load voltage ratio in a 1-phase, 50-Hz, core-type transformer is 1,200/440. Find the number of turns in each winding if the maximum flux is to be 0.075 Wb. **[24 and 74 turns]**

7. A 1-phase transformer has 500 primary and 1200 secondary turns. The net cross-sectional area of the core is 75 cm^2. If the primary winding be connected to a 400-V, 50 Hz supply, calculate

(*i*) the peak value of flux density in the core and (*ii*) voltage induced in the secondary winding.

[0.48 Wb/m^2 ; 60 V]

8. A 10-kVA, 1-phae transformer has a turn ratio of 300/23. The primary is connected to a 1500-V, 60 Hz supply. Find the secondary volts on open-circuit and the approximate values of the currents in the two windings on full-load. Find also the maximum value of the flux. **[115 V ; 6.67 A; 87 A ; 11.75 mWb]**

9. A 100-kVA, 3300/400-V, 50 Hz, 1 phase transformer has 110 turns on the secondary. Calculate the approximate values of the primary and secondary full-load currents, the maximum value of flux in the core and the number of primary turns.

How does the core flux vary with load ? **[30.3 A; 250 A; 16.4 mWb ; 907]**

10. The no-load current of a transformer is 5.0 A at 0.3 power factor when supplied at 230-V, 50-Hz. The number of turns on the primary winding is 200. Calculate (*i*) the maximum value of flux in the core (*ii*) the core loss (*iii*) the magnetising current. **[5.18 mWb ; 345 W; 4.77 A]**

11. The no-load current of a transformer is 15 A at a power factor of 0.2 when connected to a 460-V, 50-Hz supply. If the primary winding has 550 turns, calculate

(*a*) the magnetising component of no-load current (*b*) the iron loss

(*c*) the maximum value of the flux in the core. **[(*a*) 14.7 A (*b*) 1,380 W (*c*) 3.77 mWb]**

12. The no-load current of a transformer is 4.0 A at 0.25 p.f. when supplied at 250-V, 50 Hz. The number of turns on the primary winding is 200. Calculate

(*i*) the r.m.s. value of the flux in the core (assume sinusoidal flux) (*ii*) the core loss

(*iii*) the magnetising current. **[(*i*) 3.96 m Wb (*ii*) 250 W (*iii*) 3.87 A]**

30.10. Transformer on Load

When the secondary is loaded, the secondary current I_2 is set up. The magnitude and phase of I_2 with respect to V_2 is determined by the characteristics of the load. Current I_2 is in phase with V_2 if load is non-inductive, it lags if load is inductive and it leads if load is capacitive.

The secondary current sets up its own m.m.f. (=$N_2 I_2$) and hence its own flux Φ_2 which is in opposition to the main primary flux Φ which is due to I_0. The secondary ampere-turns $N_2 I_2$ are known as *demagnetising* amp-turns. The opposing secondary flux Φ_2 weakens the primary flux Φ momentarily, hence primary back e.m.f. E_1 tends to be reduced. For a moment, V_1 gains the upper hand over E_1 and hence causes more current to flow in primary.

Let the additional primary current be I_2'. It is known as *load component of primary current*. This current is antiphase with I_2'. The additional primary m.m.f. $N_1 I_2$ sets upits own flux Φ_2' which is in opposition to Φ_2 (but is in the same direction as Φ) and is equal to it in magnitude. Hence, the two cancel each other out. So, we find that the magnetic effects of secondary current I_2 are immediately neutralized by the additional primary current I_2' which is brought into existence exactly at the same instant as I_2. The whole process is illustrated in Fig. 30-17.

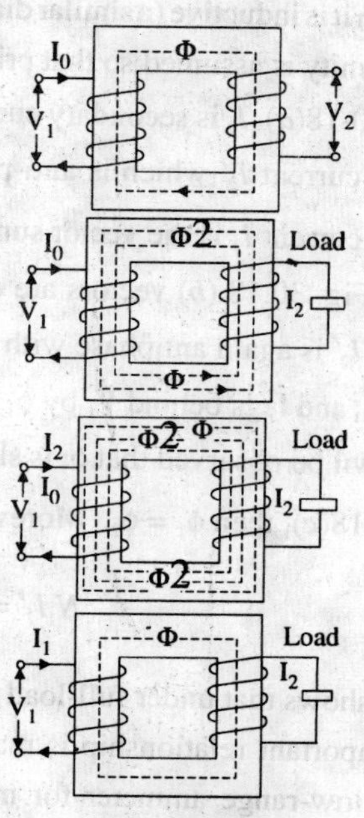

Fig. 30.17

Hence, whatever the load conditions, *the net flux passing through the core is approximately the same as at no-load.* An important deduction is that due to the constancy of core flux at all loads, *the core loss is also practically the same under all load conditions.*

$$\text{As } \Phi_2 = \Phi_2' \quad \therefore \quad N_2 I_2 = N_1 I_2' \quad \therefore \quad I_2' = \frac{N_2}{N_1} \times I_2 = K I_2$$

Hence, when transformer is on load, the primary winding has two currents in it ; one is I_0 and the other is I_2' which is anti-phase with I_2 and K times in magnitude. *The total primary current is the vector sum of I_0 and I_2'.*

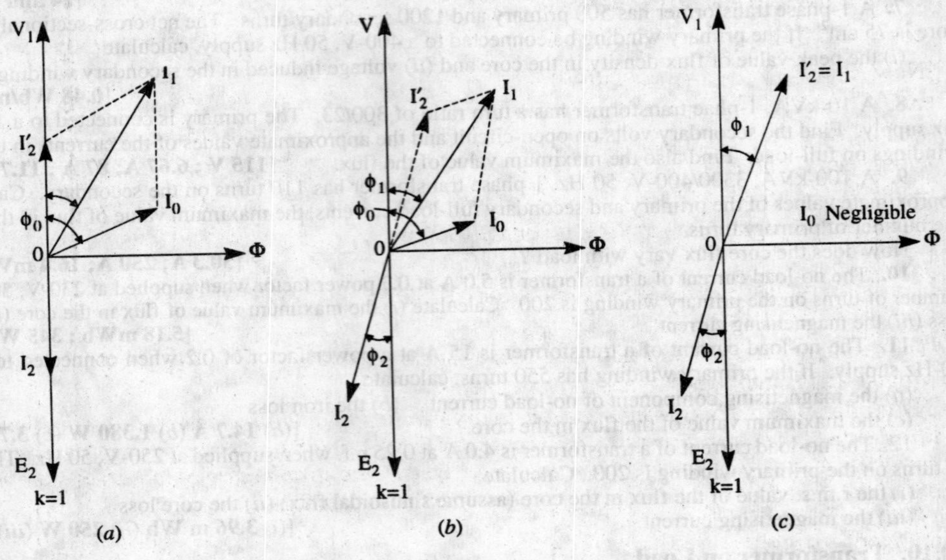

Fig. 30.18

In Fig. 30.18 are shown the vector diagrams for a load transformer when load is non-inductive and when it is inductive (a similar diagram could be drawn for capacitive load). Voltage transformation ratio of unity is assumed so that primary vectors are equal to the secondary vectors. With reference to Fig. 30-18(*a*), I_2 is secondary current in phase with E_2 (strictly speaking it should be V_2). It causes primary current I_2' which is anti-phase with it and equal to it in magnitude ($\because$ $K = 1$). Total primary current I_1 is the vector sum of I_0 and I_2' and lags behind V_1 by an angle ϕ_1.

In Fig. 30.18 (*b*) vectors are drawn for an inductive load. Here, I_2 lags E_2 (actually V_2) by ϕ_2. Current I_2' is again antiphase with I_2 and equal to it in magnitude. As before, I_1 is the vector sum of I_2' and I_0 and lags behind V_1 by ϕ_1.

It wil be observed that ϕ_1 is slightly greater than ϕ_2. But if we neglect I_0 as compared to I_2' as in Fig. 30.18(*c*), then $\phi_1 = \phi_2$. Morever, under this assumption

$$N_1 I_2' = N_2 I_1 = N_1 I_2 \qquad \therefore \quad \frac{I_2'}{I_2} = \frac{I_1}{I_2} = \frac{N_2}{N_1} = K$$

It shows that under full-load condtions, the ratio of primary and secondary currents is constant. This important relationship is made the basis of current transformer—a transformer which is used with a low-range ammeter for measuring currents in circuits where the direct connection of the ammeter is impracticable.

The different components of the primary current are shown below :

[PICTURE]

Example 30-7. *A single-phase transformer with a ratio of 440/110-V takes a no-load current of 5 A at 0.2 power factor lagging. If the secondary supplies a current of 120 A at a p.f. of 0.8 lagging, estimate the current taken by the primary.* **(Elect. Engg. Punjab Univ. 1991)**

Solution.

$\cos \phi_2 = 0.8, \phi_2 = \cos^{-1}(0.8) = 36° 54'$

$\cos \phi_0 = 0.2 \therefore \phi_0 = \cos^{-1}(0.2) = 78° 30'$

Now $K = V_2/V_1 = 110/440 = 1/4$

$\therefore I_2' = KI_2 = 120 \times 1/4 = 30$ A

$I_0 = 5$ A.

Angle between I_0 and I_2'

$= 78°30' - 36° 54' = 41° 36'$

Using parallelogram law of vectors (Fig. 30-19) we get

$$I_1 = \sqrt{(5^2 + 30^2 + 2 \times 5 \times 30 \times \cos 41° 36')}$$

$= \mathbf{34.45\ A}$

The resultant current could also have been found by resolving I_2' and I_0 into their X - and Y-components.

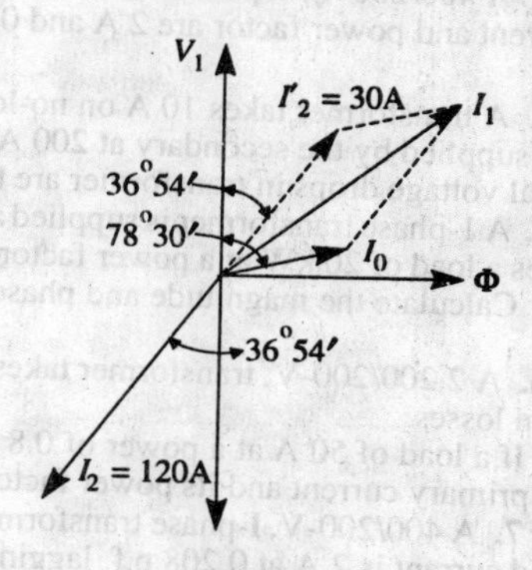

Fig. 30.19

Example 30.8. *A transformer has a primary winding of 800 turns and a secondary winding of 200 turns. When the load current on the secondary is 80 A at 0.8 power factor lagging, the primary current is 25 A at 0.707 power factor lagging. Determine graphically or otherwise the no-load current of the transformer and its phase with respect to the voltage.*

Solution. Here

$K = 200/800 = 1/4 ; I_2' = 80 \times 1/4 = 20$ A

$\phi_2 = \cos^{-1}(0.8) = 36.9° ; \phi_1 = \cos^{-1}(0.707) = 45°$

As seen from Fig. 30-20, I_1 is the vector sum of I_0 and I_2'. Let I_0 lag behind V_1 by an angle ϕ_0.

$I_0 \cos \phi_0 + 20 \cos 36.9° = 25 \cos 45°$

$\therefore I_0 \cos \phi_2 = 25 \times 0.707 - 20 \times 0.8$

$= 1.675$ A

$I_0 \sin \phi_0 + 20 \sin 36.9° = 25 \sin 45°$

$\therefore I_0 \sin \phi_0 = 25 \times 0.707 - 20 \times 0.6$

$= 5.675$ A

$\therefore \tan \phi_0 = 5.675/1.675 = 3.388$

$\therefore \phi_0 = \mathbf{73.3°}$

Now, $I_0 \sin \phi_0 = 5.675$

$\therefore I_0 = 5.675/ \sin 73.3° = \mathbf{5.93\ A}$

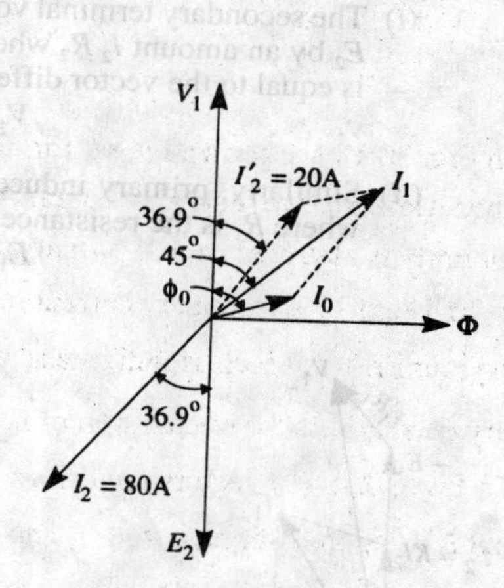

Fig. 30.20

Tutorial Problem No. 30.2

1. The primary of a certain transformer takes 1 A at a power factor of 0.4 when it is connected across a 200-V, 50-Hz supply and the secondary is on open circuit. The number of turns on the primary is twice that on the secondary. A load taking 50 A at a lagging power factor of 0.8 is now connected across the secondary. What is now the value of primary current ? **[25.9 A]**

2. The number of turns on the primary and secondary windings of a single-phase transformer are 350 and 38 respectively. If the primary winding is connected to a 2.2 kV, 50-Hz supply, determine

(*a*) the secondary voltage on no-load.

(b) the primary current when the secondary current is 200 A at 0.8 p.f. lagging, if the no-load current is 5 A at 0.2 p.f. lagging.

(c) the power factor of the primary current. **[239 V ; 25-65 A; 0.715 lag]**

3. A 400/200-V, 1-phase transformer is supplying a load of 25 A at a p.f. of 0.866 lagging. On no-load, the current and power factor are 2 A and 0.208 respectively. Calculate the current taken from the supply. **[13.9 A lagging V_1 by 36.1°]**

4. A transformer takes 10 A on no-load at a power factor of 0.1. The turn ratio is 4 : 1 (step down). If a load is supplied by the secondary at 200 A and p.f. of 0.8, find the primary current and power factor (internal voltage drops in transformer are to be ignored). **[57.2 A; 0.717 lagging]**

5. A 1-phase transformer is supplied at 1,600 V on the h.v. side and has a turn ratio of 8 : 1. The transformer supplies a load of 20 kW at a power factor of 0.8 lag and takes a magnetising current of 2.0 A at a power factor of 0.2. Calculate the magnitude and phase of the current taken from the h.v. supply.
[17.15 A ; 0.753 lag] *(Elect. Engg. Calcutta Univ. 1980)*

6. A 2,200/200-V, transformer takes 1 A at the H.T. side on no-load at a p.f. of 0.385 lagging. Calculate the iron losses.

If a load of 50 A at a power of 0.8 lagging is taken from the secondary of the transformer, calculate the actual primary current and its power factor. **[847 W; 5.44 A; 0.74 lag]**

7. A 400/200-V, I-phase transformer is supplying a load of 50 A at a power factor of 0.866 lagging. The no-load current is 2 A at 0.208 p.f. lagging. Calculate the primary current and primary power factor.
[26.4 A; 0.838 lag] *(Elect. Machines-I, Indore Univ. 1980)*

30.11. Transformer with Winding Resistance but No Magnetic Leakage

An ideal transformer was supposed to possess no resistance, but in an actual transformer, there is always present some resistance of the primary and secondary windings. Due to this resistance, there is some voltage drop in the two windings. The result is that :

(i) The secondary terminal voltage V_2 is vectorially less than the secondary induced e.m.f. E_2 by an amount $I_2 R_2$ where R_2 is the resistance of the secondary winding. Hence, V_2 is equal to the vector difference of E_2 and resistive voltage drop $I_2 R_2$.

$\therefore \quad V_2 = E_2 - I_2 R_2$ .....vector difference

(ii) Similarly, primary induced e.m.f. E_1 is equal to the vector difference of V_1 and $I_1 R_1$ where R_1 is the resistance of the primary winding.

$E_1 = V_1 - I_1 R_1$ ....vector difference

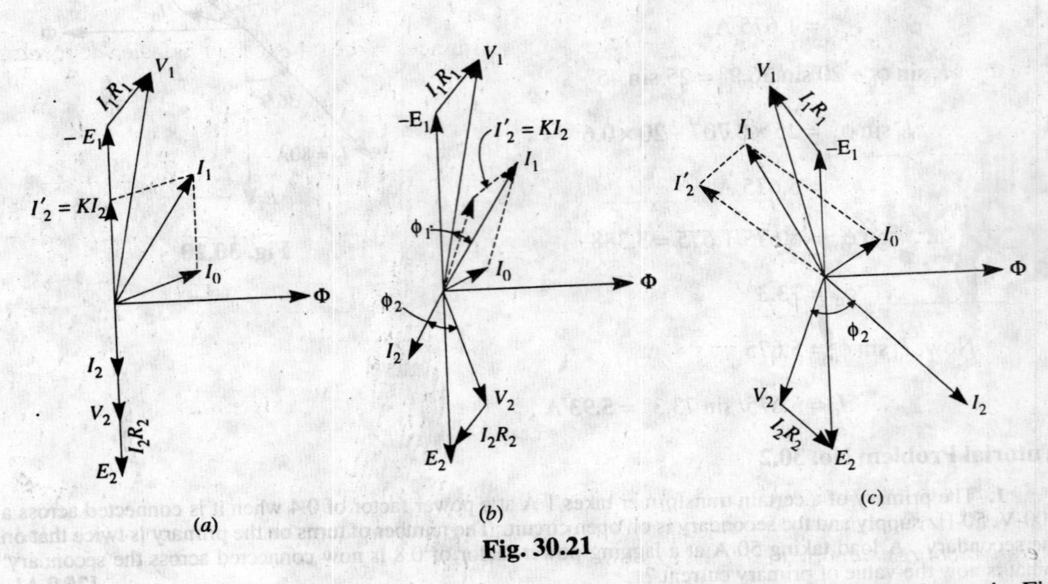

Fig. 30.21

The vector diagrams for non-inductive, inductive and capacitive loads are shown in Fig. 30-21(a), (b) and (c) respectively.

30.12. Equivalent Resistance

In Fig. 30.22 is shown a transformer whose primary and secondary windings have resistances of R_1 and R_2 respectively. The resistances have been shown external to the windings.

It would now be shown that the resistances of the two windings can be transferred to any one of the two windings. The advantage of concentrating both the resistances in one winding is that it makes calculations very simple and easy because one has then to work in one winding only. It will be proved that a resistance of R_2 in secondary is equivalent to R_2/K^2 in primary. The value R_2/K^2 will be denoted by R_2' – *the equivalent secondary resistance as referred to primary.*

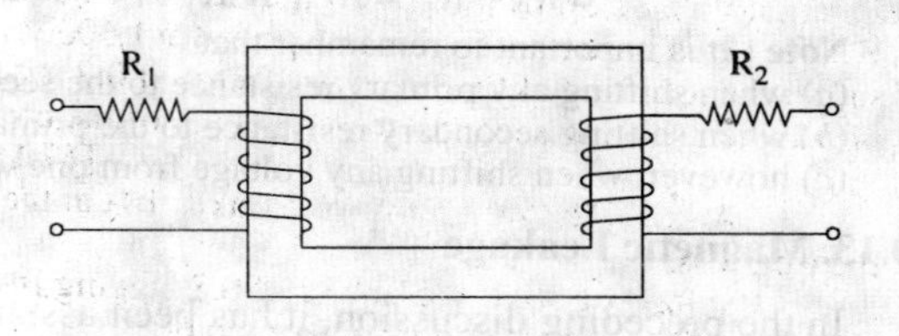

Fig. 30.22

The copper loss in secondary is $I_2^2R_2$. This loss is supplied by primary which takes a current of I_1. Hence, if R_2' is the *equivalent resistance in primary which would have caused the same loss* as R_2 in secondary, then

$$I_1^2R_2' = I_2^2R_2 \quad \text{or} \quad R_2' = (I_2/I_1)^2R_2$$

Now, if *we neglect no-load current* I_0, then $I_2/I_1 = I/K^*$. Hence, $R_2' = R_2/K^2$

Similarly, equivalent primary resistance as referred to secondary is $R_1' = K^2R_1$

In Fig. 30.23, secondary resistance has been transferred to primary side leaving secondary circuit resistanceless. The resistance $R_1 + R_2' = R_1 + R_2/K^2$ is known as the *equivalent* or *effective resistance of the transformer as referred to primary* and may be designated as R_{01}.

$$\therefore \quad R_{01} = R_1 + R_2' = R_1 + R_2K^2$$

Similarly, the *equivalent resistance of the transformer as referred to secondary is*

$$R_{02} = R_2 + R_1' = R_2 + K^2R_1$$

This fact is shown in Fig. 30.24 where all the resistances of the transformer has been concentrated in the secondary winding.

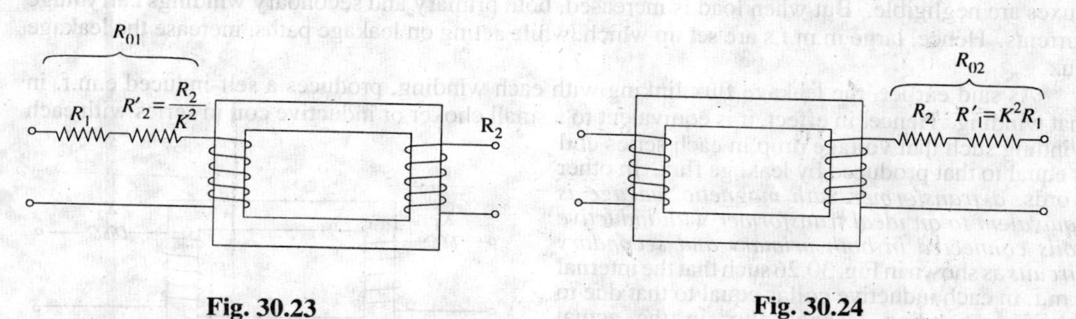

Fig. 30.23

Fig. 30.24

It is to be noted that

1. a resistance of R_1 in primary is equivalent to K^2R_1 in secondary. Hence, it is called the *equivalent resistance as referred to secondary i.e.* R_1.

*Actually I_2 #2 $/I_2' = I/K$ and not I_2 #2 $/I_1$. However, if I_0 is neglected, then $I_2' = I_1$.

2. a resistance of R_2 in secondary is equivalent to R_2/K^2 in primary. Hence, it is called the *equivalent secondary resistance as referred to primary i.e.* R_2'.

3. Total or effective resistance of the transformer as referred to primary is

R_{01} = primary resistance + equivalent secondary resistance as referred to primary

$= R_1 + R_2' = R_1 + R_2/K^2$

4. Similarly, total transformer resistance as referred to secondary is,

R_{02} = secondary resistance + equivalent primary resistance as referred to secondary

$= R_2 + R_1' = R_2 + K^2 R_1$

Note : It is important to remember that

(*a*) when shifting any primary resistance to the secondary, *multiply* it by K^2 *i.e.* (transformation ratio)^2.

(*b*) when shifting secondary resistance to the primary, *divide* it by K^2.

(*c*) however, when shifting any voltage from one winding to another only K is used.

30.13. Magnetic Leakage

In the preceding discussion, it has been assumed that all the flux linked with primary winding also links the secondary winding. But, in practice, it is impossible to realize this condition. It is found, however, that all the flux linked with primary does not link the secondary but part of it *i.e* Φ_{L_1} completes its magnetic circuit by passing through air rather than around the core, as shown in Fig. 30.25. This leakage flux is produced when the m.m.f. due to primary ampere-turns existing between points *a* and *b*, acts along the leakage paths. Hence, this flux is known as *primary leakage flux* and is proportional to the primary ampere-turns alone because the secondary turns do not link the magnetic circuit of Φ_{L_1}. The flux Φ_{L_1} is in time phase with I_1. It induces an e.m.f. e_{L1} in primary but none in secondary.

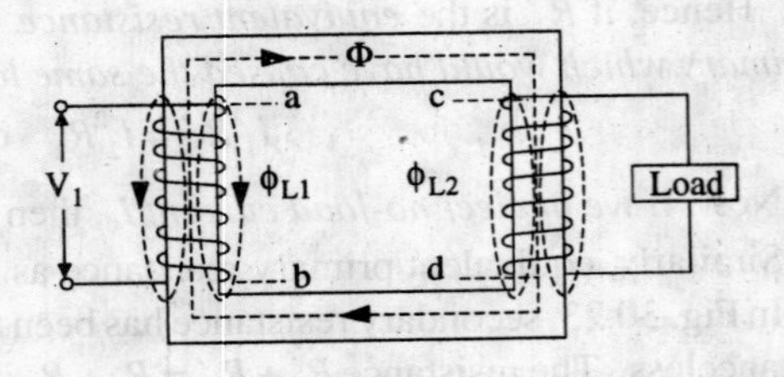

Fig. 30.25

Similarly, secondary ampere-turns (or m.m.f.) acting across points *c* and *d* set up leakage flux Φ_{L_2} which is linked with secondary winding alone (and not with primary turns). This flux Φ_{L_2} is in time phase with I_2 and produces a self-induced e.m.f. e_{L2} in secondary (but none in primary).

At no load and light loads, the primary and secondary ampere-turns are small, hence leakage fluxes are negligible. But when load is increased, both primary and secondary windings carry huge currents. Hence, large m.m.f.s are set up which, while acting on leakage paths, increase the leakage flux.

As said earlier, the leakage flux linking with each winding, produces a self-induced e.m.f. in that winding. Hence, in effect, it is equivalent to a small choker or inductive coil in series with each winding such that voltage drop in each series coil is equal to that produced by leakage flux. In other words, *a transformer with magnetic leakage is equivalent to an ideal transformer with inductive coils connected in both primary and secondary circuits* as shown in Fig. 30.26 such that the internal e.m.f. in each inductive coil is equal to that due to the corresponding leakage flux in the actual transformer.

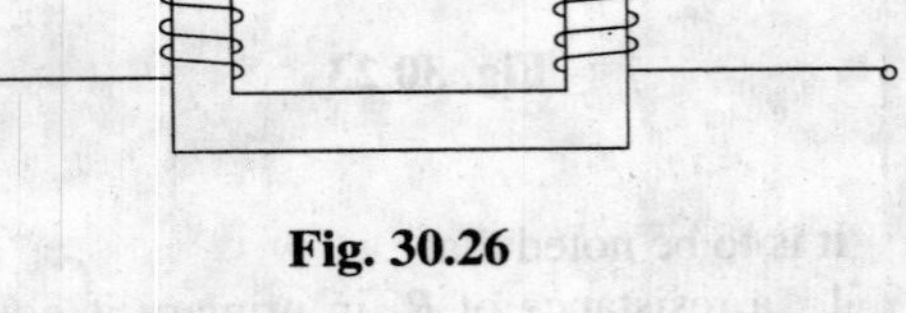

$X_1 = e_{L1}/I_1$ and $X_2 = e_{L2}/I_2$

Fig. 30.26

The terms X_1 and X_2 are known as primary and secondary *leakage* reactances respectively.

Following few points should be kept in mind:

1. The leakage flux links one or the other winding but *not both*, hence it in no way contributes to the transfer of energy from the primary to the secondary winding.

2. The primary voltage V_1 will have to supply reactive drop I_1X_1 in addition to I_1R_1. Similarly, E_2 will have to supply I_2R_2 and I_2X_2.

3. In an actual transformer, the primary and secondary windings are not placed on separate legs or limbs as shown in Fig. 30.26 because due to their being widely separated, large primary and secondary leakage fluxes would result. These leakage fluxes are minimised by sectionalizing and interleaving the primary and secondary windings as in Fig. 30.6 or Fig. 30.8.

30.14. Transformer with Resistance and Leakage Reactance

In Fig. 30.27 are shown the primary and secondary windings of a transformer with reactances taken out of the windings. The primary impedance is given by

$$Z_1 = \sqrt{(R_1^2 + X_1^2)}$$

Similarly, secondary impedance is given by

$$Z_2 = \sqrt{(R_2^2 + X_2^2)}$$

The resistance and leakage reactance of each winding is responsible for some voltage drop in each winding. In primary, the leakage reactance drop is I_1X_1 (usually 1 or 2% of V_1). Hence

$$\mathbf{V}_1 = \mathbf{E}_1 + \mathbf{I}_1 (R_1 + jX_1) = \mathbf{E}_1 + \mathbf{I}_1\mathbf{Z}_1$$

Similarly, there are I_2R_2 and I_2X_2 drops in secondary which combine with V_2 to give E_2.

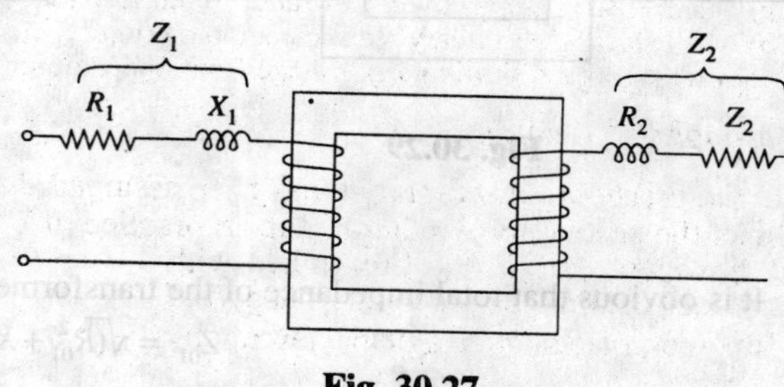

Fig. 30.27

$$\mathbf{E}_2 = \mathbf{V}_2 + \mathbf{I}_2 (R_2 + jX_2) = \mathbf{V}_2 + \mathbf{I}_2\mathbf{Z}_2$$

The vector diagram for such a transformer for different kinds of loads is shown in Fig. 30.28. In these diagrams, vectors for resistive drops are drawn parallel to current vectors whereas reactive drops are perpendicular to the current vectors. The angle ϕ_1 between V_1 and I_1 gives the power factor angle of the transformer.

It may be noted that leakage reactances can also be transferred from one winding to the other in the same way as resistance.

$$\therefore \qquad X_2' = X_2/K^2 \quad \text{and} \quad X_1' = K^2X_1$$

and $$X_{01} = X_1 + X_2' = X_1 + X_2/K^2 \text{ and } X_{02} = X_2 + X_1' = X_2 + K^2X_1$$

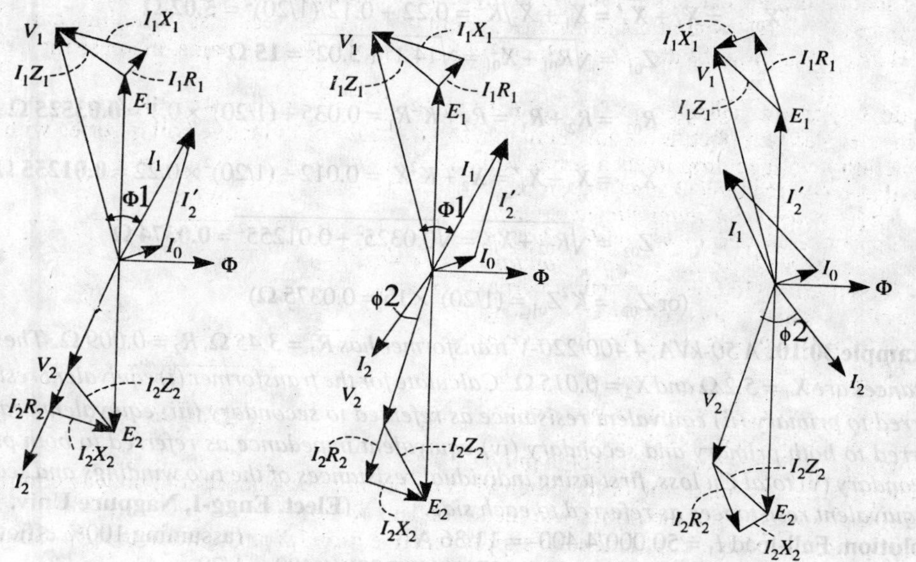

Fig. 30.28

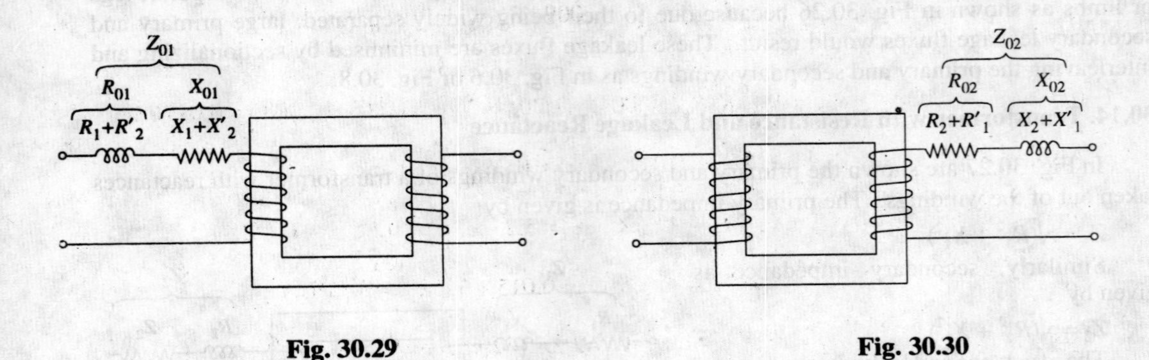

Fig. 30.29 **Fig. 30.30**

It is obvious that total impedance of the transformer as referred to primary is given by

$$Z_{01} = \sqrt{(R_{01}^2 + X_{01}^2)} \qquad \text{...Fig. 30 – 29}$$

and

$$Z_{02} = \sqrt{(R_{02}^2 + X_{02}^2)} \qquad \text{...Fig. 30.30}$$

Example 30.9. *A 30.kVA, 2400/120-V, 50-Hz transformer has a high voltage winding resistance of 0.1 Ω and a leakage reactance of 0.22 Ω. The low voltage winding resistance is 0.035 Ω and the leakage reactance is 0.012 Ω. Find the equivalent winding resistance, reactance and impedance referred to the (i) high voltage side and (ii) the low-voltage side.*

(Electrical Machines-I, Bangalore Univ. 1987)

Solution. $K = 120/2400 = 1/20$; $R_1 = 0.1\ \Omega$, $X_1 = 0.22\ \Omega$

$R_2 = 0.035\ \Omega$ and $X_2 = 0.012\ \Omega$

(*i*) Here, high-voltage side is, obviously, the primary side. Hence, values as referred to primary side are

$$R_{01} = R_1 + R_2' = R_1 + R_2/K^2 = 0.1 + 0.035/(1/20)^2 = \mathbf{14.1\ \Omega}$$

$$X_{01} = X_1 + X_2' = X_1 + X_2/K^2 = 0.22 + 0.12/(1/20)^2 = \mathbf{5.02\ \Omega}$$

$$Z_{01} = \sqrt{R_{01}^2 + X_{01}^2} = \sqrt{14.1^2 + 5.02^2} = \mathbf{15\ \Omega}$$

(*ii*)

$$R_{02} = R_2 + R_1' = R_2 + K^2 R_1 = 0.035 + (1/20)^2 \times 0.1 = \mathbf{0.03525\ \Omega}$$

$$X_{02} = X_2 + X_1' = X_2 + K^2 X_1 = 0.012 + (1/20)^2 \times 0.22 = \mathbf{0.01255\ \Omega}$$

$$Z_{02} = \sqrt{R_{02}^2 + X_{02}^2} = \sqrt{0.0325^2 + 0.01255^2} = \mathbf{0.0374\ \Omega}$$

$$(\text{or } Z_{02} = K^2 Z_{01} = (1/20)^2 \times 15 = 0.0375\ \Omega)$$

Example 30.10. *A 50-kVA, 4,400/220-V transformer has $R_1 = 3.45\ \Omega$, $R_2 = 0.009\ \Omega$. The values of reactances are $X_1 = 5.2\ \Omega$ and $X_2 = 0.015\ \Omega$. Calculate for the transformer (i) equivalent resistance as referred to primary (ii) equivalent resistance as referred to secondary (iii) equivalent reactance as referred to both primary and secondary (iv) equivalent impedance as referred to both primary and secondary (v) total Cu loss, first using individual resistances of the two windings and secondly, using equivalent resistances as referred to each side.* **(Elect. Engg-I, Nagpure Univ. 1993)**

Solution. Full-load I_1 = 50,000/4,400- = 11.36 A (assuming 100% efficiency)

Full-load I_2 = 50,000/2220 = 227 A ; K = 220/4,400 = 1/20 (*i*) $$R_{01} = R_1 + \frac{R_2}{K^2} = 3.45 + \frac{0.009}{(1/20)^2} = 3.45 + 3.6 = \mathbf{7.05\ \Omega}$$

(*ii*) $$R_{02} = R_2 + K^2R_1 = 0.009 + (1/20)^2 \times 3.45 = 0.009 + 0.0086 = \mathbf{0.0176\ \Omega}$$

Also, $$R_{02} = K^2R_{01} = (1/20)^2 \times 7.05 = 0.0176\ \Omega \text{ (check)}$$

(*iii*) $$X_{01} = X_1 + X_2' = X_1 + X_2/K^2 = 5.2 + 0.015/(1/20)^2 = \mathbf{11.2\ \Omega}$$

$$X_{02} = X_2 + X_1' = X_2 + K^2X_1 = 0.015 + 5.2/20^2 = \mathbf{0.028\ \Omega}$$

Also $$X_{02} = K^2X_{01} = 11.2/400 = 0.028\ \Omega\text{(check)}$$

(*iv*) $$\mathbf{Z}_{01} = \sqrt{(R_{01}^2 + X_{01}^2)} = \sqrt{(7.05^2 + 11.2^2)} = \mathbf{13.23\ \Omega}$$

$$\mathbf{Z}_{02} = \sqrt{(R_{02}^2 + X_{02}^2)} = \sqrt{(0.0176^2 + 0.028^2)} = \mathbf{0.0331\ \Omega}$$

Also $$\mathbf{Z}_{02} = K^2Z_{01} = 13.23/400 = 0.0331\ \Omega \text{ (check).}$$

(*v*) $$\textit{Cu}\text{ loss} = I_1^2R_1 + I_2^2R_2 = 11.36^2 \times 3.45 + 227^2 \times 0.009 = \mathbf{910\ W}$$

Also Cu loss $= I_1^2R_{01} = 11.36^2 \times 7.05 = \mathbf{910\ W}$

$= I_2^2R_{02} = 227^2 \times 0.0176 = \mathbf{910\ W}$

Example 30.11. *A transformer with a 10 : 1 ratio and rated at 50-kVA, 2400/240-V, 50-Hz is used to step down the voltage of a distribution system. The low tension voltage is to be kept constant at 240 V.*

(a) What load impedance connected to low-tension size will be loading the transformer fully at 0.8 power factor (lag) ?

(b) What is the value of this impedance referred to high tension side ?

(c) What is the value of the current referred to the high tension side ?

(Elect. Engineering-I, Bombay Univ. 1987)

Solution. (*a*) F.L. I_2 = 50,000/240 = 625/3 A; $Z_2 = \frac{240}{(625/3)} = \mathbf{1.142\ \Omega}$

(*b*) K = 240/2400 = 1/10

The secondary impedance referred to primary side is

$$Z_2' = Z_2/K^2 = 1.142/(1/10)^2 = \mathbf{114.2\ \Omega}$$

(*c*) Secondary current referred to primary side is $I_2' = KI_2 = (1/10) \times 625/3 = \mathbf{20.83\ A}$

Example 30.12. *The full-load copper loss on the h.v. side of a 100-kVA, 11000/317-V, 1-phase transformer is 0.62 kW and on the l.v. side is 0.48 kW.*

(i) *Calculate R_1, R_2, and R_3 in ohms (ii) the total reactance is 4 per cent, find X_1, X_2 and X_3 in ohms if the reactance is divided in the same proportion as resistance.*

(Elect. Machines, A.M.I.E. Sec. B, 1991)

Solution. (*i*) F.L. I_1 = 100 × 103/11000 = 9.1 A, F.L. $I_2 = 100 \times 10^3/317$ = 315.5 A

Now, $I_1^2 R_1$ = 0.62 kW or $R_1 = 620/9.1^2 = 7.5\ \Omega$

$I^2_2R_2$ = 0.48 kW, $R_2 = 480/315.5^2 = 0.00482\ \Omega$

$R'_2 = R_2/K^2 = 0.00482 \times (11{,}000/317)^2 = 5.8\ \Omega$

$$\%\text{ reactance} = \frac{I_1 \times X_{01}}{V_1} \times 100 \text{ or } 4 = \frac{9.1 \times X_{01}}{11000} \times 100,\ X_{01} = 48.4\ \Omega$$

$X_1 + X_2' = 48.4\ \Omega$. Given R_1/R'_2 = or X_1/X'_2 or $(R_1 + R'_2)/R'_2$

$= (X_1 + X'_2)/X'_2$ or $(7.5 + 5.8)/5.8 = 48.4/X'_2 \quad \therefore X'_2 = 21.1\ \Omega$

$\therefore X_1 = 48.4 - 21.1 = 27.3\ \Omega, X_2 = 21.1 \times (317/11000)^2 = 0.175\ \Omega$

Example 30.13. *The following data refer to a 1-phase transformer :*

Turn ratio 19.5 : 1 ; $R_1 = 25\ \Omega$; $X_1 = 100\ \Omega$; $R_2 = 0.06\ \Omega$; $X_2 = 0.25\ \Omega$. No-load current = 1.25 A leading the flux by 30°.

The secondary delivers 200 A at a terminal voltage of 500 V and p.f. of 0.8 lagging. Determine by the aid of a vector diagram, the primary applied voltage, the primary p.f. and the efficiency.

(Elect. Machinery-I, Madras Univ. 1989)

Solution. The vector diagram is similar to Fig. 30.28 which has been redrawn as Fig. 30.31. Let us take V_2 as the reference vector.

$\therefore \quad \mathbf{V_2} = 500 \angle 0° = 500 + j0$

$\mathbf{I_2} = 200(0.8 - j0.6) = 160 - j120$

$\mathbf{Z_2} = (0.06 + j0.25)$

$\mathbf{E_2} = \mathbf{V_2} + \mathbf{I_2 Z_2}$

$= (500 + j0) + (160 - j\,120)(0.06 + j0.25)$

$= 500 + (39.6 + j32.8) = 539.6 + j32.8 = 541 \angle 3.5°$

Obviously, $\beta = 3.5°$

$\mathbf{E_1} = \mathbf{E_2}/K = 19.5\ \mathbf{E_2} = 19.5(539.6 + j32.8)$

$= 10{,}520 + j640$

$\therefore \quad -\mathbf{E_1} = -10{,}520 - j640 = 10{,}540 \angle 183.5°$

$\mathbf{I_2'} = -\mathbf{I_2}K = (-160 + j120)/19.5 = -8.21 + j6.16$

As seen from Fig. 30.31, $\mathbf{I_0}$ leads $\mathbf{V_2}$ by an angle

$= 3.5° + 90° + 30° = 123.5°$

$\therefore \quad \mathbf{I_0} = 1.25 \angle 123.5° = 1.25 (\cos 123.5° + j \sin 123.5°)$

$= 1.25 (-\cos 56.5° + j \sin 56.5°) = -0.69 + j1.04$

$\mathbf{I_1} = \mathbf{I_2'} + \mathbf{I_0} = (-8.21 + j6.16) + (-0.69 + j1.04)$

$= -8.9 + j\,7.2 = 11.45 \angle 141°$

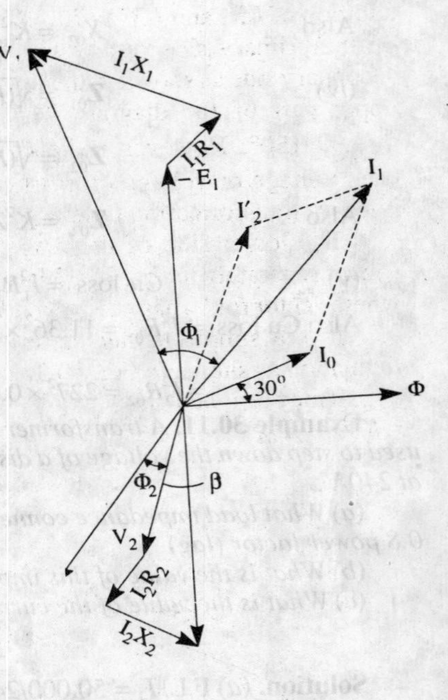

Fig. 30.31

$\mathbf{V_1} = -\mathbf{E_1} + \mathbf{I_1 Z_1} = -10{,}520 - j640 + (-8.9 + j\,7.5)(25 + j100) = -10{,}520 - j640 - 942 - j710$

$= -11{,}462 - j1350$

$= 11{,}540 \angle 186.7°$

Phase angle between $\mathbf{V_1}$ and $\mathbf{I_1}$ is $= 186.7° - 141° = 45.7°$

$\therefore$ primary p.f. = cos 45.7° = **0.698 (lag)**

No-load primary input power = $V_1 I_0 \sin \phi_0$

$= 11{,}540 \times 1.25 \times \cos 60° = 7{,}210$ **W**

$R_{02} = R_2 + K^2 R_1 = 0.06 + 25/19.5^2 = 0.1257\ \Omega$

Total Cu loss as referred to secondary $= I_2^2 R_{02} = 200^2 \times 0.1257 = 5{,}030$ W

Output $= V_2\, I_2 \cos \phi_2 = 500 \times 200 \times 0.8 = 80{,}000$ W

Total losses $= 5030 + 7210 = 12{,}240$ W

Input $= 80{,}000 + 12{,}240 = 92{,}240$ W

$\eta = 80{,}000/92{,}240 = 0.8674$ or **86.74%**

30.15. Simplified Diagram

The vector diagram of Fig. 30.28 may be considerably simplified if the no-load current I_0 is neglected. Since I_0 is 1 to 3 per cent of full-load primary current I_1, it may be neglected without serious error. Fig. 30.32 shows the diagram of Fig. 30.28 with I_0 omitted altogether.

In Fig. 30.32, V_2, V_1, ϕ_2 are known, hence E_2 can be found by adding vectorially $I_2 R_2$ and $I_2 X_2$ to V_2. Similarly, V_1 is given by the vector addition of I_1R_1 and I_1X_1 to E_1. All the voltages on the primary side can be transferred to the secondary side as shown in the figure, where the upper part of the diagram has been rotated through 180°. However, it should be noted that each voltage or voltage drop should be multiplied by transformation ratio K.

The lower side of the diagram has been shown separately in Fig. 30.33 laid horizontally where vector for V_2 has been taken along X-axis.

It is a simple matter to find transformer regulation as shown in Fig. 30.34 or Fig. 30.35.

It may be noted that $\mathbf{V}_2 = K\mathbf{V}_1 - \mathbf{I}_2(R_{02} + jX_{02}) = K\mathbf{V}_1 - \mathbf{I}_2\mathbf{Z}_{02}$*

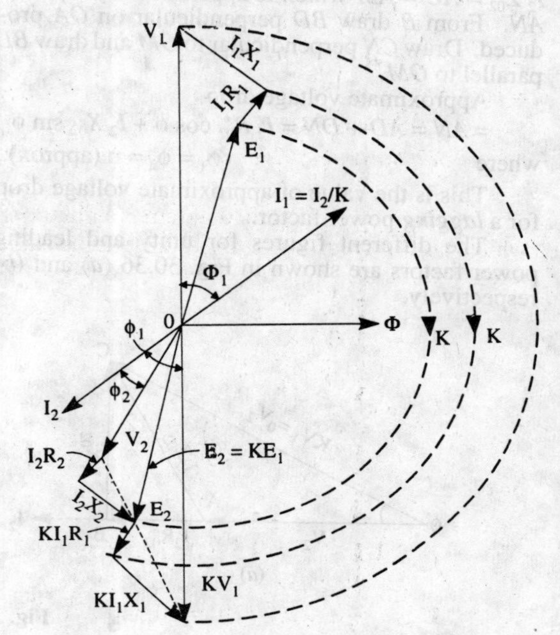

Fig. 30.32

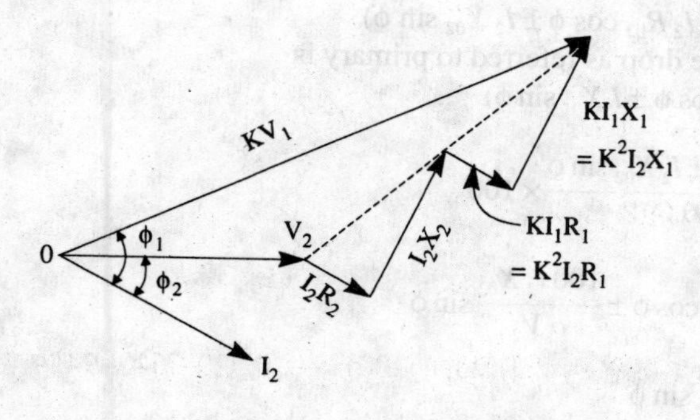

Fig. 30.33

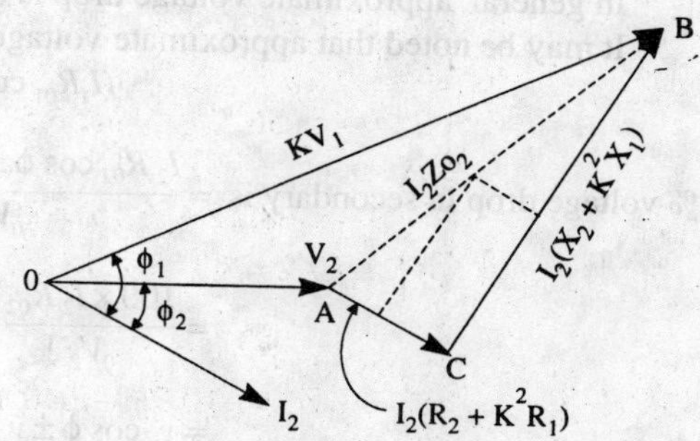

Fig. 30.34

30.16. Total Approximate Voltage Drop in a Transformer

When the transformer is on *no-load*, then V_1 is approximately equal to E_1. Hence $E_2 = KE_1 = KV_1$. Also, $E_2 = {}_0V_2$ where ${}_0V_2$ is secondary terminal voltage on *no-load*, hence no-load secondary terminal voltage is KV_1. The secondary voltage on load is V_2. The difference between the two is $I_2 Z_{02}$ as shown in Fig. 30.35. The approximate voltage drop of the transformer *as referred to secondary* is found thus :

---

*Also, $\mathbf{V}_1 = (\mathbf{V}_2 + \mathbf{I}_2\mathbf{Z}_{02})$ #3 $/K$

With O as the centre and radius OC draw an arc cutting OA produced at M. The total voltage drop $I_2 Z_{02} = AC = AM$ which is approximately equal to AN. From B draw BD perpendicular on OA produced. Draw CN perpendicular to OM and draw BL parallel to OM.

Approximate voltage drop

$= AN = AD + DN = I_2 R_{02} \cos\phi + I_2 X_{02} \sin\phi$

where $\phi_1 = \phi_2 = \phi$ (approx)

This is the value of approximate voltage drop for a *lagging* power factor.

The different figures for unity and leading power factors are shown in Fig. 30.36 (*a*) and (*b*) respectively.

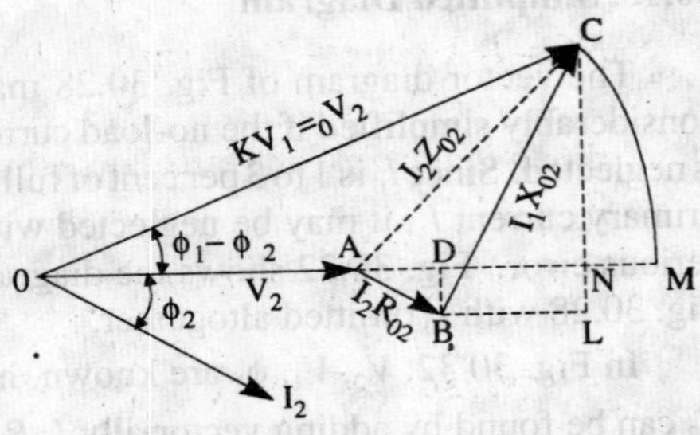

Fig. 30.35

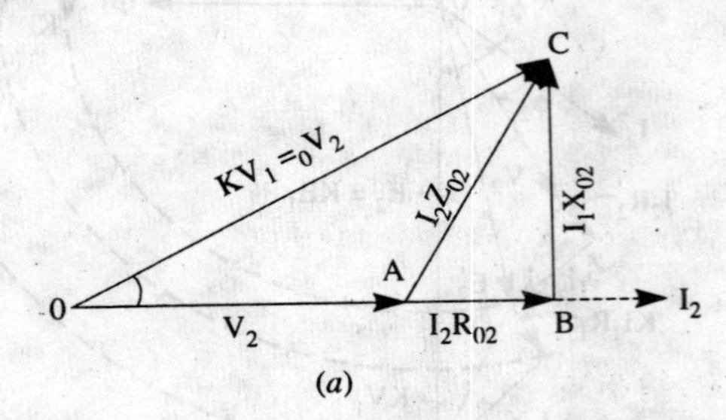

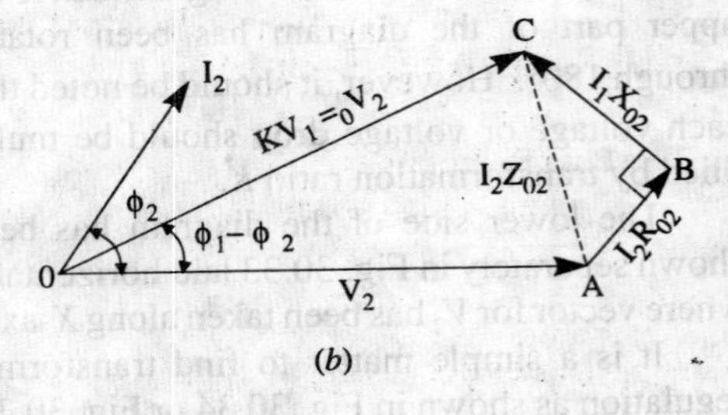

Fig. 30.36

The approximate voltage drop for *leading* power factor becomes

$$(I_2 R_{02} \cos\phi - I_2 \pm X_{02} \sin\phi)$$

In general, approximate voltage drop is $(I_2 R_{02} \cos\phi \pm I_2 X_{02} \sin\phi)$

It may be noted that approximate voltage drop as referred to primary is

$$(I_1 R_{01} \cos\phi \pm I_1 X_{01} \sin\phi)$$

$$\% \text{ voltage drop in secondary is } = \frac{I_2 R_{02} \cos\phi \pm I_2 X_{02} \sin\phi}{{}_0V_2} \times 100$$

$$= \frac{100 \times I_2 R_{02}}{{}_0V_2} \cos\phi \pm \frac{100\, I_2 X_{02}}{o\, V_2} \sin\phi$$

$$= v_r \cos\phi \pm v_x \sin\phi$$

where $v_r = \dfrac{100\, I_2 R_{02}}{{}_0V_2}$ = percentage resistive drop $= \dfrac{100\, I_1 R_{01}}{V_1}$

$v_x = \dfrac{100\, I_2 X_{02}}{{}_0V_2}$ = percentage reactive drop $= \dfrac{100\, I_1 X_{01}}{V_1}$

30.17. Exact Voltage Drop

With reference to Fig. 30.35, it is to be noted that exact voltage drop is AM and not AN. If we add the quantity NM to AN, we will get the exact value of the voltage drop.

Considering the right-angled triangle OCN, we get

$NC^2 = OC^2 - ON^2 = (OC + ON)(OC - ON) = (OC + ON)(OM - ON) = 2\, OC \times NM$

$\therefore NM = NC^2 / 2.OC$ Now, $NC = LC - LN = LC - BD$

$\therefore\ NC = I_2 X_{02} \cos\phi - I_2 R_{02} \sin\phi \quad \therefore \quad NM = \dfrac{(I_2 X_{02} \cos\phi - I_2 R_{02} \sin\phi)^2}{2_0V_2}$

$\therefore$ For a *lagging* power factor, exact voltage drop is

$$= AN + NM = (I_2 R_{02} \cos\phi + I_2 X_{02} \sin\phi) + \frac{(I_2 X_{02} \cos\phi - I_2 R_{02} \sin\phi)^2}{2_0V_2}$$

For a *leading* power factor, the expression becomes

$$= (I_2R_{02} \cos\phi - I_2 X_{02} \sin\phi) + \frac{(I_2 X_{02} \cos\phi + I_2 R_{02} \sin\phi)^2}{2_0V_2}$$

In general, the voltage drop is

$$= (I_2 R_{02} \cos\phi \pm I_2 R_{02} \sin\phi) + \frac{(I_2X_{02} \cos\phi \pm I_2R_{02} \sin\phi)^2}{2_0V_2}$$

Percentage drop is

$$= \frac{(I_2 R_{02} \cos\phi \pm I_2 X_{02} \sin\phi) \times 100}{_0V_2} + \frac{(I_2 X_{02} \cos\phi \mp I_2 R_{02} \sin\phi)^2 \times 100}{2_0V_2^2}$$

$$= (v_r \cos\phi \pm v_x \sin\phi) + (1/200)\,(v_x \cos\phi \mp v_r \sin\phi)^2$$

The upper signs are to be used for a *lagging* power factor and the lower ones for a *leading* power factor.

Example 30.14. *A 230/460-V transformer has a primary resistance of 0.2 Ω and reactance of 0.5 Ω and the corresponding values for the secondary are 0.75 Ω and 1.8 Ω respectively. Find the secondary terminal voltage when supplying 10 A at 0.8 p.f. lagging.*

(Electric. Machines-II, Bangalore Univ. 1991)

Solution. $K = 460/230 = 2\ ;\ R_{02} = R_2 + K^2R_1 = 0.75 + 2^2 \times 0.2 = 1.55\ \Omega$

$X_{02} = X_2 + K^2X_1 = 1.8 + 2^2 \times 0.5 = 3.8\ \Omega$

Voltage drop $= I_2\,(R_{02} \cos\phi + X_{02} \sin\phi) = 10(1.55 \times 0.8 + 3.8 \times 0.6) = 35.2$ V

$\therefore$ secondary terminal voltage = 460–35.2 = **424.8 V**

Example 30.15. *Calculate the regulation of a transformer in which the percentage resistance drop is 1.0% and percentage reactance drop is 5.0% when the power factor is (a) 0.8 lagging (b) unity and (c) 0.8 leading.*

(Electrical Engineering, Banaras Hindu Univ. 1988)

Solution. We will use the approximate expression of Art 30.16

(*a*) **p.f. = cos φ = 0.8 lag** $\mu = v_r \cos\phi + v_x \sin\phi = 1 \times 0.8 + 5 \times 0.6 =$ **3.8%**

(*b*) **p.f. = cos φ = 1** $\mu = 1 \times 1 + 5 \times 0 =$ **1%**

(*c*) **p.f. = cos φ = 0.8 lead** $\mu = 1 \times 0.8 - 5 \times 0.6 =$ **−2.2%**

Example 30.16. *A transformer has a reactance drop of 5% and a resistance drop of 2.5%. Find the lagging power factor at which the voltage regulation is maximum and the value of this regulation.*

(Elect. Engg. Punjab Univ. 1991)

Solution. The percentage voltage regulation (μ) is given by

$$\mu = v_r \cos\phi + v_x \sin\phi$$

where v_r is the percentage resistive drop and v_x is the percentage reactive drop.

Differentiating the above equation, we get $\dfrac{d\mu}{d\phi} = -v_r \sin\phi + v_x \cos\phi$

For regulation to be maximum, $d\mu/d\phi = 0 \quad \therefore \quad -v_r \sin\phi + v_x \cos\phi = 0$

or $\tan\phi = v_x / v_r = 5/2.5 = 2 \quad \therefore\ \phi = \tan^{-1}(2) = 63.5°$ Now, $\cos\phi = 0.45$ and $\sin\phi = 0.892$

Maximum percentage regulation is = (2.5 × 0.45) + (5 × 0.892) = 5.585

Maximum percentage regulation is **5.585** and occurs at a power factor of **0.45** (lag).

Example 30.17. *Calculate the percentage voltage drop for a transformer with a percentage resistance of 2.5% and a percentage reactance of 5% of rating 500 kVA when it is delivering 400 kVA at 0.8 p.f. lagging* **(Elect. Machinery-I, Indore Univ. 1987)**

Solution.
$$\% \text{ drop} = \frac{(\%R)\, I \cos\phi}{I_f} + \frac{(\%X)\, I \sin\phi}{I_f}$$

where I_f is the full-load current and I the actual current.

$$\therefore \qquad \% \text{ drop} = \frac{(\%R)kW}{k\text{VA rating}} + \frac{(\%X)\,\text{kVAR}}{\text{kVA rating}}$$

In the present case, kW $= 400 \times 0.8 = 320$ and kVAR $= 400 \times 0.6 = 240$

$$\therefore \qquad \% \text{ drop} = \frac{2.5 \times 320}{500} + \frac{5 \times 240}{500} = \mathbf{4\%}$$

30.18. Equivalent Circuit

The transformer shown diagrammatically in Fig. 30.37 (*a*) can be resolved into an equivalent circuit in which the resistance and leakage reactance of the transformer are imagined to be external to the winding whose only function then is to transform the voltage [Fig. 30.37 (*b*)]. The no-load

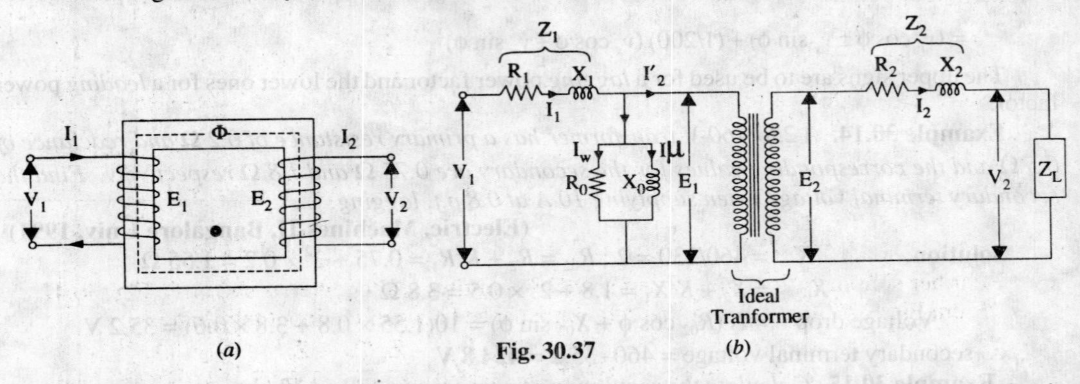

(*a*) **Fig. 30.37** (*b*)

current I_0 is simulated by pure inductance X_0 taking the magnetising component I_μ and a non-inductive resistance R_0 taking the working component I_w connected in parallel across the primary circuit. The value of E_1 is obtained by subtracting vectorially $I_1 Z_1$ from V_1. The value of $X_0 = E_1/I_0$ and of $R_0 = E_1/I_w$. It is clear that E_1 and E_2 are related to each other by expression

$$E_2/E_1 = N_2/N_1 = K$$

To make transformer calculations simpler, it is preferable to transfer voltage, current and impedance either to the primary or to the secondary. In that case, we would have to work in one winding only which is more convenient.

The primary equivalent of the secondary induced voltage is $E'_2 = E_2/K = E_1$

Similarly, primary equivalent of secondary terminal or output voltage is $V_2' = V_2/K$

Primary equivalent of the secondary current is $I'_2 = KI_2$

For transferring secondary impedance to primary, K^2 is used.

$$R'_2 = R_2/K^2, \quad X_2' = X_2/K^2, \quad Z_2' = Z_2/K^2$$

The same relationship is used for shifting an external load impedance to the primary.

The secondary circuit is shown in Fig. 30.38 (*a*) and its equivalent primary values are shown in Fig. 30.38 (*b*).

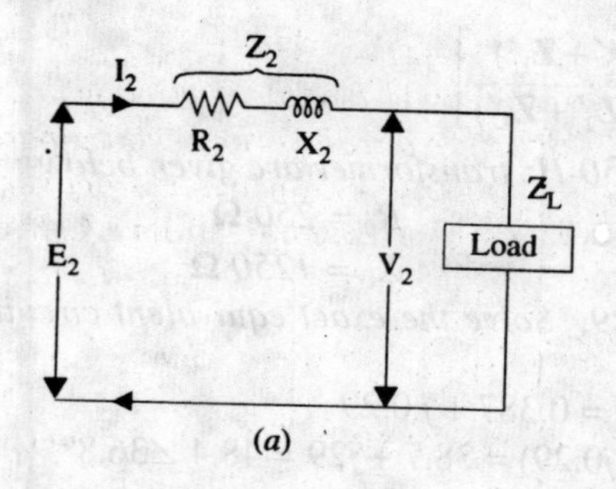

(a)

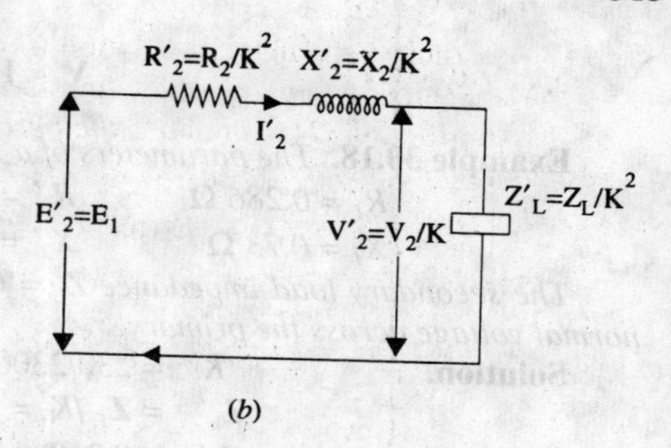

(b)

Fig. 30.38

The total equivalent circuit of the transformer is obtained by adding in the primary impedance as shown in Fig. 30.39. This is known as the exact equivalent circuit but it presents a somewhat harder circuit problem to solve. A simplification can be made by transferring the exciting circuit across the terminals as in Fig. 30.40 or in Fig. 30.41. It should be noted that in this case $X_0 = V_1/I_\mu$. The values of R_0 and X_0 are found from the open circuit test as described in Art. 30.20.

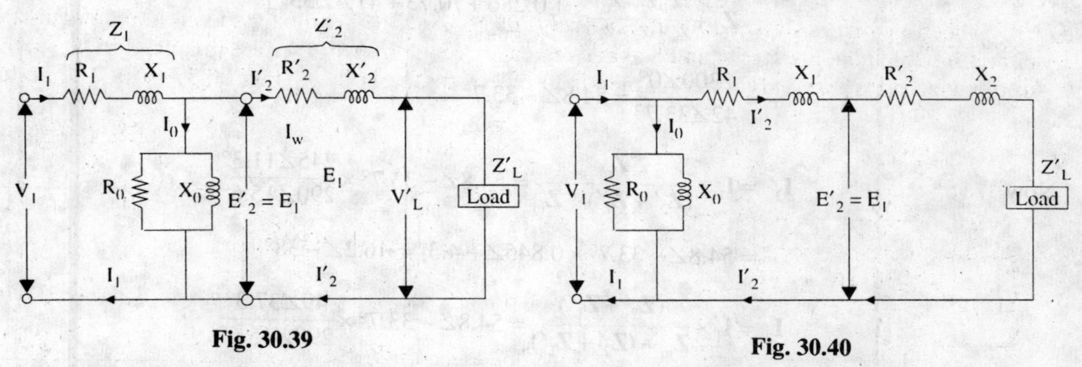

Fig. 30.39 **Fig. 30.40**

Further simplification may be achieved by omitting I_0 altogether as shown in Fig. 30.42. From Fig. 27-39, it is found that total impedance between the input terminal is

$$\mathbf{Z} = \mathbf{Z}_1 + \mathbf{Z}_m \,||\, (\mathbf{Z}_2' + \mathbf{Z}_L') = \left(\mathbf{Z}_1 + \frac{\mathbf{Z}_m(\mathbf{Z}_2' + \mathbf{Z}_L')}{\mathbf{Z}_m + (\mathbf{Z}_2' + \mathbf{Z}_L')}\right)$$

where $\mathbf{Z}_2' = R_2' + jX_2'$ and $\mathbf{Z}_m$ = impedance of the exciting circuit.

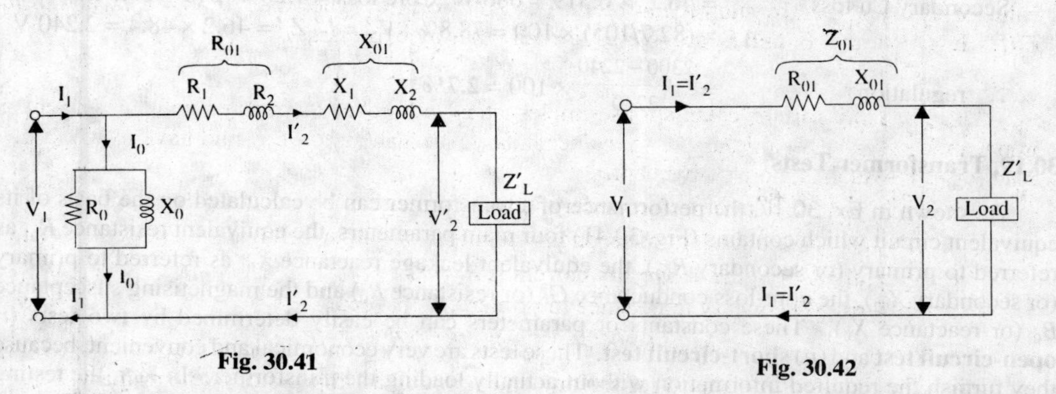

Fig. 30.41 **Fig. 30.42**

This is so because there are two parallel circuits, one having an impedance of $\mathbf{Z}_m$ and the other having $\mathbf{Z}_2'$ and $\mathbf{Z}_L'$ in series with each other.

$$\therefore \quad \mathbf{V}_1 = \mathbf{I}_1\left[\mathbf{Z}_1 + \frac{\mathbf{Z}_m(\mathbf{Z}_2' + \mathbf{Z}_L')}{\mathbf{Z}_m + (\mathbf{Z}_2' + \mathbf{Z}_L')}\right]$$

Example 30.18. *The parameters of a 2300/230.V, 50-Hz transformer are given below :*

$R_1 = 0.286\ \Omega$ $\quad R_2' = 0.319\Omega$ $\quad R_0 = 250\ \Omega$

$X_1 = 0.73\ \Omega$ $\quad X_2' = 0.73\ \Omega$ $\quad X_0 = 1250\ \Omega$

The secondary load impedance Z_L = 0.387 + j 0.29. Solve the exact equivalent circuit with normal voltage across the primary.

Solution.

$K = 230/2300 = 1/10\ ;\ \mathbf{Z}_L = 0.387 + j\,0.29$

$\mathbf{Z}_L' = \mathbf{Z}_L/K^2 = 100(0.387 + j0.29) = 38.7 + j29 = 48.4\ \angle 36.8°$

$\mathbf{Z}_2' = 0.319 + j0.73$

$\therefore\ \mathbf{Z}_2' + \mathbf{Z}_L' = (38.7 + 0.319) + j(29 + 0.73) = 39.02 + j29.73 = 49.0\ \angle 37.3°$

$\mathbf{Y}_m = (0.004 - j0.0008)\ ;\ \mathbf{Z}_m = 1/\mathbf{Y}_m = 240 + j48 = 245\ \angle 11.3°$

$\mathbf{Z}_m + (\mathbf{Z}_2' + \mathbf{Z}_L') = (240 + j48) + (39 + j29.7) = 290\ \angle 15.6°$

$$\therefore \quad \mathbf{I}_1 = \frac{\mathbf{V}_1}{\mathbf{Z}_1 + \dfrac{\mathbf{Z}_m(\mathbf{Z}_2' + \mathbf{Z}_L')}{\mathbf{Z}_m + (\mathbf{Z}_2' + \mathbf{Z}_L')}} = \left[\frac{2300\angle 0°}{0.286 + j0.73 + 41.4\angle 33°}\right]$$

$$= \frac{2300\angle 0°}{42\angle 33.7°} = 54.8\angle -33.7°$$

Now

$$\mathbf{I}_2' = \mathbf{I}_1 \times \frac{\mathbf{Z}_m}{(\mathbf{Z}_2' + \mathbf{Z}_L') + \mathbf{Z}_m} = 54.8\angle -33.7° \times \frac{245\angle 11.3°}{290\angle 15.6°}$$

$$= 54.8\angle -33.7° \times 0.845\angle -4.3° = 46.2\angle -38°$$

$$\mathbf{I}_0 = \mathbf{I}_1 \times \frac{(\mathbf{Z}_2' + \mathbf{Z}_L')}{\mathbf{Z}_m + (\mathbf{Z}_2' + \mathbf{Z}_L')} = 54.8\angle -33.7° \times \frac{49\angle 37.3°}{290\angle 15.6°}$$

$$= 54.8\angle -33.7° \times 0.169\angle 21.7° = 9.26\angle -12°$$

Input power factor $= \cos 33.7° = 0.832$ lagging

Power input $= V_1 I_1 \cos\phi_1 = 2300 \times 54.8 \times 0.832 = 105$ kW

Power output $= 46.2^2 \times 38.7 = 82.7$ kW

Primary Cu loss $= 54.8^2 \times 0.286 = 860$ W

Secondary Cu loss $= 46.2^2 \times 0.319 = 680$ W; Core loss $= 9.26^2 \times 240 = 20.6$ kW

$\eta = (82.7/105) \times 100 = 78.8\%\ ;\ V_2' = I_2' Z_L' = 46.2 \times 48.4 = 2{,}240$ V

$$\therefore \text{ regulation} = \frac{2300 - 2240}{2240} \times 100 = \mathbf{2.7\%}$$

30.19. Transformer Tests

As shown in Ex. 30.17, the performance of a transformer can be calculated on the basis of its equivalent circuit which contains (Fig. 30.41) four main parameters, the equivalent resistance R_{01} as referred to primary (or secondary R_{02}), the equivalent leakage reactance X_{01} as referred to primary (or secondary X_{02}), the core-loss conductance G_0 (or resistance R_0) and the magnetising susceptance B_0 (or reactance X_0). These constants or parameters can be easily determined by two tests (*i*) **open-circuit test** and (*ii*) **short-circuit test**. These tests are very economical and convenient, because they furnish the required information without actually loading the transformer. In fact, the testing of very large a.c. machinery consists of running two tests similar to the open and short-circuit tests of a transformer.

30.20. Open-circuit or No-load Test

The purpose of this test is to determine no-load loss or core loss and no-load I_0 which is helpful in finding X_0 and R_0.

One winding of the transformer—whichever is convenient but usually high voltage winding—is left open and the other is connected to its supply of normal voltage and frequency. A wattmeter W, voltmeter V and an ammeter A are connected in the low-voltage winding *i.e.* primary winding in the present case. With normal voltage applied to the primary, normal flux will be set up in the core, hence normal iron losses will occur which are recorded by the wattmeter. As the primary no-load current I_0 (as measured by ammeter) is small (usually 2 to 10% of rated load current), Cu loss is negligibly small in primary and nil in secondary (it being open). Hence, the wattmeter reading represents practically the core loss under no-load condition (and which is the same for all loads as pointed out in Art. 30.9).

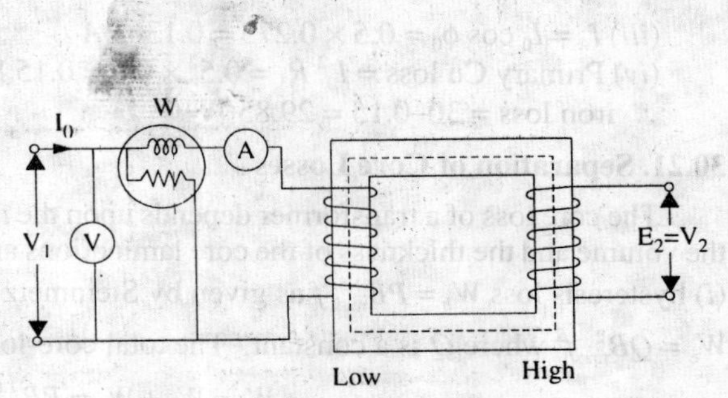

Fig. 30.43

It should be noted that since I_0 is itself very small, the pressure coils of the wattmeter and the voltmeter are connected such that the current in them does not pass through the current coil of the wattmeter.

Sometimes, a high-resistance voltmeter is connected across the secondary. The reading of the voltmeter gives the induced e.m.f. in the secondary winding. This helps to find transformation ratio K.

The no-load vector diagram is shown in Fig. 30.16. If W is the wattmeter reading (in Fig. 30.43), then

$$W = V_1 I_0 \cos\phi_0 \qquad \therefore \qquad \cos\phi_0 = W/V_1 I_0$$

$$\therefore \qquad I_\mu = I_0 \sin\phi_0, \quad I_w = I_0 \cos\phi_0 \quad \therefore X_0 = V_1/I_\mu \text{ and } R_0 = V_1/I_w$$

Or since the current is practically all-exciting current when a transformer is on no-load (*i.e.* $I_0 \cong I_\mu$) and as the voltage drop in primary leakage impedance is small*, hence the exciting admittance Y_0 of the transformer is given by $I_0 = V_1 Y_0$ or $Y_0 = I_0/V_1$

The exciting conductance G_0 is given by $W = V_1^2 G_0$ or $G_0 = W/V_1^2$

The exciting susceptance $B_0 = \sqrt{(Y_0^2 - G_0^2)}$

Example 30.19. *In no-load test of a single-phase transformer, the following test data were obtained :*

Primary voltage : 220 V ; Secondary voltage : 110 V ;
Primary current : 0.5 A ; Power input : 30 W.
Find the following :
(i) the turns ratio (ii) the magnetising component of no-load current (iii) its working (or loss) component (iv) the iron loss.
Resistance of the primary winding = 0.6 ohm.
Draw the no-load phasor diagram to scale. **(Elect. Machine A.M.I.E. 1990)**

---

*If it is not negligibly small, then $I_0 = E_1 Y_0$ *i.e.* instead of V_1 we will have to use E_1.

Solution. (*i*) Turn ratio, $N_1/N_2 = 220/110 = 2$

(*ii*) $W = V_1 I_0 \cos\phi_0$; $\cos\phi_0 = 30/220 \times 0.5 = 0.273$; $\sin\phi_0 = 0.962$

$I_\mu = I_0 \sin\phi_0 = 0.5 \times 0.962 = 0.48\ A$

(*iii*) $I_w = I_0 \cos\phi_0 = 0.5 \times 0.273 = 0.1365\ A$

(*iv*) Primary Cu loss $= I_0^2 R_1 = 0.5^2 \times 0.6 = 0.15$ W

$\therefore$ iron loss = 30–0.15 = 29.85 W

30.21. Separation of Core Losses

The core loss of a transformer depends upon the frequency and the maximum flux density when the volume and the thickness of the core laminations are given. The core loss is made up of two parts (*i*) hysteresis loss $W_h = PB_{max}^{1.6} f$ as given by Steinmetz's empirical relation and (*ii*) eddy current loss $W_e = QB_{max}^2 f^2$ where Q is a constant. The total core-loss is given by

$$W_i = W_h + W_e = PB_{max}^{1.6} f^2 + QB_{max}^2 f^2$$

If we carry out two experiments using two different frequencies but the same maximum flux density, we should be able to find the constants P and Q and hence calculate hysteresis and eddy current losses separately.

Example 30.20. *In a transformer, the core loss is found to be 52 W at 40 Hz and 90 W at 60 Hz measured at same peak flux density. Compute the hysteresis and eddy current losses at 50 Hz.* **(Elect. Machines, Nagpur Univ. 1993)**

Solution. Since the flux density is the same in both cases, we can use the relation

Total core loss $W_i = Af + Bf^2$ or $W_i/f = A + Bf$

$\therefore$ 52/40 = A + 40B and 90/60 = A + 60B ; $\therefore$ $A = 0.9$ and $B = 0.01$

At 50 Hz, the two losses are

$W_h = Af = 0.9 \times 50 = 45\ W$; $W_e = Bf^2 = 0.01 \times 50^2 = 25$ W

Example 30.21. *In a power loss test on a 10 kg specimen of sheet steel laminations, the maximum flux density and waveform factor are maintained constant and the following results were obtained :*

Frequency (Hz)	*25*	*40*	*50*	*60*	*80*
Total loss (watt)	*18.5*	*36*	*50*	*66*	*104*

Calculate the eddy current loss per kg at a frequency of 50 Hz. **(Elect. Measur. A.M.I.E. Sec B, 1991)**

Solution. When flux density and form factor remain constant, the expression for iron loss can be written as

$W_i = Af + Bf^2$ or $W_i/f = A + Bf$

The values of W_i/f for different frequencies are as under :

f	25	40	50	60	80
W_i/f	0.74	0.9	1.0	1.1	1.3

The graph between f and W_i/f has been plotted in Fig. 30.44.

As seen from it, A = 0.5 and B = 0.01

$\therefore$ eddy current loss at 50 Hz = $Bf^2 = 0.01 \times 50^2 = 25$ W

$\therefore$ eddy current loss/kg = 25 / 10 = 2.5 W

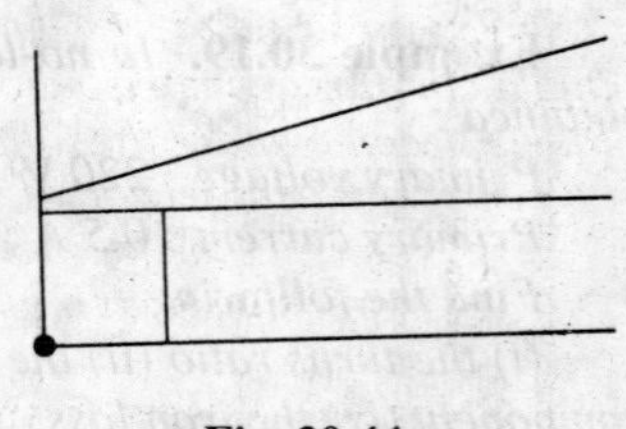

Fig. 30.44

Example 30.22. *In a test for the determination of the losses of a 440-V, 50-Hz transformer, the total iron losses were found to be 2500 W at normal voltage and frequency. When the applied voltage and frequency were 220 V and 25 Hz, the iron losses were found to be 850 W. Calculate the eddy-current loss at normal voltage and frequency.* **(Elect. Inst. and Meas. Punjab Univ. 1991)**

Solution. The flux density in both cases is the same because in second case voltage as well as frequency are halved. Flux density remaining the same, the eddy current loss is proportional to f^2 and hysteresis loss $\propto f$.

Hysteresis loss $\propto f = Af$ and eddy current loss $\propto f^2 = Bf^2$

where A and B are constants.

Total iron loss $W_i = Af + Bf^2 \quad \therefore \quad \frac{W_i}{f} = A + Bf$ ...(*i*)

Now, when $f = 50$ Hz ; $W_i = 2500$ W

and when $f = 25$ Hz ; $W_i = 850$ W

Using these values in (*i*) above, we get

$$2{,}500/50 = A + 50\,B \text{ and } 850/25 = A + 25\,B \quad \therefore \quad B = 16/25 = 0.64$$

Hence, at normal p.d. and frequency

eddy current loss = $Bf^2 = 0.64 \times 50^2 =$ **1600 W**

Hysteresis loss = 2500–1600 = 900 W

Example 30.23. *When a transformer is connected to a 1000-V, 50-Hz supply the core loss is 1000 W, of which 650 is hysteresis and 350 is eddy current loss. If the applied voltage is raised to 2,000 V and the frequency to 100 Hz, find the new core losses.*

Solution. Hysteresis loss $W_h \propto B_{max}^{1.6} f = PB_{max}^{1.6} f$

Eddy current loss $W_e \propto B^2_{max} f^2 = QB^2_{max} f^2$

From the relation $E = 4.44\, fNB_{max} A$ volt, we get $B_{max} \propto E/f$

Putting this value of B_{max} in the above equations, we have

$$W_h = P\left(\frac{E}{f}\right)^{1.6} f = PE^{1.6} f^{-0.6} \text{ and } W_e = Q\left(\frac{E}{f}\right)^2 f^2 = QE^2$$

In the first case, $E = 1000$ V, $f = 50$ Hz, $W_h = 650$ W, $W_e = 350$ W

$\therefore \quad 650 = P \times 1000^{1.6} \times 50^{-0.6} \quad \therefore P = 650 \times 1000^{-1.6} \times 50^{0.6}$

Similarly, $350 = Q \times 1000^2 \quad \therefore Q = 350 \times 1000^{-2}$

Hence, constants P and Q are known.

Using them in the second case, we get

$$W_h = (650 \times 1000^{-1.6} \times 50^{0.6}) \times 2000^{1.6} \times 100^{-0.6} = 650 \times 2 = 1{,}300\ W$$

$$W_e = (350 \times 1000^{-2}) \times 2{,}000^2 = 350 \times 4 = 1{,}400\text{W}$$

$\therefore$ core loss under new condition is = 1,300 + 1,400 = **2700 W**

Alternative Solution.

Here, both voltage and frequency are doubled, leaving the flux density unchanged.

With 1000 V at 50 Hz

$W_h = Af$ or $650 = 50\,A$; $A = 13$

$W_e = Bf^2$ or $350 = B \times 50^2$; $B = 7/50$

With 2000 V at 100 Hz

$W_h = Af = 13 \times 100 = 1300$ W and $W_e = Bf^2 = (7/50) \times 100^2 = 1400$ W

$\therefore$ new core loss = 1300 + 1400 = **2700 W**

Example 30.24. *A transformer with normal voltage impressed has a flux density of 1.4 Wb/m² and a core loss comprising of 1000 W eddy current loss and 3000 W hysteresis loss. What do these losses become under the following conditions :*

(*a*) *increasing the applied voltage by 10% at rated frequency.*

(*b*) *reducing the frequency by 10% with normal voltage impressed.*

(*c*) *increasing both impressed voltage and frequency by 10 per cent.*

(Electrical Machinery-I, Madras Univ. 1985)

Solution. As seen from Ex. 30.19.

$W_h = PE^{1.6} f^{-0.6}$ and $W_e = QE^2$

From the given data, we have $3000 = PE^{1.6} f^{-0.6}$ ...(*i*)

and $1000 = QE^2$ ...(*ii*)

where E and f are the normal values of primary voltage and frequency.

(*a*) here voltage becomes $= E + 10\%E = 1.1E$

The new hysteresis loss is $W_h = P(1.1\,E)^{1.6} f^{-0.6}$ ...(iii)

Dividing Eq. (iii) by (i), we get $\dfrac{W_h}{3000} = 1.1^{1.6}$; $W_h = 3000 \times 1.165 = \mathbf{3495\ W}$

The new eddy-current loss is

$$W_e = Q\,(1.1E)^2 \quad \therefore \quad \frac{W_e}{1000} = 1.1^2 \;\therefore\; W_e = 1000 \times 1.21 = \mathbf{1210\ W}$$

(b) As seen from Eq. (i) above, eddy-current loss would not be effected. The new hysteresis loss is $W_h = PE^{1.6}\,(0.9\,f)^{-0.6}$ ...(iv)

From (i) and (iv), we get $\dfrac{W_h}{3000} = 0.9^{-0.6}$; $W_h = 3000 \times 1.065 = \mathbf{3{,}196\ W}$

(c) In this case, both E and f are increased by 10%. The new losses are as under:-

$$W_h = P(1.1E)^{1.6}\,(1.1f)^{-0.6}$$

$$\therefore \quad \frac{W_h}{3000} = 1.1^{1.6} \times 1.1^{-.06} = 1.165 \times 0.944 \;\therefore\; W_h = 3000 \times 1.165 \times 0.944 = \mathbf{3{,}299}\ \text{W}$$

As W_e is unaffected by changes in f, its value is the same as found in (a) above i.e. **1210 W**

30.22. Short-circuit or Impedance Test

This is an economical method for determining the following :

(i) Equivalent impedance ($\mathbf{Z}_{01}$ or $\mathbf{Z}_{02}$), leakage reactance (X_{01} or X_{02}) and total resistance (R_{01} or R_{02}) of the transformer as referred to the winding in which the measuring instruments are placed.

(ii) Cu loss at full load (and at any desired load). This loss is used in calculating the efficiency of the transformer.

(iii) Knowing Z_{01} or Z_{02}, the total voltage drop in the transformer as referred to primary or secondary can be calculated and hence regulation of the transformer determined.

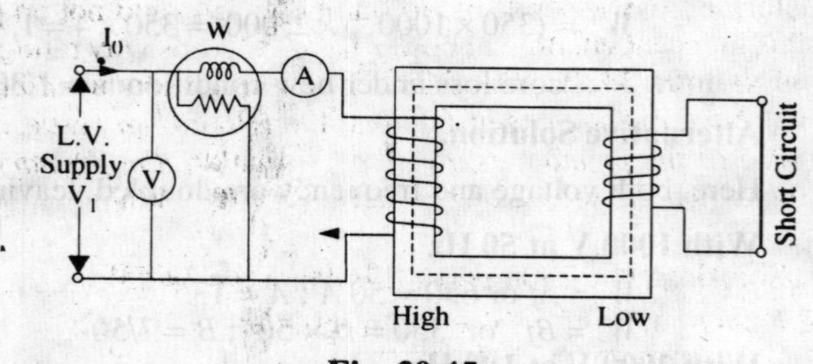

Fig. 30.45

In this test, one winding, usually the low-voltage winding, is solidly short-circuited by a thick conductor (or through an ammeter which may serve the additional purpose of indicating rated load current) as shown in Fig. 30.45.

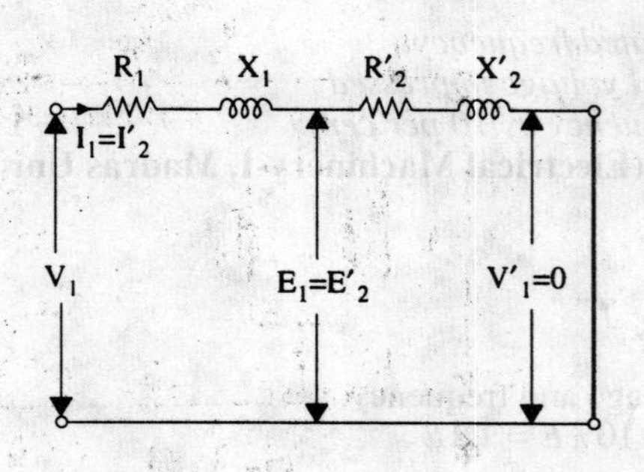

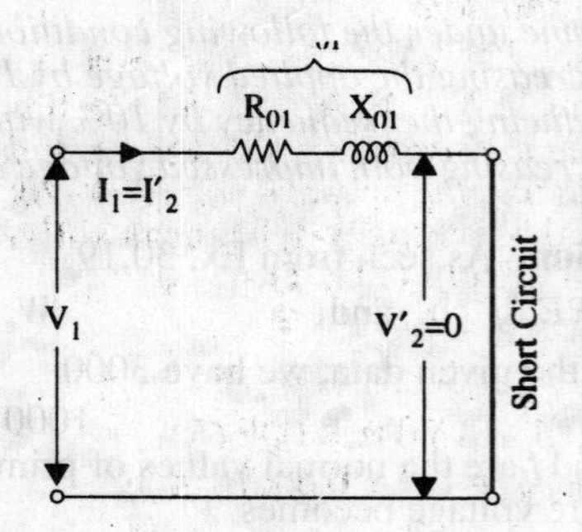

Fig. 30.46

A low voltage (usually 5 to 10% of normal primary voltage) at correct frequency (though for Cu losses it is not essential) is applied to the primary and is cautiously increased till full-load currents are flowing both in primary and secondary (as indicated by the respective ammeters).

Since, in this test, the applied voltage is a small percentage of the normal voltage, the mutual flux Φ produced is also a small percentage of its normal value (Art. 30.6). Hence, core losses are very small with the result that the wattmeter reading represents the full-load Cu loss or I^2R loss for the whole transformer *i.e.* both primary Cu loss and secondary Cu loss. The equivalent circuit of the transformer under short-circuit condition is shown in Fig. 30.46. If V_{SC} is the voltage required to circulate rated load currents, then $Z_{01} = V_{SC}/I_1$

Also $W = I_1^2 R_{01}$

$\therefore \quad R_{01} = W/I_1^2$

$\therefore \quad X_{01} = \sqrt{(Z_{01}^2 - R_{01}^2)}$

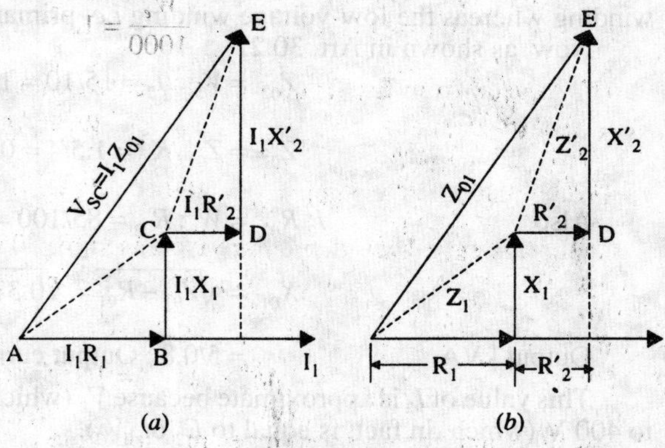

Fig. 30.47

In Fig. 30.47(*a*) is shown the equivalent circuit vector diagram for the short-circuit test. This diagram is the same as shown in Fig. 30.34 except that all the quantities are referred to the primary side. It is obvious that the entire voltage V_{SC} is consumed in the impedance drop of the two windings.

If R_1 can be measured, then knowing R_{01}, we can find $R_2' = R_{01} - R_1$. The impedance triangle can then be divided into the appropriate equivalent triangles for primary and secondary as shown in Fig. 30.47(*b*).

30.23. Why Transformer Rating in kVA ?

As seen, Cu loss of a transformer depends on current and iron loss on voltage. Hence, total transformer loss depends on volt-ampere (VA) and not on phase angle between voltage and current *i.e.* it is independent of load power factor. That is why rating of transformers is in kVA and not in kW.

Example 30.25. *The primary and secondary windings of a 30.kVA 76000/230.V, 1-ph transformer have resistances of 10 ohm and 0.016 ohm respectively. The reactance of the transformer referred to the primary is 34 ohm. Calculate the primary voltage required to circulate full-load current when the secondary is short-circuited. What is the power factor on short circuit.*

(Elect. Machines AMIE Sec. B 1991)

Solution. $K = 230/6000 = 23/600$, $X_{01} = 34\ \Omega$,

$R_{01} = R_1 + R_2/K^2 = 10 + 0.016\,(600/23)^2 = 20.9\ \Omega$

$Z_{01} = \sqrt{R_{01}^2 + X_{01}^2} = \sqrt{20.9^2 + 34^2} = 40\Omega$

F.L., $I_1 = 30{,}000/6000 = 5\ A$; $V_{SC} = I_1 Z_{01} = 5 \times 40 = 200$ V

Short circuit p.f. $= R_{01}/Z_{01} = 20.9/40 = 0.52$

Example 30.26. *Obtain the equivalent circuit of a 200/400-V, 50-Hz, 1-phase transformer from the following test data :-*

O.C. test : 200 V, 0.7 A, 70 W –on l.v. side

S.C. test : 15 V, 10 A, 85 W –on h.v. side

Calculate the secondary voltage when delivering 5 kW at 0.8 p.f. lagging, the primary voltage being 200 V.

(Electrical Machinery-I, Madras Univ. 1987)

Solution. From O.C. Test

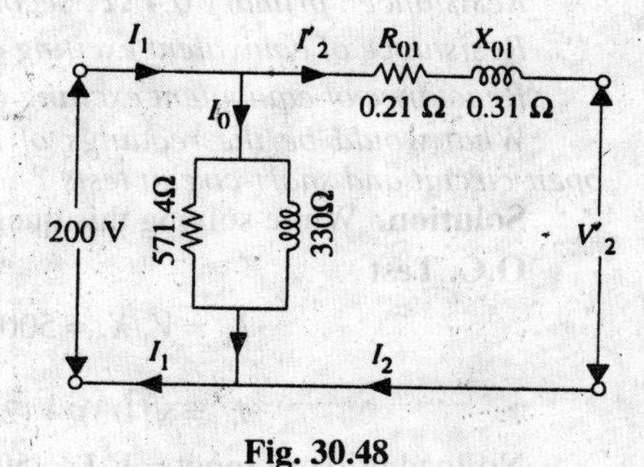

Fig. 30.48

$V_1 I_0 \cos\phi_0 = W_0$

$\therefore \quad 200 \times 0.7 \times \cos\phi_0 = 70$

$\cos\phi_0 = 0.5$ and $\sin\phi_0 = 0.866$

$I_w = I_0 \cos\phi_0 = 0.7 \times 0.5 = 0.35$ A

$I_\mu = I_0 \sin\phi_0 = 0.7 \times 0.866 = 0.606$ A

$R_0 = V_1/I_w = 200/0.35 = 571.4\ \Omega$

$X_0 = V_1/I_\mu = 200/0.606 = 330\ \Omega$

As shown in Fig. 30.48, these values refer to primary *i.e.* low-voltage side.

From S.C. Test

It may be noted that in this test, instruments have been placed in the secondary *i.e.* high-voltage winding whereas the low-voltage winding *i.e.* primary has been short-circuited.

Now, as shown in Art. 30.22

$$Z_{02} = V_{SC}/I_2 = 15/10 = 1.5\ \Omega\ ;\ K = 400/200 = 2$$

$$Z_{01} = Z_{02}/K^2 = 1.5/4 = 0.375\ \Omega$$

Also $\quad I_2^2 R_{02} = W\ ;\ R_{02} = 85/100 = 0.85\ \Omega\ ;\ R_{01} = R_{02}/K^2 = 0.85/4 = 0.21\ \Omega$

$$X_{01} = \sqrt{Z_{01}^2 - R_{01}^2} = \sqrt{0.375^2 - 0.21^2} = 0.31\ \Omega$$

Output kVA $= 5/0.8$; Output current $I_2 = 5000/0.8 \times 400 = 15.6$ A

This value of I_2 is approximate because V_2 (which is to be calculated as yet) has been taken equal to 400 V (which ,in fact, is equal to E_2 or ${}_0V_2$).

Now, $\quad Z_{02} = 1.5\ \Omega,\ R_{02} = 0.85\ \Omega\ \therefore\ X_{02} = \sqrt{1.5^2 - 0.85^2} = 1.24\ \Omega$

Total transformer drop as referred to secondary

$$= I_2 (R_{02} \cos \phi_2 + X_{02} \sin \phi_2) = 15.6\ (0.85 \times 0.8 + 1.24 \times 0.6) = 22.2\ \text{V}$$

$\therefore \quad V_2 = 400 - 22.2 = \mathbf{377.8\ V}$

Example 30.27. *Starting from the ideal transformer, obtain the approximate equivalent circuit of a commercial transformer in which all the constants are lumped and represented on one side.*

A 1-phase transformer has a turn ratio of 6. The resistance and reactance of primary winding are 0.9 Ω and 5 Ω respectively and those of the secondary are 0.03 Ω and 0.13 Ω respectively. If 330.V at 50-Hz be applied to the high voltage winding with the low-voltage winding short-circuited, find the current in the low-voltage winding and its power factor. Neglect magnetising current.

Solution. Here $K = 1/6\ ;\ R_{01} = R_1 + R'_2 = 0.9 + (0.03 \times 36) = 1.98\ \Omega$

$$X_{01} = X_1 + X_2' = 5 + (0.13 \times 36) = 9.68\ \Omega$$

$$\therefore\quad Z_{01} = \sqrt{(9.68^2 + 1.98^2)} = 9.9\ \Omega\ ;\ V_{SC} = 330\ \text{V}$$

$\therefore$ full-load primary current $I_1 = V_{SC}/Z_{01} = 330/0.9 = 100/3$ A

As I_0 is negligible, hence $I_1 = I'_2 = 100/3$ A. Now, $I_2' = KI_2$

F.L. secondary current $I_2 = I_2'/\text{K} = (100/3) \times 6 = 200$ A

Now, power input on short-circuit $= V_{SC}\, I_1 \cos \phi_{SC} =$ Cu loss $= I_1^2 R_{01}$

$$\therefore\quad (100/3)^2 \times 1.98 = 330 \times (100/3) \times \cos \phi_{SC}\ ;\ \cos \phi_{SC} = \mathbf{0.2}$$

Example 30.28. *A 1-phase, 10-kVA, 500/250-V, 50-Hz transformer has the following constants:*

Reactance : primary 0.2 Ω ; secondary 0.5 Ω

Resistance : primary 0.4 Ω ; secondary 0.1 Ω

Resistance of equivalent exciting circuit referred to primary, R_0 = 1500 Ω

Reactance of equivalent exciting circuit referred to primary, X_0 = 750 Ω

What would be the readings of the instruments when the transformer is connected for the open-circuit and short-circuit tests ?

Solution. While solving this question, reference may please be made to Art. 30.20 and 30-22.

O.C. Test

$$I_\mu = V_1/X_0 = 500/750 = 2/3\ \text{A}\ ;\ I_w = V_1/R_0 = 500/1500 = 1/3\text{A}$$

$$\therefore\quad I_0 = \sqrt{[(1/3)^2 + (2/3)^2]} = 0.745\ A$$

No-load primary input $= V_1 I_w = 500 \times 1/3 = 167$ W

Instruments used in primary circuit are : voltmeter, ammeter and wattmeter, their readings being 500 V, 0.745 A and 167 W respectively.

S.C. Test

Suppose S.C. test is performed by short-circuiting the *l.v.* winding *i.e.* the secondary so that all instruments are in primary.

$R_{01} = R_1 + R_2' = R_1 + R_2/K^2$; Here $K = 1/2$ $\therefore R_{01} = 0.2 + (4 \times 0.5) = 2.2\ \Omega$

Similarly, $X_{01} = X_1 + X_2' = X_1 + X_2/K^2 = 0.4 + (4 \times 0.1) = 0.8\ \Omega$

$$Z_{01} = \sqrt{(2.2^2 + 0.8^2)} = 2.341\ \Omega$$

Full-load primary current

$I_1 = 10,000/500 = 20$ A $\therefore V_{SC} = I_1 Z_{01} = 20 \times 2.341 = 46.8$ V

Power absorbed $= I_1^2 R_{01} = 20^2 \times 2.2 = 880$ W

Primary instruments will read : **46.8 V, 20 A, 880 W**.

Example 30.29. *The efficiency of a 1000-kVA, 110/220 V, 50-Hz, single-phase transformer, is 98.5% at half full-load at 0.8 p.f. leading and 98.8% at full-load unity p.f. Determine (i) iron loss (ii) full-load copper loss and (iii) maximum efficiency at unity p.f.*

(Elect. Engg. AMIETE Sec. A Dec. 1991)

Solution. Output at F.L. unity p.f. = 1000 × 1 = 1000 kW

F.L. input = 1000/0.988 = 1012.146 kW

F.L. losses = 1012.146 − 1000 = 12.146 kW

If F.L. Cu and iron losses are x and y respectively then

$$x + y = 12.146 \text{ kW} \qquad \ldots(i)$$

Input at half F.L. 0.8 p.f. = 500 × 0.8/0.985 = 406.091 kW

Total losses at half F.L. = 406.091 − 400 = 6.091 kW

Cu loss at half-load $= x(1/2)^2 = x/4$

$$\therefore \quad x/4 + y = 6.091 \qquad \ldots(ii)$$

From Eqn. (*i*) and (*ii*), we get (*i*) x = 8.073 kW and (*ii*) y = 4.073 kW

(*iii*) kVA for $\eta_{max} = 1000 \times \sqrt{4.073/8.073} = 710.3$ kVA

Output at u.p.f. = 710.3 × 1 = 710.3 kW

Cu loss = iron loss = 4.037 kW ; total loss = 2 × 4.037 = 8.074 kW

$\therefore \eta_{max} = 710.3/(710.3 + 8.074) = 0.989$ or 98.9%

Example 30.30. *The equivalent circuit for a 200/400-V step-up transformer has the following parameters referred to the low-voltage side.*

equivalent resistance = 0.15 Ω ; equivalent reactance = 0.37 Ω

core-loss component resistance= 600 Ω ; magnetising reactance = 300 Ω

When the transformer is supplying a load of 10 A at a power factor of 0.8 lag, calculate (i) the primary current (ii) secondary terminal voltage

(Electrical Machinery-I, Bangalore Univ. 1989)

Solution. We are given the following :

$R_{01} = 0.15\ \Omega$, $X_{01} = 0.37\ \Omega$; $R_0 = 600\ \Omega$, $X_0 = 300\ \Omega$

Using the approximate equivalent circuit of Fig. 30.41, we have,

$$I_\mu = V_1/X_0 = 200/300 = (2/3) \text{ A}$$

$$I_w = V_1/R_0 = 200/600 = (1/3) \text{ A}$$

$$I_0 = \sqrt{I_\mu^2 + I_w^2} = \sqrt{(2/3)^2 + (1/3)^2} = 0.745 \text{ A}$$

As seen from Fig. 30.49

$$\tan\theta = \frac{I_w}{I_\mu} = \frac{1/3}{2/3} = \frac{1}{2}; \theta = 26.6°$$

$\phi_0 = 90° - 26.6° = 63.4°$; Angle betwen I_0 and

$I_2' = 63.4° - 36.9° = 26.5°; K = 400/200 = 2$

$I_2' = KI_2 = 2 \times 10 = 20$ A

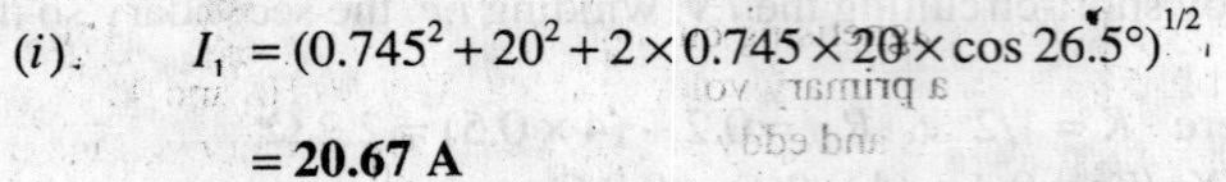

(*i*) $I_1 = (0.745^2 + 20^2 + 2 \times 0.745 \times 20 \times \cos 26.5°)^{1/2}$

$= \mathbf{20.67\ A}$

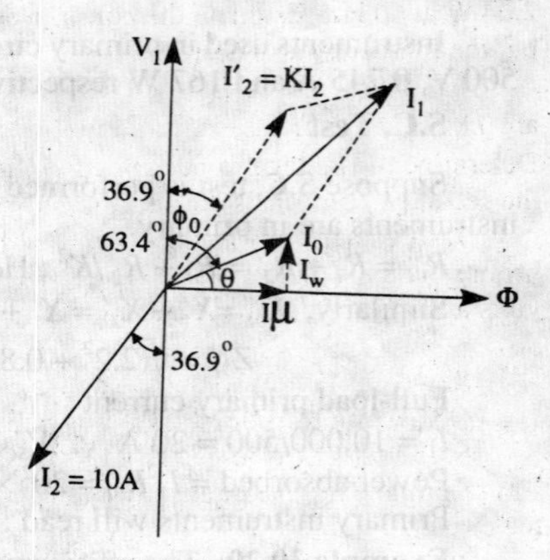

Fig. 30.49

(*ii*) $R_{02} = K^2 R_{01} = 2^2 \times 0.15 = 0.6\ \Omega$

$X_{02} = 2^2 \times 0.37 = 1.48\ \Omega$

Approximate voltage drop $= I_2 (R_{02} \cos \phi + X_{02} \sin \phi)$
$= 10 (0.6 \times 0.8 + 1.48 \times 0.6) = 13.7$ V

∴ secondary terminal voltage = 400−13.7
= **386.3 V**

Example 30.31. *The low voltage winding of a 300-kVA, 11,000/2500-V, 50-Hz transformer has 190 turns and a resistance of 0.06 Ω. The high-voltage winding has 910 turns and a resistance of 1.6 Ω. When the l.v. winding is short-circuited, the full-load current is obtained with 550-V applied to the h.v. winding. Calculate (i) the equivalent resistance and leakage reactance as referred to h.v. side and (ii) the leakage reactance of each winding.*

Solution. Assuming a full-load efficiency of 0.985, the full-load primary current is

$= 300{,}000/0.985 \times 11{,}000 = 27.7$ A

∴ $Z_{01} = 550/27.7 = 19.8\ \Omega\ ;\ R_2' = R_2/K^2 = 0.06\,(910/190)^2 = 1.38\ \Omega$

∴ $R_{01} = R_1 + R_2' = 1.6 + 1.38 = 2.98\ \Omega$

$X_{01} = \sqrt{(Z_{01}^2 - R_{01}^2)} = \sqrt{(19.8^2 - 2.98^2)} = 19.5\ \Omega$

Let us make another assumption that for each winding the ratio (reactance/resistance) is the same, then

$X_1 = 19.5 \times 1.6/2.98 = 10.5\ \Omega$

$X_2' = 19.5 \times 1.38/2.98 = 9.0\ \Omega\ ;\ X_2 = 9(190/910)^2 = 0.39\ \Omega$

(*a*) $R_{01} = \mathbf{2.98\ \Omega}\ ;\ X_{01} = \mathbf{19.5\ \Omega}$ (*b*) $X_1 = \mathbf{10.5\ \Omega}\ ;\ X_2 = \mathbf{0.39\ \Omega}$

Tutorial Problem No. 30-3

1. The S.C. test on a 1-phase transformer, with the primary winding short-circuited and 30 V applied to the secondary gave a wattmeter reading of 60 W and secondary current of 10 A. If the normal applied primary voltage is 200, the transformation ratio 1 : 2 and the full-load secondary current 10 A, calculate the secondary terminal p.d. at full-load current for (*a*) unity power factor (*b*) power factor 0.8 lagging.

If any approximations are made, they must be explained. **[394 V, 377.6 V]**

2. A single-phase transformer has a turn ratio of 6, the resistances of the primary and secondary windings are 0.9 Ω and 0.025Ω respectively and the leakage reactances of these windings are 5.4 Ω and 0.15 Ω respectively. Determine the voltage to be applied to the low-voltage winding to obtain a current of 100 A in the short-circuited high voltage winding. Ignore the magnetising current **[82 V]**

3. Draw the equivalent circuit for a 3000/400-V, I-phase transformer on which the following test results were obtained. Input to high voltage winding when l.v. winding is open-circuited: 3000 V, 0.5 A, 500 W. Input to l.v. winding when h.v. winding is short-circuited : 11 V, 100 A, 500 W. Insert the appropriate values of resistance and reactance. **[R_0 = 18,000 Ω, X_0 = 6,360 Ω, R_{01} = 2.81 Ω, X_{01} = 5.51 Ω]** (*I.E.E. London*)

4. The iron loss in a transformer core at normal flux density was measured at frequencies of 30 and 50 Hz, the results being 30 W and 54 W respectively. Calculate (*a*) the hysteresis loss and (*b*) the eddy current loss at 50 Hz. **[44 W, 10 W]**

5. An iron core was magnetised by passing an alternating current through a winding on it. The power required for a certain value of maximum flux density was measured at a number of different frequencies. Neglecting the effect of resistance of the winding, the power required per kg of iron was 0.8 W at 25 Hz and

2.04 W at 60 Hz. Estimate the power needed per kg when the iron is subject to the same maximum flux density but the frequency is 100 Hz. **[3.63 W]**

6. The ratio of turns of a 1-phase transformer is 8, the resistances of the primary and secondary windings are 0.85 Ω and 0.012 Ω respectively and leakage reactances of these windings are 4.8 Ω and 0.07 Ω respectively. Determine the voltage to be applied to the primary to obtain a current of 150 A in the secondary circuit when the secondary terminals are short-circuited. Ignore the magnetising current. **[176.4 W]**

7. A transformer has no-load losses of 55 W with a primary voltage of 250 V at 50 Hz and 41 W with a primary voltage of 200 V at 40 Hz. Compute the hysteresis and eddy current losses at a primary voltage of 300 volts at 60 Hz of the above transformer. Neglect small amount of copper loss at no-load.

[43.5 W ; 27 W] (*Elect. Machines AMIE Sec. B. (E-3) Summer 1992*)

8. A 20 kVA, 2500/250 V, 50 Hz, 1-phase transformer has the following test results :

O.C. Test (l.v.side) : 250 V, 1.4 A, 105 W

S.C. Test (h.v.side) : 104 V, 8 A, 320 W

Compute the parameters of the approximate equivalent circuit referred to the low voltage side and draw the circuit.

[R_0 = 592.5 Ω ; X_0 = 187.2 Ω ; R_{02} = 1.25Ω ; X_{12} = 3 Ω] (*Elect. Machines AMIE Sec. B Summer 1990*)

9. A 10-kVA, 2000/400-V, single-phase transformer has resistances and leakage reactances as follows : $R_1 = 5.2\ \Omega, X_1 = 12.5\ \Omega, R_2 = 0.2\ \Omega, X_2 = 0.5\ \Omega$

Determine the value of secondary terminal voltage when the transformer is operating with rated primary voltage with the secondary current at its rated value with power factor 0.8 lag. The no-load current can be neglected. Draw the phasor diagram. **[376.8 V]** (*Elect. Machines, A.M.I.E. Sec B, 1989*)

10. A 1000-V, 50-Hz supply to a transformer results in 650 W hysteresis loss and 400 W eddy current loss. If both the applied voltage and frequency are doubled, find the new core losses.

[W_h = 1300 W; W_e = 1600 W] **(Elect. Machine, A.M.I.E. Sec. B, 1993)**

30.24. Regulation of a Transformer

(1) When a transformer is loaded with a *constant primary voltage*, the secondary voltage decreases* because of its internal resistance and leakage reactance.

Let $_0V_2$ = secondary terminal voltage at *no-load*.

$= E_2 = EK_1 = KV_1$ because at no-load the impedance drop is negligible.

V_2 = secondary terminal voltage on *full-load*.

The change in secondary terminal voltage from no-load to full-load is = $_0V_2 - V_2$. This change divided by $_0V_2$ is known as regulation 'down'. If this change is divided by V_2 *i.e.* full-load secondary terminal voltage, then it is called regulation 'up'.

$$\therefore\ \%\ \text{regn 'down'} = \frac{_0V_2 - V_2}{_0V_2} \times 100 \quad \text{and} \quad \%\ \text{regn 'up'} = \frac{_0V_2 - V_2}{V_2} \times 100$$

In further treatment, unless stated otherwise, regulation is to be taken as regulation 'down'.

We have already seen in Art. 30.16 (Fig. 30.35) that the change in secondary terminal voltage from no-load to full-load, expressed as a percentage of no-load secondary voltage, is

$$= v_r \cos\phi \pm v_x \sin\phi \quad \text{(approximately)}$$

Or more accurately

$$= (v_r \cos\phi \pm v_x \sin\phi) + \frac{1}{200}(v_x \cos\phi \mp v_r \sin\phi)^2$$

$$\therefore \quad \%\ \text{regn} = v_r \cos\phi \pm v_x \sin\phi \quad \text{...approximately.}$$

The lesser this value, the better the transformer, because a good transformer should keep its secondary terminal voltage as constant as possible under all conditions of load.

(2) The regulation may also be explained in terms of primary values.

---

*Assuming lagging power factor. It will increase if power factor is leading.

In Fig. 30.50 (*a*) is shown the approximate equivalent circuit of a transformer and in Fig. 30.50 (*b*), (*c*) and (*d*) are shown the vector diagrams corresponding to different power factors.

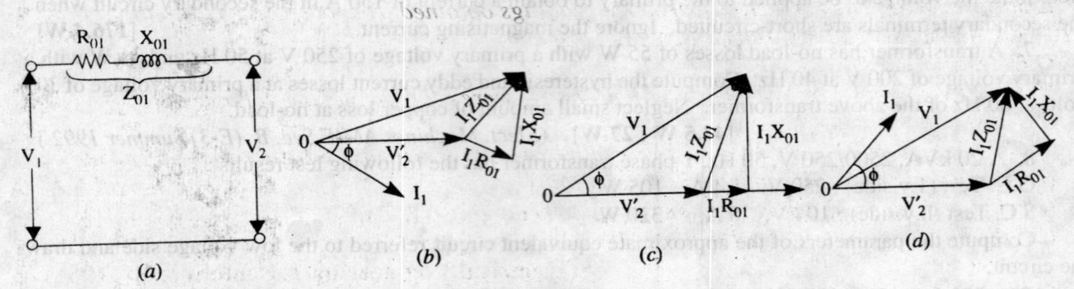

Fig. 30.50

The secondary *no-load* terminal voltage as referred to primary is $E_2' = E_2/K = E_1 = V_1$ and if the secondary full-load voltage as referred to primary is V_2' $(=V_2/K)$ then

$$\% \text{ regn}, = \frac{V_1 - V_2'}{V_1} \times 100$$

From the vector diagram, it is clear that if angle between V_1 and V_2' is neglected, then the value of numerical difference $V_1 - V_2'$ is given by $(I_1R_{01}\cos\phi + I_1X_{01}\sin\phi)$ for lagging p.f.

$$\therefore \qquad \% \text{ regn.} = \frac{I_1R_{01}\cos\phi + I_1X_{01}\sin\phi}{V_1} \times 100 = v_r\cos\phi + v_x\sin\phi$$

where $\dfrac{I_1R_{01}\times 100}{V_1} = v_r$ and $\dfrac{I_1X_{01}\times 100}{V_1} = v_x$ ...Art 30-16

As before, if angle between V_1 and V_2' is not negligible, then

$$\% \text{ regn.} = (v_r\cos\phi \pm v_x\sin\phi) + \frac{1}{200}(v_x\cos\phi \mp v_r\sin\phi)^2$$

(3) In the above definitions of regulation, *primary voltage was supposed to be kept constant* and the changes in secondary terminal voltage were considered.

As the transformer is loaded, the secondary terminal voltage falls (for a lagging p.f.). Hence, to keep the output voltage constant, the primary voltage must be increased. The rise in primary voltage required to maintain rated output voltage from no-load to full-load at a given power factor expressed as percentage of rated primary voltage gives the regulation of the transformer.

Suppose primary voltage has to be raised from its rated value V_1 to V_1', them

$$\% \text{ regn.} = \frac{V_1' - V_1}{V_1} \times 100$$

Example 30.32. *A-100 kVA transformer has 400 turns on the primary and 80 turns on the secondary. The primary and secondary resistances are 0.3 Ω and 0.01 Ω respectively and the corresponding leakage reactances are 1.1 and 0.035 Ω respectively. The supply voltage is 2200 V. Calculate (i) equivalent impedance referred to primary and (ii) the voltage regulation and the secondary terminal voltage for full load having a power factor of 0.8 leading.*

(Elect. Machines, A.M.I.E. Sec B, 1989)

Solution. $K = 80/400 = 1/5$, $R_1 = 0.3\ \Omega$, $R_{01} = R_1 + R_2/K^2 = 0.3 + 0.01/(1/5)^2 = 0.55\ \Omega$

$X_{01} = X_1 + X_2/K^2 = 1.1 + 0.035/(1/5)^2 = 1.975\ \Omega$

(*i*) $Z_{01} = 0.55 + j\,1.975 = 2.05\angle 74.44°$

(*ii*) $Z_{02} = K^2Z_{01} = (1/5)^2\,(0.55 + j\,1.975) = (0.022 + j\,0.079)$

No-load secondary voltage $= KV_1 = (1/5) \times 2200 = 440$ V, $I_2 = 10 \times 10^3/440 = 227.3$ A

Full-load voltage drop as referred to secondary
$= I_2 (R_{02} \cos\phi - X_{02} \sin\phi) = 227.3\ (0.022 \times 0.8 - 0.079 \times 0.6) = -\ 6.77$ V
% regn. $= -\ 6.77 \times 100/440 = -\ 1.54$
Secondary terminal voltage on load $= 440 - (-6.77) = 446.77$ V

Example 30.33. *The corrected instrument readings obtained from open and short-circuit tests on 10-kVA, 450/120-V, 50-Hz transformer are :*

O.C. test :
$V_1 = 120$ *V ;* $I_1 = 4.2$ *A;* $W_1 = 80$ *W;* V_1, W_1 *and* I_1 *were read on the low-voltage side.*

S.C. test :
$V_1 = 9.65$ *V ;* $I_1 = 22.2$ *A ;* $W_1 = 120$ *W –with low-voltage winding short-circuited Compute :*

(i) the equivalent circuit (approximate) constants
(ii) efficiency and voltage regulation for an 80% lagging p.f. load
(iii) the efficiency at half full-load and 80% lagging p.f. load.

(Electrical Engineering-I, Bombay Univ. 1988)

Solution. It is seen from the O.C. test, that with primary open, the secondary draws a no-load current of 4.2 A. Since $K = 120/450 = 4/15$, the corresponding no-load primary current $I_0 = 4.2 \times 4/15 = 1.12$ A.

(*i*) Now, $V_1 I_0 \cos\phi_0 = 80$ $\quad\therefore\quad$ $\cos\phi_0 = 80/450 \times 1.12 = 0.159$

$\therefore\ \phi_0 = \cos^{-1}(0.159) = 80.9°;\ \sin\phi_0 = 0.987$

$I_w = I_0 \cos\phi_0 = 1.12 \times 0.159 = 0.178$ A and $I_\mu = 1.12 \times 0.987 =$ **1.1 A**

$\therefore\ R_0 = 450/0.178 =$ **2530** Ω and $X_0 = 450/1.1 =$ **409** Ω

During S.C. test, instruments have been placed in primary.

$\therefore\quad Z_{01} = 9.65/22.2 = 0.435\ \Omega$

$R_{01} = 120/22.2^2 = 0.243\ \Omega$

$X_{01} = \sqrt{0.435^2 - 0.243^2} = 0.361\ \Omega$

The equivalent circuit is shown in Fig. 30.51.

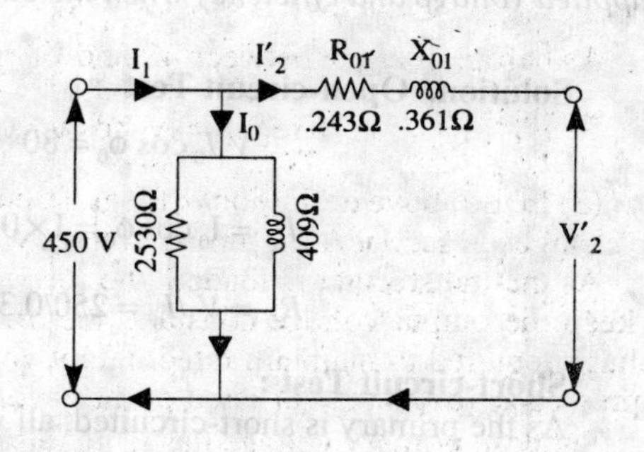

Fig. 30.51

(*ii*) Total approximate voltage drop as referred to primary is* $= I_1(R_{01} \cos\phi + X_{01} \sin\phi)$
Now, full-load $I_1 = 10{,}000/450 = 22.2$ A
$\therefore$ drop $= 22.2(0.243 \times 0.8 + 0.361 \times 0.6) = 9.2$ V
regulation $= 9.2 \times 100/450 = 2.04\%$
F.L. losses $= 80 + 120 = 200$ W ;
F.L. output $= 10{,}000 \times 0.8 = 8000$ W
$\eta = 8000/8200 = 0.9757$ or **97.57%**

(*iii*) **Half-load**
Iron loss $= 80$ W ; Cu loss $= (1/2)^2 \times 120 = 30$ W
Total losses $= 110$ W; output $= 5000 \times 0.8 = 4000$ W
$\therefore\ \eta = 4000/4110 = 0.9734$ or **97.34%**

Example 30.34. Consider a 20 KVA, 2200/220 V, 50 Hz transformer. The O.C./S.C. test results are as follows :

O.C. test : 220 V, 4.2 A, 148 W (1.v. side)
S.C. test : 86 V, 10.5 A, 360 W (h.v. side)

Determine the regulation at 0.8 p.f. lagging and at full load. What is the p.f. on short-circuit ?

(*Elect. Machines Nagpur Univ. 1993*)

Solution. It may be noted that O.C. data is not required in this question for finding the regulation.

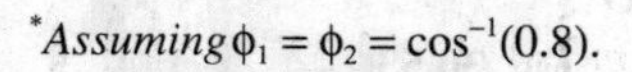

*Assuming $\phi_1 = \phi_2 = \cos^{-1}(0.8)$.

Since during S.C. test, instruments have been placed on the h.v. side i.e. primary side.

$$\therefore \quad Z_{01} = 86/10.5 = 8.19\ \Omega\ ;\ R_{01} = 360/10.5^2 = 3.26\ \Omega;$$

$$X_{01} = \sqrt{8.19^2 - 3.26^2} = 7.5\ \Omega$$

F.L. primary current, $I_1 = 20{,}000/2200 = 9.09$ A

Total voltage drop as referred to primary $= I_1\ (R_{01} \cos\phi + X_{01} \sin\phi\ )$

Drop $= 9.09\ (326 \times 0.8 + 7.5 \times 0.6) = 64.6$ V

% age regn. $= 64.6 \times 100/2200 = 2.9$ %, p.f. on short-circuit $= R_{01}/Z_{01} = 3.26/8.19 = 0.4$

Example 30.35. *A short-circuit test when performed on the h.v. side of a 10 kVA, 2000/400 V single phase transformer, gave the following data : 60 V, 4 A, 100 W*

If the l.v. side is delivering full load current at 0.8 p.f. lag and at 400 V, find the voltage applied to h.v. side. **(Elect. Machines-I Nagpur Univ. 1993)**

Solution. Here, the test has been performed on the h.v. side *i.e.* primary side.

$Z_{01} = 60/4 = 15\ \Omega$; $R_{01} = 100/4^2 = 6.25\ \Omega$; $X_{01} = \sqrt{15^2 - 6.25^2} = 13.63\ \Omega$

F.L. $I_1 = 10{,}000/2000 = 5$ A

Total transformer voltage drop as referred to primary is

$$I_1(R_{01} \cos\phi + X_{01} \sin\phi) = 5(6.25 \times 0.8 + 13.63 \times 0.6) = 67\text{ V}$$

Hence, primary voltage has to be raised from 2000 V to 2067 V in order to compensate for the total voltage drop in the transformer. In that case secondary voltage on load would remain the same as on no-load.

Example 30.36. *A 250/500-V transformer gave the following test results :*

Short-circuit test: with low-voltage winding short-circuited :

20 V; 12 A, 100 W

Open-circuit test : 250 V, 1 A, 80 W on low-voltage side.

Determine the circuit constants, insert these on the equivalent circuit diagram and calculate applied voltage and efficiency when the output is 10 A at 500 volt and 0.8 power factor lagging.

(Elect. Machines, Nagpur Univ. 1993)

Solution. Open-circuit Test :

$$V_1 I_0 \cos\phi_0 = 80 \qquad \therefore \qquad \cos\phi_0 = 80/250 \times 1 = 0.32$$

$$I_w = I_0 \cos\phi_0 = 1 \times 0.32 = 0.32\text{ A},\ I_\mu = \sqrt{(1^2 - 0.32^2)} = 0.95\text{ A}$$

$$R_0 = V_1/I_w = 250/0.32 = 781.3\ \Omega,\ X_0 = V_1/I_\mu = 250/0.95 = 263.8\Omega$$

Short-circuit Test :

As the primary is short-circuited, all values refer to secondary winding.

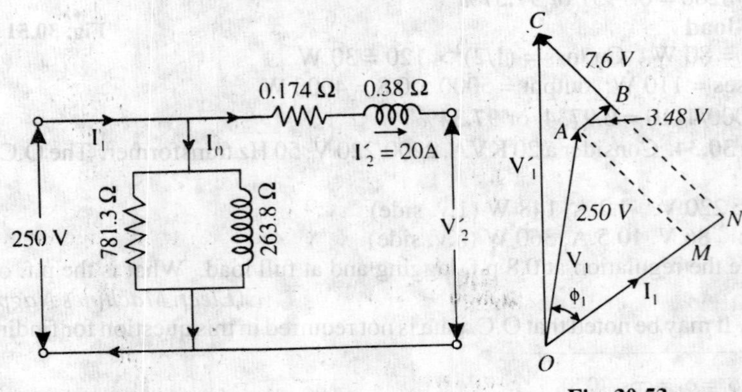

Fig. 30.52

Fig. 30.53

$$\therefore \quad R_{02} = \frac{\text{short-circuit power}}{\text{F.L. secondary current}} = \frac{100}{12^2} = 0.694\ \Omega$$

$$Z_{02} = 20/12 = 1.667\ \Omega\ ;\ X_{02} = \sqrt{(1.667^2 - 0.694^2)} = 1.518\ \Omega$$

As R_0 and X_0 refer to primary, hence we will transfer these values to primary with the help of transformation ratio.

$K = 500/250 = 2 \quad \therefore \ R_{01} = R_{02}/K^2 = 0.694/4 = 0.174\ \Omega$

$X_{01} = X_{02}/K^2 = 1.518/4 = 0.38\ \Omega\ ;\ Z_{01} = Z_{02}/K^2 = 1.667/4 = 0.417\ \Omega$

The equivalent circuit is shown in fig. 30.53.

Efficiency

Total Cu loss $= I_2^2 R_{02} = 100 \times 0.694 = 69.4$ W ; Iron loss = 80 W

Total loss = 69.4 + 80 = 149.4 W $\quad \therefore \ \eta = \dfrac{5000 \times 0.8 \times 100}{4000 + 149.4} = \mathbf{96.42\ \%}$

The applied voltage V_1' is the vector sum of V_1 and $I_1 Z_{01}$ as shown in Fig. 30.54.

$I_1 = 20$ A ; $I_1 R_{01} = 20 \times 0.174 = 3.84$ V; $I_1 X_{01} = 20 \times 0.38 = 7.6$ V

Neglecting the angle between V_1 and V_1', we have

$$V_1'^2 = OC^2 = ON^2 + NC^2 = (OM + MN)^2 + (NB + BC)^2$$

$$= (250 \times 0.8 + 3.48)^2 + (250 \times 0.6 + 7.6)^2$$

$$V_1'^2 = 203.5^2 + 157.6^2 \quad \therefore \ V_1' = \mathbf{257.4\ V}$$

30.25. Percentage Resistance, Reactance and Impedance

These quantities are usually measured by the voltage drop at full-load current expressed as a percentage of the normal voltage of the winding on which calculations are made.

(*i*) Percentage resistance at full-load

$$\%R = \frac{I_1 R_{01}}{V_1} \times 100 = \frac{I_1^2 R_{01}}{V_1 I_1} \times 100$$

$$= \frac{I_2^2 R_{02}}{V_2 I_2} \times 100 = \%\text{Cu loss at full-load}$$

$$\%R = \%\text{Cu loss} = v_r$$ ...Art. 30-16

(*ii*) Percentage reactance at full load

$$\%X = \frac{I_1 X_{01}}{V_1} \times 100 = \frac{I_2 X_{02}}{V_2} \times 100 = v_x$$

(*iii*) Percentage impedance at full-load

$$\%Z = \frac{I_1 Z_{01}}{V_1} \times 100 = \frac{I_2 Z_{02}}{V_2} \times 100$$

(*iv*) $$\%Z = \sqrt{(\%R^2 + \%X^2)}$$

It should be noted from above that the reactances and resistances in ohm can be obtained thus:

$$R_{01} = \frac{\%R \times V_1}{100 \times I_1} = \frac{\%\text{ Cu loss} \times V_1}{100 \times I_1}\ ;\ \text{Similarly } R_{02} = \frac{\%R \times V_2}{100 \times I_2} = \frac{\%\text{Cu loss} \times V_2}{100 \times I_2}$$

$$X_{01} = \frac{\%X \times V_1}{100 \times I_1} = \frac{v_x \times V_1}{100 \times I_1};\quad \text{Similarly } X_{02} = \frac{\%X \times V_2}{100 \times I_2} = \frac{v_x \times V_2}{100 \times I_2}$$

It may be noted that percentage resistance, reactance and impedance have the same value whether referred to primary or secondary.

Example 30.37. *A 3300/230.V, 50-kVA, transformer is found to have impedance of 4% and a Cu loss of 1.8% at full-load. Find its percentage reactance and also the ohmic values of resistance, reactance and impedance as referred to primary. What would be the value of primary short-circuit current if primary voltage is assumed constant?*

Solution. $\%X = \sqrt{(\%Z^2 - \%R^2)} = \sqrt{(4^2 - 1.8^2)} = 3.57\%$ ($\because$ Cu loss $= \%R$)

Full load $I_1 = 50{,}000/3300 = 15.2$ A (assuming 100% efficiency). Considering primary winding, we have

$$\%R = \frac{R_{01}I_1 \times 100}{V_L} = 1.8 \quad \therefore \quad R_{01} = \frac{1.8 \times 3300}{100 \times 15.2} = \mathbf{3.91\ \Omega}$$

$$\text{Similarly } \%X = \frac{X_{01}I_1 \times 100}{V_1} = 3.57 \quad \therefore \quad X_{01} = \frac{3.57 \times 3300}{100 \times 15.2} = \mathbf{7.76\ \Omega}$$

$$\text{Similarly } Z_{01} = \frac{4 \times 3300}{100 \times 15.2} = \mathbf{8.7\ \Omega}$$

$$\text{Now } \frac{\text{short-circuit current}^*}{\text{full load current}} = \frac{100}{4} \quad \therefore \quad \text{S.C. current} = 15.2 \times 25 = \mathbf{380\ A}$$

Example 30.38. *A 20-kVA, 2200/220-V, 50-Hz distribution transformer is tested for efficiency and regulation as follows :*

O.C. test : 220 V, 4.2 A, 148 W —l.v. side
S.C. test : 86 V, 10.5 A, 360 W —l.v. side

Determine (a) core loss (b) equivalent resistance referred to primary (c) equivalent resistance referred to secondary (d) equivalent reactance referred to primary (e) equivalent reactance referred to secondary (f) regulation of transformer at 0.8 p.f. lagging current (g) efficiency at full-load and half the full-load at 0.8 p.f. lagging current.

Solution. (*a*) As shown in Art. 30.9, no-load primary input is practically equal to the core loss. Hence, core loss as found from no-load test, is **148 W.**

(*b*) From S.C test, $R_{01} = 360/10.5^2 = \mathbf{3.26\ \Omega}$

(*c*) $R_{02} = K^2R_{01} = (220/2200)^2 \times 3.26 = \mathbf{0.0326\ \Omega}$

(*d*) $Z_{10} = \dfrac{V_{SC}}{I_{SC}} = \dfrac{86}{10.5} = 8.19\ \Omega$

$X_{01} = \sqrt{(8.19^2 - 3.26^2)} = \mathbf{7.51\ \Omega}$

(*e*) $X_{02} = K^2X_{01} = (220/2200)^2 \times 7.51 = \mathbf{0.0751\ \Omega}$.

(*f*) We will use the definition of regulation as given in Art. 30.24(3).

We will find the rise in primary voltage necessary to maintain the output terminal voltage constant from no-load to full-load.

Rated primary current = 20,000/2200 = 9.1 A

$$V_1' = \sqrt{[2200 \times 0.8 + 9.1 \times 3.26)^2 + (2200 \times 0.6 + 9.1 \times 7.51)^2]} = 2265\ \mathbf{V}$$

---

*Short circuit $\quad I_{SC} = \dfrac{V_1}{Z_{01}}$ Now, $Z_{01} = \dfrac{V_1 \times \%Z}{100 \times I_1}$

$$I_{SC} = \frac{V_1 \times 100 \times I_1}{V_1 \times \%Z} = \frac{100 \times I_1}{\%Z} \quad \therefore \quad \frac{I_{SC}}{I_1} = \frac{100}{\%Z}$$

$$\therefore \quad \% \text{ regn} = \frac{2265 - 2200}{2200} \times 100 = \mathbf{2.95\ \%}$$

We would get the same result by working in the secondary. Rated secondary current = **91** A.

$${}_0V_2 = \sqrt{[(220 \times 0.8 + 91 \times 0.0326)^2 + (220 \times 0.6 + 91 \times 0.0751)^2]} = 226.5 \ V$$

$$\therefore \quad \% \text{ regn.} = \frac{226.5 - 220}{220} \times 100 = \mathbf{2.95\ \%}$$

(*g*) Core loss = 1.48 W. It will be the same for all loads

Cu loss at full load = $I_1^2 R_{01} = 9.1^2 \times 3.26 = 270$ W

Cu loss at half full-load = $4.55^2 \times 3.26 = 67.5$ W (or F.L. Cu loss/4)

$$\therefore \eta \text{ at full-load} = \frac{20,000 \times 0.8 \times 100}{20,000 \times 0.8 + 148 + 270} = \mathbf{97.4\ \%}$$

$$\therefore \eta \text{ at half-load} = \frac{10,000 \times 0.8 \times 100}{10,000 \times 0.8 + 148 + 67.5} = \mathbf{97.3\ \%}$$

30.26. Kapp Regulation Diagram

It has been shown that secondary terminal voltage falls as the load on the transformer is increased when p.f. is lagging and it increases when the power factor is leading. In other words, secondary terminal voltage not only depends on the load but on power factor also (Art. 30.16). For finding the voltage drop (or rise) which is further used in determining the regulation of the transformer, a graphical construction is employed which was proposed by late Dr. Kappa.

For drawing Kapp regulation diagram, it is necessary to know the equivalent resistance and reactance as referred to secondary *i.e.* R_{02} and X_{02}. If I_2 is the secondary load, current, then secondary terminal voltage on load V_2, is obtained by subtracting $I_2 R_{02}$ and $I_2 X_{02}$ voltage drops vectorially from secondary no-load voltage ${}_0V_2$.

Now, ${}_0V_2$ is constant, hence it can be represented by a circle of constant radius *OA* as in Fig. 30.54. This circle is known as no-load or open-circuit e.m.f. circle. For a given load, OI_2 represents the load current and is taken as the reference vector, *CB* represents $I_2 R_{02}$ and is parallel to OI_2, *AB* represent $I_2 X_{02}$ and is drawn at right angles to *CB*. Vector *OC* obviously represents secondary terminal voltage V_2. Since I_2 is constant, the drop triangle *ABC* remains constant in size. It is seen that end point *C* of V_2 lies on another circle whose centre is *O′*. This point *O′* lies at a distance of $I_2 X_{02}$ vertically below the point *O* and a distance of $I_2 R_{02}$ to its left as shown in Fig. 30.54.

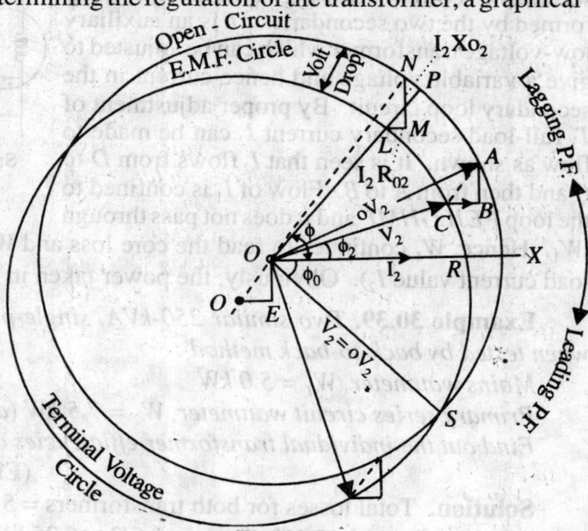

Fig. 30.54

Suppose it is required to find the voltage drop on full-load at a lagging power factor of cos φ, then a radius *OLP* is drawn inclined at an angle of φ with *OX*. $LM = I_2 R_{02}$ and is drawn horizontal. $MN = I_2 X_{02}$ and is drawn perpendicular to *LM*. Obviously, *ON* is no-load voltage ${}_0V_2$. Now, $ON = OP = {}_0V_2$. Similarly, *OL* is V_2. The voltage drop = *OP*−*OL* = *LP*.

$$\text{Hence, percentage regulation 'down' is} = \frac{OP - OL}{OP} \times 100 = \frac{LP}{OP} \times 100$$

It is seen that for finding voltage drop, the drop triangle *LMN* need not be drawn, but simply the radius *OLP*.

The diagram shows clearly how the secondary terminal voltage falls as the angle of lag increases. Conversely, for a leading power factor, the fall in secondary terminal voltage decreases till for an angle of $φ_0$ leading, the fall becomes zero ; hence $V_2 = {}_0V_2$. For angles greater than $φ_0$, the secondary terminal voltage V_2 becomes greater than ${}_0V_2$.

The Kapp diagram is very helpful in determining the variation of regulation with power factor but it has the disadvantage that since the lengths of the sides of the impedance triangle are very small as compared to the radii of the circles, the diagram has to be drawn on a very large scale, if sufficiently accurate results are desired.

30.27. Sumpner or Back-to-Back Test

This test provides data for finding the regulation, efficiency and heating under load conditions and is employed only when two similar transformers are available. One transformer is loaded on the other and both are connected to supply. The power taken from the supply is that necessary for supplying the losses of both transformers and the negligibly small loss in the control circuit.

As shown in Fig. 30.55, primaries of the two transformers are connected in parallel across the same a.c. supply. With switch S open, the wattmeter W_1 reads the core loss for the two transformers.

The secondaries are so connected that their potentials are in opposition to each other. This would be so if $V_{AB} = V_{CD}$ and A is joined to C whilst B is joined to D. In that case, there would be no secondary current flowing around the loop formed by the two secondaries. T is an auxiliary low-voltage transformer which can be adjusted to give a variable voltage and hence current in the secondary loop circuit. By proper adjustment of T, full-load secondary current I_2 can be made to flow as shown. It is seen that I_2 flows from D to C and then from A to B. Flow of I_1 is confined to the loop $FEJLGHMF$ and it does not pass through W_1. Hence, W_1 continues to read the core loss and W_2 measures full-load Cu loss (or at any other load current value I_2). Obviously, the power taken in is twice the losses of a single transformer.

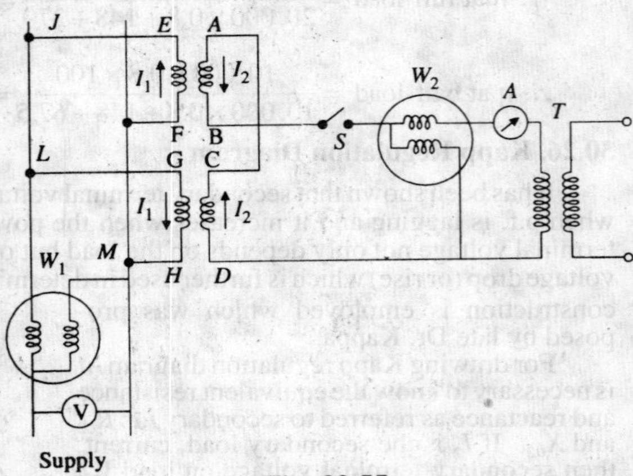

Fig. 30.55

Example 30.39. *Two similar 250-kVA, single-phase transformers gave the following results when tested by back-to-back method :*

Mains wattmeter, W_1 = 5.0 kW

Primary series circuit wattmeter, W_2 = 7.5 kW (at full-load current).

Find out the individual transformer efficiencies at 75% full-load and 0.8 p.f. lead.

(Electrical Machines-III, Gujarat Univ. 1986)

Solution. Total losses for both transformers = 5 + 7.5 = 12.5 kW

F.L. loss for each transformer = 12.5/2 = 6.25 kW

Output of each transformer at 75% F.L. and 0.8 p.f. is =(250 × 0.75) × 0.8 = 150 kW

$\therefore$ η = 150/156.25 = 0.96 or **96%**

30.28. Losses in a Transformer

In a static transformer, there are no friction or windage losses. Hence, the only losses occurring are :-

(*i*) **Core or Iron Loss.** It includes both hysteresis loss and eddy current loss. Because the core flux in a transformer remains practically constant for all loads (its variation being 1 to 3% from no-load to full-load). The core loss is practically the same at all loads.

Hysteresis loss $W_h = \eta B^{1.6}_{max} fV$ watt; eddy current loss $W_e = PB^2_{max} f^2 t^2$ watt

These losses are minimized by using steel of high silicon content for the core and by using very thin laminations. Iron or core loss is found from the O.C. test. *The input of the transformer when on no-load measures the core loss.*

(*ii*) **Copper loss.** This loss is due to the ohmic resistance of the transformer windings. Total Cu loss $= I_1^2R_1 + I_2^2R_2 = I_1^2R_{01} = I_2^2R_{02}$. It is clear that Cu loss is proportional to $(\text{current})^2$ or kVA^2. In other words, Cu loss at half the full-load is one-fourth of that at full-load.

The value of Cu loss is found from the short-circuit test (Art. 30.22).

30.29. Efficiency of a Transformer

As is the case with other types of electrical machines, the efficiency of a transformer at a particular load and power factor is defined as the output divided by the input — the two being measured in the same units (either watts or kilowatts).

$$\text{efficiency} = \frac{\text{output}}{\text{input}}$$

But a transformer being a highly efficient piece of equipment, has very small loss, hence it is impractical to try to measure transformer efficiency by measuring input and output. These quantities are nearly of the same size. A better method is to determine the losses and then to calculate the efficiency from ;

$$\text{efficiency} = \frac{\text{output}}{\text{output + losses}} = \frac{\text{output}}{\text{output + Cu loss + iron loss}}$$

or
$$\eta = \frac{\text{input - losses}}{\text{input}} = 1 - \frac{\text{losses}}{\text{input}}$$

It may be noted here that efficiency is based on power output in watts and not in volt-amperes, although losses are proportional to VA. Hence, at any volt-ampere load, the efficiency depends on power factor, being maximum at a power factor of unity.

Efficiency can be computed by determining core loss from no-load or open-circuit test and Cu loss from the short-circuit test.

30.30. Condition for Maximum Efficiency

$$\text{Cu loss} = I_1^2R_{01} \text{ or } I_2^2R_{02} = W_{cu}$$

$$\text{Iron loss} = \text{hysteresis loss + eddy current loss} = W_h + W_e = W_i$$

Considering primary side,

Primary input
$$= V_1I_1\cos\phi_1$$

$$\eta = \frac{V_1I_1\cos\phi_1 - \text{losses}}{V_1I_1\cos\phi_1} = \frac{V_1I_1\cos\phi_1 - I_1^2R_{01} - W_i}{V_1I_1\cos\phi_1}$$

$$= 1 - \frac{I_1R_{01}}{V_1\cos\phi_1} - \frac{W_i}{V_1I_1^2\cos\phi_1}$$

Differentiating both sides with respect to I_1, we get

$$\frac{d\eta}{dI_1} = 0 - \frac{R_{01}}{V_1\cos\phi_1} + \frac{W_i}{V_1I_1^2\cos\phi_1}$$

For η to be maximum, $\frac{d\eta}{dI_1} = 0$. Hence, the above equation becomes

$$\frac{R_{01}}{V_1\cos\phi_1} = \frac{W_i}{V_1I_1^2\cos\phi_1} \text{ or } W_i = I_1^2R_{01} \text{ or } I_2^2R_{02}$$

or
$$\textbf{Cu loss} = \textbf{Iron loss}$$

The output current corresponding to maximum efficiency is $I_2 = \sqrt{(W_i/R_{02})}$

It is this value of the output current which will make the Cu loss equal to the iron loss. By proper design, it is possible to make the maximum efficiency occur at any desired load.

Note. (*i*) If we are given iron loss and full-load Cu loss, then the load at which two losses would be equal (*i.e.* corresponding to maximum efficiency) is given by

$$= \text{Full load} \times \sqrt{\left(\frac{\text{Iron loss}}{\text{F.L. Cu loss}}\right)}$$

In Fig. 30.56, Cu losses are plotted as a percentage of power input and the efficiency curve as deduced from these is also shown. It is obvious that the point of intersection of the Cu and iron loss curves gives the point of maximum efficiency. It would be seen that the efficiency is high and is practically constant from 15% full-load to 25% overload.

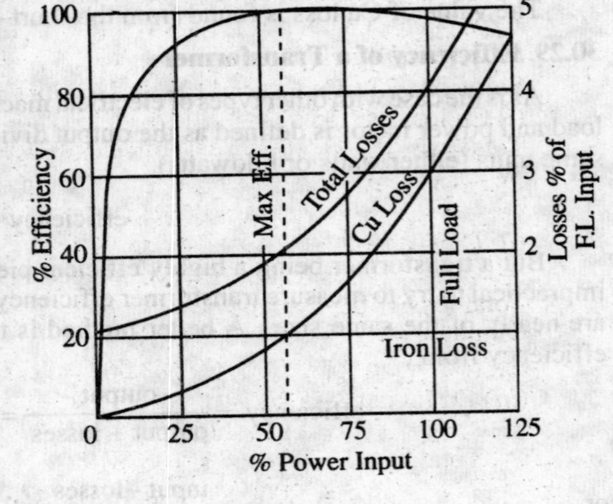

Fig. 30.56

(*ii*) The efficiency at any load is given by

$$\eta = \frac{x \times \text{full-load kVA} \times \text{p.f.}}{(x \times \text{ full-load kVA } \times \text{ p.f.}) + W_{cu} + W_i} \times 100$$

where x = ratio of actual to full-load kVA

W_i = iron loss in kW ; W_{cu} = Cu loss in kW

Example 30.40. *In a 25-kVA, 2000/200 V, single-phase transformer, the iron and full-load copper losses are 350 and 400 W respectively. Calculate the efficiency at unity power factor on (i) full load (ii) half full-load.* **(Elect. Engg. & Electronics, Bangalore Univ. 1990)**

Solution. (*i*) **Full-load Unity p.f.**

Total loss = 350 + 400 = 750 W

F.L. output at u.p.f. = 25 × 1 = 25 kW ; η = 25/25.75 = 0.97 or **97%**

(*ii*) **Half F.L. Unity p.f.**

Cu loss = 40 × $(1/2)^2$ = 100 W. Iron loss remains constant at 350 W, Total loss = 100 + 350 = 450 W

Half-load output at u.p.f. = (25/2) × 1 = 7.5 kW

∴ η = 7.5/(7.5 + 0.45) = 0.943 or **94.3%**

Example 30.41. *If P_1 and P_2 be the iron and copper losses of a transformer on full-load, find the ratio of P_1 and P_2 such that maximum efficiency occurs at 75% full-load*

(Elect. Machines AMIE Sec. B. Summer 1992)

Solution. If P_2 is the Cu loss at full-load, its value at 75% of full-load is = $P_2 \times (0.75)^2 = 9P_2/16$. At maximum efficiency, it equals the iron loss P_1 which remains constant throughout. Hence, at maximum efficiency,

$$P_1 = 9P_2/16 \quad \text{or} \quad P_1/P_2 = 9/16.$$

Example 30.42. *A 11000/230 V, 150-kVA, 1-phase, 50-Hz transformer has core loss of 1.4 kW and F.L. Cu loss of 1.6 kW. Determine*

(*i*) *the kVA load for max. efficiency and value of max. efficiency at unity p.f.*

(*ii*) *the efficiency at half F.L. 0.8 p.f. leading* **(Basic Elect. Machine, Nagpur Univ. 1993)**

Solution. (*i*) Load kVA corresponding to maximum efficiency is

$$= \text{F.L. kVA} \times \sqrt{\frac{\text{iron loss}}{\text{F.L. Cu loss}}} = 250 \times \sqrt{\frac{1.6}{1.4}} = 160 \text{ kVA}$$

Since Cu loss equals iron loss at maximum efficiency, total loss = 1.4 + 1.4 = 2.8 kW ; output = 160 × 1 = 160 kW

η_{max} = 160/162.8 = 0.982 or **98.2 %**

(*ii*) Cu loss at half full-load = 1.6 × $(1/2)^2$ = 0.4 kW ; Total loss = 1.4 + 0.4 = 1.8 kW

Half F.L. output at 0.8 p.f. = (150/2) × 0.8 = 60 kW

∴ efficiency = 60/(60 + 1.8) = 0.97 or **97%**

Example 30.43. *A 5-kVA, 2,300/230-V, 50-Hz transformer was tested for the iron losses with normal excitation and Cu losses at full-load and these were found to be 40 W and 112 W respectively. Calculate the efficiencies of the transformer at 0.8 power factor for the following kVA outputs :*

1.25 2.5 3.75 5.0 6.25 7.5

Plot efficiency vs. kVA output curve. **(Elect. Engg-I, Bombay Univ. 1987)**

Solution. F.L. Cu loss = 112 W ; Iron loss = 40 W

(*i*) Cu loss at 1.25 kVA $= 112 \times (1.25/5)^2 = 7$ W

Total loss $= 40 + 7 = 47$ W Output $= 1.25 \times 0.8 = 1$ kW $= 1{,}000$ W

$\eta = 100 \times 1{,}000/1{,}047 =$ **95.51%**

(*ii*) Cu loss at 2.5 kVA $= 112 \times (2.5/5)^2 = 28$ W

Total loss $= 40 + 28 = 68$ W

Output $= 2.5 \times 0.8 = 2$ kW

$\eta = 2{,}000 \times 100/2{,}068 =$ **96.71%**

(*iii*) Cu loss at 3.75 kVA $= 112 \times (3.75/5)^2 = 63$ W

Total loss $= 40 + 63 = 103$ W

$\eta = 3{,}000 \times 100/3{,}103 =$ **96.68 %**

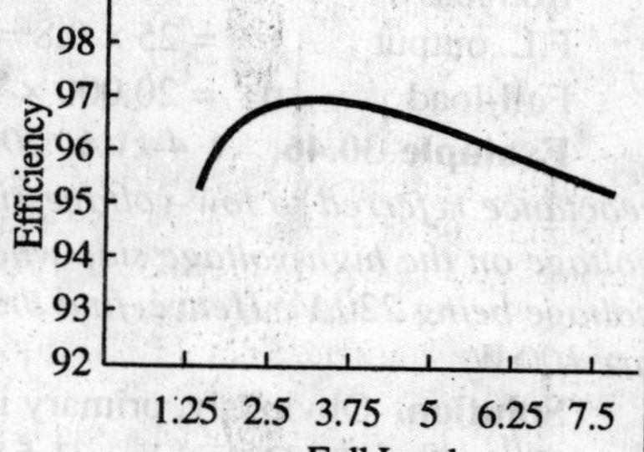

(*iv*) Cu loss at 5 kVA $= 112$ *W*

Total loss $= 152\text{W} = 0.152$ kW ; Output $= 5 \times 0.8 = 4$ kW

$\eta = 4 \times 100/4.152 =$ **96.34 %**

(*v*) Cu loss at 6.25 kVA $= 112 \times (6.25/5)^2 = 175$ W

Total loss $= 215$ kW $= 0.215$ kW ; Output $= 6.25 \times 0.8 = 5$ kW

$\eta = 5 \times 100/5.215 =$ **95.88%**

(*vi*) Cu loss at 7.5 kVA $= 112 \times (7.5/5)^2 = 252$W

Total loss $= 292\text{W} = 0.292$ kW ; Output $= 7.5 \times 0.8 = 6$ kW

$\eta = 6 \times 100/6.292 =$ **95.36 %**

Fig. 30.57

The curve is shown in Fig. 30.57.

Example 30.44. *A 200-kVA transformer has an efficiency of 98% at full load. If the max. efficiency occurs at three quarters of full-load, calculate the efficiency at half load. Assume negligible magnetizing current and p.f. of 0.8 at all loads.* **(Elect. Technology Punjab Univ. Jan. 1991)**

Solution. As given, the transformer has a F.L. efficiency of 98% at 0.8 p.f.

F.L. output $= 200 \times 0.8 = 160$ kW ; F.L. input $= 160/0.98 = 163.265$ kW

F.L. losses $= 163.265 - 160 = 3.265$ kW

This loss consists of F.L. Cu loss x and iron loss y.

$\therefore$ x + y = 3.265 kW ...(*i*)

It is also given that η_{max} occurs at three quarters of full-load when Cu loss becomes equal to iron loss.

$\therefore$ Cu loss at 75% of F.L. $= x\,(3/4)^2 = 9\,x/16$

Since y remains constant, hence $9\,x/16 = y$ ...(*ii*)

Substituting the value of *y* in Eqn. (*i*), we get $x + 9\,x/16 = 3265$ or $x = 2090$ W; $y = 1175$ W

Half-load Unity p.f.

Cu loss $= 2090 \times (1/2)^2 = 522$ W; total loss $= 522 + 1175 = 1697$ W

Output $= 100 \times 0.8 = 80$ kW ; $\eta = 80/81.697 = 0.979$ or **97.9%**

Example 30.45. *A 25-kVA, 1-phase transformer, 2,200 volts to 220 volts, has a primary resistance of 1.0 Ω and a secondary resistance of 0.01 Ω. Find the equivalent secondary resistance and the full-load efficiency at 0.8 p.f. if the iron loss of the transfomer is 80% of the full-load Cu loss.*

(Elect. Technology, Utkal Univ. 1988)

Solution. $K = 220/2,200 = 1/10$; $R_{02} = R_2 + K^2R_1 = 0.01 + 1/100 = \mathbf{0.02\ \Omega}$

Full-load $I_2 = 25,000/220 = 113.6$ A ; F.L. Cu loss $= I^2{}_2 R_{02} = 113.6^2 \times 0.02 = 258$ W.

Iron loss $= 80\%$ of $258 = 206.4$ W ; Total loss $= 258 + 206.4 = 464.4$ W

F.L. output $= 25 \times 0.8 = 20$ kW $= 20,000$ W

Full-load $\eta = 20,000 \times 100/(20,000 + 464.4) = \mathbf{97.7\ \%}$

Example 30.46. *A 4-kVA, 200/400-V, 1-phase transformer has equivalent resistance and reactance referred to low-voltage side equal to 0.5 Ω and 1.5 Ω respectively. Find the terminal voltage on the high-voltage side when it supplies 3/4th full-load at power factor of 0.8, the supply voltage being 220 V. Hence, find the output of the transformer and its efficiency if the core losses are 100 W.* **(Electrical Engineering ; Bombay Univ 1985)**

Solution. Obviously, primary is the low-voltage side and secondary, the high voltage side.

Here, $R_{01} = 0.5\ \Omega$ and $X_{01} = 1.5\ \Omega$. These can be transferred to the secondary side with the help of the transformation ratio.

$K = 400/200 = 2$; $R_{02} = K^2R_{01} = 2^2 \times 0.5 = 2\ \Omega : X_{02} = K^2X_{01} = 4 \times 1.5 = 6\ \Omega$

Secondary current when load is 3/4 the full-load is $= (1,000 \times 4 \times 3/4)/400 = 7.5$ A

Total drop as referred to transformer secondary is

$= I_2\,(R_{02} \cos\phi + X_{02} \sin\phi)^* = 7.5\,(2\times 0.8 + 6 \times 0.6) = 39$ V

$\therefore$ terminal voltage on high-voltage side under given load condition is

$= 400 - 39 = \mathbf{361\ V}$

Cu loss $= I_2^2R_{02} = 7.5^2 \times 2 = 112.5$ W Iron loss $= 100$ W

Total loss $= 212.5$ W output $= (4 \times 3/4) \times 0.8 = \mathbf{2.4\ kW}$

Input $= 2,400 + 212.5 = 2,612.5$ W $\eta = 2,400 \times 100/2,612.5 = \mathbf{91.87\ \%}$

Example 30.47. *A 20-kVA, 440/220 V, 1-φ, 50-Hz transformer has iron loss of 324 W. The cu loss is found to be 100 W when delivering half full-load current. Determine (i) efficiency when delivering full-load current at 0.8 lagging p.f. and (ii) the per cent of full-load when the efficiency will be maximum.* **(Electrotechnics-II, M.S. Univ., Baroda 1987)**

Solution. F.L. Cu loss $= 2^2 \times 100 = 400$ W; Iron loss $= 324$ W

(*i*) F.L. efficiency at 0.8 p.f. $= \dfrac{20 \times 0.8}{(20 \times 0.8) + 0.724} \times 100 = \mathbf{95.67\%}$

(*ii*) $\dfrac{\text{kVA for maximum } \eta}{\text{F.L. kVA}} = \sqrt{\dfrac{\text{iron loss}}{\text{F.L. Cu loss}}} = \sqrt{\dfrac{324}{440}} = 0.9$

Hemce. efficiency would be maximum at **90** % of F.L.

Example 30.48. *Consider a 4-kVA, 200/400 V single-phase transformer supplying full-load current at 0.8 lagging power factor. The O.C./S.C. test results are as follows : -*

O.C. test : 200 V, 0.8 A, 70 W (l.v. side)

S.C. test : 20 V, 10 A, 60 (h.v. side)

Calculate efficiency, secondary voltage and current into primary at the above load.

Calculate the load at unity power factor corresponding to maximum efficiency.

(Elect. Machines Nagpur Univ. 1993)

---

*Assuming a lagging power factor

Solution. Full-load, $I_2 = 4000/400 = 10$ A

It means that S.C. test has been carried out with full secondary current flowing. Hence, 60 W represents full-load Cu loss of the transformer.

Total F.L. losses = 60 + 70 = 130 W; F.L. output = 4 × 0.8 = 3.2 kW

F.L η = 3.2/3.33 = 0.96 or **96%**

S.C. Test

$Z_{02} = 20/10 = 2\ \Omega$; $I_2^2 R_{02} = 60$ or $R_{02} = 60/10^2 = 0.6\ \Omega$; $X_{02} = \sqrt{2^2 - 0.6^2} = 1.9\ \Omega$

Transformer voltage drop as referred to secondary

$= I_2 (R_{02} \cos\phi + X_{02} \sin\phi) = 10\,(0.6 \times 0.8 + 1.9 \times 0.6) = 16.2$ V

$\therefore\ V_2 = 400 - 16.2 = 383.8$ V

Primary current = 4000/200 = 20 A

kVA corresponding to $\eta_{max} = 4 \times \sqrt{70/60} = 4.32$ kVA

$\therefore$ load at u.p.f. corresponding to $\eta_{max} = 4.32 \times 1 =$ **4.32 kW**

Example 30.49. *A 600 kVA, 1-ph transformer has an efficiency of 92% both at full-load and half-load at unity power factor. Determine its efficiency at 60% of full-load at 0.8 power factor lag.*

(Elect. Machines, A.M.I.E. Sec. B, 1992)

Solution.

$$\frac{x \times \text{kVA} \times \cos\phi}{(x \times \text{kVA}) \times \cos\phi + W_i + x^2 W_{cu}} \times 100$$

where x represents percentage of full-load

W_i is iron loss and W_{cu} is full-load Cu loss.

At. F.L. u.p.f Here $x = 1$

$$\therefore \quad 92 = \frac{1 \times 600 \times 1}{1 \times 600 \times 1 + W_i + I^2 W_{cu}} \times 100,\ W_i + W_{cu} = 52.174\ \text{kW} \quad \ldots(i)$$

At half F.L. UPF. Here $x = 1/2$

$$92 = \frac{1/2 \times 600 \times 1}{(1/2) \times 600 \times 1 + W_i + (1/2)^2 W_{cu}} \times 100;\ \therefore\ W_i + 0.25\ W_{cu} = 26.087\ \text{kW} \quad \ldots(ii)$$

From (*i*) and (*ii*), we get, W_i = 17.39 kW, W_{cu} = 34,78 kW

60% F.L. 0.8 p.f. (lag) Here , $x = 0.6$

$$\eta = \frac{0.6 \times 600 \times 0.8 \times 100}{(0.6 \times 600 \times 0.8) + 17.39 + (0.6)^2\, 34.78} = \mathbf{85.9\ \%}$$

Example 30.50. *A 600-kVA, I-ph transformer when working at u.p.f. has an efficiency of 92% at full-load and also at half-load. Determine its efficiency when it operates at unity p.f. and 60% of full-load.*

(Electric. Machines, Kerala Univ. 1987)

Solution. The fact that efficiency is the same *i.e.* 92% at both full-load and half-load will help us to find the iron and copper losses.

At full-load

Output = 600 kW ; input = 600/0.92 = 652.2 kW ; Total loss = 652.2 − 600 = 52.2 kW

Let x = iron loss —it remains constant at all loads

y = F.L. Cu loss —it is $\propto$ (kVA)^2 $\therefore$ $x + y = 52.2$ ...(*i*)

At half-load

Output = 300 kW ; input = 300/0.92 $\therefore$ losses = (300/0.92−300) = 26.1 kW

Since Cu loss becomes one-fourth of its F.L. value, hence

$x + y/4 = 26.1$ ...(*ii*)

Solving for x and y, we get x = **17.4** kW ; y = 34.8 kW

At 60% full-load

Cu loss = $0.6^2 \times 34.8 = 12.53$ kW ; Total loss = 17.4 + 12.53 = 29.93 kW

Output = $600 \times 0.6 = 360$ kW $\therefore\ \eta = 360/389.93 = 0.965$ or **96.5%**

Example 30.51. *The maximum efficiency of a 100-kVA, single phase transformer in 98% and occurs at 80% of full load at 0.8 p.f. If the leakage impedance of the transformer is 5%. find the voltage regulation at rated load of 0.8 power factor lagging.*

(Elect. Machines-I, Nagpur Univ. 1993)

Solution. Since maximum efficiency occurs at 80 per cent of full-load at 0.8 p.f.,

Output at $\eta_{max} = (100 \times 0.8) \times 0.8 = 64$ kW ; input = 64/0.98 = 65.3 kW

$\therefore$ total loss = 65.3−64 = 1.3 kW. This loss is divided equally between Cu and iron .

$\therefore$ Cu loss at 80% of full-load = 1.3/2 = 0.65 kW

Cu loss at full-load = $0.65/0.8^2 = 1$ kW

$$\%\ R = \frac{\text{Cu loss}}{V_2 I_2} \times 100 = 1 \times \frac{100}{100} = 1\% = v_r ; v_x = 5\%$$

$$\therefore\quad \%\text{ age regn.} = (1 \times 0.8 + 5 \times 0.6) + \frac{1}{200}(5 \times 0.8 - 1 \times 0.6)^2 = \mathbf{0.166\ \%}$$

Example 30.52. *A 10 kVA, 5000/440-V, 25-Hz single-phase transformer has copper, eddy current and hysteresis losses of 1.5, 0.5 and 0.6 per cent of output on full load. What will be the percentage losses if the transformer is used on a 10-kV, 50-Hz system keeping the full-load current constant ? Assume unity power factor operation. Compare the full load efficiencies for the two cases.*

(Elect. Machines, A.M.I.E, Sec B, 1991)

Solution. We know that $E_1 = 4{,}44\, fN_1B_1A$. When both excitation voltage and frequency are doubled, flux remains unchanged

F.L. output at upf = 10 kVA × 1 = 10 kW

F.L. Cu loss = 1.5 × 10/100 = 0.15 kW ; eddy current loss

= 0.5 × 10/100 = 0.05 kW ; hysteresis loss = 0.6 × 10/100 = 0.06 kW

Now, full-load current is kept constant but voltage is increased from 5000 V to 10,000 V. Hence, output will be doubled to 20 kW. Due to constant current, Cu loss would also remain constant.

New Cu loss = 0.15 kW, % Cu loss = (0.15/20) × 100 = 0.75 %

Now, eddy current loss $\propto f^2$ and hysteresis loss $\propto f$.

New eddy current loss = $0.05\ (50/25)^2 = 0.2$ kW, % eddy current loss = (0.2/20) × 100 = 1%

Now, $W_h = 0.06 \times (50/25) = 0.12$ kW, % $W_h = (0.12/20) \times 100 = 0.6\%$

$$\eta_1 = \frac{10}{10 + 0.15 + 0.05 + 0.06} \times 100 = \mathbf{87.4\ \%}$$

$$\eta_2 = \frac{20}{20 + 0.15 + 0.2 + 0.12} \times 100 = \mathbf{97.7\%}$$

Example 30.53. *A 300-kVA, single-phase transformer is designed to have a resistance of 1.5% and maximum efficiency occurs at a load of 173.2 kVA. Find its efficiency when supplying full-load at 0.8 p.f. lagging at normal voltage and frequency.* **(Electrical Machines-I, Gujarat Univ. 1985)**

Solution. $$\%R = \frac{\text{F.L. Cu loss}}{\text{Full-load } V_2 I_2} \times 100 ;\ 1.5 = \frac{\text{F.L. Cu loss}}{300 \times 1000} \times 100$$

$$\therefore\quad \text{F.L. Cu loss} = 1.5 \times 300 \times 1000/100 = 4500 \text{ W}$$

Also, $$173.2 = 300\sqrt{\frac{\text{iron loss}}{4500}} ;\ \text{iron loss} = 1500 \text{ W}$$

Total F.L. loss $= 4500 + 1500 = 6$ kW,

F.L. η at 0.8 p.f. $= \dfrac{300 \times 0.8}{(300 \times 0.8) + 6} \times 100 = \mathbf{97.6\ \%}$

Example 30.54. *A single phase transformer is rated at 100-kVA, 2300/230-V, 50-Hz. The maximum flux density in the core is 1.2 Wb/m² and the net cross-sectional area of the core is 0.04 m². Determine*

(a) the number of primary and secondary turns needed.

(b) if the mean length of the magnetic circuit is 2.5 m and the relative permeability is 1200, determine the magnetising current. Neglect the current drawn for the core loss.

(c) on short-circuit with full-load current flowing, the powr input is 1200 W and on open-circuit with rated voltage, the power input was 400 W. Determine the efficiency of the transformer at 75% of full-load with 0.8 p.f. lag.

(d) if the same transformer is connected to a supply of similar voltage but double the frequency (i.e. 100 Hz) what is the effect on its efficiency ? **(Elect. Engg., Bombay Univ. 1988)**

Solution. (*a*) Applying e.m.f. equation of the transformer to the primary, we have

$2300 = 4.44 \times 50 \times N_1 \times (1.2 \times 0.04) \quad \therefore \quad N_1 = \mathbf{216}$

$K = 230/2300 = 1/10 \quad N_2 = KN_1 = 216/10 = 21.6$ or **22**

(*b*) $\text{AT} = H \times l = \dfrac{B}{\mu_0 \mu_r} \times l = \dfrac{1.2 \times 2.5}{4\pi \times 10^{-7} \times 1200} = 1989 \therefore I = \dfrac{1989}{216} = \mathbf{9.21\ A}$

(*c*) F.L. Cu loss $= 1200$ W – S.C. test ; Iron loss $= 400$ W – O.C. test

Cu loss at 75% of F.L. $= (0.75)^2 \times 1200 = 675$ W

Total loss $= 400 + 675 = 1.075$ kW

Output $= 100 \times (3/4) \times 0.8 = 60$ kW ; $\eta = (60/61.075) \times 100 = \mathbf{98.26\%}$

(*d*) when frequency is doubled, iron loss is increased because

(*i*) hysteresis loss is doubled – $W_h \propto f$

(*ii*) eddy current loss is quadrupled – $W_e \propto f^2$

Hence, efficiency will be decreased.

30.31. Variation of Efficiency with Power Factor

The efficiency of a transformer is given by

$$\eta = \frac{\text{output}}{\text{input}} = \frac{\text{imput - losses}}{\text{input}}$$

$$= 1 - \frac{\text{losses}}{\text{input}} = 1 - \frac{\text{losses}}{(V_2 I_2 \cos\phi + \text{losses})}$$

Let, losses/$V_2 I_2 = x$

$$\therefore \quad \eta = 1 - \frac{\text{losses}/V_2 I_2}{\cos\phi + (\text{losses}/V_2 I_2)}$$

$$= 1 - \frac{x}{(\cos\phi + x)} = 1 - \frac{x/\cos\phi}{1 + (x/\cos\phi)}$$

The variations of efficiency with power factor at different loadings on a typical transformer are shown in Fig. 30.58.

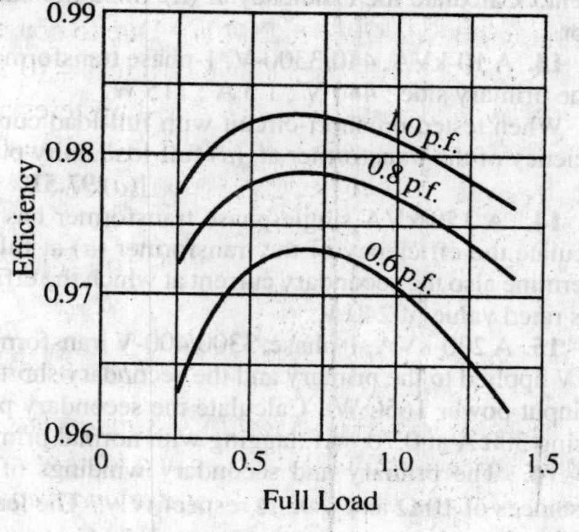

Fig. 30.58

Tutorial Problem No. 30.4

1. A 200-kVA transformer has an efficiency of 98% at full-load. If the maximum efficiency occurs at three-quarters of full-load, calculate (*a*) iron loss (*b*) Cu loss at F.L. (*c*) efficiency at half-load. Ignore magnetising current and assume a p.f. of 0.8 at all loads. [(*a*) **1.777 kW** (*b*) **2.09 kW** (*c*) **97.92** %]

2. A 600 kVA, 1-ph transformer has an efficiency of 92% both at full-load and half-load at unity power factor. Determine its efficiency at 60% of full load at 0.8 power factor lag.

[**85.9%**] (*Elect. Machines, A.M.I.E. Sec B, 1992*)

3. Find the efficiency of a 150 kVA transformer at 25% full load at 0.8 p.f. lag if the copper loss at full load is 1600 W and the iron loss is 1400 W. Ignore the effects of temperature rise and magnetising current.

[**96.15 %**] (*Elect. Machines, A.M.I.E. Sec B, 1991*)

4. The F.l. Cu loss and iron loss of a transformer are 920 W and 430 W respectively. (*i*) Calculate the loading of the transformer at which efficiency is maximum (*ii*) what would be the losses for giving maximum efficiency at 0.85 of full-load if total full-load losses are to remain unchanged ?

[(*a*) **68.4% of F.L** (*ii*)**W_i = 565 W; W_{cu} = 785 W**]

5. At full-load, the Cu and iron losses in a 100-kVA transformer are each equal to 2.5 kW. Find the efficiency at a load of 65 kVA, power factor 0.8 [**93.58 %**] (**City & Guilds London**)

6. A transformer, when tested on full-load, is found to have Cu loss 1.7% and resistance 3.8%. Calculate its full-load regulation (*i*) at unity p.f. (*ii*) 0.8 p.f. lagging (*iii*) 0.8 p.f. leading.

[(*i*) **1.77 %** (*ii*) **3.7 %** (*iii*) **−0.86%**]

7. With the help of a vector diagram, explain the significance of the following quantities in the open-circuit and short-circuit tests of a transformer (*a*) power consumed (*b*) input voltage (*c*) input current.

When a 100-kVA single-phase transformer was tested in this way, the following data were obtained : On open circuit, the power consumed was 1300 W and on short-circuit the power consumed was 1200 W. Calculate the efficiency of the transformer on (*a*) full-load (*b*) half-load when working at unity power factor.

[(*a*) **97.6%** (*b*) **96.9%**] (*London Univ.*)

8. An 11,000/230-V, 150-kVA, 50-Hz. 1-phase transformer has a core loss of 1.4 kW and full-load Cu loss of 1.6 kW. Determine (*a*) the kVA load for maximum efficiency and the maximum efficiency (*b*) the efficiency at half full-load at 0.8 power factor lagging. [**140.33 kVA, 97.6%** ; **97%**]

9. A single-phase transformer, working at unity power factor has an efficiency of 90% at both half-load and at full-load of 500 kW. Determine the efficiency at 75% of full-load. [**90.5%**] (*I.E.E. London*)

10. A 10-kVA, 500/250-V, single-phase transformer has its maximum efficiency of 94% when delivering 90% of its rated output at unity power factor. Estimate its efficiency when delivering its full-load output at p.f. of 0.8 lagging. [**92.6%**] (*Elect. Machinery, Mysore Univ. 1979*)

11. A single-phase transformer has a voltage ratio on open-circuit of 3300/660-V. The primary and secondary resistances are 0.8 Ω and 0.03 Ω respectively, the corresponding leakage reactances being 4 Ω and 0.12 Ω. The load is equivalent to a coil of resistance 4.8 Ω and inductive reactance 3.6 Ω. Determine the terminal voltage of the transformer and the output in kW. [**636 V, 54 kW**]

12. A 100-kVA, single-phase transformer has an iron loss of 600 W and a copper loss of 1.5 kW at full-load current. Calculate the efficiency at (*a*) 100 kVA output at 0.8 p.f. lagging (*b*) 50 kVA output at unity power factor [(*a*) **97.44%** (*b*) **98.09%**]

13. A 10-kVA, 440/3300-V, 1-phase transformer, when tested on open circuit, gave the following figures on the primary side : 440 V ; 1.3 A ; 115 W.

When tested on short-circuit with full-load current flowing, the power input was 140 W. Calculate the efficiency of the transformer at (*a*) full-load unity p.f. (*b*) one quarter full-load 0.8 p.f.

[(*a*) **97.51%** (*b*) **94.18%**] (*Elect. Engg-I, Sd. Patel Univ. June 1977*)

14. A 150-kVA single-phase transformer has a core loss of 1.5 kW and a full-load Cu loss of 2 kW. Calculate the efficiency of the transformer (*a*) at full-load, 0.8 p.f. lagging (*b*) at one-half full-load unity p.f. Determine also the secondary current at which the efficiency is maximum if the secondary voltage is maintained at its rated value of 240 V. [(*a*) **97.17%** (*b*) **97.4%** ; **541 A**]

15. A 200-kVA, 1-phase, 3300/400-V transformer gave the following results in the short-circuit test. With 200 V applied to the primary and the secondary short-circuited, the primary current was the full-load value and the input power 1650 W. Calculate the secondary p.d. and percentage regulation when the secondary load is passing 300 A at 0.707 p.f. lagging with normal primary voltage. [**380 V** ; **4.8%**]

16. The primary and secondary windings of a 40-kVA, 6600/250-V, single-phase transformer have resistances of 10 Ω and 0.02 Ω respectively. The leakage reactance of the transformer referred to the primary is 35 Ω. Calculate

(*a*) the primary voltage required to circulate full-load current when the secondary is short-circuited.
(*b*) the full-load regulations at (*i*) unity (*ii*) 0.8 lagging p.f. Neglect the no-load current.

[(*a*) **256 V** (*b*) (*i*) **2.2%** (*ii*) **3.7%**] (*Elect. Technology, Kerala Univ. 1979*)

17. Calculate :

(*a*) F.L. efficiency at unity p.f.
(*b*) The secondary terminal voltage when suplying full-load secondary current at p.f. (*i*) 0.8 lag (*ii*) 0.8 lead for the 4-kVA, 200/400 V, 50 Hz, 1-phase transformer of which the following are the test figures :

Open circuit with 200 V supplied to the primary winding–power 60 W. Short-circuit with 16 V applied to the h.v. winding—current 8 A, power 40 W. **(0.97 ; 383 V ; 406 V]**

18. A 100-kVA, 6600/250-V, 50-Hz transformer gave the following results :
O.C. test : 900 W, normal voltage.
S.C. test (data on h.v. side) : 12 A, 290 V, 860 W
Calculate–
(*a*) the efficiency and percentage regulation at full-load at 0.8 p.f. lagging.
(*b*) the load at which maximum efficiency occurs and the value of this efficiency at p.f. of unity, 0.8 lag and 0.8 lead. [(*a*) **97.3%, 4.32%** (*b*) **81 kVA, 97.8 %, 97.3% ; 97.3%**]

19. The primary resistance of a 440/110-V transformer is 0.5 Ω and the secondary resistance is 0.04 Ω. When 440 V is applied to the primary and secondary is left open-circuited, 200 W is drawn from the supply. Find the secondary current which will give maximum efficiency and calculate this efficiency for a load having unity power factor. **[53 A; 93.58%]** (*Basic Electricity & Electronics. Bombay Univ. 1981*)

20. Two tests were preformed on a 40-kVA transformer to predetermine its efficiency. The results were :
Open circut : 250 V at 500 W
Short circuit : 40 V at F.L. current, 750 W } both tests from primary side.
Calculate the efficiency at rated kVA and 1/2 rated kVA at (*i*) unity p.f. (*ii*) 0.8 p.f.
[96.97% ; 96.68% ; 96.24% ; 95.87%]

21. The following figures were obtained from tests on a 30-kVA, 3000/110-V transformer :
O.C. test : 3000 V 0.5 A 350 W ; S.C. test ; 150 V 10 A 500 W
Calculate the efficiency of the transformer at
(*a*) full-load, 0.8 p.f. (*b*) half-load, unity p.f.
Also, calculate the kVA output at which the efficiency is maximum. **[96.56 % ; 97% ; 25.1 kVA]**

30.32. All-day Efficiency

The ordinary or commercial efficiency of a transformer is given by the ratio

$$\frac{\text{output in watts}}{\text{input in watts}}$$

But there are certain types of transformers whose performance cannot be judged by this efficiency. Transformers used for supplying lighting and general network *i.e.* distribution transformers have their primaries energised all the twenty-four hours, although their secondaries supply little or no-load much of the time during the day except during the house lighting period. It means that whereas core loss occurs throughout the day, the Cu loss occurs only when the transformer is loaded. Hence, it is considered a good practice to design such transformers so that core losses are very low. The Cu losses are relatively less important, because they depend on the load. The performance of such a transformer shoud be judged by all-day efficiency (also known as 'operational efficiency') which is computed on the basis of *energy* consumed during a certain time period, usually a day of 24 hours.

$$\therefore \quad \eta_{\text{all - day}} = \frac{\text{output in kWh}}{\text{input in kWh}} \text{ (for 24 hours)}$$

This efficiency is always less than the commercial efficiency of a transformer.

To find this all-day efficiency or (as it is also called) energy efficiency, we have to know the load cycle on the transformer *i.e.* how much and how long the transformer is loaded during 24 hours. Practical calculations are facilitated by making use of a *load factor*.

Example 30.55. *Find the all-day efficiency of 500-kVA distribution transformer whose copper loss and iron loss at full load are 4.5 kW and 3.5 kW respectively. During a day of 24 hours, it is loaded as under :*

No. of hours	*Loading in kW*	*Power factor*
6	*400*	*0.8*
10	*300*	*0.75*
4	*100*	*0.8*
4	*0*	—

(Elect. Machines, Nagpur Univ. 1993)

Solution. It should be noted that a load of 400 kW at 0.8 pf is equal to 400/0.8 = 500 kVA. Similarly, 300 kW at 0.75 pf means 300/0.75 = 400 kVA and 100 kW at 0.8 pf means 100/0.8 = 125 kVA *i.e.* one-fourth of the full-load.

Cu loss at F.L. of 500 kVA = 4.5 kW

Cu loss at 400 kVA = $4.5 \times (400/500)^2 = 2.88$ kW

Cu loss at 125 kVA = $4.5 \times (125/500)^2 = 0.281$ kW

Total Cu loss in 24 hr. = $(6\times4.5) + (10 \times 2.88) + (4 \times 0.281) + (4\times0) = 26.9 = 56.924$ kWh

The iron loss takes place throughout the day irrespective of the load on the transformer becuae its primary is energized all the 24 hours.

∴ iron loss in 24 hours = $24 \times 3.5 = 84$ kWh

Total transformer loss = 56.924 + 84 = 140.924 kWh

Transformer output in 24 hr = $(6\times400) + (10\times300) + (4\times100) = 5800$ kWh

$$\therefore \quad \eta_{all\text{-}day} = \frac{\text{output}}{\text{output + losses}} = \frac{5800}{5800 + 140.924} = 0.976 \text{ or } \mathbf{97.6\ \%}$$

Example 30.56. *A 100-kVA lighting transformer has a full-load loss of 3 kW, the losses being equally divided between iron and copper. During a day, the transformer operates on full-load for 3 hours, one half-load for 4 hours, the output being negligible for the remainder of the day. Calculate the all-day efficiency.* **(Elect. Engg. Punjab Univ. 1990)**

Solution. It should be noted that lighting transformers are taken to have a load p.f. of unity.

Iron loss for 24 hours = $1.5 \times 24 = 36$ kWh ; F.L. Cu loss = 1.5 kW

∴ Cu loss for 3 hours on F.L = $1.5 \times 3 = 4.5$ kWh

Cu loss at half full-load = 1.5/4 kW

Cu loss for 4 hours at half the load = $(1.5/4) \times 4 = 1.5$ kWh

Total losses = 36 + 4.5 + 1.5 = 42 kWh

Total output = $(100 \times 3) + (50 \times 4)$ = 500 kWh

∴ $\eta_{all\text{-}day} = 500 \times 100/542 = \mathbf{92.26\%}$

Incidentally, ordinary or commercial efficiency of the transformer is

= 100/(100 + 3) = 0.971 or **97.1%**

Example 30.57. *Two 100-kW transformers each has a maximum efficiency of 98% but in one the maximum efficiency occurs at full-load while in the other, it occurs at half-load. Each transformer is on full-load for 4 hours, on half-load for 6 hours and on one-tenth load for 14 hours per day. Determine the all-day efficiency of each transformer.* **(Elect. Machines-I, Vikram Univ. 1988)**

Solution. Let x be the iron loss and y the full-load Cu loss. If the ordinary efficiency is a maximum at $1/m$ of full-load, then $x = y/m^2$

Now, output = 100 kW ; input = 100/0.98

∴ total losses = 100/0.98−100 = 2.04 kW

∴ $y + y/m^2 = 2.04$

Ist Transformer

Here $m = 1$; $y + y = 2.04$; $y = 1.02$ kW and $x = 1.02$ kW

Iron loss for 24 hours = $1.02 \times 24 = 24.48$ kWh

Cu loss for 24 hours = $4 \times 1.02 + 6 \times (1.02/4) + 14\,(1.02/10^2) = 5.73$ kWh

Total loss = 24.48 + 5.73 = 30.21 kWh

Output for 24 hours = $4 \times 100 + 6 \times 50 + 14 \times 10 = 840$ kWh

∴ $\eta_{all\text{-}day}$ = 840/870.21 = 0.965 or **96.5%**

2nd Transformer

Here $1/m = 1/2$ or $m = 2$ ∴ $y + y/4 = 2.04$

or $y = 1.63$ kW ; $x = 0.41$ kW

Output = 840 kWh −as above

Iron loss for 24 hours = $0.41 \times 24 = 9.84$ kWh

Cu loss for 24 hours = $4 \times 1.63 + 6(1.63/4) + 14(1.63/10^2) = 9.19$ kWh

Total loss = 9.84 + 9.19 = 19.03 kWh

∴ $\eta_{all\text{-}day}$ = 840/859.03 = 0.978 or **97.8%**

Example 30.58. *A 5-kVA distribution transformer has a full–load efficiency at unity p.f. of 95%, the copper and iron losses then being equal. Calculate its all-day efficiency if it is loaded throughout the 24 hours as follows :*

No. load for 10 hours Quarter load for 7 hours
Half load for 5 hours Full load for 2 hours
Assume load p.f. of unity. **(Power Apparatus-I, Delhi Univ. 1987)**

Solution. Let us first find out the losses from the given commercial efficiency of the transformer.

Output = 5×1 = 5 kW; input = 5/0.95 = 5.264 kW

Losses = (5.264–5.000) = 0.264 kW = 264 W

Since efficiency is maximum, the losses are divided equally between Cu and iron.

$\therefore$ Cu loss at F.L. of 5 kVA = 264/2 = 132 W ; iron loss = 132 W

Cu loss at one-fourth F.L = $(1/4)^2 \times 132 = 8.2$ W

Cu loss at one-half F.L. = $(1/2)^2 \times 132 = 33$ W

Quarter load Cu loss for 7 hours = $7 \times 8.2 = 57.4$ Wh

Half-load Cu loss for 5 hours = $5 \times 33 = 165$ Wh

F.L. Cu loss for 2 hours = $2 \times 132 = 264$ Wh

Total Cu loss during one day = 57.4 + 165 + 264 = 486.4 Wh = 0.486 kWh

Iron loss in 24 hours = 24×132 = 3168 Wh = 3.168 kWh

Total losses in 24 hours = 3.168 + 0.486 = 3.654 kWh

Since load p.f. is to be assumed as unity.

F.L. output = 5×1 = 5 kW ; Half F.L. output = $(5/2) \times 1$ = 2.5 kW

Quarter load output =$(5/4) \times 1$ = 1.25 kW

Transformer output in a day of 24 hours

$$= (7 \times 1.25) + (5 \times 2.5) + (2 \times 5) = 31.25 \text{ kWh}$$

$$\eta_{\text{all-day}} = \frac{31.25}{(31.25 + 3.654)} \times 100 = \mathbf{89.53\%}$$

Example 30.59. *Find "all day" efficiency of a transformer having maximum efficiency of 98% at 15 kVA at unity power factor and loaded as follows :*

12 hours–2 kW at 0.5 p.f. lag
6 hours–12 kW at 0.8 p.f. lag
6 hours– at no load. **(Elect. Machines-I, Nagpur Univ. 1993)**

Solution. Output = 15×1 = 15 kW, input = 15/0.98

Losses = (15/0.98–15) = 0.306 kW = 306 W

Since efficiency is maximum, the losses are divided equally between Cu and iron.

$\therefore$ Cu loss at 15 kVA = 306/2 = 153 W, Iron loss = 153 W

2 kW at 0.5 p.f. = 2/0.5 = 4 kVA, 12 kW at 0.8 p.f. = 12/0.8 = 15 kVA

Cu loss at 4 kVA = $153\ (4/15)^2 = 10.9$ W ; Cu loss at 15 kVA = 153 W.

Cu loss in 12 hr = 12×10.9 = 131 Wh ; Cu loss in 6 hr = 6×153 = 918 Wh.

Total Cu loss for 24 hr = 131 + 918 = 1050 Wh = 1.05 kWh

Iron loss for 24 hr = 24×153 = 3,672 Wh = 3.672 kWh

Output in 24 hr = $(2 \times 12) + (6 \times 12)$ = 96 kWh

Input in 24 hr = 96 + 1.05 + 3.672 = 100.72 kWh

$\therefore$ $\eta_{call\text{-}day}$ = $96 \times 100/100.72$ = **95.3 %**

Example 30.60. *A 150-kVA transformer is loaded as follows :*

Load increases from zero to 100 kVA in 3 hours from 7 a.m. to 10.00 a.m, stays at 100 kVA from 10 a.m. to 6 p.m. and then the transformer is disconnected till next day. Assuming the load to be resistive and core-loss equal to full-load copper loss of 1 kW, determine the all-day efficiency and the ordinary efficiency of the transformer. **(Electricial Machines-II, Indore Univ. 1990)**

Solution. Since load is resistive, its p.f. is unity.

Average load from 7 a.m. to 10 a.m. = (0 + 100)/2 = 50 kVA *i.e.* one-third F.L.

Load from 10 a.m. to 6 p.m. = 100 kVA *i.e.* 2/3 of F.L.

Ordinary Efficiency

In this case, load variations are not relevant.

Output = 150 × 1 = 150 kW ; Iron loss = Cu loss = 1 kW ; Total loss = 2 kW .

∴ ordinary η = 150/(150 + 2) = 0.9868 or **98.68%**

All-day Efficiency

Cu loss from 7–10 a.m.	$= 3 \times (1/3)^2 \times 1 = 0.333$ kWh
Cu loss from 10 a.m. to 6 p.m.	$= 8 \times (2/3)^2 \times 1 = 3.555$ kWh
Total Cu loss for 24 hrs	= 0.333 + 3.555 = 3.888 kWh
Total iron loss for 24 hrs	= 24 × 1 = 24 kWh
Losses for a day of 24 hours	= 27.888 kWh
Output for 24 hrs	= 3×(50 × 1) + 8(100 × 1) = 950 kWh

$$\eta_{all\text{-}day} = \frac{950 \times 100}{(950 + 27.888)} = \mathbf{97.15\ \%}$$

Tutorial Problem No. 30.5

1. A 100-kVA distribution transformer has a maximum efficiency of 98% at 50% full-load and unity power factor. Determine its iron losses and full-load copper losses.

The transformer undergoes a daily load cycle as follows :

Load	*Power factor*	*Load duration*
100 kVA	1.0	8 hr
50 kVA	0.8	6 hr
No load		10 hr

Determine its all-day efficiency. (*Electrical Engineering, MS Univ. Baroda 1979*)

2. What is meant by energy efficiency of a transformer ?

A 20-kVA transformer has a maximum efficiency of 98 per cent when delivering three-fourth full-load at u.p.f. If during the day, the transformer is loaded as follows :

12 hours	No. load
6 hours	12 kW, 0.8 p.f.
6 hours	20 kW, u.p.f.

calculate the energy efficiency of the transformer.

(*Electrical Technology-III, Gwalior Univ. 1980*)

30.33. Auto-transformer

It is a transformer with one winding only, part of this being common to both primary and secondary. Obviously, in this transformer the primary and secondary are not electrically *isolated* from each other as is the case with a 2-winding transformer. But its theory and operation are similar to those of a two-winding transformer. Because of one winding, it uses less copper and hence is cheaper. It is used where transformation ratio differs little from unity. Fig. 30.59 shows both step-down and step-up auto-transformers.

As shown in Fig. 30.59 (*a*), *AB* is primary winding having N_1 turns and *BC* is secondary winding having N_2 turns. Neglecting iron losses and no-load current,

$$\frac{V_2}{V_1} = \frac{N_2}{N_1} = \frac{I_1}{I_2} = K$$

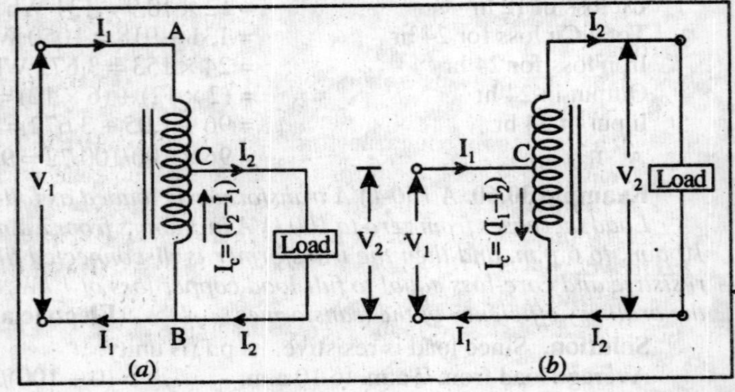

Fig. 30.59

The current in section *CB* is vector difference* of I_2 and I_1. But as the two currents are practically in phase opposition, the resultant current is $(I_2 - I_1)$ where I_2 is greater than I_1.

As compared to an ordinary 2-winding transformer of same output, an auto-transfomer has higher efficiency but smaller size. Moreover, its voltage regulation is also superior.

Saving of Cu

Volume and hence weight of Cu, is proportional to the length and area of cross-section of the conductors. Now, length of conductor is proportional to the number of turns and cross-section depends on current. Hence, weight is proportional to the product of the current and number of turns.

With reference to Fig. 30.59.

Wt. of Cu in section *AC* is $\propto (N_1 - N_2)I_1$; Wt. of Cu in section *BC* is $\propto N_2(I_2 - I_1)$

$\therefore$ total Wt. of Cu in auto-transformer $\propto (N_1 - N_2)\, I_1 + N_2(I_2 - I_1)$

If a two-winding transformer were to perform the same duty, then

Wt. of Cu on its primary $\propto N_1 I_1$; Wt. of Cu on secondary $\propto N_2\, I_2$

Total Wt. of Cu $\propto N_1 I_1 + N_2\, I_2$

$$\therefore \quad \frac{\text{Wt. of Cu in auto-transformer}}{\text{Wt of Cu in ordinary transformer}} = \frac{(N_1 - N_2)I_1 + N_2(I_2 - I_1)}{N_1 I_1 + N_2\, I_2}$$

$$= I - \frac{2\frac{N_2}{N_1}}{1 + \frac{N_2}{N_1} \times \frac{I_2}{I_1}} = 1 - \frac{2K}{2} = 1 - K \quad \left(\because \frac{N_2}{N_1} = K \, ; \frac{I_2}{I_1} = \frac{1}{K} \right)$$

Wt. of Cu in auto-transformer $(W_a) = (1 - K) \times$ Wt. of Cu in ordinary transformer W_0)

$\therefore$ saving $= W_0 - W_a =$

$W_0 - (1 - K)\, W_0 = KW_0$

$\therefore$ saving $= K \times$ (Wt. of Cu in ordinary transformer)

Hence, saving will increase as K approaches unity.

It can be proved that power transformed inductively is = input $(1 - K)$

The rest of the power = $(K \times \text{input})$ is *conducted* directly from the source to the load *i.e.* it is transferred **conductively** to the load

Uses

As said earlier, auto-transformers are used when *K* is nearly equal to unity and where there is no objection to electrical connection between primary and secondary. Hence, such transformers are used :

1. to give small boost to a distribution cable to correct for the voltage drop.
2. as auto-starter transformers to give upto 50 to 60% of full voltage to an induction motor during starting.
3. as furnace transformers for getting a convenient supply to suit the furance winding from a 230-V supply
4. as interconnecting transformers in 132 kV/330 kV system.
5. in control equipment for 1-phase and 3-phase electrical locomotives.

Example 30.61. *An auto-transformer supplies a load of 3 kW at 115 volts at a unity power factor. If the applied primary voltage is 230 volts, calculate the power transferred to the load (a) inductively and (b) conductively.* **(Basic Elect. Machines, Nagpur Univ. 1991)**

Solution. As seen from Art. 30.33

Power transferred inductively = input $(1 - K)$

*In fact, current flowing in the common winding of the auto-transformer is always equal to the difference between the primary and secondary currents of an ordinary transformer.

Power transferred conductively = input × K

Now, K = 115/230 = 1/2, input ≅ output = 3 kW

∴ inductively transferred power = 3(1−1/2) = **1.5 kW**

conductively transferred power = (1/2) × 3 = **1.5 kW**

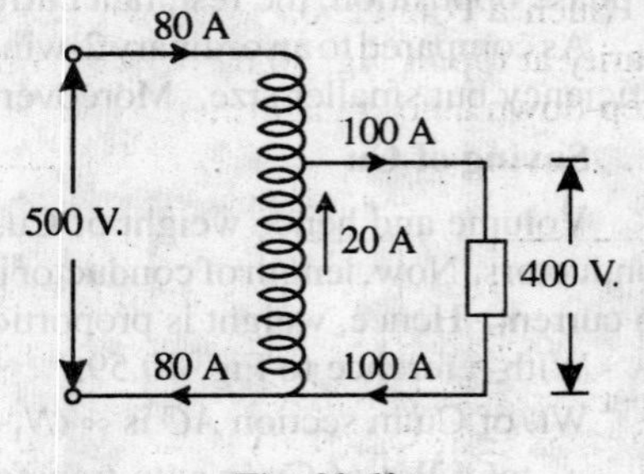

Fig. 30.60

Example 30.62. *The primary and secondary voltages of an auto-transformer are 500 V and 400 V respectively. Show with the aid of a diagram, the current distribution in the winding when the secondary current is 100 A and calculate the economy of Cu in this particular case.*

Solution. The circuit is shown in Fig. 30.60

$K = V_2/V_1 = 400/500 = 0.8$

∴ $I_1 = KI_2 = 0.8 \times 100 = 80$ A

The current distribution is shown in Fig. 30.60.

Saving = KW_0 = 0.8 W_0 −Art. 30.33

∴ percentage saivng = 0.8 × 100 = **80.**

Example 30.63. *Determine the core area, the number of turns and the position of the tapping point for a 500-kVA, 50-Hz, single-phase, 6,600/5,000-V auto-transformer, assuming the following approximate values : e.m.f. per turn 8 V. Maximum flux density 1.3 Wb/m².*

Solution. $E = 4.44 f \Phi_m N$ volt

$$\Phi_m = \frac{E/N}{4.44f} = \frac{8}{4.44 \times 50} = 0.03604 \text{ Wb}$$

Core area = 0.03604/1.3 = 0.0277 m² = **277 cm²**

Turns of h.v. side = 6600/8 = **825** ; Turns of l.v. side = 5000/8 = **625**

Hence, tapping should be 200 turns from high voltage end or 625 turns from the common end.

30.34. Conversion of 2-Winding Transformer into Auto-transformer

Any two-winding transformer can be converted into an auto-transformer either step-down or step-up. Fig. 30.61 (*a*) shows such a transformer with its polarity markings. Suppose it is a 20-kVA,

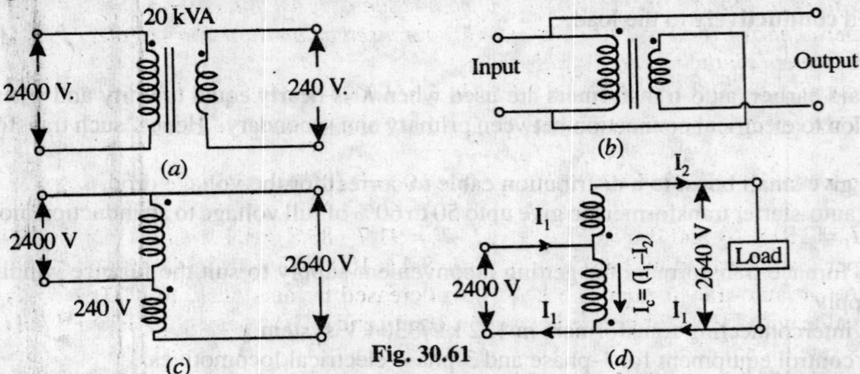

Fig. 30.61

2400/240-V transformer. If we employ *additive* polarity between the high-voltage and low-voltage sides, we get a step-up auto-transformer. If, however, we use the *subtractive* polarity, we get a step-down auto-transformer.

(*a*) Additive Polarity

Connections for such a polarity are shown in Fig. 30.61 (*b*). The circuit is re-drawn in Fig. 30.61 (*c*) showing common terminal of the transformer at the top whereas Fig. 30.61 (*d*) shows the same circuit with common terminal at the bottom. Because of additive polarity, V_2 = 2400 + 240 = 2640 V and V_1 is 2400 V. There is a marked increase in the kVA of the auto-transformer (Ex. 30.64). As shown in Fig. 30.61 (*d*), common current flows *towards* the common terminal. The transformer acts as a step-up transformer.

(*b*) **Subtracting Polarity**

Such a connection is shown in Fig. 30.62 (*a*). The circuit has been re-drawn with common polarity at top in Fig. 30.62 (*b*) and at bottom in Fig. 30.62 (*c*). In this case, the transformer acts as a step-down auto-transformer.

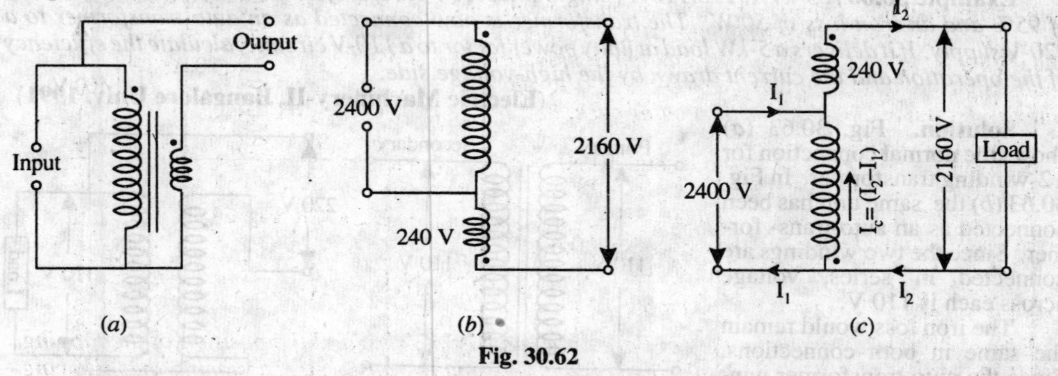

Fig. 30.62

The common current flows *away* from the common terminal. As will be shown in Example 30.65, in this case also, there is a very large increase in kVA rating of the auto-transformer though not as marked as in the previous case. Here, $V_2 = 2400 - 240 = 2160$ V.

Example 30.64. *For the 20-kVA, 2400/240-V two-winding step-down transformer shown in Fig. 30.62 (a) connected as an auto-transformer with additive polarity as shown in Fig. 30.62 (d), compute*

(*i*) *original current capacity of HV-winding.*
(*ii*) *original current capacity of LV winding.*
(*iii*) *kVA rating of auto-transformer using current capacity of LV winding as calculated in (ii) above.*
(*iv*) *per cent increase in kVA capacity of auto-transformer as compared to original two-winding transformer.*
(*v*) *values of I_1 and I_c in Fig. 30.61 (d) from value of I_2 used in (iii) above.*
(*vi*) *per cent overload of 2400-V winding when used as an auto-transformer.*
(*vii*) *comment on the results obtained.*

Solution. (*i*) $I_1 = 20 \times 10^3/2400 = \mathbf{8.33\ A}$ (*ii*) $I_2 = I_1/K = 8.33 \times 10 = \mathbf{83.3\ A}$

(*iii*) kVA rating of auto-transformer $V_2 I_2 = 2640 \times 83.3 \times 10^{-3} = \mathbf{220\ kVA}$

(*iv*) per cent increase in kVA rating $= \dfrac{220}{20} \times 100 = \mathbf{1100\%}$

(*v*) $I_1 = 220 \times 10^3/2400 = \mathbf{91.7\ A}$, $I_c = I_1 - I_2 = 91.7 - 83.3 = \mathbf{8.4\ A}$

(*vi*) per cent overload of 2400 V winding = 8.4 × 100/8.33 = **100.8 %**

(*vii*) As an auto-transformer, the kVA has increased tremendously to 1100% of its original value with *LV* coil at its rated current capacity and *HV* coil at negligible overload *i.e.* 1.008 × rated load.

Example 30.65. *Repeat Example 30.64 for subtractive polarity as shown in Fig. 30.62 (c).*

Solution. (*i*) $I_1 = \mathbf{8.33\ A}$ (*ii*) $I_2 = \mathbf{83.3\ A}$

(*iii*) New kVA rating of auto-transformer is $2160 \times 83.3 \times 10^{-3} = \mathbf{180\ kVA}$

(*iv*) Per cent increase in kVA rating $= \dfrac{180}{20} \times 100 = \mathbf{900\%}$

(*v*) $I_1 = 180 \times 10^3/2400 = \mathbf{75\ A}$ $I_c = I_2 - I_1 = 83.3 - 75 = \mathbf{8.3\ A}$

(*vi*) Per cent overload of 2400-V winding = 8.3 × 100/8.33 = **100%**

(*vii*) In this case, kVA has increased to 900% of its original value as a two-winding transformer with both low-voltage and high-voltage windings carrying their rated currents.

The above phenomenal increase in kVA capacity is due to the fact that in an auto-transformer energy transfer from primary to secondary is by both conduction as well as induction whereas in a 2-winding transformer it is by induction only. This extra conductive link is mainly responsible for the increase in kVA capacity.

Example 30.66 *A 5-kVA, 110/110-V, single-phase, 50-Hz transformer has full-load efficiency of 95% and an iron loss of 50 W. The transformer is now connected as an auto-transformer to a 220-V supply. If it delivers a 5-kW load at unity power factor to a 110-V circuit, calculate the efficiency of the operation and the current drawn by the high-voltage side.*

(Electric Machinery-II, Bangalore Univ. 1991)

Solution. Fig. 30.63 (*a*) shows the normal connection for a 2-winding transformer. In Fig. 30.63 (*b*) the same unit has been connected as an auto-transformer. Since the two windings are connected in series, voltage across each is 110 V.

The iron loss would remain the same in both connections. Since the auto-transformer windings will each carry but half the current as compared to the conventional two-winding transformer, the copper loss will be one-fourth of the previous value.

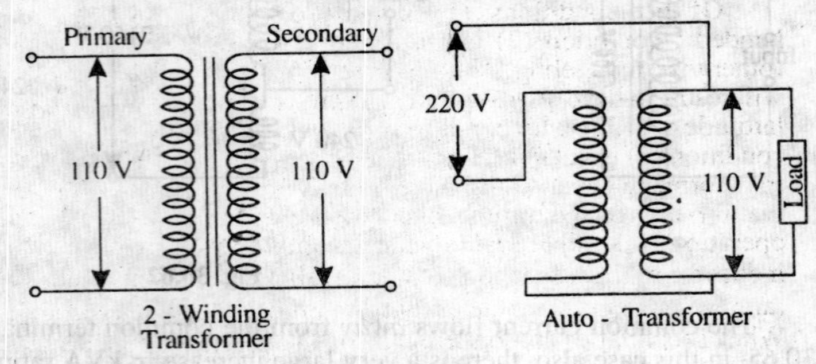

Fig. 30.63

Two-winding Transformer

$$\eta = 0.95 \quad \therefore \quad 0.95 = \frac{\text{output}}{\text{output + losses}} = \frac{5,000}{5,000 + 50 + \text{Cu loss}}$$

$\therefore$ Cu loss = 212 W

Auto-transformer

Cu lose = 212/4 = 53 W ; Iron loss = 50 W $\therefore \eta = \frac{5,000}{5,000 + 53 + 50} = 0.9797$ or **97.97%**

Current of the h.v. side = 5103/220 = **23.2 A**

30.35. Parallel Operation of Single-phase Transformers

For supplying a load in excess of the rating of an existing transformer, a second transformer may be connected in parallel with it as shown in Fig. 30.64. It is seen that primary windings are connected to the supply bus bars and secondary windings are connected to the load bus-bars. In connecting two or more than two transformers in parallel, it is essential that their terminals of similar polarities are joined to the same bus-bars as in Fig. 30.64. If this is not done, the the two e.m.fs. induced in the secondaries which are paralleled with incorrect polarities, will act together in the local secondary circuit even when supplying no load and will hence produce the equivalent of a dead short-circuit as shown in Fig. 30.65.

There are certain definite conditions which must be satisfied in order to avoid any local circulating currents and to ensure that the transformers share the common load in proportion to their kVA ratings. The conditions are :-

1. Primary windings of the transformers should be suitable for the supply system voltage and frequency.
2. The transformers should be properly connected with regard to polarity.

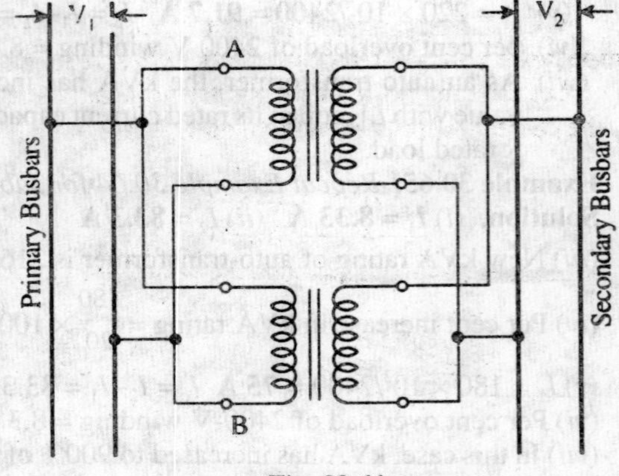

Fig. 30.64

3. The voltage ratings of both primaries and secondaries should be identical. In other words, the transformers should have the same turn ratio *i.e.* transformation ratio.
4. The percentage impedances should be equal in magnitude and have the same X/R ratio in order to avoid circulating currents and operation at different power factors.
5. With transformers having different kVA ratings, the equivalent impedances should be inversely proportional to the individual kVA rating if circulating currents are to be avoided.

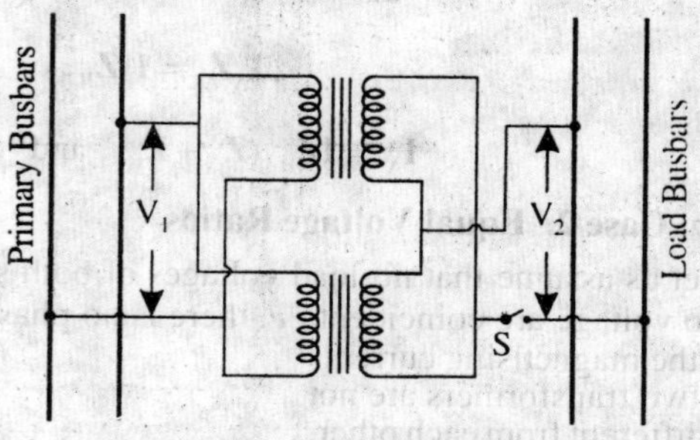

Fig. 30.65 Short Circuit

Of thse conditions, (1) is easily comprehended ; condition (2) is absolutely essential (otherwise paralleling with incorrect polarities will result in dead short-circuit). There is some lattitude possible with conditions (3) and (4). If condition (3) is not exactly satisfied *i.e.* the two transformers have slightly different transformation or voltage ratios, even then parallel operation is possible. But due to inequality of induced e.m.fs. in secondaries, there will be even on no-load, some circulating current between them (and therefore between the primary windings also) when secondary terminals are connected in parallel. When secondaries are loaded, this localized circulating current will tend to produce unequal loading condition. Hence, it may be impossible to take full kVA output from the parallel connected group without one of the transformers becoming over-heated.

If condition (4) is not exactly satisfied *i.e* impedance triangles are not identical in shape and size, parallel operation will still be possible, but the power factors at which the two transformers operate will be different from the power factor of the common load. Therefore, in this case, the two transformers will not share the load in proportion to their kVA ratings.

It should be noted that the impedances of two transformers may differ in magnitude and in quality (*i.e.* ratio of equivalent resistance to reactance). It is worthwhile to distinguish between the *percentage* and *numerical* value of an impedance. For example, consider two transformers having ratings in the ratio 1 : 2. It is obvious that to carry double the current, the latter must have half the impedance of the former for the same regulation. For parallel operation, the regulation must be the same, this condition being enforced by the very fact of their being connected in parallel. It means that the currents carried by the two transformers are proportional to their ratings provided their *numerical* impedances are inversely proportional to these ratings and their *percentage* impedances are identical.

If the *quality* of the two percentage impedances is different (*i.e.* ratio of percentage resistance to reactance is different), then this will result in divergence of phase angle of the two currents, with the result that one transformer will be operating with a higher and the other with a lower power factor than that of the combined load.

(*a*) **Case 1. Ideal Case**

We will first consider the ideal case of two transformers having the same voltage ratio and having impedance voltage triangles identical in size and shape.

Let E be the no-load secondary voltage of each transformer and V_2 the terminal voltage ; I_A and I_B the currents supplied by them and I—the total current, lagging behind V_2 by an angle ϕ (Fig. 30.66).

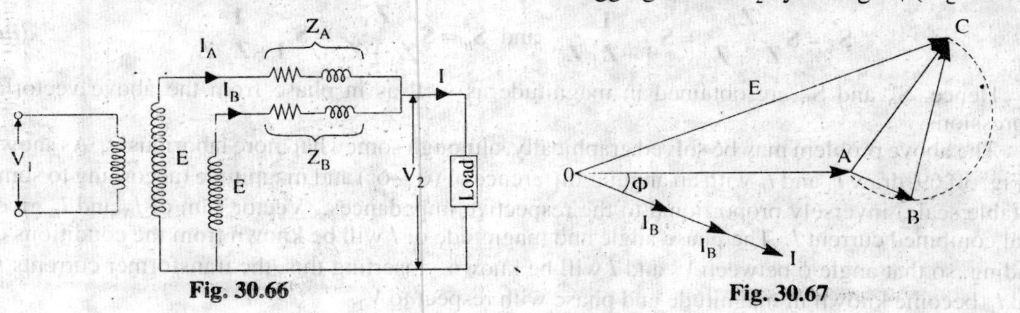

Fig. 30.66 **Fig. 30.67**

In Fig. 30.67, a single triangle *ABC* represents the identical impedance voltage triangles of both the transformers. The currents I_A and I_B of the individual transformers are in phase with the load current *I* and are inversely proportional to the respective impedances. Following relations are obvious.

$$\mathbf{I} = \mathbf{I}_A + \mathbf{I}_B; \quad \mathbf{V}_2 = \mathbf{E} - \mathbf{I}_A\mathbf{Z}_A = \mathbf{E} - \mathbf{I}_B\mathbf{Z}_B = \mathbf{E} - \mathbf{I}\mathbf{Z}_{AB}$$

Also
$$\mathbf{I}_A\mathbf{Z}_A = \mathbf{I}_B\mathbf{Z}_B \quad \text{or} \quad \mathbf{I}_A/\mathbf{I}_B = \mathbf{Z}_B/\mathbf{Z}_A$$

$\therefore$
$$\mathbf{I}_A = \mathbf{I}\mathbf{Z}_B/(\mathbf{Z}_A + \mathbf{Z}_B) \quad \text{and} \quad \mathbf{I}_B = \mathbf{I}\mathbf{Z}_A/(\mathbf{Z}_A + \mathbf{Z}_B)$$

(*b*) Case 2. Equal Voltage Ratios

Let us assume that no-load voltages of both secondaries is the same *i.e.* $E_A = E_B = E$, and that the two voltage are coincident *i.e.* there is no phase difference between E_A and E_B, which would be true if the magnetising currents of the two transformers are not much different from each other. Under these conditions, both primaries and secondaries of the two transformers can be connected in parallel and there will circulate no current between them on *no-load*.

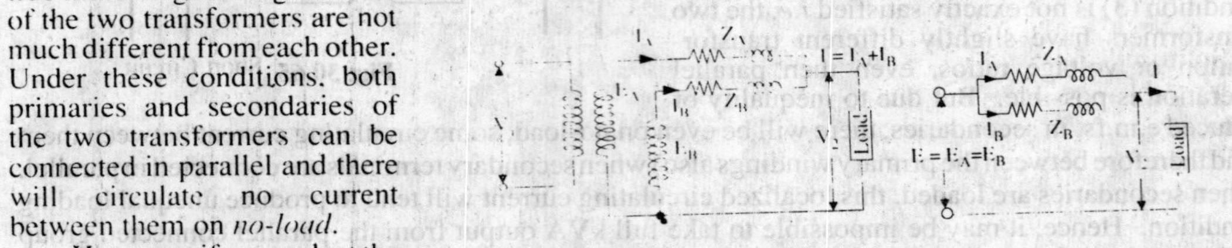

Fig. 30.68

However, if we neglect the magnetising admittances, the two transformers can be connected as shown by their equivalent circuits in Fig. 30.68. The vector diagram is shown in Fig. 30.69.

From Fig. 30.68 (*a*) or (*b*) it is seen that it represents two impedances in parallel. Considering all values *consistently with reference to secondaries*, let

$\mathbf{Z}_A, \mathbf{Z}_B$ = impedances of the transformers
$\mathbf{I}_A, \mathbf{I}_B$ = their respective currents
$\mathbf{V}_2$ = common terminal voltage
$\mathbf{I}$ = combined current

It is seen that
$$\mathbf{I}_A\mathbf{Z}_A = \mathbf{I}_B\mathbf{Z}_B = \mathbf{I}\mathbf{Z}_{AB} \qquad ...(i)$$
where $\mathbf{Z}_{AB}$ is the combined impedance of $\mathbf{Z}_A$ and $\mathbf{Z}_B$ in parallel.

$$1/\mathbf{Z}_{AB} = 1/\mathbf{Z}_A + 1/\mathbf{Z}_B$$
hence
$$\mathbf{Z}_{AB} = \mathbf{Z}_A\mathbf{Z}_B/(\mathbf{Z}_A + \mathbf{Z}_B) \qquad ...(ii)$$

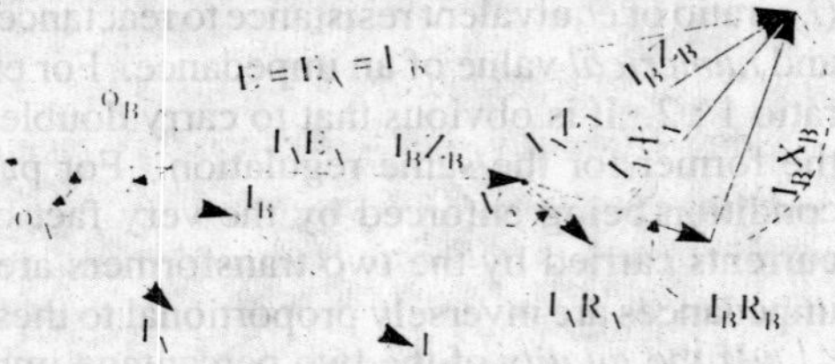

Fig. 30.69

From equation (*i*), we get

$$\mathbf{I}_A = \mathbf{I}\mathbf{Z}_{AB}/\mathbf{Z}_A = \mathbf{I}\mathbf{Z}_B/(\mathbf{Z}_A + \mathbf{Z}_B) \text{ and } \mathbf{I}_B = \mathbf{I}\mathbf{Z}_{AB}/\mathbf{Z}_B = \mathbf{I}\mathbf{Z}_A/(\mathbf{Z}_A + \mathbf{Z}_B)$$

Multiplying both sides by common terminal voltage $\mathbf{V}_2$, we have

$$\mathbf{V}_2\mathbf{I}_A = \mathbf{V}_2\mathbf{I}\frac{\mathbf{Z}_B}{\mathbf{Z}_A + \mathbf{Z}_B}; \text{ Similarly } \mathbf{V}_2\mathbf{I}_B = \mathbf{V}_2\mathbf{I}\frac{\mathbf{Z}_A}{\mathbf{Z}_A + \mathbf{Z}_B}$$

Let $\mathbf{V}_2\mathbf{I} \times 10^{-3} = \mathbf{S}$ the combined load kVA. Then, the kVA carried by each transformer is

$$\mathbf{S}_A = \mathbf{S}\frac{\mathbf{Z}_B}{\mathbf{Z}_A + \mathbf{Z}_B} = \mathbf{S}\frac{1}{1 + \mathbf{Z}_A/\mathbf{Z}_B} \text{ and } \mathbf{S}_B = \mathbf{S}\frac{\mathbf{Z}_A}{\mathbf{Z}_A + \mathbf{Z}_B} = \mathbf{S}\frac{1}{1 + \mathbf{Z}_B/\mathbf{Z}_A} \qquad ...(iii)$$

Hence, $\mathbf{S}_A$ and $\mathbf{S}_B$ are obtained in magnitude as well as in phase from the above vectorial expressions.

The above problem may be solved graphically, although somewhat more laboriously. As shown in Fig. 30.69, draw I_A and I_B with an angular difference of $(\phi_A - \phi_B)$ and magnitude (according to some suitable scale) inversely proportional to the respective impedances. Vector sum of I_A and I_B gives total combined current *I*. The phase angle and magnitude of *I* will be known from the conditions of loading, so that angle ϕ between V_2 and *I* will be known. Inserting this, the transformer currents I_A and I_B become known in magnitude and phase with respect to V_2.

Note. (*a*) In equation (*iii*) above, it is not necessary to use the ohmic values of resistances and reactances, because only impedance ratios are required.

(*b*) The two percentage impedances must be adjusted to the same kVA in the case of transformers of different ratings as in Ex. 30.72.

(*c*) From equation (*iii*) above, it is seen that if two transformers having the same rating and the same transformation ratio are to share the load equally, then their impedances should be equal *i.e.* equal resistances and reactances and not numerical equality of impedances. In general, for transformers of different ratings but same transformation ratio, their equivalent impedances must be inversely proportional to their ratings if each transformer is to assume a load in proportion to its rating. For example, as said earlier, a transformer operating in parallel with another of twice the rating, must have an impedance twice that of the large transformer in order that the load may be properly shared between them.

Example 30.67. *Two 1-phase transformers with equal turns have impedances of (0.5 + j3) ohm and (0.6 + j10) ohm with respect to the secondary. If they operate in parallel, determine how they will share a total load of 100 kW at p.f. 0.8 lagging ?*

(Electrical Technology, Madras Univ. 1987)

Solution. $\mathbf{Z}_A = 0.5 + j3 = 3.04 \angle 80.6°$ $\mathbf{Z}_B = 0.6 + j10 = 10.02 \angle 86.6°$

$\mathbf{Z}_A + \mathbf{Z}_B = 1.1 + j13 = 13.05 \angle 85.2°$

Now, a load of 100 kW at 0.8 p.f. means a kVA of 100/0.8 = 125. Hence,

$$\mathbf{S} = 125 \angle -36.9°$$

$$\mathbf{S}_A = \mathbf{S}\frac{\mathbf{Z}_B}{\mathbf{Z}_A + \mathbf{Z}_B} = \frac{125\angle -36.9° \times 10.02\angle 86.6°}{13.05 \angle 85.2°} = 96\angle -35.5°$$

$$= \text{a load of } 96 \times \cos 35.5° = 78.2\text{kW}$$

$$\mathbf{S}_B = \mathbf{S}\frac{\mathbf{S}_A}{\mathbf{Z}_A + \mathbf{Z}_B} = \frac{125\angle -36.9° \times 3.04\angle 80.6°}{13.05\angle 85.2°} = 29.1\angle -41.5°$$

$$= \text{a load of } 29.1 \times \cos 41.5° = \mathbf{21.8\ kW}$$

Note. Obviously, transformer *A* is carrying more than its due share of the common load.

Example 30.68. *Two single-phase transformers A and B are connected in parallel. They have same kVA ratings but their resistances are respectively 0.005 and 0.01 per unit and their leakage reactances 0.05 and 0.04 per unit. If A is operated on full-load at a p.f. of 0.8 lagging, what will be the load and p.f. of B.* **(A.C. Machines-I, Jadavpur Univ. 1985)**

Solution. In general, $\mathbf{S}_A = \mathbf{S}\dfrac{\mathbf{Z}_B}{\mathbf{Z}_A + \mathbf{Z}_B}$ and $\mathbf{S}_B = \mathbf{S}\dfrac{\mathbf{Z}_A}{\mathbf{Z}_A + \mathbf{Z}_B}$

where **S** is the total kVA supplied and $\mathbf{Z}_A$ and $\mathbf{Z}_B$ are the percentage impedances of these transformers.

$$\therefore \qquad \frac{\mathbf{S}_B}{\mathbf{S}_A} = \frac{\mathbf{Z}_A}{\mathbf{Z}_B}$$

Now, $\mathbf{Z}_A = 0.005 + j0.05$ per unit; $\%\mathbf{Z}_A = 0.5 + j5$ and $\%\mathbf{Z}_B = 1 + j4$

Let $\mathbf{S}_A = S_A \angle -36.87°$

where S_A represents the rating of transformer *A* (and also of *B*)

$$\mathbf{S}_B = S_A \angle -36.87° \times \frac{(0.5 + j5)}{(1 + j4)}$$

$$= \mathbf{S}_A \angle -36.87° \times \frac{20.7}{17} \angle 8.3° = 1.22\ \mathbf{S}_A \angle -28.57°$$

It is obvious that transformer *B* is working **22%** over-load and its power factor is

$\cos 28.57° = \mathbf{0.878\ (lag)}$

Example 30.69. *Two 1-phase transformers A and B rated at 250 kVA each are operated in parallel on both sides. Percentage impedances for A and B are (1 + j 6) and (1.2 + j 4.8) respectively. Compute the load shared by each when the total load is 500 kVA at 0.8 p.f. lagging.*

(Electrical Machines-II, Indore Univ. 1989)

Solution.

$$\frac{\mathbf{Z_A}}{\mathbf{Z_A}+\mathbf{Z_B}} = \frac{1+j6}{2.2+j10.8} = 0.55\angle 2.1°\ ; \quad \frac{\mathbf{Z_B}}{\mathbf{Z_A}+\mathbf{Z_B}} = \frac{1.2+j4.8}{2.2+j10.8} = 0.45\angle -2.5°$$

$$\mathbf{S_A} = \mathbf{S}\frac{\mathbf{Z_B}}{\mathbf{Z_A}+\mathbf{Z_B}} = 500\angle -36.9° \times 0.45\angle -2.5° = 225\angle -39.4°$$

$$\mathbf{S_B} = \mathbf{S}\frac{\mathbf{Z_A}}{\mathbf{Z_A}+\mathbf{Z_B}} = 500\angle -36.9° \times 0.55\angle 2.1° = 275\angle -34.8°$$

Obviously, transformer *B* is overloaded to the extent of (275–250) × 100/250 = 10%. It carries (275/500) × 100 = 55% of the total load.

Example 30.70. *Two 100-kW, single-phase transformers are connected in parallel both on the primary and secondary. One transformer has an ohmic drop of 0.5% at full-load and an inductive drop of 8% at full-load current. The other has an ohmic drop of 0.75% and inductive drop of 2%. Show how will they share a load of 180 kW at 0.9 power factor.*

(Elect. Machines-I, Calcutta Univ. 1988)

Solution. A load of 180 kW at 0.9 p.f. means a kVA of 180/0.9 = 200

∴ Load **S** = 200 –25.8°

$$\frac{\mathbf{Z_1}}{\mathbf{Z_1}+\mathbf{Z_2}} = \frac{(0.5+j8)}{(1.25+j12)} = \frac{(0.5+j8)(1.25-j12)}{1.25^2+12^2}$$

$$= \frac{96.63+j4}{145.6} = \frac{96.65\angle 2.4°}{145.6} = 0.664\angle 2.4°$$

$$\frac{\mathbf{Z_2}}{\mathbf{Z_1}+\mathbf{Z_2}} = \frac{(0.75+j4)(1.25-j12)}{145.6} = \frac{48.94-j4}{145.6} = \frac{49.1\angle -5°}{145.6} = 0.337\angle -5°$$

$$\mathbf{S_1} = \mathbf{S}\frac{\mathbf{Z_2}}{\mathbf{Z_1}+\mathbf{Z_2}} = 200\angle -25.8° \times 0.337\angle -5° = 67.4\angle -30.8°$$

$$\therefore \quad \text{kW}_1 = 67.4 \times \cos 30.8° = 67.4 \times 0.859 = 57.9 \text{ kW}$$

$$\mathbf{S_2} = 200\ \angle -25.8° \times 0.664\ \angle 2.4° = 132.8\ \angle -23.4^*$$

$$\text{kW}_2 = 132.8 \times \cos 23.4° = 132.8 \times 0.915 = 121.5 \text{ kW}$$

Note. Second transformer is working 21.5% over-load. Also, it shares 65.7% of the total load.

Example 30.71. *A load of 200 kW at 0.85 power factor lagging is to be shared by two transformers A and B having the same ratings and the same transformation ratio. For transformer A, the full-load resistive drop is 1% and reactance drop 5% of the normal terminal voltage. For transformer B the corresponding values are : 2% and 6%. Calculate the load kVA supplied by each transformer.*

---

*Or $\mathbf{S_B} = \mathbf{S} - \mathbf{S_A}$.

Solution.

$$\mathbf{Z_A} = 1 + j5;\ \mathbf{Z_B} = 2 + j6$$

$$\frac{\mathbf{Z_A}}{\mathbf{Z_B}} = \frac{1+j5}{2+j6} = 0.8 + j0.1; \quad \frac{\mathbf{Z_B}}{\mathbf{Z_A}} = \frac{2+j6}{1+j5} = 1.23 - j0.514$$

Load

$$\text{kVA} = \text{kW/p.f.} = 200/0.85 = 235$$

$$\mathbf{kVA} = \mathbf{S} = 235(0.85 - j0.527) = 200 - j123.8 (\because \cos\phi = 0.85; \sin\phi = 0.527)$$

$$\mathbf{S_A} = \mathbf{S}.\frac{\mathbf{Z_B}}{\mathbf{Z_A}+\mathbf{Z_B}} = \mathbf{S}\frac{1}{1+(\mathbf{Z_A}/\mathbf{Z_B})}$$

$$= \frac{200 - j\,123.8}{1+(0.8+j0.1)} = \frac{200 - j123.8}{1.8 + j0.1} = 107.3 - j74.9$$

$$\mathbf{S_A} = \sqrt{(107.3^2 + 74.9^2)} = 131; \cos\phi_A = 107.3/131 = 0.82(\text{lag})$$

Similarly

$$\mathbf{S_B} = \mathbf{S}\frac{\mathbf{Z_A}}{\mathbf{Z_A}+\mathbf{Z_B}} = \mathbf{S}\frac{1}{1+(\mathbf{Z_B}/\mathbf{Z_A})}$$

$$= \frac{200 - j123.8}{1+(1.23 - j0.514)} = \frac{200 - j123.8}{2.23 - j0.154} = 93 - j49$$

$$\therefore \quad S_B = \sqrt{(93^2 + 49^2)} = 105; \cos\phi_B = 90/105 = 0.888 \text{ (lag)}$$

(As a check, $S = S_A + S_B = 131 + 105 = 236$. The small error is due to approximations made in calculations.)

Example 30.72. *Two 2,200/110-V, transformers are operated in parallel to share a load of 125 kVA at 0.8 power factor lagging. Transformers are rated as below :*

A : 100 kVA ; 0.9% resistance and 10% reactance

B : 50 kVA ; 1.0% resistance and 5% reactance

Find the load carried by each transformer. **(Elect. Technology, Utkal Univ. 1989)**

Solution. It should be noted that the percentages given above refer to different ratings. As pointed out in Art. 30.35, these should be adjusted to the same basic kVA, say, 100 kVA.

$$\%\mathbf{Z_A} = 0.9 + j10; \quad \%\mathbf{Z_B} = (100/50)(1 + j5) = (2 + j10)$$

$$\frac{\mathbf{Z_A}}{\mathbf{Z_A}+\mathbf{Z_B}} = \frac{0.9+j10}{(2.9+j20)} = \frac{(0.9+j10)(2.9-j20)}{2.9^2+20^2}$$

$$= \frac{(202.6+j11)}{408.4} = \frac{202.9\angle 3.1°}{408.4} = 0.4968 \angle 3.1°$$

$$\frac{\mathbf{Z_B}}{\mathbf{Z_A}+\mathbf{Z_B}} = \frac{(2+j10)(2.9-j20)}{408.4} = \frac{206 - j11}{408.4} = \frac{206.1\angle -3.1°}{408.4} = 0.504 \angle -3.1°$$

Also $\cos\phi = 0.8, \phi = \cos^{-1}(0.8) = 36.9°$

$$\mathbf{S_A} = \mathbf{S}\frac{\mathbf{Z_B}}{\mathbf{Z_A}+\mathbf{Z_B}} = 125\angle -36.9° \times 0.504 \angle -3.1° = 63 \angle -40°$$

$$\mathbf{S_B} = \mathbf{S}\frac{\mathbf{S_A}}{\mathbf{Z_A}+\mathbf{Z_B}} = 125 \angle -36.9° \times 0.4968 \angle 3.1° = 62.1 \angle -33.8°$$

Example 30.73. *A 500-kVA transformer with 1% resistance and 5% reactance is connected in parallel with a 250-kVA transformer with 1.5% resistance and 4% reactance. The secondary voltage of each transformer is 400 V on no-load. Find how they share a load of 750-kVA at a p.f. of 0.8 lagging.* **(Electrical Machinery-I, Madras Univ. 1987)**

Solution. It may be noted that percentage drops given above refer to different ratings. These should be adjusted to the same basic kVA *i.e.* 500 kVA.

$$\% \mathbf{Z}_A = 1 + j5 = 5.1 \angle 78.7^\circ ; \% \mathbf{Z}_B = \left(\frac{500}{250}\right)(1.5 + j4) = 3 + j8 = 8.55 \angle 69.4$$

$$\% (\mathbf{Z}_A + \mathbf{Z}_B) = 4 + j13 = 13.6 \angle 72.9^\circ ; \ \mathbf{S} = 750 \angle -36.9^\circ$$

$$\mathbf{S}_A = \mathbf{S}\frac{\mathbf{Z}_B}{\mathbf{Z}_A + \mathbf{Z}_B} = \frac{750 \angle -36.9^\circ \times 8.55 \angle 69.4^\circ}{13.6 \angle 72.9^\circ} = 470 \angle -40.4^\circ$$

= 470 kVA at p.f. of 0.762 lagging

$$\mathbf{S}_B = \mathbf{S}.\frac{\mathbf{Z}_A}{\mathbf{Z}_A + \mathbf{Z}_B} = \frac{750 \angle -36.9^\circ \times 5.1 \angle 78.7^\circ}{13.6 \angle 72.9^\circ} = 280 \angle -31.1^\circ$$

= 280 kVA at p.f. of 0.856 lagging

Note. The above solution has been attempted vectorially, but in practice, the angle between I_A and I_B is so small that if instead of using vectorial expressions, arithmetic expressions were used, the answer would not be much different. In most cases, calculations by both vectorial and arithmetical methods generally yield results that do not differ sufficiently to warrant the more involved procedure by the vector solution. The above example will now be attempted arithmetically :

$$Z_A = 5.1\ \Omega, Z_B = 8.55\ \Omega ; I_A/I_B = Z_B/Z_A = 8.55/5.1 = 1.677 \ \therefore\ I_A = 1.677\, I_B$$

Total current = 750,000/400 = 1875 A

$$I = I_A + I_B ; 1875 = 1.677 I_B + I_B = 2.677\, I_B \quad \therefore \quad I_B = 1875/2.677$$

$$\therefore \quad S_B = 400 \times 1875/2.677 \times 100 = 280 \text{ kVA} \quad I_A = 1.677 \times 1875/2.677$$

$$S_A = 400 \times 1.677 \times 1875/2.677 \times 1000 = 470 \text{ kVA}$$

Example 30.74. *Two single-phase transformers A and B of equal voltage ratio are running in parallel and supply a load of 1000 A at 0.8 p.f. lag. The equivalent impedances of the two transformers are (2 + j3) and (2.5 + j5) ohms respectively. Calculate the current supplied by each transformer and the ratio of the kW output of the two transformers.* **(Electrical Machines-I, Bombay Univ. 1986)**

Solution. $\mathbf{Z}_A = (2 + j3), \ \mathbf{Z}_B = (2.5 + j5)$

Now, $\dfrac{\mathbf{I}_A}{\mathbf{I}_B} = \dfrac{\mathbf{Z}_B}{\mathbf{Z}_A} = \dfrac{2.5 + j5}{2 + j3} = (1.54 + j0.2) ; \mathbf{I}_A = \mathbf{I}_B (1.54 + j0.2)$

Taking secondary terminal voltage as reference vector, we get

$$\mathbf{I} = 1000(0.8 - j0.6) = 800 - j\,600 = 200\,(4 - j3)$$

Also, $\mathbf{I} = \mathbf{I}_A + \mathbf{I}_B = \mathbf{I}_B\,(1.54 + j0.2) + \mathbf{I}_B = \mathbf{I}_B\,(2.54 + j0.2)$

$\therefore \quad 200(4 - j3) = \mathbf{I}_B\,(2.54 + j0.2) ; \ \mathbf{I}_B = 294.6 - j\,259.5 = \mathbf{392.6 \angle -41.37^\circ}$

$\mathbf{I}_A = \mathbf{I}_B\,(1.54 + j0.2) = (294.6 - j\,259.5)\,(1.54 + j\,0.2) = 505.6 - j340.7$

$= \mathbf{609.7 \angle -33.95^\circ}$

The ratio of the kW outputs is given by the ratio of the in-phase components of the two currents.

$$\frac{\text{output of } A}{\text{output of } B} = \frac{505.6}{294.6} = \frac{1.7}{1}$$

Note. Arithmetic solution mentioned above could also be attempted.

Example 30.75. *Two transformers A and B, both of no-load ratio 1,000/500-V are connected in parallel and supplied at 1,000 V. A is rated at 100 kVA, its total resistance and reactance being 1% and 5% respectively, B is rated at 250 kVA, with 2% resistance and 2% reactance. Determine the load on each transformer and the secondary voltage when a total load of 300 kVA at 0.8 power factor lagging is supplied.*

Solution. Let the percentage impedances be adjusted to the common basic kVA of 100.

Then $\%\mathbf{Z}_A = (1 + j5)$; $\%\mathbf{Z}_B = (100/250)(2 + j2) = (0.8 + j0.8)$

$$\frac{\mathbf{Z}_A}{\mathbf{Z}_A + \mathbf{Z}_B} = \frac{1 + j5}{(1.8 + j5.8)} = 0.839 \angle 5.9^\circ$$

$$\frac{\mathbf{Z}_B}{\mathbf{Z}_A + \mathbf{Z}_B} = \frac{0.8 + j0.8}{(1.8 + j5.8)} = 0.1865 \angle -27.6^\circ$$

Now, $\mathbf{S} = 300 \angle -36.9^\circ = 240 - j\,180$

$$\mathbf{S}_A = \mathbf{S}.\frac{\mathbf{Z}_B}{\mathbf{Z}_A + \mathbf{Z}_B} = 300 \angle -36.9^\circ \times 0.1865 \angle -27.6^\circ = 55.95 \angle -64.5^\circ$$

$$\mathbf{S}_B = \mathbf{S}\frac{\mathbf{Z}_A}{\mathbf{Z}_A + \mathbf{Z}_B} = 300 \angle -36.9^\circ \times 0.839 \angle 5.9^\circ = 251.7 \angle -31^\circ$$

Since $\mathbf{Z}_A$ and $\mathbf{Z}_B$ are in parallel, their combined impedance on 100 kVA basis is

$$\mathbf{Z}_{AB} = \frac{\mathbf{Z}_A \mathbf{Z}_B}{\mathbf{Z}_A + \mathbf{Z}_B} = \frac{(1 + j5)(0.8 + j0.8)}{(1.8 + j5.8)} = 0.6 + j0.738$$

Percentage drop over $\mathbf{Z}_{AB}$ is

$$= (0.6 \times 240/100) + (0.738 \times 180/100) = 1.44 + 1.328 = 2.768\%$$

$$\therefore \quad V_2 = \left(500 - \frac{500 \times 2.768}{100}\right) = \mathbf{486.12\ V}$$

Example 30.76. *Two 1-φ transformers are connected in parallel at no-load. One has a turn ratio of 5,000/440 and a rating of 200 kVA, the other has a ratio of 5,000/480 and a rating of 350 kVA. The leakage reactance of each is 3.5%.*

What is the no-load circulation current expressed as a percentage of the nominal current of the 200 kVA transformer.

Solution. The normal currents are

$$200 \times 10^3/440 = 455 \text{ A and } 350 \times 10^3/480 = 730 \text{ A}$$

Reactances seen from the secondary side are

$$\frac{3.5}{100} \times \frac{440}{455} = 0.034\ \Omega, \quad \frac{3.5}{100} \times \frac{480}{730} = 0.023\ \Omega$$

The difference of induced voltage is 40 V. The circulating current is

$I_C = 40/0.057 = 704$ A = **1.55 times** the normal current of 200 kVA unit.

Tutorial Problem No. 30.6

1. Two single-phase transformers *A* and *B* of equal voltage ratio are running in parallel and supplying a load requiring 500 A at 0.8 power factor lagging at a terminal voltage of 400 V. The equivalent impedances of the transformers, as referred to secondary windings, are (2 + *j*3) and (2.5 + *j*5) ohm. Calculate the current supplied by each transformer.

(**Note.** The student is advised to try by arithmetic method also). [$\mathbf{I_A = 304\ A}$; $\mathbf{I_B = 197\ A}$]

2. Two single-phase transformers *A* and *B* are operating in parallel and supplying a common load of 1000 kVA at 0.8 p.f. lagging. The data regarding the transformers is as follows :

Transformer	Rating	% Resistance	% Reactance
A	750 kVA	3	5
B	500 kVA	2	4

Determine the loading of each transformer. [$S_A = 535 \angle -34.7°$; $S_B = 465 \angle -39.3°$]

3. Two transformers *A* and *B* give the following test results. With the low-tension side short-circuited, *A* takes a current of 10 A at 200 V, the power input being 1000 W. Similarly, *B* takes 30 A at 200 V ; the power input being 1,500 W. On open circuit, both transformers give a secondary voltage of 2200 when 11,000 volts are applied to the primary terminals. These transformers are connected in parallel on both high tension and low tension sides. Calculate the current and power in each transformer when supplying a load of 200 A at 0.8 power factor lagging. The no-load currents may be neglected.

[**Hint** : Calculate $\mathbf{Z}_A$ and $\mathbf{Z}_B$ from S.C. test data]

[I_A = 50.5 A, P_A = 100 kW ; I_B = 151 A, P_B = 252 kW] (*London University*)

4. Two 6600/250-V transformers have the following short-circuit characteristics : Applied voltage 200 V, current 30 A, power input 1,200 W for one of the transformers ; the corresponding data for the other transformer being 120 V, 20 A and 1,500 W. All values are measured on the H.V. side with the L.V. terminals short circuited. Find the approximate current and the power factor of each transformer when working in parallel with each other on the high and low voltage sides and taking a total load of 150 kW at a p.f. of 0.8 lagging from the high voltage bus-bars. **[I_A = 13.8 A, cos ϕ_A = 0.63 ; I_B=15.35 A, cos ϕ_B = 0.91]** (*Electrical Engg-IV, Baroda Univ. 1978*)

5. Two 11,000/2,200-V, 1-phase transformers are connected in parallel to supply a total load of 200 at 0.8 p.f. lagging at 2,200 V. One transformer has an equivalent resistance of 0.4 Ω and equivalent reactance of 0.8 Ω referred to the low-voltage side. The other has equivalent resistance of 0.1 Ω and a reactance of 0.3 Ω. Determine the current and power supplied by each transformer. **[52 A ; 148 A ; 99 A ; 252 kW]**

6. A 2,000-kVA transformer (*A*) is connected in parallel with a 4,000-kVA transformer (*B*) to supply a 3-phase load of 5,000 kVA at 0.8 p.f. lagging. Determine the kVA supplied by each transformer assuming equal no-load voltages. The percentage volt drops in the windings at the rated loads are as follows :

Transformer *A* : resistance 2% ; reactance 8%
Transformer *B* : resistance 1.6% ; reactance 3%

[A : 860 kVA, 0.661 lag ; B : 4170 kVA, 0.824 lag] (*A.C. Machines-I, Jadavpur Univ. 1979*)

7. Two single-phase transformers work in parallel on a load of 750 A at 0.8 p.f. lagging. Determine secondary voltage and the output and power factor of each transformer. Test data are :

Open circuit : 11,000/3,300 V for each transformer
Short circuit : with h.t. winding short-circuit
Transformer *A* : secondary input 200 V, 400 A, 15 kW
Transformer *B* : secondary input 100 V, 400 A, 20 kW

[3,190 V ; A : 80 kVA, 0.65 lag ; B : 1,615 kVA ; 0.86 lag]

(*c*) Case 3. Unequal Voltage Ratios

In this case, the voltage ratios (or transformation ratios) of the two transformers are different. It means that their no-load secondary voltages are unequal. Such cases can be more easily handled by phasor algebra than graphically.

Let $\mathbf{E}_A$, $\mathbf{E}_B$ = no-load secondary e.m.f.s of the two transformers.

$\mathbf{Z}_L$ = load impedance across the secondary.

The equivalent circuit and vector diagram are also shown in Fig. 30.70 and 30.71.

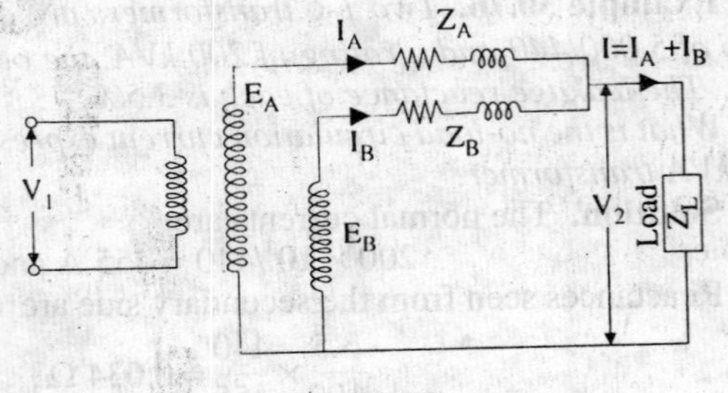

Fig. 30.70

It is seen that even when secondaries are on no-load, there will be some cross-current in them because of inequality in their induced e.m.fs. This circulating current I_C is given by

$$\mathbf{I}_C = (\mathbf{E}_A - \mathbf{E}_B)/(\mathbf{Z}_A + \mathbf{Z}_B) \quad ...(i)$$

As the induced e.m.fs. of the two transformers are equal to the total drops in their respective circuits.

$\therefore \quad \mathbf{E}_A = \mathbf{I}_A \mathbf{Z}_A + \mathbf{V}_2$; $\mathbf{E}_B = \mathbf{I}_B \mathbf{Z}_B + \mathbf{V}_2$

Now, $\quad \mathbf{V}_2 = \mathbf{I}\mathbf{Z}_L = (\mathbf{I}_A + \mathbf{I}_B)\,\mathbf{Z}_L$

where $\mathbf{Z}_L$ = load impedance

$$\mathbf{E}_A = \mathbf{I}_A \mathbf{Z}_A + (\mathbf{I}_A + \mathbf{I}_B)\,\mathbf{Z}_L \quad ...(ii)$$

$$\mathbf{E}_B = \mathbf{I}_B \mathbf{Z}_B + (\mathbf{I}_A + \mathbf{I}_B)\,\mathbf{Z}_L \quad ...(iii)$$

$$\therefore \quad \mathbf{E}_A - \mathbf{E}_B = \mathbf{I}_A \mathbf{Z}_A - \mathbf{I}_B \mathbf{Z}_B \quad ...(iv)$$

$$\mathbf{I}_A = [(\mathbf{E}_A - \mathbf{E}_B) + \mathbf{I}_B \mathbf{Z}_B]/\mathbf{Z}_A$$

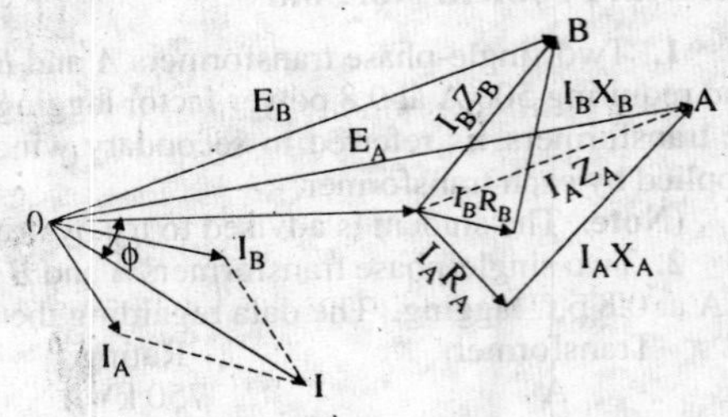

Fig. 30.71

Substituting this value of $\mathbf{I}_A$ in equation (*iii*), we get

$$\mathbf{E}_B = \mathbf{I}_B\mathbf{Z}_B + [\{(\mathbf{E}_A - \mathbf{E}_B) + \mathbf{I}_B\mathbf{Z}_B\}/\mathbf{Z}_A + \mathbf{I}_B]/\mathbf{Z}_L$$

$$\mathbf{I}_B = [\mathbf{E}_B\,\mathbf{Z}_A - (\mathbf{E}_A - \mathbf{E}_B)\mathbf{Z}_L]/[\mathbf{Z}_A\mathbf{Z}_B + \mathbf{Z}_L.(\mathbf{Z}_A + \mathbf{Z}_B)] \quad ...(iv)$$

From the symmetry of the expression, we get

$$\mathbf{I}_A = \mathbf{E}_A\mathbf{Z}_B + (\mathbf{E}_A - \mathbf{E}_B)\mathbf{Z}_L]/[\mathbf{Z}_A\mathbf{Z}_B + \mathbf{Z}_L(\mathbf{Z}_A + \mathbf{Z}_B)] \quad ...(vi)$$

Also,
$$\mathbf{I} = \mathbf{I}_A + \mathbf{I}_B = \frac{\mathbf{E}_A\mathbf{Z}_B + \mathbf{E}_B\mathbf{Z}_A}{\mathbf{Z}_A\mathbf{Z}_B + \mathbf{Z}_L(\mathbf{Z}_A + \mathbf{Z}_B)}$$

By multiplying the numerator and denominator of this equation by $1/\mathbf{Z}_A\mathbf{Z}_B$ and the result by $\mathbf{Z}_L$ we get

$$\mathbf{V}_2 = \mathbf{I}\mathbf{Z}_L = \frac{\mathbf{E}_A/\mathbf{Z}_A + \mathbf{E}_B/\mathbf{Z}_B}{1/\mathbf{Z}_A + 1/\mathbf{Z}_B + 1/\mathbf{Z}_L}$$

The two equations (*v*) and (*vi*) then give the values of secondary currents. The primary currents may be obtained by the division of transformation ratio *i.e.* *K* and by addition (if not negligible) of the no-load current. Usually, $\mathbf{E}_A$ and $\mathbf{E}_B$ have the same phase (as assumed above) but there may be some phase difference between the two due to some difference of internal connection in parallel of a star/star and a star/delta 3-phase transformers.

If Z_A and $\mathbf{Z}_B$ are small as compared to $\mathbf{Z}_L$ *i.e.* when the transformers are not operated near short-circuit conditions, then equations for $\mathbf{I}_A$ and $\mathbf{I}_B$ can be put in a simpler and more easily understandable form. Neglecting $\mathbf{Z}_A\mathbf{Z}_B$ in comparison with the expression $\mathbf{Z}_L(\mathbf{Z}_A + \mathbf{Z}_B)$, we have

$$\mathbf{I}_A = \frac{\mathbf{E}_A\mathbf{Z}_B}{\mathbf{Z}_L(\mathbf{Z}_A + \mathbf{Z}_B)} + \frac{\mathbf{E}_A - \mathbf{E}_B}{\mathbf{Z}_A + \mathbf{Z}_B} \quad ...(vii)$$

$$\mathbf{I}_B = \frac{\mathbf{E}_B\mathbf{Z}_A}{\mathbf{Z}_L(\mathbf{Z}_A + \mathbf{Z}_B)} - \frac{\mathbf{E}_A - \mathbf{E}_B}{\mathbf{Z}_A + \mathbf{Z}_B} \quad ...(viii)$$

The physical interpretation of the second term in equations (*vii*) and (*viii*) is that it represents the cross-current between the secondaries. The first term shows how the actual load current divides between the loads. The value of current circulating in transformer secondaries (even when there is no-load) is given by* $\mathbf{I}_C = (\mathbf{E}_A - \mathbf{E}_B)/(\mathbf{Z}_A + \mathbf{Z}_B)$ assuming that $\mathbf{E}_A > \mathbf{E}_B$. It lags behind $\mathbf{E}_A$ by an angle α given by $\tan\alpha = (X_A + X_B)/(R_A + R_B)$. If $E_A + E_B$ the ratios of the currents are inversely as the impedances (numerical values).

If in Eq. (*iv*) we substitute $\mathbf{I}_B = \mathbf{I} - \mathbf{I}_A$ and simplify, we get

$$\mathbf{I}_A = \frac{\mathbf{I}\mathbf{Z}_B}{\mathbf{Z}_A + \mathbf{Z}_B} + \frac{\mathbf{E}_A - \mathbf{E}_B}{\mathbf{Z}_A + \mathbf{Z}_B} \quad ...(ix)$$

Similarly, if we substitute $\mathbf{I}_A = \mathbf{I} - \mathbf{I}_B$ and simplify, then

$$\mathbf{I}_B = \frac{\mathbf{I}\mathbf{Z}_A}{\mathbf{Z}_A + \mathbf{Z}_B} + \frac{\mathbf{E}_A - \mathbf{E}_B}{\mathbf{Z}_A + \mathbf{Z}_B} \quad ...(x)$$

In a similar manner, value of terminal voltage $\mathbf{V}_2$ is given by

---

*Under load conditions, the circulating current is

$$I_C = \frac{E_A - E_b}{Z_A + Z_B + Z_A Z_B/Z_L}$$

If $Z_L = \infty$ *i.e.* on open-circuit, the expression reduces to that given above.

$$\mathbf{V}_2 = \mathbf{IZ}_L = \frac{\mathbf{E}_A\mathbf{Z}_B + \mathbf{E}_B\mathbf{Z}_A - \mathbf{IZ}_A\mathbf{Z}_B}{\mathbf{Z}_A + \mathbf{Z}_B} \quad \ldots(xi)$$

These expressions give the values of transformer currents and terminal voltage in terms of the load current. The value of $\mathbf{V}_2$ may also be found as under :

As seen from Fig. 30.70,

$$\mathbf{I}_A = (\mathbf{E}_A - \mathbf{V}_2)/\mathbf{Z}_A = (\mathbf{E}_A - \mathbf{V}_2)\mathbf{Y}_A; \mathbf{I}_B = (\mathbf{E}_B - \mathbf{V}_2)\mathbf{Y}_B$$

$$\mathbf{I} = \mathbf{V}_2\mathbf{Y}_L = \mathbf{I}_A + \mathbf{I}_B \text{ or } \mathbf{V}_2\mathbf{Y}_L = (\mathbf{E}_A - \mathbf{V}_2)\mathbf{Y}_A + (\mathbf{E}_B - \mathbf{V}_2)\mathbf{Y}_B$$

$$\therefore \mathbf{V}_2(\mathbf{Y}_L + \mathbf{Y}_A + \mathbf{Y}B) = \mathbf{E}_A\mathbf{Y}_A + \mathbf{E}_B\mathbf{Y}_B \text{ or } \mathbf{V}_2 = \frac{\mathbf{E}_A\mathbf{Y}_A + \mathbf{E}_B\mathbf{Y}_B}{\mathbf{Y}_L + \mathbf{Y}_A + \mathbf{Y}_B} \quad \ldots(xii)$$

Eq. (*xi*) gives $\mathbf{V}_2$ in terms of load current. But if only load kVA is given, the problem becomes more complicated and involves the solution of a quadratic equation in $\mathbf{V}_2$.

Now, $\mathbf{S} = \mathbf{V}_2\mathbf{I}$. When we substitute this value of **I** in Eq. (*xi*), we get

$$\mathbf{V}_2 = \frac{\mathbf{E}_A\mathbf{Z}_B + \mathbf{E}_B\mathbf{Z}_A - \mathbf{S}\mathbf{Z}_A\mathbf{Z}_B/\mathbf{V}_2}{(\mathbf{Z}_A + \mathbf{Z}_B)}$$

or

$$\mathbf{V}_2^2(\mathbf{Z}_A + \mathbf{Z}_B) - \mathbf{V}(\mathbf{E}_A\mathbf{Z}_B\mathbf{E}_B\mathbf{Z}_A) + \mathbf{S}\mathbf{Z}_A\mathbf{Z}_B = 0$$

When $\mathbf{V}_2$ becomes known, then $\mathbf{I}_A$ and $\mathbf{I}_B$ may be directly found from

$$\mathbf{V}_2 = \mathbf{E}_A - \mathbf{I}_A\mathbf{Z}_A \text{ and } \mathbf{V}_2 = \mathbf{E}_B - \mathbf{I}_B\mathbf{Z}_B$$

Note. In the case considered above, it is found more convenient to work with *numerical* values of impedances instead of % values.

Example 30.77. *Two transformers A and B are joined in parallel to the same load. Determine the current delivered by each transformer having given : open-circuit e.m.f. 6600 V for A and 6,400 V for B. Equivalent leakage impedance in terms of the secondary = 0.3 + j 3 for A and 0.2 + j1 for B. The load impedance is 8 + j6* **(Elect. Machines-I, Indore Univ. 1987)**

Solution. $$\mathbf{I}_A = \frac{\mathbf{E}_A\mathbf{Z}_B + (\mathbf{E}_A - \mathbf{E}_B)\mathbf{Z}_L}{\mathbf{Z}_A\mathbf{Z}_B + \mathbf{Z}_L(\mathbf{Z}_A + \mathbf{Z}_B}$$

Here $\mathbf{E}_A = 6,600\text{ V}, \mathbf{E}_B = 6,400 V, \mathbf{Z}_L = 8 + j6\,; \mathbf{Z}_A = 0.3 + j3\,; \mathbf{Z}_B = 0.2 + j1$

$$\therefore \quad \mathbf{I}_A = \frac{6600(0.2 + j1) + (6600 - 6400)(8 + j6)}{(0.3 + j3)(0.2j1) + (8 + j6)(0.3 + j3 + 0.2 + j1)}$$

$$117 - j156 = 195\text{A in magnitude})$$

Similarly,

$$\mathbf{I}_B = \frac{\mathbf{E}_B\mathbf{Z}_A(\mathbf{E}_A - \mathbf{E}_B)\mathbf{Z}_L}{\mathbf{Z}_A\mathbf{Z}_B + \mathbf{Z}_L(\mathbf{Z}_A + \mathbf{Z}_B)} = \frac{6400(0.3 + j3) - (6600 - 6400)(8 + j6)}{(0.3 + j3)(0.2 + j1) + (8 + j6)(0.5 + j4)}$$

$$= 349 - j\,231 = 421\text{ A (in magnitude)}$$

Example 30.78. *Two 1−φ transformers, one of 100 kVA and the other of 50 kVA are connected in parallel to the same bus-bars on the primary side, their no-load secondary voltages being 1000 V and 950 V respectively. Their resistances are 2.0 and 2.5 per cent respectively and their reactances 8 and 6 per cent respectively. Calculate no-load circulating current in the secondaries.*

(Adv. Elect Machines, A.M.I.E. Sec B, 1991)

Solution. The circuit connections are shown in Fig. 30.72.

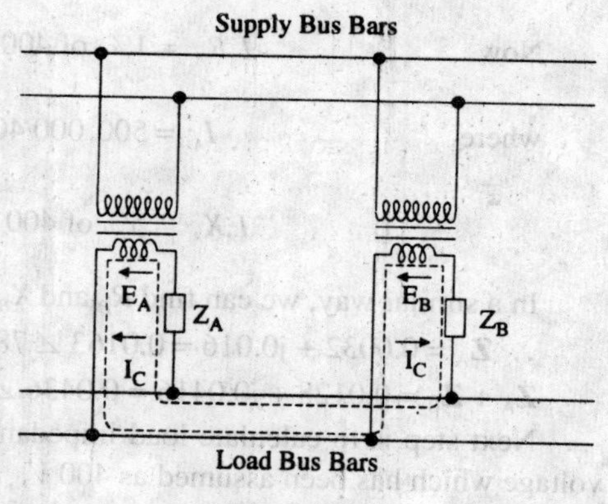

Fig. 30.72

Ist transformer

Normal secondary current

= 100,000/1000 = 100 A

$$R_A = \frac{1000 \times 2.0}{100 \times 100} = 0.2\,\Omega\,;\, X_A = \frac{1000 \times 8}{100 \times 100} = 0.8\,\Omega$$

2nd Transformer

Normal secondary current

= 50,000/950 = 52.63 A

R_B = 950 × 2.5/100 × 52.63 = 0.45 Ω

X_B = 950 × 6/100 × 52.63 = 1.08 Ω

Combined impedance of the two secondaries is

$$Z = \sqrt{(R_A + R_B)^2 + (X_A + X_B)^2}$$

$$= \sqrt{0.65^2 + 1.88^2} = 1.99\,\Omega$$

$\therefore\; I_C = (1000 - 950)/1.99 = \mathbf{25.1\ A} \quad \alpha = \tan^{-1}(1.88/0.65) = \mathbf{71°}$

Example 30.79. *Two single-phase transformers, one of 1000-kVA and the other of 500-kVA are connected in parallel to the same bus-bars on the primary side ; their no-load secondary voltages being 500 V and 510 V respectively. The impedance voltage of the first transformer is 3% and that of the second 5%. Assuming that ratio of resistance to reactance is the same and equal to 0.4 in each, what will be the cross current when the secondaries are connected in parallel ?*

(Electrical Machines-I, Madras Univ. 1985)

Solution. Let us first determine the ohmic values of the two impedances. Also, let the secondary voltage be 480 V*

Full-load $\quad I_A = 1000 \times 1000/480 = 2083$ A ; F.L. $I_B = 500 \times 1000/480 = 1042$ A

$$Z_A = \frac{\%Z_A \times E_A}{100 \times I_A} = \frac{9 \times 500}{100 \times 2083} = 0.0072\,\Omega$$

$$Z_B = \frac{5 \times 510}{100 \times 1042} = 0.0245\,\Omega$$

$$I_C = \frac{E_A - E_B}{Z_A + Z_B} = \frac{510 - 500}{(0.0072 + 0.0245)} = \mathbf{315.4\ A}$$

Note. Since the value of X/R is the same for the two transformers, their is no phase difference between E_A and E_B.

Example 30.80. *Two transformers A and B of ratings 500 kVA and 250 kVA are supplying a load kVA of 750 at 0.8 power factor lagging. Their open-circuit voltages are 405V and 415 V respectively. Transformer A has 1% resistance and 5% reactance and transformer B has 1.5% resistance and 4% reactance. Find (a) cross-current in the secondaries on no-load and (b) the load shared by each transformer.*

Solution. As said earlier, it is more convenient to work with ohmic impedances and for that purpose, we will convert percentage values into numerical values by assuming 400 volt as the terminal voltage (this value is arbitrary but this assumption will not introduce appreciable error).

---

*Though it is chosen arbitrarily, its value must be less than either of the two no-load e.m.fs.

Now $I_A R_A = 1\%$ of 400 $\therefore$ $R_A = \frac{1}{100} \times \frac{400}{1250} = 0.0032\ \Omega$

where $I_A = 500,000/400 = 1250$ A

$I_A X_A = 5\%$ of 400 ; $X_A = \frac{5}{100} \times \frac{400}{1250} = 0.016\ \Omega$ $(i.e. X_A = 5R_A)$

In a similar way, we can find R_B and X_B ; $R_B = 0.0096\ \Omega$; $X_B = 0.0256\ \Omega$

$\therefore\ \mathbf{Z_A} = 0.0032 + j0.016 = 0.0163 \angle 78.5°$; $\mathbf{Z_B} = 0.0096 + j0.0256 = 0.0275\angle 69.4°$

$\mathbf{Z_A} + \mathbf{Z_B} = 0.0128 + j0.0416 = 0.0436 \angle 72.9°$

Next step is to calculate load impedance. Let $\mathbf{Z_L}$ be the load impedance and $\mathbf{V_2}$ the terminal voltage which has been assumed as 400 V.

$\therefore\ V_2^2/\mathbf{Z_L}) \times 10^{-3} = 750\ \angle -36.9°$

$\therefore\ \mathbf{Z_L} = 400^2 \times 10^{-3}/750 \angle -36.9° = 0.214 \angle 36.9° = (0.171 + j0.128)\ \Omega$

(a) $\mathbf{I_C} = \frac{\mathbf{E_A} - \mathbf{E_B}}{\mathbf{Z_A} + \mathbf{Z_B}} = \frac{(405 - 415)}{0.0436 \angle 72.9°} = -230 \angle -72.9°$

(b) $\mathbf{I_A} = \frac{405 \times 0.0275 \angle 69.4° + (405 - 415) \times 0.214 \angle 36.9°}{0.0163 \angle 78.5° \times 0.0275 \angle 69.4° + 0.214 \angle 36.9° \times 0.0436 \angle 72.9°}$

$= 970 \angle -35°$

Similarly, $I_B = \frac{415 \times 0.0163 \angle 78.5° - (405 - 415) \times 0.214 \angle 36.9°}{0.0163 \angle 78.5° \times 0.0275 \angle 69.4° + 214 \angle 36.9° \times 0.0436 \angle 72.9°}$

$\therefore\ \mathbf{S_A} = 400 \times 970 \times 10^{-3} \angle -35° = 388 \angle -35°$ kVA ; $\cos\phi_A = \cos 35° = 0.82$ (lag)

$\mathbf{S_B} = 400 \times 875 \times 10^{-3} \angle 42.6° = 350 \angle -42.6°$ kVA

$\cos\phi_B = \cos 42.6° = 0.736$ (lag)

Example 30.81. *Two transformers A and B are connected in parallel to a load of (2 + j1.5) ohms. Their impedances in secondary terms are Z_A = (0.15 + j0.5) ohm and Z_B = (0.1 + j0.6) ohm. Their no-load terminal voltages are E_A = 207 ∠0° volt and E_B = 205 ∠0° volt. Find the power output and power factor of each transformer.* **(Elect. Machines–I, Punjab Univ. 1991)**

Solution. Using the equations derived in Art. 30.34 (c), we have

$$\mathbf{I_A} = \frac{\mathbf{E_A Z_B} + (\mathbf{E_A} - \mathbf{E_B})\mathbf{Z_L}}{\mathbf{Z_A Z_B} + \mathbf{Z_L}(\mathbf{Z_A} + \mathbf{Z_B})}$$

Here $\mathbf{E_A} = 207 \angle 0°$; $\mathbf{E_B} = 205 \angle 0°$

$\mathbf{Z_A} = (0.15 + j0.5)\ \Omega$; $\mathbf{Z_B} = (0.1 + j0.6)\ \Omega$; $\mathbf{Z_L} = (2 + j1.5) = 2.5 \angle 36.9°$

$$\therefore\ \mathbf{I_A} = \frac{207(0.1 + j0.6) + (207 - 205)(2 + j1.5)}{(0.15 + j0.5)(0.1 + j0.6) + (2 + j1.5)(0.25 + j1.1)}$$

$$= \frac{24.7 + j127.2}{-1.435 + j2.715} = \frac{129.7 \angle 79°}{3.07 \angle 117.9°} = 42.26 \angle -38.9°\ A = (32.89 - j26.55)\ A$$

$$\mathbf{I}_B = \frac{\mathbf{E}_B\mathbf{Z}_A - (\mathbf{E}_A - \mathbf{E}_B)\mathbf{Z}_L}{\mathbf{Z}_A\mathbf{Z}_B + \mathbf{Z}_L(\mathbf{Z}_A + \mathbf{Z}_B)} = \frac{205(0.15 + j0.5) - 2(2 + j1.5)}{-1.435 + j2.715} = \frac{103\angle 75°}{3.07\angle 117.9°}$$

$$= 33.56\angle -42.9° = (24.58 - j22.84)\,A$$

Now $\mathbf{V}_2' = \mathbf{I}\mathbf{Z}_L = (\mathbf{I}_A + \mathbf{I}_B)\mathbf{Z}_L$

$$= (57.47 - j49.39)(2 + j1.5) = 189 - j12.58 = 189.4\angle -3.9°$$

p.f. angle of transformer $A = -3.9° - (-38.9°) = 35°$

$\therefore$ p.f. of $A = \cos 35° = 0.818$ (lag) ; p.f. of $B = \cos[-3.9° - (-42.9°)] = 0.776$ (lag)

Power output of transformer A is $P_A = 189.4 \times 42.26 \times 0.818 =$ **6,548 W**

Similarly, $P_B = 189.4 \times 33.56 \times 0.776 =$ **4,900 W**

Example 30.82. *Two transformers have the folliwng particulars :*

	Transformer A	*Transformer B*
Rated current	*200 A*	*600 A*
Per unit resistance	*0.02*	*0.025*
Per unit reactance	*0.05*	*0.06*
No-load emf	*245 V*	*240 V*

Calculate the terminal voltage when they are connected in parallel and supply a load impedance of (0.25 + j0.1) Ω **(Elect. Machines-I, Sd. Patel Univ. 1981)**

Solution. Impedance in ohms = $\mathbf{Z}_{pu} \times$ N.L. emf/full-load current

$\therefore$ $\mathbf{Z}_A = (245/200)(0.02 + j0.05) = 0.0245 + j0.0613\ \Omega = 0.066\angle 68.2°$

$Z_B = (240/600)(0.025 + j0.06) = 0.01 + j0.024\ \Omega = 0.026\angle 67.3°$

$\mathbf{Z} = (0.25 + j0.1) = 0.269\angle 21.8°$; $\mathbf{Z}_A + \mathbf{Z}_B = 0.0345 + j0.0853 = 0.092\angle 68°$

$\mathbf{Z}_L(\mathbf{Z}_A + \mathbf{Z}_B) = 0.269 \times 0.092\angle 89.8° = 0.0247\angle 89.8° = (0 + j0.0247)$

$\mathbf{Z}_A\mathbf{Z}_B = 0.066 \times 0.026\angle 135.5° = (-0.001225 + j0.001201)$

$\therefore$ $\mathbf{Z}_A\mathbf{Z}_B + \mathbf{Z}_L(\mathbf{Z}_A + \mathbf{Z}_B) = (-0.00125 + j0.259) = 0.0259\angle 92.7°$

Let us take $\mathbf{E}_A$ as reference quantity.

Also $\mathbf{E}_b$ is in phase with $\mathbf{E}_A$ because transformers are in parallel on both sides.

$\mathbf{E}_A\mathbf{Z}_B = 245(0.01 + j0.0245) = 2.45 + j5.87$

$\mathbf{E}_B\mathbf{Z}_A = 240(0.0245 + j0.0613) = 5.88 + j14.7$

$\mathbf{E}_A\mathbf{Z}_B + \mathbf{E}_B\mathbf{Z}_A = 8.33 + j20.57 = 22.15\angle 67.9°$

Now, $\mathbf{I} = \dfrac{\mathbf{E}_A\mathbf{Z}_B + \mathbf{E}_B\mathbf{Z}_A}{\mathbf{Z}_A\mathbf{Z}_B + \mathbf{Z}_L(\mathbf{Z}_A + \mathbf{Z}_B)} = \dfrac{22.15\angle 67.9°}{0.0259\angle 92.7°} = 855\angle -24.8°$

$\therefore$ $\mathbf{V}_2 = \mathbf{I}\mathbf{Z}_L = 885\angle -24.8° \times 0.269\angle 21.8° = 230\angle -3°$

Tutorial Problem No. 30.7

1. A 1000-kV and a 500-kVA, 1-phase transformers are connected to the same bus-bars on the primary side. The secondary e.mfs. at no-load are 500 and 510 V respectively. The impedance voltage of the first transformer is 3.4% and of the second 5%. What cross-current will pass between them when the secondaries are connected together in parallel ? Assuming that the ratio of resistance to reactance is the same in each, what currents will flow in the windings of the two transformers when supplying a total load of 1200 kVA.

[(*i*) **290 A** (*ii*) **1577 and 900 A**] (*City & Guilds, London*)

2. Two transformers *A* and *B* are connected in parallel to supply a load having an impedance of (2 + *j*1.5 Ω). The equivalent impedances referred to the secondary windings are 0.15 + *j*0.5 Ω and 0.1 + *j*0.6 Ω respectively. The open-circuit e.m.f. of *A* is 207 V and of *B* is 205 **V**. Calculate (*i*) the voltage at the load (*ii*) the power supplied to the load (*iii*) the power output of each transformer and (*iv*) the kVA input to each transformer.

[(*i*) **189 ∠–3.8° V** (*ii*) **11.5 kW** (*iii*) **6.5 kW, 4.95 kW** (*iv*) **8.7 kVA, 6.87 kVA**]

QUESTIONS AND ANSWERS ON TRANSFORMERS

Q.1. How is magnetic leakage reduced to a minimum in commercial transformers ?

Ans. By interleaving the primary and secondary windings.

Q.2. Mention the factors on which hysteresis loss depends ?
Ans. (*i*) quality and amount of iron in the core (*ii*) flux density and (*iii*) frequency.
Q.3 How can eddy current loss be minimised
Ans. By laminating the core.
Q.4. In practice, what determines the thickness of the laminae or stampings ?
Ans. Frequency.
Q.5. Does the transformer draw any current when its secondary is open ?
Ans. Yes, no-load primary current.
Q.6. Why ?
Ans. For supplying no-load iron and copper losses.
Q.7. Is Cu loss affected by power factor ?
Ans. Yes, Cu loss varies inversely with power factor.
Q.8. Why ?
Ans. Cu loss depends on currents in the primary and secondary windings. It is well-known that current required is higher when power factor is lower.
Q.9. What effects are produced by change in voltage ?
Ans.
1. Iron loss — varies approximately as V^2.
2. Cu loss — it also varies as V^2 but decreases with an increase in voltage *if constant kVA output is assumed.*
3. efficiency — for distribution transformers, efficiency at fractional loads decreases with increase in voltage while at full load or overload it increases with increase in voltage and *vice versa.*
4. regulation — it varies as V^2 but decreases with increase in voltage *if constant kVA output is assumed.*
5. heating — for constant kVA output, iron temperatures increase whereas Cu temperatures decrease with increase in voltages and *vice versa.*

Q.10. How does change in frequency affect the operation of a given transformer ?
Ans.
1. Iron loss — increases with a decrease in frequency. A 60-Hz transformer will have nearly 11% higher losses when worked on 50 Hz instead of 60 Hz. However, when a 25-Hz transformer is worked on 60 Hz, iron losses are reduced by 25%.
2. Cu loss — in distribution transformers, it is independent of frequency.
3. efficiency — since Cu loss is unaffected by change in frequency, a given transformer efficiency is less at a lower frequency than at a higher one.
4. regulation — regulation at unity power factor is not affected because *IR* drop is independent of frequency. Since reactive drop is affected, regulation at low power factors decreases with a decrease in frequency and *vice versa.* For example, the regulation of a 25-Hz transformer when operated at 50-Hz and low power factor is much poorer.
5. heating — since total loss is greater at a lower frequency, the temperature is increased with decrease in frequency.

Objective Tests–30

1. A transformer transforms
(*a*) frequency
(*b*) voltage
(*c*) current
(*d*) voltage and current.

2. Which of the following is not a basic element of a transformer ?
(*a*) core
(*b*) primary winding
(*c*) secondary winding
(*d*) mutual flux.

3. In an ideal transformer,
(*a*) windings have no resistance
(*b*) core has no losses
(*c*) core has infinite permeability
(*d*) all of the above.

4. The main purpose of using core in a transformer is to
(*a*) decrease iron losses
(*b*) prevent eddy current loss
(*c*) eliminate magnetic hysteresis
(*d*) decrease reluctance of the common magnetic circuit.

5. Transformer cores are laminated in order to
(*a*) simplify its construction
(*b*) minimise eddy current loss
(*c*) reduce cost
(*d*) reduce hysteresis loss.

6. A transformer having 1000 primary turns is connected to a 250-V a.c. supply. For a secondary voltage of 400 V, the number of secondary turns should be

(*a*) 1600 (*b*) 250
(*c*) 400 (*d*) 1250

7. The primary and secondary induced e.mfs. E_1 and E_2 in a two-winding transformer are **always**
(*a*) equal in magnitude
(*b*) antiphase with each other
(*c*) in-phase with each other
(*d*) determined by load on transformer secondary.

8. A step-up transformer increases
(*a*) voltage
(*b*) current
(*c*) power
(*d*) frequency.

9. The primary and secondary windings of an ordinary 2-winding transformer **always** have
(*a*) different number of turns
(*b*) same size of copper wire
(*c*) a common magnetic circuit
(*d*) separate magnetic circuits.

10. In a transformer, the leakage flux of each winding is proportional to the current in that winding because
(*a*) Ohm's law applies to magnetic circuits
(*b*) leakage paths do not saturate
(*c*) the two windings are electrically isolated
(*d*) mutual flux is confined to the core.

11. In a two-winding transformer, the e.m.f. per turn in secondary winding is **always** —— the induced e.m.f. power turn in primary.
(*a*) equal to K times
(*b*) equal to $1/K$ times
(*c*) equal to
(*d*) greater than.

12. In relation to a transformer, the ratio 20 : 1 indicates that
(*a*) there are 20 turns on primary one turn on secondary
(*b*) secondary voltage is 1/20th of primary voltage
(*c*) primary current is 20 times greater than the secondary current.
(*d*) for every 20 turns on primary, there is one turn on secondary.

13. In performing the short circuit test of a transformer
(*a*) high voltage side is usually short circuited
(*b*) low voltage side is usually short circuited
(*c*) any side is short circuited with preference
(*d*) none of the above.
(Elect. Machines, A.M.I.E. Sec. B, 1993)

14. The equivalent resistance of the primary of a transformer having $K = 5$ and $R_1 = 0.1$ ohm when referred to secondary becomes —— ohm.
(*a*) 0.5
(*b*) 0.02
(*c*) 0.004
(*d*) 2.5

15. A transformer has negative voltage regulation when its load power factor is
(*a*) zero (*b*) unity
(*c*) leading (*d*) lagging.

16. The primary reason why open-circuit test is performed on the low-voltage winding of the transformer is that it
(*a*) draws sufficiently large no-load current for convenient reading
(*b*) requires least voltage to perform the test
(*c*) needs minimum power input
(*d*) involves less core loss.

17. No-load test on a transformer is carried out to determine
(*a*) copper loss
(*b*) magnetising current
(*c*) magnetising current and no-load loss
(*d*) efficiency of the transformer.

18. The main purpose of performing open-circuit test on a transformer is to measure its
(*a*) Cu loss
(*b*) core loss
(*c*) total loss
(*d*) insulation resistance.

19. During short-circuit test, the iron lose of a transformer is negligible because
(*a*) the entire input is just sufficient to meet Cu losses only
(*b*) flux produced is a small fraction of the normal flux
(*c*) iron core becomes fully saturated
(*d*) supply frequency is held constant.

20. The iron loss of a transformer at 400 Hz is 10 W. Assuming that eddy current and hysteresis losses vary as the square of flux density, the iron loss of the transformer at rated voltage but at 50 Hz would be —— watt
(*a*) 80 (*b*) 640
(*c*) 1.25 (*d*) 100

21. In operating a 400 Hz transformer at 50 Hz
(*a*) only voltage is reduced in the same proportion as the frequency
(*b*) only kVA rating is reduced in the same proportion as the frequency
(*c*) both voltage and kVA rating are reduced in the same proportion as the frequency
(*d*) non of the above.
(Elect. Machines, A.M.I.E. Sec. B, 1993)

22. The voltage applied to the h.v. side of a transformer during short-circuit test is 2% of its rated voltage. The core loss will be —— percent of the rated core loss
(*a*) 4 (*b*) 0.4
(*c*) 0.25 (*d*) 0.04

23. Transformers are rated in kVA instead of kW because
(*a*) load power factor is often not known
(*b*) kVA is fixed whereas kW depends on load p.f.
(*c*) total transformer loss depends on volt-ampere
(*d*) it has become customary.

24. When a 400-Hz transformer is operated at 50 Hz. its kVA rating is

(a) reduced to 1/8
(b) increased 8 times
(c) unaffected
(d) increased 64 times.

25. At relatively light loads, transformer efficiency is low because
(a) secondary output is low
(b) transformer losses are high
(c) fixed loss is high in proportion to the output
(d) Cu loss is small.

26. A 200 kVA transformer has an iron loss of 1 kW and full-load Cu loss of 2kW. Its load kVA corresponding to maximum efficiency is ——— kVA.
(a) 100 (b) 141.4
(c) 50 (d) 200

27. If Cu loss of a transformer at 7/8th full load is 4900 W, then its full-load Cu loss would be ——— watt.
(a) 5600 (b) 6400
(c) 375 (d) 429

28. The ordinary efficiency of a given transformer is maximum when
(a) it runs at half full-load
(b) it runs at full-load
(c) its Cu loss equals iron loss
(d) it runs slightly overload.

29. The output current corresponding to maximum efficiency for a transformer having core loss of 100 W and equivalent resistance referred to secondary of 0.25 Ω is ——— ampere.
(a) 20 (b) 25
(c) 5 (d) 400

30. The maximum efficiency of a 100-kVA transformer having iron loss of 900 kW and F.L. Cu loss of 1600 W occurs at ——— kVA.
(a) 56.3 (b) 133.3
(c) 75 (d) 177.7

31. The all-day efficiency of a transformer depends primarily on
(a) its copper loss (b) the amount of load
(c) the duration of load (d) both (b) and (c).

32. The marked increase in kVA capacity produced by connecting a 2 winding transformer as an autotransfomer is due to
(a) increase in turn ratio
(b) increase in secondary voltage
(c) increase in transformer efficiency
(d) establishment of conductive link between primary and secondary.

33. The kVA rating of an ordinary 2-winding transformer is increased when connected as an autotransformer because
(a) transformation ratio is increased
(b) secondary voltage is increased
(c) energy is transferred both inductively and conductivity
(d) secondary current is increased.

34. The saving in Cu achieved by converting a 2-winding trnsformer into an autotransformer is determined by
(a) voltage transformation ratio
(b) load on the secondary
(c) magnetic quality of core material
(d) size of the transformer core.

35. An autotransformer having a transformation ratio of 0.8 supplies a load of 3 kW. The power transferred conductively from primary to secondary is ——— kW.
(a) 0.6 (b) 2.4
(c) 1.5 (d) 0.27

36. The essential condition for parallel operation of two 1–ϕ transformers is that they should have the same
(a) polarity
(b) kVA rating
(c) voltage ratio
(d) percentage impedance.

37. If the impedance triangles of two transformers operating in parallel are not identical in shape and size, the two transformers will
(a) share the load unequally
(b) get heated unequally
(c) have a circulatory secondary current even when unloaded
(d) run with different power factors.

38. Two transformers A and B having equal outputs and voltage ratios but unequal percentage impedances of 4 and 2 are operating in parallel. Transformer A will be running over-load by ——— percent.
(a) 50 (b) 66
(c) 33 (d) 25

Answers

1.d 2.d 3.d 4.d 5.b 6.a 7.c 8.a 9.c 10.b 11.c 12.d 13.b 14.d 15 c 16.a 17.c 18.b
19.b 20.b 21.b 22.d 23.c 24.a 25.c 26.b 27.b 28.c 29.a 30.c 31.d 32.d 33.c 34.a
35.b 36.a 37.d 38.c

31 TRANSFORMER-THREE PHASE

31.1. Three-Phase Transformers

Large scale generation of electric power is usually 3-phase at generated voltages of 13.2 kV or somewhat higher. Transmission is generally accomplished at higher voltages of 110, 132, 275, 400 and 750 kV for which purpose 3-phase transformers are necessary to step up the generated voltage to that of the transmission line. Next, at load centres, the transmission voltages are reduced to distribution voltages of 6,600, 4,600 and 2,300 volts. Further, at most of the consumers, the distribution voltages are still reduced to utilization voltages of 440, 220 or 110 volts. Years ago, it was a common practice to use suitably interconnected three single-phase transformers instead of a single 3-phase transformer. But these days, the latter is gaining popularity because of improvement in design and manufacture but principally because of better acquaintance of operating men with the three-phase type. As compared to a bank of single-phase transformers, the main advnatages of a 3-phase transformer are that it occupies less floor space for equal rating, weighs less, costs about 15% less and further, that only one unit is to be handled and connected.

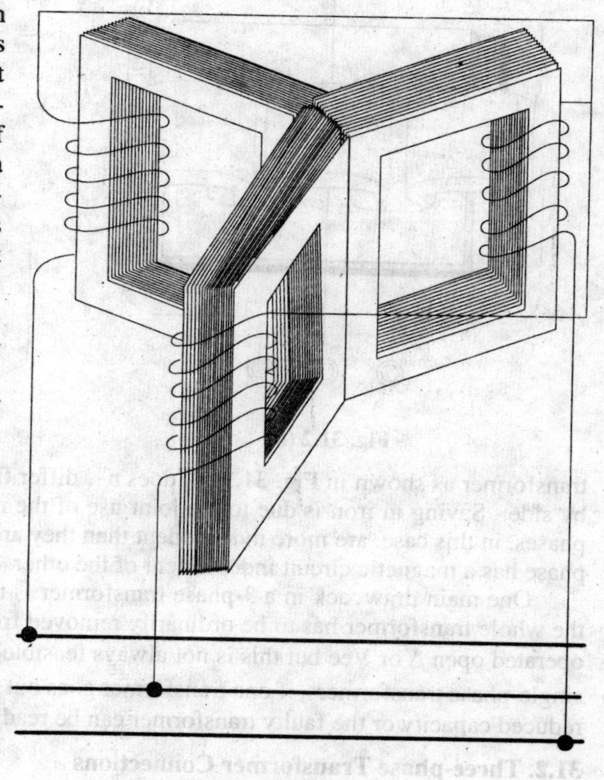

Fig. 31.1

Like single-phase transformers, the three-phase transformers are also of the core type or shell type. The basic principle of a 3-phase transformer is illustrated in Fig. 31.1 in which only primary windings have been shown interconnected in star and put across 3-phase supply. The three cores are 120° apart and their empty legs are shown in contact with each other. The centre leg, formed by these three, carries the flux produced by the three-phase currents I_R, I_Y and I_B. As at any instant $\mathbf{I}_R + \mathbf{I}_Y + \mathbf{I}_B = 0$, hence the sum of three fluxes is

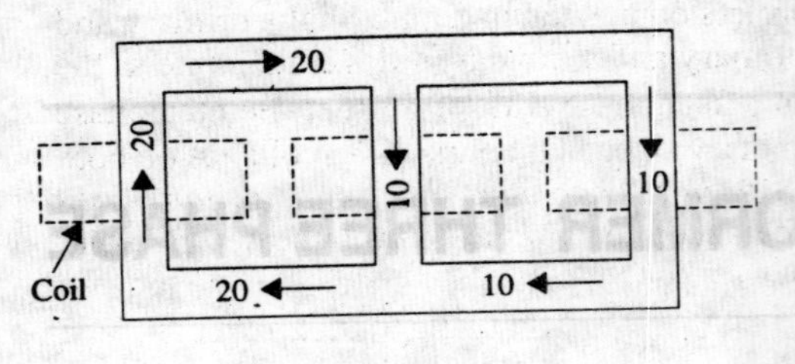

Fig. 31.2 (*a*)

also zero. Therefore, it will make no difference if the common leg is removed. In that case any two legs will act as the return for the third just as in a 3-phase system any two conductors act as the return for the current in the third conductor. This improved design is shown in Fig. 31.2(*a*) where dotted rectangles indicate the three windings and numbers in the cores and yokes represent the directions and magnitudes of fluxes at a particular instant. It will be seen that at any instant, the amount of 'up' flux in any leg is equal to the sum of 'down' fluxes in the other two legs. The core type transfomers are usually wound with circular cylindrical coils.

In a similar way, three single-phase shell type transformers can be combined together to form a 3-phase shell type unit. But some saving in iron can be achieved in constructing a single 3-phase

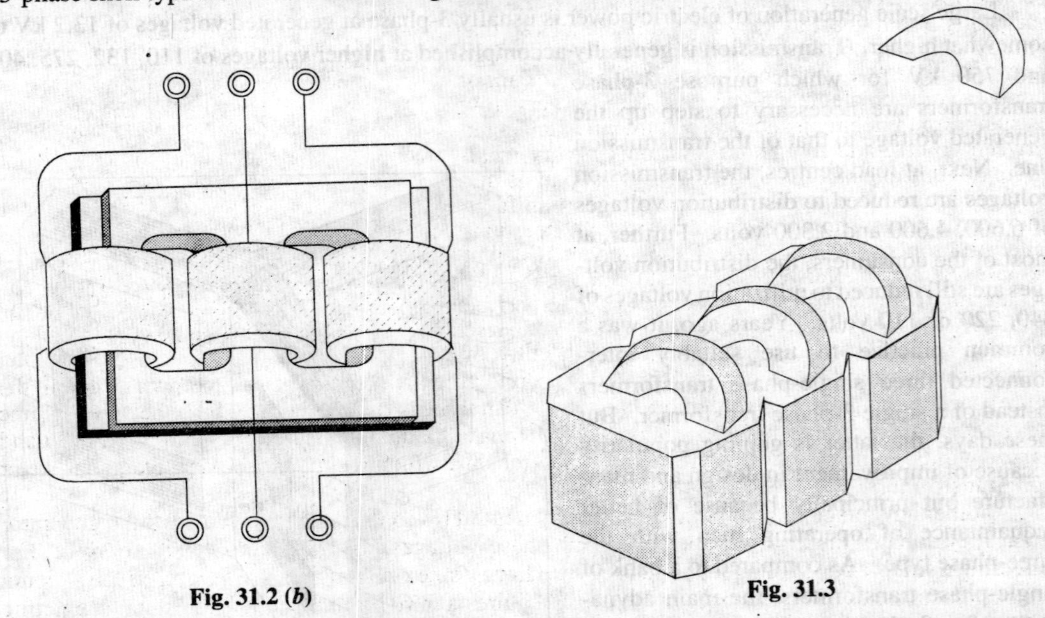

Fig. 31.2 (*b*)

Fig. 31.3

transformer as shown in Fig. 31.3. It does n t differ from three single phase transforn ers put side by side. Saving in iron is due to the joint use of the magnetic paths between the coils. The three phases, in this case, are more independent than they are in the core type transformers, because each phase has a magnetic circuit independent of the other.

One main drawback in a 3-phase transformer is that if any one phase becomes disabled, hen the whole transformer has to be ordinarily removed from service for repairs (the shell type may be operated open Δ or Vee but this is not always feasible). However, in the case of a 3-phase bank of single-phase transformers, if one transformer goes out of order, the system can still be run open-Δ at reduced capacity or the faulty transformer can be readily replaced by a single spare.

31.2. Three-phase Transformer Connections

There are various methods available for transforming 3-phase voltages to higher or lower 3-phase voltages *i.e.* for handling a considerable amount of power. The most common connections are (*i*) Y–Y (*ii*) Δ–Δ (*iii*) Y–Δ (*iv*) Δ–Y (*v*) open-delta or V–V (*vi*) Scott connection or T–T connection.

31.3. Star/Star or Y/Y Connection

This connection is most economical for small, high-voltage transformers becasue the number of turns/phase and the amount of insulation required is minimum (as phase voltage is only $1/\sqrt{3}$ of line voltage). In Fig. 31.4 is shown a bank of 3 transformers connected in Y on both the primary and the secondary sides. The ratio of line voltages on the primary and secondary sides is the same as the transformation ratio of each transformer. However, there is a phase shift of 30° between the phase voltages and line voltages both on the primary and secondary sides. Of course, line voltages on both sides as well as primary voltages are respectively in phase with each other. *This connection works satisfactorily only if the load is balanced.* With the unbalanced load to the neutral, the neutral point shifts thereby making the three line-to-neutral (*i.e.* phase) voltages unequal. The effect of unbalanced loads can be illustrated by placing a single load between phase (or coil) *a* and the neutral on the secondary side. The power to the load has to be supplied by primary phase (or coil) *A*. This primary coil *A* cannot supply the required power because it is in series with primaries *B* and *C* whose secondaries are open. Under these conditions, the primary coils *B* and *C* act as very high impedances so that primary coil can obtain but very little current through them from the line. Hence, secondary coil *a* cannot supply any appreciable power. In fact, a very low resistance approaching a short-circuit may be connected between point *a* and the neutral and only a very small amount of current will flow. This, as said above, is due to the reduction of voltage E_{an} because of neutral shift. In other words, under short-circuit conditions, the neutral is pulled too much towards coil *a*. This reduces E_{an} but increases E_{bn} and E_{in} (however line voltages E_{AB}, E_{BC} and E_{CA} are unaffected). On the primary side, E_{AN} will be practically reduced to zero whereas E_{BN} and E_{CN} will rise to nearly full primary line voltage. This difficulty of shifting (or floating) neutral can be obviated by connecting the primary neutral (shown dotted in the figure) back to the generator so that primary coil *A* can take its required power from between its line and the neutral. It should be noted that if a single phase load is connected between the lines *a* and *b*, there will be a similar but less pronounced neutral shift which results in an overvoltage on one or more transformers.

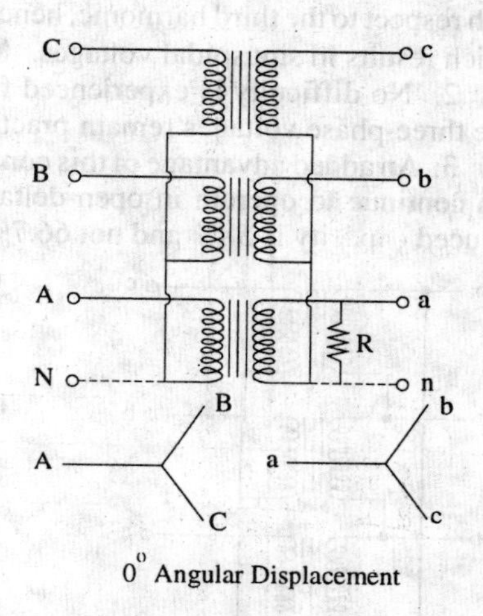

Fig. 31.4

Another advantage of stabilizing the primary neutral by connecting it to neutral of the generator is that it eliminates distortion in the secondary phase voltages. This is explained as follows. For delivering a sine wave of voltage, it is necessary to have a sine wave of flux in the core, but on account of the characteristics of iron, a sine wave of flux requires a third harmonic component in the exciting current. As the frequency of this component is thrice the frequency of the circuit, at any given instant, it tends to flow either towards or away from the neutral point in all the three transformers. If the primary neutral is isolated, the triple frequency current cannot flow. Hence, the flux in the core cannot be a sine wave and so the voltages are distorted. But if the primary neutral is earthed *i.e.* joined to the generator neutral, then this provides a path for the triple-frequency currents and e.mfs. and the difficulty is overcome. Another way of avoiding this trouble of oscillating neutral is to provide each of the transformers with a third or tertiary winding of relatively low kVA rating. This tertiary winding is connected in Δ and provides a circuit in which the triple-frequency component of the magnetising current can flow (with an isolated neutral, it could not). In that case, a sine wave of voltage applied to the primary will result in a sine wave of phase voltage in the secondary. As said above, the advantage of this connection is that insulation is stressed only to the extent of line to neutral voltage *i.e.* 58% of the line voltage.

31.4. Delta–Delta or Δ–Δ Connection

This connection is economical for large, low-voltage transformers in which insulation problem is not so urgent, because it increases the number of turns/phase. The transformer connections and voltage triangles are shown in Fig. 31.5. The ratio of transformation between primary and secondary line voltage is exactly the same as that of each transformer. Further, the secondary voltage triangle *abc* occupies the same relative position as the primary voltage triangle *ABC i.e.* there is no angular displacement between the two. Moreover, there is no internal phase shift between phase and line voltages on either side as was the case in Y–Y connection. This connection has the following advantages:

1. As explained above, in order that the output voltage be sinusoidal, it is necessary that the magnetising current of the transformer must contain a third harmonic component. In this case, the third harmonic component of the magneising current can flow in the Δ-connected transformer primaries without flowing in the line wires. The three phases are 120° apart which is 3 × 120 = 360° with respect to the third harmonic, hence it merely circulates in the Δ. Therefore, the flux is sinusoidal which results in sinusoidal voltages.

2. No difficulty is experienced from unbalanced loading as was the case in *Y–Y* connection. The three-phase voltages remain practically constant regardless of load imbalance.

3. An added advantage of this connection is that if one transformer becomes disabled, the system can continue to operate in open-delta or in V–V although with reduced available capacity. The reduced capacity is 58% and not 66.7% of the normal value, as expalined in Art. 31.7.

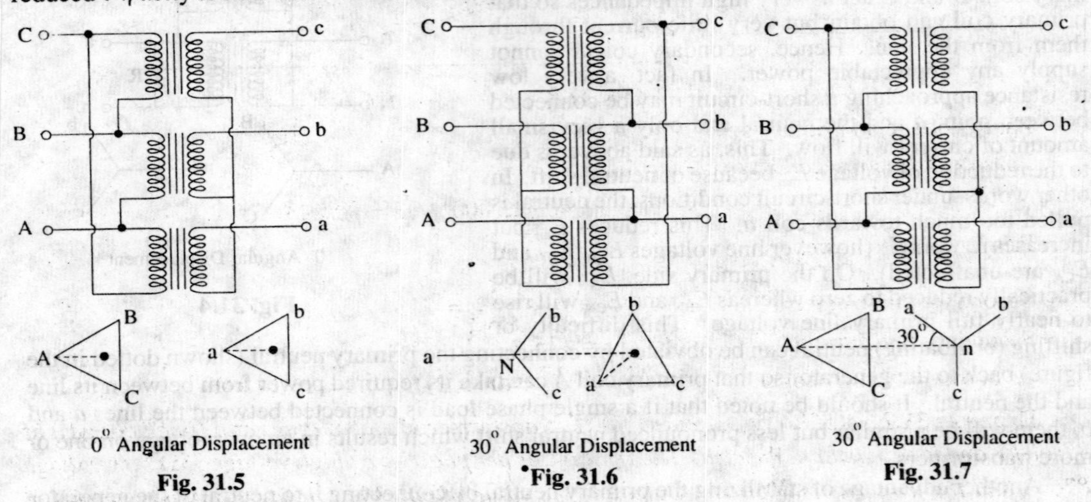

0° Angular Displacement

Fig. 31.5

30° Angular Displacement

Fig. 31.6

30° Angular Displacement

Fig. 31.7

31.5. Wye/Delta or Y/Δ Connection

The main use of this connection is at the substation end of the transmission line where the voltage is to be stepped down. The primary winding is Y-connected with grounded neutral as shown in Fig. 31.6. The ratio between the secondary and primary line voltage is $1/\sqrt{3}$ times the transformation ratio of each transformer. There is a 30° shift between the primary and secondary line voltages which means that a Y-Δ transformer bank cannot be paralleled with either a Y–Y *or* a Δ–Δ bank. Also, third harmonic currents flows in the Δ to provide a sinusoidal flux.

31.6. Delta/Wye or Δ/Y Connection

This connection is generally employed where it is necessary to step up the voltage as for example, at the beginning of high tension transmission system. The connection is shown in Fig. 31.7. The neutral of the secondary is grounded for providing 3-phase 4-wire service. In recent years, this connection has gained considerable popularity because it can be used to serve both the 3-phase power equipment and single-phase lighting circuits.

This connection is not open to the objection of a floating neutral and voltage distortion because the existence of a Δ-connection allows a path for the third-harmonic currents. It would be observed that the primary and secondary line voltages and line currents are out of phase with each other by 30°. Because of this 30° shift, it is impossible to parallel such a bank with a Δ–Δ or Y–Y bank of transformers even though the voltage ratios are correctly adjusted. The ratio of secondary of primary voltage is $\sqrt{3}$ times the transformation ratio of each transformer.

Example 31.1. *A 3-phase, 50-Hz transformer has a delta-connected primary and star-connected secondary, the line voltages being 22,000 V and 400 V respectively. The secondary has a star-connected balanced load at 0.8 power factor lagging. The line current on the primary side is 5 A. Determine the current in each coil of the primary and in each secondary line. What is the output of the transformer in kW ?*

Solution. It should be noted that in three-phase transformers, the *phase* transformation ratio is equal to the turn ratio but the terminal or the voltages depend upon the method of connection employed. The Δ/Y connection is shown in Fig. 31.8.

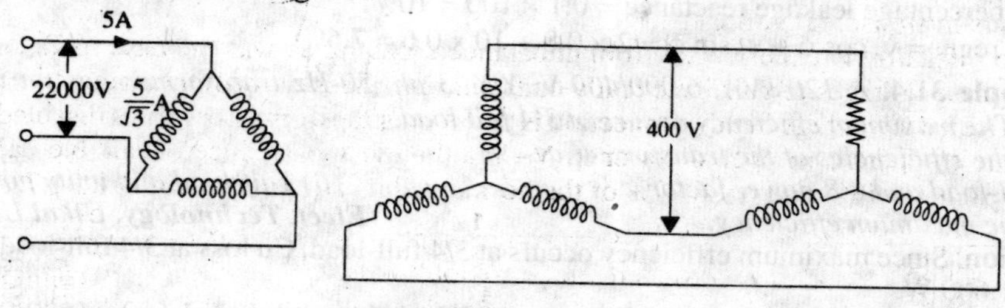

Fig. 31.8

Phase voltage on primary side $= 22{,}000$ V

Phase voltage on secondary side $= 400/\sqrt{3}$

$\therefore \quad K = 400/22{,}000 \times \sqrt{3} = 1/55\sqrt{3}$

Primary phase current $= 5/\sqrt{3}$ A

Secondary phase current $= \dfrac{5}{\sqrt{3}} \div \dfrac{1}{55\sqrt{3}} = 275$ A

Secondary line current $=$ **275 A**

$\therefore$ output $= \sqrt{3} V_L I_L \cos\phi = \sqrt{3} \times 400 \times 275 \times 0.8 =$ **15.24 kW.**

Example 31.2. *A 500-kVA, 3-phase, 50-Hz transformer has a voltage ratio (line voltages) of 33/11-kV and is delta/star connected. The resistances per phase are : high voltage 35Ω, low voltage 876 Ω and the iron loss is 3050 W. Calculate the value of efficiency at full-load and one-half of full-load respectively (a) at unity p.f. and (b) 0.8 p.f.*(**Electrical Machinery, Madras Univ. 1985**)

Solution. Transformation ratio $K = \dfrac{11{,}000}{\sqrt{3} \times 33{,}000} = \dfrac{1}{3\sqrt{3}}$

Per phase $R_{02} = 0.876 + (1/3\sqrt{3})^2 \times 35 = 2.172\ \Omega$

Secondary phase current $= \dfrac{500{,}000}{\sqrt{3} \times 11{,}000} = \dfrac{500}{11\sqrt{3}}$ A

Full-load condition

Full-load total Cu loss $= 3 \times (500/11\sqrt{3})^2 \times 2.172 = 4{,}490$ W ; Iron loss $= 3{,}050$ W

Total full-load losses $= 4{,}490 + 3{,}050 = 7{,}540$ W; Output at unity p.f. $= 500$ kW

$\therefore$ F.L. efficiency $= 500{,}000/507{,}540 = 0.9854$ or **98.54%**; Output at 0.8 p.f. $= 400$ kW

$\therefore$ efficiency $= 400{,}000/407{,}540 = 0.982$ or **98.2%**

Half-load condition

Output at unity p.f. $= 250$ kW Cu losses $= (1/2)^2 \times 4{,}490 =$ **1,222 W**

Total losses $= 3{,}050 + 1{,}122 = 4{,}172$ W

$\therefore \quad \eta = 250{,}000/254{,}172 = 0.9835 =$ **98.35%**

Output at 0.8 p.f. $= 200$ kW $\therefore \quad \eta =$ **200,000,/204,172 = 0.98 or 98%**

Example 31.3. *A 3-phae, 6,600/415-V, 2,000-kVA transformer has a per unit resistance of 0.02 and a per unit leakage reactance of 0.1. Calculate the Cu loss and regulation at full-load 0.8 p.f. lag.* **(Electrical Machines-I, Bombay Univ. 1987)**

Solution. As seen from Art. 27–16, $\% R = \%$ Cu loss $= \dfrac{\text{Cu loss}}{\text{VA}} \times 100$.

Now, $\% R = 0.02 \times 100 = 2\%$ $\quad \therefore \quad 2 = \dfrac{\text{Cu loss}}{2,000} \times 100 \; \therefore$ Cu loss = **40 kW**

Now, percentage leakage reactance = $0.1 \times 100 = 10\%$

regn. = $v_r \cos\phi + v_x \sin\phi = 2 \times 0.8 + 10 \times 0.6 =$ **7.6%**

Example 31.4. *A 120-kVA, 6,000/400-V, Y/Y, 3-ph, 50-Hz transformer has an iron loss of 1,600 W. The maximum efficiency occurs at 3/4 full load.*

Find the efficiencies of the transformer at

(i) full-load and 0.8 power factor *(ii) half-load and unity power factor*

(iii) the maximum efficiency. ***(Elect. Technology, Utkal Univ. 1987)***

Solution. Since maximum efficiency occurs at 3/4 full-load, Cu loss at 3/4 full-load equals iron of loss of 1,600 W.

Cu loss at 3/4 F.L $= 1,600$ W; Cu loss at F.L. $= 1,600 \times (4/3)^2 = 2,845$ W

(*i*) F.L. output at 0.8 p.f. $= 120 \times 0.8 = 96$ kW $= 96,000$ W

Total loss $= 1,600 + 2,845 = 4,445$ W

$\therefore \quad \eta = \dfrac{96,000}{100,445} \times 100 =$ **95.57%**

(*ii*) Cu loss at 1/2 full-load $= (1/2)^2 \times 2,845 = 710$ W

Total loss $= 710 + 1,600 = 2310$ W

Output at 1/2 F.L. and u.p.f. is = 60 kW = 60,000 W ; $\eta = \dfrac{60,000}{62,310} \times 100 =$ **96.3%**

(*iii*) Maximum efficiency occurs at 3/4 full-load when iron loss equals Cu loss.

Total loss $= 2 \times 1,600 = 3,200$ W

Output at u.p.f. $= (3/4) \times 120 = 90$ kW $= 90,000$ W

Input $= 90,000 + 3,200 = 93,200$ W $\therefore \eta = \dfrac{90,000}{93,200} \times 100 =$ **96.57%**

Example 31.5. *A 3-phase transformer, ratio 33/6.6-kVA, Δ/Y, 2-MVA has a primary resistance of 8 Ω per phase and a secondary resistance of 0.08 ohm per phase. The percentage impedance is 7%. Calculate the secondary voltage with rated primary voltage and hence the regulation for full-load 0.75 p.f. lagging conditions.* **(Elect. Machine-I, Nagpur, Univ. 1993)**

Solution. F.L. secondary current $= \dfrac{2 \times 10^6}{\sqrt{3} \times 6.6 \times 10^3} = 175$ A

$K = 6.6/\sqrt{3} \times 33 = 1/8.65$; $R_{02} = 0.08 + 8/8.65^2 = 0.1867\ \Omega$ per phase

Now, secondary impedance drop per phase $= \dfrac{7}{100} \times \dfrac{6,600}{\sqrt{3}} = 266.7$ V

$\therefore \quad Z_{02} = 266.7/175 = 1.523\ \Omega$ per phase

$$X_{02} = \sqrt{Z_{02}^2 - R_{02}^2} = \sqrt{1.523^2 - 0.1867^2} = 1.51\ \Omega\text{/phase}$$

Drop per phase $= I_2(R_{02}\cos\phi + X_{02}\sin\phi) = 175(0.1867 \times 0.75 + 1.51 \times 0.66) = 200$ V

Secondary voltage/phase $= 6,600/\sqrt{3} = 3,810$ V $\therefore \; V_2 = 3,810 - 200 = 3,610$ V

$\therefore$ secondary *line* voltage $= 36,10 \times \sqrt{3} =$ **6,250 V**

% regn. $= 200 \times 100/3,810 =$ **5.23%**

Example 31.6. *A 100-kVA, 3-phase, 50-Hz, 3,300/400 V transformer is Δ-connected on the h.v. side and Y-connected on the l.v. side. The resistance of the h.v. winding is 3.5 Ω per phase and that of the l.v. winding 0.02 Ω per phase. Calculate the iron losses of the transformer at normal voltage and frequency if its full-load efficiency be 95.8% at 0.8 p.f. (lag)*

(A.C. Machines-I, Jadavpur Univ. 1989)

Solution. F.L. output = 100 × 0.8 = 80 kW; input = 80/0.958 = 83.5 kW
Total loss = input − output = 83.5 − 80 = 3.5 kW
Let us find full-load Cu losses for which purpose, we would first calculate R_{02}.

$$K = \frac{\text{secondary voltage/phase}}{\text{primary voltage/phase}} = \frac{400/\sqrt{3}}{3.300} = \frac{4}{33\sqrt{3}}$$

$$R_{02} = R_2 + K^2R_1 = 0.02 + (4/\sqrt{3}\times 33)^2 \times 3.5 = 0.037\ \Omega$$

Full-load secondary phase current is $I_2 = 100{,}000/\sqrt{3} \times 400 = 144.1$ A
Total Cu loss $= 3I_2^2R_{02} = 3 \times 144.1^2 \times 0.037 = 2{,}305$ W
Iron loss = Total loss − F.L. Cu loss = 3,500 − 2,305 = **1,195 W**

Example 31.7. *A 5,000-kVA, 3-phase transformer, 6.6/33-kV, Δ/Y, has a no-load loss of 15 kW and a full-load loss of 50 kW. The impedance drop at full-load is 7%. Calculate the primary voltage when a load of 3,200 kW at 0.8 p.f. is delivered at 33 kV.*

Solution. Full-load $I_2 = 5 \times 10^6/\sqrt{3}\times 33{,}000 = 87.5$ A
Impedance drop/phase = 7% of $(33/\sqrt{3})$ = 7% of 19 kV = 1,330 V
∴ $Z_{02} = 1{,}330/87.5 = 15.3$ Ω/phase; F.L. Cu loss = 50 − 15 = 35 kW
∴ $3I_2R_{02} = 35{,}000$; $R_{02} = 35{,}000/3 \times 8.75^2 = 1.53$ Ω/phase
∴ $X_{02} = \sqrt{15.3^2 - 1.53^2} = \mathbf{15.23\ \Omega}$

When load is 3,200 kW at 0.8 p.f.

$I_2 = 3{,}200/\sqrt{3}\times 33 \times 0.8 = 70$ A; drop $= 70(1.53\times 0.8 + 15.23\times 0.6) = 725$ V/phase

$$\therefore \quad \%\ \text{regn.} = \frac{725\times 100}{19{,}000} = 3.8\%$$

Primary voltage will have to be increased by 3.8%.
∴ primary voltage = 6.6 + 3.8% of 6.6 = 6.85 kV = **6,850 V**

Example 31.8. *A 3-phase transformer has its primary connected in Δ and its secondary in Y. It has an equivalent resistance of 1% and an equivalent reactance of 6%. The primary applied voltage is 6,600 V. What must be the ratio of transformation in order that it will deliver 4,800 V at full-load current and 0.8 power factor (lag).* **(Elect. Technology-II, Magadh Univ. 1991)**

Solution. Percentage regulation
$= v_r \cos\phi + v_x \sin\phi$
$= 1 \times 0.8 + 6 \times 0.6 = 4.4\%$
Induced secondary e.m.f. (line value)
= 4,800 + 4.4% of 4,800 = 5,010 V
Secondary phase voltage
$= 5{,}010/\sqrt{3} = 2{,}890$ V
Transformation ratio
$K = 2{,}890/6{,}600 = \mathbf{0.437}$.

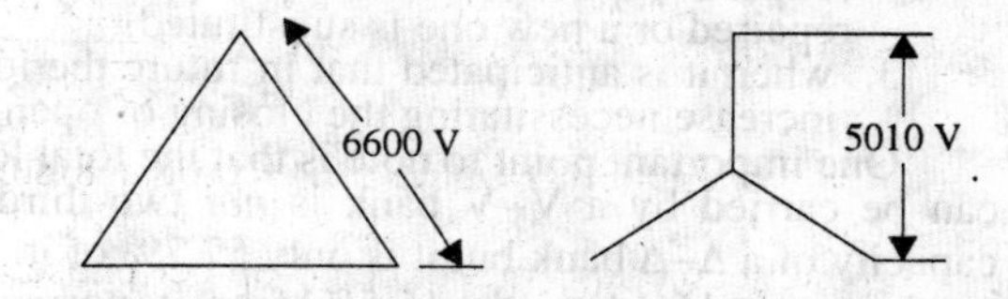

Fig. 31.9

Example 31.9. *A 2000-kVA, 6,600/400-V, 3-phase transformer is delta-connected on the high voltage side and star-connected on the low-voltage side. Determine its % resistance and % reactance drops, % efficiency and % regulation on full load 0.8 p.f. leading given the following data :*
S.C. test ; H.V. data : 400 V, 175 A and 17 kW
O.C. test ; L.V. data : 400 V, 150 A and 15 kW

[Basic Elect. Machines, Nagpur Univ. 1993]

Solution. From S.C. test data, we have

Primary voltage/phase $= 400/\sqrt{3} = 231$ V ; Primary current/phase $= 175/\sqrt{3} = 100$ A

$\therefore \quad Z_{10} = 231/100 = 2.31\ \Omega$

Now $I_1^2 R_{01} = 17{,}000/3$ or $R_{01} = 0.567\ \Omega$; $X_{01} = \sqrt{2.31^2 - 0.567^2} = 2.24\ \Omega$

$$\%R = \frac{I_1 R_{01}}{V_1} \times 100 = \frac{100 \times 0.567}{6{,}600} \times 100 = \mathbf{0.86}$$

$$\%X = \frac{I_1 X_{01}}{V_1} \times 100 = \frac{100 \times 2.24}{6{,}600} \times 100 = \mathbf{3.4}$$

%regn. $= v_r \cos\phi - v_x \sin\phi = 0.86 \times 0.8 - 3.4 \times 0.6 = -1.34\%$

Full-load primary line current can be found from

$$\sqrt{3} \times 6{,}600 \times I_1 = 2{,}000 \times 1{,}000 \ ; \ I_1 = 175 \text{ A}$$

It shows that S.C. test has been carried out under full-load conditions.

Total losses $= 17 + 15 = 32$ kW ; F.L. output $= 2{,}000 \times 0.8 = 1600$ kW

$\eta = 1{,}600/1{,}632 = 0.98$ or **98%**

Tutorial Problem No. 31.1

1. A 3-phase, star-connected alternator generates 6,360 V per phase and supplies 500 kW at a p.f. 0.9 lagging to a load through a step-down transformer of turns 40 : 1. The transformer is delta connected on the primary side and star-connected on the secondary side. Calculate the value of the line volts at the load. Calculate also the currents in (*a*) alternator windings (*b*) transforemr primary windings (*c*) transforme secondary windings. **[476 V (*a*) 29.1 A (*b*) 16.8 A (*c*) 672 A]**

2. A 11,000/6,600 V, 3-ϕ, transformer has a star-connected primary and a delta-connected secondary. It supplies a 6.6 kV motor having a star-connected stator, developing 969.8 kW at a power factor of 0.9 lagging and an efficiency of 92 per cent. Calculate (*i*) motor line and phase currents (*ii*) transformer secondary current and (*iii*) transformer primary current. **[(*a*) Motor ; $I_L = I_{ph} = 126.3$ A (*b*) phase current 73 A (*c*) 75.8 A]**

31.7. Open-Delta or V–V connection

If one of the transformers of a Δ–Δ is removed and 3-phase supply is connected to the primaries as shown in Fig. 31.10, then three equal 3-phase voltages will be available at the secondary terminals on no-load. This method of transforming 3-phase power by means of only two transformers is called the open –Δ or V–V connection.

It is employed:

1. when the three-phase load is too small to warrant the installation of full three-phase transformer bank.
2. when one of the transformers in a Δ-Δ bank is disabled, so that service is continued although at reduced capacity, till the faulty transformer is repaired or a new one is substituted.
3. when it is anticipated that in future the load will increase necessitating the closing of open delta.

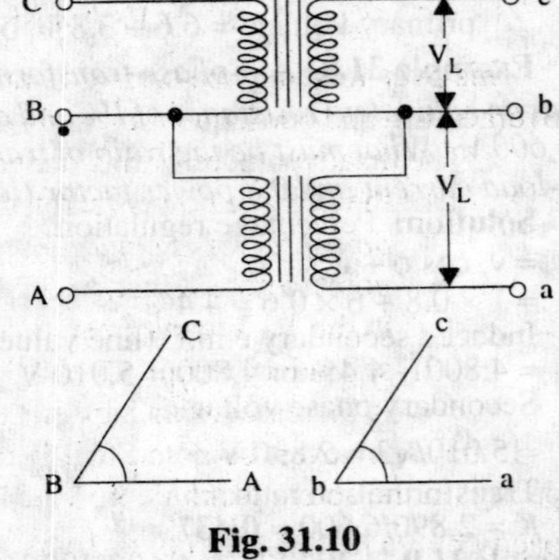

Fig. 31.10

One important point to note is that the total load that can be carried by a V–V bank is *not* two-third of the capacity of a Δ–Δ bank but it is only 57.7% of it. That is a reduction of 15% (strictly, 15.5%) from its normal rating. Suppose there is Δ – Δ bank of three 10-kVA transformers. When one transformer is removed, then it runs in V–V. The total rating of the two transformers is 20 kVA. But the capacity of the V–V bank is not the sum of the transformer kVA ratings but only 0.866 of it *i.e.* $20 \times 0.866 = 17.32$ (or $30 \times 0.577 = 17.3$ kVA). The fact that the ratio of V-capacity to Δ-capacity is $1/\sqrt{3} = 57.7\%$ (or nearly 58%) instead of $66\frac{2}{3}$ per cent can be proved as follows :

As seen from Fig. 31.11 (*a*)

Δ-Δ capacity $= \sqrt{3}.\ V_L \,.\, V_L = \sqrt{3}.\ V_L\,(\sqrt{3}.\ I_S) = 3 V_L\, I_S$

In Fig. 31.11 (*b*), it is obvious that when Δ-Δ bank becomes V–V bank, the secondary line current I_L becomes equal to the secondary phase current I_S.

$$\therefore \quad \text{V- V capacity} = \sqrt{3}.V_L I_L = \sqrt{3}\,V_L.I_S$$

$$\therefore \quad \frac{\text{V - V capacity}}{\Delta - \Delta \text{ capacity}} = \frac{\sqrt{3}.V_L I_S}{3V\,I_S} = \frac{I}{\sqrt{3}} = 0.577 \text{ or } 58 \text{ percent}$$

It means that the 3-phase load which can be carried *without exceeding the ratings* of the transformers is 57.7 percent of the original load rather than the expected 66.7%.

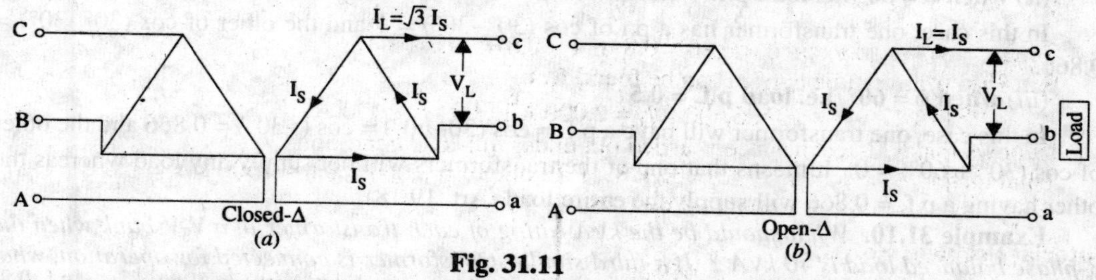

Fig. 31.11

It is obvious from above that when one transformer is removed from a Δ–Δ bank.

1. the bank capacity is reduced from 30 kVA to 30 × 0.577 = 17.3 kVA and not to 20 kVA as might be thought off-hand.
2. only 86.6% of the rated capacity of the two remaining transformers is available (*i.e.* 20 × 0.866 = 17.3 kVA). In other words, ratio of *operating* capacity to *available* capacity of an open-Δ is 0.866. This factor of 0.866 is sometimes called the *utility factor*.
3. each transformer will supply 57.7% of load and not 50% when operating in V–V (Ex. 31.11)

However, it is worth noting that if three transformers in a Δ-Δ bank are delivering their *rated load** and one transformer is removed, the overload on *each* of the two remaining transformers is 73.2% because.

$$\frac{\text{total load in } V-V}{VA/\text{transformer}} = \frac{\sqrt{3}\cdot V_L I_S}{V_L I_S} = \sqrt{3} = 1.732$$

This over-load may be carried temporarily but some provision must be made to reduce the load if overheating and consequent breakdown of the remaining two transformers is to be avoided.

The disadvantages of this connection are :

1. the average power factor at which the V-bank operates is less than that of the load. *This power factor is actually 86.6% of the balanced load power factor*. Another significant point to note is that, except for a balanced unity power factor load, the two transformers in the V–V bank operate at different power factors (Art. 31.8).
2. secondary terminal voltages tend to become unbalanced to a great extent when the load is increased, this happens even when the load is perfectly balanced.

It may, however, be noted that if two transformers are operating in V–V and loaded to rated capacity (in the above example, to 17.3 kVA), the addition of a third transformer increases the total capacity by $\sqrt{3}$ or 173.2% (*i.e.* to 30 kVA). It means that for an increase in cost of 50% for the third transformer, the increase in capacity is 73.2% when converting from a V–V system to a Δ-Δ system.

---

*In Ex. 31.11, the three transformers are not supplying their rated load of 20 × 3 = 60 kVA but only 40 kVA.

31.8. Power Supplied by V–V Bank

When a *V-V* bank of two transformers supplies a balanced 3-phase load of power factor cos ϕ, then one transformer operates at a p.f. of cos $(30°-\phi)$ and the other at cos $(30° + \phi)$. Consequently, the two transformers will not have the same voltage regulation.

$\therefore \quad P_I = \text{kVA} \cos(30°-\phi)$ and $P_2 = \text{kVA} \cos(30° + \phi)$

(*i*) when ϕ = 0 *i.e.* load p.f. = 1

Each transformer will have a p.f. = cos 30° = 0.866

(*ii*) when ϕ = 30° *i.e.* load p.f. = 0.866

In this case, one transformer has a p.f.of cos (30°–30°) = 1 and the other of cos (30° 30°) = 0.866.

(*iii*) when ϕ = 60° i.e. load p.f. = 0.5

In this case, one transformer will have a p.f. = cos (30–60°) = cos (–30°) = 0.866 and the other of cos (30° + 60°) = 0. It means that one of the transformers will not supply any load whereas the other having a p.f. = 0.866 will supply the entire load (Art. 19.18).

Example 31.10. *What should be the kVA rating of each transformer in a V-V bank when the 3-phase balanced load is 40 kVA ? If a third similar transformer is connected for operation, what is the rated capacity ? What percentage increase in rating is affected in this way ?*

Solution. As pointed out earlier, the kVA rating of each transformer has to be 15% greater.

$\therefore$ kVA/transformer = (40/2) × 1.15 = **23**

Δ-Δ bank rating = 23 × 3 = 69 ; Increase = [(69–40)/ × 100 = **72.5%**

Example 31.11. *A Δ-Δ bank consisting of three 20-kVA, 2300/230-V transformers supplies a load of 40 kVA. If one transformer is removed, find for the resulting V–V connection*

(*i*) *kVA load carried by each transformer*

(*ii*) *percent of rated load carried by each transformer*

(*iii*) *total kVA rating of the V-V bank*

(*iv*) *ratio of the V–V bank to Δ–Δ bank transformer ratings*

(*v*) *percent increase in load on each transformer when bank is converted into V–V bank.*

Solution. (*i*) As explained earlier in Art. 31.7, $\dfrac{\text{total kVA load in V-V bank}}{\text{VA/transformer}} = \sqrt{3}$

$\therefore$ kVA load supplied by *each* of the two transformers = $40/\sqrt{3}$ = **23.1 kVA**

Obviously, each transformer in V–V bank does not carry 50% of the original load but 57.7%.

(*ii*) percent of rated load $= \dfrac{\text{kVA load/transformer}}{\text{kVA rating/transformer}} = \dfrac{23.1}{20} = \mathbf{115.5\%}$

carried by each transformer.

Obviously, in this case, each transformer is overloaded to the extent of 15.5 percent.*

(*iii*) kVA *rating* of the V–V bank = (2 × 20) = 0.866 = **34.64 kVA**

(*iv*) $\dfrac{\text{V-V rating}}{\Delta-\Delta \text{ rating}} = \dfrac{34.64}{60} = 0.577$ or **57.7%**

As seen, the rating is reduced to 57.7% of the original rating.

(*v*) Load supplied by each transformer in Δ–Δ bank = 40/3 = 13.33 kVA

$\therefore$ percentage increase in load supplied by each transformer

---

*Overloading becomes 73.2% only when full rated load is supplied by the Δ–Δ bank (*i.e.*. 3 × 20 = 60 kVA in this case) before it becomes V–V bank.

$$= \frac{\text{kVA load/transformer in V-V bank}}{\text{kVA load/transformer in } \Delta-\Delta \text{ bank}} = \frac{23.1}{13.3} = 1.732 = \mathbf{173.2\ \%}$$

It is obvious that each transformer in the Δ–Δ bank supplying 40 kVA was running underloaded (13.33 *vs* 20 kVA) but runs overloaded (23.1 *vs* 20 kVA) in V–V connection.

Example 31.12. *A balanced 3-phase load of 150 kVA at 1000 V, 0.866 lagging power factor is supplied from 2000 V, 3-phase mains through single-phase transformers (assumed to be ideal) connected in (i) delta-delta (ii) Vee-Vee. Find the current in the windings of each transformer and the power factor at which they operate in each case. Explain your calculations with circuit and vector diagrams.*

Solution. (*i*) **Delta-Delta Connection**

$\sqrt{3} \times V_L I_L \cos\phi = 15{,}000$; $\sqrt{3} \times 1000 \times I_L \times 0.866 = 150{,}000$ $\therefore I_L = 100$ A

∴ secondary line current = 100 A ; secondary phase current = $100/\sqrt{3}$ = **57.7A**

Transformation ratio = 1000/2000 = 1/2

∴ primary phase current = 57.7/2 = **28.85 A**

(*ii*) **Vee-Vee Connection**

Let I be the secondary line current which is also the phase current in V–V connection. Then

$$\sqrt{3} \times 1000 \times I \times 0.866 = 150{,}000 \quad \therefore \quad I = 100 \text{ A}$$

∴ secondary phase current = **100 A** ; primary phase current = 100 × 1/2 = **50 A**

Transformer power factor = 86.6 per cent of 0.866 = **0.75 (lag).**

Example 31.13. *(a) Two identical 1-phase transformers are connected in open-delta across 3-phase mains and deliver a balanced load of 3000 kW at 11 kV and 0.8 p.f. lagging. Calculate the line and phase currents and the power factors at which the two transformers are working.*

(b) If one more identical unit is added and the open delta is converted to closed delta, calculate the additional load of the same power factor that can now be supplied for the same temperature rise. Also calculate the phase and line currents. **(Elect. Machinery-I, Madras Univ. 1987)**

Solution. (*a*) If I is the line current, then

$$\sqrt{3} \times 11{,}000 \times I \times 0.8 = 3{,}000{,}000 \qquad \mathbf{I = 197\ A}$$

Since, this also represents the phase current,

∴ secondary phase current = **197 A** : transformer p.f. = 86.6 per cent of 0.8 = **0.693.**

(*b*) Additional load = 72.5 per cent of 3000 = **2175 kW**

Total load = 3000 + 2175 = 5175 kW.

Now, $\sqrt{3} V_L I_L \cos\phi = 5{,}175{,}000$ or $\sqrt{3} \times 11{,}000 \times I_L \times 0.8 = 5{,}175{,}000$

∴ I_L = **340 A** ; phase current = $340/\sqrt{3}$ = **196 A**

Example 31.14. *Two transformers connected in open delta supply a 400-kVA balanced load operating at 0.866 p.f. (lag). The load voltage is 440 V. What is the (a) kVA supplied by each transformer ? (b) kW supplied by each transformer ?* **(Elect. Machines-I, Gwalior Univ. 1991)**

Solution. As stated in Art. 31.7, the ratio of operating capacity to available capacity in an open-Δ is 0.866. Hence, kVA of each transformer is one-half of the total kVA load divided by 0.866.

(*a*) kVA of each transformer $= \dfrac{(400/2)}{0.866}$ = **231 kVA**

(*b*) As stated in Art. 31.8, the two transformers have power factors of cos (30°–ϕ) and cos (30° + ϕ).

∴ P_1 = kVA cos (30°–ϕ) and P_2 = kVA cos (30° + ϕ)

Now, load p.f. = cos ϕ = 0.866; ϕ = $\cos^{-1}$ (0.866) = 30°

∴ P_1 = 231 × cos 0° = **231 kW** ; P_2 = 231 × cos 60° = **115.5 kW**

Obviously, $P_1 + P_2$ must equal 400 × 0.86 = 346.5 kW

Tutorial Problem No. 31–2

1. Three 1100/110-V transformers connected delta-delta supply a lighting load of 100 kW. One of th transformers is damaged and removed for repairs. Find

(*a*) What currents were flowing in each transformer when the three transformers were in service ?

(*b*) What current flows in each transformer when the third is removed ? and

(*c*) The output kVA of each transformer if the transformers connected in open Δ supply the full-load with normal heating ?

[(*a*) **primary = 30.3 A ; secondary = 303 A** (*b*) **primary = 30.3 $\sqrt{3}$A ; Secondary = 303 $\sqrt{3}$ A** (*c*) **33.33 kVA**] (*Elect. Machines-I, Gwalior Univ. Apr. 1977*)

31.9. Scott Connection or T-T Connection

This is a connection by which 3-phase to 3-phase transformation is accomplished with the help of two transformers as shown in Fig. 31.12. Since it was first proposed by Charles F. Scott, it is frequently referred to as Scott connection. This connection can also be used for 3-phase to 2-phase transformation as explained in Art. 31.10.

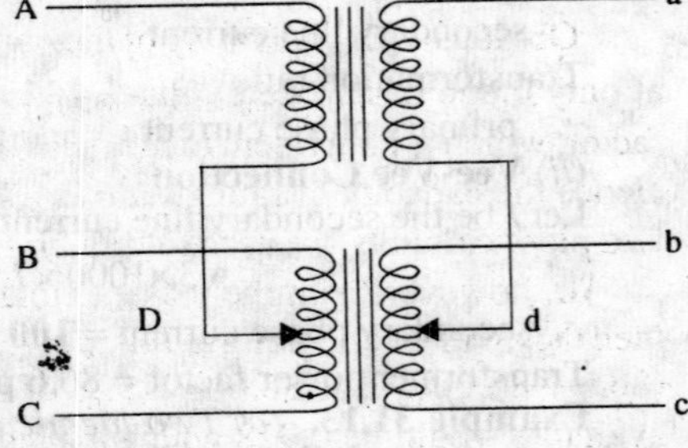

Fig. 31.12

One of the transformers has centre taps both on the primary and secondary windings (Fig. 31.12) and is known as the *main* transformer. It forms the horizontal member of the connection (Fig. 31.13).

The other transformer has a 0.866 tap and is known as *teaser* transformer. One end of both the primary and secondary of the teaser transformer is joined to the centre taps on both primary and secondary of the main transformer respectively as shown in Fig. 31.13 (*a*). The other end *A* of the teaser primary and the two ends *B* and *C* of the main transformer primary are connected to the 3-phase supply.

The voltage diagram is shown in Fig. 31.13 (*a*) where the 3-phase supply line voltage is assumed to be 100 V and a transformation ratio of unity. For understanding as to how 3-phase transformation results from this arrangement, it is desirable to think of the primary and secondary vector voltages as forming geometrical T'_s (from which this connection gets its name).

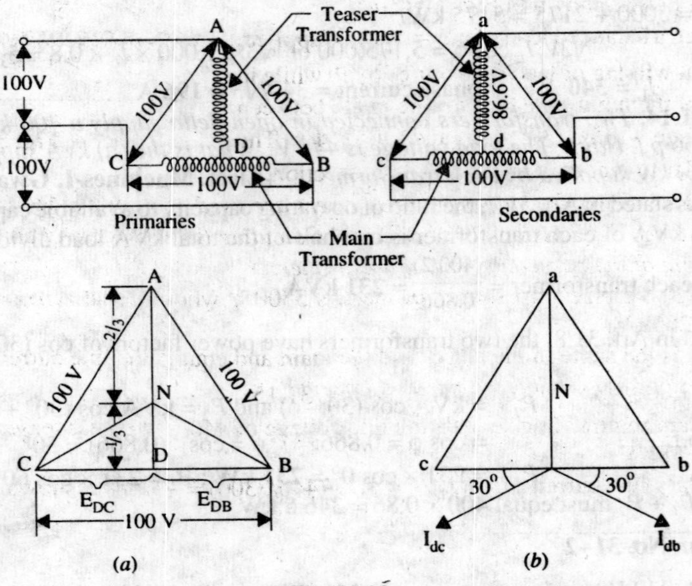

Fig. 30.13

In the primary voltage T of Fig. 31.13 (*a*), E_{DC} and E_{DB} are each 50 V and differ in phase by 180°, because both coils DB and DC are on the same magnetic circuit and are connected in opposition. Each side of the equilateral triangle represents 100 V. The voltage E_{DA} being the altitude of the equilaterial triangle is equal to $(\sqrt{3}/2) \times 100 = 86.6$ V. and lags behind the voltage across the main by 90°. The same relation holds good in the secondary winding so that *abc* is a symmetrical 3-phase system.

With reference to the secondary voltage triangle of Fig. 31.13 (*b*), it should be noted that for a load of unity power factor, current I_{db} lags behind volts E_{db} by 30° and I_{ds} leads E_{dc} by 30°. In other words, the teaser transformer and each half of the main transformer, all operate at different power factors.

Obviously, the full rating of the transformers is not being utilized. The teaser transformer operates at only 0.866 of its rated voltage and the main transformer coils operate at cos 30° = 0.866 power factor, which is equivalent to the main transformer's coils working at 86.6 per cent of this kVA rating. Hence the capacity of rating ratio in a T.T. connection is 86.6% –the same as in $V–V$ connection if two identical units are used, although heating in the two cases is not the same.

If, however, both the teaser primary and secondary windings are designed for 86.6 volts only, then they will be operating at full rating, hence the combined rating of the arrangement would become $(86.6 + 86.6)/(100 + 86.6) = 0.928$ of its total rating.* In other words, ratio of kVA utilized to that available would be 0.928 which makes this connection more economical than open-Δ with its ratio of 0.866.

Fig. 31.14 shows the secondary of the T-T connection with its different voltages based on a nominal voltage of 100 V. As seen, the neutral point *n* is one third way up from point *d*. If secondary voltage and current vector diagram is drawn for load power factor of unity, it will be found that

1. current in teaser transformer is in phase with the voltage.
2. in the main transformer, current leads the voltage by 30° across one half but lags the voltage by 30° across the other half as shown in Fig. 31.13(*b*).

Hence, when a balanced load of p.f. = cos φ, is applied, the teaser current will lag or lead the voltage by Φ while in the two halves of the main transformer, the angle between current and voltage will be (30° – Φ) and (30° + Φ). The situation is similar to that existing in a $V–V$ connection.

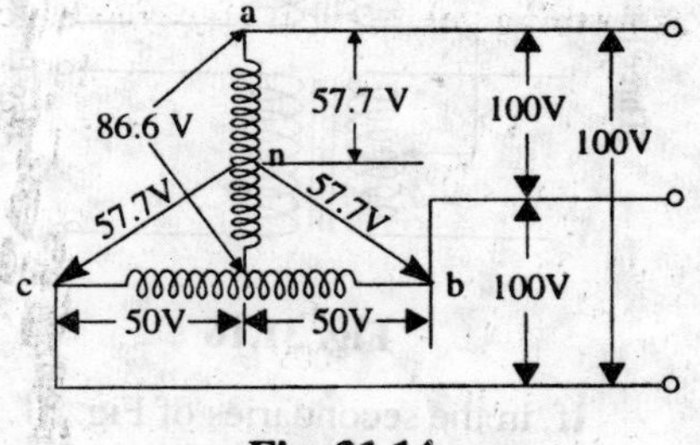

Fig. 31.14

Example **31.15.** *Two T-connected transformers are used to supply a 440-V, 33-kVA balanced load from a balanced 3-phase supply of 3300 V. Calculate (a) voltage and current rating of each coil (b) kVA rating of the main and teaser transformer.*

Solution. (*a*) Voltage across main primary is 3300 V whereas that across teaser primary is = $0.866 \times 3300 = 2858$ V.

The current is the same in the teaser and the main and equals the line current.

$\therefore \quad I_{LP} = 33{,}000/\sqrt{3} \times 3300 = 5.77$ A –Fig. 31.15

The secondary main voltage equals the line voltage of 440 V whereas teaser secondary voltage $= 0.866 \times 440 = 381$ V

The secondary line current, $I_{LS} = I_{LP}/k = 5.77/(440/3300) = 43.3$A as shown in Fig. 31.15.

---

*Alternatively, *VA* capacity available is = $V_L I_L + (0.866\ V_L) I_1 = 1.866\ V_L I_L$ where I_L is the primary line current. Since 3-phase power is supplied, volt-amperes actually utilized = $1.732\ V_L I_L$. Hence, ratio of kVA actually utilized to those avaialble is = $1.732\ V_L I_L / 1.866\ V_L I_L = 0.928$

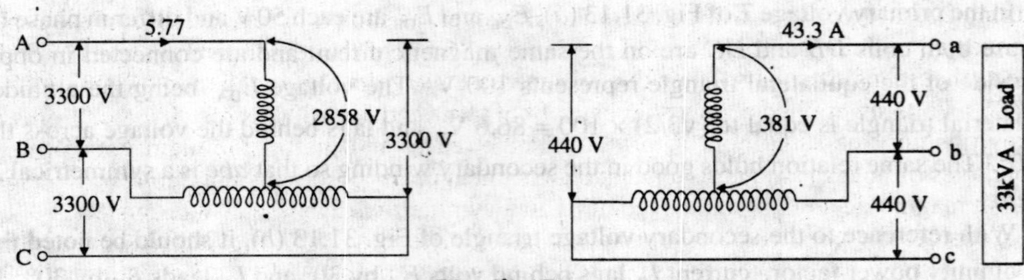

Fig. 31.15

(b) Main kVA $= 3300 \times 5.77 \times 10^{-3} =$ **19 kVA**

Teaser kVA $= 0.866 \times$ main kVA $= 0.866 \times 19 =$ **16.4 kVA**

31.10. Three-phase to Two-phase Conversion and *vice-versa*

This conversion is required to supply two-phase furnaces, to link two-phase circuit with 3-phase system and also to supply a 3-phase apparatus from a 2-phase supply source. For this purpose, Scott connection as shown in Fig. 31.16 is employed. This connection requires two transformers of different ratings although for interchangeabilty and provision of spares, both transformers may be identical but having suitable tappings.

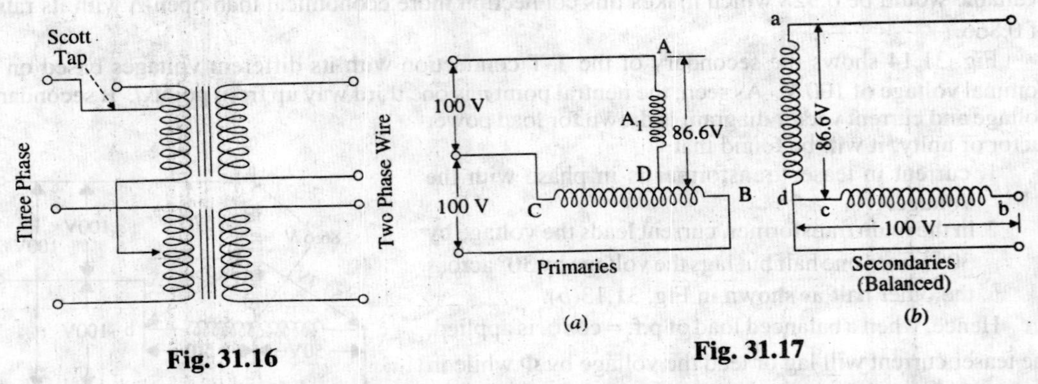

Fig. 31.16 **Fig. 31.17**

If, in the secondaries of Fig. 31.13(*b*), points *c* and *d* are connected as shown in Fig. 31.17 (*b*), then a 2-phase, 3-wire system is obtained. The voltage E_{dc} is 86.6 V but $E_{Cb} = 100$ V, hence the

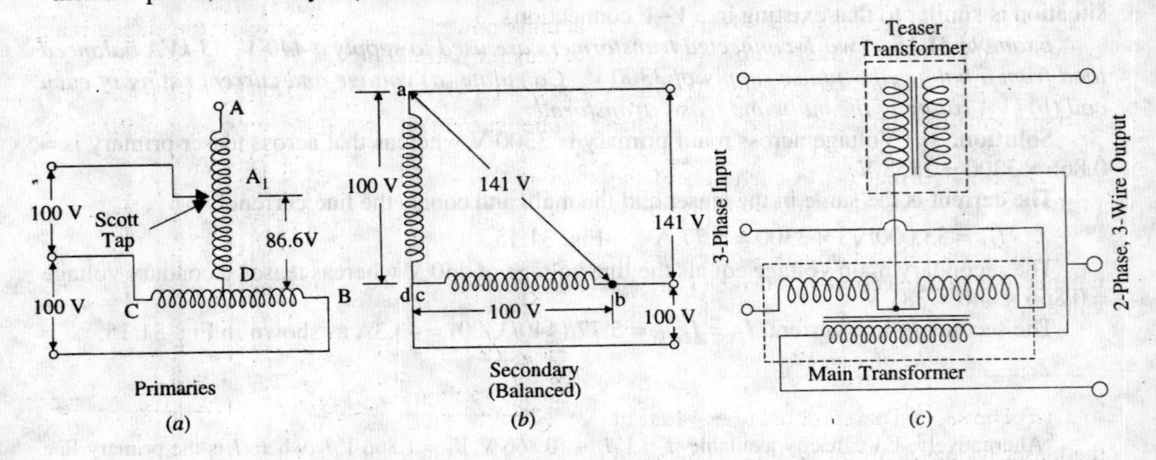

Fig. 31.18

resulting 2-phase voltages will be unequal. However, as shown in Fig. 31.18(*a*) if the 3-phase line is connected to point A_1, such that DA_1 represents 86.6% of the teaser primary turns (which are the same as that of main primary), then this will increase the volts/turn in the ratio of 100 : 86.6, because now 86.6 volts are applied across 86.6 per cent of turns and not 100% turns. In other words, this will make volts/turn the same both in primary of the teaser and that of the main transformer. If the secondaries of both the transformers have the same number of turns, then secondary voltage will be equal in magnitude as shown, thus resulting in a symmetrical 2-phase, 3-wire system.

Consider the same connection drawn slightly differently as in Fig. 31.19. The primary of the main transformer having N_1 turns is connected between terminals *CB* of a 3-phase supply. If supply line voltage is *V*, then obviously $V_{AB} = V_{BC} = V_{CA} = V$ but voltage between *A* and *D* is $V \times \sqrt{3}/2$. As said above, the number of turns between *A* and *D* should be also $(\sqrt{3}/2)\ N_1$ for making volt/turn the same in both primaries. If so, then for secondaries having equal turns, the secondary terminal voltages will be equal in magnitude although in phase quadrature.

It is to be noted that point *D* is not the neutral point of the primary supply because its voltage with respect to any line is not $V/\sqrt{3}$. Let *N* be the neutral point. Its position can be determined as follows. Voltage of *N* with respect to *A* must be $V/\sqrt{3}$ and since *D* to *A* voltage is $V \times \sqrt{3}/2$, hence *N* will be $(\sqrt{3}\ V/2 - V/\sqrt{3}) = 0.288$ V or 0.29 V from *D*. Hence, *N* is above *D* by a number of turns equal to 29% of N_1. Since 0.288 is one-third of 0.866, hence *N* divides the teaser winding *AD* in the ratio 2 : 1.

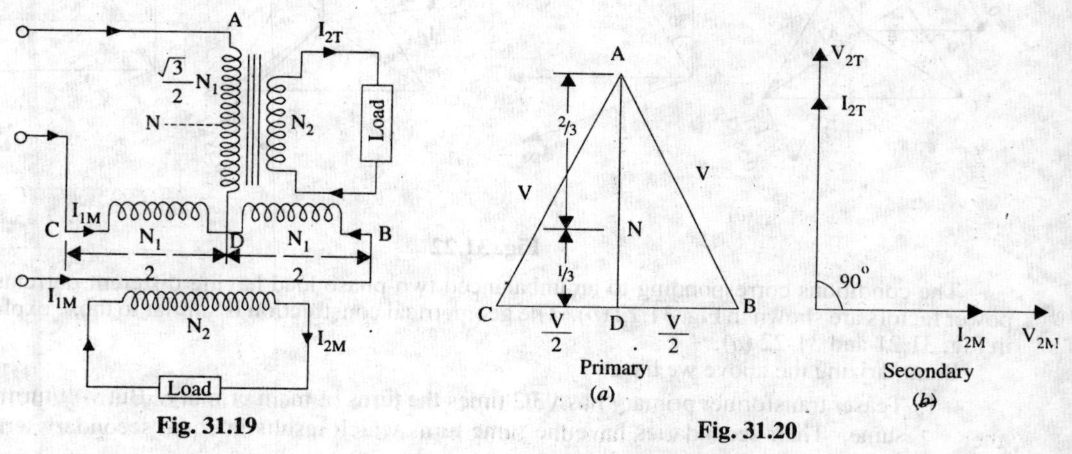

Fig. 31.19 **Fig. 31.20**

Let the teaser secondary supply a current I_{2T} at unity power factor. If we neglect the magnetizing current I_0, then teaser primary current is $I_{1T} = I_{2T} \times$ transformation ratio.

$\therefore\ I_{1T} = I_{2T} \times N_2/(\sqrt{3}N_1/2) = (2/\sqrt{3}) \times (N_2/N_1) \times I_{2T} = 1.15\ (N_2/N_1)\ I_{2T} = 1.15\ KI_{2T}$ where $K = N_2/N_1$ = transformation ratio of *main* transformer. The current is in phase with star voltage of the primary supply (Fig. 31.20).

The total current I_{1M} in each half of the primary of the *main* transformer consists of two parts :

(*i*) One part is that which is necessary to balance the main secondary current I_{2M}. Its value is

$$= I_{2M} \times \frac{N_2}{N_1} = KI_{2M}$$

(*ii*) The second part is equal to one-half of the teaser primary current *i.e* $\frac{1}{2} I_{1T}$. This is so

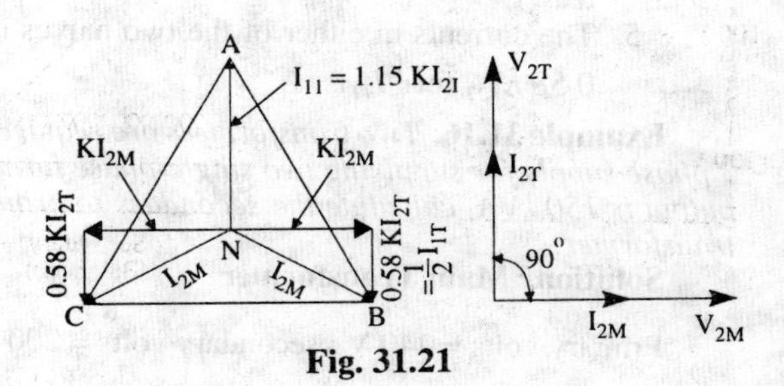

Fig. 31.21

because the main transformer primary forms a return path for the teaser primary current which divides itself into two halves at mid-point D in either direction. The value of each half is $= I_{1T}/2 = 1.15\ KI_{2T}/2 = 0.58\ KI_{2T}$.

Hence, the currents in the lines B and C are obtained vectorially as shown in Fig. 31.21. It should be noted that as the two halves of the teaser primary current flow in opposite directions from point D, they have no magnetic effect on the core and play no part at all in balancing the secondary ampere-turns of the main transformer.

The line currents thus have rectangular components of KI_{2M} and $0.58\ KI_{2T}$ and, as shown in Fig. 31.21, are in phase with the primary star voltages V_{NB} and V_{NC} and are equal to the teaser primary current. Hence, the three-phase side is balanced when the two-phase load of unity power factor is balanced.

Fig. 31.22 (*a*) illustrates the condition corresponding to a balanced two-phase load at a lagging power factor of 0.866. The construction is the same as in Fig. 31.21. It will be seen that the 3-phase side is again balanced. But under these conditions, the main transformer rating is 15% greater than that of the teaser, because its voltage is 15% greater although its current is the same.

Hence, we conclude *that if the load is balanced on one side, it would always be balanced on the other side.*

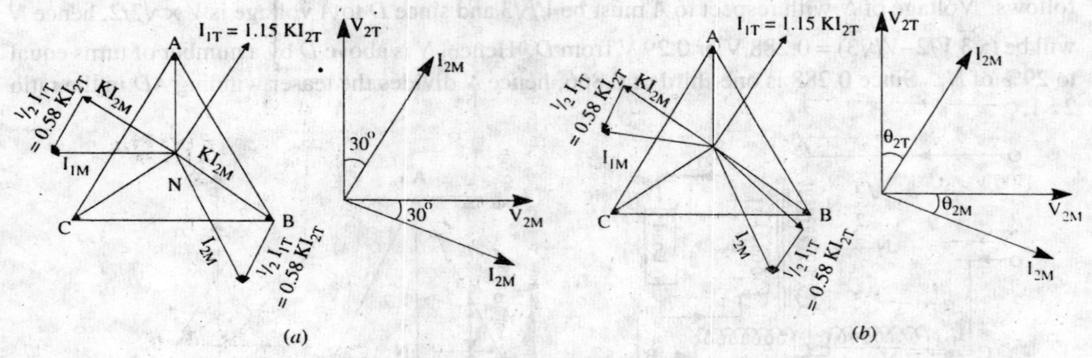

Fig. 31.22

The conditions corresponding to an unbalanced two-phase load having different currents and power factors are shown in Fig. 31.22 (*b*). The geometrical construction is similar to those explained in Fig. 31.21 and 31-22 (*a*).

Summarizing the above we have :

1. Teaser transformer primary has $\sqrt{3}/2$ times the turns of main primary. But volt/turn is the same. Their secondaries have the same turns which results in equal secondary terminal voltages.
2. If main primary has N_1 turns and main secondary has N_2 turns, then main transformation ratio is N_2/N_1 . However, the transformation ratio of teaser is
$$N_2/(\sqrt{3}N_1/2) = 1.15\ N_2/N_1 = 1.15\ K$$
3. If the load is balanced on one side, it is balanced on the other side as well.
4. Under balanced load conditions, main transformer rating is 15% greater than that of the teaser.
5. The currents in either of the two halves of main primary are the vector sum of KI_{2M} and $0.58\ KI_{2T}$ (or $\frac{1}{2} I_{1T}$).

Example 31.16. *Two transformers are required for a Scott connection operating from a 440-V, 3 phase supply for supplying two single-phase furnaces at 200 V on the two-phase side. If the total output is 150 kVA, calculate the secondary to primary turn ratio and the winding currents of each transformer.*

Solution. Main Transformer

Primary volts = 440 V ; secondary volts = 200 V $\therefore \quad \frac{N_2}{N_1} = \frac{200}{440} = \frac{1}{2.2}$

Secondary current = 150,000/2 × 200 = **375 A**

∴ primary current = 375 × 1/2.2 = **197 A**

Teaser Transformer

$$\text{primary volts } = (\sqrt{3}/2 \times 440) = 381 \text{ V: secondary volts } = 200\ V$$

$$\frac{\text{secondary turns}}{\text{primary turns}} = \frac{200}{381} = \frac{1}{1.905} \text{ (also teaser ratio } = 1.15 \times 1/2.2 = 1/1.905)$$

Example 31.17. *Two single-phase furnaces working at 100 V are connected to 3300-V, 3-phase mains through Scott-connected transformers. Calculate the current in each line of the 3-phase mains when the power taken by each furnace is 400-kW at a power factor of 0.8 lagging. Neglect losses in the transformer.* **(Elect. Machines-III, South Gujarat Univ. 1988)**

Solution. Here K = 100/3,300 = 1/33 (main transformer)

$$I_2 = \frac{400,000}{0.8 \times 100} = 5,000 \text{ A (Fig. 31-23) ; Here } I_{2T} = I_{2M} = I_2 = 5,000 \text{ A}$$

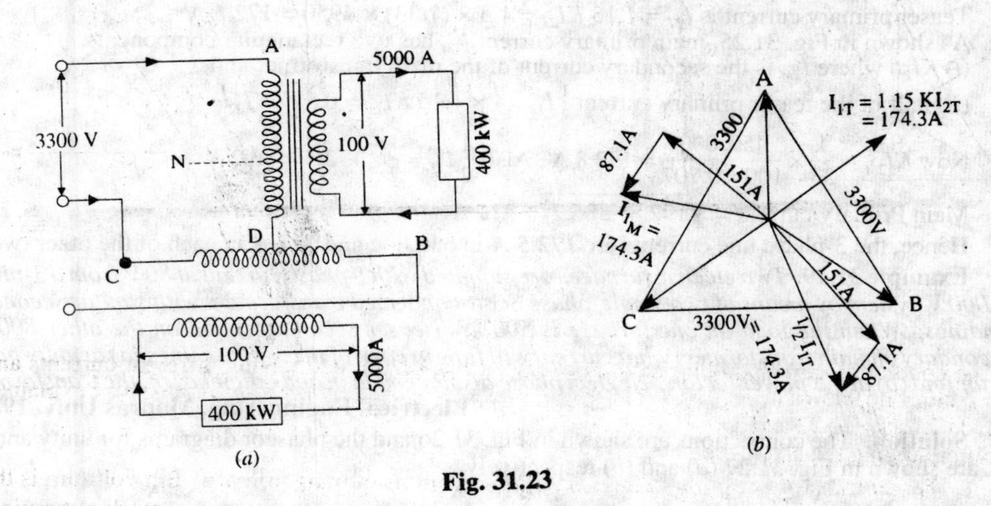

Fig. 31.23

As the two-phase load is balanced, the 3-phase side is also balanced.

Primary phase currents are = 1.15 KI_2 = 1.15 × (1/33) × 5,000 = 174.3 A

Since for a star-connection, phase current is equal to line current,

∴ line current = **174.3 A**

Note. We have made use of the fact that since secondary load is balanced, primary load is also balanced. If necessary, I_{1M} can also be found.

I_{1M} is the vector sum of (*i*) KI_{2M} and (*ii*) $\frac{1}{2} I_{1T}$ or 0.58 KI_{2T}.

Now, KI_{2M} = (1/33) × 5,000 = 151 A and 0.58 $KI_2 = \frac{1}{2} I_{1T}$ = 174.3/2 = 87.1 A

$$\therefore \qquad I_{1M} = \sqrt{151^2 + 87.1^2} = 174.3 \text{ A}$$

Example 31.18. *In a Scott-connection, calculate the values of line currents on the 3-phase side if the loads on the 2-phase side are 300 kW and 450 kW both at 100 V and 0.707 p.f. (lag) and the 3-phase line voltage is 3,300 V. The 300-kW load is on the leading phase on the 2-phase side. Neglect transformer losses.* **(Elect. Technology, Allahabad Univ. 1991)**

Solution. Connections are shown in Fig. 31.24 and phasor diagram in Fig. 31.25.

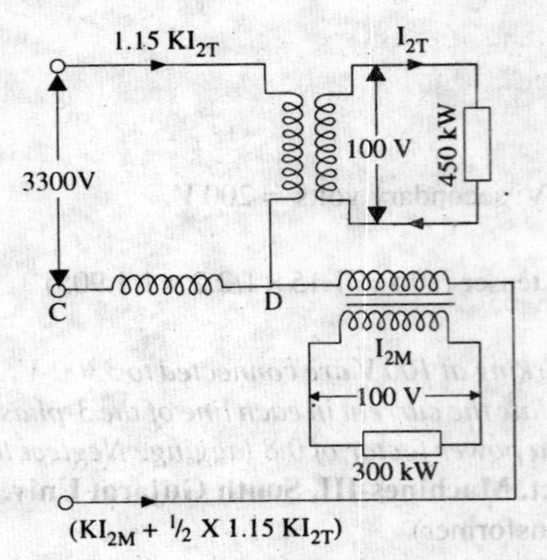

Fig. 31.24

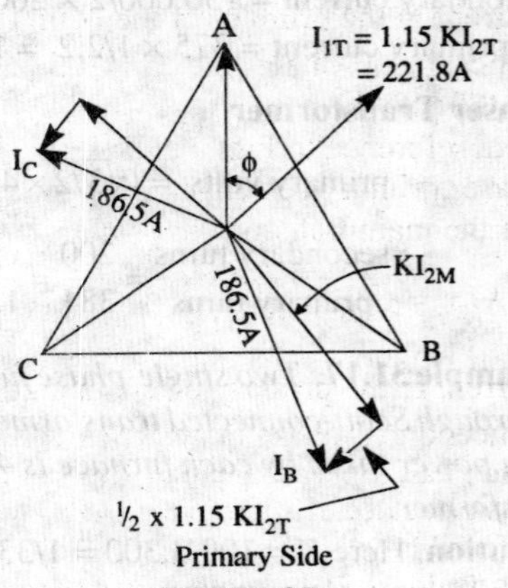

Fig. 31.25

Here, $K = 100/3{,}300 = 1/33$

Teaser secondary current is $I_{2T} = 300{,}000/100 \times 0.707 = 4950$ A

Teaser primary current is $I_{1T} = 1.15\, KI_{2T} = 1.5 \times (1/33) \times 4950 = 172.5$ A

As shown in Fig. 31.25, main primary current I_{1M} has two rectangular components.

(*i*) KI_{2M} where I_{2M} is the secondary current of the main transformer and

(*ii*) half of the teaser primary current $\frac{1}{2} I_{1T} = \frac{1}{2} \times 1.15\, KI_{2T} = 0.577\, KI_{2T}$

$$\text{Now } KI_{2M} = \frac{1}{33} \times \frac{450{,}000}{100 \times 0.707} = 192.8 \text{ A} \text{ ; Also, } \frac{1}{2} I_{1T} = \frac{1}{2} \times 172.5 = 86.2 \text{ A}$$

Main Primary current $= \sqrt{192.8^2 + 86.2^2} = 211$ A

Hence, the 3-phase line currents are **172.5 A** in one line and **211 A** in each of the other two.

Example 31.19. *Two electric furnaces are supplied with I-phase current at 80 V from a 3-phase, 11,000 V system by means of two single-phase Scott-connected transformers with similar secondary windings. When the load on one furnace is 500 kW (teaser secondary) and on the other 800 kW (secondary of main transformer) what current will flow in each of the 3-phase lines (a) at unity power factor and (b) at 0.5 power factor ? Neglect phase displacement in and efficiency of, the transformers.*

(Electrical Enginnering, Madras Univ. 1987)

Solution. The connections are shown in Fig. 31.26 and the phaseor diagrams for unity and 0.5 p.f. are shown in Fig. 31.27 (*a*) and (*b*) respectively.

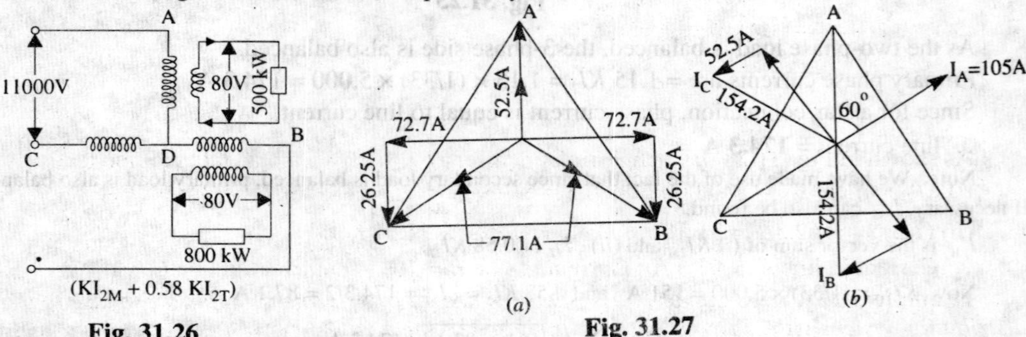

Fig. 31.26

Fig. 31.27

Here, $K = 80/11{,}000 = 2/275$

(*a*) **Unity p.f.**

With reference to Fig. 31.27 (*a*), we have $I_{2T} = 500{,}000/80 \times 1 - 6{,}250$ A

Teaser primary current $I_{1T} = 1.15\, KI_{2T} = 1.15 \times (2/275) \times 6{,}250 = 52.5$ A

For the main transformer primary

$$(i)\ KI_{2M} = \frac{2}{275} \times \frac{800{,}000}{80 \times 1} = 72.7 \text{ A} \quad (ii)\ \frac{1}{2} \times I_{1T} = 52.5/2 = 26.25 \text{ A}$$

Current in the primary of the main transformer is $= \sqrt{72.7^2 + 26.25^2} = 77.1$ A

Hence, one 3-phase line carries **52.5 A** whereas the other 2 carry **77.1 A** each [Fig. 31.27 (*a*)]

(*b*) **0.5 p.f.**

With reference to Fig. 31.27 (*b*) we have, $I_{2T} = 500{,}000/80 \times 0.5 = 12{,}500$ A

Teaser primary current $I_{IT} = 1.15 \times (2/275) \times 12{,}500 = 105$ A

For the main transformer primary

(*i*) $KI_{2M} = \dfrac{2}{275} \times \dfrac{500{,}000}{80 \times 0.5} = 145.4$ A (*ii*) $\dfrac{1}{2} I_{1T} = 105/2 = 52.5$ A

Current in the primary of the main transformer is $= \sqrt{145.4^2 + 52.5^2} = 154.2$ A.

Hence, one 3-phase line carries **105 A** and the other two carry **154.2 A** each.

Note : Part (*b*) need not be worked out in full because at 0.5 p.f., each component current and hence the resultant are doubled. Hence, in the second case, answers can be found by multiplying by a factor of 2 the line currents found in (*a*).

Example 31.20. *Two furnaces are supplied with 1-phase current at 50 V from a 3-phase, 4.6 kV system by means of two 1-phase, Scott-connected transformers with similar secondary windings. When the load on the main transformer is 350 kW and that on the other transformer is 200 kW at 0.8 p.f. lagging, what will be the current in each 3-phase line ? Neglect phase displacement and losses in transformers.* **(Electrical Machinery-II, Bangalore Univ. 1991)**

Solution. Connections and vector diagrams are shown in Fig. 31.28.

$K = 50/4{,}600 = 1/92$; $I_{2T} = 200{,}000/50 \times 0.8 = 5{,}000$ A

$I_{IT} = 1.15\, KI_{2T} = 1.1 \times (1/92) \times 5{,}000 = 62.5$ A

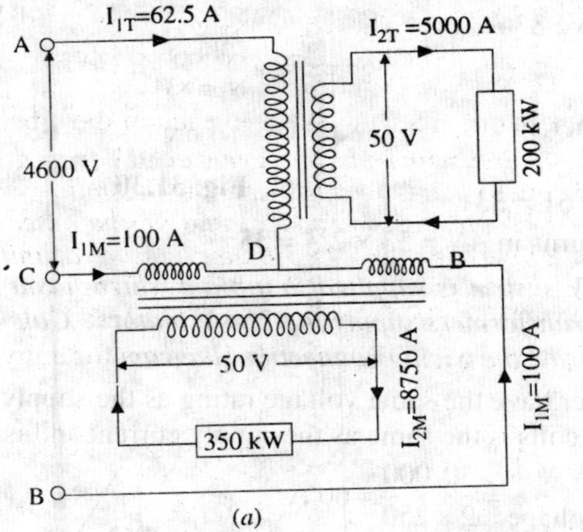

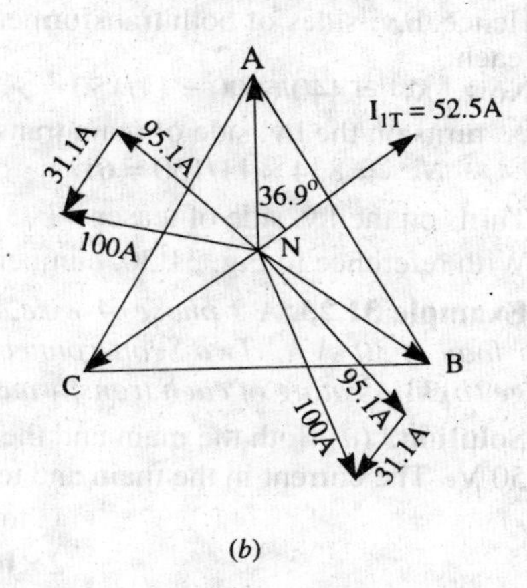

Fig. 31.28

As shown in Fig. 31.28 (*b*), main primary current I_{1M} has two rectangular components

(*i*) KI_{2M} where $I_{2M} = 350{,}000/50 \times 0.8 = 8{,}750$ A $\quad \therefore \; KI_{2M} = 8{,}750/92 = 95.1$ A

(*ii*) $(1/2)\, I_{IT} = 62.5/2 = 31.3$ A $\quad \therefore \; I_{IM} = \sqrt{95.1^2 + 31.3^2} = 100$ A

$\therefore$ Current in line A = **62.5 A** ; Current in line B = **100 A** ; Current in line C = **100 A.**

Example 31.21. *Two single-phase Scott-connected transformers supply a 3-phase four-wire distribution system with 231 volts between lines and the neutral. The h.v. windings are connected to a two-phase system with a phase voltage of 6,600 V. Determine the number of turns in each section of the h.v. and l.v. winding and the position of the neutral point if the induced voltage per turn is 8 volts.*

Solution. As the volt/turn is 8 and the h.v. side voltage is 6,600 V, the h.v. side turns are = 6,600/8 = **825** on both transformers.

Now, voltage across points B and C of main winding = line voltage = $231 \times \sqrt{3} = 400$ V

No. of turns on the l.v. side of the main transformer = 400/8 = **50**

No. of turns on the l.v. side of teaser transformer

$= (\sqrt{(3/2)} \times \text{main turns}$

$= \sqrt{3} \times 50/2 =$ **43** (whole number)

The neutral point on the 3-phase side divides teaser turns in the ratio 1 : 2.

$\therefore$ Number of turns between A and $N = (2/3) \times AD = (2/3) \times 43 =$ **29**

Hence, neutral point is located on the 29th turn from A downwards (Fig. 31.29).

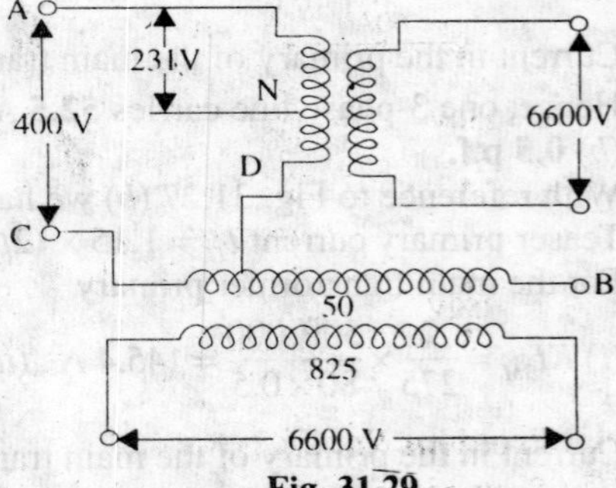

Fig. 31.29

Example 31.22. *A Scott-connected (2 to 3-phase) transformer links a 6,000 V, 2-phase system with a 440-V, 3-phase system. The frequency is 50 Hz, the gross core area is 300 cm², while the maximum flux density is to be about 1.2 Wb/m². Find the number of turns on each winding and the point to be tapped for the neutral wire on the 3-phase side. If the load is balanced on the one side of such a transformer, find whether it will also be balanced on the other side.* **(London Univ.)**

Solution. Use the transformer voltage equation.

$E = 4.44\, f N \Phi_x$ volt

Gross core area = 300 cm²

Assuming net iron = 0.9 of gross area, and considering the h.v. side, we have

$6000 = 4.44 \times 50 \times N_1 \times 1.2(300 \times 0.9 \times 10^{-4})$

$N_1 =$ **834**

Hence, h.v. sides of both transformers have 834 turns each.

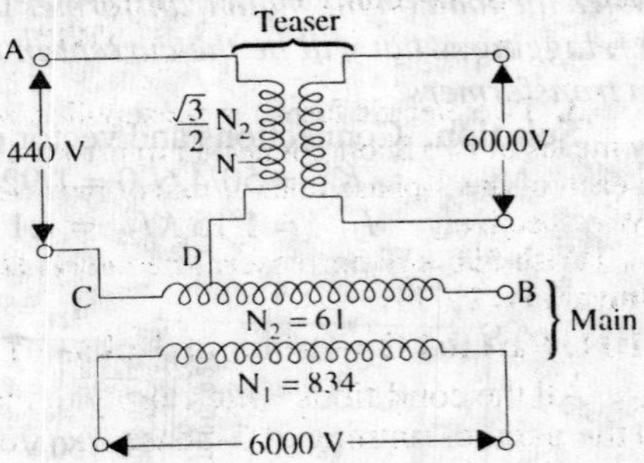

Now $K = 440/6000 = 11/150$

$\therefore$ turns on the l.v. side of main transformer

$N_2 = 834 \times 11/150 =$ **61**

Turns on the l.v. side of teaser $= (\sqrt{3}/2) \times 61 =$ **53**

Fig. 31.30

With reference to Fig. 31.30, number of turns in $AN = 53 \times 2/3 =$ **35**

Example 31.23. *A 2-phase, 4-wire, 250-V system is supplied to a plant which has a 3-phase motor load of 30 kVA. Two Scott-connected transformers supply the 250 V motors. Calculate (a) voltage (b) kVA rating of each transformer. Draw the wiring connection diagram.*

Solution. (*a*) Both the main and the teaser have the same voltage rating as the supply voltage *i.e.* 250 V. The current in the main and teaser coils is the same as the supply current and is

$$= \frac{\text{total kVA}}{2 \times \text{line voltage}} = \frac{30,000}{2 \times 250} = 60 \text{ A}$$

On the three-phase side, current is the same in all coils and is equal to the load line current = $30,000/\sqrt{3} \times 250 = 69.3$ A

Load voltage on main secondary = line voltage = 250 V

Load voltage on teaser secondary = $0.866 \times 250 = 216.5$ V

Hence, voltage rating of main transformer is **250/250** whereas that of teaser transformer is **250/216.5**

The current rating of main transformer is **60/69.3** and it is the same for the teaser transformer.

(*b*) The volt-amp rating of the teaser primary as well as secondary is the same *i.e.* $60 \times 250 \times 10^{-3} = 69.3 \times 216.5 \times 10^{-3} =$ **15 kVA**

The main volt-ampere rating is $= 250 \times 69.3 \times 10^{-3} =$ **17.3 kVA**

Incidentally, if two identical transformers are used for providing inter-changeability, then both must be rated at 17.3 kVA. In that case, a total capacity of 34.6 kVA would be required to provide a 30 kVA load.

The wiring connections are shown in Fig. 31.31.

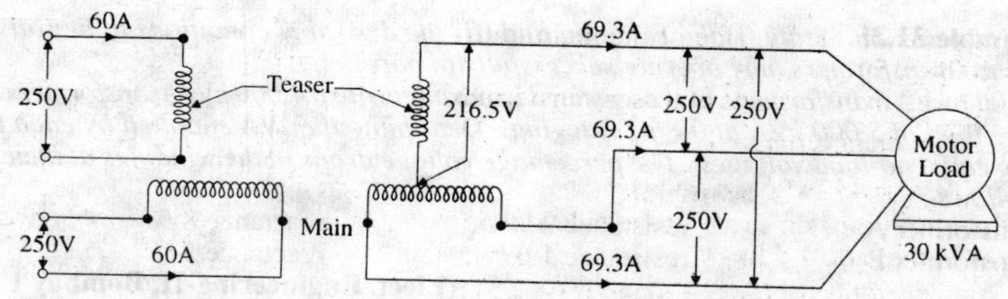

Fig. 31.31

Tutorial Problem No. 31.3

1. A Scott-connected transformer is fed from a 6,600-V, 3-phase network and supplies two single-phase furnaces at 100 V. Calculate the line currents on the 3-phase side when the furnaces take 400 kW and 700 kW respectively at 0.8 power factor lagging.

[With 400 kW on teaser, line currents are 87.2 A ; 139 A ; 139 A]
(Elect. Machines-II, Indore Univ. 1977)

2. Two 220-V, 1-phase electrical furnaces take loads of 350 kW and 500 kW respectively at a power factor of 0.8 lagging. The main supply is at 11-kV, 3-phase, 50 Hz. Calculate current in the 3-phase lines which energise a Scott-connected transformer combination.

[With 350 kW on teaser, line currents are : 45.7 A ; 61.2 A ; 61.2 A]
(Electric Machines, Madras Univ. 1978)

3. Two electric furnaces are supplied with 1-phase current at 80 V from 3-phase, 11,000-V supply mains by means of two Scott-connected transformers with similar secondary windings. Calculate the current flowing in each of the 3-phase lines at unity power factor when the loads on the two transformer are 550 kW and 800 kW respectively.

[With 550 kW on teaser, line currents are : 57.5 A ; 78.2 A ; 78.2 A] (Electric Machines-I, Madras University, 1977)

31.11. Parallel Operation of 3-phase Transformers

All the conditions which apply to the parallel operation of single-phase transformers also apply to the parallel running of 3-phase transformers but with the following additions :

1. The voltage ratio must refer to the terminal *voltage of primary and secondary*. It is obvious that this ratio may not be equal to the ratio of the number of turns per phase. For example, if V_l, V_2 are the primary and secondary terminal voltages, then for Y/Δ connection, the turn ratio is $V_2/(V_l/\sqrt{3}) = \sqrt{3}\,V_2/V_l$.

2. The phase displacement between primary and secondary voltages must be the same for all transformers which are to be connected for parallel operation.

3. The phase sequence must be the same.

4. All three transformers in the 3-phase transformer bank will be of the same construction either core or shell.

Note. (*i*) In dealing with 3-phase transformers, calculations are made for one phase only. The value of equivalent impedance used is the equivalent impedance per phase referred to secondary.

(*ii*) In case the impedances of primary and secondary windings are given separately, then primary impedance must be referred to secondary by multiplying it with (transformation ratio)^2.

(*iii*) For Y/Δ or Δ/Y transformers, it should be remembered that the voltage ratios as given in the questions, refer to terminal voltages and are quite different from turn ratio.

Example 31.24. *A load of 500 kVA at 0.8 power factor lagging is to be shared by two three-phase transformers A and B of equal ratings. If the equivalent delta impedances as referred to secondary are (2 + j6) Ω for A and (2 + j5) Ω for B, calculate the load supplied by each transformer.*

Solution. $$\mathbf{S_A} = \mathbf{S}\frac{\mathbf{Z_B}}{\mathbf{Z_A}+\mathbf{Z_B}} = \mathbf{S}\frac{1}{1+(\mathbf{Z_A}/\mathbf{Z_B})}$$

Now $\mathbf{S} = 500(0.8 - j0.6) = (400 - j300)$

$\mathbf{Z_A/Z_B} = (2 + j6)/(2 + j5) = 1.17 + j0.07$; $\mathbf{Z}_B/\mathbf{Z}_A = (2 + j5)/(2 + j6) = 0.85 - j0.05$

$\mathbf{S_A} = (400 - j300)/(2.17 + j0.07) = 180 - j144.2 = 230.7\angle{-38.7°}$

$\cos\phi_A = 0.78$ lagging

$\mathbf{S_B} = (400 - j300)/(1.85 - j0.05) = 220.1 - j156 = 270\angle - 40°28'$ ∴ $\cos\Phi_B = 0.76$ lagging.

Example 31.25. *State (i) the essential and (ii) the desirable conditions to be satisfied so that two 3-phase transformers may operate successfully in parallel.*

A 2,000-kVA transformer (A) is connected in parallel with a 4,000 kVA transformer (B) to supply a 3-phase load of 5,000 kVA at 0.8 p.f. lagging. Determine the kVA supplied by each transformer assuming equal no-load voltages. The percentage voltage drops in the windings at their rated loads are as follows :

Transformer A	resistance 2 % ;	reactance 8 %
Transformer B	resistance 1.6 % ;	reactance 3 %

(Elect. Engineering-II, Bombay Univ. 1987)

Solution. On the basis of 4,000 kVA

$$\% \mathbf{Z}_A = (4{,}000/2{,}000)(2 + j8) = (4 + j16) = 16.5 \angle 76°$$

$$\% \mathbf{Z}_B = (1.6 + j3) ; \% \mathbf{Z}_A + \% \mathbf{Z}_B = (5.6 + j16) = 19.8 \angle 73.6°$$

$$\mathbf{S} = 5{,}000 \angle -36.9° = (4{,}000 - j3{,}000)$$

Now $$\mathbf{S}_A = \mathbf{S}.\frac{\mathbf{Z}_A}{\mathbf{Z}_A + \mathbf{Z}_B} = 5{,}000 \angle -36.9° \times \frac{16.5\angle 76°}{19.8\angle 73.6°}$$

$$= 5{,}000 \angle -36.9° \times 0.832 \angle 2.4° = 4{,}160 \angle -34.5° = (3{,}425 - j2{,}355)$$

$$\mathbf{S}_B = \mathbf{S} - \mathbf{S}_A = (4{,}000 - j3{,}000) - (3{,}425 - j2355)$$

$$= (575 - j645) = 864 \angle -48.3°$$

$$\cos \phi_A = \cos 34.5° = 0.824 \text{ (lag)} ; \cos \phi_B = \cos 48.3° = 0.665 \text{ (lag)}.$$

Example 31.26. *A load of 1,400 kVA at 0.866 p.f. lagging is supplied by two 3-phase transformers of 1,000 kVA and 500 kVA capacity operating in parallel. The ratio of transformation is the same in both : 6,600/400 delta-star. If the equivalent secondary impedances are (0.001 + j0.003) ohm and (0.0028 + j0.005) ohm per phase respectively, calculate the load and power factor of each transformer.* **(Elect. Engg-I, Nagpur Univ. 1993)**

Solution. On the basis of 1000 kVA, $\mathbf{Z}_A = (0.001 + j0.003)\ \Omega$

$$\mathbf{Z}_B = (1000/500)(0.0028 + j0.005) = (0.0056 + j0.01)\ \Omega$$

$$\frac{\mathbf{Z}_A}{\mathbf{Z}_A + \mathbf{Z}_B} = \frac{(0.001 + j0.003)}{(0.0066 + j0.013)} = \frac{3.162 \times 10^{-3} \angle 71.6°}{14.57 \times 10^{-3} \angle 63.1°} = 0.2032 \angle 8.5°$$

$$\mathbf{S} = 1400 \angle \cos^{-1}(0.866) = 1400 \angle -30° = (1212 - j700)$$

$$\mathbf{S}_B = \mathbf{S}.\frac{\mathbf{Z}_A}{\mathbf{Z}_A + \mathbf{Z}_B} = 1400 \angle -30° \times 0.2032 \angle 8.5°$$

$$= 284.5 \angle -21.5° = 265 - j104$$

$$\mathbf{S}_A = \mathbf{S} - \mathbf{S}_B = (1212 - j700) - (265 - j104) = (947 - j596) = 1145 \angle -32.2°$$

$$\cos \phi_A = \cos 32.3° = 0.846 \text{ (lag)} ; \cos \phi_B = \cos 21.5° = 0.93 \text{ (lag)}.$$

Example 31.27. *Two 3-phase transformers A and B having the same no-load line voltage ratio 3,300/400-V supply a load of 750 kVA at 0.707 lagging when operating in parallel. The rating of A is 500 kVA, its resistance is 2% and reactance 3%. The corresponding values for B are 250 kVA ; 1.5% and 4% respectively. Assuming that both transformers have star-connected secondary windings, calculate*

(a) the load supplied by each transformer

(b) the power factor at which each transformer is working

(c) the secondary line voltage of the parallel circuit.

Solution. On the basis of 500 kVA,

$$\%\mathbf{Z}_A = 2 + j3 ; \%\mathbf{Z}_B = (500/200)(1.5 + j4) = (3 + j8)$$

$$\frac{\mathbf{Z_A}}{\mathbf{Z_A}+\mathbf{Z_B}}=\frac{2+j3}{5+j1}=0.3\angle-9.3°;\ \frac{\mathbf{Z_B}}{\mathbf{Z_A}+\mathbf{Z_B}}=\frac{3+j8}{5+j11}=0.711\angle 3.8°$$

Now, $\mathbf{S}=750\angle-45°$

(*a*)
$$\mathbf{S_A}=\mathbf{S}\frac{\mathbf{Z_B}}{\mathbf{Z_A}+\mathbf{Z_B}}=750\angle-45°\times0.711\angle3.8°$$
$$=533\angle-41.2°=(400-j351)$$
$$S_B=750\angle-45°\times0.3\angle09.3°=225\angle-54.3°$$

(*b*) $\cos\phi_A=\cos41.2°=0.752$ (lag) ; $\cos\phi_B=\cos54.3°=0.5835$ (lag)

(*c*) Since voltage drop of each transformer is the same, its value in the case of transformer *A* would only be calculated. Now, for transformer *A*, kW = 400 for the active component of the current and kVAR = 351 for the reactive component.

∴ % resistive drop = 2 × 400/500 = 1.6% ; % reactive drop = 3 × 351/500 = 2.1%

Total percentage drop = 1.6 + 2.1 = 3.7

Secondary line voltage = 400 – (3.7 × 400/100) = **385.2 V.**

31.12. Instrument Transformers

In d.c. circuit when large currents are to be measured, it is usual to use low-range ammeters with suitable shunts. For measuring high voltages, low-range voltmeters are used with a high resistance connected in series with them. But it is not convenient to use this method with alternating current and voltage instruments. For this purpose, specially constructed accurate ratio instrument transformers are employed in conjunction with standard low-range a.c. instruments. These instrument transformers are of two kinds (*i*) current transformers for measuring large alternating currents and (*ii*) potential transformers for measuring high alternating voltages.

31.13. Current Transformers

These transformers are used with low-range ammeters to measure currents in high-voltage alternating-current circuits where it is not practicable to connect instruments and meters directly to the lines. In addition to insulating the instrument from the high voltage line, they step down the current in a known ratio. The current (or series) transformer has a primary coil of one or more turns of thick wire connected in *series* with the line whose current is to be measured as shown in Fig. 31.32. The secondary consists of a large number of turns of fine wire and is connected across the ammeter terminals (usually of 5-ampere or 1-ampere range).

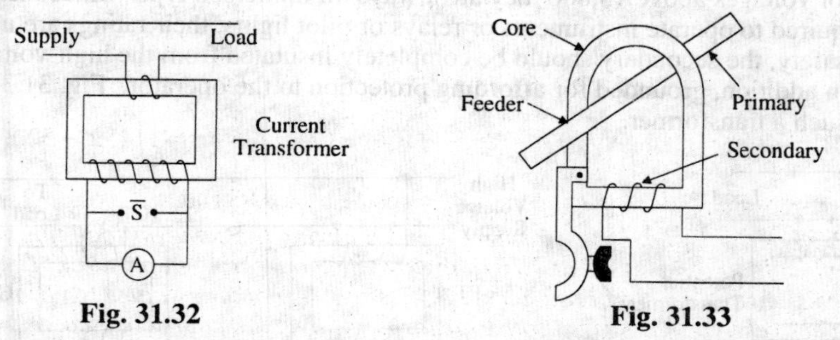

Fig. 31.32 Fig. 31.33

As regards voltage, the transformer is of step-up variety but it is obvious that current will be stepped down. Thus, if the current transformer has *primary to secondary* current ratio of 100 : 5, then it steps up the voltage 20 times whereas it steps down the current to 1/20th of its actual value. Hence, if we know current ratio (I_1/I_2) of the transformer and the reading of the a.c. ammeter, the line current can be calculated. In fact, line current is given by the current transformation ratio times the reading on the ammeter. One of the most commonly-used current transformer is the one known as *clamp-on or clip-on type*. It has a laminated core which is so arranged that it can be opened out at hinged section by merely pressing a trigger-like projection (Fig. 31.33). When the core is thus opened, it permits the admission of very heavy current-carrying bus bars or feeders whreupon the trigger is released and the core is tightly closed by a spring. The current carrying conductor or feeder acts as a single-turn primary whereas the secondary is connected across the standard ammeter conveniently mounted in the handle.

It should be noted that, since the ammeter resistance is very low, the current transformer normally works short circuited. If for any reason, the ammeter is taken out of the secondary winding, then this winding must be short-circuited with the help of short-circuiting switch *S*. If this is not done, then due to the absence of counter amp-turns of the secondary, the unopposed primary m.m.f. will set up an abnormally high flux in the core which will produce excessive core loss with subsequent heating and a high voltage across the secondary terminals. This is not the case with ordinary constant-potential transformers, because their primary current is determined by the load on their secondary whereas in a current transformer, the primary current is determined entirely by the load on the system and not by the load on its own secondary.

Hence, *the secondary of a current transformer should never be left open under any circumstances.*

Example 31.28. *A 100 : 5 transformer is used in conjunction with a 5-amp ammeter. If the latter reads 3.5 A, find the line current.*

Solution. Here, the ratio 100 : 5 stands for the ratio of primary-to-secondary currents *i.e.* $I_1/I_2 = 100/5$

$\therefore$ primary (or line) current = 3.5 × (100/5) = **70 A**

Example 31.29. *It is desired to measure a line current of the order of 2,000 A to 2,500 A. If a standard 5-amp ammeter is to be used along with a current transformer, what should be the turn ratio of the latter ? By what factor should the ammeter reading be multiplied to get the line current in each case ?*

Solution. $I_1/I_2 = 2000/5 = 400$ or $2500/5 = 500$. Since $I/I_2 = N_2/N_1$ hence $N_2/N_1 = 400$ in the first case and 500 in the second case. It means that $N_1 : N_2 :: 1 : 400$ or $1 : 500$.

Ratio or multiplication factor in the first case is 400 and in the second case 500.

31.14. Potential Transformers

These transformers are extremely accurate-ratio step-down transformers and are used in conjunction with standard low-range voltmeters (usually 150-V) whose deflection when divided by voltage transformation ratio, gives the true voltage on the high voltage side. In general, they are of the shell-type and do not differ much from the ordinary two-winding transformers discussed so far, except that their power rating is extremely small. Upto voltages of 5,000, potential transformers are usually of the dry type, between 5,000 and 13,800 volts, they may be either dry type or oil immersed type, although for voltages above 13,800 they are always oil immersed type. Since their secondary windings are required to operate instruments or relays or pilot lights, their ratings are usually of 40 to 100 W. For safety, the secondary should be completely insulated from the high-voltage primary and should be, in addition, grounded for affording protection to the operator. Fig. 31.34 shows the connections of such a transformer.

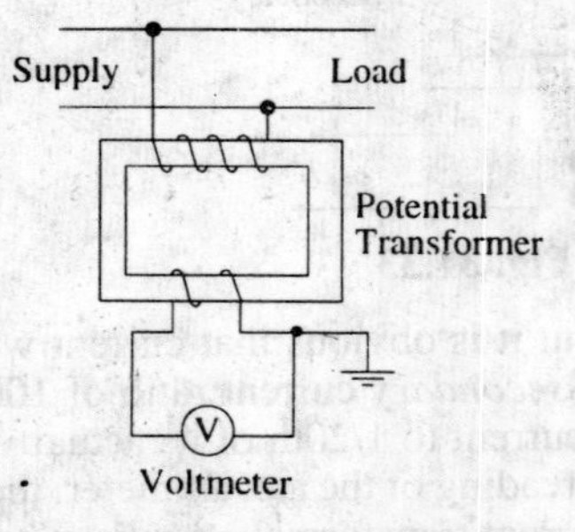

Fig. 31.34

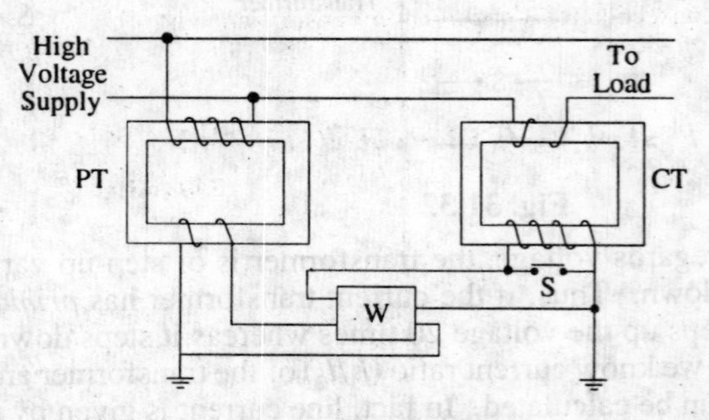

Fig. 31.35

Fig. 31.35 shows the connections of instrument transformers to a wattmeter. While connecting the wattmeter, the relative polarities of the secondary terminals of the transformers with respect to their primary terminals must be known for connections of the instruments.

Objective Tests–31

1. Which of the following connections is best suited for 3-phase, 4-wire service ?
(*a*) Δ – Δ (*b*) Y–Y
(*c*) Δ – Y (*d*) Y–Δ

2. In a three-phase Y–Y transformer connection, neutral is fundamental to the
(*a*) suppression of harmonics
(*b*) passage of unbalanced currents due to unbalanced loads
(*c*) provision of dual electric service
(*d*) balancing of phase voltages with respect to line voltages

3. As compared to Δ–Δ bank, the capacity of the V–V bank of transformers is —— percent.
(*a*) 57.7 (*b*) 66.7
(*c*) 50 (*d*) 86.6

4. If three transformers in a Δ – Δ are delivering their rated load and one transformer is removed, then overload and *each* of the remaining transformers is —— percent.
(*a*) 66.7 (*b*) 173.2
(*c*) 73.2 (*d*) 58

5. When a V–V system is converted into a Δ–Δ system, increase in capacity of the system is —— percent.
(*a*) 86.6 (*b*) 66.7
(*c*) 73.2 (*d*) 50

6. For supplying a balanced 3 – ϕ load of 40-kVA, rating of each transformer in V–V bank should be nearly —— kVA.
(*a*) 20 (*b*) 23
(*c*) 34.6 (*d*) 25

7. When a closed –Δ bank is converted into an open –Δ bank, each of the two remaining transformers supplies —— percent of the original load.
(*a*) 66.7 (*b*) 57.7
(*c*) 50 (*d*) 73.2

8. If the load p.f is 0.866, then the average p.f. of the V–bank is
(*a*) 0.886 (*b*) 0.75
(*c*) 0.51 (*d*) 0.65

9. A T–T connection has higher ratio of utilization thatn a V–V connection only when
(*a*) identical transformers are used
(*b*) load power factor is leading
(*c*) load power factor is unity
(*d*) non-identical transformers are used.

10. The biggest advantage of T–T connection over the V V connection for 3-phase power transformation is that it provides
(*a*) a set of balanced voltages under load
(*b*) a true 3-phase, 4-wire system
(*c*) a higher ratio of utilization
(*d*) more voltages

11. Of the following statements concerning parallel operation of transformers, the one which is **not** correct is
(*a*) transformers must have equal voltage ratings
(*b*) transformers must have same ratio of transformation
(*c*) transformers must be operated at the same frequency
(*d*) transformers must have equal kVA ratings

12. Statement
An auto-transformer is more efficient in transferring energy from primary to secondary circuit.
Reason
because it does so both inductively and conductively.
Key.
(*a*) statement is false, reason is correct and relevant
(*b*) statement is correct, reason is correct but irrelevant
(*c*) both statement and reason are correct and are connected to each other as cause and effect
(*d*) both statement and reason are false.

13. Out of the following given choices for poly phase transformer connections which one will you select for three-to-two phase conversion ?
(*a*) Scott
(*b*) star/star
(*c*) double Scott
(*d*) star/double-delta

14. A T–T transformer cannot be paralleled with —— transformer
(*a*) V–V (*b*) Y– Δ
(*c*) Y–Y (*d*) Δ – Δ

15. Instrument transformers are used on ac circuits for extending the range of
(*a*) ammeters
(*b*) voltmeters
(*c*) wattmeters
(*d*) all of the above

16. Before removing the ammeter from a current transformer, its secondary must be short-circuited in order to avoid
(*a*) excessive heating of the core
(*b*) high secondary emf
(*c*) increase in iron losses
(*d*) all of the above

Answers

1. *c* **2.** *a* **3.** *a* **4.** *c* **5.** *c* **6.** *b* **7.** *b* **8.** *b* **9.** *d* **10.** *b* **11.** *d* **12.** *c* **13.** *a* **14.** *b* **15.** *d* **16.** *d*

32 INDUCTION MOTOR

32.1. Classification of A.C. Motors

With the almost universal adoption of a.c. system of distribution of electric energy for light and power, the field of application of a.c. motors has widened considerably during recent years. As a result, motor manufactures have tried, over the last few decades, to perfect various types of a.c. motors suitable for all classes of industrial drives and for both single and three-phase a.c. supply. This has given rise to bewildering multiplicity of types whose proper classification often offers considerable difficulty. Different a.c. motors may, however, be classified and divided into various groups from the following different points of view :

1. AS REGARDS THEIR PRINCIPLE OF OPERATION
   - **(A) Synchronous motors**
     - (*i*) plain and (*ii*) super—
   - **(B) Asynchronous motors**
     - (*a*) **Induction motors**
       - (*i*) Squirrel cage { single / double
       - (*ii*) Slip-ring (external resistance)
     - (*b*) **Commutator motors**
       - (*i*) Series { single phase / universal
       - (*ii*) Compensated { conductively / inductively
       - (*iii*) shunt { simple / compensated
       - (*iv*) repulsion { straight / compensated
       - (*v*) repulsion-start induction
       - (*vi*) repulsion induction
2. AS REGARDS THE TYPE OF CURRENT
   (*i*) single phase (*ii*) three phase
3. AS REGARDS THEIR SPEED
   (*i*) constant speed (*ii*) variable speed (*iii*) adjustable speed
4. AS REGARDS THEIR STRUCTURAL FEATURES
   (*i*) open (*ii*) enclosed (*iii*) semi-enclosed (*iv*) ventilated
   (*v*) pipe-ventilated (*vi*) riverted frame eye etc.

29.2. Induction Motor : General Principle

As a general rule, conversion of electrical power into mechanical power takes place in the *rotating* part of an electric motor. In d.c. motors, the electric power is *conducted* directly to the armature (*i.e.* rotating part) through brushes and commutator (Art. 27.1). Hence, in this sense, a d.c. motor can be called a *conduction* motor. However, in a.c. motors, the rotor does not receive electric power by conduction but by *induction* in exactly the same way as the secondary of a 2-winding transformer receives its power from the primary. That is why such motors are known as *induction* motors. In fact, an induction motor can be treated as a *rotating transformer i.e.* one in which primary winding is stationary but the secondary is free to rotate (Art. 32.47).

Of all the a.c. motors, the polyphase induction motor is the one which is extensively used for various kinds of industrial drives. It has the following main advantages and also some disadvantages:

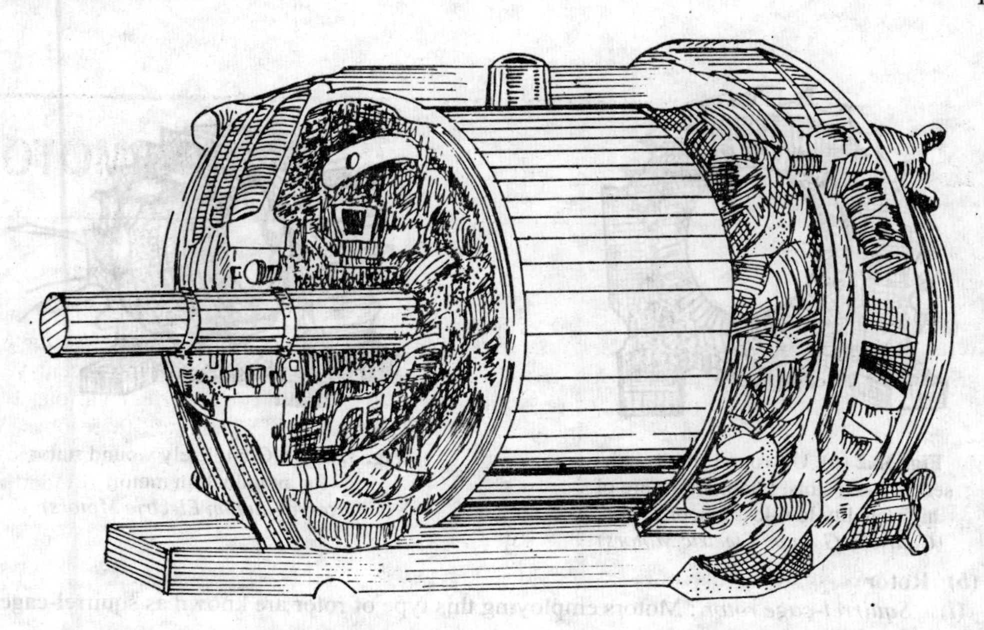

Fig. 32.1. Cut-away view of totally-enclosed, fan-cooled squirrel-cage Induction Motor. (*Courtesy:AEG, Frankfurt, Germany*)

Advantages :

1. It has very simple and extremely rugged, almost unbreakable construction (especially squirrel-cage type).
2. Its cost is low and it is very reliable.
3. It has sufficiently high efficiency. In normal running condition, no brushes are needed, hence frictional losses are reduced. It has a reasonably good power factor.
4. It requires minimum of maintenance.
5. It starts up from rest and needs no extra starting motor and has not to be synchronised. Its starting arrangement is simple especially for squirrel-cage type motor.

Disadvantages :

1. Its speed cannot be varied without sacrificing some of its efficiency.
2. Just like a d.c. shunt motor, its speed decreases with increase in load.
3. Its starting torque is somewhat inferior to that of a d.c. shunt motor.

32.3 Construction

An induction motor consists essentially of two main parts :

(*a*) a stator and (*b*) a rotor.

(*a*) Stator

The stator of an induction motor is, in principle, the same as that of a synchronous motor or generator. It is made up of a number of stampings, which are slotted to receive the windings [Fig.32.2 (*a*)]. The stator carries a 3-phase winding [Fig.32.2 (*b*)] and is fed from a 3-phase supply. It is wound for a definite number of poles*, the exact number of poles being determined by the requirements of speed. Greater the number of poles, lesser the speed and *vice versa*. It will be shown in Art. 32.6 that the stator windings, when supplied with 3-phase currents, produce a magnetic flux, which is of constant magnitude but which revolves (or rotates) at synchronous speed (given by $N_s = 120\ f/P$). This revolving magnetic flux induces an e.m.f. in the rotor by mutual induction.

*The number of poles P, produced in the rotating field is $P = 2n$ where n is the number of stator slots/pole/phase.

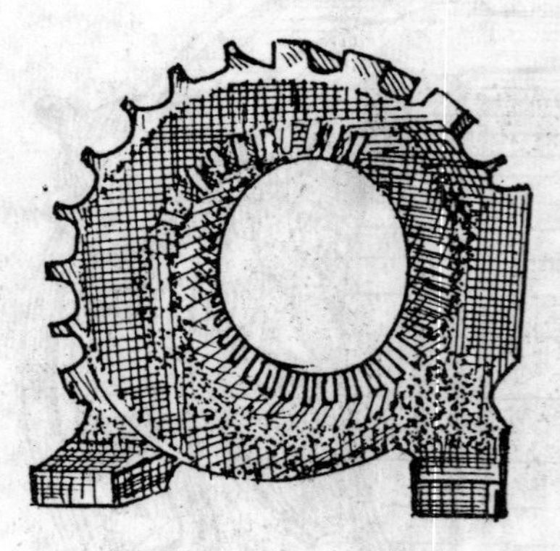

Fig. 32.2 (*a*). Unwound stator with semi-closed slots. Laminations are of high-quality low-loss silicon steel. (*Courtesy:Gautam Electric Motors*)

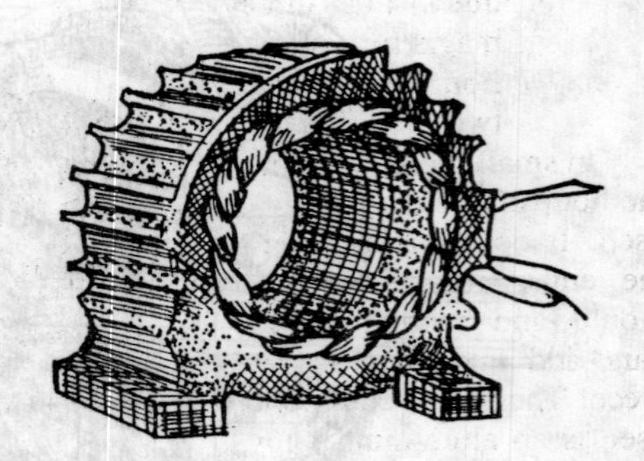

Fig. 32.2 (*b*). Completely wound stator for an induction motor. (*Courtesy: Gautam Electric Motors*)

(*b*) Rotor

(*i*) *Squirrel-cage rotor* : Motors employing this type of rotor are known as squirrel-cage induction motors.

(*ii*) *Phase-wound or wound rotor* : Motors employing this type of rotor are variously known as 'phase-wound' motors or 'wound' motors or as 'slip-ring' motors.

32.4. Squirrel-cage Rotor

Almost 90 per cent of induction motors are squirrel-cage type, because this type of rotor has the simplest and most rugged construction imaginable and is almost indestructible. The rotor consists of a cylindrical laminated core with parallel slots for carrying the rotor conductors which,

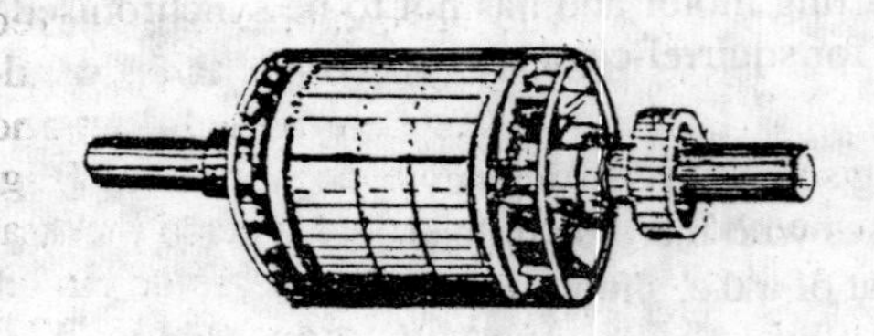

Fig.32.3(*a*) Squirrel-cage rotor with copper bars and alloy brazed end-rings. (*Courtesy:Gautam Electric Motors*)

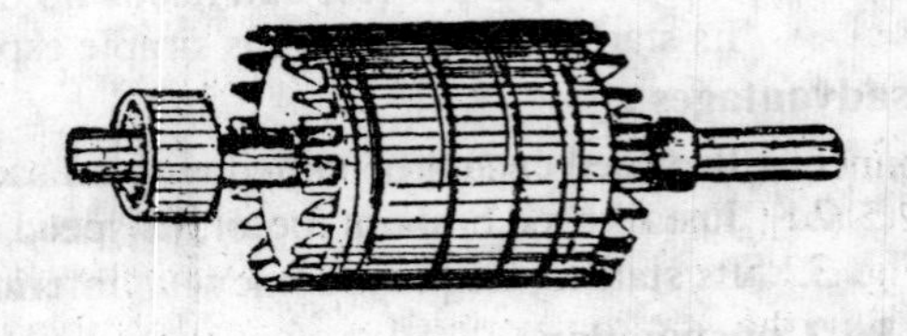

Fig. 32.3(*b*). Rotor with shaft and bearings (*Courtesy:Gautam Electric Motors*)

it should be noted clearly, are not wires but consist of heavy bars of copper, aluminium or alloys. One bar is placed in each slot, rather the bars are inserted from the end when semi-closed slots are used. The rotor bars are brazed or electrically welded or bolted to two heavy and stout short-circuiting end-rings, thus giving us, what is so picturesquely called, a squirrel-case construction (Fig. 32.3)

It should be noted that the *rotor bars are permanently short-circuited on themselves,* hence it is not possible to add any external resistance in series with the rotor circuit for starting purposes.

The rotor slots are usually not quite parallel to the shaft but are purposely given a slight skew (Fig. 32.4). This is useful in two ways :

(*i*) it helps to make the motor run quietly by reducing the magnetic hum and

(*ii*) it helps in reducing the locking tendency of the rotor *i.e.* the tendency of the rotor teeth to remain under the stator teeth due to direct magnetic attraction between the two.*

In small motors, another method of construction is used. It consists of placing the entire rotor core in a mould and casting all the bars and end-rings in one piece. The metal commonly used is an aluminium alloy.

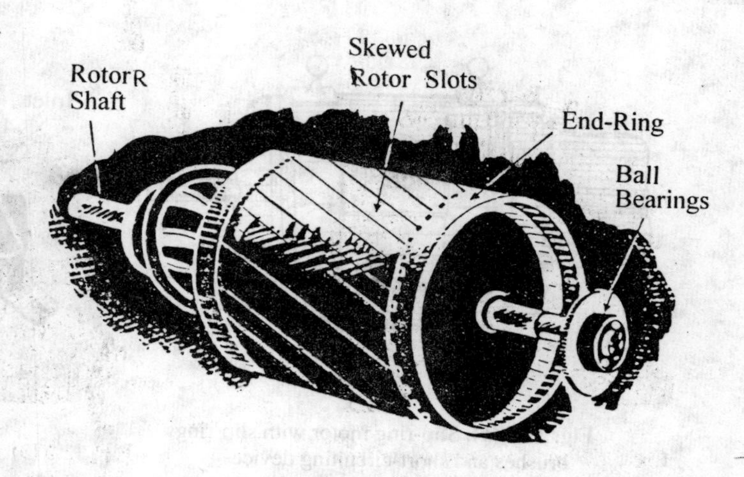

Fig. 32.4

Another form of rotor consists of a solid cylinder of steel without any conductors or slots at all ! The motor operation depends upon the production of eddy currents in the steel rotor.

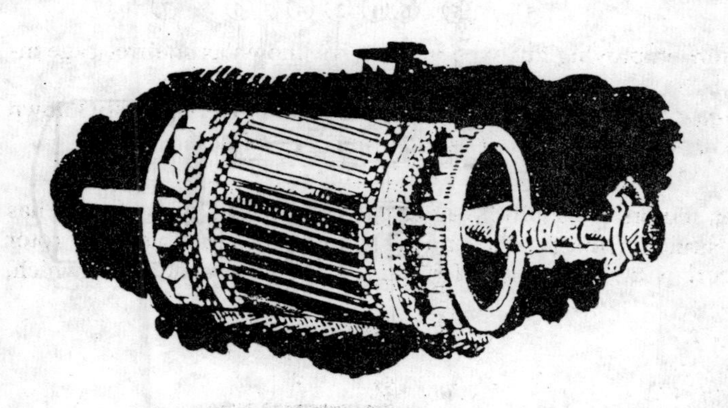

Fig. 32.5 (*a*)

32.5. Phase-wound Rotor

This type of rotor is provided with 3-phase, double-layer, distributed winding consisting of coils as used in alternators. The rotor is wound for as many poles as the number of stator poles and is always wound 3-phase even *when the stator is wound two-phase*.

The three phases are starred internally. The other three winding terminals are brought out and connected to three insulated slip-rings mounted on the shaft with brushes resting on them [Fig. 32.5 (*b*)]. These three brushes are further externally connected to a 3-phase star-connected rheostat [Fig. 32.5 (*c*)]. This makes possible the introduction of additional resistance in the rotor circuit during the starting period for increasing the starting torque of the motor, as shown in Fig. 32.6 (*a*) (Ex.32.7 and 32.10) and for changing its speed-torque/current characteristics. When running under normal conditions, the *slip-rings are automatically short-circuited* by means of a metal collar, which is pushed along the shaft and connects all the rings together. Next, the brushes are automatically lifted from the slip-rings to reduce the frictional losses and the wear and tear. Hence, it is seen that under normal running conditions, the wound rotor is short-circuited on itself just like the squirrel-case rotor.

Fig. 32.6(*b*) shows the longitudinal section of a slip-ring motor, whose structural details are as under :

1. **Frame.** Made of close-grained alloy cast iron.
2. **Stator and Rotor Core.** Built from high-quality low-loss silicon steel laminations and flash-enamelled on both sides.

*Other results of skew which may or may not be desirable are (*i*) increase in the effective ratio of transformation between stator and rotor (*ii*) increased rotor resistance due to increased length of rotor bars (*iii*) increased impedance of the machine at a given slip and (*iv*) increased slip for a given torque.

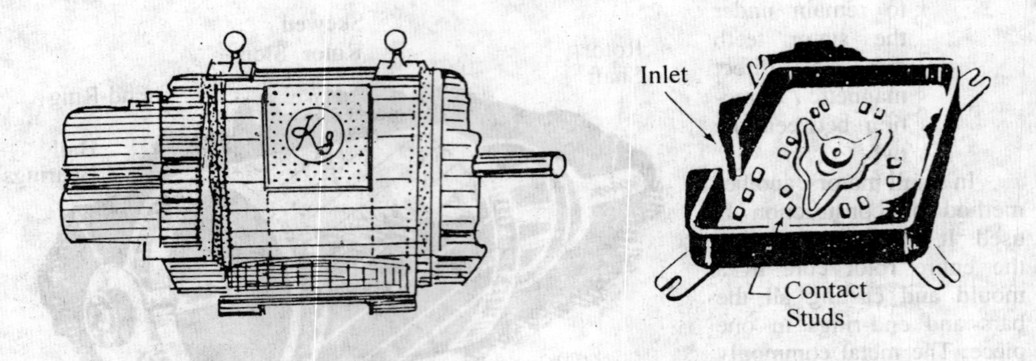

Fig. 32.5 (*b*). Slip-ring motor with slip-rings brushes and short-circuiting devices
(*Courtesy:Kirloskar Electric Company*)

Fig. 32.5 (*c*)

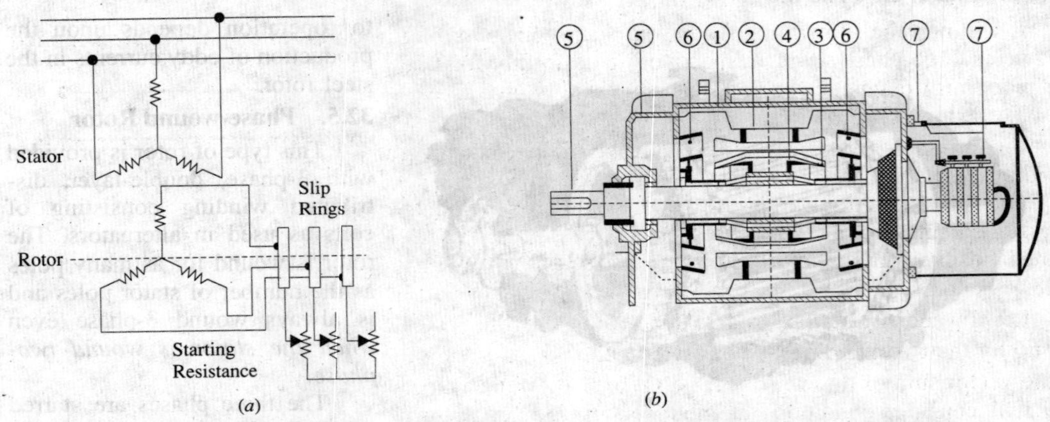

Fig.32.6 (*a*)

Fig.32.6 (*b*) Longitudinal section of a Jyoti splash-proof slip-ring motor
(*Courtesy. Jyoti Calor-Emag Ltd.*)

3. **Stator and Rotor Windings.** Have moisture proof tropical insulation embodying mica and high quality varnishes. Are carefully spaced for most effective air circulation and are rigidly braced to withstand centrifugal forces and any short-circuit stresses.
4. **Air-gap.** The stator rabbets and bore are machined carefully to ensure uniformity of air-gap.
5. **Shafts and Bearings.** Ball and roller bearings are used to suit heavy duty, toruble-free running and for enhanced service life.
6. **Fans.** Light aluminium fans are used for adequate circulation of cooling air and are securely keyed onto the rotor shaft.
7. **Slip-rings and Slip-ring Enclosures.** Slip-rings are made of high quality phosphor-bronze and are of moulded construction.

Fig. 32.6(*c*) shows the disassembled view of an induction motor with squirrel-cage rotor. According to the labelled notation (*a*) represents stator (*b*) rotor (*c*) bearing shields (*d*) fan (*e*) ventilation grill and (*f*) terminal box.

Similarly, Fix 32.6 (*d*) shows the disassembled view of a slip-ring motor where (*a*) represents stator (*b*) rotor (*c*) bearing shields (*d*) fan (*e*) ventilation grill (*f*) terminal box (*g*) slip-rings (*h*) brushes and brush holders.

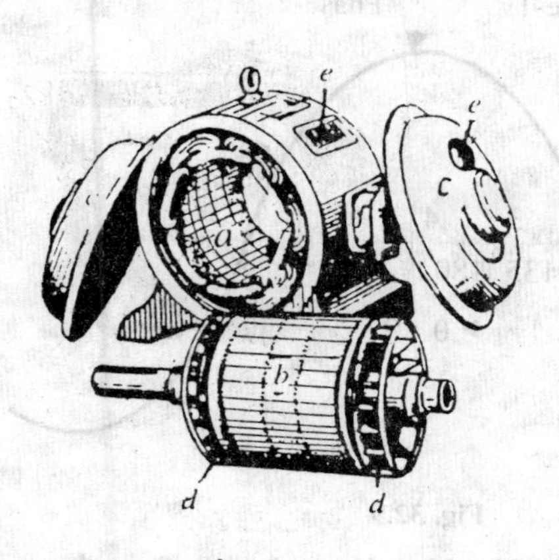

Fig. 32.6 (*c*)

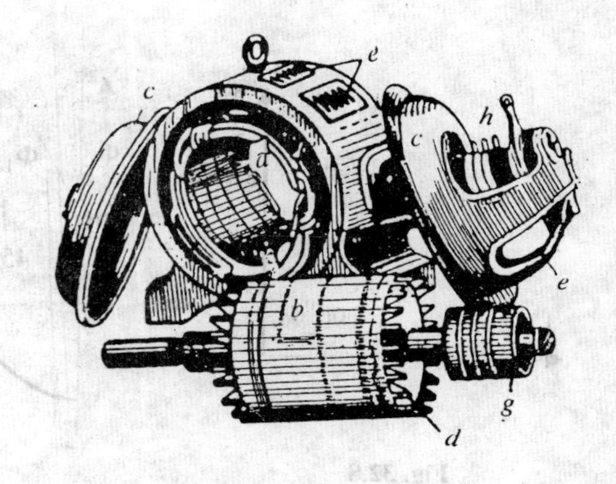

Fig. 32.6 (*d*)

32.6. Production of Rotating Field

It will now be shown that when stationary coils, wound for two or three phases, are supplied by two or three-phase supply respectively, a uniformly-rotating (or revolving) magnetic flux of constant value is produced.

Two-phase Supply

The principle of *a* 2-ϕ, 2-pole stator having two identical windings, 90 space degrees apart, is illustrated in Fig. 32.7.

The flux due to the current flowing in each phase winding is assumed sinusoidal and is represented in Fig. 32.9. The assumed positive directions of fluxes are those shown in Fig. 32.8.

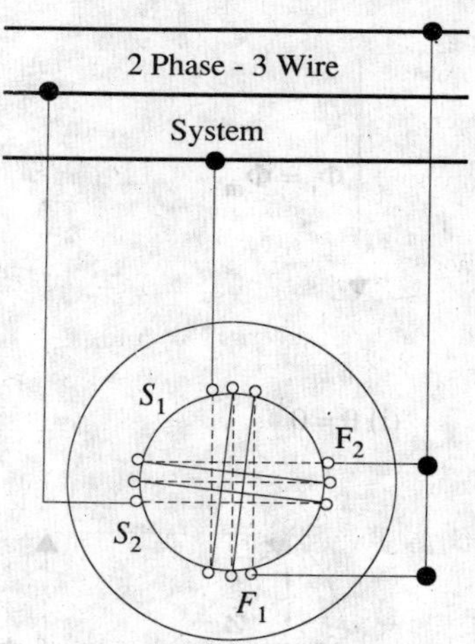

Fig. 32.7

Let Φ_1 and Φ_2 be the instantaneous values of the fluxes set up by the two windings. The resultant flux Φ_r at any time is the vector sum of these two fluxes (Φ_1 and Φ_2) at that time. We will consider conditions at intervals of 1/8th of a time period *i.e.* at intervals corresponding to angles of 0°, 45°, 90°, 135° and 180°. It will be shown that resultant flux Φ_r is constant in magnitude *i.e.* equal to Φ_m-the maximum flux due to either phase and is making one revolution/cycle. In other words, it means that the resultant flux rotates synchronously.

(*a*) When $\theta = 0°$ *i.e.* corresponding to point *O* in Fig. 32.9, $\Phi_1 = 0$, but Φ_2 is maximum *i.e.* equal to Φ_m and negative. Hence, resultant flux $\Phi_r = \Phi_m$ and, being negative, is shown by a vector pointing downwards [Fig. 32.10 (*i*)].

(*b*) When $\theta = 45°$ *i.e.* corresponding to point 1 in Fig. 32.9. At this instant, $\Phi_1 = \Phi_m / \sqrt{2}$ and is positive ; $\Phi_2 = \Phi_m / \sqrt{2}$ but is still negative. Their resultant, as shown in Fig. 32.10 (*ii*), is $\Phi_r = \sqrt{[(\Phi_m / \sqrt{2})^2 + (\Phi_m / \sqrt{2})^2]} = \Phi_m$ although this resultant has shifted 45° clockwise.

(*c*) When $\theta = 90°$ *i.e.* corresponding to point 2 in Fig. 32.9. Here $\Phi_2 = 0$, but $\Phi_1 = \Phi_m$ and is positive. Hence, $\Phi_r = \Phi_m$ and has further shifted by an angle of 45° from its position in (*b*) or by 90° from its original position in (*a*).

(*d*) When $\theta = 135°$ *i.e.* corresponding to point 3 in Fig. 32.9. Here, $\Phi_1 = \Phi_m / \sqrt{2}$ and is positive, $\Phi_2 = \Phi_m / \sqrt{2}$ and is also positive. The resultant $\Phi_r = \Phi_m$ and has further shifted clockwise by another 45°, as shown in Fig. 32.10 (*iv*).

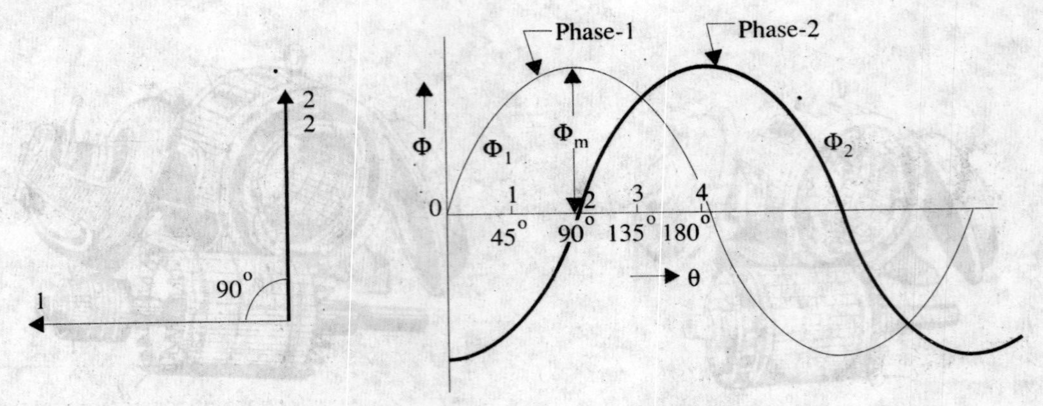

Fig. 32.8 **Fig. 32.9**

(e) When θ = 180° *i.e.* corresponding to point 4 in Fig. 32.9. Here, $\Phi_1 = 0$, $\Phi_2 = \Phi_m$ and is positive. Hence, $\Phi_r = \Phi_m$ and has shifted clockwise by another 45° or has rotated through an angle of 180° from its position at the beginning. This is shown in Fig. 32.10 (v).

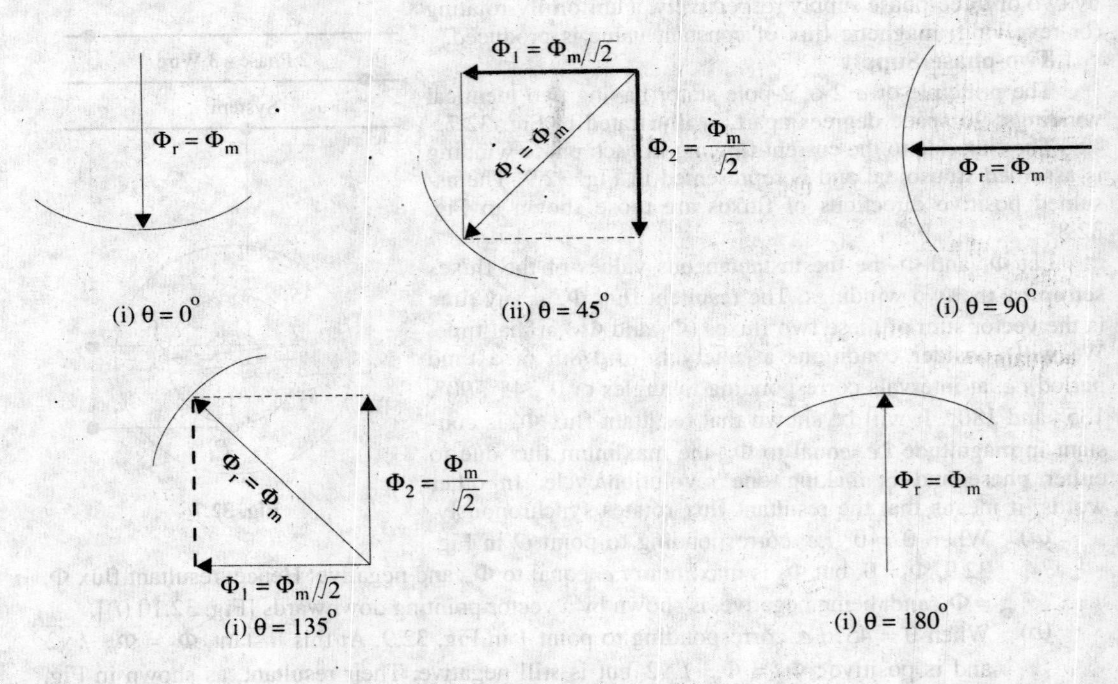

Fig. 32.10

Hence, we conclude

1. *that the magnitude of the resultant flux is constant and is equal to Φ_m — the maximum flux due to either phase.*
2. *that the resultant flux rotates at synchronous speed given by $N_s = 120\, f/P$ rpm.*

However, it should be clearly understood that in this revolving field, there is no actual revolution of the flux. The flux due to each phase changes periodically, according to the changes in the phase current, but the magnetic flux itself does not move around the stator. It is only the *seat* of the resultant flux which keeps on shifting synchronously around the stator.

Mathematical Proof

Let $\Phi_1 = \Phi_m \sin \omega t$ and $\Phi_2 = \Phi_m \sin(\omega t - 90°)$

$\therefore \quad \Phi_r^2 = \Phi_1^2 + \Phi_2^2$

$\Phi_r^2 = (\Phi_m \sin \omega t)^2 + [\Phi_m \sin(\omega t\text{-}90°)]^2 = \Phi_m^2 (\sin^2 \omega t + \cos^2 \omega t) = \Phi_m^2$

$\therefore \quad \Phi_r = \Phi_m$

It shows that the flux is of constant value and does not change with time.

32.7. Three-phase Supply

It will now be shown that when three-phase windings displaced in space by 120°, are fed by three-phase currents, displaced in time by 120°, they produce a resultant magnetic flux, which rotates in space as if actual magnetic poles were being rotated mechanically.

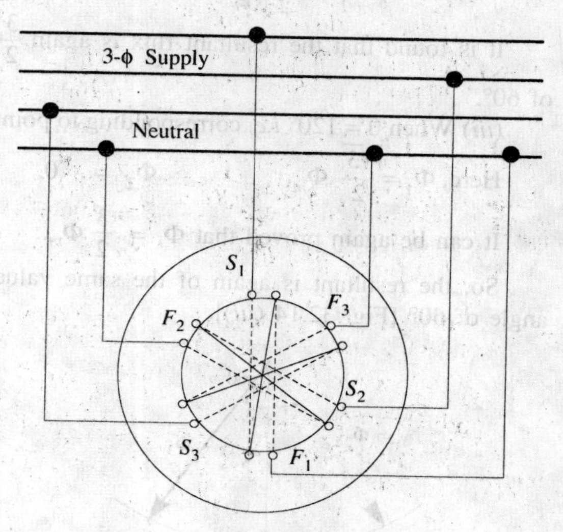

Fig. 32.11

The principle of a 3-phase, two-pole stator having three identical windings placed 120 space degrees apart is shown in fig. 32.11. The flux (assumed sinusoidal) due to three-phase windings is shown in Fig 32.12.

The assumed positive directions of the fluxes are shown in Fig 32.13. Let the maximum value of flux due to any one of the three phases be Φ_m. The resultant flux Φ_r, at any instant, is given by the vector sum of the individual fluxes, Φ_1, Φ_2 and Φ_3 due to three phases. We will consider values of Φ_r at four instants 1/6th time-period apart corresponding to points marked 0, 1, 2 and 3 in Fig. 32.12.

(*i*) When $\theta = 0°$ *i.e.* corresponding to point *O* in Fig. 32.12

Here $\Phi_1 = 0$, $\Phi_2 = -\frac{\sqrt{3}}{2}\Phi_m$, $\Phi_3 = \frac{\sqrt{3}}{2}\Phi_m$,. The vector for Φ_2 in Fig. 32.14 (*i*) is drawn in a direction opposite to the direction assumed positive in Fig. 32.13

$$\therefore \quad \Phi_r = 2 \times \frac{\sqrt{3}}{2}\Phi_m \cos\frac{60°}{2} = \sqrt{3} \times \frac{\sqrt{3}}{2}\Phi_m = \frac{3}{2}\Phi_m$$

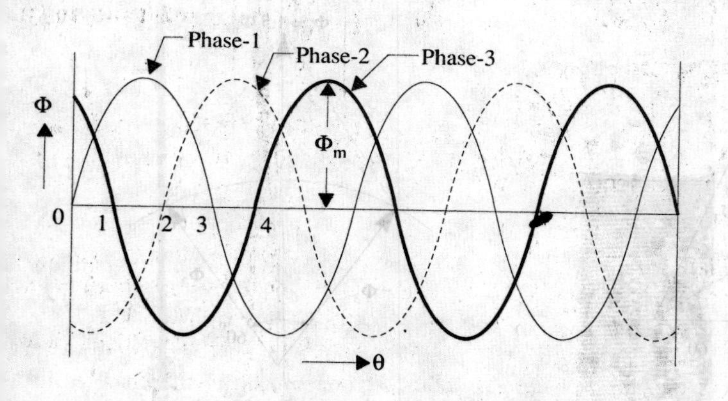

Fig. 32.12

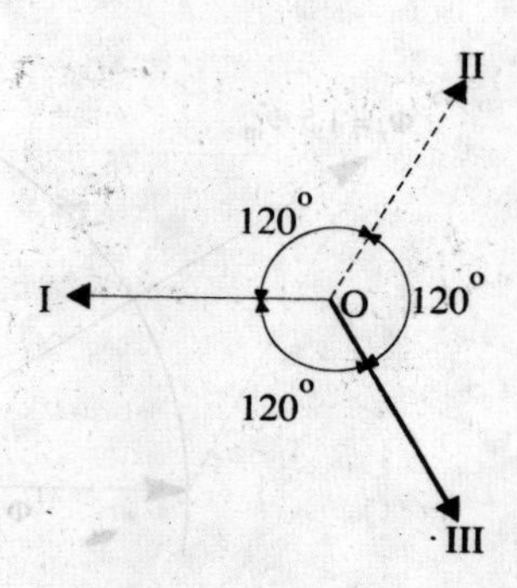

Fig. 32.13

(*ii*) when $\theta = 60°$ *i.e.* corresponding to point 1 in Fig. 32.12.

Here $\Phi_1 = \frac{\sqrt{3}}{2}\,\Phi_m$ ... drawn parallel to *OI* of Fig. 32.13 as shown in Fig. 32.14 (*ii*)

$\Phi_2 = -\frac{\sqrt{3}}{2}\,\Phi_m$ ...drawn in opposition to *OII* of Fig. 32.13.

$\Phi_3 = 0$

$$\therefore \quad \Phi_r = 2 \times \frac{\sqrt{3}}{2}\,\Phi_m \times \cos 30° = \frac{3}{2}\,\Phi_m \qquad \text{[Fig. 32.14 (ii)]}$$

It is found that the resultant flux is again $\frac{3}{2}\Phi_m$ but has rotated clockwise through an angle of 60°.

(*iii*) When $\theta = 120°$ *i.e.* corresponding to point 2 in Fig. 32.12.

Here, $\Phi_1 = \frac{\sqrt{3}}{2}\,\Phi_m$, $\quad \Phi_2 = 0, \quad \Phi_3 = -\frac{\sqrt{3}}{2}\,\Phi_m$

It can be again proved that $\Phi_r = \frac{3}{2}\,\Phi_m$.

So, the resultant is again of the same value, but has further rotated clockwise through an angle of 60° [Fig. 32.14 (*iii*)].

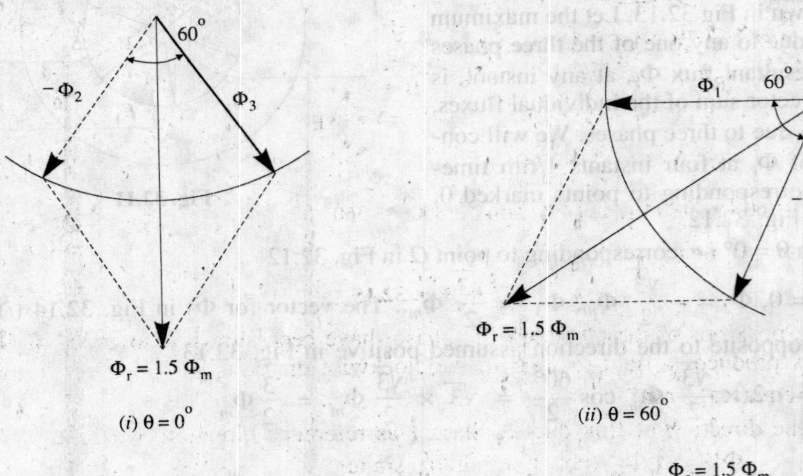

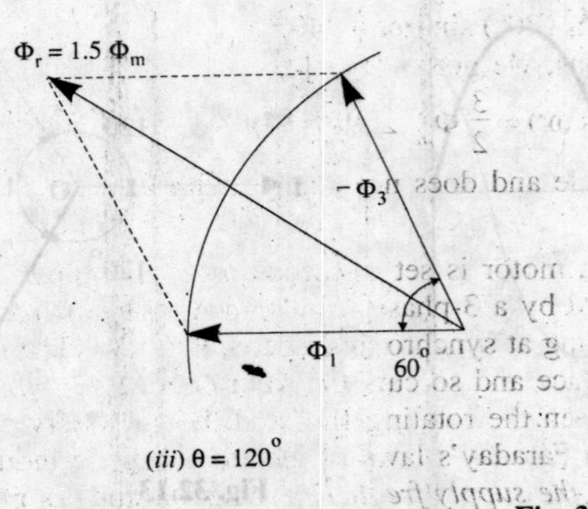

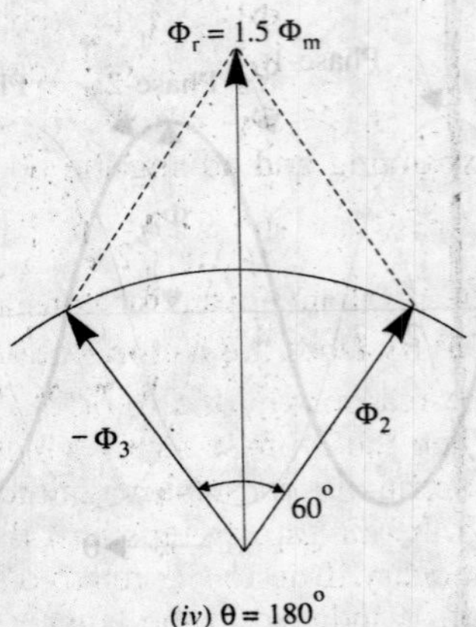

Fig. 32.14

(*iv*) When $\theta = 180°$ *i.e.* corresponding to point 3 in Fig. 32.12.

$$\Phi_1 = 0, \Phi_2 = \frac{\sqrt{3}}{2}\Phi_m, \Phi_3 = -\frac{\sqrt{3}}{2}\Phi_m$$

The resultant is $\frac{3}{2}\Phi_m$ and has rotated clockwise through an additional angle 60° or through an angle of 180° from the start.

Hence, we conclude that

1. *the resultant flux is of constant value* $= \frac{3}{2}\Phi_m$ *i.e. 1.5 times the maximum value of the flux due to any phase.*
2. *the resultant flux rotates around the stator at synchronous speed given by* $N_s = 120\, f/P$.

Fig. 32.15 (*a*) shows the graph of the rotating flux in a simple way. As before, the positive directions of the flux phasors have been shown separately in Fig. 32.15 (*b*). Arrows on these flux phasors are reversed when each phase passes through zero and becomes negative.

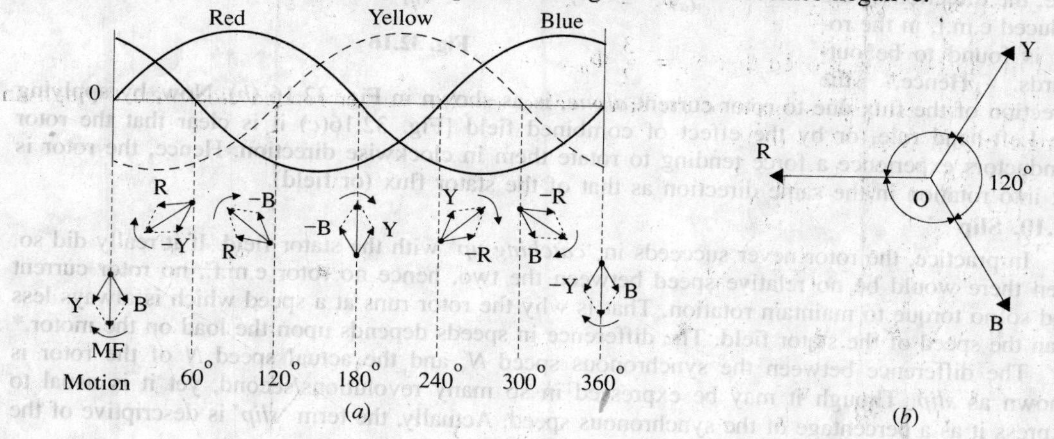

Fig. 32.15

As seen, positions of the resultant flux phasor have been shown at intervals of 60° only. The resultant flux produces a field rotating in the clockwise direction.

32.8. Mathematical Proof

Taking the direction of flux due to phase 1 as reference direction, we have

$$\Phi_1 = \Phi_m (\cos 0° + j \sin 0°) \sin \omega t$$

$$\Phi_2 = \Phi_m (\cos 240° + j \sin 240°) \sin (\omega t - 120°)$$

$$\Phi_3 = \Phi_m (\cos 120° + j \sin 120°) \sin (\omega t - 240°)$$

Expanding and adding the above equations, we get

$$\Phi_r = \frac{3}{2}\Phi_m (\sin \omega t + j \cos \omega t) = \frac{3}{2}\Phi_m \angle 90° - \omega t$$

The resultant flux is of constant magnitude and does not change with time '*t*'.

32.9. Why Does the Rotor Rotate ?

The reason why the rotor of an induction motor is set into rotation is as follow :

When the 3-phase stator windings, are fed by a 3-phase supply then, as seen from above, a magnetic flux of constant magnitude, but rotating at synchronous speed, is set up. The flux passes through the air-gap, sweeps past the rotor surface and so cuts the rotor conductors which, as yet, are stationary. Due to the relative speed between the rotating flux and the stationary conductors, an e.m.f. is induced in the latter, according to Faraday's laws of electro-magnetic induction. *The frequency of the induced e.m.f. is the same as the supply frequency.* Its magnitude is proportional to the relative velocity between the flux and the conductors and its direction is given by Fleming's

Right-hand rule. Since the rotor bars or conductors form a closed circuit, rotor current is produced whose direction, as given by Lenz's law, is such as to oppose the very cause producing it. In this case, the cause which produces the rotor current is the relative velocity between the rotating flux of the stator and the stationary rotor conductors. Hence, to reduce the relative speed, the rotor starts running in the *same* direction as that of the flux and tries to catch up with the rotating flux.

The setting up of the torque for rotating the rotor is explained below :

In Fig 32.16 (*a*) is shown the stator field which is assumed to be rotating clockwise. The relative motion of the rotor with respect to the stator is *anticlockwise*. By applying Right-hand rule, the direction of the induced e.m.f. in the rotor is found to be outwards. Hence, the direction of the flux due to rotor current *alone*, is as shown in Fig. 32.16 (*b*). Now, by applying the Left-hand rule, or by the effect of combined field [Fig. 32.16(*c*) it is clear that the rotor conductors experience a force tending to rotate them in clockwise direction. Hence, the rotor is set into rotation in the same direction as that of the stator flux (or field).

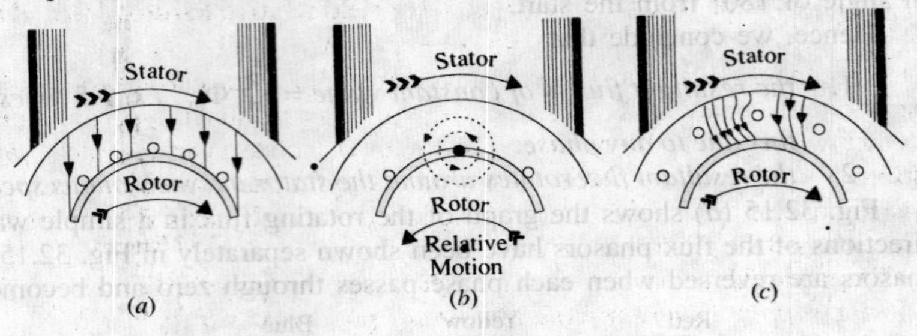

Fig. 32.16

32.10. Slip

In practice, the rotor never succeeds in '*catching up*' with the stator field. If it really did so, then there would be no relative speed between the two, hence no rotor e.m.f., no rotor current and so no torque to maintain rotation. That is why the rotor runs at a speed which is always less than the speed of the stator field. The difference in speeds depends upon the load on the motor.*

The difference between the synchronous speed N_s and the actual speed N of the rotor is known as *slip*. Though it may be expressed in so many revolutions/second, yet it is usual to express it as a percentage of the synchronous speed. Actually, the term '*slip*' is descriptive of the way in which the rotor '*slips back*' from synchronism.

$$\% \text{ slip } s = \frac{N_s - N}{N_s} \times 100$$

Sometimes, $N_s - N$ is called the *slip speed*.

Obviously, rotor (or motor) speed is $N = N_s (1 - s)$.

It may be kept in mind that revolving flux is rotating synchronously, relative to the stator (*i.e.* stationary space) but at slip speed relative to the rotor.

32.11. Frequency of Rotor Current

When the rotor is stationary, the frequency of rotor current is *the same as the supply frequency.* But when the rotor starts revolving, then the frequency depends upon the relative speed or on slip-speed. Let at any slip-speed, the frequency of the rotor current be f'. Then

$$N_s - N = \frac{120 f'}{P}. \quad \text{Also,} \quad N_s = \frac{120 f}{P}$$

Dividing one by the other, we get, $\frac{f'}{f} = \frac{N_s - N}{N_s} = s$; $\therefore f' = sf$

As seen, rotor currents have a frequency of $f' = sf$ and when flowing through the individual

*It may be noted that as the load is applied, the natural effect of the load or braking torque is to slow down the motor. Hence, slip increases and with it increases the current and torque, till the driving torque of the motor balances the retarding torque of the load. This fact determines the speed at which the motor runs on load.

phases of rotor winding, give rise to rotor magnetic fields. These individual rotor magnetic fields produce a combined rotating magnetic field, whose speed relative to rotor is

$$= \frac{120 f'}{P} = \frac{120\, sf}{P} = s N_S$$

However, the rotor itself is running at speed N with respect to *space*. Hence,

speed of rotor field in *space* = speed of field relative to rotor + speed of rotor relative to space

$$= s N_s + N = s N_s + N_s (1 - s) = N_s$$

It means that no mattar what the value of slip, rotor currents and stator currents each produce a sinusoidally distributed magnetic field of constant magnitude and constant space speed of N_s. In other words, both the rotor and stator fields rotate synchronously, which means that they are stationary *with respect to each other.* These two synchronously rotating magnetic fields, in fact, superimpose on each other and give rise to the actually existing rotating field, which corresponds to the magnetising current of the stator winding.

Example 32.1. *A slip-ring induction motor runs at 290 r.p.m. at full load, when connected to 50-Hz supply. Determine the number of poles and slip.*

(Utilisation of Electric Power AMIE Sec. E 1991)

Solution. Since N is 290 rpm; N_s has to be somewhere near it, say 300 rpm. If N_s is assumed as 300 *rpm*, then 300 = 120 × 50/*P*. Hence, $P = 20$. ∴ $s = (300 - 290)/300 = 3.33\%$

Example 32.2. *The stator of a 3–φ induction motor has 3 slots per pole per phase. If supply frequency is 50 Hz, calculate*

(*i*) *number of stator poles produced and total number of slots on the stator*

(*ii*) *speed of the rotating stator flux (or magnetic field).*

Solution. (*i*) $P = 2n = 2 \times 3 =$ **6 poles**

Total No. of slots = 3 slots/pole/phase × 6poles × 3phases = **54**

(*ii*) $N_s = 120 f/P = 120 \times 50/6 =$ **1000 r.p.m.**

Example 32.3. *A 4-pole, 3-phase induction motor operates from a supply whose frequency is 50 Hz. Calculate:*

(*i*) *the speed at which the magnetic field of the stator is rotating.*

(*ii*) *the speed of the rotor when the slip is 0.04.*

(*iii*) *the frequency of the rotor currents when the slip is 0.03.*

(*iv*) *the frequency of the rotor currents at standstill.*

(Electrical Machinery II, Banglore Univ. 1991)

Solution. (*i*) Stator field revolves at synchronous speed, given by

$N_s = 120 f/P = 120 \times 50/4 =$ **1500 r.p.m.**

(*ii*) rotor (or motor) speed, $N = N_s(1 - s) = 1500(1 - 0.04) =$ **1440 r.p.m.**

(*iii*) frequency of rotor current, $f' = sf = 0.03 \times 50 = 1.5$ r.p.s = **90 r.p.m**

(*iv*) Since at standstill, $s = 1$, $f' = sf = 1 \times f = f =$ **50Hz**

Example 32.4. *A 3-φ induction motor is wound for 4 poles and is supplied from 50-Hz system. Calculate (i) the synchronous speed (ii) the rotor speed, when slip is 4% and (iii) rotor frequency when rotor runs at 600rpm.* **(Electrical Engineering-I, Pune Univ. 1991)**

Solution. (*i*) $N_s = 120 f/P = 120 \times 50/4 = 1500$ rpm

(*ii*) rotor speed, $N = N_s(1 - s) = 1500(1 - 0.04) =$ **1440 rpm**

(*iii*) when rotor speed is 600 rpm, slip is

$$s = (N_s - N)/N_s = (1500 - 600)/1500 = 0.6$$

rotor current frequency, $f' = sf = 0.6 \times 50 =$ **30 Hz**

Example 32.5. *A 12-pole, 3-phase alternator driven at a speed of 500 r.p.m. supplies power to an 8-pole, 3-phase induction motor. If the slip of the motor, at full-load is 3%, calculate the full-load speed of the motor.*

Solution. Let N = actual motor speed; Supply frequency, $f = 12 \times 500/120 = 50$ Hz Synchronous speed $N_s = 120 \times 50/8 = 750$ r.p.m.

$$\% \text{ slip } s = \frac{N_S - N}{N} \times 100 \; ; \quad 3 = \frac{750 - N}{750} \times 100 \qquad \therefore \; N = \mathbf{727.5 \text{ r.p.m.}}$$

Note. Since slip is 3%, actual speed N is less than N_s by 3% of N_s *i.e.* by $3 \times 750/100 = 22.5$ r.p.m.

32.12. Relation Between Torque and Rotor Power Factor

In Art. 32.7, it has been shown that in the case of a d.c. motor, the torque T_a is proportional to the product of armature current and flux per pole *i.e.* $T_a \propto \phi \, I_a$. Similarly, in the case of an induction motor, the torque is also proportional to the product of flux per stator pole and the rotor current. However, there is one more factor that has to be taken into account *i.e.* the power factor of the rotor.

$\therefore \; T \propto \phi \, I_2 \cos \phi_2$ or $T = k \phi \, I_2 \cos \phi_2$

where I_2 = rotor current at *standstill*

ϕ_2 = angle between rotor e.m.f. and rotor current

k = a constant

Denoting rotor e.m.f. at *standstill* by E_2, we have that $E_2 \propto \phi$

$$\therefore \; T \propto E_2 I_2 \cos \phi$$

$$\text{or } T = k_1 E_2 I_2 \cos \phi_2$$

where k_1 is another constant.

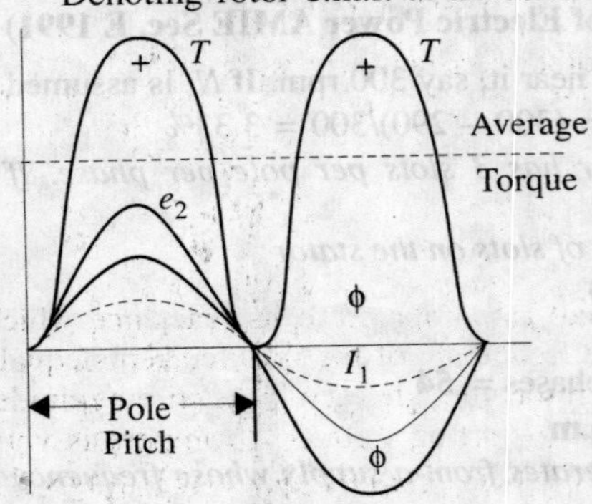

Fig. 32.17

The effect of rotor power factor on rotor torque is illustrated in Fig. 32.17 and Fig. 32.18 for various values of ϕ_2. From the above expression for torque, it is clear that as ϕ_2 increases (and hence, $\cos \phi$ decreases) the torque decreases and *vice versa*.

In the discussion to follow, the stator flux distribution is assumed sinusoidal. This revolving flux induces in each rotor conductor or bar an *e.m.f.* whose value depends on the flux density, in which the conductor is lying at the instant considered ($\because \; e = Blv$ volt). Hence, the induced e.m.f. in the rotor is also sinusoidal.

(*i*) **Rotor Assumed Non-inductive** (or $\phi_2 = 0$)

In this case, the rotor current I_2 is in phase with the e.m.f. E_2 induced in the rotor (Fig. 32.17). The instantaneous value of the torque acting on each rotor conductor is given by the product of instantaneous value of the flux and the rotor current ($\because \; F \propto BI_2 l$). Hence, torque curve is obtained by plotting the products of flux ϕ (or flux density B) and I_2. It is seen that the torque is always positive *i.e.* unidirectional.

(*ii*) **Rotor Assumed Inductive**

This case is shown in Fig. 32-18. Here, I_2 lags behind E_2 by an angle $\phi_2 = \tan^{-1} X_2/R_2$ where R_2 = rotor resistance/phase ; X_2 = rotor reactance / phase at *standstill*.

It is seen that for a portion '*ab*' of the pole pitch, the torque is negative *i.e.* reversed. Hence, the total torque which is the difference of the forward and the backward torques, is considerably reduced. If $\phi_2 = 90°$, then the total torque is zero because in that case the backward and the forward torques become equal and opposite.

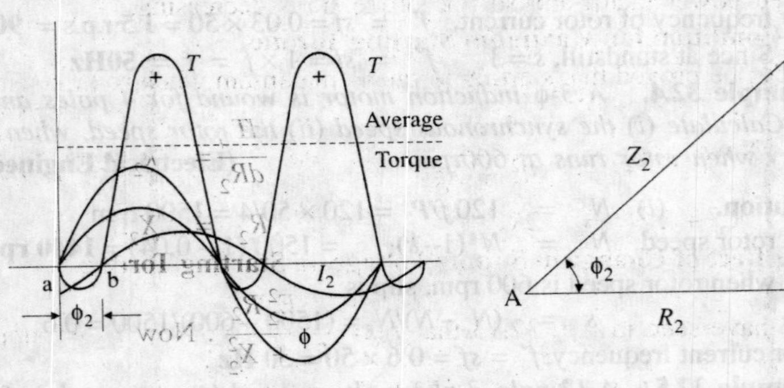

Fig. 32.18

Fig. 32.19

32.13. Starting Torque

The torque developed by the motor at the instant of starting is called starting torque. In some cases, it is greater than the normal running torque, whereas in some other cases it is somewhat less.

Let E_2 = rotor *e.m.f.* per phase at *standstill*; R_2 = rotor resistance/phase

X_2 = rotor reactance/phase at *standstill*

$\therefore\ Z_2 = \sqrt{(R_2^2 + X_2^2)}$ = rotor impedance/phase at *standstill* ...Fig. 32.19

$$\text{Then, } I_2 = \frac{E_2}{Z_2} = \frac{E_2}{\sqrt{(R_2^2 + X_2^2)}} \quad ; \quad \cos\phi_2 = \frac{R_2}{Z_2} = \frac{R_2}{\sqrt{(R_2^2 + X_2^2)}}$$

Standstill or starting torque $T_{st} = k_1 E_2 I_2 \cos\phi_2$ ... Art. 32.12

$$\text{or } T_{st} = k_1 E_2 \cdot \frac{E_2}{\sqrt{(R_2^2 + X_2^2)}} \times \frac{R_2}{\sqrt{(R_2^2 + X_2^2)}} = \frac{k_1 E_2^2 R_2}{R_2^2 + X_2^2} \quad ...(i)$$

If supply voltage V is constant, then the flux Φ and hence, E_2 both are constant.

$$\therefore\ T_{st} = k_2 \frac{R_2}{R_2^2 + X_2^2} = k_2 \frac{R_2}{Z_2^2} \text{ where } k_2 \text{ is some other constant.}$$

$$\text{Now, } k_1 = \frac{3}{2\pi N_s}, \quad \therefore\ T_{st} = \frac{3}{2\pi N_s} \cdot \frac{E_2^2 R_2}{R_2^2 + X_2^2}$$

32.14. Starting Torque of a Squirrel-cage Motor

The resistance of a squirrel-cage rotor is fixed and small as compared to its reactance which is very large especially at the start because at standstill, the frequency of the rotor currents equals the supply frequency. Hence, the starting current I_2 of the rotor, though very large in magnitude, lags by a very large angle behind E_2, with the result that the starting torque per ampere is very poor. It is roughly 1.5 times the full-load torque, although the starting current is 5 to 7 times the full-load current. Hence, such motors are not useful where the motor has to start against heavy loads.

32.15. Starting Torque of a Slip-ring Motor

The starting torque of such a motor is increased by improving its power factor by adding external resistance in the rotor circuit from the star-connected rheostat, the rheostat resistance being progressively cut out as the motor gathers speed. Addition of external resistance, however, increases the rotor impedance and so reduces the rotor current. At first, the effect of improved power factor predominates the current-decreasing effect of impedance. Hence, starting torque is increased. But after a certain point, the effect of increased impedance predominates the effect of improved power factor and so the torque starts decreasing.

32.16. Condition for Maximum Starting Torque

It can be proved that starting torque is maximum when rotor resistance equals rotor reactance.

$$\text{Now} \quad T_{st} = \frac{k_2 R_2}{R_2^2 + X_2^2} \quad \therefore\ \frac{dT_{st}}{dR_2} = k_2 \left[\frac{1}{R_2^2 + X_2^2} - \frac{R_2(2R_2)}{(R_2^2 + X_2^2)^2} \right] = 0$$

$$\text{or} \quad R_2^2 + X_2^2 = 2R_2^2 \quad \therefore\ R_2 = X_2$$

32.17. Effect of Change in Supply Voltage on Starting Torque

We have seen in Art. 32.13 that $T_{st} = \dfrac{k_1 E_2^2 R_2}{R_2^2 + X_2^2}$. Now $E_2 \propto$ supply voltage V

$$\therefore\ T_{st} = \frac{k_3 V^2 R_2}{R_2^2 + X_2^2} = \frac{k_3 V^2 R_2}{Z_2^2} \text{ where } k_3 \text{ is yet another constant. Hence } T_{st} \propto V^2.$$

Clearly, the torque is very sensitive to any changes in the supply voltage. A change of 5 per cent in supply voltage, for example, will produce a change of approximately 10% in the rotor torque. This fact is of importance in star-delta and auto transformer starters (Art. 33-11).

Example 32.6. *A 3-φ induction motor having a star-connected rotor has an induced e.m.f. of 80 volts between slip-rings at standstill on open-circuit. The rotor has a resistance and reactance per phase of 1 Ω and 4 Ω respectively. Calculate current/phase and power factor when (a) slip-rings are short-circuited (b) slip-rings are connected to a star-connected rheostat of 3 Ω per phase.* **(Electrical Technology, Bombay Univ. 1987)**

Solution. Standstill e.m.f/rotor phase $= 80/\sqrt{3} = 46.2$ V

(*a*) Rotor impedance/phase $= \sqrt{(1^2 + 4^2)} = 4.12\ \Omega$

Rotor current/phase $= 46.2/4.12 = \mathbf{11.2\ A}$

Power factor $= \cos\phi = 1/4.12 = \mathbf{0.243}$

As *p.f.* is low, the starting torque is also low.

(*b*) Rotor resistance/phase $= 3 + 1 = 4\ \Omega$

Rotor impedance/phase $= \sqrt{(4^2 + 4^2)} = 5.66\ \Omega$

$\therefore$ rotor current/phase $= 46.2/5.66 = \mathbf{8.16\ A}$; $\cos\phi = \mathbf{4/5.66 = 0.707}$.

Hence, the starting torque is increased due to the improvement in the power factor. It will also be noted that improvement in *p.f.* is much more than the decreases in current due to increased impedance.

Example 32.7. *A 3-phase, 400-V, star-connected induction motor has a star-connected rotor with a stator to rotor turn ratio of 6.5. The rotor resistance and standstill reactance per phase are 0.05 Ω and 0.25 Ω respectively. What should be the value of external resistance per phase to be inserted in the rotor circuit to obtain maximum torque at starting and what will be rotor starting current with this resistance?*

Solution. Here $K = \dfrac{1}{6.5}$ because transformation ratio K is defined as

$$= \frac{\text{rotor turns/phase}}{\text{stator turns/phase}}$$

$$\text{Standstill rotor } e.m.f./\text{phase, } E_2 = \frac{400}{\sqrt{3}} \times \frac{1}{6.5} = 35.5 \text{ volt}$$

It has been shown in Art. 32.16 that starting torque is maximum when $R_2 = X_2$ *i.e.* when $R_2 = 0.25\ \Omega$ in the present case

$\therefore$ external resistance/phase required $= 0.25 - 0.05 = \mathbf{0.2\ \Omega}$

rotor impedance/phase $= \sqrt{(0.25^2 + 0.25^2)} = 0.3535\ \Omega$

rotor current/phase, $I_2 = 35.5/0.3535 = \mathbf{100\ A\ (approx)}$

Example 32.8. *A 1100-V, 50-Hz delta-connected induction motor has a star-connected slip-ring rotor with a phase transformation ratio of 3.8. The rotor resistance and standstill leakage reactance are 0.012 ohm and 0.25 ohm per phase respectively. Neglecting stator impedance and magnetising current determine.*

(*i*) *the rotor current at start with slip-rings shorted*

(*ii*) *the rotor power factor at start with slip-rings shorted*

(*iii*) *the rotor current at 4% slip with slip-rings shorted*

(*iv*) *the rotor power factor at 4% slip with slip-rings shorted*

(*v*) *the external rotor resistance per phase required to obtain a starting current of 100 A in the stator supply lines.*

(Elect. Machines AMIE Sec. B 1992)

Solution. It should be noted that in a Δ/Y connection, primary phase voltage is the same as the line voltage. The rotor phase voltage can be found by using the phase transformation ratio of 3.8 *i.e.* $K = 1/3.8$.

Rotor phase voltage at standstill $= 1100 \times 1/3.8 = 289.5$V

(*i*) Rotor impedence / phase = $\sqrt{0.012^2 + 0.25^2} = 0.2503\Omega$
Rotor phase current at start = 289.5/0.2503 = **1157 A**
(*ii*) $p.f. = R_2/Z_2 = 0.012/0.2503 = \mathbf{0.048\ lag}$
(*iii*) at 4% slip, $X_r = sX_2 = 0.04 \times 0.25 = \mathbf{0.01\ \Omega}$
$\therefore\ Z_r = \sqrt{0.012^2 + 0.01^2} = \mathbf{0.0156\ \Omega}$
$E_r = s\,E_2 = 0.04 \times 289.5 = 11.58$ V; $I_2 = 11.58/0.0156 = \mathbf{742.3\ A}$
(*iv*) $p.f. = 0.012/0.0156 = \mathbf{0.77}$
(*v*) $I_2 = I_1/K = 100 \times 3.8 = 380$ A; E_2 at standstill = 289.5 V
$Z_2 = 289.5/380 = 0.7618\ \Omega$; $R2 = \sqrt{Z_2^2 - X_2^2} = \sqrt{0.7618^2 - 0.25^2} = 0.7196\ \Omega$
$\therefore$ external resistance reqd./phase = 0.7196 – 0.012 = **0.707 Ω**

Example 32.9. *A 150-kw, 3000-V, 50-Hz, 6-pole star-connected induction motor has a star-connected slip-ring rotor with a transformation ratio of 3.6 (stator/rotor). The rotor resistance is 0.1Ω/phase and its per phase leakage reactance is 3.61 mH. The stator impedance may be neglected. Find the starting current and starting torque on rated voltage with short-circuited slip rings.* **(Elect. Machines, A.M.I.E. Sec. B, 1989)**

Solution. $X_2 = 2\pi \times 50 \times 5.61 \times 10^{-3} = 1.13\ \Omega$

$K = 1/3.6,\quad R_2' = R_2/K^2 = (3.6)^2 \times 0.1 = 1.3\Omega$

$X_2 = 2\pi \times 50 \times 3.61 \times 10^{-3} = 1.13\ \Omega\ ;\ X_2' = (3.6)^2 \times 1.13 = 14.7\Omega$

$$I_{st} = \frac{V}{(R_2')^2 + (X_2')^2} \cdot \frac{3000/\sqrt{3}}{\sqrt{(1.3)^2 + 14.7^2}} = \mathbf{117.4\ A}$$

Now, $N_s = 120 \times 50/6 = 1000$ rpm = (50/3) rps

$$T_{st} = \frac{3}{2\pi N_s} \cdot \frac{V^2 R_2'}{(R_2')^2 + (X_2')^2} = \frac{3}{2\pi\,(50/3)} \times \frac{(3000/\sqrt{3})^2 \times 1.3}{(1.3^2 + 14.7^2)} = \mathbf{513\ N\text{-}m}$$

Tutorial Problem No : 32.1

1. In the case of an 8-pole induction motor, the supply frequency was 50-Hz and the shaft speed was 735 r.p.m. What were the magnitudes of the following:
   (*i*) synchronous speed (*ii*) speed of slip
   (*ii*) per unit slip (*iv*) percentage slip **[750 r.p.m. ; 15 r.p.m.; 0.02 ; 2%]**
2. A 6-pole, 50-Hz squirrel-cage induction motor runs on load at a shaft speed of 970 r.p.m. Calculate:-
   (*i*) the percentage slip
   (*ii*) the frequency of induced current in the rotor. **[3% ; 1.5 Hz]**
3. An 8-pole alternator runs at 750. r.p.m. and supplies power to a 6-pole induction motor which has at full-load a slip of 3%. Find the full-load speed of the induction motor and the frequency of its rotor e.m.f. **[970 r.p.m. ; 1.5 Hz]**
4. A 3-phase, 50-Hz induction motor with its rotor star-connected gives 500 *V* (r.m.s.) at standstill between the slip-rings on open-circuit. Calculate the current and power factor at standstill when the rotor winding is joined to a star-connected external circuit, each phase of which has a resistance of 10 Ω and an inductance of 0.04 H. The resistance per phase of the rotor winding is 0.2 Ω and its inductance is 0.04 H.
   Also, calculate the current and power factor when the slip-rings are short-circuited and the motor is running with a slip of 5 per cent. Assume the flux to remain constant.
   [10.67 A; 0.376; 21.95 A; 0.303]
5. Obtain an expression for the condition of maximum torque of an induction motor. Sketch the torque-slip curves for several values of rotor circuit resistance and indicate the condition for maximum torque to be obtained at starting.
   If the motor has a rotor resistance of 0.02 Ω and a standstill reactance of 0.1 Ω, what must be the value of the total resistance of a starter for the rotor circuit for maximum torque to be exerted at starting? **[0.08 Ω]** *(City and Guilds, London)*

6. The rotor of a **6**-pole, 50-Hz induction motor is rotated by some means at 1000 r.p.m. Compute (*i*) rotor voltage (*ii*) rotor frequency (*iii*) rotor slip and (*iv*) torque developed. Can the rotor rotate at this speed by itself? **[(*i*) 0 (*ii*) 0 (*iii*) 0 (*iv*) 0; No]** (*Elect. Engg.Grad I.E.T.E. June 1985*)
7. The rotor resistances per phase of a 4-pole, 50-Hz, 3-phase induction motor are 0.024 ohm and 0.12 ohm respectively. Find the speed at maximum torque. Also find the value of the additional rotor resistance per phase required to develop 80% of maximum torque at starting. **[1200 r.p.m. 0.036 Ω]** (*Elect. Machines, A.M.I.E. Sec. B, 1990*)

32.18. Rotor E.M.F. and Reactance Under Running Conditions

Let E_2 = *standstill* rotor induced e.m.f/phase
X_2 = *standstill* rotor reactance/phase, f_2 = rotor current frequency at *standstill*

When rotor is stationary *i.e.* $s = 1$, the frequency of rotor e.m.f. is the same as that of the stator supply frequency. The value of e.m.f. induced in the rotor at standstill is maximum because the relative speed between the rotor and the revolving stator flux is maximum. In fact, the motor is equivalent to a 3-phase transformer with a short-circuited rotating secondary.

When rotor starts running, the relative speed between it and the rotating stator flux is decreased. Hence, the rotor induced e.m.f. which is directly proportional to this relative speed, is also decreased (and may disappear altogether if rotor speed were to become equal to the speed of stator flux). Hence, for a slip s, the rotor induced e.m.f. will be s times the induced e.m.f. at standstill.

Therefore, under *running conditions* $E_r = s\,E_2$

The frequency of the induced e.m.f. will likewise become $f_r = s f_2$

Due to decrease in frequency of the rotor e.m.f., the rotor reactance will also decrease.

$$\therefore \quad X_r = s\,X_2$$

where E_r and X_r are rotor e.m.f. and reactance under *running* conditions.

32.19. Torque Under Running Conditions

$$T \propto E_r\, I_r \cos\phi_2 \text{ or } T \propto \phi\, I_r \cos\phi_2 \qquad (\because E_r \propto \phi)$$

where E_r = rotor e.m.f/phase under *running conditions*
I_r = rotor current/phase under *running conditions*

Now $E_r = sE_2$

$$\therefore \quad I_r = \frac{E_r}{Z_r} = \frac{s\,E_2}{\sqrt{[R_2^{\,2} + (s\,X_2)^2]}}$$

$$\cos\phi_2 = \frac{R_2}{\sqrt{[R_2^{\,2} + (s\,X_2)^2]}} \qquad \text{— Fig. 32.20}$$

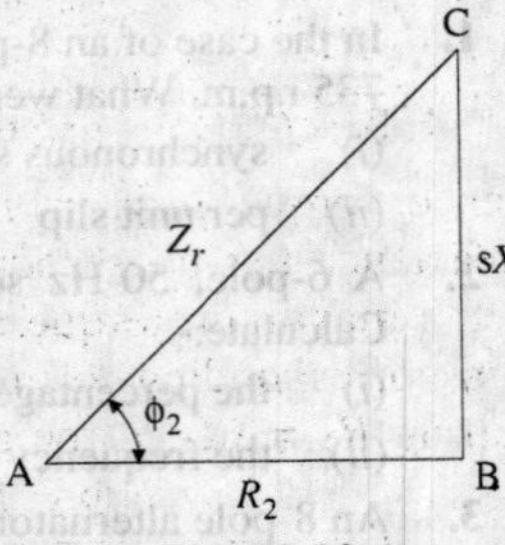

Fig. 32.20

$$\therefore \quad T \propto \frac{s\,\Phi E_2\, R_2}{R_2^{\,2} + (s\,X_2)^2} = \frac{k\,\Phi\,.\,s\,.\,E_2\,R_2}{R_2^{\,2} + (s\,X_2)^2}$$

$$\text{Also} \quad T = \frac{k\,.\,s\,E_2^{\,2}\,R_2}{R_2^{\,2} + (s\,X_2)^2} \qquad (\because E_2 \propto \phi)$$

where k_1 is another constant. Its value can be proved to be equal to $3/2\,\pi N_s$ (Art. 32.38). Hence, in that case, expression for torque becomes

$$T = \frac{3}{2\,\pi\,N_S} \cdot \frac{s\,E_2^{\,2}\,R_2}{R_2^{\,2} + (s\,X_2)^2} = \frac{3}{2\,\pi\,N_S} \cdot \frac{s\,E_2^{\,2}\,R_2}{Zr^2}$$

At standstill when $s = 1$, obviously

$$T_{st} = \frac{k_1\,E_2^{\,2}\,R_2}{R_2^{\,2} + X_2^{\,2}} \quad \text{or} = \frac{3}{2\,\pi\,N_S} \cdot \frac{E_2^{\,2}\,R_2}{R_2^{\,2} + X_2^{\,2}}$$ ... the same as in Art. 32.13.

Example 32.10. *The star connected rotor of an induction motor has a standstill impedance of* (0.4 + *j*4) *ohm per phase and the rheostat impedance per phase is* (6 + *J*2) *ohm.*

The motor has an induced emf of 80 V between slip-rings at standstill when connected to its normal supply voltage. Find

(*i*) *rotor current at standstill with the rheostat is in the circuit.*

(*ii*) *when the slip-rings are short-circuited and motor is running with a slip of 3%.*

(Elect.Engg. I, Nagpur Univ. 1993)

Solution. (1) Standstill Conditions

Voltage/rotor phase $= 80/\sqrt{3} = 46.2$. V; rotor and

starter impedance/phase $= (6.4 + j6) = 8.77 \angle 43.15°$

Rotor current/phase $= 46.2/8.77 =$ **5.27 A** ($p.f. = \cos 43.15° = 0.729$)

(2) Running Conditions. Here, starter impedance is cut out.

Rotor voltage/phase, $E_r = s\,E_2 = 0.03 \times 46.2 = 1.386$ V

Rotor reactance/phase, $X_r = 0.03 \times 4 = 0.12\ \Omega$

Rotor impedance/phase, $Z_r = 0.4 + j0.12 = 0.4176 \angle 16.7°$

Rotor current/phase $= 1.386/0.4176 =$ **3.32** *A* (p.f. $= \cos 16.7° = 0.96$)

Note. It has been assumed that flux across the air -gap remains constant

32.20. Condition for Maximum Torque Under Running Conditions

The torque of a rotor under *running* conditions is

$$T = \frac{k\,\Phi\,s\,E_2\,R_2}{R_2^{\,2} + (sX_2)^2} = k_1 \frac{s\,E_2^{\,2}\,R_2}{R_2^{\,2} + (s\,X_2)^2} \qquad ...(i)$$

The condition for maximum torque may be obtained by differentiating the above expression with respect to slip s and then putting it equal to zero. However, it is simpler to put $Y = \frac{1}{T}$ and then differentiate it.

$$\therefore \quad Y = \frac{R_2^{\,2} + (sX_2)^2}{k\,\Phi\,s\,E_2\,R_2} = \frac{R_2}{k\,\Phi\,s\,E_2} + \frac{sX_2^2}{k\,\Phi\,E_2\,R_2}; \quad \frac{dY}{ds} = \frac{-R_2}{k\,\Phi\,s^2\,E_2} + \frac{X_2^{\,2}}{k\,\Phi\,E_2\,R_2} = 0$$

$$\therefore \quad \frac{R_2}{k\,\Phi\,s^2\,E_2} = \frac{X_2^2}{k\,\Phi\,E_2\,R_2} \quad \text{or} \quad R_2^2 = s^2X_2^2 \quad \text{or} \quad R_2 = s\,X_2$$

Hence, torque *under running condition* is maximum at that value of the slip s which makes rotor reactance per phase equal to rotor resistance per phase. This slip is sometimes written as s_b and the maximum torque as T_b.

Slip corresponding to maximum torque is $s = R_2/X_2$

Putting $R_2 = sX_2$ in the above equation for the torque, we get

$$T_{max} = \frac{k\,\Phi\,s^2\,E_2\,X_2}{2\,s^2\,X_2^{\,2}} \left(\text{or } \frac{k\,\Phi\,s\,E_2\,R_2}{2R_2^{\,2}} \right) \text{ or } T_{max} = \frac{k\,\Phi\,E_2}{2\,X_2} \left(\text{or } \frac{k\,\Phi\,s\,E_2}{2\,R_2} \right) \qquad ...(ii)$$

Substituting value of $s = R_2/X_2$ in the other equation given in (*i*) above, we get

$$T_{max} = k_1 \frac{(R_2/X_2)\,.\,E_2^{\,2}\,.R_2}{R_2^{\,2} + (R_2/X_2)^2\,.\,X_2^{\,2}} = k_1 \frac{E_2^{\,2}}{2X_2}$$

Since, $k_1 = 3/2\pi\,N_S$, we have $T_{max} = \frac{3}{2\pi\,N_s} . \frac{E_2^{\,2}}{2X_2}$ N–m

From the above, it is found

1. *that the maximum torque is independent of rotor resistance as such.*
2. *however, the speed or slip at which maximum torque occurs is determined by the rotor resistance. As seen from above, torque becomes maximum when rotor reactance equals its resistance. Hence, by varying rotor resistance (possible only with slip-ring motors) maximum torque can be made to occur at any desired slip (or motor speed).*

3. *maximum torque varies inversely as standstill reactance. Hence, it should be kept as small as possible.*
4. *maximum torque varies directly as the square of the applied voltage.*
5. *for obtaining maximum torque at starting (s =1), rotor resistance must be equal to rotor reactance.*

Example 32.11. *A 3-phase, slip-ring, induction motor with star-connected rotor has an induced e.m.f. of 120 volts between slip-rings at standstill with normal voltage applied to the stator. The rotor winding has a resistance per phase of 0.3 ohm and standstill leakage reactance per phase of 1.5 ohm.*

Calculate (*i*) *rotor current/phase when running short-circuited with 4 percent slip and* (*ii*) *the slip and rotor current per phase when the rotor is developing maximum torque.*

(Elect. Engg.-II, Pune Univ. 1989)

Solution. (*i*) induced e.m.f./rotor phase, $E_r = sE_2 = 0.04 \times (120/\sqrt{3}) = \mathbf{2.77V}$

rotor reactance/phase, $X_r = sX_2 = 0.04 \times 1.5 = 0.06\ \Omega$

rotor impedance/phase $= \sqrt{0.3^2 + 0.06^2} = 0.306\ \Omega$

rotor current/phase $= 2.77/0.306 = \mathbf{9A}$

(*ii*) For developing maximum torque,

$$R_2 = sX_2 \quad \text{or} \quad s = R_2/X_2 = 0.3/1.5 = \mathbf{0.2}$$

$$X_r = 0.2 \times 1.5 = 0.3\ \Omega, \quad Z_r = \sqrt{0.3^2 + 0.3^2} = 0.42\ \Omega$$

$$E_r = sE_2 = 0.2 \times (120/\sqrt{3}) = 13.86\text{ V}$$

$\therefore$ rotor current/phase $= 13.86/0.42 = \mathbf{33\ A}$

32.21. Rotor Torque and Breakdown Torque

The rotor torque at any slip s can be expressed in terms of the maximum (or breakdown) torque T_b by the following equation

$$T = T_b \left[\frac{2}{(s_b/s) + (s/s_b)} \right]$$ where s_b is the breakdown or pull-out slip.

Example 32.12. *Calculate the torque exerted by an 8-pole, 50-Hz, 3-phase induction motor operating with a 4 per cent slip which develops a maximum torque of 150 kg-m at a speed of 660 r.p.m. The resistance per phase of the rotor is 0.5 Ω*

(Elect. Machines, A.M.I.E. Sec. B, 1989)

Solution. $N_s = 120 \times 50/8 = 750$ r.p.m.

Speed at maximum torque = 660 r.p.m. Corresponding slip $s_b = \dfrac{750 - 660}{750} = 0.12$

For maximum torque, $R_2 = s_b X_2$ $\quad\therefore\quad X_2 = R_2/s_b = 0.5/0.12 = \mathbf{4.167\ \Omega}$

As seen from Eq. (*ii*) of Art. 32.20,

$$T_{max} = k\Phi E_2 \cdot \frac{s_b}{2R_2} = k\Phi E_2 \cdot \frac{0.12}{2 \times 0.5} = 0.12\, k\Phi E_2 \quad \text{...(i)}$$

When slip is 4 per cent

As seen from Eq. (*i*) of Art. 32.20

$$T = k\Phi E_2 \frac{sR_2}{R_2^2 + (sX_2)^2} = k\Phi E_2 \frac{0.04 \times 0.5}{0.5^2 + (0.04 \times 4.167)^2} = \frac{0.02\, k\Phi E_2}{0.2778}$$

$$\therefore \quad \frac{T}{T_{max}} = \frac{T}{150} = \frac{0.02}{0.2778 \times 0.12} \quad \therefore \quad T = \mathbf{90\ kg\text{-}m}$$

Alternative Solution

$T_b = 150$ kg.m; $s_b = 0.12$, $s = 4\% = 0.04$, $T = ?$

$$T = T_b\left(\frac{2}{(s_b/s) + (s/s_b)}\right) \quad \text{...Art 32.21}$$

$$= 150\left(\frac{2}{(0.12/0.04) + (0.04/0.12)}\right) = \mathbf{90\ kg\text{-}m}$$

32.22. Relation Between Torque and Slip

A family of torque/slip curves is shown in Fig. 32.21 for a range of $s = 0$ to $s = 1$ with R_2 as the parameter. We have seen above in Art. 32.19 that

$$T = \frac{k\Phi\, s\, E_2\, R_2}{R_2^2 + (s\, X_2)^2}$$

It is clear that when $s = 0$, $T = 0$, hence the curve starts from point O.

At normal speeds, close to synchronism, the term $(s\ X_2)$ is small and hence negligible w.r.t. R_2.

$$\therefore \quad T \propto \frac{s}{R_2}$$

or $\quad T \propto s$ if R_2 is constant.

Hence, for low values of slip, the torque/slip curve is approximately a straight line. As slip increases (for increasing load on the motor), the torque also increases and becomes maximum when $s = R_2/X_2$. This torque is known as 'pull-out' or 'breakdown' torque T_b or stalling torque. As the slip further increases (*i.e.* motor speed falls) with further increase in motor load, then R_2 becomes negligible as compared to $(s\ X_2)$. Therefore, for large values of slip

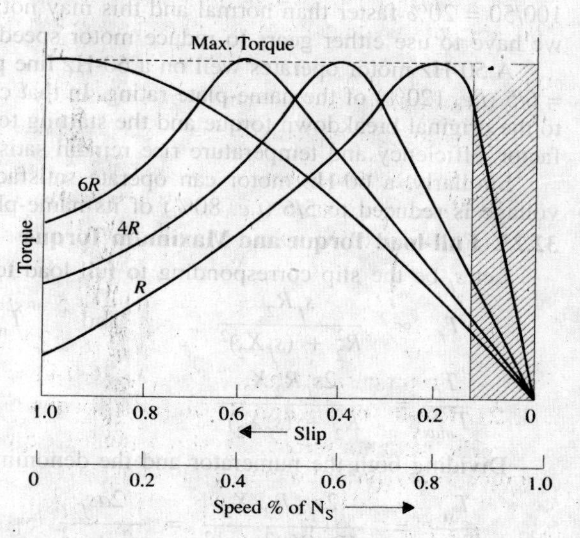

Fig. 32.21

$$T \propto \frac{s}{(sX_2)^2} \propto \frac{1}{s}$$

Hence, the torque/slip curve is a rectangular hyperbola. So, we see that beyond the point of maximum torque, any further increase in motor load results in decrease of torque developed by the motor. The result is that the motor slows down and eventually stops. The circuit-breakers will be tripped open if the circuit has been so protected. In fact, the stable operation of the motor lies between the values of $s = 0$ and that corresponding to maximum torque. The operating range is shown shaded in Fig. 32.21.

It is seen that although maximum torque does not depend on R_2, yet *the exact location of* T_{max} is dependent on it. Greater the R_2, greater is the value of slip at which the maximum torque occurs.

32.23. Effect of Change in Supply Voltage on Torque and Speed

As seen from Art. 32.19, $T = \dfrac{k\Phi\, sE_2R_2}{R_2^{\,2} + (sX_2)^2}$

As $\quad E_2 \propto \phi \propto V$ where V is supply voltage $\quad \therefore \; T \propto sV^2$

Obviously, torque at any speed is proportional to the square of the applied voltage, If stator voltage decreases by 10%, the torque decreases by 20%. Changes in supply voltage not only affect the starting torque T_{st} but torque under running conditions also. If V decreases, then T also decreases. Hence, for maintaining the same torque, slip increases *i.e.* speed falls.

Let V change to V', s to s' and T to T'; then $\dfrac{T}{T'} = \dfrac{sV^2}{s'V'^2}$

32.24. Effect of Changes in supply Frequency on Torque and Speed

Hardly any important changes in frequency take place on a large distribution system except during a major disturbance. However, large frequency changes often take place on isolated, low-power systems in which electric electric energy is generated by means of diesel engines or gas turbines. Examples of such systems are : emergency supply in a hospital and the electrical system on a ship etc.

The major effect of change in supply frequency is on motor *speed*. If frequency drops by 10%, then motor speed also drops by 10%. Machine tools and other motor-driven equipment meant for 50 Hz causes problem when connected to 60-Hz supply. Everything runs (60 – 50) × 100/50 = 20% faster than normal and this may not be acceptable in all applications. In that case, we have to use either gears to reduce motor speed or an expensive 50-Hz source.

A 50-Hz motor operates well on a 60-Hz line provided its terminal voltage is raised to 60/50 = 6/5 (*i.e.* 120%) of the name-plate rating. In that case, the new breakdown torque becomes equal to the original breakdown torque and the starting torque is only slightly reduced. However, power factor, efficiency and temperature rise remain satisfactory.

Similarly, a 60-Hz motor can operate satisfactorily on 50-Hz supply provided its terminal voltage is reduced to 5/6 (*i.e.* 80%) of its name-plate rating.

32.25. Full-load Torque and Maximum Torque

Let s_f be the slip corresponding to full-load torque, then

$$T_f \propto \frac{s_f R_2}{R_2^2 + (s_f X_2)^2} \quad \text{and} \quad T_{max} \propto \frac{1}{2 \times X_2} \qquad \text{... Art 32.20}$$

$$\therefore \frac{T_f}{T_{max}} = \frac{2 s_f R_2 X_2}{R_2^2 + (s_f X_2)^2}.$$

Dividing both the numerator and the denominator by X_2^2, we get

$$\frac{T_{st}}{T_{max}} = \frac{2 s_f \cdot R_2/X_2}{(R_2/X_2)^2 + s_f^2} = \frac{2 a s_f}{a^2 + s_f^2}$$

where $a = R_2/X_2$ = resistance/standstill reactance*

In general, $$\frac{\text{operating torque at any slip } s}{\text{maximum torque}} = \frac{2\,a\,s}{a^2 + s^2}$$

32.26. Starting Torque and Maximum Torque

$$T_{st} \propto \frac{R_2}{R_2^2 + X_2^2} \qquad \text{...Art 32.13}$$

$$T_{max} \propto \frac{1}{2 X_2} \qquad \text{... Art. 32.20}$$

$$\therefore \frac{T_{st}}{T_{max}} = \frac{2 R_2 X_2}{R_2^2 + X_2^2} = \frac{2R_2/X_2}{1 + (R_2/X_2)^2} = \frac{2a}{1 + a^2}$$

where $$a = \frac{R_2}{X_2} = \frac{\text{rotor resistance}}{\text{standstill reactance}} \text{ per phase**}$$

Example. 32.13. *A 3-φ induction motor is driving full-load torque which is independent of speed. If line voltage drops to 90% of the rated value, find the increase in motor copper losses.*

---

*In fact $a = s_m$ —slip corresponding to maximum torque. In that case, the relation becomes $\frac{T_f}{T_{max}} = \frac{2 s_m s_f}{s_m^2 + s_f^2}$ — where s_f = full-load slip.

**Similarly, the relation becomes $\frac{T_{st}}{T_{max}} = \frac{2 s_m}{1 + s_m^2}$

Solution. As seen from Art. 32-23, when T remains constant, $s_1 V_1^2 = s_2 V_2^2$

$$\therefore \quad \frac{s_2}{s_1} = \left(\frac{V_1}{V_2}\right)^2 = \left(\frac{1}{0.9}\right)^2 = 1.23$$

Again from Art. 32-19, $I_2 \propto sV \quad \therefore \quad \frac{I_2'}{I_2} = \frac{s_2 V_2}{s_1 V_1} = 1.2 \times 0.9 = 1.107$

Now, Cu losses are nearly proportional to I_2^2

$$\therefore \quad \frac{\text{Cu loss in the 2nd case}}{\text{Cu loss in the 1st case}} = \frac{I_2'^2}{I_2^2} = 1.107^2 = 1.23$$

Thus a reduction of 10% in line voltage causes about 23% increase in Cu losses.

Example. 32.14. *A 230-V, 6-pole, 3-φ, 50-Hz, 15-kW induction motor drives a constant torque load at rated frequency, rated voltage and rated kW output and has a speed of 980 rpm and an efficiency of 93%. Calculate (i) the new operating speed if there is a 10% drop in voltage and 5% drop in frequency and (ii) the new output power. Assume all losses to remain constant.*

Solution. (*i*) $V_2 = 0.9 \times 230 = 207$ V; $f_2 = 0.95 \times 50 = 47.5$ Hz; $N_{s1} = 120 \times 50/6 = 1000$ rpm; $N_{s2} = 120 \times 47.5/6 = 950$ rpm; $s_1 = (1000 - 980)/1000 = 0.02$

Since the load torque remains constant, the product (sV^2/f) remains constant.

$$\therefore \quad s_1 V_1^2/f_1 = \frac{s_2 V_2^2}{f_2} \text{or } s_2 = s_2 = s_1 \left(\frac{V_1}{V_2}\right)^2 . f_2/f_1 = 0.02(230/230 \times 0.9)^2 \times 47.5/50 = 0.2234$$

$\therefore \quad N_2 = Ns_2 (1 - s_2) = 950(1 - 0.0234) = 928$ rpm

(*ii*) $P \propto TN$. Since torque remains constant, $P \propto N$

$\therefore \quad P_1 \propto N_1 ; \; P_2 \propto N_2 ; \quad$ or $\quad P_2 = P_1 \times N_2/N_1 = 15 \times 928/980 =$ **14.2 kW**

Example 32.15. *A 3-phase, 400/200-V, Y-Y connected wound-rotor induction motor has 0.06 Ω rotor resistance and 0.3 Ω standstill reactance per phase. Find the additional resistance required in the rotor circuit to make the starting torque equal to the maximum torque of the motor.*

(Electrical Technology, Bombay Univ. 1990)

Solution. $\frac{T_{st}}{T_{max}} = \frac{2a}{1 + a^2}$; Since $T_{st} = T_{max}$

$\therefore \quad 1 = \frac{2a}{1 + a^2}$ or $a = 1$ Now, $a = \frac{R_2 + r}{X_2}$

where r = external resistance per phase added to the rotor circuit

$\therefore \quad 1 = \frac{0.06 + r}{0.3} \quad \therefore \quad r = 0.3 - 0.06 =$ **0.24 Ω**

Example 32.16. *A 746-kW, 3-phase, 50-Hz, 16-pole induction motor has a rotor impedance of (0.02 + J 0.15) Ω at standstill. Full-load torque is obtained at 360 rpm. Calculate (i) the ratio of maximum to full-load torque (ii) the speed of maximum torque and (iii) the rotor resistance to be added to get maximum starting torque.* **(Elect. Machines, Nagpur Univ. 1993)**

Solution. Let us first find out the value of full-load slip s_f

$N_s = 120 \times 50/16 = 375$ rpm.; F.L. Speed = 360 rpm.

$s_f = (375 - 360)/375 = 0.04$; $a = R_2/X_2 = 0.02/0.15 = 2/15$

(*i*) $\frac{T_f}{T_{max}} = \frac{2as_f}{a^2 + s_f^2} = \frac{2 \times (2/15) \times 0.04}{(2/15)^2 + (0.04)^2} = 0.55$ or $\frac{T_{max}}{T_f} = \frac{1}{0.55} =$ **1.818**

(*ii*) At maximum torque, $a = s_m = R_2/X_2 = 0.02/0.15 = 2/15$

$N = N_s(1 - s) = 375(1 - 2/15) =$ **325 r.p.m.**

(*iii*) For maximum starting torque, $R_2 = X_2$. Hence, total rotor resistance per phase = **0.15Ω**

$\therefore$ external resistance required/phase = 0.15 – 0.02 = **0.13 Ω**

Example 32.17. *The rotor resistance and reactance per phase of a 4-pole, 50-Hz, 3-phase induction motor are 0.025 ohm and 0.12 ohm respectively. Make simplifying assumptions, state them and :*

(i) find speed at maximum torque

(ii) find value of additional rotor resistance per phase required to give three-fourth of maximum torque at starting.

Draw the equivalent circuit of a single-phase induction motor.

(Elect. Machines, Nagpur Univ. 1993)

Solution. (*i*) At maximum torque, $s = R_2/X_2 = 0.025/0.12 = 0.208$.

$N_s = 120 \times 50/4 = 1500$ rpm ; $\therefore\ N = 1500\,(1 - 0.208) = \mathbf{1188}$ rpm

(*ii*) It is given that $T_{st} = 0.75\, T_{max}$. Now, $\dfrac{T_{st}}{T_{max}} = \dfrac{2a}{1 + a^2} = \dfrac{3}{4}$

$\therefore\ 3a^2 - 8a + 3 = 0$; $a = \dfrac{8 \pm \sqrt{64 - 36}}{6} = 0.45\ \Omega$*

Let, r = additional rotor resistance reqd., then

$a = \dfrac{R_2 + r}{R_2}$ or $0.45 = \dfrac{0.025 + r}{0.12}$ $\therefore\ r = \mathbf{0.029\Omega}$

Example. 32.18. *A 50-Hz, 8-pole induction motor has F.L. slip of 4%. The rotor resistance/phase = 0.01 ohm and standstill reactance/phase = 0.1 ohm. Find the ratio of maximum to full load torque and the speed at which the maximum torque occurs.*

(Basic Elect. Machine Nagpur Univ. 1993)

Solution. $\dfrac{T_f}{T_{max}} = \dfrac{2as_f}{a^2 + s_f^2}$

Now, $a = R_2/X_2 = 0.01/0.1 = 0.1,\ s_f = 0.04$

$\therefore\ \dfrac{T_f}{T_{max}} = \dfrac{2 \times 0.1 \times 0.04}{0.1^2 + 0.04^2} = \dfrac{0.008}{0.0116} = 0.69$ $\therefore\ \dfrac{T_{max}}{T_f} = \dfrac{1}{0.69} = \mathbf{1.45}$

$N_s = 120 \times 50/8 = 750$ rpm, $s_m = 0.1$ $N = (1 - 0.1) \times 750 = \mathbf{675}$ **rpm**

Example 32.19. *For a 3-phase slip-ring induction motor, the maximum torque is 2.5 times the full-load torque and the starting torque is 1.5 times the full-load torque. Determine the percentage reduction in rotor circuit resistance to get a full-load slip of 3%. Neglect stator impedance.*

(Elect. Machines, A.M.I.E. Sec. B, 1992)

Solution. Given, $T_{max} = 2.5\, T_f$; $T_{st} = 1.5\, T_f$; $T_{st}/T_{max} = 1.5/2.5 = 3/5$.

Now, $\dfrac{T_{st}}{T_f} = \dfrac{3}{5} = \dfrac{2a}{1 + a^2}$ or $3a^2 - 10a + 3 = 0$ or $a = 1/3$

Now, $a = R_2/X_2$ or $R_2 = X_2/3$

When F.L. slip is 0.03

$\dfrac{T_f}{T_{st}} = \dfrac{2\,as}{a^2 + s^2}$ or $\dfrac{2}{2.5} = \dfrac{2a \times 0.03}{a^2 + 0.03^2}$

$a^2 - 0.15\,a + 0.009 = 0$ or $a = 0.1437$

If R'_2 is the new rotor circuit resistance, then $0.1437 = R'_2/X_2$ or $R'_2 = 0.1437\, X_2$

% reduction in rotor resistance is

$= \dfrac{(X_2/3) - 0.1437 \times X_2}{(X_2/3)} \times 100 = \mathbf{56.8\%}$.

Example 32.20. *An 8-pole, 50-Hz, 3-phase slip-ring induction motor has effective rotor re-*

*The larger value of 2.215Ω has been rejected.

sistance of 0.08 Ω/phase. Stalling speed is 650 r.p.m. How much resistance must be inserted in the rotor phase to obtain the maximum torque at starting? Ignore the magnetising current and stator leakage impedance. **(Elect.Machines-I, Punjab Univ. 1991)**

Solution. It should be noted that stalling speed corresponds to maximum torque (also called stalling torque) and to maximum slip *under running conditions.*

$N_s = 120 \times 50/8 = 750$ r.p.m.; stalling speed is = 650 r.p.m.

$s_b = (750 - 650)/750 = 2/15 = 0.1333$ or 13.33%

Now, $s_b = R_2/X_2 \qquad \therefore X_2 = 0.08 \times 15/2 = 0.6\ \Omega$

$$\frac{T_{st}}{T_{max}} = \frac{2a}{1+a^2}. \quad \text{Since} \quad T_{st} = T_{max} \quad \therefore \quad 1 = \frac{2a}{1+a^2} \quad \text{or} \quad a = 1$$

Let r be the external resistance per phase added to the rotor circuit. Then

$$a = \frac{R_2 + r}{X_2} \quad \text{or} \quad 1 = \frac{0.08 + r}{0.6} \quad \therefore \quad r = \mathbf{0.52\ \Omega\ per\ phase.}$$

Example 32.21. *A 4-pole, 50-Hz, 3-φ induction motor develops a maximum torque of 162.8 N-m at 1365 r.p.m. The resistance of the star-connected rotor is 0.2 Ω/phase. Calculate the value of the resistance that must be inserted in series with each rotor phase to produce a starting torque equal to half the maximum torque.*

Solution. $N_s = 120 \times 50/4 = 1500$ r.p.m. $\qquad N = 1365$ r.p.m.

∴ slip corresponding to maximum torque is

$s_b = (1500 - 1365)/1500 = 0.09 \quad s_b = R_2/X_2 \therefore X_2 = 0.2/0.09 = 2.22\ \Omega$

$$\text{Now,} \quad T_{max} = \frac{k\Phi E_2}{2X_2} = \frac{K}{2X_2} \quad (\text{where } K = k\Phi E_2) \qquad \text{... Art 32.20}$$

$$= \frac{K}{2 \times 2.22} = 0.225\,K$$

Let 'r' be the external resistance introduced per phase in the rotor circuit, then

$$\text{Starting torque} \quad T_{st} = \frac{k\Phi E_2 (R_2 + r)}{(R_2 + r)^2 + (X_2)^2} = \frac{K(0.2 + r)}{(0.2 + r)^2 + (0.2/0.09)^2}$$

$$\because \quad T_{st} = \frac{1}{2} \cdot T_{max} \quad \therefore \quad \frac{K(0.2 + r)}{(0.2 + r)^2 + (2.22)^2} = \frac{0.225\,K}{2}$$

Solving the quadratic equation for 'r', we get $r = \mathbf{0.4\ \Omega}$

Example 32.22. *A 4-pole, 50-Hz, 7.46.kW motor has, at rated voltage and frequency, a starting torque of 160 per cent and a maximum torque of 200 per cent of full-load torque. Determine (i) full-load speed (ii) speed at maximum torque.*

(Electrical Technology-I, Osmania Univ. 1990)

$$\textbf{Solution.} \quad \frac{T_{st}}{T_f} = 1.6 \text{ and } \frac{T_{max}}{T_f} = 2 \quad \therefore \quad \frac{T_{st}}{T_{max}} = \frac{1.6}{2} = 0.8$$

$$\text{Now,} \quad \frac{T_{st}}{T_{max}} = \frac{2a}{1+a^2} \quad \therefore \quad \frac{2a}{1+a^2} = 0.8$$

or $0.8a^2 - 2a + 0.8 = 0 \qquad a = 0.04 \quad \therefore \quad a = R_2/X_2 = 0.04$ or $R_2 = 0.04X_2$

$$\text{Also,} \quad \frac{T_f}{T_{max}} = \frac{2as_f}{a^2 + s_f^2} = \frac{1}{2} \quad \text{or} \quad \frac{2 \times 0.04 s_f}{0.0016 + s_f^2} = \frac{1}{2} \quad \text{or} \quad s_f = 0.01$$

(*i*) full-load speed occurs at a slip of 0.01 or 1 per cent. Now,

$N_s = 120 \times 50/4 = 1500$ r.p.m.; $N = 1500\text{-}15 =$ **1485 r.p.m.**

(*ii*) Maximum torque occurs at a slip given by $s_b = R_2/X_2$. As seen from above slip corresponding to maximum torque is 0.04.

∴ $N = 1500 - 1500 \times 0.04 =$ **1440 r.p.m.**

Example 32.23. *A 3-phase induction motor having a 6-pole, star-connected stator winding runs on 240-V, 50-Hz supply. The rotor resistance and standstill reactance are 0.12 ohm and 0.85 ohm per phase. The ratio of stator to rotor turns is 1.8. Full load slip is 4%.*

Calculate the developed torque at full load, maximum torque and speed at maximum torque.

(Elect. Machines, Nagpur Univ. 1993)

Solution. Here, $K = \dfrac{\text{rotor turns/phase}}{\text{stator turns/phase}} = \dfrac{1}{1.8}$

$$E_2 = KE_1 = \frac{1}{1.8} \times \frac{240}{\sqrt{3}} = 77 \text{ V}; \quad s = 0.04;$$

$$N_s = 120 \times 50/6 = 1000 \text{ rpm} = 50/3 \text{ rps}$$

$$T_f = \frac{3}{2\pi N_s} \cdot \frac{s E_2^2 R_2}{R_2^2 + (sX_2)^2} \quad \text{... Art. 32.19}$$

$$= \frac{3}{2\pi(50/3)} \cdot \frac{0.04 \times 77^2 \times 0.12}{0.12^2 + (0.04 \times 0.85)^2} = 52.4 \text{ N-m}$$

For maximum torque, $s = R_2/X_2 = 0.12/0.85 = 0.14$

$$\therefore \quad T_{max} = \frac{3}{2\pi(50/3)} \cdot \frac{0.14 \times 77^2 \times 0.12}{0.12^2 + (0.14 \times 0.85)^2} = 99.9 \text{ N–m}$$

Alternatively, as seen from Art 32.20.

$$T_{max} = \frac{3}{2\pi N_S} \cdot \frac{E_2^2}{2X_2}$$

$$\therefore \quad T_{max} = \frac{3}{2\pi(50/3)} \cdot \frac{77^2}{2 \times 0.85} = 99.9 \text{ N–m}$$

Speed corresponding to maximum torque, $N = 1000(1 - 0.14) =$ **860 rpm**

Example 32.24. *A 440-V, 3-φ, 50-Hz, 4-pole, Y-connected induction motor has a full-load speed of 1425 rpm. The rotor has an impedance of (0.4 + J 4) ohm and rotor/stator turn ratio of 0.8. Calculate (i) full-load torque (ii) rotor current and full-load rotor Cu loss (iii) power output if windage and friction losses amount to 500 W (iv) maximum torque and the speed at which it occurs (v) starting current and (vi) starting torque.*

Solution. $N_s = 120 \times 50/4 = 1500$ rpm $= 25$ rps, $s = 75/1500 = 0.05$

$E_1 = 440/1.73 = 254$ V/phase

$$(i) \text{ Full-load } T_f = \frac{3}{2\pi \times 25} \times \frac{0.05\,(0.8 \times 254)^2 \times 0.4}{(0.4)^2 + (0.05 \times 4)^2} = \mathbf{78.87 \text{ N-m}}$$

$$(ii) \quad I_r = \frac{s E_2}{\sqrt{R_2^2 + (s X_2)^2}} = \frac{s KE_1}{\sqrt{R_2^2 + (s X_2)^2}} = \frac{0.05 \times (0.8 \times 254)}{\sqrt{(0.4)^2 + (0.05 \times 4)^2}}$$

$$= \mathbf{22.73 \text{ A}}$$

$$\text{Total Cu loss} = 3I_r^2 R = 3 \times 22.73^2 \times 0.4 = \mathbf{620 \text{ W}}$$

(*iii*) Now, $P_m = 2\pi NT = 2\pi \times (1425/60) \times 78.87 = 11{,}745$ W

$\therefore \quad P_{out} = P_m -$ windage and firiction loss $= 11{,}745 - 500 =$ **11,245 W**

(*iv*) For maximum torque, $s = R_2/X_2 = 0.4/4 = 0.1$

$$\therefore \quad T_{max} = \frac{3}{2\pi \times 25} \times \frac{0.1 \times (0.8 \times 254)^2 \times 0.4}{(0.4)^2 + (0.1 \times 4)^2} = \mathbf{98.5 \text{ N-m}}$$

Since $s = 0.1$, slip speed $= sN_s = 0.1 \times 1500 = 150$ rpm.

$\therefore$ speed for maximum torque $= 1500 - 150 =$ **1350 rpm.**

$$(v) \text{ starting current} = \frac{E_2}{\sqrt{R_2^2 + X_2^2}} = \frac{KE_1}{\sqrt{R_2^2 + X_2^2}} = \frac{0.8 \times 254}{\sqrt{0.4^2 + 4^2}} = \mathbf{50.5 \text{ A}}$$

(*vi*) At start, $s = 1$, hence

$$T_{st} = \frac{3}{2\pi \times 25} \times \frac{(0.8 \times 254)^2 \times 0.4}{(0.4)^2 + 4^2} = \mathbf{19.5\ N\text{-}m}$$

It is seen that as compared to full-load torque, the starting torque is much less-almost 25 per cent.

Example 32.25. *The rotor resistance and standstill reactance of a 3-phase induction motor are respectively 0.015 Ω and 0.09 Ω per phase. At normal voltage, the full-load slip is 3%. Estimate the percentage reduction in stator voltage to develop full-load torque at half full-load speed. Also, calculate the power factor.* **(Adv.Elect. Machines, A.M.I.E. 1989)**

Solution. Let N_s = 100 r.p.m. F.L. speed = (1 – 0.03)100 = 97 r.p.m.

Let the normal voltage be V_1 volts.

Speed in second case = 97/2 = 48.5 r.p.m.

$$\therefore \quad \text{slip} = (100 - 48.5)/100 = 0.515 \text{ or } 51.5\%$$

$$\text{Now,} \quad T = \frac{k\,\Phi\, s\, E_2\, R_2}{R_2^2 + (s\,X_2)^2} = \frac{k\, sV^2\, R_2}{R_2^2 + (s\,X_2)^2} \qquad (\because \ E_2 \propto \Phi \propto V)$$

Since torque is the same in both cases,

$$\frac{k\,V_1^2\, s_1\, R_2}{R_2^2 + (s_1\,X_2)^2} = \frac{k\,V_2^2\, s_2\, R_2}{R_2^2 + (s_2\,X_2)^2}$$

where V_2 = sector voltage in second case

$$\therefore \quad \left(\frac{V_1}{V_2}\right)^2 = \frac{s_2}{s_1} \cdot \frac{R_2^2 + (s_1 X_2)^2}{R_2^2 + (s_2 X_2)^2} = \frac{51.3}{3} \cdot \frac{0.015^2 + (0.03 \times 0.09)^2}{0.015^2 + (0.515 \times 0.09)^2} = 1.68$$

$$\therefore \quad \frac{V_1}{V_2} = \sqrt{1.68} = 1.296 \quad \text{or} \quad \frac{V_1 - V_2}{V_1} = \frac{0.296}{1.296}$$

Hence, percentage, reduction in stator (or supply voltage) is

$$= \frac{V_1 - V_2}{V_1} \times 100 = \frac{0.296 \times 100}{1.296} = \mathbf{22.84\%}$$

In the second case, $\tan \phi = s_2 X_2 / R_2 = 0.515 \times 0.09/0.015 = 3.09$

$$\therefore \quad \phi = \tan^{-1}(3.09) = 72°4' \text{ and p.f.} = \cos\phi = \cos 72°4' = \mathbf{0.31}$$

32.27. Torque/Speed Curve

The torque developed by a conventional 3-phase motor depends on its speed but the relation between the two cannot be represented by a simple equation. It is easier to show the relationship in the form of a curve (Fig. 32.22). In this diagram, T represents the nominal full-load torque of the motor. As seen, the starting torque (at N = 0) is 1.5 T and the maximum torque (also called breakdown torque) is 2.5 T.

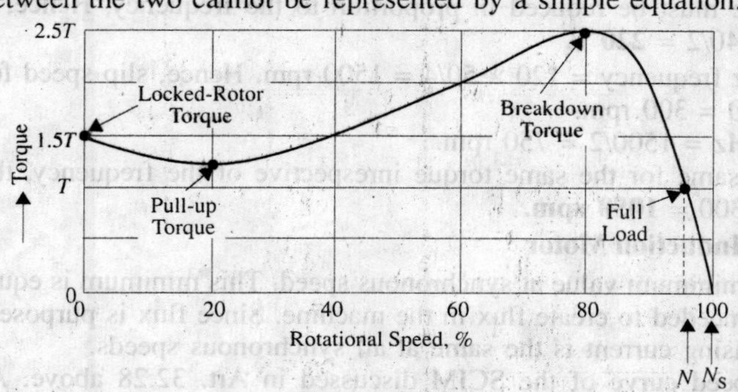

Fig. 32.22

At full-load, the motor runs at a speed of N. When mechanical load increases, motor speed decreases till the motor torque again becomes equal to the load torque. As long as the two torques are in balance, the motor will run at constant (but lower) speed. However, if the load torque exceeds 2.5 T, the motor will suddenly stop.

32.28. Shape of Torque/Speed Curve

For a squirrel-cage induction motor (SCIM), shape of its torque/speed curve depends on the voltage and frequency applied to its stator. If f is fixed, $T \propto V^2$ (Art 32.22). Also, synchronous speed depends on the supply frequency. Now, let us see what happens when *both* stator voltage and frequency are changed. In practice, supply voltage and frequency are varied in the *same proportion* in order to maintain a constant flux in the air-gap. For example, if voltage is doubled, then frequency is also doubled. Under these conditions, shape of the torque/speed curve remains the same but its position along the X-axis (*i.e.*speed axis) shifts with frequency.

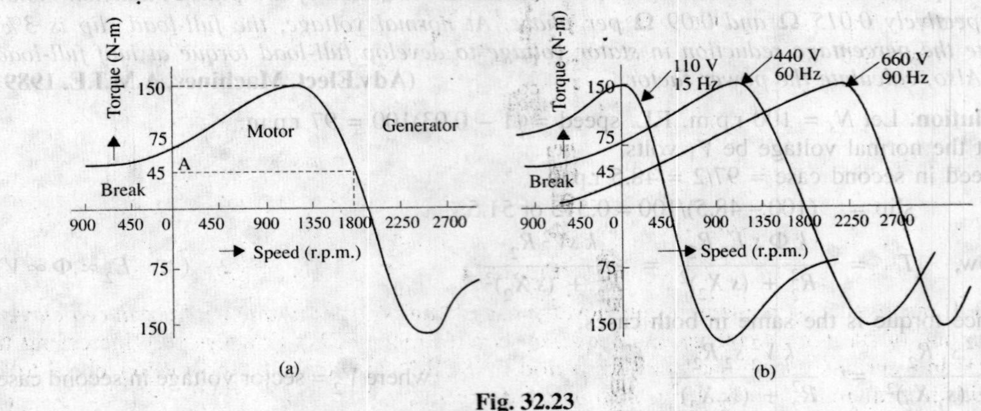

Fig. 32.23

Fig.32.23 (*a*) shows the torque/speed curve of an 11.kW, 440-V, 60-Hz 3-φ SCIM. As seen, full-load speed is **1728** rpm and full-load torque is 45 N-m (point-A) whereas breakdown torque is 150 N-m and locked-rotor current is 75A.

Suppose, we now reduce both the voltage and fequency to *one-fourth* their original values *i.e.* to 110 V and 15 Hz respectively. As seen in Fig. 32.23 (*b*), the torque/speed curve shifts to the left. Now, the curve crosses the X-axis at the synchronous speed of 120 × 15/4 = 450 rpm (*i.e.* 1800/4 = 450 rpm). Similarly, if the voltage and frequency are increased by 50% (660 V, 90 Hz), the curve shifts to the right and cuts the X-axis at the synchronous speed of 2700 rpm.

Since the *shape* of the torque/speed curve remains the same at *all frequencies*,it follows that torque developed by a SCIM is the same *whenever slip-speed is the same.*

Exampel 32.26. *A 440-V, 50-Hz, 4-pole, 3-phase SCIM develops a torque of 100 N-m at a speed of 1200 rpm. If the stator supply frequency is reduced by half, calculate*

(*a*) *the stator supply voltage required for maintaining the same flux in the machine*

(*b*) *the new speed at a torque of 100 N-m.*

Solution. (*a*) The stator voltage must be reduced in proportion to the frequency. Hence, it should also be reduced by half to 440/2 = **220 V.**

(*b*) Synchronous speed at 50 Hz frequency = 120 × 50/4 = 1500 rpm. Hence, slip speed for a torque of 100 N-m = 1500 − 1200 = 300 rpm.

Now, synchonous speed at 25 Hz = 1500/2 = 750 rpm.

Since slip-speed has to be the same for the same torque inrespective of the frequency, the new speed at 100 N-m is = 750 + 300 = **1050 rpm.**

32.29. Current/Speed Curve of an Induction Motor

It is a V-shaped curve having a minimum value at synchronous speed. This minimum is equal to the magnetising current which is needed to create flux in the machine. Since flux is purposely kept constant, it means that magnetising current is the same at all synchronous speeds.

Fig.32.24. shows the current/speed curve of the SCIM discussed in Art. 32.28 above. As seen, locked rotor current is 100 A and the corresponding torque is 75 N-m. If stator voltage and frequency are varied in the same proportion, current/speed curve has the same shape, but shifts along the speed axis. Suppose that voltage and frequency are reduced to one-fourth of their previous values *i.e.* to 110 V, 15 Hz respectively. Then, locked rotor current decreases to 75 A but corresponding torque *increases* to 150 N-m which is equal to full breakdown torque (Fig.

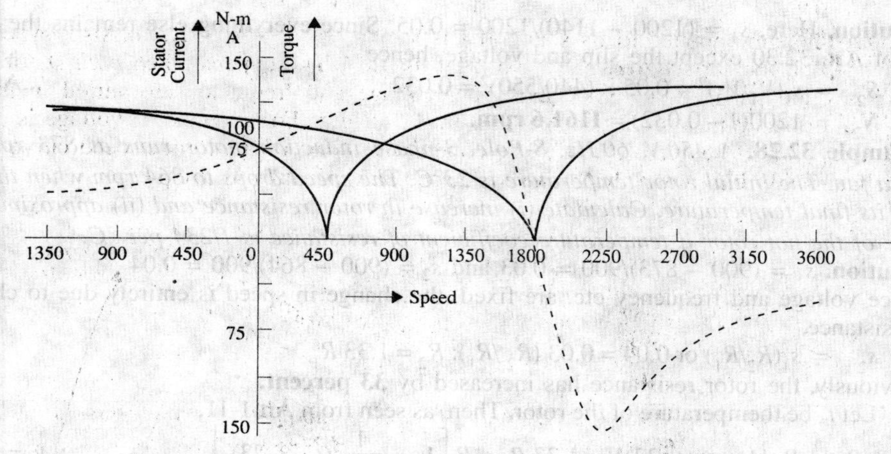

Fig. 32.24

32.25). It means that by reducing frequency, we can obtain *a larger torque with a reduced current*. This is one of the big advantages of frequency control method. By progressively increasing the voltage and current during the start-up period, a SCIM can be made to develop close to its breakdown torque all the way from zero to rated speed.

Another advantage of frequency control is that it permits regenerative braking of the motor. In fact, the main reason for the popularity of frequency-controlled induction motor drives is their ability to develop high torque from zero to full speed together with the economy of regenerative braking.

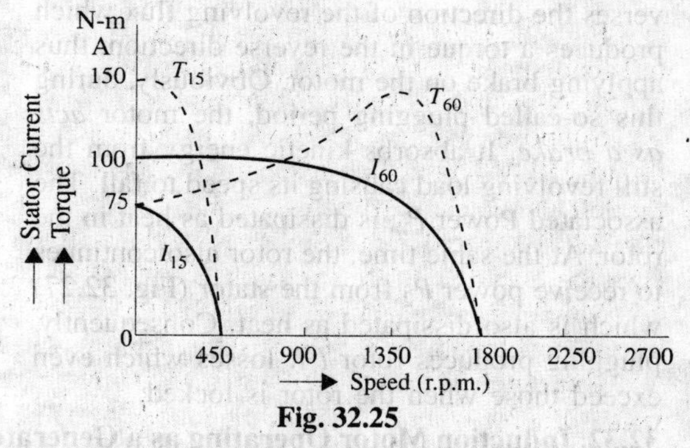

Fig. 32.25

32.30. Torque/Speed Characteristic Under Load

As stated earlier, stable operation of an induction motor lies over the linear portion of its torque/speed curve. The slope of this straight line depends mainly on the rotor resistance. Higher the resistance, sharper the slope. This linear relationship between torque and speed (Fig. 32.26) enables us to establish a very simple equation between different parameters of an induction motor. The parameters under two different load conditions are related by the equation

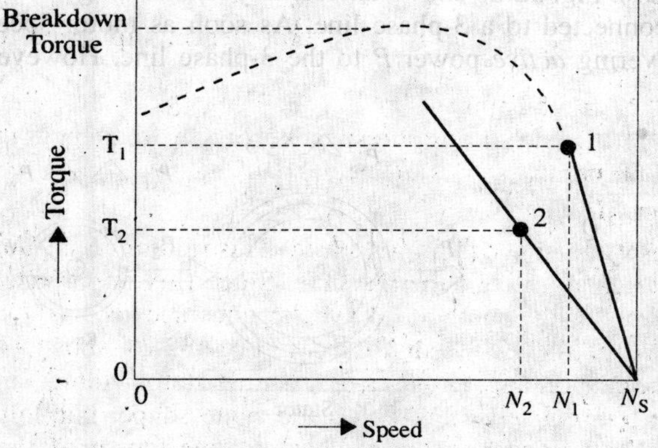

Fig. 32.26

$$s_2 = s_1 \cdot \frac{T_2}{T_1} \cdot \frac{R_2}{R_1} \left(\frac{V_1}{V_2}\right)^2 \qquad \text{... } (i)$$

The only restriction in applying the above equation is that the new torque T_2 must not be greater than $T_1 (V_2/V_1)^2$. In that case, the above equation yields an accuracy of better than 5% which is sufficient for all practical purposes.

Example 32.27. *A 400-V, 60-Hz, 8-pole, 3-φ induction motor runs at a speed of 1140 rpm when connected to a 440-V line. Calculate the speed if voltage increases to 550 V.*

Solution. Here, $s_1 = (1200 - 1140)/1200 = 0.05$. Since everything else remains the same in Eq. (*i*) of Art. 32.30 except the slip and voltage, hence

$S_2 = s_1(V_1/V_2)^2 = 0.05 \times (440/550)^2 = 0.032$ ...Art. 32.23

$\therefore \; N_2 = 1200(1 - 0.032) =$ **1161.6 rpm.**

Example 32.28. *A 450.V, 60.Hz, 8-Pole, 3-phase induction motor runs at 873 rpm when driving a fan. The initial rotor temperature is 23°C. The speed drops to 864 rpm when the motor reaches its final temperature. Calculate (i) increase in rotor resistance and (ii) approximate temperature of the hot rotor if temperature coefficient of resistance is 1/234 per °C.*

Solution. $s_1 = (900 - 873)/900 = 0.03$ and $s_2 = (900 - 864)/900 = 0.04$

Since voltage and frequency etc. are fixed, the change in speed is entirely due to change in rotor resistance.

(*i*) $s_2 = s_1(R_2/R_1)$ or $0.04 = 0.03\,(R_2/R_1)$; $R_2 = 1.33\,R_1$

Obviously, the rotor resistance has increased by **33 percent.**

(ii) Let t_2 be themperature of the rotor. Then, as seen from Art.1-11,

$$R_2 = R_1\,[1 + \alpha\,(t_2 - 23)] \text{ or } 1.33\,R_1 = R_1\left[1 + \frac{1}{234}(t_2 - 23)\right] \qquad \therefore t_2 = \mathbf{100.2°C}$$

32.31. Plugging of an Induction Motor

As fully discussed in Art. 32.13, an induction motor can be quickly stopped by simply interchanging any of its two stator leads. It reverses the direction of the revolving flux which produces a torque in the reverse direction, thus applying brake on the motor. Obviously, during this so-called plugging period, the motor *acts as a brake*. It absorbs kinetic energy from the still revolving load causing its speed to fall. The associated Power P_m is dissipated as heat in the rotor. At the same time, the rotor also continues to receive power P_2 from the stator (Fig. 32.27) which is also dissipated as heat. Consequently, plugging produces rotor I^2R losses which even exceed those when the rotor is locked.

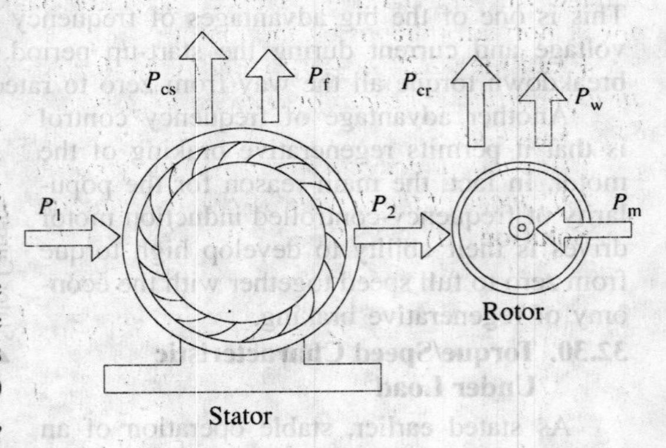

Fig. 32.27

32.32. Induction Motor Operating as a Generator

When run *faster than* its synchronous speed, an induction motor runs as a generator called a *synchronous generator*. It converts the mechanical energy it receives into electrical energy and this energy is released by the stator (Fig. 32.29). Fig. 32.28 shows an ordinary squirrel-cage motor which is driven by a petrol engine and is connected to a 3-phase line. As soon as motor speed exceeds its synchronous speed, it starts delivering *active* power P to the 3-phase line. However,

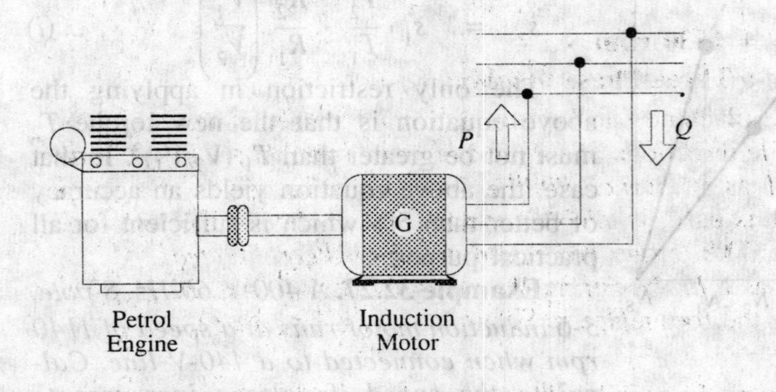

Fig. 32.28

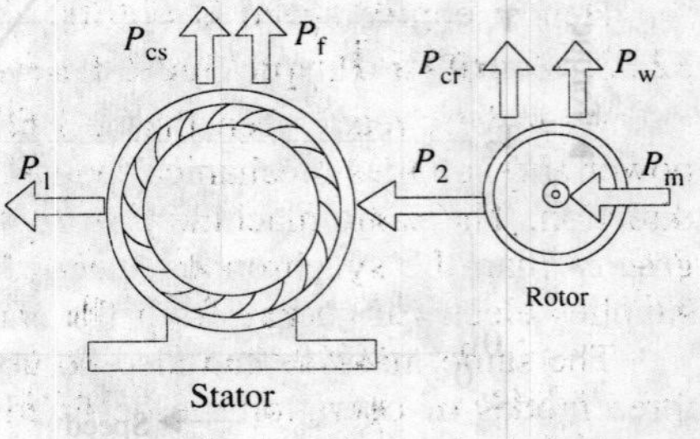

Fig. 32.29

for creating its own magnetic field, it absorbs *reactive* power Q from the line to which it is connected. As seen, Q flows in the *opposite* direction to P.

The active power is *directly proportional to the slip* above the synchronous speed. The reactive power required by the motor can also be supplied by a group of capacitors connected across its terminals (Fig. 32.30). This arrangement can be used to supply a 3-phase load without using an external source. The frequency generated is slightly less than that corresponding to the speed of rotation.

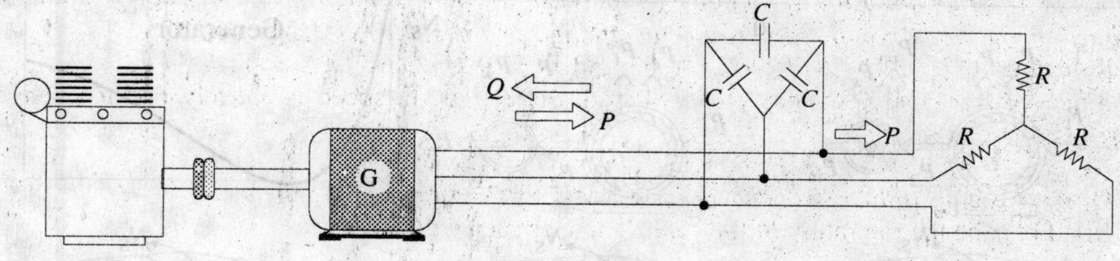

Fig. 32.30

The terminal voltage increases with capacitance. If capacitance is insufficient, the generator voltage will not build up. Hence, capacitor bank must be large enough to supply the reactive power normally drawn by the motor.

Example 32.29 *A 440-V, 4-pole, 1470 rpm. 30-kW, 3-phase induction motor is to be used as an asynchronous generator. The rated current of the motor is 40 A and full-load power factor is 85%. Calculate*

(a) *capacitance required per phase if capacitors are connected in delta.*
(b) *speed of the driving engine for generating a frequency of 50 Hz.*

Solution. (*i*) $S = \sqrt{3}VI = 1.73 \times 440 \times 40 = 30.4$ kVA

$P = S \cos \phi = 30.4 \times 0.85 = 25.8$ kW

$Q = \sqrt{S^2 - P^2} = \sqrt{30.4^2 - 25.8^2} = 16$ kVAR

Hence, the Δ-connected capacitor bank (Fig. 32.31) must provide 16 / 3 = 5.333 kVAR per phase. Capacitor current per phase is = 5,333/440 = 12 A. Hence $X_c = 440/12 = 36.6\ \Omega$. Now, $C = 1/2\pi f X_c = 1/2\pi \times 50 \times 36.6 =$ **87μF**

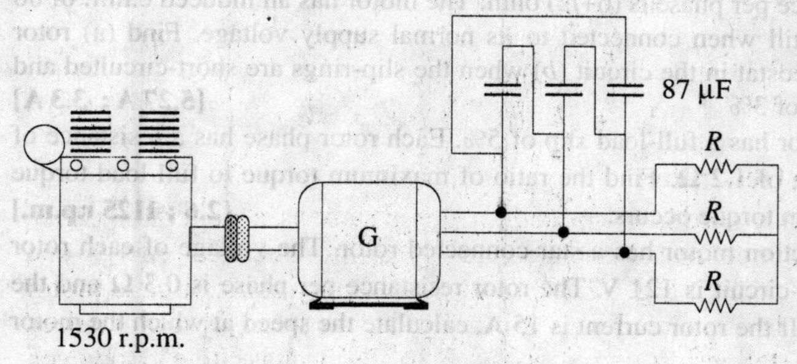

Fig. 32.31

(*ii*) The driving engine must run at slightly more than synchronous speed. The slip speed is usually the same as that when the machine runs as a motor *i.e.* 30 rpm.

Hence, engine speed is = 1500 + 30 = **1530 rpm.**

32.33. Complete Torque/Speed Curve of a Three-Phase Machine

We have already seen that a 3-phase machine can be run as a *motor,* when it takes electric power and supplies mechanical power. The directions of torque and rotor rotation are in the *same* direction. The same machine can be used as an *asynchronous generator* when driven at a speed *greater* than the synchronous speed. In this case, it receives mechanical energy in the rotor and supplies electrical energy from the stator. The torque and speed are *oppositely-directed.*

The same machine can also be used as a *brake* during the plugging period (Art. 32.31). The three modes of operation are depicted in the torque/speed curve shown in Fig. 32.32.

Tutorial Problem No. 32.2

1. In a 3-phase, slip-ring induction motor, the open-circuit voltage across slip-rings is measured to be 110 V with normal voltage applied to the stator. The rotor is star-connected and has a resistance of

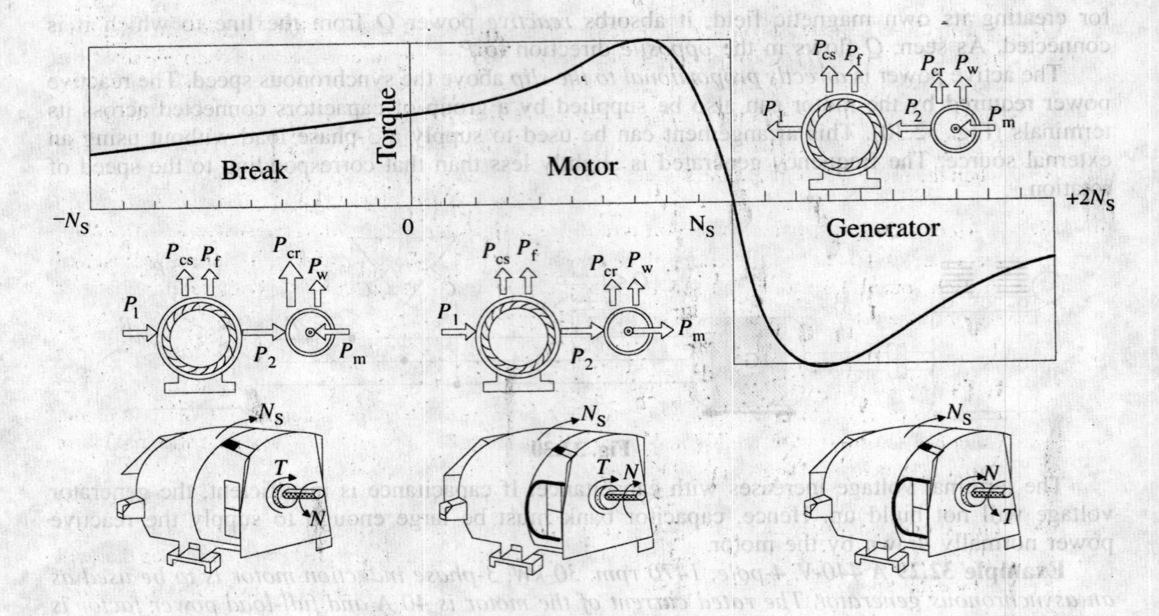

Fig. 32.32

1 Ω and reactance of 4 Ω at standstill condition. Find the rotor current when the machine is (a) at standstill with slip-rings joined to a star-connected starter with a resistance of 2Ω per phase and negligible reactance (*b*) running normally with 5% slip. State any assumptions made.

[12.7 A ; 3.11 A] (*Electrical Technology-I, Bombay Univ. 1978*)

2. The star-connected rotor of an induction motor has a standstill impedance of (0.4 + *j*4) ohm per phase and the rheostat impedance per phase is (6+*j*2) ohm. The motor has an induced e.m.f. of 80 V between slip-rings at standstill when connected to its normal supply voltage. Find (a) rotor current at standstill with the rheostat in the circuit (*b*) when the slip-rings are short-circuited and the motor is running with a slip of 3% **[5.27 A ; 3.3 A]**

3. A 4-pole, 50-Hz induction motor has a full-load slip of 5%. Each rotor phase has a resistance of 0.3 Ω and a standstill reactance of 1.2 Ω. Find the ratio of maximum torque to full-load torque and the speed at which maximum torque occurs. **[2.6 ; 1125 r.p.m.]**

4. A 3-phase, 4-pole, 50-Hz induction motor has a star-connected rotor. The voltage of each rotor phase at standstill and on open-circuit is 121 V. The rotor resistance per phase is 0.3 Ω and the reactance at standstill is 0.8 Ω. If the rotor current is 15 A, calculate the speed at which the motor is running.
Also, calculate the speed at which the torque is a maximum and the corresponding value of the input power to the motor, assuming the flux to remain constant. **[1444 r.p.m.; 937.5 r.p.m.]**

5. A 4-pole, 3-phase, 50 Hz induction motor has a voltage between slip-rings on open-circuit of 520 V. The star-connected rotor has a standstill reactance and resistance of 2.0 and 0.4 Ω per phase respectively. Determine.
(*a*) the full-load torque if full-load speed is 1,425 r.p.m.
(*b*) the ratio of starting torque to full-load torque
(*c*) the additional rotor resistance required to give maximum torque at standstill

[(*a*) 200 N-m (*b*) 0.82 (*c*) 1.6 Ω] (*Elect. Machines-II, Vikram Univ. Ujjain 1977*)

6. A 50-Hz, 8-pole induction motor has a full-load slip of 4 per cent. The rotor resistance is 0.001 Ω per phase and standstill reactance is 0.005 W per phase. Find the ratio of the maximum to the full-load torque and the speed at which the maximum torque occurs.

[2.6; 600 r.p.m.] (*City & Guilds, London*)

7. A 3-ϕ, 50-Hz induction motor with its rotor star-connected gives 500 V (r.m.s.) at standstill between slip-rings on open circuit. Calculate the current and power factor in each phase of the rotor windings at standstill when joined to a star-connected circuit, each limb of which has a resistance of 10 Ω and an inductance of 0.03 H. The resistance per phase of the rotor windings is 0.2 Ω and inductance 0.03 H. Calculate also the current and power factor in each rotor phase when the rings are short-circuited and the motor is running with a slip of 4 per cent.

[13.6 A, 0.48; 27.0 A, 0.47] (*London University*)

8. A 4-pole, 50-Hz, 3-phase induction motor has a slip-ring rotor with a resistance and standstill reactance of 0.04 Ω and 0.2 Ω per phase respectively. Find the amount of resistance to be inserted in each rotor phase to obtain full-load torque at starting. What will be the approximate power factor in the rotor at this instant? The slip at full-load is 3 per cent.

[0.084 Ω, 0.516 p.f.] (*London University*)

9. A 3-ϕ induction motor has a synchronous speed of 250 r.p.m. and 4 per cent slip at full-load. The rotor has a resistance of 0.02 Ω/phase and a standstill leakage reactance of 0.15 Ω/phase. Calculate (*a*) the ratio of maximum and full-load torque (*b*) the speed at which the maximum torque is developed. Neglect resistance and leakage of the stator winding.

[(*a*) 1.82 (*b*) 217 r.p.m.] (*London University*)

10. The rotor of an 8-pole, 50-Hz, 3-phase induction motor has a resistance of 0.2 Ω/phase and runs at 720 r.p.m. If the load torque remains unchanged. Calculate the additional rotor resistance that will reduce this speed by 10% **[0.8 Ω]** (*City & Guilds, London*)

11. A 3-phase induction motor has a rotor for which the resistance per phase is 0.1 Ω and the reactance per phase when stationary is 0.4 Ω. The rotor induced e.m.f. per phase is 100 V when stationary. Calculate the rotor current and rotor power factor (*a*) when stationary (*b*) when running with a slip of 5 per cnet. **[(*a*) 242.5 A; 0.243 (*b*) 49 A; 0.98]**

12. An induction motor with 3-phase star-connected rotor has a rotor resistance and standstill reactance of 0.1 Ω and 0.5 Ω respectively. The slip-rings are connected to a star-connected resistance of 0.2 Ω per phase. If the standstill voltage between slip-rings is 200 volts, calculate the rotor current per phase when the slip is 5%, the resistance being still in circuit. **[19.1 A]**

13. A 3-phase, 50-Hz induction motor has its rotor windings connected in star. At the moment of starting the rotor, induced e.m.f. between each pair of slip-rings is 350 V. The rotor resistance per phase is 0.2 Ω and the standstill reactance per phase is 1 Ω. Calculate the rotor starting current if the external starting resistance per phase is 8 Ω and also the rotor current when running with slip-rings short-circuited, the slip being 3 per cent. **[24.5 A ; 30.0 A]**

14. In a certain 8-pole, 50-Hz machine, the rotor resistance per phase is 0.04 Ω and the maximum torque occurs at a speed of 645 r.p.m. Assuming that the air-gap flux is constant at all loads, determine the percentage of maximum torque (*a*) at starting (*b*) when the slip is 3%

[(*a*) 0.273 (*b*) 0.41] (*London University*)

15. A 6-pole, 3-phase, 50-Hz induction motor has rotor resistance and reactance of 0.02 Ω and 0.1 Ω respectively per phase. At what speed would it develop maximum torque? Find out the value of resistance necessary to give half of maximum torque at starting.

[800 rpm; 0.007 Ω] (*Elect.Engg. Grad I.E.T.E. June 1988*)

32.34. Measurement of Slip

Following are some of the methods used for finding the slip of an induction motor whether squirrel-cage or slip-ring type.

(*i*) **By actual measurement of motor speed**

This method requires measurement of actual motor speed N and calculation of synchronous speed N_s. N is measured with the help of a speedometer and N_s calculated from the knowledge of supply frequency and the number of poles of the motor.* Then slip can be calculated by using the equation.

$$s = (N_s - N) \times 100/N_s$$

*Since an induction motor does not have salient poles, the number of poles is usually inferred from the no-load speed or from the rated speed of the motor.

(*ii*) **By comparing rotor and stator supply frequencies**

This method is based on the fact that $s = f_r / f$

Since f is generally known, s can be found if frequency of rotor current can be measured by some method. In the usual case, where f is 50 Hz, f_r is so low that individual cycles can be easily counted. For this purpose, a d.c. moving-coil millivoltmeter, preferably of centre-zero, is employed as described below :

(*a*) In the case of a slip-ring motor, the leads of the millivoltmeter are lightly pressed against the adjacent slip-rings as they revolve (Fig. 32.33). Usually, there is sufficient voltage drop in the brushes and their short-circuiting strap to provide an indication on the millivoltmeter. The current in the millivoltmeter follows the variations of the rotor current and hence the pointer oscillates about its mean zero position. The number of complete cycles made by the pointer per second can be easily counted (it is worth remembering that one cycle consists of a movement from zero to a maximum to the right, back to zero and on to a maximum to the left and then back to zero).

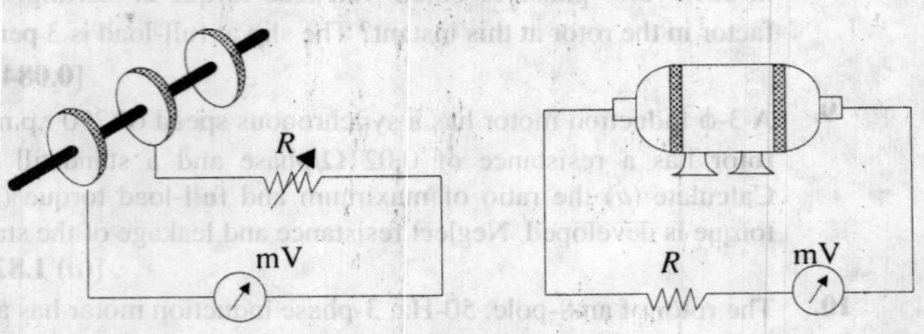

Fig. 32.33 Fig. 32.34

As an example, consider the case of a 4-pole motor fed from a 50-Hz supply and running at 1,425 r.p.m. Since N_s = 1,500 r.p.m., its slip is 5% or 0.05. The frequency of the rotor current would be $f_r = sf = 0.05 \times 50 = 2.5$ Hz which (being slow enough) can be easily counted.

(*b*) For squirrel-cage motors (which do not have slip-rings) it is not possible to employ the millivoltmeter so *directly*, althought it is sometimes possible to pick up some voltage by connecting the milivoltmeter across the ends of the motor shaft (Fig. 32.34)

Another method, sometime employed, is as follows :

A large flat search coil of many turns is placed centrally against the end plate on the non-driving end of the motor. Quite often, it is possible to pick up sufficient voltage by induction from the leakage fluxes to obtain a reading on the millivoltmeter. Obviously, a large 50-Hz voltage will also be induced in the search coil although it is too rapid to affect the millivoltmeter. Commerical slip-indicators use such a search coil and, in addition, contain a low-pass filter amplifier for eliminating fundamental frequency and a bridge circuit for comparing stator and rotor current frequencies.

(*iii*) **Stroboscopic Method**

In this method, a circular metallic disc is taken and painted with alternately black and white segments. The number of segments (both black and white) is equal to the number of poles of the motor. For a 6-pole motor, there will be six segments, three black and three white, as shown in Fig. 32.35(*a*). The painted disc is mounted on the end of the shaft and illuminated by means of a neon-filled stroboscopic lamp, which may be supplied preferably with a combined d.c. and a.c. supply although only a.c. supply will do*. The connections for combined supply are shown in Fig. 32.36 whereas Fig. 32.35 (*b*) shows the connection for a.c. supply only. It must be noted that with combined d.c. and a.c. supply, the lamp will flash once per cycle**. But with a.c. supply, it will flash twice per cycle.

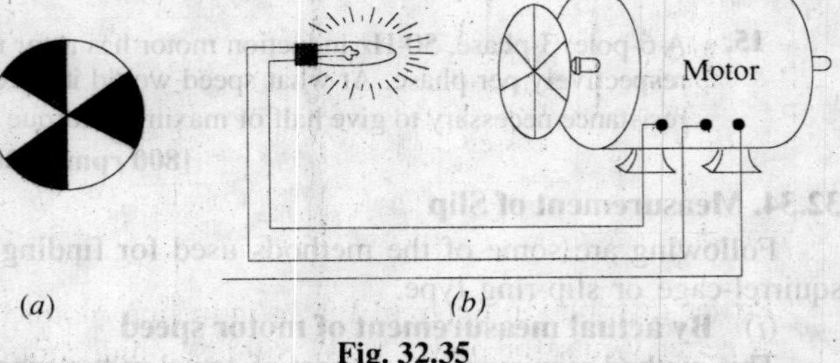

(*a*) (*b*)

Fig. 32.35

*When combined d.c. and a.c. supply is used, the lamp should be tried both ways in its socket to see which way it gives better light.

**It will flash only when the two voltages add and remain extinguished when they oppose.

Consider the case when the revolving disc is seen in the flash light of the bulb which is fed by the combined d.c. and a.c. supply.

If the disc were to rotate at synchronous speed, it would appear to be stationary. Since, in actual practice, its speed is slightly less than the synchronous speed, it appears to rotate slowly backwards. The reason for this apparent backward movement is as follows:

Let Fig. 32.37 (*a*) represent the position of the white lines when they are illuminated by the first flash. When the next flash comes, they have *nearly* reached positions 120° ahead (but not quite), as shown in Fig. 32.37 (*b*). Hence, line No. 1 has *almost* reached the position previously

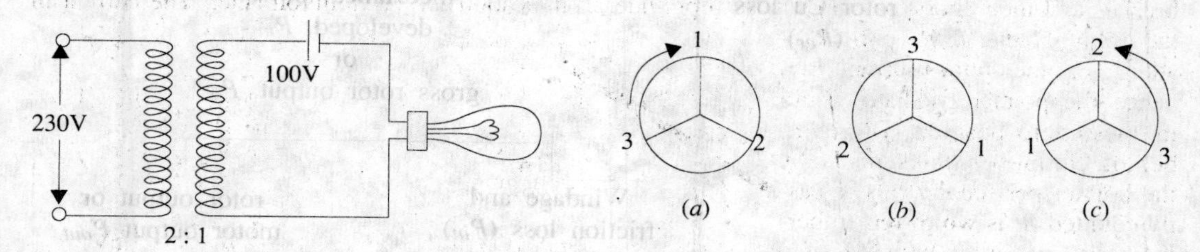

Fig. 32.36 **Fig. 32.37**

occupied by line No. 2 and one flash still later [Fig. 32.37 (*c*)] it has *nearly* reached the position previously occupied by line No. 3 in Fig. 32.37 (*a*).

By counting the number of lines passing a fixed point in, say, a minute and dividing by the number of lines seen (*i.e.* three in the case of a 6-pole motor and so on) the apparent backward speed in r.p.m. can be found. This gives slip-speed in r.p.m. *i.e.* $N_s - N$. The slip may be found from the relation $s = \dfrac{N_S - N}{N_S} \times 100$

Note. If the lamp is fed with a.c. supply alone, then it will flash twice per cycle and twice as many lines will be seen rotating as before.

32.35. Power Stages in an Induction Motor

Stator iron loss (consisting of eddy and hysteresis losses) depends on the supply frequency and the flux density in the iron core. It is practically constant. The iron loss of the rotor is, however, negligible because frequency of rotor currents under normal running conditions is always small. Total rotor Cu loss = $3\, I_2^2\, R_2$

Different stages of power development in an induction motor are as under:

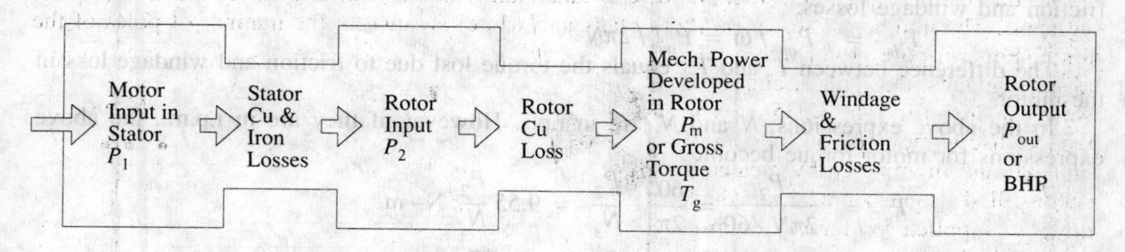

Power distribution diagram for an induction motor is given on the next page:

A better visual for power flow, within an induction motor, is given in Fig. 32.38.

32.36. Torque Developed by an Induction Motor

An induction motor develops gross torque T_s due to gross rotor output P_m (Fig 32.38). Its value can be expressed either in terms of rotor input P_2 or rotor gross output P_m as given below.

$$T_g = \frac{P_2}{\omega_s} = \frac{P_2}{2\,\pi\,N_s} \qquad \text{...in terms of rotor input}$$

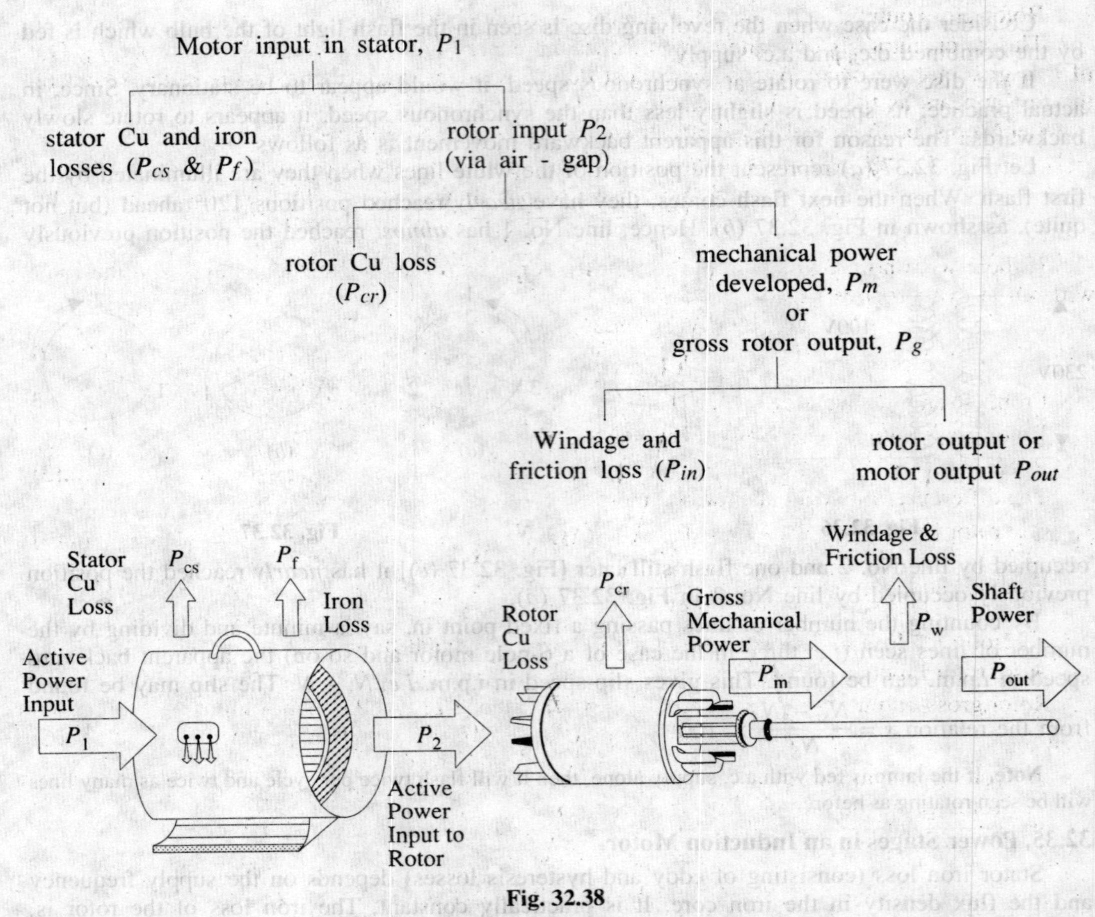

Fig. 32.38

$$= \frac{P_m}{\omega} = \frac{P_m}{2\pi N} \quad \text{... in terms of rotor output}$$

The shaft torque T_{sh} is due to output power P_{out} which is less than P_m because of rotor friction and windage losses.

$$\therefore \quad T_{sh} = P_{out} / \omega = P_{out} / 2\pi N$$

The difference between T_g and T_{sh} equals the torque lost due to friction and windage loss in the motor.

In the above expressions, N and N_s are in r.p.s. However, if they are in r.p.m., the above expressions for motor torque become

$$T_g = \frac{P_2}{2\pi N_s/60} = \frac{60}{2\pi} \cdot \frac{P_2}{N_s} = 9.55 \frac{P_2}{N_s} \text{ N–m}$$

$$= \frac{P_m}{2\pi N/60} = \frac{60}{2\pi} \cdot \frac{P_m}{N} = 9.55 \frac{P_m}{N} \text{ N–m}$$

$$T_{sh} = \frac{P_{out}}{2\pi N/60} = \frac{60}{2\pi} \cdot \frac{P_{out}}{N} = 9.55 \frac{P_{out}}{N} \text{ N–m}$$

32.37. Torque, Mechanical Power and Rotor Output

Stator input P_1 = stator output + stator losses

The stator output is transferred entirely inductively to the rotor circuit.

Obviously, rotor input P_2 = stator output

Rotor gross output, P_m = rotor input P_2 — rotor Cu losses

This rotor output is converted into mechanical energy and gives rise to gross torque T_g. Out of this gross torque developed, some is lost due to windage and friction losses in the rotor and the rest appears as the useful or shaft torque T_{sh}.

Let N r.p.s. be the actual speed of the rotor and if T_g is in N-m, then

$$T_g \times 2\pi N = \text{rotor gross output in watts, } P_m$$

$$\therefore \quad T_g = \frac{\text{rotor gross output in watts, } P_m}{2\pi N} \text{ N–m*} \qquad ...(1)$$

If there were no Cu losses in the rotor, then rotor output will equal rotor input and the rotor will run at synchronous speed.

$$\therefore \quad T_g = \frac{\text{rotor input } P_2}{2\pi N_S} \qquad ...(2)$$

From (1) and (2), we get,

Rotor gross output $P_m = T_g\,\omega = T_g \times 2\pi N$

Rotor input $P_2 = T_g\,\omega_s = T_g \times 2\pi N_s$ ...(3)

The difference of two equals rotor Cu loss.

$\therefore$ rotor Cu loss = $P_2 - P_m = T_g \times 2\pi\ (N_s - N)$ ...(4)

From (3) and (4), $\dfrac{\text{rotor Cu loss}}{\text{rotor input}} = \dfrac{N_s - N}{N_s} = s$

$\therefore$ rotor Cu loss = $s \times$ rotor input = $s \times$ power across air-gap = $s\,P_2$ ... (5)

Also, rotor input = rotor Cu loss / s

Rotor gross output, P_m = input P_2 – rotor Cu loss = input – $s \times$ rotor input

= $(1-s)$ input P_2 ...(6)

$\therefore$ rotor gross output P_m = $(1\text{-}s)$ rotor input P_2

or $\dfrac{\text{rotor gross output, } P_m}{\text{rotor input, } P_2} = 1 - s = \dfrac{N}{N_s}\,;\quad \dfrac{P_m}{P_2} = \dfrac{N}{N_s}$

$\therefore$ rotor efficiency $= \dfrac{N}{N_s}$ Also, $\dfrac{\text{rotor Cu loss}}{\text{rotor gross output}} = \dfrac{s}{1-s}$

Important Conclusion

If some power P_2 is delivered to a rotor, then a part sP_2 is lost in the rotor itself as copper loss (and appears as heat) and the remaining $(1 - s)P_2$ appears as gross mechanical power P_m (including friction and windage losses).

$\therefore P_2 : P_m : I^2R :: 1 : (1-s) : s$ or $P_2 : P_m : P_{cr} :: 1 : (1-s) : s$

The rotor input power will always divide itself in this ratio, hence it is advantageous to run the motor with as small a slip as possible.

32.38. Induction Motor Torque Equation

The gross torque T_g developed by an induction motor is given by

$T_g = P_2 / 2\pi N_s$ —N_s in r.p.s.

$= 60\,P_2 / 2\pi N_s = 9.55\,P_2 / N_s$ —N_s in r.p.m.

Now, P_2 = rotor Cu loss/s = $3I_2^2 R_2 / s$

As seen from Art. 32.19, $I_2 = \dfrac{sE_2}{\sqrt{R_2^2 + (s\,X_2)^2}} = \dfrac{sKE_1}{\sqrt{R_2^2 + (s\,X_2)^2}}$

---

*The value of gross torque in kg-m is given by

$$T_g = \frac{\text{rotor gross output in watts}}{9.81 \times 2\pi N} \text{ kg–m.} = \frac{P_m}{9.81 \times 2\pi N} \text{ kg-m}$$

where K is rotor/stator turn ratio per phase.

$$\therefore \; P_2 = 3 \times \frac{s^2 E_2^2 R_2}{R_2^2 + (s X_2)^2} \times \frac{1}{s} = \frac{3 \, s \, E_2^2 R_2}{R_2^2 + (s X_2)^2}$$

$$\text{Also, } P_2 = 3 \times \frac{s^2 K^2 E_1^2 R_2}{R_2^2 + (s X_2)^2} \times \frac{1}{s} = \frac{3 \, s \, K^2 E_1^2 R_2}{R_2^2 + (s X_2)^2}$$

$$\therefore \; T_g = \frac{P_2}{2\pi N_S} = \frac{3}{2\pi N_S} \times \frac{s E_2^2 R_2}{R_2^2 + (s X_2)^2} \qquad \text{— in terms of } E_2$$

$$\text{or} \quad = \frac{3}{2\pi N_S} \times \frac{s K^2 E_1^2 R_2}{R_2^2 + (sX_2)^2} \qquad \text{— in terms of } E_1$$

Here, E_1, E_2, R_2 and X_2 represent phase values.

In fact, $3 K^2/2 \pi N_s = k$ is called the constant of the given machine. Hence, the above torque equation may be simplified to

$$T_g = k \frac{s E_1^2 R_2}{R_2^2 + (s X_2)^2} \qquad \text{— in terms of } E_1$$

32.39. Synchronous Watt

It is clear from the above relations that torque is proportional to rotor input. By defining a new unit of torque (instead of the force-at-radius unit), we can say that the rotor torque *equals* rotor input. The new unit is synchronous watt. When we say that a motor is developing a torque of 1,000 synchronous watts, we mean that the rotor input is 1,000 watts and that the torque is such that power developed would be 1,000 watts provided the rotor were running synchronously and developing the same torque.

Or

Synchronous watt is that torque which, at the synchronous speed of the machine under consideration, would develop a power of 1 watt.

$$\text{rotor input} = T_{sw} \times 2\pi N_S \qquad \therefore \; T_{sw} = \frac{\text{rotor input}, P_2}{2\pi \times \text{synch.speed}} = \frac{1}{\omega_s} \cdot \frac{N_S}{N} \cdot P_g = \frac{1}{\omega_s} \cdot \frac{N_S}{N} \cdot P_m$$

Synchronous wattage of an induction motor equals the power transferred across the air-gap to the rotor.

$\therefore$ torque in synchronous watt

$$= \text{rotor input} = \frac{\text{rotor Cu loss}}{s} = \frac{\text{gross output power}, P_m}{(1 - s)}$$

Obviously, at $s = 1$, torque in synchronous watt equals the total rotor Cu loss because at standstill, entire rotor input is lost as Cu loss.

Suppose a 23-*kW*, 4-pole induction motor has an efficiency of 92% and a speed of 1440 r.p.m. at rated load. If mechanical losses are assumed to be about 25 per cent of the total losses, then

motor inpur = 23/0.92 = 25 kW, total loss = 25 – 23 = 2 kW.

Friction and windage loss = 2/4 = 0.5 kW

$\therefore \; P_m = 23 + 0.5 = 23.5$ kW

Power in synchronous watts $P_{sw} = P_2 = 23.5 \times 1500/1440 = 24.5$ kW

synchronous speed $w_S = 2\pi\,(1500/60) = 157$ rad / s

$\therefore$ synchronous torque, $T_{sw} = 24.5 \times 10^3/157 = 156$ N-m

$$\text{or} \quad T_{sw} = P_m \frac{N_S}{N} \cdot \frac{1}{\omega_s} = 23.5 \times \frac{1500}{1440} \times \frac{1}{2\pi\,(1500/60)}$$

$$= 156 \text{ N–m} \qquad \text{—Art 32.39}$$

32.40. Variations in Rotor Current

The magnitude of the rotor current varies with load carried by the motor.

As seen from Art. 32.37

$$\frac{\text{rotor output}}{\text{rotor input}} = \frac{N}{N_S} \quad \text{or rotor output} = \text{rotor input} \times \frac{N}{N_S}$$

$\therefore$ rotor input = rotor output $\times N_s / N$

Also, rotor output $\propto 2\pi N T = k N T$

$\therefore$ rotor input $= k N T \times N_s / N = k N_s T$

Now, $\dfrac{\text{rotor Cu loss}}{\text{rotor input}} = s$ or $\dfrac{3 I_2^2 R_2}{s}$ = rotor input

$\therefore$ $3 I_2^2 R_2/s = k N_s T$ *or* $T \propto I_2^2 R_2/s$

$T_{st} \propto I_{2st}^2 R_2$ — since $s = 1$

$T_f \propto I_{2f}^2 R_2 / s_f$ — s_f = full-load slip

$$\therefore \quad \frac{T_{st}}{T_f} = s_f \left(\frac{I_{2st}}{I_2 f}\right)^2$$

where I_{2st} and I_{2f} are the rotor currents for starting and full-load running conditions.

32.41. Analogy with a Mechanical Clutch

We have seen above that, rotor Cu loss = slip × rotor input

This fact can be further clarified by considering the working of a mechanical clutch (though it is not meant to be a proof for the above) similar to the one used in automobiles. A plate clutch is shown in Fig. 32.39. It is obvious that the torque on the driving shaft must exactly equal the torque on the driven shaft. In fact, these two torques are actually one and the same torque, because the torque is caused by friction between the discs and it is true whether the clutch is slipping or not. Let ω_1 and ω_2 be the angular velocities of the shaft when the clutch is slipping.

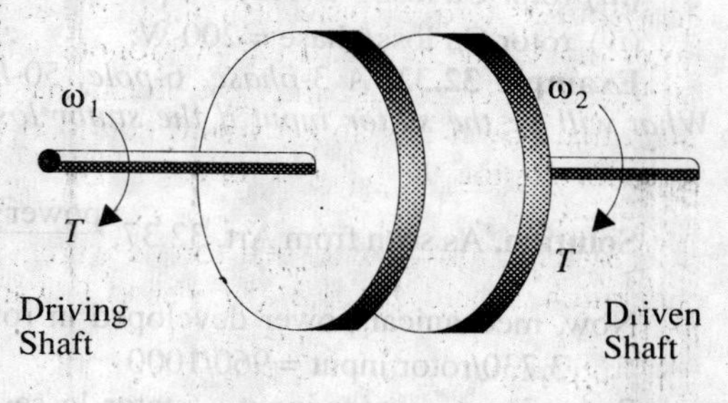

Fig. 32.39

Then, input $= T\,\omega_1$ and output $= T\,\omega_2 = T\,\omega_1(1 - s)$: [$\because \omega_2 = \omega_1(1 - s)$]

loss $= T\,\omega_1 - T\,\omega_2 = T\,\omega_1 - T\,\omega_1(1 - s) = sT\,\omega_1$ = slip × input

32.42. Analogy with a D.C. Motor

The above relations could also be derived by comparing an induction motor with a d.c. motor. As shown in Art 27.3, in a d.c. shunt motor, the applied voltage is always opposed by a back e.m.f. E_b. The power developed in the motor armature is $E_b\ I_a$ where I_a is armature current. This power, as we know, is converted into mechanical power in the armature of the motor.

Now, in in an induction motor, it is seen that the induced e.m.f in the rotor decreases from its standstill value of E_2 to sE_2 when in rotation. Obviously, the difference$(1 - s)E_2$ is the e.m.f. called forth by the rotation of the rotor similar to the back e.m.f. in a d.c. motor. Hence, gross power P_m developed in the rotor is given by the product of the back e.m.f., armature current and rotor power factor.

$$P_m = (1 - s)E_2 \times I_2 \cos\phi_2 \;; \quad \text{Now } I_2 = \frac{sE_2}{\sqrt{[R_2^2 + (sX_2)^2]}} \text{ and } \cos\phi_2 = \frac{R_2}{\sqrt{[R_2^2 + (sX_2)^2]}}$$

$$\therefore \quad P_m = (1 - s)E_2 \times \frac{sE_2}{\sqrt{[R_2^2 + (sX_2)^2]}} \times \frac{R_2}{\sqrt{[R_2^2 + (sX_2)^2]}} = \frac{s(1 - s)\,E_2^2\,R_2}{R_2^2 + (sX_2)^2}$$

Multiplying the numerator and the denominator by s, we get

$$P_m = \left(\frac{1-s}{s}\right) R_2 \times \frac{s^2E_2^2}{R_2^2 + (sX_2)^2} = \left(\frac{1-s}{s}\right) I_2^2R_2 \qquad \left(\because \; I_2 = \frac{sE_2}{\sqrt{[R_2^2 + (sX_2)^2]}}\right)$$

Now, $I_2^2R_2$ = rotor Cu loss/phase $\qquad \therefore \; \dfrac{\text{Cu loss}}{\text{rotor output}} = \dfrac{s}{1-s}$

This is the same relationship as derived in Art. 32-37.

Example 32.30. *The power input to a 3-phase induction motor is 60 kW. The stator losses total 1 kW. Find the mechanical power developed and the rotor copper loss per phase if the motor is running with a slip of 3%.* **(Elect. Machines AMIE Sec. E Summer 1991)**

Solution. Rotor input, P_2 = stator input − stator losses = 60 − 1 = 59 kW

$P_m = (1-s)P_2 = (1-0.03) \times 59 = 57.23$ kW

Total rotor Cu loss $= sP_2 = 0.03 \times 59 = 1.77$ kW = 1770 W

Rotor Cu loss/phase = 1770/3 = 590 W

Example 32.31. *The power input to the rotor of a 400 - V, 50.Hz, 6-pole, 3-phase induction motor is 20 kW. The slip is 3%. Calculate (i) the frequency of rotor currents (ii) rotor speed (iii) rotor copper losses and (iv) rotor resistance per phase if rotor current is 60 A.*

(Elect. Engg. Punjab Univ. 1991)

Solution. (*i*) Frequency of rotor current = $sf = 0.03 \times 50 = 1.5$ Hz

(ii) $N_s = 120 \times 50/6 = 1000$ rpm; $N = 1000(1 - 0.03) = 700$ rpm

(*iii*) rotor Cu loss = $s \times$ rotor input = $0.03 \times 20 = 0.6$ *kW* = 600 W

(*iv*) rotor Cu loss/phase = 200 W; $\quad \therefore 60^2R_2 = 200;\; R_2 = 0.055\,\Omega$

Example 32.32. *A 3-phase, 6-pole, 50-Hz induction motor develops 3.73 kW at 960 rpm. What will be the stator input if the stator loss is 280 W?*

(Electrical Machines.I, Indore Univ. 1987)

Solution. As seen from Art. 32.37, $\dfrac{\text{power developed in rotor}}{\text{rotor input}} = \dfrac{N}{N_S}$

Now, mechanical power developed in rotor = 3.73 *kW*., $N_s = 120 \times 50/6 = 1000$ r.p.m.

$\therefore$ 3,730/rotor input = 960/1000 $\qquad \therefore$ rotor input = 3,885 W

Stator input = rotor input + stator losses = 3885 + 280 = **4,156 W**

Example 32.33. *The power input to the rotor of a 400 - V, 50-Hz, 6-pole, 3-ϕ induction motor is 75 kW. The rotor electromotive force is observed to make 100 complete alteration per minute. Calculate:*

(i) slip (ii) rotor speed (iii) rotor copper losses per phase (iv) mechanical power developed.

(Elect. Engg. I, Nagpur Univ. 1993)

Solution. Frequency of rotor emf, f' = 100/60 = 5/3 Hz

(*i*) Now, $f' = sf$ or $5/3 = s \times 50$; $s = 1/30 =$ **3.33**

(*ii*) $N_s = 120 \times 50/6 = 1000$ rpm; $N = Ns\,(1-s) = 1000\,(1 - 1/30) = 966.7$ rpm

(*iii*) $P_2 = 75$ *kW*; total rotor Cu loss $= sP_2 = (1/30) \times 75 = 2.5$ kW

rotor Cu loss/phase = 2.5/3 = **0.833** kW

(*iv*) $P_m = (1-s)P_2 = (1 - 1/30) \times 75 =$ **72.5 kW**

Example 32.34. *The power input to a 500 - V, 50-Hz, 6-pole, 3-phase induction motor running at 975 rpm is 40 kW. The stator losses are 1 kW and the friction and windage losses total 2 kW. Calculate : (i) the slip (ii) the rotor copper loss (iii) shaft power and (iv) the efficiency.*

(Elect. Engg. - II, Pune Univ. 1989)

Solution. (*i*) $N_s = 120 \times 50/6 = 1000$ rpm ; $s = (100 - 975)/1000 = 0.025$ or 2.5%

(*ii*) Motor input = 40 kW ; stator loss = 1 kW; rotor input $P_2 = 40 - 1 = 39$ kW

$\therefore$ rotor Cu loss = $s \times$ rotor input $= 0.025 \times 39 =$ **0.975 kW** (*iii*) P_m = P_2 – rotor Cu loss = 39 – 0.975 = 38.025 kW

P_{out} = P_m – friction and windage loss = 38.025 – 2 = **36.025 kW**

(*iv*) η = P_{out}/P_1 = 36.025/40 = 0.9 or **90%**

Example 32.35. *A 100-kW (output), 3300-V, 50-Hz, 3-phase, star-connected induction motor has a synchronous speed of 500 r.p.m. The full-load slip is 1.8% and F.L. power factor 0.85. Stator copper loss = 2440 W. Iron loss = 3500 W. Rotational losses = 1200 W. Calculate (i) the rotor copper loss (ii) the line current (iii) the full-load efficiency.*

(Elect. Machines, Nagpur Univ. 1993)

Solution. P_m = output + rotational losss = 100 + 1.2 = 101.2 kW

(*i*) rotor Cu loss $= \dfrac{s}{1-s} \times P_m = \dfrac{0.018}{1-0.018} \times 101.2 =$ **1.855 kW**

(*ii*) rotor input, P_2 = P_m + rotor Cu loss = 101.2 + 1.855 = 103.055 kW

Stator input = P_2 + stator Cu and iron losses

= 103.055 + 2.44 + 3.5 = 108.995 kW

$\therefore$ 108,995 = $\sqrt{3} \times 3300 \times I_L \times 0.85$; I_L = **22.4 A**

The entire power flow in the motor is given below.

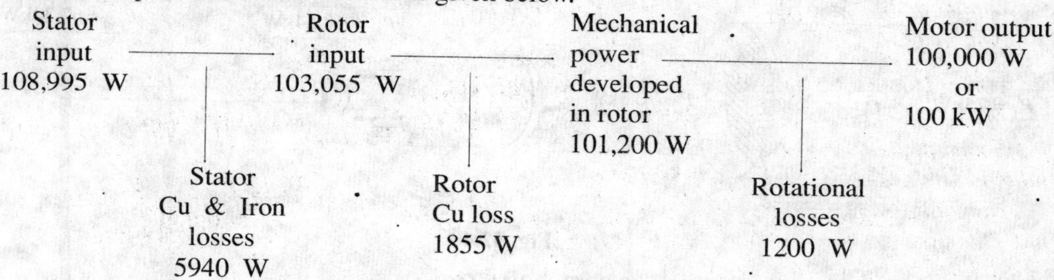

(*iii*) F.L. efficiency = 100,000/108,995 = 0.917 or **91.7%**

Example 32.36. *The power input to the rotor of a 440-V, 50-Hz, 6-pole, 3-phase induction motor is 100 kW. The rotor electromotive force is observed to make 120 cycles per minute, Calculate (i) the slip (ii) the rotor speed (iii) mechanical power developed (iv) the rotor copper loss per phase and (v) speed of stator field with respect to rotor.*

(Elect. Engg. AMIETE Sec. A June 1991)

Solution. (*i*) $f' = sf$ or (120/60) = $s \times 50$; s = **0.01**

(*ii*) N_s = 120 × 50/6 = 1000 rpm; N = 1000(1 - 0.01) = **990 rpm**

(*iii*) P_m $(1-s)P_2$ = (1 - 0.01) × 100 = **99 kW**

(*iv*) total rotor Cu loss = sP_2 = 0.01 × 100 = 1 kW ; Cu loss/phase = **1/3 kW**

(*v*) N_s = 1000 rpm; N = 990 rpm. Hence, speed of stator field with respect to rotor is = 1000 – 990 = **10 rpm.**

Example 32.37. *An induction motor has an efficiency of 0.9 when delivering an output of 37 kW. At this load, the stator Cu loss and rotor Cu loss each equals the stator iron loss. The mechanical losses are one-third of the no-load loss. Calculate the slip.*

(Adv. Elect. Machines, A.M.I.E. Sec. B Winter 1993)

Solution. Motor input = 37,000/0.9 = 41,111 W

$\therefore$ total loss = 41,111 – 37,000 = 4,111 W

This includes (*i*) stator Cu and iron losses (*ii*) rotor Cu loss (its iron loss being negligibly small) and (*iii*) rotor mechanical losses.

Now, no-load loss of an induction motor consists of (*i*) stator iron loss and (*ii*) mechanical losses provided we neglect the small amount of stator Cu loss under no-load condition. Moreover, these two losses are independent of the load on the motor.

no-load loss = $W_i + W_m = 3W_m$ $\quad\therefore W_m = W_i/2$

where W_i is the stator iron loss and W_m is the rotor mechanical losses.

Let, stator iron loss = x; then stator Cu loss = x; rotor Cu loss = x; mechanical loss = $x/2$

$\therefore$ $3x + x/2 = 4{,}111$ or $x = 1175$ W

Now, rotor input = gross output + mechanical losses + Cu loss

$= 37{,}000 + (1175/2) + 1175 = 38{,}752$ W

$$s = \frac{\text{rotor Cu loss}}{\text{rotor input}} = \frac{1175}{38{,}752} = 0.03 \text{ or } \mathbf{3\%}.$$

Example 32.38. *A 400-V, 50-Hz, 6-pole, Δ-connected, 3-φ induction motor consumes 75 kW with a line current of 75 A and runs at a slip of 2.5%. If stator iron loss is 2 kW, windage and fricition loss is 1.2 kW and resistance between two stator terminals is 0.32 Ω, calculate (i) power supplied to the rotor P_2 (ii) rotor Cu loss P_{cr} (iii) power supplied to load P_{out} (iv) efficiency and (v) shaft torque developed.*

Solution. (*i*) $P_2 = P_1 - P_{cs} - P_f$

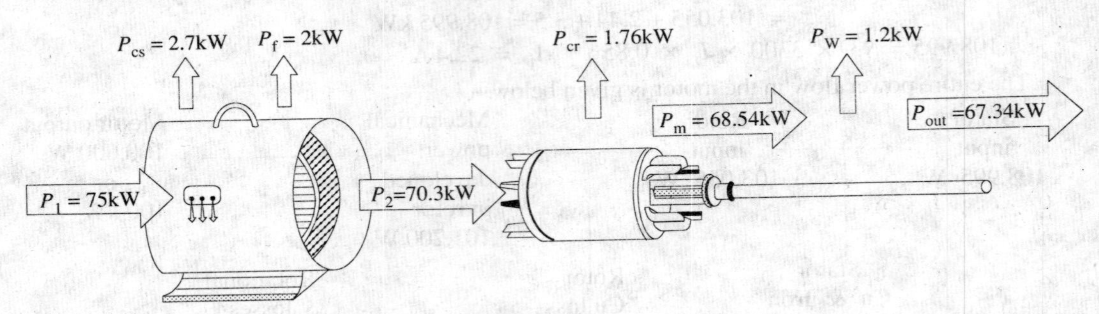

Fig. 32.40

stator resistance per phase = 0.32/2 = 0.16 Ω

stator Cu loss $P_{cs} = 3I^2R = 3 \times 75^2 \times 0.16 = 2.7$ kW

total stator loss $= P_{cs} + P_f = 2.7 + 2 = 4.7$ kW $\quad\therefore P_2 = 75 - 4.7 = \mathbf{70.3\ kW}$

(*ii*) $P_{cr} = sP_2 = 0.025 \times 70.3 = \mathbf{1.76\ kW}$

(*iii*) $P_{out} = P_2 - P_{er} - P_w = 70.3 - 1.76 - 1.2 = \mathbf{67.34\ kW}$

(*iv*) $\eta = P_{out} / P_1 = 67.34/75 = 0.898$ or **89.8%**

(*v*) $T_{sh} = 9.55\, P_{out} / N = 9.55 \times 67{,}340/975 = \mathbf{659\ N\text{-}m}$

Example 32.39. *A 3-phase induction motor has a 4-pole, star-connected stator winding and runs on a 220-V, 50-Hz supply. The rotor resistance per phase is 0.1 Ω and reactance 0.9 Ω. The ratio of stator to rotor turns is 1.75. The full-load slip is 5%. Calculate for this load:-*

(*a*) *the load torque in kg-m* (*b*) *speed at maximum torque*

(*c*) *rotor e.m.f. at maximum torque.* **(Electrical Machines-I, South Gujarat Univ. 1985)**

Solution. (*a*) K = rotor turns/stator turns = 1/1.75

stator voltage/phase, $E_1 = 220/\sqrt{3}$ V

$\therefore$ standstill rotor e.m.f./phase, $E_2 = K E_1 = \dfrac{220}{\sqrt{3}} \times \dfrac{1}{1.75} = \mathbf{72.6\ V}$

$Z_r = \sqrt{R_2^2 + (sX_2^2)} = \sqrt{0.1^2 + (0.05 \times 0.9)^2} = 0.11\ \Omega$

$I_2 = sE_2/Z_r = 0.05 \times 72.6/0.11 = 33$ A

Rotor Cu loss $P_{cr} = 3I_2^2R_2 = 3 \times 33^2 \times 0.1 = 327$ W

$$\frac{\text{rotor Cu loss}}{\text{mech. power developed}} = \frac{s}{1-s};\quad \frac{327}{P_m} = \frac{0.05}{1-0.05};\quad P_m = 6213 \text{ W}$$

$T_g = 9.55\, P_m / N;\quad N = N_s(1-s) = 1500\,(1 - 0.05) = 1425$ rpm

$\therefore$ $T_g = 9.55 \times 6213/1425 = 41.6$ N-m $= 41.6/9.9 = \mathbf{4.24\ kg.m}$

(b) For maximum torque, $s_m = R_2/X_2 = 0.1/0.81 = \mathbf{1/9}$

$\therefore \quad N = N_s(1-s) = 1500(1-1/9) = \mathbf{1333\ r.p.m.}$

(c) rotor e.m.f./phase at maximum torque $= (1/9) \times 72.6 = \mathbf{8.07\ V}$

Example 32.40. *A 3-phase, 440-V, 50-Hz, 40-pole, Y-connected induction motor has rotor resistance of 0.1 Ω and reactance 0.9 Ω per phase. The ratio of stator to rotor turns is 3.5. Calculate:*

(a) gross output at a slip of 5%

(b) the maximum torque in synchronous watts and the corresponding slip.

Solution. (a) Phase voltage $= \dfrac{440}{\sqrt{3}}$ V ; $K = \dfrac{\text{rotor turns}}{\text{stator turns}} = \dfrac{1}{3.5}$

Standstill e.m.f. per rotor phase is $E_2 = KE_1 = \dfrac{440}{\sqrt{3}} \times \dfrac{1}{3.5} = 72.6$ V

$$E_r = s.E_2 = 0.05 \times 72.6 = 3.63\text{ V};\ Z_r = \sqrt{0.1^2 + (0.05 \times 0.9)^2} = 0.1096\ \Omega$$

$$I_2 = \frac{E_r}{Z_r} = \frac{3.63}{0.1096} = 33.1\text{ A};\quad \text{Total Cu loss} = 3I_2^2R = 3 \times 33.1 \times 0.1 = 330\text{W}$$

$$\frac{\text{rotor Cu loss}}{\text{rotor gross output}} = \frac{s}{1-s} \quad \therefore \text{ rotor gross output} = \frac{330 \times 0.95}{0.05} = \mathbf{7{,}250\ W}$$

(b) **For Maximum Torque**

$$R_2 = s_m . X_2 \quad \therefore s_m = R_2/X_2 = 0.1/0.9 = \mathbf{1/9};\quad E_r = (1/9) \times 72.6 = 8.07\text{ V}$$

$$Z_r = \sqrt{0.1^2 + (0.9/9)^2} = 0.1414\ \Omega;\quad I_2 = 8.07/0.1414 = 57.1\text{ A}$$

Total rotor Cu loss $= 3 \times 57.1^2 \times 0.1 = 978$ W

$$\text{rotor input} = \frac{\text{rotor Cu loss}}{s} = \frac{978}{1/9} = 8{,}802\text{ W}$$

$\therefore$ maximum torque in synchronous watts = rotor input = **8,802 W**

Example 32.41. *An 18.65-kW, 4-pole, 50-Hz, 3-phase induction motor has friction and windage losses of 2.5 per cent of the output. The full-load slip is 4%. Compute for full load (a) the rotor Cu loss (b) the rotor input (c) the shaft torque (d) the gross electromagnetic torque.*

(Elect. Machines-II, Indore Univ. 1987)

Solution. Motor output $P_{out} = 18{,}650$ W

Friction and windage loss $P_w = 2.5\%$ of $18{,}650 = 466$ W

Rotor gross output $P_m = 18{,}650 + 466 = 19{,}116$ W

(a) $\dfrac{\text{rotor Cu loss}}{\text{rotor gross output}} = \dfrac{s}{1-s}$, rotor Cu loss $= \dfrac{0.04}{(1-0.04)} \times 19{,}116 = \mathbf{796.6\ W}$

(b) Rotor input $P_2 = \dfrac{\text{rotor Cu loss}}{s} = \dfrac{796.5}{0.04} = \mathbf{19{,}912.5\ W}$

(or rotor input = 19,116 + 796.6 = 19,913 W)

(c) $T_{sh} = 9.55\, P_{out}/N$: $N_s = 120 \times 50/4 = 1500$ r.p.m.

$N = (1 - 0.04) \times 1500 = 1440$ r.p.m

$\therefore \quad T_{sh} = 9.55 \times 18{,}650/1440 = \mathbf{123.7\ N\text{-}m}$

(d) Gross torque $T_g = 9.55\, P_m/N = 9.55 \times 19{,}116/1440 = \mathbf{126.8\ N\text{-}m}$

(or $T_g = P_2/2\pi N_s = 19{,}913/2\pi \times 25 = 127$ N-m)

Example. 32.42. *An 8-pole, 3-phase, 50 Hz, induction motor is running at a speed of 710 rpm with an input power of 35 kW. The stator losses at this operating condition are known to be 1200 W while the rotational losses are 600 W. Find (i) the rotor copper loss, (ii) the gross torque developed, (iii) the gross mechanical power developed, (iv) the net torque and (v) the mechanical power output.*

(Elect. Engg. AMIETE Sec. A 1991)

Solution. (*i*) $P_2 = 35 - 1.2 = 33.8$ kW ; $N_s = 120 \times 50/8 = 750$ rpm ; $N = 710$ rpm; $s = (750 - 710)/750 = 0.0533$

$\therefore$ Cu loss $= sP_2 =$ **1.8 kW**

(*ii*) $P_m = P_2$ - rotor Cu loss $= 33.8 - 1.8 =$ **32 kW**

(*iii*) $T_g = 9.55\, P_m/N = 9.55 \times 32000/710 = 430.42$ *N-m*

(*iv*) $P_{out} = P_m$ − rotational losses = 32000 - 600 = **31400 W**

(*v*) $T_{sh} = 9.55\, P_{out}/N = 9.55 \times 31400/710 =$ **422.35 N-m**

Example 32.43. *A 6-pole, 50-Hz, 3-phase, induction motor running on full-load with 4% slip develops a torque of 149.3 N-m at its pulley rim. The friction and windage losses are 200 W and the stator Cu and iron losses equal 1,620 W Calculate (a) output power (b) the rotor Cu loss and (c) the efficiency at full-load.* **(Elect. Technology, Mysore Univ. 1989)**

Solution. $N_s = 120 \times 50/6 = 1{,}000$ r.p.m; $N = (1 - 0.04)\, 1{,}000 = 960$ r.p.m.

(*s*) Out put power $= T_{sh} \times 2\,\pi N = 2\,\pi \times (960/60) \times 149.3 =$ **15 kW**

Now, output = 15,000 W

Friction and windage losses = 200 W ; Rotor gross output = 15,200 W

$$\frac{P_m}{P_2} = \frac{N}{N_S} \qquad \therefore \quad \text{rotor input } P_2 = 15{,}200 \times 1{,}000/960 = 15{,}833 \text{ W}$$

(*b*) $\therefore$ rotor Cu loss = 15,833 - 15,200 = **633 W**

$$\left(\text{rotor Cu loss could by using } \frac{\text{rotor Cu loss}}{\text{rotor output}} = \frac{s}{1-s}\right)$$

(*c*) stator output = rotor input = 15,833 W

Stator Cu and iron losses = 1,620 W

$\therefore$ stator input P_1 = 15,833 + 1,620 = 17,453 W

overall efficiency, η = 15,000 × 100/17,453 = **86%**

Example 32.44. *An 18.65-kW, 6-pole, 50-Hz, 3-φ slip-ring induction motor runs at 960 r.p.m. on full-load with a rotor current per phase of 35 A. Allowing 1 kW for mechanical losses, find the resistance per phase of 3-phase rotor winding.* **(Elect. Engg-I, Nagpur Univ. 1992)**

Solution. Motor output = 18.65 kW; Mechanical losses = 1 kW

$\therefore$ mechanical power developed by rotor, $P_m = 18.65 + 1 = 19.65$ kW

Now, $N_s = 120 \times 50/6 = 1000$ r.p.m. ; $s = (1000 - 960)/1000 = 0.04$

$$\text{rotor Cu loss} = \frac{s}{1-s} \times P_m = \frac{0.04}{(1-0.04)} \times 19.65 = 0.819 \text{ kW} = 819 \text{ W}$$

$\therefore \quad 3I_2^2R_2 = 819$ or $3 \times 35^2 \times R_2 = 819$ $\qquad \because R_2 =$ **0.023 Ω/phase**

Example 32.45. *A 400-V, 4-pole, 3-phase, 50-Hz induction motor has a rotor resistance and reactance per phase of 0.01 Ω and 0.1 Ω respectively. Determine* (a) *maximum torque in N-m and the corresponding slip (b) the full-load slip and power output in watts, if maximum torque is twice the full-load torque. The ratio of stator to rotor turns is 4.*

Solution. Applied voltage/phase $E_1 = 400/\sqrt{3} = 231$ V

Standstill e.m.f. induced in rotor, $E_2 = KE_1 = 231/4 = 57.75$ V

(*a*) Slip for maximum torque, $s_m = R_2/X_2 = 0.01/0.1 = 0.1$ or **10%**

$$T_{max} = \frac{3}{2\pi N_S} \times \frac{E_2^2}{2\,X_2} \qquad \text{... Art. 32.20}$$

$$N_s = 120 \times 50/4 = 1500 \text{ r.p.m.} = 25 \text{ r.p.s.}$$

$$\therefore \quad T_{max} = \frac{3}{2\pi \times 25} \times \frac{57.75^2}{2 \times 0.1} = \mathbf{320\ N\text{-}m}$$

(*b*) $\dfrac{T_f}{T} = \dfrac{2as_f}{a^2 + s_f^2} = \dfrac{1}{2}$; Now, $a = R_2/X_2 = 0.01/0.1 = 0.1$

$\therefore\ 2 \times 0.1 \times s_f/(0.1^2 + s_f^2) = 1/2 \qquad \therefore\ s_f = 0.027$ or 0.373

Since $s_f = 0.373$ is not in the operating region of the motor, we select $s_f = 0.027$. Hence,

s_f = **0.027.** $N = 1500\,(1 - 0.027) = 1459.5$ r.p.m.

Full-load torque T_f = 320/2 = 160 N-m

F.L. Motor output = $2\pi N T_f/60 = 2\pi \times 1459.5 \times 160/60 =$ **24,454 W**

Example 32.46. *A 3-phase induction motor has a 4-pole, Y-connected stator winding. The motor runs on 50-Hz supply with 200V between lines. The motor resistance and standstill reactance per phase are 0.1 Ω and 0.9 Ω respectively. Calculate*

(a) the total torque at 4% slip (b) the maximum torque

(c) the speed at maximum torque if the ratio of the rotor to stator turns is 0.67. Nelgect rotor impedance.

(Elect. Machinery, Mysore Univ. 1987)

Solution. (*a*) Voltage/phase $= \dfrac{200}{\sqrt{3}}$ V; $K = \dfrac{\text{rotor turns}}{\text{stator turns}} = 0.67$

Standstill rotor e.m.f. per phase is $E_2 = \dfrac{200}{\sqrt{3}} \times 0.67 = 77.4$ V and $s = 0.04$

$$Z_r = \sqrt{R_2^2 + (sX_2^2)} = \sqrt{0.1^2 + (0.04 \times 0.9)^2} = 0.106\ \Omega$$

$$I_2 = \frac{sE_2}{Z_r} = \frac{77.4 \times 0.04}{0.106} = 29.1\ \text{A}$$

Total Cu loss in rotor $= 3I_2^2R_2 = 3 \times 29.1^2 \times 0.1 = 255$ W

Now, $\dfrac{\text{rotor Cu loss}}{\text{rotor gross output}} = \dfrac{s}{1-s} \quad \therefore\ P_m = 255 \times 0.96/0.04 = 6{,}120$ W

$N_s = 120 \times 50/4 = 1{,}500$ r.p.m.; $N = (1 - 0.04) \times 1{,}500 = 1440$ r.p.m.

Gross torque developed, $T_g = 9.55\, P_m/N = 9.55 \times 6120/1440 =$ **40.6 N-m**

(*b*) For maximum torque, $s_m = R_2/X_2 = 0.1/0.9 = 1/9$ and $E_r = sE_2 = 77.4 \times 1/9 = 8.6$ V

$Z_r = \sqrt{0.1^2 + (0.9 \times 1/9)^2} = 0.1414\ \Omega$; $I_2 = 8.6/0.1414 = 60.8$ A

Total rotor Cu loss $= 3 \times 60.8^2 \times 0.1 = 1{,}110$ W

$$\text{Rotor gross output} = \frac{1{,}100 \times (1 - 1/9)}{1/9} = 8{,}800\ \text{W}$$

$N = (1 - 1/9) \times 1{,}500 = 1{,}333$ rpm.

$\therefore\ T_{max} = 9.55 \times 8{,}800/1333 =$ **63 N-m**

(*c*) Speed at maximum torque, as found above, is **1333.3 r.p.m.**

Example 32.47. *The rotor resistance and standstill reactance of a 3-phase induction motor are respectively 0.015 Ω and 0.09 Ω per phase.*

(i) What is the p.f. of the motor at start? (ii) What is the p.f. at a slip of 4 per cent

(iii) If the number of poles is 4, the supply frequency is 50-Hz and the standstill e.m.f. per rotor phase is 110 V, find out the full-load torque. Take full-load slip as 4 per cent.

(Electrical Technology-I, Osmania Univ. 1990)

Solution. (*i*) rotor impedance/phase $= \sqrt{0.015^2 + 0.09^2} = 0.092\ \Omega$

p.f. = 0.015/0.0912 = **0.164**

(*ii*) reactance/phase $= sX_2 = 0.04 \times 0.09 = 0.0036\ \Omega$

rotor impedance/phase $= \sqrt{0.015^2 + 0.0036^2} = 0.0154\ \Omega$

p.f. = 0.015/0.0154 = **0.974**

(*iii*) $N_s = 120 \times 50/4 = 1{,}500$ r.p.m.; $N = 1{,}500 - 0.04 \times 1{,}500 = 1{,}440$ r.p.m.

$E_r = sE_2 = 0.04 \times 110 = 4.4$ V ; $Z_r = 0.0154\ \Omega$ ...found above

$I_2 = 4.4/0.0154 = 286$ A

Total rotor Cu loss $= 3I_2^2R_2 = 3 \times 286^2 \times 0.015 = 3{,}650$ W

Now, $\dfrac{\text{rotor Cu loss}}{\text{rotor gross output}} = \dfrac{s}{1-s}$

∴ rotor gross output $P_m = 3{,}650 \times 0.96/0.04 = 87{,}600$ W

If T_g is the gross torque developed by the rotor, then

$T_g = 9.55\ P_m/N = 9.55 \times 87{,}600/1440 =$ **581 N-m**

Example 32.48. *The useful full load orque of 3-phase, 6-pole, 50-Hz induction motor is 162.84 N-m. The rotor e.m. f. is observed to make 90 cycles per minute. Calculate (a) motor output (b) Cu loss in rotor (c) motor input and (d) efficiency if mechanical torque lost in windage and friction is 20.36 N-m and stator losses are 830 W.* **(Elect. Machines-II, Indore Univ. 1988)**

Solution. $N_s = 120 \times 50/6 = 1{,}000$ r.p.m.

Frequency of rotor e.m.f. = 90/60 = 1.5 Hz; $s = f_r/f = 1.5/50 = 0.03$

Rotor speed = 1,000(1 – 0.03) = 970 r.p.m. ; Useful F.L. torque = 162.84 N-m

(*a*) motor output $= T_{sh}\dfrac{2\pi N}{60} = \dfrac{2\pi \times 970 \times 162.84}{60} =$ **16,540 W**

(*b*) *gross torque* $T_g = 162.84 + 20.36 = 183.2$ N-m

Now, $T_g = 9.55\dfrac{P_2}{N_S}$ ∴ $183.2 = 9.55 \times \dfrac{P_2}{1000}$

∴ rotor input, $P_2 = 183.2 \times 1000/9.55 = 19{,}170$ W

∴ rotor Cu loss $= s \times$ rotor input $= 0.03 \times 19{,}170 =$ **575.1 W**

(*c*) Motor input, $P_1 = 19{,}170 + 830 =$ **20,000 W**

(*d*) $\eta = (16{,}540/20{,}000) \times 100 =$ **82.27%**

Example 32.49. *Estimate in kg-m the starting torque exerted by an 18.65-kW, 420-V, 6-pole, 50-Hz, 3-phase induction motor when an external resistance of 1 Ω is inserted in each rotor phase.*

stator impedance : (0.25 + j 0.75) Ω *rotor impedance:(0.173 + J 0.52) Ω*

stato/rotor voltage ratio: 420/350 *connection: Star-Star*

Solution. $K = E_2/E_1 = 350/420 = 5/6$

Equivalent resistance of the *motor* as referred to rotor is

$R_{02} = R_2 + K^2R_1 = 0.173 + (5/6)^2 \times 0.25 = 0.346\ \Omega$ / phase

Similarly, $X_{02} = 0.52 + (5/6)^2 \times 0.75 = 1.04\ \Omega$ / phase

When an external resistance of 1 ohm/phase is added to the rotor circuit, the equivalent motor impedance as referred to rotor circuit is

$$Z_{02} = \sqrt{(1 + 0.346)^2 + 1.04^2} = 1.7\ \Omega$$

Short-circuit rotor current is $I_2 = \dfrac{350/\sqrt{3}}{1.7} = 119$ A

Rotor Cu loss per phase on short-circuit $= 119^2 \times 0.173 = 16{,}610$ W

Now, rotor power input $= \dfrac{\text{rotor Cu loss}}{s}$; On short-circuit, s = 1

∴ rotor power input = rotor Cu loss on short-circuit

= 16,610 W/phase = 49,830 W for 3 phases

$N_s = 120 \times 50/6 = 1000$ r.p.m.

If T_{st} is the starting torque in newton-metres, then

$T_{st} = 9.55\ P_2/N_s = 9.55 \times 49{,}830/1000 = 476$ N-m = 476/9.81 = **46.7 kg-m.**

Example 32.50. *An 8-pole, 37.3-kW, 3-phase induction motor has both stator and rotor windings connected in star. The supply voltage is 280 V per phase at a frequency of 50 Hz. The short-circuit*

current is 200 A per phase at a short-circuit power factor of 0.25. The stator resistance per phase is 0.15 Ω. If transformation ratio between the stator and rotor windings is 3, find

(i) *the resistance per phase of the rotor winding* (ii) *the starting torque of the motor.*

Solution. Under short circuit, all the power supplied to the motor is dissipated in the stator and rotor winding resistances. Short-circuit power supplied to the motor is

$$W_{sc} = 3V_1 I_{sc} \cos \phi_{sc} = 3(280 \times 200 \times 0.25) \text{ W}$$

Power supplied/phase $= 280 \times 200 \times 0.25 = 14{,}000$ W

Let r_2' be the rotor resistance per phase as referred to stator. If r_1 is the stator resistance per phase, then

$$I_{sc}^2 (r_1 + r_2') = 14{,}000 \quad \therefore \quad r_1 + r_2' = 14{,}000 / 200^2 = 0.35\ \Omega$$

$$\therefore \quad r_2' = 0.35 - 0.15 = 0.2\ \Omega; \quad \text{Now, } r_2' = r_2 / K^2 \text{ where } K = 1/3$$

(i) $\therefore \quad r_2 = K^2 r_2' = 0.2/9 = \mathbf{0.022\ \Omega}$ **per phase**

(ii) Power supplied to the rotor circuit is

$$= 3I_{sc}^2 r_2' = 3 \times 200^2 \times 0.2 = 24{,}000 \text{ W}$$

$$Ns = 120f/P = 120 \times 50/8 = 750 \text{ r.p.m.} = 12.5 \text{ r.p.s.}$$

$$\therefore \text{ starting torque } = 9.55\, P_2 / N_s = 9.55 \times 24{,}000/750 = \mathbf{305.6\ N\text{-}m}$$

Example 32.51. *A 3-phase induction motor, at rated voltage and frequency has a starting torque of 160 per cent and a maximum torque of 200 per cent of full-load torque. Neglecting stator resistance and rotational losses and assuming constant rotor resistance, determine*

(a) *the slip at full-load* (b) *the slip at maximum torque*

(c) *the rotor current at starting in terms of F.L. rotor current.*

(Electrical Machine - II, Bombay Univ. 1987)

Solution. As seen from Example 32.22 above,

(a) $s_f = 0.01$ or **1%**

(b) From the same example it is seen that at maximum torque, $a = s_b = 0.04$ or **4%**

(c) As seen from Art. 32.40.

$$\frac{I_{2st}}{I_{2f}} = \sqrt{\frac{T_{st}}{sf \cdot T_f}} = \sqrt{\frac{1.6}{0.01}} = 12.65$$

$\therefore$ starting rotor current = 12.65 × F.L. rotor current

32.43. Sector Induction Motor

Consider a standard 3-phase, 4-pole, 50-Hz, Y-connected induction motor. Obviously, its N_s = 1500 rpm. Suppose we cut the stator in half *i.e.* we remove half the stator winding with the result that only two complete *N* and *S* poles are left behind. Next, let us star the three phases without making any other changes in the existing connections. Finally, let us mount the original rotor above this *sector stator* leaving a small air-gap between the two. When this stator is energised from a 3-phase 50-Hz source, the rotor is found to run at almost 1500 rpm. In order to prevent saturation, the stator voltage should be reduced to half its original value because the sector stator winding has only half the original number of turns. It is found that under these conditions, this half-truncated sector motor still develops about 30% of its original rated power.

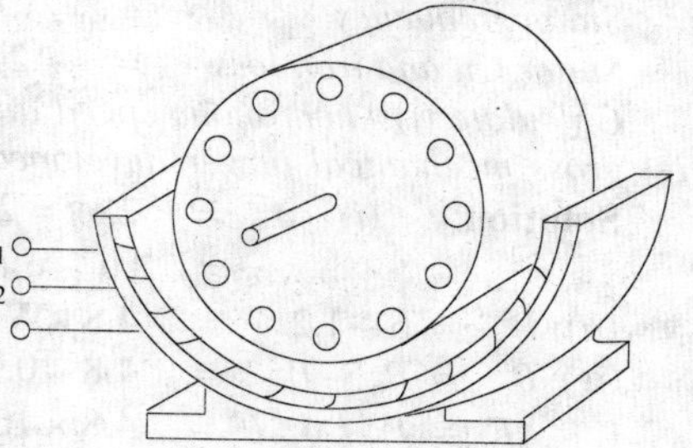

Fig. 32.41

The stator flux of the sector motor revolves at the same peripheral speed as the flux in the original motor. But instead of making a complete round, the flux in the sector motor simply travels *continuously from one end of the stator to the other.*

32.44. Linear Induction Motor

If, in a sector motor, the sector is laid out flat and a flat squirrel-cage winding is brought near to it, we get a linear induction motor (Fig. 32.42). In practice, instead of a flat squirrel-cage winding, an aluminium or copper or iron plate is used as a 'rotor'. The flat stator produces a flux that moves in a straight line from its one end to the other at a linear synchronous speed given by

$$\upsilon_s = 2w_f$$

where υ_s = linear synchronous speed (m/s)
w = width of one pole-pitch (m)
f = supply frequency (Hz)

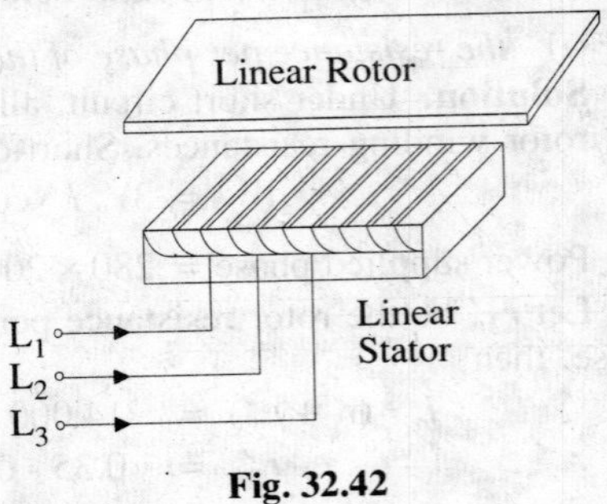

Fig. 32.42

It is worth noting that speed *does not depend on the number of poles*, but only on the pole-bitch and stator supply frequency. As the flux moves linearly, it drags to rotor plate along with it in the *same* direction. However, in many practical applications, the 'rotor' is stationary, while the stator moves. For example, in high-speed trains, which utilize magnetic levitation (Art. 32.45), the rotor is composed of thick aluminium plate that is fixed to the ground and extends over the full length of the track. The linear stator is bolted to the undercarriage of the train.

32.45. Properties of a Linear Induction Motor

These properties are almost identical to those of a standard rotating machine.

1. *Synchronous speed.* It is given by $\upsilon_s = 2wf$
2. *Slip.* It is given by $s = (v_s - v) / v_s$

where v is the actual speed.

3. *Thrust or Force.* It is given by $F = P_2/v_s$

where P_2 is the active power supplied to the rotor.

4. *Active Power Flow.* It is similar to that in a rotating motor.

(*i*) $P_{cr} = sP_2$ and (*ii*) $P_m = (1 - s)P_2$

Example 32.52. *An electric train, driven by a linear motor, moves with 200 km/h, when stator frequency is 100 Hz. Assuming negligible slip, calculate the pole-pitch of the linear motor.*

Solution. $\upsilon_s = 2\,\omega f$ $\quad \omega = \dfrac{(200 \times 5/18)}{2 \times 100} =$ **277.8 mm**

Example 32.53. *An overhead crane in a factory is driven horizontally by means of two similar linear induction motors, whose 'rotors' are the two steel I-beams, on which the crane rolls. The 3-phase, 4-pole linear stators, which are mounted on opposite sides of the crane, have a pole-pitch of 6 mm and are energised by a variable-frequency electronic source. When one of the motors was tested, it yielded the following results:*

Stator frequency = 25 Hz; Power to stator = 6 kW
Stator Cu and iron loss = 1.2 kW; crane speed = 2.4 m/s

Calculate (i)synchronous speed and slip (ii) power input to rotor (iii) Cu losses in the rotor (iv) gross mechanical power developed and (v) thrust.

Solution. (*i*) $\upsilon_s = 2\,\omega f = 2 \times 0.06 \times 25 =$ **3 m/s**

$s = (\upsilon_s - \upsilon)/\upsilon_s = (3 - 2.4)/3 = 0.2$ or **20%**

(*ii*) $P_2 = 6 - 1.2 =$ **4.8 kW** $\quad$ (*iii*) $P_{cr} = s\,P_2 = 0.2 \times 4.8 =$ **0.96 kW**

(*iv*) $P_m = P_2 - P_{cr} = 4.8 - 0.96 =$ **3.84 kW**

(*v*) $F = P_2 / \upsilon_s = 4.8 \times 10^3/3 = 1600$ N = **1.60 kN**

32.46. Magnetic Levitation

As shown in Fig. 32.43 (*a*), when a moving permanent magnet sweeps across a conducting ladder, it tends to drag the ladder along with, because it applies a horizontal tractive force $F = Bll$. It will now be shown that this horizontal force is also accompanied by a *vertical* force (particularly, at high magnet speeds), which tends to push the magnet away from the ladder in the upward direction.

A portion of the conducting ladder of Fig. 32.43 (*a*) has been shown in Fig. 32.43 (*b*). The

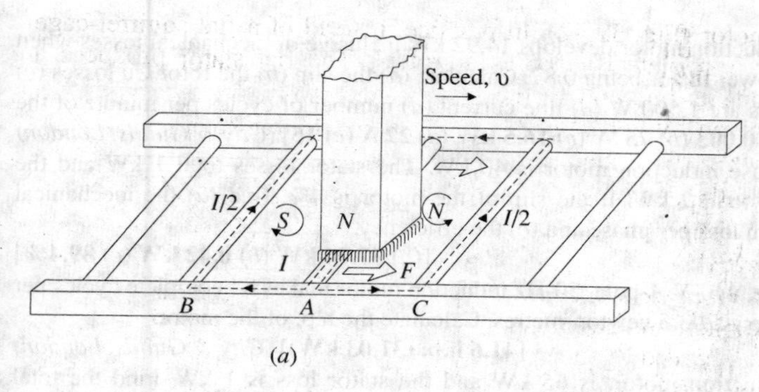

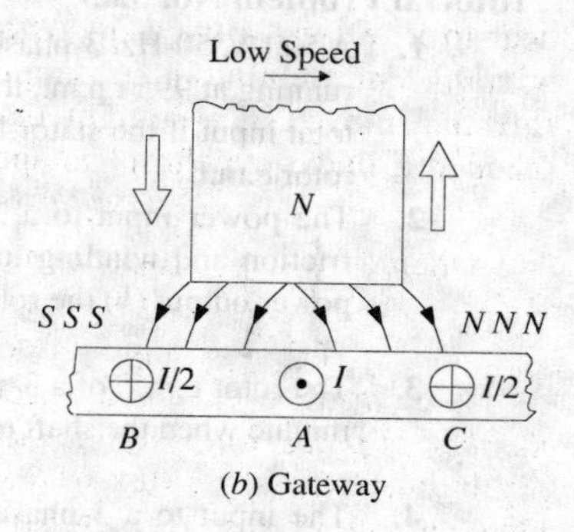

Fig. 32.43

voltage induced in conductor (or bar) A is maximum because flux is greatest at the centre of the N pole. If the magnet speed is very low, the induced current reaches its maximum value in A at virtually the *same time* (because delay due to conductor inductance is negligible). As this current flows via conductors B and C, it produces induced SSS and NNN poles, as shown. Consequently, the front half of the magnet is pushed upwards while the rear half is pulled downwards. Since distribution of SSS and NNN pole is symmetrical with respect to the centre of the magnet, the vertical forces of attraction and repulsion, being equal and opposite, cancel each other out, leaving behind only horizontal tractive force.

Now, consider the case when the magnet sweeps over the conductor A with a *very high speed*. Due to conductor inductance, current in A reaches its maximum value a fraction of a second (Δt) after voltage reaches its maximum value. Hence, by the time I in conductor A reaches its maximum value, the centre of the magnet is already ahead of A by a distance = $\upsilon . \Delta t$ where υ is the magnet velocity. The induced poles SSS and NNN are produced, as before, by the currents returning via conductors B *and* C respectively. But, by now, the N pole of the permanent magnet lies over the induced NNN pole, which pushes it upwards with a strong vertical force.* This forms the basis of *magnetic levitation* which literally means 'floating in air'.

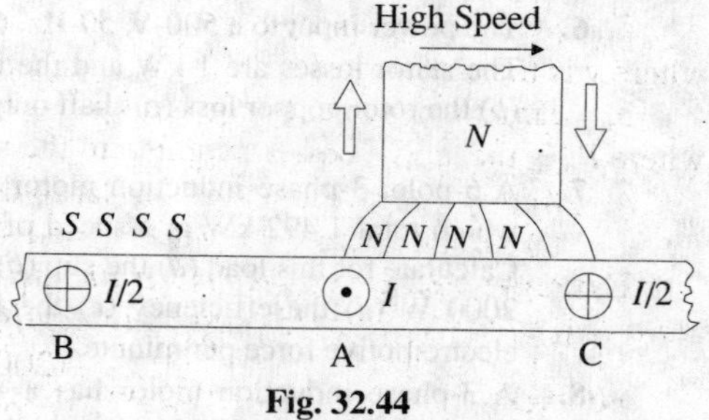

Fig. 32.44

Magnetic levitation is being used in ultrahigh speed trains (upto 300 km/h) which float in the air about 100 mm to 300 mm above the metallic track. They do not have any wheels and do not require the traditional steel rail. A powerful electromagnet (whose coils are cooled to about 4°K by liquid helium) fixed underneath the train moves across the conducting rail, thereby inducing current in the rail. This gives rise to vertical force (called force of levitation) which keeps the train pushed up in the air above the track. Linear motors are used to propel the train.

A similar magnetic levitation system of transit is being considered for connecting Vivek Vihar in East Delhi to Vikaspuri in West Delhi. The system popularly known as Magneto-Bahn (M-Bahn) completely eliminates the centuries-old 'steel-wheel-over steel rail' traction. The M-Bahn train floats in the air through the principle of magnetic levitation and propulsion is by linear induction motors. There is 50% decrease in the train weight and 60% reduction in energy consumption for propulsion purposes. The system is extraordinarily safe (even during an earthquake) and the operation is fully automatic and computer-based.

*The induced current is always delayed (even at low magnet speeds) by an interval of time Δt which depends on the L / R time-constant of the conductor circuit. This delay is so brief at slow speeds that voltage and the current reach their maximum value virtually at the same *time* and *place*. But at high speeds, the same delay the Δt is sufficient to produce large shift *in space* between the points where the voltage and current achieve their maximum values.

Tutorial Problem No. 32.3

1. A 500-V, 50-Hz, 3-phase induction motor develops 14.92 kW inclusive oı ..echanical losses when running at 995 r.p.m., the power factor being 0.87. Calculate (*a*) the slip (*b*) the rotor Cu losses (*c*) total input if the stator losses are 1,500 W (*d*) line current (*d*) number of cycles per minute of the rotor e.m.f. **[(*a*) 0.005 (*b*) 75 W (*c*) 16.5 kW (*d*) 22 A (*e*) 15]** (*City & Guilds, London*)
2. The power input to a 3-phase induction motor is 40 kW. The stator losses total 1 kW and the friction and winding losses total 2 kW. If the slip of the motor is 4%, find (*a*) the mechanical power output (*b*) the rotor Cu loss per phase and (*c*) the efficiency. **[(*a*) 37.74 kW (*b*) 0.42 kW (*c*) 89.4%]**
3. The rotor e.m.f. of a 3-phase, 440-V, 4-pole, 50-Hz induction motor makes 84 complete cycles per minute when the shaft torque is 203.5 newton-metres. Calculate the h.p. of the motor. **[41.6 h.p. (31.03 kW)]** (*City & Guilds, London*)
4. The input to a 3-phase induction motor, is 65 kW and the stator loss is 1 kW. Find the total mechanical power developed and the rotor copper loss per phase if the slip is 3%. Calculate also in terms of the mechanical power developed the input to the rotor when the motor yields full-load torque at half speed. **[83.2 h.p. (62.067 kW) : 640 W, Double the output]** (*City & Guilds. London*)
5. A 6-pole, 50-Hz, 8-phase induction motor, running on full-load, develops a useful torque of 162 N-m and it is observed that the rotor electromotive force makes 90 complete cycles per min. Calculate the shaft output. If the mechanical torque lost in friction be 13.5 Nm, find the copper loss in the rotor windings, the input to the motor and the efficiency. Stator losses total 750 W. **[16.49 kW; 550 W; 19.2 kW; 86%]**
6. The power input to a 500-V, 50-Hz, 6-pole, 3-phase induction motor running at 975 rpm is 40 kW. The stator losses are 1 kW and the friction and windage losses total 2 kW. Calculate (*a*) the slip (*b*) the rotor copper loss (*c*) shaft output (*d*) the efficiency. **[(*a*) 0.025 (*b*) 975 W (*c*) 36.1 kW (*d*) 90%]**
7. A 6-pole, 3-phase induction motor develops a power of 22.38kW, including mechanical losses which total 1.492 kW at a speed of 950 rpm on 550-V, 50-Hz mains. The power factor is 0.88. Calculate for this load (*a*) the slip (*b*) the rotor copper loss (*c*) the total input if the stator losses are 2000 W (*d*) the efficiency (*e*) the line current (*f*) the number of complete cycles of the rotor electromotive force per minute. **[(*a*) 0.05 (*b*) 1175 (*c*) 25.6 kW (*d*) 82% (*e*) 30.4 A(*f*) 150]**
8. A 3-phase induction motor has a 4-pole, star-connected stator winding. The motor runs on a 50-Hz supply with 200 V between lines. The rotor resistance and standstill reactance per phase are 0.1 Ω and 0.9 Ω respectively. The ratio of rotor to stator turns is 0.67. Calculate (a) total torque at 4% slip (*b*) total mechanical power at 4% slip (*c*) maximum torque (*d*) speed at maximum torque (*e*) maximum mechanical power. Prove the formulae employed, neglecting stator impedance. **[(*a*) 40 Nm (*b*) 6 kW (*c*) 63.7 Nm (*d*) 1335 rpm (*e*) 8.952 kW.]**
9. A 3-phase induction motor has a 4-pole, star-connected, stator winding and runs on a 220-V, 50-Hz supply. The rotor resistance is 0.1 Ω and reactance 0.9. The ratio of stator to rotor turns is 1.75. The full load slip is 5%. Calculate for this load (*a*) the total torque (*b*) the shaft output. Find also (*c*o the maximum torque (*d*) the speed at maximum torque. **[(*a*) 42 Nm (*b*) 6.266 kW (*c*) 56 Nm (*d*) 1330 rpm]**
10. A 3000-V, 24-pole, 50-Hz 3-phase, star-connected induction motor has a slip-ring rotor of resistance 0.016 Ω and standstill reactance 0.265 Ω per phase. Full-load torque is obtained at a speed of 247 rpm
    Calculate (*a*) the ratio of maximum to full-load torque (*b*) the speed at maximum torque. Neglect stator impedance. **[(*a*) 2.61 (*b*) 235 rpm]**
11. The rotor resistance and standstill reactance of a 3-phase induction motor are respectively 0.015 Ω and 0.09 Ω per phase. At normal voltage, the full-load slip is 3%. Estimate the percentage reduction in stator voltage to develop full-load torque at one-half of full-load speed. What is then the power factor? **[22.5%; 0.31]**
12. The power input to a 3-phase, 50-Hz induction motor is 60 kW. The total stator loss is 1000 W. Find the total mechanical power developed and rotor copper loss if it is observed that the rotor

e.m.f. makes 120 complete cycles per minute.

[56.64 kW; 2.36 kW] (*AMIE Sec. B Elect. Machine (E-3) Summer 1990*)

13. A balanced three phase induction motor has an efficiency of 0.85 when its output is 44.76 kW. At this load both the stator copper loss and the rotor copper loss are equal to the core losses. The mechanical losses are one-fourth of the no-load loss. Calculate the slip.

[4.94%] (*AMIE Sec. B Elect. Machines (E-3) Winter 1991*)

14. An induction motor is running at 20% slip, the output is 36.775 kW and the total mechanical losses are 1500 W. Estimate Cu losses in the rotor circuit. If the stator losses are 3 kW, estimate efficiency of the motor. **[9,569 W, 72.35%]** (*Electrical Engineering-II, Bombay Univ. 1978*)

15. A 3- ϕ, 50-Hz, 500-V, 6-pole induction motor gives an output of 37.3 kW at 955 r.p.m. The power factor is 0.86, frictional and windage losses total 1.492 kW; stator losses amount to 1.5 kW. Determine (*i*) line current (*ii*) the rotor Cu loss for this load.

[(*i*) 56.54 A (*ii*) 88.6% (*iii*) 1.828 kW] (*Electrical Technology, Kerala Univ. 1977*)

16. Determine the efficiency and the output horse-power of a 3-phase, 400-V induction motor running on load with a slip of 4 per cent and taking a current of 50 A at a power facotr of 0.86. When running light at 440 V, the motor has an input current of 15 A and the power taken is 2,000 W, of which 650 W represent the friction, windage and rotor core loss. The resistance per phase of the stator winding (delta-connected) is 0.5 Ω

[85.8 per cent; 34.2 h.p. (25.51 kW)] (*Electrical Engineering-II, M.S. Univ. Baroda 1977*)

17. The power input to the rotor of a 440-V, 50-Hz, 3-phase, 6-pole induction motor is 60 kW. It is observed that the rotor e.m.f. makes 90 complete cycles per minute. Calculate (*a*) the slip (*b*) rotor speed (*c*) rotor Cu loss per phase (*c*) the mechanical power developed and (*e*) the rotor resistance per phase if rotor current is 60 A. **[(*a*) 0.03 (*b*) 970 r.p.m. (*c*) 600 W (*d*) 58.2 kW (*e*) 0.167 Ω]**

18. An induction motor is running at 50% of the synchronous speed with a useful output of 41.03 kW and the mechanical losses total 1.492 kW. Estimate the Cu loss in the rotor circuit of the motor. If the stator losses total 3.5 kW, at what efficiency is the motor working?

[42.52 kW; 46.34%] (*Electrical Engineering-II, Bombay Univ. 1975*)

19. Plot the torque/speed curve of a 6-pole, 50-Hz, 3-phase induction motor. The rotor resistance and reactance per phase are 0.02 Ω and 0.1 Ω respectively. At what speed is the torque a maximum? What must be the value of the external rotor resistance per phase to give two-thirds of maximum torque at starting? **[(*a*) 800 rpm (*b*) 0.242 Ω or 0.018 Ω]**

32.47. Induction Motor as a Generalized Transformer

The transfer of energy from stator to the rotor of an induction motor takes place entirely *inductively,* with the help of a flux mutually linking the two. Hence, an induction motor is essentially a transformer with stator forming the primary and rotor forming (the short-circuited) rotating secondary (Fig. 32.45). The vector diagram is similar to that of a transformer (Art. 30.14).

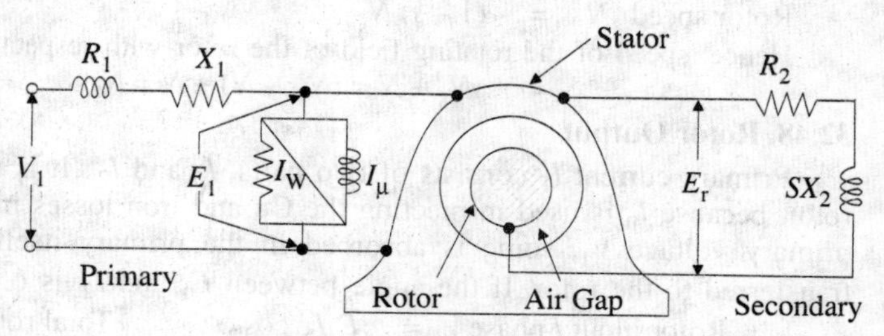

Fig. 32.45

In the vector diagram of Fig. 32.46, V_1 is the applied voltage per stator phase, R_1 and X_1 are stator resistance and leakage reactance per phase respectively, shown external to the stator winding in Fig. 32.45. The applied voltage V_1 produces a magnetic flux which links both primary and secondary thereby producing a counter e.m.f of self-induction E_1 in primary (*i.e.* stator) and a mutually-induced e.m.f. E_r (= $s\,E_2$) in secondary (*i.e.* rotor). There is no secondary terminal voltage V_2 in secondary because whole of the induced e.m.f. E_r is used up in circulating the rotor current as the rotor is closed upon itself (which is equivalent to its being short-circuited).

Obviously $\quad \mathbf{V_1} = \mathbf{E_1} + I_1R_1 + j\,I_1Z_1$

The magnitude of $\mathbf{E_r}$ depends on voltage transformation ratio K between stator and rotor and the slip. As it is wholly absorbed in the rotor impedance.

$\therefore \quad \mathbf{E_r} = \mathbf{I_2 Z_2} = I_2 (R_2 + j\, s\, X_2)$

In the vector diagram, I_0 is the no-load primary current. It has two components (*i*) the working or iron loss components I_ω which supplies the no-load motor losses and (*ii*) the magnetising component I_μ which sets up magnetic flux in the core and the air-gap.

Obviously $I_0 = \sqrt{(I_\omega^2 + I_\mu^2)}$ In Fig. 32.45, I_ω and I_μ are taken care of by an exciting circuit containing $R_0 = E_1/I_\omega$ and $X_0 = E_1/I_\mu$ respectively.

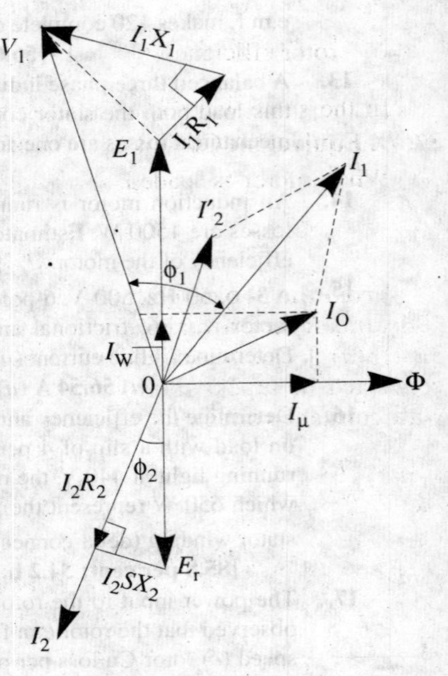

Fig. 32.46

It should be noted here, in passing that in the usual two-winding transformer, I_0 is quite small (about **1**% of the full-load current). The reason is that the magnetic flux path lies almost completely in the steel core of low reluctance, hence I_μ is small, with the result that I_0 is small But in an induction motor, the presence of an air-gap (of high reluctance) necessitates a large I_μ hence I_0 is very large (approximately 40 to 50% of the full-load current).

In the vector diagram, I_2' is the equivalent load current in primary and is equal to KI_2. Total primary current is the vecotr sum of I_0 and I_2'.

At this place, a few words may be said to justify the representation of the stator and rotor quantities on the same vector diagram, even though the frequency of rotor current and e.m.f is only a fraction of that of the stator. We will now show that even though the frequencies of stator and rotor currents are different, yet magnetic fields due to them are synchronous with each other, when seen by an observer stationed in space—both fields rotate at synchronous speed Ns (Art. 32.11).

The current flowing in the short-circuited rotor produces a magnetic field, which revolves round the rotor in the same direction as the stator field. The speed of rotation of the rotor field is

$$= \frac{120 f_r}{P} = \frac{120\, s f}{P} = s N_S = N_S \times \frac{N_S - N}{N_S} = (N_S - N)$$

Rotor speed $N = (1 - s) N_s$

Hence, speed of the rotating field of the rotor with respect to the stationary stator or space is

$$= sN_S + N = (N_S - N) + N = N_s$$

32.48. Rotor Output

Primary current I_1 consists of two parts, I_0 and I_2'. It is the latter which is transferred to the rotor, because I_0 is used in meeting the Cu and iron losses in the stator itself. Out of the applied primary voltage V_1, some is absorbed in the primary itself ($= I_1 Z_1$) and the remaining E_1 is transferred to the rotor. If the angle between E_1 and I_2' is ϕ, then

Rotor input / phase $= E_1 I_2' \cos\phi$; Total rotor input $= 3E_1 I_2 \cos\phi$

The electrical input to the rotor which is wasted in the form of heat is

$$= 3I_2 E_r \cos\phi \qquad (\text{or} = 3\, I_2^{\,2} R_2)$$

Now $I_2' = K I_2$ or $I_2 = I_2'/K$

$E_r = s\,E_2$ or $E_2 = KE_1$ $\quad \therefore E_r = sK E_1$

$\therefore$ electrical input wasted as heat

$$= 3 \times (I_2'/K) \times sK\, E_1 \times \cos\phi = 3E_1 I_2' \cos\phi \times s = \text{rotor input} \times s$$

Now, rotor output $=$ rotor input $-$ losses $= 3E_1\, I_2' \cos\phi - 3E_1 I_2' \cos\phi \times s$

$$= (1 - s) 3E_1 I_2' \cos\phi = (1 - s) \times \text{rotor input}$$

$$\therefore \quad \frac{\text{rotor output}}{\text{rotor input}} = 1 - s \qquad \therefore \quad \text{rotor Cu loss} = s \times \text{rotor input}$$

$$\text{rotor efficiency} = 1 - s = \frac{N}{N_S} = \frac{\text{actual speed}}{\text{synchronous speed}}$$

In the same way, other relation similar to those derived in Art. 32.37 can be found.

32.49. Equivalent Circuit of the Rotor

When motor is loaded, the rotor current I_2 is given by

$$I_2 = s\frac{E_2}{\sqrt{[R_2^2 + (sX_2)^2]}} = \frac{E_2}{\sqrt{[(R_2/s)^2 + X_2^2]}}$$

From the above relation it appears that the rotor circuit which actually consists of a fixed resistance R_2 and a variable reactance sX_2 (proportional to slip) connected across $E_r = sE_2$ [Fig. 32.47 (*a*)] can be looked upon as equivalent to a rotor circuit having a fixed reactance X_2 connected in series with a variable resistance R_2/s (inversely proportional to slip) and supplied with constant voltage E_2, as shown in Fig. 32.47 (*b*).

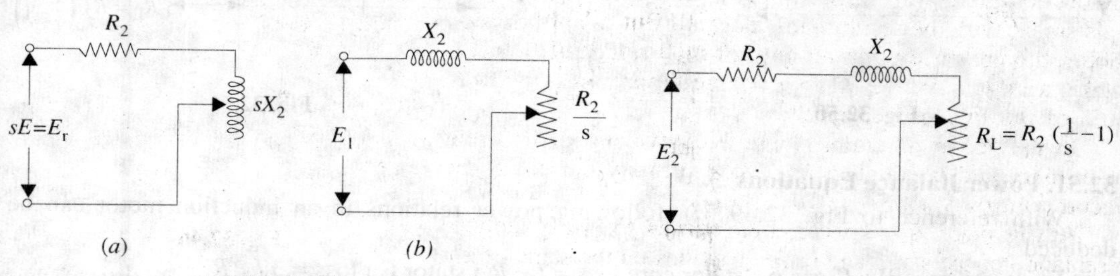

Fig. 32.47

Fig. 32.48

Now, the resistance $\frac{R_2}{s} = R_2 + R_2\left(\frac{1}{s} - 1\right)$. It consists of two parts :

(*i*) the first part R_2 is the rotor resistance itself and represents the rotor Cu loss.

(*ii*) the second part is $R_2\left(\frac{1}{s} - 1\right)$

This is known as the load resistance R_L and is the electrical equivalent of the mechanical load on the motor. In other words, the mechanical load on an induction motor can be represented by a non-inductive resistance of the value $R_2\left(\frac{1}{s} - 1\right)$. The equivalent rotor circuit along with the load resistance R_L may be drawn as in Fig. 32.48.

32.50. Equivalent Circuit of an Induction Motor

As in the case of a transformer (Fig. 30.14), in this case also, the secondary values may be transferred to the primary and *vice versa*. As before, it should be remembered that when shifting impedance or resistance from secondary to primary, it should be *divided* by K^2 whereas current

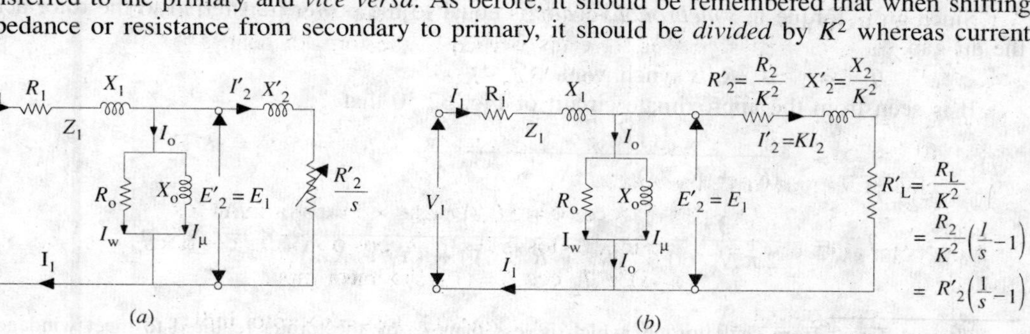
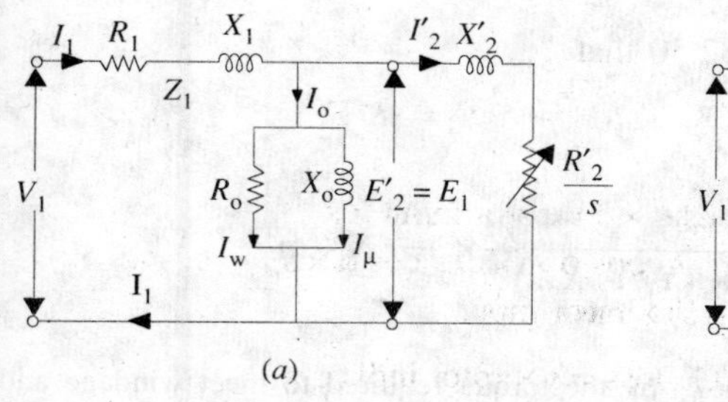

Fig. 32.49

should be *multiplied* by K. The equivalent circuit of an induction motor where all values have been referred to primary *i.e.* stator is shown in Fig. 32.49.

As shown in Fig. 32.50, the exciting circuit may be transferred to the left, because inaccuracy involved is negligible but the circuit and hence the calculations are very much simplified. This is known as the *approximate equivalent* circuit of the induction motor.

If transformation ratio is assumed unity *i.e.* $E_2/E_1 = 1$, then the equivalent circuit is as shown in Fig. 32.51 instead of that in Fig. 32.49.

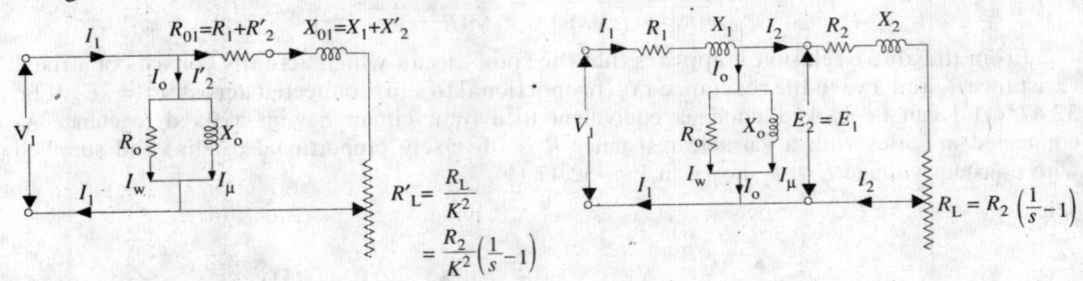

Fig. 32.50 **Fig. 32.51**

32.51. Power Balance Equations

With reference to Fig. 32.49 (*a*), following power relations in an induction motor can be deduced :

Input power = $3V_1 I_1 \cos\phi_1$; stator core loss = $I_\omega^2 R_0$; stator Cu loss = $3 I_1^2 R_1$

Power transferred to rotor = $3 I_2'^2 R_2'/s$; Rotor Cu loss = $3 I_2'^2 R_2'$

Mechanical power developed by rotor (P_m) or gross power developed by rotor (P_g)

$$= \text{rotor input} - \text{rotor Cu losses}$$

$$= 3 I_2'^2 R_2'/s - 3 I_2'^2 R_2' = 3 I_2'^2 R_2'\left(\frac{1-s}{s}\right) \text{ watt}$$

If T_g is the gross torque* developed by the rotor, then

$$T_g \times \omega = T_g \times 2\pi \frac{N}{60} = 3 I_2'^2 R_2'\left(\frac{1-s}{s}\right)$$

$$\therefore \quad T_g = \frac{3 I_2' R_2'\left(\frac{1-s}{s}\right)}{2\pi N/60} \text{ N-m}$$

Now, $N = N_S(1-s)$. Hence gross torque becomes

$$T_g = \frac{3 I_2' R_2'/s}{2\pi N_S/60} \text{ N-m} = 9.55 \times \frac{3 I_2'^2 R_2'/s}{N_S} \text{ N-m}$$

Since gross torque in *synchronous watts* is equal to the power transferred to the rotor across the air-gap.

$$\therefore \quad T_g = 3 I_2'^2 R_2'/s \text{ synch. watt.}$$

It is seen from the approximate circuit of Fig. 32.50 that

$$I_2' = \frac{V_1}{(R_1 + R_2'/s) + j(X_1 + X_2')}$$

$$T_g = \frac{3}{2\pi N_S/60} \times \frac{V_1^2}{(R_1 + R_2'/s)^2 + (X_1 + X_2')^2} \times \frac{R_2'}{s} \text{ N-m}$$

*It is different from shaft torque, which is less than T_g by the torque required to meet windage and frictional losses.

32.52. Maximum Power Output

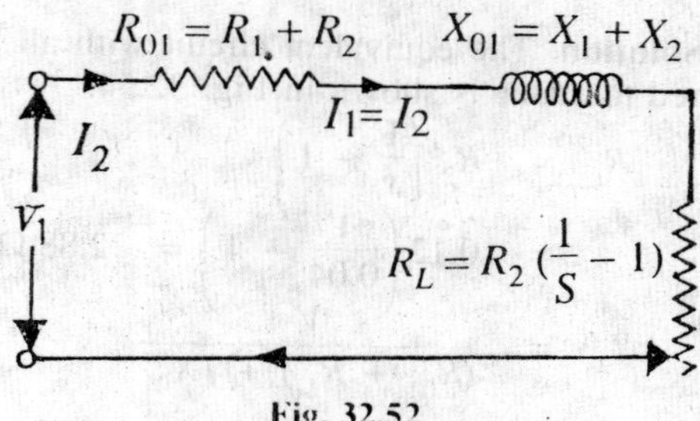

Fig. 32.52

Fig. 32.52. shows the approximate equivalent circuit of an induction motor with the simplification that—

(*i*) exciting circuit is omitted *i.e.* I_0 is neglected and

(*ii*) K is assumed unity.

As seen, gross power output for 3-phase induction motor is

$$P_g = 3 I_1^2 R_L$$

Now, $I_1 = \dfrac{V_1}{\sqrt{[(R_{01} + R_L)^2 + X_{01}^2]}}$ $\quad \therefore \quad P_g = \dfrac{3V_1^2 R_L}{(R_{01} + R_L)^2 + X_{01}^2}$

The condition for maximum power output can be found by differentiating the above equation and by equating the first derivative to zero. If it is done, it will be found that

$$R_L^2 = R_{01}^2 + X_{01}^2 = Z_{01}^2$$

where Z_{01} = leakage impedance of the motor as referred to primary

$$\therefore \quad R_L = Z_{01}$$

Hence, the power output is maximum *when the equivalent load resistance is equal to the standstill leakage impedance of the motor.*

32.53. Corresponding Slip

Now $R_L = R_2[(1/s) - 1]$ $\quad \therefore \quad Z_{01} = R_L = R_2 [(1/s) - 1]$ or $s = \dfrac{R_2}{R_2 + Z_{01}}$

This is the slip corresponding to maximum gross power output. The vlaue of P_{gmax} is obtained by substituting R_L by Z_{01} in the above equation.

$$\therefore \quad P_{gmax} = \frac{3 V_1^2 Z_{01}}{(R_{01} + Z_{01})^2 + X_{01}^2} = \frac{3 V_1^2 Z_{01}}{R_{01}^2 + Z_{01}^2 + 2 R_{01} Z_{01} + X_{01}^2} = \frac{3 V_1^2}{2 (R_{01} + Z_{01})}$$

It should be noted that V_1 is voltage/phase of the motor and K has been taken as unity.

Example. 32.54. *The maximum torque of a 3-phase induction motor occurs at a slip of 12%. The motor has an equivalent secondary resistance of 0.08 Ω/phase. Calculate the equivalent load resistance R_L, the equivalent load voltage V_L and the current at this slip if the gross power output is 9,000 watts.*

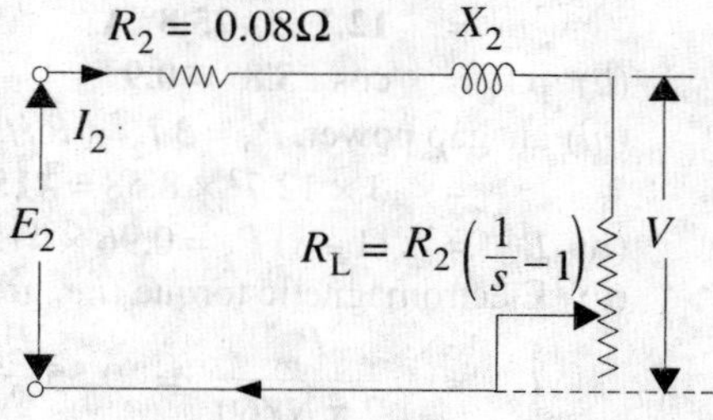

Fig. 32.53

Solution. $R_L = R_2[(1/s) - 1] = 0.08 [(1/0.12) - 1]$ $= \mathbf{0.587\ \Omega/phase}$

As shown in the equivalent circuit of the rotor in Fig. 32.53, V is a fictitious voltage drop equivalent to that consumed in the load connected to the secondary *i.e.* rotor. The value of $V = I_2 R_L$

Now, gross power $P_g = 3 I_2^2 R_L = 3V^2/R_L$

$$V = \sqrt{(P_g \times R_L/3)}$$

$$= \sqrt{(0.587 \times 9000/3)} = \mathbf{42\ V}$$

Equivalent load current

$$= V/R_L = 42/0.587 = \mathbf{71.6A}$$

Example 32.55. *A 3-phase, star-connected 400-V, 50-Hz, 4-pole induction motor has the following per phase parameters in ohms, referred to the stators.*

$R_1 = 0.15$, $X_1 = 0.45$, $R_2' = 0.12$, $X_2' = 0.45$, $X_m = 28.5$

Compute the stator current and power factor when the motor is operated at rated voltage and frequency with s = 0.04. **(Elect. Machines, A.M.I.E. Sec. B, 1990)**

Solution. The equivalent circuit with all values referred to stator is shown in Fig. 32.54.

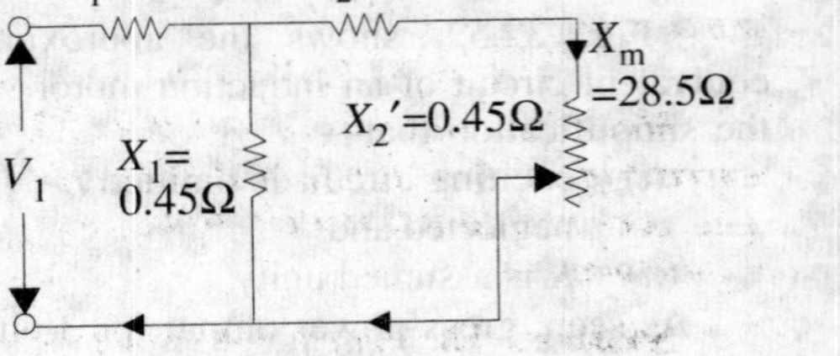

$$R_L' = R_2'\left(\frac{1}{s} - 1\right) = 0.12\left(\frac{1}{0.04} - 1\right) = 2.88\ \Omega$$

$$I_2' = \frac{V}{(R_{01} + R_L') + jX_{01}} = \frac{400/\sqrt{3}}{(0.15 + 0.12 + 2.88) + (0.45 + 0.05)} = 67.78 - j19.36$$

$$I_0 = \frac{400/\sqrt{3}}{X_m} = \frac{400}{\sqrt{3} \times j\,28.5} = j\,8.1$$

Stator current, $I_1 = I_0 + I_2' = 67.78 - j\,19.36 - j\,8.1 = $ **73.13 ∠–22°**

p.f. $= \cos\phi = \cos 22° = 0.927$ (lag)

Fig. 32.54

Example 32.56. *A 220-V, 3-φ, 4-pole, 50-Hz, Y-connected induction motor is rated 3.73 kW. The equivalent circuit parameters are:*

$R_1 = 0.45\ \Omega$, $X_1 = 0.8\ \Omega$; $R_2' = 0.4\ \Omega$, $X_2' = 0.8\ \Omega$, $B_0 = -1/30$ mho

The stator core loss is 50 W and rotational loss is 150 W. For a slip of 0.04, find (i) input current (ii) p.f. (iii) air-gap power (iv) mechanical power (v) electro-magnetic torque (vi) output power and (vii) efficiency.

Solution. The exact equivalent circuit is shown in Fig. 32.55. Since R_0 (or G_0) is negligible in determining I_1, we will consider B_0 (or X_0) only

$$Z_{AB} = \frac{jX_m[(R_2'/s) + jX_2']}{(R_2'/s) + j(X_2' + X_m)} = \frac{j\,30\,(10 + j\,0.8)}{10 + j\,30.8} = 8.58 + j\,3.56 = 9.29\angle 22.5°$$

$$\mathbf{Z}_{01} = \mathbf{Z}_1 + \mathbf{Z}_{AB} = (0.45 + j\,0.8) + (8.58 + j\,3.56) = 9.03 + j\,4.36 = 10\angle 25.8°$$

$$V_{ph} = \frac{220}{\sqrt{3}}\angle 0° = 127\angle 0°$$

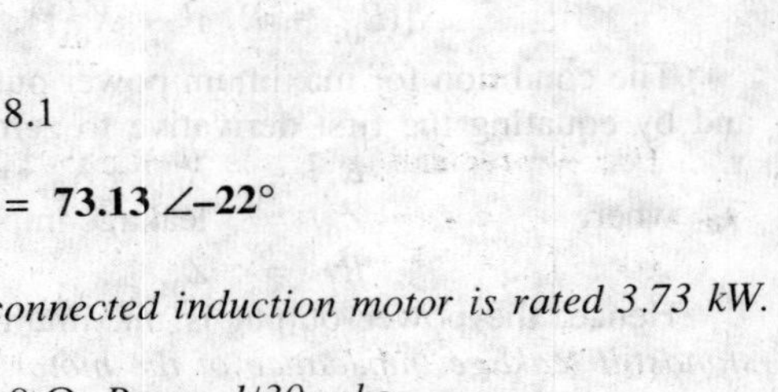

(i) ∴ $\mathbf{I}_1 = V_1/Z_{01} = 127\angle 0°/10\angle 25.8°$ = **12.7 ∠–25.8° A**

(ii) p.f. = cos 25.8° = **0.9**

(iii) air-gap power, $P_2 = 3\,I_2'^2\,(R_2'/s) = 3I_1^2 R_{AB}$ $= 3 \times 12.7^2 \times 8.58 =$ **4152 W**

(iv) $P_m = (1 - s)\,P_2 = 0.96 \times 4152 =$ **3,986 W**

(v) Electromagnetic torque (*i.e.* gross torque)

$$T_g = \frac{P_m}{2\pi N/60} = 9.55\frac{P_m}{N}\ \text{N–m}$$

Fig. 32.55

Now, N_s = 1500 r.p.m., $N = 1500\,(1 - 0.04) = 1440$ r.p.m.

$T_g = 9.55 \times 3986/1440 =$ **26.4 N-m**

(or $T_g = 9.55\dfrac{P_2}{N_S} = 9.55 \times \dfrac{4152}{1500} =$ **26.4 N–m)**

(vi) output power = 3986 – 150 = **3836 W**

(vii) stator core loss = 50 W; stator Cu loss $= 3\,I_1^2 R_1 = 3 \times 12.7^2 \times 0.45 = 218$ W

Rotor Cu loss $= 3\,I_2'^2\,R_2' = sP_2 = 0.04 \times 4152 = 166$ W; Rotational losses = 150 W

Total loss = 50 + 218 + 166 + 150 = 584 W; η = 3836/(3836 + 584) = 0.868 or **86.8%**

Example 32.57. *A 440-V, 3-ϕ 50-Hz, 37.3kW, Y-connected induction motor has the following parameters:*

$$R_1 = 0.1\ \Omega,\ X_1 = 0.4\ \Omega,\ R_2' = 0.15\ \Omega,\ X_2' = 0.44\ \Omega$$

Motor has stator core loss of 1250 W and rotational loss of 1000 W. It draws a no-load line current of 20 A at a p.f. of 0.09 (lag). When motor operates at a slip of 3%, calculate (i) input line current and p.f. (ii) electromagnetic torque developed in N-m (iii) output and (iv) efficiency of the motor. **(Elect. Machines -II, Nagpur Univ. 1992)**

Solution. The equivalent circuit of the motor is shown in Fig. 32.49(*a*). Applied voltage per phase = 440/√3 = 254 V

$$I_2' = \frac{V_1}{(R_1 + R_2'/s) + j(X_1 + X_2')} = \frac{254\angle 0^\circ}{(0.1 + 0.15/0.03) + j(0.4 + 0.44)}$$

$$= \frac{254\angle 0^\circ}{5.1 + j0.84} = \frac{254\angle 0^\circ}{5.17\angle 9.3^\circ} = 49.1\angle -9.3^\circ = 48.4 - j\,7.9$$

For all practical purposes, no-load motor current may be taken as equal to magnetising current I_0. Hence, $I_0 = 20\angle -84.9^\circ = 1.78 - j\,19.9$.

(*i*) $\mathbf{I}_1 = \mathbf{I}_0 + \mathbf{I}_2' = (48.4 - j\,7.9) + (1.78 - j\,19.9) = 50.2 - j\,27.8 = \mathbf{57.4\angle -29^\circ}$

∴ p.f. = cos 29° = **0.875 (lag).**

(*ii*) $P_2 = 3I_2'^2(R_2'/s) = 3 \times 49.1^2 \times (0.15/0.03) = 36{,}160$ W

N_s = 1500 r.p.m. ∴ $T_g = 9.55 \times 36{,}160/1500 =$ **230 N-m**

(*iii*) $P_m = (1 - s)P_2 = 0.97 \times 36{,}160 = 35{,}075$ W

output power = 35,075 - 1000 = **34,075 W**

Obviously, motor is delivering less than its rated output at this slip.

(*iv*) Let us total up the losses.

Core loss = 1250 W, stator Cu loss = $3\,I_1^2R_1 = 3 \times 57.4^2 \times 0.1 = 988$ W

Rotor Cu loss = $3\,I_2'^2R_2' = s\,P_2 = 0.03 \times 36{,}160 = 1085$ W

rotational *i.e.* friction and windage losses = 1000 W

Total losses = 1250 + 988 + 1085 + 1000 = 4323 W

η = 34,075/(34,075 + 4323) = 0.887 or **88.7%**

or input = √3 × 440 × 57.4 × 0.875 = 38,275 W

∴ η = 1 – (4323/38,275) = 0.887 or **88.7%**

Example 32.58. *A 400-V, 3-ϕ, star-connected induction motor has a stator impedance of (0.06 + j 0.2) Ω and an equivalent rotor impedance of (0.06 + j 0.22) Ω. Neglecting current, find the maximum gross power and the slip at which it occurs.*

(Elect.Engg.-II, Bombay Univ. 1987)

Solution. The equivalent circuit is shown in Fig. 32.56.

$R_{01} = R_1 + R_2' = 0.06 + 0.06 = 0.12\ \Omega$

$X_{01} = X_1 + X_2' = 0.2 + 0.22 = 0.42\ \Omega$

∴ $Z_{01} = \sqrt{(0.12^3 + 0.42^2)} = 0.44\ \Omega$

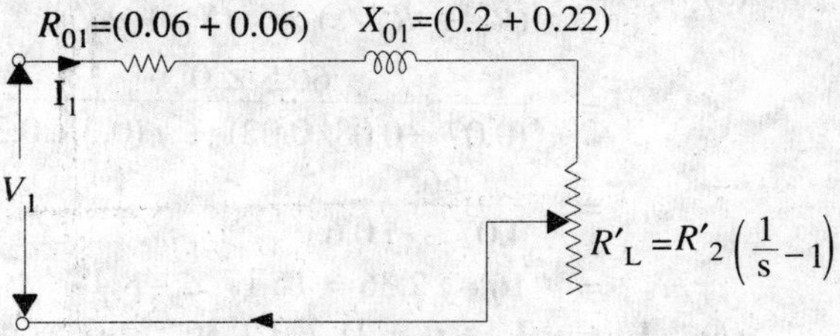

R_{01}=(0.06 + 0.06) X_{01}=(0.2 + 0.22) I_1 V_1 $R'_L = R'_2\left(\frac{1}{s} - 1\right)$

Fig. 32.56

As shown in Art. 32.53, slip corresponding to maximum gross power output is given by

$$s = \frac{R_2}{R_2 + Z_{01}} = \frac{0.06}{0.06 + 0.44}$$

= 0.12 or **12%**

Voltage/phase, $V_1 = 400/\sqrt{3}$ V

$$P_{g\,max} = \frac{3V_1^2}{2(R_{01} + Z_{01})} = \frac{3(400/\sqrt{3})^2}{2(0.12 + 0.44)} = \mathbf{142{,}900\ W}$$

Example 32.59. *A 115-V, 60-Hz, 3-phase, Y-Connected, 6-pole induction motor has an equivalent T-circuit consisting of stator impedance of (0.07 + j 0.3) Ω and an equivalent rotor impedance at standstill of (0.08 + j 0.3) Ω. Magnetising branch has G_o = 0.022 mho, B_o = 0.158 mho. Find (a) secondary current (b) primary current (c) primary p.f. (d) gross power output (e) gross torque (f) input (g) gross efficiency by using approximate equivalent circuit. Assume a slip of 2%.*

Solution. The equivalent circuit is shown in Fig. 32.57 $R_L' = R_2'\,[\,(1/s) - 1]$

$$= 0.88\left(\frac{1}{0.02} - 1\right) = 3.92\ \Omega\ /\ \text{phase}$$

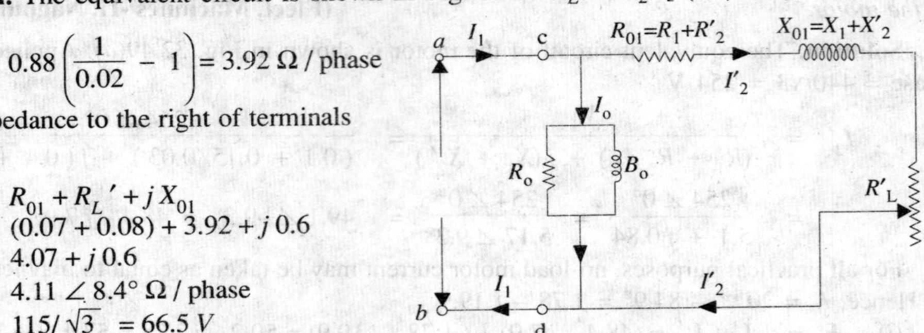

Fig. 32.57

The impedance to the right of terminals *c* and *d* is

$$Z_{cd} = R_{01} + R_L' + jX_{01}$$
$$= (0.07 + 0.08) + 3.92 + j\,0.6$$
$$= 4.07 + j\,0.6$$
$$= 4.11\ \angle 8.4°\ \Omega\ /\ \text{phase}$$
$$V = 115/\sqrt{3} = 66.5\ V$$

(*a*) Secondary current $\mathbf{I}_2' = \mathbf{I}_2$

$$= \frac{66.5}{4.11\ \angle 8.4°} = 16.17\ \angle -j\,8.4°$$
$$= 16 - j\,2.36\ \text{A}$$

The exciting current $\mathbf{I}_0 = V\,(G_0 - jB_0) = 66.5\,(0.022 - j\,0.158) = 1.46 - j\,10.5$ A

(*b*) $\mathbf{I}_1 = \mathbf{I}_0 + \mathbf{I}_2' = \mathbf{I}_0 + \mathbf{I}_2 = (1.46 - j\,10.5) + (16 - j\,2.36)$

$$= 17.46 - j\,12.86 = 21.7\ \angle -36.5°$$

(*c*) Primary p.f. = cos 36.5° = **0.804**

(*d*) $P_g = 3\,I_2^2 R_L' = 3 \times 16.17^2 \times 3.92 =$ **3,075 W**

(*e*) Synchronous speed N_s = 120 × 60/6 = 1,200 r.p.m.

Actual rotor speed $N = (1 - s)\,N_s = (1 - 0.02) \times 1200 = 1{,}176$ r.p.m.

$$\therefore\quad T_g = 9.55.\frac{P_m}{N} = 9.55 \times \frac{3075}{1176} = \mathbf{24.97\ N\text{-}m}$$

(*f*) Primary power input $= \sqrt{3}\,V_1 I_1 \cos\phi = \sqrt{3} \times 115 \times 21.7 \times 0.804 =$ **3,450 W**

(*g*) Gross efficiency = 3,075 × 100/3,450 = **89.5%**

Alternative Solution

Instead of using the equivalent circuit of Fig. 32.57, we could use that shown in Fig. 32.49 which is reproduced in Fig. 32.58.

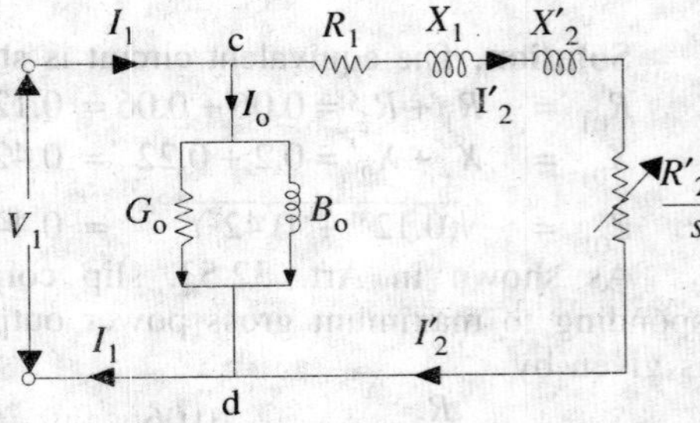

Fig. 32.58

(*a*) $$I_2' = \frac{V_1}{(R_1 + R_2'/s) + j\,(X_1 + X_2')}$$

$$= \frac{66.5\ \angle 0°}{(0.07 + 0.08/0.02) + j\,(0.3 + 0.3)}$$

$$= \frac{66.5}{4.07 + j\,0.6}$$

$$= 16 - j\,2.36 = 16.17\ \angle -8.4°$$

(*b*) $\mathbf{I}_1 = \mathbf{I}_o + \mathbf{I}_2 = 21.7\ \angle -36.5°$ ...as before

(*c*) primary p.f. = **0.804** ...as before

(*d*) gross power developed, $P_g = 3\,I_2'^2\,R_2'\left(\frac{1-s}{s}\right)$

$$= 3 \times 16.17^2 \times 0.08\left(\frac{1 - 0.02}{0.02}\right) = \mathbf{3075\ W}$$

The rest of the solution is the same as above.

Example 32.60. *The equivalent circuit of a 400-V, 3-phase induction motor with a star-connected winding has the following impedances per phase referred to the stator at standstill:*

Stator : (0.4 + j 1) ohm; Rotor : (0.6 + j 1) ohm; Magnetising branch : (10 + j 50) ohm.

Find (i) maximum torque developed (ii) slip at maximum torque and (iii) p.f. at a slip of 5%. Use approximate equivalent circuit. **(Elect. Machinery-III, Bangalore Univ. 1987)**

Solution. (ii) Gap power transferred and hence the mechanical torque developed by rotor would be maximum when there is maximum transfer of power to the resistor R_2'/s shown in the

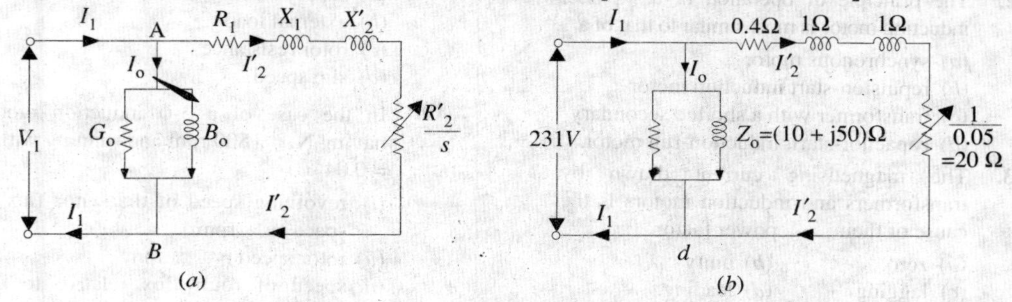

Fig. 32.59

approximate equivalent circuit of the motor in Fig. 32.59. It will happen when R_2'/s equals the impedance looking back into the supply source. Hence,

$$\frac{R_2'}{s_m} = \sqrt{R_1^2 + (X_1 + X_2')^2}$$

or

$$s_m = \frac{R_2'}{\sqrt{R_1^2 + (X_1 + X_2')^2}} = \frac{0.6}{\sqrt{0.4^2 + 2^2}} = 0.29 \text{ or } \mathbf{29\%}$$

(*i*) Maximum value of gross torque developed by rotor

$$T_{g\,max} = \frac{P_{g.max}}{2\pi N_S/60} = \frac{3\, I_2'^2 R_2'/s_m}{2\pi N_S/60} \text{ N-m}$$

Now,

$$I_2' = \frac{V_1}{\sqrt{(R_1+R_2')^2 + (X_1 + X_2')^2}} = \frac{400/\sqrt{3}}{\sqrt{(0.4+0.6)^2 + (1+1)^2}} = 103.3 \text{ A}$$

$$\therefore \quad T_{g\text{-}max} = \frac{3 \times 103.3^2 \times 1/0.29}{2\pi \times 1500/60} = \mathbf{351 \text{ N-m}} \quad \text{... assuming } N_S = 1500 \text{ r.p.m.}$$

(*iii*) The equivalent circuit for one phase for a slip of 0.05 is shown in Fig. 32.59 (*b*).

$$\mathbf{I}_2' = 231/[(20 + 0.4) + j\,2] = 11.2 - j\,1.1$$

$$\mathbf{I}_0 = 231/(10 + j\,50) = 0.89 - j\,4.4$$

$$\mathbf{I}_1 = \mathbf{I}_0 + \mathbf{I}_2' = 12.09 - j\,5.5 = 13.28\angle -24.4° \text{ ; p.f.} = \cos 24.4° = \mathbf{0.91 \text{ (lag)}}$$

Tutorial Problem No. 32.4.

1. A 3-phase, 115-volt induction motor has the following constants : $R_2 = 0.07\ \Omega$; $R_2' = 0.08\ \Omega$, $X_1 = 0.4\ \Omega$ and $X_2' = 0.2\ \Omega$. All the values are for one phase only. At which slip the gross power output will be maximum and the value of the gross power output? **[11.4% ; 8.6 kW]**
2. A 3-phase, 400-V, Y-connected induction motor has an equivalent *T*-circuit consisting of $R_1 = 1\ \Omega$, $X_1 = 2\ \Omega$, equivalent rotor values are $R_2' = 1.2\ \Omega$, $X_2' = 1.5\ \Omega$. The exciting branch has an impedance of $(4 + j\,40)\ \Omega$. If slip is 5% find (*i*) current (*ii*) efficiency (*iii*) power factor (*iv*) output. Assume friction loss to be 250 W. **[(*i*) 10.8 A (*ii*) 81% (*iii*) 0.82 (*iv*) 5 kW]**

OBJECTIVE TESTS - 32

1. Regarding skewing of motor bars in a squirrel-cage induction motor, (SCIM) which statement is **false**?
   (*a*) it prevents cogging
   (*b*) it increases starting torque
   (*c*) it produces more uniform torque
   (*d*) it reduces motor 'hum' during its operation.
2. The principle of operation of a 3-phase. induction motor is most similar to that of a
   (*a*) synchronous motor
   (*b*) repulsion-start induction motor
   (*c*) transformer with a shorted secondary
   (*d*) capacitor-start, induction-run motor.
3. The magnetising current drawn by transformers and induction motors is the cause of their ....power factor.
   (*a*) zero (*b*) unity
   (*c*) lagging (*d*) leading.
4. The effect of increasing the length of air-gap in an induction motor will be to increase the
   (*a*) power factor
   (*b*) speed
   (*c*) magnetising current
   (*d*) air-gap flux.
   (Power App-II, Delhi Univ. Jan. 1987)
5. In a 3-phase induction motor, the relative speed of stator flux with respect to ....is zero.
   (*a*) stator winding (*b*) rotor
   (*c*) rotor flux (*d*) space.
6. An eight-pole wound rotor induction motor operating on 60 Hz supply is driven at 1800 r.p.m. by a prime mover in the opposite direction of revolving magnetic field. The frequency of rotor current is
   (*a*) 60 Hz (*b*) 120 Hz
   (*c*) 180 Hz (*d*) none of the above.
   (Elect. Machines, A.M.I.E. Sec. B, 1993)
7. A 3-phase, 4-pole, 50-Hz induction motor runs at a speed of 1440 r.p.m. The rotating field produced by the rotor rotates at a speed of ....r.p.m. with respect to the rotor.
   (*a*) 1500 (*b*) 1440
   (*c*) 60 (*d*) 0.
8. In a 3-ϕ induction motor, the rotor field rotates at synchronous speed with respect to
   (*a*) stator
   (*b*) rotor
   (*c*) stator flux
   (*d*) none of the above.
9. Irrespective of the supply frequency, the torque developed by a SCIM is the same whenever .... is the same.
   (*a*) supply voltage
   (*b*) external load
   (*c*) rotor resistance
   (*d*) slip speed.
10. In the case of a 3-ϕ induction motor having N_S = 1500 rpm and running with s = 0.04
    (*i*) revolving speed of the stator flux is space is .....rpm
    (*ii*) rotor speed is ....rpm
    (*iii*) speed of rotor flux relative to the rotor is ....rpm
    (*iv*) speed of the rotor flux with respect to the stator is ....rpm.
11. The number of stator poles produced in the rotating magnetic field of a 3-ϕ induction motor having 3 slots per pole per phase is
    (*a*) 3 (*b*) 6
    (*c*) 2 (*d*) 12.
12. The power factor of a squirrel-cage induction motor is
    (*a*) low at light loads only
    (*b*) low at heavy loads only
    (*c*) low at light and heavy loads both
    (*d*) low at rated load only.
    (Elect. Machines, A.M.I.E. Sec.B, 1993)
13. Which of the following rotor quantity in *a* SCIM does NOT depend on its slip?
    (*a*) reactance
    (*b*) speed
    (*c*) induced emf
    (*d*) frequency.
14. A 6-pole, 50-Hz, 3-ϕ induction motor is running at 950 rpm and has rotor Cu loss of 5 kW. Its rotor input is ......kW.
    (*a*) 100 (*b*) 10
    (*c*) 95 (*d*) 5.3.
15. The efficiency of a 3-phase induction motor is approximately proportional to
    (*a*) $(1 - s)$ (*b*) s
    (*c*) N (*d*) N_S.

16. A 6-pole, 50-Hz, 3-ϕ induction motor has a full-load speed of 950 rpm. At half-load, its speed would be ......rpm.

(*a*) 475 (*b*) 500
(*c*) 975 (*d*) 1000.

17. If rotor input of a SCIM running with a slip of 10% is 100 kW, gross power developed by its rotor is ...... kW.

(*a*) 10 (*b*) 90
(*c*) 99 (*d*) 80.

18. Pull-out torque of a SCIM occurs at that value of slip where rotor power factor equals

(*a*) unity (*b*) 0.707
(*c*) 0.866 (*d*) 0.5.

19. Fill in the blanks.

When load is placed on a 3-phase induction motor, its

(*i*) speed ....
(*ii*) slip ......
(*iii*) rotor induced emf ......
(*iv*) rotor current ......
(*v*) rotor torque ......
(*vi*) rotor continues to rotate at that value of slip at which developed torque equals ...... torque.

20. When applied rated voltage per phase is reduced by one-half, the starting torque of a SCIM becomes ...... of the starting torque with full voltage.

(*a*) 1/2 (*b*) 1/4
(*c*) $1/\sqrt{2}$ (*d*) $\sqrt{3}/2$.

21. If maximum torque of an induction motor is 200 kg-m at a slip of 12%, the torque at 6% slip would be ...... kg-m.

(*a*) 100 (*b*) 160
(*c*) 50 (*d*) 40.

22. The fractional slip of an induction motor is the ratio

(*a*) rotor Cu loss/rotor input
(*b*) stator Cu loss/stator input
(*c*) rotor Cu loss/rotor output
(*d*) rotor Cu loss/stator Cu loss.

23. The torque developed by a 3-phase induction motor depends on the following three factors:

(*a*) speed, frequency, number of poles
(*b*) voltage, current and stator impedance
(*c*) synchronous speed, rotor speed and frequency
(*d*) rotor emf, rotor current and rotor p.f.

24. If the stator voltage and frequency of an induction motor are reduced proportionately, its

(*a*) locked rotor current is reduced
(*b*) torue developed is increased
(*c*) magnetising current is decreased
(*d*) both (*a*) and (*b*) .

25. The efficiency and p.f. of a SCIM increases in proportion to its

(*a*) speed
(*b*) mechanical load
(*c*) voltage
(*d*) rotor torque.

26. A SCIM runs at *constant* speed only so long as

(*a*) torque developed by it remains constant
(*b*) its supply voltage remains constant
(*c*) its torque exactly equals the mechanical load
(*d*) stator flux remains constant.

27. The synchronous speed of a linear induction motor does NOT depend on

(*a*) width of pole pitch
(*b*) number of poles
(*c*) supply frequency
(*d*) any of the above.

28. Thrust developed by a linear induction motor depends on

(*a*) synchronous speed
(*b*) rotor input
(*c*) number of poles
(*d*) both (*a*) and (*b*).

ANSWERS

1. *b* **2.** *c* **3.** *c* **4.** *c* **5.** *c* **6.** *c* **7.** *c* **8.** *a* **9.** *d* **10.** (*i*) 1500 (*ii*) 1440 (*iii*) 60 (*iv*) 1500 **11.** *b* **12.** *a* **13.** *b* **14.** *a* **15.** *a* **16.** *c* **17.** *b* **18.** *b* **19.** (*i*) decreases (*ii*) increases (*iii*) increases (*iv*) increases (*v*) increases (*vi*) applied **20.** *b* **21.** *b* **22.** *a* **23.** *d* **24.** *d* **25.** *b* **26.** *c* **27.** *b* **28.** *d*

33 COMPUTATIONS AND CIRCLE DIAGRAMS

33.1. General

In this chapter, it will be shown that the performance characteristics of an induction motor are derivable from a circular locus. The data necessary to draw the circle diagram may be found from no-load and blocked-rotor tests, corresponding to the open-circuit and short-circuit tests of a transformer. The stator and rotor Cu losses can be separated by drawing a torque line. The parameters of the motor, in the equivalent circuit, can be found from the above tests, as shown below.

33.2. Circle Diagram for a Series Circuit

It will be shown that the end of the current vector for a series circuit with constant reactance and voltage, but with a variable resistance is a circle. With reference to Fig. 33.1, it is clear that

$$I = \frac{V}{Z} = \frac{V}{\sqrt{(R^2 + X^2)}}$$

$$= \frac{V}{X} \times \frac{X}{\sqrt{(R^2 + X^2)}} = \frac{V}{X} \sin \phi$$

$$\because \sin \phi = \frac{X}{\sqrt{(R^2 + X^2)}} \qquad \text{—Fig. 33.2}$$

$$\therefore \quad I = (V/X) \sin \phi$$

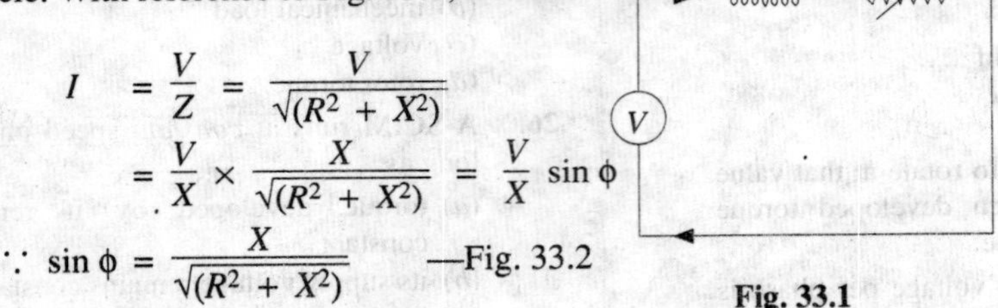

Fig. 33.1 Fig. 33.2

It is the equation of a circle in polar co-ordinates, with diameter equal to V/X. Such a circle is drawn in Fig. 33.3, using the magnitude of the current and power factor angle ϕ as polar co-ordinates of the point A. In other words, as resistance R is varied (which means, in fact, ϕ is changed), the end of the current vector lies on a circle with diameter equal to V/X. For a lagging current, it is usual to orientate the circle of Fig. 33.3 (*a*) such that its diameter is horizontal and the voltage vector takes a vertical position, as shown in Fig. 33.3 (*b*). There is no difference between the two so far as the magnitude and phase relationships are concerned.

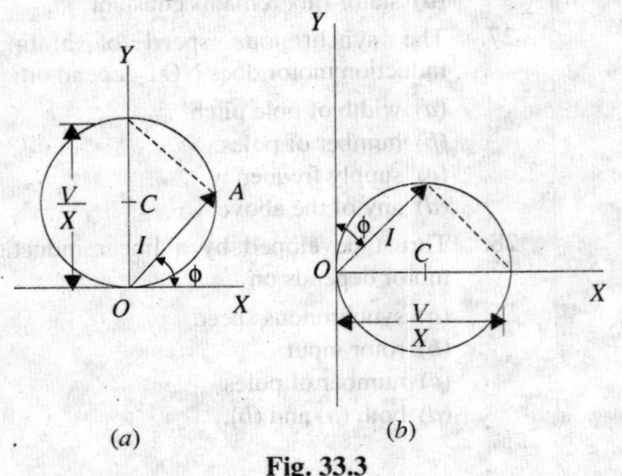

Fig. 33.3

33.3. Circle Diagram for the Approximate Equivalent Circuit

The approximate equivalent diagram is redrawn in Fig. 33.4. It is clear that the circuit to the right of points *ab* is similar to a series circuit, having a constant voltage V_1 and reactance X_{01} but variable resistance (corresponding to different values of slip *s*).

Hence, the end of current vector for I_2' will lie on a circle with a diameter of V/X_{01}. In Fig. 33.5, I_2' is the rotor current referred to stator, I_0 is no-load current (or exciting current) and I_1 is the total stator current and is the vector sum of the first two. When I_2' is lagging and ϕ_2 =

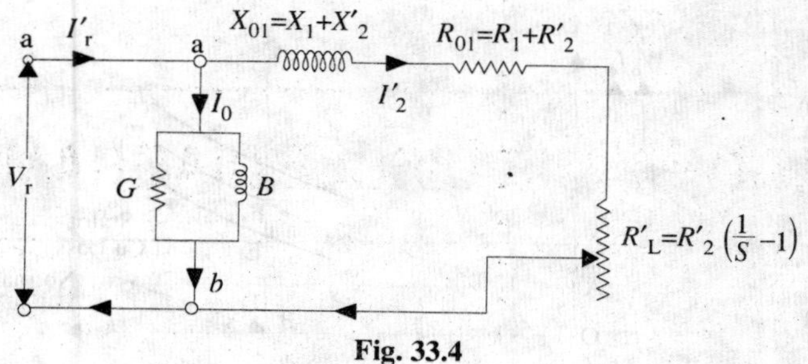

Fig. 33.4

90°, then the position of vector for I_2' will be along OC *i.e.* at right angles to the voltage vector OE. For any other value of ϕ_2, point A will move along the circle shown dotted. The exciting current I_0 is drawn lagging V by an angle ϕ_0. If conductance G_0 and susceptance B_0 of the exciting circuit are assumed constant, then I_0 and ϕ_0 are also constant. The end of current vector for I_1 is also seen to lie on another circle which is displaced from the dotted circle by an amount I_0. Its diameter is still V/X_{01} and is parallel to the horizontal axis OC. Hence, we find that if an induction motor is tested at various loads, the locus of the end of the vector for the current (drawn by it) is a circle.

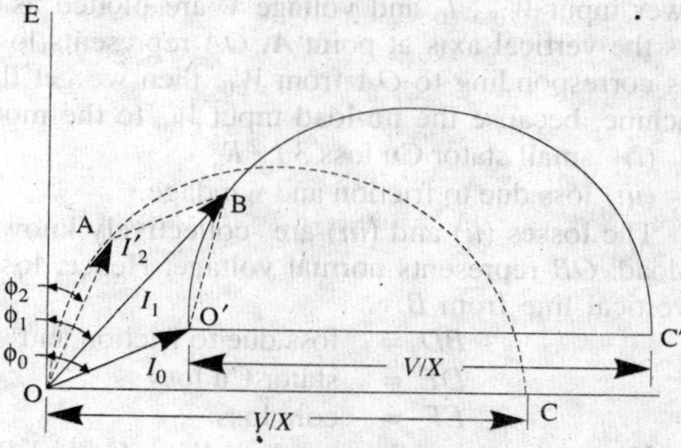

Fig. 33.5

33.4. Determination of G_0 and B_0

It the total leakage reactance X_{01} of the motor, exciting conductance G_0 and exciting susceptance B_0 are found, then the position of the circle $O'BC'$ is determined uniquely. One method of finding G_0 and B_0 consists in running the motor synchronously so that slip $s = 0$. In practice, it is impossible for an induction motor to run at synchronous speed, due to the inevitable presence of friction and windage losses. However, the induction motor may be run at synchronous speed by another machine which supplies the friction and windage losses. In that case, the circuit to the right of points ab behaves like an open circuit, because with $s = 0$, $R_L = \infty$ (Fig. 33.6). Hence, the current drawn by the motor is I_0 only. Let

V = applied voltage/phase; I_0 = motor current / phase

W = wattmeter reading *i.e.* input in watt ; Y_0 = exciting admittance of the motor. Then, for a 3-phase induction motor

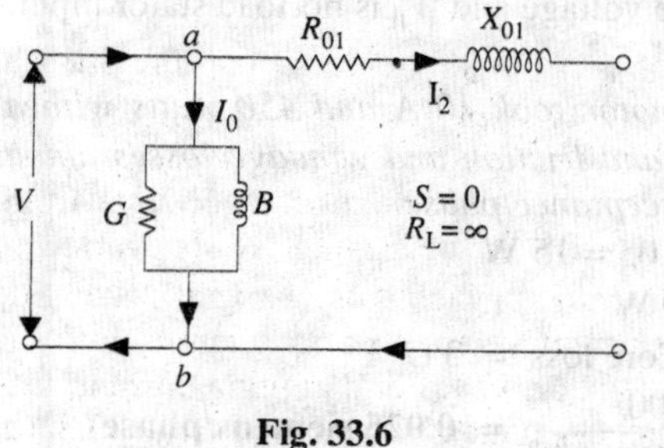

Fig. 33.6

$$W = 3G_0V^2 \quad \text{or} \quad G_0 = \frac{W}{3V^2} \qquad \text{Also,} \quad I_0 = VY_0 \quad \text{or} \quad Y_0 = I_0/V$$

$$B_0 = \sqrt{(Y_0^2 - G_0^2)} = \sqrt{[(I_0/V)^2 - G_0^2]}$$

Hence, G_0 and B_0 can be found.

33.5. No-load Test

In practice, it is neither necessary nor feasible to run the induction motor synchronously for getting G_0 and B_0. Instead, the motor is run without any external mechanical load on it. The speed of the rotor would not be synchronous, but very much near to it ; so that, for all practical purposes, the speed may be assumed synchronous. The no load test is carried out with different values of applied voltage, below and above the value of normal voltage. The power input is measured by two wattmeters, I_0 by an ammeter and V by a voltmeter, which are included in the circuit of Fig. 33.7. As motor is running on light load, the p.f. would be low *i.e.* less than 0.5, hence total power input will be the difference of the two wattmeter readings W_1 and W_2. The readings of the total

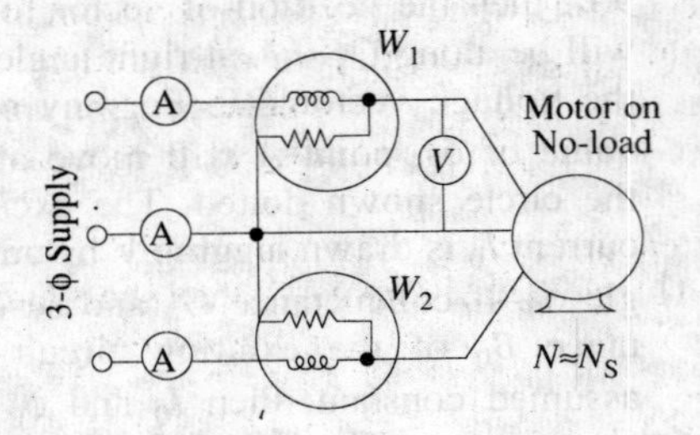

Fig. 33.7

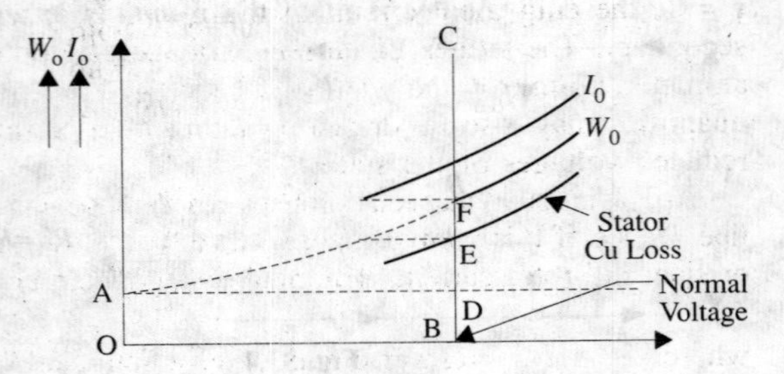

Fig. 33.8

power input W_0 , I_0 and voltage V are plotted as in Fig. 33.8. If we extend the curve for W_0, it cuts the vertical axis at point **A.** OA represents losses due to friction and windage. If we subtract loss corresponding to OA from W_0, then we get the no-load electrical and magnetic losses in the machine, because the no-load input W_0 to the motor consists of

(*i*) small stator Cu loss $3\,I_0^2\,R_1$ (*ii*) stator core loss $W_{CL} = 3G_0\,V_{1d}^{\,2}$

(*iii*) loss due to friction and windage.

The losses (*ii*) and (*iii*) are collectively known as fixed losses, because they are independent of load. OB represents normal voltage. Hence, losses at normal voltage can be found by drawing a vertical line from B.

BD = loss due to friction and windage

DE = stator Cu loss

EF = core loss

Hence, knowing the core loss W_{CL}, G_0 and B_0 can be found, as discussed in Art. 33.4.

Additionaly, ϕ_0 can also be found from the relation $W_0 = \sqrt{3}V_L I_0 \cos\phi_0$

$$\therefore \quad \cos\phi_0 = \frac{W_0}{\sqrt{3}\,V_L\,I_0}$$ where V_L = line voltage and W_0 is no-load stator input.

Example 33.1. *In a no-load test, an induction motor took 10 A and 450 watts with a line voltage of 110 V. If stator resistance/phase is 0.05 Ω and friction and windage losses amount to 135 watts, calculate the exciting conductance and susceptance/phase.*

Solution. stator Cu loss $= 3\,I_0^2\,R_1 = 3 \times 10^2 \times 0.05 = 15$ W

$\therefore$ stator core loss $= 450 - 135 - 15 = 300$ W

Voltage/phase $V = 110/\sqrt{3}$ V ; Core loss $= 3\,G_0 V^2$

$$300 = 3G_0 \times (110/\sqrt{3}\,)^2 \quad ; \quad G_0 = \frac{300}{3 \times (110/\sqrt{3})^2} = \mathbf{0.025\ siemens/phase}$$

$$Y_0 = I_0/V = (10 \times \sqrt{3})/110 = 0.158 \text{ siemens/phase}$$

$$B_0 = \sqrt{(Y_0^2 - G_0^2)} = \sqrt{(0.158^2 - 0.025^2)} = \mathbf{0.156\ siemens/phase.}$$

33.6. Blocked Rotor Test

It is also known as locked-rotor or short-circuit test. This test is used to find—

1. short-circuit with *normal* voltage applied to stator
2. power factor on short-circuit

Both the values are used in the construction of circle diagram

3. total leakage reactance X_{01} of the motor as referred to primary (*i.e.* stator)
4. total resistance of the motor R_{01} as referred to primary.

In this test, the rotor is locked (or allowed very slow rotation) and the rotor windings are short-circuited at slip-rings, if the motor has a wound rotor. Just as in the case of a short-circuit test on a transformer, a reduced voltage (up to 15 or 20 per cent of normal value) is applied to the stator terminals and is so adjusted that full-load current flows in the stator. As in this case

$s = 1$, the equivalent circuit of the motor is exactly like a transformer, having a short-circuited secondary. The values of current, voltage and power input on short-circuit are measured by the ammeter, voltmeter and wattmeter connected in the circuits as before. Curves connecting the above quantities may also be drawn by taking two or three additional sets of readings at progressively reduced voltages of the stator.

(*a*) It is found that relation between the short-circuit current and voltage is approximately a straight line. Hence, if V is normal stator voltage, V_s the short-circuit voltage (a fraction of V), then short-circuit or standstill rotor current, if normal voltage were applied to stator, is found from the relation

$$I_{SN} = I_s \times V/V_s$$

where I_{SN} = short-circuit current obtainable with normal voltage

I_s = short-circuit current with voltage V_S

(*b*) Power factor on short-circuit is found from

$$W_S = \sqrt{3}\, I_{SL} \cos \phi_S \,; \qquad \therefore \quad \cos \phi_S = W_S / (\sqrt{3}\, V_{SL}\, I_{SL})$$

where W_S = total power input on short-circuit

V_{SL} = line voltage on short-circuit

I_{SL} = line current on short-circuit.

(*c*) Now, the motor input on short-circuit consists of

(*i*) mainly stator and rotor Cu losses

(*ii*) core-loss, which is small due to the fact that applied voltage is only a small percentage of the normal voltage. This core-loss (if found appreciable) can be calculated from the curves of Fig. 33.8.

$$\therefore \quad \text{Total Cu loss} = W_S - W_{CL}$$

$$3 I_s^2 R_{01} = W_s - W_{CL} : \quad R_{01} = (W_s - W_{CL}) / 3 I_s^2$$

(*d*) With reference to the approximate equivalent circuit of an induction motor (Fig. 33.4), motor leakage reactance per phase X_{01} as referred to the stator may be calculated as follows:

$$Z_{01} = V_S / I_S \qquad X_{01} = \sqrt{(Z_{01}^2 - R_{01}^2)}$$

Usually, X_1 is assumed equal to X_2' where X_1 and X_2 are stator and rotor reactances per phase respectively as referred to stator. $X_1 = X_2' = X_{01} / 2$

If the motor has a wound rotor, then stator and rotor resistances are separated by dividing R_{01} in the ratio of the d.c. resistances of stator and rotor windings.

In the case of squirrel-cage rotor, R_1 is determined as usual and after allowing for 'skin effect' is subtracted from R_{01} to give R_2'—the effective rotor resistance as referred to stator.

$$\therefore \quad R_2' = R_{01} - R_1$$

Example 33.2. *A 110-V, 3-ϕ, star-connected induction motor takes 25 A at a line voltage of 30 V with rotor locked. With this line voltage, power input to motor is 440 W and core loss is 40 W. The d.c. resistance between a pair of stator terminals is 0.1 Ω. If the ratio of a.c. to d.c. resistance is 1.6, find the equivalent leakage reactance/phase of the motor and the stator and rotor resistance per phase.* **(Electrical Technology, Madras Univ. 1987)**

Solution. S.C. voltage/phase, $V_s = 30 / \sqrt{3} = 17.3 \text{ V} : I_s = 25$ A per phase

$Z_{01} = 17.3/25 = 0.7\ \Omega$ (approx.) per phase

Stator and rotor Cu losses = input – core loss = 440 – 40 = 400 W

$\therefore \quad 3 \times 25^2 \times R_{01} = 400 \qquad \therefore R_{01} = 400/3 \times 625 = \mathbf{0.21\ \Omega}$

where R_{01} is equivalent resistance/phase of motor as referred to stator.

Leakage reactance/phase $X_{01} = \sqrt{(0.7^2 - 0.21^2)} = 0.668\ \Omega$

d.c. resistance/phase of stator = 0.1/2 = 0.05 Ω

a.c. resistance/phase $R_1 = 0.05 \times 1.6 = 0.08\ \Omega$

Hence, effective resistance/phase of rotor as referred to stator

$R_2' = 0.21 - 0.08 = \mathbf{0.13\ \Omega}$

33.7. Construction of the Circle Diagram

Circle diagram of an induction motor can be drawn by using the data obtained from (1) **no-load** (2) **short-circuit test** and (3) **stator resistance test**, as shown below.

Step No. 1

From no-load test, I_0 and ϕ_0 can be calculated. Hence, as shown in Fig. 33.9, vector for I_0 can be laid off lagging ϕ_0 behind the applied voltage *V*.

Step No. 2

Next, from blocked rotor test or short-circuit test, short-circuit current I_{SN} *corresponding to normal voltage* and ϕ_S are found. The vector ***OA*** represents $I_{SN} = (I_S V/V_S)$ in magnitude and phase. Vector *O′A* represents rotor current I_2' as referred to stator.

Clearly, the two points *O′* and *A* lie on the required circle. For finding the centre *C* of this circle, chord *O′A* is bisected at right angles—its bisector giving point *C*. The diameter *O′D* is drawn perpendicular to the voltage vector.

As a matter of practical contingency, it is recommended that the scale of current vectors should be so chosen that the diameter is more than 25 cm, in order that the performance data of the motor may be read with reasonable accuracy from the circle diagram. With centre *C* and radius = *CO′*, the circle can be drawn. The line *O′A* is known as **out-put line.**

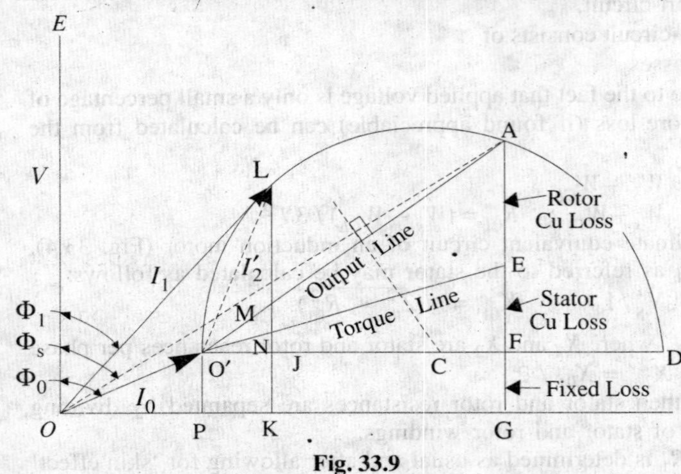

Fig. 33.9

It should be noted that as the voltage vector is drawn vertically, all vertical distances represent the active or power or energy components of the currents. For example, the vertical component *O′P* of no-load current *OO′* represents the no-load input, which supplies core loss, friction and windage loss and a negligibly small amount of stator I^2R loss. Similarly, the vertical component *AG* of short-circuit current *OA* is proportional to the motor input on short-circuit or if measured to a proper scale, may be said to equal power input.

Step No. 3

Torque line. *This is the line which separates the stator and the rotor copper losses.* When the rotor is locked, then all the power supplied to the motor goes to meet core losses and Cu losses in the stator and rotor windings. The power input is proportional to *AG*. Out of this, *FG* (= *O′P*) represents fixed losses *i.e.* stator core loss and friction and windage losses. *AF* is proportional to the sum of the stator and rotor Cu losses. The point *E* is such that

$$\frac{AE}{EF} = \frac{\text{rotor Cu loss}}{\text{stator Cu loss}}$$

As said earlier, line *O′E* is known as *torque line.*

How to locate point E ?

(*i*) *Squirrel-cage Rotor.* Stator resistance/phase *i.e.* R_1 is found from stator-resistance test. Now, the short-circuit motor input W_s is approximately equal to motor Cu losses (neglecting iron losses).

$$\text{Stator Cu loss} = 3I_S{}^2R_1 \qquad \therefore \quad \text{rotor Cu loss} = W_S - 3I_S{}^2R_1 \qquad \therefore \quad \frac{AE}{EF} = \frac{W_S - 3I_S^2R_1}{3I_S^2R_1}$$

(*ii*) *Wound Rotor.* In this case, rotor and stator resistances per phase r_2 and r_1 can be easily computed. For any values of stator and rotor currents I_1 and I_2 respectively, we can write

$$\frac{AE}{EF} = \frac{I_2^2 r_2}{I_1^2 r_1} = \frac{r_2}{r_1}\left(\frac{I_2}{I_1}\right)^2; \quad \text{Now,} \quad \frac{I_1}{I_2} = K = \text{transformation ratio}$$

$$\frac{AE}{EF} = \frac{r_2}{r_1} \times \frac{1}{K^2} = \frac{r_2 / K^2}{r_1} = \frac{r_2'}{r_1} = \frac{\text{equivalent rotor resistance per phase}}{\text{stator resistance per phase}}$$

Value of K may be found from short-circuit test itself by using two ammeters, both in stator and rotor circuits.

Let us assume that the motor is running and taking a current OL (Fig. 33.9). Then, the perpendicular JK represents fixed losses, JN is stator Cu loss, NL is the rotor input, NM is rotor Cu loss, ML is rotor output and LK is the total motor input.

From our knowledge of the relations between the above-given various quantities, we can write:

$\sqrt{3} \cdot V_L \cdot LK$ = motor input $\sqrt{3} \cdot V_L \cdot JK$ = fixed losses

$\sqrt{3} \cdot V_L \cdot JN$ = stator copper loss $\sqrt{3} \cdot V_L \cdot MN$ = rotor copper loss

$\sqrt{3} \cdot V_L \cdot MK$ = total loss $\sqrt{3} \cdot V_L \cdot ML$ = mechanical output

$\sqrt{3} \cdot V_L \cdot NL$ = rotor input $\propto$ torque

1. ML / LK = output/input = efficiency
2. MN / NL = (rotor Cu loss)/(rotor input) = slip, s.
3. $\dfrac{ML}{NL} = \dfrac{\text{rotor output}}{\text{rotor input}} = 1 - s = \dfrac{N}{N_S} = \dfrac{\text{actual speed}}{\text{synchronous speed}}$
4. $\dfrac{LK}{OL}$ = power factor

Hence, it is seen that, at least, theoretically, it is possible to obtain all the characteristics of an induction motor from its circle diagram. As said earlier, for drawing the circle diagram, we need (*a*) stator-resistance test for separating stator and rotor Cu losses and (*b*) the data obtained from (*i*) no-load test and (*ii*) short-circuit test.

33.8. Maximum Quantities

It will now be shown from the circle diagram (Fig. 33.10) that the maximum values occur at the positions stated below:

(*i*) **Maximum Output**

It occurs at point M where the tangent is parallel to output line $O'A$. Point M may be located by drawing a line CM from point C such that it is perpendicular to the output line $O'A$. Maximum output is represented by the vertical MP.

(*ii*) **Maximum Torque or Rotor Input**

It occurs at point N where the tangent is parallel to torque line $O'E$. Again, point N may be found by drawing CN perpendicular to the torque line. Its value is represented by NQ. Maximum torque is also known as stall-ing or pull-out torque.

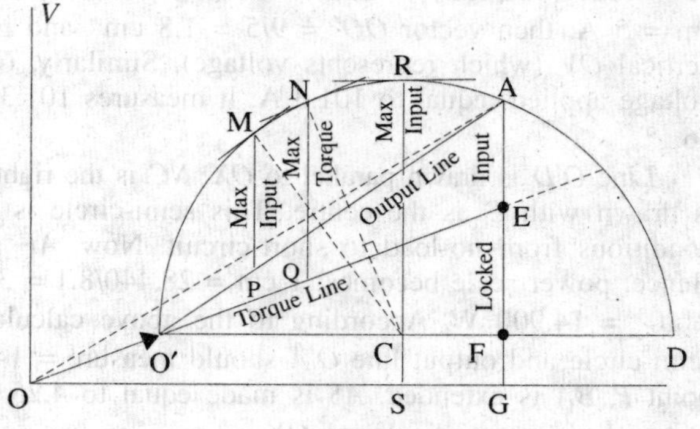

Fig. 33.10

(*iii*) **Maximum Input Power**

It occurs at the highest point of the circle *i.e.* at point R where the tangent to the circle is horizontal. It is proportional to RS. As the point R is beyond the point of maximum torque, the induction motor will be unstable here. However, the maximum input is a measure of the size of the circle and is an indication of the ability of the motor to carry short-time over-loads. Generally, RS is twice or thrice the motor input at rated load.

Example 33.3. *A 3-ph, 400-V induction motor gave the following test readings;*
No-load : 400 V, 1250 W, 9 A, Short-circuit : 150 V, 4 kW, 38 A
Draw the circle diagram.
If the normal rating is 14.9 kW, find from the circle diagram, the full-load value of current, p.f. and slip. **(Electrical Machines-I, Gujarat Univ. 1985)**

Solution. $\cos\phi_0 = \dfrac{1250}{\sqrt{3} \times 400 \times 9} = 0.2004\,; \quad \phi_0 = 78.5°$

$\cos\phi_S = \dfrac{4000}{\sqrt{3} \times 150 \times 38} = 0.405\,; \quad \phi_S = 66.1°$

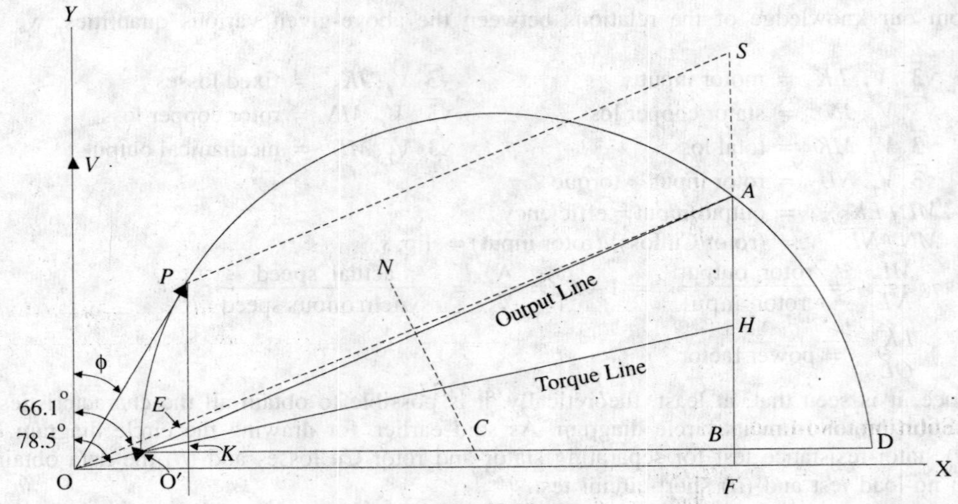

Fig. 33.11

Short-circuit current with normal voltage is I_{SN} = 38(400/150) = 101.3 A. Power taken would be = 4000 (400/150)^2 = 28,440 W. In Fig. 33.11, *OO′* represents I_0 of 9 A. If current scale is 1 cm = 5 A, then vector *OO′* = 9/5 = 1.8 cm* and is drawn at an angle of ϕ_0 = 78.5° with the vertical *OV* (which represents voltage). Similarly, *OA* represents I_{SN} (S.C. current with normal voltage applied) equal to 101.3 A. It measures 101.3/5 = 20.26* cm and is drawn at an angle of 66.1°.

Line *O′D* is drawn parallel to *OX*. *NC* is the right-angle bisector of *O′A*. The semi-circle *O′AD* is drawn with *C* as the centre. This semi-circle is the locus of the current vector for all load conditions from no-load to short-circuit. Now, *AF* represents 28,440 W and measures 8.1 cm. Hence, power scale becomes: 1 cm = 28,440/8.1 = 3,510 W. Now, full-load motor output = 14.9 × 10^3 = 14,900 W. According to the above calculated power scale, the intercept between the semi-circle and output line *O′A* should measure = 14,900/3510 = 4.25 cm. For locating full-load point *P, BA* is extended. *AS* is made equal to 4.25 cm and *SP* is drawn parallel to output line *O′*A. *PL* is perpendicular to *OX*.

Line current = *OP* = 6 cm = 6 × 5 = **30 A;** φ = 30° (by measurement)

p.f. = cos 30° = **0.886** (or cos φ = *PL*/*OP* = 5.2/6 = 0.865)

Now, $\quad \text{slip} = \dfrac{\text{rotor Cu loss}}{\text{rotor input}}$

In fig. 33.11, *EK* represents rotor Cu loss and *PK* represents rotor input.

$\therefore \quad \text{slip} = \dfrac{EK}{PK} = \dfrac{0.3}{4.5} = 0.067$ or **6.7%**

*The actual lengths are different from these values, due to reduction in block making.

Example 33.4. *Draw the circle diagram for a 3.73 kW, 200-V, 50-HZ, 4-pole, 3-ϕ star-connected induction motor from the following test data:*

No-load : Line voltage 200 V, line current 5 A; total input 350 W

Blocked rotor: Line voltage 100 V, line current 26 A; total input 1700 W

Estimate from the diagram for full-load condition, the line current, power factor and also the maximum torque in terms of the full-load torque. The rotor Cu loss at standstill is half the total Cu loss. **(Electrical Engineering, Bombay Univ. 1987)**

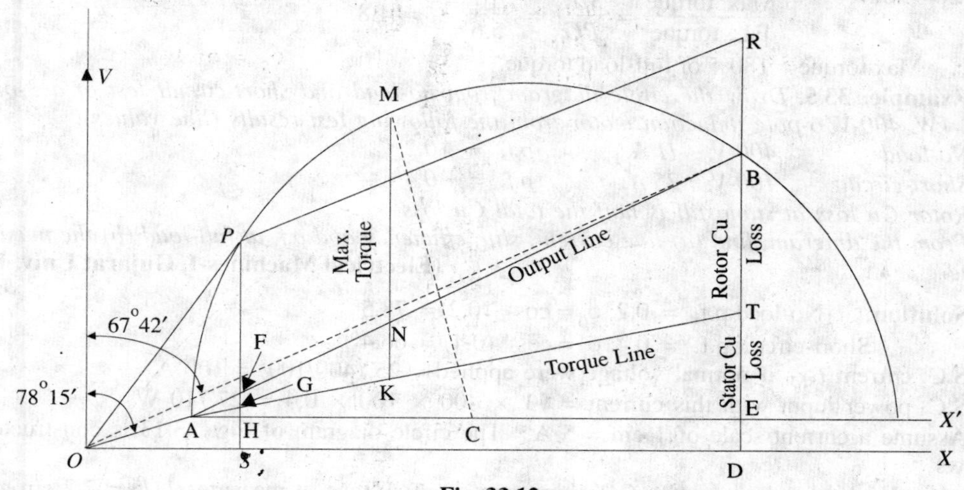

Fig. 33.12

Solution. No-load test

$$I_0 = 5\text{ A}, \cos\phi_0 = \frac{350}{\sqrt{3} \times 200 \times 5} = 0.202\,; \quad \phi_0 = 78°15'$$

Blocked-rotor test :

$$\cos\phi_s = \frac{1700}{\sqrt{3} \times 100 \times 26} = 0.378\,; \quad \phi_s = 67°42'$$

Short-circuit current with normal voltage, I_{SN} = 26 × 200/100 = 52 A

Short-circuit/blocked rotor input with normal voltage = 1700(52/26)^2 = 6,800 W

In the circle diagram of Fig. 33.12, voltage is represented along *OV* which is drawn perpendicular to *OX*. Current scale is 1 cm = 2 A

Line *OA* is drawn at an angle of ϕ_0 = 78°15′ with *OV* and 2.5 cm in length. Line *AX′* is drawn parallel to *OX*. Line *OB* represents short-circuit current with normal voltage *i.e.* 52 A and measures 52/2 = 26 cm. *AB* represents output line. Perpendicular bisector of *AB* is drawn to locate the centre *C* of the circle. With *C* as centre and radius = *CA*, a circle is drawn which passes through points *A* and *B*. From point *B*, a perpendicular is drawn to the base. *BD* represents total input of 6,800 W for blocked rotor test. Out of this, *ED* represents no-load loss of 350 W and *BE* represents 6,800 − 350 = 6,450 W. Now *BD* = 9.8 cm and represents 6,800 W

∴ power scale = 6,800/9.8 = 700 watt/cm or 1 cm = **700 W**

BE which represents total copper loss in rotor and stator, is bisected at point *T* to separate the two losses. *AT* represents torque line.

Now, motor output = 3,730 watt. It will be represented by a line = 3,730/700 = 5.33 cm

The output point *P* on the circle is located thus:-

DB is extended and *BR* is cut = 5.33 cm. Line *RP* is drawn parallel to output line *AB* and cuts the circle at point *P*. Perpendicular *PS* is drawn and *P* is joined to origin *O*.

Point *M* corresponding to maximum torque is obtained thus:

From centre *C*, *a* line *CM* is drawn such that it is perpendicular to torque line *AT*. It cuts the circle at M which is the required point. Point M could also have been located by drawing a line parallel to the torque line. MK is drawn vertical and it represents maximum torque.

Now, in the circle diagram, OP = line current on full-load = 7.6 cm. Hence, OP represents 7.6 × 2 = 15.2 A

$$\text{Power factor on full-load} = \frac{SP}{OP} = \frac{6.45}{7.6} = 0.86$$

$$\frac{\text{Max. torque}}{\text{F.L. torque}} = \frac{MK}{PG} = \frac{10}{5.6} = 1.8$$

∴ Max.torque = 180% of full-load torque.

Example. 33.5. *Draw the circle diagram from no-load and short-circuit test of a 3-phase. 14.92 kW, 400-V, 6-pole induction motor from the following test results (line values).*

No-load : 400-V, 11 A, p.f. = 0.2
Short-circuit : 100-V, 25 A, p.f. = 0.4

Rotor Cu loss at standstill is half the total Cu loss.

From the diagram, find (a) line current, slip, efficiency and p.f. at full-load (b) the maximum torque.

(Electrical Machines-I, Gujarat Univ. 1985)

Solution. No-load p.f. = 0.2; $\phi_0 = \cos^{-1}(0.2) = 78.5°$

Short-circuit p.f. = 0.4: $\phi_s = \cos^{-1}(0.4) = 66.4°$

S.C. current I_{SN} if normal voltage were applied = 25 (400/100) = 100 A

S.C. power input with this current = $\sqrt{3}$ × 400 × 100 × 0.4 = 27,710 W

Assume a current scale of 1 cm = 5 A.* The circle diagram of Fig. 33.13 is constructed as follows:

(*i*) No-load current vector OO' represents 11 A. Hence, it measures 11/5 = 2.2 cm and is drawn at an angle of 78.5° with OY.

(*ii*) Vector OA represents 100 A and measures 100/5 = 20 cm. It is drawn at an angle of 66.4° with OY.

(*iii*) $O'D$ is drawn parallel to OX. NC is the right angle bisector of $O'A$.

(*iv*) With C as the centre and CO' as radius, a semicircle is drawn as shown.

(*v*) AF represents power input on short-circuit with normal voltage applied. It measures 8 cm and (as calculated above) represents 27,710 W. Hence, power scale becomes

1 cm = 27,710/8 = 3,465 W

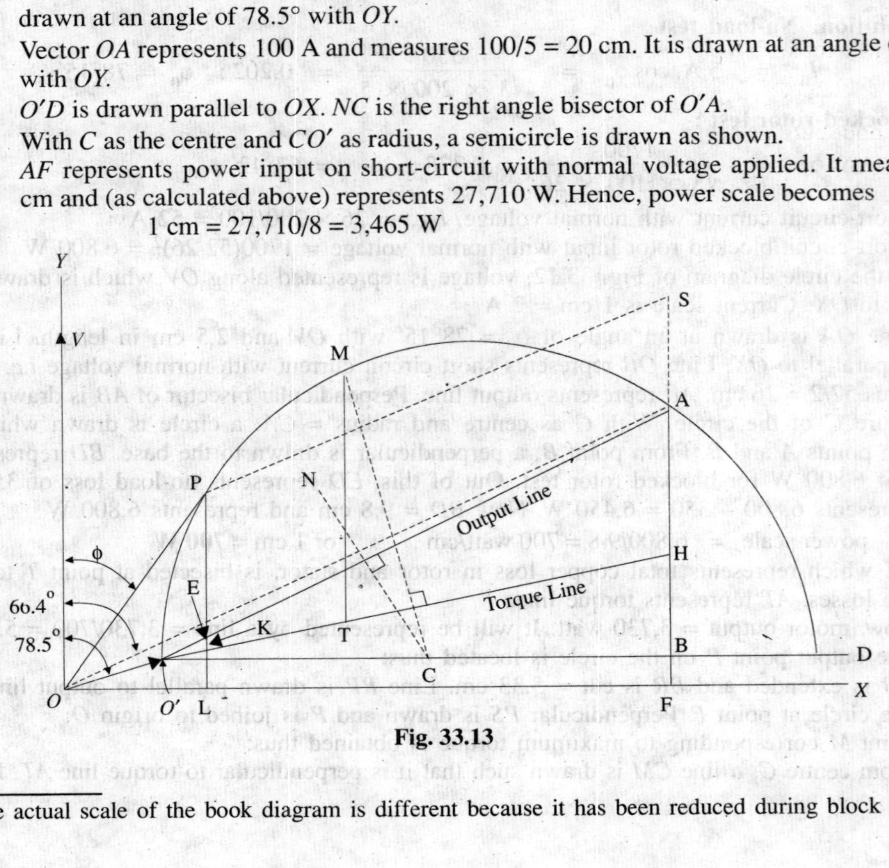

Fig. 33.13

*The actual scale of the book diagram is different because it has been reduced during block making.

(*a*) F.L motor output = 14,920 W. According to the above power scale, the intercept between the semicircle and the output line $O'A$ should measure = 14,920/3,465 = 4.31 cm. Hence, vertical line PL is found which measures 4.31 cm. Point P represents the full-load operating point.*

(*a*) Line current = OP = 6.5 cm which means that full-load line current

$= 6.5 \times 5 = \mathbf{32.5\ A.}$ $\qquad \phi = 32.9°$ (by measurement)

$\therefore \quad \cos 32.9° = \mathbf{0.84}$ (or $\cos \phi = PL/OP = 5.4/6.5 = 0.84$)

$$\text{slip} = \frac{EK}{PK} = \frac{0.3}{5.35} = 0.056 \text{ or } \mathbf{5.6\%}\ ; \quad \eta = \frac{PE}{PL} = \frac{4.3}{5.4} = 0.8 \text{ or } \mathbf{80\%}$$

(*b*) For finding maximum torque, line CM is drawn ⊥ to torque line $O'H$. MT is the vertical intercept between the semicircle and the torque line and represents the maximum torque of the motor in synchronous watts.

Now. MT = 7.8 cm (by measurement) $\quad \therefore T_{max} = 7.8 \times 3465 = \mathbf{27{,}030\ synch.\ watt}$

Example 33.6. *A 415-V, 29.84 kW, 50-Hz, delta-connected motor gave the following test data :*

No-load test	*: 415 V,*	*21 A,*	*1,250 W*
Locked rotor test	*: 100 V,*	*45 A,*	*2,730 W*

Construct the circle diagram and determine

(a) the line current and power factor for rated output (b) the maximum torque.

Assume stator and rotor Cu losses equal at standstill.

(A.C. Machines-I, Jadavpur Univ. 1990)

Solution. Power factor on no-load is $= \dfrac{1{,}250}{\sqrt{3} \times 415 \times 21} = 0.0918$

$\therefore \quad \phi_0 = \cos^{-1}(0.0918) = 84° 44'$

Power factor with locked rotor is $= \dfrac{2{,}730}{\sqrt{3} \times 100 \times 45} = 0.3503$

$\therefore \quad \phi_S = \cos^{-1}(0.3503) = 69°30'$

The input current I_{SN} on short-circuit if normal voltage were applied = 45(415/100) = 186.75 A and power taken would be = $2{,}730(415/100)^2$ = 47,000 W.

Let the current scale be 1 cm = 10 A. The circle diagram of Fig. 33.14 is constructed as follows :

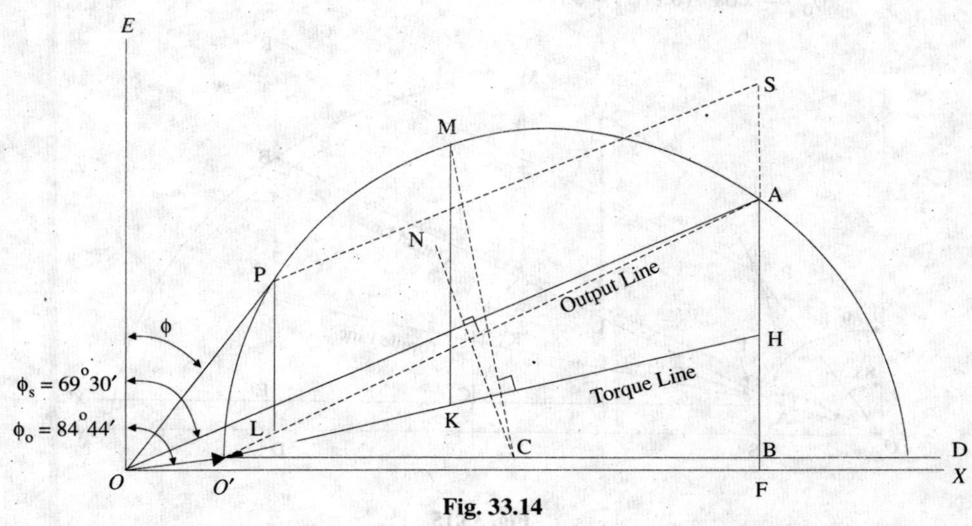

Fig. 33.14

*The operating point may also be found by making AS = 4.31 cm and drawing SP parallel to $O'A$.

(i) Vector OO' represents 21 A so that it measures 2.1 cm and is laid at an angle of 84°44′ with OE (which is vertical *i.e.* along *Y*-axis).

(ii) Vector OA measures 186.75/10 = 18.675 cm and is drawn at an angle of 69°30′ with OE.

(iii) $O'D$ is drawn parallel to OX. NC is the right-angle bisector of $O'A$

(iv) With C as the centre and CO' as radius, a semi-circle is drawn as shown. This semi-circle is the locus of the current vector for all load conditions from no-load to short-circuit.

(v) The vertical AF represents power input on short-circuit with normal voltage applied. AF measures 6.6 cm and (as calculated above) represents 47,000 W. Hence, power scale becomes,1 cm = 47,000/6.6 = 7,120 W

(a) Full-load output = 29,840 W. According to the above power scale, the intercept between the semicircle and output line $O'A$ should measure 29,840/7,120 = 4.19 cm. Hence, line PL is found which measures 4.19 cm. Point P represents the full-load operating point.*

Phase current = OP = 6 cm = 6 × 10 = 60 A; Line current = $\sqrt{3}$ × 60 = **104 A**

Power factor = cos ∠ POE = cos 35° = **0.819**

(b) for finding the maximum torque, line CM is drawn ⊥ to the torque line $O'H$. Point H is such that

$$\frac{AH}{BH} = \frac{\text{rotor Cu loss}}{\text{stator Cu loss}}$$

Since the two Cu losses are equal, point H is the mid-point of AB.

Line MK represents the maximum torque of the motor in synchronous watts

MK = 7.3 cm (by measurement) = 7.3 × 7,120 = **51,980 synch. watt.**

Example 33.7. *Draw the circle diagram for a 5.6 kW, 400-V, 3-ϕ, 4-pole, 50-Hz, slip-ring induction motor from the following data:*

No-load readings : 400 V, 6 A, cos ϕ_0 = *0.087 : Short-circuit test : 100 V, 12 A, 720 W.*

The ratio of primary to secondary turns = 2.62, stator resistance per phase is 0.67 Ω and of the rotor is 0.185 Ω. Calculate

(i) *full-load current* (ii) *full-load slip* (iii) *full-load power factor*

(iv) $\dfrac{\textit{maximum torque}}{\textit{full-load torque}}$ (v) *maximum power.*

Solution. No-load condition

$\phi_0 = \cos^{-1}(0.087) = 85°$

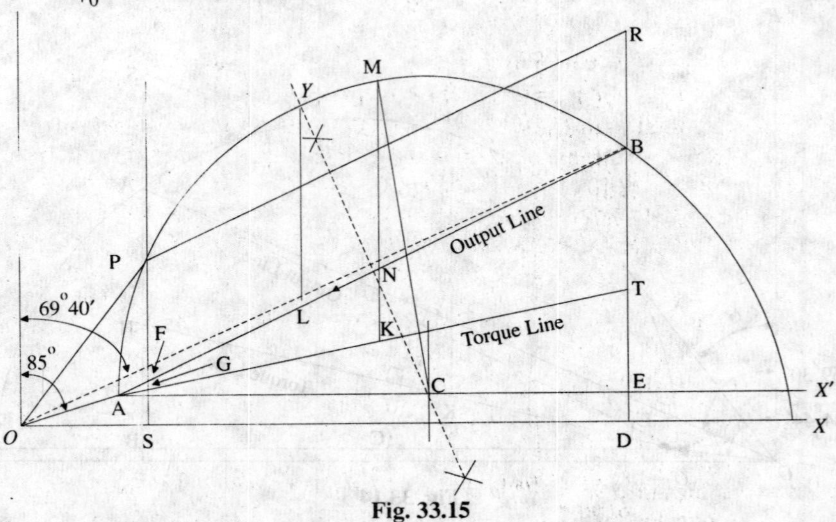

Fig. 33.15

*The operating point may also be found be making AS = 4.19 cm and drawing SP parallel to $O'A$.

Short-circuit condition

Short-circuit current with normal voltage = 12 × 400/100 = 48 A

Total input = $720 \times (48/12)^2 = 11.52$ kW

$$\cos \phi_s = \frac{720}{\sqrt{3} \times 100 \times 12} = 0.347 \quad \text{or} \quad \phi_s = 69° 40'$$

Current scale is, 1 cm = 2 A

In the circle diagram of Fig. 33.15, *OA* = 3 cm and inclined at 85° with *OV*. Line *OB* represents short-circuit current with normal voltage. It measures 48/2 = 24 cm and represent 48 A. *BD* is perpendicular to *OX*.

For Drawing Torque Line

$K = 2.62$ $\quad R_1 = 0.67\ \Omega$ $\quad R_2 = 0.185\ \Omega$

(in practice, an allowance of 10% is made for skin effect)

$$\therefore \frac{\text{rotor Cu loss}}{\text{stator Cu loss}} = 2.62^2 \times \frac{0.185}{0.67} = 1.9 \qquad \therefore \frac{\text{rotor Cu loss}}{\text{total Cu loss}} = \frac{1.9}{2.9} = 0.655$$

Now *BD* = 8.25 cm and represents 11.52 kW

$\because$ power scale = 11.52/8.25 = 1.4 kW/cm $\quad \therefore$ 1 cm = 1.4 kW

BE represents total Cu loss and is divided at point *T* in the ratio 1.9 : 1.

$BT = BE \times 1.9/2.9 = 0.655 \times 8 = 5.24$ cm

AT is the torque line

Full-load output = 5.6 kW

It is represented by a line = 5.6/1.4 =4 cm

DB is produced to *R* such that *BR* = 4 cm. Line *RP* is parallel to output line and cuts the circle at *P*. *OP* represents full-load current.

PS is drawn vertically. Points *M* and *Y* represent points of maximum torque and maximum output respectively.

(*i*) F.L. current $= OP = 5.75 \text{ cm} = 5.75 \times 2 = \mathbf{11.5\ A}$

(*ii*) F.L. slip $= \frac{FG}{PG} = \frac{0.2}{4.25} = 0.047$ or **4.7%**

(*iii*) p.f. $= \frac{SP}{OP} = \frac{4.6}{5.75} = \mathbf{0.8}$

(*iv*) $\frac{\text{max. torque}}{\text{full–load torque}} = \frac{MK}{PG} = \frac{10.05}{4.25} = \mathbf{2.37}$

(*v*) Maximum output is represented by *YL* = 7.75 cm.

$\therefore$ Max. output $= 7.75 \times 1.4 = \mathbf{10.8\ kW}$

Example 33.8. *A 440-V, 3-φ, 4-pole, 50-Hz slip-ring motor gave the following test results :*

No-load reading : 440 V, 9 A, p.f. = 0.2

Blocked rotor test : 110 V, 22 A, p.f. = 0.3

The ratio of stator to rotor turns per phase is 3.5/1. The stator and rotor Cu losses are divided equally in the blocked rotor test. The full-load current is 20 A. Draw the circle diagram and obtain the following :-

(*a*) *power factor, output power, efficiency and slip at full-load*

(*b*) *standstill torque or starting torque.*

(*c*) *resistance to be inserted in the rotor circuit for giving a starting torque 20 % of the full-load torque. Also, find the current and power factor under these conditions.*

Solution. No-load p.f. = 0.2 $\quad \therefore \phi_0 = \cos^{-1}(0.2) = 78.5°$

Short-circuit p.f. = 0.3 $\quad \therefore \phi_S = 72.5°$

Short-circuit current at normal voltage = 22 × 440/110 = 88 A

S.C. input = $\sqrt{3} \times 440 \times 88 \times 0.3 = 20{,}120$ W = 20.12 kW

Take a current scale of 1 cm = 4 A

In the circle diagram of Fig. 33.16, OA = 2.25 cm drawn at an angle of 78.5° behind OV. The semi-circle is drawn as usual. Point T is such that $BT = TD$. Hence, torque line AT can be drawn. BC represents 20.12 kW. By measurement BC = 6.6 cm

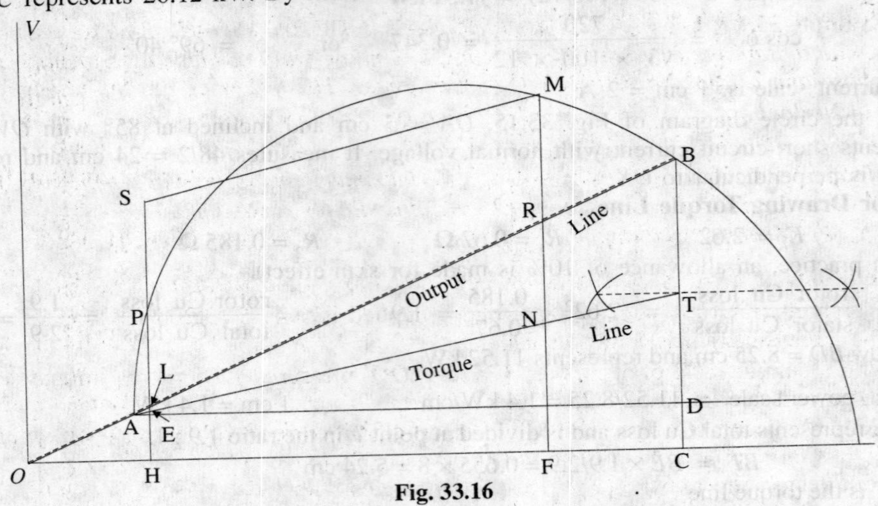

Fig. 33.16

∴ power scale = 20.12/6.6 = 3.05 ∴ 1 cm = 3.05 kW

Full-load current = 20 A. Hence, it is represented by a length of 20/4 = 5 cm. With O as centre and 5 cm as radius, an are is drawn which cuts the semi-circle at point P. This point represents full-load condition. PH is drawn perpendicular to the base OC.

(a) (i) p.f. = $\cos \phi = PH/OP$ = 4.05/5 = **0.81**

(ii) Torque can be found by measuring the input.

Rotor input = PE = 3.5 cm = 3.5 × 3.05 = 10.67 kW

Now N_s = 120 × 50/4 = 1500 r.p.m.

∴ $T_g = 9.55\, P_2/N_s = 9.55 \times 10{,}670/1500 =$ **61 N-m**

(iii) output = PL = 3.35 × 3.05 = **10.21 kW**

(iv) efficiency $= \dfrac{\text{output}}{\text{input}} = \dfrac{PL}{PH} = \dfrac{3.35}{4.05} = 0.83$ or **83%**

(v) slip $s = \dfrac{\text{rotor Cu loss}}{\text{rotor input}} = \dfrac{LE}{PE} = \dfrac{0.1025}{3.5} = 0.03$ or **3%**

(b) Standstill torque is represented by BT.

BT = 3.1 cm = 3.1 × 3.05 = 9.45 kW ∴ $T_{st} = 9.55 \times \dfrac{9.45 \times 10^3}{1500} =$ **60.25 N-m**

(c) We will now locate point M on the semi-circle which corresponds to a starting torque twice the full-load torque *i.e.* 200% of F.L. torque.

Full-load torque = PE. Produce EP to point S such that $PS = PE$. From point S draw a line parallel to torque line AT cutting the semi-circle at M. Draw MN perpendicular to the base.

At starting when rotor is stationary, MN represents total rotor copper losses.

NR = Cu loss in rotor itself as before ; RM = Cu loss in external resistance

RM = 4.5 cm = 4.5 × 3.05 = 13.716 kW = 13,716 watt.

Cu loss/phase = 13,716/3 = 4,572 watt

Rotor current AM = 17.5 cm = 17.5 × 4 = 70 A

Let r_2' be the additional external resistance in the rotor circuit (as referred to stator) then

$r_2' \times 70^2 = 4{,}572$ or $r_2' = 4{,}572/4{,}900 = 0.93\ \Omega$

Now K = 1/3.5

$\therefore$ rotor resistance/phase, $r_2 = r_2' \times K^2 = 0.93/3.5^2 = \mathbf{0.076\ \Omega}$

Stator current $= OM = 19.6 \times 4 = \mathbf{78.4\ A}$; power factor $= \dfrac{MF}{OM} = \dfrac{9.75}{18.7} = \mathbf{0.498}$

Example 33.9. *Draw the circle diagram of a 7.46 kW, 200-V, 50-Hz, 3-phase slip-ring induction motor with a star-connected stator and rotor, a winding ratio of unity, a stator resistance of 0.38 ohm/phase and a rotor resistance of 0.24 ohm/phase. The following are the test readings ;*

No-load : 200 V, 7.7 A, $\cos\phi_0 = 0.195$

Short-circuit : 100 V, 47.6 A, $\cos\phi_s = 0.454$

Find (a) starting torque and (b) maximum torque, both in synchronous watts

(c) the maximum power factor (d) the slip for maximum torque

(e) the maximum output **(Elect. Tech.-II, Madras Univ. 1989)**

Solution. $\phi_0 = \cos^{-1}(0.195) = 78°45'$; $\phi_S = \cos^{-1}(0.454) = 63°$

The short-circuit I_{SN} with normal voltage applied is $= 47.6 \times (200/100) = 95.2$ A

The circle diagram is drawn as usual

With a current scale of 1 cm = 5 A, vector OO' measures 7.7/5 =1.54 cm and represents the no-load current of 7.7 A.

Similarly, vector OA represents I_{SN} *i.e.* short-circuit current with normal voltage and measures 95.2/5 = 19.04 cm

Both vectors are drawn at their respective angles with OE

The vertical line AF measures the power input on short-circuit with normal voltage and is $= \sqrt{3} \times 200 \times 95.2 \times 0.454 = 14{,}970$ W

Since AF measures 8.6 cm, the power scale is 1 cm = 14,970/8.6 = 1740 W

The point H is such that

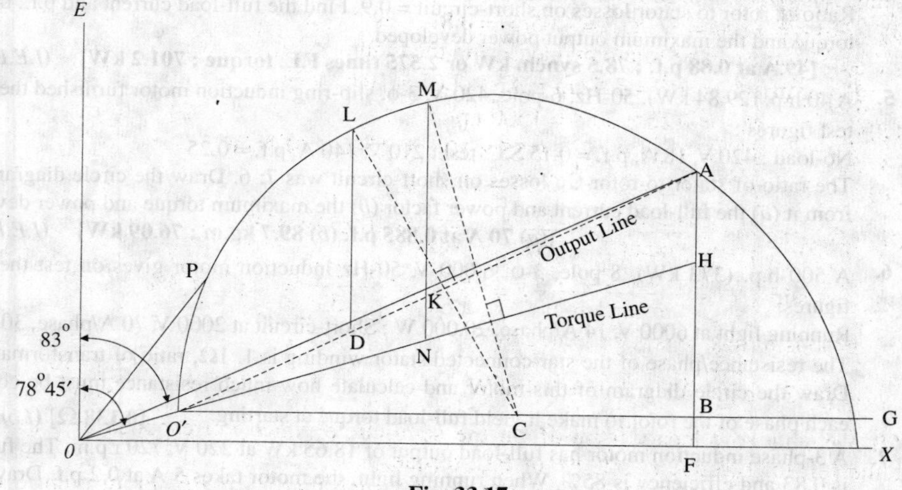

Fig. 33.17

$$\frac{AH}{AB} = \frac{\text{rotor Cu loss}}{\text{total Cu loss}} = \frac{\text{rotor resistance}^*}{\text{rotor + stator resistance}} = \frac{0.24}{0.62}$$

Now AB = 8.2 cm (by measurement) $\therefore AH = 8.2 \times 0.24/0.62 = 3.2$ cm

(*a*) Starting torque $= AH = 3.2$ cm $= 3.2 \times 1740 =$ **5,570 synch. watt.**

(*b*) Line CM is drawn perpendicular to the torque line $O'H$. The intercept MN represents the maximum torque in synchronous watts.

Maximum torque $= MN = 7.15$ cm $= 7.15 \times 1740 =$ **12,440 synch. watts.**

---

*Because $K = 1$, otherwise it should be $R_2' = R_2/K^2$.

(c) For finding the maximum power, line OP is drawn tangential to the semi-circle.

$\angle POE = 28.5°$ $\therefore$ maximum p.f. = cos 28.5° = **0.879**

(d) The slip for maximum torque is = KN/MN= 1.4/7.15 = **0.195**

(e) Line CL is drawn perpendicular to the output line $O'A$. From L is drawn the vertical line LD. It measures 5.9 cm and represents the maximum output.

$\therefore$ maximum output = 5.9 × 1740 = **10,270 W**

Tutorial Problem No. 33.1

1. A 300 h.p. (223.8 kW), 3000-V, 3-φ, induction motor has a magnetising current of 20 A at 0.10 p.f. and a short-circuit (or locked) current of 240 A at 0.25 p.f. Draw the circuit diagram, determine the p.f. at full-load and the maximum horse-power. **[0.85 p.f. 621 h.p. (463.27 kW)]** *(I.E.E. London)*
2. The following are test results for a 18.65 kW, 3-φ, 440.V slip-ring induction motor :
Light load : 440-V, 7.5 A, 1350 W (including 650 W friction loss).
S.C. test : 100 V, 32 A, 1800 W
Draw the locus diagram of the stator current and hence obtain the current, p.f. and slip on full-load. On short-circuit, the rotor and stator copper losses are equal.
[30 A, 0.915, 0.035] *(London Univ)*
3. Draw the circle diagram for 20 h.p. (14.92 kW), 440-V, 50-Hz, 3-φ induction motor from the following test figures (line values) :
No-load : 440 V, 10A, p.f. 0.2 Short-circuit :200 V, 50 A, p.f. 0.4
From the diagram, estimate (*a*) the line current and p.f. at full-load (*b*) the maximum power developed (*c*) the starting torque. Assume the rotor and stator I^2R losses on short-circuit to be equal.
[(*a*) 28.1 A at 0.844 p.f. (*b*) 27.75 kW (*c*) 11.6 synchronous kW/phase] *(London Univ.)*
4. A 40 h.p. (29.84.kW), 440-V, 50-Hz, 3-phase induction motor gave the following test results
No.load : 440 V, 16 A, p.f. = 0.15 S.C. test : 100 V, 55 A, p.f. = 0.225
Ratio of rotor to stator losses on short-circuit = 0.9. Find the full-load current and p.f., the pull-out torque and the maximum output power developed.
[49 A at 0.88 p.f. ; 78.5 synch. kW or 2.575 times F.L. torque ; 701.2 kW] *(I.E.E. London)*
5. A 40 h.p. (29.84 kW), 50-Hz, 6-pole, 420-V, 3-φ, slip-ring induction motor furnished the following test figures :
No-load : 420 V, 18 A, p.f. = 0.15 S.C. test : 210 V, 140 A, p.f. = 0.25
The ratio of stator to rotor Cu losses on short-circuit was 7: 6. Draw the circle diagram and find from it (*a*) the full-load current and power factor (*b*) the maximum torque and power developed.
[(*a*) 70 A at 0.885 p.f. (*b*) 89.7 kg.m ; 76.09 kW] *(I.E.E. London)*
6. A 500 h.p. (373 kW), 8-pole, 3-φ, 6,000-V, 50-Hz induction motor gives on test the following figures:
Running light at 6000 V, 14 A/phase, 20,000 W ; Short-circuit at 2000 V, 70 A/phase, 30,500 W
The resistance/phase of the star-connected stator winding is 1. 1Ω, ratio of transformation is 4:1. Draw the circle diagram of this motor and calculate how much resistance must be connected in each phase of the rotor to make it yield full-load torque at starting. **[0.138 Ω]** *(London Univ.)*
7. A 3-phase induction motor has full-load output of 18.65 kW at 220 V, 720 r.p.m. The full-load p.f. is 0.83 and efficiency is 85%. When running light, the motor takes 5 A at 0.2 p.f. Draw the circle diagram and use it to determine the maximum torque which the motor can exert (*a*) in N-m (*b*) in terms of full-load torque and (*c*) in terms of the starting torque.
[(*a*) 268.7 N-m (*b*) 1.08 (*c*) 7.2 approx.] *(London Univ.)*
8. A 415-V, 40 h.p: (29.84 kW), 50.Hz, Δ-connected motor gave the following test data :
No-load test : 415 V, 21 A, 1250 W ; Locked rotor test : 100 V, 45 A, 2,730 W
Construct the circle diagram and determine
(*a*) the line current and power factor for rated output (*b*) the maximum torque. Assume stator and rotor Cu losses equal at standstill.
[(*a*) 104 A : 0.819 (*b*) 51,980 synch watt] *(A.C. Machines-I, Jadavpur Univ. 1978)*
9. Draw the no-load and short circuit diagram for a 14.92 kW., 400-V, 50-Hz, 3-phase star-connected induction motor from the following data (line values):

No load test : 400 V, 9 A, cos ϕ = 0.2

Short circuit test : 200 V, 50 A, cos ϕ = 0.4

From the diagram find (a) the line current and power factor at full load, and (b) the maximum output power. **[(a) 32.0 A, 0.85 (b) 21.634 kW]**

33.9. Starting of Induction Motors

It has been shown earlier that a plain induction motor is similar in action to a polyphase transformer with a short-circuited rotating secondary. Therefore, if normal supply voltage is applied to the stationary motor, then, as in the case of a transformer, a very large initial current is taken by the primary, at least, for a short while. It would be remembered that exactly similar conditions exist in the case of a d.c. motor, if it is thrown directly across the supply lines, because at the time of starting it, there is no back e.m.f. to oppose the initial inrush of current.

Induction motors, when direct-switched, take five to seven times their full-load current and develop only 1.5 to 2.5 times their full-load torque. This initial excessive current is objectionable because it will produce large line-voltage drop that, in turn, will affect the operation of other electrical equipment connected to the same lines. Hence, it is not advisable to line-start motors of rating above 25 kW to 40 kW.

It was seen in Art. 32.15 that the starting torque of an induction motor can be improved by increasing the resistance of the rotor circuit. This is easily feasible in the case of slip-ring motors but not in the case of squirrel-cage motors. However, in their case, the initial in-rush of current is controlled by applying a reduced voltage to the stator during the starting period, full normal voltage being applied when the motor has run up to speed.

33.10. Direct-switching or Line starting of Induction Motors

It has been shown earlier that

Rotor input $= 2\pi N_s T = kT$ —Art. 32.36

Also, rotor Cu loss $= s \times$ rotor input

$\therefore \quad 3 I_2^2 R_2 = s \times kT \qquad \therefore \quad T \propto I_2^2/s$ (if R_2 is the same)

Now $\quad I_2 \propto I_1 \qquad \therefore \quad T \propto I_1^2/s$ or $T = K I_1^2/s$

At starting moment $s = 1 \qquad \therefore \quad T_{st} = K I_{st}^2$ where I_{st} = starting current

If $\quad I_f$ = normal full-load current and $\quad s_f$ = full-load slip

then $\quad T_f = K I_f^2 / s_f \qquad \therefore \quad \dfrac{T_{st}}{T_f} = \left(\dfrac{I_{st}}{I_f}\right)^2 . s_f$

When motor is direct-switched onto normal voltage, then starting current is the short-circuit current I_{sc}.

$$\therefore \quad \frac{T_{st}}{T_f} = \left(\frac{I_{sc}}{I_f}\right)^2 . s_f = a^2 . s_f \quad \text{where } a = I_{sc} / I_f$$

Suppose in a case, $I_{sc} = 7 I_f$, s_f = 4% = 0.04, the $T_{st} / T_f = 7^2 \times 0.04 = 1.96$

$\therefore$ starting torque $= 1.96 \times$ full-load torque

Hence, we find that with a current as great as seven times the full-load current, the motor develops a starting torque which is only 1.96 times the full-load value.

Some of the methods for starting induction motors are discussed below:

Squirrel-cage Motors

(a) Primary resistors (or rheostat) or reactors

(b) Auto-transformer (or autostarter)

(c) Star-delta switches

In all these methods, terminal voltage of the squirrel-cage motor is reduced during starting.

Slip-ring Motors

(a) Rotor rheostat

33.11. Squirrel-cage Motors

(a) Primary resistors

Their purpose is to drop some voltage and hence reduce the voltage applied across the motor terminals. In this way, the initial current drawn by the motor is reduced. However, it should be noted that whereas current varies directly as the voltage, the torque varies as square of applied voltage* (Art 32.17). If the voltage applied across the motor terminals is reduced by 50%, starting current is reduced by 50%, but torque is reduced to 25% of the full-voltage value.

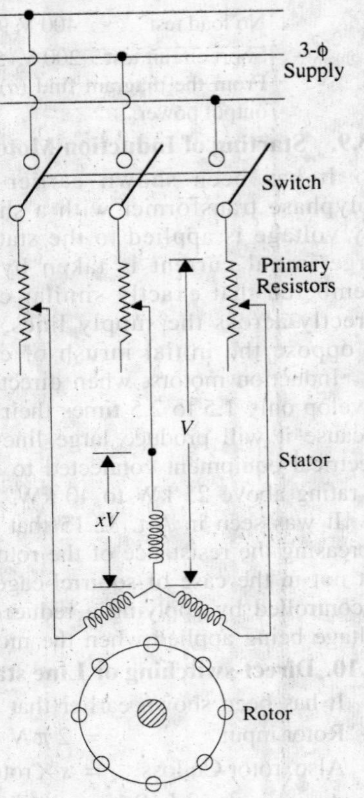

Fig. 33.18

By using primary resistors (Fig. 33.18), the applied voltage/phase can be reduced by a fraction '*x*' (and it additionally improves the power factor of the line slightly).

$$I_{st} = x\,I_{sc} \quad \text{and} \quad T_{st} = x^2\,T_{sc}$$

As seen from Art 33.10, above,

$$\frac{T_{st}}{T_f} = \left(\frac{I_{st}}{I_f}\right)^2 s_f = \left(\frac{x\,I_{sc}}{I_f}\right)^2 s_f$$

$$= x^2 \left(\frac{I_{sc}}{I_f}\right)^2 s_f = x^2 \,.\, a^2 \,.\, s_f$$

It is obvious that the ratio of the starting torque to full-load torque is x^2 of that obtained with direct switching or across-the-line starting. This method is useful for the smooth starting of small machines only.

(b) Auto-transformers

Such starters, known variously as *auto-starters* or *compensators,* consist of an auto-transformer, with necessary switches. We may use either two auto-transformers connected as usual [Fig. 33.19 (*b*)] or 3 auto-transformers connected in open delta [Fig. 33.19 (*a*)]. *This method can be used both for star-and delta-connected motors.* As shown in Fig. 33.20 with starting connections, a reduced voltage is applied across the motor terminals. When the motor has ran up to say, 80% of its normal speed, connections are so changed that auto-transformers are cut out and full supply voltage is applied across the motor. The switch making these changes from 'start' to 'run' may be airbreak (for small motors) or may be oil-immersed (for large motors) to reduce sparking. There is also provision for no-voltage and over-load protection, along with a time-delay device, so that momentary interruption of voltage or momentary over-load do not disconnect the motor from supply line. Most of the auto-starters are provided with 3 sets of taps, so as to reduce voltage to 80, 65 or 50 per cent of the line voltage, to suit the local conditions of supply. The *V*-connected auto-transformer is commonly used, because it is cheaper, although the currents are unbalanced during starting period. This is, however, not much objectionable firstly, because the current imbalance is about 15 per cent and secondly, because balance is restored as soon as running conditions are attained.

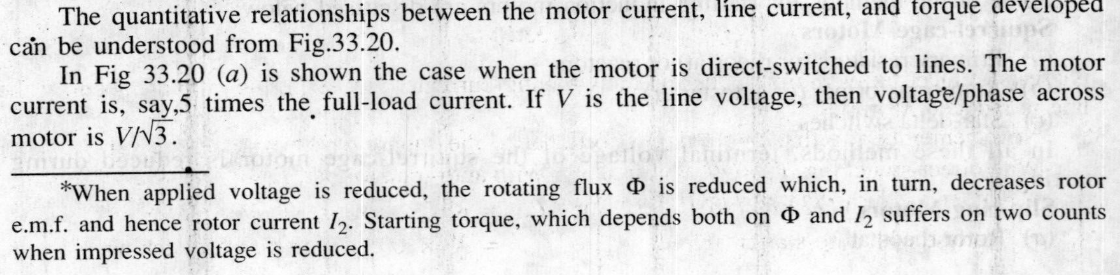

The quantitative relationships between the motor current, line current, and torque developed can be understood from Fig.33.20.

In Fig 33.20 (*a*) is shown the case when the motor is direct-switched to lines. The motor current is, say,5 times the full-load current. If *V* is the line voltage, then voltage/phase across motor is $V/\sqrt{3}$.

*When applied voltage is reduced, the rotating flux Φ is reduced which, in turn, decreases rotor e.m.f. and hence rotor current I_2. Starting torque, which depends both on Φ and I_2 suffers on two counts when impressed voltage is reduced.

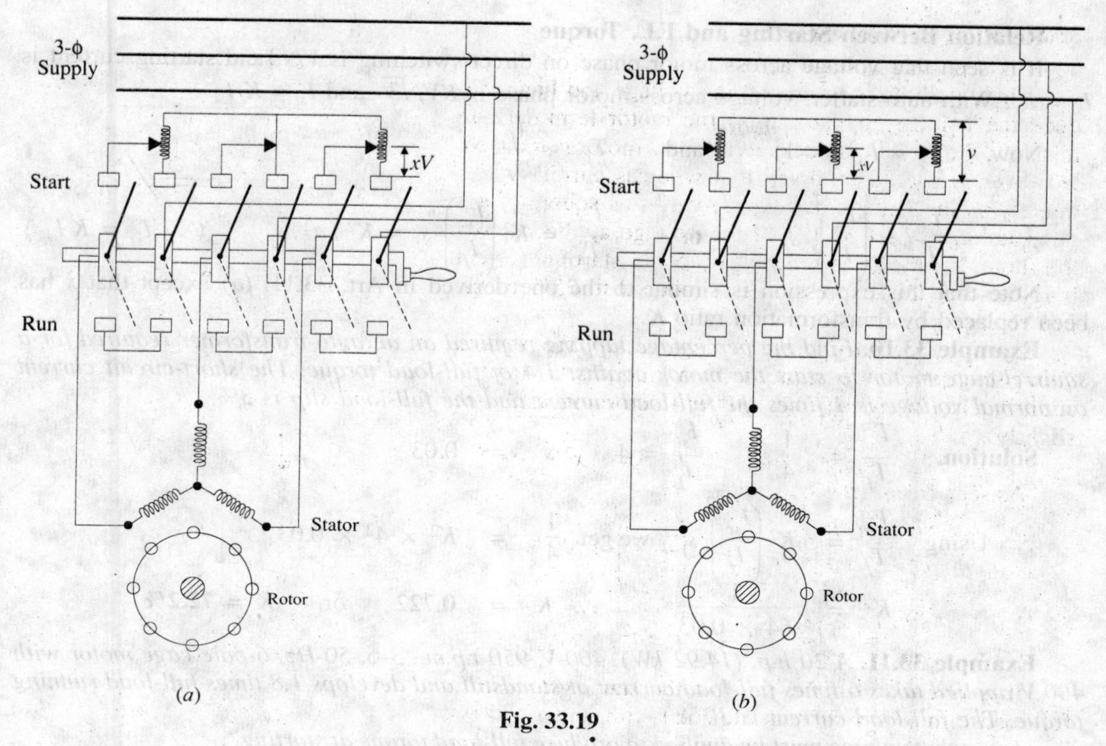

Fig. 33.19

$\therefore \quad I_{sc} \;=\; 5\,I_f = \dfrac{V}{\sqrt{3}Z}$ where Z is stator impedance/phase.

In the case of auto-transformer, if a tapping of transformation ratio K is used, then phase voltage across motor is $KV/\sqrt{3}$.

$\therefore$ motor current at starting $I_2 = \dfrac{KV}{\sqrt{3}Z} = K \cdot \dfrac{V}{\sqrt{3}Z} = K \cdot I_{sc} = K \cdot 5\,I_f$

The current taken from supply or by auto-transformer is $I_1 = K\,I_2 = K^2 \times 5\,I_f = K^2\,I_{sc}$ if magnetising current of the transformer is ignored. Hence, we find that although motor current per

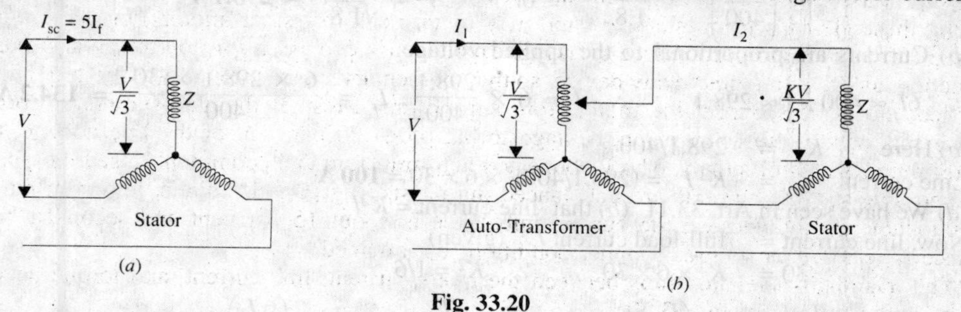

Fig. 33.20

phase is reduced only K times, the direct-switching current ($\because K < 1$), the current taken by the line is reduced K^2 times.

Now, remembering that torque is proportional to the square of the voltage, we get

With direct-switching, $T_1 \propto (V/\sqrt{3})^2$; With auto-transformer, $T_2 \propto (KV/\sqrt{3})^2$

$\therefore \quad T_2/T_1 = (K\,V/\sqrt{3})^2/(V/\sqrt{3})^2$ or $T_2 = K^2 T_1$ or $T_{st} = K^2 . T_{sc}$

$\therefore$ torque with auto-starter $= K^2 \times$ torque with direct-switching.

Relation Between Starting and F.L. Torque

It is seen that voltage across motor phase on direct-switching is $V/\sqrt{3}$ and starting current is $I_{st} = I_{sc}$. With auto-starter, voltage across motor phase is $KV/\sqrt{3}$ and $I_{st} = K\,I_{sc}$

Now, $T_{st} \propto I_{st}^2\ (s = 1)$ and $T_f \propto \dfrac{I_f^2}{s_f^2}$

$$\therefore \quad \frac{T_{st}}{T_f} = \left(\frac{I_{st}}{I_f}\right)^2 s_f \quad \text{or} \quad \frac{T_{st}}{T_f} = K^2\left(\frac{I_{sc}}{I_f}\right)^2 s_f = K^2 . a^2 . s_f \qquad (\because\ I_{st} = K\,I_{sc})$$

Note that this expression is similar to the one derived in Art. 33.11. (*a*) except that x has been replaced by transformation ratio K.

Example 33.10. *Find the percentage tapping required on an auto-transformer required for a squirrel-cage motor to start the motor against 1/4 of full-load torque. The short-circuit current on normal voltage is 4 times the full-load current and the full-load slip is 3%.*

Solution. $\dfrac{T_{st}}{T_f} = \dfrac{1}{4}, \quad \dfrac{I_{sc}}{I_f} = 4, \quad s_f = 0.03$

$$\therefore \text{ Using } \frac{T_{st}}{T_f} = K^2\left(\frac{I_{sc}}{I_f}\right)^2 s_f, \text{ we get } \frac{1}{4} = K^2 \times 4^2 \times 0.03$$

$$\therefore \quad K^2 = \frac{1}{64 \times 0.03} \qquad \therefore\ K = 0.722 \quad \text{or} \quad K = \mathbf{72.2\%}$$

Example 33.11. *A 20 h.p. (14.92 kW), 400-V, 950 r.p.m., 3-φ, 50-Hz, 6-pole cage motor with 400 V applied takes 6 times full-load current at standstill and develops 1.8 times full-load running torque. The full-load current is 30 A.*

(a) what voltage must be applied to produce full-load torque at starting?

(b) what current will this voltage produce?

(c) if the voltage is obtained by an auto-transformer, what will be the line current?

(d) if starting current is limited to full-load current by an auto-transformer, what will be the starting torque as a percentage of full-load torque?

Ignore the magnetising current and stator impedance drops.

Solution. (*a*) Remembering that $T \propto V^2$, we have

In the first case, $1.8\,T_f \propto 400^2$, ; In the second case, $T_f \propto V^2$

$$\therefore \quad \left(\frac{V}{400}\right)^2 = \frac{1}{1.8} \quad \text{or} \quad V = \frac{400}{\sqrt{1.8}} = \mathbf{298.1\ V}$$

(*b*) Currents are proportional to the applied voltage.

$$\therefore \quad 6I_f \propto 400\ ;\ I \propto 298.1 \qquad \therefore\ I = 6 \times \frac{298.1}{400},\ I_f = \frac{6 \times 298.1 \times 30}{400} = \mathbf{134.2\ A}$$

(*c*) Here $K = 298.1/400$

Line current $= K^2 I_{sc} = (298.1/400)^2 \times 6 \times 30 = \mathbf{100\ A}$

(*d*) We have seen in Art. 33.11 (*b*) that line current $= K^2 I_{sc}$

Now, line current = full-load current I_f (given)

$\therefore \quad 30 = K^2 \times 6 \times 30 \qquad \therefore\ K^2 = 1/6$

$$\text{Now, using } \frac{T_{st}}{T_f} = K^2\left(\frac{I_{sc}}{I_f}\right)^2 \times s_f \quad \text{we get } \frac{T_{st}}{T_f} = \frac{1}{6} \times \left(\frac{6\,I_f}{I_f}\right)^2 \times 0.05 = 0.3$$

Here $N_s = 120 \times 50/6 = 1000$ r.p.m. $\qquad N = 950$ r.p.m.; $s_f = 50/1000 = 0.05$

$\therefore \quad T_{st} = 0.3\,T_f$ or **30%** F.L. torque

Example 33.12. *Determine the suitable auto-transformation ratio for starting a 3-phase induction motor with line current not exceeding three times the full-load current. The short-circuit current is 5 times the full-load current and full-load slip is 5%.*

Estimate also the starting torque in terms of the full-load torque.

(Elect. Engg.II, Bombay Univ. 1987)

Solution. Supply *line* current = $K^2 I_{sc}$

It is given that supply line current at start equals $3I_f$ and short-circuit current $I_{sc} = 5I_f$ where I_f is the full-load current

$\therefore \quad 3 I_f = K^2 \times 5 I_f \quad$ or $K^2 = 0.6 \qquad \therefore \quad K = 0.775$ or **77.5%**

In the case of an auto-load starter,

$$\frac{T_{st}}{T_f} = K^2 \left(\frac{I_{sc}}{I_f}\right)^2 \times s_f \qquad \therefore \quad \frac{T_{st}}{T_f} = 0.6 \times \left(\frac{5I_f}{I_f}\right)^2 \times 0.05 = 0.75\%$$

$\therefore \quad T_{st} = 0.75\, T_f =$ **75%** of full-load torque.

Example 33.13. *The full-load slip of a 400-V, 3-phase cage induction motor is 3.5% and with locked rotor, full-load current is circulated when 92 volt is applied between lines. Find necessary tapping on an auto-transformer to limit the starting current to twice the full-load current of the motor. Determine also the starting torque in terms of the full-load torque.*

(Elect. Machines, Banglore Univ. 1991)

Solution. Short-circuit current with full normal voltage applied is

$$I_{sc} = (400/92)\, I_f = (100/23)\, I_f$$

Supply *line* current $= I_{st} = 2\, I_f$

Now, line current $I_{st} = K^2 I_{sc}$

$\therefore \quad 2I_f = K^2 \times (100/23)\, I_f \qquad \therefore K^2 = 0.46\ ;\ K = 0.678$ or **67.8%**

Also, $$\frac{T_{st}}{T_f} = K^2 \left(\frac{I_{sc}}{I_f}\right)^2 \times s_f = 0.46 \times (100/23)^2 \times 0.035 = 0.304$$

$\therefore \quad T_{st} =$ **30.4%** of full-load torque

Tutorial Problem No. 33.2

1. A 3-ϕ motor is designed to run at 5% slip on full-load. If motor draws 6 times the full-load current at starting at the rated voltage, estimate the ratio of starting torque to the full-load torque.
[1.8] (*Electrical Engineering Grad, I.E.T.E. Dec. 1986*)
2. A squirrel-cage induction motor has a short-circuit current of 4 times the full-load value and has a full-load slip of 5%. Determine a suitable auto-transformer ratio if the supply line current is not to exceed twice the full-load current. Also, express the starting torque in terms of the full-load torque. Neglect magnetising current. **[70.7%, 0.4]**
3. A 3-ϕ, 400-V, 50-Hz induction motor takes 4 times the full-load current and develops twice the full-load torque when direct-switched to 400-V supply. Calculate in terms of full-load values (*a*) the line current, the motor current and starting torque when started by an auto-starter with 50% tap and (*b*) the voltage that has to be applied and the motor current, if it is desired to obtain full-load torque on starting. **[(*a*) 100%, 200%, 50% (b) 228 V, 282%]**

(*c*) Star-delta Starter

This method is used in the case of motors which are built to run normally with a delta-connected stator winding. It consists of a two-way switch which connects the motor in star for starting and then in delta for normal running. The usual connections are shown in Fig. 33.21. When star-connected, the applied voltage over each motor phase is reduced by a factor of $1/\sqrt{3}$ and hence the torque developed becomes 1/3 of that which would have been developed if motor were directly connected in delta. The line current is reduced to 1/3. Hence, during starting period when motor *Y*-is connected, it takes 1/3rd as much starting current and develops 1/3rd as much torque as would have been developed were it directly connected in delta.

Relation Between Starting and F.L. Torque

$$I_{st} \text{ per phase} = \frac{1}{\sqrt{3}} I_{sc} \text{ per phase}$$

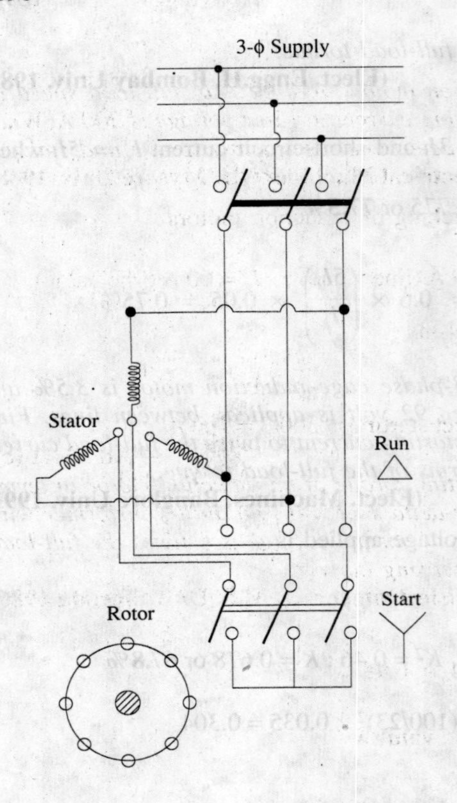

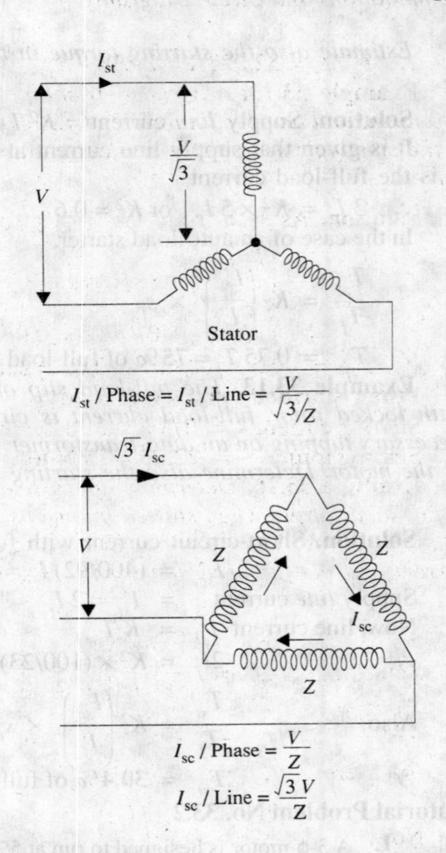

Fig. 33.21

where I_{sc} is the current/phase which Δ-connected motor would have taken if switched on to the supply directly (however, line current at start = 1/3 of line I_{sc})

Now $T_{st} \propto I_{st}^2$ $(s = 1)$ —Art. 33.10

$T_f \propto I_f^2/s_f$

$$\therefore \quad \frac{T_{st}}{T_f} = \left(\frac{I_{st}}{I_f}\right)^2 s_f = \left(\frac{I_{sc}}{\sqrt{3}I_f}\right)^2 s_f = \frac{1}{3}\left(\frac{I_{sc}}{I_f}\right)^2 s_f = \frac{1}{3}a^2\, s_f$$

Here, I_{st} and I_{sc} represent phase values.

It is clear that the star-delta swith is equivalent to an auto-transformer of ratio 1/ $\sqrt{3}$ or 58% approximately.*

This method is cheap and effective provided the starting torque is required not to be more than 1.5 times the full-load torque. Hence, it is used for machine tools, pumps and motor-generators etc.

Example 33.14. *The full-load efficiency and power factor of a 12-kW, 440-V, 3-phase induction motor are 85% and 0.8 lag respectively. The blocked rotor line current is 45 A at 220 V. Calculate the ratio of starting to full-load current, if the motor is provided with a star-delta starter. Neglect magnetising current.* **(Elect. Machines, A.M.I.E. Sec.B, 1991)**

Solution. Blocked rotor current with full voltage applied

$I_{sc} = 45 \times 440/220 = 90$ A

Now, $\sqrt{3} \times 440 \times I_f \times 0.8 = 12{,}000/0.85$, $\therefore \quad I_f = 23.1$ A

In star-delta starter, $I_{st} = I_{sc} / \sqrt{3} = 90 / \sqrt{3} = 52$ A

*By comparing it with the expression given in Art. 33.11 (*b*)

$\therefore \quad I_{st}/I_f = 52/23.1 = \mathbf{2.256}$

Example 33.15. *A 3-phase, 6-pole, 50-Hz induction motor takes 60 A at full-load speed of 940 r.p.m. and develops a torque of 150 N-m. The starting current at rated voltage is 300 A. What is the starting torque? If a star/delta starter is used, determine the starting torque and starting current.*

(Electrical Machinery-II, Mysore Univ. 1988)

Solution. As seen from Art. 33.10, for direct-switching of induction motors

$$\frac{T_{st}}{T_f} = \left(\frac{I_{sc}}{I_f}\right)^2 s_f \qquad \text{Here, } I_{st} = I_{sc} = 300 \text{ A (line value)}; \quad I_f = 60 \text{ A (line value)},$$

$$s_f = (1000 - 940)/1000 = 0.06 \; ; \; T_f = 150 \text{ N-m}$$

$$\therefore \quad T_{st} = 150(300/60)^2 \times 0.06 = \mathbf{225 \text{ N-m}}$$

When star/delta starter is used

Starting current = 1/3 × starting current with direct starting = 300/3 = **100 A**

Starting torque = 225/3 = **75 N-m** —Art 33-11 (*c*)

Example 33.16. *Determine approximately the starting torque of an induction motor in terms of full-load torque when started by means of (a) a star-delta switch (b) an auto-transformer with 70.7 % tapping. The short-circuit current of the motor at normal voltage is 6 times the full-load current and the full-load slip is 4%. Neglect the magnetising current.*

(Electrotechnics, M.S. Univ. Baroda 1986)

Solution. (*a*) $\dfrac{T_{st}}{T_f} = \dfrac{1}{3}\left(\dfrac{I_{sc}}{I_f}\right)^2 s_f = \dfrac{1}{3} \times 6^2 \times 0.04 = 0.48$

$\therefore \quad T_{st} = 0.48\, T_f$ or **48% of F.L. value**

(*b*) Here $K = 0.707 = 1/\sqrt{2}\; ; \; K^2 = 1/2$

Now, $\dfrac{T_{st}}{T_f} = K^2\left(\dfrac{I_{sc}}{I_f}\right)^2 s_f = \dfrac{1}{2} \times 6^2 \times 0.04 = 0.72$

$\therefore \quad T_{st} = 0.72\, T_f$ or **72% of $\mathbf{T_f}$**

Example 33.17. *A 15 h.p. (11.2 kW), 3-φ, 6-pole, 5-Hz, 400-V, Δ-connected induction motor runs at 960 r.p.m. on full-load. If it takes 86.4 A on direct starting, find the ratio of starting torque to full-load torque with a star-delta starter. Full-load efficiency and power factor are 88% and 0.85 respectively.*

Solution. Here, I_{sc}/phase $= 86.4/\sqrt{3}$ A

$$I_{st} \text{ per phase} = \frac{1}{\sqrt{3}} \cdot I_{sc} \text{ per phase} = \frac{86.4}{\sqrt{3} \times \sqrt{3}} = 28.8 \text{ A}$$

Full-load input line current may be found from

$$\sqrt{3} \times 400 \times I_L \times 0.85 = 11.2 \times 10^3/0.88 \qquad \therefore \text{ Full-load } I_L = 21.59 \text{ A}$$

$$\text{F.L. } I_{ph} = 21.59/\sqrt{3} \text{ A}; \qquad I_f = 21.59/\sqrt{3} \text{ A per phase}$$

$$N_s = 120 \times 50/6 = 1000 \text{ r.p.m.}, \qquad N = 950\; ; \; s_f = 0.05$$

$$\frac{T_{st}}{T_f} = \left(\frac{I_{st}}{I_f}\right)^2 s_f = \left(\frac{28.8 \times \sqrt{3}}{21.59}\right)^2 \times 0.05 \quad \therefore \; T_{st} = 0.267\, T_f \text{ or } \mathbf{26.7\% \text{ F.L. torque}}$$

Example 33.18. *Find the ratio of starting to full-load current in a 10 kW (output), 400-V, 3-phase induction motor with star/delta starter, given that full-load p.f. is 0.85, the full-load efficiency is 0.88 and the blocked rotor current at 200 V is 40 A. Ignore magnetising current.*

(Electrical Engineering, Madras Univ. 1985)

Solution. F.L. line current drawn by the Δ-connected motor may be found from

$$\sqrt{3} \times 400 \times I_L \times 0.85 = 10 \times 1000/0.88 \qquad \therefore \; I_L = 19.3 \text{ A}$$

Now, with 200 V, the line value of S.C. current of the Δ-connected motor is 40 A. If full normal voltage were applied, the *line* value of S.C. current would be = 40 × (400/200) = 80 A.

∴ I_{sc} (line value) = 80 A ; I_{sc} (phase value) = $80/\sqrt{3}$ A

When connected in star across 400 V, the starting current per phase drawn by the motor stator during starting is

$$I_{st} \text{ per phase} = \frac{1}{\sqrt{3}} \times I_{sc} \text{ per phase} = \frac{1}{\sqrt{3}} \times \frac{80}{\sqrt{3}} = \frac{80}{3} \text{ A} \qquad \text{—Art. 33.10}$$

Sjince during starting, motor is star-connected, I_{st} per phase = line value of I_{sc} = 80/3A

$$\therefore \quad \frac{\text{line value of starting current}}{\text{line value of F.L. current}} = \frac{80/3}{19.3} = \mathbf{1.38}$$

Example 33.19. *A 5 h.p. (3.73 kW), 400-V, 3-φ, 50-Hz cage motor has a full-load slip of 4.5%. The motor develops 250% of the rated torque and draws 650% of the rated current when thrown directly on the line. What would be the line current, motor current and the starting torque if the motor were started (i) be means of a star/delta starter and (ii) by connecting across 60% taps of a starting compensator.* **(Elect. Machines-II, Indore Univ. 1989)**

Solution. (*i*) Line current = (1/3) × 650 = **216.7%**

Motor being star-connected, line current is equal to phase current.

∴ motor current = 650/3 = **216.7%**

As shown earlier, starting torque developed for star-connection is one-third of that developed on direct switching with delta-connection ∴ T_{st} = 250/3 = **83.3%**

(*ii*) Line current = $K^2 \times I_{sc}$ = $(60/100)^2 \times 650$ = **234%**

Motor current = $K \times I_{sc}$ = (60/100) × 650 = **390%**

$T_{st} = K^2 \times T_{sc} = (60/100)^2 \times 250$ = **90%**

Example. 33.20. *A squirrel-cage type induction motor when started by means of a star/delta starter takes 180% of full-load line current and develops 35% of full-load torque at starting. Calculated the starting torque and current in terms of full-load values, if an auto-transformer with 75% tapping were employed.* **(Utilization of Elect. Power, A.M.I.E. 1987)**

Solution. With star-delta starter, $\dfrac{T_{st}}{T_f} = \dfrac{1}{3}\left(\dfrac{I_{sc}}{I_f}\right)^2 s_f$

Line current on line-start I_{sc} = 3 × 180% of I_f = $3 \times 1.8\, I_f = 5.4\, I_f$

Now, T_{st}/T_f = 0.35 (given) ; I_{sc}/I_f = 5.4

∴ 0.35 = $(1/3) \times 5.4^2\, s_f$ or $5.4^2\, s_f = 1.05$

Autostarter : Here, K = 0.75

Line starting current = $K^2 I_{sc} = (0.75)^2 \times 5.4\, I_f$ = **3.04 I_f = 304% of F.L. current**

$$\frac{T_{st}}{T_f} = K^2\left(\frac{I_{sc}}{I_f}\right)^2 s_f \; ; \quad \frac{T_{st}}{T_f} = (0.75)^2 \times 5.4^2\, s_f$$

$$= (0.75)^2 \times 1.05 = 0.59$$

T_{st} = 0.59 T_f = **59% F.L. torque**

Example. 33.21. *A 10 h.p. (7.46 kW) motor when started at normal voltage with a star-delta switch in the star position is found to take an initial current of 1.7 × full-load current and gave an initial starting torque of 35% of full-load torque. Explain what happens when the motor is started under the following conditions : (a) an auto-transformer giving 60% of normal voltage (b) a resistance in series with the stator reducing the voltage to 60% of the normal and calculate in each case the value of starting current and torque in terms of the corresponding quantities at full-load.* **(Elect. Machinery-III, Kerala Univ. 1987)**

Solution. If the motor were connected in delta and direct-switched to the line, then it would take a line current three times that which it takes when star-connected.

$\therefore$ line current on line start or $I_{sc} = 3 \times 1.7\, I_f = 5.1\, I_f$

We know $\dfrac{T_{st}}{T_f} = \dfrac{1}{3}\left(\dfrac{I_{sc}}{I_f}\right)^2 s_f$

Now $\dfrac{T_{st}}{T_f} = 0.35$ ... given ; $\dfrac{I_{sc}}{I_f} = 5.1$ ...calculated

$\therefore$ $0.35 = (1/3) \times 5.1^2 \times s_f$

We can find s_f from $5.1^2 \times s_f = 1.05$

(*a*) When it is started with an auto-starter, then $K = 0.6$

Line starting current $= K^2 \times I_{sc} = 0.6^2 \times 5.1\, I_f = \mathbf{0.836\, I_f}$

$T_{st}/T_f = 0.6^2 \times 5.1^2 \times s_f = 0.6^2 \times 1.05 = 0.378 \quad \therefore T_{st} = \mathbf{37.8\%}$ of F.L. torque

(*b*) Here, voltage across motor is reduced to 60% of normal value. In this case *motor* current is the same as *line* current but it decreases in proportion to the decrease in voltage.

As voltage across motor $=$ 0.6 of normal voltage

$\therefore$ line starting current $= 0.6 \times 5.1\, I_f = \mathbf{3.06\, I_f}$

Torque at starting would be the same as before.

$T_{st}/T_f = 0.6^2 \times 5.1^2 \times S_f = 0.378 \quad \therefore \quad T_{st} = \mathbf{37.8\%}$ of F.L. torque.

Tutorial Problem No. 33.3.

1. A 3-phase induction motor whose full-load slip is 4 per cent, takes six times full-load current when switched directly on to the supply. Calculate the approximate starting torque in terms of the full-load torque when started by means of an auto-transformer starter, having a 70 percent voltage tap. **[0.7 T_f]**
2. A 3-phase, cage induction motor takes a starting current at normal voltage of 5 times the full-load value and its full-load slip is 4 per cent. What auto-transformer ratio would enable the motor to be started with not more than twice full-load current drawn from the supply ?
   What would be the starting torque under these conditions and how would it compare with that obtained by using a stator resistance starter under the same limitations of line current ? **[63.3% tap ; 0.4 T_f ; 0.16 T_f]**
3. A 3-phase, 4-pole, 50-Hz induction motor takes 40 A at a full-load speed of 1440 r.p.m. and develops a torque of 100 N-m at full-load. The starting current at rated voltage is 200 A. What is the starting torque ? If a star-delta starter is used, what is the starting torque and starting current? Neglect magnetising current.
   [100 N-m; 33.3 N-m; 66.7 A] (*Electrical Machines-IV, Bangalore Univ. Aug. 1978*)
4. Determine approximately the starting torque of an induction motor in terms of full-load torque when started by means of (*a*) a star-delta switch (*b*) an auto-transformer with 50% tapping. Ignore magnetising current. The short-circuit current of the motor at normal voltage is 5 times the full-load current and the full-load slip is 4 per cent.
   [(*a*) 0.33 (*b*) 0.25] (*A.C. Machines, Madras Univ. 1976*)
5. Find the ratio of starting to full-load current for a 7.46 kW, 400-V, 3-phase induction motor with star/delta starter, given that the full-load efficiency is 0.87, the full-load p.f. is 0.85 and the short-circuit current is 15 A at 100 V. **[1.37]** (*Electric Machinery-II, Madras Univ. April 1978*)
6. A four-pole, 3-phase, 50-Hz, induction motor has a starting current which is 5 times its full-load current when directly switched on. What will be the percentage reduction in starting torque if (*a*) star-delta switch is used for starting (*b*) auto-transformer with a 60 per cent tapping is used for starting? (*Electrical Technology-III, Gwalior Univ. Nov. 1917*)
7. Explain how the performance of induction motor can be predicted by circle diagram. Draw the circle diagram for a 3-phase, mesh-connected, 22.38 kW, 500-V, 4-pole, 50-Hz induction motor. The data below give the measurements of line current, voltage and reading of two wattmeters connected to measure the input :—

No load	500 V	8.3 A	2.85 kW	−1.35 kW
Short circuit	100 V	32 A	− 0.75 kW	2.35 kW

From the diagram, find the line current, power factor, efficiency and the maximum output.

[83 A, 0.9, 88%, 50.73 kW] (*Electrical Machines-II, Vikram Univ. Ujjan 1977*)

33.12. Starting of Slip-ring Motors

These motors are practically always started with full line voltage, applied across the stator terminals. The value of starting current is adjusted by introducing a variable resistance in the rotor circuit. The controlling resistance is in the form of a rheostat, connected in star (Fig. 33.22), the resistance being gradually cut-out of the rotor circuit, as the motor gathers speed. It has been already shown that by increasing the rotor resistance, not only is the rotor (and hence stator) current reduced at starting, but at the same time, the starting torque is also increased due to improvement in power factor.

The controlling rheostat is either of stud or contactor type and may be hand-operated or automatic. The starter unit usually includes a line switching contactor for the stator along with no-voltage (or low- voltage) and over-current protective devices. There is some form of interlocking to ensure proper sequential operation of the line contactor and the starter. This interlocking prevents the closing of stator contactor unless the starter is 'all in'.

As said earlier, the introduction of additional external resistance in the rotor circuit enables a slip-ring motor to develop a high starting torque with reasonably moderate starting current. Hence, such motors can be started under load. This additional resistance is for starting purpose only. It is gradually cut out as the motor comes up to speed.

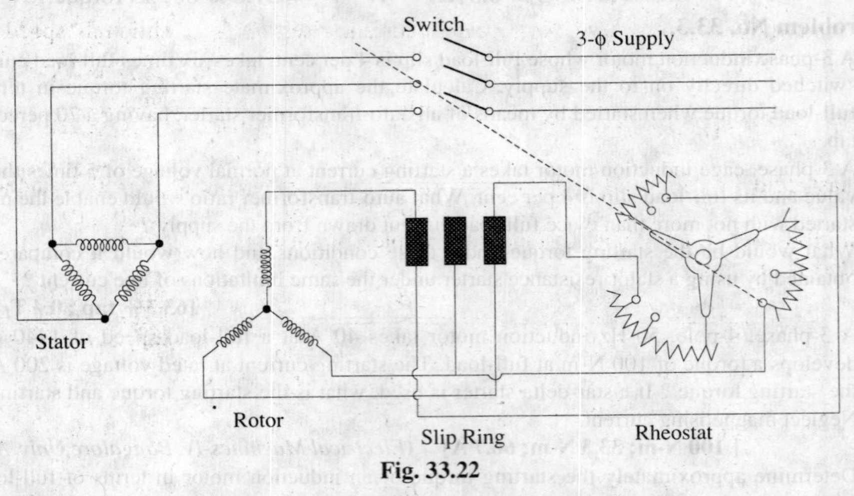

Fig. 33.22

The rings are, later on, short-circuited and brushes lifted from them when motor runs under normal conditions.

33.13. Starter Steps

Let it be assumed, as usually it is in the case of starters, that (*i*) the motor starts against a constant torque and (*ii*) that the rotor current fluctuates between fixed maximum and minimum values of I_{2max} and I_{2min} respectively.

In Fig. 33.23 is shown one phase of the 3-phase rheostat *AB* having *n* steps and the rotor circuit. Let R_1, R_2 ....etc. be the *total* resistances of the rotor circuit on the first, second step...etc. respectively. The resistances R_1, R_2 ..., etc. consist of rotor resistance per phase r_2 and the external resistances ρ_1, ρ_2 .... etc. Let the corresponding values of slips be s_1, s_2 ...etc. at stud No.1, 2...etc. At the commencement of each step, the current is I_{2max} and at the instant of leaving it, the current is I_{2min} . Let E_2 be the standstill e.m.f. induced in each phase of the rotor. When the handle touches first stud, the current rises to a maximum value I_{2max}, so that

$$I_{2\,max} = \frac{s_1 E_2}{\sqrt{[R_1^2 + (s_1 X_2)^2]}} = \frac{E_2}{\sqrt{[(R_1/s_1)^2 + X_2^2]}}$$

where s_1 = slip at starting *i.e.* unity and X_2 = rotor reactance/phase

Then, *before* moving to stud No. 2, the current is reduced to $I_{2\,min}$ and slip changes to s_2 such that $I_{2\,min} = \dfrac{E_2}{\sqrt{[(R_1/s_2)^2 + X_2^2]}}$

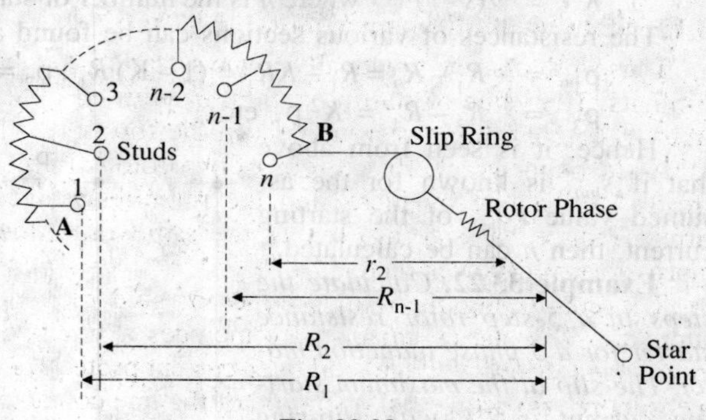

As we now move to stud No. 2, the speed momentarily remains the same, but current rises to $I_{2\,max}$ because some resistance is cut out.

Hence, $I_{2max} = \dfrac{E_2}{\sqrt{[(R_2/s_2)^2 + X_2^2]}}$

After some time, the current is again reduced to I_{2min} and the slip changes to s_3 such that

Fig. 33.23

$$I_{2\,min} = \frac{E_2}{\sqrt{[(R_2/s_3)^2 + X_2^2]}}$$

As we next move over to stud No.3, again current rises to I_{2max} although speed remains momentarily the same.

$$\therefore \quad I_{2\,max} = \frac{E_2}{\sqrt{[(R_2/s_3)^2 + X_2^2]}} \qquad \text{Similarly} \quad I_{2\,min} = \frac{E_2}{[(R_3/s_4)^2 + X_2^2]}$$

At the last stud *i.e.* *n*th stud, $I_{2max} = \dfrac{E_2}{\sqrt{[(r_2/s_{max})^2 + X_2^2]}}$ where s_{max} = slip under normal running conditions, when external resistance is completely cut out.

It is found from above that

$$I_{2max} = \frac{E_2}{\sqrt{[(R_1/s_1)^2 + X_2^2]}} = \frac{E_2}{\sqrt{[(R_2/s_2)^2 + X_2^2]}} = ... + \frac{E_2}{\sqrt{[(r_2/s_{max})^2 + X_2^2]}}$$

$$\text{or} \quad \frac{R_1}{s_1} = \frac{R_2}{s_2} = \frac{R_3}{s_3} = ... = \frac{R_{n-1}}{s_{n-1}} = \frac{R_n}{s_n} = \frac{r_2}{s_{max}} \quad ...(i)$$

Similarly,

$$I_{2min} = \frac{E_2}{\sqrt{[(R_1/s_2)^2 + X_2^2]}} = \frac{E_2}{\sqrt{[(R_2/s_3)^2 + X_2^2]}} = ... = \frac{E_2}{\sqrt{[(R_{n-1}/s_{max})^2 + X_2^2]}}$$

$$\text{or} \quad \frac{R_1}{s_2} = \frac{R_2}{s_3} = \frac{R_3}{s_4} = ... = \frac{R_{n-1}}{s_{max}} \quad ...(ii)$$

From (*i*) and (*ii*), we get

$$\frac{s_2}{s_1} = \frac{s_3}{s_2} = \frac{s_4}{s_3} = ... = \frac{R_2}{R_1} = \frac{R_3}{R_2} = \frac{R_4}{R_3} = ... = \frac{r_2}{R_{n-1}} = \text{K (say)} \quad ...(iii)$$

Now, from (*i*) it is seen that $R_1 = \dfrac{s_1 \times r_2}{s_{max}}$.

Now, s_1 = 1 at starting, when rotor is stationary.

$\therefore$ $R_1 = r_2/s_{max}$. Hence, R_1 becomes known in terms of rotor resistance/phase and normal slip.

From (*iii*), we obtain

$$R_2 = KR_1\,;\ R_3 = KR_2 = K^2R_1\,;\ R_4 = KR_3 = K^3R_1 \text{ and } r_2 = KR_{n-1} = K^{n-1}.R_1$$

or $$r_2 = K^{n-1}\,.\,\frac{r_2}{s_{max}} \quad \text{(putting the value of } R_1\text{)}$$

$$K = (s_{max})^{1/n-1} \text{ where } n \text{ is the number of starter studs.}$$

The resistances of various sections can be found as given below:-

$$\rho_1 = R_1 - R_2 = R_1 - KR_1 = (1 - K)\,R_1\,;\ \rho_2 = R_2 - R_3 = KR_1 - K^2R_1 = K\rho_1$$

$$\rho_3 = R_3 - R_4 = K^2\,\rho_1 \text{ etc.}$$

Hence, it is seen from above that if s_{max} is known for the assumed value I_{2max} of the starting current, then n can be calculated.

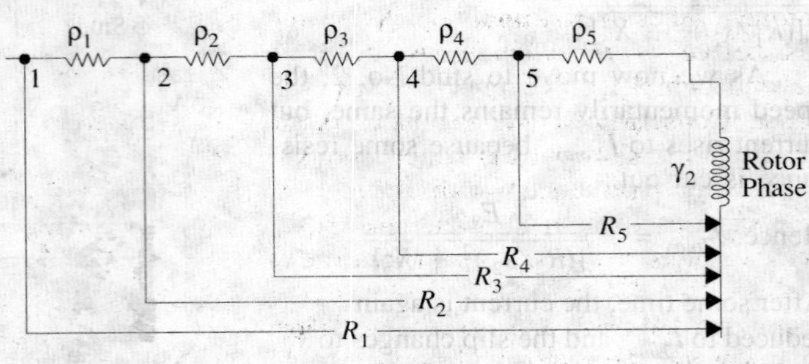

Fig. 33.24

Example 33.22. *Calculate the steps in a 5-step rotor resistance starter for a 3 phase induction motor. The slip at the maximum starting current is 2% with slip-ring short-circuited and the resistance per rotor phase is 0.02 Ω*

Solution. Here, $s_{max} = 2\%$

$= 0.02;\ r_2 = 0.02\ \Omega,\ n = 6$

$R_1 =$ total resistance in rotor circuit/phase on first stud

$= r_2/s_{max} = 0.02/0.02 = 1\ \Omega$

Now, $K = (s_{max})^{1/n-1} = (0.02)^{1/5} = 0.4573$

$R_1 = 1\Omega;\ R_2 = KR_1 = 0.4573 \times 1 = 0.4573\ \Omega$

$R_3 = KR_2 = 0.4573 \times 0.4573 = 0.2091\ \Omega;\ R_4 = KR_3 = 0.4573 \times 0.2091 = 0.0956\ \Omega$

$R_5 = KR_4 = 0.4573 \times 0.0956 = 0.0437\ \Omega;\ r_2 = KR_5 = 0.4573 \times 0.0437 = 0.02\ \Omega$ (as given)

The resistances of various starter sections are as found below.

$\rho_1 = R_1 - R_2 = 1 - 0.4573 = \mathbf{0.5427}\ \Omega;\quad \rho_2 = R_2 - R_3 = 0.4573 - 0.2091 = \mathbf{0.2482}\ \Omega$

$\rho_3 = R_3 - R_4 = 0.2091 - 0.0956 = \mathbf{0.1135}\ \Omega;\quad \rho_4 = R_4 - R_5 = 0.0956 - 0.0437 = \mathbf{0.0519}\ \Omega$

$\rho_5 = R_5 - r_2 = 0.0437 - 0.02 = \mathbf{0.0237}\ \Omega$

The resistances of various sections are shown in Fig. 33.24.

33.14. Crawling

It has been found that induction motors, particularly the squirrel-cage type, sometimes exhibit a tendency to run stably at speeds as low as one-seventh of their synchronous speed N_s. This phenomenon is known as crawling of an induction motor.

This action is due to the fact that the a.c. winding of the stator produces a flux wave, which is not a pure sine wave. It is a complex wave consisting of a fundamental wave, which revolves synchronously and odd harmonics like 3rd, 5th, and 7th etc. which rotate either in the forward or backward direction at $N_s / 3$, $N_s / 5$ and $N_s / 7$ speeds respectively. As a result, in addition to the fundamental torque, harmonic torques are also developed, whose synchronous speeds are $1/n$th of the speed for the fundamental torque *i.e.* N_s / n, where n is the order of the harmonic torque. Since 3rd harmonic currents are absent in a balanced 3-phase system, they produce no rotating field and, therefore, no torque. Hence, total motor torque has three components : (*i*) the fundamental torque, rotating with the synchronous speed N_s (*ii*) 5th harmonic torque* rotating at $N_s / 5$ speed and (*iii*) 7th harmonic torque, having a speed of $N_s / 7$.

Now, the 5th harmonic currents have a phase difference of 5 × 120° = 600° = –120° in three stator windings. The revolving field, set up by them, rotates in the *reverse* direction at $N_s / 5$. The forward speed of the rotor corresponds to a slip greater than 100%. The small amount of 5th harmonic reverse torque produces a braking action and may be neglected.

*The magnitude of the harmonic torques is $1/n^2$ of the fundamental torque.

The 7th harmonic currents in the three stator windings have a phase difference of 7 × 120° = 2 × 360° + 120° = 120°. They set up a forward rotating field, with a synchronous speed equal to 1/7th of the synchronous speed of the fundamental torque.

If we neglect all higher harmonics, the resultant torque can be taken as equal to the sum of the fundamental torque and the 7th harmonic torque, as shown in Fig. 33.25. It is seen that the 7th harmonic torque reaches its maximum positive value just before 1 / 7th synchronous speed N_s, beyond which it becomes negative in value. Consequently, the resultant torque characteristic shows a dip which may become very pronounced with certain slot combinations. If the mechanical load on the shaft involves a constant load torque, it is possible that the torque developed by the motor may fall below this load torque. When this happens, the motor will not accelerate upto its normal speed but will remain running at a speed, which is nearly 1 / 7th of its full-speed. This is referred to as crawling of the motor.

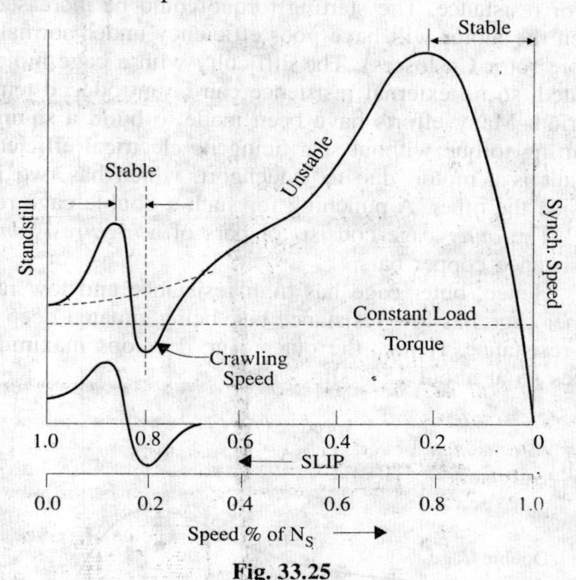

Fig. 33.25

33.15. Cogging or Magnetic Locking

The rotor of a squirrel-cage motor sometimes refuses to start at all, particularly when the voltage is low. This happens when the number of stator teeth S_1 is equal to the number of rotor teeth S_2 and is due to the magnetic locking between the stator and rotor teeth. That is why this phenomenon is sometimes referred to as teeth-locking.

It is found that the reluctance of the magnetic path is minimum when the stator and rotor teeth face each other rather than when the teeth of one element are opposite to the slots on the other. It is in such positions of minimum reluctance, that the rotor tends to remain fixed and thus cause serious trouble during starting. Cogging of squirrel cage motors can be easily overcome by making the number of rotor slots prime to the number of stator slots.

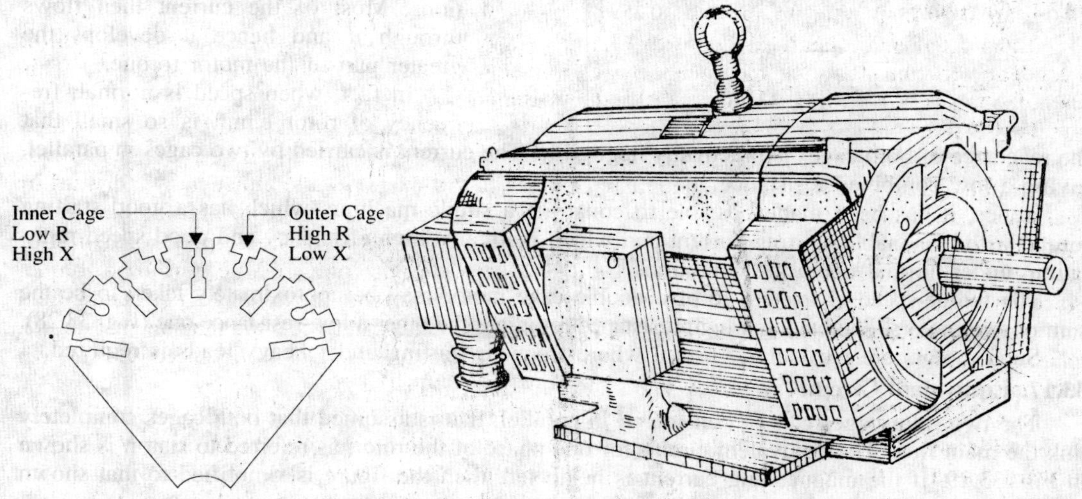

Fig. 33.26

Fig. 33.27. Double-cage, 30-kW, 400/440-V, 3-φ, 960 r.p.m. squirrel-cage motor. (*Courtesy : Jyoti Ltd., Baroda*).

33.16. Double Squirrel Cage Motor

The main disadvantage of a squirrel-cage motor is its poor starting torque, because of its low rotor resistance. The starting torque could be increased by having a cage of high resistance, but then the motor will have poor efficiency under normal running conditions (because there will be more rotor Cu losses). The difficulty with a cage motor is that its cage is permanently short-circuited, so no external resistance can be introduced temporarily in its rotor circuit during starting period. Many efforts have been made to build a squirrel-cages motor which should have a high starting torque without scarificing its electrical efficiency, under normal running conditions. The result is a motor, due to Boucheort, which has two independent cages on the same rotor, one inside the other. A punching for such a double cage rotor is hown in Fig. 33.26.

The *outer* cage consists of bars of a *high-resistance* metal, whereas the inner cage has low-resistance copper bars.

Hence, outer cage has high resistance and low ratio of reactance-to-resistance, whereas the inner cage has low resistance but, being situated deep in the rotor, has a large ratio of reactance-to-resistance. Hence, the outer cage develops maximum torque at starting, while the inner cage does so at about 15% slip.

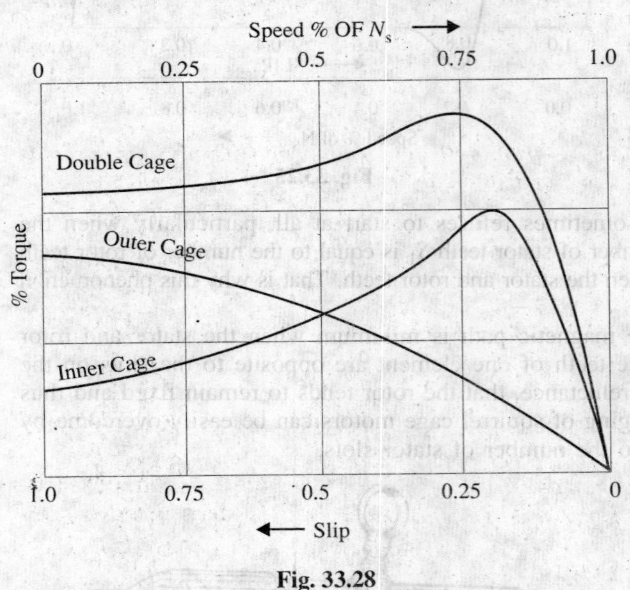

Fig. 33.28

As said earlier, at starting and at large slip values, frequency of induced e.m.f in the rotor is high. So the reactance of the inner cage ($=2\pi fL$) and therefore, its impedance are both high. Hence, very little current flows in it. Most of the starting current is confined to outer cage, despite its high resistance. Hence, the motor develops a high starting torque due to high-resistance outer cage. Double cage squirrel-cage motor is shown in Fig. 33.27.

As the speed increases, the frequency of the rotor e.m.f. decreases, so that the reactance and hence the impedance of inner cage deceases and becomes very small, under normal running conditions. Most of the current then flows through it and hence it develops the greater part of the motor torque.

In fact, when speed is normal, frequency of rotor e.m.f. is so small that the reactance of both cages is practically negligible. The current is carried by two cages in parallel, giving a low combined resistance.

Hence, it has been made possible to construct a single machine, which has a good starting torque with reasonable starting current and which maintains high efficiency and good speed regulation, under normal operating conditions.

The torque-speed characteristic of a double cage motor may be approximately taken to be the sum of two motors, one having a high-resistance rotor and the other a low-resistance one (Fig. 33.28).

Such motors are particularly useful where frequent starting under heavy loads is required.

33.17. Equivalent Circuit

The two rotor cages can be considered in parallel, if it is assumed that both cages completely link the main flux. The equivalent circuit for one phase of the rotor, as referred to stator, is shown in Fig. 33.29. If the magnetising current is neglected, then the figure is simplified to that shown in Fig. 33.30. Hence, R_0'/s and R_i'/s are resistances of outer and inner rotors as referred to stator respectively and X_0' and X_i' their reactances

Total impedance as referred to primary is given by

$$Z_{01} = R_1 + jX_1 + \frac{1}{1/Z_1' + 1/Z_0'} = R_1 + jX_1 + \frac{Z_i' Z_0'}{Z_0' + Z_i'}$$

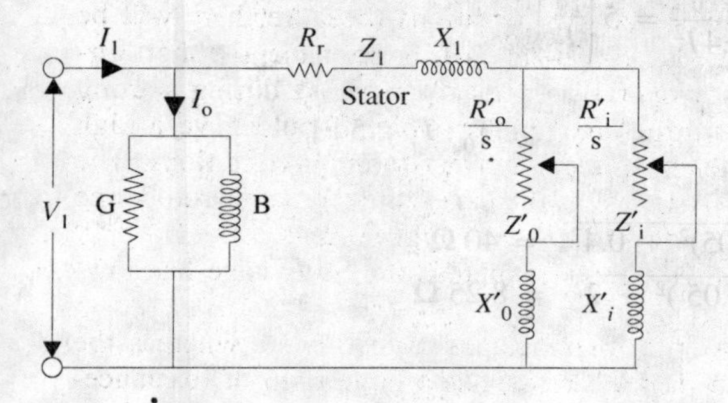

Fig. 33.29

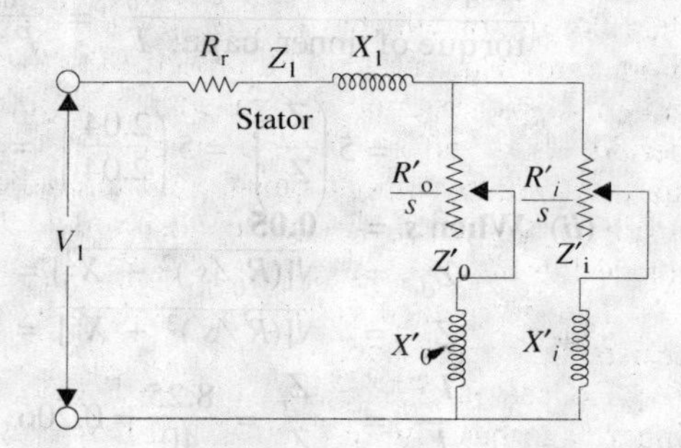

Fig. 33.30

Example 33.23. *A double-cage induction motor has the following equivalent circuit parameters, all of which are phase values referred to the primary :*

Primary	$R_1 = 1\ \Omega$	$X_1 = 3\ \Omega$
Outer cage	$R_0' = 3\ \Omega$	$X_0' = 1\Omega$
Inner cage	$R_i' = 0.6\ \Omega$	$X_i' = 5\ \Omega$

The primary is delta-connected and supplied from 440 V. Calculate the starting torque and the torque when running at a slip of 4%. The magnetising current may be neglected.

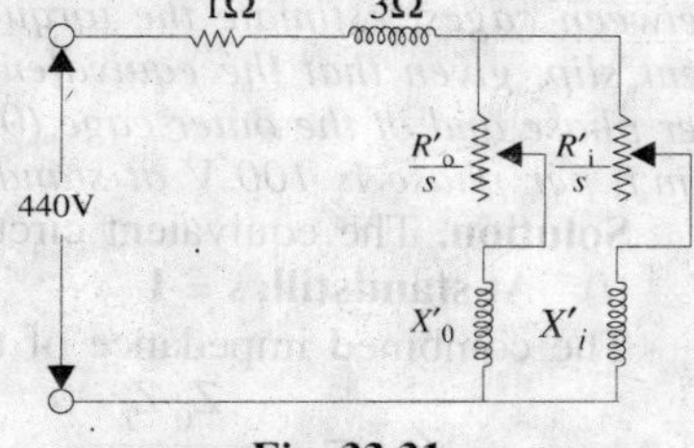

Fig. 33.31

Solution.

(*i*) **At start,** **s = 1**

$$\mathbf{Z_{01}} = 1 + j\,3 + \frac{1}{1/(3+j\,1) + 1/(0.6 + j\,5)}$$

$$= 2.68 + j\,4.538\ \Omega$$

Current / phase = $440 / (2.68^2 + 4.538^2)^{1/2} = 83.43$ A

Torque = $3 \times 83.43^2 \times (2.68 - 1)$ = **35,000 synch watt.**

(*ii*) **when s = 4 per cent**

In this case, approximate torque may be found by neglecting the outer cage impedance altogether. However, it does carry some current, which is almost entirely determined by its resistance.

$$\mathbf{Z_{01}} = 1 + j\,3 + \frac{1}{1/(3/0.04) + 1/(0.6/0.04 + j\,5)} = 13.65 + j\,6.45$$

Current/phase = $440/(13.65^2 + 6.45^2)^{1/2}$ = 29.14 A

Torque = $3 \times 29.14^2 \times (13.65 - 1)$ = **32,000 synch watt.**

Example 33.24. *At standstill, the equivalent impedance of inner and outer cages of a double-cage rotor are (0.4 + j2)Ω and (2+ j0.4) Ω respectively. Calculate the ratio of torques produced by the two cages (i) at standstill (ii) at 5% slip.*

(Elect. Machines-II, Punjab Univ. 1989)

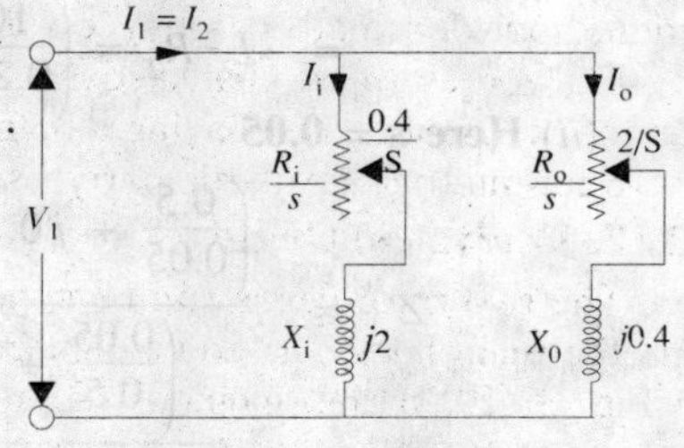

Fig. 33.32

Solution. The equivalent circuit for one phase is shown in Fig. 33.32.

(*i*) **At standstill, s = 1**

Impedance of inner cage = $Z_i = \sqrt{0.4^2 + 2^2} = 2.04\ \Omega$

Impedance of outer cage = $Z_0 = \sqrt{2^2 + 0.4^2} = 2.04\ \Omega$

If I_0 and I_i are the current inputs of the two cages, then

power input of inner cage, $P_i = I_i^2 R_i = 0.4\, I_i^2$ watt
power input of outer cage, $P_0 = I_0^2 R_0 = 2\, I_0^2$

$$\therefore \quad \frac{\text{torque of outer cage, } T_0}{\text{torque of inner cage, } T_i} = \frac{P_0}{P_i} = \frac{2I_0^2}{0.4 I_i^2} = 5\left(\frac{I_0}{I_i}\right)^2$$

$$= 5\left(\frac{Z_i}{Z_0}\right)^2 = 5\left(\frac{2.04}{2.04}\right)^2 = 5 \qquad \therefore\ T_0 : T_i :: 5 : 1$$

(*ii*) When s = 0.05

$$Z_0 = \sqrt{[(R_0/s)^2 + X_0^2]} = \sqrt{(2/0.05)^2 + 0.4^2} = 40\ \Omega$$

$$Z_i = \sqrt{[(R_i/s)^2 + X_i^2]} = \sqrt{0.4/0.05)^2 + 2^2} = 8.25\ \Omega$$

$$\therefore \quad \frac{I_o}{I_i} = \frac{Z_i}{Z_o} = \frac{8.25}{40} = 0.206$$

$$P_0 = I_0^2 R_0 / s = 40\, I_0^2; \quad P_i = I_i^2 R_i / s = 8\, I_i^2$$

$$\therefore \quad \frac{T_o}{T_i} = \frac{P_o}{P_i} = \frac{40 I_o^2}{8 I_i^2} = 5\left(\frac{I_o}{I_i}\right)^2 = 5(0.206)^2 = 0.21$$

$$\therefore \quad T_0 : T_i :: 0.21 : 1$$

It is seen from above, that the outer cage provides maximum torque at starting, whereas inner cage does so later.

Example 33.25. *A double-cage rotor has two independent cages. Ignoring mutual coupling between cages, estimate the torque in synchronous watts per phase(i) at standstill and at 5 per cent slip, given that the equivalent standstill impedance of the inner cage is (0.05 + j 0.4) ohm per phase and of the outer cage (0.5 + j 0.1) ohm per phase and that the rotor equivalent induced e.m.f. per phase is 100 V at standstill.*

Solution. The equivalent circuit of the double-cage rotor is shown in Fig. 33.33.

(*i*) **At standstill, $s = 1$**

The combined impedance of the two cages is

$$Z = \frac{Z_0 Z_i}{Z_0 + Z_i}$$

where Z_0 = impedance of the outer cage
Z_i = impedance of the inner cage

$$Z = \frac{(0.5 + j\,0.1)\,(0.05 + j\,0.4)}{(0.55 + j\,0.5)}$$

$$= 0.1705 + j\,0.191 \text{ ohm}$$

$$\therefore \quad Z = \sqrt{0.1705^2 + 0.191^2} = 0.256\ \Omega$$

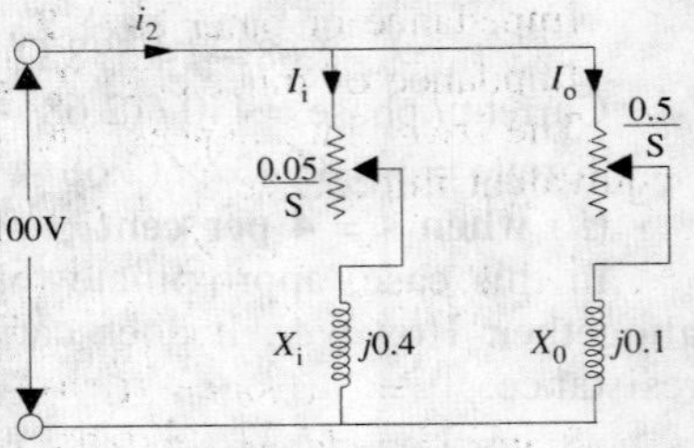

Fig. 33.33

Rotor current $I_2 = 100/0.256$ A; Combined resistance $R_2 = 0.1705\ \Omega$

Torque at standstill in synchronous watts per phase is

$$= I_2^2 R_2 = \left(\frac{100}{0.256}\right)^2 \times 0.1705 = \mathbf{26{,}000 \text{ synch watts}}$$

(*ii*) **Here s = 0.05**

$$Z = \frac{\left(\dfrac{0.5}{0.05} + j\,0.1\right)\left(\dfrac{0.05}{0.05} + j\,0.4\right)}{\left(\dfrac{0.05}{0.5} + \dfrac{0.5}{0.05} + j\,0.5\right)} = 1.01 + j\,0.326 \text{ ohm}$$

$$Z = \sqrt{(1.01)^2 + 0.326^2} = 1.06\ \Omega$$

Combined resistance $R_2 = 1.01\ \Omega$; rotor current = 100/1.06 A

Torque in synchronous watts per phase is

$$= I_2^2 R_2 = (100/1.06)^2 \times 1.1 = \mathbf{9,000\ synch.watt.}$$

Example 33.26. *In a double-cage induction motor, if the outer cage has an impedance at standstill of (2 + j 1.2) ohm, determine the slip at which the two cages develop equal torques if the inner cage has an impedance of (0.5 + j 3.5) ohm at standstill.*

(Electric Machines, Osmania Univ. 1991)

Solution. Let s be the slip at which two cage develop equal torques.

Then $Z_1 = \sqrt{(2/s)^2 + 1.2^2}$ and $Z_2 = \sqrt{(0.5/s)^2 + 3.5^2}$

$$\therefore \quad \left(\frac{I_1}{I_2}\right)^2 = \left(\frac{Z_2}{Z_1}\right)^2 = \frac{(0.25/s^2) + 3.5^2}{(4/s^2) + 1.44}$$

Power input to outer cage $P_1 = I_1^2 R_1/s$

$$\therefore \quad P_1 = I_1^2 \times \frac{2}{s}; \qquad P_2 = I_2^2 \times \frac{0.5}{s}$$

$$\therefore \quad \frac{T_1}{T_2} = \frac{P_1}{P_2} = \left(\frac{I_1}{I_2}\right)^2 \times 4 = \frac{(0.25/s^2) + 3.5^2}{(4/s^2) + 1.44} \times 4$$

As $T_1 = T_2$

$$\therefore \quad \frac{4}{s^2} + 1.44 = \left(\frac{0.25}{s^2} + 12.25\right) \times 4 \qquad \therefore \quad s = 0.251 = \mathbf{25.1\%}$$

Example 33.27. *The resistance and reactance (equivalent) values of a double-cage induction motor for stator, outer and inner cage are 0.25, 1.0 and 0.15 ohm resistance and 3.5, zero and 3.0 ohm reactance respectively. Find the starting torque if the phase voltage is 250 V and the synchronous speed is 1000 r.p.m.*

(I.E.E. London)

Solution. The equivalent circuit is shown in Fig. 33.34 where magnetising current has been neglected. At starting, $s = 1$

Impedance of outer cage $\mathbf{Z}_0' = (1 + j\,0)$

Impdance of inner cage $\mathbf{Z}_i' = (0.15 + j\,3)$.

The two impedances are in parallel. Hence, their equivalent impedance

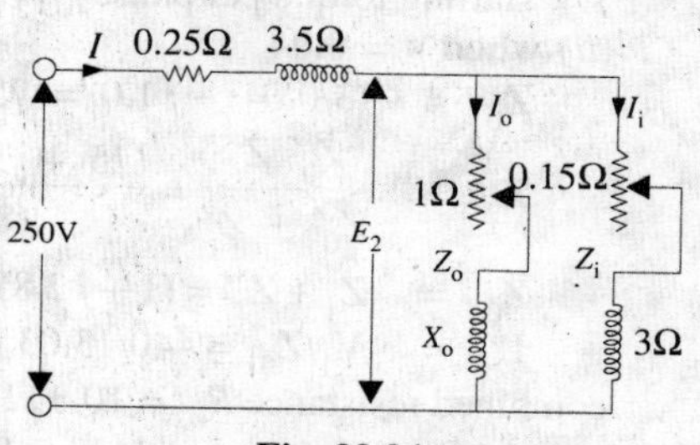

Fig. 33.34

$$\mathbf{Z}_2' = \frac{Z_0' Z_i'}{Z_0' + Z_i'} = \frac{(1 + j\,0)(0.15 + j\,3)}{(1 + j\,0) + (0.15 + j\,3)}$$

$$= 0.889 + j\,0.29$$

Stator impedance $= (0.25 + j3.5)$

$\therefore$ total impedance $Z_{01} = (0.889 + j\,0.29) + (0.25 + j\,3.5) = (1.14 + j\,3.79)\ \Omega$ (approx.)

$$\text{Current } I = \frac{\text{phase voltage}}{\text{total phase impedance}} = \frac{250 + j\,0}{1.14 + j\,3.79} = 18.2 - j\,60.5 = 66.15 \text{ A}$$

Rotor Cu loss/phase = (current)^2 × total resistance/phase of two rotors

$$= 66.15^2 \times 0.889 = 3{,}890 \text{ W}$$

Total Cu loss in 3-phases = 3 × 3890 = 11,670 W

Now, rotor input $= \dfrac{\text{rotor Cu loss}}{s}$; At starting, $s = 1$ $\therefore$ rotor input = 11,670 W

$$\therefore \quad T_{start} = 11{,}670 \text{ synchronous watts}$$

Also $T_{start} \times 2\pi N_S = 11{,}670$ or $T_{start} = \dfrac{11{,}670}{2\pi \times (1000/60)} = \mathbf{111.6\ N\text{-}m\ (approx.)}$

Note. If torques developed by the two rotors separately are required, then find E_2 (Fig. 33.34), then I_1 and I_2. Knowing these values, T_1 and T_2 can be found as given in previous example.

Example 33.28. *A double-cage induction motor has the following equivalent circuit parameters all of which are phase values referred to the primary:*

Primary :	$R_1 = 1$ *ohm*	$X_1 = 2.8$ *ohm*
Outer cage;	$R_0' = 3$ *ohm*	$X_0' = 1.0$ *ohm*
Inner cage:	$R_i' = 0.5$ *ohm*	$X_i' = 5$ *ohm*

The primary is delta-connected and supplied from 440 V. Calculate the starting torque and the torque when running at a slip of 4 per cent. The magnetizing branch can be assumed connected across the primary terminals. **(Electrical Machines-II, South Gujarat Univ. 1987)**

Solution. The equivalent circuit for one phase is shown in Fig. 33.35. It should be noted that magnetising impedance Z_0 has no bearing on the torque and speed and hence, can be neglected so far as these two quantities are concerned-

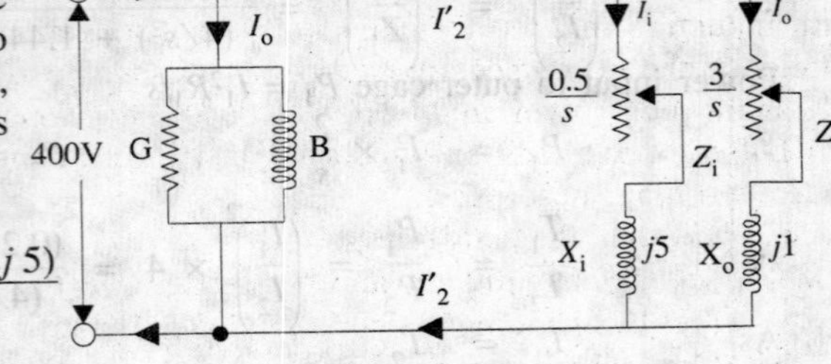

Fig. 33.35

(*i*) **At standstill s = 1**

$$Z_2' = \frac{Z_0'\,Z_i'}{Z_0' + Z_i'} = \frac{(3 + j\,1.0)\,(0.5 + j\,5)}{(3.5 + j\,6)}$$

$$= 1.67 + j\,1.56$$

$$\mathbf{Z_{01}} = Z_1 + Z_2' = (1 + j\,2.8) + (1.67 + j\,1.56)$$

$$= (2.67 + j\,4.36) = 5.1 \angle 58.5°$$

Voltage per phase, $V_1 = 440$ V. Since stator is delta-connected.

$$\mathbf{I_2'} = \mathbf{V_1/Z_{01}} = 440/5.1 \angle 58.5° = 86.27 \angle -58.5°$$

Combined resistance $R_2 = 1.6\ \Omega$

$\therefore$ starting torque per phase $= I_2'^2 R_2 = 86.27^2 \times 1.67 =$ **12,430 synch. watt.**

(*ii*) **when s = 0.04**

$$\mathbf{Z_0'} = (3/0.04) + j\,1.0 = 75 + j\,1.0; \qquad \mathbf{Z_i'} = (0.5/0.04) + j\,5 = 12.5 + j\,5$$

$$\mathbf{Z_2'} = \frac{Z_0'\,Z_i'}{Z_0' + Z_i'} = \frac{(75 + j\,1.0)\,(12.5 + j5)}{(87.5 + j\,6)} = 10.3 + j\,3.67$$

$$\mathbf{Z_{01}} = Z_1 + Z_2' = (1 + j\,2.8) + (10.3 + j\,6.47) = 11.3 + j\,6.47 = 13.03 \angle 29.8°$$

$$\mathbf{I_2'} = \mathbf{V_1/Z_{01}} = 440/13.03 \angle 29.8° = 33.76 \angle -29.8°$$

combined resistance $R_{02} = 10.3\ \Omega$

$\therefore$ full-load torque per phase $= I_2' R_{02} = 33.76^2 \times 10.3 =$ **11.740 synch. watts**

Obviously, starting torque is higher than full-load torque.

Tutorial Problem No. 33.4.

1. Calculate the steps in a 5-section rotor starter of a 3-phase induction motor for which the starting current should not exceed the full-load current, the full-load slip is 0.018 and the rotor resistance is 0.015 Ω per phase

   $\rho_1 =$ **0.46 Ω** ; $\rho_2 =$ **0.206 Ω** ; $\rho_3 =$ **0.092 Ω** ; $\rho_4 =$ **0.042 Ω** ; $\rho_5 =$ **0.0185 Ω**

   (Electrical Machinery-III, Kerala Univ. Apr. 1976)

2. The fulll-load slip of a 3-phase double-cage induction motor is 6% and the two cages have impedances of (3.5 + *j* 1.5) Ω and (0.6 + *j* 7.0) Ω respectively. Neglecting stator impedances and magnetising current, calculate the starting torque in terms of full-load torque. **[79%]**

3. In a double-cage induction motor, if the outer cage has an impedance at standstill of (2 + *j* 2) ohm and the inner cage an impedance of (0.5 + *j* 5) Ω, determine the slip at which the two cages develop equal torques. **[17.7%]**

4. The two independent cages of a rotor have the respective standstill impedance of (3 + *j* 1) ohm and (1 + *j* 4) ohm. What proportion of the total torque is due to the outer cage (*a*) at starting and (*b*) at

a fractional slip of 0.05?

[(*a*) **83.6%** (*b*) **25.8%**] (*Principle of Elect. Engg.I, Jadavpur Univ. 1975*)

5. An induction motor has a double cage rotor with equivalent impedance at standstill of $(1.0 + j\,1.0)$ and $(0.2 + j\,4.0)$ ohm. Find the relative value of torque given by each cage at a slip of 5%.

[(*a*) **40.1** (*b*) **0.4 : 1**] (*Electrical Machines-I, Gwalior Univ. Nov. 1977*)

33.18. Speed Control of Induction Motors*

A 3-phase induction motor is practically a constant-speed machine, more or less like a d.c. shunt motor. The speed regulation of an induction motor (having low resistnace) is usually less than 5% at full-load. However, there is one difference of practical importance between the two. Whereas d.c. shunt motors can be made to run at any speed within wide limits, with good efficiency and speed regulation, merely by manipulating a simple field rheostat, the same is not possible with induction motors. In their case, speed reduction is accompanied by a corresponding loss of efficiency and good speed regulation. That is why it is much easier to build a good adjustable-speed d.c. shunt motor than an adjustable speed induction motor.

Different methods by which speed control of induction motors is achieved, may be grouped under two main headings :

1. **Control from stator side**
   (*a*) by changing the applied voltage (*b*) by changing the applied frequency
   (*c*) by changing the number of stator poles
2. **Control from rotor side**
   (*d*) rotor rheostat control
   (*e*) by operating two motors in concatenation or cascade
   (*f*) by injection an e.m.f. in the rotor circuit.

A brief description of these methods would be given below :

(*a*) Changing Applied Voltage

This method, though the cheapest and the easiest, is rarely used because

(*i*) a large change in voltage is required for a relatively small change in speed

(*ii*) this large change in voltage will result in a large change in the flux density thereby seriously disturbing the magnetic conditions of the motor.

(*b*) Changing the Applied Frequency

This method is also used very rarely. We have seen that the synchronous speed of an induction motor is given by $N_s = 120f/P$. Clearly, the synchronous speed (and hence the running speed) of an induction motor can be changed by changing the supply frequency *f*. However, this method could only be used in cases where the induction motor happens to be the only load on the generators, in which case, the supply frequency could be controlled by controlling the speed of the prime movers of the generators. But, here again the range over which the motor speed may be vardied is limited by the economical speeds of the prime movers. This method has been used to some extent on electrically-driven ships.

(*c*) Changing the Number of Stator Poles

This method is *easily applicable to squirrel-cage motors* because the squirrel-cage rotor adopts itself to any reasonable number of stator poles.

From the above equation it is also clear that the synchronous (and hence the running)speed of an induction motor could also be changed by changing the number of stator poles. This change of number of poles is achieved by having two or more entirely independent stator windings in the same slots. Each winding gives a different number of poles and hence different synchronous speed. For example, a 36-slot stator may have two 3-φ windings, one with 4 poles and the other with 6-poles. With a supply frequency of 50-Hz, 4-pole winding will give $N_s = 120 \times 50/4 = 1500$ r.p.m. and the 6-pole winding will give $N_s = 120 \times 50/6 = 1000$ r.p.m. Motors with four independent stator winding are also in use and they give four different synchronous (and hence running) speeds. Of course, one winding is used at a time, the others being entirely disconnected.

*For Electronic Control of AC Motors, please consult the relevant chapter 43 of this book.

This method has been used for elevator motors, traction motors and also for small motors driving machine tools.

Speeds in the ratio of 2:1 can be produced by a single winding if wound on the consequent-pole principle. In that case, each of the two stator windings can be connected by a simple switch to give two speeds, each, which means four speeds in all. For example, one stator winding may give 4 or 8-poles and the other 6 or 12-poles. For a supply frequency of 50-Hz, the four speeds will be 1500, 750, 1000 and 500 r.p.m. Another combination, commonly used, is to group 2- and 4-pole winding with a 6- and 12-pole winding, which gives four synchronous speeds of 3000, 1500, 1000 and 500 r.p.m.

(*d*) Rotor Rheostat Control

In this method (Fig. 33.36), which is applicable to slip-ring motors alone, the motor speed is reduced by introducing an external resistance in the rotor circuit. For this purpose, the rotor starter may be used, provided it is continuously rated. *This method is, in fact, similar to the armature rheostat control method of d.c. shunt motors.*

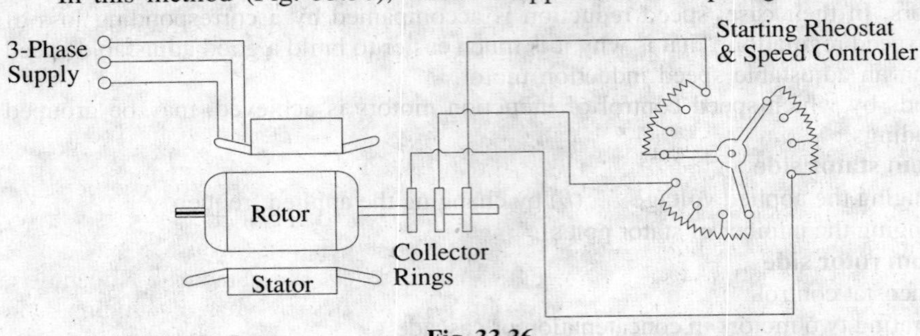

Fig. 33.36

It has been shown in Art 32.22 that near synchronous speed (*i.e.* for very small slip value), $T \propto s/R_2$.

It is obvious that for a given torque, slip can be increased *i.e.* speed can be decreased by increasing the rotor resistance R_2.

One serious disadvantage of this method is that with increase in rotor resistance, I^2R losses also increase which decrease the operating efficiency of the motor. In fact, the loss is directly proportional to the reduction in the speed.

The second disadvantage is the double dependence of speed, not only on R_2 but on load as well.

Because of the wastefulness of this method, it is used where speed changes are needed for short periods only.

Example 33.29. *The rotor of a 4-pole, 50-Hz slip-ring induction motor has a resistance of 0.30 Ω per phase and runs at 1440 rpm. at full load. Calculate the external resistance per phase which must be added to lower the speed to 1320 rpm, the torque being the same as before.*

(Advanced Elect. Machines AMIE Sec.E1992)

Solution. The motor torque is given by $T = \dfrac{K\,s\,R_2}{R_2^2 + (s\,X_2)^2}$

Since, X_2 is not given, $T = \dfrac{K\,s\,R_2}{R_2{}^2} = \dfrac{K\,s}{R_2}$

In the first case, $T_1 = Ks_1/R_2$; in the second case, $T_2 = Ks_2/(R_2 + r)$

where r is the external resistance per phase, added to the rotor circuit

Since $T_1 = T_2$ $\quad\therefore\; K\,s_1/R_2 = K\,s_2/(R_2 + r)$ or $(R_2 + r)/R_2 = s_2/s_1$

Now, $N_s = 120 \times 50/4 = 1500$ rpm; $N_1 = 1440$ rpm; $N_2 = 1320$ rpm

$\therefore\; s_1 = (1500 - 1440)/1500 = 0.04$; $s_2 = (1500 - 1320)/1500 = 0.12$

$\therefore\; \dfrac{0.3 + r}{0.3} = \dfrac{0.12}{0.04} \qquad \therefore\; r = \mathbf{0.6\ \Omega}$

Example 33.30. *A certain 3-phase, 6-pole, 50-Hz induction motor when fully-loaded, runs with a slip of 3%. Find the value of the resistance necessary in series per phase of the rotor to reduce the speed by 10%. Assume that the resistance of the rotor per phase is 0.2 ohm.*

(Electrical Engineering-II (M), Bangalore Univ. 1989)

Solution. $T = \dfrac{Ks\,R_2}{R_2^2 + (s\,X_2)^2} = \dfrac{K_s\,R_2}{R_2^2} = \dfrac{K\,s}{R_2}$ —neglecting $(s\,X_2)$

$\therefore$ $T_1 = Ks_1 / R_2$ and $T_2 = K\,s_2 / (R_2 + r)$ where r is the external resistance per phase added to the rotor circuit.

Since $T_1 = T_2$, $K\,s_1 / R_2 = K\,s_2 / (R_2 + r)$ or $(R_2 + r)/R_2 = s_2 / s_1$

Now, $N_s = 120 \times 50/6 = 1000$ rpm., $s_1 = 0.03$, $N_1 = 1000(1 - 0.030) = 970$ rpm.

$N_2 = 970 - 10\%$ of $970 = 873$ rpm., $s_2 = (1000 - 873)/1000 = 0.127$

$$\therefore \quad \frac{0.2 + r}{0.2} = \frac{0.127}{0.03}\ ; \qquad r = \mathbf{0.65\ \Omega}.$$

(*e*) Cascade or Concatenation or Tandem Operation

In this method, two motors are used (Fig. 33.37) and are ordinarily mounted on the same shaft, *so that both run at the same speed* (or else they may be geared together).

The stator winding of the main motor A is connected to the mains in the usual way, while that of the auxiliary motor B is fed from the rotor circuit of motor A. For satisfactory operation, the main motor A should be phase-wound *i.e.* of slip-ring type with stator to rotor winding ratio of 1 : 1, so that, in addition to concatenation, each motor may be run from the supply mains separately.

There are at least three ways (and sometimes four ways) in which the combination may be run.

1. Main motor A may be run separately from the supply. In that case, the synchronous speed is $N_{sa} = 120\ f / P_a$ where P_a = Number of stator poles of motor A.
2. Auxiliary motor B may be run separately from the mains (with motor A being disconnected). In that case, synchronous speed is $N_{sb} = 120 \times f / P_b$ where P_b = Number of stator poles of motor B.
3. The combination may be connected in cumulative cascade *i.e.* in such a way that the phase rotation of the stator fields of both motors is in the same direction. The synchronous speed of the cascaded set, in this case, is $N_{sc} = 120\,f/(P_a + P_b)$.

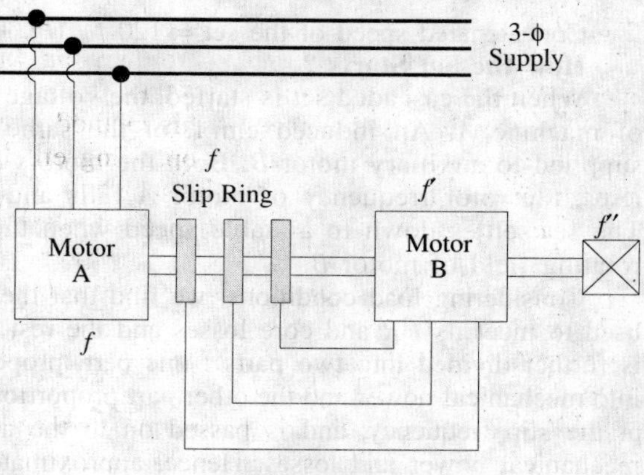

Fig. 33.37

Proof

Let N = actual speed of concatenated set ;

N_{sa} = synchronous speed of motor A, it being independent of N.

Clearly, the relative speed of rotor A with respect to its stator field is $(N_{sa} - N)$. Hence, the frequency f' of the induced e.m.f. in rotor A is given by

$$f' = \frac{N_{sa} - N}{N_{sa}} \times f$$

This is also the frequency of the e.m.f. applied to the stator of motor B. Hence, the synchronous speed of motor B with this input frequency is

$$N' = 120\frac{f'}{P_b} = \frac{120\,(N_{sa} - N)f}{P_b \times N_{sa}} \quad ...(i)$$

(Note that N' is not equal to N_{sb} which is the synchronous speed of motor B with supply frequency).

This will induce an e.m.f. of frequency, say, f'' in the rotor B. Its value is found from the fact that the stator and rotor frequencies are proportional to the speeds of stator field and the rotor

$$\therefore \quad f'' = \frac{N' - N}{N'} f$$

Now, on no-load, the speed of rotor B is almost equal to its synchronous speed, so that the frequency of induced e.m.f. is, to a first approximation, zero.

$$f'' = 0, \text{ or } \frac{N' - N}{N'} f = 0 \text{ or } N' = N \quad ...(ii)$$

From (i) above $$N' = \frac{120f\,(N_{sa} - N)}{P_b \times N_{sa}} = \frac{120f}{P_b}\left(1 - \frac{N}{N_{sa}}\right)$$

Hence, from (ii) above,

$$\frac{120f}{P_b}\left(1 - \frac{N}{N_{sa}}\right) = N \text{ or } \frac{120f}{P_b} = N\left(1 + \frac{1}{N_{sa}} \times \frac{120f}{P_b}\right)$$

Putting $N_{sa} = 120f/P_a$, we get

$$\frac{120f}{P_b} = N\left(1 + \frac{P_a}{120f} \times \frac{120f}{P_b}\right) = N\left(1 + \frac{P_a}{P_b}\right) \therefore N = \frac{120f}{(P_a + P_b)}$$

Concatenated speed of the set $= 120\,f\,/\,(P_a + P_b)$

How the Set Starts ?

When the cascaded set is started, the voltage at frequency f is applied to the stator winding of machine A. An induced e.m.f. of the same frequency is produced in rotor A which is supplied to auxiliary motor B. Both the motors develop a forward torque. As the shaft speed rises, the rotor frequency of motor A falls and so does the *synchronous* speed of motor B. The set settles down to a stable speed when the shaft speed becomes equal to the speed of rotating field of motor B.

Considering load conditions, we find that the electrical power taken in by stator A is partly used to meet its I^2R and core losses and the rest is given to its rotor. The power given to rotor is further divided into two parts : one part, proportional to the speed of set *i.e.* N is converted into mechanical power and the other part proportional to $(N_{sa} - N)$ is developed as electrical power at the slip frequency, and is passed on to the auxilary motor B, which uses it for producing mechanical power and losses. Hence, approximately, the machanical outputs of the two motors are in the ratio $N : (N_{sa} - N)$. *In fact, it comes to that the mechanical outputs are in the ratio of the number of poles of the motors.*

It may be of interest to the reader to know that it can be proved that

i. $s = f'/f$ where s = slip of the set referred to its synchronous speed N_{sc}.

$= (N_{sc} - N)/N_{sc}$

ii. $s = s_a\,s_b$

where s_a and s_b are slips of two motors, referred to their respective stators *i.e*

$$s_a = \frac{N_{sa} - N}{N_{sa}} \quad \text{and} \quad s_b = \frac{N' - N}{N}.$$

Conclusion

We can briefly note the main conclusions drawn from the above discussion :

(*a*) the mechanical outputs of the two motors are in the ratio of their number of poles.

(*b*) $s = f''/f$ (*c*) $s = s_a.s_b$

4. The fourth possible connection is the *differential cascade*. In this method, the phase rotation of stator field of the motor B is opposite to that of the stator of motor A. This reversal of phase rotation of stator of motor B is obtained by interchanging any of its two leads. It can be proved in the same way as above, that for this method of connection, the synchronous speed of the set is

$$N_{sc} = 120f/(P_a - P_b)$$

As the differentially-cascaded set has a very small or zero starting torque, this method is rarely used. Moreover, the above expression for synchronous speed becomes meaningless for $P_a = P_b$.

Example 33.31. *Two 50-Hz, 3-ϕ induction motors having six and four poles respectively are cumulatively cascaded, the 6-pole motor being connected to the main supply. Determine the frequencies of the rotor currents and the slips referred to each stator field if the set has a slip of 2 per cent.* **(Elect. Machinery-II Madras Univ. 1987)**

Solution. Synchronous speed of set $N_{sc} = 120 \times 50/10 = 600$ r.p.m.

Actual rotor speed $N = (1 - s)N_{sc} = (1 - 0.02)600 = 588$ r.p.m.

Synchronous speed of the stator field of 6-pole motor, $N_{sa} = 120 \times 50/6 = 1000$ r.p.m.

Slip referred to this stator field is $s_a = \dfrac{N_{sa} - N}{N_{sa}} = \dfrac{1000 - 588}{1000} = 0.412$ or **41.2%**

Frequency of the rotor currents of 6-pole motor $f' = s_a f = 0.412 \times 50 =$ **20.6 Hz**

This is also the frequency of stator currents of the four pole motor. The synchronous speed of the stator 4-pole motor is

$$N' = 120 \times 20.6/4 = 618 \text{ r.p.m.}$$

This slip, as referred to the 4-pole motor, is $s_b = \dfrac{N' - N}{N'} = \dfrac{618 - 588}{618}$

$= 0.0485$ or **4.85%**

The frequency of rotor current of 4-pole motor is

$$f'' = s_b f' = 0.0485 \times 20.6 = \mathbf{1.0\ Hz\ (approx)}$$

As a check, $f'' = sf = 0.02 \times 50 = 1.0$ Hz

Example 33.32. *A 4-pole induction motor and a 6-pole induction motor are connected in cumulative cascade. The frequency in the secondary circuit of the 6-pole motor is observed to be 1.0 Hz. Determine the slip in each machine and the combined speed of the set. Take supply frequency as 50 Hz.* **(Electrical Machinery-II, Madras Univ. 1986)**

Solution. With reference to Art. 33.18 (*e*) and Fig. 33.38

$N_{sc} = 120 \times 50/(4 + 6) = 600$ r.p.m.

$s = f''/f = 1/50 = 0.02$

$N =$ actual speed of the concatenated set

$\therefore\ 0.02 = \dfrac{600 - N}{600}$ or $N =$ **588 r.p.m.**

$N_{sa} = 120 \times 50/4 = 1500$ r.p.m.

$s_a = (1500 - 588)/1500 = 0.608$ or **60.8%**

$f' = s_a f = 0.608 \times 50 = 30.4$ Hz

$N' =$ synchronous speed of 6-pole motor with frequency f'

$= 120 \times 30.4/6 = 608$ r.p.m.

$s_b = \dfrac{N' - N}{N'} = \dfrac{608 - 588}{608} = 0.033$ or **3.3%**

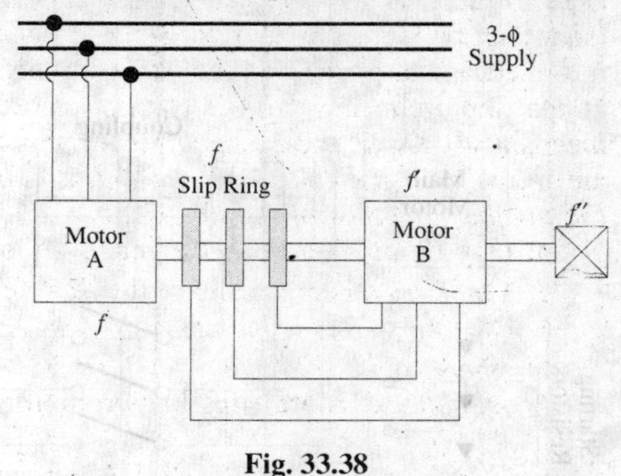

Fig. 33.38

Example 33.33. *The stator of a 6-pole motor is joined to a 50-Hz supply and the machine* *is mechanically coupled and joined in cascade with a 4-pole motor, Neglecting all losses, determine the speed and output of the 4-pole motor when the total load on the combination is 74.6 kW.*

Solution. As all losses are neglected, the actual speed of the rotor is assumed to be equal to the synchronous speed of the set.

Now, $N_{sc} = 120 \times 50/10 =$ **600 r.p.m.**

As said earlier, mechanical outputs are in the ratio of the number of poles of the motors.

$\therefore$ output of 4-pole motor = $74.6 \times 4/10 =$ **29.84 kW**

Example 33.34. *A cascaded set consists of two motors A and B with 4 poles and 6 poles respectively. The motor A is connected to a 50-Hz supply. Find*

(i) the speed of the set

(ii) the electric power transferred to motor B when the input to motor A is 25 kW. Neglect losses.

(Electric Machines-I, Utkal Univ. 1990/

Solution. Synchronous speed of the set is*

$N_{sc} = 120\,f/(P_a + P_b) = 120 \times 50/(6 + 4) =$ **600 r.p.m.**

(*ii*) The outputs of the two motors are proportional to the number of their poles.

$\therefore$ output of 4-pole motor $B = 25 \times 4/10 =$ **10 kW**

(*f*) Injecting an e.m.f. in the Rotor Circuit

In this method, the speed of an induction motor is controlled by injecting a voltage in the rotor circuit, it being of course, necessary for the injected voltage to have the same frequency as the slip frequency. There is, however, no restriction as to the phase of the injected e.m.f.

When we insert a voltage which is in *phase opposition* to the induced rotor e.m.f., it amounts to increasing the rotor resistance, whereas inserting a voltage which is in phase with the induced rotor e.m.f., is equivalent to *decreasing* its resistance. Hence, by changing the phase of the injected e.m.f. and hence the rotor resistance, the speed can be controlled.

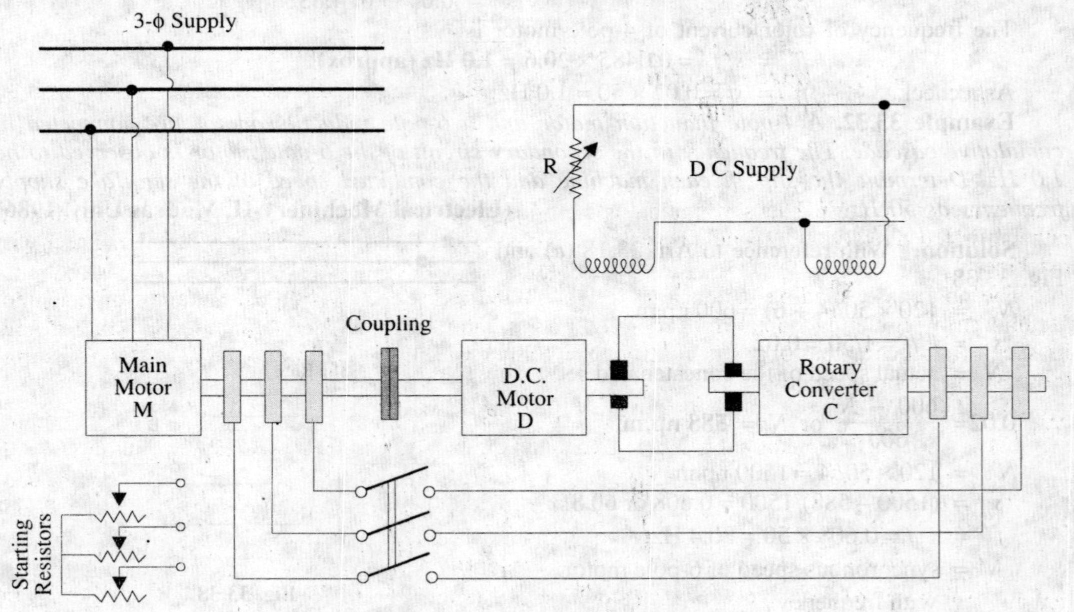

Fig. 33.39

One such practical method of this type of speed control is Kramer system, as shwon in Fig. 33.39, which is used in the case of large motors of 4000 kW or above. It consists of a rotary

*It is assumed that the two motors are connected in cumulative cascade.

converter *C* which converts the low-slip frequency a.c. power into d.c. power, which is used to drive a d.c. shunt motor *D*, mechanically coupled to the main motor *M*.

The main motor is coupled to the shaft of the d.c. shunt motor *D*. The slip-rings of *M* are connected to those of the rotary converter *C*. The d.c. output of *C* is used to drive *D*. Both *C* and *D* are excited from the d.c. bus-bars or from an exciter. There is a field regulator which governs the back e.m.f. E_b of *D* and hence the d.c. potential at the commutator of *C* which further controls the slip-ring voltage and therefore, the speed of *M*.

One big advantage of this method is that any speed, within the working range, can be obtained instead of only two or three, as with other methods of speed control.

Yet another advantage is that if the rotary converter is over-excited, it will take a leading current which compensates for the lagging current drawn by main motor *M* and hence improves the power factor of the system.

In Fig. 33.40 is shown another method, known as Scherbius system, for controlling the speed of

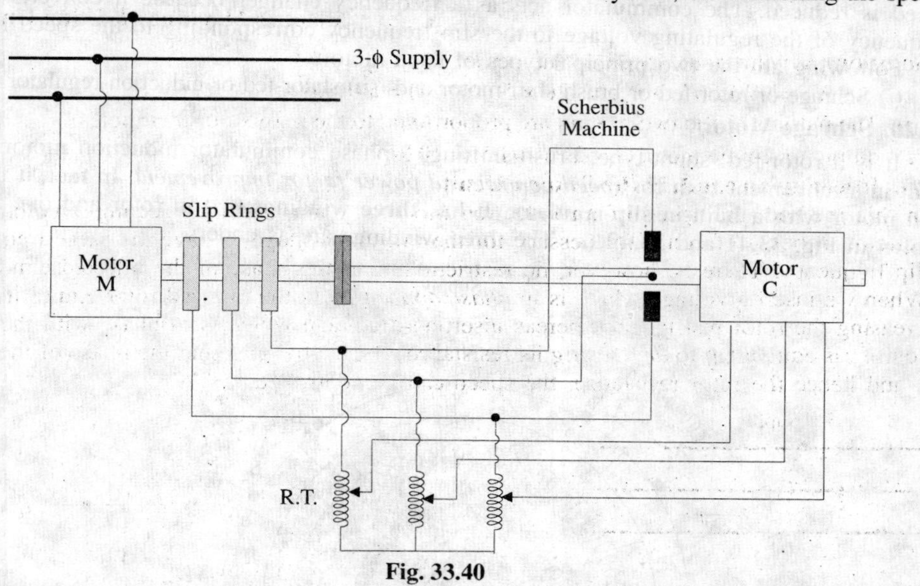

Fig. 33.40

large induction motors. The slip energy is not converted into d.c. and then fed to a d.c. motor, rather it is fed directly to a special 3-phase (or 6-phase) a.c. commutator motor—called a, Scherbius machine.

The polyphase winding of machine *C* is supplied with the low-frequency output of machine *M* through a regulating transformer *RT*. The commutator motor *C* is a variable-speed motor and its speed (and hence that of *M*) is controlled by either varying the tappings on *RT* or by adjusting the position of brushes on *C*.

Tutorial Problems 33. 5

1. An induction motor has a double-cage rotor with equivalent impedances at standstill of $(1.0 + j\ 1.0)$ and $(0.2 + j\ 4.0)\ \Omega$. Find the relative values of torque given by each cage (*a*) at starting and (*b*) at 5% slip. **[(*a*) 40:1 (*b*) 0.4:1]** (*Adv. Elect. Machines AMIE Sec. B 1991*)

2. The cages of a double-cage induction motor have standstill impedances of $(3.5 + j\ 1.5)\ \Omega$ and $(0.6 + j\ 7.0)\ \Omega$ respectively. The full-load slip is 6%. Find the starting torque at normal voltage in terms of full-load torque. Neglect stator impedance and magnetizing current.
[300%] (*Elect. Machines-I, Nagput Univ. 1993*)

3. The rotor of a 4 pole, 50 Hz, slip ring induction motor has a resistance of 0.25 ohm per phase and runs at 1440 rpm at full-load. Calculate the external resistance per phase, which must be added to lower the speed to 1200 rpm, the torque being same as before.
[1 Ω] (*Utilisation of Elctric Power (E-8) AMIE Sec. B Summer 1992*)

33.19. Three-phase A.C. Commutator Motors

Such motors have shunt speed characteristics *i.e.* change in their speed is only moderate, as compared to the change in the load. They are ideally suited for drives, requiring a uniform accelerating torque and continuously variable speed characteristics over a wide range. Hence, they find wide use in high-speed lifts, fans and pumps and in the drives for cement kilns, printing presses, pulverised fuel plants, stokers and many textile machines. Being more complicated, they are also more expensive than single-speed motors. Their efficiency is high over the whole speed range and their power factor varies from low value at synchronous speed to unity at maximum (supersynchronous) speed.

The speed control is obtained by injecting a variable voltage at *correct* frequency into the secondary winding on the motor via its commutator. If injected voltage assists the voltage induced in the secondary winding, the speed is increased but if it is in the opposing direction, then motor speed is reduced. The commutator acts as a frequency changer because it converts the supply frequency of the regulating voltage to the slip frequency corresponding to the speed required.

Following are the two principal types of such motors :

(*i*) Schrage or rotor-fed or brush shift motor and (*ii*) stator-fed or induction-regulator type motor.

33.20. Schrage Motor*

It is a rotor-fed, shunt-type, brush-shifting, 3-phase commutator induction motor which has built-in arrangement both for *speed control and power factor improvement*. In fact, it is an induction motor with a built-in slip-regulator. It has **three** windings:two in rotor and one in stator as shown in Fig. 33.41 and 33.42 (*a*) The three windings are as under:

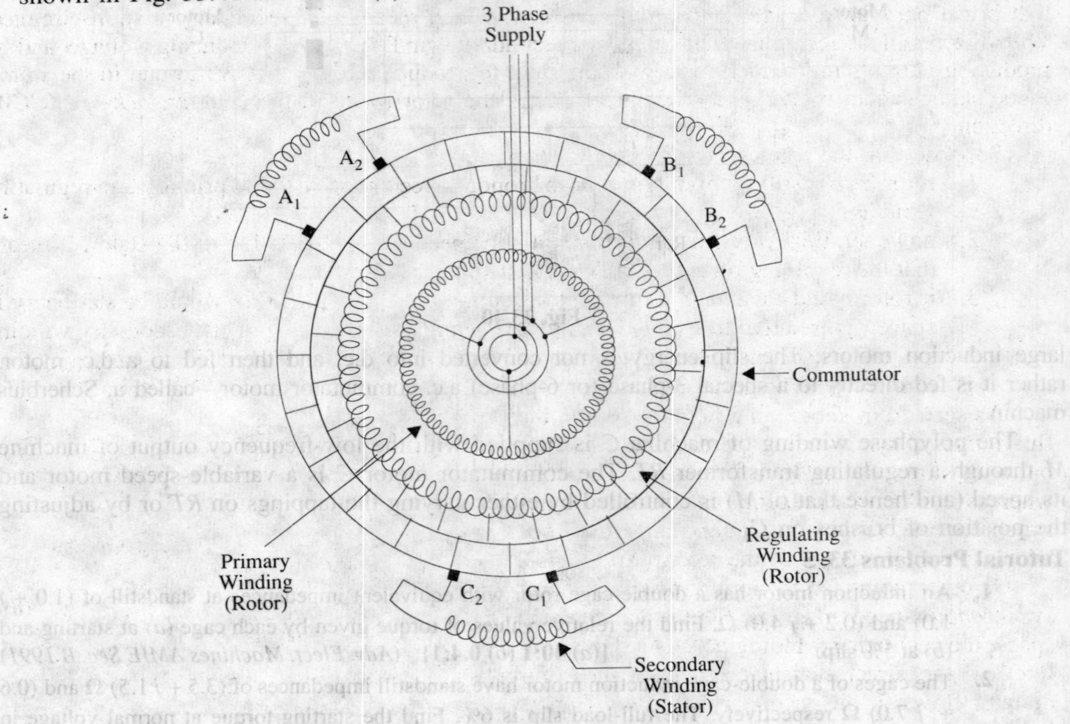

Fig. 33.41

(*i*) **Primary winding**—It is housed in the lower part of the rotor slots (not stator) and is supplied through slip-rings and brushes at *line* frequency. It generates the working flux in the machine.

*After the name of its inventor K.H. Schrage of Sweden.

(*ii*) **Regulating winding**—It is variously known as *compensating* winding or *tertiary* winding. It is also housed in *rotor slots* (in the upper part) and is connected to the commutator in a manner similar to the armature of a d.c. motor.

(*iii*) **Secondary winding**—It is contained in the *stator slots*, but end of each phase winding is connected to one of a pair of brushes arranged on the commutator. These brushes are mounted on two separate brush rockers, which are designed to move in opposite directions relative to the centre line of the corresponding stator phase (usually by a rack and pinion mechanism). Brushes A_1, B_1, and C_1 move together and are 120 electrical degrees apart. Similarly, brushes A_2, B_2 and C_2 move together and are also 120 electrical degrees apart. A sectional drawing of the motor is shown in Fig. 33.44.

(*a*) **Working**

When primary is supplied at line frequency, there is transformer action between primary and regulating winding and normal induction motor action between primary and secondary winding. Hence, voltage at line frequency is induced in the regulating winding by transformer action. The commutator, acting as a frequency changer, converts this line-frequency voltage of the regulating winding to the slip frequency for feeding it into the secondary winding on the stator. The voltage across brush pairs A_1A_2, B_1B_2 and C_1C_2 increases as brushes are separated. In fact, magnitude of the voltage injected into the secondary winding depends on the angle of separation of the brushes A_1 and A_2, B_1 and B_2 and C_1 and C_2. How slip-frequency e.m.f. is induced in secondary winding is detailed below :

When 3-ϕ power is connected to slip-rings, synchronously rotating field is set up in the rotor core. Let us suppose that this field revolves in the clockwise direction. Let us further suppose that brush pairs are on *one commutator segment*, which means that secondary is short-circuited. With rotor still at rest, this field cuts the secondary winding, thereby inducing voltage and so producing currents in it which react with the field to produce *clockwise* (*CW*) torque in the *stator*. Since stator cannot rotate, as a reaction, it makes the rotor rotate in the *counterclockwise* (*CCW*) direction.

Suppose that the rotor speed is N rpm. Then

1. rotor flux is still revolving with synchronous speed relative to the primary and regulating winding.
2. however, this rotor flux will rotate at slip speed ($N_s - N$) relative to the stator. It means that the revolving rotor flux will rotate at slip speed in *space*.
3. if rotor could rotate at synchronous speed *i.e.* if $N = N_s$, then flux would be stationary in space (*i.e.* relative to stator) so that there would be no cutting of the secondary winding by the flux and, consequently, no torque would be developed in it.

As seen from above, in a Schrage motor, the flux rotates at synchronous speed, relative to rotor but with slip speed relative to space (*i.e.* stator), whereas in a normal induction motor, flux rotates synchronously relative to stator (*i.e.* space) but with slip speed relative to the rotor. (Art. 32.11).

Another point worth noting is that since at *synchronous* speed, magnetic field is stationary in space, the regulating winding acts as a d.c. armature and the direct current taken from the commutator flows in the secondary winding. Hence, Schrage motor then operates like a synchronous motor.

(*b*) **Speed Control**

It is quite easy to obtain speeds above as well as below synchronism in a Schrage motor as shown in Fig. 33.42 (*b*) (*i*).

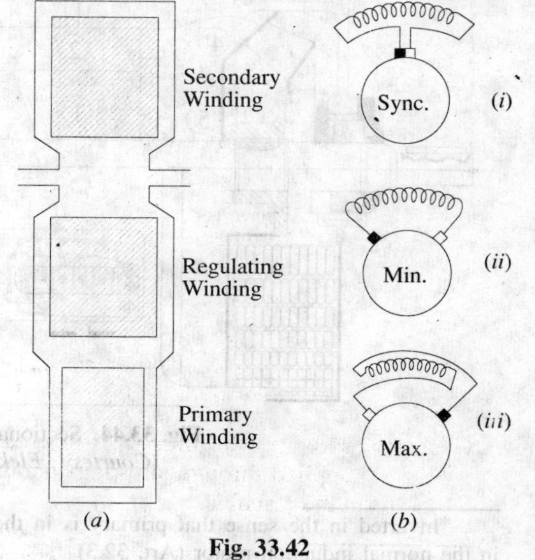

Fig. 33.42

When brush pairs are together on the same commutator segment (*i.e.* are electrically connected via commutator), the secondary winding is short-circuited and the machine operates as an inverted* plain squirrel-cage induction motor, running with a small positive slip. Parting the brushes in one direction, as shown in Fig. 33.42 (*b*) (*ii*) produces subsynchronous speeds, because in this case, regulating voltage injected into the secondary winding *opposes* the voltage induced in it from primary winding. However, when movement of brushes is reversed and they are parted in opposite directions, the direction of the regulating voltage is reversed and so motor speed *increases* to super synchronous (maximum) value, as shown in Fig. 33.41 (*b*) (*iii*) The commutator provides maximum voltage when brushes are separated by one pole pitch.

No-load motor speed is given by $N \cong N_s(1 - K \sin 0.5\beta)$ where β is brush separation in electrical degrees and K is a constant whose value depends on turn ratio of the secondary and regulating windings.

Maximum and minimum speeds are obtained by changing the magnitude of the regulating voltage. Schrage motors are capable of speed variations from zero to nearly twice the synchronous speed, though a speed range of 3:1 is sufficient for most applications. It is worth noting that Schrage motor is essentially a shunt machine, because for a particular brush separation, speed remains approximately constant as the load torque is increased as happens with dc shunt motors (Art 27.14).

(*c*) **Power Factor Improvement**

Power factor improvement can be brought about by changing the phase angle of the voltage injected into the secondary winding. As shown in Fig. 33.43, when one set of brushes is advanced more rapidly than the other is retarted, then injected voltage has a quadrature component which leads the rotor induced voltage. Hence, it results in the improvement of motor power factor. This differential movement of brush sets is obtained by coupling the racks driving the brush rockers to the hand wheel with gears having differing ratios. In Schrage motor, speed depends on angular distance between the individual brush sets (A_1 and A_2 in Fig. 33.41) but p.f. depends on the angular positions of the brushes as a whole.

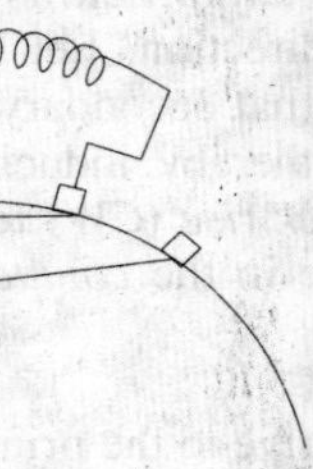

Fig. 33.43

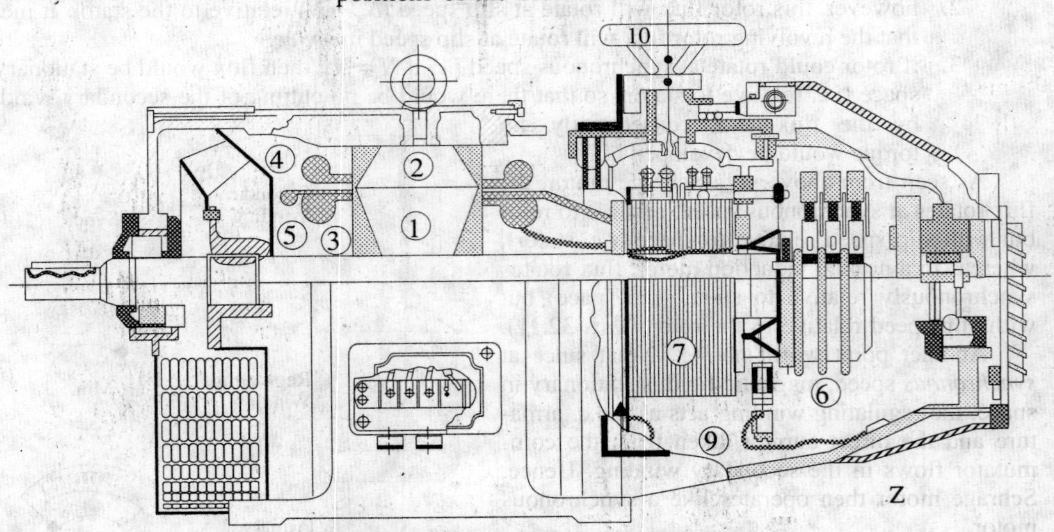

Fig. 33.44. Sectional drawing of a Scharge motor
(*Courtesy : Elekrta Faurndou, Germany*)

*Inverted in the sense that primary is in the rotor and secondary in the stator—just opposite of that in the normal induction motor (Art. 32.3).

(*d*) **Starting**

Schrage motors are usually started with brushes in the lowest speed position by direct-on contactor starters. Usually, interlocks are provided to prevent the contactor getting closed on the line when brushes are in any other position. One major disadvantage of this motor is that its operating voltage is limited to about 700 V because a.c. power has to be fed through slip-rings. It is available in sizes upto 40 kW and is designed to operate on 220, 440 and 550 V. It is ordinarily wound for four or six poles.

Fig. 33.44 shows a sectional drawing of a Schrage motor. The details of different parts labelled in the diagram are as under:

1. rotor laminations 2. stator laminations 3. primary winding 4. secondary winding 5. regulating winding 6. slip-ring unit 7. commutator 8. cable feed for outer brush yoke 9. cable feed for inner brush yoke 10. hand wheel.

33.21. Motor Enclosures

Enclosed and semi-enclosed motors are practically identical with open motors in mechanical construction and in their operating characteristics. Many different types of frames or enclosures are available to suit particular requirements. Some of the common type enclosures are described below:

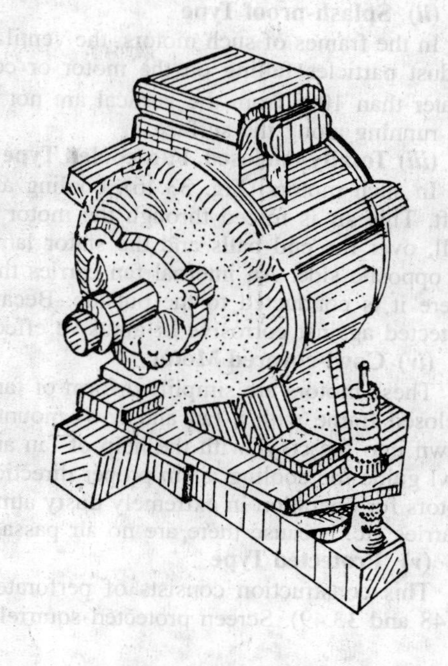

Fig. 33.45. Totally-enclosed surface cooled 750 W, 750 r.p.m. high torque induction motor (*Courtesy : Jyoti Limited*)

(*i*) **Totally-enclosed, Non-ventilated Type**

Such motors have solid frames and end-shields, but no openings for ventilation. They get cooled by surface radiation only (Fig. 33.45).

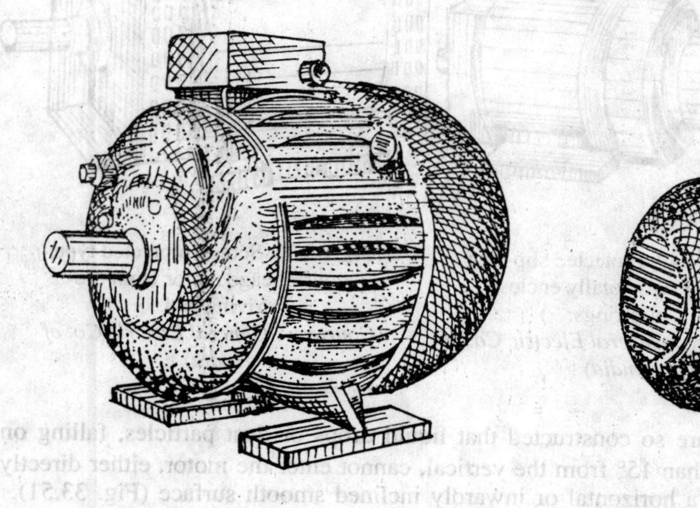

Fig. 33.46. Totally-enclosed, fan-cooled 10-kW 440/400-V, 1000 r.p.m. 50-Hz induction motor (*Courtesy : Jyoti Limited*)

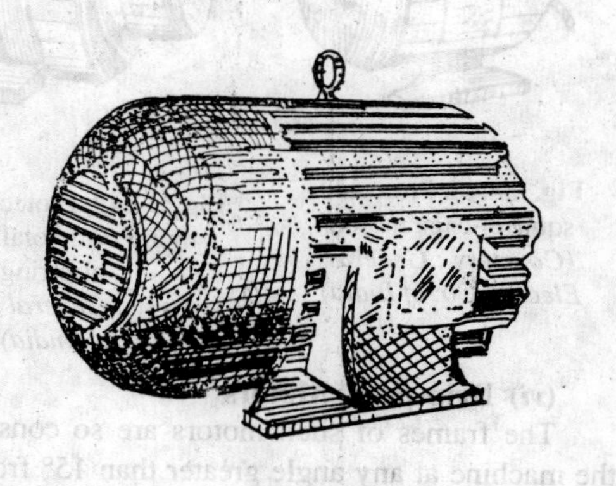

Fig. 33.47. Squirrel-cage motor, showing cowl over the external fan. (*Courtesy : General Electric Co. of India*)

Such surface-cooled motors are seldom furnished in sizes above two or three kW, because higher ratings require frames of much larger sizes than fan-cooled motors of corresponding rating.

(ii) Splash-proof Type

In the frames of such motors, the ventilating openings are so constructed that the liquid drops or dust particles falling on the motor or coming towards it in a straight line at any angle not greater than 100° from the vertical are not able to enter the motor either directly or by striking and running along the surface.

(iii) Totally-enclosed, Fan-cooled Type

In such motors (Fig. 33.46), cooling air is drawn into the motor by a fan mounted on the shaft. This air is forced through the motor between the inner fully-enclosed frame and an outer shell, over the end balls and the stator laminations and is then discharged through openings in the opposite side. An internal fan carries the generated heat to the totally enclosing frame, from where it is conducted to the outside. Because of totally enclosing frame, all working parts are protected against corrosive or abrasive effects of fumes, dust, and moisture.

(iv) Cowl-covered Motor

These motors are simplified form of fan-cooled motors (Fig. 33.47). These consist of totally-enclosed frame with a fan and cowl mounted at the end opposite to the driving end. The air is drawn into the cowl with the help of fan and is then forced over the frame. The contours of the cowl guide the cooling air in proper directions. These motors are superior to the usual fan-cooled motors for operation in extremely dusty atmosphere *i.e.* gas works, chemical works, collieries and quarries etc. because there are no air passages which will become clogged with dust.

(v) Protected Type

This construction consists of perforated covers for the openings in both end shields (Fig. 33.48 and 33.49). Screen protected squirrel cage motor is shown in Fig. 33.50.

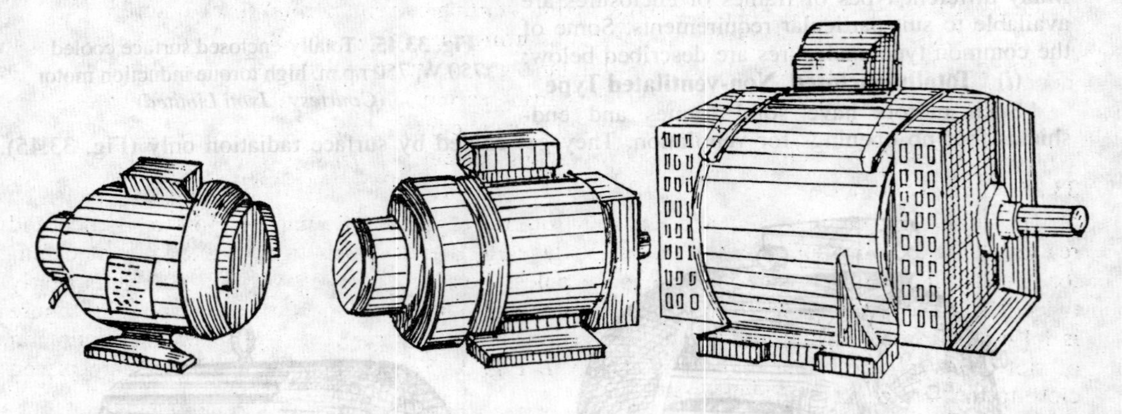

Fig. 33.48. Protected squirrel-cage motor. (*Courtesy : General Electric Co. of India*)

Fig. 33.49. Protected slip-ring motor with totally enclose slip-rings. (*Courtesy : General Electric Co. of India*)

Fig. 33.50. Screen protected type squirrel cage motor from the driving end. (*Courtesy : General Electric Co. of India*)

(vi) Drip-proof Motors

The frames of such motors are so constructed that liquid drops or dust particles, falling on the machine at any angle greater than 15° from the vertical, cannot enter the motor, either directly or by striking and running along a horizontal or inwardly inclined smooth surface (Fig. 33.51).

(vii) Self (Pipe) Ventilated Type

The construction of such motors consists of enclosed shields with provision for pipe connection on both the shields. The motor fan circulates sufficient air through pipes which are of ample section.

(*viii*) Separately (Forced) Ventilated Type

These motors are similar to the self-ventilated type except that ventilation is provided by a separated blower.

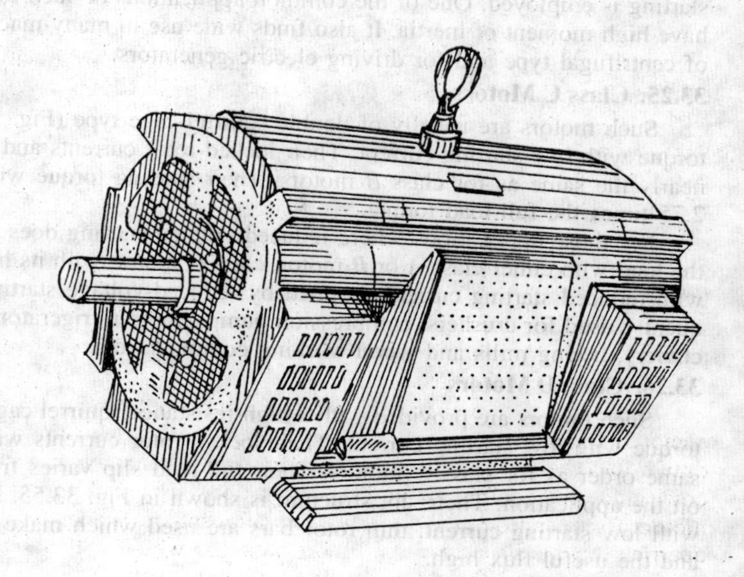

Fig. 33.51

33.22. Standard Types of Squirrel-cage Motors

Different types of 3-phase squirrel-cage motors have been standardized, according to their electric characterstics, into six types, designated as design *A, B, C, D, E* and *F* respectively. The original commercial squirrel-cage induction motors which were of shallow-slot type are designated as class *A*. For this reason, Class *A* motors are used as a reference and are referred to as 'normal starting-torque, normal starting-current, normal slip' motors.

(*i*) Class *A*—Normal starting troque, normal starting current, normal slip
(*ii*) Class *B*—Normal starting torque, low starting current, normal slip
(*iii*) Class *C*—High starting torque, low starting current, normal slip
(*iv*) Class *D*—High starting torque, low starting current, high slip
(*v*) Class *E*—Low starting torque, normal starting current, low slip
(*vi*) Class *F*—Low starting torque, low starting current, normal slip

33.23. Class A Motors

It is the most popular type and employs squirrel cage having relatively low resistance and reactance. Its locked-rotor current with full voltage is generally more than 6 times the rated full-load current. For smaller sizes and number of poles, the starting torque with full voltage is nearly twice the full-load torque whereas for larger sizes and number of poles, the corresponding figure is 1.1 times the full-load torque. The full-load slip is less than 5 per cent. The general configuration of slot construction of such motors is shown in Fig. 33.52. As seen, the rotor bars are placed close to the surface so as to reduce rotor reactance.

Such motors are used for fans, pumps, compressors and conveyors etc. which are started and stopped infrequently and have low inertia loads so that the motor can accelerate in a few seconds.

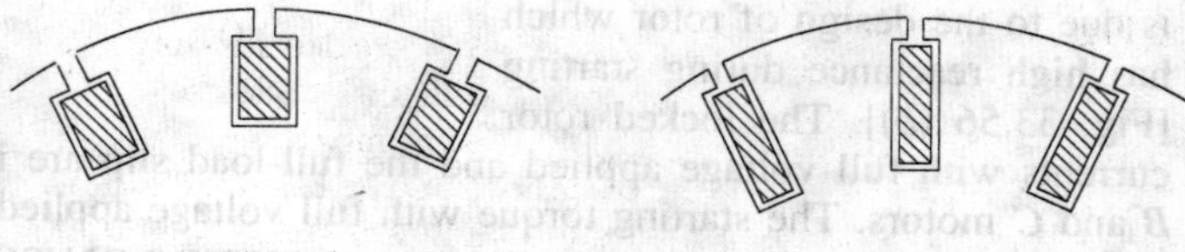

Fig. 33.52 Fig. 33.53

33.24. Class B Motors

These motors are so built that they can be started at full-load while developing normal starting torque with relatively low starting current. Their locked-rotor current with full voltage applied is generally 5 to 5 1/2 times the full-load current. Their cages are of high reactance as seen from Fig. 33.53. The rotor is constructed with deep and narrow bars so as to obtain high reactance during starting.

Such motors are well-suited for those applications where there is limitation on the starting current or if the starting current is still in excess of what can be permitted, then reduced voltage starting is employed. One of the common applications of such motors is large fans most of which have high moment of inertia. It also finds wide use in many machine tool applications, for pumps of centrifugal type and for driving electric generators.

33.25. Class C Motors

Such motors are usually of double squirrel-cage type (Fig. 33.54) and combine high starting torque with low starting current. Their locked-rotor currents and slip with full voltage applied are nearly the same as for class *B* motors. Their starting torque with full voltage applied is usually 2.75 times the full-load torque.

For those applications where reduced voltage starting does not give sufficient torque to start the load with either class *A* or *B* motor, class *C* motor, with its high inherent starting torque along with reduced starting current supplied by reduced-voltage starting may be used. Hence, it is frequently used for crushers, compression pumps, large refrigerators, coveyor equipment, textile machinery, boring mills and wood-working equipment etc.

33.26. Class D Motors

Such motors are provided with a high-resistance squirrel cage giving the motor a high starting torque with low starting current. Their locked-rotor currents with full voltage applied are of the same order as for class *C* motors. Their full-load slip varies from 5% to 20 per cent depending on the application. Their slot structure is shown in Fig. 33.55. For obtaining high starting torque with low starting current, thin rotor bars are used which make the leakage flux of the rotor low and the useful flux high.

Since these motors are used where extremely high starting torque is essential, they are usually used for bulldozers, shearing machines, punch presses, foundry equipment, stamping machines, hoists, laundry equipment and metal drawing equipment etc.

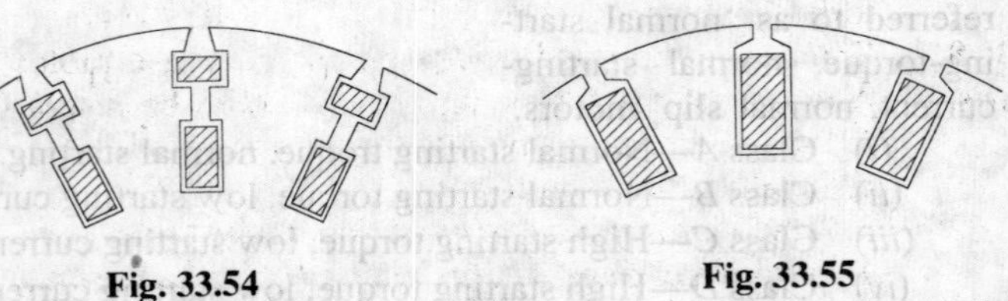

Fig. 33.54 Fig. 33.55

33.27. Class E Motors

These motors have a relatively low slip at rated load. For motors above 5 kW rating, the starting current may be sufficiently high as to require a compensator or resistance starter. Their slot structure is shown in Fig. 33.56 (*a*).

33.28. Class F Motors

Such motors combine a low starting current with a low starting torque and may be started on full voltage. Their low starting current is due to the design of rotor which has high reactance during starting [Fig. 33.56 (*b*)]. The locked rotor currents with full voltage applied and the full-load slip are in the same range as those for class *B* and *C* motors. The starting torque with full voltage applied is nearly 1.25 times the full-torque.

() (*b*)

Fig. 33.56

QUESTIONS AND ANSWERS ON THREE-PHASE INDUCTION MOTORS

Q. 1. How do changes in supply voltage and frequency affect the performance of an induction motor?

Ans. High voltage decreases both power factor and slip, but increases torque. Low voltage does just the opposite. Increase in frequency increases power factor but decreases the torque. However, per cent slip remains unchanged. Decrease in frequency decreases power factor but increases torque leaving per cent sli unaffe ted as before.

Q. 2. What is, in brief, the basis of operation of a 3-phase induction motor ?

Ans. The revolving magnetic field which is produced when a 3-phase stator winding is fed from a 3-phase supply.

Q. 3. What factors determine the direction of rotation of the motor ?

Ans. The phase sequence of the supply lines and the order in which these lines are connected to the stator winding.

Q. 4. How can the direction of rotation of the motor be reversed ?

Ans. By transposing or changing over any two line leads, as shown in Fig. 33.57.

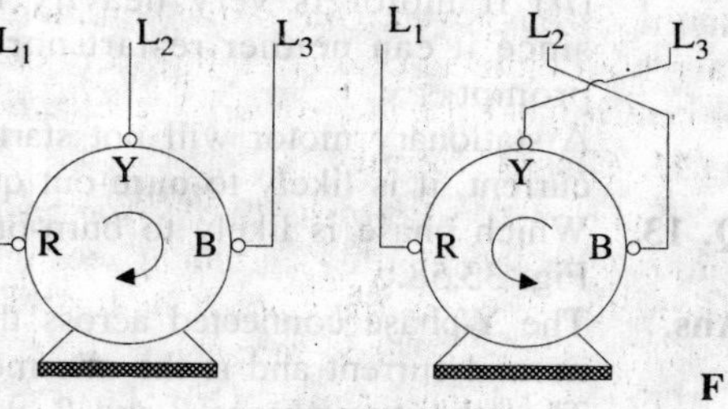

L_1 L_2 L_3 Y R B L_1 L_2 L_3 Y R B

ig. 33.57

Q. 5. Why are induction motors called asynchronous ?

Ans. Because their rotors can never run with the synchronous speed.

Q. 6. How does the slip vary with load ?

Ans. The greater the load, greater is the slip or slower is the rotor speed.

Q. 7. What modifications would be necessary if a motor is required to operate on voltage different from that for which it was originally designed ?

Ans. The number of conductors per slot will have to be changed in the same ratio as the change in voltage. If the voltage is doubled, the number of conductors per slot will have to be doubled.

Q. 8. Enumerate the possible reasons if a 3-phase motor fails to start.

Ans. Any one of the following reasons could be responsible :

1. one or more fuses may be blown.
2. voltage may be too low.
3. the starting load may be too heavy.
4. worn bearings due to which the armature may be touching field laminae, thus introducing excessive friction.

Q. 9. A motor stops after starting *i.e.* it fails to carry load. What could be the causes ?

Ans. Any one of the following:

1. hot bearings, which increase the load by excessive friction.
2. excessive tension on belt, which causes the bearings to heat.
3. failure of short cut-out switch.
4. single-phasing on the running position of the starter.

Q. 10. Which is the usual cause of blow-outs in induction motors ?

Ans. The commonest cause is single-phasing.

Q. 11. What is meant by 'single-phasing' and what are its causes ?

Ans. By single-phasing is meant the opening of one wire (or leg) of a three-phase circuit whereupon the remaining leg at once becomes single-phase. When a three-phase circuit functions normally, there are three distinct currents flowing in the circuit. As is known, any two of these currents use the third wire as the return path *i.e.* one of the three phases acts as a return path for the other two. Obviously, an open circuit in one leg kills two of the phases and there will be only one current or phase working, even though two wires are left intact. The remaining phase attempts to carry all the load. The usual cause of single-phasing is, what is generally referred to as *running fuse,* which is a fuse whose current-carrying capacity is equal to the full-load current of the motor connected in the circuit. This fuse will blow-out whenever there is overload (either momentary or sustained) on the motor.

Q. 12. What happens if single-phasing occurs when the motor is running ? And when it is stationary?

Ans. (*i*) If already running and carrying half load or less, the motor will continue running as a single-phase motor on the remaining single-phase supply, without damage because half loads do not blow normal fuses.

(*ii*) If motor is very heavily loaded, then it will stop under single-phasing and since it can neither restart nor blow out the remaining fuses, the burn-out is very prompt.

A stationary motor will not start with one line broken. In fact, due to heavy standstill current, it is likely to burn-out quickly unless immediately disconnected.

Q. 13. Which phase is likely to burn-out in a single-phasing delta-connected motor, shown in Fig. 33.58.

Ans. The Y-phase connected across the live or operative lines carries nearly three times its normal current and is the one most likely to burn-out.

The other two phases *R* and *B*, which are in series across L_2 and L_3 carry more than their full-load currents.

Q. 14. What currents flow in single-phasing star-connected motor of Fig. 33.59.

Ans. With L_1 disabled, the currents flowing in L_2 and L_3 and through phases *Y* and *B* in series will be of the order of 250 per cent of the normal full-load current, 160 per cent on 3/4 load and 100 per cent on 1/2 load.

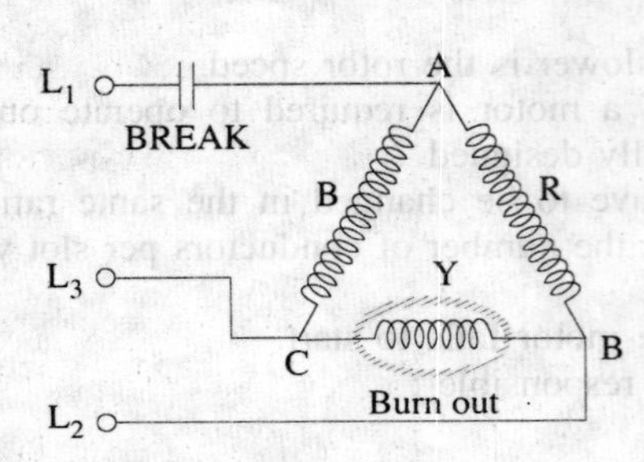

Fig. 33.58

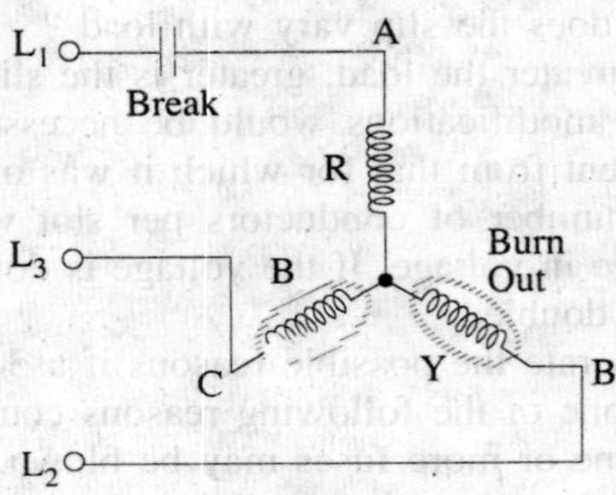

Fig. 33.59

Q. 15. How can the motors be protected against single-phasing ?

Ans. (*i*) By incorporating a combined overload and single-phasing relay in the control gear.

(*ii*) by incorporating a phase-failure relay in the control gear. The relay may be either voltage or current-operated.

Q. 16. Can a 3-phase motor be run on a single-phase line ?

Ans. Yes, it can be. But a phase-splitter is essential.

Q. 17. What is a meant by a phase-splitter ?

Ans. It is a device consisting of a number of capacitors so connected in the motor circuit that it produces, from a single input wave, three output waves which differ in phase from each other.

Q. 18. What is the standard direction of rotation of an induction motor ?

Ans. Counterclockwise, when looking from the front end *i.e.* non-driving end of the motor.

Q. 19. Can a wound-motor be reversed by transposing any two leads from the slip-rings ?

Ans. No. There is only one way of doing so *i.e.* by transposing any two line leads.

Q. 20. What is jogging ?

Ans. It means inching a motor *i.e.* make it move a little at a time by constant starting and stopping.

Q. 21. What is meant by plugging ?

Ans. It means stopping a motor by instantaneously reversing it till it stops.

Q. 22. What are the indications of winding faults in an induction motor ?

Ans. Some of the indications are as under:

(*i*) excessive and unbalanced starting currents

(*ii*) some peculiar noises and (*iii*) overheating.

OBJECTIVE TEST—33

1. In the circle diagram for a 3-ϕ induction motor, the diameter of the circle is determined by
(*a*) rotor current
(*b*) exciting current
(*c*) total stator current
(*d*) rotor current referred to stator.

2. Point out the WRONG statement.
Blocked rotor test on a 3-ϕ induction motor helps to find
(*a*) short-circuit current with normal voltage
(*b*) short-circuit power factor
(*c*) fixed losses
(*d*) motor resistance as referred to stator.

3. In the circle diagram of an induction motor, point of maximum input lies on the tangent drawn parallel to
(*a*) output line
(*b*) torque line
(*c*) vertical axis
(*d*) horizontal axis.

4. An induction motor has a short-circuit current 7 times the full-load current and a full-load slip of 4 per cent. Its line-starting torque is .... times the full-load torque.
(*a*) 7. (*b*) 1.96
(*c*) 4 (*d*) 49

5. In a SCIM, torque with autostarter is .... times the torque with direct-switching.
(*a*) K^2 (*b*) K
(*c*) $1/K^2$ (*d*) $1/K$
where K is the transformation ratio of the autostarter.

6. If stator voltage of a SCIM is reduced to 50 per cent of its rated value, torque developed is reduced by .... per cent of its full-load value.
(*a*) 50 (*b*) 25
(*c*) 75 (*d*) 57.7

7. For the purpose of starting an induction motor, a Y-Δ switch is equivalent to an auto-starter of ratio.......per cent.
(*a*) 33.3 (*b*) 57.7
(*c*) 73.2 (*d*) 60.

8. A double squirrel-cage motor (DSCM) scores over SCIM in the matter of
(*a*) starting torque
(*b*) high efficiency under running conditions
(*c*) speed regulation under normal operating conditions
(*d*) all of the above.

9. In a DSCM, outer cage is made of high resistance metal bars primarily for the purpose of increasing its
(*a*) speed regulation
(*b*) starting torque
(*c*) efficiency
(*d*) starting current.

10. A SCIM with 36-slot stator has two separate windings: one with 3 coil groups/phase/pole and the other with 2 coil groups/phase/pole. The obtainable two motor speeds would be in the ratio of
(*a*) 3 : 2 (*b*) 2 : 3
(*c*) 2 : 1 (*d*) 1 : 2

11. A 6-pole 3-ϕ induction motor taking 25 kW from a 50-Hz supply is cumulatively-cascaded to a 4-pole motor. Neglecting all losses, speed of the 4-pole motor would be .... r.p.m.
(*a*) 1500 (*b*) 1000
(*c*) 600 (*d*) 3000.
and its output would be .... kW.
(*e*) 15 (*f*) 10
(*g*) 50/3 (*h*) 2.5.

12. Which class of induction motor will be well suited for large refrigerators?
(*a*) Class E (*b*) Class B
(*c*) Class F (*d*) Class C

13. In a Schrage motor operating at supersynchronous speed, the injected emf and the standstill secondary induced emf
(*a*) are in phase with each other
(*b*) are at 90° in time phase with each other
(*c*) are in phase opposition
(*d*) none of the above.
(Power App.-III, Delhi Univ. July 1987)

14. For starting a Schrage motor, 3-ϕ supply is connected to
(*a*) stator
(*b*) rotor via slip-rings
(*c*) regulating winding
(*d*) secondary winding via brushes.

15. Two separate induction motors, having 6 poles and 5 poles respectively and their cascade combination from 60 Hz, 3-phase supply can give the following synchronous speeds in rpm

(*a*) 720, 1200, 1500 and 3600

(*b*) 720, 1200 1800

(*c*) 600, 1000, 15000

(*d*) 720 and 3000

(Power App.-II, Delhi Univ.Jan 1987)

16. Mark the WRONG statement.

A Chrage motor is capable of behaving as a/an

(*a*) inverted induction motor

(*b*) slip-ring induction motor

(*c*) shunt motor

(*d*) series motor

(*e*) synchronous motor.

17. When a **stationary** 3-phase induction motor is switched on with one phase disconnected

(*a*) it is likely to burn out quickly unless immediately disconnected

(*b*) it will start but very slowly

(*c*) it will make jerky start with loud growing noise

(*d*) remaining intact fuses will be blown out due to heavy inrush of current

18. If single-phasing of a 3-phase induction motor occurs under running conditions, it

(*a*) will stall immediately

(*b*) will keep running though with slightly increased slip

(*c*) may either stall or keep running depending on the load carried by it

(*d*) will become noisy while it still keeps running.

ANSWERS

1. *c* **2.** *c* **3.** *d* **4.** *b* **5.** *a* **6.** *c* **7.** *b* **8.** *d* **9.** *b* **10.** *a* **11.** *c,f* **12.** *d* **13.** *a* **14.** *b*
15. *a* **16.** *d* **17.** *a* **18.** *c*

34 SINGLE-PHASE MOTORS

34.1. Types of Single-Phase Motors

Such motors, which are designed to operate from a single-phase supply, are manufactured in a large number of types to perform a wide variety of useful services in home, offices, factories, workshops and in business establishments etc. Small motors, particularly in the fractional kilo watt sizes are better known than any other. In fact, most of the new products of the manufacturers of space vehicles, aircrafts, business machines and power tools etc. have been possible due to the advances made in the design of fractional-kilowatt motors. Since the performance requirements of the various applications differ so widely, the motor-manufacturing industry has developed many different types of such motors, each being designed to meet specific demands.

Single-phase motors may be classified as under, depending on their construction and method of starting :

1. *Induction Motors* (split-phase, capacitor and shaded-pole etc.)
2. *Repulsion Motors* (sometime called inductive-series motors)
3. *A.C. Series Motor*
4. *Un-excited Synchronous Motors*

34.2. Single-phase Induction Motor

Constructionally, this motor is, more or less, similar to a polyphase induction motor, except that (*i*) its stator is provided with a single-phase winding and (*ii*) a centrifugal switch is used in some types of motors, in order to cut out a winding, used only for starting purposes. It has distributed stator winding and a squirrel-cage rotor. When fed from a single-phase supply, its stator winding produces a flux (or field) which is only *alternating i.e.* one which alternates along one space axis only. It is not a synchronously revolving (or rotating) flux, as in the case of a two- or a three-phase stator winding, fed from a 2-or 3-phase supply. Now, an alternating or pulsating flux acting on a *stationary* squirrel-cage rotor cannot produce rotation (only a revolving flux can). That is why a single-phase motor is *not self-starting*.

However, if the rotor of such a machine is given an initial start by hand (or small motor) or otherwise, *in either* direction, then immediately a torque arises and the motor accelerates to its final speed (unless the applied torque is too high).

This peculiar behaviour of the motor has been explained in two ways: (*i*) by two -field or double-field revolving theory and (*ii*) by cross-field theory. Only the first theory will be discussed briefly.

34.3. Double-field Revolving Theory

This theory makes use of the idea that an alternating uni-axial quantity can be represented by two oppositely-rotating vectors of half magnitude. Accordingly, an *alternating* sinusoidal flux can be represented by two *revolving* fluxes, each equal to half the value of the alternating flux and each rotating synchronously ($N_s = 120f / P$) in opposite direction*.

As shown in Fig. 34.1 (*a*), let the alternating flux have a maximum value of Φ_m. Its component fluxes *A* and *B* will each be equal to $\Phi_m /2$ revolving in anticlockwise and clockwise directions respectivdely.

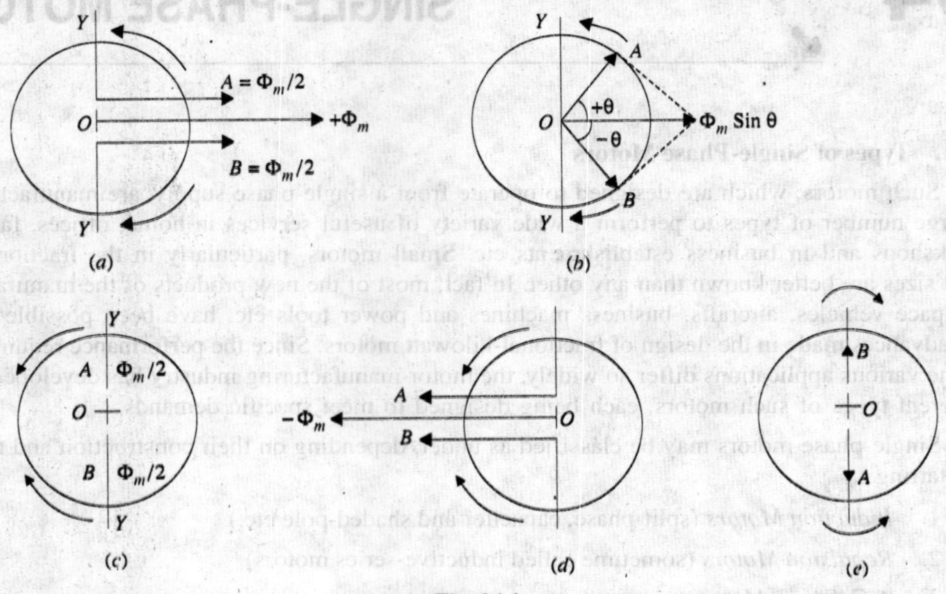

Fig. 34.1

After some time, when *A* and *B* would have rotated through angle + θ and − θ, as in Fig. 34.1 (*b*), the resultant flux would be

$$= 2 \times \frac{\Phi_m}{2} \sin \frac{2\theta}{2} = \Phi_m \sin \theta$$

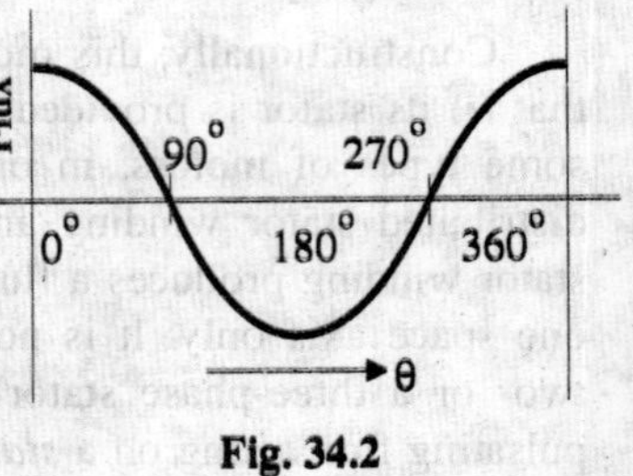

Fig. 34.2

After a quarter cycle of rotation, fluxes *A* and *B* will be oppositely-directed as shown in Fig. 34.1 (*c*) so that the resultant flux would be zero.

After half a cycle, fluxes *A* and *B* will have a resultant of $-2 \times \Phi_m /2 = -\Phi_m$. After three-quarters of a cycle, again the resultant is zero, as shown in Fig. 34.1 (*e*) and so on. If we plot the values of resultant flux against θ between limits θ = 0° to

---

* For example, a flux given by $\Phi = \Phi_m \cos 2\pi ft$ is equivalent to two fluxes revolving in opposite directions, each with a magnitude of 1/2Φ and an angular velocity of $2\pi f$. It may be noted that Euler's expressions for cos θ provides interesting justification for the decomposition of a pulsating flux. His expression is

$$\cos \theta = \frac{e^{j\theta} + e^{-j\theta}}{2}$$

The term $e^{j\theta}$ represents a vector rotated clockwise through an angle θ whereas $e^{-j\theta}$ represents rotation in anticlockwise direction. Now, the above given flux can be expressed as

$$\phi_m \cos 2\pi f t = \frac{\phi_m}{2}\left(e^{j^2 \pi f t} + e^{-j^2 \pi f t}\right)$$

The right-hand expression represents two oppositely-rotating vectors of half magnitude.

θ = 360°, then a curve similar to the one shown in Fig. 34.2 is obtained. That is why an alternating flux can be looked upon as composed of two revolving fluxes, each of half the value and revolving synchronously in opposite directions.

It may be noted that if the slip of the rotor is s with respect to the forward rotating flux (*i.e.* one which rotates in the same direction as rotor) then its slip with respect to the backward rotating flux is $(2-s)$*.

Each of the two component fluxes, while revolving round the stator, cuts the rotor, induces an e.m.f. and this produces its own torque. Obviously, the two torques (called forward and backward torques) are oppositely-directed, so that the net or resultant torques is equal to their difference as shown in Fig. 34.3.

Now, power developed by a rotor is $P_g = \left(\frac{1-s}{s}\right) I_2^{\,2} R_2$

If N is the rotor r.p.s., then torque is given by $T_g = \frac{1}{2\pi N} \cdot \left(\frac{1-s}{s}\right) I_2^{\,2} R_2$

Now, $N = N_S(1-s)$ $\therefore$ $T_g = \frac{1}{2\pi N_S} \cdot \frac{I_2^{\,2} R_2}{s} = k \cdot \frac{I_2^{\,2} R_2}{s}$

Hence, the forward and backward torques are given by

$$T_f = K\frac{I_2^{\,2} R_2}{s} \quad \text{and} \quad T_b = -K \cdot \frac{I_2^{\,2} R_2}{(2-s)}$$

or

$$T_f = \frac{I_2^{\,2} R_2}{s} \text{ synch.watt} \quad \text{and} \quad T_b = -\frac{I_2^{\,2} R_2}{(2-s)} \text{ synch. watt}$$

Total torque $T = T_f + T_b$

Fig. 34.3 shows both torques and the resultant torque for slips between zero and +2. At standstill, $s = 1$ and $(2 - s) = 1$. Hence, T_f and T_b are numerically equal but, being oppositely directed, produce no resultant torque. That explains why there is no *starting* torque in a single-phase motor.

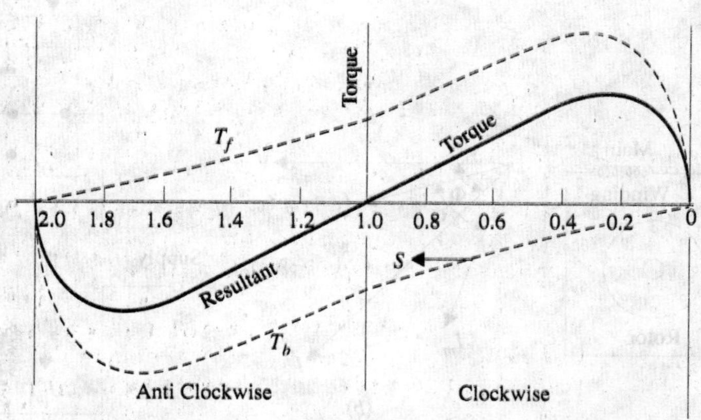

Fig. 34.3

*It may be proved thus : If N is the r.p.m. of the rotor, then its slip with respect to forward rotating flux is

$$s = \frac{N_S - N}{N_S} = 1 - \frac{N}{N_S} \text{ or } \frac{N}{N_s} = 1 - s$$

Keeping in mind the fact that the backward rotating flux rotates opposite to the rotor, the rotor slip with respect to this flux is

$$s_b = \frac{N_S - (-N)}{N_S} = 1 + \frac{N}{N_S} = 1 + (1-s) = (2-s)$$

However, if the rotor is started somehow, say, in the clockwise direction, the clockwise torque starts increasing and, at the same time, the anticlockwise torque starts decreasing. Hence, there is a certain amount of net torque in the clockwise direction which accelerates the motor to full speed.

34.4. Making Single-phase Induction Motor Self-starting

As discussed above, a single-phase induction motor is not self-starting. To overcome this drawback and make the motor self-starting, it is temporarily converted into a two-phase motor during starting period. For this purpose, the stator of a single-phase motor is provided with an extra winding, known as *starting* (or auxiliary) winding, in addition to the *main* or *running* winding. The two windings are spaced 90° electrically apart and are connected in *parallel* across the single-phase supply as shown in Fig. 34.4.

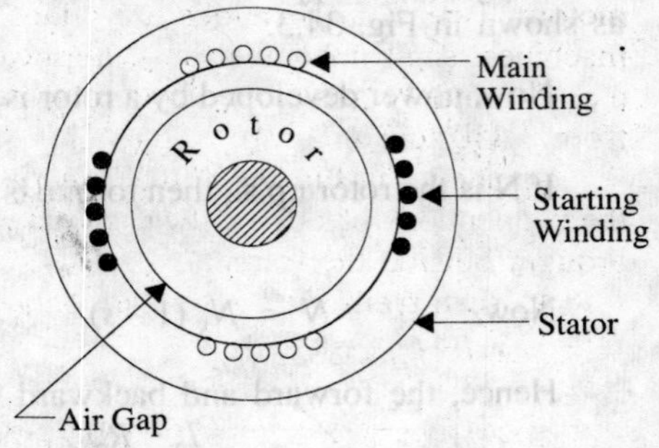

Fig. 34.4

It is so arranged that the phase-difference between the currents in the two stator windings is very large (ideal value being 90°). Hence, the motor behaves like a two-phase motor. These two currents produce a revolving flux and hence make the motor self-starting.

There are many methods by which the necessary phase-difference between the two currents can be created.

(*i*) In *split*-phase machine, shown in Fig. 34.5 (*a*), the main winding has low resistance but high reactance whereas the starting winding has a high resistance, but low reactance. The resistance of the starting winding may be increased either by connecting a high resistance R in series with it or by choosing a high-resistance fine copper wire for winding purposes.

Hence, as shown in Fig. 34.5 (*b*), the current I_s drawn by the starting winding lags behind the applied voltage by a small angle whereas current I_m taken by the main winding lags behind

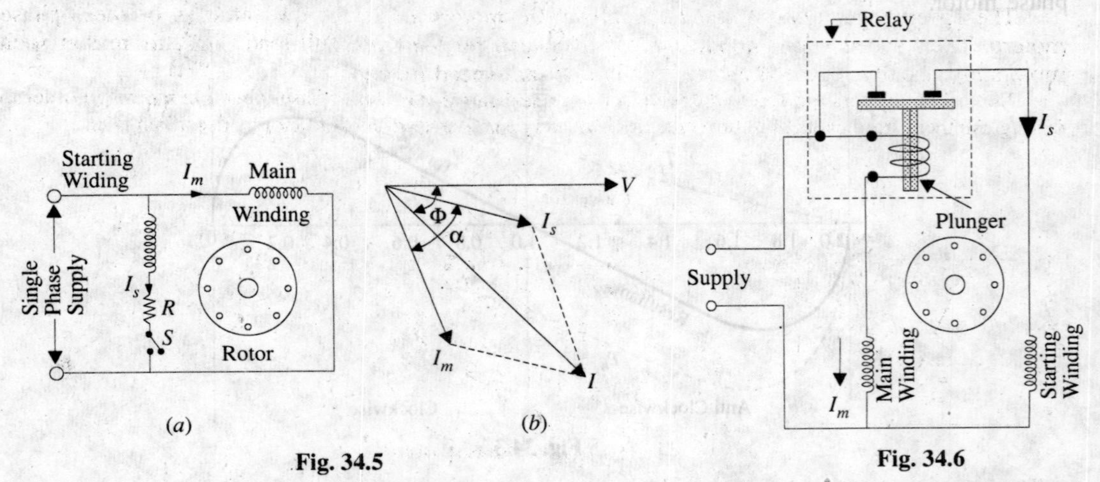

Fig. 34.5 **Fig. 34.6**

V by a very large angle. Phase angle between I_s and I_m is made as large as possible because the starting torque of a split-phase motor is proportional to sin α. A centrifugul switch S is connected in series with the starting winding and is located inside the motor. Its function is to automatically disconnect the starting winding from the supply when the motor has reached 70 to 80 per cent of its full-load speed.

In the case of split-phase motors that are hermetically sealed in refrigeration units, instead of internally-mounted centrifugal switch, an electromagnetic type of relay is used. As shown in

Fig. 34.6, the relay coil is connected in series with main winding and the pair of contacts which are normally open, is included in the starting winding. During starting period, when I_m is large, relay contacts close thereby allowing I_s to flow and the motor starts as usual. After motor speeds up to 75 per cent of full-load speed, I_m drops to a value that is low enough to cause the contacts to open.

A typical torque/speed characteristic of such a motor is shown in Fig. 34.7. As seen, the starting torque is 150 to 200 per cent of the full-load torque with a starting current of 6 to 8 times the full-load current. These motors are often used in preference to the costlier capacitor-start motors. Typical applications are : fans and blowers, centrifugal pumps and separators, washing machines, small machine tools, duplicating machines and domestic refrigerators and oil burners etc. Commonly available sizes range from 1/20 to 1/3 h.p. (40 to 250 W) with speeds ranging from 3,450 to 865 r.p.m.

As shown in Fig. 34.8, the direction of rotation of such motors can be reversed by reversing the connections of one of the two stator windings (not both). For this purpose, the four leads are brought outside the frame.

As seen from Fig. 34.9, the connections of the starting winding have been reversed.

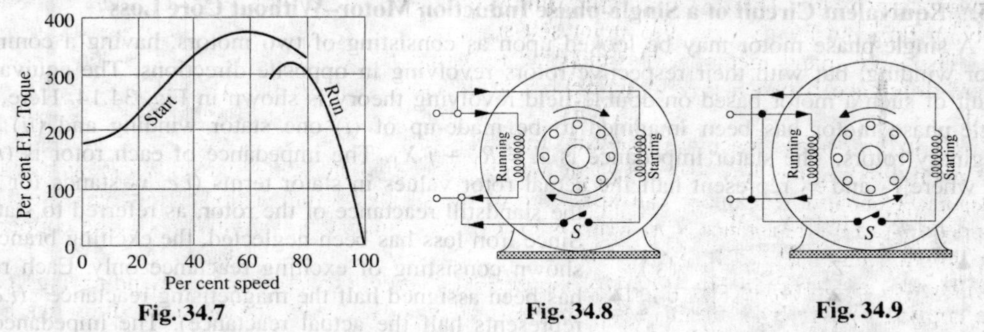

Fig. 34.7 Fig. 34.8 Fig. 34.9

The speed regulation of standard split-phase motors is nearly the same as of the 3-phase motors. Their speed varies about 2 to 5% between no load and full-load. For this reason such motors are usually regarded as practically constant-speed motors.

Note. Such motors are sometimes referred to as resistance-start split-phase induction motors in order to distinguish them from capacitor-start induction run and capacitor start-and-run motors described later.

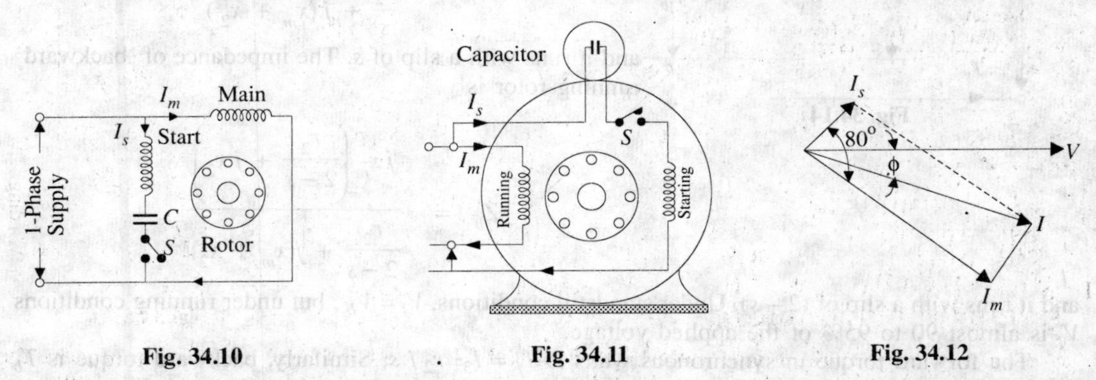

Fig. 34.10 Fig. 34.11 Fig. 34.12

(*ii*) **Capacitor-start Induction-run motors.** In these motors, the necessary phase difference between I_s and I_m is produced by connecting a capacitor in series with the starting winding as shown in Fig. 31.10. The capacitor is generally of the electrolytic type and is usually mounted on the outside of the motor as a separate unit (Fig. 34.11).

The capacitor is designed for extremely short-duty service and is guaranteed for not more than 20 periods of operation per hour, each period not to exceed 3 seconds. When the motor reaches about 75 per cent of full speed, the centrifugal switch S opens and cuts out both the starting winding and the capacitor from the supply, thus leaving only the running winding across the lines.

As shown in Fig. 33.12, current I_m drawn by the main winding lags the supply voltage V by a large angle whereas I_s leads V by a certain angle. The two currents are out of phase with each other by about 80° (for a 200-W 50-Hz motor) as compared to nearly 30° for a split-phase motor. Their resultant current I is small and is almost in phase with V as shown in Fig. 33.12.

Since the torque developed by a split-phase motor is proportional to the sine of the angle between I_s and I_m, it is obvious that the increase in the angle (from 30° to 80°) alone increases the starting torque to nearly twice the value developed by a standard split-phase induction motor. Other improvements in motor design have made it possible to increase the starting torque to a value as high as 350 to 450 per cent.

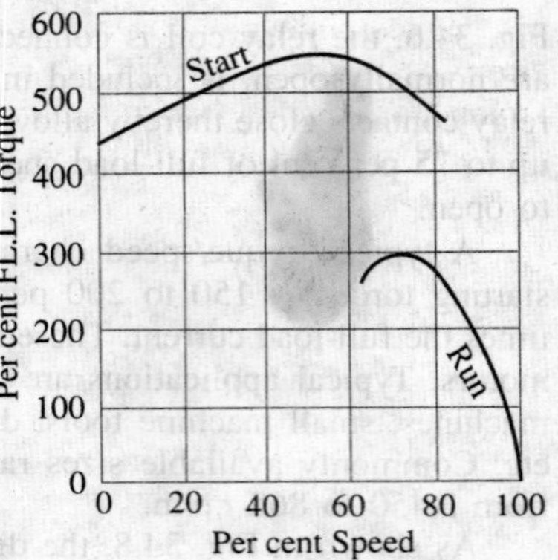

Fig. 34.13

Typical performance curve of such a motor is shown in Fig. 34.13.

34.5. Equivalent Circuit of a Single-phase Induction Motor–Without Core Loss

A single-phase motor may be looked upon as consisting of two motors, having a common stator winding, but with their respective rotors revolving in opposite directions. The equivalent circuit of such a motor based on double-field revolving theory is shown in Fig. 34.14. Here, the single-phase motor has been imagined to be made-up of (*i*) one stator winding and (*ii*) two imaginary rotors. The stator impedance is $Z = R_1 + j X_1$. The impedance of each rotor is $(r_2 + jx_2)$ where r_2 and x_2 represent half the actual rotor values in stator terms (*i.e.* x_2 stands for half the standstill reactance of the rotor, as referred to stator). Since iron loss has been neglected, the exciting branch is shown consisting of exciting reactance only. Each rotor has been assigned half the magnetising reactance* (*i.e* x_m represents half the actual reactance). The impedance of 'forward running' rotor is

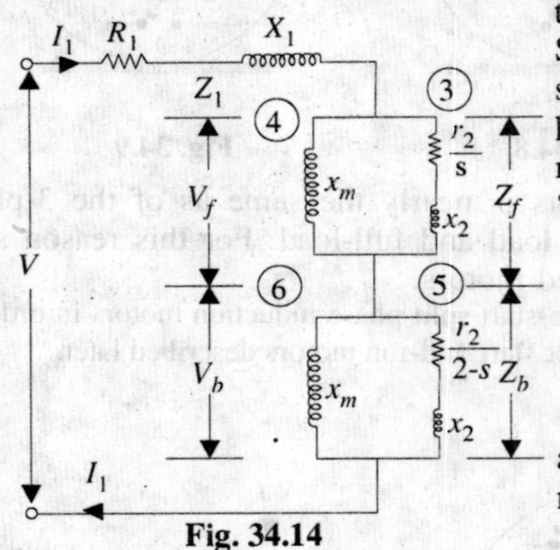

Fig. 34.14

$$Z_f = \frac{j x_m \left(\frac{r_2}{s} + j x_2\right)}{\frac{r_2}{s} + j(x_m + x_2)}$$

and it runs with a slip of s. The impedance of 'backward' running rotor is

$$Z_b = \frac{j x_m \left(\frac{r_2}{2-s} + j x_2\right)}{\frac{r_2}{2-s} + j(x_m + x_2)}$$

and it runs with a slip of $(2 - s)$. Under standstill conditions, $V_f = V_b$, but under running conditions V_f is almost 90 to 95% of the applied voltage.

The forward torque in synchronous watts is $T_f = I_3^2\, r_2 / s$. Similarly, backward torque is $T_b = I_5^2\, r_2 / (2 - s)$

The total torque is $T = T_f - T_b$.

33.6. Equivalent Circuit—With Core Loss

The core loss can be represented by an equivalent resistance which may be connected either in parallel or in series with the magnetising reactance as shown in Fig. 34.15.

*In fact, full values are shown by capital letters and half values by small letters.

Since under running conditions V_f is very high (and V_b is correspondingly, low) most of the iron loss takes place in the 'forward motor' consisting of the common stator and forward-running rotor. Core-loss current I_w = core loss / V_f. Hence, half value of core-loss equivalent resistance is $r_c = V_f / I_w$. As shown in Fig. 34.15 (*a*), r_c has been connected in parallel with x_m in each rotor.

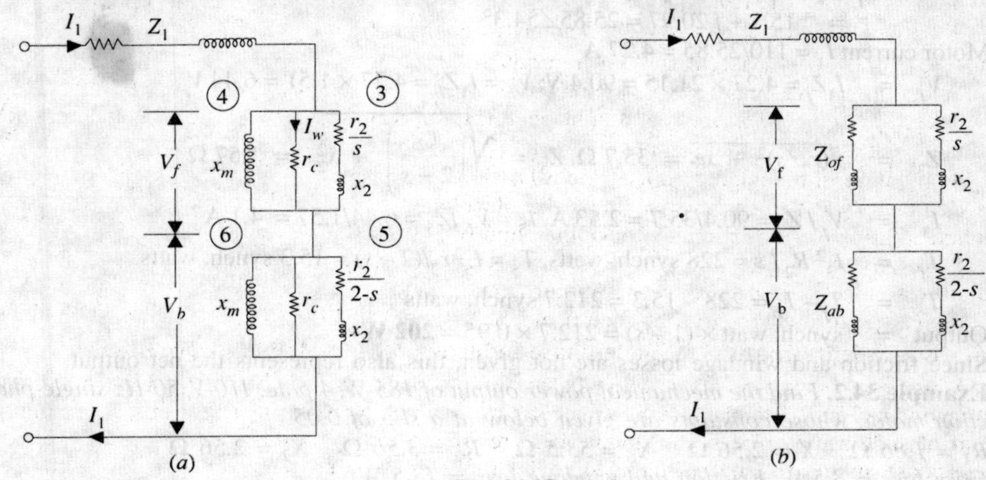

Fig. 34.15

Example 34.1. *Discuss the revolving field theory of single-phase induction motors. Find the mechanical power output at a slip of 0.05 of the 185-W, 4-pole, 110-V, 60-Hz single-phase induction motor, whose constants are given below:*

Resistance of the stator main winding	R_1	=	*1.86 ohm*
Reactance of the stator main winding	X_1	=	*2.56 ohm*
Magnetizing reactance of the stator main winding	X_m	=	*53.5 ohm*
Rotor resistance at standstill	R_2	=	*3.56 ohm*
Rotor reactance at standstill	X_2	=	*2.56 ohm*

(Elect. Machines, Nagpur Univ. 1991)

Solution. Here, $X_m = 53.5\ \Omega$, hence $x_m = 53.5/2 = 26.7\ \Omega$

Similary, $r_2 = R_2 / 2 = 3.56 / 2 = 1.78\ \Omega$ and $x_2 = X_2 / 2 = 2.56 / 2 = 1.28\ \Omega$

$$\therefore\quad Z_f = \frac{j\,x_m\left(\dfrac{r_2}{s} + j\,x_2\right)}{\dfrac{r_2}{s} + j(x_2 + x_m)} = x_m \frac{\dfrac{r_2}{s} \cdot x_m + j\left[(r_2/s)^2 + x_2 x_0\right]}{(r_2/s)^2 + x_2^2} \quad \text{where } x_0 = (x_m + x_2)$$

$$\therefore\quad Z_f = 26.7\,\frac{(1.78/0.05) \times 26.7 + j[(1.78/0.05)^2 + 1.28 \times 27.98]}{(1.78/0.05)^2 + (27.98)^2}$$

$$= 12.4 + j\,17.15 = 21.15\angle 54.2°$$

$$\text{Similarly,}\quad Z_b = \frac{j\,x_m\left(\dfrac{r_2}{2-s} + jx_2\right)}{\dfrac{r_2}{2-s} + j(x_2 + x_m)} = x_m \frac{\left(\dfrac{r_2}{2-s}\right)x_m + j\left[\left(\dfrac{r_2}{2-s}\right)^2 + x_0 x_2\right]}{\left(\dfrac{r_2}{2-s}\right)^2 + x_0^2}$$

$$= 26.7\,\frac{(1.78/1.95) \times 26.7 + j[(1.78/1.95)^2 + 1.28 \times 27.98]}{(1.78/1.95)^2 + (27.98)^2}$$

$$= \quad 0.84 + j\,1.26 = 1.51\angle 56.3°$$

$$\mathbf{Z}_1 = R_1 + jX_1 = 1.86 + j\,2.56 = 3.16\angle 54°$$

Total circuit impedance is

$$\mathbf{Z}_{01} = \mathbf{Z}_1 + \mathbf{Z}_f + \mathbf{Z}_b = (1.86 + j\,2.56) + (12.4 + j\,17.15) + (0.84 + j\,1.26)$$

$$= 15.1 + j\,20.97 = 25.85\angle 54.3°$$

Motor current $I_1 = 110/25.85 = 4.27$ A

$$V_f = I_1 Z_f = 4.27 \times 21.15 = 90.4 \text{ V};\ V_b = I_1 Z_b = 4.27 \times 1.51 = 6.44\ V$$

$$Z_3 = \sqrt{(r_2/s)^2 + x_2^2} = 35.7\ \Omega,\ Z_5 = \sqrt{\left(\frac{r_2}{2-s}\right)^2 + x_2^2} = 1.57\ \Omega$$

$$I_3 = V_f / Z_3 = 90.4/35.7 = 2.53 \text{ A},\ I_5 = V_b/Z_5 = 6.44/1.57 = 4.1 \text{ A}$$

$$T_f = I_3^2 R_2 / s = 228 \text{ synch. watts},\ T_5 = I_5^2 r_2/(2 - s) = 15.3 \text{ synch. watts.}$$

$$T = T_f - T_b = 228 - 15.3 = 212.7 \text{ synch. watts}$$

Output $=$ synch. watt $\times (1 - s) = 212.7 \times 0.95 =$ **202 W**

Since friction and windage losses are not given, this also represents the net output.

Example 34.2. *Find the mechanical power output of 185-W, 4 pole, 110-V, 50-Hz single-phase induction motor, whose constants are given below at a slip of 0.05.*

$R_1 = 1.86\ \Omega \quad X_1 = 2.56\ \Omega \quad X_\phi = 53.5\ \Omega \quad R_2 = 3.56\ \Omega \quad X_2 = 2.56\ \Omega$

Core loss = 3.5 W, Friction and windage loss = 13.5 W.

(Electrical Machines-III, Indore Univ. 1987)

Solution. It would be seen that major part of this problem has already been solved in Example 34.1. Let us, now, assume that V_f = 82.5% of 110 V = 90.7 V. Then the core-loss current I_c = 35/90.7 = 0.386 A ; r_c = 90.7/0.386 = 235 Ω

Motor I

conductance of core-loss branch $= 1/r_c = 1/235 = 0.00426$ S

susceptance of magnetising branch $= -j/x_m = -j/26.7 = -j\,0.0374$ S

$$\text{admittance of branch 3} = \frac{(r_2/s) - j\,x_2}{(r_2/s)^2 + x_2^2} = 0.028 - j\,0.00101 \text{ S}$$

admittance of 'motor' I is $\mathbf{Y}_f = 0.00426 - j\,0.0374 + 0.028 - j\,0.00101$

$= 0.03226 - j0.03841$ S

impedance $\mathbf{Z}_f = \mathbf{1/Y}_f = 12.96 + j\,15.2$ or 19.9 Ω

Motor II

$$\text{admitance of branch 5} = \frac{\dfrac{r_2}{2-s} - j\,x_2}{\left(\dfrac{r_2}{2-s}\right)^2 + x_2^2} = \frac{0.91 - j\,1.28}{2.469} = 0.369 - j\,0.517$$

admittance of 'motor' **II,** $\mathbf{Y}_b = 0.00426 - j\,0.0374 + 0.369 - j\,0.517$

$= 0.3733 - j\,0.555$ S

Impedance of 'motor' II, $\mathbf{Z}_b = \mathbf{1/Y}_b = 0.836 + j\,1.242$ or 1.5 Ω

Impedance of entire motor (Fig. 31.16) $\mathbf{Z}_{01} = \mathbf{Z}_1 + \mathbf{Z}_f + \mathbf{Z}_b = 15.66 + j19$ or 24.7 Ω

$$I_1 = V/Z_{01} = 110/24.7 = 4.46 \text{ A}$$

$$V_f = I_1 Z_f = 4.46 \times 19.9 = 88.8 \text{ V}$$

$$V_b = 4.46 \times 1.5 = 6.69 \text{ V}$$

$$I_3 = 88.8/35.62 = 2.5 \text{ A}$$

$$I_5 = 6.69/1.57 = 4.25 \text{ A}$$

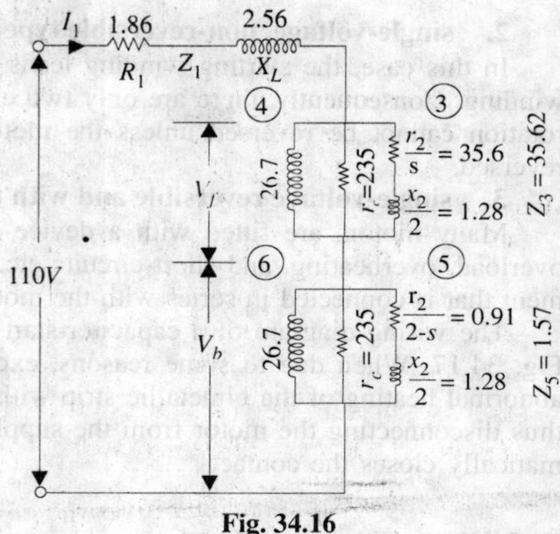

Fig. 34.16

$T_f \;=\; I_3^2(r_2/s) = 222$ synch. watt

$T_b \;=\; I_5^2\left(\frac{r_2}{2-s}\right) \;=\; 16.5$ synch. watt

$T \;=\; T_f - T_b = 205.5$ synch. watt

Watts converted $=$ synch. watt $(1 - s)$

$= 205.5 \times 0.95 = 195$ W

Net output $= 195 - 13.5 =$ **181.5 W.**

Example 34.3. *A 250-W, 230-V, 50-Hz capacitor-start motor has the following constants for the main and auxiliary windings: Main winding, Z_m = (4.5 + j 3.7) ohm. Auxiliary winding Z_a = (9.5 + j 3.5)ohm.* Determine *the value of the starting capacitor that will place the main and auxiliary winding currents in quadrature at starting.*

(Electrical Machines-III, South Gujarat Univ. 1985)

Solution. Let X_C be the reactance of the capacitor connected in the auxiliary winding.

Then $\mathbf{Z}_a = 9.5 + J\,3.5 - jX_C = (9.5 + jX)$

where X is the net reactance.

Now, $\mathbf{Z}_m = 4.5 + j\,3.5 = 5.82 \angle 39.4°$ ohm

Obviously, I_m lags behind V by 39.4°

Since, time phase angle between I_m and I_a has to be 90°, I_a must lead V by

$(90° - 39.4°) = 50.6°$.

For auxiliary winding, $\tan \phi_a = X/R$ or $\tan 50.6° = X/9.5$

or $X = 9.5 \times 1.217 = 11.56\ \Omega$ (capacitive)

$\therefore\ X_C = 11.56 + 3.5 = 15.06\ \Omega$ $\qquad \therefore\ 15.06 = 1/314\,C;\quad C =$ **211 μF.**

Tutorial Problem No. 34.1.

1. A 1-φ, induction motor has stator windings in space quadrature and is supplied with a single-phase voltage of 200 V at 50 Hz. The standstill impedance of the main winding is (5.2 + *j* 10.1) and of the auxiliary winding is (19.7 + *j* 14.2). Find the value of capacitane to be inserted in the auxiliary winding for maximum starting torque.
   (Electrical Machines-III, Indore Univ. July, 1977)
2. A 230-V, 50-Hz, 6-pole, single-phase induction motor has the following constants.
   $r_1 = 0.12\ \Omega,\ r_2 = 0.14\ \Omega,\ x_1 = x_2 = 0.25\ \Omega,\ x_m = 15\ \Omega.$
   If the core loss is 250 W and friction and windage losses are 500 W, determine the efficiency and torque at $s = 0.05$. *(Electrical Machines-IV, Bangalore Univ. Aug. 1978)*
3. Explain how the pulsating mmf of a singl-phase induction motor may be considered equivalent to two oppositely-rotating fields. Develop an expression for the torque of the motor.
   A 125-W, 4-pole, 110-V, 50-Hz single-phase induction motor has the no-load rotational loss of 25 watts and total rotor copper loss at rated load of 25 watts at a slip of 0.06. The rotor I^2R loss may be neglected.
   At a slip $s = 0.06$, what is the power input to the machine?
   (Electrical Machines-III, Indore Univ. July 1977)

34.7. Types of Capacitor–start Motors

Some of the important types of such motors are given below :

1. single-voltage, externally-reversible type

In this motor, four leads are brought outside its housing; two from the main winding and two from the starting-winding circuit. These four leads are necessary for external reversing. As usual, internally, the starting winding is connected in series with the electrolytic capacitor and a centrifugal switch. The direction of rotation of the motor can be easily reversed externally by reversing the starting winding leads with respect to the running winding leads.

2. single-voltage, non-reversible type

In this case, the starting winding leads are connected internally to the leads of the running winding. Consequently, there are only two external leads in such motors. Obviously, direction of rotation cannot be reversed unless the motor is taken apart and leads of the starting winding reversed.

3. single-voltage reversible and with thermostat type

Many motors are fitted with a device called thermostat which provides protection against overload, overheating, and short-circuits etc. The thermostat usually consists of a bimetallic element that is connected in series with the motor and is often mounted on the outside of the motor.

The wiring diagram of a capacitor-start motor fitted with this protective device is shown in Fig. 34.17. When due to some reasons, excessive current flows through the motor, it produces abnormal heating of the bimetallic strip with the result that it bends and opens the contact points thus disconnecting the motor from the supply lines. When the thermostat element cools, it automatically closes the contacts*.

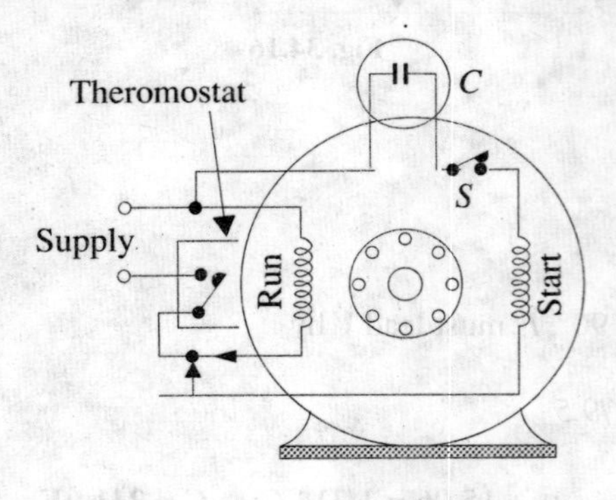

Fig. 34.17

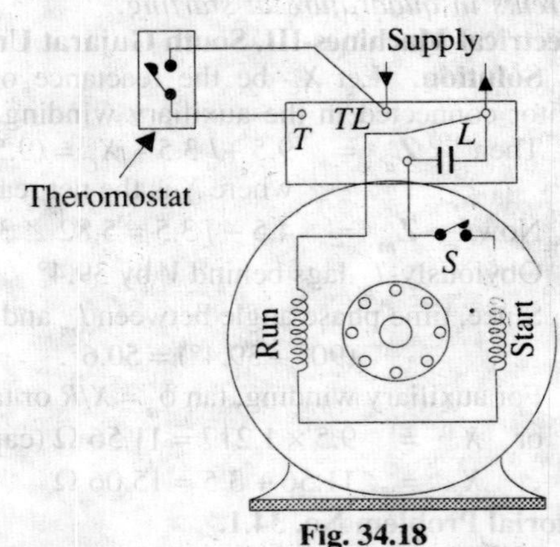

Fig. 34.18

In the case of capacitor-start motors used for refrigerators, generally a terminal block is attached to the motor. Three out of the four block terminals are marked *T*, *TL* and *L* as shown in Fig. 34.18. Thermostat is connected to *T* and *TL*, capacitor between *L* and the unmarked terminal and the supply lines to *TL* and *L*.

4. single-voltage, non-reversible with magnetic switch type

Such motors are commonly used in refrigerators where it is not possible to use a centrifugal switch. The circuit diagram is similar to that shown in Fig. 31.16. Since their application requires just one direction of rotation, these motors are not connected for reversing.

One disadvantage of a capacitor-start motor having magnetic switch lies in the possibility that slight overloads may operate the plunger thereby connecting the starting winding circuit to the supply. Since this winding is designed to operate for very short periods (3 seconds or less) it is likely to be burnt out.

5. two-voltage, non-reversible Type

These motors can be operated from two a.c. voltage either 110 V and 220 V or 220 V and 440 V. Such motors have two main windings (or one main winding in two sections) and one starting winding with suitable number of leads brought out to permit changeover from one voltage to another.

When the motor is to operate from lower voltage, the two main windings are connected in

---

*However, in some thermal units, a reset button has to be operated manually to restore the motor to operation.

In certain types of thermal units, a heating element is used for heating the bimetallic strip. In that case, the heating element is connected in the line and the element or bimetallic strip is placed either inside the heating unit or besides it.

parallel (Fig. 34.19). whereas for higher voltage, they are connected in series (Fig. 34.20). As will be seen from the above circuit diagrams, the starting winding is always operated on the low-voltage for which purpose it is connected across one of the main windings.

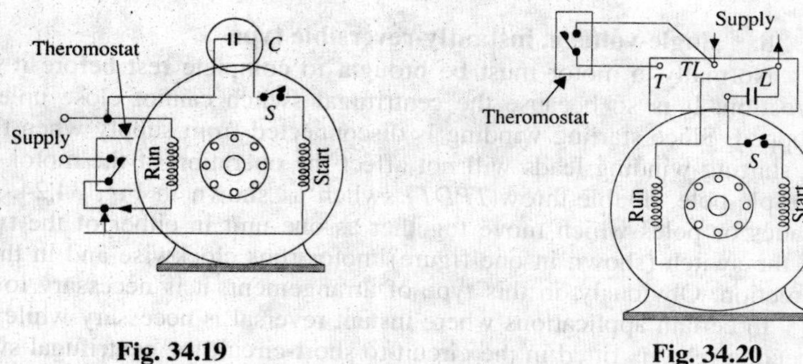

Fig. 34.19 Fig. 34.20

6. Two-voltage, reversible type

External reversing is made possible by means of two additional leads that are brought out from the starting winding.

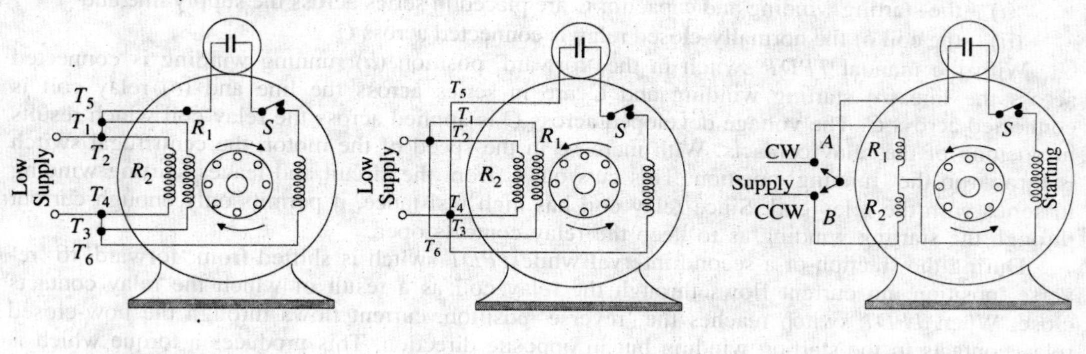

Fig. 34.21 Fig. 34.22 Fig. 34.23

Fig. 34.21 and 34.22 show connections for clockwise and anticlockwise rotations respectively when motor is operated from lower voltage. Similar wiring diagram can be drawn for higher voltage supply.

7. single-voltage, three-lead reversible type

In such motors, a two-section running winding is used. The two sections R_1 and R_2 are internally connected in series and one lead of the starting winding is connected to the mid-point of R_1 and R_2. The second lead of the starting winding and both leads of the running winding are brought outside as shown in Fig. 34.23. When the external lead of the starting winding is connected to point *A*, the winding is connected across R_1 and the motor runs clockwise. When the lead of starter winding is connected to point *B*, it is connected across R_2. Since current flowing through starting winding is reversed, the motor runs in counter-clockwise direction.

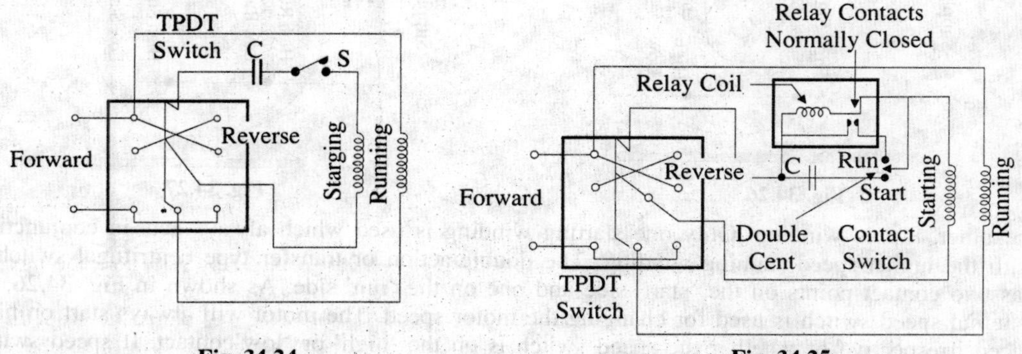

Fig. 34.24 Fig. 34.25

8. single-voltage, instantly-reversible type

Normally, a motor must be brought to complete rest before it can be started in the reverse direction. It is so because the centrifugal switch cannot close unless the motor has practically stopped. Since starting winding is disconnected from supply when the motor is running, reversal of starting winding leads will not affect the operation of the motor. This reversal is achieved by a triple-pole, double-throw(*TPDT*) switch as shown in Fig. 34.24. The switch consists of three blades or poles which move together as one unit in either of the two positions. In one position of the switch (shown in one figure) motor runs clockwise and in the other, in counter-clockwise direction. Obviously, in this type of arrangement, it is necessary to wait till motor stops.

In certain applications where instant reversal is necessary while the motor is operating at full speed, a relay is fitted in the circuit to short-circuit the centrifugal switch and connect the starting winding in the circuit in the reversed direction(Fig. 34.25).

It will be seen that when at rest, the double-contact centrifugal switch is in the 'start' position. In this position, two connections are made

(*i*) the starting winding and capacitor *C* are placed in series across the supply line and

(*ii*) the coil of the normally-closed relay is connected across *C*

With the manual *TPDT* switch in the 'forward' position (*a*) running winding is connected across the line (*b*) starting winding and *C* are in series across the line and (*c*) relay coil is connected across *C*. The voltage developed across *C* is applied across the relay coil which results in opening of the relay contacts. With increase in the speed of the motor, the centrifugal switch is thrown in the 'running' position. This cuts out *C* from the circuit and leaves starting winding in series with the relay coil. Since relay coil has high resistance, it permits only enough current through the starting winding as to keep the relay contacts open.

During the fraction-of-a-second interval while *TPDT* switch is shifted from 'forward' to 'reverse' position, no current flows through the relay coil as a result of which the relay contacts close. When *TPDT* switch reaches the 'reverse' position, current flows through the now-closed relay contacts to the starting winding but in opposite direction. This produces a torque which is applied in a direction opposite to the rotation. Hence, (*i*) rotor is immediately brought to rest and (*ii*) centrifugal switch falls to the 'start' position. As before, *C* is put in series with the starter winding and the motor starts rotating in the opposite direction.

9. two-speed type

Speed can be changed by changing the number of poles in the winding for which purpose two separate running windings are placed in the slots of the stator, one being 6-pole winding and

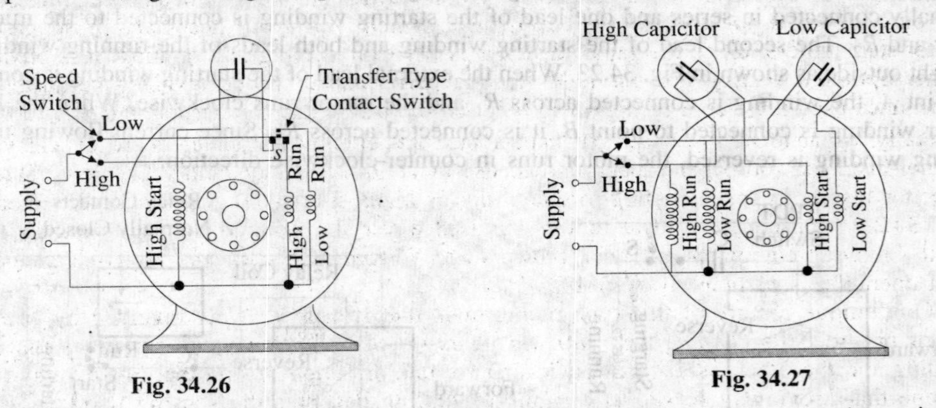

Fig. 34.26

Fig. 34.27

the other, 8-pole winding. Only one starting winding is used which always acts in conjunction with the higher-speed running winding. The double-action or transfer type centrifugal switch *S* has two contact points on the 'start' side and one on the 'run' side. As shown in Fig. 34.26, an external speed switch is used for changing the motor speed. The motor will always start on high speed irrespective of whether the speed switch is on the 'high' or 'low' contact. If speed switch is set on 'low', then as soon as the motor comes up to speed, the centrifugal switch

(*a*) cuts out the starting winding and high-speed running winding and
(*b*) cuts in the low-speed running winding.

10. two-speed with two-capacitor type

As shown in Fig. 34.27, this motor has two running windings, two starting windings and two capacitors. One capacitor is used for high-speed operation and the other for low-speed operation. A double centrifugal switch *S* is employed for cutting out the starting winding after start.

34.8. Capacitor Starts-and-Run Motor

This motor is similar to the capacitor-start motor [Art.34.4 (*ii*)] except that the starting winding and capacitor are connected in the circuit at *all times*. The advantages of leaving the capacitor permanently in circuit are (*i*) improvement of over-load capacity of the motor (*ii*) a higher power factor (*iii*) higher efficiency and (*iv*) quieter running of the motor which is so much desirable for small power drivers in offices and laboratories. Some of these motors which start and run with one value of capacitance in the circuit are called *single-value* capacitor-run motors. Other which start with high value of capacitance but run with a low value of capacitance are known as *two-value* capacitor-run motors.

(*i*) single-value capacitor-run motor

It has one running winding and one starting winding in series with a capacitor as shown in Fig. 34.28. Since capacitor remains in the circuit permanently, this motor is often referred to as

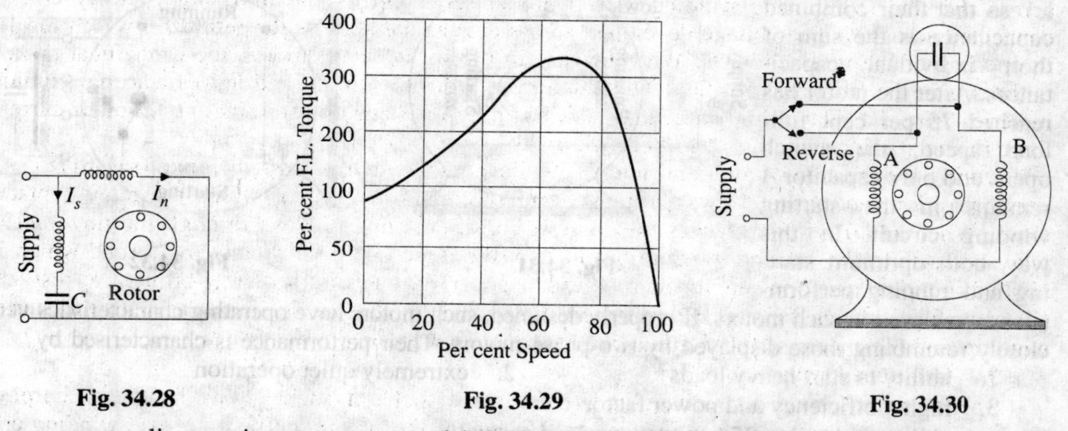

Fig. 34.28 **Fig. 34.29** **Fig. 34.30**

permanent-split capacitor-run motor and behaves practically like an unbalanced 2-phase motor. Obviously, there is no need to use a centrifugal switch which was necessary in the case of capacitor-start motors. Since the same capacitor is used for starting and running, it is obvious that neither optimum starting nor optimum running performance can be obtained because value of capacitance used must be a compromise between the best value for starting and that for running. Generally, capacitors of 2 to 20 μF capacitance are employed and are more expensive oil or pyranol-insulated foil-paper capacitors because of continuous-duty rating. The low value of the capacitor results in small starting torque which is about 50 to 100 per cent of the rated torque (Fig. 34.29). Consequently, these motors are used where the required starting torque is low such as air - moving equipment *i.e.* fans, blowers and voltage regulators and also oil burners where quiet operation is particularly desirable.

One unique feature of this type of motor is that it can be easily reversed by an external switch provided its *running and starting windings are identical*. One serves as the running winding and the other as a starting winding for one direction of rotation. For reverse rotation, the one that previously served as a running winding becomes the starting winding while the former starting winding serves as the running winding. As seen from Fig. 34.30 when the switch is in the forward position, winding *B* serves as running winding and *A* as starting winding. When switch is in 'reverse' position, winding *A* becomes the running winding and *B* the starting winding.

Such reversible motors are often used for operating devices that must be moved back and forth very frequently such as rheostats, induction regulations, furnace controls, valves and arc-welding controls.

(*ii*) two-value capacitor-run motor

This motor starts with a high capacitor in series with the starting winding so that the starting torque is high. For running, a lower capacitor is substituted by the centrifugal switch. Both the running and starting windings remain in circuit.

The two values of capacitance can be obtained as follows:

1. by using two capacitors in parallel at the start and then switching out one for low-value run (Fig. 34.31) or
2. by using a step-up auto-transformer in conjunction with one capacitor so that effective capacitance value is increased for starting purposes.

In Fig. 34.31, *B* is an electrolytic capacitor of high capacity (short duty) and *A* is an oil capacitor of low value (continuous duty). Generally, starting capacitor *B* is 10 to 15 times the running capacitor *A*. At the start, when the centrifugal switch is closed, the two capacitors are put in parallel, so that their combined capacitance is the sum of their individual capacitances. After the motor has reached 75 per cent full-load speed, the switch opens and only capacitor *A* remains in the starting winding circuit. In this way, both optimum starting and running performance is achieved in such motors. If properly designed, such motors have operating characteristics very closely resembling those displayed by two-phase motors. Their performance is characterised by

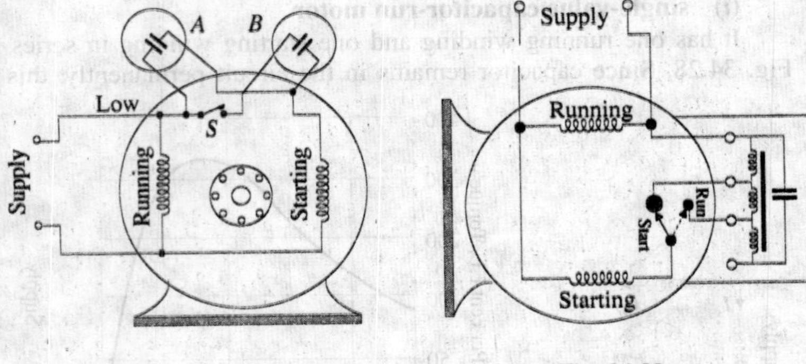

Fig. 34.31

Fig. 34.32

1. ability to start heavy loads
2. extremely quiet operation
3. higher efficiency and power factor
4. ability to develop 25 per cent overload capacity

Hence, such motors are ideally suited where load requirements are severe as in the case of compressors and fire strokers etc.

The use of an auto-transformer and single oil-type capacitor is illustrated in Fig.34.32. The transformer and capacitor are sealed in a rectangular iron box and mounted on top of the motor. The idea behind using this combination is that a capacitor of value *C* connected to the secondary of a step-up transformer, appears to the primary as though it had a value of K^2C where *K* is voltage transformation ratio. For example, if actual value of $C = 4\ \mu F$ and $K = 6$, then low-voltage primary acts as if it had a $144\ \mu F(= 6^2 \times 4)$ capacitor connected across its terminals. Obviously, effective value of capacitance has increased 36 times. In the 'start' position of the switch, the connection is made to the mid-tap of the auto-transformer so that $K = 2$. Hence, effective value of capacitance at start is 4 times the running value and is sufficient to give a high starting torque. As the motor speeds up, the centrifugal switch shifts the capacitor from one voltage tap to another so that the voltage transformation ratio changes from higher value at starting to a lower value for running. The capacitor which is actually of the paper-tinfoil construction is immersed in a high grade insulation like wax or mineral oil.

34.9. Shaded-pole Single-phase Motor

In such motors, the necessary phase-splitting is produced by induction. These motors have salient poles on the stator and a squirrel-cage type rotor Fig.34.33. shows a four-pole motor with the field poles connected in series for alternate polarity. One pole of such a motor is shown separately in Fig. 34.34. The laminated pole has a slot cut across the laminations approximately one-third distance from one edge. Around the small part of the pole is placed a short-circuited

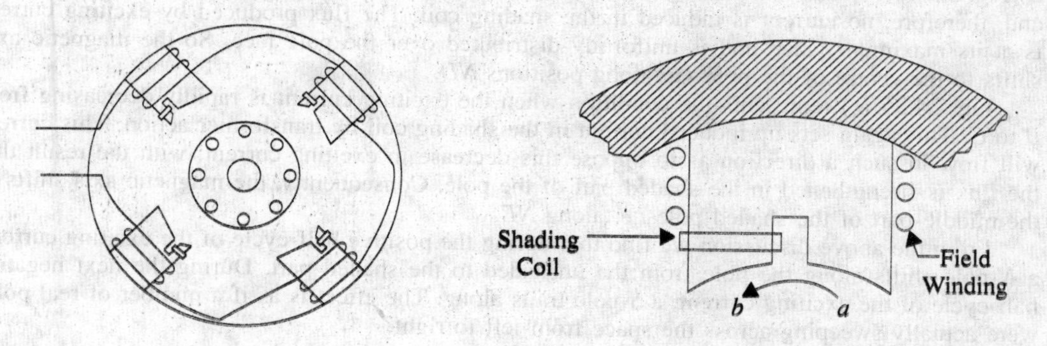

Fig. 34.33 **Fig. 34.34**

Cu coil known as *shading coil*. This part of the pole is known as *shaded* part and the other as *unshaded* part. When an alternating current is passed through the exciting (or field) winding surrounding the whole pole, the axis of the pole shifts from the unshaded part *a* to the shaded part *b*. This shifting of the magnetic axis is, in effect, equivalent to the actual physical movement of the pole. Hence, the rotor starts rotating in the direction of this shift *i.e.* from unshaded part to the shaded part.

Let us now discuss why shifting of the magnetic axis takes place. It is helpful to remember that the shading coil is highly inductive. When the alternating current through exciting coil tends to increase, it induces a current in the shading coil by transformer action in such a direction as to oppose its growth. Hence, flux density decreases in the shaded part when exciting current increases. However, flux density increases in the shaded part when exciting current starts decreasing (it being assumed that exciting current is sinusoidal)

In Fig. 34.35 (*a*) exciting current is rapidly increasing along *OA* (shown by dots). This will produce an e.m.f. in the shading coil. As shading coil is of low resistance, a large current will

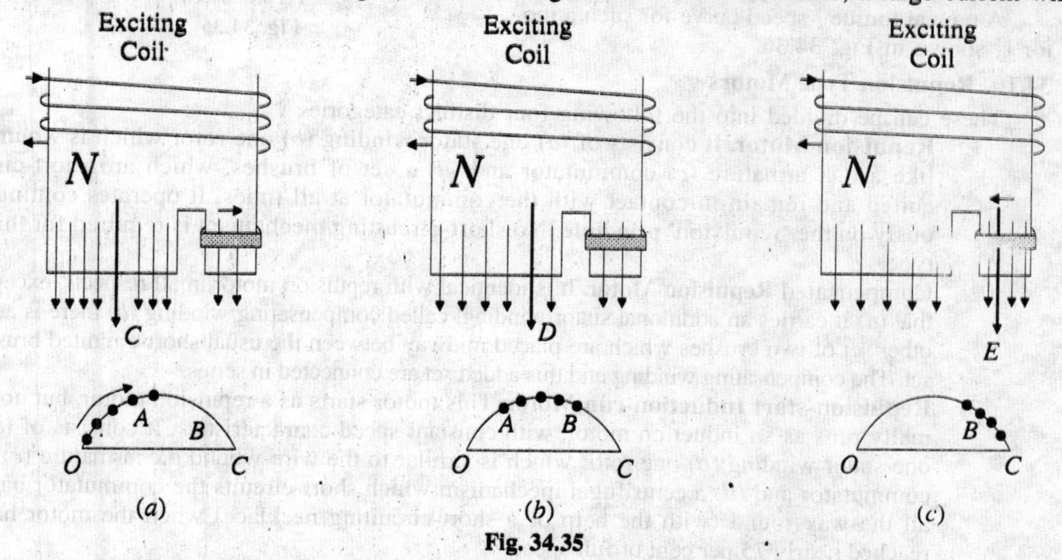

Fig. 34.35

be set up in such a direction (according to Lenz's law) as to oppose the rise of exciting current (which is responsible for its production). Hence, the flux mostly shifts to the unshaded part and the magnetic axis lies along the middle of this part *i.e.* along *NC*.

Next, consider the moment when exciting current is near its peak value *i.e.* from point A to B [Fig. 34.35 (*b*)]. Here, the change in exciting current is very slow. Hence, practically no voltage and, therefore, no current is induced in the shading coil. The flux produced by exciting current is at its maximum value and is uniformly distributed over the pole face. So the magnetic axis shifts to the centre of the pole *i.e.* along positions ND.

Fig. 34.35 (*c*) represents the condition when the exciting current is rapidly decreasing from B to C. This again sets up induced current in the shading coil by transformer action. This current will flow in such a direction as to oppose this decrease in exciting current, with the result that the flux is strengthened in the shaded part of the pole. Consequently, the magnetic axis shifts to the middle part of the shaded pole *i.e.* along NE.

From the above discussion we find that during the positive half-cycle of the exciting current, a N-pole shifts along the pole from the unshaded to the shaded part. During the next negative half-cycle of the exciting current, a S-pole trails along. The effect is as if a number of real poles were actually sweeping across the space from left to right.

Shaded pole motors are built commercially in very small sizes, varying approximately from 1/250 h.p. (3W) to 1/6 h.p. (125 W). Although such motors are simple in construction, extremely rugged, reliable and cheap, they suffer from the disadvantages of (*i*) low starting torque (*ii*) very little overload capacity and (*iii*) low efficiency. Efficiencies vary from 5% (for tiny sizes) to 35 (for higher ratings). Because of its low starting torque, the shaded-pole motor is generally used for small fans, toys, instruments, hair dryers, ventilators, circulators and electric clocks. It is also frequently used for such devices as churns, phonograph turntables and advertising displays etc. The direction of rotation of this motor cannot be changed, because it is fixed by the position of copper rings.

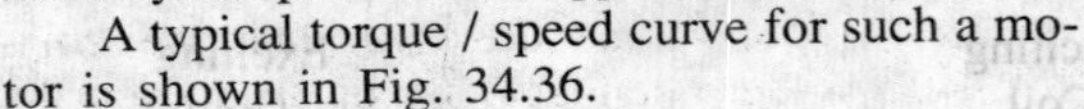

A typical torque / speed curve for such a motor is shown in Fig. 34.36.

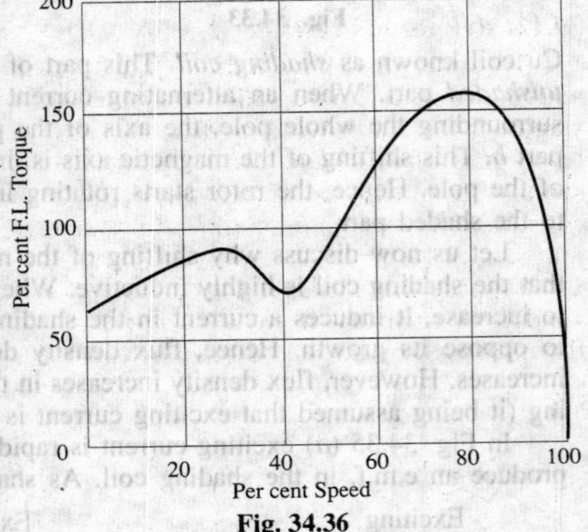

Fig. 34.36

34.10. Repulsion Type Motors

These can be divided into the following four distinct categories :

1. **Repulsion Motor.** It consists of (*a*) one stator winding (*b*) one rotor which is wound like a d.c. armature (*c*) commutator and (*d*) a set of brushes, which are short-circuited and remain in contact with the commutator at all times. It operates continuously on the 'repulsion' principle. No short-circuiting mechanism is required for this type.
2. **Compensated Repulsion Motor.** It is identical with repulsion motor in all respects, except that (*a*) it carries an additional stator winding, called compensating winding (*b*) there is another set of two brushes which are placed midway between the usual short-circuited brush set. The compensating winding and this added set are connected in series.
3. **Repulsion-start Induction-run Motor.** This motor starts as a repulsion motor, but normally runs as an induction motor, with constant speed characteristics. It consists of (*a*) one stator winding (*b*) one rotor which is similar to the wire-wound d.c. armature (*c*) a commutator and (*d*) a centrifugal mechanism which short-circuits the commutator bars all the way round (with the help of a short-circuiting necklace) when the motor has reached nearly 75 per cent of full speed.
4. **Repulsion-induction Motor.** It works on the combined principle of repulsion and induction. It consists of (*a*) stator winding (*b*) two rotor windings : one squirrel cage and the other usual d.c. winding connected to the commutator and (*c*) a short-circuited set of two brushes.

It may be noted that repulsion motors have excellent characterstics, but are expensive and require more attention and maintenance than single-phase motors. Hence, they are being replaced by two-value capacitor motors for nearly all applications.

34.11. Repulsion Motor

Constructionally, it consists of the following :

1. Stator winding of the distributed non-salient pole type housed in the slots of a smooth-cored stator (just as in the case of split-phase motors). The stator is generally wound for four, six or eight poles.
2. A rotor (slotted core type) carrying a distributed winding (either lap or wave) which is connected to the commutator. The rotor is identical in construction to the d.c. armature.
3. A commutator, which may be one of the two types : an axial commutator with bars parallel to the shaft or a radial or vertical commutator having radial bars on which brushes press horizontally.
4. Carbon brushes (fitted in brush holders) which ride against the commutator and are used for conducting current through the armature (*i.e.* rotor) winding.

34.12. Repulsion Principle

To understand how torque is developed by the repulsion principle, consider Fig. 34.37 which shows a 2-pole salient pole motor with the magnetic axis vertical. For easy understanding, the stator winding has been shown with concentrated salient-pole construction (actually it is of distributed non-salient type). The basic functioning of the machine will be the same with either type of construction. As mentioned before, the armature is of standard d.c. construction with commutator and brushes (which are short-circuited with a low-resistance jumper).

Suppose that the direction of flow of the alternating current in the exciting or field (stator) winding is such that it creates a *N*-pole at the top and a *S*-pole at the bottom. The alternating flux produced by the stator winding will induce e.m.f. in the armature conductors by transformer action. The direction of the induced e.m.f. can be found by using Lenz's law and is as shown in Fig. 34.37 (*a*). However, the direction of the induced *currents* in the armature conductors will depend on the *positions of the short-circuited brushes*. If brush axis is colinear with magnetic axis of the main poles, the directions of the induced currents (shown by dots and arrows) will be as indicated in Fig. 34.37 (*a*)*. As a result, the armature will become an electromagnet with a *N*-pole on its top, directly under the main *N*-pole and with a *S*-pole at the bottom, directly over the main *S*-pole. Because of this face-to-face positioning of the main and induced magnetic poles, no torque will be developed. The two forces of repulsion on top and bottom act along *YY′* in direct opposition to each other.**

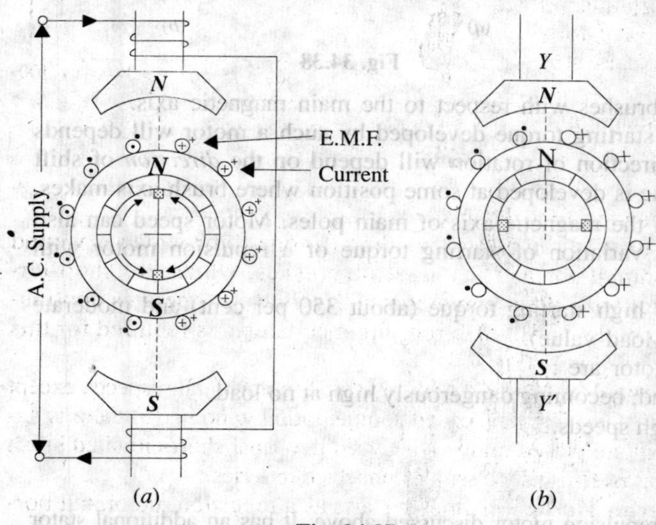

Fig. 34.37

*It should be noted that during the next half-cycle of the supply current, the directions of the respective voltages will be in the opposite directions.

**Alternatively, the absence of torque may be explained by arguing that the torques developed in the four quadrants neutralize each other.

If brushes are shifted through 90° to the position shown in Fig. 34.37 (*b*) so that the brush axis is at right angles to the magnetic axis of the main poles, the directions of the induced voltages at any time in the respective armature conductors *are exactly the same* as they were for the brush position of Fig. 34.37 (*a*). However, with brush positions of Fig. 33.37 (*b*), the voltages induced in the armature conductors in each path between the brush terminals will neutralize each other, hence there will be no net voltage across brushes to produce armature current. If there is no armature current, obviously, no torque will be developed.

If the brushes are set in position shown in Fig. 34.38 (*a*) so that the brush axis is neither in line with nor 90° from the magnetic axis *YY′* of the main poles, a net voltage* will be induced between the brush terminals which will produce armature current. The armature will again act as an electromagnet and develop its own *N*-and *S*-poles which, in this case, will not directly face the respective main poles. As shown in Fig. 34.38 (*a*), the armature poles lie along *AA′* making an angle of α with *YY′*.

Hence, *rotor N*-pole will be repelled by the *main N*-pole and the rotor *S*-pole will, similarly, be repelled by the main *S*-pole. Consequently, the rotor will rotate in clockwise direction [Fig.34.38 (*b*)]. Since the forces are those of *repulsion,* it is appropriate to call the motor as repulsion motor.

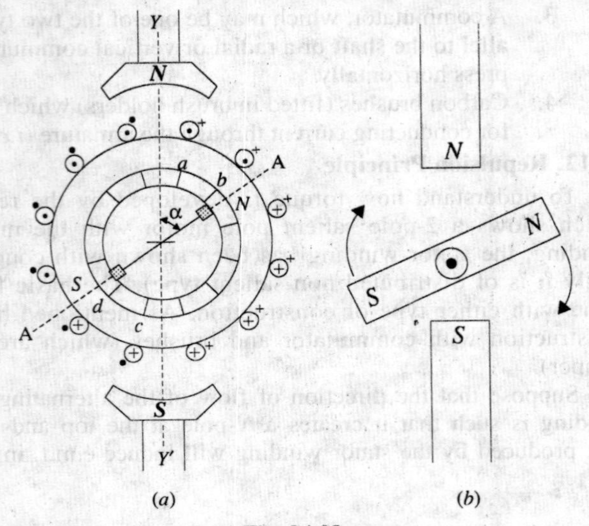

Fig. 34.38

It should be noted that if the brushes are shifted counter-clockwise from *YY′*, rotation will also be counter-clockwise. Obviously, direction of rotation of the motor is determined by the position of brushes with respect to the main magnetic axis.

It is worth noting that the value of starting torque developed by such a motor will depends on the *amount* of brush-shift whereas direction of rotation will depend on the *direction* of shift [Fig. 34.39 (*a*)]. Maximum starting torque is developed at some position where brush axis makes, an angle lying between 0° and 45° with the magnetic axis of main poles. Motor speed can also be controlled by means of brush shift. Variation of starting torque of a repulsion motor with brush-shift is shown in Fig. 33.39 (*b*).

A straight repulsion type motor has high starting torque (about 350 per cent) and moderate starting current (about 3 to 4 times full-load value).

Principal shortcomings of such a motor are :

1. speed varies with changing load, becoming dangerously high at no load.
2. low power factor, except at high speeds.
3. tendency to spark at brushes.

34.13. Compensated Repulsion Motor

It is a modified form of the straight repulsion motor discussed above, It has an additional stator winding, called compensating winding whose purpose is (*i*) to improve power-factor and (*ii*) to provide better speed regulation. This winding is much smaller than the stator winding and is usually wound in the inner slots of each main pole and is connected in series with the armature (Fig. 34.40) through an additional set of brushes placed mid-way between the usual short-circuited brushes.

*It will be seen from Fig. 34.38 (*a*) that the induced voltages in conductors *a* and *b* oppose the voltages in other conductors lying above brush-axis. Similarly, induced voltages in conductors *c* and *d* oppose the voltages in other conductors, lying below the brush-axis. Yet the net voltage across brush terminals will be sufficient to produce current which will make the armature a powerful magnet.

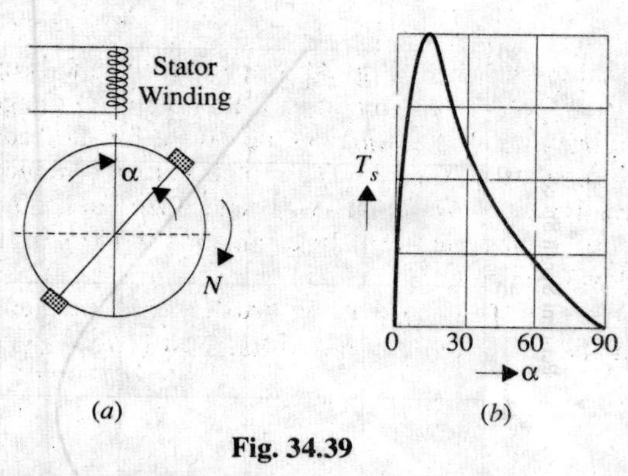

Fig. 34.39

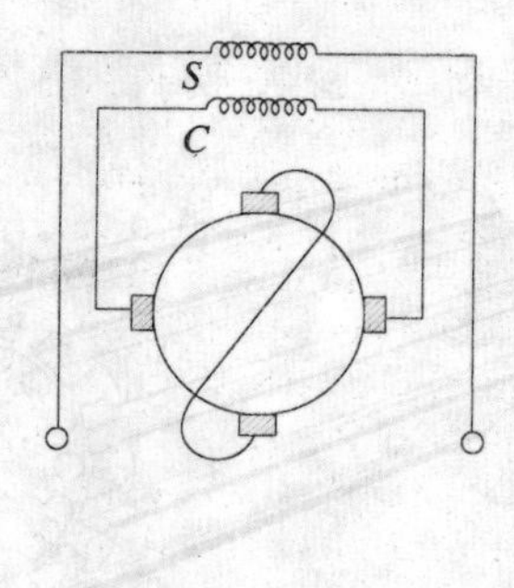

Fig. 34.40

34.14. Repulsion-start Induction-run Motor

As mentioned earlier, this motor starts as an ordinary repulsion motor, but after it reaches about 75 per cent of its full speed, centrifugal short-circuiting device short-circuits its commutator. From then on, it runs as an induction motor, with a short-circuited squirrel-cage rotor. After the commutator is short-circuited, brushes do not carry any current, hence they may also be lifted from the commutator, in order to avoid unnecessary wear and tear and friction losses.

Repulsion-start motors are of two different designs :

1. Brush-lifting type in which the brushes are automatically lifted from the commutator when it is short-circuited. These motors generally employ radial form of commutator and are built both in small and large sizes.
2. Brush-riding type in which brushes ride on the commutator at all times. These motors use axial form of commutator and are always built in small sizes.

The starting torque of such a motor is in excess of 350 per cent with moderate starting current. It is particularly useful where starting period is of comparatively long duration, because of high inertia loads. Applications of such motors include machine tools, commercial refrigerators, compressors, pumps, hoists, floor-polishing and grinding devices etc.

34.15. Repulsion Induction Motor

In the field of repulsion motor, this type is becoming very popular, because of its good all-round characteristics which are comparable to those of a compund d.c. motor. It is particularly suitable for those applications where the load can be removed entirely by de-clutching or by a loose pulley.

This motor is a combination of the repulsion and induction types and is sometimes referred to as *squirrel-cage repulsion motor*. It possesses the desirable characteristics of a repulsion motor and the constant-speed characteristics of an induction motor.

It has the usual stator winding as in all repulsion motors. But there are two separate and independent windings in the rotor (Fig. 33.41).

(*i*) a squirrel-cage winding and

(*ii*) commutated winding similar to that of a d.c. armature.

Both these windings function during the *entire period* of operation of the motor. The commutated winding lies in the outer slots while squirrel-cage winding is located in the inner slots*. At start, the commutated winding supplies most of the torque, the squirrel-cage winding being practically inactive because of its high reactance. When the rotor accelerates, the squirrel-cage winding takes up a larger portion of the load.

The brushes are short-circuited and ride on the commutator continuously. One of the advantages of this motor is that it requires no centrifugal short-circuiting mechanism. Sometimes such motors are also made with compensating winding for improving the power factor.

As shown in Fig. 34.42, its starting torque is high, being in excess of 300 per cent. Moreover,

*Hence, commutated winding has low reactance, whereas the squirrel-cage winding has inherently a high reactance.

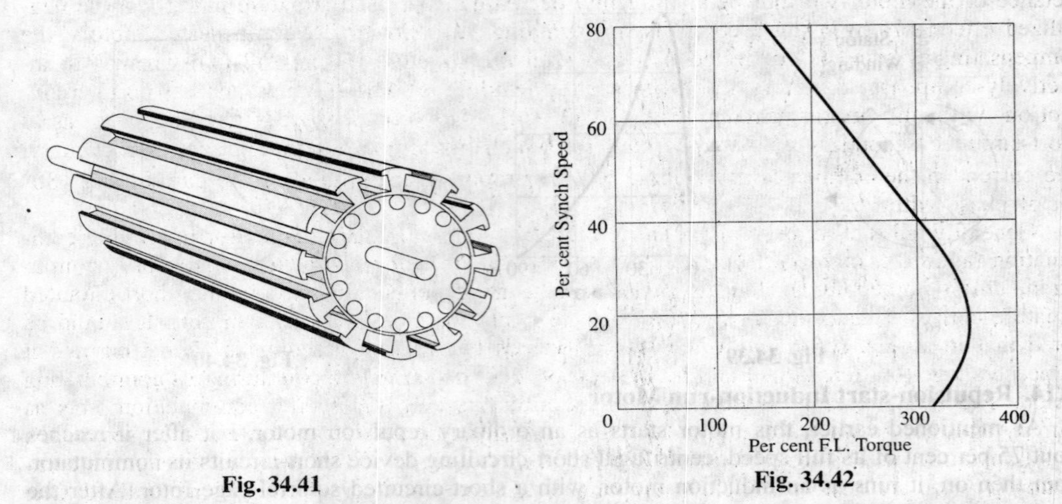

Fig. 34.41 **Fig. 34.42**

it has a fairly constant speed regulation. Its field of application includes house-hold refrigerators, garage air pumps, petrol pumps, compressors, machine tools, mixing machines, lifts and hoists etc.

These motors can be reversed by the usual brush-shifting arrangement.

34.16. A.C. Series Motors

If an ordinary d.c. series motor were connected to an a.c. supply, it will rotate and exert unidirectional torque* because the current flowing both in the armature and field reverses at the same time. But the performance of such a motor will not be satisfactory for the following reasons :-

1. the alternating flux would cause excessive eddy current loss in the yoke and field cores which will become extremely heated.
2. vicious sparking will occur at brushes because of the huge voltage and current induced in the short-circuited armature coils during their commutation period.
3. power factor is low because of high inductance of the field and armature circuits.

However, by proper modification of design and other refinements, a satisfactory single-phase motor has been produced.

The eddy current loss has been reduced by laminating the entire iron structure of the field cores and yoke.

Power factor improvement is possible only by reducing the magnitudes of the reactances of the field and armature windings. Field reactance is reduced by reducing the number of turns on the field windings. For a given current, it will reduce the field m.m.f. which will result in reduced air-gap flux. This will tend to increase the speed but reduce motor torque. To obtain the same torque, it will now be necessary to increase the number of armature turns proportionately. This will, however, result in increased inductive reactance of the armature, so that the overall

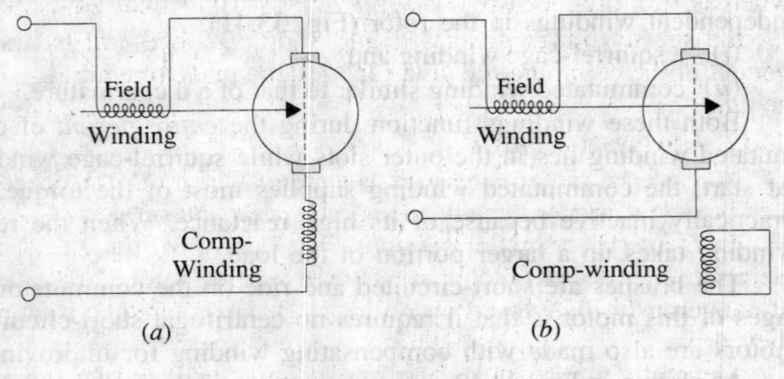

Fig. 34.43

*However, torque developed is not of constant magnitude (as in d.c. series motors) but pulsates between zero and maximum value each half-cycle.

rectance of the motor will not be significantly decreased. Increased armature m.m.f. can be neutralized effectively by using a compensating winding. In conductively-compensated motors, the compensating winding is connected in series with the armature [Fig. 34.43 (*a*)] whereas in inductively-compensated motors, the compensating winding is short-circuited and has no interconnection with the motor circuit [Fig. 34.43 (*b*)]. The compensating winding acts as a short-circuited secondary of a transformer, for which the armature winding acts as a primary. The current in the compensating winding will be proportional to the armature current and 180° out of phase with it.

Generally, all d.c. series motors are 'provided' with commutating poles for improving commutation (as in d.c. motors). But commutating poles alone will not produce satisfactory commutation, unless something is done to neutralize the huge voltage induced in the short-circuited armature coil by transformer action (this voltage is not there in d.c. series motor). It should be noted that in an a.c. series motor, the flux produced by the field winding is alternating and it induces voltage (by transformer action) in the short-circuited armature coil during its commutating period. The field winding, associated with the armature coil undergoing commutation, acts as primary and the armature coil during its commutating period acts as a short-circuited secondary. This transformer action produces heavy current in the armature coil as it passes through its commutating period and results in vicious sparking, unless the transformer voltage is neutralized. One method, which is often used for large motors, consists of shunting the winding of each commutating pole with a non-inductive resistance, as shown in Fig. 34.44 (*a*).

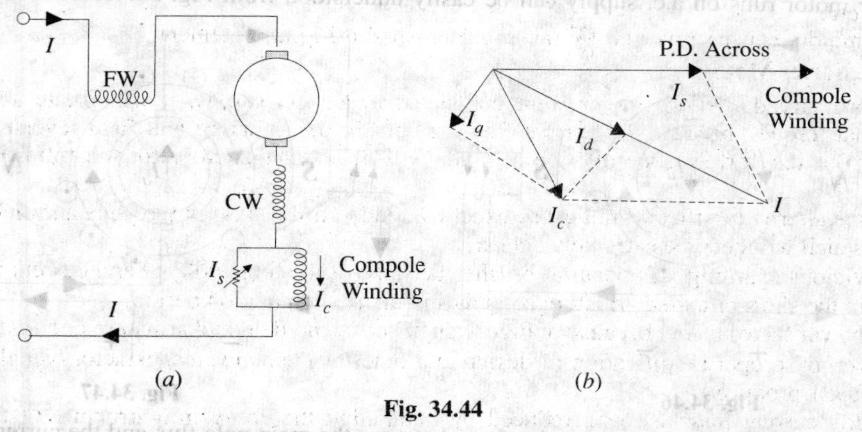

Fig. 34.44

Fig. 34.44. (*b*) shows the vector diagram of a shunted commutator pole. The current I_c through the commutating pole (which lags the total motor current) can be resolved into two reactangular components I_d and I_q as shown. I_d produces a flux which is in phase with total motor current I whereas flux produced by I_q lags I by 90°. By proper adjustment of shunt resistance (and hence I_s), the speed voltage generated in a short-circuited coil by the cutting of the 90° lagging component of the commutating pole flux may be made to neutralize the voltage induced by transformer action.

34.17. Universal Motor

A universal motor is defined as a motor which may be operated either on direct or single-phase a.c. supply at approximately the same speed and output.

In fact, it is a smaller version (5 to 150 W) of the a.c. series motor described in Art. 34.16. Being a series-wound motor, it has high starting torque and a variable speed characteristic. It runs at dangerously high speed on no-load, That is why such motors are usually built into the device they drive.

Generally, universal motors are manufactured in two types:-

1. *concentrated-pole, non-compensated type* (low power rating)
2. *distributed-field compensated type* high power rating)

The non-compensated motor has two salient poles and is just like a 2-pole series d.c. motor except that whole of its magnetic path is laminated (Fig. 34.45). The laminated stator is necessary because the flux is alternating when motor is operated from a.c. supply. The armature is of wound type and similar to that of a small d.c. motor. It consists essentially of a laminated core having either straight or skewed slots and a commutator to which the leads of the armature winding are connected. The distributed-field compensated type motor has a stator core similar to that of a split-phase motor and a wound armature similar to that of a small d.c. motor. The compensating winding is used to reduce the reactance voltage present in the armature when motor runs on a.c. supply. This voltage is caused by the alternating flux by transformer action (Art.34.16).

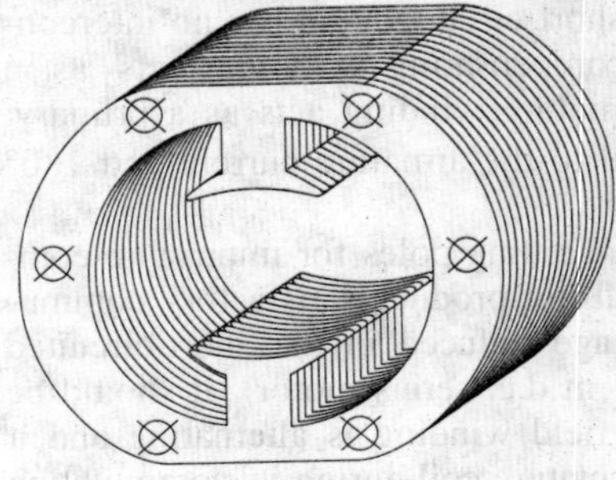

Fig. 34.45

In a 2-pole non-compensated motor, the voltage induced by transformer action in a coil during its commutation period is not sufficient to cause any serious commutation trouble. Moreover, high-resistance brushes are used to aid commutation.

(*a*) **Operation.** As explained in Art. 34.16, such motors develop unidirectional torque, regardless of whether they operate on d.c. or a.c. supply. The production of unidirectional torque, when the motor runs on a.c. supply can be easily understood from Fig. 34.46. The motor works

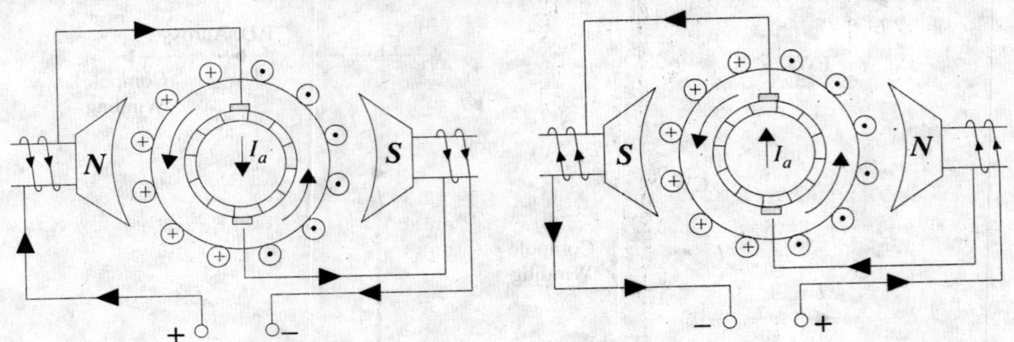

Fig. 34.46 **Fig. 34.47**

on the same principle as a d.c. motor *i.e.* force between the main pole flux and the current-carrying armature conductors. This is true regardless of whether the current is alternating or direct (Fig. 34.47).

(*b*) **Speed/Load Characteristic.** The speed of a universal motor varies just like that of a d.c. series motor *i.e.* low at full-load and high on no-load (about 20,000 r.p.m. in some cases). In fact, on no-load the speed is limited only by its own friction and windage load. Fig. 34.48 shows typical torque characteristics of a universal motor both for d.c. and a.c. supply. Usually, gear trains are used to reduce the actual load speeds to proper values.

(*c*) **Applications.** Universal motors are used in vacuum cleaners where actual motor speed is the load speed. Other applications where motor speed is reduced by a gear train are : drink and food mixers, portable drills and domestic sewing machine etc.

(*d*) **Reversal of Rotation.** The concentrated-pole (or salient-pole) type universal motor may be reversed by reversing the flow of current through either the armature or field windings. The usual method is to interchange the leads on the brush holders (Fig.34.49).

The distributed-field compensated type universal motor may be reversed by interchanging either the armature or field leads and shifting the brushes against the direction in which the motor will rotate. The extent of brush shift usually amounts amounts to several commutator bars.

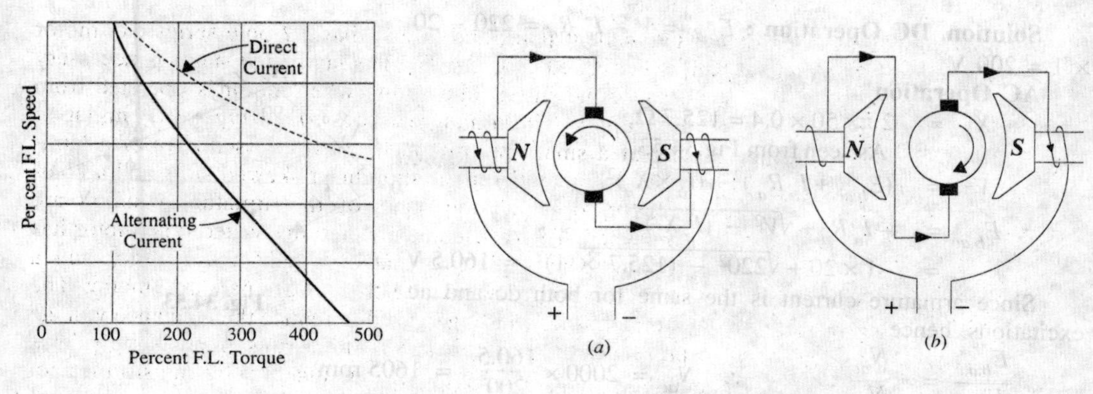

Fig. 34.48

Fig. 34.49

34.18. Speed Control of Universal Motors

The following methods are usually employed for speed-control purposes :

(*i*) **Resistance Method.** As shown in Fig. 34.50, the motor speed is controlled by connecting a variable resistance *R* in series with the motor. This method is employed for motors used in sewing machines. The amount of resistance in the circuit is changed by means of a foot-pedal.

(*ii*) **Tapping-field Method.** In this method, a field pole is tapped at various points and speed is controlled by varying the field strength (Fig. 34.51). For this purpose, either of the following two arrangements may be used :-

(*a*) The field pole is wound in various sections with different sizes of wire and taps are brought out from each section.

(*b*) Nichrome resistance wire is wound over one field pole and taps are brought out from this wire.

(*iii*) **Centrifugal Mechanism**. Universal motors, particularly those used for home food and drink mixers, have a number of speeds.Selection is made by a centrifugal device located inside

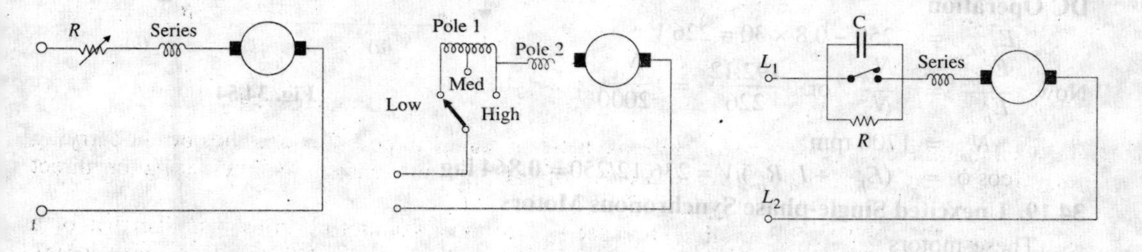

Fig. 34.50 Fig. 34.51 Fig. 34.52

the motor and connected, as shown in Fig. 34.52. The switch is adjustable by means of an external lever. If the motor speed rises above that set by the lever, the centrifugal device opens two contacts and inserts resistance *R* in the circuit, which causes the motor speed to decrease. When motor runs slow, the two contacts close and short-circuit the resistance, so that the motor speed rises. This process is repeated so rapidly that variations in speed are not noticeable.

The resistance *R* is connected across the governor points as shown in Fig. 34.52. A capacitor *C* is used across the contact points in order to reduce sparking produced due to the opening and closing of these points. Moreover, it prevents the pitting of contacts.

Example 34.4. *A 250-W, single-phase, 50-Hz, 220-V universal motor runs at 2000 rpm and takes 1.0 A when supplied from a 220-V dc. supply. If the motor is connected to 220-V ac supply and takes 1.0 A (r.m.s), calculate the speed, torque and power factor, Assume* $R_a = 20\ \Omega$ *and* $L_a = 0.4\ H$.

Solution. DC Operation : $E_{b.dc} = V - I_a R_a = 220 - 20 \times 1 = 200$ V

AC Operation

$X_a = 2\pi \times 50 \times 0.4 = 125.7\ \Omega.$

As seen from Fig.34.53.

$V^2 = (E_{b.ac} + I_a R_a)^2 + (I_a \times X_a)^2$

$\therefore E_{b.ac} = -I_a R_a + \sqrt{V^2 - (I_a X_a)^2}$

$= -1 \times 20 + \sqrt{220^2 - (125.7 \times 1)^2} = 160.5$ V

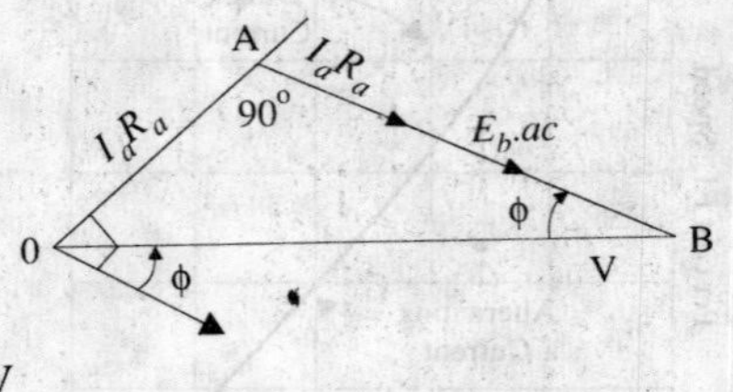

Fig. 34.53

Since armature current is the same for both dc and ac excitations, hence

$\frac{E_{b.d.c}}{E_{b.ac}} = \frac{N_{dc}}{N_{ac}}$; $\therefore N_{ac} = 2000 \times \frac{160.5}{200} = 1605$ rpm

$\cos\phi = AB/OB = (E_{b.ac} + I_a R_a)/V = (160.5 + 20)/220 = 0.82$ lag

$P_{mech} = E_{b.ac} I_a = 160.5 \times 1 = 160.5$ W

$T = 9.55 \times 160.5/1605 =$ **0.955 *N-m***

Example 34.5. *A universal series motor has resistance of 30 Ω and an inductance of 0.5 H. When connected to a 250 V d.c. supply and loaded to take 0.8 A, it runs at 2000 r.p.m. Estimate its speed and power factor, when connected to a 250-V, 50-Hz a.c. supply and loaded to take the same current.* **(Elect. Machine, A.M.I.E. Sec. B, 1992)**

Solution. Ac Operation

$X_a = 2\pi \times 50 \times 0.5 = 157\ \Omega,\ R_a = 30\ \Omega$

$I_a R_a = 0.8 \times 30 = 24\ \Omega$

$I_a X_a = 0.8 \times 157 = 125.6$ V

The phasor diagram is shown in Fig. 34.54 (*b*)

$V^2 = (E_{b.ac} + I_a R_a)^2 + (I_a X_a)^2$

$250^2 = (E_{b.ac} + 24)^2 + 125.6^2 = 192.12$V

30Ω 0.5 H R_a X_a 250 V (dc) (*a*)

A $I_aR_a + E_b.ac$ 90° θ 0 θ V B (*b*)

Fig. 34.54

DC Operation

$E_{b.dc} = 250 - 0.8 \times 30 = 226\ V$

Now, $\frac{E_{b.ac}}{E_{b.dc}} = \frac{N_{ac}}{N_{dc}}$ or $\frac{192.12}{226} = \frac{N_{ac}}{2000}$;

$N_{ac} = 1700$ rpm

$\cos\phi = (E_{b.ac} + I_a R_a)/V = 236.12/250 =$ **0.864 lag**

34.19. Unexcited Single-phase Synchronous Motors

These motors

1. operate from a single-phase a.c. supply
2. run at a constant speed—the synchronous speed of the revolving flux
3. need no d.c. excitation for their rotors (that is why they are called *unexcited*)
4. are self-starting.

These are of two types (*a*) reluctance motor and (*b*) hysteresis motor.

34.20. Reluctance Motor

It has either the conventional split-phase stator and a centrifugal switch for cutting out the auxiliary winding (split-phase type reluctance motor) or a stator similar to that of a permanent-split capacitor-run motor (capacitor-type reluctance motor). The stator produces the revolving field.

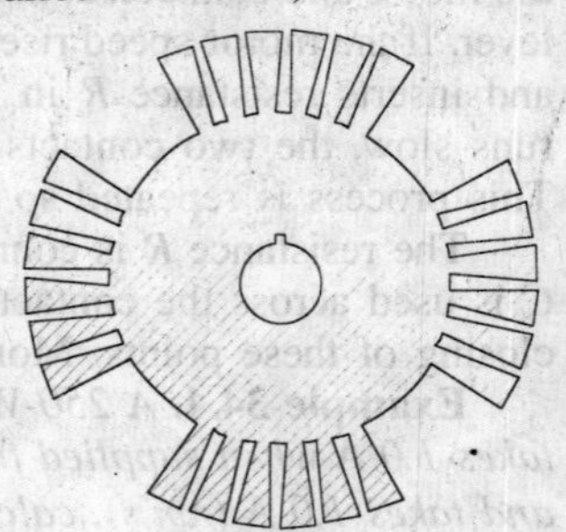

Fig. 34.55

The squirrel-cage rotor is of unsymmetrical magnetic construction. This type of unsymmetrical construction can be achieved

by removing some of the teeth of a symmetrical squirrel-cage rotor punching. For example, in a 48-teeth, four-pole rotor following teeth may be cut away :

1, 2, 3, 4, 5, 6—13, 14, 15, 16, 17, 18—25, 26, 27, 28, 29, 30—37, 38, 39, 40, 41, 42.

This would leave four projecting or salient poles (Fig. 34.55) consisting of the following sets of teeth:7.12; 19.24; 31.36 and 43—48. In this way, the rotor offers variable magnetic reluctance to the stator flux, the reluctance varying with the position of the rotor.

Working

For understanding the working of such a motor one basic fact must be kept in mind. And it is that when a piece of magnetic material is located in a magnetic field, a force acts on the material, tending to bring it into the most dense portion of the field. The force tends to align the specimen of material in such a way that the reluctance of the magnetic path that lies through the material will be minimum.

When the stator winding is energised, the revolving magnetic field exerts reluctance torque on the unsymmetrical rotor tending to align the salient pole axis of the rotor with the axis of the revolving magnetic field (because in this position, the reluctance of the magnetic path is minimum). If the reluctance torque is sufficient to start the motor and its load, the rotor will pull into step with the revolving field and continue to run at the speed of the revolving field.*

However, even though the rotor revolves synchronously, its poles lag behind the stator poles by a certain angle known as torque angle, (something similar to that in a synchronous motor). The reluctance torque increases with increase in torque angle, attaining maximum value when $\alpha = 45°$. If α increases beyond 45°, the rotor falls out of synchronism. The average value of the reluctance torque is given by $T = K\,(V/f)^2 \sin 2\,\delta$ where K is a motor constant.

It may be noted that the amount of load which a reluctance motor could carry at *its constant speed* would only be a fraction of the load that the motor could normally carry when functioning as an induction motor. If the load is increased beyond a value under which the reluctance torque cannot maintain synchronous speed, the rotor drops out of step with the field. The speed, then, drops to some value at which the slip is sufficient to develop necessary torque to drive the load by induction-motor action.

The constant-speed characteristic of a reluctance motor makes it very suitable fo such applications as signalling devices, recording instruments, many kinds of timers and phonographs etc.

31.21. Hysteresis Motor

The operation of this motor depends on the presence of a *continuously-revolving* magnetic flux. Hence, for the split-phase operation, its stator has two windings which remain connected to the single-phase supply continuously both at starting as well as during the running of the motor. Usually, shaded-pole principle is employed for this purpose giving shaded-pole hysteresis motor. Alternatively, stator winding of the type used in capacitor-type motor may be used giving capacitor-type shaded-pole motor. Obviously, in either type, no centrifugal device is used.

The rotor is a smooth chrome-steel cylinder** having high retentivity so that the hysteresis loss is high. It has no winding. Because of high retentivity of the rotor material, it is very difficult to change the magnetic polarities once they are induced in the rotor by the revolving flux. The rotor revolves synchronously because the rotor poles magnetically lock up with the revolving stator poles of opposite polarity. However, the rotor poles always lag behind the stator poles by an angle α. Mechanical power developed by rotor is given by $P_m = P_h \left(\frac{1-s}{s}\right)$ where P_h is hysteres loss in rotor. Also $T_h = 9.55\ P_m / N_s$. It is seen that hysteresis torque depends solely on the area of rotor's hysteresis loop.

*Actually, the motor starts as an induction motor and after it has reached its maximum speed as an induction motor, the reluctance torque pulls its rotor into step with the revolving field so that the motor now runs as a synchronous motor by virtue of its saliency.

**Rotors of ceramic permanant magnet material are used whose resistivity approaches that of an insulator. Consequently, it is impossible to set up eddy currents in such a rotor. Hence, there is no eddy current loss but only hystesesis loss.

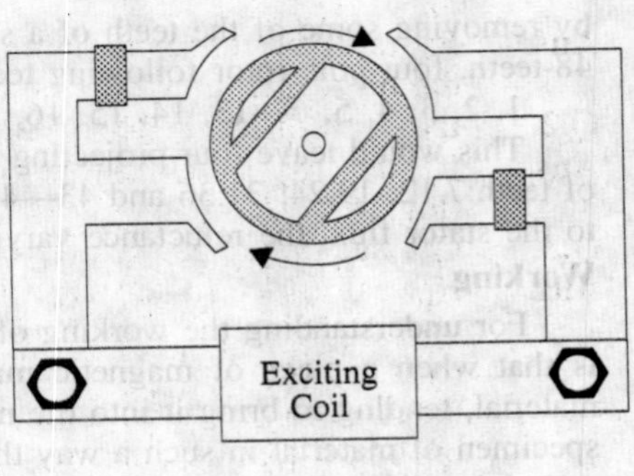

Fig. 34.56

The fact that the rotor has no teeth or winding of any sort, results in making the motor extremely quiet in operation and free from mechanical and magnetic vibrations. This makes the motor particularly useful for driving tape-decks, tape-decks, turn-tables and other precision audio equipment. Since, commercial motors usually have two poles, they run at 3,000 r.p.m. at 50-Hz single-phase supply. In order to adopt such a motor for driving an electric clock and other indicating devices, gear train is connected to the motor shaft for reducing the load speed. The unit accelerates rapidly, changing from rest to full speed almost instantaneously. It must do so because it cannot accelerate gradually as an ordinary motor it is either operating at synchronous speed or not at all.

Some unique features of a hysteresis motor are as under :—

(*i*) since its hysteresis torque remains practically constant from locked rotor to synchronous speed, a hysteresis motor is able to synchronise any load it can accelerate—something no other motor does.

(*ii*) due to its smooth rotor, the motor operates quietly and does not suffer from magnetic pulsations caused by slots/salient-poles that are present in the rotors of other motors.

In Fig. 34.56. is shown a two-pole shaded-pole type hysteresis motor used for driving ordinary household electric clocks. The rotor is a thin metal cylinder and the shaft drives a gear train.

Example 34.6. *A 8-kW, 4-pole, 220-V, 50-Hz reluctance motor has a torque angle of 30° when operating under rated load conditions. Calculate (i) load torque (ii) torque angle if the voltage drops to 205 V and (iii) will the rotor pulled out of synchronism?*

Solution. (*i*) $N_s = 120 \times 50/4 = 1500$ rpm; $T_{sh} = 9.55 \times$ output$/N = 9.55 \times 8000/1500 =$ 51 *N-m*

(*ii*) With the same load torque and constant frequency,

$V_1 \sin 2\alpha_1 = V_2^2$ sing $2\,\alpha_2$

$\therefore\quad 220^2 \times \sin(2 \times 30°) = 205^2 \times \sin 2\alpha; \qquad \therefore\ \alpha = 42.9°$

(*iii*) since the new load angle is less than 45°, the rotor will not pull out of synchronous.

Tutorial Problems-34.2.

1. A 230-V, 50-Hz, 4-pole, class-A, single-phase induction motor has the following parameters at an operating temperature 63°*C*:

$r_{1m} = 2.51$ ohms, $r_2^1 = 7.81$ ohm, $X_m = 150.88$ ohm, $X_{im} = 4.62$ ohm, $X_2' = 4.62$ ohms

Determine stator main winding current and power factor when the motor is running at a slip of 0.05 at the specified temperature of 63°*C*.

[3.74 ∠48.24°, 0.666] *(AMIE Sec. B Elect. Machines (E-B) Summer 1991)*

2. A fractional horse-power universal motor has armature circuit resistance of 20 ohm and inductance of 0.4 H. On being connected to a 220-V d.c. supply, it draws 1.0 A from the mains and runs at 2000 r.p.m. Estimate the speed and power factor of the motor, when connected to a 230-V, 50-Hz supply drawing the same armature current. Draw relevant phasor diagram.

[1726 rpm, 0.84] *(AMIE Sec. B Elect. Machines 1991)*

3. A universal series motor, when operating on 220 V d.c. draws 10 A and runs at 1400 r.p.m Find the new speed and power factor, when connected to 220 V, 25 Hz supply, the motor current remaining the same. The motor has total resistance of 1 ohm and total inductance of 0.1 H.

[961 rpm; 0.7] *(AMIE Sec. B Elect. Machines 1990)*

QUESTIONS AND ANSWERS ON SINGLE-PHASE MOTORS

Q.1. How would you reverse the direction of, rotation of a capacitor start-induction-run motor ?

Ans. By reversing either the running or starting-winding leads where they are connected to the lines. Both must not be reversed.

Q.2. In which direction does a shaded-pole motor run ?

Ans. It runs from the unshaded to the shaded pole (Fig.34.57)

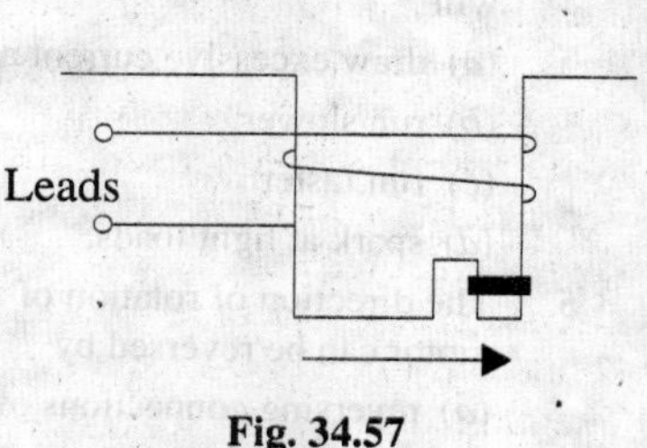

Fig. 34.57

Q.3. Can such a motor be reversed ?

Ans. Normally, such motors are not reversible because that would involve mechanical dismantling and re-assembly. However, special motors are made having two rotors on a common shaft, each having one stator assembly for rotation in opposite direction.

Q.4. What is a universal motor ?

Ans. It is built like a series d.c. motor with the difference that both its stator and armature are laminated. They can be used either on d.c. or a.c. supply although the speed and power are greater on direct current. They cannot be satisfactorily made to run at less than about 2000 r.p.m.

Q.5. How can a universal motor be reversed ?

Ans. By reversing either the field leads or armature leads but not both.

Q.6. How can we reverse the direction of rotation of repulsion, repulsion-induction and repulsion-induction and repulsion-start-induction-run motors ?

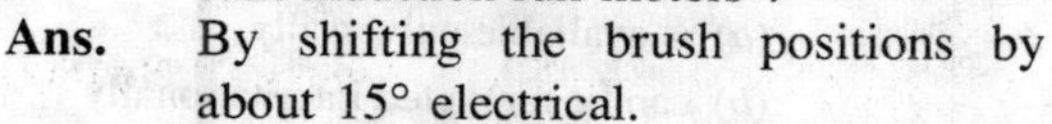

Ans. By shifting the brush positions by about 15° electrical.

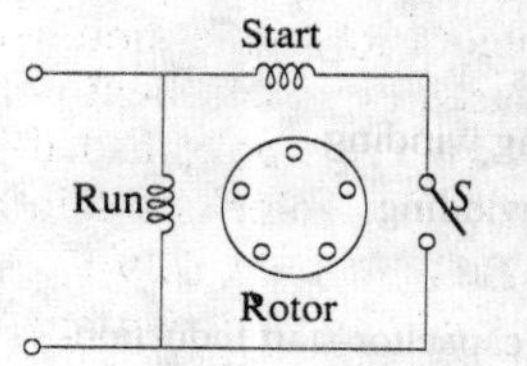

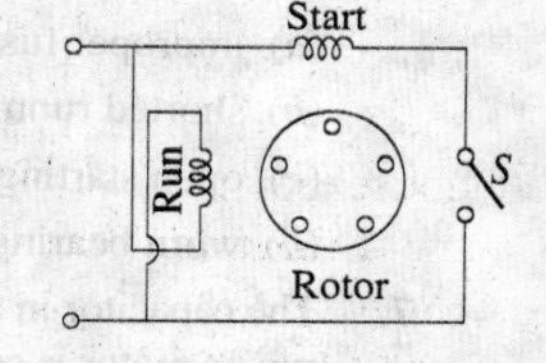

Fig. 34.58

Q.7. How can we reverse the rotation of a 1-phase, split-phase motor ?

Ans. By reversing the leads to either the running or starter winding (Fig. 34.58) but not both.

Q.8. What could be the reasons if a repulsion-induction motor fails to start ?

Ans. Any one of the following:-

1. no supply voltage 2. low voltage 3. excessive overload
4. the bearing lining may be stuck or 'frozen' to the shaft
5. armature may be rubbing 6. brush yoke may be incorrectly located
7. brush spacing may be wrong.

Q.9. What could be the reasons if a split-phase motor fails to start and hums loudly ?

Ans. It could be due to the starting winding being open or grounded or burnt out.

Q.10. What could be the reasons if a split-phase motor runs too slow ?

Ans. Any one of the following factors could be responsible:

1. wrong supply voltage and frequency
2. overload 3. grounded starting and running windings
4. short-circuited or open winding in field circuit.

OBJECTIVE TESTS-34

1. The starting winding of a single-phase motor is placed in the
(*a*) rotor (*b*) stator
(*c*) armature (*d*) field.

2. One of the characteristics of a single-phase motor is that it
(*a*) is self-starting
(*b*) is not self-starting
(*c*) requires only one winding
(*d*) can rotate in one direction only.

3. After the starting winding of a single-phase induction motor is disconnected from supply, it continues to run only on ....winding.
(*a*) rotor (*b*) compensating
(*c*) field (*d*) running

4. If starting winding of a single-phase induction motor is left in the circuit, it will

(*a*) draw excessive current and overheat

(*b*) run slower

(*c*) run faster

(*d*) spark at light loads.

5. The direction of rotation of a single-phase motor can be reversed by

(*a*) reversing connections of both windings

(*b*) reversing connections of starting winding

(*c*) using a reversing switch

(*d*) reversing supply connections.

6. If a single-phase induction motor runs slower than normal, the more likely defect is

(*a*) improper fuses

(*b*) shorted running winding

(*c*) open starting winding

(*d*) worn bearings.

7. The capacitor in a capacitor-start induction-run ac motor is connected in series with ...... winding.

(*a*) starting (*b*) running

(*c*) squirrel-cage (*d*) compensating

8. A permanent-split single-phase capacitor motor does **not** have

(*a*) centrifugal switch

(*b*) starting winding

(*c*) squirrel-cage rotor

(*d*) high power factor.

9. The starting torque of a capacitor-start induction-run motor is directly related to the angle α between its two winding currents by the relation

(*a*) $\cos \alpha$ (*b*) $\sin \alpha$

(*c*) $\tan \alpha$ (*d*) $\sin \alpha/2$.

10. In a two-value capacitor motor, the capacitor used for running purposes is a/an

(*a*) dry-type ac electrolytic capacitor

(*b*) paper-spaced oil-filled type

(*c*) air-capacitor

(*d*) ceramic type.

11. If the centrifugal switch of a two-value capacitor motor using two capacitors fails to open, then

(*a*) electrolytic capacitor will, in all probability, suffer breakdown

(*b*) motor will not carry the load

(*c*) motor will draw excessively high current

(*d*) motor will not come upto the rated speed.

12. Each of the following statements regarding a shaded-pole motor is true **except**

(*a*) its direction of rotation is from un-shaded to shaded portion of the poles

(*b*) it has very poor efficiency

(*c*) it has very poor p.f.

(*d*) it has high starting torque.

13. Compensating winding is employed in an ac series motor in order to

(*a*) compensate for decrease in field flux

(*b*) increase the total torque

(*c*) reduce the sparking at brushes

(*d*) reduce effects of armature reaction.

14. A universal motor is one which

(*a*) is avaliable universally

(*b*) can be marketed internationally

(*c*) can be operated either on dc or ac supply

(*d*) runs at dangerously high speed on no-load.

15. In a single-phase series motor the main purpose of inductively-wound compensating winding is to reduce the

(*a*) reactance emf of commutation

(*b*) rotational emf of commutation

(*c*) transformer emf of commutation

(*d*) none of the above.

(Power App.-II, Delhi Univ. Jan. 1987)

16. A repulsion motor is equipped with

(*a*) a commutator

(*b*) slip-rings

(*c*) a repeller

(*d*) neither (*a*) nor (*b*).

17. A repulsion-start induction-run single-phase motor runs as an induction motor only when

(*a*) brushes are shifted to neutral plane

(*b*) short-circuiter is disconnected

(*c*) commutator segments are short-circuited

(*d*) stator winding is reversed.

18. If a dc series motor is operated on ac supply, it will
(*a*) have poor efficiency
(*b*) have poor power factor
(*c*) spark excessively
(*d*) all of the above
(*e*) none of the above.

19. An outstanding feature of a universal motor is its
(*a*) best performance at 50 Hz supply
(*b*) slow speed at all loads
(*c*) excellent performance on dc. supply
(*d*) highest output kW/kg ratio.

20. The direction of rotation of a hysteresis motor is determined by the
(*a*) retentivity of the rotor material
(*b*) amount of hysteresis loss
(*c*) permeability of rotor material
(*d*) position of shaded pole with respect to the main pole.

21. Speed of the universal motor is
(*a*) dependent on frequency of supply
(*b*) proportional to frequency of supply
(*c*) independent of frequency of supply
(*d*) none of the above.

(Elect. Machines, A.M.I.E. Sec. B, 1993)

22. In the shaded pole squirrel cage induction motor the flux in the shaded part always
(*a*) leads the flux in the unshaded pole segment
(*b*) is in phase with the flux in the unshaded pole segment
(*c*) lags the flux in the unshaded pole segment
(*d*) none of the above.

(Elect. Machines, A.M.I.E. Sec. B, 1993)

23. Which of the following motor is an interesting example of beneficially utilizing a phenomenon that is often considered undesirable?
(*a*) hysteresis motor
(*b*) reluctance motor
(*c*) stepper motor
(*d*) shaded-pole motor.

24. Usually, large motors are more efficient than small ones. The efficiency of the tiny motor used in a wrist watch is approximately........... per cent.
(*a*) 1 (*b*) 10
(*c*) 50 (*d*) 80

ANSWERS

1. *b* **2.** *b* **3.** *d* **4.** *a* **5.** *b* **6.** *d* **7.** *a* **8.** *a* **9.** *b* **10.** *b* **11.** *a* **12.** *d* **13.** *d* **14.** *c*
15. *d* **16.** *a* **17** *c* **18.** *d* **19.** *d* **20.** *d* **21.** *a* **22.** *c* **23.** *a* **24.** *a*

35 ALTERNATORS

35.1. Basic Principle

A.C. generators or alternators (as they are usually called) operate on the same fundamental principles of electromagnetic induction as d.c. generators. They also consist of of an armature winding and a magnetic field. But there is one important difference between the two. Whereas in d.c. generators, the *armature rotates* and the field system is *stationary,* the arrangement in alternators is just the reverse of it. In their case, standard construction consists of armature winding mounted on a stationary element called *stator* and field windings on a rotating element called *rotor*. The details of construction are shown in Fig. 35.1.

The stator consists of a cast-iron frame, which supports the armature core, having slots on its inner periphery for housing the armature conductors. The rotor is like a flywheel having alternate N and S poles fixed to its outer rim. The magnetic poles are excited (or magnetised) from direct current supplied by a d.c. source at 125 to 600 volts. In most cases, necessary exciting

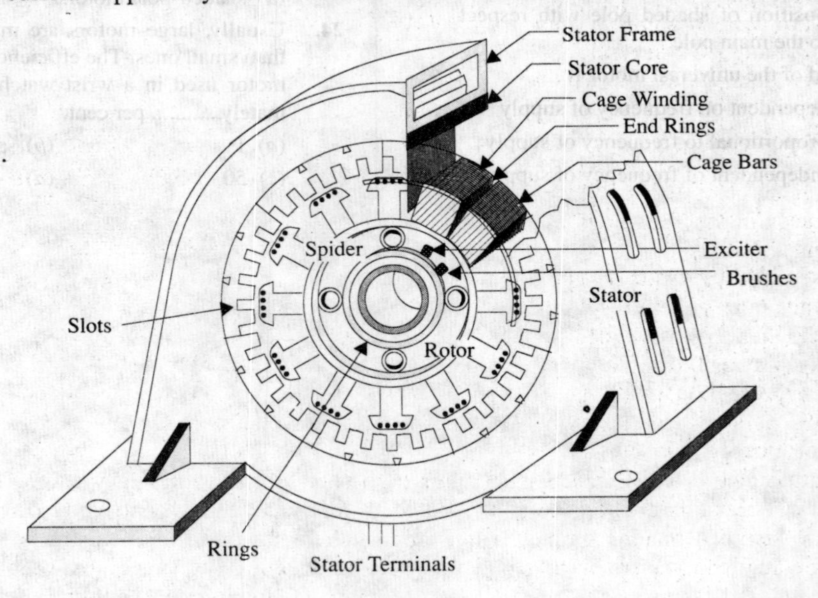

Fig. 35.1

(or magnetising) current is obtained from a small d.c. shunt generator which is belted or mounted on the shaft of the alternator itself. Because the field magnets are rotating, this current is supplied through two slip-rings. As the exciting voltage is relatively small, the slip-rings and brush gear are of light construction. Recently, brushless excitation systems have been developed in which a 3-phase a.c. exciter and a group of rectifiers supply d.c. to the alternator. Hence, brushes, slip-rings and commutator are eliminated.

When the rotor rotates, the stator conductors (being stationary) are cut by the magnetic flux, hence they have induced e.m.f. produced in them. Because the magnetic poles are alternately N

and *S*, they induce an e.m.f. and hence current in armature conductors, which first flows in one direction and then in the other. Hence, an alternating e.m.f. is produced in the stator conductors (*i*) whose frequency depends on the number of *N* and *S* poles moving past a conductor in one second and (*ii*) whose direction is given by Fleming's Right-hand rule.

35.2. Stationary Armature

Advantages of having stationary armature (and a rotating field system) are :

1. The output current can be led directly from fixed terminals on the stator (or armature windings) to the load circuit, without having to pass it through brush-contacts.
2. It is easier to insulate stationary armature winding for high a.c. voltages, which may have as high a value as 30 kV or more.
3. The sliding contacts *i.e.* slip-rings are transferred to the low-voltage, low-power d.c. field circuit which can, therefore, be easily insulated.
4. The armature windings can be more easily braced to prevent any deformation, which could be produced by the mechanical stresses set up as a result of short-circuit current and the high centrifugal forces brought into play.

35.3. Details of Construction

1. Stator Frame

In d.c. machines, the outer frame (or yoke) serves to carry the magnetic flux but in alternators, it is not meant for that purpose. Here, it is used for holding the armature stampings and windings in position. Low-speed large-diameter alternators have frames which because of ease of manufacture, are cast in sections. Ventilation is maintained with the help of holes cast in the frame itself. The provision of radial ventilating spaces in the stampings assists in cooling the machine.

But, these days, instead of using castings, frames are generally fabricated from mild steel plates welded together in such a way as to form a frame having a box type section.

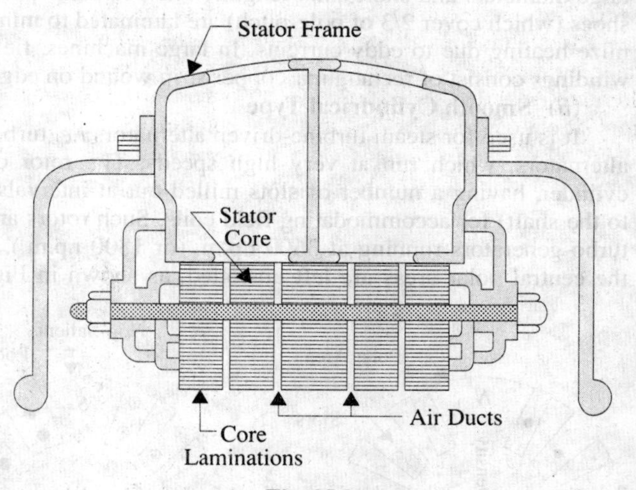

Fig. 35.2

In Fig. 35.2 is shown the section through the top of a typical stator.

2. Stator Core

The armature core is supported by the stator frame and is built up of laminations of special magnetic iron or steel alloy. The core is laminated to minimise loss due to eddy currents. The laminations are stamped out in complete rings (for smaller machine) or in segments (for larger machines). The laminations are insulated from each other and have spaces between them for allowing the cooling air to pass through. The slots for housing the armature conductors lie along the inner periphery of the core and are stamped out at the same time when laminations are formed. Different shapes of the armature slots are shown in Fig. 35.3.

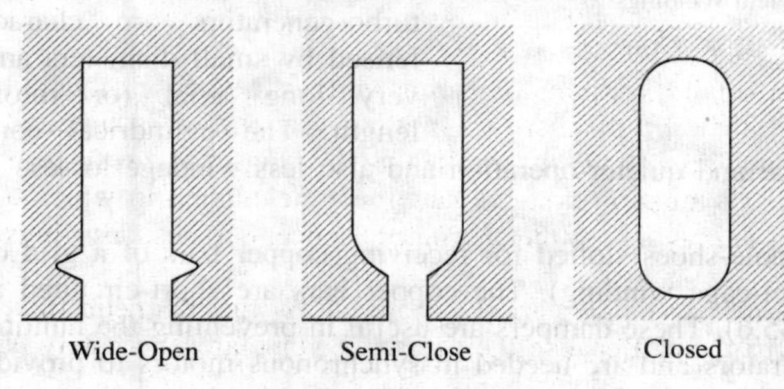

Fig. 35.3

The wide-open type slot (also used in d.c. machines) has the advantage of permitting easy installation of form-wound coils and their easy removal in case of repair. But it has the disadvantage of distributing the air-gap flux into bunches or tufts, that produce ripples in the wave of the generated e.m.f. The semi-closed type slots are better in this respect, but do not allow the use of form-wound coils. The wholly-closed type slots or tunnels do not disturb the air-gap flux but (*i*) they tend to increase the inductance of the windings (*ii*) the armature conductors have to be threaded through, thereby increasing initial labour and cost of winding and (*iii*) they present a complicated problem of end-connections. Hence, they are rarely used.

35.4. Rotor

Two types of rotors are used in alternators (i) *salient-pole type and (ii) smooth-cylindrical type*.

(*i*) Salient (or projecting) Pole Type

It is used in low-and medium-speed (engine driven) alternators. It has a large number of projecting (salient) poles, having their cores bolted or dovetailed onto a heavy magnetic wheel of cast-iron, or steel of good magnetic quality (Fig. 35.4). Such generators are characterised by their large diameters and short axial lengths. The poles and pole-shoes (which cover 2/3 of pole-pitch) are laminated to minimize heating due to eddy currents. In large machines, field windings consist of rectangular copper strip wound on edge.

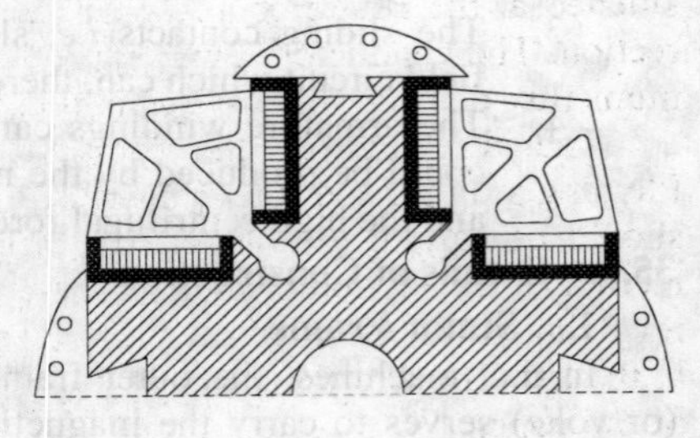

Fig. 35.4

(*ii*) Smooth Cylindrical Type

It is used for steam turbine-driven alternators *i.e.* turbo-alternators, which run at very high speeds. The rotor consists of a smooth solid forged steel cylinder, having a number of slots milled out at intervals along the outer periphery (and parallel to the shaft) for accommodating field coils. Such rotors are designed mostly for 2-pole (or 4-pole) turbo-generators running at 3600 r.p.m. (or 1800 r.p.m.). Two (or four) regions corresponding to the central polar areas are left unslotted, as shown in Fig. 35.5 (*a*) and (*b*).

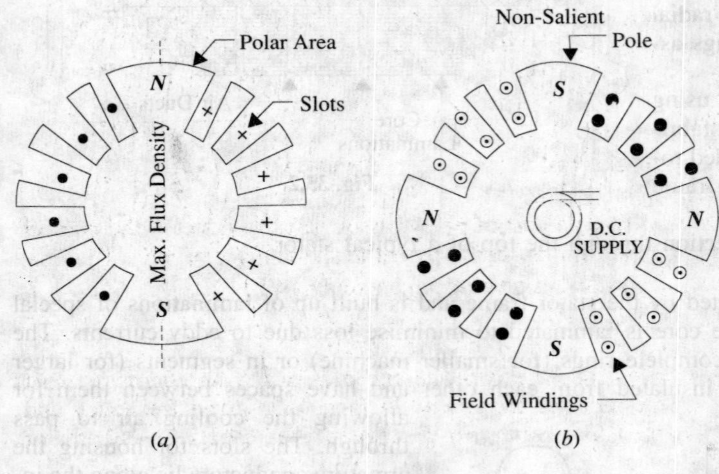

Fig. 35.5

The central polar areas are surrounded by the field windings placed in slots. The field coils are so arranged around these polar areas that flux density is maximum on the polar central line and gradually falls away on either side. It should be noted that in this case, poles are non-salient *i.e.* they do not project out from the surface of the rotor. To avoid excessive perpheral velocity, such rotors have very small diameters (about 1 metre or so). Hence, turbo-generators are characterised by small diameters and very long axial (or rotor) length. The cylindrical construction of the rotor gives better balance and quieter-operation and also less windage losses.

35.5. Damper Windings

Most of the alternators have their pole-shoes slotted for receiving copper bars of a grid or damper winding (also known as squirrel-cage winding). The copper bars are short-circuited at both ends by heavy copper rings (Fig. 35.6). These dampers are useful in preventing the hunting (momentary speed fluctuations) in generators and are needed in synchronous motors to provide the starting torque. Turbo-generators usually do not have these damper windings (except in special case to assist in synchronizing) because the solid field-poles themselves act as efficient dampers. It should be clearly understood that under normal running conditions, damper winding does not carry any current because rotor runs at synchronous speed.

The damper winding also tends to maintain balanced 3-ϕ voltage under unbalanced load conditions.

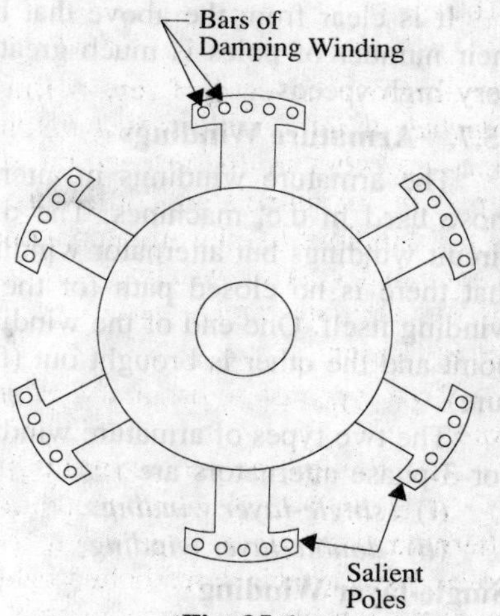

Fig. 35.6

35.6. Speed and Frequency

In an alternator, there exists a definite relationship between the rotational speed (N) of the rotor, the frequency (f) of the generated e.m.f. and the number of poles P.

Consider the armature conductor marked X in Fig. 35.7 situated at the centre of a N-pole rotating in clockwise direction. The conductor being, situated at the place of maximum flux density will have maximum e.m.f. induced in it.

The direction of the induced e.m.f. is given by Fleming's right'hand rule. But while applying this rule, one should be careful to note that the thumb indicates the direction of the motion of the *conductor* relative to the *field*. To an observer stationed on the clockwise revolving poles, the conductor would seem to be rotating anti-clockwise. Hence, thumb should point to the left. The direction of the induced e.m.f. is downwards, in a direction at right angles to the plane of the paper.

When the conductor is in the interpolar gap, as at A in Fig. 35.7, it has minimum e.m.f. induced in it, because flux density is minimum there. Again, when it is at the centre of a S-pole, it has maximum e.m.f. induced in it, because flux density at B is maximum. But the direction of the e.m.f. when conductor is over a N-pole is opposite to that when it is over a S-pole.

Obviously, one cycle of emf is induced in a conductor when one pair of poles passes over it. In other words, the e.m.f. in an armature conductor goes through one cycle in angular distance equal to twice the pole-pitch, as shown in Fig. 35.7.

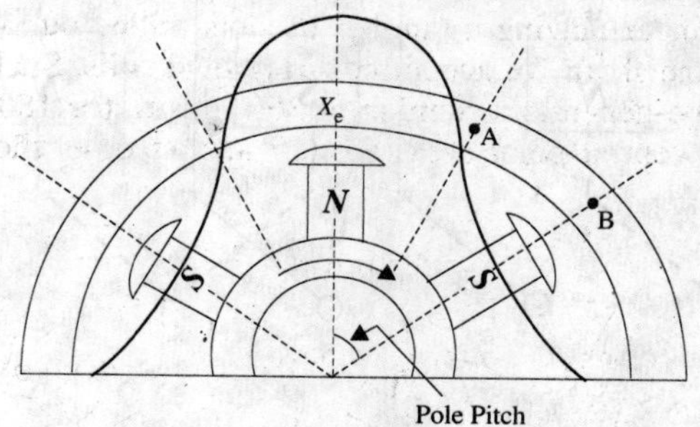

Fig. 35.7

Let P = total number of magnetic poles

N = rotative speed of the rotor in r.p.m.

f = frequency of generated e.m.f. in Hz.

Since one cycle of e.m.f. is produced when a pair of poles passes past a conductor, the number of cycles of e.m.f. produced in one revolution of the rotor is equal to the number of *pair* of poles.

$\therefore$ No. of cycles/revolution = $P/2$ and No. of revolutions/second = $N/60$

$$\therefore \quad \text{frequency} = \frac{P}{2} \times \frac{N}{60} = \frac{PN}{120} \text{ Hz} \quad \text{or} \quad f = \frac{PN}{120} \text{ Hz}$$

N is known as the synchronous speed, because it is the speed at which an alternator must run, in order to generate an e.m.f. of the required frequency. In fact, for a given frequency and given number of poles, the speed is fixed. For producing a frequency of 60 Hz, the alternator will have to run at the following speeds :

No. of poles	2	4	6	12	24	36
Speed (r.p.m.)	3600	1800	1200	600	300	200

Referring to the above equation, we get $P = 120f/N$

It is clear from the above that because of slow rotative speeds of engine-driven alternators, their number of poles is much greater as compared to that of the turbo-generators which run at very high speeds.

35.7. Armature Windings

The armature windings in alternators are different from those used in d.c. machines. The d.c. machines have closed circuit windings but alternator windings are open, in the sense that there is no closed path for the armature currents in the winding itself. One end of the winding is joined to the neutral point and the other is brought out (for a star-connected armature).

The two types of armature windings most commonly used for 3-phase alternators are :

(*i*) *single-layer winding*

(*ii*) *double-layer winding*

Single-layer Winding

Fig. 35.8

It is variously referred to as concentric or chain winding. Sometimes, it is of simple bar type or wave winding.

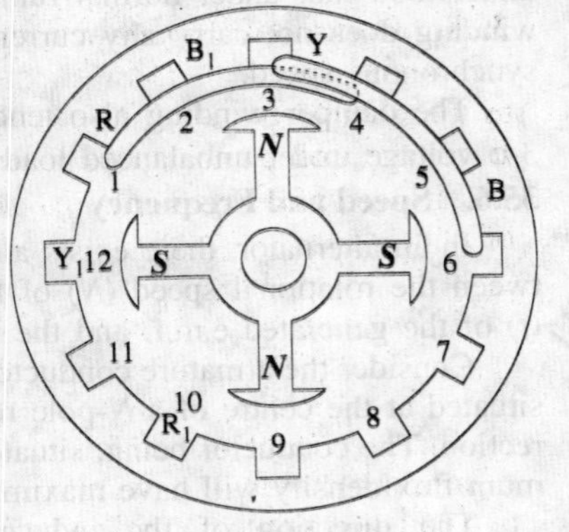

The fundamental principle of such a winding is illustrated in Fig. 35.8 which shows a single-layer, one-turn, full-pitch winding for a four-pole generator. There are 12 slots in all, giving

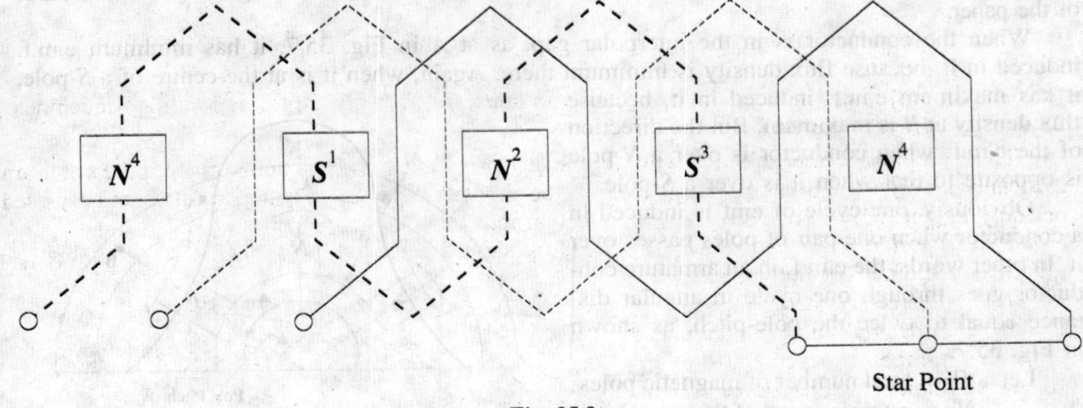

Fig. 35.9

3 slots per pole or 1 slot/phase/pole. The pole pitch is obviously 3. To get maximum e.m.f., two sides of a coil should be one pole-pitch apart *i.e.* coil span should be equal to one pole pitch. In other words, if one side of the coil is under the centre of a *N*-pole, then the, other side of the same *coil* should be under the centre of *S*-pole *i.e.* 180° (electrical) apart. In that case, the e.m.fs. induced in the two sides of the coil are added together. It is seen from the above figure, that *R* phase starts at slot No. 1, passes through slots 4, 7 and finishes at 10. The *Y*-phase starts 120° afterwards *i.e.* from slot No. 3 which is two slots away from the start of *R*-phase (because when 3 slots correspond to 180° electrical degrees, two slots correspond to an angular displacement of 120° electrical). It passes through slots 6, 9 and finishes at 12.

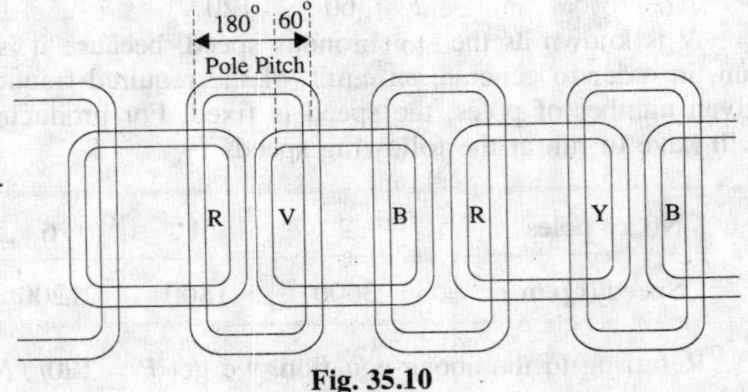

Fig. 35.10

Similarly, *B*-phase starts from slot No.5 *i.e.* two slots away from the start of *Y*-phase. It passes through slots 8, 11 and finishes at slot No. 2, The developed diagram is shown in Fig. 35.9. The ends of the windings are joined to form a star point for a *Y*-connection.

35.8. Concentric or Chain Windings

For this type of winding, the number of slots is equal to twice the number of coils or equal to the number of coil *sides*. In Fig. 35.10 is shown a concentric winding for 3-phase alternator. It has one coil per pair of poles per phase.

It would be noted that the polar group of each phase is 360° (electrical) apart. in this type of winding

1. It is necessary to use two different shapes of coils to avoid fouling of end connections.
2. Since polar groups of each phase are 360 electrical degrees apart, all such groups are connected in the same direction.
3. The disadvantage is that short-pitched coils cannot be used.

In Fig. 35.11 is shown a concentric winding with two coils per group per pole. Different shapes of coils are required for this winding.

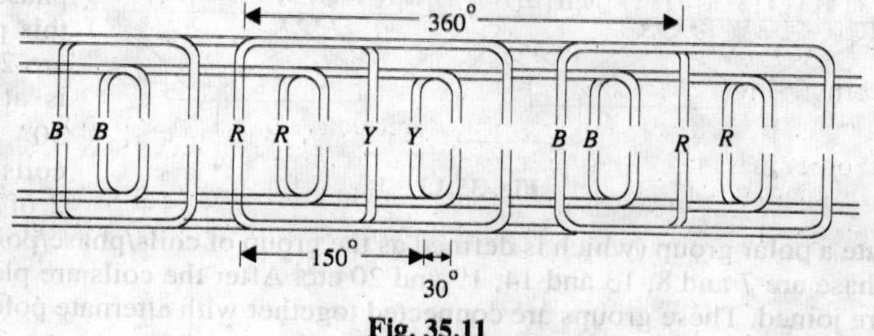

Fig. 35.11

All coil groups of phase *R* are connected in the same direction. It is seen that in each group, one coil has a pitch of 5/6 and the other has a pitch of 7/6 so that pitch factor (explained later) is 0.966. Such windings are used for large high-voltage machines.

35.9. Two-Layer Winding

This winding is either of wave-wound type or lap-wound type (this being much more common especially for high-speed turbo-generators). It is the simplest and, as said above, most commonly-used not only in synchronous machines but in induction motors as well.

Two important points regarding this winding should be noted :

(*a*) Oridinarily, the number of slots in stator (armature) is a multiple of the number of poles and the number of phases. Thus, the stator of a 4-pole, 3-phase alternator may have 12, 24, 36, 48 etc. slots all of which are seen to be multiple of 12 (*i.e.* 4 × 3).

(*b*) The number of stator slots is equal to the number of coils (which are all of the same shape). In other words, each slot contains two coil sides, one at the bottom of the slot and the other at the top. The coils overlap each other, just like shingles on a roof top.

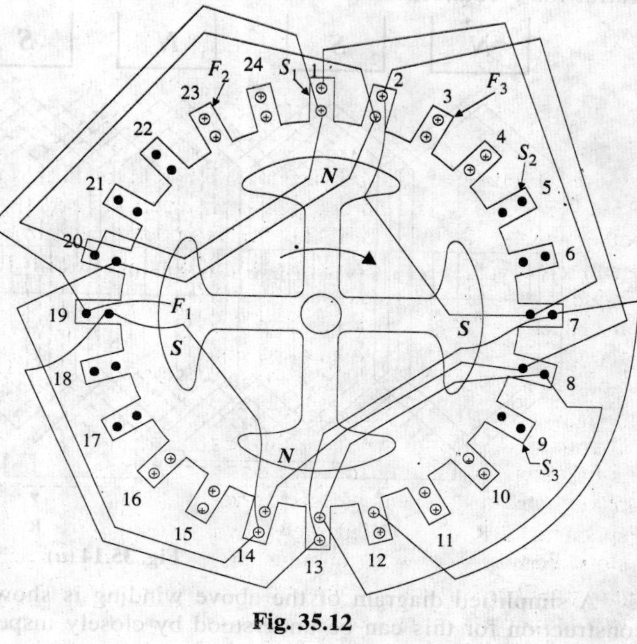

Fig. 35.12

For the 4-pole, 24-slot stator machine shown in Fig. 35.12, the pole-pitch is 24/4 = 6. For maximum voltage, the coils should be full-pitched. It means that if one side of the coil is in slot No.1, the other side should be in slot No.7, the two slots 1 and 7 being one pole-pitch or 180° (electrical) apart. To make matters simple, coils have been labelled as 1, 2, 3 and 4 etc. In the developed diagram of Fig. 35.14, the coil number is the number of the slot in which the left-hand side of the coil is placed.

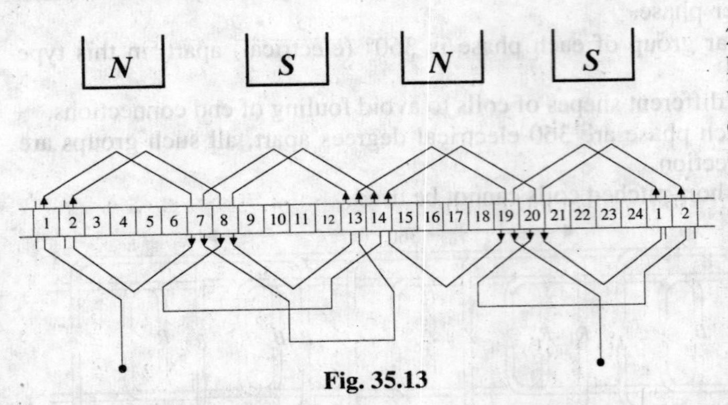

Fig. 35.13

Each of the three phases has 24/3 = 8 coils, these being so selected as to give maximum voltage when connected in series. When connected properly, coils 1, 7, 13 and 19 will add directly in phase. Hence, we get 4 coils for this phase. To complete eight coils for this phase, the other four selected are 2, 8, 14 and 20 each of which is at an angular displacement of 30° (elect.) from the adjacent coils of the first. The coils 1 and 2 of this phase are said to constitute a polar group (which is defined as the group of coils/phase/pole). Other polar groups for this phase are 7 and 8, 13 and 14, 19 and 20 etc. After the coils are placed in slots, the polar groups are joined. These groups are connected together with alternate poles reversed (Fig. 35.13) which shows winding for one phase only.

Now, phase Y is to be so placed as to be 120° (elect.) away from phase R. Hence, it is started from slot 5 *i.e.* 4 slots away (Fig. 35.14). It should be noted that angular displacement between slot No. 1 and 5 is 4 × 30 = 120° (elect). Starting from coil 5, each of the other eight coils of phase Y will be placed 4 slots to the right of corresponding coils for phase R. In the same way, B phase will start from coil 9. The complete wiring diagram for three phases is shown in Fig. 35.14. The terminals R_2, Y_2 and B_2 may be connected together to form a neutral for Y-connection.

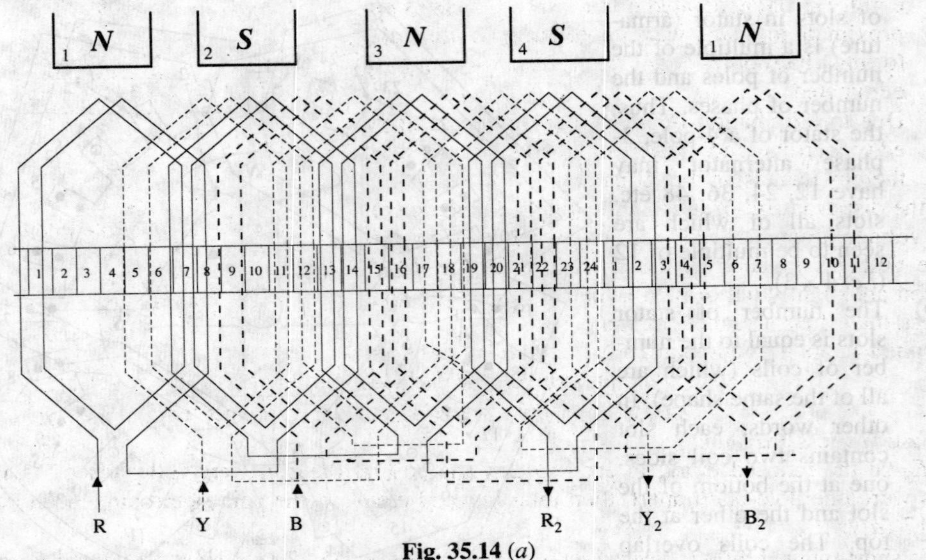

Fig. 35.14 (*a*)

A simplified diagram of the above winding is shown below Fig. 35.14. The method of construction for this can be understood by closely inspecting the developed diagram.

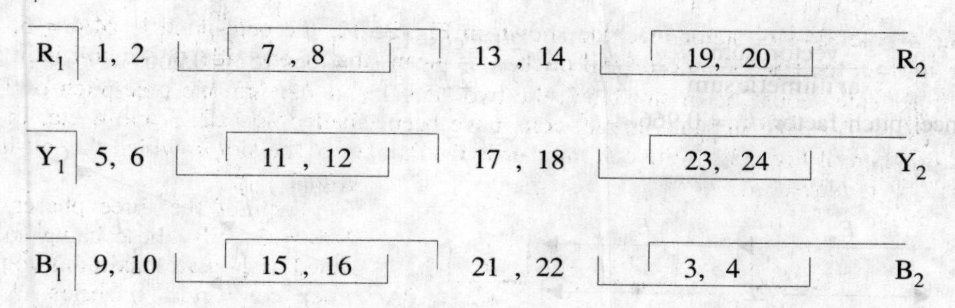

Fig. 35.14 (*b*)

35.10. Wye and DeltaConnections

For *Y*-connection, R_1, Y_1 and B_1 are joined together to form the star-point. Then, ends R_2, Y_2 and B_2 are connected to the terminals. For delta connection, R_2 and Y_1, Y_2 and B_1 B_2 and R_1 are connected together and terminal leads are brought out from their junctions as shown in Fig. 35.15 (*a*). and (*b*).

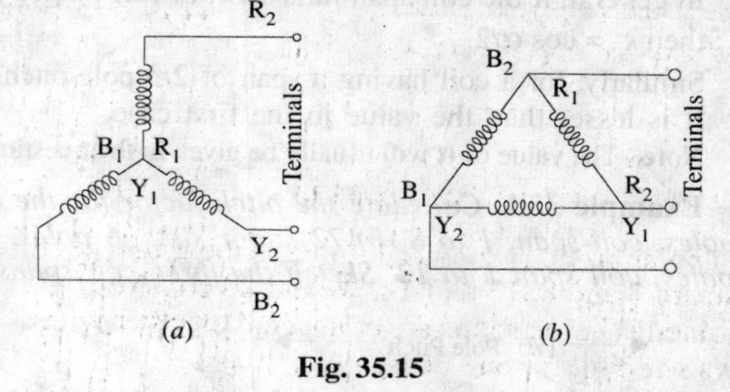

Fig. 35.15

35.11. Short-pitch Winding : Pitch factor/chording factor

So far we have discussed full-pitched coils *i.e.* coils having span which is equal to one pole-pitch *i.e.* spanning over 180° (electrical).

As shown in Fig. 35.16, if the coil sides are placed in slots 1 and 7, then it is full-pitched. If the coil sides are placed in slots 1 and 6, then it is short-pitched or fractional-pitched because coil span is equal to 5/6 of a pole-pitch It falls short by 1/6 pole-pitch or by 180°/6 = 30°. Short-pitched coils are deliberately used because of the following advantages:

1. They save copper of end connections.
2. They improve the wave-form of the generated e.m.f. *i.e.* the generated e.m.f. can be made to approximate to a sine wave more easily and the distorting harmonics can be reduced or totally eliminated.
3. Due to elimination of high frequency harmonics, eddy current and hysteresis losses are reduced thereby increasing the efficiency.

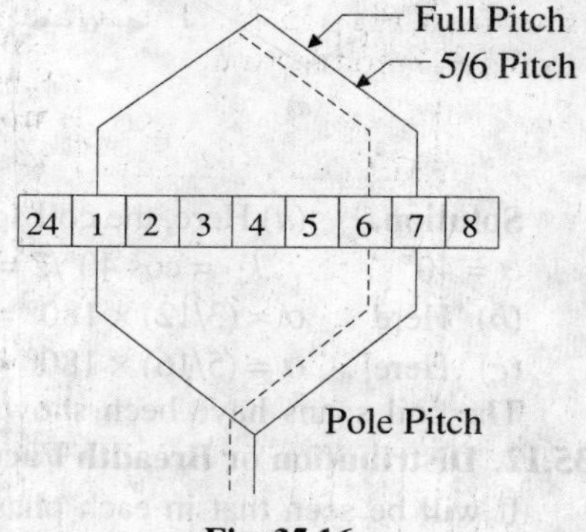

Fig. 35.16

But the disadvantage of using short-pitched coils is that the total voltage around the coils is somewhat reduced. Because the voltages induced in the two sides of the short-pitched coil are slightly out of phase, their resultant vectorial sum is less than their arithmetical sum.

The pitch factor or coil-span factor k_p or k_c is defined as

$$= \frac{\text{vector sum of the induced e.m.fs. per coil}}{\text{arithmetic sum of the induced e.m.fs. per coil}}$$

It is always less than unity.

Let E_S be the induced e.m.f. in each side of the coil. If the coil were full-pitched *i.e.* if its two sides were one pole-pitch apart, then total induced e.m.f. in the coil would have been = $2E_S$ [Fig. 35.17 (*a*)].

If it is short-pitched by 30° (elect.) then as shown in Fig. 35.17 (*b*), their resultant is *E* which is the vector sum of two voltage 30° (electrical) apart.

$$\therefore \quad E = 2\,E_S \cos 30°/2 = 2E_S \cos 15°$$

$$k_c = \frac{\text{vector sum}}{\text{arithmetic sum}} = \frac{E}{2E_S} = \frac{2E_s \cos 15°}{2E_S} = \cos 15° = 0.966$$

Hence, pitch factor, $k_c = 0.966$.

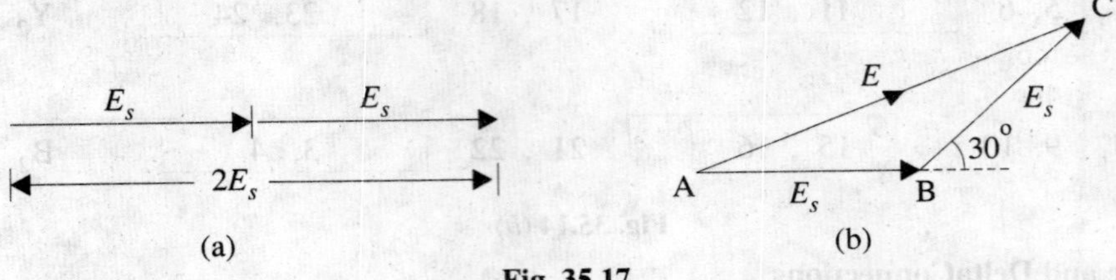

Fig. 35.17

In general, if the coil span falls short of full-pitch by an angle α (electrical)*, then $k_c = \cos \alpha/2$.

Similarly, for a coil having a span of 2/3 pole-pitch, $k_c = \cos 60°/2 = \cos 30° = 0.866$

It is lesser than the value in the first case.

Note. The value of α will usually be given in the question, if not, then assume $k_c = 1$

Example 35.1. *Calculate the pitch factor for the under-given windings : (a) 36 stator slots, 4-poles, coil-span, 1 to 8 (b) 72 stator slots, 6 poles, coils span 1 to 10 and (c) 96 stator slots, 6 poles, coil span 1 to 12. Sketch the three coil spans.*

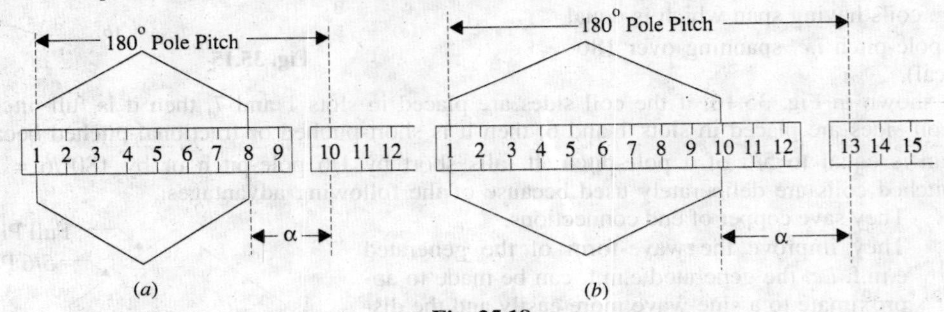

Fig. 35.18

Solution. (*a*) Here, the coil span falls short by (2/9) × 180° = 40°

α = 40° ∴ $k_c = \cos 40°/2 = \cos 20° = \mathbf{0.94}$

(*b*) Here α = (3/12) × 180° = 45° ∴ $k_c = \cos 45°/2 = \cos 22.5° = \mathbf{0.924}$

(*c*) Here α = (5/16) × 180° = 56° 16′ ∴ $k_c = \cos 28°8' = \mathbf{0.882}$

The coil spans have been shown in Fig. 32.18.

35.12. Distribution or Breadth Factor or Winding Factor or Spread Factor

It will be seen that in each phase, coils are not concentrated or bunched in one slot, but are distributed in a number of slots to form *polar groups* under each pole. These coils / phase are displaced from each other by a certain angle. The result is that the e.m.fs. induced in coil sides constituting a polar group are not in phase with each other but differ by an angle equal to angular displacement of the slots.

In Fig. 35.19 are shown the end connections of a 3-phase single-layer winding for a 4-pole alternator. It has a total of 36 slots *i.e.* 9 slots/pole. Obviously, there are 3 slots / phase / pole. For example, coils 1, 2 and 3 belong to *R* phase. Now, these three coils which constitute one polar group are not bunched in one slot but in three different slots. Angular displacement between any two adjacent slots = 180° / 9 = 20° (elect.)

*This angle is known as chording angle and the winding employing short-pitched coils is called chorded winding.

If the three coils were bunched in one slot, then total e.m.f. induced in the three sides of the coil would be the arithmetic sum of the three e.m.f.s. *i.e.* = 3 E_S, where E_S is the e.m.f. induced in one coil *side* [Fig.35.20(*a*)].

Since the coils are distributed, the individual e.m.fs. have a phase difference of 20° with each other. Their vector sum as seen from Fig. 35.20 (*b*) is

$$E = E_S \cos 20° + E_S + E_S \cos 20°$$
$$= 2E_S \cos 20° + E_S$$
$$= 2E_S \times 0.9397 + E_S = 2.88\, E_S$$

The distribution factor (k_d) is defined as

$$= \frac{\text{e.m.f. with distributed winding}}{\text{e.m.f. with concentrated winding}}$$

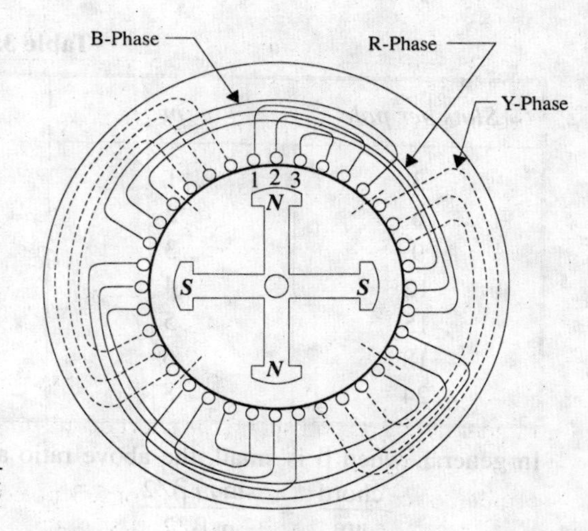

Fig. 35.19

In the present case

$$k_d = \frac{\text{e.m.f. with winding in 3 slots/pole/phase}}{\text{e.m.f. with winding in 1 slot/pole/phase}} = \frac{E}{3\,E_S} = \frac{2.88\,E_S}{3\,E_S} = 0.96$$

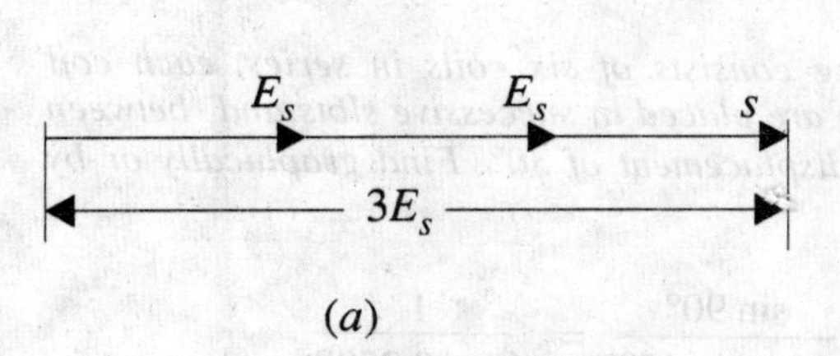

Fig. 35.20

General Case

Let β be the value of angular displacement between the slots. Its value is

$$\beta = \frac{180°}{\text{No. of slots/pole}} = \frac{180°}{n}$$

Let m = No. of slots/phase/pole

$m\beta$ = phase spread angle

Then, the resultant voltage induced in one polar group would be mE_S where E_S is the voltage induced in one coil side. Fig. 32.21 illustrates the method for finding the vector sum of m voltages each of value E_S and having a mutual phase difference of β (if m is large, then the curve *ABCDE* will become part of a circle of radius r).

$$AB = E_S = 2r \sin \beta/2$$

Arithmetic sum is= $mE_S = m \times 2r \sin \beta/2$

Their vector sum = $AE = E_r = 2r \sin m\beta/2$

$$k_d = \frac{\text{vector sum of coil e.m.f.s}}{\text{arithmetic sum of coil e.m.f.s.}}$$

$$= \frac{2r \sin m\,\beta/2}{m \times 2r \sin \beta/2} = \frac{\sin m\,\beta/2}{m \sin \beta/2}$$

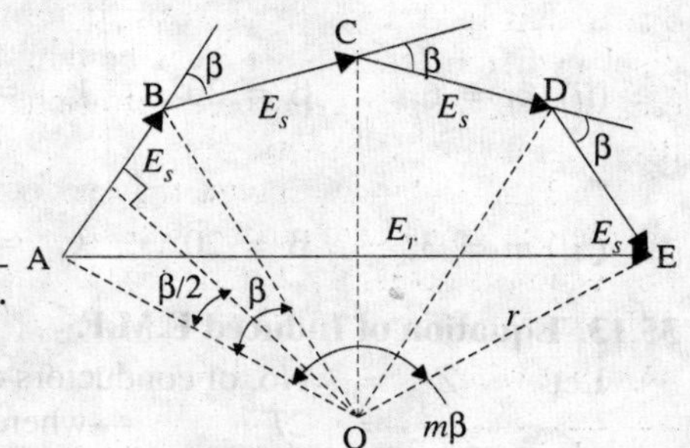

Fig. 35.21

The value of distribution factor of a 3-phase alternator for different number of slots pole phase is given in table No. 35.1.

Table 35.1

Slots per pole	*m*	β°	*Distribution factor* k_d
3	1	60	1.000
6	2	30	0.966
9	3	20	0.960
12	4	15	0.958
15	5	12	0.957
18	6	10	0.956
24	8	7.5	0.955

In general, when β is small, the above ratio approaches

$$\frac{\text{chord}}{\text{arc}} = \frac{\sin m\,\beta/2}{m\,\beta/2} \qquad \text{— angle } m\beta/2 \text{ in radians.}$$

Example 35.2. *Calculate the distribution factor for a 36-slots, 4-pole, single-layer three-phase winding.* **(Elect. Machine-I Nagpur Univ. 1993)**

Solution. $n = 36/4 = 9;\quad \beta = 180°/9 = 20°;\quad m = 36/4 \times 3 = 3$

$$k_d = \frac{\sin m\,\beta/2}{m \sin \beta/2} = \frac{\sin 3 \times 20°/2}{3 \sin 20°/2} = \mathbf{0.96}$$

Example. 35.3. *A part of an alternator winding consists of six coils in series, each coil having an e.m.f. of 10 V r.m.s. induced in it. The coils are placed in successive slots and between each slot and the next, there is an electrical phase displacement of 30°. Find graphically or by calculation, the e.m.f. of the six coils in series.*

Solution. By calculation

Here $\beta = 30° : m = 6 \quad \therefore\ k_d = \dfrac{\sin m\,\beta/2}{m \sin \beta/2} = \dfrac{\sin 90°}{6 \times \sin 15°} = \dfrac{1}{6 \times 0.2588}$

Arithmetic sum of voltage induced in 6 coils = 6 × 10 = 60 V

Vector sum = k_d × arithmetic sum = 60 × 1/6 × 0.2538 = **38.64V**

Example 35.4. *Find the value of k_d for an alternator with 9 slots per pole for the following cases:*

(i) One winding in all the slots (ii) one winding using only the first 2/3 of the slots/pole (iii) three equal windings placed sequentialy in 60° group.

Solution. Here, β = 180°/9 = 20° and values of *m i.e.* number of slots in a group are 9, 6 and 3 respectively

(*i*) $m = 9,\quad \beta = 20°,\quad k_d = \dfrac{\sin 9 \times 20°/2}{9 \sin 20°/2} = \mathbf{0.64} \quad \left[\text{or } k_d = \dfrac{\sin \pi/2}{\pi/2} = 0.637\right]$

(*ii*) $m = 6,\quad \beta = 20°,\quad k_d = \dfrac{\sin 6 \times 20°/2}{6 \sin 20°/2} = \mathbf{0.83} \quad \left[\text{or } k_d = \dfrac{\sin \pi/3}{\pi/3} = 0.827\right]$

(*iii*) $m = 3,\quad \beta = 20°,\quad k_d = \dfrac{\sin 3 \times 20°/2}{3 \sin 20°/2} = \mathbf{0.96} \quad \left[\text{or } k_d = \dfrac{\sin \pi/6}{\pi/6} = 0.955\right]$

35.13. Equation of Induced E.M.F.

Let Z = No. of conductors or coil sides in series/phase

= $2T$ —where T is the No. of coils or turns per phase

(remember one turn or coil has two sides)

P = No. of poles
f = frequency of induced e.m.f. in Hz
Φ = flux pole in webers
k_d = distribution factor $= \dfrac{\sin m\beta/2}{m \sin \beta/2}$
k_c or k_p = pitch or coil span factor = $\cos \alpha/2$
k_f = from factor = **1.11** —if e.m.f. is assumed sinusoidal
N = rotor r.p.m.

In one revolution of the rotor (*i.e.* in 60./N second) each stator conductor is cut by a flux of ΦP webers.

$\therefore \quad d\Phi = \Phi P$ and $dt = 60/N$ second

$\therefore$ average e.m.f. induced per conductor $= \dfrac{d\Phi}{dt} = \dfrac{\Phi P}{60/N} = \dfrac{\Phi NP}{60}$

Now, we know that $f = PN/120$ or $N = 120\,f/P$

Substituting this value of N above, we get

Average e.m.f. per conductor $= \dfrac{\Phi P}{60} \times \dfrac{120 f}{P} = 2f\Phi$ volt

If there are Z conductors in series/phase, then Average e.m.f./phase = $2f\,\Phi Z$ volt = $4\,f\,\Phi T$ volt

R.M.S. value of e.m.f./phase = $1.11 \times 4f\,\Phi T = 4.44\,f\,\Phi\,T$ volt*

This would have been the actual value of the induced voltage if all the coils in a phase were (*i*) full-pitched and (*ii*) concentrated or bunched in one slot (instead of being distributed in several slots under poles). But this not being so, the actually available voltage is reduced in the ratio of these two factors.

$\therefore$ actually available voltage/phase $= 4.44\,k_c\,k_d\,f\,\Phi\,T = 4\,k_f k_c k_d\,f\,\Phi\,T$ volt

If the alternator is star-connected (as is usually the case) then the line voltage is $\sqrt{3}$ times the phase voltage (as found from the above formula).

35.14. Effect of Harmonics on Pitch and Distribution Factors

(*a*) If the short-pitch angle or chording angle is α degrees (electrical) for the fundamental flux wave, then its values for different harmonics are

for 3rd harmonic = 3α ; for 5th harmonic = 5α and so on.

$\therefore$ pitch-factor, $k_c = \cos \alpha/2$ —for fundamental
$= \cos 3\alpha/2$ —for 3rd harmonic
$= \cos 5\alpha/2$ —for 5th harmonic etc.

(*b*) Similarly, the distribution factor is also different for different harmonics. Its value becomes

$$k_d = \frac{\sin mn\,\beta/2}{m \sin n\beta/2}$$ where n is the order of the harmonic

for fundamental, $n = 1$ $\quad k_{d1} = \dfrac{\sin\, m\,\beta/2}{m\,\sin \beta/2}$

for 3rd harmonic, $n = 3$ $\quad k_{d3} = \dfrac{\sin 3\,m\,\beta/2}{m \sin 3\,\beta/2}$

for 5th harmonic, $n = 5$ $\quad k_{d5} = \dfrac{\sin 5\,m\,\beta/2}{m \sin 5\,\beta/2}$

(*c*) Frequency is also changed. If fundamental frequency is 50 Hz *i.e.* f_1 = 50 Hz then other frequencies are :

3rd harmonic, $f_3 = 3 \times 50 = 150$ Hz, 5th harmonic, $f_5 = 5 \times 50 = 250$ Hz etc.

*It is exactly the same equation as the e.m.f. equation of a transformer. (Art 30.6)

Example 35.5. *An alternator has 18 slots/pole and the first coil lies in slots 1 and 16. Calculate the pitch factor for (i) fundamental (ii) 3rd harmonic (iii) 5th harmonic and (iv) 7th harmonic.*

Solution. Here, coil span is = (16 – 1) = 15 slots, which falls short by 3 slots.

Hence, $\alpha = 180° \times 3/18 = 30°$

(*i*) $k_{c1} = \cos 30°/2 = \cos 15° = \mathbf{0.966}$ (*ii*) $k_{c3} = \cos 3 \times 30°/2 = \mathbf{0.707}$

(*iii*) $k_{c5} = \cos 5 \times 30°/2 = \cos 75° = \mathbf{0.259}$ (*iv*) $k_{c7} = \cos 7 \times 30°/2 = \cos 105° = \cos 75° = \mathbf{0.259}$

Example 35.6. *A 3-phase, 16-pole alternator has a star-connected winding with 144 slots and 10 conductors per slot. The flux per pole is 0.03 Wb, Sinusoidally distributed and the speed is 375 r.p.m. Find the frequency rpm and the phase and line e.m.f. Assume full-pitched coil.*

(Elect. Machines, AMIE Sec. B, 1991)

Solution. $f = PN/120 = 16 \times 375/120 = \mathbf{50\ Hz}$

Since k_c is not given, it would be taken as unity.

$n = 144/16 = 9; \beta = 180°/9 = 20°; m = 144/16 \times 3 = 3$

$k_d = \sin 3 \times (20°/2) / 3 \sin (20°/2) = 0.96$

$Z = 144 \times 10 / 3 = 480; T = 480/2 = 240$ / phase

$E_{ph} = 4.44 \times 1 \times 0.96 \times 50 \times 0.03 \times 240 = \mathbf{1534\ V}$

Line voltage, $E_L = \sqrt{3}\, E_{ph} = \sqrt{3} \times 1534 = \mathbf{2658\ V}$

Example 35.7. *Find the no-load phase and line voltage of a star-connected 3-phase, 6-pole alternator which runs at 1200 rpm, having flux per pole of 0.1 Wb sinusoidally distributed. Its stator has 54 slots having double layer winding. Each coil has 8 turns and the coil is chorded by 1 slot.*

(Elect.Machines-I, Nagput Univ. 1993)

Solution. Since winding is chorded by one slot, it is short-pitched by 1/9 or 180°/9 = 20°

$\therefore k_c = \cos 20°/2 = 0.98$; $f = 6 \times 1200/120 = 60$ Hz

$n = 54/6 = 9$; $\beta = 180°/9 = 20°$, $m = 54/6 \times 3 = 3$

$k_d = \sin 3 \times (20°/2) / 3 \sin (20°/2) = 0.96$

$Z = 54 \times 8/3 = 144$; $T = 144/2 = 72$, $f = 6 \times 1200/120 = 60$ Hz

$E_{ph} = 4.44 \times 0.98 \times 0.96 \times 60 \times 0.1 \times 72 = 1805$ V

Line voltage, $E_L = \sqrt{3} \times 1805 = 3125$ V

Example 35.8. *The stator of a 3-phase, 16-pole alternator has 144 slots and there are 4 conductors per slot connected in two layers and the conductors of each phase are connected in series. If the speed of the alternator is 375 r.p.m., calculate the e.m.f. inducted per phase. Resultant flux in the air-gap is 5 × 10^{-2} webers per pole sinusoidally distributed. Assume the coil span as 150° electrical.*

(Elect. Machine, Nagpur Univ. 1993)

Solution. For sinusoidal flux distribution, $k_f = 1.11$; $\alpha = (180° - 150°) = 30°$ (elect)

$k_c = \cos 30°/2 = 0.966$*

No. of slots / pole, $n = 144/16 = 9$; $\beta = 180°/9 = 20°$

m = No. of slots/pole/phase = $144/16 \times 3 = 3$

$\therefore k_d = \dfrac{\sin m\beta/2}{m \sin \beta/2} = \dfrac{\sin 3 \times 20°/2}{3 \sin 20°/2} = 0.96$; $f = 16 \times 375/120 = 50$ Hz

No. of slots / phase = 144/3 = 48; No of conductors / slot = 4

$\therefore$ No. of conductors in series/phase = 48 × 4 = 192

$\therefore$ turns / phase = conductiors per phase / 2 = 192/2 = 96

$E_{ph} = 4\, k_f\, k_c\, k_d\, f\, \Phi T$

$= 4 \times 1.11 \times 0.966 \times 0.96 \times 50 \times 5 \times 10^{-2} \times 96 = \mathbf{988\ V}$

*Also, $k_d = \sin 150°/2 = \sin 75° = 0.966$

Example 35.9. *A 10-pole, 50-Hz, 600 r.p.m. alternator has flux density distribution given by the following expression*

$$B = \sin\theta + 0.4 \sin 3\theta + 0.2 \sin 5\theta$$

The alternator has 180 slots wound with 2-layer 3-turn coils having a span of 15 slots. The coils are connected in 60° groups. If armature diameter is = 1.2 m and core length = 0.4 m, calculate

(i) the expression for instantaneous e.m.f./conductor
(ii) the expression for instantaneous e.m.f./coil
(iii) the r.m.s. phase and line voltages

Solution. For find voltage/conductor, we may either use the relation *Blv* or use the relation of Art.35-13.

Area of pole pitch = $(1.2\,\pi/10) \times 0.4 = 0.1508\ m^2$

Fundamental flux/pole, ϕ_1 = av. flux density × area = $0.637 \times 1 \times 0.1508 = 0.096$ Wb

(*a*) RMS value of fundamental voltage per conductor,

$$= 1.1 \times 2 f\phi_1 = 1.1 \times 2 \times 50 \times 0.096 = \mathbf{10.56\ V}$$

Peak value $= \sqrt{2} \times 10.56 = 14.93$ V

Since harmonic conductor voltages are in proportion to their flux densities,

3rd harmonic voltage = $0.4 \times 14.93 = 5.97$ V

5th harmonic voltage = $0.2 \times 14.93 = 2.98$ V

Hence, equation of the instantaneous e.m.f./conductor is

$$\mathbf{e = 14.93 \sin\theta + 5.97 \sin 3\theta + 2.98 \sin 5\theta}$$

(*b*) Obviously, there are 6 conductors in a 3-turn coil. Using the values of k_c found in solved Ex. 35.5, we get

fundamental coil voltage = $6 \times 14.93 \times 0.966 = 86.5$ V

3rd harmonic coil voltage = $6 \times 5.97 \times 0.707 = 25.3$ V

5th harmonic coil voltage = $6 \times 2.98 \times 0.259 = 4.63$ V

Hence, coil voltage expression is*

$$\mathbf{e = 86.5 \sin\theta + 25.3 \sin 3\theta + 4.63 \sin 5\theta}$$

(*c*) Here, $m = 6$, $\beta = 180°/18 = 10°$; $k_{d1} = \dfrac{\sin 6 \times 10°/2}{6 \sin 10°/2} = 0.956$

$$kd_3 = \frac{\sin 3 \times 6 \times 10°/2}{6 \sin 3 \times 10°/2} = 0.644 \quad k_{d5} = \frac{\sin 5 \times 6 \times 10°/2}{6 \sin 5 \times 10°/2} = 0.197$$

It should be noted that number of coils per phase = 180/3 = 60

Fundamental phase e.m.f. = $(86.5/\sqrt{2}) \times 60 \times 0.956 = 3510$ V

3rd harmonic phase e.m.f. = $(25.3/\sqrt{2}) \times 60 \times 0.644 = 691$ V

5th harmonic phase e.m.f. = $(4.63/\sqrt{2}) \times 60 \times 0.197 = 39$ V

RMS value of phase voltage = $(3510^2 + 691^2 + 39^2)^{1/2}$ = **3577 V**

RMS value of line voltage = $\sqrt{3} \times (3510^2 + 39^2)^{1/2}$ = **6080 V**

Example 35.10. *A 4-pole, 3-phase, 50-Hz, star-connected alternator has 60 slots, with 4 conductors per slot. Coils are short-pitched by 3 slots. If the phase spread is 60°, find the line voltage induced for a flux per pole of 0.943 Wb distributed sinusoidally in space. All the turns per phase are in series.* **(Electrical Machinery, Mysore Univ.1987)**

Solution. As explained in Art. 35.12, phase spread = $m\beta = 60°$ —given

Now, $m = 60/4 \times 3 = 5$ $\therefore$ $5\beta = 60°, \beta = 12°$

*Since they are not of much interest, the relative phase angles of the voltages have not been included in the expression.

$$k_d = \frac{\sin 5 \times 12°/2}{5 \sin 12°/2} = 0.957\,; \quad \alpha = (3/15) \times 180° = 36°\,;\; k_c = \cos 18° = 0.95$$

$$Z = 60 \times 4/3 = 80;\; T = 80/2 = 40;\; \Phi = 0.943 \text{ Wb};\; k_f = 1.11$$

$$\therefore \quad E_{ph} = 4 \times 1.11 \times 0.95° \times 0.975 \times 50 \times 0.943 \times 40 = 7613 \text{ V}$$

$$E_L = \sqrt{3} \times 7613 = \mathbf{13{,}185\ V}$$

Example 35.11. *A 4-pole, 50-Hz, star-connected alternator has 15 slots per pole and each slot has 10 con luctors. All the conductors of each phase are connected in series' the winding factor being 0.95. When running on no-load for a certain flux per pole the terminal e.m.f. was 1825 volt. If the windings are lap-connected as in a d.c. machine, what would be the e.m.f. between the brushes for the same speed and the same flux/pole. Assume sinusoidal distribution of flux.*

Solution. Here k_f = 1.11, k_d = 0.95, k_c = 1 (assumed)

f = 50 Hz; e.m.f./phase = 1825/$\sqrt{3}$ V

Total No. of slots = 15×4 = 60

$\therefore$ No. of slots/phase= 60/3 = 20; No. of turns/phase = 20 × 10/2 = 100

$\therefore$ $1825/\sqrt{3} = 4 \times 1.11 \times 1 \times 0.95 \times \Phi \times 50 \times 100$ $\quad\therefore\; \Phi = 49.97$ mWb

When connected as a d.c. generator

$$E_g = (\Phi ZN/60) \times (P/A) \text{ volt}$$

$$Z = 60 \times 10 = 600,\; N = 120\, f/P = 120 \times 50/4 = 1500 \text{ r.p.m}$$

$$\therefore \quad E_g = \frac{49.97 \times 10^{-3} \times 600 \times 1500}{60} \times \frac{4}{4} = \mathbf{750\ V}$$

Example 35.12. *An alternator on open-circuit generates 360 V at 60 Hz when the field current is 3.6 A. Neglecting saturation, determine the open-circuit e.m.f. when the frequency is 40 Hz and the field current is 2.4 A.*

Solution. As seen from the e.m.f. equation of an alternator,

$$E \propto \Phi f \qquad \therefore \quad \frac{E_1}{E_2} = \frac{\Phi_1 f_1}{\Phi_2 f_2}$$

Since saturation is neglected, $\Phi \propto I_f$ is the field current.

$$\therefore \quad \frac{E_1}{E_2} = \frac{I_{f1} \cdot f_1}{I_{f2} \cdot f_2} \quad \text{or} \quad \frac{360}{E_2} = \frac{3.6 \times 60}{2.4 \times 40}\,; \quad E_2 = \mathbf{160\ V}$$

Example 35.13. *Calculate the R.M.S. value of the induced e.m.f. per phase of a 10-pole, 3-phase, 50-Hz alternator with 2 slots per pole per phase and 4 conductors per slot in two layers. The coil span is 150°. The flux per pole has a fundamental component of 0.12 Wb and a 20% third component.* **(Elect. Machines-III, Punjab Univ. 1991)**

Solution. Fundamental E.M.F.

$$\alpha = (180° - 150°) = 30°; \quad k_{c1} = \cos \alpha/2 = \cos 15° = 0.966$$

$$m = 2\,; \text{ No. of slots / pole} = 6\,; \quad \beta = 180°/6 = 30°$$

$$\therefore \quad k_{d1} = \frac{\sin m\,\beta/2}{m \sin \beta/2} = \frac{\sin 2 \times 30°/2}{2 \sin 30°/2} = 0.966$$

$$Z = 10 \times 2 \times 4 = 80\,; \quad \text{turn/phase}, \quad T = 80/2 = 40$$

$$\therefore \quad \text{fundamental E.M.F./phase} = 4.44\, k_c k_d\, f\, \Phi\, T$$

$$\therefore \quad E_1 = 4.44 \times 0.966 \times 0.966 \times 50 \times 0.12 \times 40 = 995\text{V}$$

Hormonic E.M.F.

$$K_{c3} = \text{Cos } 3\,\alpha/2 = \text{Cos } 3 \times 30°/2 = \text{Cos } 45° = 0.707$$

$$k_{d3} = \frac{\sin mn\,\beta/2}{m \sin n\,\beta/2} \quad \text{where } n \text{ is the order of the harmonic } i.e.\ n = 3$$

$$\therefore \quad k_{d3} = \frac{\sin 2 \times 3 \times 30°/2}{2 \sin 3 \times 30°/2} = \frac{\sin 90°}{2 \sin 45°} = 0.707,\; f_2 = 50 \times 3 = 150 \text{ Hz}$$

$$\Phi_3 = (1/3) \times 20\% \text{ of fundamental flux} = (1/3) \times 0.02 \times 0.12 = 0.008 \text{ Wb}$$

$$\therefore \quad E_3 = 4.44 \times 0.707 \times 0.707 \times 150 \times 0.008 \times 40 = 106 \text{ V}$$

$$\therefore \quad E \text{ per phase} = \sqrt{E_1^2 + E_3^2} = \sqrt{995^2 + 106^2} = \mathbf{1000\ V}$$

Note. Since phase e.m.fs. induced by the 3rd, 9th and 15th harmonics etc. are eliminated from the line voltages, the line voltage for a Y-connection would be = $995 \times \sqrt{3}$ volt.

Example 35.14. *A 3-phase alternator has generated e.m.f. per phase of 230 V with 10 per cent third harmonic and 6 per cent fifth harmonic content. Calculate the r.m.s. line voltage for* (a) *star connection* (b) *delta-connection. Find also the circulating current in delta connection if the reactance per phase of the machine at 50-Hz is 10 Ω.*

(Elect. Machines-III, Osmania Univ. 1988)

Solution. It should be noted that in both star and delta-connections, the third harmonic components of the three phases cancel out at the line terminals because they are co-phased. Hence, the line e.m.f. is composed of the fundamental and the fifth harmonic only.

(*a*) **Star-connection**

$$E_1 = 230 \text{ V} ; \quad E_5 = 0.06 \times 230 = 13.8 \text{ V}$$

$$\text{E.M.F./phase} = \sqrt{E_1^2 + E_5^2} = \sqrt{230^2 + 13.8^2} = 230.2 \text{ V}$$

$$\text{R.M.S. value of line e.m.f.} = \sqrt{3} \times 230.2 = \mathbf{399\ V}$$

(*b*) **Delta-connection**

Since for delta-connection, line e.m.f. is the same as the phase e.m.f.

R.M.S. value of line e.m.f. = **230.2 V**

In delta-connection, third harmonic components are additive round the mesh, hence a circulating current is set up whose magnitude depends on the reactance per phase at the third harmonic frequency.

R.M.S. value of third harmonic e.m.f. per phase = 0.1 × 230 = 23 V

Reactance at triple frequency = 10 × 3 = 30 Ω

Circulating current = 23/30 = **0.77 A**

Tutorial Problem No. 35.1.

1. Find the no-load phase and line voltage of a star-connected, 4-pole alternator having flux per pole of 0.1 Wb sinusoidally distributed; 4 slots per pole per phase, 4 conductors per slot, double-layer winding with a coil span of 150°.
   [Assuming f = 50 Hz; 789 V; 1366 V] (*Elect. Technology-I, Bombay Univ. 1978*)
2. A 3-φ, 10-pole, Y-connected alternator runs at 600 r.p.m. It has 120 stator slots with 8 conductors per slot and the conductors of each phase are connected in series. Determine the phase and line e.m.fs. if the flux per pole is 56 mWb. Assume full-pitch coils.
   [1910 V; 3300 V] (*Electrical Technology-II, Madras Univ. April 1977*)
3. Calculate the speed and open-circuit line and phase voltages of a 4-pole, 3-phase, 50-Hz, star-connected alternator with 36 slots and 30 conductors per slot. The flux per pole is 0.0496 Wb and is sinusoidally distributed. **[1500 r.p.m.; 3,300 V; 1,905 V]** (*Elect. Engg-II, Bombay Univ. 1979*)
4. A 4-pole, 3-phase, star-connected alternator armature has 12 slots with 24 conductors per slot and the flux per pole is 0.1 Wb sinusoidally distributed.
   Calculate the line e.m.f. generated at 50 Hz. **[1850 V]**
5. A 3-phase, 16-pole alternator has a star-connected winding with 144 slots and 10 conductors per slot. The flux per pole is 30 mWb sinusoidally distributed. Find the frequency, the phase and line voltage if the speed is 375 rpm.
   [50 Hz; 1530 V; 2650 V] (*Electrical Machines-I, Indore Univ. April 1977*)
6. A synchronous generator has 9 slots per pole. If each coil spans 8 slot pitches, what is the value of the pitch factor? **[0.985]** (*Elect. Machines, A.M.I.E. Sec. B. 1989*)
7. A 3-phase, Y-connected, 2-pole alternator runs at 3,600 r.p.m. If there are 500 conductors per phase in series on the armature winding and the sinusoidal flux per pole is 0.1 Wb, calculate the magnitude and frequency of the generated voltage from first principles. **[60 Hz; 11.5 kV]**

8. One phase of a 3-phase alternator consists of twelve coils in series. Each coil has an r.m.s. voltage of 10 V induced in it and the coils are arranged in slots so that there is a successive phase displacement of 10 electrical degrees between the e.m.f. in each coil and the next. Find graphically or by calculation, the r.m.s. value of the total phase voltage developed by the winding. If the alternator has six pole and is driven at 100 r.p.m., calculate the frequency of the e.m.f. generated. **[108 V; 50 Hz]**

9. A 4-pole, 50-Hz, 3-phase, *Y*-connected alternator has a single-layer, full-pitch winding with 21 slots per pole and two conductors per slot. The fundamental flux is 0.6 Wb and air-gap flux contains a third harmonic of 5% amplitude. Find the r.m.s. values of the phase e.m.f. due to the fundamental and the 3rd harmonic flux and the total induced e.m.f. **[3,550 V; 119.5 V; 3,553 V]** (*Elect. Machines-III, Osmania Univ. 1977*)

10. A 3-phase, 10-pole alternator has 90 slots, each containing 12 conductors. If the speed is 600 r.p.m. and the flux per pole is 0.1 Wb, calculate the line e.m.f. when the phases are (*i*) star connected (*ii*) delta connected. Assume the winding factor to be 0.96 and the flux sinusoidally distributed. **[(*i*) 6.93 kV (*ii*) 4 kV]** (*Elect. Engg-II, Kerala Univ. 1979*)

11. A star-connected 3-phase, 6-pole synchronous generator has a stator with 90 slots and 8 conductors per slot. The rotor revolves at 1000 r.p.m. The flux per pole is 4×10^{-2} weber. Calculate the e.m.f. generated, if all the conductors in each phase are in series. Assume sinusoidal flux distribution and full-pitched coils. **[Eph = 1,066 V]** (*Elect. Machines, A.M.I.E. Summer, 1979*)

12. A six-pole machine has an armature of 90 slots and 8 conductors per slot and revolves at 1000 r.p.m. the flux per pole being 50 milli weber. Calculate the e.m.f. generated as a three-phase star-connected machine if the winding factor is 0.96 and all the conductors in each phase are in series. **[1280 V]** (*Elect. Machines, AMIE, Sec. B, (E-3), Summer 1992*)

35.15. Factors Affecting Alternator Size

The efficiency of an alternator always increases as its power increases. For example, if an alternator of 1 kW has an efficiency of 50%, then one of 10 MW will inevitably have an efficiency of about 90%. It is because of this improvement in efficiency with size that alternators of 1000 MW and above possess efficiencies of the order of 99%.

Another advantage of large machines is that power output per kilogram increases as the alternator power increases. If 1 kW alternator weighs 20 kg (*i.e.* 50 W/kg), then 10MW alternator weighing 20,000 kg yields 500 W/kg. In other words, larger alternators weigh relatively less than smaller ones and are, consequently, cheaper.

However, as alternator size increases, cooling problem becomes more serious. Since large machines inherently produce high power loss per unit surface area (W/m^2), they tend to overheat. To keep the temperature rise within acceptable limits, we have to design efficient cooling system which becomes ever more elaborate as the power increases. For cooling alternators of rating upto 50 MW, circulating cold-air system is adequate but for those of rating between 50 and 300 MW, we have to resort to hydrogen cooling. Very big machines in 1000 MW range have to be equipped with hollow water-cooled conductors. Ultimately, a point is reached where increased cost of cooling exceeds the saving made elsewhere and this fixes the upper limit of the alternator size.

So for as the speed is concerned, low-speed alternators are always bigger than high speed alternators of the same power. Bigness always simplifies the cooling problem. For example, the large 200-rpm, 500-MVA alternators installed in a typical hydropower plant are air-cooled whereas much smaller 1800-rpm,, 500-MVA alternators installed in a steam plant are hydrogen cooled.

35.16. Alternator on Load

As the load on an alternator is varied, its terminal voltage is also found to vary as in d.c. generators. This variation in terminal voltage *V* is due to the following reasons:

1. voltage drop due to armature resistance R_a
2. voltage drop due to armature leakage reactance X_L
3. voltage drop due to armature reaction

(*a*) Armature Resistance

The armature resistance/phase R_a causes a voltage drop/phase of IR_a which is in phase with the armature current I. However, this voltage drop is practically negligible.

(*b*) **Armature Leakage Reactance**

When current flows through the armature conductors, fluxes are set up which do not cross the air-gap, but take different paths. Such fluxes are known as *leakage fluxes*. Various types of leakage fluxes are shown in Fig. 35.22.

The leakage flux is practically independent of saturation, but is dependent on I and its phase angle with terminal voltage V. This leakage flux sets up an e.m.f. of self-inductance which is known as *reactance e.m.f.* and which is ahead of I by 90°. Hence, armature winding is assumed

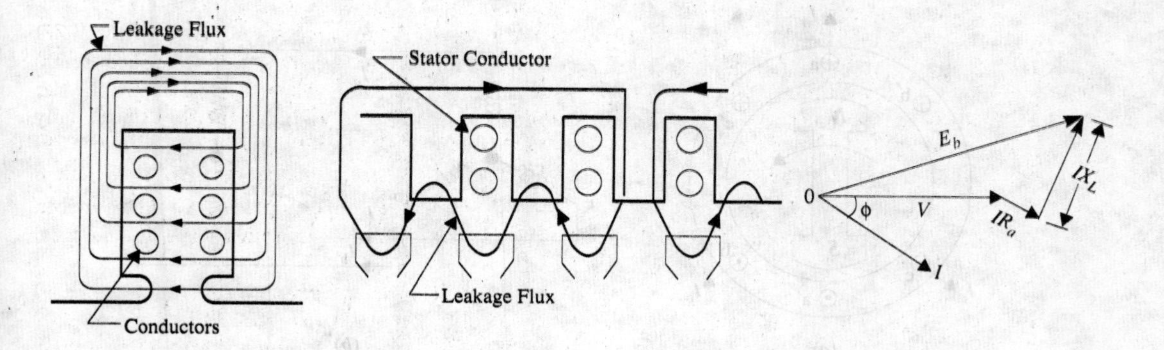

Fig. 35.22 Fig. 35.23

to possess leakage reactance X_L (also known as Potier rectance X_P) such that voltage drop due to this equals IX_L. A part of the generated e.m.f. is used up in overcoming this reactance e.m.f.

$$\therefore \quad \mathbf{E} = \mathbf{V} + \mathbf{I}(R + jX_L)$$

This fact is illustrated in the vector diagram of Fig. 35.23.

(*c*) **Armature Reaction**

As in d.c. generators, armature reaction is the effect of armature flux on the main field flux. *In the case of alternators, the power factor of the load has a considerable effect on the armature reaction.* We will consider three cases: (*i*) when load of p.f. is unity (*ii*) when p.f. is zero lagging and (*iii*) when p.f. is zero leading.

Before discussing this, it should be noted that in a 3-phase machine the combined ampere-turn wave (or m.m.f. wave) is sinusoidal which moves synchronously. This amp-turn or m.m.f. wave is fixed relative to the poles, its amplitude is proportional to the load current, but its position depends on the p.f. of the load.

Consider a 3-phase, 2-pole alternator having a single-layer winding, as shown in Fig. 35.24 (*a*). For the sake of simplicity, assume that winding of each phase is concentrated (instead of being distributed) and that the number of turns per phase is N. Further suppose that the alternator is loaded with a resistive load of unity power factor, so that phase currents I_a, I_b and I_c are in phase with their respective phase voltages. Maximum current I_a will flow when the poles are in position shown in Fig. 35.24 (*a*) or at a time t_1 in Fig. 35.24 (*c*). When I_a has a maximum value, I_b and I_c have one-half their maximum values (the arrows attached to I_a, I_b and I_c are only polarity marks and are not meant to give the instantaneous directions of these currents at time t_1). The instantaneous directions of currents are shown in Fig. 35.24 (*a*). At the instant t_1, I_a flows in conductor a whereas I_b and I_c flow out.

As seen from Fig. 35.24 (*d*), the m.m.f. (= NI_m) produced by phase *a-a′* is horizontal, whereas that produced by other two phases is $(I_m/2)\,N$ each at 60° to the horizontal. The total armature m.m.f. is equal to the vector sum of these three m.m.fs.

$$\therefore \quad \text{armature m.m.f.} = NI_m + 2.(1/2\, NI_m) \cos 60° = 1.5\, NI_m$$

As seen, at this instant t_1, the m.m.f. of the main field is upwards and the armature m.m.f.

is behind it by 90 electrical degrees.

Next, let us investigate the armature m.m.f. at instant t_2. At this instant, the poles are in the horizontal position. Also $I_a = 0$, but I_b and I_c are each equal to 0.866 of their maximum values. Since I_c has not changed in direction during the interval t_1 to t_2, the direction of its m.m.f. vector remains unchanged. But I_b has changed direction, hence, its m.m.f. vector will now be in the position shown in Fig. 35.24 (*d*). Total armature m.m.f. is again the vector sum of these two m.m.fs.

$\therefore$ armature m.m.f. $= 2 \times (0.866\ NI_m) \times \cos 30° = 1.5\ NI_m$

If further investigations are made, it will be found that:

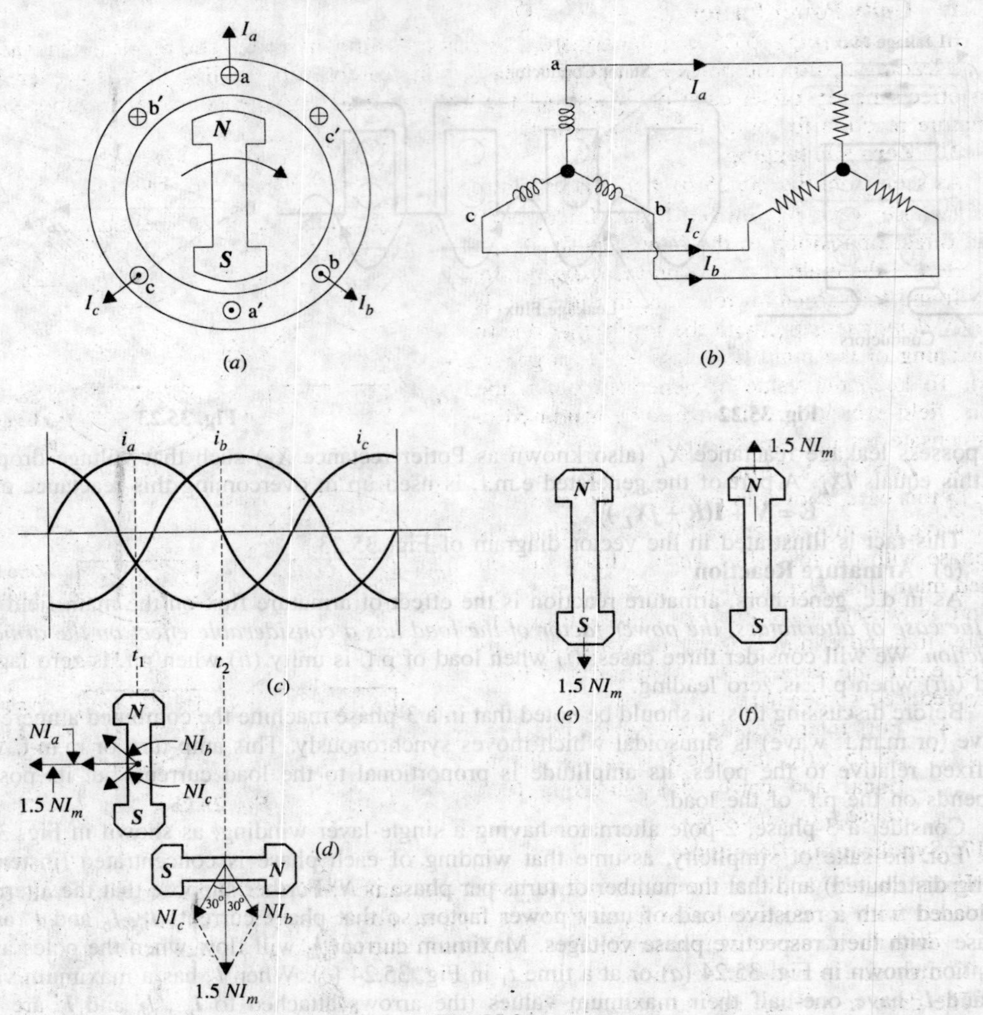

Fig. 35.24

1. armature m.m.f. remains constant with time
2. it is 90 space degrees behind the main field m.m.f., so that it is only distortional in nature.
3. it rotates synchronously round the armature *i.e.* stator.

For a lagging load of zero power factor, all currents would be delayed in time 90° and armature m.m.f. would be shifted 90° with respect to the poles as shown in Fig. 35.24 (*e*). Obviously, armature m.m.f. would demagnetise the poles and cause a reduction in the induced e.m.f. and hence the terminal voltage.

For leading loads of zero power factor, the armature m.m.f. is advanced 90° with respect to the position shown in Fig. 35.24 (*d*). As shown in Fig. 35.24 (*f*), the armature m.m.f. strengthens the main m.m.f. In this case, armature reaction is wholly magnetising and causes an increase in the terminal voltage.

The above facts have been summarized briefly in the following paragraphs where the matter is discussed in terms of 'flux' rather than m.m.f. waves.

1. Unity Power Factor

In this case [Fig. 35.25 (*a*)] the armature flux is cross-magnetising. The result is that the flux at the leading tips of the poles is reduced while it is increased at the trailing tips. However, these two effects nearly offset each other leaving the average field strength constant. In other words, armature reaction for unity p.f. is distortional.

2. Zero P.F. lagging

As seen from Fig. 35.25(*b*), here the armature flux (whose wave has moved backward by 90°) is in direct opposition to the main flux.

Hence, the main flux is decreased. Therefore, it is found that armature reaction, in this case, is wholly *demagnetising*, with the result, that due to weakening of the main flux, less e.m.f. is generated. To keep the value of generated e.m.f. the same, field excitation will have to be increased to compensate for this weakening.

3. Zero P.F. leading

In this case, shown in Fig. 35.25 (*c*) armature flux wave has moved forward by 90° so that it is in phase with the main flux wave. This results in added main flux. Hence, in this case, armature reaction is wholly *magnetising, which* results in greater induced e.m.f. To keep the value of generated e.m.f. the same, field excitation will have to be reduced somewhat.

4. For intermediate power factor [Fig. 35.25 (*d*)], the effect is partly distortional and partly demagnetising (because p.f. is lagging).

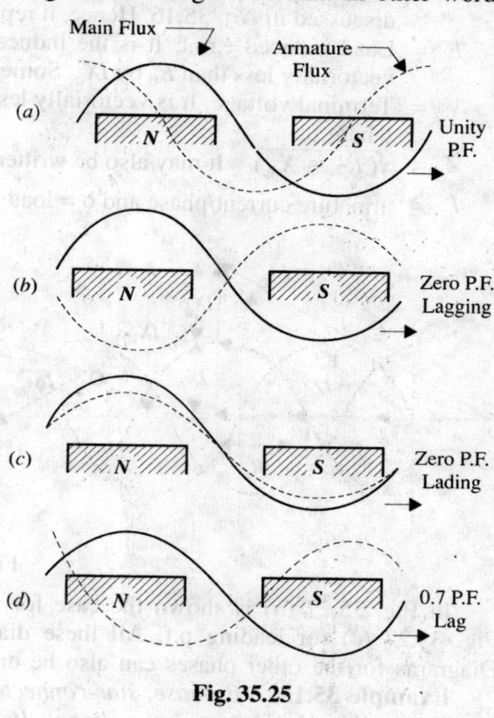

Fig. 35.25

35.17. Synchronous Reactance

From the above discussion, it is clear that for the same field excitation, terminal voltage is decreased from its no-load value E_0 to V (for a lagging power factor). This is because of

1. drop due to armature resistance, IR_a
2. drop due to leakage reactance, $I X_L$
3. drop due to armature reaction.

The drop in voltage due to armature reaction may be accounted for by assuming the presence of a fictitious reactance X_a in the armature winding. The value of X_a is such that IX_a represents the voltage drop due to armature reaction.

The leakage reactance X_L (or X_P) and the armature reactance X_a may be combined to give synchronous reactance X_S.

Hence $X_S = X_L + X_a$ *

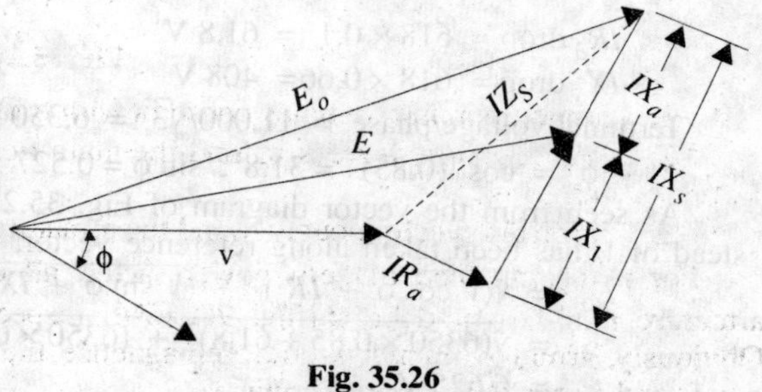

Fig. 35.26

*The ohmic value of Xa varies with the p.f. of the load because armature reaction depends on load p.f.

Therefore, total voltage drop in an alternator under load is

$$= IR_a + jIX_S = I(R_a + jX_S) = \mathbf{IZ_S}$$

where Z_S is known as *synchronous impedance* of the armature, the word 'synchronous' being used merely as an indication that it refers to the working conditions.

Hence, we learn that the vector difference between no-load voltage E_0 and terminal voltage V is equal to IZ_S, as shown in Fig. 35.26.

35.18. Vector Diagrams of a Loaded Alternator

Before discussing the diagrams, following symbols should be clearly kept in mind.

E_0 = No-load e.m.f. This being the voltage induced in armature in the absence of three factors discussed in Art. 35.16. Hence, it represents the maximum value of the induced e.m.f.

E = Load induced e.m.f. It is the induced e.m.f. after allowing for armature reaction. E is vectorially less than E_0 by IX_a. Sometimes, it is written as E_a (Ex 35.16).

V = Terminal voltage, It is vectorially less than E_0 by IZ_S or it is vectorially less than E by I_Z where

$Z = \sqrt{(R_a^2 + X_L^2)}$. It may also be written as Z_a.

I = armature current/phase and ϕ = load p.f. angle.

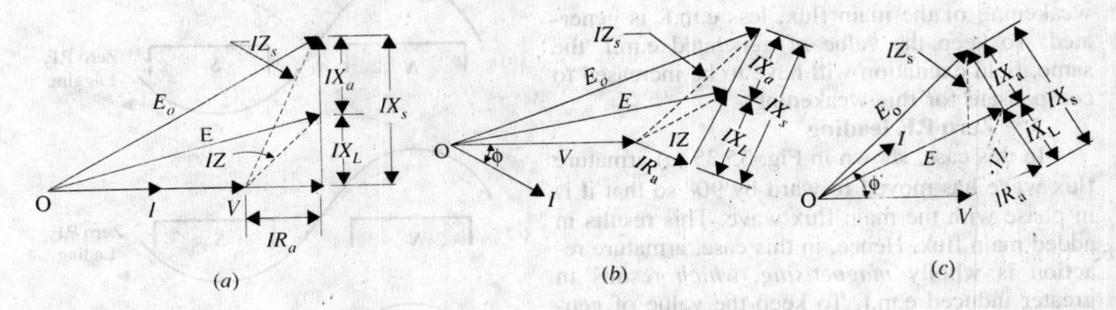

Fig. 35.27

In Fig. 35.27 (*a*) is shown the case for unity p.f., in Fig. 35.27 (*b*) for lagging p.f. and in Fig. 35.27 (c) for leading p.f. All these diagrams apply to one phase of a 3-phase machine. Diagrams for the other phases can also be drawn similary.

Example 35.15. *A 3-phase, star-connected alternator supplies a load of 10 MW at p.f. 0.85 lagging and at 11 kV (terminal voltage). Its resistance is 0.1 ohm per phase and synchronous reactance 0.66 ohm per phase. Calculate the line value of e.m.f. generated.*

(Electrical Technology, Aligarh Muslim Univ. 1988)

Solution. F.L. output current $= \dfrac{10 \times 10^6}{\sqrt{3} \times 11,000 \times 0.85} = 618$ A

IR_a drop = $618 \times 0.1 = 61.8$ V

IX_S drop = $618 \times 0.66 = 408$ V

Terminal voltage/phase $= 11,000/\sqrt{3} = 6,350$ V

$\phi = \cos^{-1}(0.85) = 31.8°$; $\sin\phi = 0.527$

As seen from the vector diagram of Fig. 35.28 where I instead of V has been taken along reference vector,

$$E_0 = \sqrt{(V\cos\phi + IR_a)^2 + (V\sin\phi + IX_S)^2}$$

$$= \sqrt{(6350 \times 0.85 + 61.8)^2 + (6350 \times 0.527 + 408)^2}$$

$$= 6,625 \text{ V}$$

Line e.m.f. $= \sqrt{3} \times 6,625 =$ **11,486 volt**

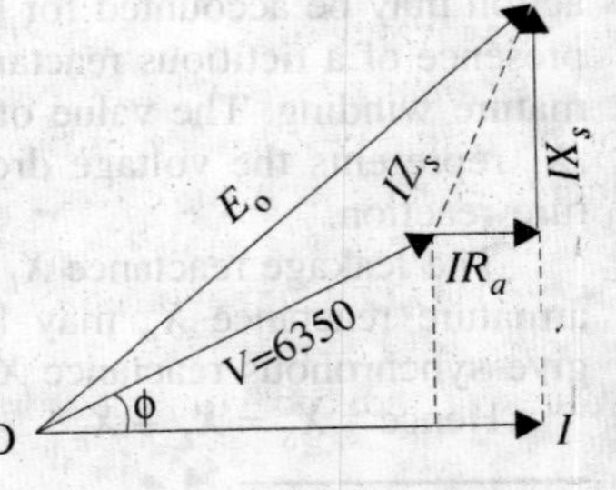

Fig. 35.28

35.19. Voltage Regulation

It is clear that with change in load, there is a change in terminal voltage of an alternator. The magnitude of this change depends *not only on the load but also on the load power factor.*

The voltage regulation of an alternator is defined as "the *rise* in voltage when full-load is removed (field excitation and speed remaining the same) divided by the rated terminal voltage."

$$\therefore \quad \% \text{ regulation 'up'} = \frac{E_0 - V}{V} \times 100$$

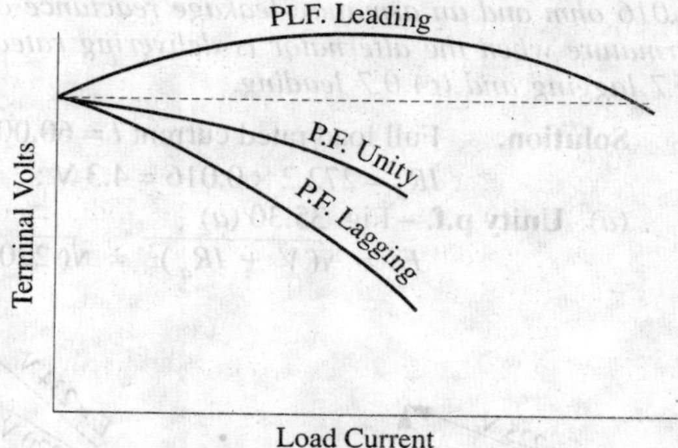

Fig. 35.29

Note. (*i*) $E_0 - V$ is the arithmetical difference and not the vectorial one.

(*ii*) In the case of *leading* load p.f., terminal voltage will fall on removing the full-load. Hence, regulation is negative in that case.

(*iii*) The rise in voltage when full-load is thrown off is not the same as the fall in voltage when full-load is applied.

Voltage characterstics of an alternator are shown in Fig. 35.29.

35.20. Determination of Voltage Regulation

In the case of small machines, the regulation may be found by direct loading. The procedure is as follows :

The alternator is driven at synchronous speed and the terminal voltage is adjusted to its rated value V. The load is varied until the wattmeter and ammeter (connected for the purpose) indicate the rated values at desired p.f. Then the entire load is thrown off while the speed and field excitation are kept constant. The open-circuit or no-load voltage E_0 is read. Hence, regulation can be found from

$$\% \text{ regn} = \frac{E_0 - V}{V} \times 100$$

In the case of large machines, the cost of finding the regulation by direct loading becomes prohibitive. Hence, other indirect methods are used as discussed below. It will be found that all these methods differ chiefly in the way the no-load voltage E_0 is found in each case.

1. *Synchronous Impedance or E.M.F. Method.* It is due to Behn Eschenberg.
2. *The Ampere-turn or M.M.F. Method.* This method is due to Rothert.
3. *Zero Power Factor or Potier Method.* As the name indicates, it is due to Potier.

All these methods require —

1. Armature (or stator) resistance R_a
2. Open-circuit/No-load characteristic.
3. Short-circuit characteristic (but zero power factor lagging characteristic for Potier method).

Now, let us take up each of these methods one by one.

(*i*) Value of R_a

Armature resistance R_a per phase can be measure directly be voltmeter and ammeter method or by using Wheatstone bridge. However, under working conditions, the effective value of R_a is increased due to 'skin effect'*. The value of R_a so obtained is increased by 60% or so to allow for this effect. Generally, a value 1.6 times the d.c. value is taken.

(*ii*) O.C. Characteristic

As in d.c. machines, this is plotted by running the machine on no-load and by noting the values of induced voltage and field excitation current. It is just like the *B-H* curve.

---

*The 'skin effect' may sometimes increase the effective resistance of armature conductors as high as 6 times its d.c. value.

***(iii)* S.C. Characteristic**

It is obtained by short-circuiting the armature (*i.e.* stator) windings through a low-resistance ammeter. The excitation is so adjusted as to give 1.5 to 2 times the value of full-load current. During this test, the speed which is not necessarily synchronous, is kept constant.

Example 35.16. *A 60-kVa, 220 V, 50-Hz, 1-ϕ alternator has effective armature resistance of 0.016 ohm and an armature leakage reactance of 0.07 ohm. Compute the voltage induced in the armature when the alternator is delivering rated current at a load power factor of (a) unity (b) 0.7 lagging and (c) 0.7 leading.* **(Elect. Machines-I, Indore Univ. 1981)**

Solution. Full load rated current $I = 60{,}000/220 = 272.2$ A

$IR_a = 272.2 \times 0.016 = 4.3$ V ; $IX_L = 272.2 \times 0.07 = 19$ V

(*a*) **Unity p.f.** – Fig. 35.30 (*a*)

$$E = \sqrt{(V + IR_a)^2} = \sqrt{(220 + 4.3)^2 + 19^2} = \mathbf{225\ V}$$

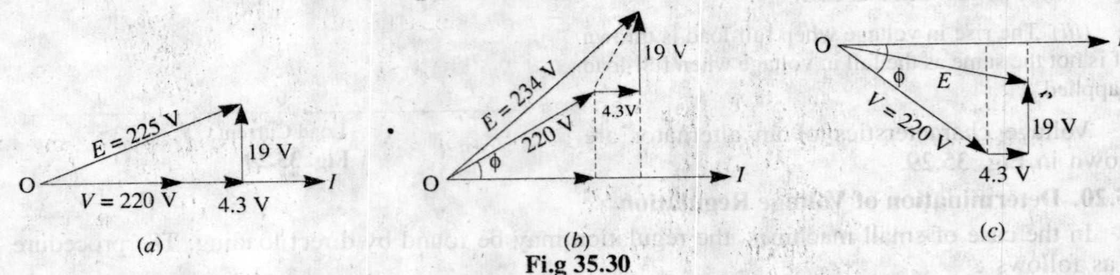

Fi.g 35.30

(*b*) **p.f. 0.7 (lag)** – Fig. 35.30 (*b*)

$$E = [V \cos \phi + IR_a)^2 + (V \sin \phi + IX_L)^2]^{1/2}$$
$$= [(220 \times 0.7 + 4.3)2 + (220 \times 0.7 + 19)^2]^{1/2} = \mathbf{234V}$$

(*c*) **p.f. = 0.7 (lead)**–Fig. 35.30 (*c*)

$$E = [(V \cos \phi + IR_a)^2 + (V \sin \phi - IX_L)^2]^{1/2}$$
$$= [(220 \times 0.7 + 4.3)^2 + (220 \times 0.7 - 19)^2]^{1/2} = \mathbf{208\ V}$$

Example 35.17. *In a 50-kVA, star-connected, 440-V, 3-phase, 50-Hz alternator, the effective armature resistance is 0.25 ohm per phase. The synchronous reactance is 3.2 ohm per phase and leakage reactance is 0.5 ohm per phase. Determine at rated load and unity power factor:*

(a) internal e.m.f E_a (b) no-load e.m.f. E_0 (c) percentage regulation on full-load (d) value of synchronous reactance which replaces armature reaction. **(Electrical Engg. Bombay Univ. 1987)**

Solution. (*a*) The e.m.f. E_a is the vector sum of (*i*) terminal voltage V (*ii*) IR_a and (*iii*) IX_L as detailed in Art. 35.17. Here,

$$V = 440/\sqrt{3} = 254 \text{ V}$$

F.L. output current at u.p.f. is

$$= 50{,}000/\sqrt{3} \times 440 = 65.6 \text{ A}$$

Resistive drop $= 65.6 \times 0.25 = 16.4$ V

Leakage reactance drop $IX_L = 65.6 \times 0.5 = 32.8$ V

$$\therefore \quad E_a = \sqrt{(V + IR_a)^2 + (IX_L)^2}$$
$$= \sqrt{(254 + 16.4)^2 + 32.8^2} = 272 \text{ volt}$$

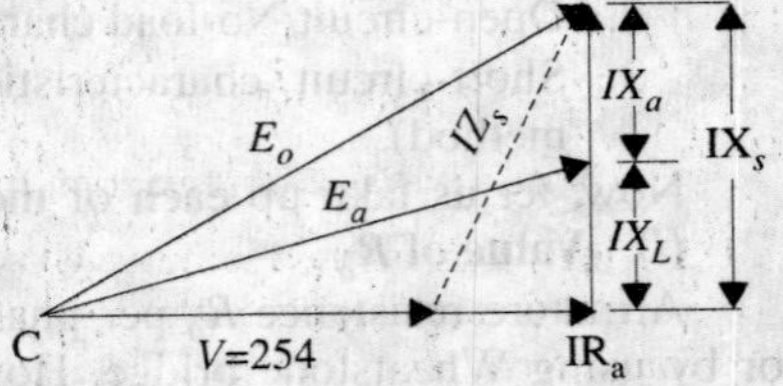

Fig. 35.31

Line value $= \sqrt{3} \times 272 = \mathbf{471\ volt.}$

(*b*) The no-load e.m.f. E_0 is the vector sum of (*i*) V (*ii*) IR_a and (*iii*) IX_S or is the vector sum of V and IZ_S (Fig. 35.31).

$$\therefore \quad E_0 = \sqrt{(V + IR_a)^2 + (IX_S)^2} = \sqrt{(254 + 16.4)^2 + (65.6 \times 3.2)^2} = 342 \text{ volt}$$

Line value $= \sqrt{3} \times 342 =$ **592 volt**

(*c*) %age regulation 'up' $= \dfrac{E_0 - V}{V} \times 100 = \dfrac{342 - 254}{254} \times 100 =$ **34.65 per cent**

(*d*) $X_a = X_S - X_L = 3.2 - 0.5 = \mathbf{2.7\ \Omega}$ ... Art. 35.17

35.21. Synchronous Impedance Method

Following procedural steps are involved in this method:

1. O.C.C is plotted from the given data as shown in Fig. 35.32.

2. Similarly, S.C.C. is drawn from the data given by the short-circuit test. It is a straight line passing through the origin. Both these curves are drawn on a common field-current base.

Consider a field current I_f. The O.C. voltage corresponding to this field current is E_1. When winding is short-circuited, the terminal voltage is zero. Hence, it may be assumed that the whole of this voltage E_1 is being used to circulate the armature short-circuit current I_1 against the synchronous impedance Z_S.

$$\therefore \quad E_1 = I_1 Z_S \qquad \therefore \quad Z_S = \frac{E_1 \text{ (open-circuit)}}{I_1 \text{ (short-circuit)}}$$

3. Since R_a can be found as discussed earlier, $X_S = \sqrt{(Z_S^2 - R_a^2)}$

4. Knowing R_a and X_S, vector diagram as shown in Fig. 35.33 can be drawn for any load and any power factor.

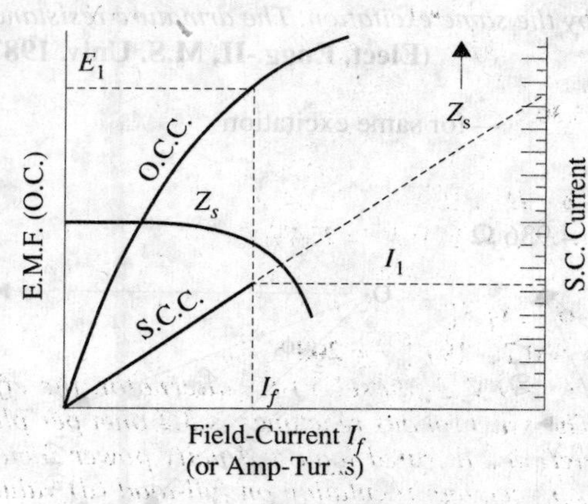

Fig. 35.32

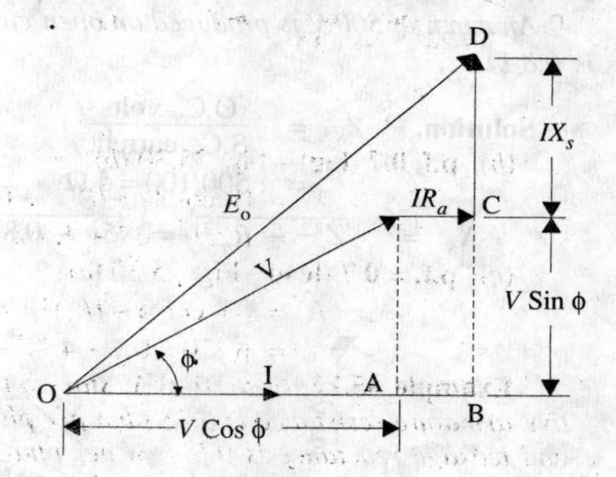

Fig. 35.33

Here $\quad OD = E_0 \qquad \therefore \quad E_0 = \sqrt{(OB^2 + BD^2)}$

or $\quad E_0 = \sqrt{[(V \cos \phi + IR_a)^2 + (V \sin \phi + IX_S)^2]}$

$$\therefore \quad \% \text{ regn. 'up'} = \frac{E_0 - V}{V} \times 100$$

Note. (*i*) Value of regulation for unity power factor or leading p.f. can also be found in a similar way.

(*ii*) This method is not accurate because the value of Z_S so found is always more than its value under normal voltage conditions and saturation. Hence, the value of regulation so obtained is always more than that found from an actual test. That is why it is called pessimistic method. The value of Z_S is not constant but varies with saturation. At low saturation, its value is larger because then the effect of a given armature ampere-turns is much more than at high saturation. Now, under short-circuit conditions, saturation is very low, because armature m.m.f is directly demagnetising. Different values of Z_S corresponding to different values of field current are also plotted in Fig. 35.32.

(*iii*) The value of Z_S usually taken is that obtained from full-load current in the short-circuit test.

(*iv*) Here, armature reactance X_a has not been treated separately but along with leakage reactance X_L.

Example 35.18. *Find the synchronous impedance and reactance of an alternator in which* *a given field current produces an armature current of 200 A on short-circuit and a generated e.m.f. of 50 V on open-circuit. The armature resistance is 0.1 ohm. To what induced voltage must the alternator be excited if it is to deliver a load of 100 A at a p.f. of 0.8 lagging, with a terminal voltage of 200 V.* **(Elect. Machinery, Banglore Univ. 1991)**

Solution. It will be assumed that alternator is a single phase one. Now, for same field current,

$$Z_S = \frac{\text{O.C. volts}}{\text{S.C. current}} = \frac{50}{200} = \mathbf{0.25\ \Omega}$$

$$X_S = \sqrt{Z_S^2 - R_a^2} = \sqrt{0.25^2 - 0.1^2} = \mathbf{0.23\ \Omega}$$

Now, $IR_a = 100 \times 0.1 = 10$ V, $IX_S = 100 \times 0.23 = 23$ V;

$\cos\phi = 0.8$, $\sin\phi = 0.6$. As seen from Fig. 35.34.

$$E_0 = \sqrt{(V\cos\phi + IR_a)^2 + (V\sin\phi + IX_S)^2}$$

$$= [(200 \times 0.8 + 10)^2 + (200 \times 0.6 + 23)^2]^{1/2} = \mathbf{222\ V}$$

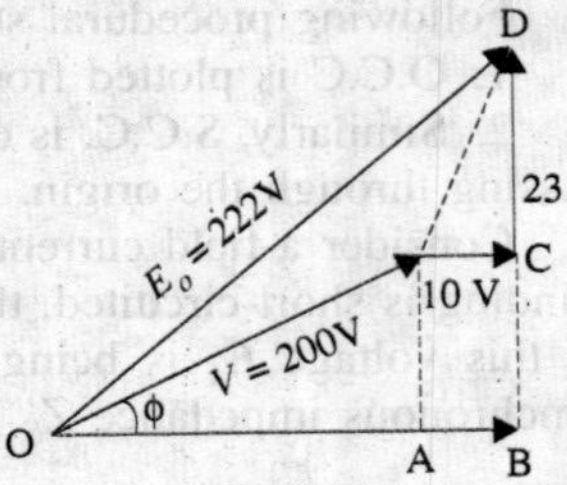

Fig. 35.34

Example 35.19. *From the following test results, determine the voltage regulation of a 2000-V, 1-phase alternator delivering a current of 100 A at (i) unity p.f. (ii) 0.8 leading p.f. and (iii) 0.71 lagging p.f.*

Test results : Full-load current of 100 A is produced on short-circuit by a field excitation of 2.5 A.

An e.m.f. of 500 V is produced on open-circuit by the same excitation. The armature resistance is 0.8 Ω. **(Elect. Engg.-II, M.S. Univ. 1987)**

Solution. $Z_S = \frac{\text{O.C. volts}}{\text{S.C. current}}$ —for same excitation

$= 500/100 = 5\ \Omega$

$$X_S = \sqrt{Z_s^2 - R_a^2} = \sqrt{5^2 - 0.8^2} = 4.936\ \Omega$$

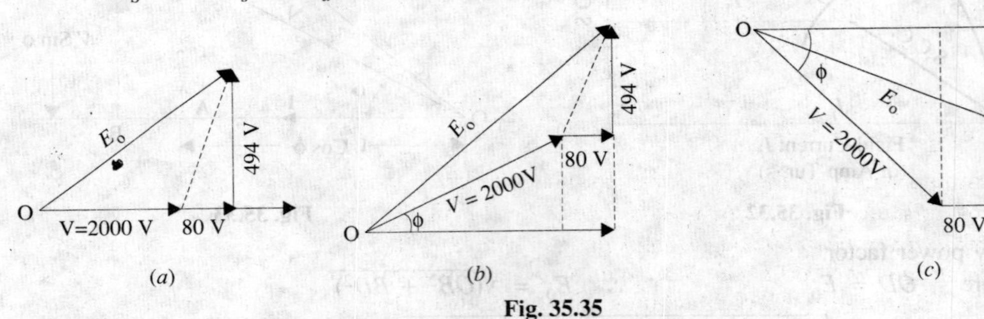

Fig. 35.35

***(i)* Unity p.f.** (Fig. 35.35 (*a*)]

$IR_a = 100 \times 0.8 = 80$ V; $IX_S = 100 \times 4.936 = 494$ V

$$\therefore \quad E_0 = \sqrt{(2000 + 80)^2 + 494^2} = 2140\text{ V}$$

$$\%\text{ regn} = \frac{2140 - 2000}{2000} \times 100 = \mathbf{7\%}$$

***(ii)* p.f. = 0.8 (lead)** [Fig. 35.35 (*c*)]

$$E_0 = [(2000 \times 0.8 + 80)^2 + (2000 \times 0.6 - 494)^2]^{1/2} = 1820\text{ V}$$

$$\%\text{ regn} = \frac{1820 - 2000}{2000} \times 100 = \mathbf{9\%}$$

***(iii)* p.f. = 0.71(lag)** [Fig.35.35 (*b*)]

$$E_0 = [(2000 \times 0.71 + 80)^2 + (2000 \times 0.71 + 494)^2]^{1/2} = 2432\text{ V}$$

$$\%\text{ regn} = \frac{2432 - 2000}{2000} \times 100 = \mathbf{21.6\ \%}$$

Example 35.20. *A 100-kVA, 3000-V, 50-Hz 3-phase star-connected alternator has effective armature resistance of 0.2 ohm. The field current of 40 A produces short-circuit current of 200 A and an open-circuit emf of 1040 V (line value). Calculate the full-load voltage regulation at 0.8 p.f. lagging and 0.8 p.f. leading. Draw phasor diagrams.*

(Basic Elect. Machines, Nagpur Univ. 1993)

Solution. $Z_S = \dfrac{\text{O.C. voltage/phase}}{\text{S.C. current/phase}}$ — for same excitation

$= \dfrac{1040/\sqrt{3}}{200} = 3\Omega$

$X_S = \sqrt{Z_S^2 - R_a^2} = \sqrt{3^2 - 0.2^2}$

$= 2.99\ \Omega$

F.L. current,

$I = 100{,}000/\sqrt{3} \times 3000 = 19.2$ A

$IR_a = 19.2 \times 0.2 = 3.84\ V$

$IX_S = 19.2 \times 2.99 = 57.4\ V$

Voltage/phase $= 3000/\sqrt{3} = 1730$ V

$\cos\phi = 0.8$; $\sin\phi = 0.6$

(*i*) **p.f. = 0.8 lagging** -Fig. 35.36 (*a*)

$E_0 = [(V\cos\phi + IRa)^2 + (V\sin\phi + IXs)^2]^{1/2}$

$= (1730 \times 0.8 + 3.84)^2 + (1730 \times 0.6 + 57.4)^2]^{1/2} = 1768$ V

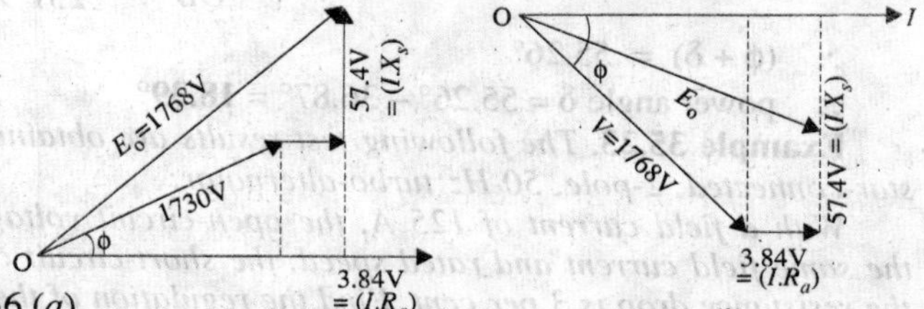

% regn. 'up' $= \dfrac{(1768 - 1730)}{1730} \times 100 =$ **2.2%**

(*ii*) **0.8 p.f. leading** —Fig. 35.36 (*b*)

$E_0 = [(V\cos\phi + IR_a)^2 + (V\sin\phi - IX_S)^2]^{1/2}$

$= [(1730 \times 0.8 + 3.84)^2 + (1730 \times 0.6 - 57.4)^2]^{1/2}$

$= 1699$ V

% regn. $= \dfrac{1699 - 1730}{1730} \times 100 =$ **−1.8%**

Fig. 35.36

Example 35.21. *A 3-phase, star-connected alternator is rated at 1600 kVA, 13,500 V. The armature resistance and synchronous reactance are 1.5 Ω and 30 Ω respectively per phase. Calculate the percentage regulation for a load of 1280 kW at 0.8 leading power factor.*

(Advanced Elect. Machines AMIE Sec B. 1991)

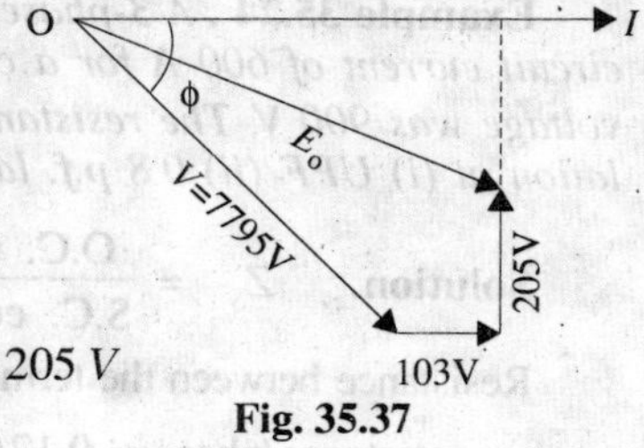

Solution. $1280{,}000 = \sqrt{3} \times 13,500 \times I \times 0.8$;

$\therefore\ I = 68.4$ A

$IR_a = 68.4 \times 1.5 = 103\ V$; $IX_S = 68.4 \times 30 = 205\ V$

Voltage/phase $= 13{,}500/\sqrt{3} = 7795$ V

As seen from Fig. 35.37.

$E_0 = [(7795 \times 0.8 + 103)^2 + (7795 \times 0.6 - 205)^2]^{1/2} = 7758$ V

% regn. $= (7758 - 7795)/7795$

$= -0.0047$ or **− 0.47%**

Fig. 35.37

Example 35.22. *A 3-phase, 10-kVA, 400-V, 50-Hz, Y-connected alternator supplies the rated load at 0.8 p.f. lag. If arm. resistance is 0.5 ohm and syn. reactance is 10 ohms, find the power angle and voltage regulation.* **(Elect. Machines-I Nagpur Univ. 1993)**

Solution. F.L. current, $I = 10{,}000/\sqrt{3} \times 400 = 14.4$ A

$IR_a = 14.4 \times 0.5 = 7.2$ V

$IX_S = 14.4 \times 10 = 144$ V

Voltage/phase = $400/\sqrt{3} = 231$ V

$\phi = \cos^{-1} 0.8 = 36.87°$

$E_0 = [(V \cos \phi + IR_a)^2 + (V \sin \phi + IX_S)^2]^{1/2}$

$= (231 \times 0.8 + 7.2)^2 + (231 \times 0.6 + 144)^2]^{1/2} = 342$ V

$$\% \text{ regn.} = \frac{342 - 231}{231} \times 100 = 0.48 \text{ or } 48\%$$

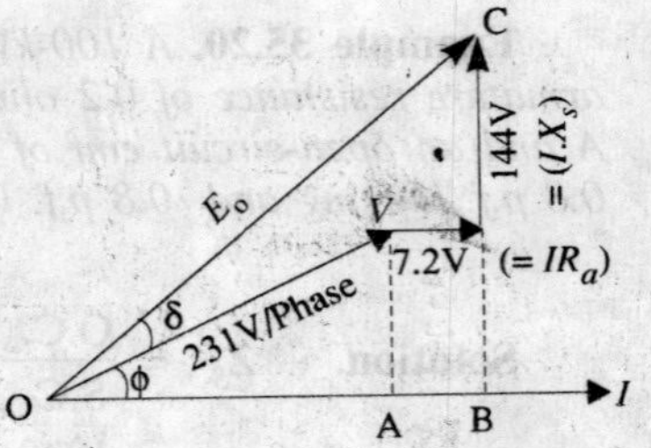

Fig. 35.38

The power angle of the machine is defined as the angle between V and E_0 *i.e.* angle δ

As seen from Fig. 35.38, $\tan(\phi + \delta) = \frac{BC}{OB} = \frac{231 \times 0.6 + 144}{231 \times 0.8 + 7.2} = \frac{282.6}{192} = 1.4419;$

$\therefore \quad (\phi + \delta) = 55.26°$

$\therefore \quad$ power angle $\delta = 55.26° - 36.87° = \mathbf{18.39°}$

Example 35.23. *The following test results are obtained from a 3-phase, 6,000-kVA, 6,600 V, star-connected, 2-pole, 50-Hz turbo-alternator:*

With a field current of 125 A, the open-circuit voltage is 8,000 V at the rated speed; with the same field current and rated speed, the short-circuit current is 800 A. At the rated full-load, the resistance drop is 3 per cent. Find the regulation of the alternator on full-load and at a power factor of 0.8 lagging. **(Electrical Technology, Utkal Univ. 1987)**

Solution. $Z_S = \frac{\text{O.C. voltage/phase}}{\text{S.C. current/phase}} = \frac{8,000/\sqrt{3}}{800} = 5.77\ \Omega$

Voltage/phase $= 6,600/\sqrt{3} = 3,810$ V

Resistive drop $= 3\%$ of $3,810$ V $= 0.03 \times 3,810 = 114.3$ V

Full-load current $= 6,000 \times 10^2/\sqrt{3} \times 6,600 = 525$ A

Now $IR_a = 114.3$V $\quad \therefore \quad R_a = 114.3/525 = 0.218\ \Omega$

$X_S = \sqrt{Z_s^2 - R_a^2} = \sqrt{5.77^2 - 0.218^2} = 5.74\ \Omega$ (approx.)

As seen from the vector diagram of Fig. 35.33,

$E_0 = \sqrt{[(3,810 \times 0.8 + 114.3)^2 + (3,810 \times 0.6 + 525 \times 5.74)^2]} = 6,180$ V

$\therefore$ regulation $= (6,180 - 3,810) \times 100/3,810 = \mathbf{62.2\%}$

Example 35.24 . *A 3-phase 50-Hz star-connected 2000-kVA, 2300 V alternator gives a short-circuit current of 600 A for a certain field excitation. With the same excitation, the open circuit voltage was 900 V. The resistance between a pair of terminals was 0.12 Ω. Find full-load regulation at (i) UPF (ii) 0.8 p.f. lagging.* **(Elect. Machines, Nagpur Univ. 1993)**

Solution. $Z_S = \frac{\text{O.C. volts/phase}}{\text{S.C. current/phase}} = \frac{900/\sqrt{3}}{600} = 0.866\ \Omega$

Resistance between the terminals is 0.12 Ω. It is the resistance of two phases connected in series.

$\therefore$ resistance/phase $= 0.12/2 = 0.06\ \Omega;$

effective resistance/phase $= 0.06 \times 1.5 = 0.09\ \Omega;$

$X_S = \sqrt{0.866^2 - 0.09^2} = 0.86\ \Omega$

F.L. $I = 2000,000/\sqrt{3} \times 2300 = 500$ A

$IR_a = 500 \times 0.06 = 30$ V ;

$IX_S = 500 \times 0.86 = 430$ V

rated voltage/phase $= 2300/\sqrt{3} = 1328$ V

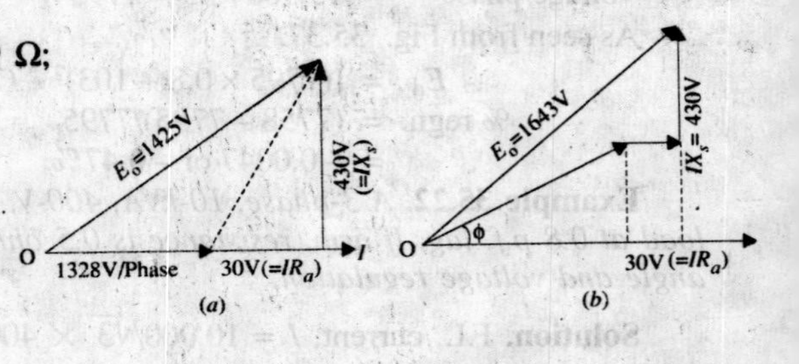

Fig. 35.39

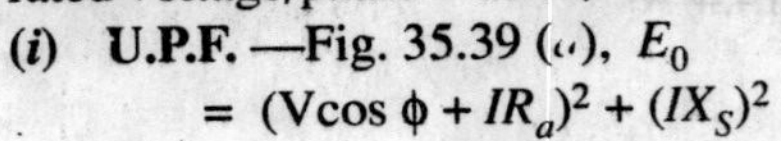

(*i*) **U.P.F.** —Fig. 35.39 (*a*), E_0

$= (V\cos \phi + IR_a)^2 + (IX_S)^2$

$= \sqrt{(1328 + 30)^2 + 430^2} = 1425$ V

% regn. = (1425 – 1328)/1328 = 0.073 or **7.3%**

(*ii*) 0.8 p.f. lagging —Fig. 35.39 (*b*)

$E_0 = [(V \cos\phi + IR_a)^2 + (V \sin\phi + IXs)^2]^{1/2}$

$= [(1328 \times 0.8 + 30)^2 + (1328 \times 0.6 + 430)^2]^{1/2} = 1643$ V

$\therefore$ % regn. = (1643 - 1328)/1328 = 0.237 or **23.7%**

Example 35.25. *A 2000-kVA, 11-kV, 3-phase, star-connected alternator has a resistance of 0.3 ohm and reactance of 5 ohm per phase. It delivers full-load current at 0.8 lagging power factor at rated voltage. Compute the terminal voltage for the same excitation and load current at 0.8 power factor leading.*

(Elect. Machines, Nagpur Univ. 1993)

Solution. (*i*) **At 0.8 p.f. lagging**

F.L. $I = 2000{,}000/\sqrt{3} \times 11{,}000 = 105$ A

Terminal voltage = $11{,}000/\sqrt{3} = 6350$ V

$IR_a = 105 \times 0.3 = 31.5$ V; $IX_s = 105 \times 5 = 525$ V

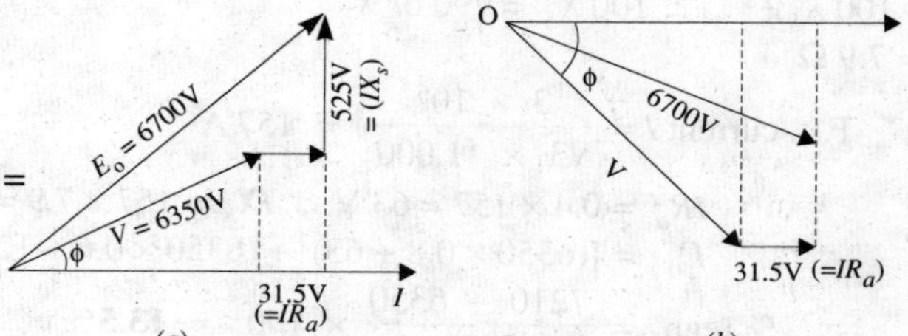

Fig. 35.40

As seen from Fig. 35.40 (*a*)

$E_0 = [6350 \times 0.8 + 31.5)^2 + (6350 \times 0.6 + 525)^2]^{1/2} = 6700$ V

As seen from Fig. 35.40 (*b*), now, we are given $E_0 = 6700$ V and we are required to find the terminal voltage V at 0.8 p.f. leading.

$6700^2 = (0.8V + 31.5)^2 + (0.6V - 525)^2$; V = **6975 V**

Example 35.26. *The effective resistance of a 1200-kVA, 3.3-kV, 50-Hz, 3-phase, Y-connected alternator is 0.25 Ω/phase. A field current of 35 A produces a current of 200 A on short-circuit and 1.1 kV (line to line) on open circuit. Calculate the power angle and p.u. change in magnitude of the terminal voltage when the full load of 1200 kVA at 0.8 p.f. (lag) is thrown off. Draw the corresponding phasor diagram.* **(Elect. Machines, A.M.I.E. Sec. B, 1993)**

Solution. $Z_s = \dfrac{\text{O.C. voltage}}{\text{S.C. voltage}}$ — same excitation

$= \dfrac{1.1 \times 10^3/3}{200} = 3.175\,\Omega$

$X_S = \sqrt{3.175^2 - 0.25^2} = 3.165\,\Omega$

$V = 3.3 \times 10^3/\sqrt{3} = 1905$ V

$\tan\theta = X_s / R_a = 3.165/0.25,\ \theta = 85.48°$

$\therefore\ Z_s = 3.175\ \angle 85.48°$

Rated $I_a = 1200 \times 10^3 / \sqrt{3} \times 3.3 \times 10^3 = 210$ A

Fig. 35.41

Let, $V = 1905\angle 0°$, $I_a = 210/-36.87°$

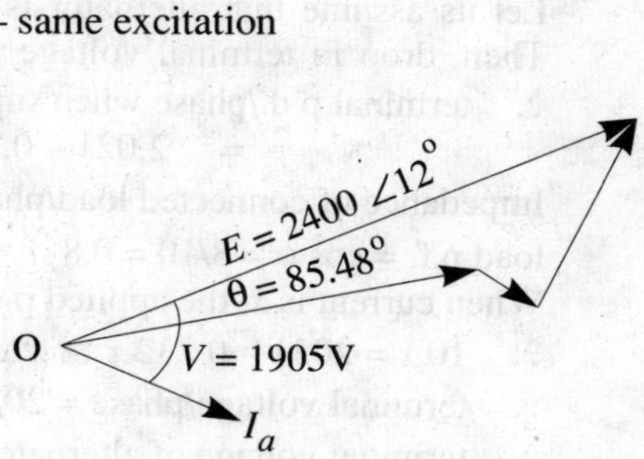

As seen from Fig. 35.41, $E = V + I_a Z_s = 1905 + 210\ \angle{-36.87°} \times 3.175\ \angle 85.48° = 2400\angle 12°$

Per unit change in terminal voltage is

$= (2400 - 1905)/1905 = \mathbf{0.26}$

Example 35.27. *A given 3-MVA, 50-Hz, 11-kV, 3-φ, Y-connected alternator when supplying 100 A at zero p.f. leading has a line-to-lin voltage of 12,370 V; when the load is removed, the terminal voltage falls down to 11,000 V. Predict the regulation of the alternator when supplying* *full-load at 0.8 p.f. lag. Assume an effective resistance of 0.4 Ω per phase.*

(Elect Machines, Nagpur Univ. 1993)

Solution. As seen from Fig. 35.42 (*a*), at zero p.f. leading

$E_0^2 = (V \cos\phi + IR_a)^2 + (V \sin\phi - IX_S)^2$

Now, $E_0 = 11{,}000/\sqrt{3} = 6350$ V

$V = 12{,}370/\sqrt{3} = 7{,}142$ V, $\cos\phi = 0$, $\sin\phi = 1$

$\therefore \quad 6350^2 = (0 + 100 \times 0.4)^2 + (7142 - 100\,X_S)^2 \quad \therefore\ 100\,X_S = 790$ or $X_S = 7.9\ \Omega$

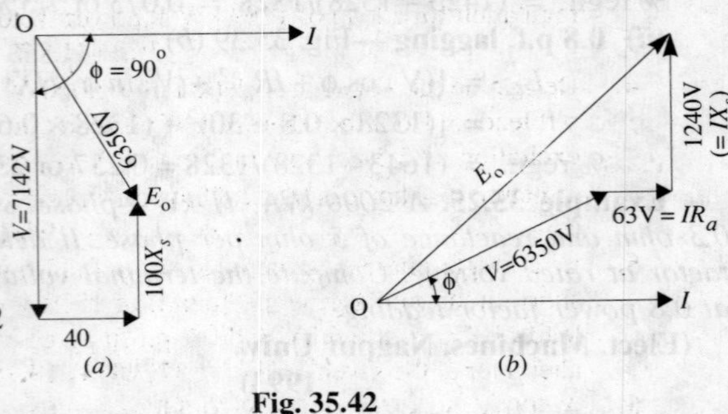

Fig. 35.42

$$\text{F.L. current } I = \frac{3 \times 10^6}{\sqrt{3} \times 11{,}000} = 157 \text{ A}$$

$IR_a = 0.4 \times 157 = 63$ V ; $IX_S = 157 \times 7.9 = 1240$ V

$\therefore \quad E_0 = [(6350 \times 0.8 + 63)^2 + (6350 \times 0.6 + 1240)^2]^{1/2} = 7210$ V/phase

$$\therefore \quad \%\text{ regn} = \frac{7210 - 6350}{6350} \times 100 = \mathbf{13.5\%}$$

Example 35.28. *A straight line law connects terminal voltage and load of a 3-phase star-connected alternator delivering current at 0.8 power factor lagging. At no-load, the terminal voltage is 3,500 V and at full-load of 2,280 kW, it is 3,300 V. Calculate the terminal voltage when delivering current to a 3-φ, star-connected load having a resistance of 8 Ω and a reactance of 6 Ω per phase. Assume constant speed and field excitation.* **(London Univ.)**

Solution. No-load phase voltage = $3{,}500/\sqrt{3} = 2{,}021$ V

Phase voltage on full-load and 0.8 power factor = $3{,}300/\sqrt{3}$ = 1.905 V

Full-load current is given by

$$\sqrt{3} V_L I_L \cos\phi = 2{,}280 \times 1000 \qquad \therefore \quad I_L = \frac{2{,}280 \times 1000}{\sqrt{3} \times 3{,}300 \times 0.8} = 500 \text{ A}$$

drop in terminal voltage/phase for 500 A = 2,021 – 1,905 = 116 V

Let us assume that alternator is supplying a current of x ampere.

Then, drop in terminal voltage per phase for x ampere is = $116\,x / 500 = 0.232\,x$ volt

$\therefore$ terminal p.d./phase when supplying x amperes at a p.f. of 0.8 lagging is

$= 2{,}021 - 0.232\,x$ volt

Impedance of connected load/phase = $\sqrt{(8^2 + 6^2)} = 10\ \Omega$

load p.f. = $\cos\phi = 8/10 = 0.8$

When current is x, the applied p.d. is = $10\,x$

$\therefore \quad 10\,x = 2021 - 0.232\,x$ or $x = 197.5\ A$

$\therefore$ terminal voltage/phase = 2021 – (0.232 × 197.5) = 1975.2 V

$\therefore$ terminal voltage of alternator = $1975.2 \times \sqrt{3}$ = **3,421 V**

Tutorial Problem No. 35.2

1. If a field excitation of 10 A in a certain alternator gives a current of 150 A on short-circuit and a terminal voltage of 900 V on open-circuit, find the internal voltage drop with a load current of 60A. **[360 V]**
2. A 500-V, 50-kVA, 1-φ alternator has an effective resistance of 0.2 Ω. A field current of 10 A produces an armature current of 200 A on short-circuit and an e.m.f. of 450 V on open-circuit. Calculate the full-load regulation at p.f. 0.8 lag **[34.4%]** *(Electrical Technology, Bombay Univ. 1978)*

3. A 3-ϕ star-connected alternator is rated at 1600 kVA, 13,500 V. The armature effective resistance and synchronous reactance are 1.5 Ω and 30 Ω respectively per phase. Calculate the percentage regulation for a load of 1280 kW at power factors of (*a*) 0.8 leading and (*b*) 0.8 lagging.
[(a) –11.8% (b) 18.6%] (*Elect. Engg.-II, Bombay Univ. 1977*)

4. Determine the voltage regulation of a 2,000-V, 1-phase alternator giving a current of 100 A at 0.8 p.f. leading from the test results. Full-load current of 100 A is produced on short-circuit by a field excitation of 2.5 A. An e.m.f. of 500 V is produced on open-circuit by the same excitation. The armature resistance is 0.8 Ω. Draw the vector diagram.
[– 8.9%] (*Electrical Machines-I, Gujarat Univ. Apr. 1976*)

5. In a single-phase alternator, a given field current produces an armature current of 250 A on short-circuit and a generated e.m.f. of 1500 V on open-circuit. Calculate the terminal p.d. when a load of 250 A at 6.6 kV and 0.8 p.f. lagging is switched off. Calculate also the regulation of the alternator at the given load. **[7,898 V; 19.7%]** (*Elect. Machines-II, Indore Univ. Dec. 1977*)

6. A 500-V, 50-kVA, single-phase alternator has an effective resistance of 0.2 Ω. A field current of 10 A produces an armature current of 200 A on short-circuit and e.m.f. of 450 V on open circuit. Calculate (*a*) the synchronous impedance and reactance and (*b*) the full-load regulation with 0.8 p.f. lagging. **[(a) 2.25 Ω, 2.24 Ω, (b) 34.4%]** (*Elect. Technology, Mysore Univ. 1979*)

7. A 100-kVA, 3,000-V, 50-Hz, 3-phase star-connected alternator has effective armature resistance of 0.2 Ω. A field current of 40 A produces short-circuit current of 200 A and an open-circuit e.m.f. of 1040 V (line value). Calculate the full-load percentage regulation at a power factor of 0.8 lagging. How will the regulation be affected if the alternator delivers its full-load output at a power factor of 0.8 leading? **[24.4% - 13.5%]** (*Elect. Machines-II, Indore Univ. July 1977*)

8. A 3-ϕ, 50-Hz, star-connected, 2,000 kVA, 2,300-V alternator gives a short-circuit current of 600 A for a certain field excitation. With the same excitation, the O.C. voltage was 900 V. The resistance between a pair of terminals was 0.12 Ω. Find full-load regulation at (*a*) u.p.f. (*b*) 0.8 p.f.lagging (*c*) 0.8 p.f. leading.
[(*a*) 7.3% (*b*) 23.8% (*c*) –13.2%] (*Elect. Machinery-III, Bangalore Univ. Aug. 1979*)

35.22. Rothert's M.M.F. or Ampere-turn Method

This method also utilizes O.C. and S.C. data, but is the converse of the E.M.F. method in the sense that *armature leakage reactance is treated as an additional armature reaction.* In other words, it is assumed that the change in terminal p.d. on load is due entirely to armature reaction (and due to the ohmic resistance drop which, in most cases, is negligible). This fact is shown in Fig. 35.43.

Now, field A.T. required to produce a voltage of V on *full-load* is the vector sum of the following :

(*i*) Field A.T. required to produce V (or if R_a is to be taken into account, then $V + I\,Ra \cos \phi$) on *no-load.* This can be found from O.C.C. and

(*ii*) Field A.T. required to overcome the demagnetising effect of armature reaction on full-load. This value is found from short-circuit test. The field A.T. required to produce full-load current on short-circuit balances the armature reaction and *the impedance drop*.

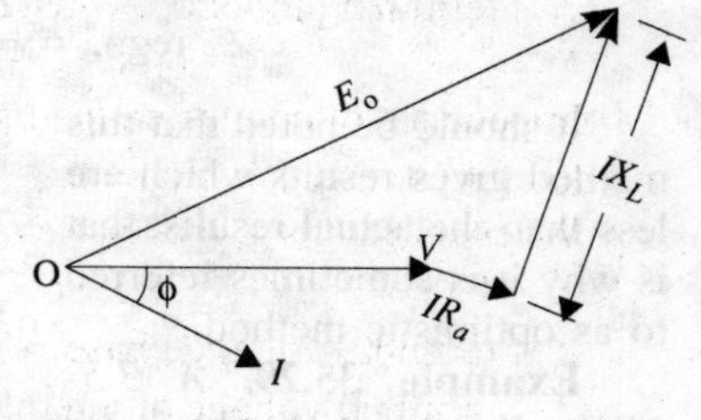

Fig. 35.43

The impedance drop can be neglected because R_a is usually very small and X_S is also small under short-circuit conditions. Hence, p.f. on short-circuit is almost zero lagging and the field A.T. are used entirely to overcome the armature reaction which is wholly demagnetising (Art. 35.15). *In other words, the demagnetising armature A.T. on full-load are equal and opposite to the field A.T. required to produce full-load current on short-circuit.*

Now, if the alternator, instead of being on short-circuit, is supplying full-load current at its normal voltage and zero p.f. lagging, then total field A.T. required are the vector sum of

(*i*) the field A.T. = OA necessary to produce normal voltage (as obtained from O.C.C.) and (*ii*) the field A.T. necessary to neutralize the armature reaction AB_1. The total field A.T. are represented by OB_1 in Fig. 35.44 (*a*) and equals the vector sum of OA and AB_1

If the p.f. is zero leading, the armature reaction is *wholly magnetising*. Hence, in that case, the field A.T. required is OB_2 which is less than OA by the field A.T. = AB_2 required to produce full-load current on short-circuit [Fig. 35.44 (*b*)]

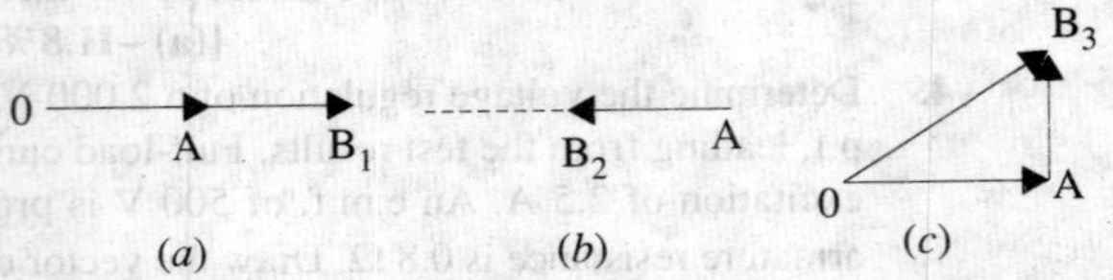

If p.f. is unity, them armature reaction is *cross-magnetising i.e.* its effect is distortional only. Hence, field A.T. required is OB_3 *i.e.* vector sum of OA and AB_3 which is drawn at right angles to OA as in Fig. 35.44 (*c*).

Fig. 35.44

35.23. General Case

Let us consider the general case when the p.f. has any value between zero (lagging or leading) and unity. Field ampere-turns OA corresponding to V(or $V + IR_a \cos \phi$) is laid off horizontally. Then AB_1, representing full-load short-circuit field A.T. is drawn at an angle of (90° + ϕ) for a lagging p.f. The total field A.T. are given by OB_1 as in Fig. 35.45. *(a)*. For a leading *p.f.*, short-circuit A.T. = AB_2 *is* drawn at an angle of *(*90° – ϕ) as shown in Fig. 35.45 (*b*) *and* for unity p.f., AB_3 is drawn at right angles as shown in Fig. 35.45 *(c)*.

In those cases where the number of turns on the field coils is not known, it is usual to work in terms of the field current as shown in Fig. 35.46.

In Fig. 35.47. is shown the complete diagram along with O.C. and S.C. characteristics. OA represents field current for normal voltage V. OC represents field current required for producing

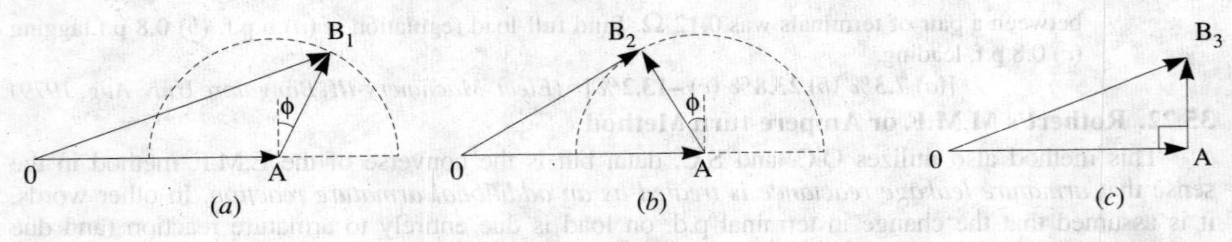

Fig. 35.45

full-load current on *short-circuit*. Vector AB = OC is drawn at an angle of (90° + ϕ) to OA (if the p.f. is lagging). The total field current is OB for which the corresponding O.C. voltage is E_0

$$\therefore \quad \%\ \text{regn.} = \frac{E_0 - V}{V} \times 100$$

It should be noted that this method gives results which are less than the actual results, that is why it is sometimes referred to as optimistic method.

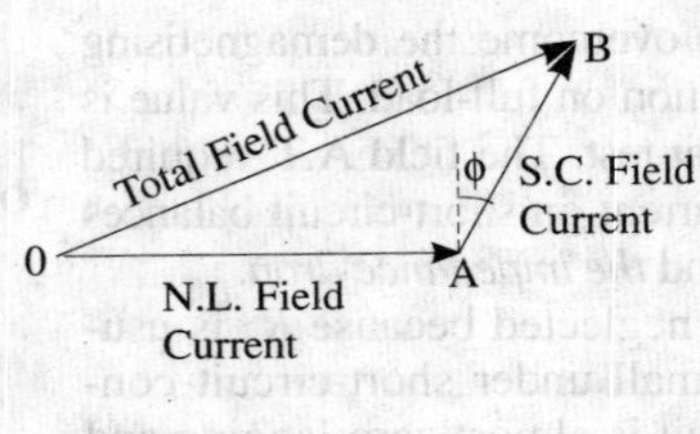

Fig. 35.46

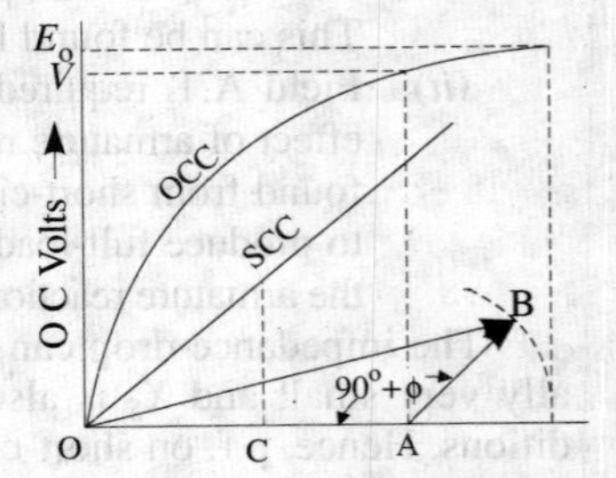

Fig. 35.47

Example 35.29. *A 3.5-MVA, Y-connected alternator rated at 4160 volts at 50-Hz has the open-circuit characteristic given by the following data:-*

Field Current (Amps)	50	100	150	200	250	300	350	400	450
E.M.F. (Volts)	1620	3150	4160	4750	5130	5370	5550	5650	5750

A field current of 200 A is found necessar to circulate full-load current on short-circuit of the alternator. Calculate by (i) synchronous impedance method and (ii) ampere-turn method the full-load voltage regulation at 0.8 p.f. lagging. Neglect resistance. Comment on the results obtained. **(Electrical Machines-II, Indore Univ. 1984)**

Solution. (*i*) As seen from the given data, a field current of 200 A produces *O.C.* voltage of 4750 (line value) and full-load current on short-circuit which is

$$= 3.5 \times 10^6/\sqrt{3} \times 4160 = 486 \text{ A}$$

$$Z_S = \frac{\text{O.C. volt/phase}}{\text{S.C. current/phase}} = \frac{4750/\sqrt{3}}{486} = \frac{2740}{486} = 5.64\ \Omega/\text{phase}$$

Since $R_a = 0,\ X_S = Z_S \quad \therefore\ IR_a = 0,\ IX_S = IZ_S = 486 \times 5.64 = 2740$ V

F.L. Voltage/phase $= 4160/\sqrt{3} = 2400$ V, $\cos\phi = 0.8$, $\sin\phi = 0.6$

$$E_0 = (V\cos\phi + IR_a)^2 + (V\sin\phi + I, X_S)^2]^{1/2}$$

$$= [(2400 \times 0.8 + 0)^2 + (2400 \times 0.6 + 2740)^2]^{1/2} = 4600 \text{ V}$$

$$\%\text{ regn. up} = \frac{4600 - 2400}{2400} \times 100 = \mathbf{92.5\%}$$

(*ii*) It is seen from the given data that for normal voltage of 4160 V, field current needed is 150 A. Field current necessary to circulate F.L. current on short-circuit is 200 A.

In Fig. 35.48, *OA* represents 150 A. The vector *AB* which represents 200 A is vectorially added to *OA* at $(90° + \phi) = (90° + 36°52') = 126°52'$. Vector *OB* represents excitation necessary to produce a terminal p.d. of 4160 V at 0.8 p.f. lagging at *full-load.*

$OB = [150^2 + 200^2 + 2 \times 150 \times 200 \times \cos(180° - 126°, 52')]^{1/2}$
$= 313.8$ A

The generated phase e.m.f. E_0, corresponding to this excitation as found from *OCC* (if drawn) is 3140 V. Line value is $3140 \times \sqrt{3} = 5440$ V.

$$\%\text{ regn.} = \frac{5440 - 4160}{4160} \times 100 = \mathbf{30.7\%}$$

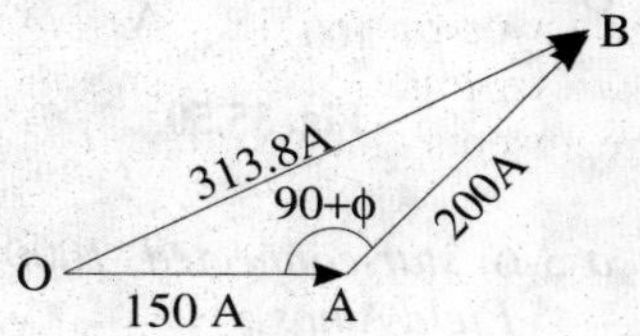

Fig. 35.48

Example 35.30. *The following test results are obtained on a 6,600-V alternator:*

Open-circuit voltage :	*3,100*	*4,900*	*6,600*	*7,500*	*8,300*
Field current (amps) :	*16*	*25*	*37.5*	*50*	*70*

A field current of 20 A is found necessary to circulate full-load current on short-circuit of the armature. Calculate by (i) the ampere-turn method and (ii) the synchronous impedance method the full-load regulation at 0.8 p.f. (lag). Neglect resistance and leakage reactance. State the drawbacks of each of these methods.

(Elect. Machinery-II, Bangalore Univ. 1992)

Solution. (*i*) Ampere-turn Method

It is seen from the given data that for the normal voltage of 6,600 V, the field current needed is 37.5 A.

Field-current for full-load current, on short-circuit, is given as 20 A.

In Fig. 35.49, *OA* represents 37.5 A. The vector *AB*, which represents 20 A, is vectorially added to *OA* at $(90° + 36°52') = 126°52'$. Vector *OB* represents the excitation necessary to produce a terminal p.d. of 6,600 V at 0.8 p.f. lagging on *full-load*

$$OB = \sqrt{37.5^2 + 20^2 + 2 \times 3.75 \times 20 \times \cos 53°8'} = 52 \text{ A}$$

The generated e.m.f. E_0 corresponding to this excitation, as found from O.C.C. of Fig. 35.49 is 7,600V.

$$\text{Percentage regulation} = \frac{E_o - V}{V} \times 100 = \frac{7{,}600 - 6{,}600}{6{,}600} \times 100 = \mathbf{15.16\%}$$

(*ii*) Synchronous Impedance Method

Let the voltage of 6,600 V be taken as 100 per cent and also let 100 per cent excitation be that which is required to produce 6,600 V on open-circuit, that is, the excitation of 37.5 A.

Full-load or 100 per cent armature current is produced on short-circuit by a field current of 20 A. If 100 per cent field current were applied on short-circuit, then S.C. current would be 100 × 37.5/20 = 187.5 per cent.

$$\therefore \quad Z_S = \frac{\text{O.C. voltage}}{\text{S.C. current}} \Bigg| \text{same excitation}$$

$$= 100/187.5 \text{ or } 0.533 \text{ or } 53.3\%$$

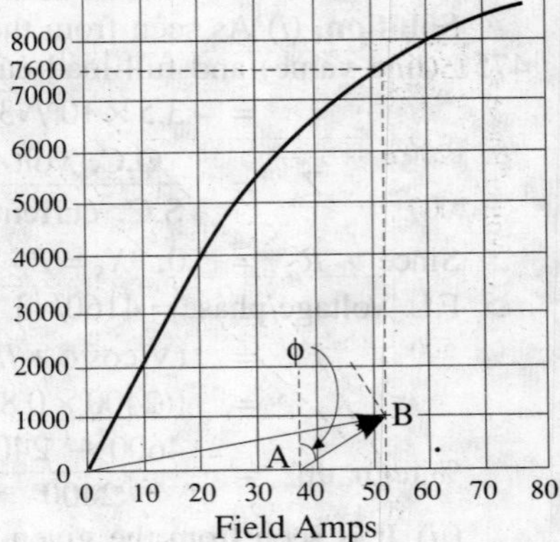

Fig. 35.49

The impedance drop IZ_S is equal to 53.3% of the normal voltage. When the two are added vectorially (Fig. 35.50), the value of voltage is

$$E_0 = \sqrt{[100 + 53.3 \cos(90 - \phi)]^2 + [53.3 \sin(90 - \phi)]^2}$$

$$= \sqrt{(100 + 53.3 \times 0.6)^2 + (53.3 \times 0.8)^2} = 138.7\%$$

$$\% \text{ regn.} = \frac{138.7 - 100}{100} \times 100 = \mathbf{38.7\%}$$

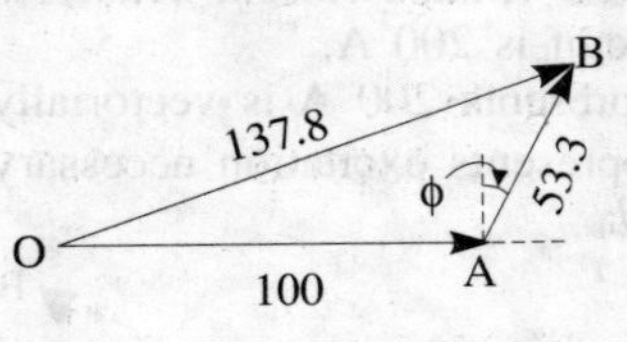

Fig. 35.50

∴ The two value of regulation, found by the two methods, are found to differ widely from each other. The first method gives somewhat lesser value, while the other method gives a little higher value as compared to the actual value. However, the first value is more likely to be nearer the actual value, because the second method employs Z_S, which does not have a constant value. Its value depends on the field excitation.

Example 35.31. *The open-and short-circuit test readings for a 3-φ, star-connected, 1000-kVA, 2000 V, 50-Hz, synchronous generator are :*

Field Amps ;	*10*	*20*	*25*	*30*	*40*	*50*
O.C. Terminal V	*800*	*1500*	*1760*	*2000*	*2350*	*2600*
S.C. armature current in A:	—	*200*	*250*	*300*	—	—

The armature effective resistance is 0.2 Ω per phase. Draw the characteristic curves and estimate the full-load percentage regulation at (a) 0.8 p.f. lagging (b) 0.8 p.f. leading.

Solution. The O.C.C. and S.C.C. are plotted in Fig. 35.51

The phase voltages are : 462, 866, 1016, 1155, 1357, 1502.

Full-load phase voltage = $2000/\sqrt{3}$ = 1155 V

Full-load current = $1{,}000{,}000/2000 \times \sqrt{3}$ = 288.7 A

Voltage/phase at full-load at 0.8 p.f. = $V + IR_a \cos\phi = 1155 + (288.7 \times 0.2 \times 0.8) = 1200$ volt

Form open-circuit curve, it is found that field current necessary to produce this voltage = 32 A.

From short-circuit characteristic, it is found that field current necessary to produce full-load current of 288.7 A is = 29 A.

(*a*) $\cos\phi = 0.8, \phi = 36°52'$ (lagging)

In Fig. 35.52, OA = 32 A, *AB* = 29 A and is at an angle of (90° + 36°52′) = 126°52′ with *OA*. The total field current at full-load 0.8 p.f. lagging is *OB* = 54.6 A

O.C. volt corresponding to a field current of 54.6 A is = 1555 V

% regn. = (1555 – 1155) × 100/1155 = **34.6%**

(*b*) In this case, as p.f. is leading, *AB* is drawn with *OA* (Fig. 35.53) at an angle of 90° – 36° 52′ = 53°8′. *OB* = 27.4 A.

O.C. voltage corresponding to 27.4A of field excitation is 1080 V.

$$\% \text{ regn.} = \frac{1080 - 1155}{1155} \times 100 = \mathbf{-6.4\%}$$

Example 35.32. *A 3-phase, 800-kV A, 3,300-V, 50-Hz alternator gave the following results:*

Exciting current (A)	*50*	*60*	*70*	*80*	*90*	*100*

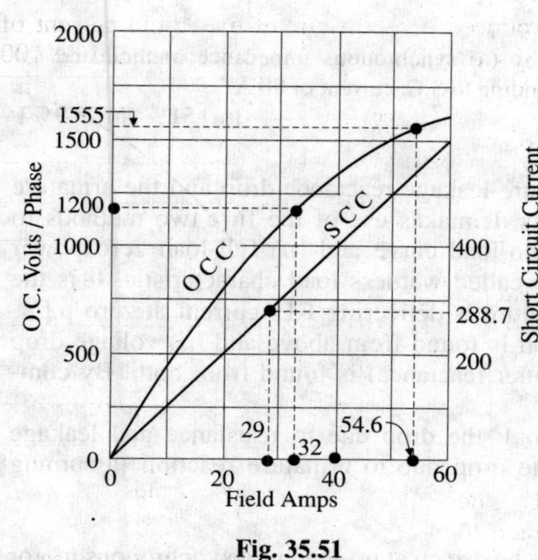

Fig. 35.51

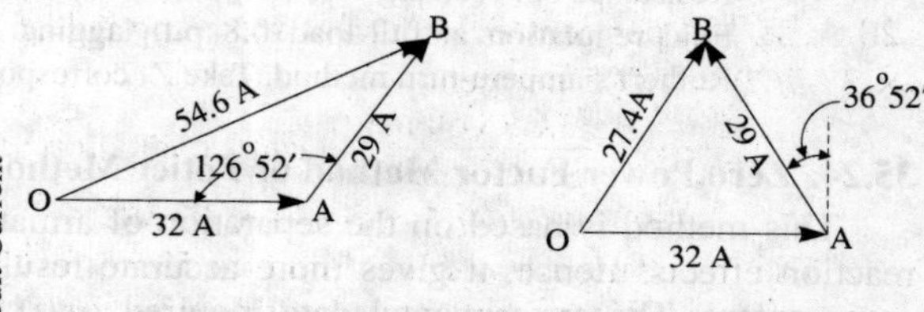

Fig. 35.52 Fig. 35.53

O.C. volt (line)	*2560*	*3000*	*3300*	*3600*	*3800*	*3960*
S.C current	*190*					—

The armature leakage reactance drop is 10% and the resistance drop is 2% of the normal voltage. Determine the excitation at full-load 0.8 power factor lagging by the m.m.f. method.

Solution. The phase voltages are : 1478, 1732, 1905, 2080, 2195, 2287

The O.C.C. is drawn in Fig. 35.54.

Normal phase voltage $= 3300/\sqrt{3} = 1905$ V ;

$I.R_a$ drop = 2% of 1905 = 38.1 volt

Leakage reactance drop = 10% of 1905 = 190.5 Volt

$$\therefore \quad E = \sqrt{[(1905 \times 0.8 + 38.1)^2 + (1905 \times 0.6 + 190.5)^2]} = 2{,}068 \text{ V}$$

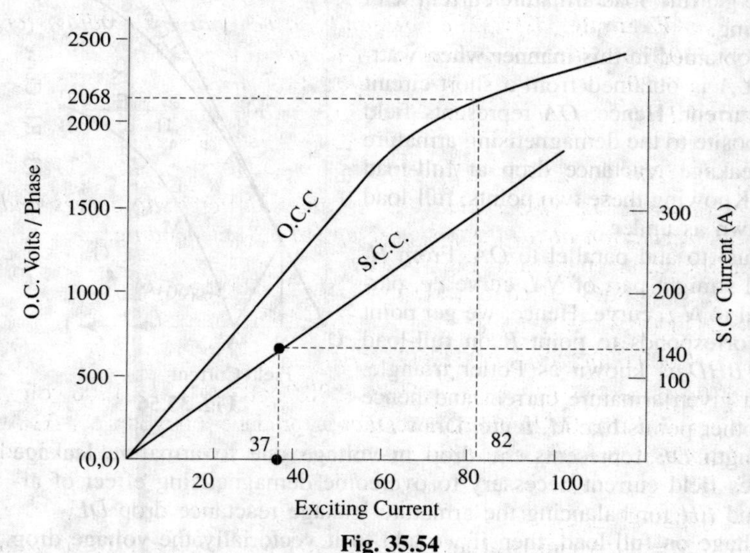

Fig. 35.54

The exciting current required to produce this voltage (as found from O.C.C.) is 82 A.

Full load current = $800.000/\sqrt{3} \times 3300 = 140$A

As seen from S.C.C., the exciting current required to produce this full-load current of 140 A on short-circuit is 32 A.

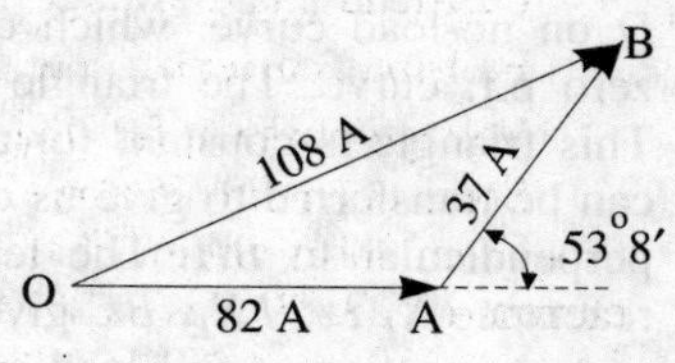

Fig. 35.55

In Fig. 35.55, *OB* gives the excitation required on full-load to give a terminal phase voltage of 1905 V (or line voltage of 3300 V) at 0.8 p.f. lagging and its value is

$$= \sqrt{82^2 + 37^2 + 2 \times 82 \times 37 \times \cos 53°8'} = \mathbf{108\ A}$$

Tutorial Problem No. 35.3

1. A 30-kVA, 440-V, 50-Hz, 3-ϕ, star-connected synchronous generator gave the following test data:

Field current (A) :	2	4	6	7	8	10	12	14
Terminal volts :	155	287	395	440	475	530	570	592
S.C. current :	11	22	34	40	46	57	69	80

Resistance between any two terminals is 0.3 Ω

Find regulation at full-load 0.8 p.f. lagging by (a) synchronous impedance method and (b) Rothert's ampere-turn method. Take Z_s corresponding to S.C. current of 80 A.

[(a) **51%** (b) **29.9%**]

35.24. Zero Power Factor Method or Potier Method

This method is based on the separation of armature-leakage reactance drop and the armature reaction effects. Hence, it gives more accurate results. It makes use of the first two methods to some extent. The experimental data required is (i) no-load curve and (ii) full-load zero power factor curve (not the short-circuit characteristic) also called wattless load characteristic. It is the curve of *terminal* volts against excitation when armature is delivering F.L. current at zero p.f.

The reduction in voltage due to armature reaction is found from above and (ii) voltage drop due to armature leakage reactance X_L (also called Potier reactance) is found from both. By combining these two, E_0 can be calculated.

It should be noted that if we vectorially add to V the drop due to resistance and leakage reactance X_L, we get E. If to E is further added the drop due to armature reaction (assuming lagging p.f.), then we get E_0 (Art. 35.18).

The zero p.f. lagging curve can be obtained.

(a) if a similar machine is available which may be driven at no-load as a synchronous motor at practically zero p.f. or

(b) by loading the alternator with pure reactors

(c) by connecting the alternator to a 3-ϕ line with ammeters and wattmeters connected for measuring current and power and by so adjusting the field current that we get full- load armature current with zero wattmeter reading.

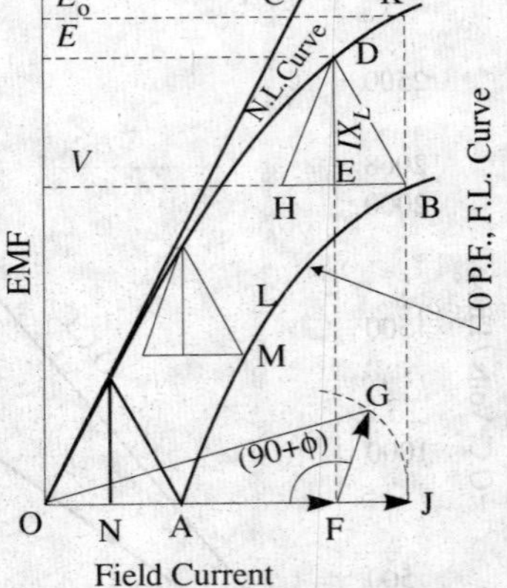

Fig. 35.56

Point B (Fig. 35.56) was obtained in this manner when wattmeter was reading zero. Point A is obtained from a short-circuit test with full-load armature current. Hence, OA represents field current which is equal and opposite to the demagnetising armature reaction and for balancing leakage reactance drop at full-load (please refer to A.T. method). Knowing these two points, full-load zero p.f. curve AB can be drawn as under.

From B, BH is drawn equal to and parallel to OA. From H, HD is drawn parallel to initial straight part of N-L curve *i.e.* parallel to OC, which is tangential to N-L curve. Hence, we get point D on no-load curve, which corresponds to point B on full-load zero p.f. curve. The triangle BHD is known as Potier triangle. This triangle is constant for a given armature current and hence can be transferred to give us other points like M, L etc. Draw DE perpendicular to BH. The length DE represents the drop in voltage due to armature leakage reactance X_L *i.e.* $I.X_L$. BE gives field current necessary to overcome demagnetising effect of armature reaction at full-load and EH for balancing the armature leakage reactance drop DE.

Let V be the terminal voltage on full-load, then if we add to it vectorially the voltage drop due to armature leakage reactance alone (neglecting R_a), then we get voltage $E = DF$ (and not E_0). Obviously, field excitation corresponding to E is given by OF. NA (= BE) represents the field current needed to overcome armature reaction. Hence, if we add NA vectorially to OF (as in Rothert's A.T. method) we get excitation for E_0 whose value can be read from N-L curve.

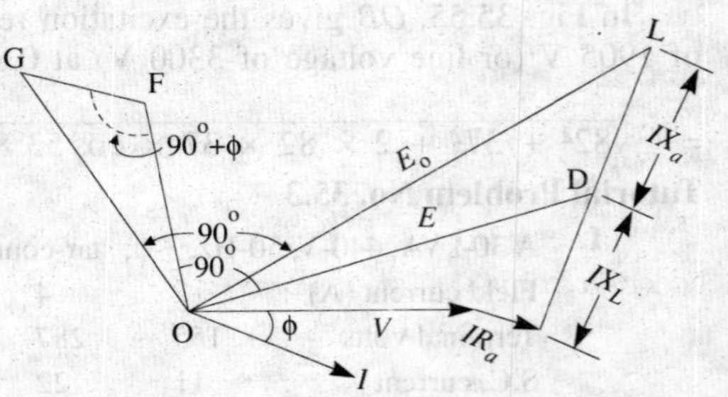

Fig. 35.57

In Fig. 35.56, FG (= NA) is drawn at an angle of (90° + ϕ) for a lagging p.f. (or it is drawn at an angle of 90° – ϕ for a leading p.f.). The voltage corresponding to this excitation is $JK = E_0$

$$\therefore \quad \% \text{ regn.} = \frac{E_0 - V}{V} \times 100$$

The vector diagram is also shown separately in Fig. 35.57.

Assuming a lagging p.f. with angle ϕ, vector for I is drawn at an angle of ϕ to V. IR_a is drawn parallel to current vector and IX_L is drawn perpendicular to it. OD represents voltage E. The excitation corresponding to it *i.e.*. OF is drawn at 90° ahead of it. FG (= NA = BE in Fig. 35.56) representing field current equivalent of full-load armature reaction, is drawn parallel to current vector OI. The closing side OG gives field excitation for E_0. Vector for E_0 *is 90°* lagging behind OG. DL represents voltage drop due to armature reaction.

35.25. Procedural Steps for Potier Method

1. Suppose we are given V-the terminal voltage/phase.
2. We will be given or else we can calculate armature leakage reactance X_L and hence can calculate IX_L.
3. Adding IX_L (and IR_a if given) vectorially to V, we get voltage E.
4. We will next find from N-L curve, field excitation for voltage E. Let it be i_{f1}.
5. Further, field current i_{f2} necessary for balancing armature reaction is found from Potier triangle.
6. Combine i_{f1} and i_{f2} vertorially (as in A.T. method) to get i_f.
7. Read from N-L curve, the e.m.f. corresponding to i_f. This gives us E_0. Hence, regulation can be found.

Example 35.33. *A 3-phase, 6,00-V alternator has the following O.C.C. at normal speed :*

Field amperes :	*14*	*18*	*23*	*30*	*43*
Terminal volts:	*4000*	*5000*	*6000*	*7000*	*8000*

With armature short-circuited and full-load current flowing the field current is 17 A and when the machine is supplying full-load of 2,000 kVA at zero power factor, the field current is 42.5 A and the terminal voltage is 6,000 V.

Determine the field current required when the machine is supplying the full-load at 0.8 p.f. lagging. **(A.C. Machines-I, Jadavpur Univ. 1988)**

Solution. The O.C.C. is drawn in Fig. 35.58 with phase voltages which are

2310, 2828, 3465 4042 4620

The full-load zero p.f. characteristic can be drawn because two points are known *i.e.* (17, 0) and (42.5, 3465).

In the Potier ΔBDH, line DE represents the leakage reactance drop (= IX_L) and is (by measurement) equal to 450 V. As seen from Fig. 35.59.

$$E = \sqrt{(V \cos \phi)^2 + (V \sin \phi + IX_L)^2}$$

$$= \sqrt{(3465 \times 0.8)^2 + (3465 \times 0.6 + 450)^2}$$

$$= 3750 \text{ V}$$

From O.C.C. of Fig. 35.58, it is found that field amperes required for this voltage = 26.5 A.

Field amperes required for balancing armature reaction = BE = 14.5 A (by measurement from Potier triangle BDH).

As seen from Fig. 35.60, the field currents are added vectorially at an angle of (90° + ϕ) = 126° 52′.

Resultant field current is

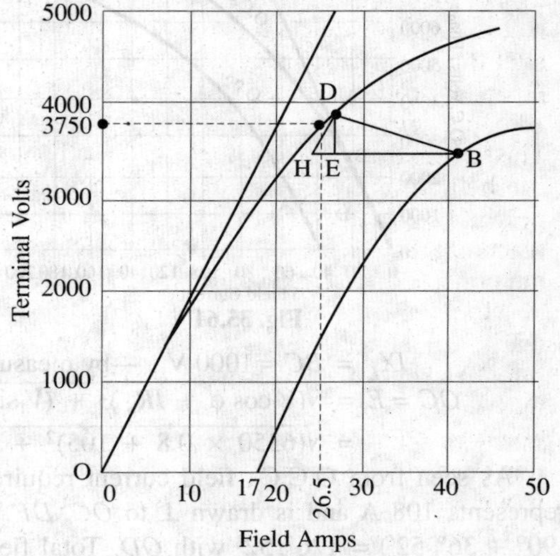

Fig. 35.58

$$OB = \sqrt{26.5^2 + 14.5^2 + 2 \times 26.5 \times 14.4 \cos 53°8'} = \mathbf{37.2 \text{ A}}$$

Example 35.34. *An 11-kV, 1000-kVA, 3-phase, Y-connected alternator has a resistance of 2 Ω per phase. The open-circuit and full-load zero power factor characteristics are given below. Find the voltage regulation of the alternator for full load current at 0.8 p.f.lagging by Potier method.*

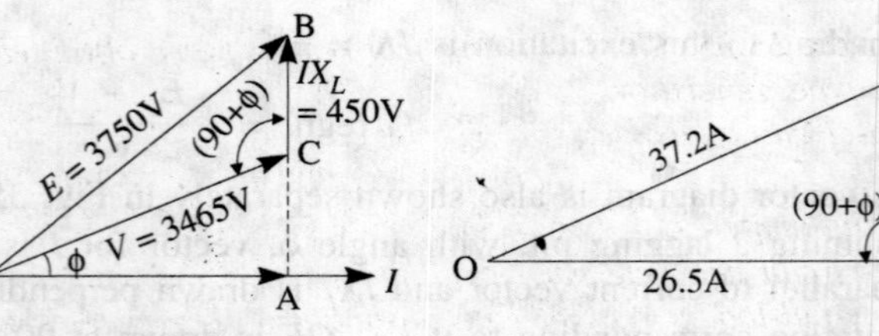

Fig. 35.59 Fig. 35.60

Field current (A):	*40*	*50*	*110*	*140*	*180*
O.C.C. line voltage:	*5,800*	*7,000*	*12,500*	*13,750*	*15,000*
Line volts zero p.f.	*0*	*1500*	*8500*	*10,500*	*12,500*

(Electrical Machines-II, Calcutta Univ. 1987)

Solution. The O.C.C. and full-load zero p.f. curve for phase voltage are drawn in Fig. 35.61. The corresponding phase voltages are :

O.C.C. phase voltage	3350	4040	7220	7940	8660
Phase voltage zero p.f.	0	866	4900	6060	7220

Full-load current $= 1000 \times 1000/\sqrt{3} \times 11,000 \;=\; 52.5$ A

Phase voltage $= 11,000/\sqrt{3} = 6,350$ A

In the Potier Δ *ABC*, *AC = 40 A*, *CB* is parallel to the tangent to the initial portion of the *O.C.C.* and *BD* is ⊥ to *AC*.

BD = leakage reactance drop IX_L = 1000 V – by measurement

AD = 30 A — field current required to overcome demagnetising effect of armature reaction on *full-load*.

As shown in Fig. 35.62,

$OA = 6,350$ V; $AB = IR_a = 52.5 \times 2 = 105$ V

Fig. 35.61

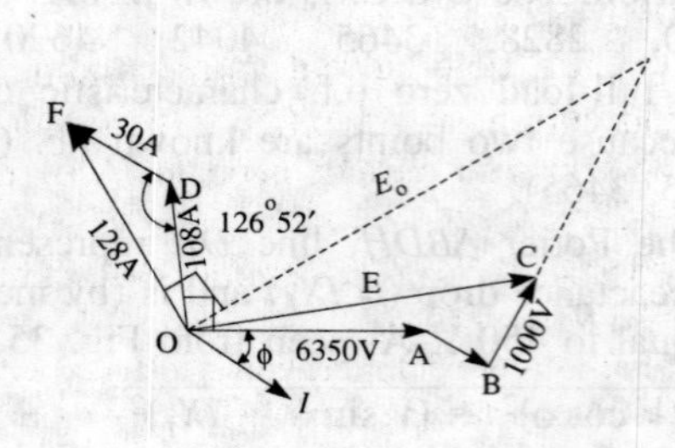

Fig. 35.62

$IX_L = BC = 1000$ V —by measurement

$$OC = E = \sqrt{(V\cos\phi + IR_a)^2 + (V\sin\phi + IX_L)^2}$$

$$= \sqrt{(6350 \times 0.8 + 105)^2 + (6350 \times 0.6 + 1000)^2}\;;\; I. = 7,080 \text{ V}$$

As seen from O.C.C., field current required for 7,080 V is 108 A. Vector *OD* (Fig. 35.62) represents 108 A and is drawn ⊥ to *OC*. *DF* represents 30 A and is drawn parallel to *OI* or at (90° + 36° 52′) = 126° 52′ with *OD*. Total field current is *OF*.

$$OF = \sqrt{108^2 + 30^2 + 2 \times 108 \times 30 \cos 53° 8'} = 128 \text{ A}$$

From O.C.C., it is found that the e.m.f. corresponding to this field current is 7,700V

$$\therefore \quad E_0 = 7,700 \text{ V}\;;\; \text{regulation} = \frac{7,700 - 6,350}{6,350} \times 100 = \textbf{21.3 per cent}$$

Example 35.35. *The following test results were obtained on a 275-kW, 3-φ, 6,600-V non-salient pole type generator.*

Open-circuit characteristic :

Volts :	5600	6600	7240	8100
Exciting amperes:	46.5	58	67.5	96

Short-circuit characteristic : Stator current 35 A with an exciting current of 50 A. Leakage reactance on full-load = 8%. Neglect armature resistance. Calculate as accurately as possible the exciting current (for full-load) at power factor 0.8 lagging and at unity. **(City & Guilds, London)**

Solution. First convert the O.C. line volts into phase volts by dividing the given terminal values by √3.

∴ O.C. volts (phase) : 3233, 3810, 4180, 4677.

O.C.C. is plotted in Fig. 35.63. For plotting S.C.C., we need two points. One is (0, 0) and the other is (50 A, 35 A). In fact, we can do without plotting the S.C.C. because it being a straight line, values of field currents corresponding to any armature current can be found by direct ratio.

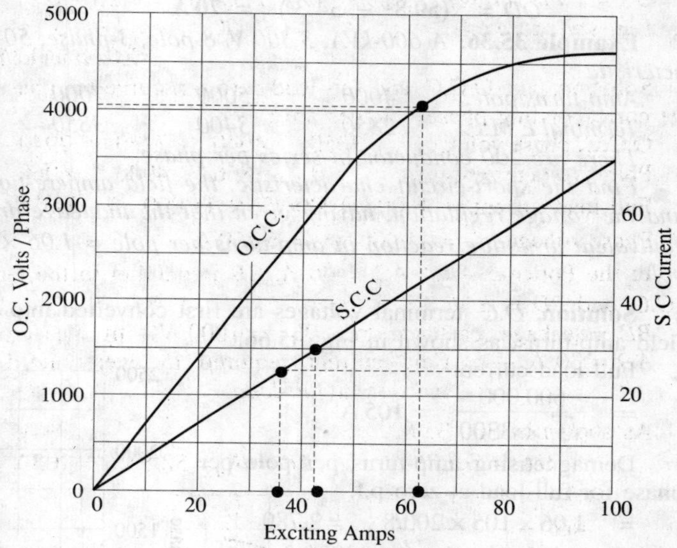

Fig. 35.63

Leakage reactance drop

$$= \frac{3810 \times 8}{100} = 304.8 \text{ V}$$

Normal phase voltage

$= 6{,}600/\sqrt{3} \quad = 3{,}810$ V

In Fig. 35.64, OA = 3810 V and at an angle φ ahead of current vector OI.

AB = 304.8 V is drawn at right angles to OI. Resultant of the two is OB = 4010 V.

From O.C.C., field current corresponding to 4,010 V is 62 A.

Full-load current at 0.8 p.f. = $275{,}000/V_3 \times 6600 \times 0.8 = 30$ A

35 A of armature current need 50 A of field current, hence 30 A of armature current need 30 × 50/35 = 43 A.

In Fig. 35.64, OC = 62 A is drawn at right angles to OB. Vector CD = 43 A is drawn parallel to OI. Then, OD = **94.3 A**

Note. Here. 43 A pf field excitation is assumed as having all been used for balancing armature reaction. In fact, a part of it is used for balancing armature leakage drop of 304.8 V. This fact has been clarified in the next example.

At Unity p.f.

In Fig. 35.65, OA again represents V = 3810 V, AB = 304.8 V and at right angles to OA.

The resultant $OB = \sqrt{(3810^2 + 304.8^2)} \quad = 3830$ V

Field current from O.C.C. corresponding to this voltage = 59.8 A.

Hence, OC = 59.8 A is drawn perpendicular to OB (as before)

Full-load current at u.p.f. = $275{,}000/\sqrt{3} \times 6600 \times 1 = 24$ A

Now, 35 A armature current corresponds to a field current of 50 A, hence 24 A of armature current corresponds to 50 × 24/35 = 34.3 A.

Hence, CD = 34.3 A is drawn ∥ to OA (and ⊥ to OC approximately).*

*It is so because angle between OA and OB is negligibly small. If not, then CD should be drawn at an angle of (90 + α) where α is the angle between OA and OB.

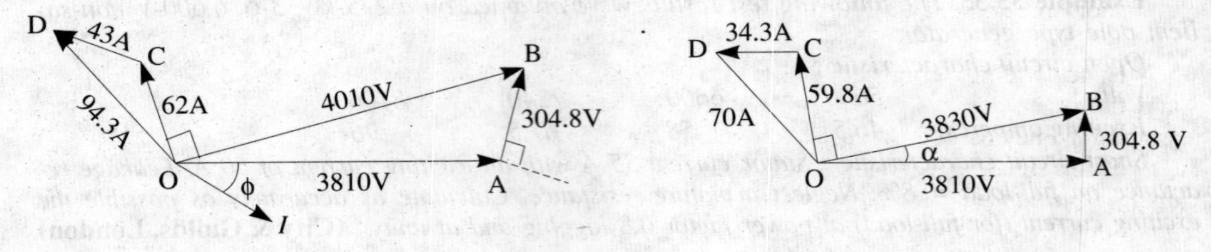

Fig. 35.64

Fig. 35.65

$$\therefore \quad OD = \sqrt{(59.8^2 + 34.3^2)} = \mathbf{70\ A}$$

Example 35.36. *A 600-kVA, 3,300-V, 8-pole, 3-phase, 50-Hz alternator has following characteristic :—*

Amp-turns/pole:	*4000*	*5000*	*7000*	*10,000*
Terminal E.M.F.:	*2850*	*3400*	*3850*	*4400*

There are 200 conductor in series per phase.

Find the short-circuit characteristic, the field ampere-turns for full-load 0.8 p.f. (lagging) and the voltage regulation, having given that the inductive drop at full-load is 7% and that the equivalent armature reaction in amp-turns per pole = 1.06 × ampere-conductors per phase per pole. **(London Univ.)**

Solution. *O.C.* terminal voltages are first converted into phase voltages and plotted against field amp-turns, as shown in Fig. 35.66.

Full-load current

$$= \frac{600,000}{\sqrt{3} \times 3300} = 105\ \text{A}$$

Demagnetising amp-turns per pole per phase for full-load at zero p.f.

$$= 1.06 \times 105 \times 200/8 = 2,780$$

Normal phase voltage $= 3300/\sqrt{3}$
$= 1910$ volt

Leakage reactance drop

$$= \frac{3300 \times 7}{\sqrt{3} \times 100} = 133\ \text{V}$$

In Fig. 35.67, *OA* represents 1910 V. $AB = 133$ V is drawn $\perp$ *OI*, *OB* is the resultant voltage *E* (not E_0).

$\therefore$ $OB = E = 1987$ *volt*

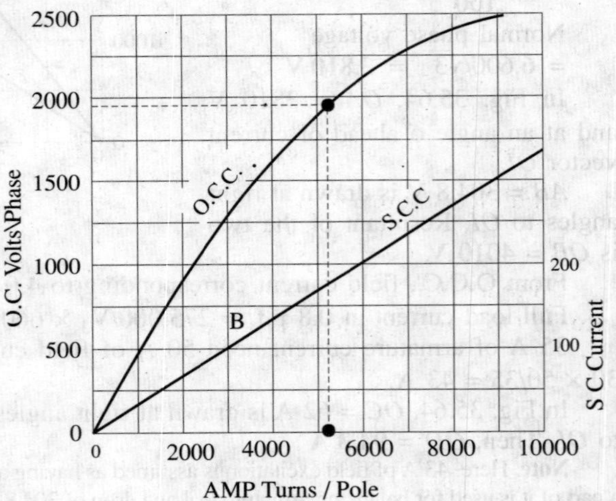

Fig. 35.66

From O.C.C., we find that 1987 V correspond to 5100 field amp-turns. Hence, *OC* = 5100 is drawn $\perp$ to *OB*. *CD* = 2780 is $\parallel$ to *OI*. Hence, *OD* = 7240 (approx). From O.C.C. it is found that this corresponds to an O.C. voltage of 2242 volt. Hence, when load is thrown off, the voltage will rise to 2242 V.

$$\therefore \ \% \text{ regn.} = \frac{2242 - 1910}{1910} \times 100 = \mathbf{17.6\%}$$

How to deduce S.C.C.?

We have found that field amp-turns for balancing armature reaction only are 2,780. To this should be added field amp-turns required for balancing the leakage reactance voltage drop of 133 V.

Field amp-turns corresponding to 133 volt on O.C. are 300 approximately. Hence, with reference to Fig. 35.56, *NA* = 2780, *ON* = 300

∴ short-circuit field amp-turns = OA = 2780 + 300
= 3080

Hence, we get a point B on S.C.C. *i.e.* (3080, 105) and the other point is the origin. So S.C.C. (which is a straight line) can be drawn as shown in Fig. 35.66.

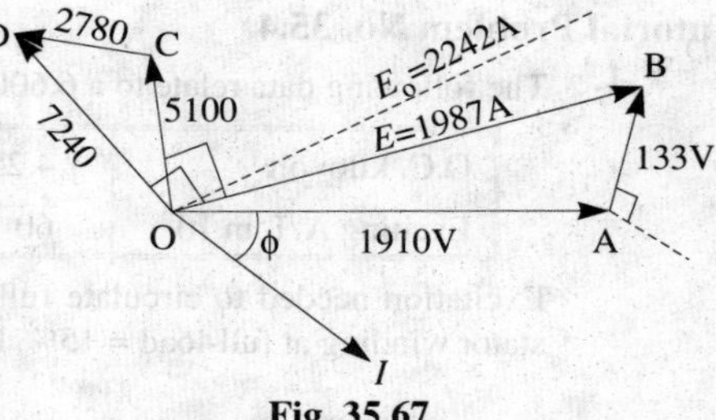

Fig. 35.67

Example 35.37. *The following figures give the open-circuit and full-load zero p.f saturation curves for a 15,000-kVA. 11,000 V, 3-φ, 50-Hz, star-connected turbo-alternator:*

Field AT in 10^3:	*10*	*18*	*24*	*30*	*40*	*45*	*50*
O.C. line kV:	*4.9*	*8.4*	*10.1*	*11.5*	*12.8*	*13.3*	*13.65*
Zero p.f. full-load line kV:	—	*0*	—	—	—	*10.2*	—

Find the armature reaction, the armature reactance and the synchronous reactance. Deduce the regulation for full-load at 0.8 power lagging.

Solution. First, O.C.C. is drawn between phase voltages and field amp-turns, as shown in Fig. 35.68.

Full-load, zero p.f. line can be drawn, because two points are known *i.e.* A (18, 0) and C (45, 5890). Other points on this curve can be found by transferring the Potier triangle. At point C, draw CD ∥ to and equal to OA and from D draw DE ∥ to ON. Join EC. Hence, CDE is the Potier triangle.

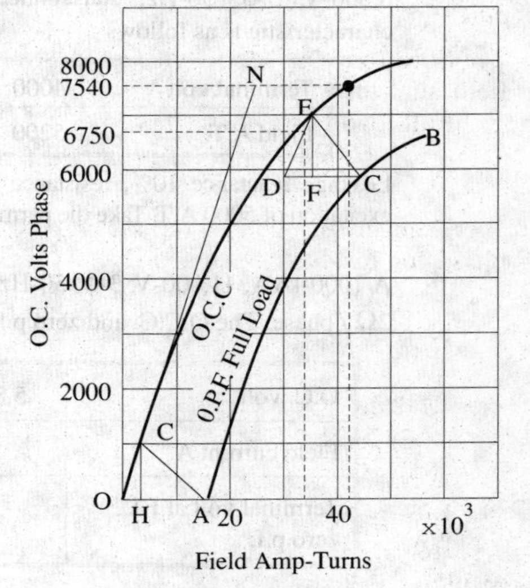

Fig. 35.68

Line EF is ⊥ to DC

CF = field amp-turns for balancing armature reaction only = 15,700

EF = GH = 640 volt

= leakage reactance drop / phase

Short-circuit A.T. required = OA = 18,000

$$\text{Full-load current} = \frac{15,000 \times 1000}{\sqrt{3} \times 11,000} = 788 \text{ A}$$

$\therefore\ 640 = I \times X_L \quad \therefore X_L = 640 / 788 = 0.812\ \Omega$

From O.C.C., we find that 18,000 A.T. correspond to an O.C. voltage of 8,400 / $\sqrt{3}$ = 4,850 V.

$$\therefore Z_S = \frac{\text{O.C. volt}}{\text{S.C. current}} = \frac{4,850}{788}$$

$$= 6.16\ \Omega \qquad \text{(Art. 35.21)}$$

As R_a is negligible, hence Z_S equals X_S.

Regulation

In Fig. 35.69, OA = phase voltage = 11,000/$\sqrt{3}$ = 6,350 V

AB = 640 V and is drawn at right angles to OI or at (90° + φ) to OA.

Resultant is OB = 6,750 V

Field A.T. coresponding to O.C voltage of 6,750 V is = OC = 30,800 and is drawn ⊥ to OB.

CD = armature reaction at F.L. = 15,700 and is drawn ∥ to OI or at (90° + φ) to OC.

Hence, OD = 42,800.

From O.C.C., e.m.f. corresponding to 42,800 A.T. of rotor = 7,540 V

∴ % regn. up = (7,540 – 6,350)/6,350 = **0.187 or 18.7%**

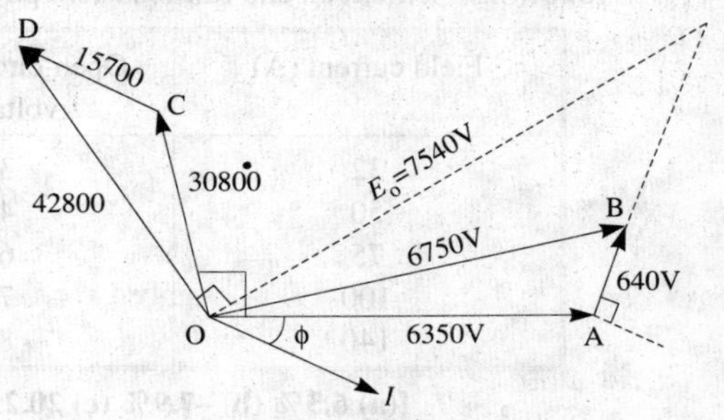

Fig. 35.69

Tutorial Problem No. 35.4

1. The following data relate to a 6,600-V, 10,000-kVA, 50-Hz, 3-ϕ, turbo- alternator:

O.C. kilovolt	4.25	5.45	6.6	7.3	8	9
Exciting A.T. in 10^3	60	80	100	120	145	220

Excitation needed to circulate full-load current on short circuit : 117,000 A.T. Inductive drop in stator winding at full-load = 15%. Find the voltage regulation at full-load 0.8 power factor.

[34.4%] *(City & Guilds, London)*

2. Deduce the exciting current for a 3-ϕ, 3300-V generator when supplying 100 kW at 0.8 power factor lagging, given magnetisation curve on open-circuit :

Line voltage :	3300	3600	3900
Exciting current :	80	96	118

There are 16 poles, 144 slots, 5 conductors/slot, single-circuit, full-pitched winding, star-connected. The stator winding has a resistance per phase of 0.15 Ω and a leakage reactance of 1.2 Ω. The field coils have each 108 turns. **[124 A]** *(London Univ.)*

3. Estimate the percentage regulation at full-load and power factor 0.8 lagging of a 1000-kVA, 6,600-V, 3-ϕ, 50-Hz, star-connected salient-pole synchronous generator. The open-circuit characteristic is as follows :

Terminal volt	4000	6000	6600	7200	8000
Field A.T.	5200	8500	10,000	12,500	17,500

Leakage reactance 10%, resistance 2%. Short-circuit characteristic : full-load current with a field excitation of 5000 A.T. Take the permeance to cross armature reaction as 35% of that to direct reaction.

[20% up]

4. A 1000-kVA, 11,000-V, 3-ϕ, 50-Hz, star-connected turbo-generator has an effective resistance of 2Ω / phase. The O.C.C. and zero p.f. full-load data is as follows :

O.C. volt	5,805	7,000	12,550	13,755	15,000
Field current A	40	50	110	140	180
Terminal volt at F.L. zero p.f.	0	1500	8,500	10,500	12,400

Estimate the % regulation for F.L. at 0.8 p.f. lagging. **[22 %]**

5. A 5-MVA, 6.6 kV, 3-ϕ, star-connected alternator has a resistance of 0.075 Ω per phase. Estimate the regulation for a load of 500 A at p.f. *(a)* unity and *(b)* 0.9 leading *(c)* 0.71 lagging from the following open-circuit and full-load zero power factor curve.

Field current (A)	Open-circuit terminal voltage (V)	Saturation curve zero p.f.
32	3100	0
50	4900	1850
75	6600	4250
100	7500	5800
140	8300	7000

[(a) 6.3% (b) –7.9% (c) 20.2%] *(Electrical Machines-II, Indore Univ. Feb. 1978)*

35.26. Operation of a Salient Pole Synchronous Machine

A multipolar machine with cylindrical rotor has a uniform air-gap, because of which its reactance remains the same, irrespective of the spatial position of the rotor. However, a synchronous machine with salient or projecting poles has non-uniform air-gap due to which its reactance varies with the rotor position. Consequently, a cylindrical rotor machine possesses one axis of symmetry (pole axis or direct axis) whereas salient-pole machine possesses two axes of geometric symmetry (*i*) field poles axis, called direct axis or d-axis and (*ii*) axis passing through the centre of the interpolar space, called the quadrature axis or q-axis, as shown in Fig. 35.70.

Obviously, two mmfs act on the d-axis of a salient-pole synchronous machine *i.e.* field mmf and armature mmf whereas only one mmf, *i.e.* armature mmf acts on the q-axis, because field mmf has no component in the q-axis. The magnetic reluctance is low along the poles and high between the poles. The above facts form the basis of the two-reaction theory proposed by Blondel, according to which

(*i*) armature current I_a can be resolved into two components *i.e.* I_d perpendicular to E_0 and I_q along E_0 as shown in Fig. 35.71 (*b*)

(*ii*) armature reactance has two components *i.e.* d-axis armature reactance X_{ad} associated with I_d and *q*-axis armature reactance X_{aq} linked with I_q.

If we include the armature leakage reactance X_l which is the same on both axes, we get

$$X_d = X_{ad} + X_l \text{ and } X_q = X_{aq} + X_l$$

Since reluctance on the q-axis is higher, owing to the larger air-gap, hence,

$$X_{aq} < X_{ad} \text{ or } X_q < X_d \text{ or } X_d > X_q$$

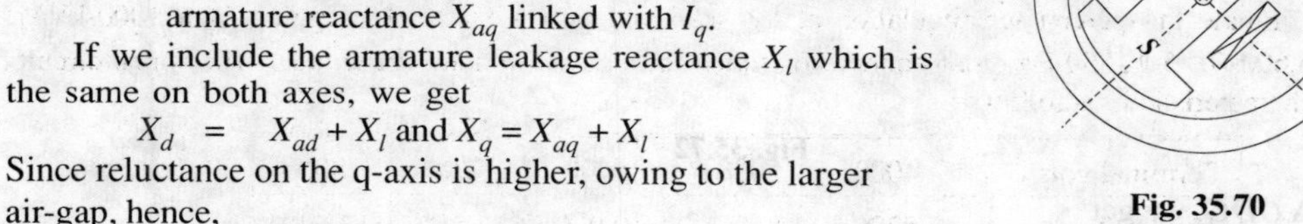

Fig. 35.70

35.27. Phasor Diagram for a Salient Pole Synchronous Machine

The equivalent circuit of a salient-pole synchronous generator is shown in Fig. 35.71 (*a*). The component currents I_d and I_q provide component voltage drops $j\,I_d\,X_d$ and $j\,I_q\,X_q$ as shown in Fig. 35.71 (*b*) for a lagging load power factor.

The armature current I_a has been resolved into its rectangular components with respect to the axis for excitation voltage E_0. The angle ϕ between E_0 and I_a is known as the internal power factor angle. The vector for the armature resistance drop $I_a\,R_a$ is drawn parallel to I_a. Vector for the drop $I_d\,X_d$ is drawn perpendicular to I_a whereas that for $I_q \,.\, X_q$ is drawn perpendicular to I_q. The angle δ between E_0 and V is called the power angle. Following phasor relationships are obvious from Fig. 35.71 (*b*)

$$E_0 = V + I_a R_a + jI_d X_d + jI_q X_q \text{ and } I_a = I_d + I_q$$

If R_a is neglected the phasor diagram becomes as shown in Fig. 35.72 (*a*). In this case,

$$E_0 = V + jI_d X_d + jI_q X_q$$

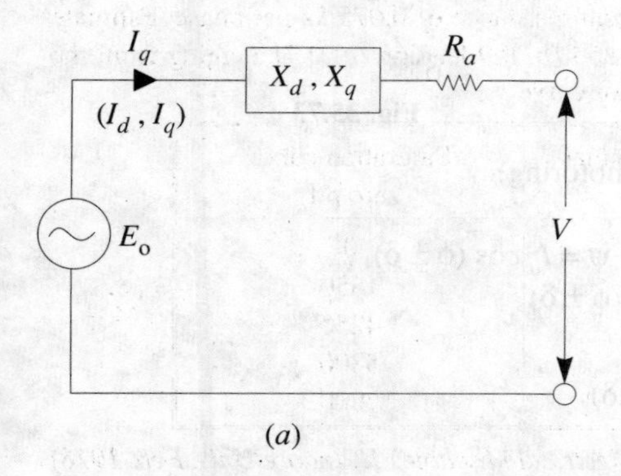

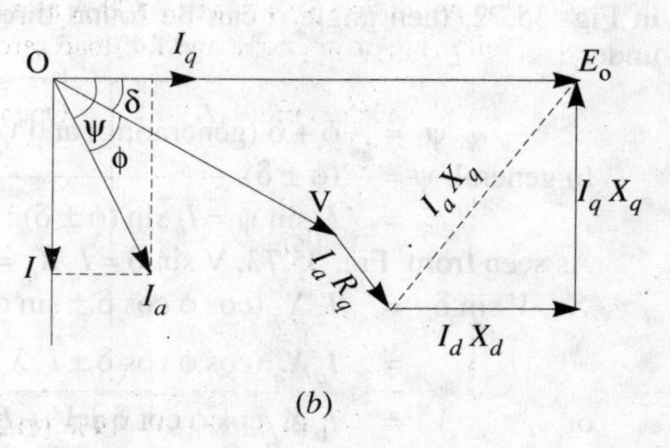

Fig. 35.71

Incidentally, we may also draw the phasor diagram with terminal voltage V lying in the horizontal direction as shown in Fig. 35-72 (*b*). Here, again drop $I_a R_a$ is $\parallel I_a$ and $I_d X_d$ is $\perp$ to I_d and drop $I_q X_q$ is $\perp$ to I_q as usual.

35.28. Calculations from Phasor Diagram

In Fig. 35.73, dotted line AC has been drawn perpendicular to I_a and CB is perpendicular to the phasor for E_0. The angle $ACB = \psi$ because angle between two lines is the same as between their perpendiculars. It is also seen that

$I_d = I_a \sin \psi$; $I_q = I_a \cos \psi$; hence, $I_a = I_q / \cos \psi$

In $\Delta\, ABC$, $BC/AC = \cos \psi$ or $AC = BC/\cos \psi = I_q X_q / \cos \psi = I_a X_q$

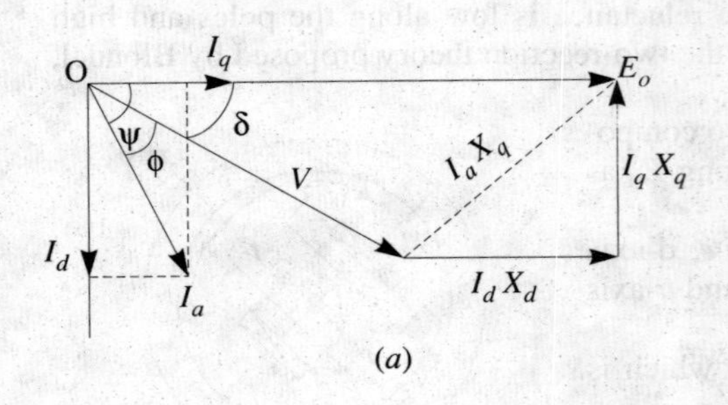

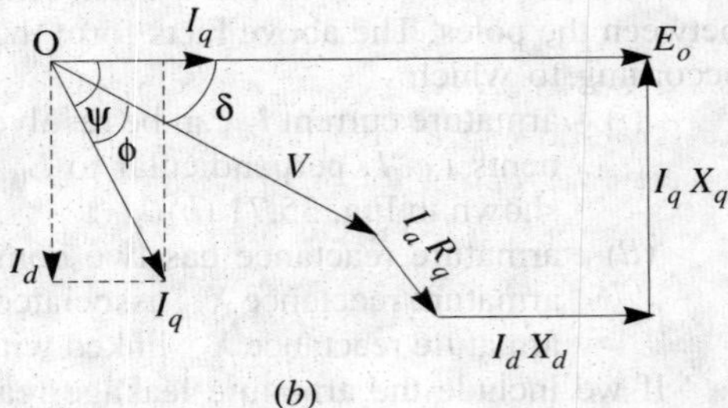

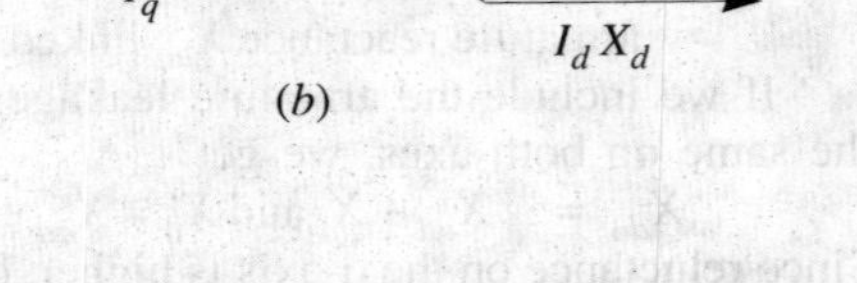

Fig. 35.72

From $\Delta\, ODC$, we get

$$\tan \psi = \frac{AD + AC}{OE + ED} = \frac{V \sin \phi + I_a X_q}{V \cos \phi + I_a R_a} \quad \text{.... generating}$$

$$= \frac{V \sin \phi - I_a X_q}{V \sin \phi - I_a R_a} \quad \text{... motoring}$$

The angle ψ can be found from the above equation. Then, $\delta = \psi - \phi$ (generating) and $\delta = \phi - \psi$(motoring)

As seen from Fig. 35.73, the excitation voltage is given by

$$E_0 = V \cos \delta + I_q R_a + I_d X_d \quad \text{—generating}$$

$$= V \cos \delta - I_q R_a - I_d X_d \quad \text{— motoring}$$

Note. Since angle ϕ is taken positive for lagging p.f., it will be taken negative for leading p.f.

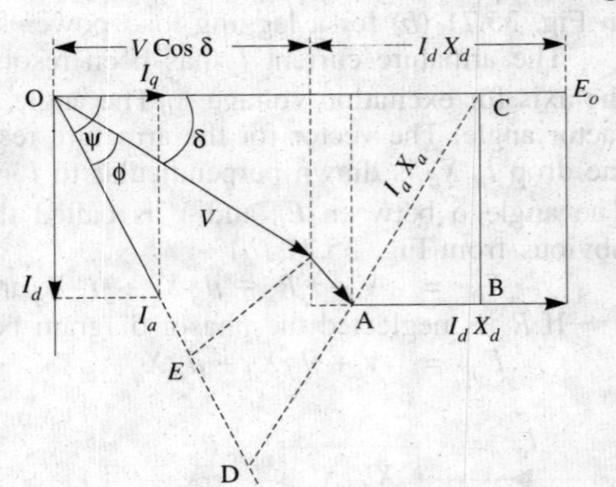

Fig. 35.73

If we neglect the armature resistance as shown in Fig. 35.72, then angle δ can be found directly as under :

$$\psi = \phi + \delta \text{ (generating) and } \psi = \phi - \delta \text{ (motoring).}$$

In general, $\psi = (\phi \pm \delta)$.

$$I_d = I_a \sin \psi = I_a \sin (\phi \pm \delta) \; ; \; I_q = I_a \cos \psi = I_a \cos (\phi \pm \delta)$$

As seen from Fig. 35.73, $V \sin \delta = I_q X_q = I_a X_q \cos(\phi \pm \delta)$

$$\therefore \quad V \sin \delta = I_a X_q (\cos \phi \cos \delta \pm \sin \phi \sin \delta)$$

$$= I_a X_q (\cos \phi \cos \delta \pm I_a X_q \sin \psi \sin \delta)$$

$$\text{or} \quad V = I_a X_q \cos \phi \cot \delta \pm V + I_a X_q \sin \phi$$

$$\therefore \quad I_a X_q \cos\phi \cot\delta = V \pm I_a X_q \sin\phi$$

$$\therefore \quad \tan\delta = \frac{I_a X_q \cos\phi}{V \pm I_a X_q \sin\phi}$$

In the above expression, plus sign is for synchronous generators and minus sign for synchronous motors.

Similarly, when R_a is neglected, then,

$$E_0 = V\cos\delta \pm I_d X_d$$

However, if R_a and hence $I_a R_a$ drop is not negligible then,

$$E_0 = V\cos\delta + I_q R_a + I_d X_d \qquad \text{—generating}$$

$$= V\cos\delta - I_q R_a - I_d X_d \qquad \text{—motoring}$$

35.29. Power Developed by a Synchronous Generator

If we neglect R_a and hence Cu loss, then the power developed (P_d) by an alternator is equal to the power output (P_{out}). Hence, the per phase power output of an alternator is

$$P_{out} = VI_a \cos\phi = \text{power developed } (p_d) \qquad \text{—(i)}$$

Now, as seen from Fig., 35.72(*a*), $I_q X_q = V\sin\delta$; $I_d X_d = E_0 - V\cos\delta$ —(*ii*)

Also, $I_d = I_a \sin(\phi + \delta)$; $I_q = I_a \cos(\phi + \delta)$ —(*iii*)

Substituting Eqn. (*iii*) in Eqn. (*ii*) and solving for $I_a \cos\phi$, we get

$$I_a \cos\phi = \frac{V}{X_d}\sin\delta + \frac{V}{2X_q}\sin 2\delta - \frac{V}{2X_d}\sin 2\delta$$

Finally, substituting the above in Eqn. (*i*), we get

$$P_d = \frac{E_0 V}{X_d}\sin\delta + \frac{1}{2}V_2\left(\frac{1}{X_q} - \frac{1}{X_d}\right)\sin 2\delta = \frac{E_0 V}{X_d}\sin\delta + \frac{V^2(X_d - X_q)}{2X_d X_q}\sin 2\delta$$

The total power developed would be three times the above power.

As seen from the above expression, the power developed consists of two components, the first term represents power due to field excitation and the second term gives the reluctance power *i.e.* power due to saliency. If $X_d = X_q$ *i.e.* the machine has a cylindrical rotor, then the second term becomes zero and the power is given by the first term only. If, on the other hand, there is no field excitation *i.e.* $E_0 = 0$, then the first term in the above expression becomes zero and the power developed is given by the second term. It may be noted that value of δ is positive for a generator and negative for a motor.

Example 35.38. *A 3-phase alternator has a direct-axis synchronous reactance of 0.7 p.u and a quadrature axis synchronous reactance of 0.4 p.u. Draw the vector diagram for full-load 0.8 p.f. lagging and obtain therefrom (i) the load angle and (ii) the no-load per unit voltage .*

(Advanced Elect. Machines, AMIE Sec. B 1991)

Solution. $V = 1$ p.u.; $X_d = 0.7$ p.u.; $X_q = 0.4$ p.u.;

$\cos\phi = 0.8$; $\sin\phi = 0.6$; $\phi = \cos^{-1} 0.8 = 36.9°$; $I_a = 1$ p.u.

$$(i) \quad \tan\delta = \frac{I_a X_q \cos\phi}{V + I_a X_q \sin\phi} = \frac{1 \times 0.4 \times 0.8}{1 + 1 \times 0.4 \times 0.6} = 0.258, \quad \delta = 16.5°$$

$$(ii) \quad I_d = I_a \sin(\phi + \delta) = 1 \sin(36.9° + 14.9°) = 0.78 \text{ A}$$

$$E_0 = V\cos\delta + I_d X_d = 1 \times 0.966 + 0.78 \times 0.75 = \mathbf{1.553}$$

Example 35.39. *A 3-phase, star-connected, 50-Hz synchronous generator has direct-axis synchronous reactance of 0.6 p.u. and quadrature-axis synchronous reactance of 0.45 p.u. The generator delivers rated kVA at rated voltage. Draw the phasor diagram at full-load 0.8 p.f. lagging and hence calculate the open-circuit voltage and voltage regulation. Resistive drop at full-load is 0.015 p.u.* **(Elect. Machines-II, Nagpur Univ. 1993)**

Solution. $I_a = 1$ p.u.; $V = 1$ p.u.; $X_d = 0.6$ p.u.; $X_q = 0.45$ p.u. ; $R_a = 0.015$ p.u.

$$\tan\psi = \frac{V\sin\phi + I_a X_q}{V\cos\phi + I_a R_a} = \frac{1\times 0.6 + 1\times 0.45}{1\times 0.8 + 1\times 0.015} = 1.288\ ; \quad \psi = 52.2°$$

$$\delta = \psi - \phi = 52.2° - 36.9° = 15.3°$$

$$I_d = I_a \sin\psi = 1\times 0.79 = 0.79\text{ A}\ ; \quad I_q = I_a\cos\psi = 1\times 0.61 = 0.61\text{A}$$

$$E_0 = V\cos\delta + I_q R_a + I_d X_d$$

$$= 1\times 0.965 + 0.61\times 0.015 + 0.79\times 0.6 = 1.448$$

$$\therefore \quad \%\text{ regn.} = \frac{1.448 - 1}{1}\times 100 = \mathbf{44.8\%}$$

Example 35.40. *A 3-phase, Y-connected syn. generator supplies current of 10 A having phase angle of 20° lagging at 400 V. Find the load angle and the components of armature current I_d and I_q if X_d = 10 ohm and X_q = 6.5 ohm. Assume arm. resistance to be negligible.* **(Elect. Machines-I, Nagpur Univ. 1993)**

Solution. $\cos\phi = \cos 20° = 0.94$; $\sin\phi = 0.342$; $I_a = 10$ A

$$\tan\delta = \frac{I_a X_q\cos\phi}{V + I_a X_q\sin\phi} = \frac{10\times 6.5\times 0.94}{400 + 10\times 6.5\times 0.342} = 0.1447$$

$$\delta = 8.23°$$

$$I_d = I_a\sin(\phi+\delta) = 10\sin(20° + 8.23°) = 4.73\text{ A}$$

$$I_q = I_a\cos(\phi+\delta) = 10\cos(20° + 8.23°) = 8.81\text{ A}$$

Incidentally, if required, voltage regulation of the above generator can be found as under:

$$I_d X_d = 4.73\times 10 = 47.3\text{ V}$$

$$E_0 = V\cos\delta + I_d X_d = 400\cos 8.23° + 47.3 = 443\text{ V}$$

$$\%\text{ regn.} = \frac{E_o - V}{V}\times 100$$

$$= \frac{443 - 400}{400}\times 100 = \mathbf{10.75\%}$$

Tutorial Problem No. 35.5.

1. A 20 MVA, 3-phase, star-connected, 50-Hz, salient-pole has X_d = 1 p.u.; X_q = 0.65 p.u. and R_a = 0.01 p.u. The generator delivers 15 MW at 0.8 p.f. lagging to an 11-kV, 50-Hz system. What is the load angle and excitation e.m.f. under these conditions? **[18°; 1.73 p.u]**
2. A salient-pole synchronous generator delivers rated kVA at 0.8 p.f. lagging at rated terminal voltage. It has X_d = 1.0 p.u. and X_q = 0.6 p.u. If its armature resistance is negligible, compute the excitation voltage under these conditions. **[1.77 p.u]**
3. A 20-kVA, 220-V, 50-Hz, star-connected, 3-phase salient-pole synchronous generator supplies load at a lagging power factor angle of 45°. The phase constants of the generator are X_d = 4.0 Ω ; X_q = 2 Ω and R_a = *0.5* Ω. Calculate (*i*) power angle and (*ii*) voltage regulation under the given load conditions. **[(i) 20.6° (*ii*) 142%]**
4. A 3-phase salient-pole synchronous generator has X_d = 0.8 p.u.; X_q = 0.5 p.u. and R_a = 0. Generator supplies full-load at 0.8 p.f. lagging at rated terminal voltage. Compute (i) power angle and (ii) no-load voltage if excitation remains constant. **[(i) 17.1° (*ii*) 1.6 p.u]**

35.30. Parallel Operation of Alternators

The operation of connecting an alternator in parallel with another alternator or with common bus-bars is known as *synchronizing*. Generally, alternators are used in a power system where they are in parallel with many other alternators. It means that the alternator is connected to a *live* system of constant voltage and constant frequency. Often the electrical system to which the alternator is connected, has already so many alternators and loads connected to it that no matter what power is delivered by the incoming alternator, the voltage and frequency of the system remain the same. In that case, the alternator is said to be conncted to *infinite* bus-bars.

It is never advisable to connect a stationary alternator to live bus-bars, because, stator induced e.m.f. being zero, a short-circuit will result. For proper synchronization of alternators, the following three conditions must be satisfied :-

1. The terminal voltage (effective) of the incoming alternator must be the same as bus-bar voltage.
2. The speed of the incoming machine must be such that its frequency (= $PN/120$) equals bus-bar frequency.
3. The phase of the alternator voltage must be identical with the phase of the bus-bar voltage. It means that the switch must be closed at (or very near) the instant the two voltages have correct phase relationship.

Condition (1) is indicated by a voltmeter, conditions (2) and (3) are indicated by synchronizing lamps or a synchronoscope.

35.31. Synchronizing of Alternators

(*a*) Single-phase Alternators

Suppose machine 2 is to be synchronized with or 'put on' the bus-bars to which machine 1 is already connected. This is done with the help of two lamps L_1 and L_2 (known as synchronizing lamps) connected as shown in Fig. 35.74.

It should be noted that E_1 and E_2 are in-phase relative to the external circuit but are in direct phase opposition in the local circuit (shown dotted).

If the speed of the incoming machine 2 is not brought up to that of machine 1, then its frequency will also be different, hence there will be a phase-difference between their voltages (even when they are equal in magnitude, which is determined by field excitation). This phase-difference will be continously changing with the changes in their frequencies. The result is that their resultant voltage will undergo changes similar to the frequency changes of beats produced, when two sound sources of *nearly* equal frequency are sounded together, as shown in Fig. 35.75.

Sometimes the resultant voltage is maximum and some other times minimum. Hence, the current is alternatingly maximum and minimum. Due to this changing current through the lamps, a flicker will be produced, the frequency of flicker being $(f_2 \sim f_1)$. Lamps will dark out and glow up alternately. Darkness indicates that the two voltages E_1 and E_2 are in exact phase opposition

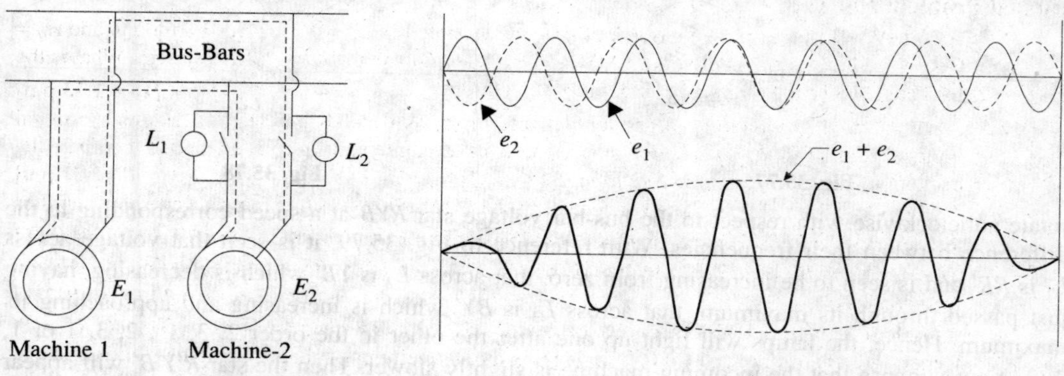

Fig. 35.74 **Fig. 35.75**

relative to the local circuit and hence there is no resultant current throught the lamps. Synchronizing is done at the middle of the dark period. That is why, sometimes, it is known as 'lamps dark' synchronizing. Some engineers prefer 'lamps bright' synchronization because of the fact the lamps are much more sensitive to changes in voltage at their maximum brightness than when they are dark. Hence, a sharper and more accurate synchronization is obtained. In that case, the lamps are connected as shown in Fig. 35.76. Now, the lamps will glow brightest when the two voltages are in-phase with the bus-bar voltage because then voltage across them is twice the voltage of each machine.

(b) Three-phase Alternators

In 3-φ alternators, it is necessary to synchronize one phase only, the other two phases will then be synchronized automatically. However, first it is necessary that the incoming alternator is correctly 'phased out' *i.e.* the phases are connected in the proper order of *R, Y, B* and not *R, B, Y* etc.

In this case, three lamps are used. But they are deliberately connected asymmetrically, as shown in Fig. 35.77 and 35.78.

This transposition of two lamps, suggested by Siemens and Halske, helps to indicate whether the incoming machine is running too slow. If lamps were connected symmetrically, they would dark out or glow up simultaneously (if the phase rotation is the same as that of the bus-bars).

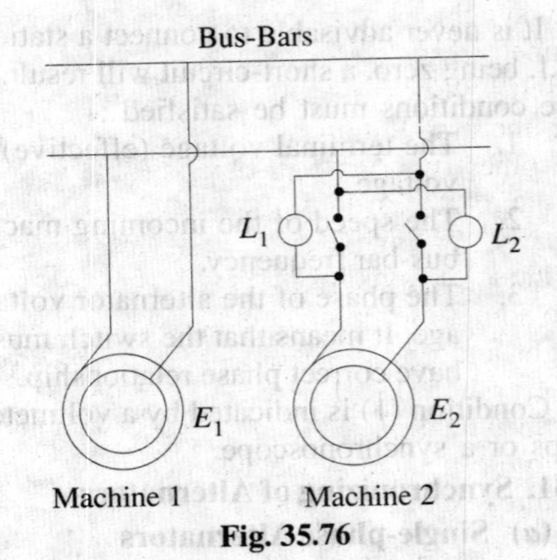

Fig. 35.76

Lamp L_1 is connected between *R* and R′, L_2 between *Y* and *B′* (not *Y* and *Y′*) and L_3 *between* *B* and *Y′* (and not *B* and *B′*), as shown in Fig. 35.78.

Voltage stars of two machines are shown superimposed on each other in Fig. 35.79.

Two sets of star vectors will rotate at unequal speeds if the frequencies of the two machines are different. If the incoming alternator is running faster, then voltage star R′*Y′B′* will appear to

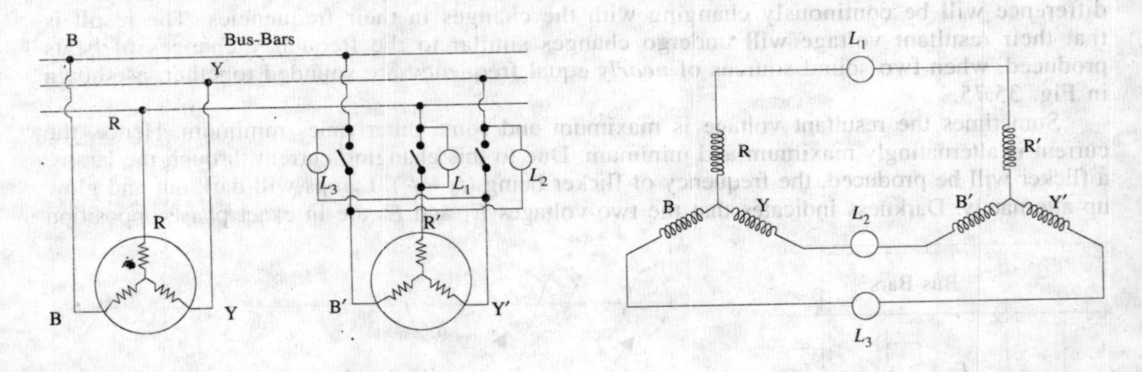

Fig. 35.77

Fig. 35.78

rotate anticlockwise with respect to the bus-bar voltage star *RYB* at a speed corresponding to the difference between their frequencies. With reference to, Fig. 35.79, it is seen that voltage across L_1 is *RR′* and is seen to be increasing from zero, that across L_2 is *YB′* which is decreasing, having just passed through its maximum, that across L_3 is *BY′* which is increasing and approaching its maximum. Hence, the lamps will light up one after the other in the order 2, 3, 1 ; 2, 3, 1 or 1, 2, 3. Now, suppose that the incoming machine is slightly slower. Then the star *R′Y′B′* will appear to be rotating clockwise relative to voltage star *RYB* (Fig. 35.80). Here, we find that voltage across L_3 *i.e.* *Y′B* is decreasing having just passed through its maximum, that across L_2 *i.e.* *YB′* is increasing and approaching its maximum, that across L_1 is decreasing having passed through its maximum earlier. Hence, the lamps will light up one after the other in the order 3, 2, 1 ; 3, 2, 1, etc. which is just the reverse of the first order. Usually, the three lamps are mounted at the three corners of a triangle and the apparent direction of rotation of light indicates whether the incoming alternator is running too fast or too slow (Fig. 35.81). Synchronization is done at the moment the uncrossed lamp L_1 is in the middle of the dark period. When the alternator voltage is too high for the lamps to be used directly, then usually step-down transformers are used and the synchronizing lamps are connected to the secondaries.

It will be noted that when the uncrossed lamp L_1 is dark, the other two 'crossed' lamps L_2 and L_3 are dimly but equally bright. Hence, this method of synchronizing is also sometimes known as 'two bright and one dark' method.

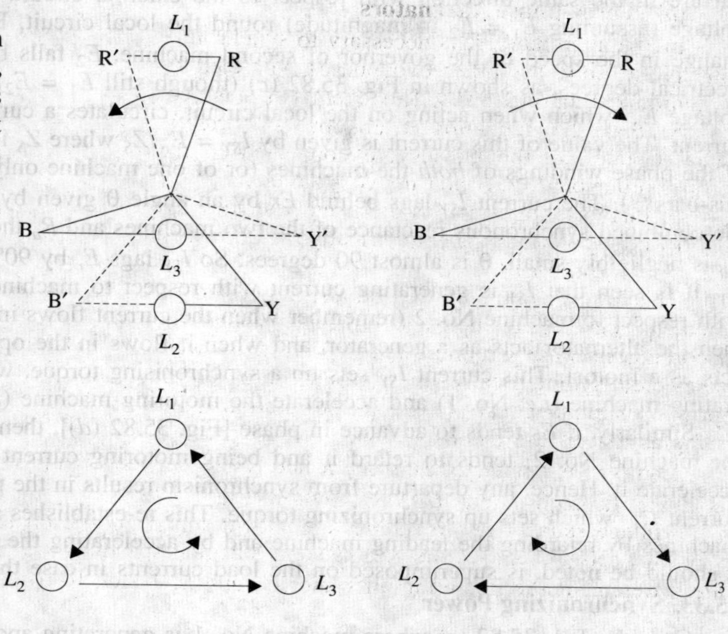

Fig. 35.79 **Fig. 35.80**

It should be noted that synchronization by lamps is not quite accurate, because to a large extent, it depends on the sense of correct judgement of the operator. Hence, to eliminate the element of personal judgment in routine operation of alternators, the machines are synchronized by a more accurate device called a synchronoscope. It consists of 3 stationary coils and a rotating iron vane which is attached to a pointer. Out of three coils, a pair is connected to one phase of the line and the other to the corresponding machine terminals, potential transformer being usually used. The pointer moves to one side or the other from its vertical position depending on whether the incoming machine is too fast or too slow. For correct speed, the pointer points vertically up.

Fast Slow

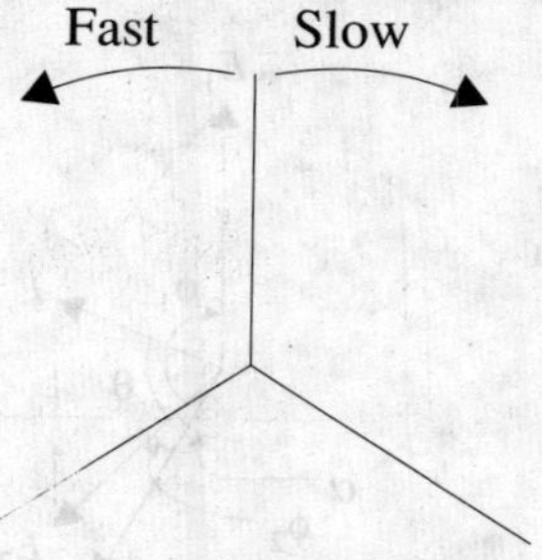

Fig. 35.81

Example 35.41. *In Fig. 35.74, E_1 = 220 V and f_1 = 60 Hz, whereas E_2 = 222 V and f_2 = 59 Hz. With the switch open; calculate*

(i) maximum and minimum voltage across each lamp.

(ii) frequency of voltage across the lamps.

(iii) peak value of voltage across each lamp.

(iv) phase relations at the instants maximum and minimum voltages occur.

(v) the number of maximum light pulsations/minute.

Solution. (*i*) E_{max} /lamp = (220 + 222)/2 = **221 V**

E_{min} /lamp = (222 – 220)/2 = **1.0 V**

(*ii*) $f = (f_1 - f_2)$ = (60 – 59) = **1.0 Hz** (*iii*) E_{peak} = 221/0.707 = **313 V**

(*iv*) in-phase and anti-phase respectively in the local circuit.

(*v*) No. of pulsation/min = (60 – 59) × 60 = **60.**

35.32. Synchronizing Current

Once synchronized properly, two alternators continue to run in synchronism. Any tendency on the part of one to drop out of synchronism is immediately counteracted by the production of a synchronizing torque, which brings it back to synchronism.

When in exact synchronism, the two alternators have equal terminal p.d.'s and are in exact phase opposition, so far as the local circuit (consisting of their armatures) is concerned. Hence, there is no current circulating round the local circuit. As shown in Fig. 35.82 (*b*) e.m.f. E_1 of machine No. 1 is in exact phase opposition to the e.m.f. of machine No. 2 *i.e.* E_2. It should be clearly understood that the two e.m.f.s. are in opposition, so far as their local circuit is concerned but are in the same direction with respect to the external circuit. Hence, there is no resultant voltage (assuming $E_1 = E_2$ in magnitude) round the local circuit. But now suppose that due to change in the speed of the governor of second machine, E_2 falls back* by a phase angle of α electrical degrees, as shown in Fig. 35.82 (*c*) (though still $E_1 = E_2$). Now, they have a resultant voltage E_r, which when acting on the local circuit, circulates a current known as synchronizing current. The value of this current is given by $I_{SY} = E_r / Z_S$ where Z_S is the synchronous impedance of the phase windings of *both* the machines (or of one machine only if it is connected to infinite bus-bars**). The current I_{SY} lags behind Er by an angle θ given by $\tan \theta = X_S / R_a$ where X_S is the combined synchronous reactance of the two machines and R_a their armature resistance. Since R_a is negligibly small, θ is almost 90 degrees. So I_{SY} lags E_r by 90° and is almost in phase with E_1. It is seen that I_{SY} is generating current with respect to machine No.1 and motoring current with respect to machine No. 2 (remember when the current flows in the *same* direction as e.m.f., then the alternator acts as a generator, and when it flows in the opposite direction, the machine acts as a motor). This current I_{SY} sets up a synchronising torque, which tends to retard the generating machine (*i.e.* No. 1) and accelerate the motoring machine (*i.e.* No. 2).

Similarly, if E_2 tends to advance in phase [Fig. 35.82 (*d*)], then I_{SY}, being generating current for machine No. 2, tends to retard it and being motoring current for machine No. 1 tends to accelerate it. Hence, any departure from synchronism results in the production of a synchronizing current I_{SY} which sets up synchronizing torque. This re-establishes synchronism between the two machines by retarding the leading machine and by accelerating the lagging one. This current I_{SY}, it should be noted, is superimposed on the load currents in case the machines are loaded.

35.33. Synchronizing Power

Consider Fig. 35.82 (*c*) where machine No. 1 is generating and supplying the synchronizing power $= E_1 I_{SY} \cos \phi_1$ which is approximately equal to $E_1 I_{SY}$ ($\because \phi_1$ is small). Since $\phi_1 = (90° - \theta)$, synchronizing power $= E_1 I_{SY} \cos \phi_1 = E_1 I_{SY} \cos (90° - \theta) = E_1 I_{SY} \sin \theta \cong E_1 I_{SY}$ because θ

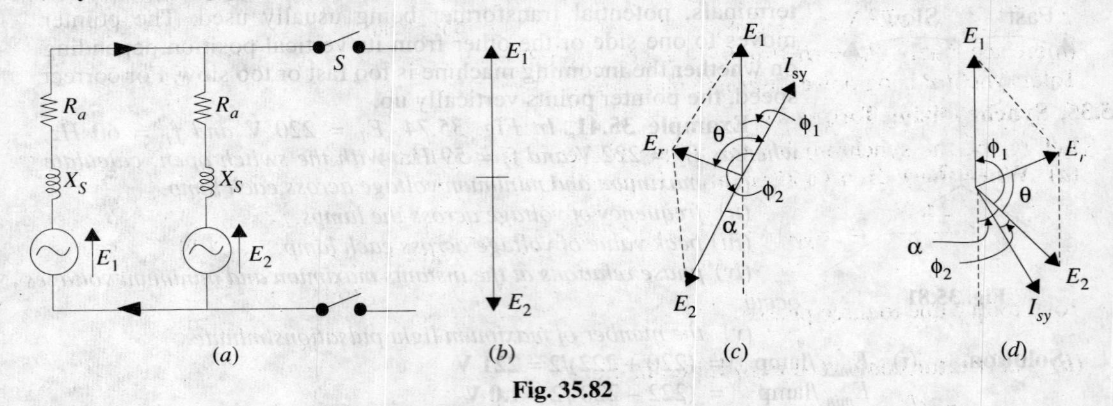

Fig. 35.82

$\cong 90°$ so that $\sin \theta \cong 1$. This power output from machine No. 1 goes to supply (*a*) power input to machine No. 2 (which is motoring) and (*b*) the Cu losses in the local armature circuit of the two machines. Power input to machine No. 2 is $E_2 I_{SY} \cos \phi_2$ which is approximately equal to $E_2 I_{SY}$.

$\therefore \quad E_1 I_{SY} = E_2 I_{SY}$ + Cu losses

Now, let $E_1 = E_2 = E$ (say)

Then, $E_r = 2E \cos [(180° - \alpha)/2]$*** $= 2E \cos [90° - (\alpha/2)]$

---

*Please remember that vectors are supposed to be rotating anticlockwise.

**Infinite bus-bars are those whose frequency and the phase of p.d.'s are not affected by changes in the conditions of any one machine connected in parallel to it. In other words, they are constant-frequency, constant-voltage bus-bars.

***Strictly speaking, $E_r = 2E \sin \theta . \sin \alpha / 2 \cong 2E \sin \alpha/2$.

$= 2\,E \sin \alpha/2 = 2\,E \times \alpha/2 = \alpha E$ ($\because$ α is small)

Here, the angle α is in electrical radians.

Now, $I_{SY} = \dfrac{E_r}{\text{synch. impedance } Z_S} \cong \dfrac{E_r}{2X_S} = \dfrac{\alpha E}{2\,X_S}$ —if R_a of both machines is negligible

Here, X_S represents synchronous reactance of one machine and not of both as in Art. 35.31

Synchronizing power (supplied by machine No. 1) is

$P_{SY} = E_1 I_{SY} \cos \phi_1 = E\,I_{SY} \cos (90° - \theta) = EI_{SY} \sin \theta \cong EI_{SY}$

Substituting the value of I_{SY} from above,

$P_{SY} = E.\alpha E / 2\,Z_S = \alpha E^2 / 2Z_S \cong \alpha E^2 / 2X_S$ — per phase

(more accurately, $P_{SY} = \alpha E_2 \sin \theta/2X_S$)

Total synchronizing power for three phases

$= 3P_{SY} = 3\,\alpha E^2 / 2X_S$ (or $3\,\alpha E^2 \sin \theta/2X_S$)

This is the value of the synchronizing power when two alternators are connected in parallel and are on no-load.

35.34. Alternators Connected to Infinite Bus-bars

Now, consider the case of an alternator which is connected to infinite bus-bars. The expression for P_{SY} given above is still applicable but with one important difference *i.e.* impedance (or reactance) of only that one alternator is considered (and not of two as done above). Hence, expression for synchronizing power in this case becomes

$E_r = \alpha E$ —as before

$I_{SY} = E_r / Z_S \cong E_r / X_S = \alpha_E / X_S$ —*if R_a is negligible*

$\therefore$ synchronizing power $P_{SY} = E\,I_{SY} = E.\alpha E/Z_S = \alpha E^2 / Z_S \cong \alpha E^2 / X_S$ — per phase

Now, $E / Z_S \cong E/X_S = S.C.$ current I_{SC}

$\therefore$ $P_{SY} = \alpha E^2/X_S = \alpha E.E/X_S = \alpha E.I_{SY}$ —per phase

(more accurately, $P_{SY} = \alpha E^2 \sin \theta/X_S = \alpha E.I_{SC}.\sin \theta$)

Total synchronizing power for three phases = 3 P_{SY}

35.35. Synchronizing Torque T_{SY}

Let T_{SY} be the synchronizing torque per phase in newton-metre (N-m)

(*a*) When there are two alternators in parallel

$$T_{SY} \times \frac{2\pi N_S}{60} = P_{SY} \quad \therefore \quad T_{SY} = \frac{P_{SY}}{2\pi N_S/60} = \frac{\alpha E^2/2X_S}{2\pi N_S/60} \text{ N-m}$$

Total torque due to three phases. $= \dfrac{3P_{SY}}{2\pi N_S/60} = \dfrac{3\,\alpha E^2/2X_S}{2\pi N_S/60}$ N-m

(*b*) Alternator connected to infinite bus-bars

$$T_{SY} \times \frac{2\pi N_S}{60} = P_{SY} \quad \text{or} \quad T_{SY} = \frac{P_{SY}}{2\pi N_S/60} = \frac{\alpha E^2/X_S}{2\pi N_S/60} \text{ N-m}$$

Again, torque due to 3 phase $= \dfrac{3P_{SY}}{2\pi N_S/60} = \dfrac{3\,\alpha\, E^2/X_S}{2\pi N_S/60}$ N-m

where N_S = synchronous speed in r.p.m. = $120\,f/P$

35.36. Effect of Load on Synchronizing Power

In this case, instead of $P_{SY} = \alpha\,E^2 / X_S$, the approximate value of synchronizing power w[illegible] be $\cong \alpha EV/X_S$ where V is bus-bar voltage and E is the alternator induced e.m.f. per phas[illegible] value of $E = V + IZ_S$

As seen from Fig. 35.83, for a lagging p.f.,

$E = [(V \cos \phi + IR_a)^2 + (V \sin \phi + IX_S)^2]^{1/2}$

Example 35.42. *Find the power angle when a 1500-kVA, 6.6 kV, 3-phase, Y*[illegible] *ternator having a resistance of 0.4 ohm and a reactance of 6 ohm per phase delivers full-load current at normal rated voltage and 0.8 p.f. lag. Draw the phasor diagram.*

(Electrical Machinery-II, Bangalore Univ. 1981)

Solution. It should be remembered that angle α between V and E known as power angle (Fig. 35.84)

Full-load $I = 15 \times 10^5/\sqrt{3} \times 6600 = 131$ A

$IR_a = 131 \times 0.4 = 52.4$ V, $IX_S = 131 \times 6 = 786$ V

V/phase $= 6600/\sqrt{3} = 3810$ V;

$\phi = \cos^{-1}(0.8) = 36°50'$.

As seen from Fig. 35.84

$$\tan(\phi + \alpha) = \frac{AB}{OA} = \frac{V\sin\phi + IX_S}{V\cos\phi + IR_a}$$

$$= \frac{3810 \times 0.6 + 786}{3810 \times 0.8 + 52.4} = 0.991$$

$\therefore \quad (\phi + \alpha) = 44° \quad \therefore \ \alpha = 44° - 36°50' = \mathbf{7°10'}$

The angle α is also known as load angle or torque angle.

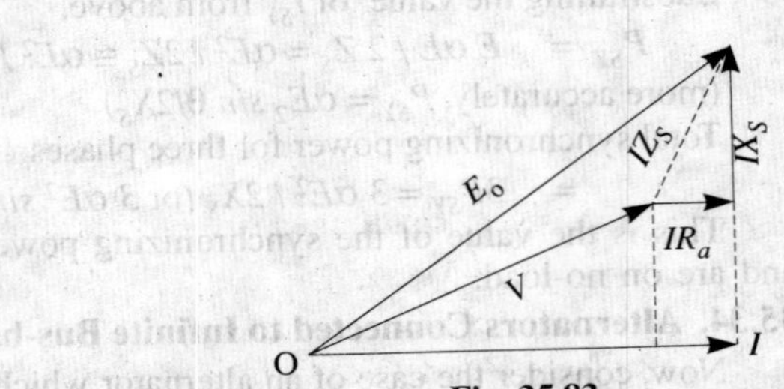

Fig. 35.83

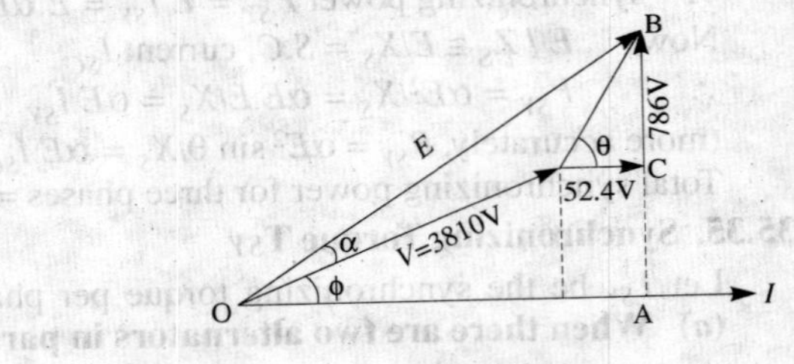

Fig. 35.84

35.37. Alternative Expression for Synchronizing Power

As shown in Fig. 35.85, let V and E (or E_0) be the terminal voltage and induced e.m.f. per phase of the rotor. Then, taking $\mathbf{V} = V\angle 0°$, the load current supplied by the alternator is

$$I = \frac{E - V}{Z_S} = \frac{E\angle\alpha - V\angle 0°}{Z_S\angle\theta}$$

$$= \frac{E}{Z_S}\angle\alpha - \theta - \frac{V}{Z_S}\angle -\theta$$

$$= \frac{E}{Z_S}[\cos(\theta - \alpha) - j\sin(\theta - \alpha)]$$

$$-\frac{V}{Z_S}(\cos\theta - j\sin\theta)$$

$$= \left[\frac{E}{Z_S}\cos(\theta - \alpha) - \frac{V}{Z_S}\cos\theta\right] - j\left[\frac{E}{Z_S}\sin(\theta - \alpha) - \frac{V}{Z_S}\sin\theta\right]$$

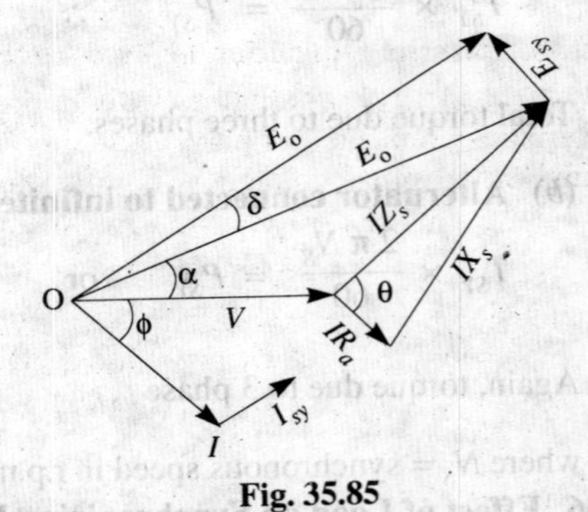

Fig. 35.85

These components represent the $I\cos\phi$ and $I\sin\phi$ respectively. The power P converted internally is given be the sum of the product of corresponding components of the current with $E\cos\alpha$ and $E\sin\alpha$.

$$\therefore \quad P = E\cos\alpha\left[\frac{E}{Z_S}\cos(\theta - \alpha) - \frac{V}{Z_S}\cos\theta\right] - E\sin\alpha\left[\frac{E}{Z_S}\sin(\theta - \alpha) - \frac{V}{Z_S}\sin\theta\right]$$

$$= E\left[\frac{E}{Z_S}\cos\theta\right] - E\left[\frac{V}{Z_S}.\cos(\theta+\alpha)\right] = \frac{E}{Z_S}[E\cos\theta - V(\cos\theta+\alpha)] \text{— per phase*}$$

Now, let, for some reason, angle α be changed to $(\alpha \pm \delta)$. Since V is held rigidly constant, due to displacement $\pm\,\delta$, an additional e.m.f. of divergence *i.e.* $E_{SY} = 2E.\sin\alpha/2$ will be produced, which will set up an additional current I_{SY} given by $I_{SY} = E_{SY}/Z_S$. The internal power will become

$$P' = \frac{E}{Z_S}[E\cos\theta - V\cos(\theta+\alpha\pm\delta)]$$

The difference between P' and P gives the synchronizing power.

$$\therefore\ P_{SY} = P' - P = \frac{EV}{Z_S}[\cos(\theta+\alpha) - \cos(\theta+\alpha\pm\delta)]$$

$$= \frac{EV}{Z_S}[\sin\delta\,.\sin(\theta+\alpha) \pm 2\cos(\theta+\alpha)\sin^2\delta/2]$$

If δ is very small, then $\sin^2(\delta/2)$ is zero, hence P_{SY} per phase is

$$R_{SY} = \frac{EV}{Z_S}.\sin(\theta+\alpha)\sin\delta \qquad \text{(i)}$$

(*i*) In large alternators, R_a is negligible, hence $\tan\theta = X_S/R_a = \infty$, so that $\theta \cong 90°$. Therefore, $\sin(\theta+\alpha) = \cos\alpha$.

$$\therefore\ P_{SY} = \frac{EV}{Z_S}.\cos\alpha\sin\delta \quad \text{— per phase} \qquad ...(ii)$$

$$= \frac{EV}{X_S}.\cos\alpha\sin\delta \quad \text{— per phase} \qquad ...(iii)$$

(*ii*) Consider the case of synchronizing an unloaded machine on to a constant-voltage bus-bars. For proper operation, $\alpha = 0$ so that E coincides with V. In that case, $\sin(\theta+\alpha) = \sin\theta$.

$$\therefore\ P_{SY} = \frac{EV}{Z_S}\sin\theta\sin\delta \quad \text{— from (i) above.}$$

Since δ is very small, $\sin\delta = \delta$,

$$\therefore\ P_{SY} = \frac{EV}{Z_S}\delta\sin\theta = \frac{EV}{X_S}\delta\sin\theta$$

Usually, $\sin\theta \cong 1$, hence

$$\therefore\ P_{SY} = \frac{EV}{Z_S}.\delta^{**} = V\left(\frac{E}{Z_S}\right)\delta = V\left(\frac{E}{X_S}\right)\delta$$

$$= V.I_{SC}.\delta \quad \text{—per phase}$$

35.38. Parallel Operation of Two Alternators

Consider two alternators with identical speed/load characteristics connected in parallel as shown in Fig. 35.86. The common terminal voltage **V** is given by

$$\mathbf{V} = \mathbf{E}_1 - \mathbf{I}_1\mathbf{Z}_1 = \mathbf{E}_2 - \mathbf{I}_2\mathbf{Z}_2$$

$$\therefore\ \mathbf{E}_1 - \mathbf{E}_2 = \mathbf{I}_1\mathbf{Z}_1 - \mathbf{I}_2\mathbf{Z}_2$$

Also $\mathbf{I} = \mathbf{I}_1 + \mathbf{I}_2$ and $\mathbf{V} = \mathbf{IZ}$

$$\therefore\ \mathbf{E}_1 = \mathbf{I}_1\mathbf{Z}_1 + \mathbf{IZ} = \mathbf{I}_1(\mathbf{Z}+\mathbf{Z}_1) + \mathbf{I}_2\mathbf{Z}$$

$$\mathbf{E}_2 = \mathbf{I}_2\mathbf{Z}_2 + \mathbf{IZ} = \mathbf{I}_3(\mathbf{Z}+\mathbf{Z}_2) + \mathbf{I}_1\mathbf{Z}$$

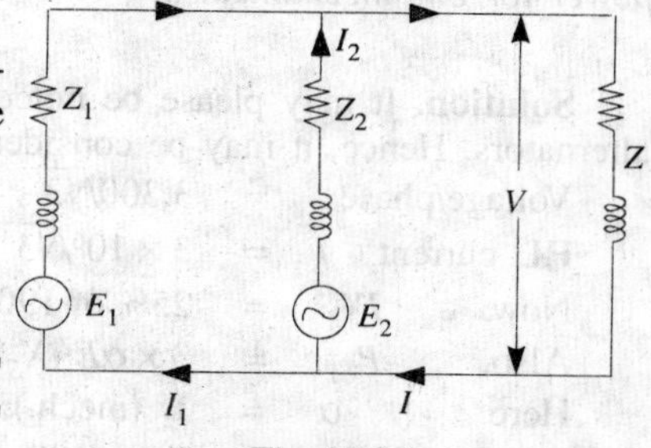

Fig. 35.86

* In large machines, R_a is very small so that $\theta = 90°$, hence

$$P = \frac{E}{Z_S}V\cos(90°+\alpha) = \frac{E}{Z_S}V\sin\alpha = \alpha\,EV/Z_S \text{—if } \alpha \text{ is so small that } \sin\alpha = \alpha.$$

** With $E = V$, the expression becomes $P_{SY} = \frac{V^2}{Z_S}\delta = \frac{\delta V^2}{X_S}$ — per phase

It is the same as in Art. 35.33.

$$\therefore \quad \mathbf{I_1} = \frac{(\mathbf{E_1} - \mathbf{E_2})\,\mathbf{Z} + \mathbf{E_1 Z_2}}{\mathbf{Z(Z_1 + Z_2) + Z_1 Z_2}}$$

$$\mathbf{I_2} = \frac{(\mathbf{E_2} - \mathbf{E_1})\,\mathbf{Z} + \mathbf{E_2 Z_1}}{\mathbf{Z(Z_1 + Z_2) + Z_1 Z_2}}\;;\quad \mathbf{I} = \frac{\mathbf{E_1 Z_2 + E_2 Z_1}}{\mathbf{Z(Z_1 + Z_2) + Z_1 Z_2}}$$

$$\mathbf{V} = \mathbf{IZ} = \frac{\mathbf{E_1 Z_2 + E_2 Z_1}}{\mathbf{Z_1 + Z_2 + (Z_1 Z_2/Z)}}\;;\quad \mathbf{I_1} = \frac{\mathbf{E_1 - V}}{\mathbf{Z_1}}\;;\quad \mathbf{I_2} = \frac{\mathbf{E_2 - V}}{\mathbf{Z_2}}$$

The circulating current under no-load condition is $\mathbf{I_C} = (\mathbf{E_1 - E_2})/(\mathbf{Z_1 + Z_2})$.

Using Admittances

The terminal Voltage may also be expressed in terms of admittances as shown below:

$$\mathbf{V} = \mathbf{IZ} = (\mathbf{I_1 + I_2})\mathbf{Z} \qquad \therefore \quad \mathbf{I_1 + I_2} = \mathbf{V/Z} = \mathbf{VY} \qquad ...(i)$$

Also $\quad \mathbf{I_1} = (\mathbf{E_1 - V})/\mathbf{Z_1} = (\mathbf{E_1 - V})\mathbf{Y_1}\;;\quad \mathbf{I_2} = (\mathbf{E_2 - V})/\mathbf{Z_2} = (\mathbf{E_2 - V})\mathbf{Y_2}$

$$\therefore \quad \mathbf{I_1 + I_2} = (\mathbf{E_1 - V})\,\mathbf{Y_1} + (\mathbf{E_2 - V})\mathbf{Y_2} \qquad ...(ii)$$

From Eq. (*i*) and (*ii*), we get

$$\mathbf{VY} = (\mathbf{E_1 - V})\mathbf{Y_1} + (\mathbf{E_2 - V})\mathbf{Y_2} \quad \text{or} \quad \mathbf{V} = \frac{\mathbf{E_1 Y_1 + E_2 Y_2}}{\mathbf{Y_1 + Y_2 + Y}}$$

Using Parallel Generator Theorem

$$\mathbf{V} = \mathbf{IZ} = (\mathbf{I_1 + I_2})\mathbf{Z} = \left(\frac{\mathbf{E_1 - V}}{\mathbf{Z_1}} + \frac{\mathbf{E_2 - V}}{\mathbf{Z_2}}\right)\mathbf{Z}$$

$$= \left(\frac{\mathbf{E_1}}{\mathbf{Z_1}} + \frac{\mathbf{E_2}}{\mathbf{Z_2}}\right)\mathbf{Z} - \mathbf{V}\left(\frac{1}{\mathbf{Z_1}} + \frac{1}{\mathbf{Z_2}}\right)\mathbf{Z}$$

$$\therefore \quad \mathbf{V}\left(\frac{1}{\mathbf{Z}} + \frac{1}{\mathbf{Z_1}} + \frac{1}{\mathbf{Z_2}}\right) = \frac{\mathbf{E_1}}{\mathbf{Z_1}} + \frac{\mathbf{E_2}}{\mathbf{Z_2}} = \mathbf{I_{SC1} + I_{SC2}} = \mathbf{I_{SC}}$$

where $\mathbf{I_{SC1}}$ and $\mathbf{I_{SC2}}$ are the short-circuit currents of the two alternators.

If $\quad \dfrac{1}{\mathbf{Z_0}} = \left(\dfrac{1}{\mathbf{Z}} + \dfrac{1}{\mathbf{Z_1}} + \dfrac{1}{\mathbf{Z_2}}\right)$; then $\mathbf{V} \times \dfrac{1}{\mathbf{Z_0}} = \mathbf{I_{SC}}$ or $\mathbf{V} = \mathbf{Z_0 I_{SC}}$

Example 35.43. *A 3,000-kVA, 6-pole alternator runs at 1000 r.p.m. in parallel with other machines on 3,300-V bus-bars. The synchronous reactance is 25%. Calculate the synchronizing power for one mechanical degree of displacement and the corresponding synchronizing torque.* **(Elect. Machines-I, Gwalior Univ. 1984)**

Solution. It may please be noted that here the alternator is working in parallel with many alternators. Hence, it may be considered to be connected to infinite bus-bars.

Voltage/phase $= 3,300/\sqrt{3} = 1905$ V

F.L. current $I = 3 \times 10^6/\sqrt{3} \times 3,300 = 525$ A

Now, $IX_S = 25\%$ of 1905 $\quad \therefore \quad X_S = 0.25 \times 1905/525 = 0.9075\ \Omega$

Also, $P_{SY} = 3 \times \alpha E^2/X_S$

Here $\alpha = 1°$ (mech.); α (elect.) $= 1 \times (6/2) = 3°$

$\therefore \quad \alpha = 3 \times \pi/180 = \pi/60$ elect. radian.

$$\therefore \quad P_{SY} = \frac{3 \times \pi \times 1905^2}{60 \times 0.9075 \times 1000} = 628.4 \text{ kW}$$

$$T_{SY} = \frac{60\,.\,P_{SY}}{2\pi N_S} = 9.55\,\frac{P_{SY}}{N_S} = 9.55\,\frac{628.4 \times 10^3}{1000} = \mathbf{6,000\ N\text{-}m}$$

Example 35.44. *A 3-MVA, 6-pole alternator runs at 1000 r.p.m on 3.3-kV bus-bars. The synchronous reactance is 25 percent. Calculate the synchronising power and torque per mechanical degree of displacement when the alternator is supplying full-load at 0.8 lag.* **(Electrical Machines-1, Bombay Univ. 1987)**

Solution. $V = 3{,}300/\sqrt{3} = 1905$ V/phase, F.L. $I = 3 \times 10^6/\sqrt{3} \times 3{,}300 = 525$ A

$IX_S = 25\%$ of $1905 = 476$ V; $X_S = 476/525 = 0.9075\ \Omega$

Let, $\mathbf{I} = 525 \angle 0°$, then, $\mathbf{V} = 1905\,(0.8 + j0.6) = 1524 + j1143$

$\mathbf{E}_0 = \mathbf{V} + \mathbf{IX}_S = (1524 + j1143) + (0 + j476) = (1524 + j1619) = 2220 \angle 46°44'$

Obviously, E_0 leads I by 46°44′. However, V leads I by $\cos^{-1}(0.8) = 36°50'$.

Hence, $\alpha = 46°44' - 36°50' = 9°54'$

$\alpha = 1°$ (mech.), No. of pair of poles $= 6/2 = 3$ $\quad\therefore\ \alpha = 1 \times 3 = 3°$ (elect.)

$$P_{SY} \text{ per phase} = \frac{EV}{X_S}\cos\alpha\sin\delta = \frac{2220 \times 1905}{0.9075} \times \cos 9°54' \sin 3° = 218 \text{ kW}$$

P_{SY} for three phases $= 3 \times 218 =$ **654 kW**

$T_{SY} = 9.55 \times P_{SY}/N_S = 9.55 \times 654 \times 10^2/1000 =$ **6.245 N-m**

Example 35.45. *A 750-kVA, 11-kV, 4-pole, 3-φ, star-connected alternator has percentage resistance and reactance of 1 and 15 ohm respectively. Calculate the synchronising power per mechanical degree of displacement at (a) no-load (b) at full-load 0.8 p.f. lag. The terminal voltage in each case is 11 kV.* **(Electrical Machines-II, Indore Univ. 1985)**

Solution. F.L. Current $I = 75 \times 10^3/\sqrt{3} \times 11 \times 10^3 = 40$ A

$V_{ph} = 11{,}000/\sqrt{3} = 6{,}350$ V, $IR_a = 1\%$ of $6{,}350 = 63.5$

or $40\,R_a = 63.5$, $R_a = 1.6\Omega$; $40 \times X_S = 15\%$ of $6{,}350 = 952.5$ V

$\therefore\ X_S = 23.8\ \Omega$; $Z_S = \sqrt{1.6^2 + 23.8^2} \cong 23.8\ \Omega$

(*a*) No-load

α(mech) $= 1°$: α(elect) $= 1 \times (4/2) = 2° = 2 \times \pi/180 = \pi/90$ elect. radian.

$$P_{SY} = \frac{\alpha E^2}{Z_S} \cong \frac{\alpha E^2}{X_S} = \frac{(\pi/90) \times 6350^2}{23.8} = 59{,}140\text{W} = \textbf{59.14 kW/phase.}$$

On no-load, V has been taken to be equal to E.

(*b*) F.L. 0.8 p.f.

As indicated in Art. 35.35, $P_{SY} = \alpha EV/X_S$. The value of E (or E_0) can be found from Fig. 35-87.

$E = [(V\cos\phi + IR_a)^2 + (V\sin\phi + IX_S)^2]^{1/2}$

$= [(6350 \times 0.8 + 63.5)^2 + (6350 \times 0.6 + 952.5)^2]^{1/2}$

$= 7010$ V

$$P_{SY} = \frac{\alpha EV}{X_S} = \frac{(\pi/90) \times 7010 \times 6350}{23.8} = 65{,}290 \text{ W}$$

$=$ **65.29 kW/phase**

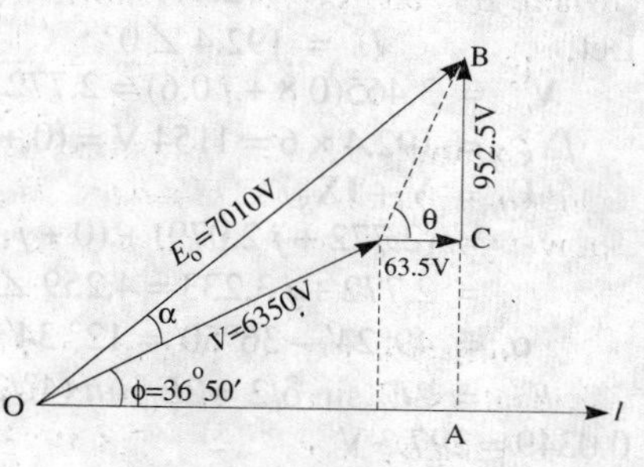

Fig. 35.87

More Accurate Method [Art. 35.35]

$$P_{SY} = \frac{EV}{X_S}\cos\alpha\sin\delta$$

Now, $E = 7010$ V, V = 6350 V, $\delta = 1° \times (4/2) = 2°$ (elect)

As seen from Fig. 35.87, $\sin(\phi + \alpha) = AB/OB = (6350 \times 0.6 + 952.5)/7010 = 0.6794$

$\therefore\ (\phi + \alpha) = 42°30'$; $\alpha = 42°30' - 36°50' = 5°40'$

$$\therefore\ P_{SY} = \frac{7010 \times 6350}{23.8} \times \cos 5°40' \times \sin 2°$$

$= 7010 \times 6350 \times 0.9953 \times 0.0349/23.8 = 64{,}970$ W $=$ **64.97 kW/phase**

Note. It would be instructive to link this example with Ex. 36.1 since both are concerned with synchronous machines, one generating and the other motoring.

Example 35.46. *A 2,000-kVA, 3-phase, 8-pole alternator runs at 750 r.p.m. in parallel with* *other machines on 6,000 V bus-bars. Find synchronizing power on full-load 0.8 p.f. lagging per mechanical degree of displacement and the corresponding synchronizing torque. The synchronous reactance is 6 ohm per phase.* **(Elect. Machines-II, Bombay Univ. 1987)**

Solution. Approximate Method

As seen from Art. 35.37 and 38, $P_{SY} = \alpha EV/X_S$ —per phase

Now α = 1°(mech); No. of pair of poles = 8/2 = 4;

$\alpha = 1 \times 4 = 4°$(elect)

$= 4\pi/180 = \pi/45$ elect. radian

$V = 6000/\sqrt{3} = 3{,}465$ —assuming Y-connection

F.L. current $I = 2000 \times 10^3/\sqrt{3} \times 6000 = 192.4$ A

As seen from Fig. 35.88,

$E_0 = [(V\cos\phi)^2 + (V\sin\phi + IX_S)^2]^{1/2} = 4295$ V

$= [(3465 \times 0.8)^2 + (3465 \times 0.6 + 192.4 \times 6)^2]^{1/2}$

$= 4295$ V

$P_{SY} = (\pi/45) \times 4295 \times 3465/6 = 173{,}160$ W = **173.16 kW/phase**

P_{SY} for three phases $= 3 \times 173.16 = 519.5$ kW

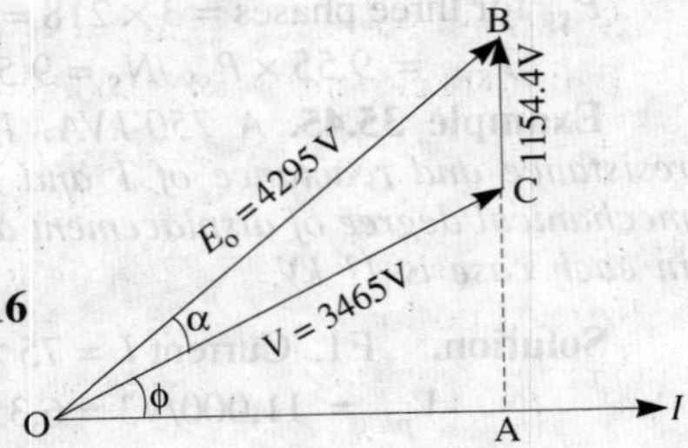

If T_{SY} is the total synchronizing torque for three phases in N-m, then

$T_{SY} = 9.55\, P_{SY}/N_S = 9.55 \times 519{,}500/750 =$ **6,614 N-m**

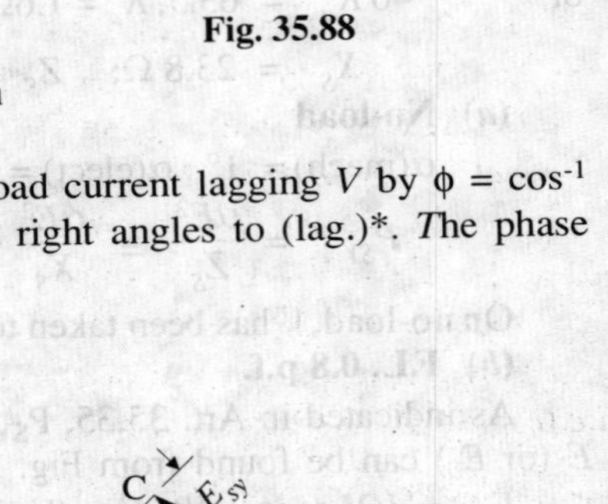

Fig. 35.88

Exact Method

As shown in the vector diagram of Fig. 35.89, I is full-load current lagging V by $\phi = \cos^{-1}(0.8) = 36°50'$. The reactance drop is IX_S and its vector is at right angles to (lag.)*. The phase angle between E_0 and V is α.

F.L. current $I = 2{,}000{,}000/\sqrt{3} \times 6{,}000$

$= 192.4$ A

Let, $I = 192.4 \angle 0°$

$\mathbf{V} = 3{,}465(0.8 + j\,0.6) = 2{,}772 + j\,2{,}079$

$IX_S = 192.4 \times 6 = 1154$ V $= (0 + j\,1154)$ V

$\mathbf{E_0} = \mathbf{V} + \mathbf{IX_S}$

$= (2{,}772 + j\,2{,}079) + (0 + j\,1154)$

$= 2{,}772 + j\,3{,}233 = 4{,}259 \angle 49°24'$

$\alpha = 49°24' - 36°50' = 12°\,34'$

$E_{SY} = 2E_0 \sin\delta/2 = 2E_0 \sin(4°/2) = 2 \times 4{,}259 \times 0.0349 = 297.3$ V

$I_{SY} = 297.3/6 = 49.55$ A

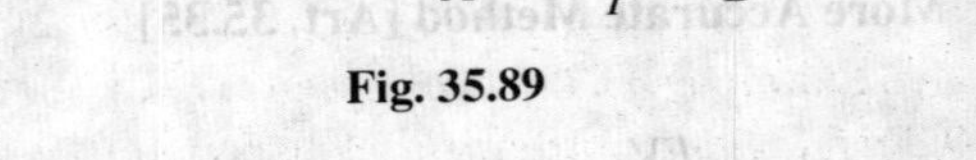

Fig. 35.89

As seen, V leads I by ϕ and I_{SY} leads I by $(\phi + \alpha + \delta/2)$, hence I_{SY} leads V by $(\alpha + \delta/2) = 12°34' + (4°/2) = 14°34'$.

$\therefore$ P_{SY}/phase $= VI_{SY} \cos 14°34' = 3465 \times 49.55 \times \cos 14°34' = 166{,}200$ W $= 166.2$ kW

Synchronising power for three phases is $= 3 \times 166.2 =$ **498.6 kW**

If T_{SY} is the total synchronizing torque, then $T_{SY} \times 2\pi \times 750/60 = 498{,}600$

$\therefore$ $T_{SY} = 9.55 \times 498{,}600/750 =$ **6,348 N-m**

Alternative Method

We may use Eq. (*iii*) of Art. 35.36 to find the total synchronizing power.

$$P_{SY} = \frac{EV}{E_S}\cos\alpha\sin\delta \quad \text{—per phase}$$

*Earlier, we had called this e.m.f. as E when discussing regulation.

Here, E = 4,259 V ; V = 3,465 V; $\alpha = 12° 34'$; $\delta = 4°$ (elect.)

$\therefore$ P_{SY} / phase = $4{,}259 \times 3{,}465 \times \cos 12° 34' \times \sin 4° / 6$

= $4{,}259 \times 3{,}465 \times 0.976 \times 0.0698/6 = 167{,}500$ W = 167.5 kW

P_{SY} for 3 phases = 3×167.5 = **502.5 kW**

Next, T_{SY} may be found as above.

Example 35.47. *A 5,000-kV A, 10,000 V, 1500-r.p.m., 50-Hz alternator runs in parallel with other machines. Its synchronous reactance is 20%. Find for (a) no-load (b) full-load at power factor 0.8 lagging, synchronizing power per unit mechanical angle of phase displacement and calculate the synchronizing torque, if the mechanical displacement is 0.5°.*

(Elect. Engg. V, M.S. Univ. Baroda, 1986)

Solution. Voltage / phase = $10{,}000 / \sqrt{3}$ = 5,775 V

Full -load current = $5{,}000{,}000 / \sqrt{3} \times 10{,}000$ = 288.7 A

$$X_S = \frac{20}{100} \times \frac{5{,}775}{288.7} = 4\ \Omega, \quad P = \frac{120 f}{N_S} = \frac{120 \times 50}{1500} = 4$$

$\alpha = 1°$ (mech.) ; No. of pair of poles = 2 $\therefore$ $\alpha = 1 \times 2 = 2°$ (elect.) $= 2\pi / 180 = \pi / 90$ radian

(*a*) At no-load

$$P_{SY} = \frac{3\alpha E^2}{X_S} = 3 \times \frac{\pi}{90} \times \frac{5{,}775^2}{4 \times 1000} = \mathbf{873.4\ kW}$$

$\therefore$ $T_{SY} = 9.55 \times (873.4 \times 10^3)/1500 = 5{,}564$ N-m

$\therefore$ T_{SY} for 0.5° = 5564/2 = **2,782 N-m**

(*b*) At F.L. p.f. 0.8 lagging

Let $\mathbf{I} = 288.7 \angle 0°$. Then $\mathbf{V} = 5775\,(0.8 + j\,0.6) = 4620 + j\,3465$

$\mathbf{I.X_S} = 288.7 \angle 0° \times 4 \angle 90° = (0 + j\,1155)$

$\mathbf{E_0} = \mathbf{V} + \mathbf{IX_S} = (4620 + j3465) + (0 + j1155) = 4620 + j4620 = 6533 \angle 45°$

$\cos\phi = 0.8, \phi = \cos^{-1}(0.8) = 36°50'$

Now, E_0 leads I by 45° and V leads I by 36° 50′. Hence, E_0 leads V by (45° - 36°50′) = 8°10′ *i.e.*α = 8°10′. As before, δ = 2° (elect).

As seen from Art. 35.36, $P_{SY} = \dfrac{EV}{X_S} \cos\alpha \sin\delta$ —per phase

$= 6533 \times 5775 \times \cos 8° 10' \times \sin 2°/4 = 326$ kW

P_{SY} for three phases $= 3 \times 326 = 978$ kW

T_{SY} / unit displacement $= 9.55 \times 978 \times 10^3/1500 = 6{,}237$ N-m

T_{SY} for 0.5° displacement = 6,237/2 = **3118.5 N-m**

(*c*) We could also use the approximate expression of Art. 35.36

P_{SY} per phase $= \alpha E V / X_S = (\pi / 90) \times 6533 \times 5775 / 4 = \mathbf{329.3\ kW}$

Example 35.48. *Two 3-phase, 6.6-kW, star-connected alternators supply a load of 3000 kW at 0.8 p.f. lagging. The synchronous impedance per phase of machine A is (0.5 + j10) Ω and of machine B is (0.4 + j 12) Ω. The xcitation of machine A is adjusted so that it delivers 150 A at a lagging power factor and the governors are so set that load is shared equally between the machines. Determine the current, power factor, induced emf. and load angle of each machine.*

(Electrical Machines-II, South Gujarat Univ. 1985)

Solution. It is given that each machine carries a load of 1500 kW. Also, $V = 6600 / \sqrt{3}$ = 3810 V. Let $V = 3810 \angle 0° = (3810 + j\,0)$.

For machine No. 1

$\sqrt{3} \times 6600 \times 150 \times \cos\phi_1 = 1500 \times 10^3$; $\cos\phi_1 = 0.874$, $\phi_1 = 29°$; $\sin\phi_1 = 0.485$

Total current $I = 3000 / \sqrt{3} \times 6.6 \times 0.8 = 328$ A

or $I = 828(0.8 - j\,0.6) = 262 - j195$

Now, $\mathbf{I}_1 = 150\,(0.874 - j\,0.485) = 131 - j\,72.6$

$\therefore$ $\mathbf{I}_2 = (262 - j\,195) - (131 - j\,72.6)$

$= (131 - j\,124.4)$

or $I_2 = \mathbf{181\ A}$, $\cos\phi_2 = 131/181 = \mathbf{0.723\ (lag)}$.

$\mathbf{E}_A = \mathbf{V} + \mathbf{I}_1\mathbf{Z}_1 = 3810 + (131 - j72.6)(0.5 + j10)$

$= 4600 + j1270$

Line value of e.m.f. $= \sqrt{3}\ \sqrt{(4600^2 + 1270^2)} = \mathbf{8{,}260\ V}$

Load angle $\alpha_1 = (1270/4600) = \mathbf{15.4°}$

$\mathbf{E}_B = \mathbf{V} + \mathbf{I}_2\mathbf{Z}_2 = 3810 + (131 - j124.4)\,(0.4 + j12)$

$= 5350 + j1520$

Line value of e.m.f $= \sqrt{3}\ \sqrt{5350^2 + 1520^2} = \mathbf{9600\ V}$

Load angle $\alpha_2 = \tan^{-1}(1520/5350) = \mathbf{15.9°}$

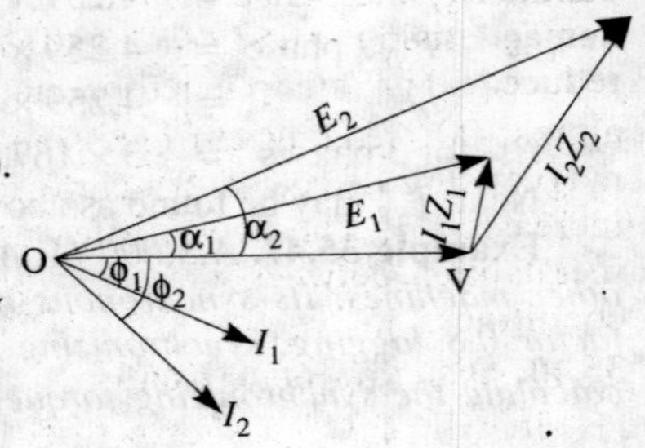

Fig. 35.90

Example 35.49. *Two single-phase alternator operating in parallel have induced e.m.fs on open circuit of 230 ∠ 0° and 230 ∠ 10° volts and respective reactances of j2 Ω and j3 Ω. Calculate (i) terminal voltage (ii) currents and (iii) power delivered by each of the alternators to a load of impedance 6 Ω (resistive).* **(Electrical Machines-II, Indore Univ. 1987)**

Solution. Here, $\mathbf{Z}_1 = j2$, $\mathbf{Z}_2 = j3$, $\mathbf{Z} = 6$; $\mathbf{E}_1 = 230\angle 0°$ and

$\mathbf{E}_2 = 230\angle 10° = 230(0.985 + j\,0.174) = (226.5 + j\,39.9)$

$$(ii)\ \mathbf{I}_1 = \frac{(E_1 - E_2)Z + E_1Z_2}{Z(Z_1 + Z_2) + Z_1Z_2} = \frac{[(230 + j0) - (226.5 + j\,39.9)] \times 6 + 230 \times j3}{6(j2 + j3) + j2 \times j3}$$

— Art. 35.38

$= 14.3 - j3.56 = 14.73\angle -14°$

$$\mathbf{I}_2 = \frac{(E_2 - E_1)Z + E_2Z_1}{Z(Z_1 + Z_2) + Z_1Z_2} = \frac{(-3.5 + j39.9) + (222.5 + j39.9) \times j2}{6(j2 + j3) + j2 \times j3}$$

$= 22.6 - j\,1.15 = 22.63\angle -3.4°$

$(i)\ \mathbf{I} = \mathbf{I}_1 + \mathbf{I}_2 = 36.9 - j\,4.71 = 37.2\angle -7.3°$

$\mathbf{V} = \mathbf{IZ} = (36.9 - j4.71) \times 6 = 221.4 - j28.3 = 223.2\angle -7.3°$

$(iii)\ P_1 = VI_1\cos\phi_1 = 223.2 \times 14.73 \times \cos 14° = \mathbf{3190\ W}$

$P_2 = VI_2\cos\phi_1 = 223.2 \times 22.63 \times \cos 3.4° = \mathbf{5040\ W}$

Tutorial Problem No. 35.6.

1. Calculate the synchronizing torque for unit mechanical angle of phase displacement for a 5,000-kVA, 3-ϕ alternator running at 1,500 r.p.m. when connected to 6,600-volt, 50-Hz bus-bars. The armature has a short-circuit reactance of 15%. **[43,370 kg-m]** *(City & Guilds, London)*
2. Calculate the synchronizing torque for one mechanical degree of phase displacement in a 6.000-kVA, 50-Hz, alternator when running at 1,500 r.p.m with a generated e.m.f. of 10,000 volt. The machine has a synchronous impedance of 25%.
   [544 kg.m] *(Electrical Engineering-III, Madras Univ. April 1978; Osmania Univ.May 1976)*
3. A 10,000-kVA, 6,600-V, 16-pole, 50-Hz, 3-phase alternator has a synchronous reactance of 15%. Calculate the synchronous power per mechanical degree of phase displacement from the full load position at power factor 0.8 lagging **[10 MW]** *(Elect. Machines-I, Gwalior Univ. 1977)*
4. A 6.6 kV, 3-phase, star-connected turbo-alternator of synchronous reactance 0.5 ohm/phase is applying 40 MVA at 0.8 lagging p.f. to a large system. If the steam supply is suddently cut off, explain what takes place and determine the current the machine will then carry. Neglect losses. **[2100 A]** *(Elect. Machines (E-3) AMIE Sec. B Summer 1990)*

35.39. Effect of Unequal Voltages

Let us consider two alternators, which are running exactly in-phase (relative to the external circuit) but which have slightly unequal voltages, as shown in Fig. 35.91. If E_1 is greater than E_2, then their resultant is $E_r = (E_1 - E_2)$ and is in-phase with E_1. This E_r or E_{SY} sets up a local synchronizing current I_{SY} which (as discussed earlier) is almost 90°

behind E_{SY} and hence behind E_1 also. This lagging current produces demagnetising effect (Art. 35.16) on the first machine, hence E_1 is reduced. The other machine runs as a synchronous motor, taking almost 90° leading current. Hence, its field is strengthened due to magnetising effect of armature reaction (Art. 35.16). This tends to increase E_2. These two effects act together and hence lessen the inequalities between the two voltages and tend to establish stable conditions.

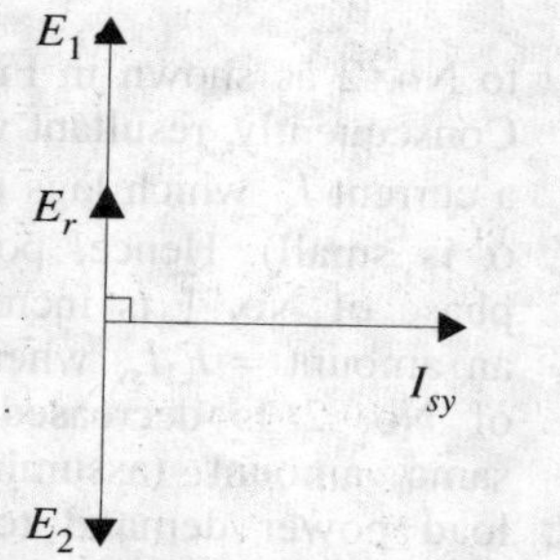

Fig. 35. 91

35.40. Distribution of Load

It will, now be shown that the amount of load taken up by an alternator running, in parallel with other machines, is solely determined by its driving torque *i.e.* by the power input to its prime mover (by giving it more or less steam, in the case of steam drive). Any alternation in its excitation merely changes its kVA output, but not its kW output. In other words, it merely changes the power factor at which the load is delivered.

(*a*) Effect of Change in Excitation

Suppose the initial operating conditions of the two parallel alternators are identical *i.e.* each alternator supplies one half of the active load (kW) and one-half of the reactive load (kVAR), the operating power factors thus being equal to the load p.f. In other words, both active and reactive powers are divided equally thereby giving equal apparent power triangles for the two machines as shown in Fig. 35.92 (*b*). As shown in Fig. 35.92 (*a*), each alternator supplies a load current I so that total output current is 2 I.

Now, let excitation of alternator No. 1 be increased, so that E_1 becomes greater than E_2. The difference between the two e.m.fs. sets up a circulating current $I_C = I_{SY} = (E_1 - E_2) / 2Z_S$ which is confined to the local path through the armatures and round the bus-bars. This current is superimposed on the original current distribution. As seen, I_C is vectorially added to the load current of alternator No. 1 and subtracted from that of No. 2. The two machines now deliver load currents I_1 and I_2 at respective power factors of cos ϕ_1 and cos ϕ_2. These changes in load currents lead to changes in power factors, such that cos ϕ_1 is reduced, whereas cos ϕ_2 is increased. However,

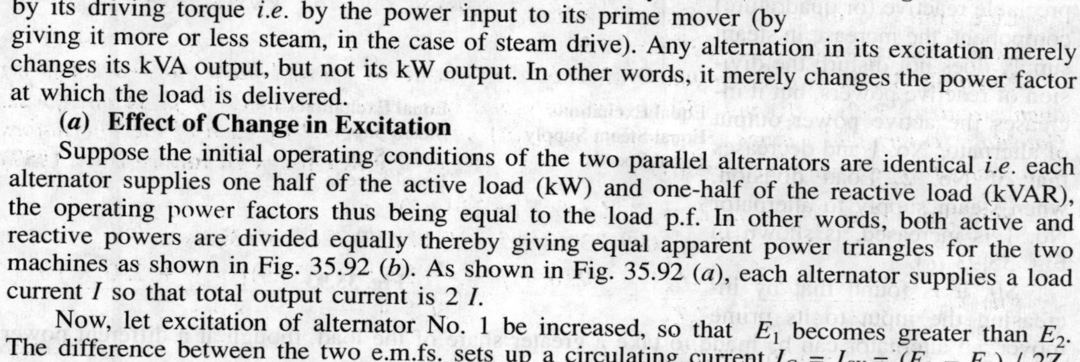

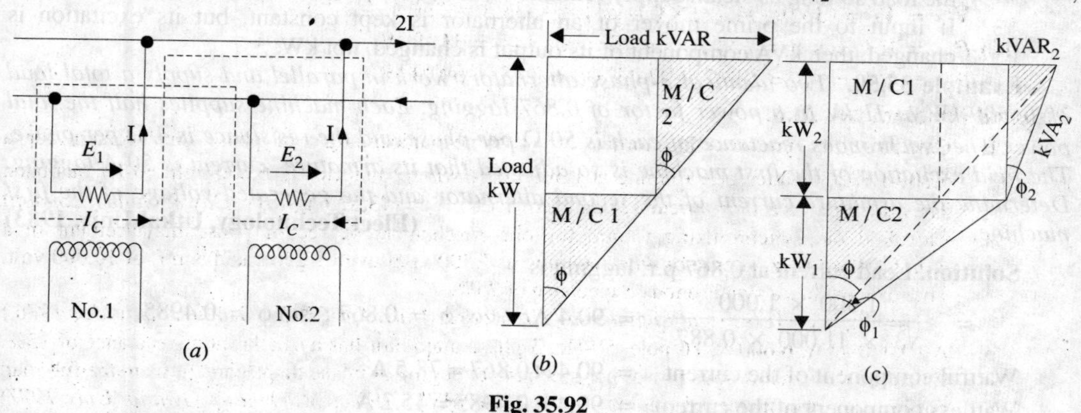

Fig. 35.92

effect on the kW loading of the two alternators is negligible, but kVAR$_1$ supplied by alternator No. 1 is increased, whereas kVAR$_2$ supplied by alternator No. 2 is correspondingly decreased, as shown by the kVA triangles of Fig. 35.92 (*c*).

(*b*) Effect of Change in Steam Supply

Now, suppose that excitations of the two alternators are kept the same but steam supply to alternator No. 1 is increased *i.e.* power input to its prime mover is increased. Since the speeds of the two machines are tied together by their synchronous bond, machine No. 1 cannot overrun machine No 2. Alternatively, it utilizes its increased power input for carrying more load than No. 2. This can be made possible only when rotor No. 1 advances its angular position with respect to No. 2 as shown in Fig. 35.93 (*b*) where E_1 is shown advanced ahead of E_2 by an angle α. Consequently, resultant voltage E_r (or E_{sy}) is produced which, acting on the local circuit, sets up a current I_{sy} which lags by almost 90° behind E_r but is almost in phase with E_1 (so long as angle α is small). Hence, power per phase of No. 1 is increased by an amount = $E_1 I_{sy}$ whereas that of No. 2 is decreased by the same amount (assuming total load power demand to remain unchanged). Since I_{sy} has no appreciable reactive (or quadrature) component, the increase in steam supply does not disturb the division of reactive powers, but it increases the active power output of alternator No. 1 and decreases that of No. 2. Load division, when steam supply to alternator No. 1 is increased, is shown in Fig. 35.93 (*c*).

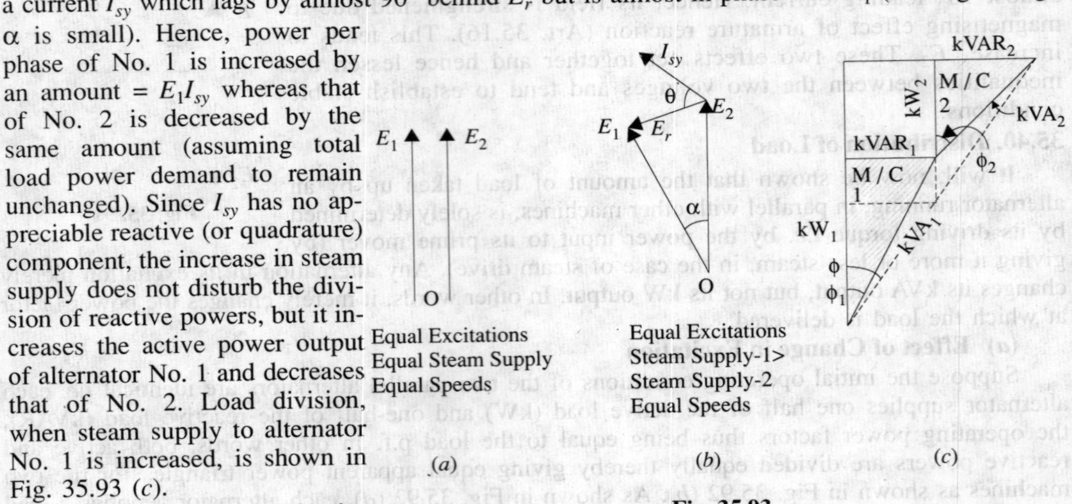

Fig. 35.93

So, it is found that by increasing the input to its prime mover, an alternator can be made to take a greater share of the load, though at a different power factor.

The points worth remembering are :

1. The load taken up by an alternators directly depends upon its driving torque or in other words, upon the angular advance of its rotor.
2. The excitation merely changes the p.f. at which the load is delivered without affecting the load so long as steam supply remains unchanged.
3. If input to the prime mover of an alternator is kept constant, but its excitation is changed, then kVA component of its output is changed, not kW.

Example 35.50. *Two identical 3-phase alternators work in parallel and supply a total load of 1, 500 kW at 11 kV at a power factor of 0.867 lagging. Each machine supplies half the total power. The synchronous reactance of each is 50 Ω per phase and the resistance is 4 Ω per phase. The field excitation of the first machine is so adjusted that its armature current is 50 A lagging. Determine the armature current of the second alternator and the generated voltage of the first machine.* **(Elect.Technology, Utkal Univ. 1983)**

Solution. Load current at 0.867 p.f. lagging is

$$= \frac{1{,}500 \times 1{,}000}{\sqrt{3} \times 11{,}000 \times 0.887} = 90.4 \text{ A}; \quad \cos\phi = 0.867 \,;\ \sin\phi = 0.4985$$

Wattful component of the current $= 90.4 \times 0.867 = 78.5$ A

Wattless component of the current $= 90.4 \times 0.4985 = 45.2$ A

Each alternator supplies half of each of the above two component when conditions are identical (Fig. 35.94).

Current supplied by each machine = 90.4/2 = 45.2 A

Since the steam supply of first machine is not changed, the working components of both machines would remain the same at 78.5 / 2 = 39.25 A. But the wattless or reactive components would be redivided due to change in excitation. The armature current of the first machine is changed from 45.2 A to 50 A.

∴ Wattless component of the 1st machine $= \sqrt{50^2 - 39.52^2} = 31$ A

Wattless component of the 2nd machine = 45.2 – 31 = 14.1 A

The new current diagram is shown in Fig. 35.95 (*a*)

(*i*) Armative current of the 2nd alternator, $I_2 = \sqrt{39.25^2 + 14.1^2} = 41.75$ A

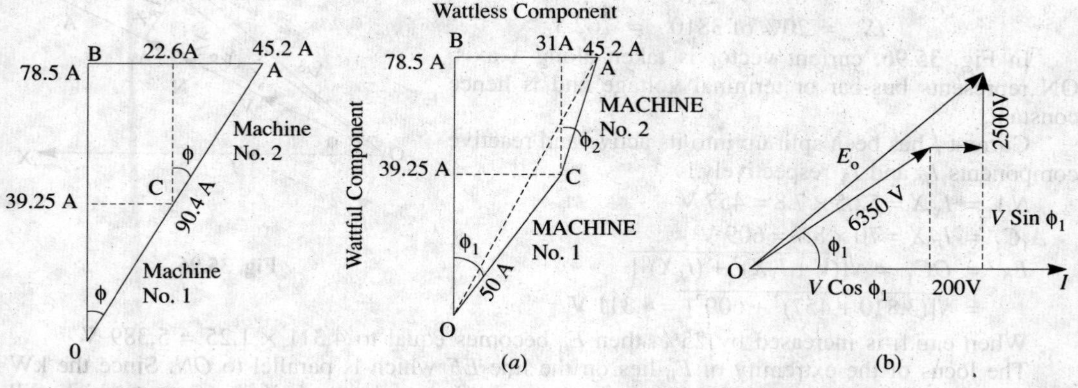

Fig. 35.94 **Fig. 35.95**

(*ii*) Terminal voltage / phase $= 11{,}000 / \sqrt{3} = 6{,}350$

Considering the first alternator,

IR drop $= 4 \times 50 = 200$ V ; IX drop $= 50 \times 50 = 2{,}500$ V

$\cos \phi_1 = 39.25/50 = 0.785$; $\sin \phi_1 = 0.62$

Then, as seen from Fig. 35.95 (*b*)

$$E = \sqrt{(V \cos \phi_1 + IR)^2 + (V \sin \phi_1 + IX)^2}$$

$$= \sqrt{(6{,}350 \times 0.785 + 200)^2 + (6{,}350 \times 0.62 + 2{,}500)^2} = 8{,}350 \text{ V}$$

Line voltage $= 8{,}350 \times \sqrt{3} =$ **14,450 V**

Example 35.51. *Two alternators A and B operate in parallel and supply a load of 10 MW at 0.8 p.f. lagging (a) By adjusting steam supply of A, its power output is adjusted to 6,000 kW and by changing its excitation, its p.f. is adjusted to 0.92 lag. Find the p.f. of alternator B.*

(b) If steam supply of both machines is left unchanged, but excitation of B is reduced so that its p.f. becomes 0.92 lead, find new p.f. of A.

Solution. (*a*) $\cos \phi = 0.8$, $\phi = 36.9°$, $\tan \phi = 0.7508$; $\cos \phi_A = 0.92$, $\phi_A = 23°$; $\tan \phi_A = 0.4245$

load kW $=$ 10,000, load kVAR $= 10{,}000 \times 0.7508 = 7508$ (lag)

kW of A $=$ 6,000, kVAR of A $= 6{,}000 \times 0.4245 = 2547$ (lag)

Keeping in mind the convention that lagging kVAR is taken as negative we have,

kW of B $= (10{,}000 - 6{,}000) = 4{,}000$: kVAR of $B = (7508 - 2547) = 4961$ (lag)

$\therefore$ kVA of $B = 4{,}000 - j4961 = 6373 \angle -51.1°$; $\cos \phi_B = \cos 51.1° =$ **0.628**

(*b*) Since steam supply remains unchanged, load kW of each machine remains as before but due to change in excitation, kVARs of the two machines are changed.

kW of $B =$ 4,000, new kVAR of $B = 4000 \times 0.4245 = 1698$ (lead)

kW of $A =$ 6,000, new kVAR of $A = -7508 - (+1698) = -9206$ (lag.)

$\therefore$ new kVA of $A = 6{,}000 - j9206 = 10{,}988 \angle -56.9°$; $\cos \phi_A =$ **0.546 (lag)**

Example 35.52. *A 6,000-V, 1,000-kVA, 3-φ alternator is delivering full-load at 0.8 p.f. lagging. Its reactance is 20% and resistance negligible. By changing the excitation, the e.m.f. is increased by 25% at this load. Calculate the new current and the power factor. The machine is connected to infinite bus-bars.*

Solution. Full-load current

$$I = \frac{1{,}000{,}000}{\sqrt{3} \times 6{,}600} = 87.5 \text{ A}$$

Voltage / phase $= 6{,}600/\sqrt{3} \quad = 3{,}810$ V

Reactance $= \dfrac{3810 \times 20}{87.5 \times 100} = 8.7\ \Omega$

$IX = 20\%$ of $3810 = 762$ V

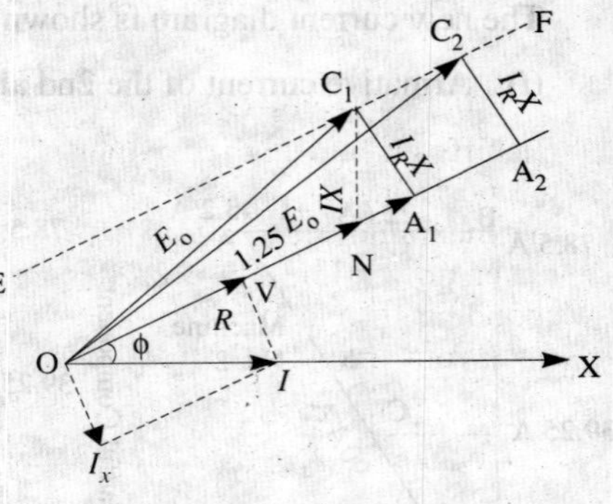

Fig. 35.96

In Fig. 35.96, current vector is taken along X-axis. ON represents bus-bar or terminal voltage and is hence constant.

Current I has been split up into its active and reactive components I_R and I_X respectively.

$NA_1 = I_X.X = 52.5 \times 7.8 = 457$ V

$A_1C_1 = I_R.X = 70 \times 8.7 = 609$ V

$E_0 = OC_1 = \sqrt{[(V + I_XX)^2 + (I_RX)^2]}$

$= \sqrt{[(3{,}810 + 457)^2 + 609^2]} = 4{,}311$ V

When e.m.f. is increased by 25%, then E_0 becomes equal to $4{,}311 \times 1.25 = 5{,}389$ V

The locus of the extremity of E_0 lies on the line EF which is parallel to ON. Since the kW is unchanged, I_R and hence $I_R.X$ will remain the same. It is only the $I_X.X$ component which will be changed. Let OC_2 be the new value of E_0. Then $A_2C_2 = A_1C_1 = I_RX$ as before. But the $I_X X$ component will change. Let I' be the new line current having active component I_R (the same as before) and the new reactive component I_X'. Then, $I_X'X = NA_2$

From right-angled triangle OC_2A_2

$OC_2^2 = OA_2^2 + A_2C_2^2$; $5{,}389^2 = (3810 + VA_2)^2 + 609^2$

$\therefore \quad VA_2 = 1546$ V or $I_X'X = 1546 \qquad \therefore I_X' = 1546/8.7 = 177.7$ A

$\therefore$ new line current $I' = \sqrt{(70^2 + 177.7^2)} =$ **191 A**

New angle of lag, $\phi' = \tan^{-1}(177.7/70) = 68°\ 30'$; $\cos\phi' = \cos 68°\ 30' =$ **0.3665**

As a check, the new power $= \sqrt{3} \times 6{,}600 \times 191 \times 0.3665 = 800$ kW

It is the same as before $= 1000 \times 0.8 = 800$ kW

Example 35.53. *A 6,600-V, 1000-kVA alternator has a reactance of 20% and is delivering full-load at 0.8 p.f. lagging. It is connected to constant-frequency bus-bars. If steam supply is gradually increased, calculate (i) at what output will the power factor become unity (ii) the maximum load which it can supply without dropping out of synchronism and the corresponding power factor.*

Solution. We have found in Example 35.52 that

$I = 87.5$ A, $X = 8.7\ \Omega$, V/phase $= 3{,}810$ V

$E_0 = 4{,}311$ V, $I_R = 70$ A, $I_X = 52.5$ A and $I.X = 87.5 \times 8.7 = 762$ V

Using this data, vector diagram of Fig. 35.97 can be constructed.

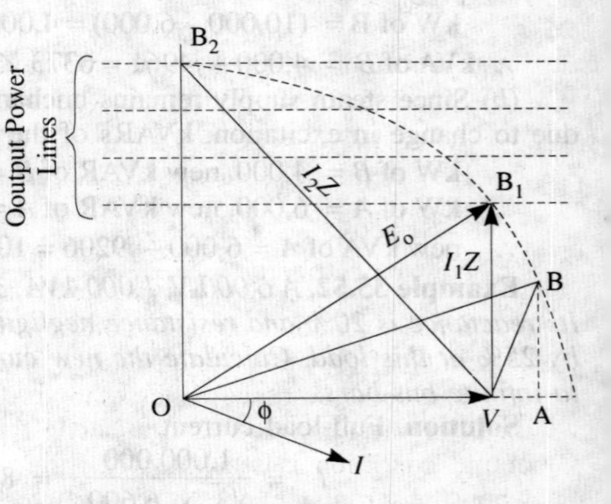

Fig. 35.97

Since excitation is constant, E_0 remains constant, the extremity of E_0 lies on the arc of a circle of radius E_0 and centre O. Constant power lines have been shown dotted and they are all parallel to OV. Zero power output line coincides with OV. When p.f. is unity, the current vector lies along OV, I_1Z is ⊥ to OV and cuts the arc at B_1. Obviously

$VB_1 = \sqrt{(OB_1^2 - OV^2)}$

$= \sqrt{(4{,}311^2 - 3{,}810^2)} = 2018$ V

Now $Z = X \qquad \therefore \quad I_1X = 2{,}018$ V

$\therefore \quad I_1 = 2018/8.7 = 232$ A

(*i*) $\therefore$ power output at u.p.f.

$$= \frac{\sqrt{3} \times 6,600 \times 232}{1000} = \mathbf{2,652\ kW}$$

(*ii*) As vector *OB* moves upwards along the arc, output power goes on increasing *i.e.*, point *B* shifts on to a higher output power line. Maximum output power is reached when *OB* reaches the position OB_2 where it is vertical to *OV*. The output power line passing through B_2 represents the maximum output for that excitation. If OB is further rotated, the point B_2 shifts down to a lower power line *i.e.* power is decreased. Hence, $B_2V = I_2Z$ where I_2 is the new current corresponding to maximum output.

From triangle OB_2V, it is seen that

$$B_2V = \sqrt{(OV^2 + OB_2^2)} = \sqrt{(3,810^2 + 4,311^2)} = 5,753\text{V}$$

$$\therefore \quad I_2 Z = 5,753 \text{ V} \qquad \therefore \quad I_2 = 5753/8.7 = 661\ A$$

Let I_{2R} and I_{2X} be the power and wattless components of I_2, then

$$I_{2R}\ X = OB_2 = 4311 \text{ and } \quad I_{2R} = 4311/8.7 = 495.6 \text{ A}$$

Similarly $\quad I_{2X} = 3810/8.7 = 438 \text{ A}; \quad \tan\phi_2 = 438/495.6 = 0.884$

$$\therefore \quad \phi_2 = 41°28'; \cos\phi_2 = \mathbf{0.749}$$

$$\therefore \quad \text{Max. power output} = \frac{\sqrt{3} \times 6,600 \times 661 \times 0.749}{1000} = \mathbf{5,658\ kW}$$

Example 35.54. *A 3-phase, star-connected turbo-alternator, having a synchronous reactance of 10 Ω per phase and negligible armature resistance, has an armature current of 220 A at unity p.f. The supply voltage is constant at 11 kV at constant frequency. If the steam admission is unchanged and the e.m.f. raised by 25%, determine the current and power factor.*

If the higher value of excitation is maintained and the steam supply is slowly increased, at what power output will the alternator break away from synchronism ?

Draw the vector diagram under maximum power condition.

(Elect.Machinery-III, Banglore Univ. 1992)

Solution. The vector diagram for unity power factor is shown in Fig. 35.98. Here, the current is wholly active.

$$OA_1 = 11,000/\sqrt{3} = 6,350 \text{ V}$$

$$A_1C_1 = 220 \times 10 = 2,200 \text{ V}$$

$$E_0 = \sqrt{(6,350^2 + 2,200^2)} = 6,810 \text{ V}$$

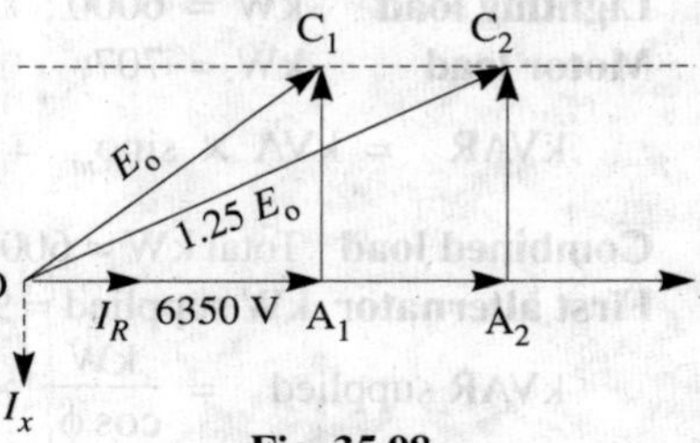

When e.m.f. is increased by 25%, the e.m.f. becomes $1.25 \times 6,810 = 8,512$ V and is represented by OC_2. Since the kW remains unchanged, $A_1C_1 = A_2C_2$. If I' is the new current, then its active component I_R would be the same as before and equal to 220 A. Let its reactive component be I_X. Then

Fig. 35.98

$$A_1A_2 = I_X X_S = 10\, I_X$$

From right-angled $\Delta\ OA_2C_2$, we have

$$8,512^2 = (6350 + A_1A_2)^2 + 2,200^2$$

$$\therefore \quad A_1A_2 = 1870 \text{ V} \qquad \therefore\ 10\,I_X = 1870 \qquad I_X = 187 \text{ A}$$

Hence, the new current has active component of 220 A and a reactive component of 187 A.

$$\text{New current} = \sqrt{220^2 + 187^2} = \mathbf{288.6\ A}$$

$$\text{New power factor} = \frac{\text{active component}}{\text{total current}} = \frac{220}{288.6} = \mathbf{0.762\ (lag)}$$

Since excitation remains constant, E_0 is constant. But as the steam supply is increased, the extremity of E_0 lies on a circle of radius E_0 and centre *O* as shown in Fig. 35.99.

The constant-power lines (shown dotted) are drawn parallel to *OV* and each represents the locus of the e.m.f. vector for a constant power output at varying excitation. Maximum power output condition is reached when the vector E_0 becomes perpendicular to *OV*. In other words, when the circular e.m.f. locus becomes tangential to the constant-power lines *i.e.* at point *B*. If the steam supply is increased further, the alternator will break away from synchronism.

B.V.= $\sqrt{6{,}350^2 + 8{,}512^2} = 10{,}620$ V

$\therefore \quad I_{max} \times 10 = 10{,}620$ or $I_{max} = 1{,}062$ A

If I_R and I_X are the active and reactive components of I_{max}, then

$10\, I_R = 8{,}512 \quad \therefore I_R = 851.2$ A; $\quad 10\, I_X = 6{,}350 \quad \therefore I_X = 635$ A

Power factor at maximum power output = 851.2/1062 = 0.8 (lead)

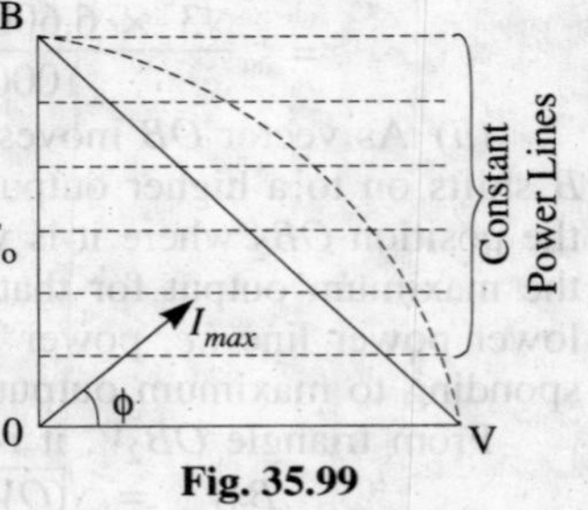

Fig. 35.99

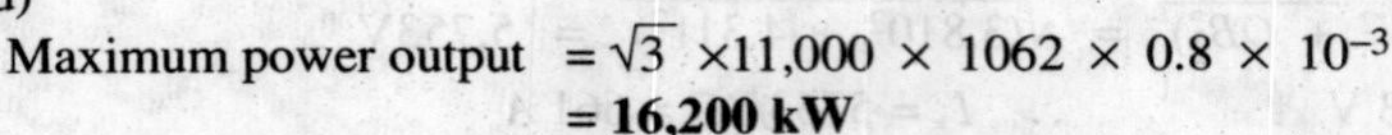

Maximum power output $= \sqrt{3} \times 11{,}000 \times 1062 \times 0.8 \times 10^{-3}$

$= \mathbf{16{,}200\ kW}$

Example 35.55. *Two 20-MVA, 3-φ alternators operate in parallel to supply a load of 35MVA at 0.8 p.f. lagging. If the output of one machine is 25 MVA at 0.9 lagging, what is the output and p.f. of the other machine?* **(Elect.Machines, Punjab Univ. 1990)**

Solution. Load MW = 35 × 0.8 = 28; load MVAR = 35 × 0.6 = 21

First Machine $\cos \phi_1 = 0.9$, $\sin \phi_1 = 0.436$; $MVA_1 = 25$, $MW_1 = 25 \times 0.9 = 22.5$

$MVAR_1 = 25 \times 0.436 = 10.9$

Second Machine $MW_2 = MW - MW_1 = 28 - 22.5 = 5.5$

$MVAR_2 = MVAR - MVAR_1 = 21 - 10.9 = 10.1$

$\therefore \quad MVA_2 = \sqrt{MW_2^2 + MVAR_2^2} = \sqrt{5.5^2 + 10.1^2} = \mathbf{11.5}$

$\cos \phi_2 = 5.5/11.5 = \mathbf{0.478}$ **(lag)**

Example 35.56. *A lighting load of 6,000 kW and a motor load of 707 kW at 0.707 p.f. are supplied by two alternators running in parallel. One of the machines supplies 900 kW at 0.9 p.f. lagging. Find the load and p.f. of the second machine.*

(Electrical Technology, Bombay Univ. 1988}

Solution. We will split the two loads into their kW and kVAR components.

Lighting load kW = 6000 ; kVAR = 0

Motor load kW = 707; $\cos \phi_m = 0.707$; $\sin \phi_m = 0.707$

$$\text{kVAR} = \text{kVA} \times \sin\phi_m = \frac{\text{kW}}{\cos \phi_m} \times \sin\phi_m = \frac{707}{0.707} \times 0.707 = 707$$

Combined load Total kW = 600 + 707 = 1307; Total kVAR = 707

First alternator kW supplied = 900; $\cos \phi_1 = 0.9$; $\sin \phi_1 = 0.4357$

$$\text{kVAR supplied} = \frac{\text{kW}}{\cos \phi_1} \times \sin \phi_1 = \frac{900}{0.9} \times 0.4357 = 392$$

Second Alternator

kW supplied = 1307 – 900 = **407** ; kVAR supplied = 707 – 392 = 315

$\tan \phi_2 = 315/407 = 0.7739$; $\phi_2 = \tan^{-1}(0.7739) = 37°44'$; $\cos \phi_2 = \mathbf{0.7916}$ **(lag)**

Example 35.57. *Two alternators, working in parallel, supply the following loads :*

(i) Lighting load of 500 kW *(ii) 1000 kW at p.f. 0.9 lagging*

(iii) 800 kW at p.f. 0.8 lagging *(iv) 500 kW at p.f. 0.9 leading*

One alternator is supplying 1500 kW at 0.95 p.f. lagging. Calculate the kW output and p.f. of the other machine.

Solution. We will tabulate the kW and kVAR components of each load separately :

Load	*kW*	*kVAR*
(i)	500	

(*ii*)	1000	$\frac{1000}{0.9} \times 0.436 = 485$
(*iii*)	800	$\frac{800 \times 0.6}{0.8} = 600$
(*iv*)	500	$\frac{500 \times 0.436}{0.9} = -242$
Total	2800	+843

For 1st machine, it is given : kW = 1500, kVAR = (1500/0.95) × 0.3123 = 493

$\therefore$ kW supplied by other machine = 2800 – 1500 = **1300**

kVAR supplied = 843 – 493 = 350 $\therefore$ tan ϕ 350/1300 = 0.27 $\therefore$ cos ϕ = **0.966**

Example 35.58. *Two 3-ϕ synchronous mechanically-coupled generators operate in parallel on the same load. Determine the kW output and p.f. of each machine under the following conditions: synchronous impedance of each generator : 0.2 + j2 ohm/phase. Equivalent impedance of the load : 3 + j4 ohm/phase. Induced e.m.f. per phase, 2000 + j0 volt for machine I and 2,2000 + j 100 for II.* **[London Univ.]**

Solution. Current of 1st machine $= \mathbf{I}_1 = \frac{\mathbf{E}_1 - \mathbf{V}}{0.2 + j2}$ or $\mathbf{E}_1 - \mathbf{V} = \mathbf{I}_1 (0.2 + j2)$

Similarly $\mathbf{E}_2 - \mathbf{V} = \mathbf{I}_2(0.2 + j2)$

Also $\mathbf{V} = (\mathbf{I}_1 + \mathbf{I}_2)(3 + j4)$ where $3 + j4$ = load impedance

Now $\mathbf{E}_1 = 2{,}000 + j0$, $\mathbf{E}_2 = 2{,}200 + j100$

Solving from above, we get $\mathbf{I}_1 = 68.2 - j\,102.5$

Similarly $\mathbf{I}_2 = 127 - j\,196.4$; $\mathbf{I} = \mathbf{I}_1 + \mathbf{I}_2 = 195.2 - j299$

Now $\mathbf{V} = \mathbf{IZ} = (192.2 - j\,299)\,(3 + j\,4) = 1781 - j\,115.9$

Using the method of conjugate for power calculating, we have for the first machine

$P_{VA1} = (1781 - j\,115.9)\,(68.2 + j\,102.5) = 133{,}344 + j\,174{,}648$

$\therefore$ kW_1 = 133.344 kW/phase = 3 × 133.344 = **400 kW** —for 3 phases

Now $\tan^{-1}(102.5/68.2) = 56°24'$; $\tan^{-1}(115.9/1781) = 3°\,43'$

$\therefore$ for 1st machine ; cos(56°24′ - 3°43′) = **0.6062**

$P_{VA2} = (1781 - j115.9)\,(127 + j\,196.4) = 248{,}950 + j\,335{,}069$

$\therefore$ kW_2 = 248.95 kW/phase = **746.85 kW** —for 3 phases

$\tan^{-1}(196.4/127) = 57°6'$; cos ϕ = cos (57°6′ – 3°43′) = **0.596**

Example 35.59. *The speed regulations of two 800-kW alternators A and B, running in parallel, are 100% to 104% and 100% to 105% from full-load to no-load respectively. How will the two alternators share a load of 1000 kW? Also, find the load at which one machine ceases to supply any portion of the load.* **(Power Systems-I, A.M.I.E. 1989)**

Solution. The speed / load characteristics (assumed straight) for the alternators are shown in Fig. 35.100. Out of the combined load *AB* = 1000 kW, *A*'s share is *AM* and *B*'s share is *BM*. Hence, *AM* + *BM* = 1000 kW. *PQ* is the horizontal line drawn through point *C*, which is the point of intersection.

From similar ∆s *GDA* and *CDP*, we have

$$\frac{CP}{GA} = \frac{PD}{AD} \text{ or}$$

$$CP = GA.\frac{PD}{AD}$$

Since, $PD = (4 - h)$

$\therefore \quad CP = 800(4 - h)/4$
$= 200(4 - h)$

Similarly, from similar Δs BEF and QEC, we get

$$\frac{QC}{BF} = \frac{QE}{BE} \quad \text{or}$$

$$QC = 800(5 - h)/5 = 160\,(5 - h)$$

$\therefore \; CP + QC = 1000$ or $200\,(4 - h) + 160\,(5 - h) = 1000$ or $h = 5/3$

$\therefore \quad CP = 200(4 - 5/3) = 467$ kW, $QC = 160\,(5 - 5/3) = 533$ kW

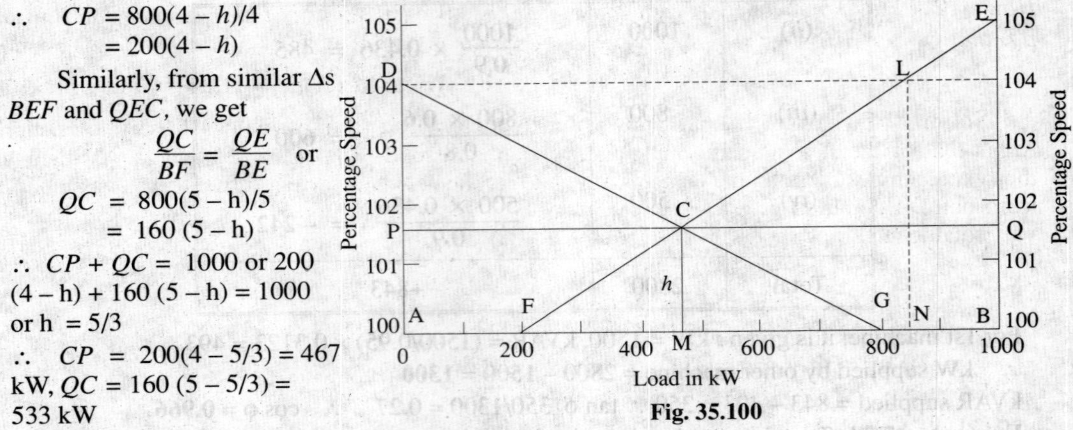

Fig. 35.100

Hence, alternator A supplies **467 kW** and B supplies **533 kW.**

Alternator A will cease supplying any load when line PQ is shifted to point D. Then, load supplied by alternator B (= BN) is such that the speed variation is from 105% to 104%.

Knowing that when its speed varies from 105% to 100%, alternator B supplies a load of 800 kW, hence load supplied for speed variation from 105% to 100% is (by proportion)

$= 800 \times 1/5 =$ **160 kW (=** BN**)**

Hence, when load drops from 1000 kW to 160 kW, alternator A will cease supplying any portion of this load.

Example 35.60. *Two 50-MVA, 3-φ alternators operate in parallel. The settings of the governors are such that the rise in speed from full-load to no-load is 2 per cent in one machine and 3 per cent in the other, the characteristics being straight lines in both cases. If each machine is fully loaded when the total load is 100 MW, what would be the load on each machine when the total load is 60 MW?* **(Electrical Machines-II, Punjab Univ. 1991)**

Solution. Fig. 35.101 shows the speed/load characteristics of the two machines, NB is of the first machine and MA is that of the second. Base AB shows equal load division at full-load and speed. As the machines are running in parallel, their frequencies must be the same. Let CD be drawn throught x% speed where total load is 60 MW.

$$CE = 50 - AP = 50 - \frac{50}{3}x$$

$$ED = 50 - QB = 50 - \frac{50}{2}x$$

$$\therefore \quad CD = 50 - (50/3)\,x + 50 - 25\,x$$

$$\therefore \quad 60 = 50 - (50/3)\,x + 50 - 25\,x$$

$$x = \frac{24}{25}\,; \therefore \text{LE} = 100\frac{24}{25}\%$$

Load supplied by 1st machine

$$= ED = 50 - 25 \times \frac{24}{25} = \mathbf{26\ MW}$$

Load supplied by 2nd machine.

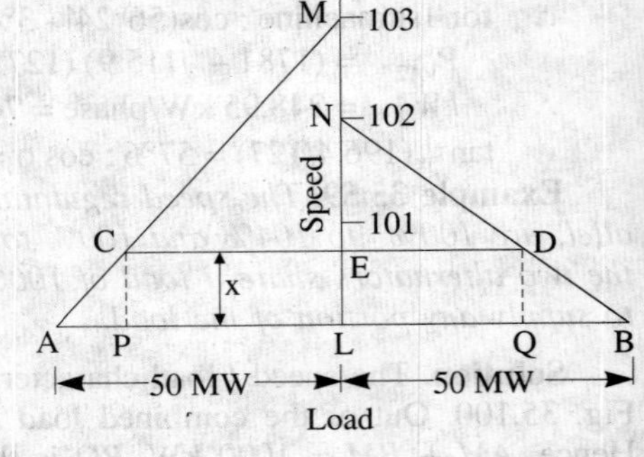

Fig. 35.101

$$= CE = 50 - \left(\frac{50}{3}\right) \times \frac{24}{25} = \mathbf{34\ MW}$$

Example 35.61. *Two identical 2,000 -kVA alternators operate in parallel. The governor of the first machine is such that the frequency drops uniformly from 50-Hz on no-load to 48-Hz on full-load. The corresponding uniform speed drop of the second machines is 50 to 47.5 Hz (a)*

How will the two machines share a load of 3,000 kW? (b) What is the maximum load at unity p.f. that can be delivered without overloading either machine ?

(Electrical Machinery-II, Osmania Univ. 1989)

Solution. In Fig. 35.102 are shown the frequency/load characteristics of the two machines, *AB* is that of the first machine and *AD* that of the second. Remembering that the frequency of the two machines must be the same at any load, a line *MN* is drawn at a frequency *x* as measured from point *A*(common point).

Total load at that frequency is $NL + ML = 3000$ kW

From Δs *ABC* and *ANL*, $NL / 2000 = x / 2.5$

$\therefore \quad NL = 2000\, x / 2.5 = 800\, x$

Similarly, $ML = 2000\, x / 2 = 1000x$

$\therefore \quad 1800\, x = 3000$ or $x = 5/3$

Frequency $= 50 - 5/3 = 145/3$ Hz.

(*a*) $NL = 800 \times 5/3 =$ **1333 kW (assuming u.p.f.)**

$ML = 1000 \times 5/3 =$ **1667 kW (assuming u.p.f.)**

(*b*) For getting maximum load, *DE* is extended to cut *AB* at *F*. Max. load = *DF*.

Now, $EF = 2000 \times 2/3.5 = 1600$ kW

$\therefore \quad$ Max. load $= DF = 2{,}000 + 1{,}600 =$ **3,600 kW.**

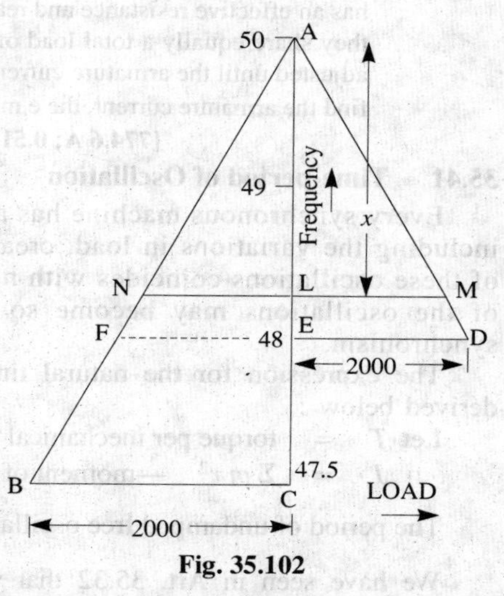

Fig. 35.102

Tutorial Problem No. 35.7

1. Two similar 6,600-V, 3-ϕ, generators are running in parallel on constant-voltage and freequency bus-bars. Each has an equivalent resistance and reactance of 0.05 Ω and 0.5 Ω respectively and supplies one half of a total load of 10,000 kW at a lagging p.f. of 0.8, the two machines being similarly excited. If the excitation of one machine be adjusted until the armature current is 438 A and the steam supply to the turbine remains unchanged, find the armature current, the e.m.f. and the p.f. of the other alternator.
[789 A, 7200 V, 0.556] (*City & Guilds, London*)

2. A single-phase alternator connected to 6,600-V bus-bars has a synchronous impedance of 10Ω and a resistance of 1 Ω. If its excitation is such that on open circuit the p.d. would be 5000 V, calculate the maximum load the machine can supply to the external circuit before dropping out of step and the corresponding armature current and p.f. **[2864 kW, 787 A, 0.551]** (*London Univ.*)

3. A turbo-alternator having a reactance of 10 Ω has an armature current of 220 A at unity power factor when running on 11,000 V, constant-frequency bus-bars. If the steam admission is unchanged and the e.m.f. raised by 25%, determine graphically or otherwise the new value of the machine current and power factor. If this higher value of excitation were kept constant and the steam supply gradually increased, at what power output would the alternator break from synchronism? Find also the current and power factor to which this maximum load corresponds. State whether this p.f. is lagging or leadin
[360 A at 0.611 p.f. ; 15 427 kW ; 1785 A at 0.7865 leading] (*City & Guilds, London*)

4. Two single-phase alternators are connected to a 50-Hz bus-bars having a constant voltage of 10^8 [176 0° kV. Generator *A* has an induced e.m.f. of 13 $\angle$ 22.6° kV and a reactance of 2 Ω; generator *B* has an e.m.f. of 12.5 $\angle$ 36.9° kV and a reactance of 3 Ω. Find the current, kW and kVAR supplied by each generator. (*Electrical Machine-II, Indore Univ. July 1977*)

5. Two 15-kVA, 400-V, 3-ph alternators in parallel supply a total load of 25 kVA at 0.8 p.f. lagging. If one alternator shares half the power at unity p.f., determine the p.f. and kVA shared by the other alternator. **[0.5548; 18.03 kVA]** (*Electrical Technology-II, Madras Univ. Apr. 1977*)

6. Two 3-ϕ, 6,600-V, star-connected alternators working in parallel supply the following loads:
(*i*) Lighting load of 400 kW (*ii*) 300 kW at p.f. 0.9 lagging

(*iii*) 400 kW at p.f. 0.8 lagging (*iv*) 1000 kW at p.f. 0.71 lagging

Find the output, armature current and the p.f. of the other machine if the armature current of one machine is 110 A at 0.9 p.f. lagging. **[970 kW, 116 A, 0.73 lagging]**

7. A 3-φ, star-connected, 11,000-V turbo-generator has an equivalent resistance and reactance of 0.5 Ω and 8 Ω respectively. It is delivering 200 A at u.p.f. when running on a constant-voltage and constant-frequency bus-bars. Assuming constant steam supply and unchanged efficiency, find the current and p.f. if the induced e.m.f. is raised by 25%. **[296 A, 0.67 lagging]**
8. Two similar 13,000-V, 3-ph alternators are operated in parallel on infinite bus-bars. Each machine has an effective resistance and reactance of 0.05 Ω and 0.5 Ω respectively. When equally excited, they share equally a total load of 18 MW at 0.8 p.f. lagging. If the excitation of one generator is adjusted until the armature current is 400 A and the steam supply to its turbine remains unaltered, find the armature current, the e.m.f. and the p.f. of the other generator.
[774.6 A; 0.5165 : 13,470 V] (*Electric Machinery-II, Madras Univ. Nov. 1977*)

35.41 Time-period of Oscillation

Every synchronous machine has a natural time period of free oscillation. Many causes, including the variations in load, create phase-swinging of the machine. If the time period of these oscillations coincides with natural time period of the machine, then the amplitude of the oscillations may become so greatly developed as to swing the machine out of synchronism.

The expression for the natural time period of oscillations of a synchronous machine is derived below :

Let T = torque per mechanical radian (in N-m/mech. radian)

J = $\Sigma m r^2$ —moment of inertia in kg-m^2.

The period of undamped free oscillations is given by $t = 2\pi \sqrt{\dfrac{J}{T}}$.

We have seen in Art. 35.32 that when an alternator swings out of phase by an angle α (electrical radian), then synchronizing power developed is

$$P_{SY} = \alpha E^2/Z \quad \text{—}\alpha \text{ in elect. radian}$$

$$= \frac{E^2}{Z} \text{ per electrical radian per phase.}$$

Now, 1 electrical radian = $\dfrac{P}{2}$ × mechanical radian—where P is the number of poles.

∴ P_{SY} per mechanical radian displacement = $\dfrac{E^2 P}{2Z}$.

The synchronizing or restoring torque is given by

$$T_{SY} = \frac{P_{SY}}{2\pi N_S} = \frac{E^2 P}{4\pi Z N_S} \quad \text{—}N_S \text{ in r.p.s.}$$

Torque for three phases is $T = 3T_{SY} = \dfrac{3\,E^2\,P}{4\,\pi\,Z\,N_S}$ where E is e.m.f. per phase ...(*i*)

Now E/Z = short-circuit current = I_{SC}

f = $PN_S/2$; hence $P/N_S = 2f/N_S^2$

Substituting these values in (*i*) above, we have

$$T_{SY} = \frac{3}{4\pi}\cdot\left(\frac{E}{Z}\right)\cdot E\cdot\frac{P}{N_S} = \frac{3}{4\,\pi}\cdot I_{SC}\cdot E\cdot\frac{2f}{N_S^2} = 0.477\,\frac{E\,I_{SC} f}{N_S^2}$$

Now, $$t = 2\pi\sqrt{\frac{J}{0.477\,E\,I_{SC}\,f/N_S^2}} = 9.1\,N_S\sqrt{\frac{J}{E\cdot I_{SC}\cdot f}} \text{ second}$$

$$= 9.1\,N_S\sqrt{\frac{J}{\frac{1}{\sqrt{3}}.E_L.I.(I_{SC}/I).f}} \qquad \left(\because E = \frac{E_L}{\sqrt{3}}\right)$$

$$= 9.1\,N_S\sqrt{\frac{J}{\frac{1}{2}.\sqrt{3}.E_L.I(I_{SC}/I).f}}$$

$$= 9.1\,N_S\sqrt{\frac{J}{\frac{1000}{3}.\frac{\sqrt{3}.E_L.I}{1000}\left(\frac{I_{SC}}{I}\right).f}}$$

$$= \frac{9.1\times\sqrt{3}}{\sqrt{1000}}.N_S.\sqrt{\frac{J}{\text{kVA}.(I_{SC}/I)f}}$$

$$\therefore \quad t = 0.4984\,N_S\sqrt{\frac{J}{\text{kVA}.(I_{SC}/I).f}}$$

where kVA = full-load kVA of the alternator; N_S = r.p.s. of the rotating system

If N_S represents the speed in r.p.m., then

$$t = \frac{0.4984}{60}N_S\sqrt{\frac{J}{\text{kVA}.(I_{SC}/I).f}} = 0.0083\,N_S\sqrt{\frac{J}{\text{kVA}\,(I_{SC}/I).f}} \text{ second.}$$

Note. It may be proved that I_{SC}/I = 100/percentage reactance = 100/% X_S.

Proof. Reactance drop $= I.X_S = \frac{V\times\%\,X_S}{100}$ $\quad\therefore\quad X_S = \frac{\text{reactance drop}}{\text{full–load current}} = \frac{V\times\%\,X_S}{100\times I}$

Now $\quad I_{SC} = \frac{V}{X_S} = \frac{V\times 100\times I}{V\times\%X_S} = \frac{100}{\%X_S}\times I$; or $\frac{I_{SC}}{I} = \frac{100}{\%\,X_S}$

For example, if synchronous reactances is 25 per cent, then

I_{SC}/I = 100/25 = 4 (please see Ex. 35.64)

Example 35.62. *A 5,000-kVA, 3-phase, 10,000-V, 50-Hz alternate runs at 1500 r.p.m. connected to constant-frequency, constant-voltage bus-bars. If the moment of inertia of entire rotating system is 1.5 × 10⁴ kg.m² and the steady short-circuit current is 5 times the normal full-load current, find the natural time period of oscillation.* **(Elect. Engg. Grad. I.E.T.E. 1991)**

Solution. The time of oscillation is given by

$$t = 0.0083\,N_S.\sqrt{\frac{J}{kVA.(I_{SC}/I)f}} \text{ second}$$

Here, N_S = 1500 r.p.m. ; $I_{SC}/I = 5$* ; $j = 1.5\times10^4$ kg-m² ; f = 50 Hz

$$t = 0.0083\times1500\sqrt{\frac{1.5\times10^4}{5000\times5\times50}} = \textbf{1.364 s.}$$

Example 35.63. *A 10,000-kVA, 4-pole, 6,600-V, 50-Hz, 3-phase star-connected alternator has a synchronous reactance of 25% and operates on constant-voltage, constant-frequency bus-bars. If the natural period of oscillation while operating at full-load and unity power factor is to be limited to 1.5 second, calculate the moment of inertia of the rotating system.* **(Electric Machinery-II, Andhra Univ. 1990)**

*It means that synchronous reactance of the alternator is 20%.

Solution. $t = 0.0083 \text{ NS} \sqrt{\dfrac{J}{kVA\,(I_{SC}/I)\,f}}$ second.

Here $I_{SC}/I = 100/25 = 4$; $N_S = 120 \times 50/4 = 1500$ r.p.m.

$\therefore \quad 1.5 = 0.0083 \times 1500 \sqrt{\dfrac{J}{10{,}000 \times 4 \times 50}} = 12.45 \times \dfrac{\sqrt{J}}{10^3 \times \sqrt{2}}$

$\therefore \quad J = (1.5 \times 10^3 \times \sqrt{2}/12.45)^2 = \mathbf{2.9 \times 10^4}$ **kg-m²**

Example 35.64. *A 10-MVA, 10-kV, 3-phase, 50-Hz, 1500 r.p.m. alternator is paralleled with others of much greater capacity. The moment of inertia of the totor is 2×10^5 kg-m² and the synchronous reactance of the machine is 40%. Calculate the frequency of oscillation of the rotor.*

(Elect. Machinery-III, Bangalore Univ. 1992)

Solution. Here, $I_{SC}/I = 100/40 = 2.5$

$$t = 0.0083 \times 1500 \sqrt{\frac{2 \times 10^5}{10^4 \times 2.5 \times 50}} = 5 \text{ second}$$

Frequency $= 1/5 =$ **0.2 Hz.**

Tutorial Problem No. 35-8

1. Show that an alternator running in parallel on constant-voltage and frequency bus-bars has a natural time period of oscillation. Deduce a formula for the time of one complete oscillation and calculate its value for a 5000-kVA, 3-phase, 10,000 V machine running at 1500 r.p.m. on constant 50-Hz bus-bars.
   The moment of inertia of the whole moving system is 14, 112 kg-m² and the steady short-circuit is five times the normal full-load value. **[1.33 second]**
2. A 10,000-kVA, 5-kV, 3-phase, 4-pole, 50-Hz alternator is connected to infinite bus-bars. The short-circuit current is 3.5 times the normal full-load current and the moment of inertia of the rotating system is 21,000 kg-m². Calculate its normal period of oscillation. **[1.365 second]**
3. Calculate for full-load and unity p.f., the natural period of oscillation of a 50-Hz, 10,000-kVA, 11-kV alternator driven at 1500 r.p.m. and connected to an infinite bus-bar. The steady short-circuit current is four times the full-load current and the moment of the inertia of the rotating masses is 17,000 kg-m². **[1.148 s.]** (*Electrical Machinery-II, Madras Univ. Apr. 1976*)
4. Calculate the rotational inertia in kg-m² units of the moving system of 10,000 kVA, 6,600-V, 4-pole, turbo-alternator driven at 1500 r.p.m. for the set to have a natural period of 1 second when running in parallel with a number of other machines. The steady short-circuit current of the alternator is five times the full-load value. **[16,828 kg-m²]** (*City & Guilds, London*)
5. A 3-φ, 4-pole, 6,000 kVA, 5,000-V, 50-Hz star-connected alternator is running on constant-voltage and constant-frequency bus-bars. It has a short-circuit reactance of 25% and its rotor has a moment of inertia of 16,800 kg-m². Calculate its natural time period of oscillation. **[1.48 second]**

35.42. Maximum Power Output

For given values of terminal voltage, excitation and frequency, there is a maximum power that the alternator is capable of delivering. Fig. 35.103 (*a*) shows full-load conditions for a cylindrical rotor where IR_a drop has been neglected*.

The power output per phase is

$$P = VI\cos\phi = \frac{VIX_S\cos\phi}{X_S}$$

Now, from Δ *OBC*, we get

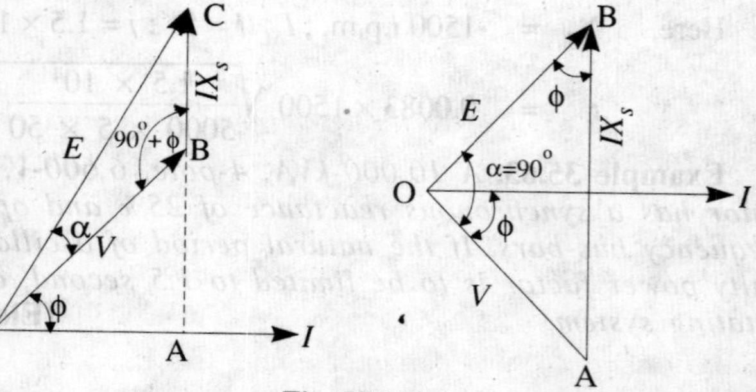

Fig. 35.103

*In fact, this drop can generally be neglected without sacrificing much accuracy of results.

$$\frac{IX_S}{\sin\alpha} = \frac{E}{\sin(90+\phi)} = \frac{E}{\cos\phi}$$

$$IX_S\cos\phi = E\sin\alpha$$

$$\therefore \quad P = \frac{EV\sin\alpha}{X_S}$$

Power becomes maximum when $\alpha = 90°$, if V, E and X_S are regarded as constant (of course, E is fixed by excitation).

$$\therefore \quad P_{max} = EV/X_S$$

It will be seen from Fig. 35.103 (*b*) that under maximum power output conditions, I leads V by ϕ and since IX_S leads I by 90°, angle ϕ and hence $\cos\phi$ is fixed $= E/IX_S$.

Now, from right-angled $\Delta\, AOB$, we have that $IX_S = \sqrt{E^2 + V^2}$. Hence, p.f. corresponding to maximum power output is

$$\cos\phi = \frac{E}{\sqrt{E^2 + V^2}}$$

The maximum power output per phase may also be written as

$$P_{max} = VI_{max}\cos\phi = VI_{max}\frac{E}{\sqrt{E^2 + V^2}}$$

where I_{max} represents the current/phase for maximum power output.

If I_f is the full-load current and $\% X_S$ is the percentage synchronous reactance, then

$$\% X_S = \frac{I_f X_S}{V} \times 100 \qquad \therefore \quad \frac{V}{X_S} = \frac{I_f \times 100}{\% X_S}$$

$$\text{Now,} \quad P_{max} = VI_{max}\frac{E}{\sqrt{E^2 + V^2}} = \frac{EV}{X_S} = \frac{EI_f \times 100}{\% X_S}$$

Two things are obvious from the above equations.

$$(i) \quad I_{max} = \frac{100\, I_f}{\% X_S} \times \frac{\sqrt{E^2 + V^2}}{V}$$

Substituting the value of $\%X_S$ from above,

$$I_{max} = \frac{100 I_f}{100\, I_f X_S} \times V \times \frac{\sqrt{E^2 + V^2}}{V} = \frac{\sqrt{E^2 + V^2}}{X_S}$$

$$(ii) \quad P_{max} = \frac{100\, EI_f}{\% X_S} = \frac{E}{V} \cdot \frac{100}{\%X_S} \times V I_f \text{ per phase}$$

$$= \frac{E}{V} \cdot \frac{100}{\% X_S} \times \text{F.L. power output at u.p.f.}$$

Total maximum power output of the alternator is

$$= \frac{E}{V} \cdot \frac{100}{\% X_S} \times \text{F.L. power output at u.p.f.}$$

Example 35.65. *Derive the condition for the maximum output of a synchronous generator connected to infinite bus-bars and working at constant excitation.*

A 3-ϕ, 11-kV, 5-MVA, Y-connected alternator has a synchronous impedance of (1 + j 10) ohm per phase. Its excitation is such that the generated line e.m.f. is 14 kV. If the alternator is connected to infinite bus-bars, determine the maximum output at the given excitation.

(Electrical Machines-III, Gujarat Univ. 1984)

Solution. For the first part, please refer to Art. 35.41

$$P_{max} \text{ per phase} = \frac{EV}{X_S} \text{ — if } R_a \text{ is neglected} = \frac{V}{Z_S}(E - V\cos\theta) \text{ — if } R_a \text{ is considered}$$

Now, $E = 14{,}000/\sqrt{3} = 8{,}083$ V ; $V = 11{,}000/\sqrt{3} = 6352$ V

$\cos\theta = R_a/Z_S = 1\sqrt{1^2 + 10^2} = 1/10.05$

$$\therefore P_{max} \text{ per phase } = \frac{8083 \times 6352}{10 \times 1000} = 5{,}135 \text{ kW}$$

Total $P_{max} = 3 \times 5{,}135 =$ **15,405 kW**

$$\text{More accurately, } P_{max}\text{/phase} = \frac{6352}{10.05}\left(8083 - \frac{6352}{10.05}\right) = \frac{6352}{10.05} \times \frac{7451}{1000} = 4{,}711 \text{ kW}$$

Total $P_{max} = 4{,}711 \times 3 =$ **14,133 kW.**

Example 35.66. *A 3-phase, 11-kVA, 10-MW, Y-connected synchronous generator has synchronous impedance of (0.8 + j 8.0) ohm per phase. If the excitation is such that the open circuit voltage is 14 kV, determine (i) the maximum output of the generator (ii) the current and p.f. at the maximum output.* **(Electrical Machines-III, Gujarat Univ. 1987)**

Solution. (*i*) If we neglect R_a*, the P_{max} per phase $= EV/X_S$

where V is the terminal voltage (or bus-bar voltage in general) and E the e.m.f. of the machine.

$$\therefore P_{max} = \frac{(11{,}000/\sqrt{3}) \times (14{,}000/\sqrt{3})}{8} = \frac{154{,}000}{24} \text{ kW/phase}$$

Total $P_{max} = 3 \times 154{,}000/24 = 19{,}250$ kW = **19.25 MW**

Incidentally, this output is nearly twice the normal output.

$$(ii) \quad I_{max} = \frac{\sqrt{E^2 + V^2}}{X_S} = \frac{\sqrt{[(14{,}000/\sqrt{3})^2 + (11{,}000/\sqrt{3})^2]}}{8} = \mathbf{1287\ A}$$

$$(iii) \quad \text{p.f.} = \frac{E}{\sqrt{E^2 + V^2}} = \frac{14000/\sqrt{3}}{\sqrt{(14{,}000/\sqrt{3})^2 + (11{,}000/\sqrt{3})^2}} = \mathbf{0.786\ (lead)}.$$

QUESTIONS AND ANSWERS ON ALTERNATORS

Q. 1. What are the two types of turbo-alternators?
Ans. Vertical and horizontal

Q. 2. How do you compare the two?
Ans. Vertical type requires less floor space and while step bearing is necessary to carry the weight of the moving element, there is very little friction in the main bearings. The horizontal type requires no step bearing, but occupies more space.

Q. 3. What is step bearing?
Ans. It consists of two cylindrical cast iron plates which bear upon each other and have a central recess between them. Suitable oil is pumped into this recess under considerable pressure.

Q. 4. What is direct-connected alternator?
Ans. One in which the alternator and engine are directly connected. In other words, there is no intermediate gearing such as belt, chain etc. between the driving engine and alternator.

Q. 5. What is the difference between direct-connected and direct-coupled units?
Ans. In the former, alternator and driving engine are directly and permanently connected. In the latter case, engine and alternator are each complete in itself and are connected by some device such as friction clutch, jaw clutch or shaft coupling.

Q. 6. Can a d.c. generator be converted into an alternator?
Ans. Yes.

Q. 7. How?
Ans. By providing two collector rings on one end of the armature and connecting these two rings to two points in the armature winding 180° apart.

Q. 8. Would this arrangement result in a desirable alternator?
Ans. No.

---

*If R_a is not neglected, then $P_{max} = \dfrac{V}{Z_S}(E - V\cos\theta)$ where $\cos\theta = R_a/Z_S$ (Ex. 35.65).

Q. 9. How is a direct-connected exciter arranged in an alternator?
Ans. The armature of the exciter is mounted on the shaft of the alternator close to the spider hub. In some cases, it is mounted at a distance sufficient to permit a pedestal and bearing to be placed between the exciter and the hub.
Q. 10. Any advantage of a direct-connected exciter?
Ans. Yes, economy of space.
Q. 11. Any disadvantage?
Ans. The exciter has to run at the same speed as the alternator which is slower than desirable. Hence, it must be larger for a given output than the gear-driven type, because it can be run at high speed and so made proportionately smaller.

OBJECTIVE TESTS - 35

1. The frequency of voltage generated by an alternator having 4-poles and rotating at 1800 r.p.m. is ....hertz.
(*a*) 60 (*b*) 7200
(*c*) 120 (*d*) 450.

2. A 50-Hz alternator will run at the greatest possible speed if it is wound for .... poles.
(*a*) 8 (*b*) 6
(*c*) 4 (*d*) 2.

3. The main disadvantage of using short-pitch winding in alterators is that it
(*a*) reduces harmonics in the generated voltage
(*b*) reduces the total voltage around the armature coils
(*c*) produces asymmetry in the three phase windings
(*d*) increases Cu of end connections.

4. Three-phase alternators are invariably Y-connected because
(*a*) magnetic losses are minimised
(*b*) less turns of wire are required
(*c*) smaller conductors can be used
(*d*) higher terminal voltage is obtained.

5. The winding of a 4-pole alternator having 36 slots and a coil span of 1 to 8 is short-pitched by .... degrees.
(*a*) 140 (*b*) 80
(*c*) 20 (*d*) 40.

6. If an alternator winding has a fractional pitch of 5/6, the coil span is .... degrees.
(*a*) 300 (*b*) 150
(*c*) 30 (*d*) 60.

7. The harmonic which would be totally eliminated from the alternator e.m.f. using a fractional pitch of 4/5 is
(*a*) 3rd (*b*) 7th
(*c*) 5th (*d*) 9th.

8. For eliminating 7th harmonic from the e.m.f. wave of an alternator, the fractional-pitch must be
(*a*) 2/3 (*b*) 5/6
(*c*) 7/8 (*d*) 6/7.

9. If, in an alternator, chording angle for fundamental flux wave is α, its value for 5th harmonic is
(*a*) 5α (*b*) $\alpha/5$
(*c*) 25α (*d*) $\alpha/25$.

10. Regarding distribution factor of an armature winding of an alternator which statement is false?
(*a*) it decreases as the distribution of coils (slots/pole) increases
(*b*) higher its value, higher the induced e.m.f. per phase
(*c*) it is not affected by the type of winding either lap. or wave
(*d*) it is not affected by the number of turns per coil.

11. When speed of an alternator is changed from 3600 r.p.m. to 1800 r.p.m., the generated e.m.f./phases will become
(*a*) one-half (*b*) twice
(*c*) four times (*d*) one-fourth.

12. The magnitude of the three voltage drops in an alternator due to armature resistance, leakage reactance and armature reaction is solely determined by
(*a*) load current, I_a
(*b*) p.f. of the load
(*c*) whether it is a lagging or leading p.f. load
(*d*) field construction of the alternator.

13. Armature reacction in an alternator primarily affects

(*a*) rotor speed
(*b*) terminal voltage per phase
(*c*) frequency of armature current
(*d*) generated voltage per phase.

14. Under no-load condition, power drawn by the prime mover of an alternator goes to
(*a*) produce induced e.m.f. in armature winding
(*b*) meet no-load losses
(*c*) produce power in the armature
(*d*) meet Cu losses both in armature and rotor windings.

15. As load p.f. of an alternator becomes more leading, the value of generated voltage required to give rated terminal voltage
(*a*) increases
(*b*) remains unchanged
(*c*) decreases
(*d*) varies with rotor speed.

16. With a load p.f. of unity, the effect of armature reaction on the main-field flux of an alternator is
(*a*) distortional (*b*) magnetising
(*c*) demagnetising (*d*) nominal.

17. At lagging loads, armature reaction in an alternator is
(*a*) cross-magnetising
(*b*) demagnetising
(*c*) non-effective
(*d*) magnetising.

18. At leading p.f., the armature flux in an alternator .... the rotor flux.
(*a*) opposes (*b*) aids
(*c*) distorts (*d*) does not affect.

19. The voltage regulation of an alternator having 0.75 leading p.f. load, no-load induced e.m.f. of 2400V and rated terminal voltage of 3000V is .... per sent.
(*a*) 20 (*b*) – 20
(*c*) 150 (*d*) – 26.7.

20. If, in a 3-ϕ alternator, a field current of 50A produces a full-load armature current of 200 A on short-circuit and 1730 V on open circuit, then its synchronous impedance is .... ohm.
(*a*) 8.66 (*b*) 4
(*c*) 5 (*d*) 34.6.

21. The power factor of an alternator is determined by its
(*a*) speed (*b*) load
(*c*) excitation (*d*) prime mover.

22. For proper parallel operation, a.c. polyphase alternators must have the same
(*a*) speed (*b*) voltage rating
(*c*) kVA rating (*d*) excitation.

23. Of the following conditions, the one which does NOT have to be met by alternators working in parallel is
(*a*) terminal voltage of each machine must be the same
(*b*) the machines must have the same phase rotation
(*c*) the machines must operate at the same frequency
(*d*) the machines must have equal ratings .

24. After wiring up two 3-ϕ alternators, you checked their frequency and voltage and found them to be equal. Before connecting them in paralled, you would
(*a*) check turbine speed
(*b*) check phase rotation
(*c*) lubricate everything
(*d*) check steam pressure.

25. Zero power factor method of an alternator is used to find its
(*a*) efficiency
(*b*) voltage regulation
(*c*) armature resistance
(*d*) synchronous impedance.

26. Some engineers prefer 'lamps bright' synchronization to 'lamps dark' synchronization because
(*a*) brightness of lamps can be judged easily
(*b*) it gives sharper and more accurate synchronization
(*c*) flicker is more pronounced
(*d*) it can be performed quickly.

27. It is never advisable to connect a *stationary* alternator to live bus-bars because it
(*a*) is likely to run as synchronous motor
(*b*) will get short-circuited
(*c*) will decrease bus-bar voltage though momentarily

(*d*) will disturb generated e.m.fs. of other alternators connected in parallel.

28. Two identical alternators are running in parallel and carry equal loads. If excitation of one alternator is increased without changing its steam supply, then

(*a*) it will keep supplying almost the same load

(*b*) kVAR supplied by it would decrease

(*c*) its p.f. will increase

(*d*) kVA supplied by it would decrease.

29. Keeping its excitation constant, if steam supply of an alternator running in parallel with another identical alternator is increased, then

(*a*) it would over-run the other alternator

(*b*) its rotor will fall back in phase with respect to the other machine

(*c*) it will supply greater portion of the load

(*d*) its power factor would be decreased.

30. The load sharing between two steam-driven alternators operating in parallel may be adjusted by varying the

(*a*) field strenhths of the alternators

(*b*) power factors of the alternators

(*c*) steam supply to their prime movers

(*d*) speed of the alternators.

31. Squirrel-cage bars placed in the rotor pole faces of an alternator help reduce hunting

(*a*) above synchronous speed only

(*b*) below synchronous speed only

(*c*) above and below synchronous speeds both

(*d*) none of the above.

(Elect. Machines, A.M.I.E. Sec. B, 1993)

32. For a machine on infinite bus active power can be varied by

(*a*) changing field excitation

(*b*) changing of prime cover speed

(*c*) both (*a*)and (*b*) above

(*d*) none of the above .

(Elect. Machines, A.M.I.E. Sec. B, 1993)

ANSWERS

1. *a* **2.** *d* **3.** *b* **4.** *d* **5.** *d* **6.** *b* **7.** *c* **8.** *d* **9.** *a* **10.** *b* **11.** *a* **12.** *a* **13.** *d* **14.** *b*
15. *c* **16.** *a* **17.** *d* **18.** *b* **19.** *b* **20.** *c* **21.** *b* **22.** *b* **23.** *d* **24.** *b* **25.** *b* **26.** *b*
27. *b* **28.** *a* **29.** *c* **30.** *c* **31.** *c* **32.** *b*

36 SYNCHRONOUS MOTOR

36.1. Synchronous Motor – General

A synchronous motor (Fig. 36.1) is electrically identical with an alternator or a.c. generator. In fact, a given synchronous machine may be used, at least theoretically, as an alternator, when driven mechanically or as a motor, when driven electrically, just as in the case of d.c. machines. Most synchronous motors are rated between 150 kW and 15 MW and run at speeds ranging from 150 to 1800 r.p.m.

Some characteristic features of a synchronous motor are worth noting :

1. It runs either at synchronous speed or not at all *i.e.* while running it maintains a constant speed. The only way to change its speed is to vary the supply frequency (because $N_S = 120\ f / P$).

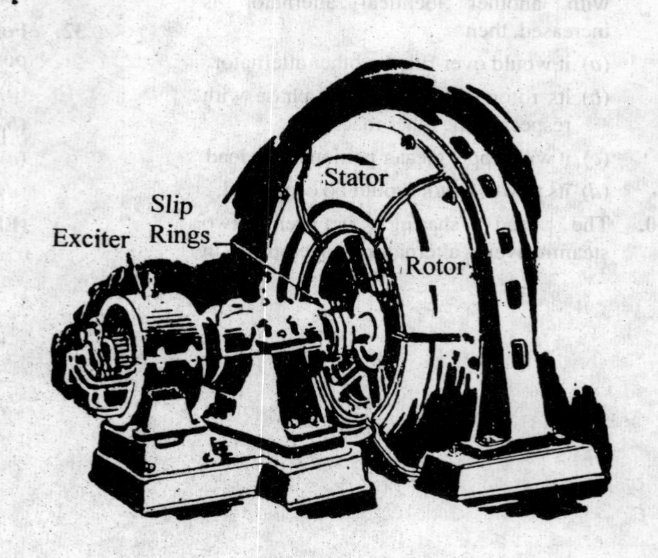

Fig. 36.1

2. It is not inherently self-starting. It has to be run upto synchronous (or near synchronous) speed by some means, before it can be synchronized to the supply.
3. It is capable of being operated under a wide range of power factors, both lagging and leading. Hence, it can be used for power correction purposes, in addition to supplying torque to drive loads.

36.2. Principle of Operation

As shown in Art. 32.7, when a 3-ϕ winding is fed by a 3-ϕ supply, then a magnetic flux of constant magnitude but *rotating* at synchronous speed, is produced. Consider a two-pole stator of Fig. 36.2, in which are shown two stator poles (marked N_S and S_S) rotating at synchronous speed, say, in clockwise direction. With the rotor position as shown, suppose the stator poles are at that instant situated at points *A* and *B*. The two similar poles, *N* (of rotor) and N_S (of stator) as well as *S* and S_S will repel each other, with the result that the rotor tends to rotate in the anti-clockwise direction.

But half a period later, stator poles, having rotated around, interchange their positions *i.e.* N_S

is at point B and S_S at point A. Under these conditions, N_S attracts S and S_S attracts N. Hence, rotor tends to rotate clockwise (which is just the reverse of the first direction). Hence, we find that due to continuous and rapid rotation of stator poles, the rotor is subjected to a torque which is rapidly reversing *i.e.* in quick succession, the rotor is subjected to torque which tends to move it first in one direction and then in the opposite direction. Owing to its large inertia, the rotor cannot instantaneously respond to such quickly-reversing torque, with the result that it remains stationary.

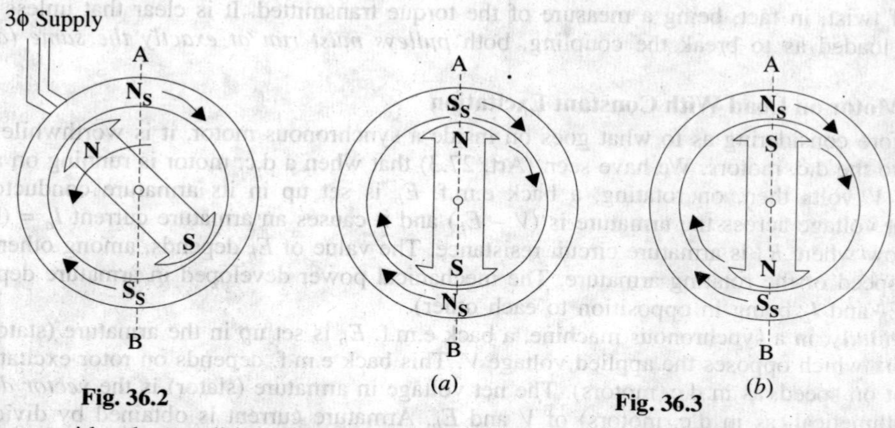

Fig. 36.2 **Fig. 36.3**

Now, consider the condition shown in Fig. 36.3 (*a*). The stator and rotor poles are attracting each other. Suppose that the rotor is not stationary, but is rotating clockwise, with such a speed that it turns through one pole-pitch by the time the stator poles interchange their positions, as shown in Fig. 36.3 (*b*). Here, again the stator and rotor poles attract each other. It means that if the rotor poles also shift their positions along with the stator poles, then they will continuously experience a unidirectional torque *i.e.* clockwise torque, as shown in Fig. 36.3.

36.3. Method of Starting

The rotor (which is as yet unexcited) is speeded up to synchronous / near synchronous speed by some arrangement and then excited by the d.c. source. The moment this (near) synchronously rotating rotor is excited, it is magnetically locked into position with the stator *i.e.* the rotor poles are engaged with the stator poles and both run synchronously in the same direction. It is because of this interlocking of stator and rotor poles that the motor has either to run synchronously or not at all. The synchronous speed is given by the usual relation $N_S = 120\,f\,/\,P$.

However, it is important to understand that the arrangement between the stator and rotor poles is *not an absolutely rigid one*. As the load on the motor is increased, the rotor progressively tends to fall back *in phase* (but *not* in speed as in d.c. motors) by some angle (Fig. 36.4) *but it still continues to run synchronously.* The value of this load angle or coupling angle (as it is called)

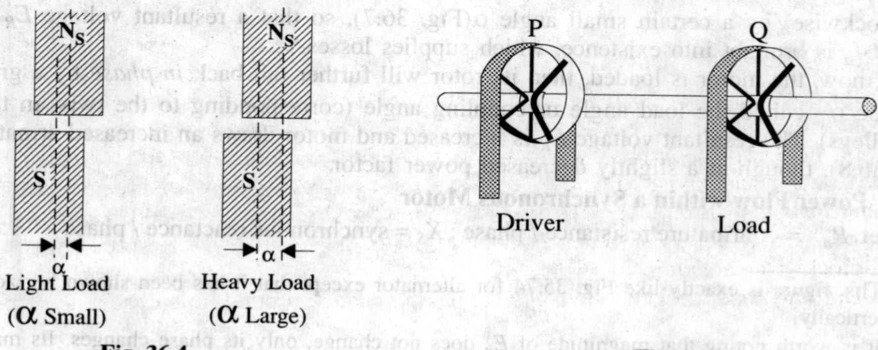

Fig. 36.4 **Fig. 36.5**

depends on the amount of load to be met by the motor. In other words, the torque developed by the motor depends on this angle, say, α.

The working of a synchronous motor is, in many ways, similar to the transmission of mechanical power by a shaft. In Fig. 36.5 are shown two pulleys P and Q transmitting power from the driver to the load. The two pulleys are assumed to be keyed together (just as stator and rotor poles are interlocked) hence they run at exactly the same (average) speed. When Q is loaded, it slightly falls behind owing to the twist in the shaft (twist angle corresponds to α in motor), the angle of twist, in fact, being a measure of the torque transmitted. It is clear that unless Q is so heavily loaded as to break the coupling, both *pulleys must run at exactly the same (average) speed.*

36.4 Motor on Load With Constant Excitation

Before considering as to what goes on inside a synchronous motor, it is worthwhile to refer briefly to the d.c. motors. We have seen (Art. 27.3) that when a d.c. motor is running on a supply of, say, V volts then, on rotating, a back e.m.f. E_b is set up in its armature conductors. The resultant voltage across the armature is $(V - E_b)$ and it causes an armature current $I_a = (V - E_b)/R_a$ to flow where R_a is armature circuit resistance. The value of E_b depends, among other factors, on the speed of the rotating armature. The mechanical power developed in armature depends on $E_b I_a$ (E_b and I_a being in opposition to each other).

Similarly, in a synchronous machine, a back e.m.f. E_b is set up in the armature (stator)by the rotor flux which opposes the applied voltage V. This back e.m.f. depends on rotor excitation only (and not on speed, as in d.c. motors). The net voltage in armature (stator) is the *vector difference* (not arithmetical, as in d.c. motors) of V and E_b. Armature current is obtained by dividing this *vector* difference of voltages by armature impedance (not resistance as in d.c. machines).

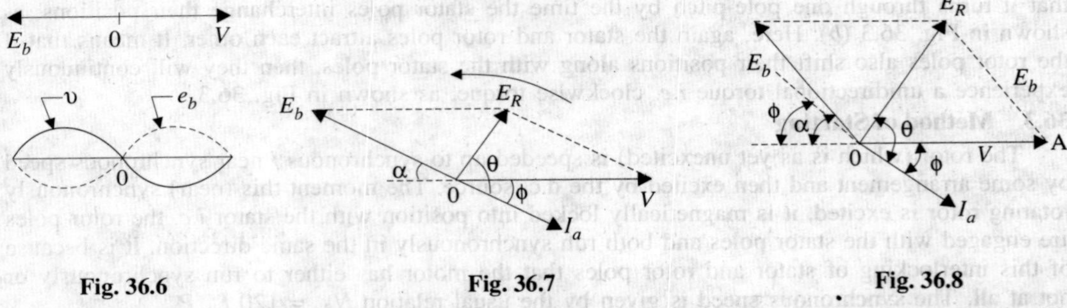

Fig. 36.6 **Fig. 36.7** **Fig. 36.8**

Fig. 36-6 shows the condition when the motor (properly synchronized to the supply) is running on *no-load* and has *no losses.** and is having field excitation which makes $E_b = V$. It is seen that vector difference of E_b and V is zero and so is the armature current. Motor intake is zero, as there is neither load nor losses to be met by it. In other words, the motor just floats.

If motor is on no-load, but it has losses, then the vector for E_b falls back (vectors are rotating anti-clockwise) by a certain small angle α(Fig. 36.7), so that a resultant voltage E_R and hence current I_a is brought into existence, which supplies losses.**

If, now, the motor is loaded, then its rotor will further fall back *in phase* by a greater value of angle α– called the load angle or coupling angle (corresponding to the twist in the shaft of the pulleys). The resultant voltage E_R is increased and motor draws an increased armature current (Fig. 36.8), though at a slightly decreased power factor.

36.5. Power Flow within a Synchronous Motor

Let R_a = armature resistance / phase ; X_S = synchronous reactance / phase

---

* This figure is exactly like Fig. 35.74 for alternator except that it has been shown horizontally rather than vertically.

** It is worth noting that magnitude of E_b does not change, only its phase changes. Its magnitude will change only when rotor dc excitation is changed *i.e.* when magnetic strength of rotor poles is changed.

then $\mathbf{Z} = \mathbf{R}_a + j X_S$; $\quad \mathbf{I}_a = \dfrac{\mathbf{E}_R}{\mathbf{Z}_S} = \dfrac{\mathbf{V} - \mathbf{E}_b}{\mathbf{Z}_s}$; Obviously, $\mathbf{V} = \mathbf{E}_b + \mathbf{I}_a \mathbf{Z}_S$

The angle θ (known as internal angle) by which I_a lags behind E_R is given by $\tan\theta = X_S / R_a$.

If R_a is negligible, then $\theta = 90°$. Motor input $= V L_a \cos\phi$ —per phase

Here, V is applied voltage / phase.

Total input for a star-connected, 3-phase machine is, $P = \sqrt{3}\, V_L \,.\, I_L \cos\phi$.

The mechanical power developed in the rotor is

P_m = back e.m.f. × armature current × cosine of the angle between the two *i.e.* angle between I_a and E_b reversed.

$= E_b I_a \cos(\alpha - \phi)$ per phase ...Fig. 36.8

Out of this power developed, some would go to meet iron and friction and excitation losses Hence, the power available at the shaft would be less than the developed power by this amount.

Out of the input power / phase $V I_a \cos\phi$, and amount $I_a^2 R_a$ is wasted in armature*, the rest $(V \,.\, I_a \cos\phi - I_a^2 R_a)$ appears as mechanical power in rotor; out of it, iron, friction and excitation losses are met and the rest is available at the shaft. If power input / phase of the motor is P, then

$$P = P_m + I_a^2 R_a$$

or mechanical power in rotor $\quad P_m = P - I_a^2 R_a$ —per phase

For three phases $\quad P_m = \sqrt{3}\, V_L I_L \cos\phi - 3 I_a^2 R_a$

The per phase power development in a synchronous machine is as under :

Power input/phase in stator
$P = VI_a \cos\phi$

Armature (*i.e.* stator) Cu loss $= I_a^2 R_a$

Mechanical power in armature $P_m = E_b I_a \cos(\alpha - \phi)$

Iron, excitation & friction losses

Output power P_{out}

Different power stages in a synchronous motor are as under :

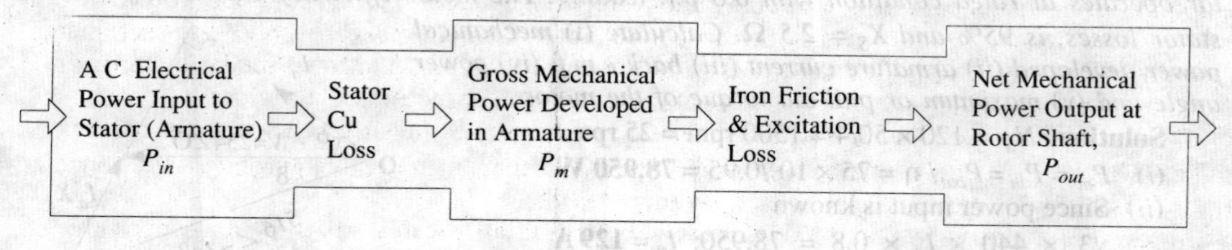

36.6. Equivalent Circuit of a Synchronous Motor

Fig. 36.9 (*a*) shows the equivalent circuit model for one armature phase of a cylindrical rotor synchronous motor.

It is seen from Fig. 36.9(*b*) that the phase applied voltage V is the vector sum of reversed back e.m.f. *i.e.* $-E_b$ and the impedance drop $I_a Z_s$. In other words, $V = (-E_b + I_a Z_s)$. The angle α* between the phasor for V and E_b is called the load angle or power angle of the synchronous motor.

---

* The Cu loss in rotor is not met by motor ac input, but by the dc source used for rotor excitation.

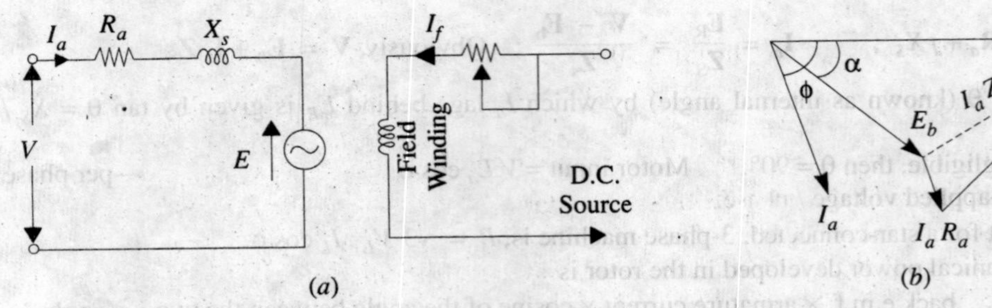

Fig. 36.9

36.7 Power Developed by a Synchronous Motor

Except for very small machines, the armature resistance of a synchronous motor is negligible as compared to its synchronous reactance. Hence, the equivalent circuit for the motor becomes as shown in Fig. 36.10 (*a*). From the phasor diagram of Fig. 36.10 (*b*), it is seen that

$AB = E_b \sin \alpha = I_a X_S \cos \phi$

or $VI_a \cos \phi = E_b V \sin \alpha$

Now, $VI_a \cos \phi$ = motor power input / phase

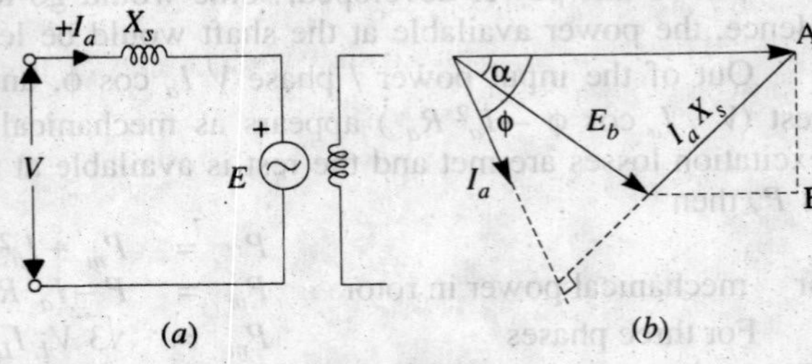

Fig. 36.10

$$\therefore \quad P_{in} = \frac{E_b V}{X_s} \sin \alpha \quad \text{...per phase**}$$

$$= 3 \frac{E_b V}{X_s} \sin \alpha \quad \text{... for three phases}$$

Since stator Cu losses have been neglected, P_{in} also represents the gross mechanical power $\{P_m\}$ developed by the motor.

$$\therefore \quad P_m = \frac{3E_b V}{X_s} \sin \alpha$$

The gross torque developed by the motor is $T_g = 9.55 \, P_m / N_s$ N–m ... N_s in rpm.

Example 36.1. *A 75-kW, 3-φ, Y-connected, 50-Hz, 440-V cylindrical rotor synchronous motor operates at rated condition with 0.8 p.f. leading. The motor efficiency excluding field and stator losses, is 95% and X_S = 2.5 Ω. Calculate* (i) *mechanical power developed (ii) armature current (iii) back e.m.f. (iv) power angle and (v) maximum or pull-out torque of the motor.*

Solution. $N_S = 120 \times 50/4 = 1500$ rpm = 25 rps

(*i*) $P_m = P_{in} = P_{out} / \eta = 75 \times 10^3/0.95 =$ **78,950 W**

(*ii*) Since power input is known

$\therefore \quad \sqrt{3} \times 440 \times I_a \times 0.8 = 78{,}950$; I_a = **129 A**

(*iii*) Applied voltage/phase = $440/\sqrt{3}$ = 254 V. Let V = 254 ∠0° as shown in Fig. 36.11.

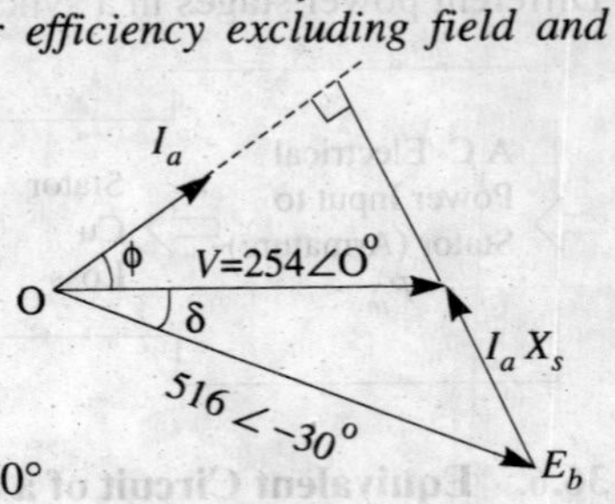

Now, $V = E_b + j\, IX_S$ or $E_b = V - j\, I_a X_s = 254 \angle 0° - 129 \angle 36.9° \times 2.5 \angle 90° = 250 \angle 0° - 322 \angle 126.9° = 254 - 322 (\cos 126.9° + j \sin 126.9°) = 254 - 322 (-0.6 + j\, 0.8) =$ **516 ∠ – 30°**

Fig. 36.11

* This angle was designated as δ when discussing synchronous generators.

** Strictly speaking, it should be $P_{in} = \frac{-E_b V}{X_s} \sin \alpha$.

(iv) $\therefore \quad \alpha = -\mathbf{30°}$

(v) pull-out torque occurs when $\alpha = 90°$

$$\text{maximum } P_m = 3\frac{E_bV}{X_s}\sin\delta = 3\frac{256 \times 516}{2.5} = \sin 90° = 157{,}275 \text{ W}$$

$\therefore$ pull-out torque= $9.55 \times 157,275 / 1500 =$ **1,000 N-m**

36.8. Synchronous Motor with Different Excitations

A synchronous motor is said to have normal excitation when its $E_b = V$. If field excitation is such that $E_b < V$, the motor is said to be *under-excited*. In both these conditions, it has a lagging power factor as shown in Fig. 36.12.

On the other hand if d.c. field excitation is such that $E_b > V$, then motor is said to be *over-excited* and draws a leading current, as shown in Fig. 36.13 *(a)*. There will be some value of excitation for which armature current will be in phase with V, so that power factor will become unity, as shown in Fig. 36.13 *(b)*.

The value of α and back e.m.f. E_b can be found with the help of vector diagrams for various power factors, shown in Fig. 36.14.

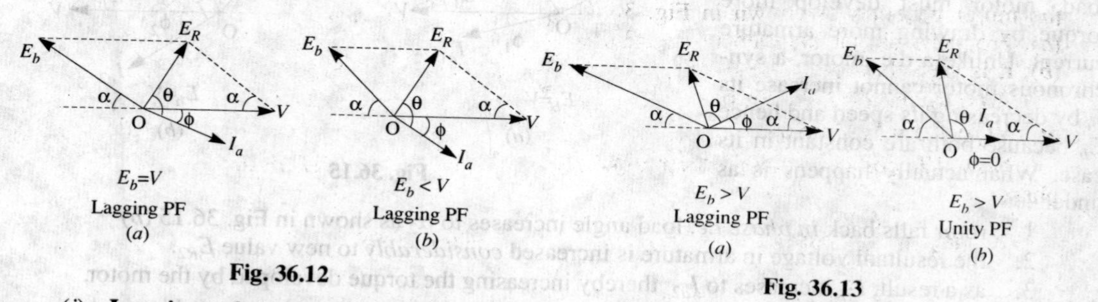

Fig. 36.12 Fig. 36.13

(i) **Lagging p.f.** As seen from Fig. 36.14 *(a)*

$$AC^2 = AB^2 + BC^2 = [V - E_R\cos(\theta - \phi)]^2 + [E_R \sin(\theta - \phi)]^2$$

$$\therefore \quad E_b = \sqrt{[V - I_a Z_s \cos(\theta - \phi]^2 + [I_a Z_S \sin(\theta - \phi]^2}$$

$$\text{Load angle } \alpha = \tan^{-1}\left(\frac{BC}{AB}\right) = \tan^{-1}\left[\frac{I_a Z_S \sin(\theta - \phi)}{V - I_a Z_S \cos(\theta - \phi)}\right]$$

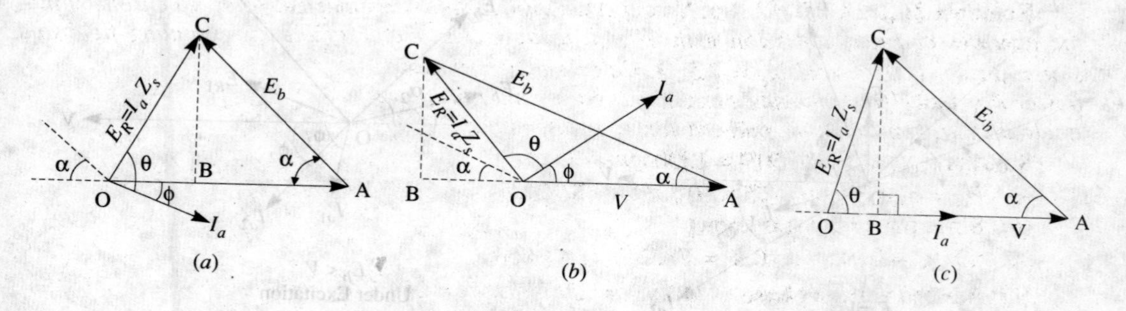

Fig. 36.14

(ii) **Leading p.f.** [36.14 *(b)*]

$$E_b = V + I_a Z_S \cos[180° - (\theta + \phi)] + j\, I_a Z_S \sin[180° - (\theta + \phi)]$$

$$\alpha = \tan^{-1}\frac{I_a Z_S \sin[180° - (\theta + \phi)]}{V + I_a Z_S \cos[180° - (\theta + \phi)]}$$

(iii) **Unity p.f.** [Fig. 36.14 *(c)*]

Here, $OB = I_a R_a$ and $BC = I_a X_S$

$$\therefore \quad E_b = (V - I_a R_a) + j I_a X_S \; ; \; \alpha = \tan^{-1}\left(\frac{I_a X_S}{V - I_a R_a}\right)$$

36.9. Effect of Increased Load with Constant Excitation

We will study the effect of increased load on a synchronous motor under conditions of normal, under and over-excitation (ignoring the effects of armature reaction). With normal excitation, $E_b = V$, with under excitation, $E_b < V$ and with over-excitation, $E_b > V$. Whatever the value of excitation, it would be kept *constant* during our discussion. It would also be assumed that R_a is negligible as compared to X_S so that phase angle between E_R and I_a *i.e.* $\theta = 90°$.

(i) Normal Excitation

Fig. 36.15. (*a*) shows the condition when motor is running with light load so that (*i*) torque angle α_1 is small (*ii*) so E_{R1} is small (*iii*) hence I_{a1} is small and (*iv*) ϕ_1 is small so that $\cos\phi_1$ is large.

Now, suppose that load on the motor is *increased* as shown in Fig. 36.15 (*b*). For meeting this extra load, motor must develop more torque by drawing more armature current. Unlike a d.c. motor, a synchronous motor cannot increase its I_a by decreasing its speed and hence E_b because both are constant in its case. What actually happens is as under :

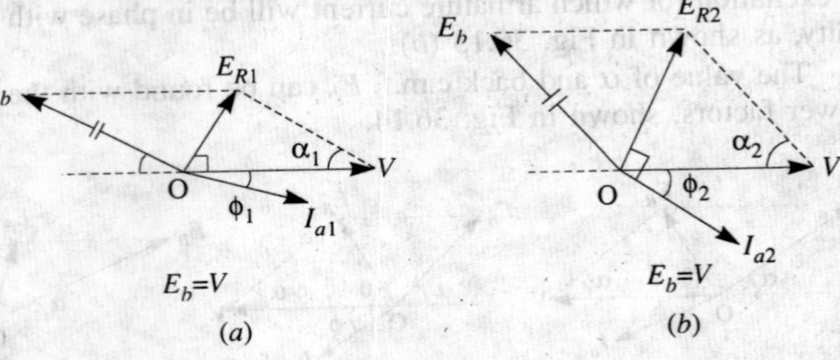

Fig. 36.15

1. rotor falls back *in phase i.e.* load angle increases to α_2 as shown in Fig. 36.15 (*b*)
2. the resultant voltage in armature is increased *considerably* to new value E_{R2}.
3. as a result, I_{a1} increases to I_{a2}, thereby increasing the torque developed by the motor.
4. ϕ_1 increases to ϕ_2, so that power factor *decreases* from $\cos\phi_1$ to the new value $\cos\phi_2$

Since increase in I_a is much greater than the *slight* decrease in power factor, the torque developed by the motor is *increased* (on the whole) to a new value sufficient to meet the extra load put on the motor. It will be seen that essentially it is by increasing its I_a that the motor is able to carry the extra load put on it.

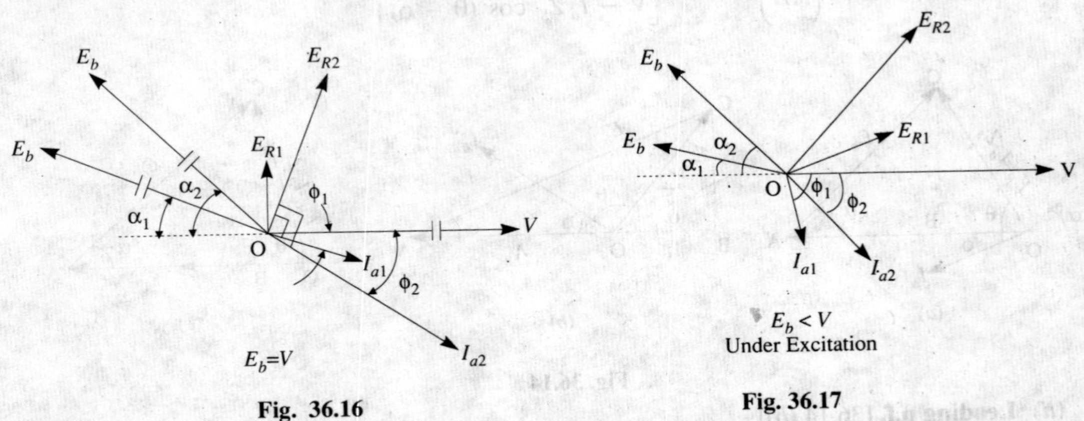

Fig. 36.16 Fig. 36.17

A phase summary of the effect of increased load on a synchronous motor at *normal excitation* is shown in Fig. 36.16. It is seen that there is a comparatively much greater *increase* in I_a than in ϕ.

(ii) Under-excitation

As shown in Fig. 36.17, with a small load and hence, small torque angle α_1, I_{a1} lags behind

V by a *large* phase angle ϕ_1 which means poor power factor. Unlike normal excitation, a much larger armature current must flow for developing the same power because of poor power factor. That is why I_{a1} of Fig. 36.17 is larger than I_{a1} of Fig. 36.15 (*a*).

As load increases, E_{R1} increases to E_{R2}, consequently I_{a1} increases to I_{a2} and p.f. angle *decreases* from ϕ_1 to ϕ_2 or p.f. *increases* from cos ϕ_1 to cos ϕ_2. Due to increase both in I_a and p.f., power generated by the armature increases to meet the increased load. As seen, in this case, *change in power factor is more than the change in* I_a.

(*iii*) Over-excitation

When running on light load, α_1 is small but I_{a1} is comparatively larger and *leads* V by a larger angle ϕ_1. Like the under-excited motor, as more load is applied, the power factor improves and *approaches unity.* The armature current also increases thereby producing the necessary increased armature power to meet the increased applied load (Fig. 36.18). However, it should be noted that in this case, power factor angle ϕ decreases (or p.f. increases) at a faster rate than the armature current thereby producing the necessary increased power to meet the increased load applied to the motor.

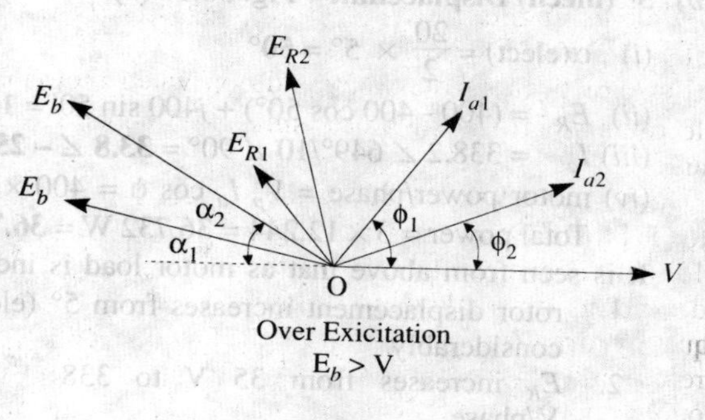

Fig. 36.18

Summary

The main points regarding the above three cases can be summarized as under :

1. As load on the motor *increases,* I_a *increases regardless of excitation.*
2. For under-and over-excited motors, p.f. tends to approach unity with increase in load.
3. Both with under-and over-excitation, change in p.f. is greater than in I_a with increase in load.
4. With normal excitation, when load is increased change in I_a is greater than in p.f. which tends to become increasingly lagging.

Example 36.2. *A 20-pole, 693-V, 50-Hz, 3-ϕ, Δ-connected synchronous motor is operating at no-load with normal excitation. It has armature ressistance per phase of 10 Ω and negligible synchronous reactance. If rotor is retarded by 0.5° (mechanical) from its synchronous position, compute.*

(*i*) *rotor displacement in electrical degrees*
(*ii*) *armature emf / phase*
(*iii*) *armature current / phase*
(*iv*) *power drawn by the motor.*
(*v*) *power developed by armature*

How will these quantities change when motor is loaded and the rotor displacement increases to 5° (mechanical)?

(Elect.Machines, AMIE Sec. B, 1993)

Solution. (*a*) **0.5° (mech) Displacement** [Fig 36.19 (a)]

(*i*) α (elect.) $= \dfrac{P}{2} \times \alpha$ (mech)

$\therefore$ α (elect)

$= \dfrac{20}{2} \times 0.5 = \mathbf{5° \text{ (elect)}}$

(*ii*) $V_p = V_L / \sqrt{3} = 693/\sqrt{3}$

$= 400$ V,

$E_b = V_p = 400$ V

$\therefore$ $E_R = (V_p - E_b \cos\alpha) + j\,E_b \sin\alpha = (400 - 400 \cos 5° + j\,400 \sin 5°)$

$= 1.5 + j35 = \mathbf{35 \angle 87.5\ V/phase}$

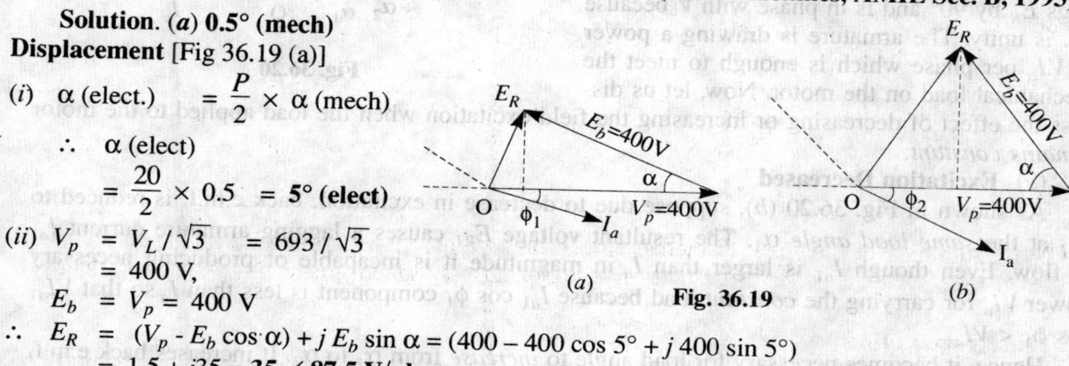

Fig. 36.19

(*iii*) $Z_S = 10 + j\,0 = 10 \angle 90°$; $I_a = E_R / Z_S = 35 \angle 87.5°/10 \angle 90° =$ **3.5 ∠ –2.5° A / phase**

Obviously, I_a lags behind V_p by 2.5°

(*iv*) Power input/phase $V_p\, I_a \cos\phi = 400 \times 3.5 \times \cos 2.5°. = 1399$ W

Total input power = 3 × 1399 = **4197 W**

(*v*) Since R_a is negligible, armature Cu loss is also negligible. Hence 4197 W also represent power developed by armature.

(*b*) 5° (mech) Displacemnt – Fig. 36.19 (*b*)

(*i*) $\alpha(\text{elect}) = \dfrac{20}{2} \times 5° = \mathbf{50°}$

(*ii*) $E_R = (400 - 400 \cos 50°) + j400 \sin 50° = 143 + j\,306.4 =$ **338.2 ∠ 64.9°**

(*iii*) $I_a = 338.2 \angle 649°/10 \angle 90° =$ **33.8 ∠ – 25.1° A/phase**

(*iv*) motor power/phase = $V_p\, I_a \cos\phi = 400 \times 33.8 \cos 25.1° = 12{,}244$ W

Total power = 3 × 12,244 = 36,732 W = **36.732 kW**

It is seen from above that as motor load is increased

1. rotor displacement increases from 5° (elect) to 50° (elect) *i.e.* E_b falls back in phase considerably.
2. E_R increases from 35 V to 338 V/phase
3. I_a increases from 3.5 A to 33.8 A
4. angle ϕ increases from 2.5° to 25.1° so that p.f. *decreases* from 0.999 (lag) to 0.906 (lag).
5. increase in power is almost *directly proportional to increase in load angle.*

Obviously, increase in I_a is much more than decrease in power factor.

It is interesting to note that not only power but even I_a, E_R and ϕ also increase almost as many times as α.

36-10. Effect of Changing Excitation of Constant load

As shown in Fig. 36.20 (*a*), suppose a synchronous motor is operating with normal excitation ($E_b = V$) at unity p.f. with a given load. If R_a is negligible as compared to X_S, then I_a lags E_R by 90° and is in phase with V because p.f. is unity. The armature is drawing a power of $V.I_a$ per phase which is enough to meet the mechanical load on the motor. Now, let us discuss the effect of decreasing or increasing the field excitation when the load applied to the motor *remains constant*.

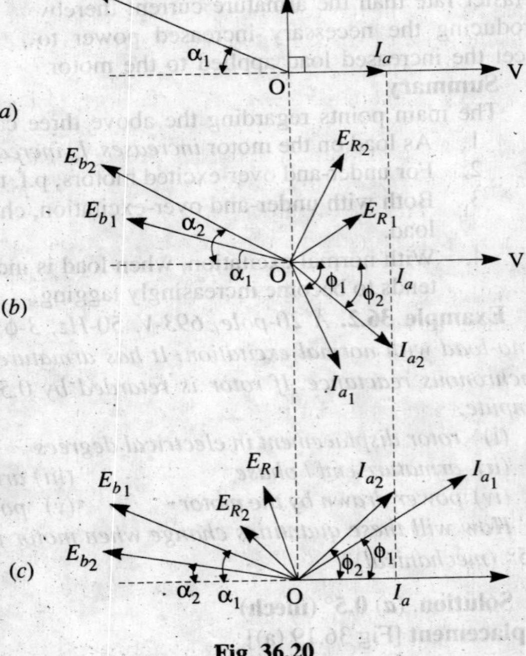

Fig. 36.20

(*a*) Excitation Decreased

As shown in Fig. 36.20 (*b*), suppose due to decrease in excitation, back e.m.f. is reduced to E_{b1} at the *same load angle* α_1. The resultant voltage E_{R1} causes a lagging armature current I_{a1} to flow. Even though I_{a1} is larger than I_a in magnitude it is incapable of producing necessary power VI_a for carrying the *constant* load because $I_{a1} \cos\phi_1$ component is less than I_a so that $VI_{a1} \cos\phi_1 < VI_a$.

Hence, it becomes necessary for load angle to *increase* from α_1 to α_2. It increases back e.m.f. from E_{b1} to E_{b2} which, in turn, increases resultant voltage from E_{R1} to E_{R2}. Consequently, armature current increases to I_{a2} whose in-phase component produces enough power (VI_{a2} cos ϕ_2) to meet the constant load on the motor.

(*b*) **Excitation Increased**

The effect of increasing field excitation is shown in Fig. 36.20 (*c*) where increased E_{b1} is shown at the original load angle α_1. The resultant voltage E_{R1} causes a *leading* current I_{a1} whose in-phase component is larger than I_a. Hence, armature develops more power than the load on the motor. Accordingly, load angle *decreases* from α_1 to α_2 which decreases resultant voltage from E_{R1} to E_{R2}. Consequently, armature current decreases from I_{a1} to I_{a2} whose in-phase component I_{a2} cos $\phi_2 = I_a$. In that case, armature develops power sufficient to carry the constant load on the motor.

Hence, we find that variations in the excitation of a synchronous motor running with a *given* load produce variations *in its load angle only*.

36.11. Different Torques of a Synchronous Motor

Various torques associated with a synchronous motor are as follows:

1. starting torque
2. running torque
3. pull-in torque and
4. pull-out torque

(*a*) **Starting Torque**

It is the torque (or turning effort) developed by the motor when full voltage is applied to its stator (armature) winding. It is also sometimes called *breakaway* torque. Its value may be as low as 10% as in the case of centrifugal pumps and as high as 200 to 250% of full-load torque as in the case of loaded reciprocating two-cylinder compressors.

(*b*) **Running Torque**

As its name indicates, it is the torque developed by the motor under running conditions. It is determined by the horse-power and speed of the *driven* machine. The peak horsepower determines the maximum torque that would be required by the driven machine. The motor must have a breakdown or a maximum running torque greater than this value in order to avoid stalling.

(*c*) **Pull-in Torque**

A synchronous motor is started as induction motor till it runs 2 to 5% below the synchronous speed. Afterwards, excitation is switched on and the rotor pulls into step with the synchronously-rotating stator field. The amount of torque at which the motor will pull into step is called the pull-in torque.

(*d*) **Pull-out Torque**

The maximum torque which the motor can develop without pulling out of step or synchronism is called the pull-out torque.

Normally, when load on the motor is increased, its rotor progressively tends to fall back *in phase* by some angle (called load angle) behind the synchronously-revolving stator magnetic field though it keeps running synchronously. Motor develops maximum torque when its rotor is retarded by an angle of 90° (or in other words, it has shifted backward by a distance equal to half the distance between adjacent poles). Any further increase in load will cause the motor to pull out of step (or synchronism) and stop.

36.12. Power Developed by a Synchronous Motor

In Fig. 36.21, *OA* represents supply voltage/phase and $I_a = I$ is the armature current, *AB* is back e.m.f. at a load angle of α. *OB* gives the resultant voltage $E_R = I Z_S$ (or $I X_S$ if R_a is negligible). *I* leads *V* by ϕ and lags behind E_R by an angle $\theta = \tan^{-1}(X_S / R_a)$. Line *CD* is drawn at an angle of θ to *AB*. *AC* and *ED* are $\perp$ to *CD* (and hence to *AE* also).

Mechanical power per phase developed in the rotor is

$$P_m = E_b I \cos \psi \qquad ...(i)$$

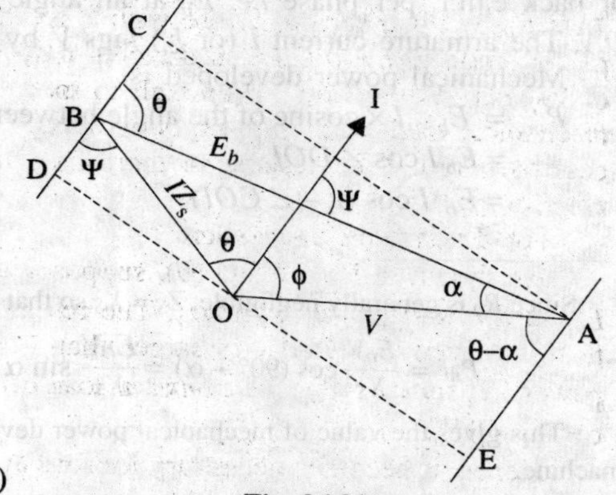

Fig. 36.21

In ΔOBD, $BD = I Z_S \cos \psi$ Now, $BD = CD - BC = AE - BC$

$I Z_S \cos \psi = V \cos (\theta - \alpha) - E_b \cos \theta$

$$\therefore \; I \cos \psi = \frac{V}{Z_S} \cos (\theta - \alpha) - \frac{E_b}{Z_S} \cos \theta$$

Substituting this value in (*i*), we get

$$P_m \text{ per phase} = E_b \left[\frac{V}{Z_S} \cos (\theta - \alpha) - \frac{E_b}{Z_S} \cos \theta \right] = \frac{E_b V}{Z_S} \cos(\theta - \alpha) - \frac{E_b^2}{Z_S} \cos\theta^* \quad ...(ii)$$

This is the expression for the mechanical power developed in terms of the load angle (α) and the internal angle (θ) of the motor for a constant voltage V and E_b (or excitation because E_b depends on excitation only)

If T_g is the gross armature torque developed by the motor, then

$$T_g \times 2 \pi N_S = P_m \text{ or } T_g = P_m / \omega_s = P_m / 2 \pi N_S \quad \text{—}N_S \text{ in rps}$$

$$T_g = \frac{P_m}{2\pi N_S / 60} = \frac{60}{2\pi} \cdot \frac{P_m}{N_S} = 9.55 \frac{P_m}{N_S} \quad \text{—}N_S \text{ in rpm}$$

Condition for maximum power developed can be found by differentiating the above expression with respect to load angle and then equating it to zero.

$$\therefore \; \frac{dP_m}{d\alpha} = -\frac{E_b V}{Z_S} \sin (\theta - \alpha) = 0 \quad \text{or} \quad \sin (\theta - \alpha) = 0 \qquad \therefore \; \theta = \alpha$$

$$\therefore \text{ value of maximum power } (P_m)_{max} = \frac{E_b V}{Z_S} - \frac{E_b^2}{Z_S} \cos \alpha \text{ or } (P_m)_{max} = \frac{E_b V}{Z_S} - \frac{E_b^2}{Z_S} \cos \theta. \quad ...(iii)$$

This shows that the maximum power and hence torque ($\because$ speed is constant) depends on V and E_b *i.e.* excitation. Maximum value of θ (and hence α) is 90°. For all values of V and E_b, this limiting value of α is the same but maximum torque will be proportional to the maximum power developed as given in equation (*iii*) Equation (*ii*) is plotted in Fig. 36.22.

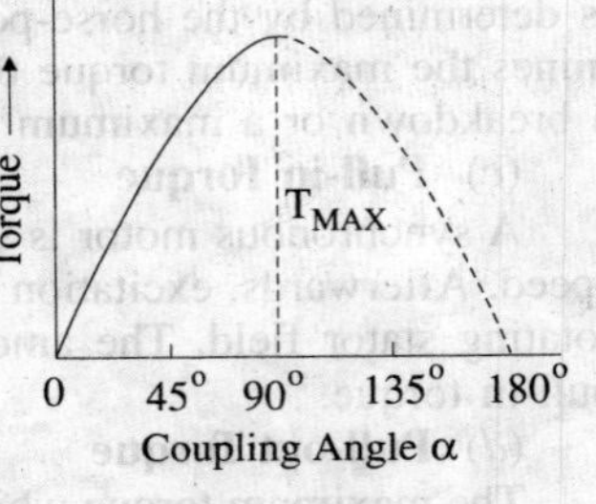

Fig. 36.22

If R_a is neglected, then $Z_S \cong X_S$ and $\theta = 90°$ $\therefore$ $\cos \theta = 0$

$$P_m = \frac{E_b V}{X_S} \sin \alpha \; ...(iv) \quad (P_m)_{max} = \frac{E_b V}{X_S} \; \text{... from equation } (iii)^{**}$$

The same value can be otained by putting $\alpha = 90°$ in equation (*iv*). This corresponds to the 'pull-out' torque.

36.13. Alternative Expression for Power Developed

In Fig. 36.23, as usual, OA represents the supply voltage per phase *i.e.* V and AB (= OC) is the induced or back e.m.f. per phase *i.e.* E_b at an angle α with OA. The armature current I (or I_a) lags V by ϕ.

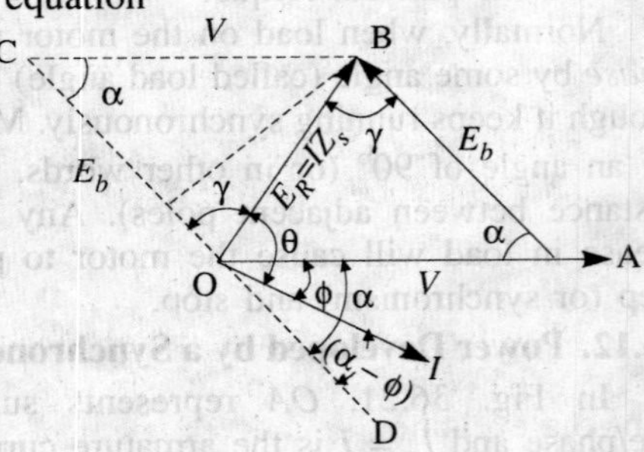

Fig. 36.23

Mechanical power developed is,

$P_m = E_b \cdot I \times$ cosine of the angle between E_b and I

$= E_b I \cos \angle DOI$

$= E_b I \cos (\pi - \angle COI)$

---

* Since R_a is generally negligible, $Z_S = X_S$ so that $\theta \cong 90°$. Hence

$$P_m = \frac{E_b V}{X_S} \cos (90° - \alpha) = \frac{E_b V}{X_S} \sin \alpha$$

This gives the value of mechanical power developed in terms of α – the basic variable of a synchronous machine.

** It is the same expression as found for an alternator or synchronous generator in Art. 35.37.

$$= -E_b I \cos(\theta + \gamma)$$

$$= -E_b\left(\frac{E_R}{Z_S}\right)(\cos\theta\cos\gamma - \sin\theta\sin\gamma) \quad ...(i)$$

Now, E_R and functions of angles θ and γ will be eliminated as follows:

From $\Delta\, OAB$; $V/\sin\gamma = E_R / \sin\alpha$ $\quad\therefore\quad \sin\gamma = V\sin\alpha / E_R$

From $\Delta\, OBC$; $E_R\cos\gamma + V\cos\alpha = E_b$ $\quad\therefore\quad \cos\gamma = (E_b - V\cos\alpha)/E_R$

Also $\cos\theta = R_a / Z_S$ and $\sin\theta = X_S / Z_S$

Substituting these values in Eq. (*i*) above, we get

$$P_m = -\frac{E_b \,.\, E_R}{Z_S}\left(\frac{R_a}{Z_S}\,.\,\frac{E_b - V\cos\alpha}{E_R} - \frac{X_S}{Z_S}\,.\,\frac{V\sin\alpha}{E_R}\right)$$

$$= \frac{E_b V}{Z_S{}^2}(R_a\cos\alpha + X_S\sin\alpha) - \frac{E_b{}^2 R_a}{Z_S{}^2} \quad ...(ii)$$

It is seen that P_m varies with E_b (which depends on excitation) and angle α (which depends on the motor load).

Note. If we substitute $R_a = Z_S\cos\theta$ and $X_S = Z_S\sin\theta$ in Eq. (*ii*), we get

$$P_m = \frac{E_b V}{Z_S{}^2}(Z_S\cos\theta\cos\alpha + Z_S\sin\alpha) - \frac{E_b^2 Z_S\cos\theta}{Z_S^2} = \frac{E_b V}{Z_S}\cos(\theta - \alpha) - \frac{E_b^2}{Z_S}\cos\theta$$

It is the same expression as found in Art. 36.10.

36.14. Various Conditions of Maxima

The following two cases may be considered :

(*i*) **Fixed E_b, V, R_a and X_S.** Under these conditions, P_m will vary with load angle α and will be maximum when $dP_m / d\alpha = 0$. Differentiating Eq. (*ii*) in Art. 36.11, we have

$$\frac{dP_m}{d\alpha} = \frac{E_b V}{Z_s^2}(X_S\cos\alpha - R_a\sin\alpha) = 0 \quad \text{or} \quad \tan\alpha = X_S / R_a = \tan\theta \text{ or } \alpha = \theta$$

Putting $\alpha = \theta$ in the same Eq. (*ii*), we get

$$(P_m)_{max} = \frac{E_b V}{Z_s^2}(R_a\cos\theta + X_S\sin\theta) - \frac{E_b^2 R_a}{Z_s^2}$$

$$= \frac{E_b V}{Z_s^2}\left(R_a\,.\,\frac{R_a}{Z_S} + X_S\,.\,\frac{X_S}{Z_S}\right) - \frac{E_b^2 R_a}{Z_s^2}$$

$$= \frac{E_b V}{Z_S^2}\left(\frac{R_a{}^2 + X_S^2}{Z_S}\right) - \frac{E_b^2 R_a}{Z_S^2} = \frac{E_b V}{Z_S} - \frac{E_b^2 R_a}{Z_S^2} \quad ...(i)$$

This gives the value of power at which the motor falls out of step.

Solving for E_b from Eq. (*i*) above, we get

$$E_b = \frac{Z_S}{2R_a}\left[V \pm \sqrt{V^2 - 4R_a\,.\,(P_m)_{max}}\right]$$

The two values of E_b so obtained represent the excitation limits for any load.

(*ii*) **Fixed V, R_a and X_s.** In this case, P_m varies with excitation or E_b. Let us find the value of the excitation or induced e.m.f. E_b which is necessary for maximum power possible. For this purpose, Eq. (*i*) above may be differentiated with respect to E_b and equated to zero.

$$\therefore \quad \frac{d(P_m)_{max}}{dE_b} = \frac{V}{Z_S} - \frac{2R_a E_b}{Z_S^2} = 0\,; \qquad E_b = \frac{VZ_S}{2R_a}{}^* \quad ...(ii)$$

Putting this value of E_b in Eq. (*i*) above, maximum power developed becomes

$$(P_m)_{max} = \frac{V^2}{2R_a} - \frac{V^2}{4R_a} = \frac{V^2}{4R_a}$$

*This is the value of induced e.m.f. to give maximum power, but it is not the maximum possible value of the generated voltage, at which the motor will operate.

36.15. Salient Pole Synchronous Motor

Cylindrical-rotor synchronous motors are much easier to analyse than those having salient-pole rotors. It is due to the fact that cylindrical-rotor motors have a uniform air-gap, whereas in salient-pole motors, air-gap is much greater between the poles than along the poles. Fortunately, cylindrical rotor theory is reasonably accurate in predicting the steady-state performance of salient-pole motors. Hence, salient-pole theory is required only when very high degree of accuracy is needed or when problems concerning transients or power system stability are to be handled.

The d-q currents and reactances for a salient-pole synchronous motor are exactly the same as discussed for salient-pole synchronous generator. The motor has d-axis reactance X_d and q-axis reactance X_q. Similarly, motor armature current I_a has two components : I_d and I_q. The complete phasor diagram of a salient-pole synchronous motor, for a lagging power factor is shown in Fig. 36.24 (*a*).

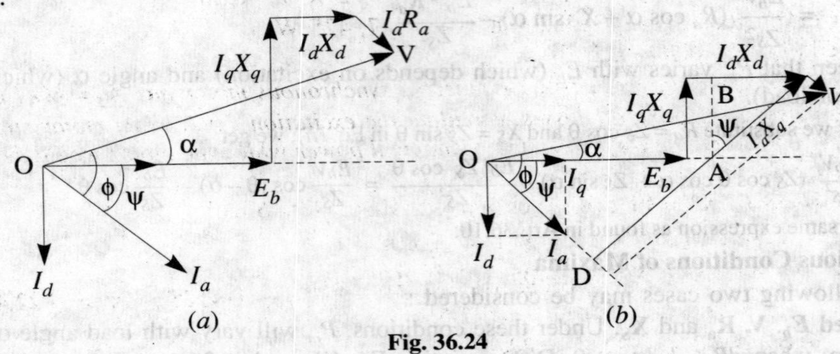

Fig. 36.24

With the help of Fig. 36.24 (*b*), it can be proved that $\tan\psi = \dfrac{V\sin\phi - I_q X_q}{V\cos\phi - I_a R_a}$

If R_a is negligible, then $\tan\psi = (V\sin\phi + I_a X_q) / V\cos\phi$

For an overexcited motor *i.e.* when motor has leading power factor,

$\tan\psi = (V\sin\phi + I_a X_q) / V\cos\phi$

The power angle α is given by $\alpha = \phi - \psi$

The magnitude of the excitation or the back e.m.f. E_b is given by

$E_b = V\cos\alpha - I_q R_a - I_d X_d$

Similarly, as proved earlier for a synchronous generator, it can also be proved from Fig. 36.24 (*b*) for a synchronous motor with $R_a = 0$ that

$$\tan\alpha = \frac{I_a X_q \cos\phi}{V - I_a X_q \sin\phi}$$

In case R_a is not negligible, it can be proved that

$$\tan\alpha = \frac{I_a X_q \cos\phi - I_a R_a \sin\phi}{V - I_a X_q \sin\phi - I_a R_a \cos\phi}$$

36.16. Power Developed by a Salient Pole Synchronous Motor

The expression for the power developed by a salient-pole synchronous generator derived in Chapter 35 also applies to a salient-pole synchronous motor.

$$\therefore \quad P_m = \frac{E_b V}{X_d}\sin\alpha + \frac{V^2 (X_d - X_q)}{2 X_d X_q}\sin 2\alpha \quad \text{... per phase}$$

$$= 3 \times \left[\frac{E_b V}{X_d}\sin\alpha + \frac{V^2 (X_d - X_q)}{2 X_d X_q}\sin 2\alpha\right] \quad \text{... per three phases}$$

$T_g = 9.55\, P_m / N_s$ ...N_s in rps.

As explained earlier, the power consists of two components, the first component is called excitation power or magnet power and the second is called reluctance power (because when excitation is removed, the motor runs as a reluctance motor).

Example 36.3. *A 3-ϕ, 150-kW, 2300-V, 50-Hz, 1000-rpm salient-pole synchronous motor has X_d = 32, Ω / phase and X_q = 20 Ω/phase. Neglecting losses, calculate the torque developed by the motor if field excitation is so adjusted as to make the back e.m.f. twice the applied voltage and α = 16°.*

Solution. $V = 2300/\sqrt{3} = 1328$ V; $E_b = 2 \times 1328 = 2656$ V

$$\text{Excitation power / phase} = \frac{E_b V}{X_d} \sin \alpha = \frac{2656 \times 1328}{32} \sin 16° = 30,382 \text{ W}$$

$$\text{Reluctance power / phase} = \frac{V^2 (X_d - X_q)}{2 X_d X_q} \sin 2\alpha = \frac{1328^2 (32 - 20)}{2 \times 32 \times 20} \sin 32° = 8760 \text{ W}$$

Total power developed, $P_m = 3(30382 + 8760) = 117,425$ W

$T_g = 9.55 \times 117,425/1000 =$ **1120 N-m**

Example 36.4 *A 3300-V, 1.5-MW, 3-ϕ, Y-connected synchronous motor has X_d = 4Ω / phase and X_q = 3 Ω / phase. Neglecting all losses, calculate the excitation e.m.f. when motor supplies rated load at unity p.f. Calculate the maximum mechanical power which the motor would develop for this field excitation.*

Solution. $V = 3300/\sqrt{3} = 1905$ V ; $\cos\phi = 1$; $\sin\phi = 0$; $\phi = 0°$

$I_a = 1.5 \times 10^6/\sqrt{3} \times 3300 \times 1 = 262$A

$$\tan\psi = \frac{V \sin\phi - I_a X_q}{V \cos\phi} = \frac{1905 \times 0 - 262 \times 3}{1905} = -0.4125 \,; \quad \psi = -22.4°$$

$\alpha = \phi - \psi = 0 - (-22.4°) = 22.4°$

$I_d = 262 \times \sin(-22.4°) = -100$ A; $I_q = 262 \cos(-22.4°) = 242$A

$E_b = V \cos\alpha - I_d X_d = 1905 \cos(-22.4°) - (-100° \times 4) = 2160V$

$= 1029 \sin\alpha + 151 \sin 2\alpha$

$$P_m = \frac{E_b V}{X_d} \sin\alpha + \frac{V^2 (X_d - X_q)}{2 X_d X_q} \sin 2\alpha \quad \text{... per phase}$$

$$= \frac{2160 \times 1905}{4 \times 1000} + \frac{1905^2 (4 - 3)}{2 \times 4 \times 3 \times 1000} \sin 2\alpha \quad \text{...kW/phase}$$

$= 1029 \sin\alpha + 151 \sin 2\alpha$ ..kW/phase

If developed power has to achieve maximum value, then

$$\frac{dP_m}{d\alpha} = 1029 \cos\alpha + 2 \times 151 \cos 2\alpha = 0$$

∴ $1029 \cos\alpha + 302 (2\cos^2\alpha - 1) = 0$ or $604 \cos^2\alpha + 1029 \cos\alpha - 302 = 0$

$$\therefore \quad \cos\alpha = \frac{-1029 \pm \sqrt{1029^2 + 4 \times 604 \times 302}}{2 \times 604} = 0.285 \,; \quad \alpha = 73.4°$$

∴ maximum $P_m = 1029 \sin 73.4° + 151 \sin 2 \times 73.4° = 1070$ kW/phase

Hence, maximum power developed for three phases = 3 × 1070 = **3210 kW**

Example 36.5. *The input to an 11000-V, 3-phase, star-connected synchronous motor is 60 A. The effective resistance and synchronous reactance per phase are respectively 1 ohm and 30 ohm. Find (i) the power supplied to the motor (ii) mechanical power developed and (iii) induced emf for a power factor of 0.8 leading.* **(Elect. Engg. AMIETE (New Scheme) June 1990)**

Solution. (*i*) Motor power input = $\sqrt{3} \times 11000 \times 60 \times 0.8 = 915$ kW

(*ii*) stator Cu loss/phase = $60^2 \times 1 = 3600$ W; Cu loss for three phases = 3 × 3600 = 10.8 kW

$P_m = P_2 -$ rotor Cu loss = 915 − 10.8 = 904.2 kW

$V_p = 11000/\sqrt{3} = 6350$ V ; $\phi = \cos^{-1} 0.8 = 36.9°$;

$\theta = \tan^{-1}(30/1) = 88.1°$

$Z_s \cong 30\ \Omega$; stator impedance drop / phase $= I_a Z_s$

$= 60 \times 30 = 1800$ V

As seen from Fig. 36.25,

$E_b^2 = 6350^2 + 1800^2 - 2 \times 6350 \times 1800 \times \cos(88.1° + 36.9°)$

$= 6350^2 + 1800^2 - 2 \times 6350 \times 1800 \times -0.572$

$\therefore\ E_b = 7528\ V$; line value of $E_b = 7528 \times \sqrt{3} =$ **13042**

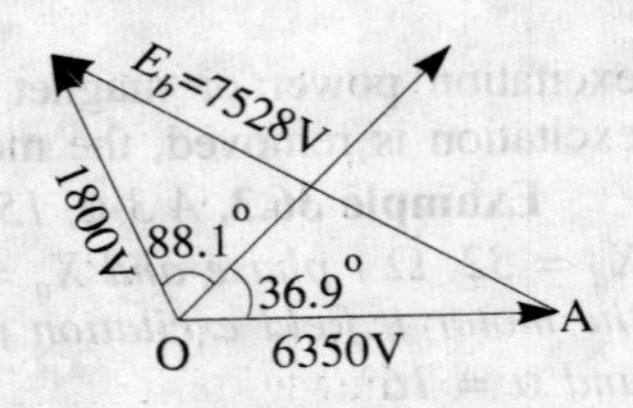

Fig. 36.25

Example 36.6. *A 500-V, 1-phase synchronous motor gives a net output mechanical power of 7.46 kW and operates at 0.9 p.f. lagging. Its effective resistance is 0.8 Ω. If the iron and friction losses are 500 W and excitation losses are 800 W, estimate the armature current. Calculate the commercial efficiency.* **(Electrical Machines-I, Gujarat Univ. 1988)**

Solution. Motor input $= VI_a \cos\phi$; Armature Cu loss $= I_a R_a^2$

Power developed in *armature* is $P_m = VI_a \cos\phi - I_a^2 R_a$

$\therefore\ I_a^2 R_a - VI_a \cos\phi + P_m = 0$ or $I_a = \dfrac{V\cos\phi \pm \sqrt{V^2 \cos^2\phi - 4R_a P_m}}{2R_a}$

Now, $P_{out} = 7.46\text{ kW} = 7{,}460$ W

$P_m = P_{out}$ + iron and friction losses + excitation losses

$= 7460 + 500 + 800 = 8760$ W ... Art. 36.5

$$I_a = \frac{500 \times 0.9 \pm \sqrt{(500 \times 0.9)^2 - 4 \times 0.8 \times 8760}}{2 \times 0.8}$$

$$= \frac{450 \pm \sqrt{202{,}500 - 28{,}030}}{1.6} = \frac{450 \pm 417.7}{1.6} = \frac{32.3}{1.6} = \mathbf{20.2\ A}$$

Motor input $= 500 \times 20.2 \times 0.9 = 9090$ W

η_c = net output / input = 7460 / 9090 = 0.8206 or **82.06%.**

Example 36.7. *A 2,300-V, 3-phase, star-connected synchronous motor has a resistance of 0.2 ohm per phase and a synchronous reactance of 2.2 ohm per phase. The motor is operating at 0.5 power factor leading with a line current of 200 A. Determine the value of the generated e.m.f. per phase.* **(Elect. Engg.-I, Nagpur Univ. 1993)**

Solution. Here, $\phi = \cos^{-1}(0.5) = 60°$ (lead)

$\theta = \tan^{-1}(2.2/0.2) = 84.8°$

$\therefore\ (\theta + \phi) = 84.8° + 60° = 144.8°$

$\cos 144.8° = -\cos 35.2°$

$V = 2300/\sqrt{3} = 1328$ volt

$Zs = \sqrt{0.2^2 + 2.2^2} = 2.209\ \Omega$

$IZ_S = 200 \times 2.209 = 442\ V$

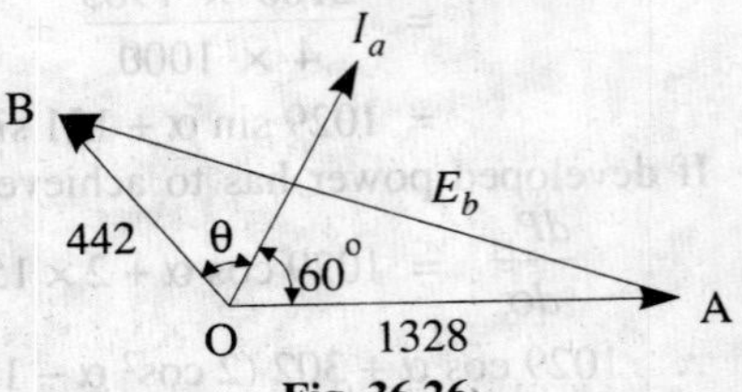

Fig. 36.26

The vector diagram is shown in Fig. 36.26.

$Eb = \sqrt{V^2 + E_{R^2} - 2\,V.\,E_R \,Cos\,(\theta + \phi)}$

$= \sqrt{1328^2 + 442^2 + 2 \times 1328 \times 442 \times Cos\,35.2°}$ = **1708 Volt / Phase**

Exaple 36.8. *A 3-phase, 6,600-volts, 50-Hz, star-connected synchronous motor takes 50 A current. The resistance and synchronous reactance per phase are 1 ohm and 20 ohm respectively. Find the power supplied to the motor and induced emf for a power factor of (i) 0.8 lagging and (ii) 0.8 leading.* **(Eect. Engg. II pune Univ. 1988)**

Solution. (*i*) **p.f.** = **0.8 lag** (Fig. 36.27).

Power input $= \sqrt{3} \times 6600 \times 50 \times 0.8 = 457{,}248$ W

Supply voltage / phase $= 6600 / \sqrt{3} = 3810$ V

$\phi = \cos^{-1}(0.8) = 36°52'$; $\theta = \tan^{-1}(X_S / R_a) = (20/1) = 87.8'$

$Z_S = \sqrt{20^2 + 1^2} = 20\ \Omega$ (approx.)

Impedance drop = $I_a Z_S = 50 \times 20 = 1000$ V/phase

$\therefore \quad E_b^2 = 3810^2 + 1000^2 - 2 \times 3810 \times 1000 \times \cos(87°8' - 36°52') \therefore \quad E_b = 3263$ V / phase

Line induced e.m.f.

$= 3263 \times \sqrt{3} = \mathbf{5651\ V}$

(*ii*) Power input would remain the same

As shown in Fig. 36.28, the current vector is drawn at a *leading* angle of

$\phi = 36°52'$

Now, $(\theta + \phi) = 87°8' + 36°52' = 124°$,

$\cos 124° = -\cos 56°$

$\therefore \quad E_b^2 = 3810^2 + 1000^2 - 2 \times 3810 \times 1000 \times - \cos 56° \qquad \therefore E_b = 4447$ V / phase

Line induced e.m.f. = $\sqrt{3} \times 4447 = \mathbf{7{,}700\ V}$

Fig. 36.27

Fig. 36.28

Note. It may be noted that if $E_b > V$, then motor has a leading power factor and if $E_b < V$,

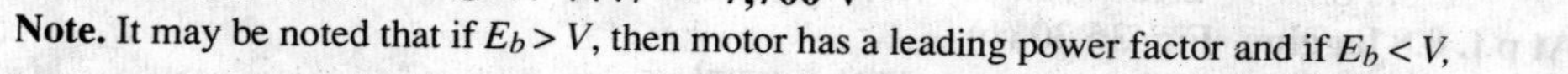

Example 36.9. *A synchronous motor having 40% reactance and a negligible resistance is to be operated at rated load at (i) u.p.f. (ii) 0.8 p.f. lag (iii) 0.8 p.f. lead. What are the values of induced e.m.f.? Indicate assumptions made if any.* **(Electrical Machines-II, Indore Univ. 1990)**

Solution. Let more $V = 100$ V, then reactance drop = $I_a X_S = 40$ V

(*i*) **At unity p.f.**

Here, $\theta = 90°$, $\quad E_b = \sqrt{100^2 + 40^2} = 108$ V ...Fig. 36.29 (*a*)

(*ii*) **At p.f. 0.8 (lag.)** Here $\angle BOA = \theta - \phi = 90° - 36°54' = 53°6'$

$$E_b^2 = 100^2 + 40^2 - 2 \times 100 \times 40 \times \cos 53°6'; \quad E_b = \mathbf{82.5\ V}$$

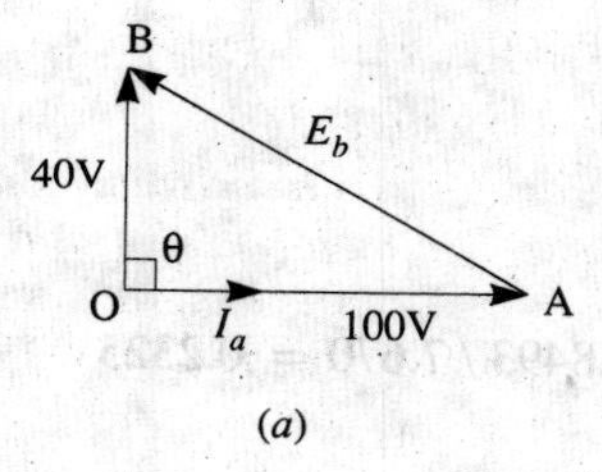

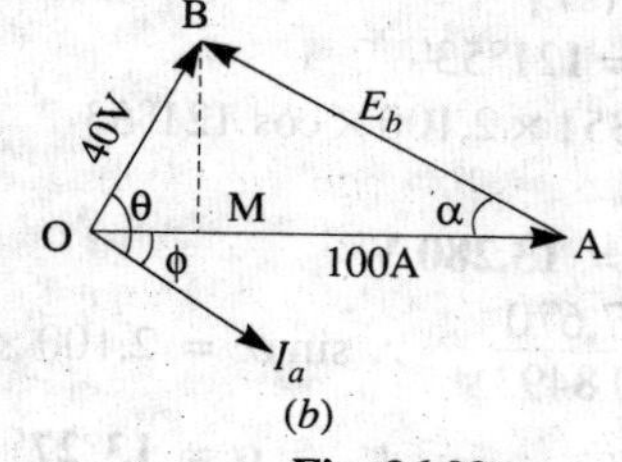

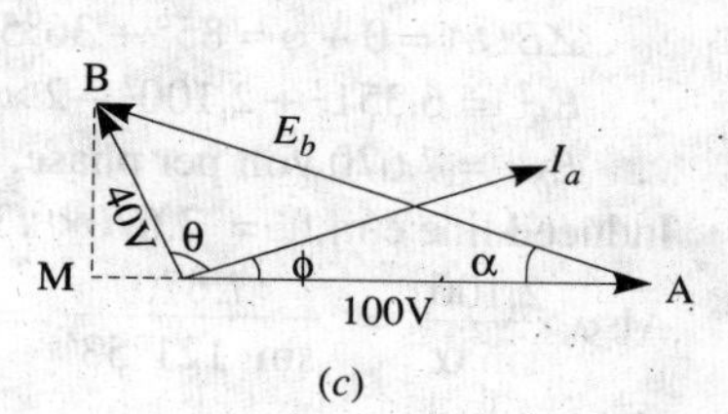

Fig. 36.29

Alternatively, $\quad E_b = AB = \sqrt{AM^2 + MB^2} = \sqrt{76^2 + 32^2} = 82.5$ V

(*iii*) **At p.f. 0.8 (lead.)** Here, $(\theta + \phi) = 90° + 36.9° = 126.9°$

$$E_b^2 = 100^2 + 40^2 - 2 \times 40 \times \cos 126.9° = \mathbf{128\ V}$$

Again from Fig. 36.29 (*c*), $E_b^2 = (OM + OA)^2 + MB^2 = 124^2 + 32^2$; $E_b = \mathbf{128\ V.}$

Example 36.10. *A 1,000-kVA, 11,000-V, 3-φ, star-connected synchronous motor has an armature resistance and reactance per phase of 3.5 Ω and 40 Ω respectively. Determine the induced e.m.f. and angular retardation of the rotor when fully loaded at (a) unity p.f. (b) 0.8 p.f. lagging (c) 0.8 p.f. leading.* **(Elect. Engineering-II, Bangalore Univ. 1992)**

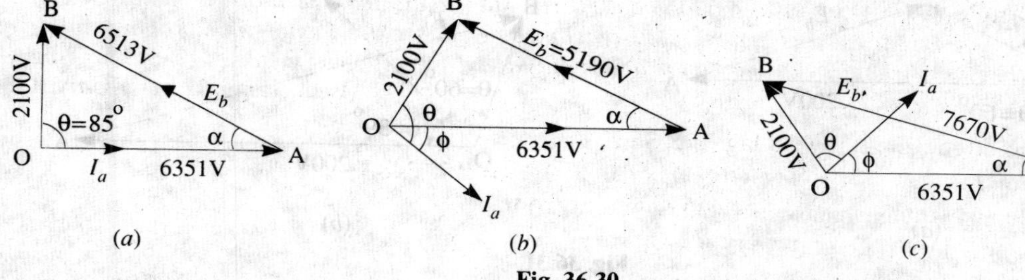

Fig. 36.30

Solution. Full-load armature current $= 1{,}000 \times 1000 / \sqrt{3} \times 11{,}000 = 52.5$ A

Voltage / phase $= 11{,}000 / \sqrt{3} = 6{,}351$ V ; $\cos\phi = 0.8$ $\quad\therefore\ \phi = 36°53'$

Armature resistance drop / phase $= I_a R_a = 3.5 \times 52.5 = 184$ V

reactance drop / phase $= I_a X_S = 40 \times 52.5 = 2{,}100$ V

$\therefore$ impedance drop / phase $= I_a Z_S = \sqrt{(184^2 + 2100^2)} = 2{,}100$ V (approx.)

$\tan\theta = X_S / R_a \quad \therefore\ \theta = \tan^{-1}(40/3.5) = 85°$

(*a*) **At unity p.f.** Vector diagram is shown in Fig. 36.30 (*a*)

$E_b^2 = 6{,}351^2 + 2{,}100^2 - 2 \times 6{,}351 \times 2{,}100 \cos 85°$; $E_b = 6{,}513$ *V* per phase

Induced line voltage $= 6{,}513 \times \sqrt{3} =$ **11,280 V**

From $\Delta\, OAB$, $\dfrac{2100}{\sin\alpha} = \dfrac{6513}{\sin 85°} = \dfrac{6513}{0.9961}$

$\sin\alpha = 2{,}100 \times 0.9961 / 6{,}513 = 0.3212$ $\quad\therefore\ \alpha =$ **18°44′**

(*b*) **At p.f. 0.8 lagging** -Fig. 36.30 (*b*)

$\angle BOA = \theta - \phi = 85° - 36°53' = 48°7'$

$E_b^2 = 6{,}351^2 + 2{,}100^2 - 2 \times 6{,}351 \times 2{,}100 \times \cos 48°7'$

$E_b = 5{,}190$ V per phase

Induced line voltage $= 5{,}190 \times \sqrt{3} =$ **8,989 V**

Again from the $\Delta\, OAB$ of Fig. 36.30 (*b*)

$$\frac{2100}{Sin\,\alpha} = \frac{5190}{Sin\,48°\,7'} = \frac{5.190}{0.7443}$$

$\therefore\ \sin\alpha = 2100 \times 0.7443/5190 = 0.3012$ $\quad\therefore\ \alpha =$ **17°32′**

(*c*) **At p.f. 0.8 leading** [Fig. 33-30 (*c*)]

$\angle BOA = \theta + \phi = 85° + 36°53' = 121°53'$

$\therefore\ E_b^2 = 6{,}351^2 + 2{,}100^2 - 2 \times 6{,}351 \times 2{,}100 \times \cos 121°53$

$\therefore\ E_b = 7{,}670$ volt per phase.

Induced line e.m.f. $= 7{,}670 \times \sqrt{3} =$ **13,280 V**

Also, $\dfrac{2{,}100}{\sin\alpha} = \dfrac{7{,}670}{\sin\ 121°53'} = \dfrac{7{,}670}{0.8493}$ $\quad\therefore\ \sin\alpha = 2{,}100 \times 0.8493 / 7{,}670 = 0.2325$

$\therefore\ \alpha =$ **13°27′**

Example 36.11. *A 1-φ alternator has armature impedance of (0.5 + j0.866). When running as a synchronous motor on 200-V supply, it provides a net output of 6 kW. The iron and friction losses amount to 500 W. If current drawn by the motor is 50 A, find the two possible phase angles of current and two possible induced e.m.f.s* **(Elec. Machines-I, Nagpur Univ. 1990)**

Solution. Arm. Cu loss/phase $= I_a^2 R_a = 50^2 \times 0.5 = 1250$ W

Motor intake $= 6000 + 500 + 1250 = 7750$ W

p.f. $= \cos\phi =$ Watts / VA $= 7750 / 200 \times 50 = 0.775$ $\therefore\ \phi = 39°$ **lag or lead.**

$\theta = \tan^{-1}(X_S/R_a) = \tan^{-1}(0.866/0.5) = 60°$; $\angle BOA = 60° - 39° = 21°$ -Fig. 36.31 (*a*)

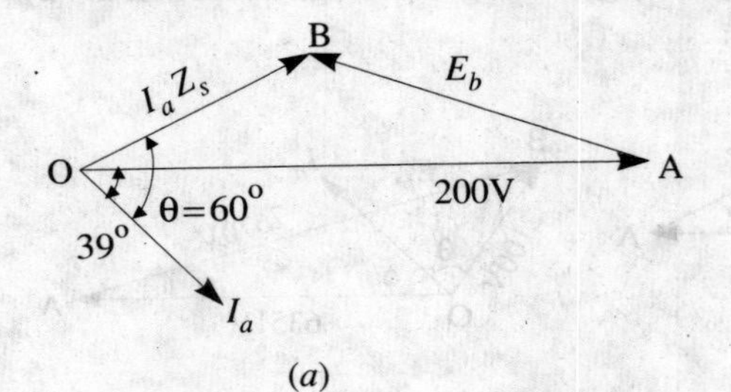

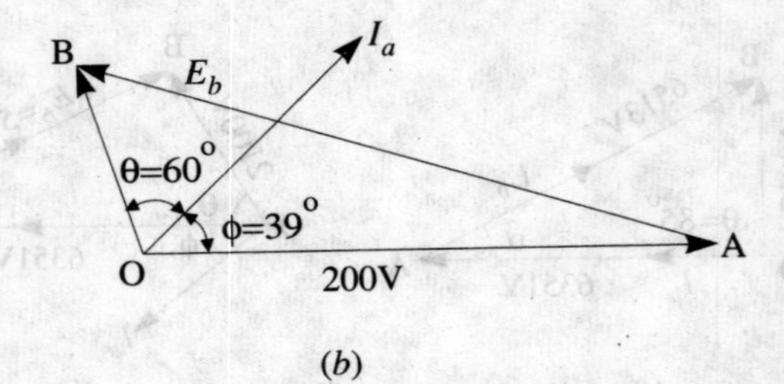

Fig. 36.31

$Z_S = \sqrt{0.5^2 + 0.866^2} = 1\ \Omega\ ;\ I_a Z_S = 50 \times 1 = 50$ V

$AB = E_b = \sqrt{200^2 + 50^2 - 2 \times 200 \times 50 \cos 21°}$; $E_b =$ **154 V.**

In Fig. 36.31 (*b*), $\angle BOA = 60° + 39° = 99°$

$\therefore\ AB = E_b = \sqrt{(200^2 + 50^2 - 2 \times 200 \times 50 \cos 99°}$; $E_b =$ **214 V.**

Example 36.12. *A 2200-V, 3-φ, Y-connected, 50-Hz, 8-pole synchronous motor has Z_S = (0.4 + j 6) ohm/phase. When the motor runs at no-load, the field excitation is adjusted so that E is made equal to V. When the motor is loaded, the rotor is retarded by 3° mechanical.*

Draw the phasor diagram and calculate the armature current, power factor and power of the motor. What is the maximum power the motor can supply without falling out of step?

(Power Apparatus-II, Delhi Univ. 1988)

Solution. Per phase $E_b = V = 2200/\sqrt{3} = 1270$ V

$\alpha = 3°$ (mech) $= 3° \times (8/2) = 12°$ (elect).

As seen from Fig 36.32.

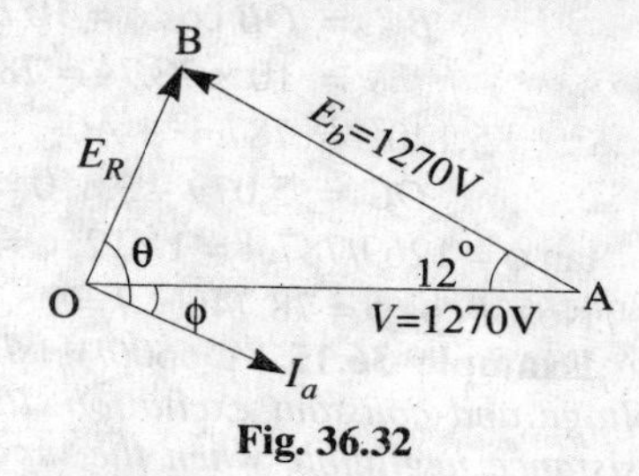

Fig. 36.32

$E_R = (1270^2 + 1270^2 - 2 \times 1270 \times 1270 \times \cos 12°)^{1/2}$

$= 266$ V; $Z_S = \sqrt{0.4^2 + 6^2} = 6.013\ \Omega$

$I_a = E_R / Z_S = 266 / 6.013 =$ **44.2 A.** From ΔOAB,

we get, $\dfrac{1270}{\sin(\theta - \phi)} = \dfrac{266}{\sin 12°}$

$\therefore\ \sin(\theta - \phi) = 1270 \times 0.2079/266 = 0.9926 \qquad \therefore (\theta - \phi) = 83°$

Now, $\theta = \tan^{-1}(X_S / R_a) = \tan^{-1}(6/0.4) = 86.18°$

$\phi = 86.18° - 83° = 3.18° \qquad \therefore$ p.f. $= \cos 3.18° =$ **0.998(lag)**

Total motor power input $= 3\ VI_a \cos\phi = 3 \times 1270 \times 44.2 \times 0.998 = 168$ kW

Total Cu loss $= 3\ I_a^2 R_a = 3 \times 44.2^2 \times 0.4 = 2.34$ kW

Power developed by motor $= 168 - 2.34 =$ **165.66 kW**

$$P_{m(max)} = \frac{E_b V}{Z_S} - \frac{E_b{}^2 R_a}{Z S^2} = \frac{1270 \times 1270}{6.013} - \frac{1270^2 \times 0.4}{6.013^2} = \textbf{250 kW}$$

Example 36.13. *A 1-φ, synchronous motor has a back e.m.f. of 250 V, leading by 150 electrical degrees over the applied voltage of 200 volts. The synchronous reactance of the armature is 2.5 times its resistance. Find the power factor at which the motor is operating and state whether the current drawn by the motor is leading or lagging.*

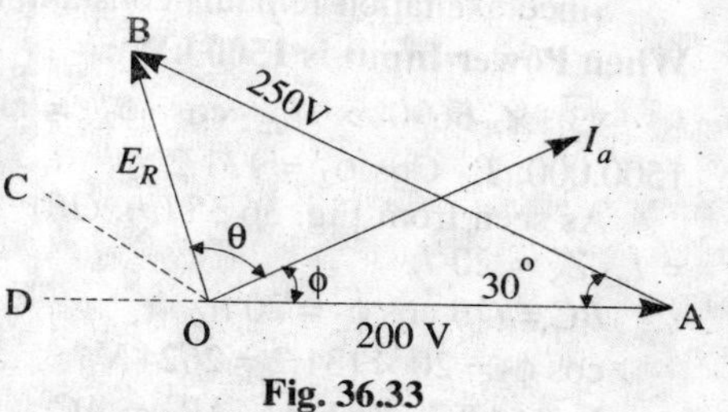

Fig. 36.33

Solution. As induced e.m.f. of 250 V is greater than the applied voltage of 200 V, it is clear that the motor is over-excited, hence it must be working with a leading power factor.

In the vector diagram of Fig. 36.33, *OA* represents applied voltage, *AB* is back e.m.f. at an angle of 30° because $\angle AOC = 150°$ and $\angle COD = \angle BAO = 30°$. *OB* represents resultant of voltage *V* and E_b *i.e.* E_R

In Δ *OBA*,

$E_R = \sqrt{(V^2 + E_b{}^2 - 2\,V E_b \cos 30°)}$

$= \sqrt{(200^2 + 250^2 - 2 \times 200 \times 250 \times 0.866)} = 126$ V

Now, $\dfrac{E_R}{\sin 30°} = \dfrac{E_b}{\sin(\theta + \phi)}$ or $\dfrac{126}{0.5} = \dfrac{250}{\sin(\theta + \phi)}$

$\therefore\ \sin(\theta + \phi) = 125/126$ (approx.) $\qquad \therefore (\theta + \phi) = 90°$

Now $\quad \tan\theta = 2.5 \qquad \therefore \theta = 68°12' \qquad \therefore \phi = 90° - 68°12' = 21°48'$

$\therefore$ p.f. of motor $= \cos 21°48' =$ **0.9285 (leading)**

Example 36.14. *The synchronous reactance per phase of a 3-phase star-connected 6,600 V synchronous motor is 10 Ω. For a certain load, the input is 900 kW and the induced line e.m.f. is 8,900 V. (line value) . Evaluate the line current. Neglect resistance.*

(Basic Elect. Machines, Nagpur Univ. (1993)

Solution. Applied voltage / phase $= 6,600/\sqrt{3} = 3,810$ V

Back e.m.f. / phase $= 8,900/\sqrt{3} = 5,140$ V

Input $= \sqrt{3}\, V_L . I \cos\phi = 900,000$

$\therefore\ I\cos\phi = 9\times 105/\sqrt{3} \times 6,600 = 78.74$ A

In Δ *ABC* of vector diagram in Fig. 36.34, we have $AB^2 = AC^2 + BC^2$

Now $OB = I.X_S = 10\,I$

$\therefore\ BC = OB\cos\phi = 10\,I\cos\phi$

$= 10\times 78.74 = 787.4$ V

$\therefore\ 5,140^2 = 787.4^2 + AC^2 \qquad \therefore\ AC = 5,079$ V

$\therefore\ OC = 5,079 - 3,810 = 1,269$ V

$\tan\phi = 1269/787.4 = 1.612;\ \phi = 58.2°,\ \cos\phi = 0.527$

Now $I\cos\phi = 78.74;\quad I = 78.74/0.527 =$ **149.4 A**

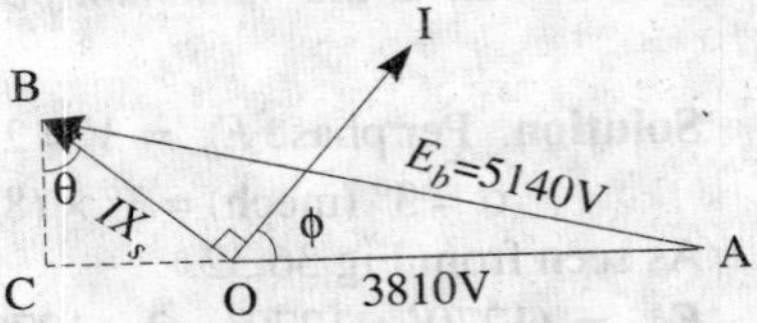

Fig. 36.34

Example 36.15. *A 6600-V, star-connected, 3-phase synchronous motor works at constant voltage and constant excitation. Its synchronous reactance is 20 ohms per phase and armature resistance negligible when the input power is 1000kW, the power factor is 0.8 leading. Find the power angle and the power factor when the input is increased to 1500 kW.*

(Elect. Machines, AMIE Sec. B 1991)

Solution. When Power Input is 1000 kW [Fig. 36.35 (*a*)]

$\sqrt{3}\times 6600\times I_{a1}\times 0.8 = 1000,000;\ I_{a1} = 109.3$ A

$Z_S = X_S = 20\ \Omega;\ I_{a1} Z_S = 109.3\times 20 = 2186\ V;\ \phi_1 = \cos^{-1} 0.8 = 36.9°;\quad \theta = 90°$

$E_b^{\,2} = 3810^2 + 2186^2 - 2\times 3810\times 2186\times\cos(90° + 36.9°)$

$= 3810^2 + 2186^2 - 2\times 3810\times 2186\times -\cos 53.1°;\ \therefore\ E_b = 5410\ V$

Since excitation remains constant, E_b in the second case would remain the same *i.e.* 5410 V.

When Power Input is 1500 kW :

$\sqrt{3}\times 6600\times I_{a2}\cos\phi_2 = 1500,000;\ I_{a2}\ \text{Cos}\ \phi_2 = 131.2$ A

As seen from Fig. 36.35 (*b*), $OB = I_{a2} Z_S = 20\, I_a$

$BC = OB\cos\phi_2 = 20\, I_{a2}$

$\cos\phi_2 = 20\times 131.2 = 2624$ V

In Δ *ABC*, we have, $AB^2 = AC^2 + BC^2$ or $5410^2 = AC^2 + 2624^2$

$\therefore\ AC = 4730\ V;\ OC = 4730 - 3810 = 920$ V

$\tan\phi_2 = 920/4730;\ \phi_2 = 11°;$ p.f. $= \cos\phi_2 = \cos 11° = 0.982$ (lead)

$\tan\alpha_2 = BC/AC = 2624/4730;\ \alpha_2 =$ **29°**

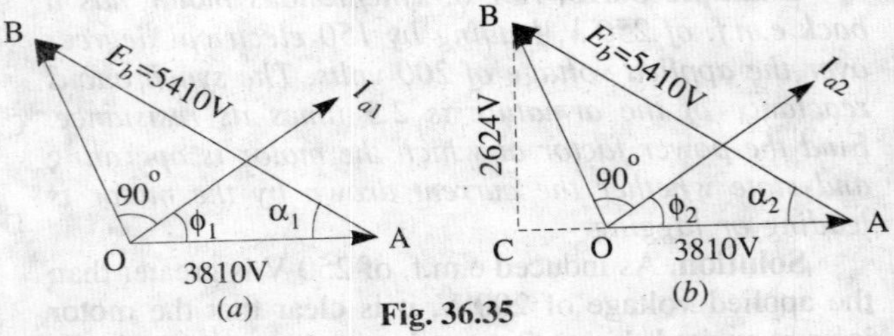

Fig. 36.35

Example 36.16. *A 3-phase star-connected 400-V synchronous motor takes a power input of 5472 watts at rated voltage. Its synchronous reactance is 10Ω per phase and resistance is negligible. If its excitation voltage is adjusted equal to the rated voltage of 400 V, calculate the load angle, power factor and the armature current.*

(Elect. Machines AMIE Sec.B 1990)

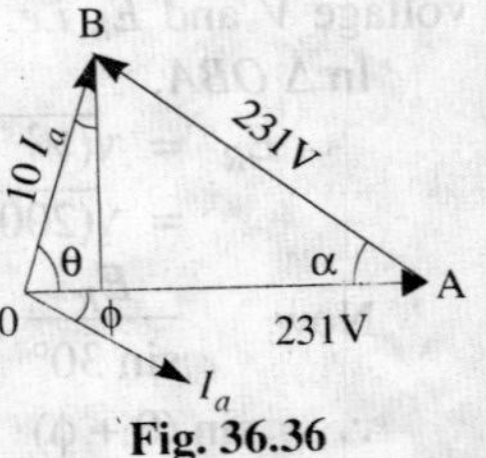

Fig. 36.36

Solution. $\sqrt{3}\times 400\times I_a\cos\phi = 5472;\ I_a\cos\phi = 7.9$ A

$Z_S = 10\ \Omega;\ E_R = I_a Z_S = 10\, I_a$

As seen from Fig. 36.36, $BC = OB \cos\phi = 10\, I_a \cos\phi = 79$ V

$AC = \sqrt{231^2 - 79^2} = 217$ V; $OC = 231 - 217 = 14$ V

$\tan\phi = 14/79$; $\phi = 10°$; $\cos\phi = 0.985$ (lag)

$I_a \cos\phi = 7.9$; $I_a = 7.9/0.985 = 8$A; $\tan\alpha = BC/AC = 79/217$; $\alpha = \mathbf{20°}$

Example 36.17. *A 2,000-V, 3-phase, star-connected synchronous motor has an effective resistance and synchronous reactance of 0.2 Ω and 2.2 Ω respectively. The input is 800 kW at normal voltage and the induced e.m.f. is 2,500 V. Calculate the line current and power factor.*

(Elect. Engg. A.M.I.E.T.E., June 1992)

Solution. Since the induced e.m.f. is greater than the applied voltage, the motor must be running with a leading p.f. If the motor current is I, then its in-phase or power component is $I \cos\phi$ and reactive component is $I \sin\phi$

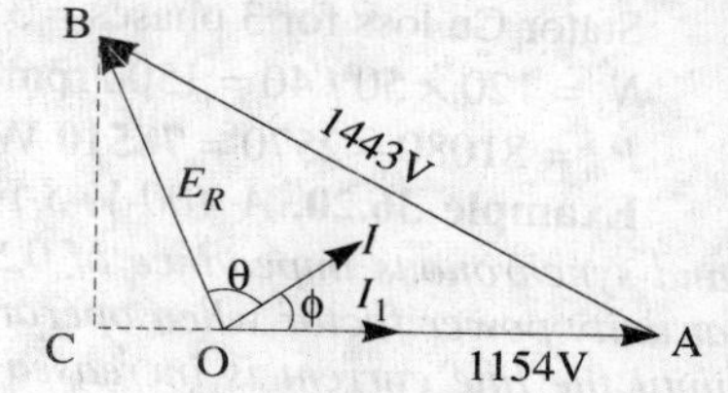

Let $I\cos\phi = I_1$ and $I \sin\phi = I_2$ so that $\mathbf{I} = (I_1 + jI_2)$

$I\cos\phi = I_1 = 800{,}00/\sqrt{3} \times 2000 = 231$ A

Applied voltage / phase = $2{,}000/\sqrt{3} = 1{,}154$ V

Induced e.m.f. / phase = $2500/\sqrt{3} = 1{,}443$ V

In Fig. 33.37 $OA = 1154$ V and

$AB = 1443$ V, OI leads OA by ϕ

$E_R = I Z_S$ and $\theta = \tan^{-1}(2.2/0.2) = 84.8°$

BC is $\perp AO$ produced.

Fig. 36.37

Now, $\mathbf{E_R} = \mathbf{I\, Z_S} = (I_1 + jI_2)(0.2 + j2.2)$

$= (231 + jI_2)(0.2 + j2.2) = (46.2 - 2.2I_2) + j(508.2 + 0.2\, I_2)$

Obviously, $OC = (46.2 - 2.2I_2)$; $BC = j(508.2 + 0.2\, I_2)$

From the right-angled Δ ABC, we have

$AB^2 = BC^2 + AC^2 = BC^2 + (AO + OC)^2$

or $1443^2 = (508.2 + 0.2\, I_2)^2 + (1154 + 46.2 - 2.2\, I_2)^2$

Solving the above quadratic equation, we get $I_2 = 71$ A

$I = \sqrt{I_1^2 + I_2^2} = \sqrt{231^2 + 71^2} = \mathbf{242\ A}$

p.f. $= I_1/I = 231/242 = \mathbf{0.95\ (lead)}$

Example 36.18. *Consider a 3300 V delta connected synchronous motor having a synchronous reactance per phase of 18 ohm. It operates at a leading pf of 0.707 when drawing 800 kW from mains. Calculate its excitation emf and the rotor angle (= delta), explaining the latter term.* **(Eleçt. Machines Nagpur Univ. 1993)**

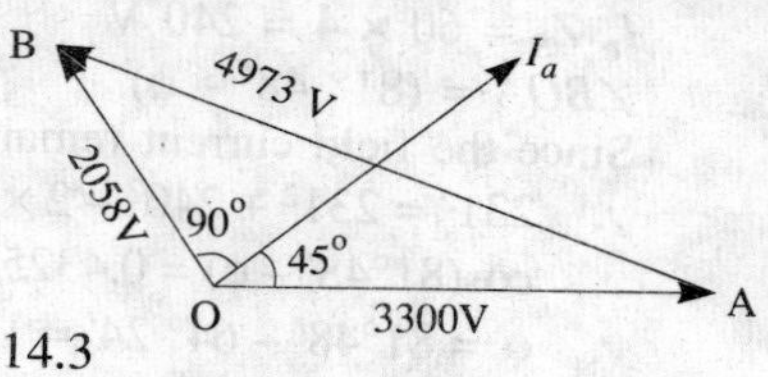

Solution. $\sqrt{3} \times 3300 \times I_a \times 0.707 = 800{,}000$

∴ Line current = 198 A, phase current, $I_a = 198/\sqrt{3} = 114.3$ A;

Fig. 36.38

$Z_S = 18\ \Omega$; $I_a Z_S = 114.3 \times 18 = 2058\ V$

$\phi = \cos^{-1} 0.707$; $\phi = 45°$; $\theta = 90°$;

$\cos(\theta + \phi) = \cos 135° = -\cos 45° = -0.707$

From Fig. 36.38, we find

$E_b^2 = 3300^2 + 2058^2 - 2 \times 3300 \times 2058 \times -0.707$

∴ $E_b = \mathbf{4973\ V}$

From Δ OAB, we get $2058/\sin\alpha = 4973/\sin 135°$. Hence, $\alpha = \mathbf{17°}$

Example 36.19. *A 75-kW, 400-V, 4-pole, 3-phase star connected synchronous motor has a resistance and synchronous reactance per phase of 0.04 ohm and 0.4 ohm respectively. Compute for full-load 0.8 p.f. lead the open circuit e.m.f. per phase and mechanical power developed.*

Assume an efficiency of 92.5%. **(Elect. Machines AMIE Sec. B 1991)**

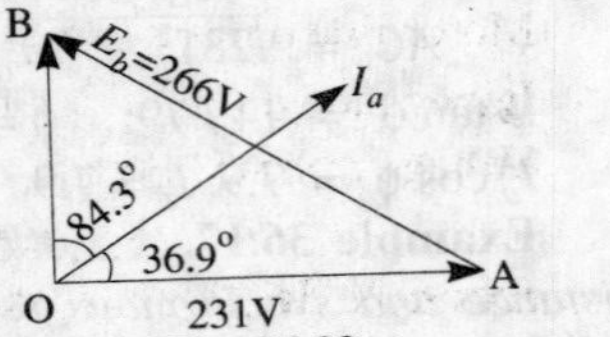

Solution. Motor input = 75,000 / 0.925 = 81,080 W

$I_a = 81{,}080 / \sqrt{3} \times 400 \times 0.8 = 146.3$ A ;

$Z_S = \sqrt{0.04^2 + 0.4^2} = 0.402\ \Omega$

$I_a Z_S = 146.3 \times 0.402 = 58.8\ V$; $\tan\phi = 0.4 / 0.04 = 10$;

$\theta = 84.3°$; $\phi = \cos^{-1} 0.8$; $\phi = 36.9°$; $(\theta + \phi) = 121.8°$;

$V_{ph} = 400 / \sqrt{3} = 231$ V

Fig. 36.39

As seen from Fig. 36.39,

$E_b^2 = 231^2 + 58.8^2 - 2 \times 231 \times 58.8 \times \cos 121.2$; E_b / phase = 266 V

Stator Cu loss for 3 phases = $3 \times 146.3^2 \times 0.04 = 2570$ W;

$N_s = 120 \times 50 / 40 = 1500$ rpm

$P_m = 81080 - 2570 =$ **78510 W** ; $T_g = 9.55 \times 78510/1500 =$ **500 N-m.**

Example 36.20. *A 400-V, 3-phase, 50-Hz, Y-connected synchronous motor has a resistance and synchronous impedance of 0.5 Ω and 4 Ω per phase respectively. It takes a current of 15 A at unity power factor when operating with a certain field current. If the load torque is increased until the line current is increased to 60 A, the field current remaining unchanged, calculate the gross torque developed and the new power factor.* **(Elect.machines, AMIE Sec. B 1992)**

Solution. The conditions corresponding to the first case are shown in Fig. 36.40.

Voltage/phase $= 400/\sqrt{3} = 231$ V ; $I_a Z_S = OB = 15 \times 4 = 60$ V

$X_S = \sqrt{Z_S^2 - R_a^2} = \sqrt{4^2 - 0.5^2} = 3.968\ \Omega$

$\theta = \tan^{-1}(3.968/0.5) = \tan^{-1}(7.936) = 81.8°$

$E_b^2 = 231^2 + 60^2 - 2 \times 231 \times 60 \times \cos 81° 48'$; $E_b = 231$ V

It is obvious that motor is running with normal excitation because $E_b = V$

When the motor load is increased, the phase angle α between the applied voltage and the induced (or back) e.m.f. is increased. Art (36.7). The vector diagram is as shown in Fig. 36.41.

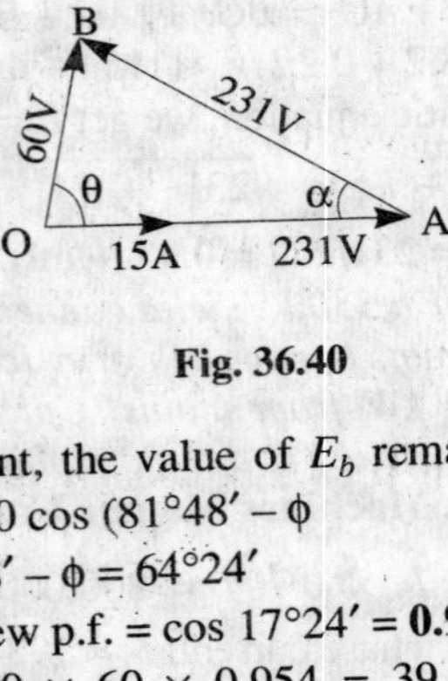

Fig. 36.40

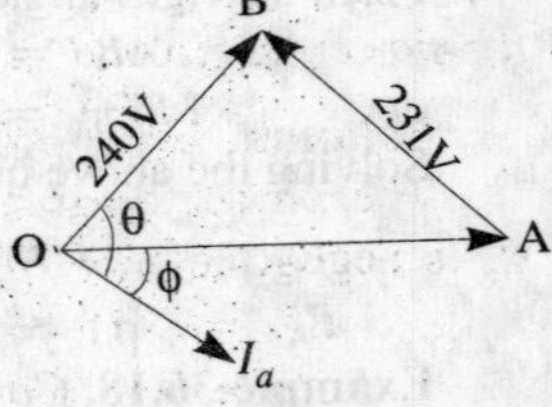

Fig. 36.41

Let φ be the new phase angle.

$I_a Z_S = 60 \times 4 = 240$ V

$\angle BOA = (81° 48' - \phi)$.

Since the field current remains constant, the value of E_b remains the same.

$\therefore\ 231^2 = 231^2 + 240^2 - 2 \times 231 \times 240 \cos(81°48' - \phi$

$\therefore\ \cos(81°48' - \phi) = 0.4325$ or $81° 48' - \phi = 64°24'$

$\therefore\ \phi = 81°48' - 64° 24' = 17° 24'$. New p.f. = cos 17°24′ = **0.954(lag)**

Motor input $= \sqrt{3} \times 400 \times 60 \times 0.954 = 39{,}660$ W

Total armature Cu loss $= 3 \times 60^2 \times 0.5 = 5{,}400$ W

Electrical power converted into mechanical power = 39,660 − 5,400 − 34,260 W

$N_S = 120 \times 50/6 = 1000$ r.p.m. $T_g = 9.55 \times 34{,}260/1000 =$ **327 N-m**

Example 36.21. *A 400-V, 10 h.p. (7.46 kW), 3-phase synchronous motor has negligible armature resistance and a synchronous reactance of 10 Ω/phase. Determine the minimum current and the corresponding induced e.m.f. for full-load conditions. Assume an efficiency of 85%.* **(A.C. Machines-I, Jadavpur Univ. 1987)**

Solution. The current is minimum when the power factor is unity *i.e.* when cos φ = 1. The vector diagram is as shown in Fig. 36.42.

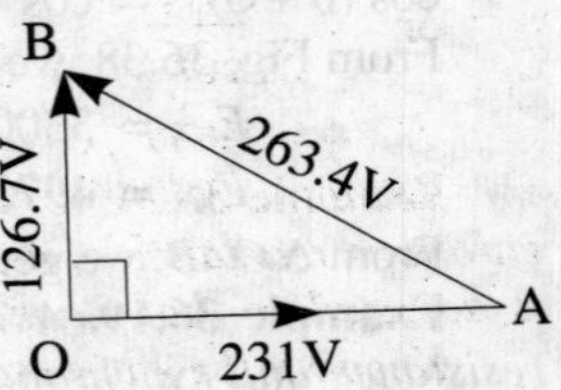

Fig. 36.42

Motor input	$= 7460 / 0.85 = 8{,}775$ W
Motor line current	$= 8{,}775 / \sqrt{3} \times 400 \times 1 =$ **12.67 A**
Impedance drop	$= I_a X_S = 10 \times 12.67 = 126.7$ V
Voltage / phase	$= 400 / \sqrt{3} = 231$ V
E_b	$= \sqrt{231^2 + 126.7^2} =$ **263.4 V**

Example 36.22. *A 400-V, 50-Hz, 3-phase, 37.5 kW, star-connected synchronous motor has a full-load efficiency of 88%. The synchronous impedance of the motor is (0.2 + j 1.6) ohm per phase. If the excitation of the motor is adjusted to give a leading power factor of 0.9, calculate the following for full load:*

(i) the excitation e.m.f.

(ii) the total mechanical power developed **(Elect.Machines, A.M.I.E. Sec. B, 1989)**

Solution. Motor input = 37.5/0.88 = 42.61 kW; $I_a = 42{,}610 / \sqrt{3} \times 400 \times 0.9 = 68.3$ A

$V = 400 / \sqrt{3} = 231$ V; $Z_s = 0.2 + j\,1.6 = 1.612 \angle 82.87°$

$E_R = I_a Z_S = 68.3 \times 1.612 = 110$ V; $\phi = \cos^{-1}(0.8) = 25.84°$

Now, $(\phi + \theta) = 25.84° + 82.87° = 108.71°$

$\cos(\phi + \theta) = \cos 108.71° = -0.32$

(a) $\therefore \quad E_b^2 = V^2 + E_R^2 - 2\,VE_R \cos 108.71°$ or $E_b = 286$ V

Line value of excitation voltage $= \sqrt{3} \times 286 =$ **495 V**

(b) From $\Delta\,OAB$, (Fig. 36.43) $E_R / \sin\alpha = E_b / \sin(\phi + \theta)$, $\alpha = 21.4°$

$$P_m = 3\frac{E_b V}{Z_s}\sin\alpha = 3\,\frac{286 \times 231}{1.612}\sin 21.4° = \mathbf{14{,}954\ W}$$

Fig. 36.43

Example 36.23. *A 6600-V, star-connected, 3-phase synchronous motor works at constant voltage and constant excitation. Its synchronous reactance is 20 ohm per phase and armature resistance negligible. When the input power is 1000 kW, the power factor is 0.8 leading. Find the power angle and the power factor when the input is 1500 kW.* **(Elect.Machines, A.M.I.E., Sec B, 1991)**

Solution. $V = 6600 / \sqrt{3} = 3810$ V, $I_a = 1000 \times 10^3 / \sqrt{3} \times 6600 \times 0.8 = 109.3$ A

The phasor diagram is shown in Fig. 36.44. Since R_a is negligible, $\theta = 90°$

$E_R = I_a X_S = 109.3 \times 20 = 2186$ V

$\cos = 0.8$, $\phi = 36.87°$

$E_b^2 = V^2 + E_R^2 - 2E_b V \cos(90° + 36.87°)$;

$E_b = 5410$ V

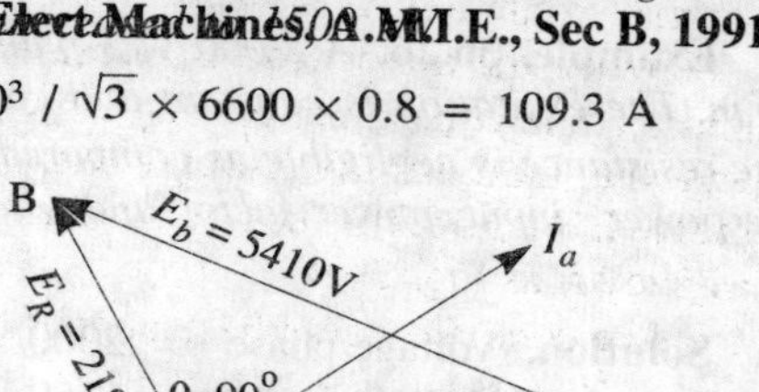

Fig. 36.44

Now, excitation has been kept constant but power has been increased to 1500 kW

$$\therefore\ 3\,\frac{E_b V}{X_a}\sin\alpha = P\ ;\quad 3 \times \frac{5410 \times 3810}{20}\sin\alpha = 1500 \times 10^3\ ;\quad \alpha = 29°$$

$$\text{Also,}\quad \frac{E_b}{\sin(90° + \phi)} = \frac{V}{\sin[180° - (\alpha + 90 + \phi)]} = \frac{V}{\cos(\alpha + \phi)}$$

$$\text{or,}\quad \frac{E_b}{\cos\phi} = \frac{V}{\cos(\alpha + \phi)}\quad \text{or}\quad \frac{V}{E_b} = \frac{\cos(29° + \phi)}{\cos\phi} = 0.3521$$

$\therefore \quad \phi = 19.39°$, $\cos\phi = \cos 19.39° =$ **0.94 (lead)**

Example 36.24. *A 400-V, 50-Hz, 6-pole, 3-phase, Y-connected synchronous motor has a synchronous reactance of 4 ohm/phase and a resistance of 0.5 ohm/phase. On full-load, the excitation is adjusted so that machine takes an armature current of 60 ampere at 0.866 p.f. leading.*

Keeping the excitation unchanged, find the maximum power output. Excitation, friction, windage and iron losses total 2 kW. **(Electrical Machinery-III, Bangalore Univ. 1990)**

Solution. $V = 400/\sqrt{3}$ = 231 V/phase; Z_S = 0.5 + j4 = 4.03 ∠ 82.9°; θ = 82.9°

$I_a Z_S$ = 60 × 4.03 = 242 V; cos φ = 0.866

φ = 30° (lead)

As seen from Fig. 36.45,

$E_b^2 = 231^2 + 242^2 - 2 \times 231 \times 242 \cos 112.9°$

E_b = **394 V**

$$(P_m)_{max} = \frac{E_b V}{Z_S} - \frac{E_b^2 R_a}{Z_S^2} = \frac{394 \times 231}{4.03} - \frac{394^2 \times 0.5}{4.03^2}$$

= 17,804 W/phase. —Art. 36.12

Maximum power developed in armature for 3 phases

= 3 × 17,804 = 52,412 W

Net output = 52,412 – 2,000 = 50,412 W = **50.4 kW**

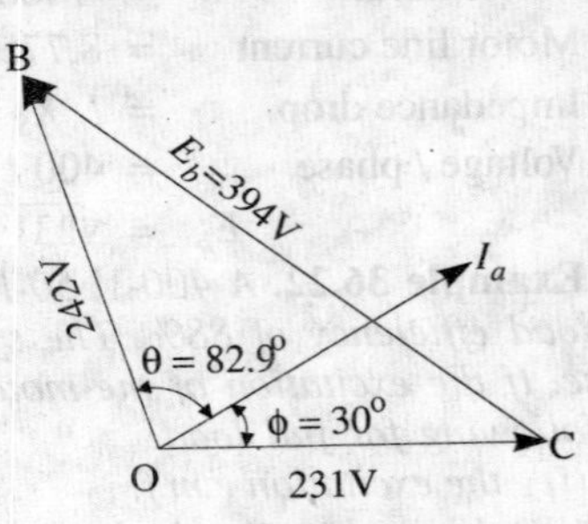

Fig. 36.45

Example 36.25. *A 6-pole synchronous motor has an armature impedance of 10 Ω and a resistance of 0.5 Ω. When running on 2,000 volts, 25-Hz supply mains, its field excitation is such that the e.m.f. induced in the machine is 1600 V. Calculate the maximum total torque in N-m developed before the machine drops out of synchronism.*

Solution. Assuming a three-phase motor,

V = 2000 V, E_b = 1600 V ; R_a = 0.5 Ω; Z_S = 10 Ω; cos θ = 0.5/10 = 1 / 20

Using equation (*iii*) of Art. 35-10, the total max. power for 3 phases is

$$(P_m)_{max} = \frac{E_b V}{Z_S} - \frac{E_b^2}{Z_S} \cos\theta = \frac{2000 \times 1600}{10} - \frac{1600^2 \times 1}{10 \times 20} = 307{,}200 \text{ watt}$$

Now, N_S = 120 f/P = 120 × 25/6 = 500 r.p.m.

Let $T_{g\cdot max}$ be the maximum gross torque, then

$$T_{max} = 9.55 \frac{(P_m)_{max}}{N_S} = 9.55 \frac{307{,}200}{500} = \mathbf{5{,}868 \text{ N-m}}$$

Example. 36.26. *A 2,000-V, 3-phase, 4-pole, Y-connected synchronous motor runs at 1500 r.p.m. The excitation is constant and corresponds to an open-circuit terminal voltage of 2,000 V. The resistance is negligible as compared with synchronous reactance of 3 Ω per phase. Determine the power input, power factor and torque developed for an armature current of 200 A.*

(Elet. Engg.-I, Nagpur Univ. 1993)

Solution. Voltage/phase = $2000/\sqrt{3}$ = 1150 V

Induced e.m.f. = 1150 V —given

Impedance drop = 200 × 3 = 600 V

As shown in Fig. 36.46, the armature current is assumed to lag behind V by an angle φ. Since R_a is negligible, θ = 90°.

∠BOA = (90° – φ)

Considering Δ BOA, we have

$1150^2 = 1150^2 + 600^2 - 2 \times 600 \times 1150 \cos(90 - \phi°)$

sin φ = 0.2605; φ = 16.2°; p.f. = cos 16.2° = **0.965 (lag)**

Power input = $\sqrt{3}$ × 2,000 × 200 × 0.965 = **668.5 kW**

N_S = 1500 r.p.m. ∴ T_g = 9.55 × 66,850/1500 = **4,255 N-m.**

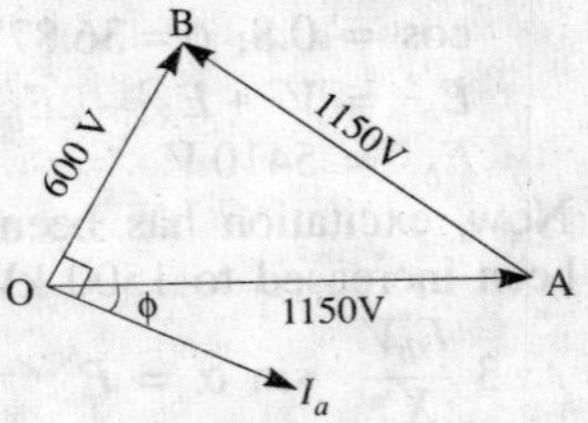

Fig. 36.46

Example 36.27. *A 3-φ, 3300-V, Y-connected synchronous motor has an effective resistance and synchronous reactance of 2.0 Ω and 18.0 Ω per phase respectively. If the open-circuit generated e.m.f. is 3800 V between lines. calculate (i) the maximum total mechanical power that the motor can develop and (ii) the current and p.f. at the maximum mechanical power.*

(Electrical Machines-III. Gujarat Univ. 1988)

Solution. θ = tan⁻¹ (18/2) = 83.7°; V_{ph} = 3300 / $\sqrt{3}$ = 1905 V; E_b = 3800 / $\sqrt{3}$ = 2195 V

Remembering that $\alpha = \theta$ for maximum power development (Ar. 36-10)

$$E_R = (1905^2 + 2195^2 - 2 \times 1905 \times 2195 \times \cos 83.7°)^{1/2} = 2744 \text{ volt per phase}$$

$$\therefore I_a Z_S = 2,744 ; \quad \text{Now, } Z_S = \sqrt{2^2 + 18^2} = 18.11\,\Omega$$

$$\therefore I_a = 2744/18.11 = 152 \text{ A/phase ; line current} = 152 \times \sqrt{3} = \mathbf{262\ A.}$$

$$(P_m)_{max} \text{ per phase} = \frac{E_b V}{Z_S} - \frac{E_b{}^2 R_a}{Z_S{}^2} = \frac{2195 \times 1905}{18.11} - \frac{2195^2 \times 2}{18.11^2}$$

$$= 230,900 - 29,380 = 201,520 \text{ W per phase}$$

Maximum power for three phases that the motor can develop *in its armature*

$$= 201,520 \times 3 = \mathbf{604,560\ W}$$

Total Cu losses = $3 \times 152^2 \times 2 = 138,700$ W

Motor input = 604,560 + 138,700 = 743,260 W

$$\therefore \sqrt{3} \times 3300 \times 262 \times \cos\phi = 743,260 \qquad \therefore \cos\phi = \mathbf{0.496\ (lead).}$$

Example 36.28. *The excitation of a 415-V, 3-phase, mesh-connected synchronous motor is such that the induced e.m.f. is 520 V. The impedance per phase is (0.5 + j4.0) ohm. If the friction and iron losses are constant at 1000 W, calculate the power output, line current, power factor and efficiency for maximum power output.* **(Elect. Machines-I, Madras Univ. 1987)**

Solution. As seen from Art. 36-12, for fixed E_b, V, R_a and X_S, maximum power is developed when $\alpha = \theta$.

Now, $\theta = \tan^{-1}(4/0.5) = \tan^{-1}(8) = 8290° = \alpha$

$$E_R = \sqrt{415^2 + 520^2 - 2 \times 415 \times 520 \times \cos 82.9°} = 625 \text{ V per phase}$$

Now, $IZ_S = 625$; $Z_S = \sqrt{4^2 + 0.5^2} = 4.03\,\Omega \qquad \therefore I = 625/4.03 = 155$ A

Line current = $\sqrt{3} \times 155 = 268.5$ A

$$(P_m)_{max} = \frac{Eb_V}{Z_S} - \frac{E_b{}^2 R_a}{Z_S{}^2} = \frac{520 \times 415}{4.03} - \frac{520^2 \times 0.5}{16.25} = 45,230 \text{ W}$$

Max. power for 3 phases $= 3 \times 45,230 = 135,690$ W

Power output = power developed – iron and friction losses

= 135,690 - 1000 = 134,690 W = **134.69 kW**

Total Cu loss = $3 \times 155^2 \times 0.5 = 36,080$ W

Total motor input = 135,690 + 36,080 = 171,770 W

$$\therefore \sqrt{3} \times 415 \times 268.5 \times \cos\phi = 171,770 ; \cos\phi = \mathbf{0.89\ (lead)}$$

Efficiency = 134,690/171,770 = 0.7845 or **78.45%**

Tutorial Problem No. 36.1.

1. A 3-phase, 400-V, synchronous motor takes 52.5 A at a power factor of 0.8 leading. Calculate the power supplied and the induced e.m.f. The motor impedance per phase is (0.25 + *j*3.2) ohm. **[29.1 kW; 670V]**
2. The input to a 11-kV, 3ϕ, Y-connected synchronous motor is 60 A. The effective resistance and synchronous reactance per phase are *I* Ω and 30Ω respectively. Find (*a*) power supplied to the motor and (*b*) the induced e.m.f. for a p.f. of 0.8 leading. **[(*a*) 915 kW (*b*) 13kV]** (*Grad. I.E.T.E. Dec. 1978*)
3. A 2,200-V, 3-phase, star-connected synchronous motor has a resistance of 0.6 Ω and a synchronous reactance of 6Ω. Find the generated e.m.f. and the angular retardation of the motor when the input is 200 kW at (*a*) power factor unity and (b) power factor 0.8 leading. **[(*a*) 2.21 kV; 14.3° (*b*) 2.62 kV; 12.8°]**
4. A 3-phase, 220-V, 50-Hz, 1500 r.p.m., mesh-connected synchronous motor has a synchronous impedance of 4 ohm per phase. it receives an input line current of 30 A at a leading power factor of 0.8. Find the line value of the induced e.m.f. and the load angle expressed in mechanical degrees.

If the mechanical load is thrown off without change of excitation, determine the magnitude of the current under the new conditions. Neglect losses. **[268 V; 6°, 20.6 A]**

5. A 400-V, 3-phase, *Y*-connected synchronous motor takes 3.73 kW at normal voltage and has an impedance of (1 + j8) ohm per phase. Calculate the current and p.f. if the induced e.m.f. is 460 V. **[6.28 A; 0.86 lead]** (*Electrical Engineering, Madras Univ. April 1979*)

6. The input to 6600-V, 3-phase, star-connected synchronous motor is 900 kW. The synchronous reactance per phase is 20 Ω and the effective resistance is negligible. If the generated voltage is 8,900 V (line), calculate the motor current and its power factor. **[Hint.** See solved Ex. 36.14 (*Electrotechnics, M.S. Univ. April 1979*)

7. A 3-phase synchronous motor connected to 6,600-V mains has a star-connected armature with an impedance of (2.5 + *j*15) ohm per phase. The excitation of machine gives 7000 V. The iron, friction and excitation losses are 12 kW. Find the maximum output of the motor **[153.68 kW]**

8. A 3300-V, 3-phase, 50-Hz, star-connected synchronous motor has a synchronous impedance of (2 + *j*15) ohm. Operating with an excitation corresponding to an e.m.f. of 2,500 V between lines, it just falls out of step at full-load. To what open-circuit e.m.f. will it have to be excited to stand a 50% excess torque. **[4 kV]**

9. A 6.6 kV, star-connected, 3-phase, synchronous motor works at constant voltage and constant excitation. Its synchronous impedance is (2.0 + *j* 20) per phase. When the input is 1000 kW, its power factor is 0.8 leading. Find the power factor when the input is increased to 1500 kW (solve graphically or otherwise) **[0.925 lead]** (*AMIE Sec. B Advanced Elect. Machines (E-9) Summer 1991*)

10. A 2200-V, 373 kW, 3-phase, star-connected synchronous motor has a resistance of 0.3 Ω and a synchronous reactance of 3.0 Ω per phase respectively. Determine the induced e.m.f. per phase if the motor works on full-load with an efficiency of 94 per cent and a p.f. of 0.8 leading. **[1510 V]** (*Electrical Machinery, Mysore Univ. 1992*)

11. The synchronous reactance per phase of a 3-phase star-connected 6600 V synchronous motor is 20 Ω. For a certain load, the input is 915 kW at normal voltage and the induced line e.m.f. is 8,942 V. Evaluate the line current and the p.f. Neglect resistance. **[97 A; 0.8258 (lead)]**

12. A synchronous motor has an equivalent armature reactance of 3.3 Ω. The exciting current is adjusted to such a value that the generated e.m.f. is 950 V. Find the power factor at which the motor would operate when taking 80 kW from a 800-V supply mains. **[0.965 leading]** (*City & Guilds, London*)

13. The input to an 11000 V, 3-phase star-connected synchronous motor is 60 A. The effective resistance and synchronous reactance per phase are respectively 1 ohm and 30 ohms. Find the power supplied to the motor and the induced electromotive force for a power factor of 0.8 leading. **[914.5 kW, 13 kV** (*Elect. Machines, A.M.I.E. Sec. B, 1990*)

14. A 400-V, 6-pole, 3-phase, 50-Hz, star-connected synchronous motor has a resistance and synchronous reactance of 0.5 ohm per phase and 4-ohm per phase respectively. It takes a current of 15 A at unity power factor when operating with a certain field current. If the load torque is increased until the line current is 60 A, the field current remaining unchanged, find the gross torque developed, and the new power factor. **[354 Nm; 0.93]** (*Elect. Engg. AMIETE Dec. 1990*)

15. The input to a 11,000-V, 3-phase, star-connected synchronous motor is 60 amperes. The effective resistance and synchronous reactance per phase are respectively 1 ohm and 30 ohm. Find the power supplied to the motor and the induced e.m.f. for power factor of 0.8 (*a*) leading and (*b*) lagging. **[915 kW (*a*) 13 kV (*b*) 9.36 kV]** (*Elect. Machines-II, South Gujarat Univ. 1981*)

16. Describe with the aid of a phasor diagram the behaviour of a synchronous motor starting from no-load to the pull-out point.

What is the output corresponding to a maximum input to a 3-ϕ delta-connected 250-V, 14.92 kW synchronous motor when the generated e.m.f. is 320 V ? The effective resistance and synchronous reactance per phase are 0.3 Ω and 4.5 Ω respectively. The friction, windage, iron and excitation losses total 800 watts and are assumed to remain constant. Give values for (*i*) output (*ii*) line current (*iii*) p.f. **[(*i*) 47.52 kW (*ii*) 161 A (*iii*) 0.804]** (*Elect. Machines, Indore Univ. Feb. 1982*)

36.17. Effect of Excitation on Armature Current and Power Factor

The value of excitation for which back em.f. E_b is equal (in magnitude) to applied voltage V is known as 100% excitation. We will now discuss what happens when motor is either over-excited or under-exicted although we have already touched this point in Art. 36-8.

Consider a synchronous motor in which the mechanical load is *constant* (and hence output is also constant if losses are neglected).

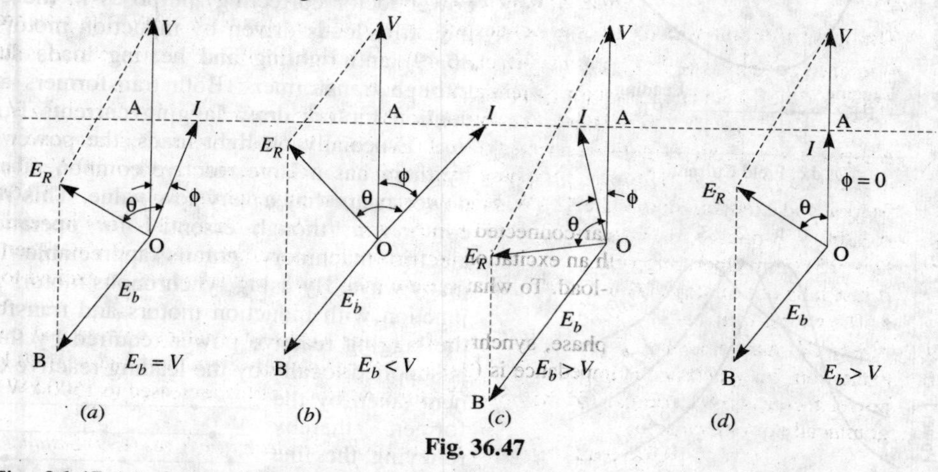

Fig. 36.47

Fig. 36.47 (*a*) shows the case for 100% excitation *i.e.* when $E_b = V$. The armature current I lags behind V by a small angle ϕ. Its angle θ with E_R is fixed by stator constants *i.e.* $\tan\theta = X_S / R_a$.

In Fig. 36.47 (*b*)* excitation is less than 100% *i.e.* $E_b < V$. Here, E_R is advanced clockwise and so is armature current (because it lags behind E_R by fixed angle θ). We note that the magnitude of I is increased but its power factor is decreased (ϕ has increased). Because input as well as V are constant, hence the power component of I *i.e.* $I\cos\phi$ remains the same as before, but wattless component $I\sin\phi$ is increased. Hence, as excitation is decreased, I will increase but p.f. will decrease so that power component of I *i.e.* $I\cos\phi = OA$ will remain constant. In fact, the locus of the extremity of current vector would be a straight horizontal line as shown.

Incidentally, it may be noted that when *field* current is reduced, the motor pull-out torque is also reduced in proportion.

Fig. 36.47 (*c*) represents the condition for overexcited motor *i.e.* when $E_b > V$. Here, the resultant voltage vector E_R is pulled anti-clockwise and so is I. It is seen that now motor is drawing a leading current. It may also happen for some value of excitation, that I may be in phase with V *i.e.* p.f. is unity [Fig. 36.47 (*d*)]. At that time, the current drawn by the motor would be *minimum*.

Two important points stand out clearly from the above discussion :

(*i*) The magnitude of armature current varies with excitation. The current has large value both for low and high values of excitation (though it is lagging for low excitation and leading for higher excitation). In between, it has minimum value corresponding to a certain excitation. The variations of I with excitation are shown in Fig. 36.48 (*a*) which are known as '*V*' curves because of their shape.

(*ii*) For the same input, armature current varies over a wide range and so causes the power factor also to vary accordingly. When over-excited, motor runs with leading p.f. and with lagging p.f. when under-excited. In between, the p.f. is unity. The variations of p.f. with excitation are shown in Fig. 36.48 (*b*). The curve for p.f. looks like inverted '*V*'

---

*These are the same diagrams as given in Fig. 36.9 and 10 expcept that vector for *V* has been shown vertical.

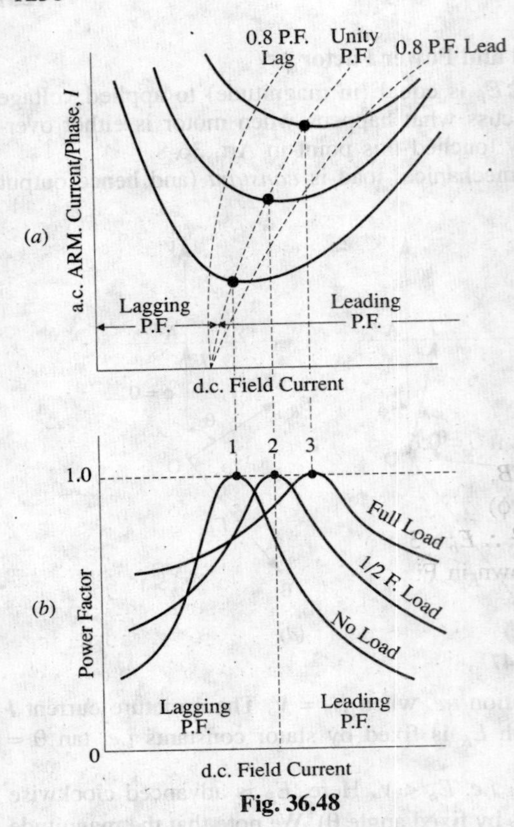

Fig. 36.48

curve. It would be noted that *minimum armature current corresponds to unity power factor.*

It is seen (and it was pointed out in Art. 36.1) that an over-excited motor can be run with leading power factor. This property of the motor renders it extremely useful for phase advancing (and so power factor correcting) purposes in the case of industrial loads driven by induction motors (Fig. 36.49) and lighting and heating loads supplied through transformers. Both transformers and induction motors draw lagging currents from the line. Especially on light loads, the power drawn by them has a large reactive component and the power factor has a very low value. This reactive component, though essential for operating the electric machinery, entails appreciable loss in many ways. By using synchronous motors in conjunction with induction motors and transformers, the lagging reactive power required by the latter is supplied locally by the leading reactive component taken by the former, thereby relieving the line and generators of much of the reactive component. Hence, they now supply only the active component of the load current. When used in this way, a synchronous motor is called a *synchronous capacitor*, because it draws, like a capacitor, leading current from the line. Most synchronous capacitors are rated between 20 MVAR and 200 MVAR and many are hydrogen-cooled.

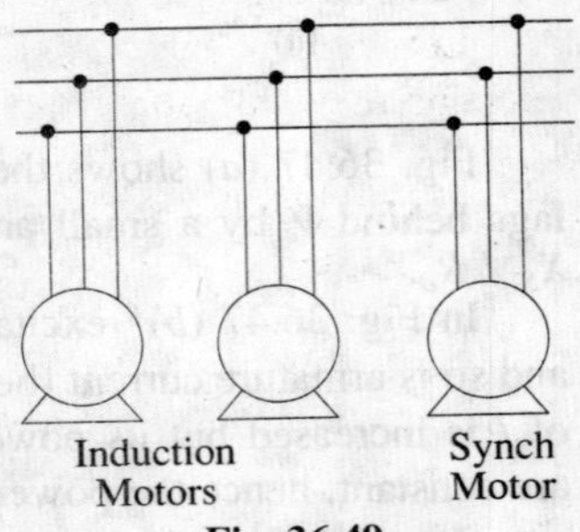

Fig. 36.49

Example 36.29. *Describe briefly the effect of varying excitation upon the armature current and p.f. of a synchronous motor when input power to the motor is maintained constant.*

A 400-V, 50-Hz, 3-φ, 37.3 kW, star-connected synchronous motor has a full-load efficiency of 88%. The synchronous impedance of the motor is (0.2 + j 1.6) Ω per phase. If the excitation of the motor is adjusted to give a leading p.f. of 0.9, calculate for full-load (a) the induced e.m.f. (b) the total mechanical power developed.

Solution. Voltage / phase $= 400/\sqrt{3} = 231$ V;

$Z_S = \sqrt{(1.6^2 + 0.2^2)} = 1.61\,\Omega$

$$\text{Full-load current} = \frac{37{,}300}{\sqrt{3} \times 400 \times 0.88 \times 0.9} = 68 \text{ A}$$

$\therefore \quad I Z_S = 1.61 \times 68 = 109.5$ V

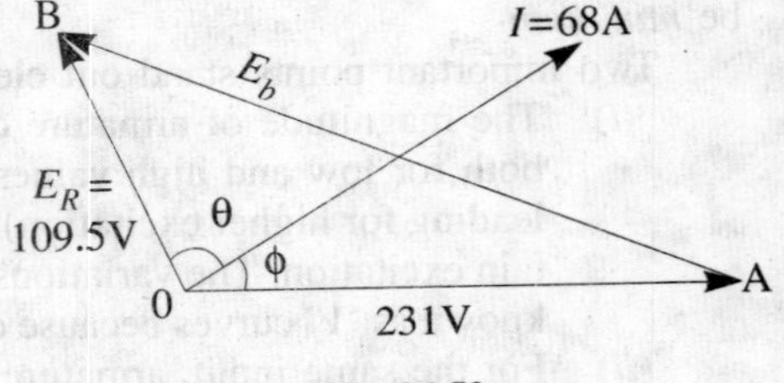

Fig. 36.50

With reference to Fig. 36.50

$\tan\theta = 1.6/0.2 = 8, \theta = 82°54'$

$\cos\phi = 0.9, \phi = 25°50'$.

$\therefore \quad \theta + \phi) = 82°54' + 25°50' = 108°44'$

Now $\cos 108°44' = -0.3212$

(a) In $\Delta\, OAB$, $E_b^2 = 231^2 + 109.5^2 - 2 \times 231 \times 109.5 \times (-0.3212) = 285.6^2$; $E_b = 285.6$ V

Line value of $E_b = \sqrt{3} \times 285.6$ = **495 V**

(b) Total motor input = 37,300 / 0.88 = 42,380 W

Total Cu losses = $3 \times I^2R_a = 3 \times 68^2 \times 0.2$ = 2,774 W

∴ electric power converted into mechanical power = 42,380 − 2,774 = **39.3 kW.**

Example 36.30. *A 3-φ, star-connected synchronous motor takes 48 kW at 693 V, the power factor being 0.8 lagging. The induced e.m.f. is increased by 30%, the power taken remaining the same. Find the current and the p.f. The machine has a synchronous reactance of 2 Ω per phase and negligible resistance.*

Solution. Full-load current

$= 48.000/\sqrt{3} \times 693 \times 0.8 = 50A$

Voltage/phase = $693/\sqrt{3}$ = 400 V

$Z_S = X_S = 2\ \Omega$ ∴ $IZ_S = 50 \times 2 = 100$ V

$\tan\theta = 2/0 = \infty$ ∴ $\theta = 90°$; $\cos\phi = 0.8, \sin\phi = 0.6$

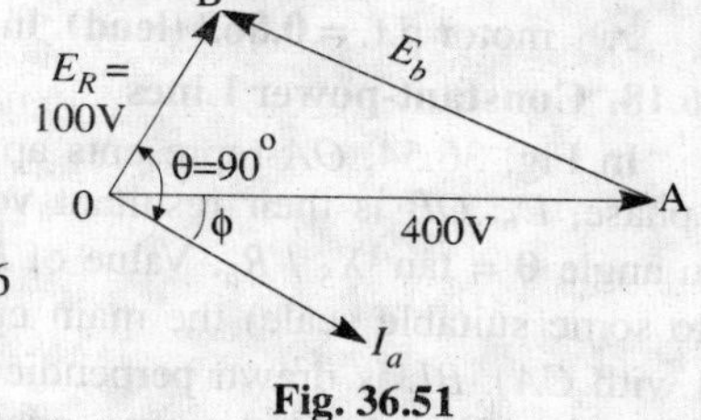

Fig. 36.51

The vector diagram is shown in Fig. 36.51. In Δ OAB,

$E_b^2 = 400^2 + 100^2 - 2 \times 400 \times 100 \times \cos(90° - \phi)$

$= 400^2 + 100^2 - 2 \times 400 \times 100 \times 0.6 = 349^2$ ∴ E_b = 349 V.

The vector diagram for increased e.m.f. is shown in Fig. 36.52. Now, $E_b = 1.3 \times 349 = 454$ V. It can be safely assumed that in the second case, current is leading V by some angle φ′.

Let the new current and the leading angle of current by I′ and φ′ respectively. As power input remains the same and V is also constant, I cos φ should be the same far the same input.

∴ $I\cos\phi = 50 \times 0.8 = 40 = I'\cos\phi'$

In $\Delta ABC, AB^2 = AC^2 + BC^2$

Now $BC = I'X_S\cos\phi'$ (∵ $OB = I'X_S$)

$= 40 \times 2 = 80$ V

∴ $454^2 = AC^2 + 80^2$ or $AC = 447$ V

∴ $OC = 447 - 400 = 47$ V

∴ $\tan\phi' = 47/80, \phi' = 30°26'$

∴ New p.f. = cos 30°26′ = **0.8623 (leading)**

Also, $I'\cos\phi' = 40$ ∴ $I' = 40/0.8623$ = **46.4 A.**

Fig. 36.52

Example 36.31. *A synchronous motor absorbing 60 kW is connected in parallel with a factory load of 240 kW having a lagging p.f. of 0.8. If the combined load has a p.f. of 0.9, what is the value of the leading kVAR supplied by the motor and at what p.f. is it working?*

[Electrical Engineering-II, Banglore Univ. 1990]

Solution. Load connections and phase relationships are shown in Fig. 36.53.

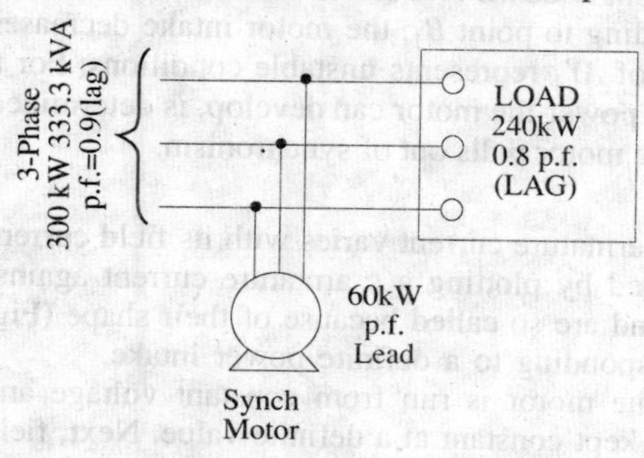

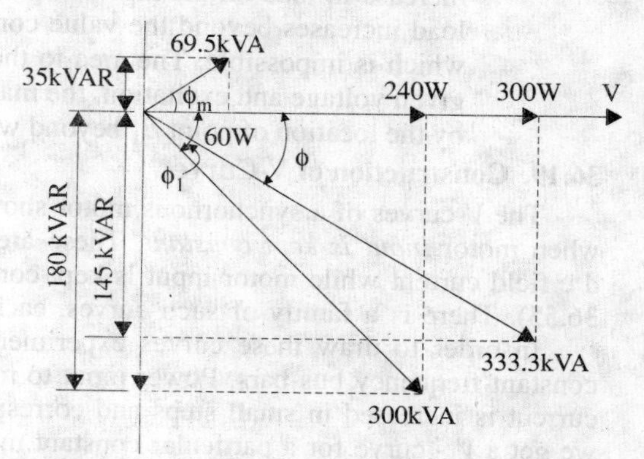

Fig. 36.53

Total load $= 240 + 60 = 300$ kW; combined p.f. = 0.9 (lag)

$\phi = 25.8°$, $\tan\phi = 0.4834$, combined kVAR $= 300 \times 0.4834 = 145$(lag)

Factory Load

$\cos\phi_L = 0.8$, $\phi_L = 36.9°$, $\tan\phi_L = 0.75$, load kVAR $= 240 \times 0.75 = 180$ (lag)

[or load kVA $= 240/0.8 = 300$, kVAR $= 300 \times \sin\phi_L = 300 \times 0.6 = 180$]

$\therefore$ *leading* kVAR supplied by synchronous motor $= 180 - 145 =$ **35.**

For Synchronous Motor

kW = 60, leading kVAR = 35, $\tan\phi_m = 35/60$; $\phi_m = 30.3°$; $\cos 30.3° = 0.863$

$\therefore$ motor p.f. = **0.863 (lead).** Incidentally, motor kVA $= \sqrt{60^2 + 35^2}$ = **69.5.**

36.18. Constant-power Lines

In Fig. 36.54, *OA* represents applied voltage / phase of the motor and *AB* is the back e.m.f. / phase, E_b. *OB* is their resultant voltage E_R. The armature current is *OI* lagging behind E_R by an angle $\theta = \tan^{-1} X_S / R_a$. Value of $I = E_R / Z_S$. Since Z_S is constant, E_R or vector *OB* represents (to some suitable scale) the main current *I*. *OY* is drawn at an angle ϕ with *OB* (or at an angle θ with *CA*). *BL* is drawn perpendicular to *OX* which is at right angles to *OY*. Vector *OB*, when referred to *OY*, also represents, on a different scale, the current both in magnitude and phase.

Hence, $OB \cos\phi = I \cos\phi = BL$

The power input / phase of the motor

$= VI \cos\phi = V \times BL.$

As *V* is constant, power input is dependent on *BL*. If motor is working with a constant intake, then locus of *B* is a straight line || to *OX* and ⊥ to *OY i.e.* line *EF* for which *BL* is constant. Hence, *EF*, represents a constant-power input line for a given voltage but varying excitation. Similarly, a series of such parallel lines can be drawn each representing a definite power intake of the motor. As regards these constant-power lines, it is to be noted that

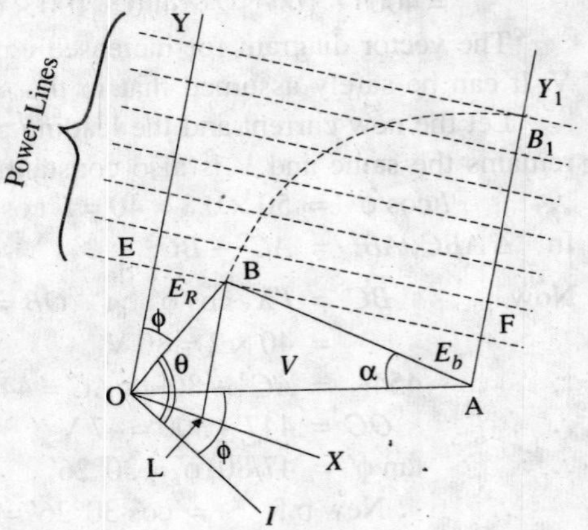

1. for equal increase in intake, the power lines are parallel and equally-spaced
2. zero power line runs along *OX*
3. the perpendicular distance from *B* to *OX* (or zero power line) represents the motor intake
4. If excitation is fixed *i.e.* *AB* is constant in length, then as the load on motor is increased, α increases. In other words, locus of *B* is a circle with radius= *AB* and centre at *A*. With increasing load, *B* goes on to lines of higher power till point B_1 is reached. Any further increase in load on the motor will bring point *B* down to a lower line. It means that as load increases beyond the value corresponding to point B_1, the motor intake decreases which is impossible. The area to the right of AY_1 represents unstable conditions. For a given voltage and excitation, the maximum power the motor can develop, is determined by the location of point B_1 beyond which the motor pulls out of synchronism.

Fig. 36.54

36.19. Construction of V-Curves

The *V*-curves of a synchronous motor show how armature current varies with its field current when motor *input is kept constant*. These are obtained by plotting a c armature current against d c field current while motor input is kept constant and are so called because of their shape (Fig. 36.55). There is a family of such curves, each corresponding to a definite power intake.

In order to draw these curves experimentally, the motor is run from constant voltage and constant-frequency bus-bars. Power input to motor is kept constant at a definite value. Next, field current is increased in small steps and corresponding armature currents are noted. When plotted, we get a *V* - curve for a particular constant motor input. Similar curves can be drawn by keeping motor input constant at different values. A family of such curves is shown in Fig. 36.55.

Detailed procedure for graphic construction of V - curves is given below:

1. First, constant-power lines are drawn as discussed in Art. 36.14.
2. Then, with A as the centre, concentric circles of different radii AB, AB_1, AB_2, etc. are drawn where AB, AB_1, AB_2, etc. are the back e.m.fs. corresponding to different excitations. The intersections of these circles with lines of constant power give positions of the working points for specific loads and excitations (hence back e.m.fs). The vectors OB, OB_1, OB_2 etc. represent different values of E_R (and hence currents) for different excitations. Back e.m.f. vectors AB, AB_1 etc. have not been drawn purposely in order to avoid confusion (Fig. 36.56)

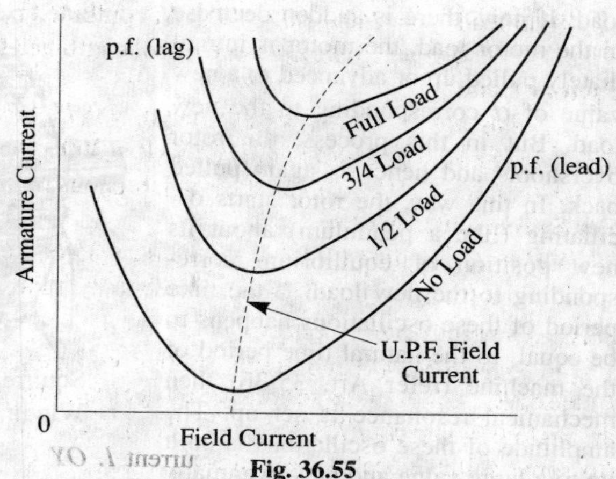

Fig. 36.55

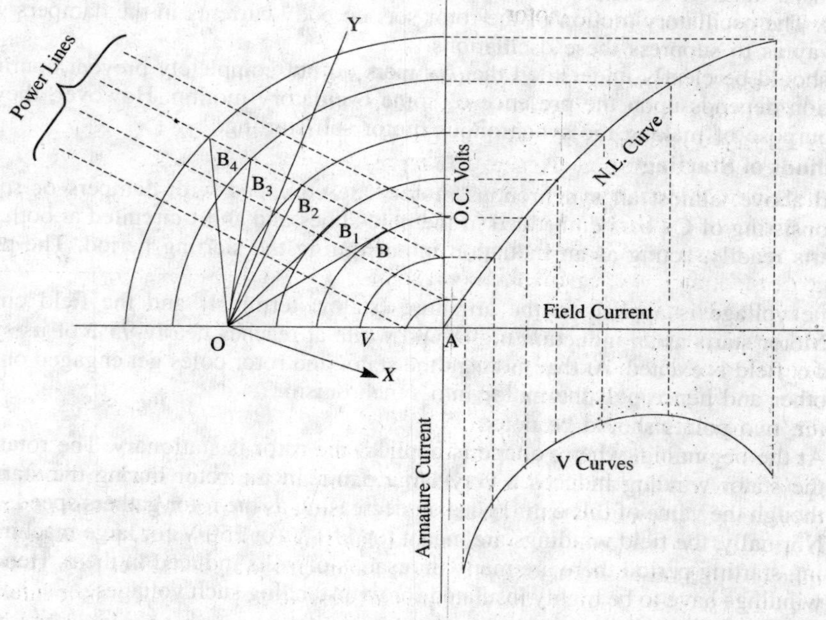

Fig. 36.56

3. The different values of back e.m.fs. like AB, AB_1, AB_2, etc. are projected on the magnetisation and corresponding values of the field (or exciting) amperes are read from it.
4. The field amperes are plotted against the corresponding armature currents, giving us 'V' curves.

36.20. Hunting or Surging or Phase Swinging

When a synchronous motor is used for driving a varying load, then a condition known as hunting is produced. Hunting may also be caused if supply frequency is pulsating (as in the case of generators driven by reciprocating internal combustion engines).

We know that when a synchronous motor is loaded (such as punch presses, shears, compressors and pumps etc.), its rotor falls back in phase by the coupling angle α. As load is progressively increased, this angle also increases so as to produce more torque for coping with the increased load. If now, there is sudden decrease in the motor load, the motor is immediately pulled up or advanced to a new value of α corresponding to the new load. But in this process, the rotor overshoots and hence is again pulled back. In this way, the rotor starts oscillating (like a pendulum) about its new position of equilibrium corresponding to the new load. If the time period of these oscillations happens to be equal to the natural time period of the machine (refer Art. 35-36) then mechanical resonance is set up. The amplitude of these oscillations is built up to a large value and may eventually become so great as to throw the machine out of synchronism. To stop the build-up of these oscillations, dampers or damping grids (also known as squirrel-cage winding) are employed. These dampers consist of short-circuited Cu bars embedded in the faces of the field poles of the motor (Fig. 36.57). The oscillatory motion of the rotor sets up eddy currents in the dampers which flow in such a way as to suppress these oscillations.

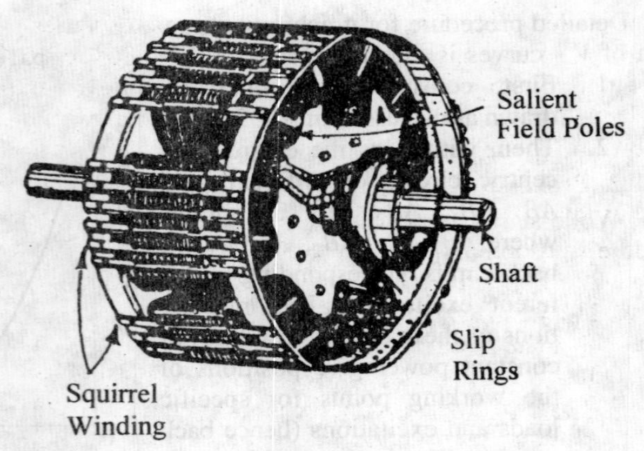

Fig. 36.57

But it should be clearly understood that dampers do not completely prevent hunting because their operation depends upon the presence of some oscillatory motion. However, they serve the additional purpose of making the synchronous motor self-starting.

36.21. Methods of Starting

As said above, almost all synchronous motors are equipped with dampers or squirrel cage windings consisting of Cu bars embedded in the pole-shoes and short-circuited at both ends. Scuh a motor starts readily, acting as an induction motor during the starting period. The procedure is as follows :

The line voltage is applied to the armature (stator) terminals and the field circuit is left *unexcited*. Motor starts as an induction motor and while at reaches nearly 95% of its synchronous speed, the d.c. field is excited. At that moment the stator and rotor poles get engaged or interlocked with each other and hence pull the motor into synchronism.

However, two points should be noted :

1. At the beginning, when voltage is applied, the rotor is stationary. The rotating field of the stator winding induces a very large e.m.f. in the rotor during the starting period, though the value of this e.m.f. goes on decreasing as the rotor gathers speed. Normally, the field windings are meant for 110-V (or 250 V for large machines) but during starting period there are many thousands of volts induced in them. Hence, the rotor windings have to be highly insulated for withstanding such voltages.
2. When full line voltage is switched on to the armature at rest, a very large current, usually 5 to 7 times the full-load armature current is drawn by the motor. In some cases, this may not be objectionable but where it is, the applied voltage at starting, is reduced by using auto-transformers (Fig. 36.58). However, the voltage should not be reduced to a very low value because the

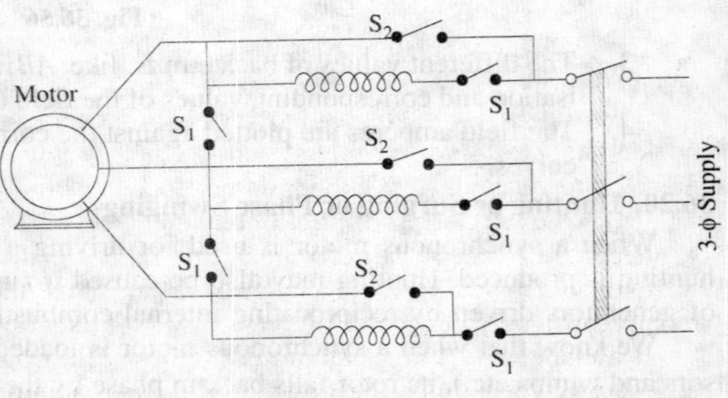

Fig. 36.58

starting torque of an induction motor varies approximately as the square of the applied voltage. Usually, a value of 50% to 80% of the full-line voltage is satisfactory.

Auto-transformer connections are shown in Fig. 36.58. For reducing the supply voltage, the switches S_1 are closed and S_2 are kept open. When the motor has been speeded-up, S_2 are closed and S_1 opened to cut out the transformers.

36.22. Procedure for Starting a Synchronous Motor

While starting a modern synchronous motor provided with damper windings, following procedure is adopted.

1. First, main field winding is short-circuited.
2. Reduced voltage with the help of auto-transformers is applied across stator terminals. The motor starts up.
3. When it reaches a steady speed (as judged by its sound), a weak d.c. excitation is applied by removing the short-circuit on the main field winding. If excitation is sufficient, then the machine will be pulled into synchronism.
4. Full supply voltage is applied across stator terminals by cutting out the auto-transformers.
5. The motor may be operated at any desired power factor by changing the d.c. excitation.

36.23. Comparison Between Synchronous and Induction Motors

1. For a given frequency, the synchronous motor runs at a constant average speed whatever the load, while the speed of an induction motor falls somewhat with increase in load.
2. The synchronous motor can be operated over a wide range of power factors, both lagging and leading, but induction motor always runs with a lagging p.f. which may become very low at light loads.
3. A synchronous motor is inherently not self-starting.
4. The changes in applied voltage do not affect synchronous motor torque as much as they affect the induction motor torque. The breakdown torque of a synchronous motor varies approximately as the first power of applied voltage whereas that of an induction motor depends on the square of this voltage.
5. A d.c excitation is required by synchronous motor but not by induction motor.
6. Synchronous motors are usually more costly and complicated than induction motors, but they are particularly attractive for low-speed drives (below 300 r.p.m.) because their power factor can always be adjusted to 1.0 and their efficiency is high. However, induction motors are excellent for speeds above 600 r.p.m.
7. Synchronous motors can be run at ultra-low speeds by using high power electronic converters which generate very low frequencies. Such motors of 10 MW range are used for driving crushers, rotary kilns and variable-speed ball mills etc.

36.24. Synchronous Motor Applications

Synchronous motors find extensive application for the following classes of service :

1. Power factor correction
2. Constant-speed, constant-load drives
3. Voltage regulation

(*a*) **power factor correction**

Overexcited synchronous motors having leading power factor are widely used for improving power factor of those power systems which employ a large number of induction motors (Fig. 36-49) and other devices having lagging p.f. such as welders and flourescent lights etc.

(*b*) **constant-speed applications**

Because of their high efficiency and high-speed, synchronous motors (above 600 r.p.m.) are well-suited for loads where constant speed is required such as centrifugal pumps, belt-driven reciprocating compressors, blowers, line shafts, rubber and paper mills etc.

Low-speed synchronous motors (below 600 r.p.m.) are used for drives such as centrifugal and screw-type pumps, ball and tube mills, vacuum pumps, chippers and metal rolling mills etc.

(*c*) **Voltage regulation**

The voltage at the end of along transmission line varies greatly especially when large inductive loads are present. When an inductive load is disconnected suddenly, voltage tends to rises considerably above its normal value because of the line capacitance. By installing a synchronous motor with a field regulator (for varying its excitation), this voltage rise can be controlled.

When line voltage decreases due to inductive load, motor excitation is increased thereby raising its p.f. which compensates for the line drop. If, on the other hand, line voltage rise due to line capacitive effect, motor excitation is decreased, thereby making its p.f. lagging which helps to maintain the line voltage at its normal value.

QUESTIONS AND ANSWERS ON SYNCHRONOUS MOTORS

Q.1. Does change in excitation affect the synchronous motor speed?
Ans. No.

Q.2. The power factor?
Ans. Yes.

Q.3. How?
Ans. When over-excited, synchronous motor has leading power factor. However, when under-excited, it has lagging power factor.

Q.4. For what service are synchronous motors especially suited?
Ans. For high voltage service.

Q.5. Which has more efficiency; synchronous or induction motor?
Ans. Synchronous motor.

Q.6. Mention some specific applications of synchronous motor?
Ans. 1. constant speed load service 2. reciprocating compressor drives 3. power factor correction 4. voltage regulation of transmission lines.

Q.7. What is a synchronous capacitor?
Ans. An overexcited synchronous motor is called synchronous capacitor, because, like a capacitor, it takes a leading current.

Q.8. What are the causes of faulty starting of a synchronous motor?
Ans. It could be due to the following causes:

1. voltage may be too low-at least half voltage is required for starting
2. there may be open-circuit in one phase—due to which motor may heat up
3. static friction may be large—either due to high belt tension or too tight bearings
4. stator windings may be incorrectly connected
5. field excitation may be too strong.

Q.9. What could be the reasons if a synchronous motor fails to start?
Ans. It is usually due to the following reasons:

1. voltage may be too low
2. some faulty connection in auxiliary apparatus
3. too much starting load
4. open-circuit in one phase or short-circuit
5. field excitation may be excessive.

Q.10. A synchronous motor starts as usual but fails to develop its full torque. What could it be due to?
Ans. 1. exciter voltage may be too low 2. field spool may be reversed 3. there may be either open-circuit or short-circuit in the field.

Q.11. Will the motor start with the field excited ?
Ans. No.

Q.12. Under which conditions a synchronous motor will fail to pull into step?
Ans.

1. no field excitation
2. excessive load
3. excessive load inertia

OBJECTIVE TESTS-36

1. In a synchronous motor, damper winding is provided in order to
(*a*) stabilize rotor motion
(*b*) suppress rotor oscillations
(*c*) develop necessary starting torque
(*d*) both (*b*) and (*c*).

2. In a synchronous motor, the magnitude of stator back e.m.f. E_b depends on
(*a*) speed of the motor
(b) load on the motor
(c) both the speed and rotor flux
(d) d.c. excitation only.

3. An electric motor in which both the rotor and stator fields rotates with the same speed is called a/an ....motor.
(*a*) d.c.
(*b*) Schrage
(*c*) synchronous
(*d*) universal

4. While running, a synchronous motor is compelled to run at synchronous speed because of
(*a*) damper winding in its pole faces
(*b*) magnetic locking between stator and rotor poles
(*c*) induced e.m.f. in rotor field winding by stator flux
(*d*) compulsion due to Lenz's law.

5. The direction of rotation of a synchronous motor can be reversed by reversing
(*a*) current to the field winding
(*b*) supply phase sequence
(*c*) polarity of rotor poles
(*d*) none of the above

6. When running under no-load condition and with normal excitation, armature current I_a drawn by a synchronous motor
(*a*) leads the back e.m.f. E_b by a small angle
(*b*) is large
(*c*) lags the applied voltage *V* by a small angle
(*d*) lags the resultant voltage E_R by 90°.

7. The angle between the synchronously-rotating stator flux and rotor poles of a synchronous motor is called....... angle.
(*a*) synchronizing
(*b*) torque
(*c*) power factor
(*d*) slip.

8. If load angle of a 4-pole synchronous motor is 8° (elect), its value in mechanical degrees is ....
(*a*) 4 (*b*) 2
(*c*) 0.5 (*d*) 0.25.

9. The maximum value of torque angle α in a synchronous motor is ....degrees electrical.
(*a*) 45 (*b*) 90
(*c*) between 45 and 90
(*d*) below 60.

10. A synchronous motor running with normal excitation adjusts to load increases *essentially* by increase in its
(*a*) power factor
(*b*) torque angle
(*c*) back e.m.f.
(*d*) armature current.

11. When load on a synchronous motor running with normal excitation is increased, armature current drawn by it increases because
(*a*) back e.m.f. E_b becomes less than applied voltage *V*.
(*b*) power factor is decreased
(*c*) net resultant voltage E_R in armature is increased
(*d*) motor speed is reduced.

12. When load on a normally-excited synchronous motor is increased, its power factor tends to
(*a*) approach unity
(*b*) become increasingly lagging
(*c*) become increasingly leading
(*d*) remain unchanged.

13. The effect of increasing load on a synchronous motor running with normal excitation is to
(*a*) increase both its I_a and p.f.
(*b*) decrease I_a but increase p.f.
(*c*) increase I_a but decrease p.f.

(*d*) decrease both I_a and p.f.

14. Ignoring the effects of armature reaction, if excitation of a synchronous motor running with constant load is increased, its torque angle must necessarily

(*a*) decrease
(*b*) increase
(*c*) remain constant
(*d*) become twice the no-load value.

15. If the field of a synchronous motor is under-excited, the power factor will be

(*a*) lagging
(*b*) leading
(*c*) unity
(*d*) more than unity.

16. Ignoring the effects of armature reaction, if excitation of a synchronous motor running with constant load is decreased from its normal value, it leads to

(*a*) increase in α but decrease in E_b
(*b*) increase in E_b but decrease in I_a
(*c*) increase in both I_a and p.f. which is lagging
(*d*) increase in both I_a and ϕ

17. A synchronous motor connected to infinite bus-bars has at *constant* full-load, 100% excitation and unity p.f. On changing the excitation only, the armature current will have

(*a*) leading p.f. with under-excitation
(*b*) leading p.f. with over-excitation
(*c*) lagging p.f. with over-excitation
(*d*) no change of p.f.

(Power App.-II, Delhi Univ. Jan 1987)

18. The *V*-curves of a synchronous motor show relationship between

(*a*) excitation current and back emf
(*b*) field current and p.f.
(*c*) dc field current and ac armature current
(*d*) armature current and supply voltage.

19. When load on a synchronous motor is increased, its armature currents is increased provided it is

(*a*) normally-excited
(*b*) over-excited
(*c*) under-excited
(*d*) all of the above.

20. If main field current of a salient-pole synchronous motor fed from an infinite bus and running at no-load is reduced to zero, it would

(*a*) come to a stop
(*b*) continue running at synchronous speed
(*c*) run at sub-synchronous speed
(*d*) run at super-synchronous speed

21. In a synchronous machine when the rotor speed becomes more than the synchronous speed during hunting, the damping bars develop

(*a*) synchronous motor torque
(*b*) d.c. motor torque
(*c*) induction motor torque
(*d*) induction generator torque.

(Power App.-II, Delhi Univ. Jan. 1987)

22. In a synchronous motor, the rotor Cu losses are met by

(*a*) motor input
(*b*) armature input
(*c*) supply lines
(*d*) d.c. source.

23. A synchronous machine is called a doubly-excited machine because

(*a*) it can be overexcited
(*b*) it has two sets of rotor poles
(*c*) both its rotor and stator are excited
(*d*) it needs twice the normal exciting current.

24. Synchronous capacitor is

(*a*) an ordinary static capacitor bank
(*b*) an over-excited synchronous motor driving mechanical load
(*c*) an over-excited synchronous motor running without mechanical load
(*d*) none of the above .

(Elect. Machines, A.M.I.E. Sec. B, 1993)

ANSWERS

1. *d* **2.** *d* **3.** *c* **4.** *b* **5.** *b* **6.** *c* **7.** *b* **8.** *a* **9.** *b* **10.** *d* **11.** *c* **12.** *b* **13.** *c* **14.** *a*
15. *a* **16.** *d* **17.** *b* **18.** *c* **19.** *d* **20.** *b* **21.** *d* **22.** *d* **23.** *c* **24.** *c*

37 CONVERSION FROM A.C. TO D.C.

37.1. General

These days most of the electric power is generated, transmitted and distributed in the form of alternating current and voltage. This is due to the reason that when generation, transmission and distribution of electric power is required on a very large scale, an a.c. plant presents relatively simpler engineering and economic problems than a d.c. system.

Nevertheless, there are several cases where the use of d.c. current is either essential or else is advantageous to a degree sufficient to make it preferable to alternating current.

A few cases are lifted below :

1. Because of their excellent starting and overload characteristics, d.c. motors are most suitable for electric traction on tramways, suburban and underground railways.
2. Direct current is absolutely essential for electrolytic and electro-chemical processes such as electroplating, electro-refining, production of aluminium, Cu and othe metals by electrolysis and fertilizers.
3. It is necessary for running arc-lamps for cinema projectors, searchlights an l fcr some types of arc-welding.
4. Direct current is required for operating relays, telephones, time-switches, and circuit-breakers etc.
5. It is essential for battery-charging work.
6. D.C. motors are superior to a.c. motors for duties where frequent starting against heavy torque, close speed adjustment and frequent reversals are required *i.e.* for elevators, printing presses, colliery winders, large machine tool drives and certain steel mill equipment etc.

Hence, it is necessary to convert alternating current into direct current by some suitable arrangement. Methods for such conversion are give below :

1. Motor-generator sets 2. Motor converters 3. Rotary converters 4. Mercury-arc rectifiers and 5. Metal rectifiers.

37.2. Motor-Generator Set

As shown in Fig. 37.1, it consists of two separate units–a 3-phase a.c. motor directly coupled to a d.c. generator. In the case of larger units, the a.c. motor is invariably synchronous and the d.c. generator is usually compound.

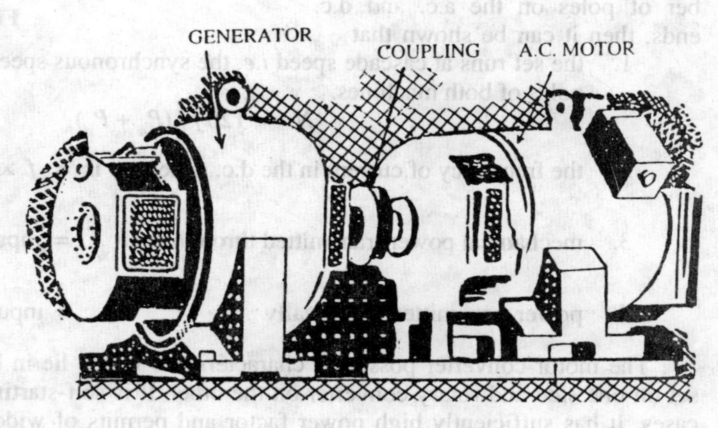

Fig. 37.1

Advantages

1. The d.c. output voltage is practically constant and is not affected by the ordinary changes in a.c. supply voltage. It is

so, because the speed of synchronous motor is constant, being independent of the load and voltage changes of the ordinary magnitude (if the magnitude is large, motor will fall out of synchronism).

2. D.C. output voltage can be controlled easily by adjusting the shunt field regulator.
3. Provided the supply voltage is not high, the synchronous motor can be directly switched to the supply line, without requiring a transformer, which is necessary for rotary converters.
4. The set is self-starting.
5. It can also be used for power factor correction purposes.

Disadvantages

1. The set is much heavier than a rotary converter. In fact, the d.c. generator of the set alone is larger than the synchronous converter. Hence, it costs more.
2. It has a comparatively low efficiency.
3. It requires more floor space.

37.3. Motor Converter

It essentially consists of an ordinary slip-ring induction motor, coupled both electrically and mechanically to a d.c. generater. Irrespective of the number of stator poles, the rotor usually has 12-phase winding, which is connected in star, under normal operating conditions. As shown in Fig. 37.2, the rotor is connected to the equidistant points on the d.c. armature winding (for simplicity, rotor has been shown wound 3-phase).

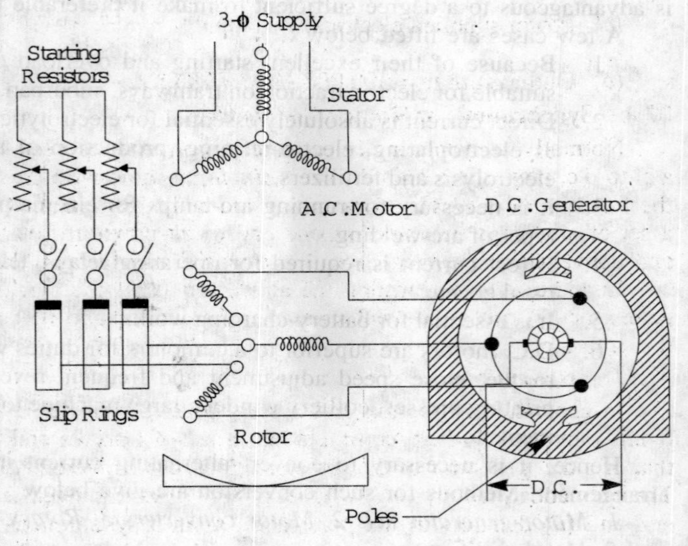

Fig. 37.2

The motor input energy is partly transmitted mechanically (through the shaft) and partly electrically *i.e.* as slip-ring energy from the rotor of the induction motor.

If P_a and P_d represent the number of poles on the a.c. and d.c. ends, then it can be shown that

1. the set runs at cascade speed *i.e.* the synchronous speed corresponding to the sum of the poles of both machines.

$$N = 120f/(P_a + P_d)$$

2. the frequency of current in the d.c. generator is $= f \times \dfrac{P_a}{P_a + P_d}$
3. mechanical power transmitted through shaft $= \text{input} \times \dfrac{P_a}{P_a + P_d}$.
4. power transmitted electrically $= \text{input} \times \dfrac{P_d}{P_a + P_d}$

The motor converter possesses characteristics which lie in between those of motor-generator set on one side and rotary converter on the other. It is self-starting and can be line-started in most cases. It has sufficiently high power factor and permits of wide regulation in d.c. voltage.

However, because of its low speed, it is more expensive and is hence not much popular.

Example 37.1. *In a 700-kW motor-generator set the induction motor has six poles and the d.c. generator has eight poles. If supply frequency is 50 Hz, calculate (a) the speed of the* *set (b) the frequency of currents in the generator and (c) power transmitted mechanically through the shaft.*

Solution. (*a*) Cascade speed $N = 120 \times 50/(6+8) = \mathbf{428.6\ r.p.m.}$

(*b*) Frequency $= 50 \times 8/14 = \mathbf{28.6\ Hz}$

(*c*) Power transmitted through shaft $= 700 \times 6/14 = \mathbf{300\ kW.}$

Example 37.2. *A motor converter consisting of a 6-pole induction motor, driving a 4-pole d.c. machine, is fed from a 3-phase, 50-Hz, 6.6-kV supply and supplies a direct current network at 380 V. If the machine is loaded to 300 kW, calculated (a) the speed of the set (b) the input current (c) the power supplied electrically to the d.c. side from the a.c. machine and (d) output current.*

Assume no loss and unity power factor. **(Elect. Technology, Utkal Univ. 1978)**

Solution. (*a*) speed of the set is $N = \dfrac{120f}{(P_a + P_d)} = \dfrac{120 \times 50}{(6+4)} = \mathbf{600\ r.p.m.}$

(*b*) If I is the input current to the a.c. machine, then

$\sqrt{3} \times 6{,}600 \times I = 300 \times 1000; \quad I = \mathbf{26.25\ A}$

(*c*) Power supplied electrically $= \text{input} \times \dfrac{P_d}{P_a \times P_d} = \dfrac{300 \times 4}{(6+4)} = \mathbf{120\ kW}$

(*d*) Output current = 300,000 / 380 = **789.5 A**

37.4. Synchronous or Rotary Converter

Normally, a synchronous (or rotary) converter is used when a large-scale conversion from a.c. to d.c. power is required. It is a *single* machine with *one* armature and *one* field. It combines the function of a synchronous motor and a d.c. generator. It receives alternating current through a set of slip-rings mounted on one end of an armature rotating synchronously ($N_S = 120\, f / P$) and delivers direct current from the opposite end through the commutator and brushes as usual.

In addition to performing the above function, this machine may be (*i*) driven as d.c. or a.c. generator (*ii*) run as a synchronous or d.c. motor and (*iii*) run as an inverter *i.e.* for changing d.c. into a.c.

37.5. Construction

In general, construction and design, rotary converter is more or less like any d.c. machine. It has interpoles to help commutation, a set of brushes and a commutator (which is larger than that of a d.c. generator of the same size, because converter has to handle larger amount of power). The only added features are (*i*) a set of slip-rings, mounted at the end opposite to the commutator end and (*ii*) dampers in the pole faces, as in a synchronous motor. A simple sketch illustrating main parts of a synchronous converter is shown in Fig. 37.3.

Now, it is well-known that the e.m.f. induced in the armature conductors of a d.c. generator is, in reality, alternating and that it becomes direct due only to the rectifying action of the commutator. Therefore, it is only necessary to connect the slip-rings to some suitable points on the armature winding to use this machine as an a.c. generator or alternator. In that case, the voltage across the slip-rings would be alternating while that across the brushes (placed) on the commutator would be direct.

Following is the rule for tapping the armature winding when it is wave-wound and also when it is lap-wound.

(*i*) Wave-wound Armature

Whatever the number of poles, there are only two parallel paths in the armature. There will be one tapping taken to each slip-ring. For a single-phase supply, the number of slip-rings is two and there will be an equal number of armature conductors—half the total number of armature conductors, in each path from one tapping point to the other. For a 3-phase supply, there will be three slip-rings, three equidistant tappings dividing the armature into three equal parts and so on. Wave winding is, however, suitable for small machines and is rarely used.

(*ii*) Lap-wound Armature

In this case, the number of parallel paths is equal to the number of poles. Therefore, the

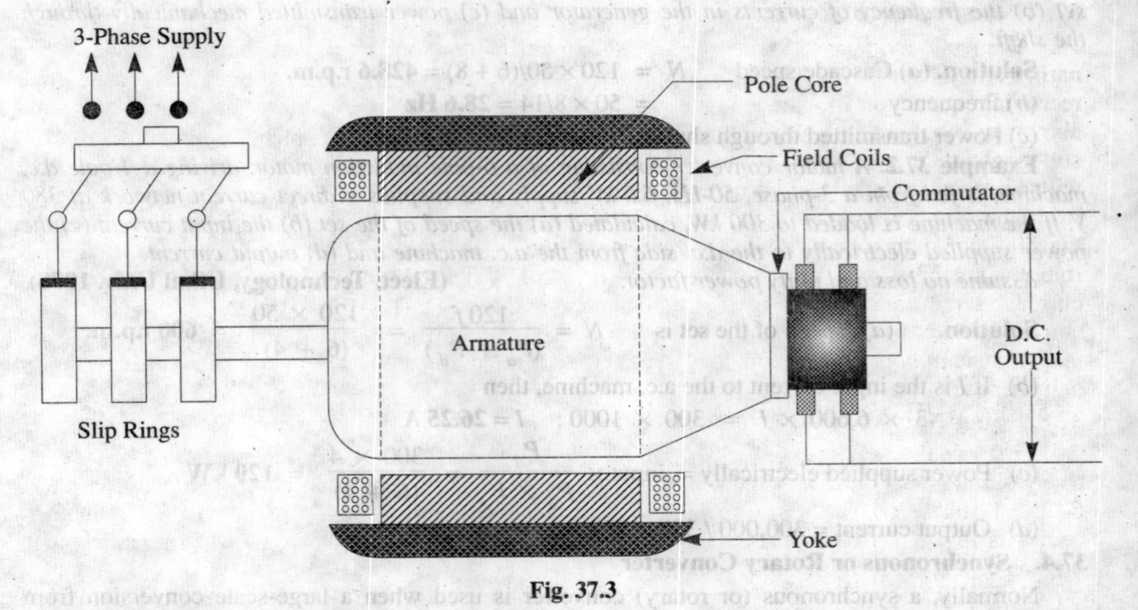

Fig. 37.3

number of equi-potential points on the armature is equal to the number of pair of poles. The number of tappings taken to each slip-ring is, therefore, equal to the number of pair of poles. Obviously, the *total* number of tappings is

= number of slip-ring × number of pairs of poles

It is, therefore, necessary that the number of armature conductors should be such that it is possible to find suitable points at the proper potential for being connected to the slip-rings. For example, in the case of a 3-phase lap-wound rotary converter, it is essential that number of armature conductors/pole should be divisible by 3. Consider a 10-pole three-ring lap-wound armature. Each slip-ring would be connected to 5 (= 10/2) equipotential points. Hence, the number of conductors should be a multiple of 3 × 5 = 15. Suppose the number is 180. The same will be the number of commutator bars.

If one slip-ring is connected to conductor No.1, then it should also be connected to conductor Nos. 37, 73, 109 and 145. All these points will be at the same potential at any given instant, because they will be located similarly in relation to the five North poles.

The second slip-ring will be connected to conductors which are located 36/3 = 12 conductors away from the first *i.e.* to conductor Nos. 13, 49, 85, 121 and 157. Similarly, the third ring would be connected to conductor Nos. 25, 61, 97, 133, and 199.

37.6. Number of Slip-rings

The above conclusions are summarized in the form of a table, which gives the relation between the number of supply phases, the number of slip-rings and the angle between adjacent tappings.

Table No. 37.1

Type	*No. of slip-rings*	*Electrical angle between adjacent tappings*
1 – φ	2	180°
3 – φ	3	120°
6 – φ	6	60°
9 – φ	9	40°
12 – φ	12	30°

37.7. Description of Operation

As mentioned above, the e.m.f. generated in armature conductors of a d.c. generator is alternating and that it flows continuously in one direction through the external circuit because of the rectifying action of the commutator. If armature conductors are suitably tapped, as described above, and joined to the slip-rings, then this machine becomes an alternator, when seen from the slip-ring end. When the armature is mechanically driven from some external source, then machine will act simultaneously as an alternator and as d.c. generator. Alternating current may be taken from the slip rings and direct current from the brushes on the commutator. In that case, the machine works as a double current generator, though it is seldom used in this role. In its normal role, the machine is connected to a suitable a.c. supply through the slip-rings and it delivers direct current at the commutator. In this application, the machine is said to work as a synchronous converter. Seen from the slip-ring side, the machine receives a.c. power and runs as a *mesh-connected* synchronous motor and as viewed from the commutator end, it runs as a d.c. generator delivering d.c. power.

37.8. Single-phase Rotary Converter

First of all, a single-phase converter will be discussed because the understanding of its relatively simple behaviour will serve to simplify the basic principles underlying the operation of 3-, 6- and 12-phase converters. In practice, single-phase converter is rarely used because (*i*) its efficiency is low (*ii*) it heats excessively (*iii*) it has tendency to hunt and (*iv*) it is physically much larger per kW than the polyphase converters.

In Fig. 37.4 is shown a diagrammatic sketch of a single-phase 2-ring, two-pole converter. Ring winding is taken for simplicity. The armature winding has twelve coils *A, B, C...X*. Tappings to the slip-rings are taken off at the back end from two diametrically opposite points like *A* and *G*.

Let us now consider the values of the d.c. voltage and a.c. voltage and see if there is any relationship between the two.

(*i*) **D.C. Brush Voltage**

As shown, d.c. brushes are placed at the neutral points. Voltage between these points is maximum and is maintained by commutation. This voltage is constant in magnitude, because its value is fixed by flux/pole, the armature speed and the number of armature conductors per path. These points being fixed in *space,* will have a constant voltage between them because whatever the position of the armature, the *number* of conductors between them is the same (though not the *same* conductors).

(*ii*) **Slip-ring A.C. Voltage**

The a.c. voltage between the two slip-rings *i.e.* between points *A* and *G* will not remain constant either in magnitude or direction, but will vary sinusoidally (assumning sinusoidal flux distribution). During the rotation of the armature, whenever tappings come directly under the d.c.

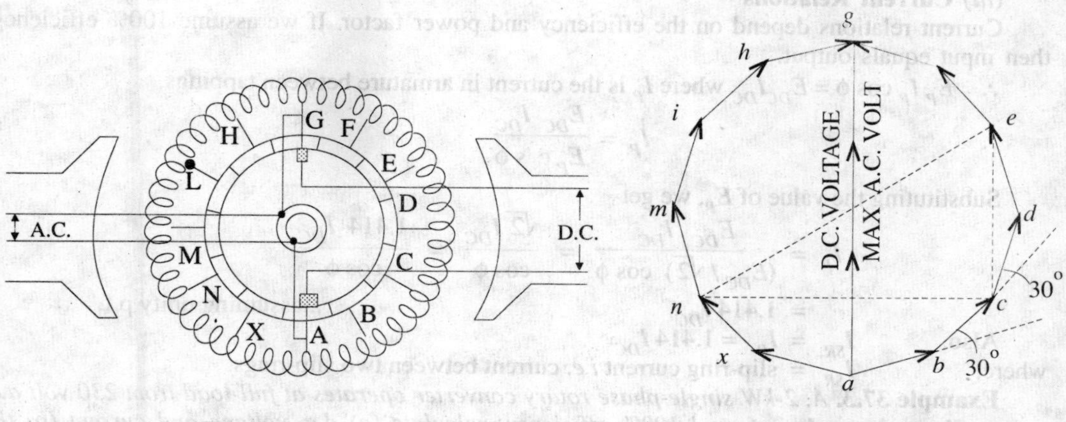

Fig. 37.4

Fig. 37.5

brushes (Fig. 37.4), the slip-ring a.c. voltage will reach maximum value, which is equal to the vector sum of e.m.fs. induced in coils *AB,BC...FG*.

It will be seen that there is phase-difference between the e.m.f. of each coil and that of its neighbouring coil. In Fig. 37.5, vector '*ab*' represents instantaneous e.m.f. induced in coil *AB* '*bc*' represents that of the coil *BC* and so on. E.M.F. in coil *BC* leads that in *AB* by 360°/12' = 30°. The resultant voltage vector is *ag* and has maximum value.

When armature has rotated through 60° (Fig. 37.6), the voltage between two tappings is equal to dotted vector *ce* (Fig. 37.5), because the voltages *ab* and *bc* are cancelled out by *ax* and *xn*. It is seen that now the slip-ring voltage has decreased to half its value when under brushes. In the same way, the slip-ring a.c. voltage will become zero, when tappings come directly under the midpoints of the two poles. Voltage relations for various positions of tapped coils with respect to the brush positions are shown in Fig. 36.7. Here, E_{AC} represents the maximum value of the a.c. volage.

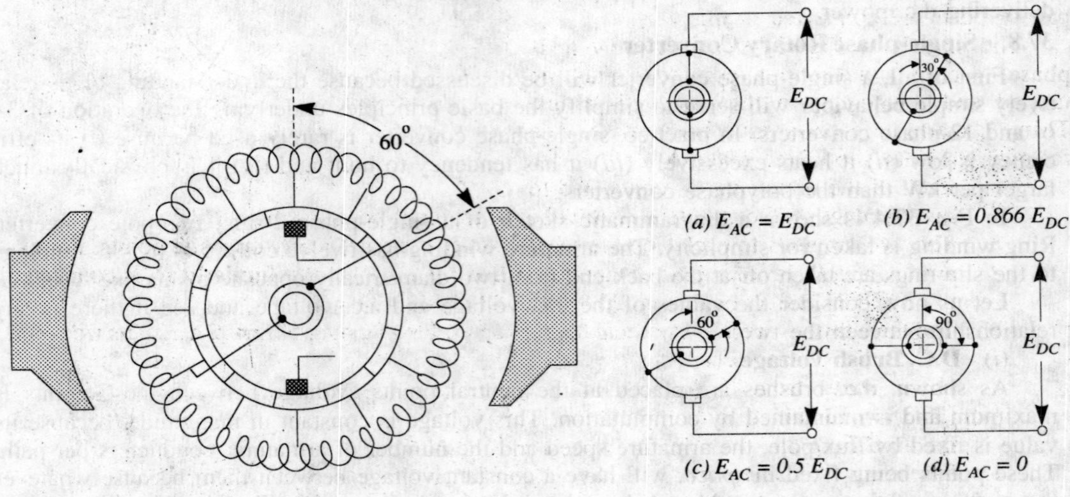

Fig. 37.6

Fig. 37.7

From the above analysis, it is clear that the maximum value of slip-ring a.c. voltage occurs when tappings come directly under d.c. brushes and in that position is equal to d.c. brush voltage.

$$\therefore \quad E_{P.Max} = E_{DC} \quad \text{or} \quad E_P = E_{DC}/\sqrt{2} = 0.707\ E_{DC}$$

where E_P is the r.m.s. value of phase voltage on a.c. side.

(*iii*) Current Relations

Current relations depend on the efficiency and power factor. If we assume 100% efficiency, then input equals output.

$\therefore \quad E_P I_P \cos\phi = E_{DC} I_{DC}$ where I_P is the current in armature between tappings.

$$\therefore \quad I_P = \frac{E_{DC}\ I_{DC}}{E_P \cos\phi}$$

Substituting the value of E_P, we get

$$I_P = \frac{E_{DC} \cdot I_{DC}}{(E_{DC}/\sqrt{2})\cos\phi} = \frac{\sqrt{2}\ I_{DC}}{\cos\phi} = \frac{1.414\ I_{DC}}{\cos\phi}$$

$$= 1.414\ I_{DC} \qquad \text{...assuming unity p.f.}$$

Also, $\quad I_{SR} = I_P = 1.414\ I_{DC}$

where, $\quad I_{SR}$ = slip-ring current *i.e.* current between two slip-rings.

Example 37.3. *A. 2-kW single-phase rotary converter operates at full-load from 230 volt a.c. source. Assuming unity p.f. and 100% efficiency, calculate* (*a*) *d.c. voltage and current* (*b*) *the a.c. input current at slip-rings.*

Solution. (*a*) $E_{DC} = \sqrt{2}\ E_P = \sqrt{2} \times 230 = \mathbf{325\ V}$

(**Note.** 230 V a.c. represents r.m.s. value)

$$I_{DC} = \frac{\text{d.c. power}}{E_{DC}} = \frac{2{,}000}{325} = \mathbf{6.15\ A} \qquad (b)\ I_P = \frac{\text{power input}}{E_P} = \frac{2000}{230} = \mathbf{8.7\ A.}$$

37.9. Three-phase Converter

As shown in Fig. 37.8, in a 2-pole, 3-φ converter, each ring is connected to one tapping*, different tappings being taken from points 120 electrical degrees apart (Table No. 37.1). For lap-wound multipolar armatures, three tappings per pair of poles are connected in this way.

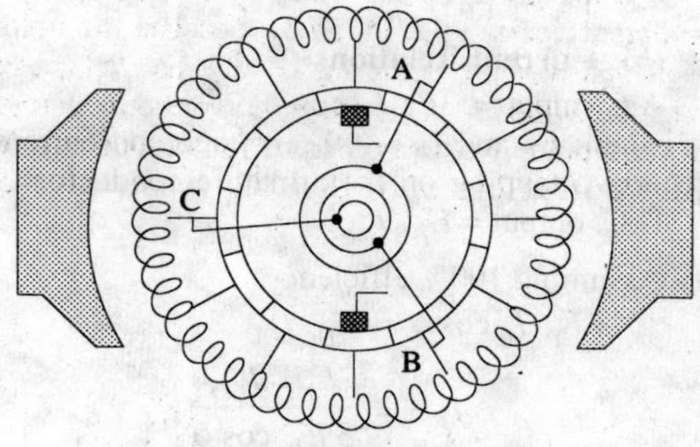

Fig. 37.8

The sketch of Fig. 37-9 shows how a 3-phase, 4-pole synchronous converter may be tapped and connected to the three slip-rings. Thirty six armature coils have been taken. The following points should be noted :

1. Each of the three rings is connected to *two* tappings on the armature *i.e.* number of tappings per slip-ring is $P / 2$.
2. The two tappings to each ring are located on two diametrically opposite points. For example, ring R_1 is connected to tappings 1 and 1 + (36/2) = 19. Similarly, R_2 is connected to 13 and 31 and R_3 to 25 and (25 + 18) = 43 *i.e.* 43 - 36 = 7.
3. The three phases are 120 electrical degrees apart. First tapping of first phase starts from coil No. 1, that of the second phase starts from 1 + (36/3) = 13 and that of the third from 25.
4. Since taps *m, n* and *p* are brought out exactly 120° apart, they represent the equivalent of a delta connection, when considered from the standpoint of a.c. winding.

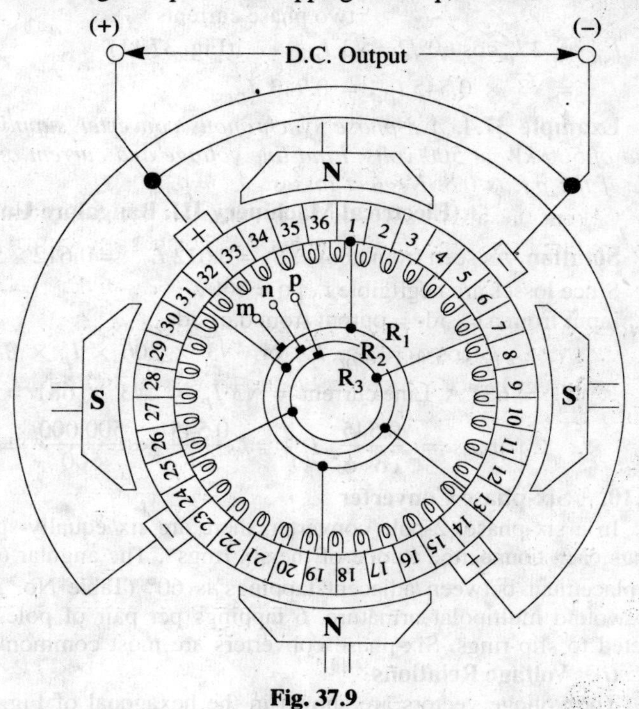

Fig. 37.9

It may be noted that in the sketch (Fig. 36.9), the commutator has been omitted for convenience and the brushes have been located diametrically opposite.

(*a*) **Voltage-Relations**

In Fig. 37.10, the p.d. between two armature tappings (or slip-rings) *i.e.* phase e.m.fs. are represented by voltage vectors *ab, bc* and *ca* and they take their maximum value in succession when in *vertical position. Vector EF* represents E_{DC}.

From the voltage vector diagram, we find that

*If the number of poles were 4, there would be two tappings connected to each ring. In general for *P* poles, each ring will be connected to *P*/2 tappings on the armature.

$E_{P.max} = ab = 2ad = 2\,ob \sin 60° = E_{DC} \sin 60° = 0.866\,E_{DC}$

$\therefore \quad E_P = 0.866\,E_{DC}/\sqrt{2} = 0.612\,E_{DC}$

or $\quad E_P = \left(\dfrac{\sqrt{3}}{2\sqrt{2}}\right) . E_{DC}$

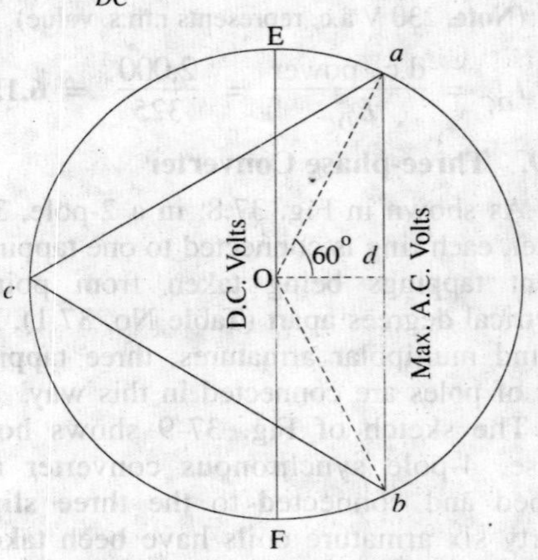

Fig. 37.10

(*b*) **Current Relations**

A.C. input $= 3E_P\,I_P \cos \phi$

where E_P and I_P are the voltage and current, between two tapping on the armature conductors.

D.C. output $= E_{DC}\,I_{DC}$

Assuming 100% efficiency

$3\,E_P\,I_P \cos\phi = E_{DC}\,I_{DC}$

$\therefore \quad I_P = \dfrac{E_{DC}\;I_{CD}}{3\,E_P\;\cos\phi}$

Substituting the value of E_P from above,

$I_P = \dfrac{E_{DC}\,I_{DC}}{3 \times 0.612\;E_{DC}\;\cos\phi} = \dfrac{0.545}{\cos\phi}\,.\,I_{DC}$

$= 0.545\,I_{DC}$ ...for unity power factor

Slip-ring current I_{SR} = vector difference between two phase currents

$\therefore \quad I_{SR} = 2\,I_P \cos 60°/2 = \sqrt{3}\;I_P$ (Fig. 37.11)

$= \sqrt{3} \times 0.545\,I_{DC} = 0.943\,I_{DC}$

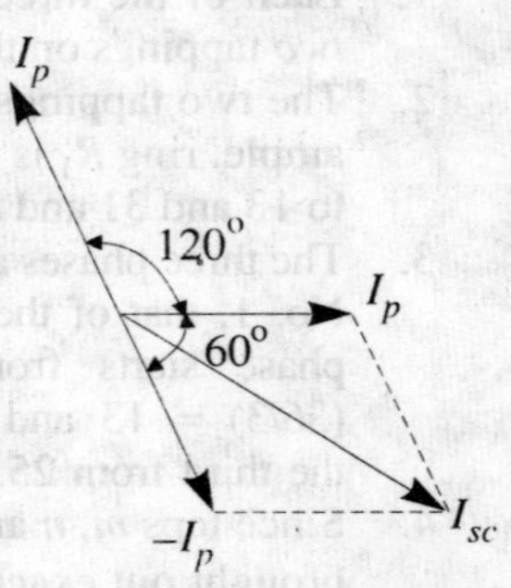

Fig. 37.11

Example 37.4. *A 3-phase synchronous converter supplies a d.c. load of 500 kW at 500 volts. Find line voltage and current on the a.c. side if the p.f. is 0.8. Neglect losses.*

(Electrical Machinery-III, Bangalore Univ. 1980)

Solution. As seen from above, $E_P = 0.612\,E_{DC} = 0.612 \times 500 =$ **306 V**

Since losses are negligible *i.e.* $\eta = 100\%$

input from a.c. side = output from d.c. side

$\therefore \quad \sqrt{3}\;E_P\,I_P \cos\phi = E_{DC}\,I_{DC}$ or $\sqrt{3} \times 306 \times I_P \times 0.8 = 500{,}000$

$\therefore \quad I_P = 681$ A, Line current $= \sqrt{3}\;I_P = \sqrt{3} \times 681 =$ **1179 A.**

$\left(\text{or } I_P = \dfrac{0.545}{\cos\phi}\,.\,I_{DC} = \dfrac{0.545}{0.8} \times \dfrac{500{,}000}{500} = 681 \text{ A}\right)$

37.10. Six-phase Converter

In a six-phase, 2-pole converter, there are six equally-spaced tappings each connectted to one of the six rings*. The angular (electrical) displacement between adjacent tappings is 60° (Table No. 37.1). For lap-wound multipolar armature, 6 tappings per pair of poles are connected to slip-rings. Six-phase converters are most commonly used.

(*i*) **Voltage Relations**

The voltage vectors are shown in the hexagonal of Fig. 37.12.

$E_{P.max} = bc = 2\,bL = 2\,Ob \sin 30° = 0.5\,E_{DC}$

($\because$ diameter represents E_{DC})

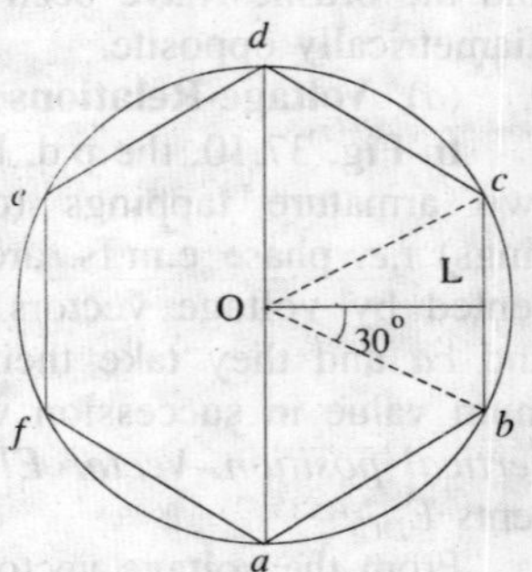

Fig. 37.12

*If the number of poles is P, then each of the six-rings would be connected to P / 2 tappings on the armature.

$$\therefore \quad E_P = \frac{0.5\ E_{DC}}{\sqrt{2}} = 0.354\ E_{DC} \quad \text{or} \quad E_P = E_{DC}\ /\ 2\sqrt{2}$$

(ii) Current Relations

A.C. input = $6\ E_P\ I_P \cos\phi$ and D.C. output = $E_{DC}\ I_{DC}$

Assuming 100% efficiency, $6\ E_P I_P \cos\phi = E_{DC} I_{DC}$

$$I_P = \frac{E_{DC}\ I_{DC}}{6\ E_P\ \cos\phi} = \frac{E_{DC}\ I_{DC}}{6 \times 0.354\ E_{DC}\ \cos\phi} = \frac{0.47}{\cos\phi}\ .\ I_{DC}$$

or $I_P = 0.47\ I_{DC} = (\sqrt{2}/3)\ I_{DC}$ ... for unity power factor.

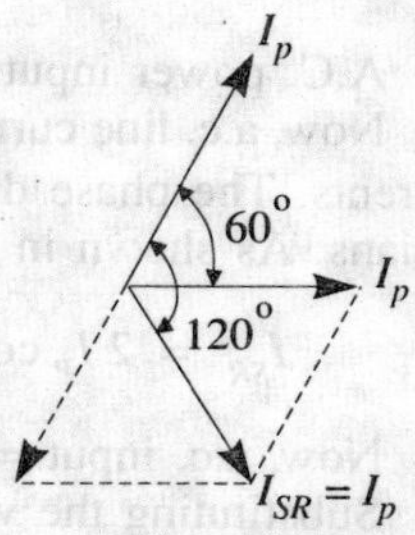

Fig. 37.13

In a 6-phase mesh connection, the displacement between phase currents is 60°.

I_{SR} = vector differences between phase currents (Fig. 37.13)

$= 2\ I_P \cos 120°/2 = I_P \qquad \therefore I_{SH} = I_P$

37.11. General Case

We will now derive an expression for an *m*-phase, 2-pole converter having m-rings. Values for any number of phases can be found by giving desired values to *m*. In Fig. 37.14, circle representing the armature also represents the voltage vector diagram.

(*i*) **Voltage Relations**

The angle between two adjacent tappings is $2\pi\ /\ m$ radians. The voltage between two adjacent tappings *a* and *b* is represented by chord *ab*. This also represents the maximum a.c. e.m.f. between the adjacent rings. The d.c.brush voltage is represented by the diameter $= 2\ r$.

$$E_{P.max} = ab = 2\ ac = 2\ r \sin 2\pi\ /\ 2m$$

$$= 2r \sin \pi\ /\ m = E_{DC} \sin \pi\ /\ m$$

$$\therefore \quad E_P = \frac{E_{DC}\ .\ \sin \pi/m}{\sqrt{2}}$$

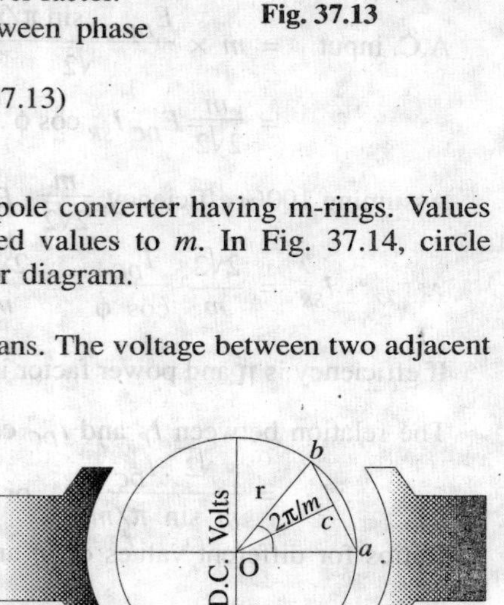

Fig. 37.14

$$\therefore \quad \frac{E_P}{E_{DC}} = \frac{\text{r.m.s. value of armature phase voltage}}{\text{d.c. output voltage}}$$

$$= \frac{\sin\ \pi/m}{\sqrt{2}}$$

$$\therefore \quad E_P = \frac{E_{DC}\ .\ \sin \pi/m}{\sqrt{2}}$$

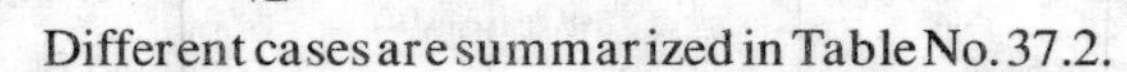

Different cases are summarized in Table No. 37.2.

Table No. 37.2

(η = 100%; p.f. = 1)

	Formula	*1-phase 2-ring*	*3-phase 3-ring*	*6-phase 6-ring*	*12-phase 12-ring*
E_P or E_{SR}	$\frac{E_{DC}\ \sin\ \pi/m}{\sqrt{2}}$	$0.707\ E_{DC}$	$0.612\ E_{DC}$	$0.354\ E_{DC}$	$0.182\ E_{DC}$

(*ii*) **Current Relations**

Let E_P = r.m.s. phase voltage *i.e.* p.d. across two armature tappings or E_{SR}.

I_P = r.m.s. phase current *i.e.* a.c. current flowing through the armature between adjacent tappings. It is not the same as I_{SR} *i.e.* lead current*.

*I_{SR} would be equal to the vector difference of two phase currents.

A.C. power input = $m\,E_P\,I_P \cos\phi$.

Now, a.c. line current at slip-rings, I_{SR}, is the vector difference of phase currents. The phase difference between these two phase currents is $2\pi/m$ radians. As shown in Fig. 37.15,

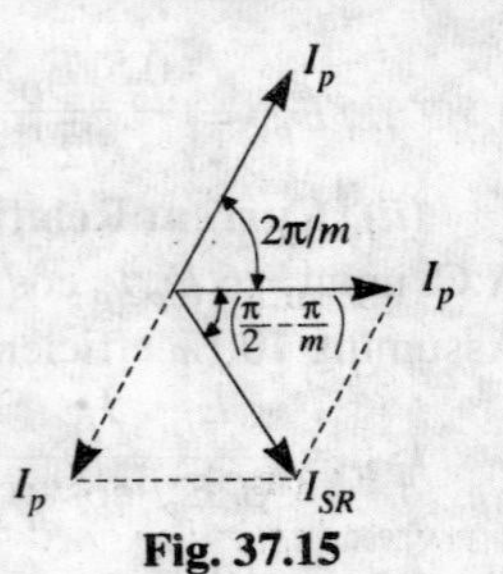

Fig. 37.15

$$I_{SR} = 2\,I_P \cos\left(\frac{\pi}{2} - \frac{\pi}{m}\right) = 2\,I_P \sin\frac{\pi}{m} \qquad \therefore\ I_P = \frac{I_{SR}}{2\sin\pi/m}$$

Now, a.c. input = $m \times E_P\,I_P \cos\phi$

Substituting the values of E_P and I_P as found above; we get

$$\text{A.C. input} = m \times \frac{E_{DC}\cdot\sin\pi/m}{\sqrt{2}} \times \frac{I_{SR}}{2\sin\pi/m} \times \cos\phi$$

$$= \frac{m}{2\sqrt{2}} E_{DC}\,I_{SR} \cos\phi \text{ . Also D.C. output} = E_{DC}\,I_{DC}$$

Assuming 100% efficiency, $\frac{m}{2\sqrt{2}} \cdot E_{DC}\,I_{SR} \cos\phi = E_{DC}\,I_{DC}$

$$\therefore\quad I_{SR} = \frac{2\sqrt{2}}{m}\,\frac{I_{DC}}{\cos\phi} = \frac{2\sqrt{2}}{m}\cdot I_{DC} = \frac{2.83}{m}\cdot I_{DC} \qquad \text{... assuming unity p.f.}$$

If efficiency is η and power factor is $\cos\phi$, then $I_{SR} = \frac{2\sqrt{2}}{m}\cdot\frac{I_{DC}}{\eta\cos\phi}$.

The relation between I_P and I_{DC} can be found to be,

$$I_P = \frac{\sqrt{2}\cdot I_{DC}}{m\sin\pi/m} \quad \text{or} \quad \frac{\sqrt{2}}{m}\cdot\frac{I_{DC}}{\eta\cos\phi}.$$

Ratios for different values of m are given in table No. 37.3.

Table No. 37.3

(η = 100%; p.f. = 1)

	Formula	*1-phase 2-ring*	*3-phase 3-ring*	*6-phase 6-ring*	*12-phase 12-ring*
I_{SR} current between leads	$\frac{2\sqrt{2}}{m}\,I_{DC}$	$1.414\,I_{DC}$	$0.943\,I_{DC}$	$0.472\,I_{DC}$	$0.236\,I_{DC}$
I_P current between two armature taps	$\frac{\sqrt{2}}{m\sin\pi/m}\,I_{DC}$	$0.707\,I_{DC}$	$0.544\,I_{DC}$	$0.472\,I_{DC}$	$0.455\,I_{DC}$

In general, it is seen from above that $I_P = I_{SR} / 2\sin\frac{\pi}{m}$

In practice, the above relations are modified to account for (*i*) efficiency (*ii*) power factor and (*iii*) the ratio, the pole arc/pole pitch.

37.12. Current Between a Slip-ring and a Tapping Point

If the converter has P poles and is lap-wound, then each slip-ring is connected to as many tappings on the armature as the *number of pair of poles*. Obviously, the current flowing in from each slip-ring is carried by $P/2$ conductors to $P/2$ lappings on the armature. Hence, current flowing from each slip-ring to its different tapping points is

$$= \frac{\text{total current}}{\text{No. of tapping points}}$$

$\therefore$ if I_{ST} is the current between a slip-ring and one of its $P/2$ tapping points, then

$$I_{ST} = \frac{I_{SR}}{P/2}$$

It should be clear that in the case of a 2-pole converter $I_{ST} = I_{SR}$

Similarly, $I_{ST} = I_{SR}/2$ ...for a 4-pole converter

$= I_{SR}/3$ ...for a 6-pole converter.

$= I_{SR}/4$ ...for an 8-pole converter and so on.

Example 37.5. *A 25-kW, 3-ϕ, 2-pole rotary converter delivers a d.c. output voltage of 230 volts. Assuming unity power factor and 100% efficiency, calculate (a) the full-load d.c. output current at brushes (b) a.c. input voltage between slip-rings and (c) a.c. input current per line at full-load.* **(Elect. Engineering, Grad. I.E.T.E. June 1985)**

Solution. If desired, table Nos. 37.2 and 37.3 may please be referred to

(*a*) $I_{DC} = \frac{\text{power}}{E_{DC}} = \frac{25{,}000}{230} = \mathbf{108.7\ A}$

(*b*) $E_P = E_{SR} = 0.612\ E_{DC} = 0.612 \times 230 = \mathbf{141\ V}$

(*c*) $I_{SR} = 0.943\ I_{DC} = 0.943 \times 108.7 = \mathbf{102.3\ A}$

Example 37.6. *A 3-phase rotary converter supplies a load of 100 kW at 550 V on the d.c. side. Find the current supplied from the a.c. side and the slip-ring voltage if the converter works at 0.9 power factor and 0.92 efficiency.* **(Elect. Engg.-II, Bombay Univ. 1979)**

Solution. $E_{SR} = \frac{E_{DC} \cdot \sin \pi/m}{\sqrt{2}} = \frac{550 \times \sin \pi/3}{\sqrt{2}} = 0.612 \times 550 = 336.6\ V$

The a.c. current at slip-rings is $I_{SR} = I_{DC}\left(\frac{2\sqrt{2}}{m}\right) \times \left(\frac{1}{\eta \times \cos\phi}\right)$

Now, $I_{DC} = 100{,}000/550 = 182\ A$

$$I_{SR} = 182 \times \left(\frac{2\sqrt{2}}{3}\right) \times \left(\frac{1}{0.9 \times 0.92}\right) = \mathbf{207.5\ A}$$

Example 37.7. *An 1,800-kW, 6-phase, 2-pole synchronous converter delivers a full-load d.c. voltage of 600 volts. Find (a) the a.c. voltage between adjacent slip-rings (b) a.c. input current at slip-rings and (c) a.c. phase current.*

Solution. (*a*) $E_P = E_{SR} = 0.354 \times 600 = \mathbf{212.4\ V}$ (*b*) $I_{SR} = 0.472\ I_{DC}$

Now $I_{DC} = 1{,}800 \times 1{,}000/600 = 3{,}000\ A$ $\therefore\ I_{SR} = 0.472 \times 3{,}000 = \mathbf{1416\ A}$

(*c*) $I_P = 0.472 \times 3{,}000 = \mathbf{1416\ A}$; Being 6-phase converter, $I_{SR} = I_P$.

Example 37.8. *A 12-pole, 6-phase, lap-wound, synchronous converter delivers a current of 1,600 A on the d.c. side. It has 1,000 armature conductors, a flux of 60 mWb per pole and runs at 600 r.p.m. Calculate (i) the voltage between slip-rings on the a.c. side (ii) the current per slip-ring and (iii) the current in the connection between slip-rings and tapping points.*

Assume unity p.f. on the a.c. side and an armature effciency of 95 per cent.

(Electrical Machinery-II, Kerala Univ. 1979)

Solution. $E_{DC} = \frac{\Phi ZN}{60}\left(\frac{P}{A}\right) = \frac{6 \times 10^{-2} \times 1{,}000 \times 600}{60}\left(\frac{12}{12}\right) = 600\ V$

(*i*) $E_{SR} = 0.354\ E_{DC} = 0.354 \times 600 = \mathbf{212.4\ V}$

(*ii*) $I_{SR} = 0.472 \times 1{,}600/0.95 = \mathbf{795\ A}$

(*iii*) $I_{ST} = 795/6 = \mathbf{132.5\ A}$ ...Art. 37.12

Example 37.9. *A 5,000-kW, 1.2 kV, 12-phase, 4-pole rotary converter has a full-load efficiency of 96% and power factor of 0.9. Calculate (a) the a.c. slip-ring voltage (b) full-load d.c. output current and (c) a.c. current at slip-rings (d) a.c. phase current (e) current between a slip-ring and tapping point.*

Solution. (*a*) $E_P = E_{SR} = 0.182 \times 1{,}200 = \mathbf{218.4\ V}$

(*b*) $I_{DC} = 5000 \times 1{,}000/1{,}200 = \mathbf{4167\ A}$

(*c*) $I_{SR} = 0.236 \times 4{,}167 / 0.96 \times 0.9 = \mathbf{1{,}140\ A}$

(*d*) $I_p = 0.455\% \times 4{,}167 / 0.096 \times 0.9 = \mathbf{2{,}195\ A}$

(*e*) $I_{ST} = I_{SR}/2 = 1{,}140/2 = \mathbf{570\ A}$

37.13. Transformer Connections for Converters

In the operation of polyphase rotary converters, it is always necessary to introduce transformers between the a.c. source and the slip-rings. There are several reasons for doing so :

1. There exists a definite theoretical relationship between d.c. output voltage and the input a.c. voltage—the latter usually having an odd value which, in most cases, does not match with any standard distribution network.
2. The a.c. voltage required at slip-rings happens to be too low for economic transmission of power over any distance from the supply.
3. In the case of 6-and 12-phase converters, it is necessary to change 3-phase supply into 6-and 12-phase respectively. This can be best accomplished by properly-connected transformers.

37.14. Three-phase Converters for Two-wire D.C. Service

For such installations, generally a bank of 3-phase transformers is used with primaries and secondaries connected in any one of the four ways (*i*) Y/Y (*ii*) Δ/Δ (*iii*) Δ/Y (*iv*) Y/Δ. The choice of any one connection depends upon the requirement of a.c. voltage which is further determined by a.c. voltage rating of the converter itself.

Example 37.10. *A 200-kW, 3-φ, rotary converter delivers d.c. output voltage of 200 volts. It receives the a.c. power from 2,300 V source through a bank of transformers connected delta/star. The machine works with a full-load efficiency of 95% and a p.f. of 0.92. Calculate (i) d.c. output current (ii) three-phase a.c. voltage at slip-rings (iii) the a.c. current delivered by transformer secondaries to the slip-rings (iv) ratio of transformation (v) current in the transformer primary coils (iv) the line current on the primary side.*

Solution. Wiring diagram along with the values of various currents and voltages in shown in Fig. 37.16.

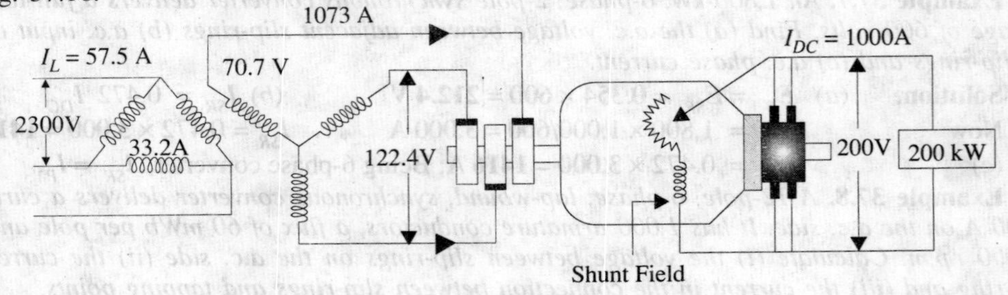

Fig. 37.16

(*i*) $I_{DC} = 200{,}000/200 = \mathbf{1{,}000\ A}$ (*ii*) $E_P = E_{SR} = 0.612 \times 200 = \mathbf{122.4\ volt.}$

(*iii*) $$I_{SR} = \frac{0.943 \times 1{,}000}{0.95 \times 0.92} = \mathbf{1{,}079\ A}$$

This current may also be calculated thus

$$\text{ISR} = \frac{\text{input power}}{\sqrt{3} \times E_p \times \cos\phi} = \frac{200 \times 1{,}000}{\sqrt{3} \times 122.4 \times 0.95 \times 0.92} = \mathbf{1{,}079\ A}$$

(*iv*) Voltage / phase across transformer secondary $= 122.4/\sqrt{3} = 70.7$ volt

∴ voltage transformation ratio $K = 70.7/2{,}300 = \mathbf{1/32.5}$

(*v*) Current in each coil of transformer primary $= 1{,}079 \times 1/32.5 = \mathbf{33.2\ A}$

(*vi*) ∴ $I_L = 33.2 \times \sqrt{3} = \mathbf{57.5\ A}$

We could also find it from $I_L = \dfrac{200,000}{\sqrt{3} \times 2,300 \times 0.95 \times 0.92} = \mathbf{57.5\ A}$

Example 37.11. *A 3-phase, 250-kW synchronous converter develops 250 V on the commutator and is supplied from a 2,200-V, 3-ϕ network through three I-phase transformers Δ-connected on the h.v. side and Y-connected on the l.v. side. The converter operates on full-load at p.f. 0.9 lag and with an efficiency of 91%. Determine the voltage and current rating of both sides of the transformer, neglecting its losses.* **(Electrical Engg.-IV, M.S. Univ. Baroda, 1978)**

Solution. The diagram of connections is similar to that shown in Fig. 37.16.

$I_{DC} = 250,000/250 = 1,000$ A; $E_{SR} = E_P = 0.612 \times 250 = 153$ V

$$I_{SR} = \frac{0.943 \times 1,000}{0.91 \times 0.9} = \mathbf{1,152\ A}$$

Voltage/phase across transformer secondary $= 153/\sqrt{3} = \mathbf{88.4\ V}$

$\therefore \quad K = 88.4/2,200 = 1/24.9$

Phase current on the primary side $= 1,152 \times 1/24.9 = \mathbf{46.3A}$

Line current on the primary side $I_L = 46.3 \times \sqrt{3} = 80.1$ A.

37.15. Three-phase Converters for 3-wire d.c. service

In this case, zig-zag connection is used on secondary side. This type of connection prevents core-saturation of transformer due to d.c. neutral current. Wiring diagram is shown in Fig. 37.17.

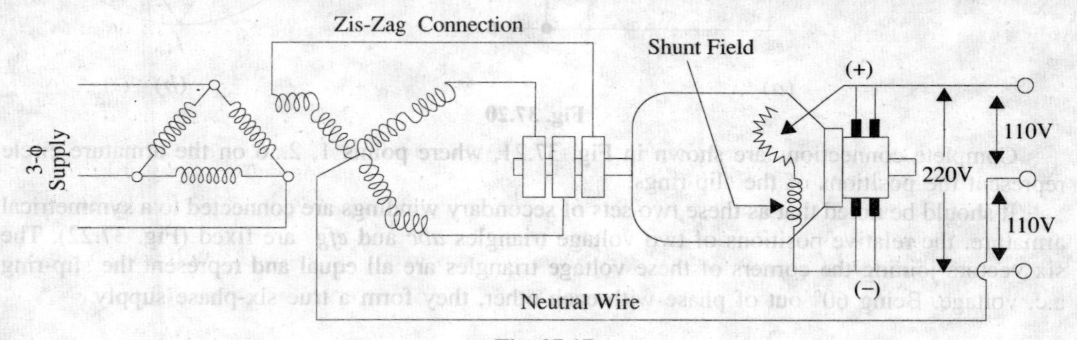

Fig. 37.17

37.16. Six-phase Connection

Conversion from 3-phase to 6-phase can be easily achieved by having two similar secondary

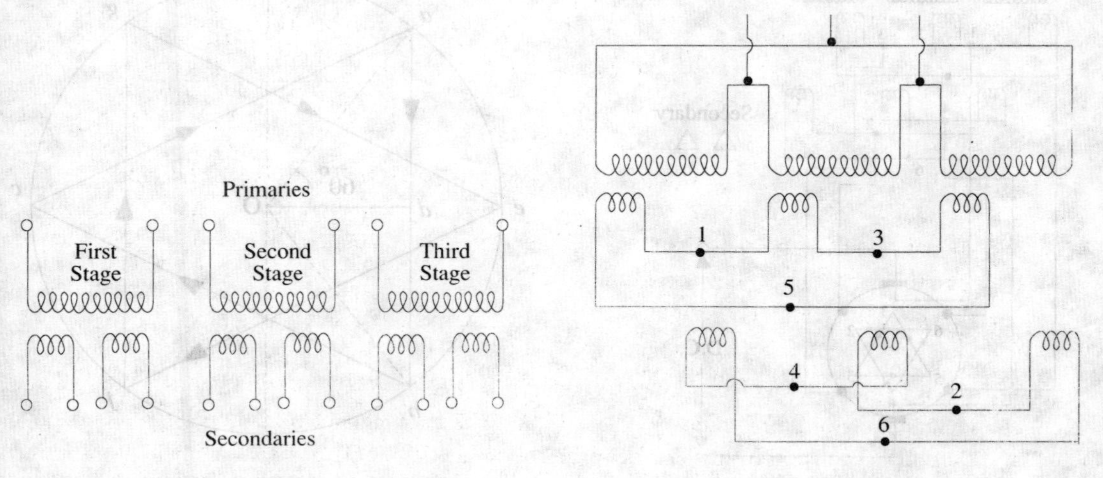

Fig. 37.18 Fig. 37.19

windings for each of the primaries of the three-phase transformer (Fig. 37.18). When primaries are fed from a 3-phase source, six-phase output can be drawn from the six secondaries. There are many ways of connecting the six secondaries for this purpose such as (*i*) double-delta (*ii*) double-star and (*iii*) diametrical. The last connection is the one most commonly used in practice.

37.17. Double-delta Connection

As shown in Fig. 37.19, the secondaries are arranged to form two independent sets of mesh or delta-connected circuits. On the first set, connections are taken from points 1, 3, 5, and on the second set, from points 2, 4 and 6, thereby reversing one of the two voltage triangles with respect to each other [Fig. 37.20 (*b*)].

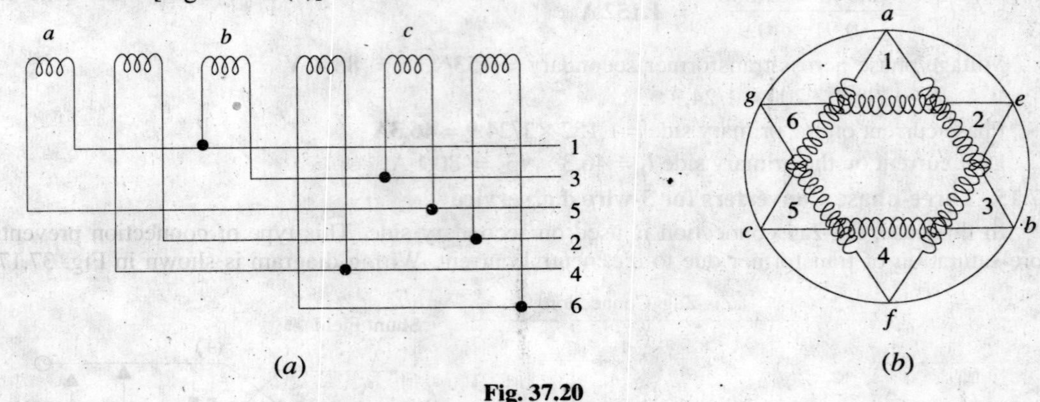

Fig. 37.20

Complete connections are shown in Fig. 37.21, where points 1, 2...6 on the armature circle represent the positions of the slip-rings.

It should be noted that as these two sets of secondary windings are connected to a symmetrical armature, the relative positions of two voltage triangles *abc* and *efg* are fixed (Fig. 37.22). The six vectors joining the corners of these voltage triangles are all equal and represent the slip-ring a.c. voltage. Being 60° out of phase with each other, they form a true six-phase supply.

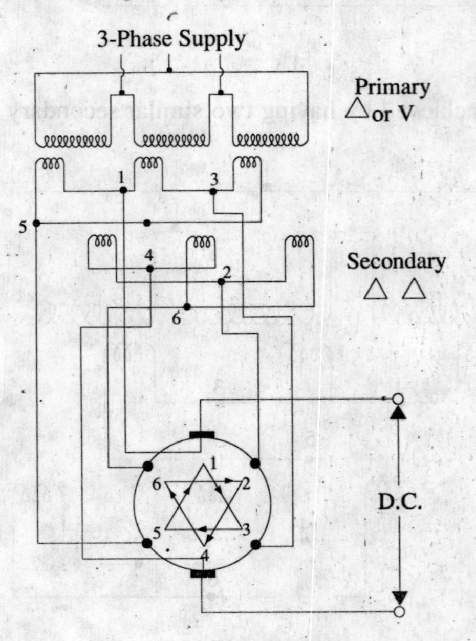

Fig. 37.21

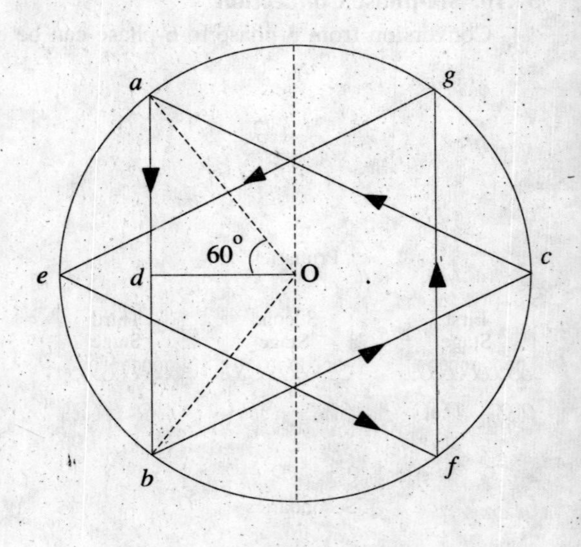

Fig. 37.22

Remembering that the diameter of the circle represents E_{DC}, it is seen that

(*i*) P.D. across each secondary $= ab = \dfrac{\sqrt{3}}{2\sqrt{2}} \cdot E_{DC} = 0.612\, E_{DC}$ —r.m.s. value

(*ii*) Phase e.m.f. $= ae = E_{DC}/2\sqrt{2} = 0.354\, E_{DC}$ (r.m.s. value)

Example 37.12. *A 1000-kW, 600-V, 6-phase synchronous converter is fed by a bank of transformers connected in star on primary side and double-delta on the secondary side, the primary line voltage is 22 kV. Assuming full-load efficiency of 95% and a power factor of 0.9, calculate (i) secondary voltage per phase and (ii) slip-ring current.*

(Elect. Engineering-II, Bombay Univ. 1979)

Solution. (*i*) Secondary voltage/phase

$= 0.612\, E_{DC}$ —as for 3-phase in Table No. 37.2

$= 0.612 \times 600 = \mathbf{367.2\ V}$

(*ii*) Slip-ring current which is also the secondary line current is

$I_{SR} = 0.472\, I_{DC} / \eta \times \cos\phi$

Now $I_{DC} = 1{,}000{,}000/600 = 1666.7$ A $\quad\therefore I_{SR} = 0.472 \times 1666.7/0.95 \times 0.9 = \mathbf{920\ A.}$

Example 37.13. *A 1500-kW, 750-V, d.c., 6-phase rotary converter is fed by a bank of transformers connected in star on primary side and double-delta on the secondary side. The three-phase primary line voltage is 23 kV. Assuming a full-load efficiency of 0.96 and power factor of 0.92, calculate (a) secondary voltage per phase (b) a.c. voltage between adjacent rings (c) slip-ring current (d) secondary phase current (e) transformation ratio (f) primary line current.*

Solution (*a*) As seen from table No. 37.2.

Voltage across each secondary is $= 0.612\, E_{DC}$ —as for 3-phase

$= 0.612 \times 750 = \mathbf{459.5\ V}$

(*b*) Similarly, referring to the same table

$E_{SR} = 0.354\, E_{DC}$ —for 6-phase $= 0.354 \times 750 = \mathbf{265\ V}$

(*c*) Slip-ring current which is secondary *line* current is given by

$$I_{SR} = 0.472\, I_{DC} \times \left(\frac{1}{\eta \times \cos\phi}\right)$$

Now $I_{DC} = 15 \times 10^5 / 750 = 2{,}000$ A $\therefore I_{SR} = 0.472 \times 2000 / 0.96 \times 0.92 = \mathbf{1068\ A}$

(*d*) Secondary phase current is $= 1068/\sqrt{3} = \mathbf{616.6\ A}$

(*e*) Since there are two Δ-connected secondary windings for each primary winding.

$$K = \frac{2 \times 459.5}{2300/\sqrt{3}} = \frac{\mathbf{1}}{\mathbf{14.45}}$$

(*f*) Primary phase current (which is also the line current) is

$= K \times$ secondary phase current $= 616.6/14.45 = \mathbf{42.7\ A}$

37.18. Double-Star Connection

In this case, the two secondaries are connected in two star sets (Fig. 37.24). The first gives us ordinary star 1,3, 5 and the other gives a reversed star 2, 4, 6. Because of the symmetry of the armature winding, the two star points coincide. The vectors joining the star ends (Fig. 37.23) give the six-phase supply.

It is seen that transformer secondary voltage per phase is

$$\frac{0.612}{\sqrt{3}} \cdot E_{DC} = 0.354\, E_{DC}$$

This connection is often used for 3-wire d.c. service because neutral point of the star supplies neutral wire for the arrangement.

Example 37.14. *A 6-ring, 200-kW, 500-V, d.c. synchronous converter has an efficiency of 96 per cent at full-load and operates at a power factor of 0.94. The primary of the transformer is delta and secondary double-star. If primary is supplied by 3-phase, 3.3 kV line, calculate*

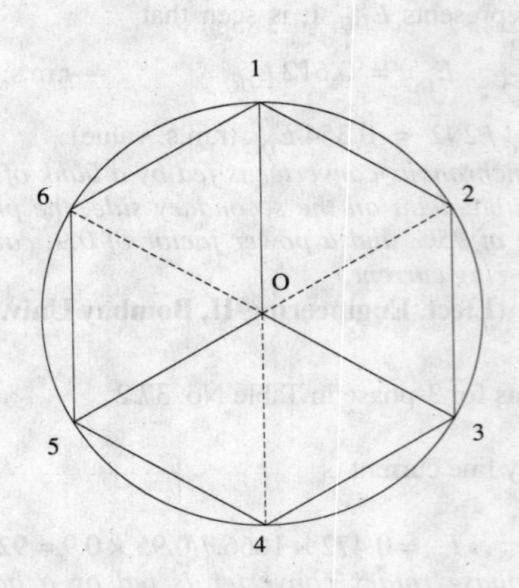

Fig. 37.23

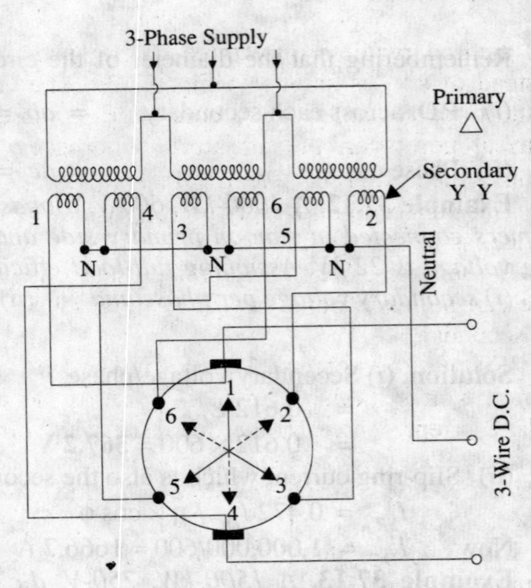

Fig. 37.24

(*a*) *slip-ring current* (*b*) *slip-ring voltage to neutral*
(*c*) *diametrical transformer secondary voltage* (*d*) *primary phase current.*
(*e*) *primary line current* (*f*) *kVA rating of each transformer.*
Neglect magnetising current and losses in transformer.

Solution (*a*) The connections are shown in Fig. 37.25.

$$I_{SR} = 0.472\, I_{DC} / \eta \times \cos \phi$$

Now, $I_{DC} = 200{,}000 / 500 = 400$ A

$\therefore \quad I_{SR} = 0.472 \times 400 / 0.96 \times 0.94 = \mathbf{209\ A}$

(*b*) The voltage to neutral is the same as between two adjacent slip-rings.

$$E_{SR} = 0.354\, E_{DC} = 0.354 \times 500 = \mathbf{177V}$$

(*c*) Diametrical transformer secondary voltage, say, between points 1 and 4 is = 2 × 177 = **354 V.**

(***d***) Voltage transformation ratio is $K = 2 \times 177/3300 = 1 / 9.32$

$\therefore$ primary phase current

= $K \times$ secondary phase current

= 209 / 9.32 = **22.4 A**

(*e*) Primary line current

= $\sqrt{3} \times 22.4$ = **37.8 A.**

(*f*) The transformer rating is

= 3.3 × 22.4 = **73.9 kVA**

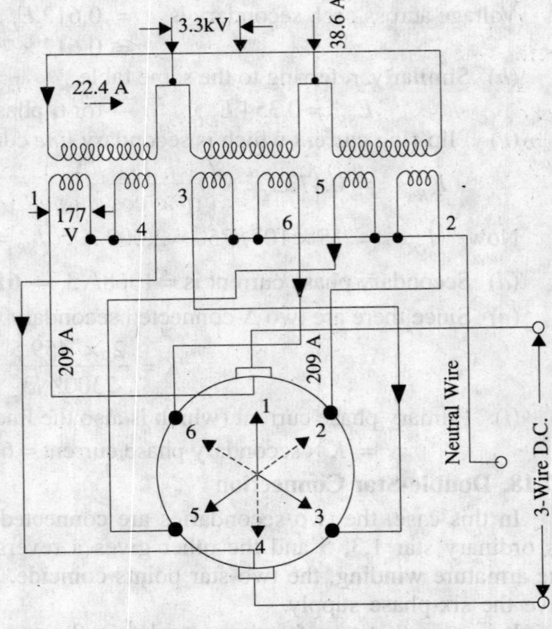

Fig. 37.25

37.19. Diametrical Connection

This connection is actually the development of the double-delta connection and is used extensively. Referring to Fig. 37.21 it is seen that each pair of secondaries can be replaced by a single winding. The mid-points like *N*, can be taken to a common star point, though it is not necessary to do so, because the potential of middle-points is fixed by the armature itself. Hence, this method requires only three single secondries (instead of six) as shown in Fig. 37.26. The ends of the three secondaries are connected to the diametrical points on the armature winding *i.e.* to points which are 180° (electrical) apart.

This connection is mostly used for two-wire d.c. service, as in street-railway system.

If used for 3-wire d.c. service, then a neutral wire can be taken out from the middle point *N* of the secondaries.

It should be noted that in this method of connection, voltage induced in each phase of the secondary is represented by the diameter of the circle *i.e. twice the value obtained in double-star connection because a single secondary takes the place of two.*

In Fig 37.28 same connection is repeated with primary star-connected.

From voltage hexagon of Fig. 37.27, it is clear that

(*i*) Voltage across each secondary $= E_{DC} / \sqrt{2} = 0.707\ E_{DC}$ (r.m.s.)

This also represents the value (r.m.s.) of the voltage between opposite slip-rings (Ex- 37.15)

(*ii*) Voltage across two armature tappings = vector 1– 2 $= E_{DC} / 2\sqrt{2} = 0.354\ E_{DC}$ (r.m.s.)

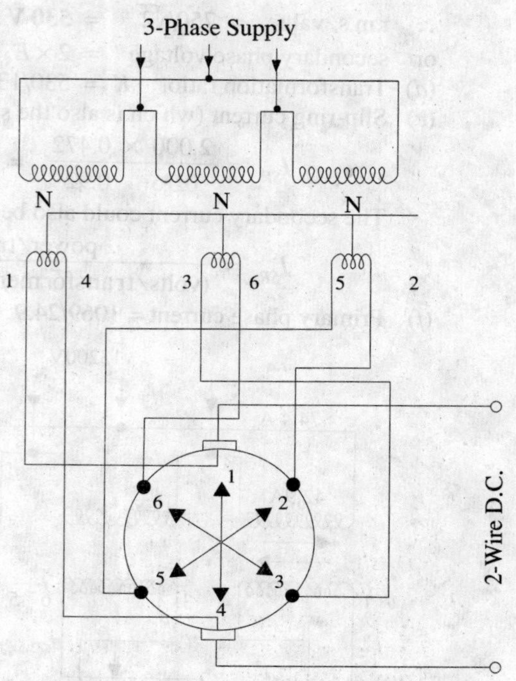

Fig. 37.26

Example 37.15. *A 1500-kW, 750-V, d.c., 6-phase synchronous converter has a full-load efficiency of 96% and a power factor of 0.92. It is fed by a bank of transformers connected in delta on primary and diametrical on the secondary side. The 3-phase primary supply line voltage is 13.2 kV. Calculate (a) d.c. output current (b) a.c. voltage between slip-rings (c) voltage across each secondary phase (d) ratio of transformation (e) a.c. current delivered by each transformer secondary (f) current in each primary coil (g) line current on primary side.*

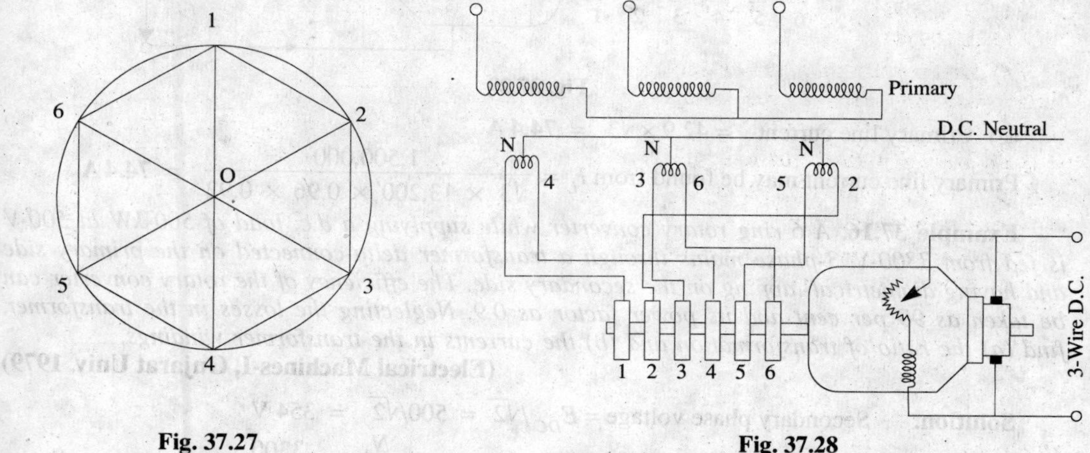

Fig. 37.27

Fig. 37.28

Solution. (*a*) I_{DC} = 1,500,000 / 750 = **2,000 A**

(*b*) Voltage between adjacent rings as per Table No. 37.2 is

E_{SR} = 0.354 × 750 = **265 V**

(*c*) Max. value of voltage across each secondary phase = 750 V

∴ r.m.s. value $= 750/\sqrt{2} = $ **530 V**

or secondary phase voltage $= 2 \times E_{SR} = 2 \times 265 = 530$ V

(*d*) Transformation ratio, $K = 530/13{,}200 =$ **1/24.9**

(*e*) Slip-ring current (which is also the secondary phase current) is

$$I_{SR} = \frac{2{,}000 \times 0.472}{0.96 \times 0.92} = \textbf{1,069 A}$$

The secondary current could also be calculated thus :

$$I_{SR} = \frac{\text{power/transformer}}{\text{(volts/transformer)} \times \text{p.f.} \times \text{efficiency}} = \frac{1{,}500{,}000/3}{530 \times 0.96 \times 0.92} = \textbf{1069A}$$

(*f*) Primary phase current = 1069/24.9 = **42.9 A**

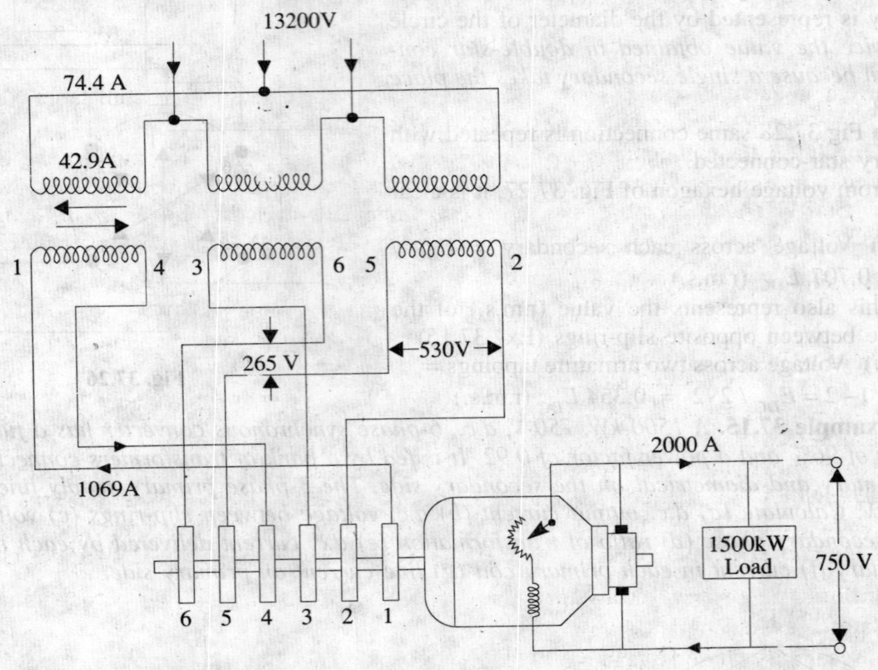

Fig. 37.29

(*g*) Primary line current $= 42.9 \times \sqrt{3} =$ **74.4 A**

Primary line current may be found from $I_L = \dfrac{1{,}500{,}000}{\sqrt{3} \times 13{,}200 \times 0.96 \times 0.92} = \textbf{74.4 A}$

Example 37.16. *A 6-ring rotary converter while supplying a d.c. load of 500 kW at 500-V is fed from 3300-V, 3-phase mains through a transformer delta-connected on the primary side and having diametrical tapping on the secondary side. The efficiency of the rotary converter can be taken as 90 per cent and its power factor as 0.9. Neglecting the losses in the transformer, find (a) the ratio of transformation and (b) the currents in the transformer windings.*

(Electrical Machines-I, Gujarat Univ. 1979)

Solution. Secondary phase voltage $= E_{DC}/\sqrt{2} = 500/\sqrt{2} = 354$ V

Primary phase voltage = 3,300 V ∴ turn ratio $= \dfrac{N_1}{N_2} = \dfrac{3300}{354} =$ **9.32**

$I_{DC} = 500{,}000/500 = 1000$ A; $I_{SR} = 0.472 \times 1000/0.9 \times 0.9 = 568$ A.

∴ secondary phase current = **568 A;** primary phase current = 568/9.32 = **62.5 A**

Example 37.17. *(a) Draw diagram of connections for a 6-ring converter supplied from a 3-phase transformer.*

(a) A synchronous converter supplies 450 kW at 500 V from the commutator side. Assuming diametrical tappings, find the voltage and current of one secondary phase of 3-phase transformer. Take the efficiency as 90 per cent and the power factor as 0.8.

(Electrica Technology, M.S. Univ. Baroda 1978)

Solution. (*a*) Connection diagram is similar to that shown in Fig. 37.29.

(*b*) Maximum value of the secondary phase voltage

$= E_{DC}$ — the diameter of the circle around the vector hexagonal of Fig. 37.27.

R.M.S. value of secondary phase voltage $= E_{DC}/\sqrt{2} = 500/\sqrt{2} =$ **354 V.**

Aslo, secondary phase voltage $= 2 \times E_{SR}$

$= 2 \times 0.354\, E_{DC} = 2 \times 0.354 \times 500 = 354$ V.

$I_{DC} = 450{,}000/500 = 900$ A

Current in each secondary is the same as slip-ring current. As seen from table No. 37.3.

$I_{SR} = 0.472 \times 900/0.9 \times 0.8 =$ **590 A**

Example 37.18. *Calculate (i) the turn ratio of the transformer winding (ii) the voltage between adjacent slip-rings of a synchronous converter having d.c. side volts = 480 V, a.c. supply volts = 3,300 V, 3-phase transformer connection:delta-diametrical, number of slip-rings = 6*

(Electrical Technology, Kerala Univ. 1978)

Solution (*i*) Secondary phase voltage $= E_{DC}/\sqrt{2} = 480/\sqrt{2} = 339.4$ V

[or $= 2 \times 0.354 \times 480 = 339.4$ V]

$$\text{Turn ratio/phase} = \frac{\text{primary phase voltage}}{\text{secondary phase voltage}} = \frac{3300}{339.4} = \mathbf{9.72.}$$

(*ii*) $E_{SR} = 0.354\, E_{DC} = 0.354 \times 480 =$ **170 V** [or E_{SR} = sec. phase voltage / 2].

Example 37.19. *A transformer Y-connected on the primary side to 6.6-kV, 3-phase mains supplies a 6-ring synchronous converter having diametral tappings and connected to a 3-wire d.c. system with 500 V between the outers. Calculate the ratio of the number of turns of primary and secondary sides of the transformer. Calculate also the voltage between the successive slip-rings of the converter.* **(Electrical Technology-II, Madras Univ, 1978)**

Solution. Secondary voltage per phase $= E_{DC}/\sqrt{2}$ where E_{DC} is the output voltage across commutator brushes.

$\therefore$ secondary phase voltage $= 500/\sqrt{2} = 354$ V

Primary phase voltage $= 6600/\sqrt{3} = 3810$ V

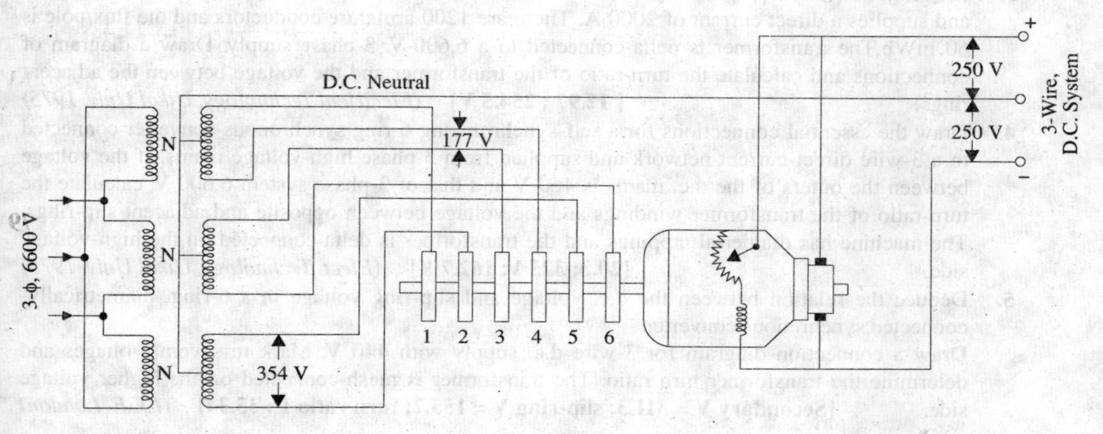

Fig. 37.30

$$\therefore \quad \frac{\text{primary turns}}{\text{secondary turns}} = \frac{\text{primary phase voltage}}{\text{secondary phase voltage}} = \frac{3810}{354} = \mathbf{10.75.}$$

Voltage between successive slip-ring is $E_{SR} = 0.354\ E_{DC} = 0.354 \times 500 = \mathbf{177\ V}$.

Note. (*i*) Voltage between opposite slip-rings = secondary voltage/phase = **354 V.**

(*ii*) If the transformer secondary is regarded as a 6-phase winding, the turn ratio would be

$$10.75 \times 2 : 1 \quad \text{or} \quad 21.5 : 1$$

Example 37.20. *A 6-ring converter delivers 500 kW at 500 V on the d.c. side, the efficiency and power factor at this load being 92% and 95% respectively. Calculate (a) the slip-ring current and (b) voltage across (i) consecutive (ii) alternate (iii) diametral slip-rings on the input side.*

Solution. (*a*) $I_{DC} = 500{,}000/500 = \mathbf{1000\ A}$

$$I_{SR} = \frac{0.472 \times 1000}{0.95 \times 0.92} = \mathbf{540\ A}$$

(*i*) Max. value of a.c. voltage between consecutive slip-rings = $bc = E_{DC}/2$ Volt

r.m.s. value $\quad E_{SR} = E_{DC}/2\sqrt{2} = \mathbf{177\ V}$

(or $= 500 \times 0.354 = 177$ V)

(*ii*) Voltage between alternate rings = bd

Now, slip-ring voltages have a mutual phase difference of 60°.

$\therefore$ r.m.s. value of voltage between alternate rings

$= 2 \times 177 \times \cos 60°/2 = 2 \times 177 \times \sqrt{3}/2 = \mathbf{306V}$

(*iii*) Max. voltage between diametral rings (which is also equal to secondry voltage/phase) = $ad = E_{DC} = \mathbf{500\ V}$

$\therefore$ r.m.s. value $= 500/\sqrt{2} = \mathbf{354\ volt.}$

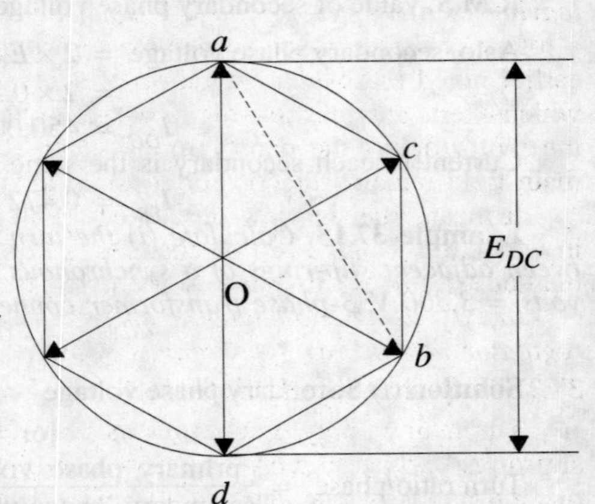

Fig. 37.31

Tutorial Problem No. 37.1

1. A 6-ring synchronous converter is supplying a load of 10 kW at 400 V. If the efficiency of the converter be 90% and the input p.f. 0.9 lagging, calculate the voltage across the slip-rings and the slip-ring current. **[141.4 V; 14.55A]**
2. A 1500-kW, 750-V, d.c., 6-phase synchronous converter is fed by a bank of transformers connected in star on primary side and double-delta on the secondary side. The primary line voltage is 23 kV. Assuming a full-load efficiency of 96% and a power factor of 0.92, calculate (*i*) secondary voltage per phase (*ii*) slip-ring current. **[459.5 V; 1068 A]** (*Elect. Engg.-II Bombay Univ. 1976*)
3. A 6-ring, 12-pole, lap-wound synchronous converter with diametrical tappings runs at 600 r.p.m. and supplies a direct current of 2000 A. There are 1200 armature conductors and the flux/pole is 60 mWb.The transformer is delta-connected to a 6,600-V, 3-phase supply. Draw a diagram of connections and calculate the turn-ratio of the transformer and the voltage between the adjacent rings. **[12.97 ; 254.5 V]** (*Electrical Technology, Utkal Univ. 1975*)
4. Draw the essential connections for a self-synchronizing 6-ring synchronous converter connected to a 3-wire direct-current network and supplied from 3-phase high-voltage mains. If the voltage between the outers of the d.c. mains is 460 V and that of 3-phase system 6,600 V, calculate the turn-ratio of the transformer windings and the voltage between opposite and adjacent slip-rings. The machine has diametral tappings and the transformer is delta-connected on the high-voltage side. **[20.3; 325 V; 162.7 V]** (*Elect. Technology, Utkal Univ. 1975*)
5. Deduce the relation between the d.c. voltage and slip-ring voltage in a 6-ring, diametrically-connected synchronous converter.
   Draw a connection diagram for 3-wire d.c. supply with 440 V. Mark in several voltages and determine the transformer turn ratio. The transformer is mesh-connected on the higher voltage side. **[Secondary V = 311.3; slip-ring V = 155.7; turn ratio 1 : 35.34]** (*I.E.E. London*)

6. Determine the alternating current voltage (r.m.s.) corresponding to 100 volt direct current in the following rotary converters (*a*) single-phase (b) 3-phase (*c*) 6-phase. State for each case the alternating current per line corresponding to direct current of 100 A when the power factor is 0.9 and the efficiency 0.95.

[(*a*) 70.7 V, 165 A (*b*) 61.2 V, 110.4 A (*c*) 35.4 volt, 55.2 A] *(I.E.E. Londan)*

37.20. Voltage Control

The d.c. output voltage of a rotary converter depends on the applied a.c. voltage at slip-rings and it drops slightly as the load increases. In d.c. generators, this voltage drop is easily adjusted by compounding *i.e.* by using a series winding whose flux helps the main flux in direct proportion to the load current. But in a rotary converter, changes in *excitation merely change its power factor, leaving its main flux practically constant*. The reason for this becomes clear if it is recalled that a rotary converter when seen from slip-ring side, operates like a synchronous motor. It has been earlier noted that when excitation of synchronous motor is increased, it draws a leading current, which exerts a demagnetising or weakening effect on the main field. Similarly, when underexcited, the motor draws the main current which exerts a magnetising effect. Hence, in both cases, the main field remains practically constant although the excitation is changed.

It means that in view of a fixed theoretical ratio between d.c. and a.c. values, the only way in which d.c. voltage can be changed is to change the applied a.c. voltage. Some of the important methods used for this purpose are given below :

(*a*) *By using tap-changing transformer (b) series reactance control method (c) Induction regulator method (d) Synchronous booster control method.*

37.21. Tap-changing Transformer

The main supply transformer is provided with tappings by which the voltage applied to the slip-rings can be suitably changed. Remote control by push buttons can also be arranged. This method has the advantages of (*i*) being cheap and (*ii*) causing no interference with power factor or commutation.

But it suffers from the following disadvantages:

1. the voltage can be changed only in jumps, hence gradual voltage regulation is not possible.
2. there is excessive wear and tear at the tappings from which heavy currents are taken out.

37.22. Series Reactance Control Method

The wiring diagram for this method is shown in Fig. 37.32. The converter is compound wound. A reactor or a choking coil is connected in series with each supply to the slip-rings. There is a voltage drop V_C across the coils which leads the current by 90° (neglecting resistance).

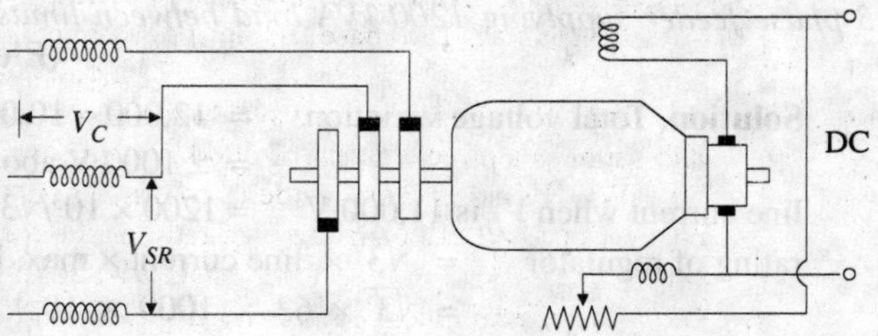

The value of this drop will vary with variations in the load. As seen, the voltage across slip-rings *i.e.* V_{SR} is the vector difference of V_C and the voltage across transformer secondary *i.e.* V_t. If V_t is assumed constant, then V_{SR} will depend on the magnitude and phase of V_C.

Fig. 37.32

It will now be shown that d.c. voltage can be kept practically constant (on varying load) by automatically adjusting the slip-ring voltage.

If load on the converter is increased, then E_{DC} tends to drop slightly. However, if V_{SR} is increased, then it will compensate for the drop in E_{DC} thereby keeping it constant. Similarly, when load is decreased, then E_{DC} tends to rise. But by decreasing V_{SR} simultaneously, this rise in E_{DC} can be neutralized, again keeping it practically constant.

In Fig. 37.33 (*a*) is shown the case when due to light load, excitation is weak and hence converter draws a lagging current. Here, V_{SR} becomes lesser than V_t.

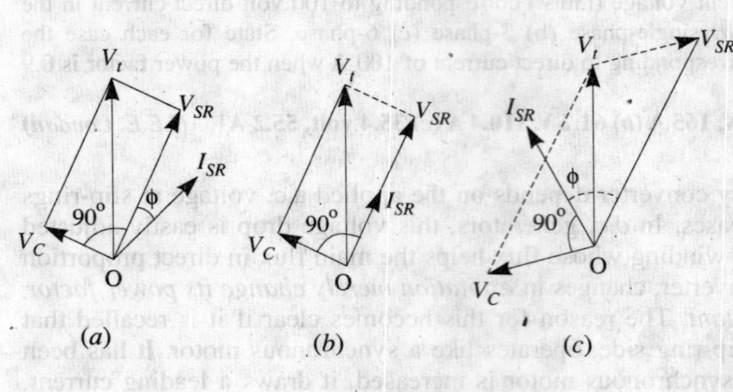

Fig. 37.33

Now, when load is increased, then excitation and hence p.f. is increased. Suppose excitation is so increased as to make power factor unity. Then as seen from Fig. 37.33 (b), V_{SR} is somewhat increased. This increase in V_{SR} is such as to counteract the increased drop in E_{DC} due to increased load.

When the converter is overloaded, the excitation is increased to such a value as to make the power factor leading as shown in Fig. 37.33 (c). Here, V_{SR} is seen to have increased further and is greater than V_t. This further increase in V_{SR} will be sufficient to counteract the increased drop in E_{DC} due to increase in load.

Hence, it is seen that E_{DC} is practically constant on all loads due to automatic and simultaneous adjustments in V_{SR} on different loads. In fact, it is possible to over-compound the converter by increasing the value of reactance in the supply lines.

The required reactance may be obtained in the following two ways :

1. separate coils may be used or
2. the reactance may be supplied by the transformer itself. For this purpose, transformer may be specially built, so as to have a high reactance. Such transformers are cheaper than those having low reactance.

Advantages

1. This method is extremely simple and relatively cheaper.
2. It responds almost instantaneously to any changes in load. Hence, this method has been widely used for supplying current for electric traction especially where service is infrequent and hence fluctuations in current are violent.
3. It gives changes in V_{SR} of ± 10 per cent, so giving a total change of 20%.

One main disadvantage of this method is that voltage changes cannot be produced without changing the power factor.

Example 37.21. *Calculate the rating of an induction regulator to regulate the voltage on a 3-phase feeder supplying 1200 kVA load between limits of 10 kV and 12 kV.*

(Electrical Machines-I, Gwalior Univ. 1979)

Solution. Total voltage variation $= 12{,}000 - 10{,}000 = 2000$ V

$= \pm 1000$ V about a mean value of 11,000 V

line current when V_L is 11,000 V $= 1200 \times 10^3/\sqrt{3} \times 11{,}000 = 63$ A

rating of regulator $= \sqrt{3} \times$ line current × max. line voltage variation up or down

$= \sqrt{3} \times 63 \times 1000 \times 10^{-3} =$ **109 kVA**

Example 37.22. *A 50-Hz, 3-ring synchronous converter has in each slip-ring lead a 2 ohm reactor of negligible resistance. The line voltage is 500 V on the a.c. side. Find the change in the voltage on the d.c. side when the operating power factor of the converter is changed from unity to 0.866 leading when the d.c. current remains unchanged at 25 A. Neglect losses.*

(Elect. Machinery-II, Kerala Univ. 1978)

Solution. It will be assumed that the given power factors are at the transformer terminals and not at slip-rings. The circuit connections and the vector diagrams are shown in Fig. 37.34.

$I_{SR} = 25$ A

Drop across reactor

$V_x = 25 \times 2 = 50$ V per phase

Line value = $50 \times \sqrt{3}$ = 86.6 V

Unity power factor [Fig. 37.34 (*b*)]

$V_{SR} = \sqrt{500^2 + 86.2^2}$

= 507.5 V

$E_{DC} = V_{SR} / 0.612 = 507.5 / 0.612$

= 830 V

Power factor 0.866 leading [Fig. 37.34 (*c*)]

$\phi = \cos^{-1}(0.866) = 30°$

$AC = AB \sin \phi = 86.6 \times \sin 30°$

= 43.3 V

$BC = AB \cos \phi = 86.6 \times \cos 30°$

= 75 V

$OB = \sqrt{(500 + 43.3)^2 + 75^2}$

= 548.3 V; E_{DC} = 548.3 / 0.612 = 896 V

(*a*) (*b*) (*c*)

Fig. 37.34

Change in voltage = 896 – 830 = **66 V.**

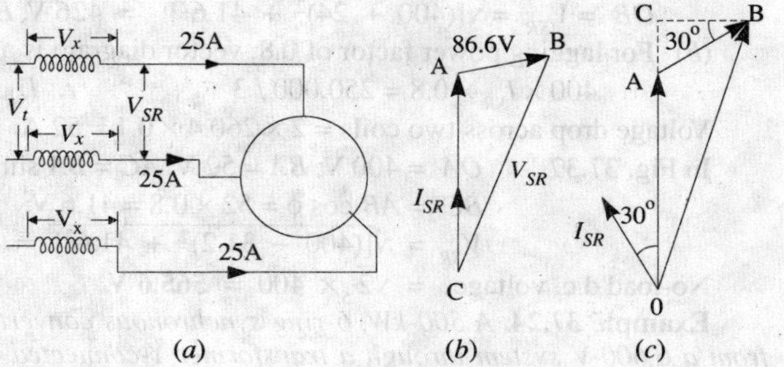

Example 37.23. *A 6-ring converter has diametral connections, with a constant p.d. 400 V across each transformer secondary winding. Reactors of 0.1 Ω each are connected in series with each of the slip-ring connections. Neglecting losses, find the d.c. output voltages for a load of 250 kW if the input power factors are adjusted to (a) 0.866 leading (b) 0.8 lagging at the transformer secondary terminals. Find also the no-load d.c. voltage.*

Solution. The connections are shown in Fig. 37.35.

Power supplied by each secondary phase = 250 / 3 kW

(*a*) ∴ $400 \times I_{SR} \times 0.866 = 250{,}000 / 3$; I_{SR} = 240.5 A

Since there are two reactors across each secondary,

∴ $V_C = 2 \times 240.5 \times 0.1 = 48$ volt*; $\cos \phi = 0.866$ (lead) ∴ $\phi = 30°$

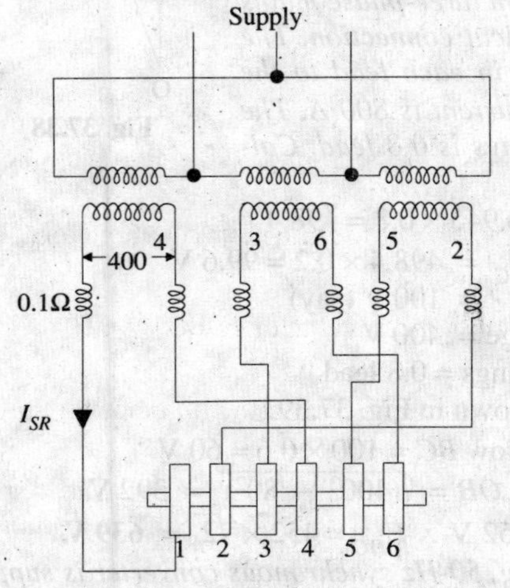

Fig. 37.35

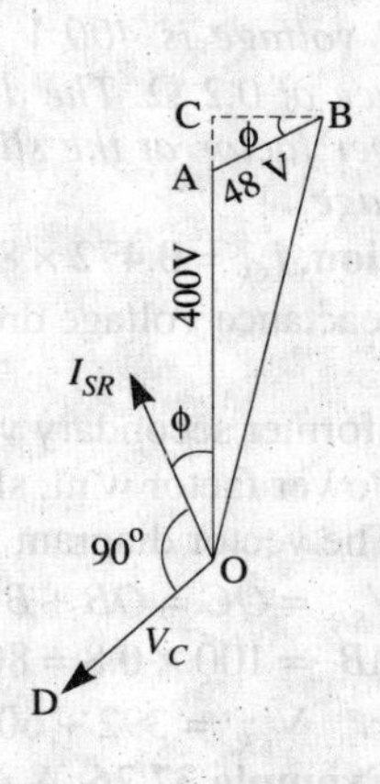

Fig. 36.36

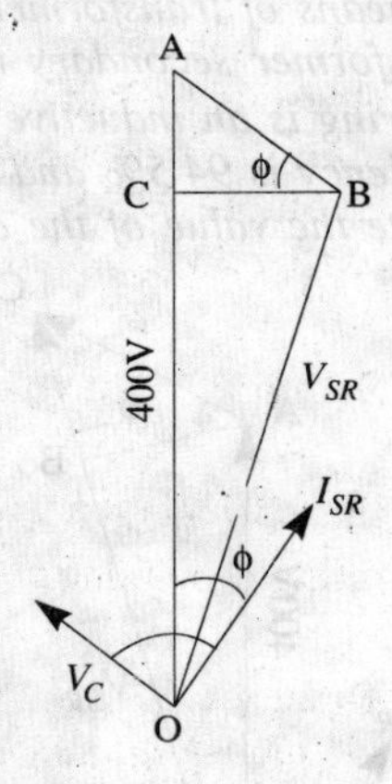

Fig. 37.37

In vector diagram of Fig. 37.36, the current I_{SR} leads the transformer secondary voltage V_t by 30°. The reactance voltage drop leads I_{SR} by 90°. Vector *AB* is parallel to *OD*. Hence, vector *OB* represents V_{SR} corresponding to which $E_{DC} = \sqrt{2} \times V_{SR}$. Produce *OA* and drop *BC* perpendicular on it.

*The reactance drop *per phase* is twice that across each reactor because there is a reactor in each slip-ring lead.

$AC = AB \sin 30° = 48 \times 0.5 = 24; \quad BC = AR \cos 30° = 48 \times 0.866 = 41.6$ V

$OB = V_{SR} = \sqrt{[(400 + 24)^2 + 41.6^2]} = 426$ V, $E_{DC} = 426 \times \sqrt{2} =$ **602 V**

(*b*) For lagging power factor of 0.8, vector diagram is as shown in Fig. 37.37

$400 \times I_{SR} \times 0.8 = 250{,}000 / 3 \qquad \therefore \ I_{SR} = 260.4$ A

Voltage drop across two coils $= 2 \times 260.4 \times 0.1 = 52$ A

In Fig. 37.37, $OA = 400$ V; $BA = 52$ V; $AC = BA \sin \phi = 52 \times 0.6 = 31.2$ V

$BC = AB \cos \phi = 52 \times 0.8 = 41.6$ V; $\quad OB^2 = OC^2 + BC^2$

$V_{SR} = \sqrt{[(400 - 31.2)^2 + 41.6^2]} = 371$ V $\therefore \ E_{DC} = 371 \times \sqrt{2} =$ **525 V**

No-load d.c. voltage $= \sqrt{2} \times 400 =$ **565.6 V**

Example 37.24. *A 500-kW, 6-ring synchronous converter with diametral tappings is supplied from a 6,600-V system through a transformer Y-connected on its primary side and having a turn ratio of 8.5. Estimate the equivalent reactance referred to the secondary in order that the converter may develop 600 V on full-load with the excitation adjusted to give unity p.f. at the slip-rings. Assume a constant normal primary transformer voltage and a constant efficiency of 90%.*

(Electrical Machinery-II, Madras Univ. 1979)

Solution. Primary phase voltage $= 6{,}600/\sqrt{3} = 3{,}810$ V

Secondary voltage $V_t = 3810/8.5 = 448$ V

$V_{SR} = E_{DC} / \sqrt{2} = 600 / \sqrt{2} = 424$ V

As seen from vector diagram of Fig. 37.38

reactive drop $AB = \sqrt{448^2 - 424^2} = 145$ V

Now, $I_{DC} = 500{,}000/600 = 833$ A

$I_{SR} = 0.472 \times 833 / 0.9 = 437$ A

If X is the reactance in each line, then remembering that there are two reactors across each secondary, we have

$2 \times 437\, X = 145 ; \qquad X =$ **0.166 Ω**

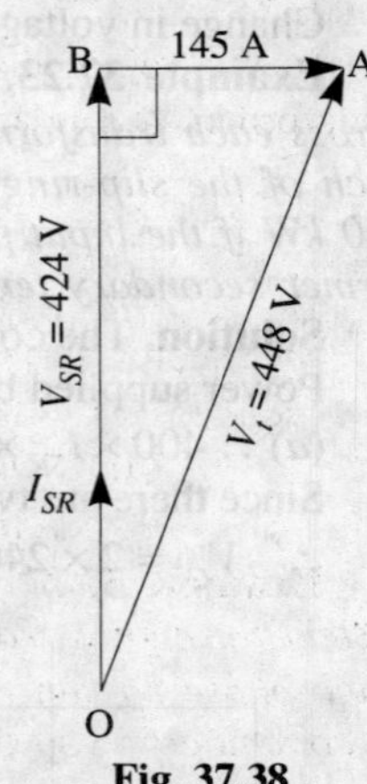

Fig. 37.38

Example 37.25. *A 6-ring rotary converter is fed from three-phase mains by means of transformers with secondaries in diametrical connection. The transformer secondary terminal voltage is 400 V and in each lead to the slip-ring is an inductive reactance of 0.2 Ω. The d.c. current is 800 A. The efficiency is 94.5% and the power factor at the slip-rings is 0.8 lead. Calculate the value of the d.c. voltage.*

Solution. $I_{SR} = 0.472 \times 800/0.945 \times 0.8 = 498.4$

$\therefore$ reactance voltage drop $V_C = 498.4 \times 0.2 = 99.6$ V $= 100$ V (say)

Transformer secondary voltage = 400 V

Power factor w.r.t. slip-rings = 0.8 lead

The vector diagram is shown in Fig. 37.39.

$V_{SR} = OC = OB + BC$; Now $BC = 100 \times 0.6 = 60$ V

$AB = 100 \times 0.8 = 80$ V; $OB = \sqrt{(400^2 - 80^2)} = 392$ V

$\therefore \ V_{SR} = 392 + 60 = 452$ V $\therefore \ E_{DC} = 452 \times \sqrt{2} =$ **639 V.**

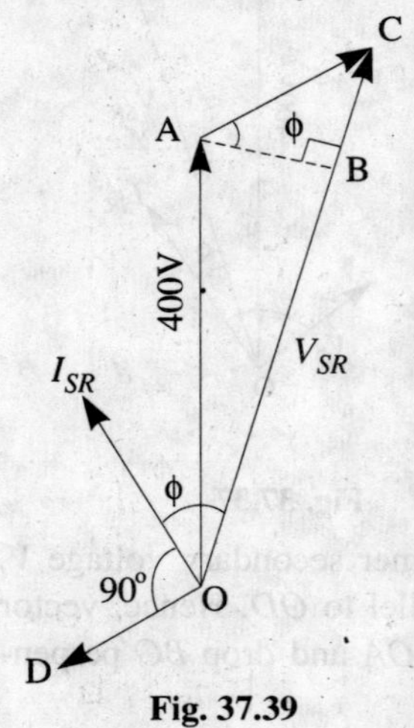

Fig. 37.39

Example 37.26. *A 3-ring, 50-Hz synchronous converter is supplied from a transformer with a constant voltage of 300 V (line value) across its star-connected low-voltage winding terminals. A choking coil with a reactance of 2 Ω and negligible resistance is inserted in each slip-ring lead. Calculate the voltage on the d.c. side when the converter draws a line current of 30 A on a.c. side and excitation is adjusted to give power factor of (a) unity (b) 0.9 leading at the transformer low-voltage winding terminals.*

Solution. The wiring diagram is shown in Fig. 37.40.

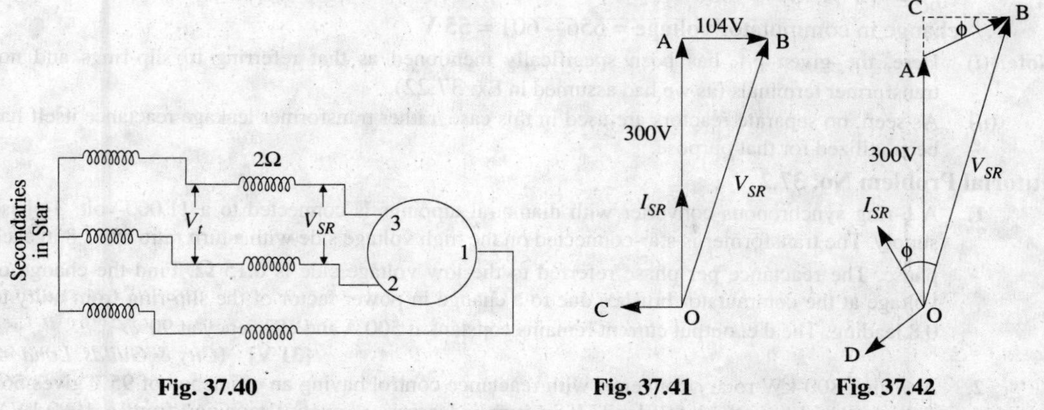

Fig. 37.40 **Fig. 37.41** **Fig. 37.42**

Here, I_{SR} = 30 A; V_C = 30 × 2 = 60 V (phase value)

Line value = 60 × √3 = 104 V

In the vector diagram, I_{SR} is in phase with V_t, because p.f. is unity as measured from the transformer secondary side. Voltage drop V_C leads I_{SR} by 90°. The voltage across slip-rings is V_{SR} (Fig. 37.41)

$$OA = 300 \text{ V}; V_C = AB = 104 \text{ V}$$

$$\therefore \quad OB = V_{SR} = \sqrt{(300^2 + 104^2)} = 317.5 \text{ V} \quad \therefore \quad E_{DC} = 317.5/0.612 = \mathbf{519 \text{ V}}$$

(*b*) Vector diagram for 0.9 power factor leading is shown in Fig. 37.42.

$$\cos\phi = 0.9, \phi = 25°51'; \sin\phi = 0.436, OA = 300 \text{ V}, AB = 104 \text{ V}$$

$$AC = BA \sin\phi = 104 \times 0.436 = 45.3 \text{ V}; BC = BA\cos\phi = 104 \times 0.9 = 93.6 \text{ V}$$

$$V_{SR} = [(300 + 45.3)^2 + 93.6^2] = 358 \text{ V} \qquad \therefore E_{DC} = 358/0.612 = \mathbf{585 \text{ V}}$$

Example 37.27. *A 6-ring rotary converter with diametral tappings is supplied from a 33-kV system through a transformer Y-connected on its primary side and having a turn ratio of 44.5. Each phase has an equivalent reactance of 0.2 Ω, considered all in secondary. Find the change in commutator voltage due to a change in the p.f. at the converter slip-rings from unity to 0.8 leading. The d.c. load is constant at 500 A. Assume a constant efficiency of 90%.*

(Electrical Machinery-II, Madras Univ. 1978)

Solution. Primary phase voltage = 33,000/√3 = 1905 V

Secondary phase voltage V_t = 1905/44.5 = 428 V

At Unity Power Factor (Fig. 37.43)

$$I_{SR} = 0.472 \times 500/0.9 = 262 \text{ A}$$

Reactance drop V_C = 0.2 × 262 = 52.4 V

With reference to Fig. 37.43, we have

$$V_{SR} = \sqrt{428^2 - 52.4^2} = 425 \text{ V}$$

$$\therefore \quad E_{DC} = \sqrt{2} \times 425 = 601 \text{ V}$$

At 0.8 p.f. leading (Fig. 37.44)

$$I_{SR} = 0.427 \times 500 / 0.9 \times 0.8 = 328 \text{ A}$$

$$V_C = 328 \times 0.2 = 65.6 \text{ V}$$

With reference to the vector diagram of Fig. 37.44,

we have $V_{SR} = OB = OC + BC$

Now $AC = AB\cos\phi = 65.6 \times 0.8 = 52.5$ V

$BC = 65.6 \times 0.6 = 39.4$ V

$OC = \sqrt{428^2 - 52.5^2} = 424.8$ V $\quad \therefore O_B = 424.8 + 39.4 = 464$ V

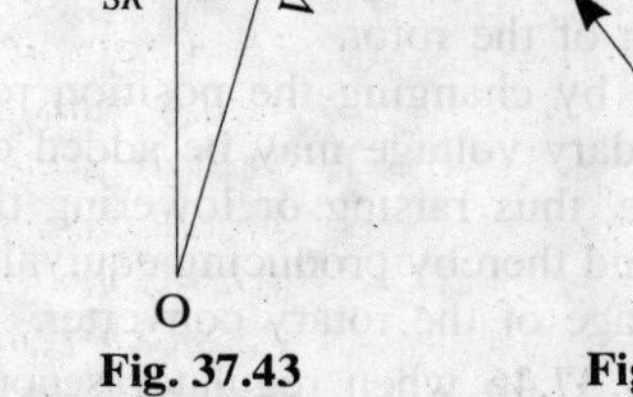

Fig. 37.43

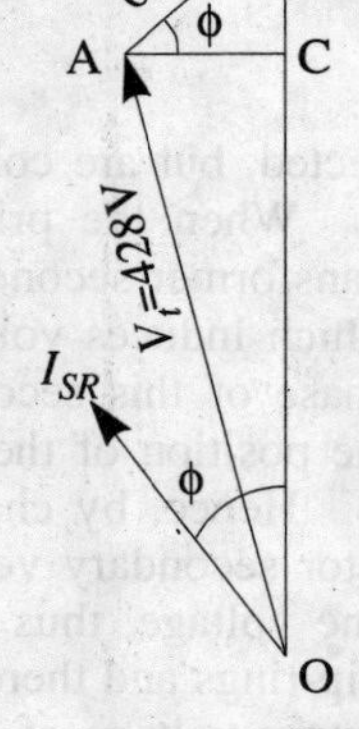

Fig. 37.44

$E_{DC} = \sqrt{2} \times 484 = 656$ V

$\therefore$ change in commutator voltage = 656 – 601 = **55 V**

Note. (*i*) Here, the given p.f. has been specifically mentioned as that referring to slip-rings and not transformer terminals (as we had assumed in Ex. 37.22).

(*ii*) As seen, no separate reactors are used in this case, rather transformer leakage reactance itself has been utilized for that purpose.

Tutorial Problem No. 37.2

1. A 6-ring synchronous converter with diametral tappings is connected to a 11,000-volt, 3-phase supply. The transformer is star-connected on the high voltage side with a turn ratio of 17.8 in each phase. The reactance per phase referred to the low voltage side is 0.15 Ω. Find the change of voltage at the commutator brushes due to a change in power factor of the slip-ring from unity to 0.8 leading. The d.c. output current remains constant at 500 A and efficiency at 90%.

[31 V] (*City & Guilds, London*)

2. A 6-ring, 800-kW rotary converter with reactance control having an efficiency of 95% gives 565 volt on the d.c. side at full-load. It is supplied across diametral tappings from a 1000-kVA, 3-phase transformer having a turn ratio of 6600 to 370. The high voltage supply is 6600 V, the reactance per phase is 20% and transformer magnetising current is 6%. Calculate the current taken by the converter at full-load and the reactive kVA on the primary side of transformer.

[821 A, 286 kVAR] (*London Univ.*)

37.23. Induction Regulator Method

Induction regulator is, in fact, a 3-phase transformer, with a rotating primary and is used for varying the voltage of an a.c. circuit through a large range. In construction, it resembles a 2-pole induction motor with a wound rotor. The rotor does not revolve but can be rotated through a desired amount by a hand crank or by using a small motor geared to the rotor shaft. The induction regulator may be placed either between the slip-rings and the transformer secondaries or it may be placed on the primary side of the transformers. In Fig. 37.45 is shown the induction regulator placed between the secondaries and the slip-rings. The rotor is connected directly to the incoming lines from the secondaries. The primary windings are connected either in star or delta fashion. The secondary windings on the stator are not interconnected, but are connected in series with the incoming lines and the slip-ring taps.

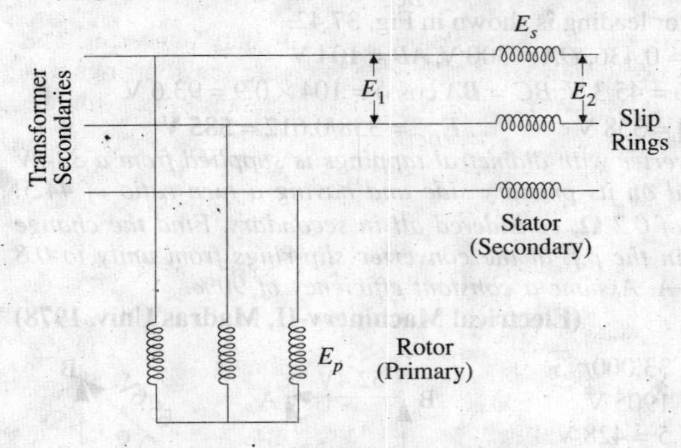

Fig. 37.45

When the primary (rotor winding) is connected to the transformer secondaries, it produces a rotating magnetic field, which induces voltage in the stator winding (secondary). The phase of this secondary voltage of the regulator depends on the position of the rotor.

Hence, by changing the position relative to stator, regulator secondary voltage may be added or subtracted from the line voltage, thus raising or lowering the a.c. voltage on the slip-rings and thereby producing equivalent changes in the d.c. output voltage of the rotary converter.

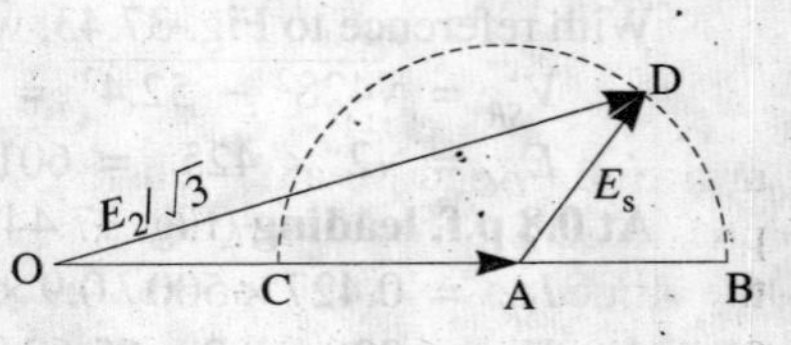

Fig. 37.46

In Fig. 37.46 when regulator secondary voltage E_S is added to the primary phase voltage, then the maximum output voltage to neutral is given by the vector *OB* and when the two are subtracted, then it has minimum value given by vector *OC*. Intermediate values of output voltage like *OD* lie in between these two extreme positions.

This method of voltage control has the following advantages :

1. it gives variations of up to 30% in voltage
2. the voltage variations follow a smooth curve over the operating range
3. remote operation and automatic working can be easily arranged
4. no additional switches are necessary in the main circuit and, constructiona ly, the regulator is robust and sim ɔle
5. can be used for converters running inverted (*i.e.* inverters).

Example 37.28. *The supply voltage to a 3-φ, 200-kVA load is fluctuating between limits of 370 V and 430 V, It is proposed to maintain this voltage constant at 400 V, Calculate the kVA capacity of a 3-φ induction regulator necessary for the same.*
(Electric Machines-II, Gujarat Univ. 1979)

Solution. F.L. line current = $200{,}000 / \sqrt{3} \times 400 = 289$ A

It is given that the regulator is required to increase or decrease the voltage by 30 V,

∴ kVA rating of the induction regulator = $\sqrt{3} \times 289 \times 30 \times 10^{-3}$ = **15 kVA**

Example 37.29. *A 3-phase induction regulator is employed on a 3300-V system to give ± 10 per cent variation of line voltage. Determine the kVA rating of the induction regulator if there is a load of 1000 kVA on the system.* **(Electrical Machinery-II, Madras Univ. 1978)**

Solution. F.L. line current = $1{,}000 \times 1000 / \sqrt{3} \times 3300 = 175$ A

line voltage variation = 10% of 3300 = 330 V

∴ regulator rating = $\sqrt{3} \times 175 \times 330 \times 10^{-3}$ = **100 kVA**

37.24. Synchronous Booster Control

This method is used in case where control of d.c. voltage output is extremely important. The booster consists of a small auxiliary synchronous generator, placed in between the slip-rings and the converter armature. This synchronous generator is of a revolving armature type with the

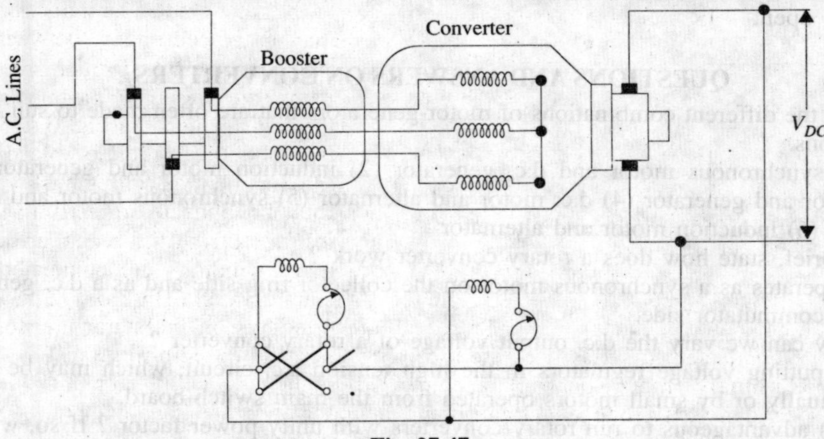

Fig. 37.47

armature placed on an extension on the converter armature. It has the same number of poles and phases as the converter. The boooster armature winding is connected electrically in series with the armature winding of the converter, as shown in Fig. 37.47. The booster receives its d.c. excitation from the d.c. output of the converter and is usually compound-wound.

The function of the series booster is to generate a small voltage proportional to the d.c. load (because its excitation depends on d.c. load), this small voltage being added to the slip-ring voltage, in order to counteract the natural drop in d.c. voltage output as the load increases.

By reversing the excitation of the series booster, its generated voltage can be made to oppose the converter voltage, thereby decreasing the d.c. voltage. In this way, voltage variations of ±15% may be achieved. This method has the following merits :

1. it gives a wide range of voltage variations by the simple adjustment of the booster field rheostat.
2. power factor is independent of load and can be made unity or even leading.
3. automatic boosting can be achieved by compound winding of booster.

However, there is some commutation trouble in this method. Moreover, synchronous booster converters are somewhat costly.

37.25. Starting of Rotary Converters

Converters may be started in any one of the following ways or any combinations of these (1) from the d.c. side, as a d.c. motor (2) from the a.c. side, as an induction motor (3) by using a small auxiliary motor mounted on the same shaft as the converter shaft.

In case d.c. is available when machine is shut down, the converter may be started from d.c. end and then synchronized on the a.c. lines.

When starting with the use of an auxiliary motor, the converter is brought up to speed, synchronized onto the a.c. line like a generator and then connected to the d.c. load. This method is no longer used these days.

Starting it from a.c. side as a polyphase induction motor is a standard practice these days. This self-starting property of the converter is due to the presence of dampers in pole faces.

37.26. Parallel Operation

Generally, there is no difficulty in operating rotary converters in parallel. For successful operation, drooping external characteristics are required on the d.c. side. On the a.c. side, it is preferable to have separate transformer banks for different converters for providing flexibility in voltage control and load division. For guarding against accidental interruption of a.c. supply and for preventing the converter from racing to dangerously high speeds by operating as a d.c. motor with reduced field excitation, the d.c. circuit-breakers must be provided with reverse current relays and speed limiting devices should be fitted. It is advantageous to interlock a.c. and d.c. circuit breakers, so that whenever the a.c. side is open, the d.c. side will also become open.

QUESTIONS AND ANSWERS ON CONVERTERS

Q.1. List the different combinations of motor-generators that are often made to suit load conditions.

Ans. (1) synchronous motor and d.c. generator (2) induction motor and generator (3) d.c. motor and generator (4) d.c. motor and alternator (5) synchronous motor and alternator and (6) induction motor and alternator.

Q.2. In brief, state how does a rotary converter work ?

Ans. It operates as a synchronous motor on the collector ring side and as a d.c. generator on the commutator side.

Q.3. How can we vary the d.c. output voltage of a rotary converter ?

Ans. By putting voltage regulators in the high tension a.c. circuit, which may be regulated manually or by small motors operated from the main switch-board.

Q.4. Is is advantageous to run rotary converters with unity power factor ? If so, why ?

Ans. Yes, because it prevents over-heating when rotary is delivering its full-load kW output.

Q.5. Which factor has greatest effect on power factor on the high-tension line ?

Ans. Strength of magnetic field.

Q.6. What is the result of a field which is either too strong ro too weak ?

Ans. A leading current is produced if the field is too strong and a lagging one if it is too weak.

Q.7. What is the usual range of rotaries ?

Ans. From 3 to 3,000 kW.

Q.8. On what factor does the ratio of conversion from a.c. to d.c. voltage depend in a rotary converter?

Ans. On the number of phases and the method of connecting the windings.

Q,9. Since ratio of voltage conversion is practically constant, what methods are employed to compensate for voltage variation due to changes in load, in order to maintain the d.c. voltage constant ?

Ans. Some of the methods are:-

1. split pole method–in which each field pole is split into two or three parts.
2. regulating pole method–regulating poles perform the same function as the interpoles in d.c. machines.
3. reactance method–generally used where there is a rapidly fluctuating load.
4. Multi-tap transformer method–being a non-automatic method, it is employed where load remains fairly constant over considerable period of time.'
5. Synchronous booster method–in which a revolving-armature alternator, having the same number of poles is combined with the convertor. It is best suited when a relatively wide variation in d.c. voltage is necessary.

OBJECTIVE TESTS — 37

1. The speed of a motor-generator set consisting of a 6-pole induction motor and a 4-pole dc generator fed from a 3-ϕ, 50-Hz supply is ...... rpm.

(*a*) 1000 (*b*) 600
(*c*) 1500 (*d*) 3000.

2. A rotary converter generally

(*a*) combines the functions of an induction motor and a d.c. generator
(*b*) has a set of slip-rings at both ends
(*c*) has one armature and two fields
(*d*) is a synchronous motor and a dc generator combined.

3. The ac line current at slip-rings in a 6-phase, 6-ring rotary converter having 100% efficiency and unity p.f. is ......times the dc current.

(*a*) 0.943 (*b*) 0.472
(*c*) 0.236 (*d*) 1.414

4. A 3-phase supply can be converted into a 6-phase supply by joining the six secondaries of the 3-ϕ transformer in

(*a*) double delta
(*b*) double star
(*c*) diametrical
(*d*) any of the above.

5. One unique advantage of employing induction regulator method for controlling the dc output voltage of a rotary converter is that it

(*a*) is extremely simple and relatively cheaper
(*b*) responds instantaneously to changes in load
(*c*) can be used for inverters
(*d*) gives voltage changes in exact jumps.

6. The most commonly used connection for joining the six secondaries of a transformer used for 3-phase to 6-phase conversion is

(*a*) diametrical
(*b*) zig-zag
(*c*) double-star
(*d*) double-delta.

ANSWERS

1. *b* **2.** *d* **3.** *b* **4.** *d* **5.** *c* **6.** *a*

38 SPECIAL MACHINES

38.1. Introduction

This chapter provides a brief introduction to electrical machines which have special applications. It includes machines whose stator coils are energized by electronically switched currents. The examples are: various types of stepper motors, brushless dc motor and switched re ıctance motor etc. There is also a brief description of dc/ac servomotors, synchro motors and resolvers. These motors are designed and built primarily for use in feedback control systems.

38.2. Stepper Motors

These motors are also called stepping motors or step motors. The name stepper is used because this motor rotates through a fixed angular step in response to each input current pulse received by its controller. In recent years, there has been wide-spread demand of stepping motors because of the explosive growth of the computer industry. Their popularity is due to the fact that they can be co:trolled directly by computers, microprocessors and programmable controllers.

As we know, industrial motors are used to convert electric energy into mechanical energy but they cannot be used for precision positioning of an object or precision control of speed without using closed-loop feedback. Stepping motors are ideally suited for situations where either precise positioning or precise speed control or both are required in automation systems.

Apart from stepping motors, other devices used for the above purposes are synchros and resolvers as well as dc/ac servomotors.(discussed later).

The unique feature of a stepper motor is that its output shaft rotates in a series of discrete angular intervals or steps, one step being taken each time a command pulse is received. When a definite number of pulses are supplied, the shaft turns through a definite known angle. This fact makes the motor well-suited for open-loop position control because no feedback need be taken from the output shaft.

Such motors develop torques ranging from 1 μN-m (in a tiny wrist watch motor of 3 mm diameter) upto 40 N-m in a motor of 15 cm diameter suitable for machine tool applications. Their power output ranges from about 1 W to a maximum of 2500 W. The only moving part in a stepping motor is its rotor which has no windings, commutator or brushes. This feature makes the motor quite robust and reliable.

Step Angle

The angle through which the motor shaft rotates for each command pulse is called the step angle β. Smaller the step angle, greater the number of steps per revolution and higher the resolution or accuracy of positioning obtained. The step angles can be as small as 0.72° or as large as 90°. But the most common step sizes are 1.8°, 2.5°, 7.5° and 15°.

The value of step angle can be expressed either in terms of the rotor and stator poles (teeth) N_r and N_s respectively or in terms of the number of stator phases (m) and the number of rotor teeth.

$$\beta = \frac{(N_s \sim N_r)}{N_s \cdot N_r} \times 360°$$

or $$\beta = \frac{360°}{mN_r} = \frac{360°}{\text{No. of stator phases} \times \text{No. of rotor teeth}}$$

For example, if $N_s = 8$ and $N_r = 6$, $\beta = (8 - 6) \times 360 / 8 \times 6 = 15°$

Resolution is given by the number of steps needed to complete one revolution of the rotor shaft. Higher the resolution, greater the accuracy of positioning of objects by the motor

$\therefore$ resolution $=$ No. of steps / revolution $= 360° / \beta$

A stepping motor has the extraordinary ability to operate at very high stepping rates (upto 20,000 steps per second in some motors) and yet to remain fully in synchronism with the command pulses. When the pulse rate is high, the shaft rotation seems continuous. Operation at high speeds is called 'slewing'. When in the slewing range, the motor generally emits an audible whine having a fundamental frequency equal to the stepping rate. If f is the stepping frequency (or pulse rate) in pulses per second (pps) and β is the step angle, then motor shaft speed is given by

$$n = \beta \times f / 360 \text{ rps} = \text{pulse frequency resolution}$$

If the stepping rate is increased too quickly, the motor loses synchronism and stops. Same thing happens if when the motor is slewing, command pulses are suddenly stopped instead of being progressively slowed.

Stepping motors are designed to operate for long periods with the rotor held in a fixed position and with rated current flowing in the stator windings. It means that stalling is no problem for such motors whereas for most of the other motors, stalling results in the collapse of back emf(E_b) and a very high current which can lead to a quick burn-out.

Applications

Such motors are used for operation control in computer peripherals, textile industry, IC fabrications and robotics etc. Applications requiring incremental motion are typewriters, line printers, tape drives, floppy disk drives, numerically-controlled machine tools, process control systems and X-Y plotters. Usually, position information can be obtained simply by keeping count of the pulses sent to the motor thereby eliminating the need for expensive position sensors and feedback controls. Stepper motors also perform countless tasks outside the computer industry. It includes commercial, military and medical applications where these motors perform such functions as mixing, cutting, striking, metering, blending and purging. They also take part in the manufacture of packed food stuffs, commercial end-products and even the production of science fiction movies.

Example 38.1. *A hybrid VR stepping has motor 8 main poles which have been castleated to have 5 teeth each. If rotor has 50 teeth, calculate the stepping angle.*

Solution. $N_s = 8 \times 5 = 40$; $N_r = 50$

$\therefore$ $\beta = (50 - 40) \times 360 / 50 \times 40 = 1.8°$.

Example 38.2. *A stepper motor has a step angle of 2.5°. Determine (a) resolution (b) number of steps required for the shaft to make 25 revolutions and (c) shaft speed, if the stepping frequency is 3600 pps.*

Solution. (*a*) Resolution = 360° / β = 360° / 2.5° = 144 steps / revolution.

(*b*) Now, steps / revolution = 144. Hence, steps required for making 25 revolutions = 144 × 25 = 3600.

(c) $n = \beta \times f / 360° = 2.5 \times 3600 / 360° = 25$ rps

38.3. Types of Stepper Motors

There is a large variety of stepper motors which can be divided into the following three basic categories:

(*i*) Variable Reluctance Stepper Motor

It has wound stator poles but the rotor poles are made of a ferromagnetic material as shown in Fig.38.1 (*a*). It can be of the single stack type (Fig.38.2) or multi-stack type (Fig.38.5) which gives smaller step angles. Direction of motor rotation independent of the polarity of the stator current. It is called variable reluctance motor because the reluctance of the magnetic circuit formed by the rotor and stator teeth varies with the angular position of the rotor.

(*ii*) Permanent Magnet Stepper Motor

It also has wound stator poles but its rotor poles are permanently magnetized. It has a cylindrical rotor as shown in Fig.38.1(*b*). Its direction of rotation depends on the polarity of the stator current.

(*iii*) Hybrid Stepper Motor

It has wound stator poles and permanently-magnetized rotor poles as shown in Fig.38.1(c). It is best suited when small step angles of 1.8°, 2.5° etc. are required.

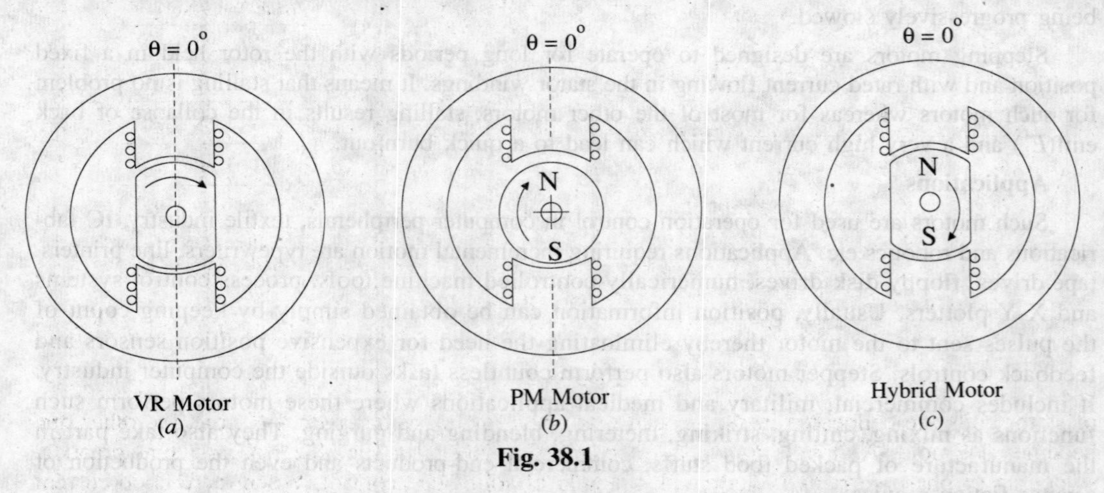

Fig. 38.1

As a variable speed machine, VR motor is sometime designed as a switched-reluctance motor. Similarly, PM stepper motor is also called variable speed brushless dc motor. The hybrid motor combines the features of VR stepper motor and PM stepper motor. Its stator construction is similar to the single-stack VR motor but the rotor is cylindrical and is composed of radially magnetized permanent magnets. A recent type uses a disc rotor which is magnetized axially to give a small stepping angle and low inertia.

38.4. Variable Reluctance Stepper Motors

Construction : A variable-reluctance motor is constructed from ferromagnetic material with salient poles as shown in Fig. 38.2. The stator is made from a stack of steel laminations and has six equally-spaced projecting poles (or teeth) each wound with an exciting coil. The rotor which may be solid or laminated has four projecting teeth of the same width as the stator teeth. As seen, there are three independent stator circuits or phases *A*, *B* and *C* and each one can be energised by a direct current pulse from the drive circuit (not shown in the figure).

A simple circuit arrangement for supplying current to the stator coils in proper sequence is shown in Fig.38.2 (*e*). The six stator coils are connected in 2-coil groups to form three separate circuits called phases. Each phase has its own independent switch. Diametrically opposite pairs

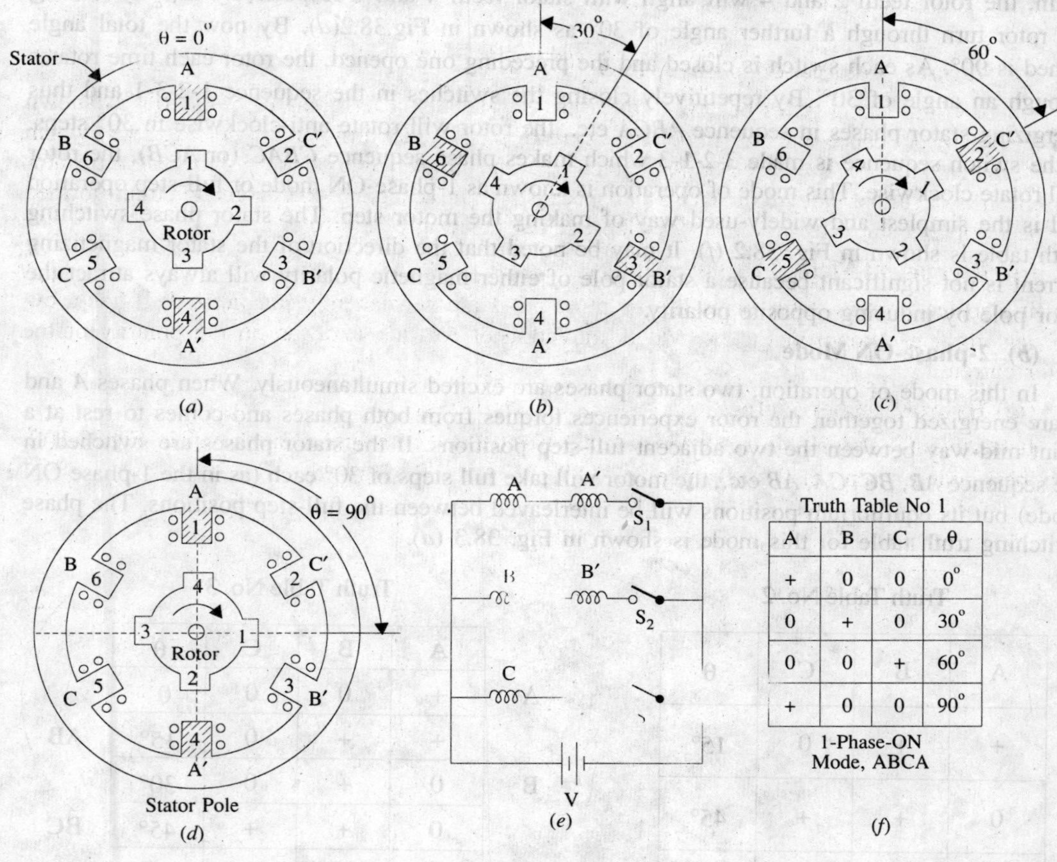

Fig. 38.2

of stator coils are connected in series such that when one tooth becomes a N-pole, the other one becomes a S-pole. Although shown as mechanical switches in Fig.38.2 (*e*), in actual practice, switching of phase currents is done with the help of solid-state control. When there is no current in the stator coils, the rotor is completely free to rotate. Energising one or more stator coils causes the rotor to step forward (or backward) to a position that forms a path of least reluctance with the magnetized stator teeth. The step angle of this three-phase, four rotor teeth motor is $\beta = 360 / 4 \times 3 = 30°$.

Working. The motor has following modes of operation:

(*a*) 1-phase-ON or Full-step Operation

Fig.38.2 (*a*) shows the position of the rotor when switch S_1 has been closed for energising phase A. A magnetic field with its axis along the stator poles of phase A is created. The rotor is therefore, attracted into a position of minimum reluctance with diametrically opposite rotor teeth 1 and 3 lining up with stator teeth 1 and 4 respectively. Closing S_2 and opening S_1 energizes phase *B* causing rotor teeth 2 and 4 to align with stator teeth 3 and 6 respectively as shown in Fig.38.2 (*b*). The rotor rotates through full-step of 30° in the counterclockwise (CCW) direction.

Similarly, when S_3 is closed after opening S_2, phase C is energized which causes rotor teeth 1 and 3 to line up with stator teeth 2 and 5 respectively as shown in Fig.38.2 (*c*). The rotor rotates through an additional angle of 30° in the CCW direction. Next if S_3 is opened and S_1 is closed again, the rotor teeth 2 and 4 will align with stator teeth 4 and 1 respectively thereby making the rotor turn through a further angle of 30° as shown in Fig.38.2(*d*). By now the total angle turned is 90°. As each switch is closed and the preceding one opened, the rotor each time rotates through an angle of 30°. By repetitively closing the switches in the sequence 1-2-3-1 and thus energizing stator phases in sequence *ABCA* etc., the rotor will rotate anti-clockwise in 30° steps. If the switch sequence is made 3-2-1-3 which makes phase sequence *CBAC* (or *ACB*), the rotor will rotate clockwise. This mode of operation is known as 1-phase-ON mode or full-step operation and is the simplest and widely-used way of making the motor step. The stator phase switching truth table is shown in Fig. 38.2 (*f*). It may be noted that the direction of the stator magnetizing current is not significant because a stator pole of either magnetic polarity will always attract the rotor pole by inducing opposite polarity.

(*b*) 2-phase-ON Mode

In this mode of operation, two stator phases are excited simultaneously. When phases *A* and *B* are energized together, the rotor experiences torques from both phases and comes to rest at a point mid-way between the two adjacent full-step positions. If the stator phases are switched in the sequence *AB*, *BC*, *CA*, *AB* etc., the motor will take full steps of 30° each (as in the 1-phase-ON mode) but its equilibrium positions will be interleaved between the full-step positions. The phase switching truth table for this mode is shown in Fig. 38.3 (*a*).

Truth Table No. 2

A	B	C	θ
+	+	0	15°
0	+	+	45°
+	0	+	75°
+	0	0	105°

2 Phase-ON Mode
AB, BC, CA, AB

Truth Table No. 3

	A	B	C	θ	
A	+	0	0	0	
	+	+	0	15°	AB
B	0	+	0	30°	
	0	+	+	45°	BC
C	0	0	+	60°	
	+	0	+	75°	CA
A	+	0	0	90°	

Half-Stepping Alternate
1-Phase-On &
2- Phase-on Mode
A, AB, B, BC, C, CA, A

Fig. 38.3

The 2-phase-ON mode provides greater holding torque and a much better damped single-stack response than the 1-phase-ON mode of operation.

(*c*) Half-step Operation

Half-step operation or 'half-stepping' can be obtained by exciting the three phases in the sequence *A*, *AB*, *B*, *BC*, *C* etc. *i.e.* alternately in the 1-phase-ON and 2-phase-ON modes. It is sometime known as 'wave' excitation and it causes the rotor to advance in steps of 15° *i.e.* half the full-step angle. The truth table for the phase pulsing sequence in half-stepping is shown in Fig. 38.3 (*b*).

Half-stepping can be illustrated with the help of Fig. 38.4 where only three successive pulses have been considered. Energizing only phase A causes the rotor position shown in Fig. 38.4 (*a*). Energising phases *A* and *B* simultaneously moves the rotor to the position shown in Fig. 38.4 (*b*) where rotor has moved through half a step only. Energising only phase *B* moves the rotor through another half-step as shown in Fig. 38.4 (*c*). With each pulse, the rotor moves 30 / 2 = 15° in the CCW direction.

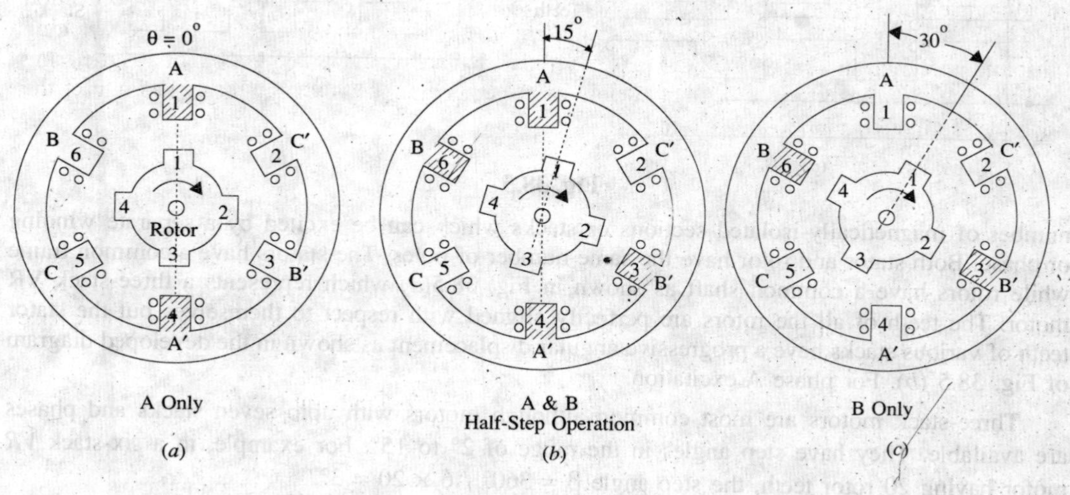

Fig. 38.4

It will be seen that in half-stepping mode, the step angle is halved thereby doubling the resolution. Moreover, continuous half-stepping produces a smoother shaft rotation.

(*d*) Microstepping

It is also known as mini-stepping. It utilizes two phases simultaneously as in 2-phase-ON mode but with the two currents deliberately made unequal (unlike in half-stepping where the two phase currents have to be kept equal). The current in phase A is held constant while that in phase *B* is increased in very small increments until maximum current is reached. The current in phase *A* is then reduced to zero using the same very small increments. In this way, the resultant step becomes very small and is called a microstep. For example, a *VR* stepper motor with a resolution of 200 steps / rev (β = 1.8°) can with microstepping have a resolution of 20,000 steps / rev (β = 0.018°). Stepper motors employing microstepping technique are used in printing and phototypesetting where very fine resolution is called for. As seen, microstepping provides smooth low-speed operation and high resolution.

Torque If I_a is the dc current pulse passing through phase *A*, the torque produced by it is given by $T = (1 / 2)\, I_a^2\, dL / d\theta$. *VR* stepper motors have a high (torque / inertia) ratio giving high rates of acceleration and fast response. A possible disadvantage is the absence of detent torque, which is necessary to retain the rotor at the step position in the event of a power failure.

38.5. Multi-stack VR Stepper Motor

So, far, we have discussed single-stack *VR* motors though multi-stack motors are also available which provide smaller step angles. The multi-stack motor is divided along its axial length into a

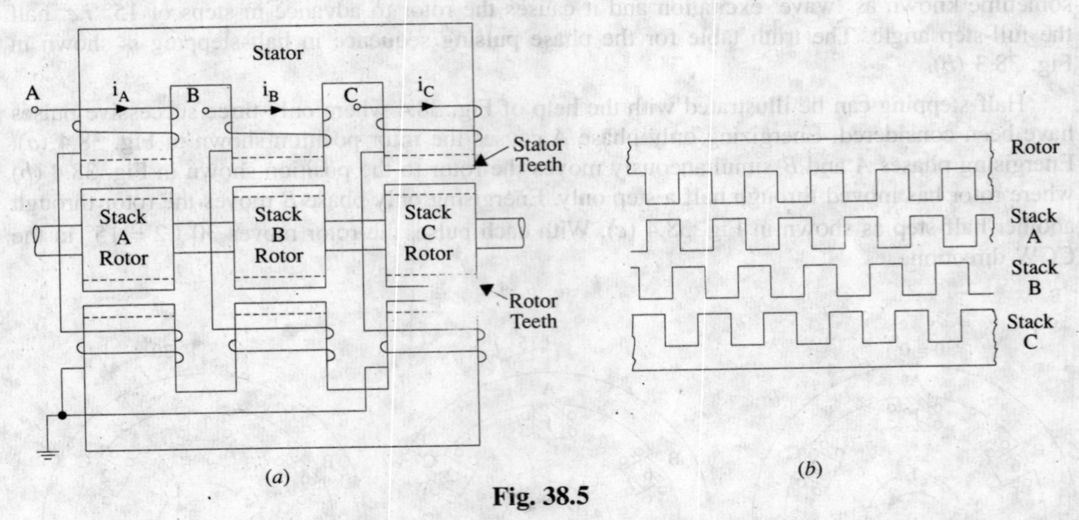

(*a*) (*b*)

Fig. 38.5

number of magnetically-isolated sections or stacks which can be excited by a separate winding or phase. Both stator and rotor have the same number of poles. The stators have a common frame while rotors have a common shaft as shown in Fig. 38.5(*a*) which represents a three-stack VR motor. The teeth of all the rotors are perfectly aligned with respect to themselves but the stator teeth of various stacks have a progressive angular displacement as shown in the developed diagram of Fig. 38.5 (*b*). For phase A excitation.

Three-stack motors are most common although motors with upto seven stacks and phases are available. They have step angles in the range of 2° to 15°. For example, in a six-stack *VR* motor having 20 rotor teeth, the step angle $\beta = 360° / 6 \times 20 = 3°$.

38.6. Permanent-Magnet Stepping Motor

(*a*) Construction. Its stator construction is similar to that of the single-stack *VR* motor discussed above but the rotor is made of a permanent-magnet material like magnetically 'hard' ferrite. As shown in the Fig. 38.6 (*a*), the stator has projecting poles but the rotor is cylindrical and has

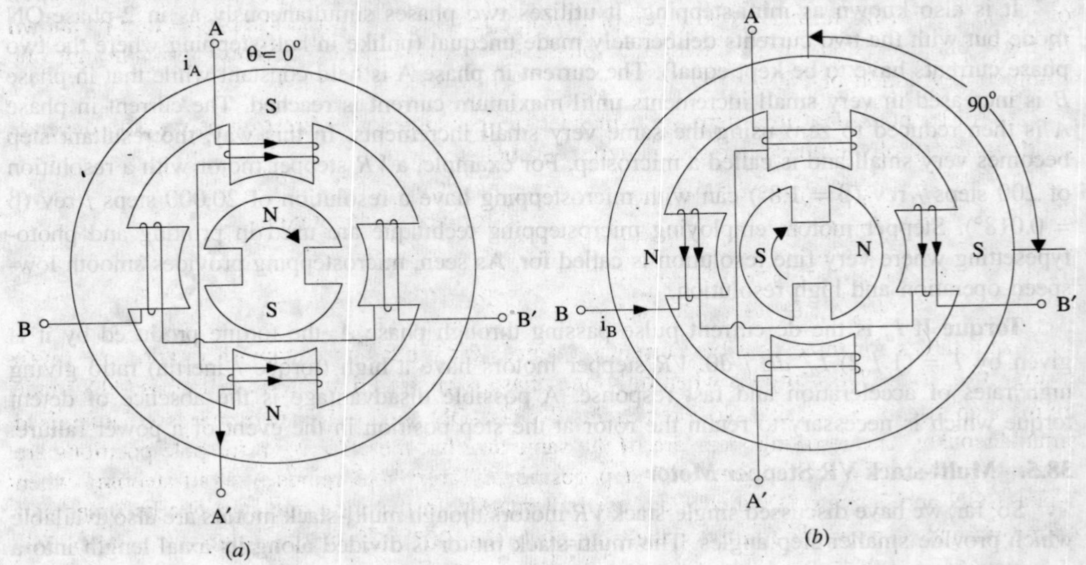

(*a*) (*b*)

Fig. 38.6

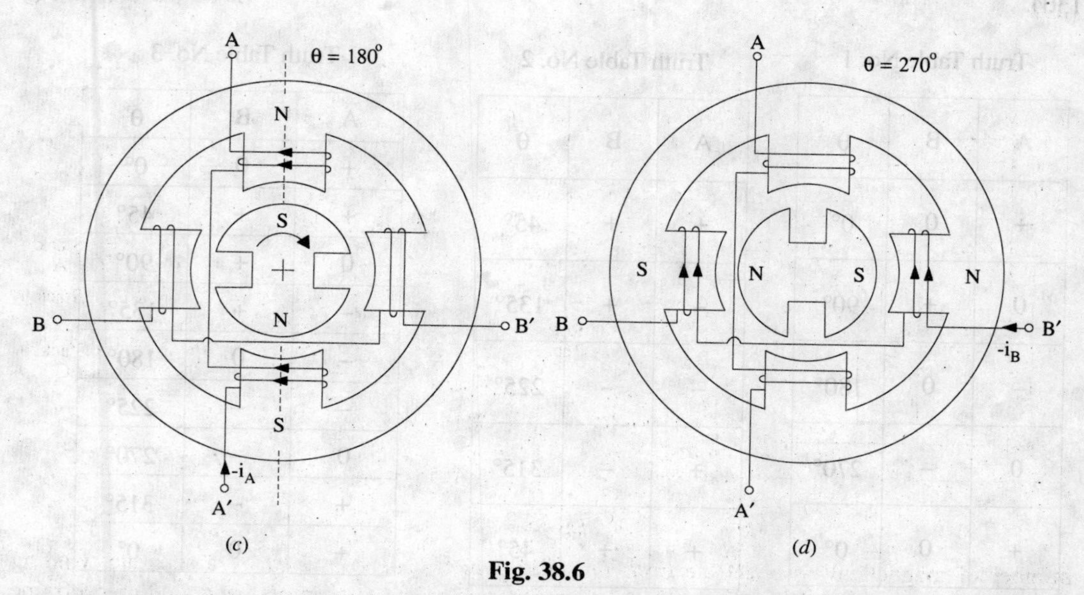

Fig. 38.6

radially magnetized permanent magnets. The operating principle of such a motor can be understood with the help of Fig. 38.6 (*a*) where the rotor has two poles and the stator has four poles. Since two stator poles are energized by one winding, the motor has two windings or phases marked *A* and *B*. The step angle of this motor $\beta = 360° / mN_r = 360° / 2 \times 2 = 90°$ or $\beta = (4 - 2) \times 360° / 2 \times 4 = 90°$.

(b) Working. When a particular stator phase is energized, the rotor magnetic poles move into alignment with the excited stator poles. The stator windings *A* and *B* can be excited with either polarity current (A^+ refers to positive current i_A^+ in the phase *A* and A^- *to negative current* i_A-). Fig.38.6(a) shows the condition when phase A is excited with positive current i_A+. Here, $\theta = 0°$. If excitation is now switched to phase *B* as in Fig. 38.6 (*b*), the rotor rotates by a full step of 90° in the clockwise direction. Next, when phase *A* is excited with negative current i_A-, the rotor turns through another 90° in CW direction as shown in Fig. 38.6 (*c*). Similarly, excitation of phase *B* with i_B- further turns the rotor through another 90° in the same direction as shown in Fig. 38.6 (*d*). After this, excitation of phase *A* with i_A+ makes the rotor turn through one complete revolution of 360°.

It will be noted that in a permanent-magnet stepper motor, the direction of rotation depends on the polarity of the phase currents as tabulated below:

$i_A^+; i_B^+; i_A^-; i_B^-; i_A^+$, ...
$A^+; B^+; A^-; B^-; A^+$; .... for clockwise rotation

$i_A+; i_B-; i_A-; i_B+; i_A+$; ....
$A^+; B^-; A^-; B^+; A^+$; .... for CCW rotation

Truth tables for three possible current sequences for producing clockwise rotation are given in Fig. 38.7. Table No.1 applies when only one phase is energized at a time in 1-phase-ON mode giving step size of 90°. Table No.2 represents 2-phase-ON mode when two phases are energised simultaneously. The resulting steps are of the same size but the effective rotor pole positions are midway between the two adjacent full-step positions. Table No.3 represents half-stepping when 1-phase-ON and 2-phase-ON modes are used alternately. In this case, the step size becomes half of the normal step or one-fourth of the pole-pitch (i.e. 90° / 2 = 45° or 180° / 4 = 45°). Mi-

Truth Table No. 1

A	B	θ
+	0	0°
0	+	90°
–	0	180°
0	–	270°
+	0	0°

1-Phase-ON Mode

Truth Table No. 2

A	B	θ
+	+	45°
–	+	135°
–	–	225°
+	–	315°
+	+	45°

2-Phase-ON Mode

Truth Table No. 3

A	B	θ
+	0	0°
+	+	45°
0	+	90°
–	+	135°
–	0	180°
–	–	225°
0	–	270°
+	–	315°
+	0	0°

Alternate
1-Phase-On &
2-Phase-On Modes

Fig. 38.7

crostepping can also be employed which will give further reduced step sizes thereby increasing the resolution.

(c) Advantages and Disadvantages. Since the permanent magnets of the motor do not require external exciting current, it has a low power requirement but possesses a high detent torque as compared to a *VR* stepper motor. This motor has higher inertia and hence slower acceleration. However, it produces more torque per ampere stator current than a *VR* motor. Since it is difficult to manufacture a small permanent-magnet rotor with large number of poles, the step size in such motors is relatively large ranging from 30° to 90°. However, recently disc rotors have been manufactured which are magnetized axially to give a small step size and low inertia.

Example 38.3. *A single-stack, 3-phase VR motor has a step angle of 15°. Find the number of its rotor and stator poles.*

Solution. Now, $\beta = 360° / mN_r$ or $15° = 360° / 3 \times N_r$; $\therefore N_r = 8$.

For finding the value of N_s, we will use the relation $\beta = (N_s \sim N_r) \times 360° / N_s.N_r$

(*i*) **When $N_s > N_r$** Here, $\beta = (N_s - N_r) \times 360° / N_s.N_r$

or $15° = (N_s - 8) \times 360° / 8\,N_s$; $\therefore N_s = \mathbf{12}$

(*ii*) **When $N_s < N_r$** Here, $15° = (8 - N_s) \times 360° / 8\,N_s$; $\therefore N_s = \mathbf{6}$.

Example 38.4. *A four-stack VR stepper motor has a step angle of 1.8°. Find the number of its rotor and stator teeth.*

Solution. A four-stack motor has four phases. Hence, $m = 4$.

$\therefore$ $1.8° = 360° / 4 \times N_r$; $\therefore N_r = \mathbf{50}$.

Since in multi-stack motors, rotor teeth equal the stator teeth, hence $N_s = \mathbf{50}$.

38.7. Hybrid Stepper Motor

(*a*) **Construction.** It combines the features of the variable reluctance and permanent-magnet stepper motors. The rotor consists of a permanent-magnet that is magnetized axially to create a pair or poles marked *N* and *S* in Fig. 38.8 (*b*). Two end-caps are fitted at both ends of this axial

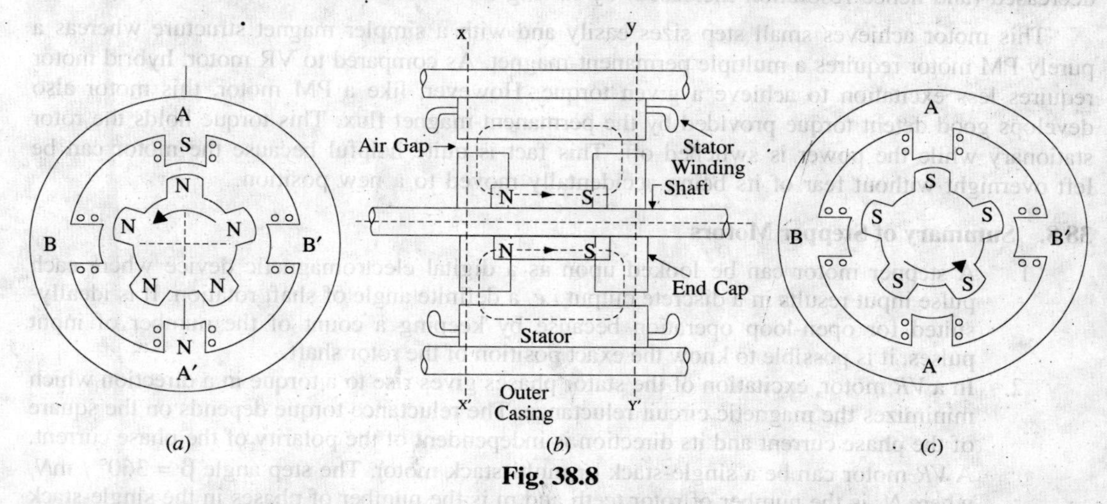

(*a*) (*b*) (*c*)

Fig. 38.8

megnet. These end-caps consist of equal number of teeth which are magnetized by the respective polarities of the axial magnet. The rotor teeth of one end-cap are offset by a half tooth pitch so that a tooth at one end-cap coincides with a slot at the other. The cross-sectional views perpendicular to the shaft along X-X′ and Y-Y′ axes are shown in Fig. 38.8 (*a*) and (*c*) respectively. As seen, the stator consists of four stator poles which are excited by two stator windings in pairs. The rotor has five N-poles at one end and five S-poles at the other end of the axial magnet. The step angle of such a motor is $= (5 - 4) \times 360° \ / \ 5 \times 4 = 18°$.

(*b*) **Working.** In Fig.38.8 (*a*), phase *A* is shown excited such that the top stator pole is a *S*-pole so that it attracts the top *N*-pole of the rotor and brings it in line with the *A*-*A*′ axis. To turn the rotor, phase A is denergized and phase *B* is excited positively. The rotor will turn in the CCW direction by a full step of 18°.

Next, phase *A* and *B* are energized negatively one after the other to produce further rotations of 18° each in the same direction. The truth table is shown in Fig. 38.9 (*a*). For producing clockwise rotation, the phase sequence should be A^+; B^-; A^-; B^+; A^+ etc.

Practical hybrid stepping motors are built with more rotor poles than shown in Fig. 38.9 in order to give higher angular resolution. Hence, the stator poles are often slotted

Truth Table

A	B	θ
+	0	0°
0	+	18°
–	0	36°
0	–	54°
+	0	72°

1-Phase ON
Full-Step Mode

(*a*) (*b*)

Fig. 38.9

or castleated to increase the number of stator teeth. As shown in Fig. 38.9 (*b*), each of the eight stator poles has been alloted or castleated into five smaller poles making N_s = 8 × 5 = 40°. If rotor has 50 teeth, then step angle = (50 – 40) × 360° / 50 × 40 = 1.8°. Step angle can also be decreased (and hence resolution increased) by having more than two stacks on the rotor.

This motor achieves small step sizes easily and with a simpler magnet structure whereas a purely PM motor requires a multiple permanent-magnet. As compared to VR motor, hybrid motor requires less excitation to achieve a given torque. However, like a PM motor, this motor also develops good detent torque provided by the permanent-magnet flux. This torque holds the rotor stationary while the power is switched off. This fact is quite helpful because the motor can be left overnight without fear of its being accidentally moved to a new position.

38.8. Summary of Stepper Motors

1. A stepper motor can be looked upon as a digital electromagnetic device where each pulse input results in a discrete output *i.e.* a definite angle of shaft rotation. It is ideally-suited for open-loop operation because by keeping a count of the number of input pulses, it is possible to know the exact position of the rotor shaft.
2. In a *VR* motor, excitation of the stator phases gives rise to a torque in a direction which minimizes the magnetic circuit reluctance. The reluctance torque depends on the square of the phase current and its direction is independent of the polarity of the phase current. A *VR* motor can be a single-stack or multi-stack motor. The step angle β = 360° / mN_r where N_r is the number of rotor teeth and m is the number of phases in the single-stack motor or the number of stacks in the multi-stack motor.
3. A permanent-magnet stepper motor has a permanently-magnetized cylindrical rotor. The direction of the torque produced depends on the polarity of the stator current.
4. A hybrid motor combines the features of *VR* and *PM* stepper motors. The direction of its torque also depends on the polarity of the stator current. Its step angle β = 360° / mN_r.
5. In the 1-phase ON mode of excitation, the rotor moves by one full-step for each change of excitation. In the 2-phase-ON mode, the rotor moves in full steps although it comes to rest at a point midway between the two adjacent full-step positions.
6. Half-stepping can be achieved by alternating between the 1-phase-ON and 2-phase-ON modes. Step angle is reduced by half.
7. Microstepping is obtained by deliberately making two phase currents unequal in the 2-phase-ON mode.

Tutorial Problems No. 38.1

1. A stepper motor has a step angle of 1.8°. What number should be loaded into the encoder of its drive system if it is desired to turn the shaft ten complete revolutions ? **[2000]**
2. Calculates the step angle of a single-stack, 4-phase, 8/6-pole VR stepper motor. What is its resolution. ? **[15°; 24 steps/rev]**
3. A stepper motor has a step angle of 1.8° and is driven at 4000 pps. Determine (*a*) resolution (*b*) motor speed (*c*) number of pulses required to rotate the shaft through 54°. **[(a) 200 steps/rev (b) 1200 rpm (c) 30]**
4. Calculate the pulse rate required to obtain a rotor speed of 2400 rpm for a stepper motor having a resolution of 200 steps/rev. **[4000 pps]**
5. A stepper motor has a resolution of 500 steps/rev in the 1-phase-ON mode of operation. If it is operated in half-step mode, determine (*a*) resolution (*b*) number of steps required to turn the rotor through 72°. **[(*a*) 1000 steps/rev (*b*) 200]**
6. What is the required resolution for a stepper motor that is to operate at a pulse frequency of 6000 pps and a travel 180° in 0.025 s. **[300 steps/rev]**

38.9. Permanent-Magnet DC Motor

A permanent-magnet dc (PMDC) motor is similar to an ordinary dc shunt motor except that its field is provided by permanent magnets instead of salient-pole wound-field structure. Fig. 38.10 (*a*) shows 2-pole PMDC motor whereas Fig. 38.10(*b*) shows a 4-pole wound-field dc motor for comparison purposes.

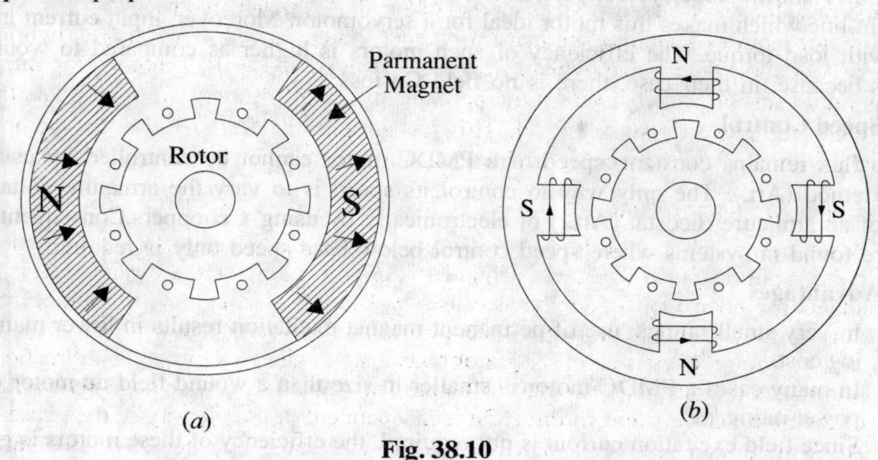

Fig. 38.10

(*a*) Construction

As shown in Fig. 38.10 (*a*), the permanent magnets of the PMDC motor are supported by a cylindrical steel stator which also serves as a return path for the magnetic flux. The rotor (*i.e.* armature) has winding slots, commutator segments and brushes as in conventional dc machines.

There are three types of permanent magnets used for such motors. The materials used have residual flux density and high coercivity.

(*i*) Alnico magnets — they are used in motors having ratings in the range of 1 kW to 150 kW.

(*ii*) ceramic (ferrite) magnets — they are much economical in fractional kilowatt motors.

(*iii*) rare-earth magnets — made of samarium cobalt and neodymium iron cobalt which have the highest energy product. Such magnetic materials are costly but are best economic choice for small as well as large motors.

Another form of the stator construction is the one in which permanent-magnet material is cast in the form of a continuous ring instead of in two pieces as shown in Fig. 38.10(*a*).

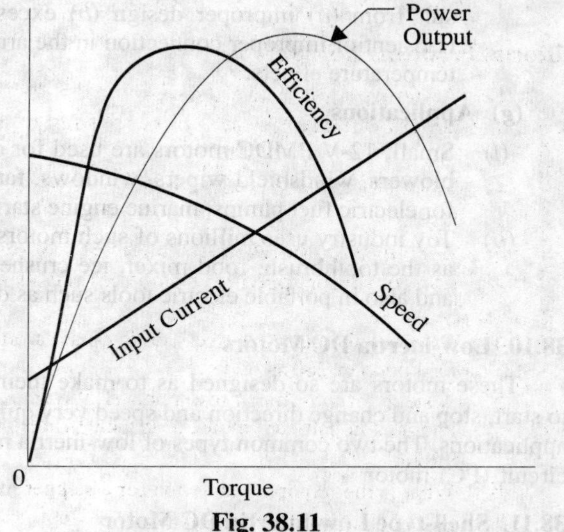

Fig. 38.11

(*b*) Working

Most of these motors usually run on 6 V, 12 V or 24 V dc supply obtained either from batteries or rectified alternating current. In such motors, torque is produced by interaction between the axial current-carrying rotor conductors and the magnetic flux produced by the permanent magnets.

(c) Performance

Fig. 38.11 shows some typical performance curves for such a motor. Its speed-torque curve is a straight line which makes this motor ideal for a servomotor. Moreover, input current increases linearly with load torque. The efficiency of such motors is higher as compared to wound-field dc motors because, in their case, there is no field Cu loss.

(d) Speed Control

Since flux remains constant, speed of a PMDC motor cannot be controlled by using Fulx Control Method (Art.). The only way to control its speed is to vary the armature voltage with the help of an armature rheostat (Art.) or electronically by using x choppers.Consequently, such motors are found in systems where speed control below base speed only is required.

(e) Advantages

(i) In very small ratings, use of permanent-magnet excitation results in lower manufacturing cost.
(ii) In many cases a PMDC motor is smaller in size than a wound-field dc motor of equal power rating.
(iii) Since field excitation current is not required, the efficiency of these motors is generally higher than that of the wound-field motors.
(iv) Low-voltage PMDC motors produce less air noise.
(v) When designed for low-voltage (12 V or less) these motors produced very little radio and TV interference.

(f) Disadvantages

(i) Since their magnetic field is active at all times even when motor is not being used, these motors are made totally enclosed to prevent their magnets from collecting magnetic junk from neighbourhood. Hence, as compared to wound-field motors, their temperature tends to be higher. However, it may not be much of a disadvantage in situations where motor is used for short intervals.
(ii) A more serious disadvantage is that the permanent magnets can be demagnetized by armature reaction mmf causing the motor to become inoperative. Demagnetization can result from (a) improper design (b) excessive armature current caused by a fault or transient or improper connection in the armature circuit (c) improper brush shift and (d) temperature effects.

(g) Applications

(i) Small, 12-V PMDC motors are used for driving automobile heater and air conditioner blowers, windshield wipers, windows, fans and radio antennas etc. They are also used for electric fuel pumps, marine engine starters, wheelchairs and cordless power tools.
(ii) Toy industry uses millions of such motors which are also used in other appliances such as the toothbrush, food mixer, ice crusher, portable vacuum cleaner and shoe polisher and also in portable electric tools such as drills, saber saws and hedge trimmers etc.

38.10. Low-inertia DC Motors

These motors are so designed as to make their armature mass very low. This permits them to start, stop and change direction and speed very quickly making them suitable for instrumentation applications. The two common types of low-inertia motors are (i) shell-type motor and (ii) printed-circuit (PC) motor

38.11. Shell-type Low-intertia DC Motor

Its armature is made up of flat aluminium or copper coils bonded together to form a hollow cylinder as shown in Fig. 38.12. This hollow cylinder is not attached physically to its iron core which is stationary and is located inside the shell-type rotor. Since iron does not form part of the rotor, the rotor inertia is very small.

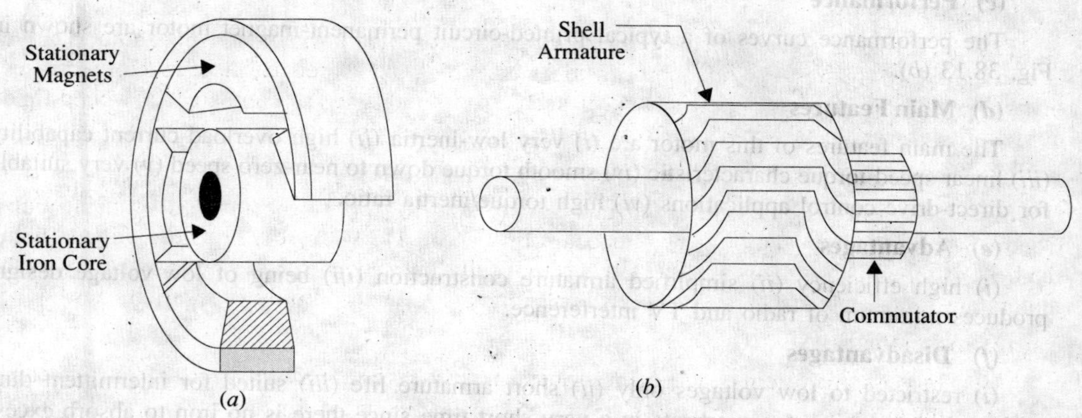

Fig. 38.12

38.12 Printed-circuit(Disc) DC Motor

(*a*) Constructional Details

It is a low-voltage dc motor which has its armature (rotor) winding and commutator printed on a thin disk of non-magnetic insulating material. This disk-shaped armature contains no iron and etched-copper conductors are printed on its both sides. It uses permanent magnets to produce the necessary magnetic field. The magnetic circuit is completed through the flux-return plate which also supports the brushes. Fig.38.13(a) shows an 8-pole motor having wave-wound armature. Brushes mounted in an axial direction bear directly on the inner parts of the armature conductors which thus serve as a commutator. Since the number of armature conductors is very large, the torque produced is uniform even at low speeds. Typical sizes of these motors are in the fractional and subfractional horsepower ranges. In many applications, acceleration from zero to a few thousand rpm can be obtained within 10 ms.

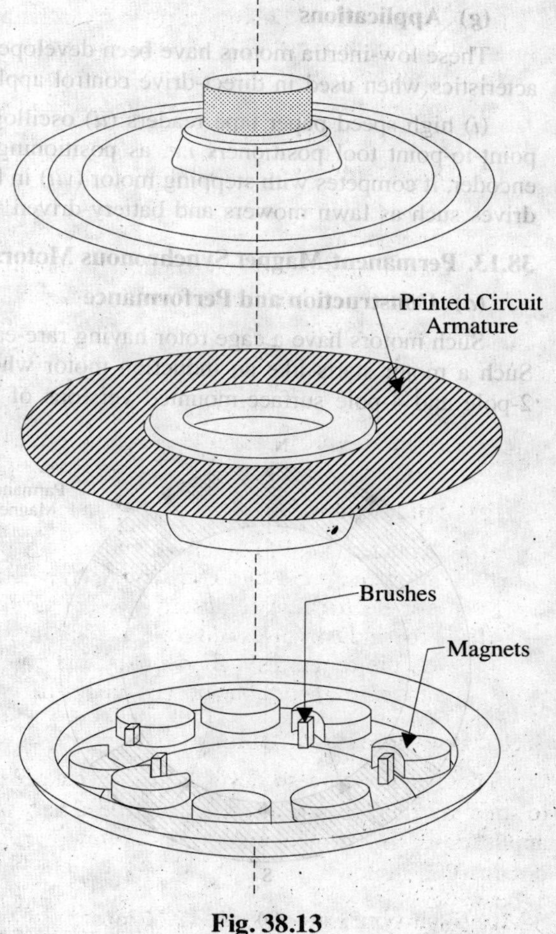

Fig. 38.13

(*b*) Speed Control

The speed can be controlled by varying either the applied armature voltage or current. Because of their high efficiency, fan cooling is not required in many applications. The motor brushes require periodic inspection and replacement. The rotor disk which carriers the conductors and commutator, being very thin, has a limited life. Hence, it requires replacing after some time.

(*c*) Performance

The performance curves of a typical printed-circuit permanent-magnet motor are shown in Fig. 38.13 (*b*).

(*d*) Main Features

The main features of this motor are (*i*) very low-inertia (*ii*) high overload current capability (*iii*) linear speed-torque characteristic (*iv*) smooth torque down to near-zero speed (*v*) very suitable for direct-drive control applications (*vi*) high torque/inertia ratio.

(*e*) Advantages

(*i*) high efficiency (*ii*) simplified armature construction (*iii*) being of low-voltage design, produces minimum of radio and TV interference.

(*f*) Disadvantages

(*i*) restricted to low voltages only (*ii*) short armature life (*iii*) suited for intermittent duty cycle only because motor overheats in a very short time since there is no iron to absorb excess heat (*v*) liable to burn out if stalled or operated with the wrong supply voltage.

(*g*) Applications

These low-inertia motors have been developed specifically to provide high performance characteristics when used in direct-drive control applications. Examples are:

(*i*) high speed paper tape readers (*ii*) oscillographs (*iii*) X-Y recorders (*iv*) layer winders (*v*) point-to-point tool positioners *i.e.* as positioning servomotors (*vi*) with in-built optical position encoder, it competes with stepping motor (*vii*) in high rating is being manufactured for heavy-duty drives such as lawn mowers and battery-driven vehicles etc.

38.13. Permanent-Magnet Synchronous Motors

(*a*) Construction and Performance

Such motors have a cage rotor having rare-earth permanent magnets instead of a wound field. Such a motor starts like an induction motor when fed from a fixed-frequency supply. A typical 2-pole and 4-pole surface-mounted versions of the rotor are shown in Fig. 38.14. Since no dc

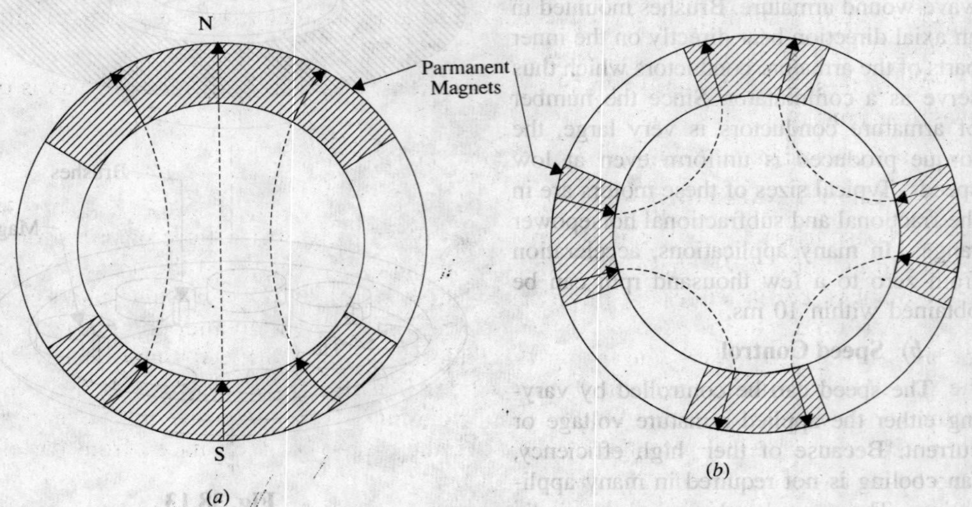

Fig. 38.14

supply is needed for exciting the rotor, it can be made more robust and reliable. These motors have outputs ranging from about 100 W upto 100 kW. The maximum synchronous torque is designed to be around 150 per cent of the rated torque. If loaded beyond this point, the motor loses synchronism and will run either as an induction motor or stall.

These motors are usually designed for direct-on-line (DOL) starting. The efficiency and power factor of the permanent-magnet excited synchronous motors are each 5 to 10 points better than their reluctance motor counterparts.

(*b*) Advantages

Since there are no brushes or slip-rings, there is no sparking. Also, brush maintenance is eliminated. Such motors can pull into synchronism with inertia loads of many times their rotor inertia.

(*c*) Applications

These motors are used where precise speed must be maintained to ensure a consistent product. With a constant load, the motor maintains a constant speed. Hence, these motors are used for synthetic-fibre drawing where constant speeds are absolutely essential.

38.14. Synchros

It is a general name for self-synchronizing machines which, when electrically energized and electrically interconnected, exert torques which cause two mechanically independent shafts either to run in synchronism or to make the rotor of one unit follow the rotor position of the other. They are also known by the trade names of selsyns and autosyns. Synchros, in fact, are small cylindrical motors varying in diameter from 1.5 cm to 10 cm depending on their power output. They are low-torque devices and are widely used in control systems for transmitting shaft position information or for making two or more shafts to run in synchronism. If a large device like a robot arm is to be positioned, synchros will not work. Usually, a servomotor is needed for a higher torque.

38.15. Types of Synchros

There are many types of synchros but the four basic types used for position and error-voltage applications are as under :

(*i*) Control Transmitter (denoted by CX) — earlier called generator (*ii*) Control Receiver (CR) — earlier called motor (*iii*) Control-Transformer (CT) and (iv) Control Differential (CD). It may be further subdivided into control differential transmitter (CDX) and control differential receiver (CDR).

All of these synchros are single-phase units except the control differential which is of three-phase construction.

(*a*) Constructional Features

1. **Control Transmitter**

Its constructional details are shown in Fig. 38.15(*a*). It has a three-phase stator winding similar to that of a three-phase synchronous generator. The rotor is of the projecting-pole type using dumbell construction and has a single-phase winding. When a single-phase ac voltage is applied to the rotor through a pair of slip rings, it produces an alternating flux field along the axes of the rotor. This alternating flux induces three unbalanced single phase/voltage in the three stator windings by transformer action. If the rotor is aligned with the axis of the stator winding 2, flux linkage of this stator winding is maximum and this rotor position is defined as the electrical zero. In Fig.38.15(b), the rotor axis is displaced from the electrical zero by an angle displaced 120° apart.

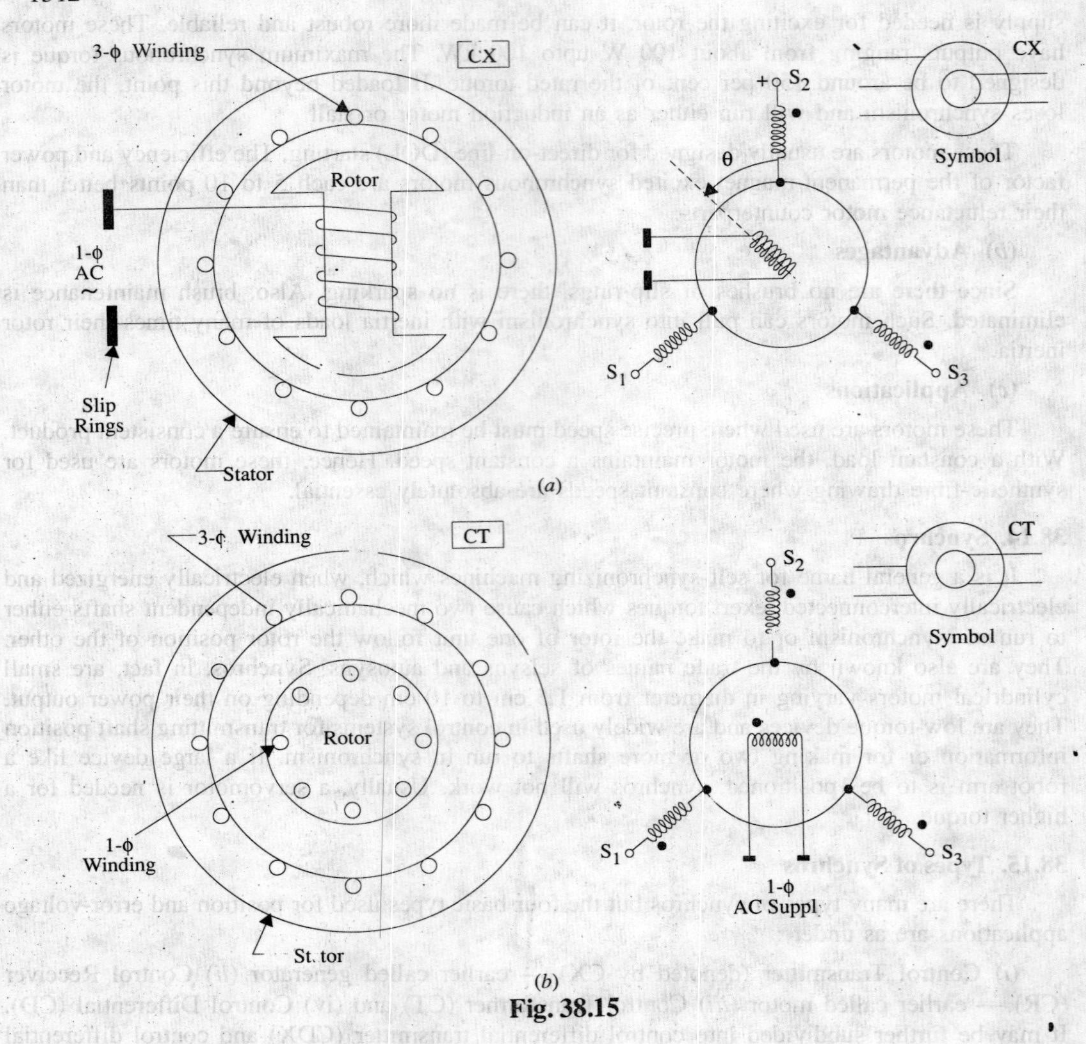

Fig. 38.15

(b) Control Receiver (CR)

Its construction is essentially the same as that of the control transmitter show in Fig. 38.15(a). It has three stator windings and a single-phase salient-pole rotor. However, unlike a CX, a CR has a mechnical viscous damper on the shaft which permits CR rotor to respond without overshooting its mark. In normal se, both the rotor and s ator windings are excited with single-phase currents. When the field of the rotor conductors interacts with the field of the stator conductors, a torque is developed which produces rotation.

(c) Control Transformer (CT)

As shown in Fig. 38.15 (b) its stator has a three-phase winding whereas the cylindrical rotor l as a single-phase windirg. In this case, the electrical zero is defined as that position of the rotor that makes the flux linkage with winding 2 of the stator zero. This rotor position has been shown in Fig. 38.15 (b) and is different from that of a control transmitter.

(d) Control Differential (CD)

The differential synchro has a balanced three-phase distributed winding in both the stator and the rotor. Moreover, it has a cylindrical rotor as shown in Fig. 38.16 (a). Although three-phase

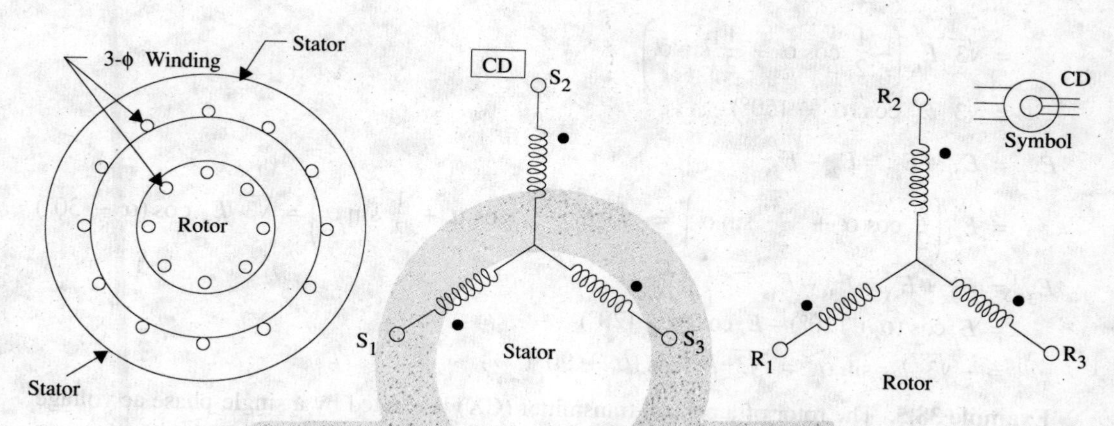

Fig. 38.16

windings are involved, it must be kept in mind that these units deal solely with single-phase voltages. The three winding voltages are not polyphase voltages. Normally, the three-phase voltages are identical in magnitude but are separated in phase by 120°. In synchros, these voltages are in phase but differ in magnitude because of their physical orientation.

(e) Voltage Relations

Consider the control transmitter shown in Fig. 38.17. Suppose that its rotor winding is excited by a single-phase sinusoidal ac voltage of rms value E_r and that rotor is held fast in its displaced position from the electrical zero. If K = stator turns / rotor turns, the rms voltage induced in the stator winding is $E = KE_r$. However, if we assume K = 1, then $E = E_r$.

The rms value of the induced emf in stator winding 2 when the rotor displacement is α is given by

$$E_{2s} = E_r \cos \alpha$$

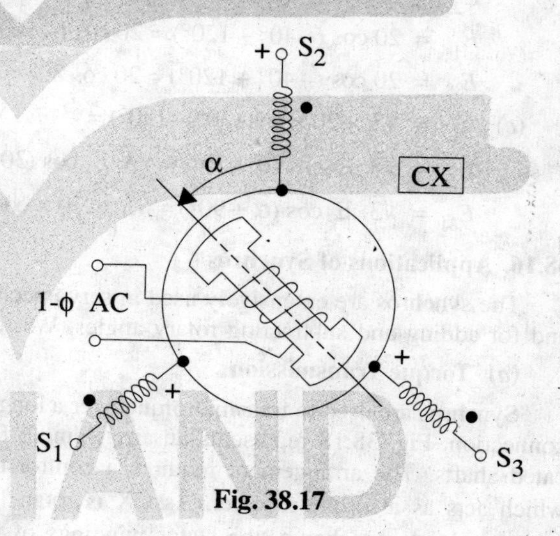

Fig. 38.17

Since the axis of the stator winding 1 is located 120° ahead of the axis of winding 2, the rms value of the induced emf in this winding is

$$E_{1s} = E_r \cos (\alpha - 120°)$$

In the same way, since winding 3 is located behind the axis of winding 2 by 120°, the expression for the induced emf in winding 3 becomes

$$E_{3s} = E_r \cos (\alpha + 120°)$$

We can also find the values of terminal induced voltages as

$$E_{12} = E_{1s} + E_{s2} = E_{1s} - E_{2s}$$

$$= E_r \cos \alpha \cos 120° + E_r \sin \alpha \sin 120° - E_r \cos \alpha$$

$$= E_r \left(-\frac{3}{2} \cos \alpha + \frac{\sqrt{3}}{2} \sin \alpha \right)$$

$$= \sqrt{3}\ E_r \left(-\frac{1}{2}\cos\alpha + \frac{1}{2}\sin\alpha\right)$$

$$= \sqrt{3}\ E_r \cos(\alpha - 150°)$$

$$E_{23} = E_{2s} + E_{s3} = E_{2s} - E_{3s}$$

$$= E_r\left(\frac{3}{2}\cos\alpha + \frac{\sqrt{3}}{2}\sin\alpha\right) = \sqrt{3}\ E_r\left(\frac{\sqrt{3}}{2}\cos\alpha + \frac{1}{2}\sin\alpha\right) = \sqrt{3}\cdot E_r \cos(\alpha - 30°)$$

$$E_{31} = E_{3s} + E_{s1} = E_{3s} - E_{1s}$$

$$= E_r \cos(\alpha + 120°) - E_r \cos(\alpha - 120°)$$

$$= -\sqrt{3}\ E_r \sin\alpha = \sqrt{3}\ E_r \cos(\alpha + 90°)$$

Example 38.5. The rotor of a control transmitter (CX) is excited by a single-phase ac voltage of rms value 20 V. Find the value of E_{1s}, E_{2s} and E_{3s} for rotor angle $\alpha = +40°$. and $-40°$. Assume the stator/rotor turn ratio as unity. Also, find the values of terminal voltages when $\alpha = +30°$.

Solution. Since K = 1, the voltage relations derived in will be used.

(*a*) $\alpha = +40°$

$$E_{2s} = E_r \cos\alpha = 20 \cos 40° = 15.3 \text{ V}$$

$$E_{1s} = E_r \cos(\alpha - 120°) = 20\cos(40° - 120°) = 3.5 \text{ V}$$

$$E_{3s} = E_r \cos(\alpha + 120°) = 20 \cos 160° = -18.8 \text{ V}$$

(*b*) $\alpha = -40°$

$$E_{2s} = 20\cos(-40°) = 15.3 \text{ V}$$

$$E_{1s} = 20\cos(-40° - 120°) = 20\cos(-160°) = -18.8 \text{ V}$$

$$E_{3s} = 20\cos(-40° + 120°) = 20\cos 80° = 3.5\text{V}$$

(*c*) $E_{12} = \sqrt{3} \times 20 \times \cos(30° - 150°) = -17.3 \text{ V}$

$$E_{23} = \sqrt{3}\ E_r \cos(\alpha - 30°) = \sqrt{3}\ E_r \cos(30° - 30°) = 34.6 \text{ V}$$

$$E_{31} = \sqrt{3}\ E_r \cos(\alpha + 90°) = \sqrt{3} \times 20 \times \cos(30° + 90°) = -17.3 \text{ V}$$

38.16. Applications of Synchros

The synchros are extensively used in servomechanism for torque transmission, error detection and for adding and subtracting rotary angles. We will consider these applications one by one.

(*a*) Torque Transmission

Synchros are used to transmit torque over a long distance without the use of a rigid mechanical connection. Fig. 38.18 represents an arrangement for maintaining alignment of two distantly-located shafts. The arrangement requires a control transmitter (CX) and a control receiver (CR) which acts as a torque receiver. As CX is rotated by an angle α, CR also rotates through the same angle α. As shown, the stator windings of the two synchros are connected together and their rotors are connected to the same single-phase ac supply.

Working. Let us suppose that CX rotor is displaced by an angle α and switch SW1 is closed to energize the rotor winding. The rotor winding flux will induce an unbalanced set of three single-phase voltages (in time phase with the rotor voltage) in the CX stator phase windings which will circulate currents in the CR stator windings. These currents produce the CR stator flux field whose axis is fixed by the angle α. If the CR rotor winding is now energized by closing switch SW2, its flux field will interact with the flux field of the stator winding and thereby produce a torque. This torque will rotate the freely-moving CR rotor to a position which exactly corresponds with the CT rotor i.e. it will be displaced by the same angle α as shown in Fig. 38.18. It should be noted that if the two rotors are in the same relative positions, the stator

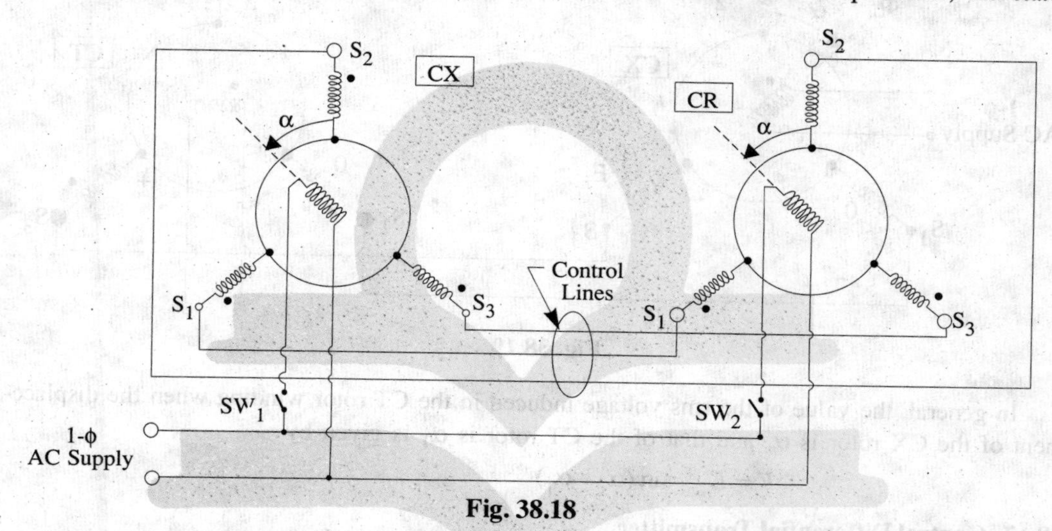

Fig. 38.18

voltages in the two synchros will be exactly equal and opposite. Hence, there will be no current flow in the two stator windings and so no torque will be produced and the system will achieve equilibrium. If now, the transmitter rotor angle changes to a new value, then new set of voltages would be induced in the transmitter stator windings which will again drive currents through the receiver stator windings. Hence, necessary torque will be produced which will turn the CR rotor through an angle corresponding to that of the CT rotor. That is why the transmitter rotor is called the master and the receiver rotor as the slave, because it follows its master. It is worth noting that this master-slave relationship is reversible because when the receiver rotor is displaced through a certain angle, it causes the transmitter rotor to turn through the same angle.

(*b*) Error Detection

Synchros are also used for error detection in a servo control system. In this case, a command in the form of a mechanical displacement of the CX rotor is converted to an electrical voltage which appears at the CT rotor winding terminals which can be further amplified by an amplifier.

For this purpose, we require a CX synchro and a CT synchro as shown in Fig. 38.19. Only the CX rotor is energized from the single-phase ac voltage supply which produces an alternating air-gap flux field. This time-varying flux field induces voltages in the stator windings whose values for $\alpha = 30°$ are as indicated in the Fig. 38.19. The CX stator voltages supply magnetizing currents in the CT stator windings which, in turn, create an alternating flux field in their own air-gap. The values of the CT stator phase currents are such that the air-gap flux produced by them induces voltages that are equal and opposite to those existing in the CX stator. Hence, the direction of the resultant flux produced by the CX stator phase currents is forced to take a position which is exactly identical to that of the rotor axis of the CT.

If the CT rotor is assumed to be held fast in its electrical zero position as shown in Fig.38.19, then the rms voltage induced in the rotor is given by $E = E_{max} \sin \alpha$ where E_{max} is the maximum voltage induced by the CT air-gap flux when coupling with the rotor windings is maximum and α is the displacement angle of the CT rotor.

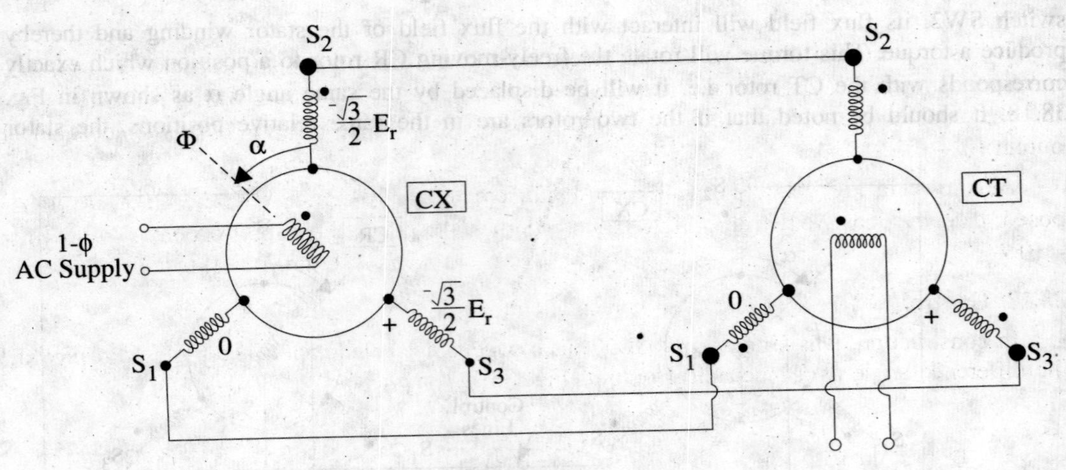

Fig. 38.19

In general, the value of the rms voltage induced in the CT rotor winding when the displacement of the CX rotor is α_x and that of the CT rotor is α_T is given by

$$E = E_{max} \sin(\alpha_x - \alpha_T)$$

38.17. Control Differential Transmitter

It can be used to produce a rotation equal to the sum of difference of the rotations of two shafts. The arrangement for this purpose is shown in Fig. 38.20(*a*). Here, a CDX is coupled to a control transmitter on one side and a control receiver on the other. The CX and CR rotor windings are engrized from the same single-phase voltage supply.

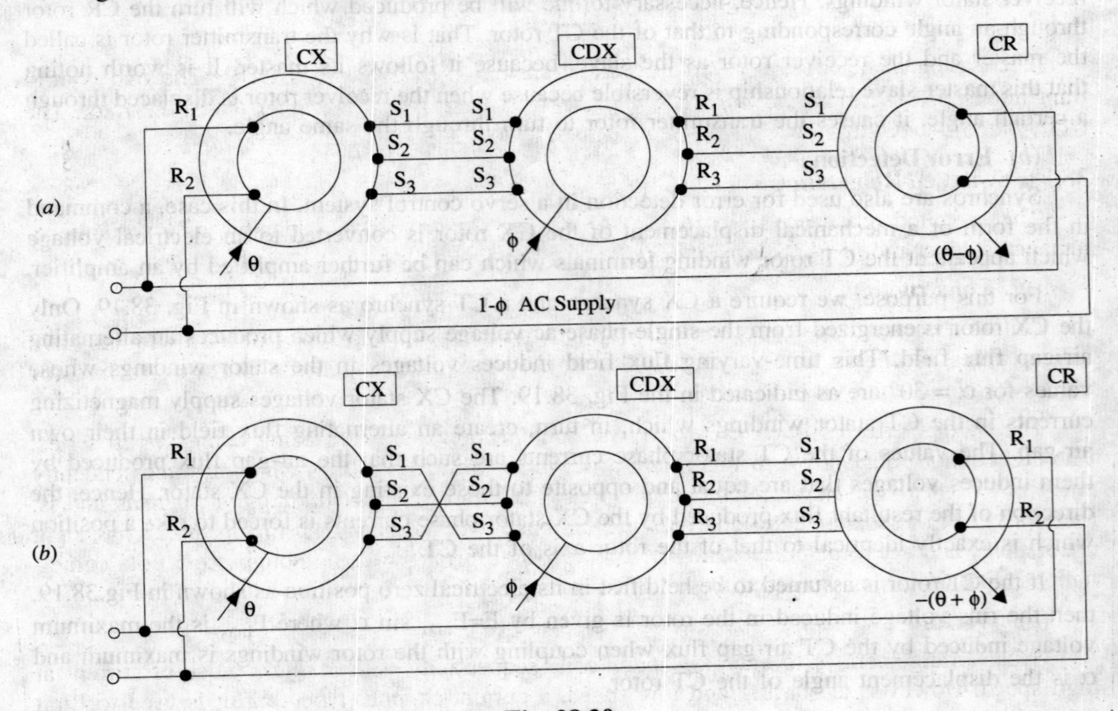

Fig. 38.20

It has two inputs: electrical θ and mechanical ϕ and the output is electrical (θ – ϕ). The mechanical input (θ) to CX is converted and applied to the CDX stator. With a rotor input (ϕ), the electrical output of the CDX is applied to the CR stator which provides the mechanical output (θ – ϕ).

As shown in Fig. 38.20(*b*), if any two stator connections between CX and CDX are transposed, the electrical input from CX to CDX becomes –θ, hence the output becomes (–θ – ϕ) = – (θ + ϕ)

38.18. Control Differential Receiver

In construction, it is similar to a CDX but it accepts two electrical input angles and provide the difference angle as a mechanical output (Fig. 38.21).

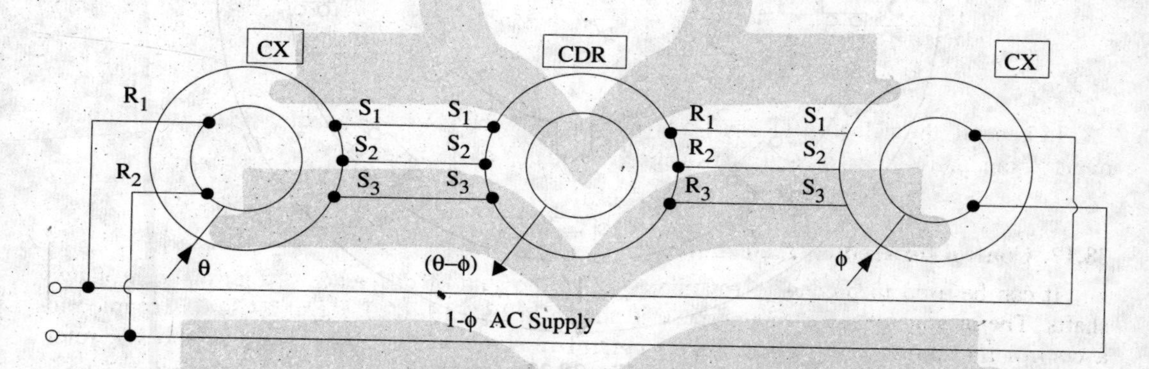

Fig. 38.21

The arrangement consists of two control transmitters coupled to a CDR. The two control transmitters provide inputs to the CDX, one (θ) to the stator and the other (ϕ) to the rotor. The CDX output is the difference of the two inputs *i.e.* (θ - ϕ).

38.19. Switched Reluctance Motor

The switched reluctance (SR) motor operates on the same basic principle as a variable reluctance stepper motor (Art. 38.4).

(*a*) Construction

Unlike a conventional synchronous motor, both the rotor and stator of an SR motor have salient poles as shown in Fig. 38.22. This doubly-salient arrangement is very effective for electromagnetic energy conversion.

The stator carries coils on each pole, the coils on opposite poles being connected in series. The eight stator coils shown in Figure are grouped to form four phases which are independently energized from a four-phase converter. The laminated rotor has no windings or magnets and is, therefore cheap to manufacture and extremely robust. The motor shown in Fig. 38.22 has eight stator poles and six rotor poles which is a widely-used arrangement although other pole combinations (like 6/4 poles) are used to suit different applications.

(*b*) Working

Usual arrangement is to energize stator coils sequentially with a single pulse of current at high speed. However, at starting and low speed, a current-chopper type control is used to limit the coil current.

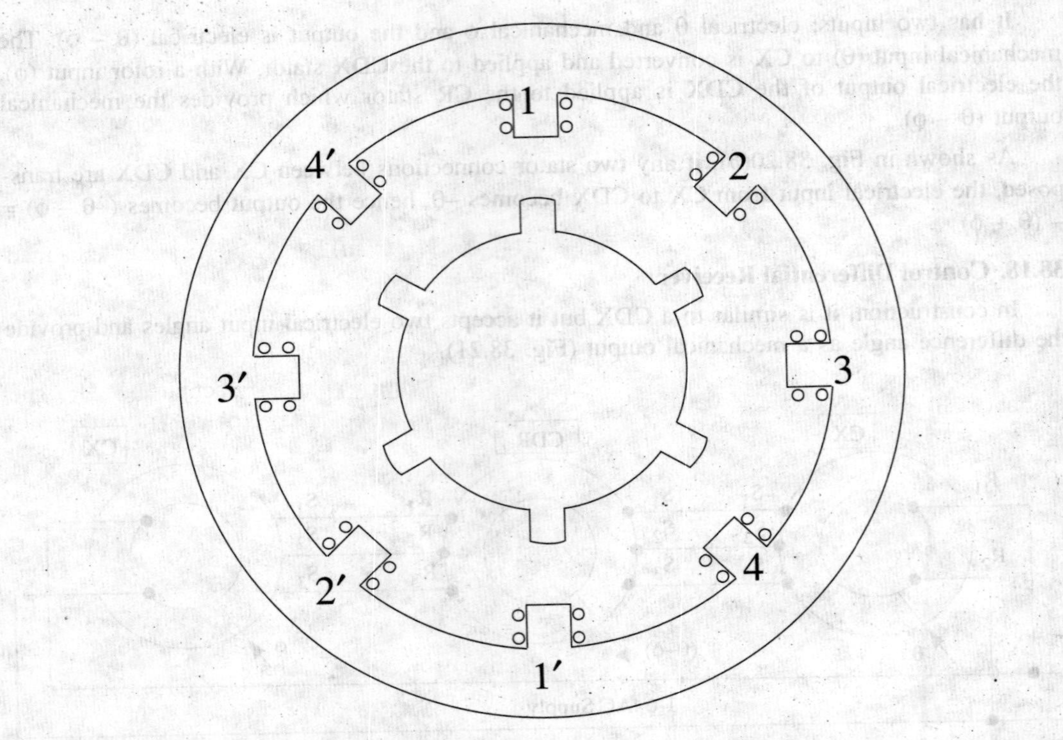

Fig. 38.22

The motor rotates in the anticlockwise direction when the stator phases are energized in the sequence 1,2,3,4 and in clockwise direction when energized in the sequence 1,4,3,2. When the stator coils are energized, the nearest pair of rotor poles is pulled into alignment with the appropriate stator poles by reluctance torque.

Closed-loop control is essential to optimize the switching angles of the applied coil voltages. The stator phases are switched by signals derived from a shaft-mounted rotor position detectors such as Hall-effect devices or optical sensors Fig. (38.23). This causes the behaviour of the SR motor to resemble that of a dc motor.

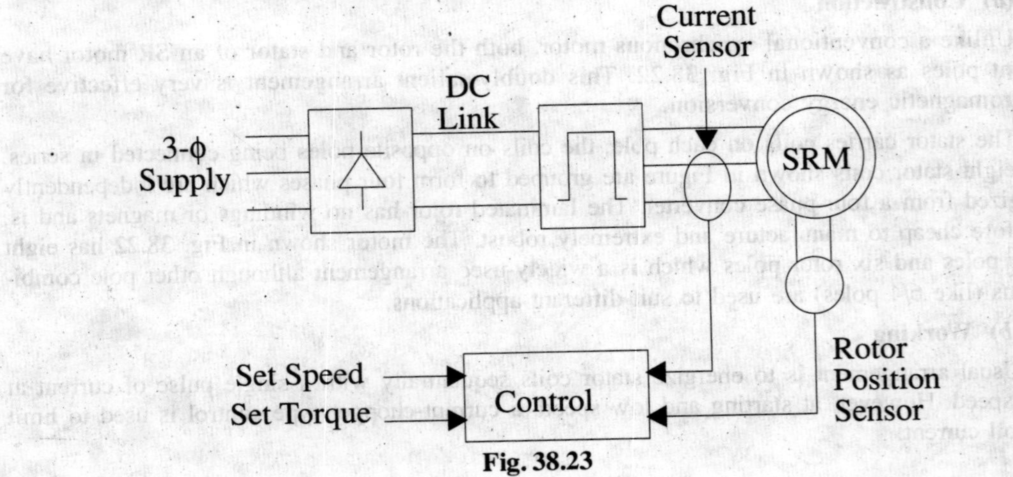

Fig. 38.23

(c) Advantages and Disadvantages

Although the newest arrival on the drives scene, the SR motor offers the following advantages:

(*i*) higher efficiency (*ii*) more power per unit weight and volume (*iii*) very robust because rotor has no windings or slip rings (*iv*) can run at very high speed (upto 30,000 rpm) in hazardous atmospheres (*v*) has versatile and flexible drive features and (*vi*) four-quadrant operation is possible with appropriate drive circuitry.

However, the drawbacks are that it is (*i*) relatively unproven (*ii*) noisy and (*iii*) not well-suited for smooth torque production.

(d) Applications

Even though the SR technology is still in its infancy, it has been successfully applied to a wide range of applications sucha as (*i*) general purpose industrial drives (*ii*) traction (*iii*) domestic appliances like food processors, vacuum cleaners and washing machines etc, and (*iv*) office and business equipment.

38.20. Comparison between VR Stepper Motor and SR Motor

VR Stepper Motor	*SR Motor*
1. It rotates in steps.	It is meant for continuous rotation.
2. It is designed first and foremost for open-loop operation.	Closed-loop control is essential for its optimal working.
3. Its rotor poles are made of ferromagnetic material	Its rotor poles are also made of ferromagnetic material.
4. It is capable of half-step operation and microstepping.	It is not designed for this purpose.
5. Has low power rating.	Has power ratings upto 75 kW (100 hp).
6. Has lower efficiency.	Has higher overall efficiency.

38.21. The Resolver

In many ways, it is similar to a synchro but differs from it in the following respects: (*i*) electrical displacement between stator windings is 90° and not 120° (*ii*) it has two stator windings and two rotor windings (Fig. 38.24) (*iii*) its input can be either to the stator or to the rotor (*iv*) they are usually not used as followers because their output voltage is put to further use.

(a) Construction

The main constructional features and the symbol for a resolver are shown in Fig. 38.24. There are two stator windings which are wound 90° apart. In most applications, only one stator winding is used, the other being short-circuited. The two rotor winding connections are brought out through slip rings and brushes.

(b) Applications

Resolvers find many applications in navigation and height determination as shown in Fig.38.25 (*a*) and (*c*) where Fig. 38.25(*b*) provides the key.

(i) Navigation Application

As shown in Fig. 38.25(*a*), the purpose is to determine the distance D to the destination. Suppose the range R to a base station as found by a radar ranging device is 369 km. The angle θ is also determined directly. If the amplifier scale is 4.5 V per 100 km, the range would be represented by 369 × (4.5 / 100) = 16.6 V. Further suppose that angle θ is foun[illegible] be 52.5°. Now, set the resolver at 52.5° and apply 16.6 V to rotor terminals R_3 R_4. Th[illegible] which

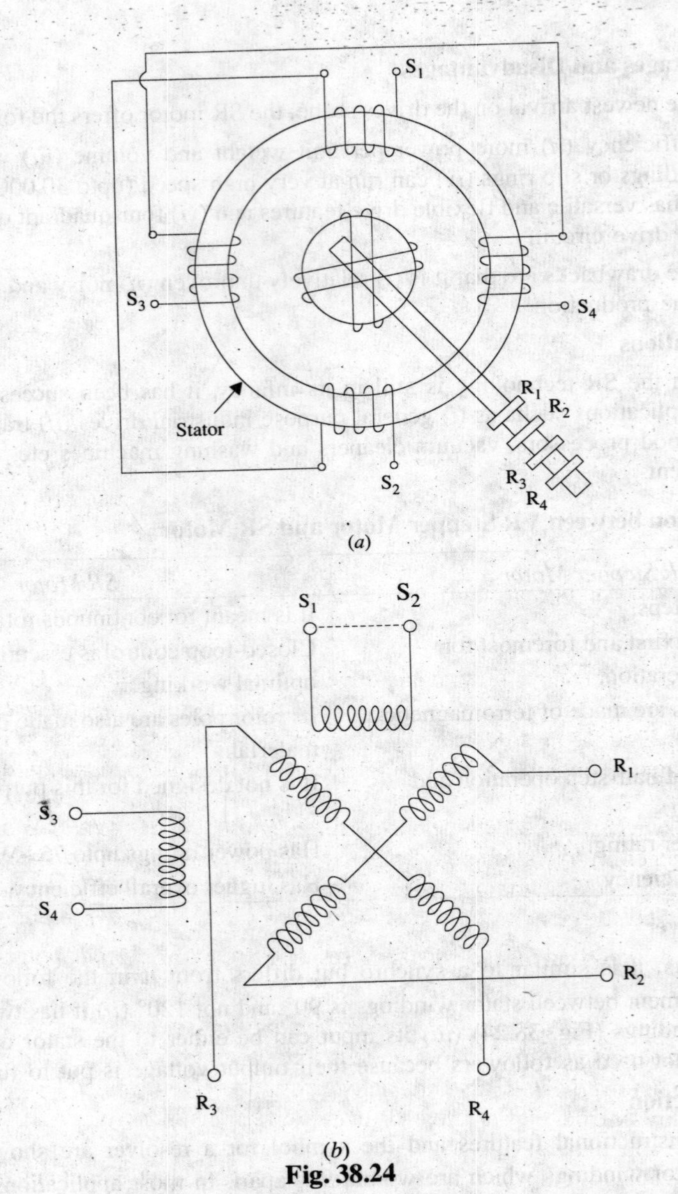

Fig. 38.24

appears at terminals S_1 S_2 represents D. If we assume K = stator turns / rotor turns = 1, the voltage available at S_1 S_2 will be = 16.6 / cos 52.5° = 16.6 / 0.6088 = 27.3 V. Since 4.5 V represent 100 km, 27.3 V represents 27.3 × 100 / 4.5 = 607 km.

(ii) Height Determination

Suppose the height H of a building is to be found. First of all, the oblique distance D to the top of the building is found by a range finder. Let D = 210 m and the scale of the amplifier to the resolver stator be 9 V per 100 m. The equivalent voltage is 9 × 210 / 100 = 18.9 V. This voltage is applied to stator terminals is S_1 S_2 of the resolver. Suppose the angle θ read from the resolver scale is 61.3°. The height of the building is given in the form of voltage which appears across the rotor terminals R_1 R_2. Assuming stator / rotor turn ratio as unity and the same amplifier ratio for the rotor output, the voltage across R_1 R_2 = 18.9 × sin 61.3° = 16.6 V. Hence, H = 16.6

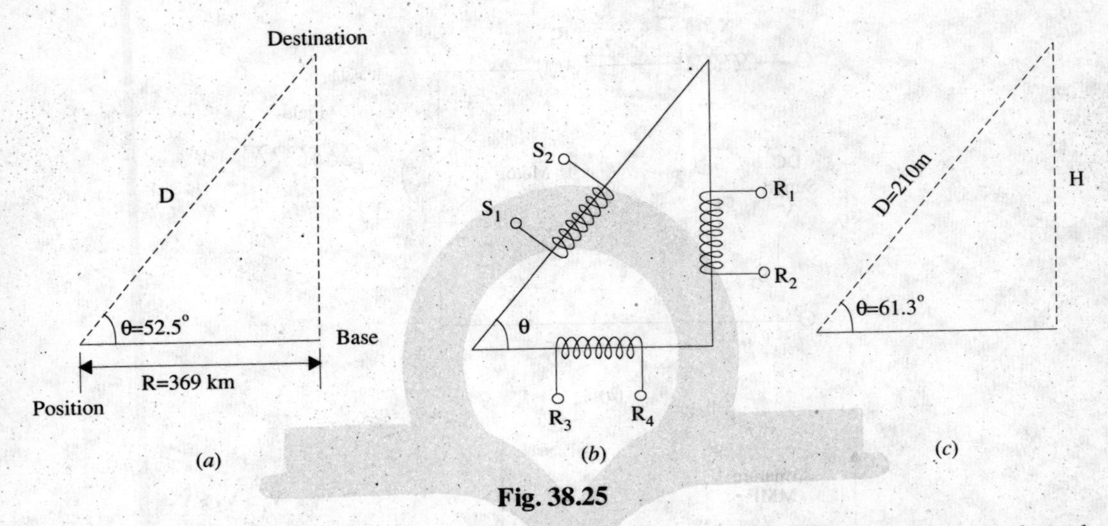

Fig. 38.25

× 100 / 9 = 184 m. It would be seen that in using the resolver, there is no need to go through trigonometric calculations because the answers come out directly.

38.22. Servomotors

They are also called control motors and have high-torque capabilities. Unlike large industrial motors, they are not used for continuous energy conversion but only for precise speed and precise position contorl at high torques. Of course, their basic principle of operation is the same as that of other electromagnetic motors. However, their construction, design and mode of operation are different. Their power ratings very from a fraction of a watt upto a few 100 W. Due to their low-inertia, they have high speed of response. That is why they are smaller in diameter but longer in length. They generally operate at very low speeds or sometime zero speed. They find wide applications in radar, tracking and guidance systems, process controllers, computers and machine tools. Both dc and ac (2-phase and 3-phase) servomotors are used at present.

Servomotors differ in application capabilities from large industrial motors in the following respects:

1. they produce high torque at all speeds including zero speed
2. they are capable of holding a static (i.e. no motion) position
3. they do not overheat at standstill or lower speeds
4. due to low-inertia, they are able to reverse directions quickly
5. they are able to accelerate and deaccelerate quickly
6. they are able to return to a given position time after time without any drift

These motors look like the usual electric motors. Their main difference from industrial motors is that more electric wires come out of them for power as well as for control. The servomotor wires go to a controller and not to the electrical line through contactors. Usually, a tachometer (speed-indicating device) is mechanically connected to the motor shaft. Sometimes, blower or fans may also be attached for motor cooling at low speeds.

38.23. DC Servomotors

These motors are either separately-excited dc motors or permanent-magnet dc motors. The schematic diagram of a separately-excited dc motor alongwith its armature and field MMFs and torque/speed characteristics is shown in Fig. 38.26.The speed of dc servomotors is normally controlled by varying the armature voltage. Their armature is deliberately designed to have large

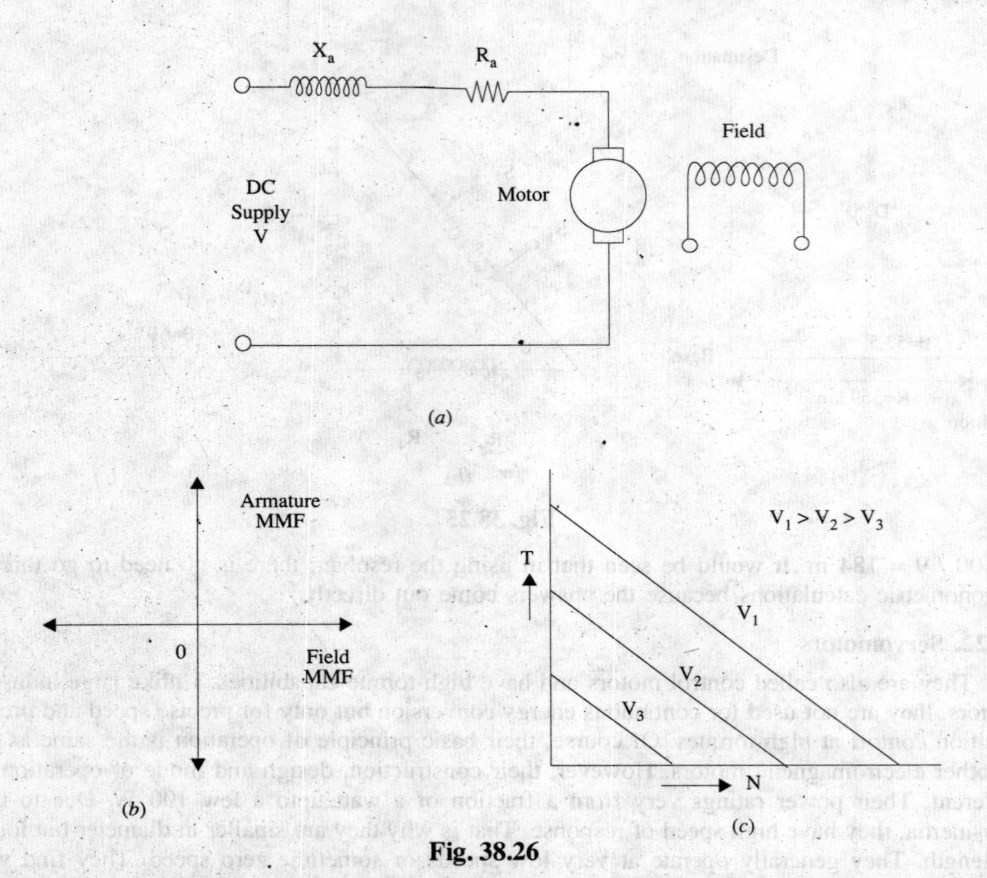

Fig. 38.26

resistance so that torque-speed characteristics are linear and have a large negative slope as shown in Fig. 38.26(*c*). The negative slope serves the purpose of providing the viscous damping for the servo drive system. As shown in Fig. 38.26(*b*), the armature mmf and excitation field mmf are in quadrature. This fact provides a fast torque response because torque and flux become decoupled. Accordingly, a step change in the armature voltage or current produces a quick change in the position or speed of the rotor.

38.24. AC Servomotors

Presently, most of the ac servomotors are of the two-phase squirrel-cage induction type and are used for low power applications. However, recently three-phase induction motors have been modified for high power servo systems which had so far been using high power dc servomotors.

(*a*) Two-phase AC Servomotor

Such motors normally run on a frequency of 60 Hz or 400 Hz (for airborne systems). The stator has two distributed windings which are displaced from each other by 90° (electrical). The main winding (also called the reference or fixed phase) is supplied from a constant voltage source, $V_m \angle 0°$ (Fig. 38.27). The other winding (also called the control phase) is supplied with a variable voltage of the same frequency as the reference phase but is phase-displaced by 90° (electrical). The control-phase voltage is controlled by an electronic controller. The speed and torque of the rotor are controlled by the phase difference between the main and control windings. Reversing the phase difference from leading to lagging (or *vice versa*) reverses the motor direction.

Since the rotor bars have high resistance, the torque-speed characteristics for various armature

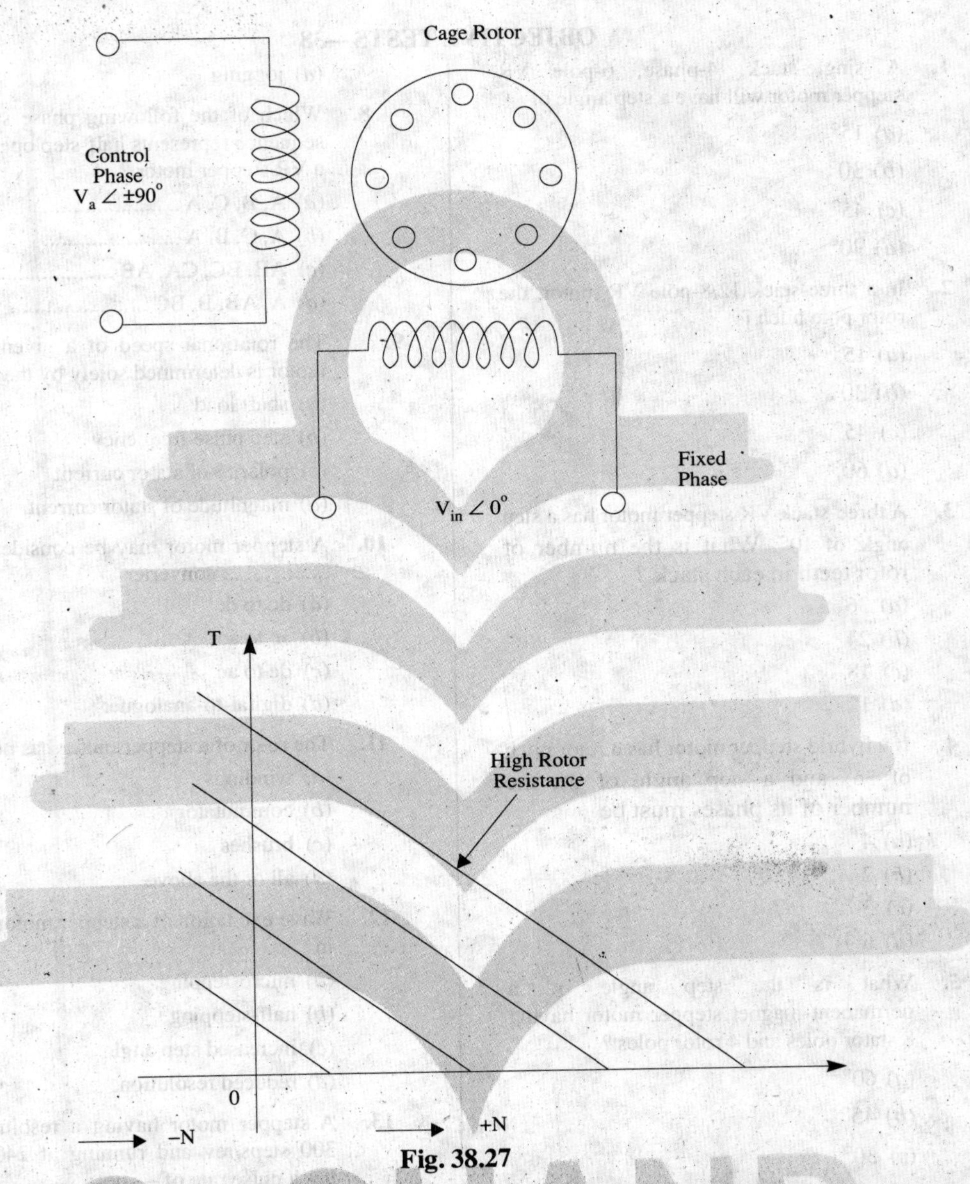

Fig. 38.27

voltages are almost linear over a wide speed range particularly near the zero speed. The motor operation can be controlled by varying the voltage of the main phase while keeping that of the reference phase constant.

(b) Three-phase AC Servomotors

A great deal of research has been to modify a three-phase squirrel-cage induction motor for use in high power servo systems. Normally, such a motor is a highly non-linear coupled-circuit device. Recently, this machine has been operated successfully as a linear decoupled machine (like a dc machine) by using a control method called vector control or field oriented control. In this method, the currents fed to the machine are controlled in such a way that its torque and flux become decoupled as in a dc machine. This results in a high speed and a high torque response.

OBJECTIVE TESTS—38

1. A single-stack, 4-phase, 6-pole VR stepper motor will have a step angle of
(*a*) 15°
(*b*) 30°
(*c*) 45°
(*d*) 90°

2. In a three-stack 12/8-pole VR motor, the rotor pole pitch is
(*a*) 15°
(*b*) 30°
(*c*) 45°
(*d*) 60°

3. A three-stack VR stepper motor has a step angle of 10°. What is the number of rotor teeth in each stack ?
(*a*) 36
(*b*) 24
(*c*) 18
(*d*) 12

4. If a hybrid stepper motor has a rotor pitch of 36° and a step angle of 9°, the number of its phases must be
(*a*) 4
(*b*) 2
(*c*) 3
(*d*) 6

5. What is the step angle of a permanent-magnet stepper motor having 8 stator poles and 4 rotor poles ?
(*a*) 60°
(*b*) 45°
(*c*) 30°
(*d*) 15°

6. A stepping motor is a .... device.
(*a*) mechanical
(*b*) electrical
(*c*) analogue
(*d*) incremental

7. Operation of stepping motors at high speeds is referred to as
(*a*) fast forward
(*b*) slewing
(*c*) inching
(*d*) jogging

8. Which of the following phase switching sequence represents half-step operation of a VR stepper motor ?
(*a*) A, B, C, A ....
(*b*) A, C, B, A ....
(*c*) AB, BC, CA, AB ....
(*d*) A, AB, B, BC ....

9. The rotational speed of a given stepper motor is determined solely by the
(*a*) shaft load
(*b*) step pulse frequency
(*c*) polarity of stator current
(*d*) magnitude of stator current.

10. A stepper motor may be considered as a .... converter
(*a*) dc to dc
(*b*) ac to ac
(*c*) dc to ac
(*d*) digital-to-analogue.

11. The rotor of a stepper motor has no
(*a*) windings
(*b*) commutator
(*c*) brushes
(*d*) all of the above.

12. Wave excitation of a stepper motor results in
(*a*) microstepping
(*b*) half-stepping
(*c*) increased step angle
(*d*) reduced resolution.

13. A stepper motor having a resolution of 300 steps/rev and running at 2400 rpm has a pulse rate of — pps.
(*a*) 4000
(*b*) 8000
(*c*) 6000
(*d*) 10,000

14. The torque exerted by the rotor magnetic field of a PM stepping motor with unexcited stator is called .... torque.
(*a*) reluctance
(*b*) detent
(*c*) holding

(*d*) either (*b*) or (*c*)

15. A variable reluctance stepper motor is constructed of .... material with salient poles.

(*a*) paramagnetic

(*b*) ferromagnetic

(*c*) diamagnetic

(*d*) non-magnetic

16. Though structurally similar to a control transmitter, a control receiver differs from it in the following way

(*a*) it has three-phase stator winding

(*b*) it has a rotor of dumbell construction

(*c*) it has a mechanical damper on its shaft

(*d*) it has single-phase rotor excitation.

17. The control .... synchro has three-phase winding both on its stator and rotor.

(*a*) differential

(*b*) transformer

(*c*) receiver

(*d*) transmitter

18. Regarding voltages induced in the three stator windings of a synchro, which statement is false ?

(*a*) they depend on rotor position

(*b*) they are in phase

(*c*) they differ in magnitude

(*d*) they are polyphase voltages

19. The low-torque synchros cannot be used for

(*a*) torque transmission

(*b*) error detection

(*c*) instrument servos

(*d*) robot arm positioning.

20. Which of the following synchros are used for error detection in a servo control system ?

(*a*) control transmitter

(*b*) control transformer

(*c*) control receiver

(*d*) both (*a*) and (*b*).

21. For torque transmission over a long distance with the help of electrical wires only, which of the following two synchros are used ?

(*a*) CX and CT

(*b*) CX and CR

(*c*) CX and CD

(*d*) CT and CD.

22. The arrangement required for producing a rotation equal to the sum or difference of the rotation of two shafts consists of the following coupled synchros.

(*a*) control transmitter

(*b*) control receiver

(*c*) control differential transmitter

(*d*) all of the above.

23. Which of the following motor would suit applications where constant speed is absolutely essential to ensure a consistent product ?

(*a*) brushless dc motor

(*b*) disk motor

(*c*) permanent-magnet synchronous motor

(*d*) stepper motor.

24. A switched reluctance motor differs from a VR stepper motor in the sense that it

(*a*) has rotor poles of ferromagnetic material

(*b*) rotates continuously

(*c*) is designed for open-loop operation only

(*d*) has lower efficiency.

25. The electrical displacement between the two stator windings of a resolver is

(*a*) 120°

(*b*) 90°

(*c*) 60°

(*d*) 45°.

26. Which of the following motor runs from a low dc supply and has permanently magnetized salient poles on its rotor ?

(*a*) permanent-magnet dc motor

(*b*) disk dc motor

(*c*) permanent-magnet synchronous motor

(*d*) brushless dc motor.

27. A dc servomotor is similar to a regular dc motor except that its design is modified to cope with

(*a*) electronic switching

(*b*) slow speeds

(*c*) static conditions
(*d*) both (*b*) and (*c*).

28. One of the basic requirements of a servomotor is that it must produce high torque at all
(*a*) loads
(*b*) frequencies
(*c*) speeds
(*d*) voltages.

29. The most common two-phase ac servomotor differs from the standard ac induction motor because it has
(*a*) higher rotor resistance
(*b*) higher power rating
(*c*) motor stator windings
(*d*) greater inertia.

30. Squirrel-cage induction motor is finding increasing application in high-power servo systems because new methods have been found to
(*a*) increase its rotor resistance
(*b*) control its torque
(*c*) decrease its intertia
(*d*) decouple its torque and flux.

ANSWERS

1.*a* 2.*c* 3.*d* 4.*a* 5.*b* 6.*d* 7.*b* 8.*d* 9.*b* 10.*d* 11.*d*
12.*b* 13.*c* 14.*d* 15.*b* 16.*c* 17.*a* 18.*d* 19.*d* 20.*d* 21.*b* 22.*d*
23.*c* 24.*b* 25.*b* 26.*a* 27.*d* 28.*c* 29.*a* 30.*d*

QUESTIONS AND ANSWERS ON SPECIAL MACHINES

Q. Do stepper motors have internal or external fans ?

Ans. No. Because the heat generated in the stator winding is conducted through the stator iron to the case which is cooled by natural conduction, convection and radiation.

Q. Why do hybrid stepping motors have many phases sometime more than six?

Ans. In order to obtain smaller step angles.

Q. Any disadvantage(s) of having more phases?

Ans. Minor ones are: more leads have to be brought out from the motor, more interconnections are required to the drive circuit and more switching devices are needed.

Q. What is the main attraction of a multi-stack VR stepper motor?

Ans. It is well-suited to high stepping rates.

Q. You are given a VR motor and a hybrid stepper motor which look exactly similar. How would you tell which is which?

Ans. Spin the rotor after short-circuiting the stator winding. If there is no mechanical resistance to rotation, it is a VR motor and if there is resistance, then it is a hybrid motor.

Q. How do you explain it ?

Ans. Since VR motor has magnetically neutral rotor, it will not induce any emf in the short-circuited winding *i.e.* the machine will not act as a generator and hence experience no drage on its rotation. However, the rotor of a hybrid motor has magnetic poles, hence it will act as a generator and so experience a drag.

Q. Will there be any harm if the rotor of a hybrid stepper motor is pulled out of its stator?

Ans. Yes. The rotor will probably become partially demagnetized and, on reassembling, will give less holding torque.